Encyclopedia of
CHROMATOGRAPHY

ENCYCLOPEDIA OF
CHROMATOGRAPHY

SECOND EDITION

VOLUME ONE

EDITED BY
JACK CAZES

Florida Atlantic University
Boca Raton, Florida, U.S.A.

Taylor & Francis
Taylor & Francis Group

Boca Raton   London   New York   Singapore

Published in 2005 by
Taylor & Francis Group
6000 Broken Sound Parkway NW, Suite 300
Boca Raton, FL 33487-2742

© 2005 by Taylor & Francis Group, LLC

No claim to original U.S. Government works
Printed in the United States of America on acid-free paper
10 9 8 7 6 5 4 3 2 1

International Standard Book Number-10: 0-8247-2785-1 (Hardcover - Set)
International Standard Book Number-10: 0-8247-2786-X (Hardcover - Volume 1)
International Standard Book Number-10: 0-8247-2787-8 (Hardcover - Volume 2)
International Standard Book Number-13: 978-0-8247-2785-7 (Hardcover - Set)
International Standard Book Number-13: 978-0-8247-2786-4 (Hardcover - Volume 1)
International Standard Book Number-13: 978-0-8247-2787-1 (Hardcover - Volume 2)

I dedicate this work to my lovely grandchildren,
Matthew, Monica, Brett, and Evan,
the shining lights of my life.

Contributors

Hassan Y. Aboul-Enein / *Pharmaceutical Analysis Laboratory, Biological and Medical Research Department, King Faisal Specialist Hospital and Research Center, Riyadh, Saudi Arabia*

Manuel Acosta / *Department of Plant Biology (Plant Physiology), University of Murcia, Murcia, Spain*

Ibrahim A. Al-Duraibi / *Pharmaceutical Analysis Laboratory, King Faisal Specialist Hospital and Research Centre, Riyadh, Saudi Arabia*

Serge Alex / *Centre d'Etudes des Procédés Chimiques du Québec, Montreal, Quebec, Canada*

Imran Ali / *National Institute of Hydrology, Roorkee, India and Pharmaceutical Analysis Laboratory, Biological and Medical Research Department, King Faisal Specialist Hospital and Research Center, Riyadh, Saudi Arabia*

Juan G. Alvarez / *Beth Israel Deaconess Medical Center, Harvard Medical School, Boston, Massachusetts, U.S.A.*

P. B. Andrade / *CEQUP/Serviço de Farmacognosia, Faculdade de Farmácia da Universidade do Porto, R. Aníbal Cunha, Porto, Portugal*

Victor P. Andreev / *Institute for Analytical Instrumentation, Russian Academy of Sciences, St. Petersburg, Russia*

M. J. Arin / *Departamento de Bioquímica y Biología Molecular, Facultad de Ciencias Biológicas y Ambientales, Universidad de León, León, Spain*

Marino B. Arnao / *Department of Plant Biology (Plant Physiology), University of Murcia, Murcia, Spain*

Christine M. Aurigemma / *Pfizer Global Research and Development, La Jolla, California, U.S.A.*

Yoshinobu Baba / *Department of Medicinal Chemistry, Faculty of Pharmaceutical Sciences, The 21st Century COE Program, The University of Tokushima, CRESTJST, Shomachi, Tokushima, Japan and Single-Molecule Bioanalysis Laboratory, National Institute of Advanced Industrial Science and Technology, Takamatsu, Japan*

M. A. Bagool / *Wockhardt Research Centre, Aurangabad, Maharashtra State, India*

James J. Bao / *Advanced Medicine, Inc., South San Francisco, California, U.S.A.*

M. A. Barbirato / *Instituto de Quimica de São Carlos, Universidade de São Paulo, São Carlos, SP, Brazil*

Csaba Barta / *Institute of Medical Chemistry, Molecular Biology, and Pathobiochemistry, Semmelweis University, Budapest, Hungary*

M. Barut / *BIA Separations d.o.o., Teslova, Ljubljana, Slovenia*

I. Bataille / *Institut Galilee, Université Paris Nord, Villetaneuse, France*

Mária Báthori / *Department of Pharmacognosy, University of Szeged, Szeged, Hungary*

Serge Battu / *Laboratoire de Cellulomique Neurale and UMR CNRS 6101, Faculté de Médecine & Pharmacie, Université de Limoges, Limoges, France*

Philippe J. Berny / *Unité de Toxicologie, Ecole Nationale Veterinaire de Lyon, Marcy l'Etoile, France*

Alain Berthod / *Laboratoire des Sciences Analytiques, CNRS UMR5180, Université de Lyon I, CPE, Villeurbanne, France*

Clayton B'Hymer / *University of Cincinnati, Cincinnati, Ohio, U.S.A.*

Jacques Bodennec / *Laboratory of Tumor Glycobiology, Université Claude Bernard Lyon I, Oullins, France*

F. Bonfils / *CIRAD-CP, TA 80/16, Montpellier, France*

Nikolaos A. Botsoglou / *Laboratory of Nutrition, Faculty of Veterinary Medicine, Aristotle University, Thessaloniki, Greece*

Nataša Brajenović / *Ruđer Bošković Institute, Zagreb, Croatia*

E. Brandšteterová / *Department of Analytical Chemistry, Slovak Technical University, Bratislava, Slovakia*

Michael Breslav / *Johnson & Johnson Pharmaceutical Research and Development, L.L.C., Spring House, Pennsylvania, U.S.A.*

Yefim Brun / *Waters Corporation, Milford, Massachusetts, U.S.A.*

Christopher E. Bunker / *Air Force Research Laboratory, Wright-Patterson Air Force Base, Ohio, U.S.A.*

Jean-Pierre Busnel / *Université Du Main, Le Mans, France*

Yong Cai / *Department of Chemistry and Biochemistry, Southeast Environmental Research Center, Florida International University, University Park, Miami, Florida, U.S.A.*

Antonio Cano / *Department of Plant Biology (Plant Physiology), University of Murcia, Murcia, Spain*

Ping Cao / *Tularik, Inc., South San Francisco, California, U.S.A.*

Wenjie Cao / *Huntsman Polymers Corporation, Odessa, Texas, U.S.A.*

Philippe J. P. Cardot / *Laboratoire de Cellulomique Neurale and UMR CNRS 6101, Faculté de Médecine & Pharmacie, Université de Limoges, Limoges, France*

M. Caude / *ESPCI, Paris, France*

Teresa Cecchi / *Dipartimento Scienze Chimiche, Università degli Studi di Camerino, Camerino, Italy and Italy Chemistry Department, ITIS Montani, Fermo (AP), Italy*

Zhikuan Chai / *Research Center for Eco-Environmental Sciences, Chinese Academy of Sciences, Beijing, P.R. China*

Jeffrey J. Chalmers / *The Ohio State University, Columbus, Ohio, U.S.A.*

Huan-Tsung Chang / *Department of Chemistry, National Taiwan University, Taipei, Taiwan*

C. Char / *CIRAD-CP, TA 80/16, Montpellier, France*

Sarah Chen / *Analytical Research Department, Merck Research Laboratories, Rahway, New Jersey, U.S.A.*

Oscar Chiantore / *Università degli Studi di Torino, Torino, Italy*

Tai-Chia Chiu / *Department of Chemistry, National Taiwan University, Taipei, Taiwan*

Josef Chmelík / *Institute of Analytical Chemistry, Academy of Sciences of the Czech Republic, Brno, Czech Republic*

Du Young Choi / *Center for Advanced Bioseparation Technology, Department of Chemical Engineering, Inha University, Nam-Ku, Incheon, South Korea*

Irena Choma / *Department of Chromatographic Methods, M. Curie-Skłodowska University, Lublin, Poland*

Gabriela Cimpan / *Sirius Analytical Instruments Ltd., Forest Row, East Sussex, U.K. and Tonbridge, Kent, U.K.*

A. Cincinelli / *Department of Chemistry, University of Florence, Florence, Italy*

Christa L. Colyer / *Department of Chemistry, Wake Forest University, Winston-Salem, North Carolina, U.S.A.*

Danilo Corradini / *Institute of Chromatography, Rome, Italy*

Tibor Cserháti / *Department of Environmental Analysis, Institute of Chemistry, Chemical Research Center, Hungarian Academy of Sciences, Budapest, Hungary*

James Curry / *International Specialty Products, Wayne, New Jersey, U.S.A.*

S.-L. Da / *Department of Chemistry, Wuhan University, Wuhan, P.R. China*

Maria de Fatima Alpendurada / *Faculdade de Farmacia, Universidade do Porto, Rua Anibal Cunha, Porto, Portugal*

M. de Moraes / *Laboratorio de CromatografiaInstituto de Quimica de São Carlos, Universidade de São Paulo, São Carlos, SP, Brazil*

Yulin Deng / *University of Saskatchewan, Saskatoon, Saskatchewan, Canada*

Yves Denizot / *Laboratoire de Cellulomique Neurale and UMR CNRS 6101, Faculté de Médecine & Pharmacie, Université de Limoges, Limoges, France*

A. A. Deo / *Wockhardt Research Centre, Aurangabad, Maharashtra State, India*

M. T. Diez / *Departamento de Bioquímica y Biología Molecular, Facultad de Ciencias Biológicas y Ambientales, Universidad de León, León, Spain*

N. Dimov / *NIHFI, Sofia, Bulgaria*

N. M. Edwards / *Canadian Grain Commission, Grain Research Laboratory, Winnipeg, Manitoba, Canada*

Jahangir Emrani / *Novartis Crop Protection, Inc., Greensboro, North Carolina, U.S.A.*

William P. Farrell / *Pfizer Global Research and Development, La Jolla, California, U.S.A.*

Petr S. Fedotov / *Vernadsky Institute of Geochemistry and Analytical Chemistry, Russian Academy of Sciences, Moscow, Russia*

Y.-Q. Feng / *Department of Chemistry, Wuhan University, Wuhan, P.R. China*

I. M. P. L. V. O. Ferreira / *REQUIMTE/Serviço de Bromatologia, Faculdade de Farmácia, Universidade do Porto, Porto, Portugal*

Sam J. Ferrito / *Analytical Services Department, Cooper Power Systems, Franksville, Wisconsin, U.S.A.*

John C. Ford / *Department of Chemistry, Indiana University of Pennsylvania, Indiana, Pennsylvania, U.S.A.*

Esther Forgács / *Department of Environmental Analysis, Institute of Chemistry, Chemical Research Center, Hungarian Academy of Sciences, Budapest, Hungary*

George M. Frame, II / *Wadsworth Laboratory, New York State Department of Health, Albany, New York, U.S.A.*

M. C. García-Alvarez-Coque / *Departamento de Química Analítica, Universidad de Valencia, Valencia Burjassot, Spain*

P. Garcia-del Moral / *Departamento de Bioquímica y Biología Molecular, Facultad de Ciencias Biológicas y Ambientales, Universidad de León, León, Spain*

J. C. García-Glez / *Physical Chemistry Department, University of León, Campus de Vegazana s/n, León, Spain*

Dimitrios Gavril / *Physical Chemistry Laboratory, Department of Chemistry, University of Patras, Patras, Greece*

Kalliopi A. Georga / *Aristotle University of Thessaloniki, Thessaloniki, Greece*

Árpád Gerstner / *Institute of Medical Chemistry, Molecular Biology, and Pathobiochemistry, Semmelweis University, Budapest, Hungary*

H. G. Gika / *Laboratory of Analytical Chemistry, Department of Chemistry, Aristotle University of Thessaloniki, Thessaloniki, Greece*

Michel Girard / *Centre for Biologics Research, Biologics and Genetic Therapies Directorate, Health Canada, Sir F.G. Banting Research Centre, Ottawa, Ontario, Canada*

Ivan Gitsov / *College of Environmental Science and Forestry, State University of New York, Syracuse, New York, U.S.A.*

Kazimierz Glowniak / *Department of Pharmacognosy with Medicinal Plant Laboratory, Medical University of Lublin, Lublin, Poland*

Simion Gocan / *Department of Analytical Chemistry, "Babes-Bolyai" University, Cluj-Napoca, Romania*

Karen M. Gooding / *Eli Lilly and Company, Indianapolis, Indiana, U.S.A.*

Tomomi Goto / *Aichi Prefectural Institute of Public Health, Tsuji-machi, Kita-ku, Nagoya, Japan*

Mohan Gownder / *Huntsman Polymers Corporation, Odessa, Texas, U.S.A.*

Susan V. Greene / *Ethyl Petroleum Additives Corporation, Richmond, Virginia, U.S.A.*

Nelu Grinberg / *Merck Research Laboratories, Rahway, New Jersey, U.S.A.*

V. K. Gupta / *Department of Chemistry, Indian Institute of Technology, Roorkee, India*

András Guttman / *Torrey Mesa Research Institute, La Jolla, California, U.S.A.*

David S. Hage / *Department of Chemistry, University of Nebraska—Lincoln, Lincoln, Nebraska, U.S.A.*

J. E. Haky / *Florida Atlantic University, Boca Raton, Florida, U.S.A.*

Susana Maria Halpine / *STArt! teaching Science Through Art, Playa del Rey, California, U.S.A.*

Jamel S. Hamada / *Southern Regional Research Center, USDA-ARS, New Orleans, Louisiana, U.S.A.*

Martin Hassellöv / *Analytical and Marine Chemistry, Göteborg University, Göteborg, Sweden*

Michael P. Henry / *Beckman Coulter, Inc., Fullerton, California, U.S.A.*

Chuichi Hirayama / *Department of Applied Chemistry and Biochemistry, Kumamoto University, Kumamoto, Japan*

Y.-L. Hu / *Department of Chemistry, Wuhan University, Wuhan, P.R. China*

Chih-Ching Huang / *Department of Chemistry, National Taiwan University, Taipei, Taiwan*

W. Jeffrey Hurst / *Hershey Foods Technical Center, Hershey, Pennsylvania, U.S.A.*

Robert J. Hurtubise / *University of Wyoming, Laramie, Wyoming, U.S.A.*

Christine Hürzeler / *Postnova Analytics, Munich, Germany*

Radovan Hynek / *Institute of Chemical Technology, Prague, Czech Republic*

Hirotaka Ihara / *Department of Applied Chemistry and Biochemistry, Kumamoto University, Kumamoto, Japan*

Gunawan Indrayanto / *Laboratory of Pharmaceutical Biotechnology, Assessment Service Unit, Faculty of Pharmacy, Airlangga University, Jl Dharmawangsa dalam Surabaya, East Java, Indonesia*

Haleem J. Issaq / *NCI-Frederick Cancer Research and Development Center, Frederick, Maryland, U.S.A.*

Yoichiro Ito / *Center of Biochemistry and Biophysics, National Heart, Lung, and Blood Institute, National Institutes of Health, Bethesda, Maryland, U.S.A. and Laboratory of Biophysical Chemistry, National Heart, Lung, and Blood Institute–National Institutes of Health (NHLBI–NIH), Bethesda, Maryland, U.S.A.*

Yuko Ito / *Aichi Prefectural Institute of Public Health, Tsuji-machi, Kita-ku, Nagoya, Japan*

Josef Janča / *Department of Chemistry, Université de La Rochelle, La Rochelle, France*

J. Jančar / *BIA Separations d.o.o., Ljubljana, Slovenia*

Pavel Jandera / *Department of Analytical Chemistry, University of Pardubice, Pardubice CZ, Czech Republic*

A. Jardy / *ESPCI, Paris, France*

Dennis R. Jenke / *Technology Resources Division, Baxter Healthcare Corporation, Round Lake, Illinois, U.S.A.*

Alfonso Jiménez / *Department of Analytical Chemistry, Nutrition and Food Sciences, University of Alicante, Alicante, Spain*

Kiyokatsu Jinno / *Department of Materials Science, Toyohashi University of Technology, Tempaku-cho, Toyohashi, Japan and School of Materials Science, Toyohashi University of Technology, Toyohashi, Japan*

Harald John / *IPF Pharmaceuticals GmbH, Hannover, Germany*

Brian Jones / *Selerity Technologies, Inc., Salt Lake City, Utah, U.S.A.*

Krzysztof Kaczmarski / *Technical University of Rzeszów, Rzeszów, Poland*

Huba Kalász / *Department of Pharmacology and Pharmacotherapy, Semmelweis University, Budapest, Hungary*

George Karaiskakis / *Department of Chemistry, University of Patras, Patras, Greece*

Jan Káš / *Institute of Chemical Technology, Prague, Czech Republic*

Galina Kassalainen / *Colorado School of Mines, Golden, Colorado, U.S.A.*

Sindy Kayillo / *Center for Biostructural and Biomolecular Research, University of Western Sydney, Richmond, New South Wales, Australia*

Sarah Kazmi / *Northeastern University, Boston, Massachusetts, U.S.A.*

Ernst Kenndler / *Institute for Analytical Chemistry, University of Vienna, Vienna, Austria*

Eileen Kennedy / *Novartis Crop Protection, Inc., Greensboro, North Carolina, U.S.A.*

Yuriko Kiba / *Department of Medicinal Chemistry, Faculty of Pharmaceutical Sciences, The 21st Century COE Program, The University of Tokushima, Tokushima, Japan*

Peter Kilz / *PSS Polymer Standards Service GmbH, Mainz, Germany*

Peter T. Kissinger / *Bioanalytical Systems, Inc. and Purdue University, West Lafayette, Indiana, U.S.A.*

E. Kitazume / *Faculty of Humanities and Social Sciences, Iwate University, Morioka, Japan*

Thorsten Klein / *Postnova Analytics, Munich, Germany*

Oliver Klett / *Institute of Chemistry, Uppsala University, Uppsala, Sweden*

Athanasia Koliadima / *University of Patras, Patras, Greece*

B. L. Kolte / *Department of Chemical Technology, Dr. Babasaheb Ambedkar Marathwada University, Wockhardt Research Centre, Aurangabad, Maharashtra State, India*

Fumio Kondo / *Department of Toxicology, Aichi Prefectural Institute of Public Health, Tsuji-machi, Kita-ku, Nagoya, Japan*

Vadim L. Kononenko / *N.M. Emanuel Institute of Biochemical Physics, Russian Academy of Sciences, Moscow, Russia*

Teresa Kowalska / *Institute of Chemistry, Silesian University, Katowice, Poland*

Anna Kozak / *Institute of Chemical Technology, Prague, Czech Republic*

Ira S. Krull / *Northeastern University, Boston, Massachusetts, U.S.A.*

Svetlana Kulevanova / *Faculty of Pharmacy, Institute of Pharmacognosy, Sts. Cyril and Methodius University, Republic of Macedonia*

Silvia Lacorte / *Department of Environmental Chemistry, IIQAB-CSIC, Barcelona, Catalonia, Spain*

Vaishali Soneji Lafita / *Abbott Laboratories, Inc., Abbott Park, Illinois, U.S.A.*

Fernando M. Lanças / *Laboratorio de Cromatografia, Instituto de Quimica de São Carlos, Universidade de São Paulo, São Carlos, SP, Brazil*

James P. Landers / *Department of Chemistry, University of Virginia, Charlottesville, Virginia, U.S.A.*

Seungho Lee / *Department of Chemistry, Hannam University, Taejon, Korea*

L. Lepri / *Department of Chemistry, University of Florence, Florence, Italy*

James Lesec / *Laboratoire Physique et Chimie Macromoleculaire, CNRS-ESPCI, Paris, France*

Vera Leshchinskaya / *Bristol-Myers Squibb Co., Princeton, New Jersey, U.S.A.*

Chenchen Li / *Institute of Analytical Chemistry, College of Chemistry and Molecular Engineering, Peking University, Beijing, P.R. China*

Guangliang Liu / *Department of Chemistry and Biochemistry, Southeast Environmental Research Center, Florida International University, University Park, Miami, Florida, U.S.A.*

Huwei Liu / *College of Chemistry and Molecular Engineering, Institute of Analytical Chemistry, Peking University, Beijing, P.R. China*

Rosario LoBrutto / *Merck Research Laboratories, Rahway, New Jersey, U.S.A. and Seton Hall University, South Orange*

E. S. M. Lutz / *AstraZeneca R&D Mölndal, Mölndal, Sweden*

Ying Ma / *Laboratory of Biophysical Chemistry, National Heart, Lung, and Blood Institute–National Institutes of Health (NHLBI–NIH), Bethesda, Maryland, U.S.A.*

Edward Malawer / *International Specialty Products, Wayne, New Jersey, U.S.A.*

Wojciech Markowski / *Department of Inorganic and Analytical Chemistry, Department of Physical Chemistry, Medical University, Lublin, Poland*

J. Martín-Villacorta / *Physical Chemistry Department, University of León, Campus de Vegazana s/n, León, Spain*

T. Maryutina / *Vernadsky Institute of Geochemistry and Analytical Chemistry, Russian Academy of Sciences, Moscow, Russia*

Kazuhiro Matsuda / *Pharmacology Division, National Cancer Center Research Institute, Tokyo, Japan*

Sachie Matsuda / *Department of Dermatology, Horikiri Central Hospital, Tokyo, Japan*

Kiichi Matsuhisa / *Asahikawa National College of Technology, Asahikawa, Hokkaido, Japan*

Maria T. Matyska / *San Jose State University, San Jose, California, U.S.A.*

Gregorio R. Meira / *INTEC (Universidad Nacional del Litoral and CONICET), Santa Fe, Argentina*

R. Méndez / *Physical Chemistry Department, University of León, Campus de Vegazana s/n, León, Spain*

Raniero Mendichi / *Insituto di Chimica delle Macromolecole (CNR), Milano, Italy*

Jean-Michel Menet / *Process Development Chemistry, Aventis Pharma, Vitry-sur-Seine, France*

Toshiaki Miura / *College of Medical Technology, Hokkaido University, Sapporo, Hokkaido, Japan*

Emi Miyamoto / *Department of Health Science, Kochi Women's University, Kochi, Japan*

N. Montes / *Physical Chemistry Department, University of León, Campus de Vegazana s/n, León, Spain*

Myeong Hee Moon / *Pusan National University, Pusan, South Korea*

J. J. S. Moreira / *Laboratorio de Cromatografia, Instituto de Quimica de São Carlos, Universidade de São Paulo, São Carlos/SP, Brazil*

Sadao Mori / *PAC Research Institute, Mie University, Nagoya, Japan*

Mark Moskovitz / *Scientific Adsorbents, Inc., Atlanta, Georgia, U.S.A.*

Tomasz Mroczek / *Department of Pharmacognosy with Medicinal Plants Laboratory, Medical University, Lublin, Poland*

Muhammad Mulja / *Laboratory of Pharmaceutical Biotechnology, Airlangga University, Surabaya, Indonesia*

D. Muller / *Institut Galilee, Université Paris Nord, Villetaneuse, France*

Roy A. Musil / *Althea Technologies, Inc., San Diego, California, U.S.A.*

Ron Myers / *Wyatt Technology Corporation, Santa Barbara, California, U.S.A.*

Noh-Hong Myoung / *Institute of Health and Environment, Seoul, Korea*

Shoji Nagaoka / *Kumamoto Industrial Research Institute, Kumamoto, Japan*

A. Negro / *Analytical Chemistry Section, Faculty of Biological and Environmental Sciences, University of León, Leon, Spain*

Tuan Q. Nguyen / *Swiss Federal Institute of Technology, Lausanne, Switzerland*

Boryana Nikolova-Damyanova / *Institute of Organic Chemistry, Bulgarian Academy of Sciences, Sofia, Bulgaria*

Hisao Oka / *Aichi Prefectural Institute of Public Health, Tsuji-machi, Kita-ku, Nagoya, Japan*

Koji Otsuka / *Himeji Institute of Technology, Hyogo, Japan*

Anders Palm / *Cell and Molecular Biology, AstraZeneca, Lund, Sweden*

Irene Panderi / *Division of Pharmaceutical Chemistry, School of Pharmacy, University of Athens, Panepistimiopolis Zografou, Athens, Greece*

Ioannis N. Papadoyannis / *Laboratory of Analytical Chemistry, Department of Chemistry, Aristotle University of Thessaloniki, Thessaloniki, Greece*

Joseph J. Pesek / *San Jose State University, San Jose, California, U.S.A.*

Terry M. Phillips / *Ultramicro Analytical Immunochemistry Resource, DBEPS, ORS, OD, National Institute of Health, Bethesda, Maryland, U.S.A.*

A. Podgornik / *BIA Separations d.o.o., Teslova, Ljubljana, Slovenia*

Jacques Portoukalian / *Laboratory of Tumor Glycobiology, Université Claude Bernard Lyon I, Oullins, France*

M. Soledad Prats / *Department of Analytical Chemistry, University of Alicante, Alicante, Spain*

K. R. Preston / *Canadian Grain Commission, Grain Research Laboratory, Winnipeg, Manitoba, Canada*

Wojciech Prus / *University of Bielsko-Biała, Bielsko-Biała, Poland and School of Technology and the Arts in Bielsko-Biała, Technical University of Lodz, Bielsko-Biala, Poland*

Waraporn Putalun / *Graduate School of Pharmaceutical Sciences, Kyushu University, Fukuoka, Japan*

Alina Pyka / *Faculty of Pharmacy, Department of Analytical Chemistry, Silesian Academy of Medicine, Sosnowiec, Poland*

B. Rabanal / *Analytical Chemistry Section, Faculty of Biological and Environmental Sciences, University of León, León, Spain*

Fred M. Rabel / *EMD Chemicals, Inc., Gibbstown, New Jersey, U.S.A.*

S. Kim Ratanathanawongs Williams / *Colorado School of Mines, Golden, Colorado, U.S.A.*

Chitra K. Ratnayake / *Beckman Coulter, Inc., Fullerton, California, U.S.A.*

B. B. Raut / *Department of Chemical Technology, Dr. Babasaheb Ambedkar Marathwada University, Wockhardt Research Centre, Aurangabad, Maharashtra State, India*

Jetse C. Reijenga / *Eindhoven University of Technology, Eindhoven, The Netherlands*

Pierluigi Reschiglian / *University of Bologna, Bologna, Italy*

J. A. Resines / *Departamento de Física, Química y Expresión Gráfica, Facultad de Ciencias Biológicas y Ambientales, Universidad de León, León, Spain*

Mark P. Richards / *Growth Biology Laboratory, USDA-ARS-ANRI, Beltsville, Maryland, U.S.A.*

Anna Rigol / *Department of Analytical Chemistry, University of Barcelona, Barcelona, Catalonia, Spain*

M.-C. Rolet-Menet / *Laboratoire de Chimie Analytique, UFR de Sciences Pharmaceutiques et Biologiques, Paris Cedex, France*

Kyung Ho Row / *Center for Advanced Bioseparation Technology, Department of Chemical Engineering, Inha University, Nam-Ku, Incheon, South Korea*

Jan K. Różyło / *Marie Curie-Sklodowska University, Lublin, Poland*

Roxana A. Ruseckaite / *Research Institute of Material Science and Technology (INTEMA), and Chemistry Department, Sciences Faculty, University of Mar del Plata, Mar del Plata, Argentina*

Takashi Sagawa / *Institute of Advanced Energy, Kyoto University, Uji, Japan*

Jirí Sajdok / *Institute of Chemical Technology, Prague, Czech Republic*

Mieczysław Sajewicz / *Institute of Chemistry, Silesian University, Katowice, Poland*

Peter Sajonz / *Merck Research Laboratories, Rahway, New Jersey, U.S.A.*

Masayo Sakata / *Department of Applied Chemistry and Biochemistry, Kumamoto University, Kumamoto, Japan*

Toshihiko Sakurai / *Department of Applied Chemistry and Biochemistry, Kumamoto University, Kumamoto, Japan*

Victoria F. Samanidou / *Laboratory of Analytical Chemistry, Department of Chemistry, Aristotle University of Thessaloniki, Thessaloniki, Greece*

Mária Sasvári-Székely / *Institute of Medical Chemistry, Molecular Biology, and Pathobiochemistry, Semmelweis University, Budapest, Hungary*

Wes Schafer / *Merck Research Laboratories, Rahway, New Jersey, U.S.A.*

Martin E. Schimpf / *Boise State University, Boise, Idaho, U.S.A.*

Oliver Schmitz / *Division of Molecular Toxicology, German Cancer Research Center, Heidelberg, Germany*

Raymond P. W. Scott / *Scientific Detectors Ltd., Banbury, Oxfordshire, U.K.*

R. M. Seabra / *CEQUP/Serviço de Farmacognosia, Faculdade de Farmácia da Universidade do Porto, R. Aníbal Cunha, Porto, Portugal*

Stephen L. Secreast / *Pharmaceutical Sciences, Pharmacia Corporation, Kalamazoo, Michigan, U.S.A.*

S. N. Semenov / *Institute of Biochemical Physics RAS, Moscow, Russia*

Larry Senak / *International Specialty Products, Wayne, New Jersey, U.S.A.*

Vince Serignese / *Pharmaceutical Analysis Laboratory, King Faisal Specialist Hospital and Research Centre, Riyadh, Saudi Arabia*

Joanne Severs / *Bayer Pharmaceuticals, Berkeley, California, U.S.A.*

R. Andrew Shalliker / *Center for Biostructural and Biomolecular Research, University of Western Sydney, Richmond, New South Wales, Australia*

Joseph Sherma / *Department of Chemistry, Lafayette College, Easton, Pennsylvania, U.S.A.*

Yoichi Shibusawa / *Department of Analytical Chemistry, Division of Structural Biology and Analytical Science, School of Pharmacy, Tokyo University of Pharmacy and Life Science, Hachioji, Tokyo, Japan*

Zak K. Shihabi / *Department of Pathology, Wake Forest University Baptist Medical Center, Winston-Salem, North Carolina, U.S.A.*

D. B. Shinde / *Department of Chemical Technology, Dr. Babasaheb Ambedkar Marathwada University, Aurangabad, Maharashtra State, India*

Kazufusa Shinomiya / *College of Pharmacy, Nihon University, Narashinodai, Funabashi-shi, Chiba, Japan*

Yukihiro Shoyama / *Graduate School of Pharmaceutical Sciences, Kyushu University, Fukuoka, Japan*

Maria Victoria Silva Elipe / *Department of Drug Metabolism, Merck Research Laboratories, Rahway, New Jersey, U.S.A.*

Edward Soczewinski / *Medical University, Lublin, Poland*

Boris Ya. Spivakov / *Vernadsky Institute of Geochemistry and Analytical Chemistry, Russian Academy of Sciences, Moscow, Russia*

Trajče Stafilov / *Faculty of Science, Institute of Chemistry, Sts. Cyril and Methodius University, Republic of Macedonia*

Raluca-Ioana Stefan / *University of Pretoria, Pretoria, South Africa*

Marina Stefova / *Faculty of Science, Institute of Chemistry, Sts. Cyril and Methodius University, Republic of Macedonia*

S. G. Stevenson / *Canadian Grain Commission, Grain Research Laboratory, Winnipeg, Manitoba, Canada*

A. Štrancar / *BIA Separations d.o.o., Teslova, Ljubljana, Slovenia*

Richard C. Striebich / *University of Dayton Research Institute, Dayton, Ohio, U.S.A.*

André M. Striegel / *Department of Chemistry and Biochemistry, Florida State University, Tallahassee, Florida, U.S.A.*

Makoto Takafuji / *Department of Applied Chemistry and Biochemistry, Kumamoto University, Kumamoto, Japan*

Hiroyuki Tanaka / *Graduate School of Pharmaceutical Sciences, Kyushu University, Fukuoka, Japan*

Naohiro Tateda / *Asahikawa National College of Technology, Asahikawa, Hokkaido, Japan*

M. C. H. Tavares / *Instituto de Quimica de São Carlos, Universidade de São Paulo, São Carlos, SP, Brazil*

Shigeru Terabe / *Himeji Institute of Technology, Hyogo, Japan*

Gerald Terfloth / *GlaxoSmithKline, King of Prussia, Pennsylvania, U.S.A.*

Georgios A. Theodoridis / *Laboratory of Analytical Chemistry, Department of Chemistry, Aristotle University of Thessaloniki, Thessaloniki, Greece*

Richard Thompson / *Merck Research Laboratories, Rahway, New Jersey, U.S.A.*

J. R. Torres-Lapasio / *Departamento de Química Analítica, Universidad de Valencia, Valencia Burjassot, Spain*

Anna Tsantili-Kakoulidou / *Department of Pharmaceutical Chemistry, School of Pharmacy, University of Athens, Panepistimiopolis, Zografou, Athens, Greece*

Anant Vailaya / *Merck Research Laboratories, Rahway, New Jersey, U.S.A.*

Jacobus F. van Staden / *University of Pretoria, Pretoria, South Africa*

Jorge R. Vega / *INTEC (Universidad Nacional del Litoral and CONICET), Santa Fe, Argentina*

Manuel C. Ventura / *Pfizer Global Research and Development, La Jolla, California, U.S.A.*

J. Vial / *ESPCI, Paris, France*

Pertti J. Viskari / *Department of Chemistry, University of Virginia, Charlottesville, Virginia, U.S.A.*

Nikolay Vladimirov / *Hercules, Inc., Wilmington, Delaware, U.S.A.*

Frank von der Kammer / *Environmental Geoscience, Institute for Geoscience, Vienna University, Vienna, Austria*

Monika Waksmundzka-Hajnos / *Department of Inorganic Chemistry, Medical University, Lublin, Poland*

Qin-Sun Wang / *National Key Laboratory of Elemento-Organic Chemistry, Nankai University, Tianjin, P.R. China*

Tao Wang / *Merck Research Laboratories, Rahway, New Jersey, U.S.A.*

Fumio Watanabe / *Department of Health Science, Kochi Women's University, Kochi, Japan*

Teresa Wawrzynowicz / *Medical University, Lublin, Poland*

Robert Weinberger / *CE Technologies, Inc., Chappaqua, New York, U.S.A.*

Jaroslaw Widelski / *Department of Pharmacognosy with Medicinal Plant Laboratory, Medical University of Lublin, Lublin, Poland*

P. Stephen Williams / *The Cleveland Clinic Foundation, Cleveland, Ohio, U.S.A.*

Chi-san Wu / *International Specialty Products, Wayne, New Jersey, U.S.A.*

Philip J. Wyatt / *Wyatt Technology Corporation, Santa Barbara, California, U.S.A.*

Feng Xu / *Department of Medicinal Chemistry, Faculty of Pharmaceutical Sciences, The 21st Century COE Program, The University of Tokushima, Tokushima, Core Research for Evolutional Science and Technology (CREST), Japan Science and Technology Agency (JST), and Analytical Instruments Division, Shimadzu Corporation, Kyoto, Japan*

Fuquan Yang / *Laboratory of Biophysical Chemistry, National Heart, Lung, and Blood Institute, National Institutes of Health, Bethesda, Maryland, U.S.A.*

Xia Yang / *Institute of Analytical Chemistry, College of Chemistry and Molecular Engineering, Peking University, Beijing, P.R. China*

Yu Yang / *East Carolina University, Greenville, North Carolina, U.S.A.*

Bing Yu / *College of Chemistry and Molecular Engineering, Institute of Analytical Chemistry, Peking University, Beijing, P.R. China*

L. M. Yuan / *Department of Chemistry, Yunnan Normal University, Kumming, P.R. China*

Mochammad Yuwono / *Laboratory of Pharmaceutical Biotechnology, Assessment Service Unit, Faculty of Pharmacy, Airlangga University, Jl. Dharmawangsa dalam, Surabaya, East Java, Indonesia*

Maciej Zborowski / *The Cleveland Clinic Foundation, Cleveland, Ohio, U.S.A.*

Igor G. Zenkevich / *Chemical Research Institute, St. Petersburg State University, St. Petersburg, Russia*

Ji-Feng Zhang / *Massachusetts Institute of Technology, Cambridge, Massachusetts, U.S.A.*

L. Zhang / *National Key Laboratory of Elemento-Organic Chemistry, Nankai University, Tianjin, P.R. China*

Lifeng Zhang / *Environmental Technology Institute, Innovation Centre (NTU), Singapore*

Weihua Zhang / *Department of Chemistry and Biochemistry, Southeast Environmental Research Center, Florida International University, University Park, Miami, Florida, U.S.A.*

Xi-Chun Zhou / *Cambridge University, Cambridge, U.K.*

Wenshan Zhuang / *Taro Pharmaceuticals, Inc., Brampton, Canada*

A. Žiaková-Čaniová / *Department of Analytical Chemistry, Slovak Technical University, Bratislava, Slovakia*

Anastasia Zotou / *Aristotle University of Thessaloniki, Thessaloniki, Greece*

Contents

xx

Preface for the Second Edition

The *Encyclopedia of Chromatography*, which was first published in 2001, has become an invaluable source of up-to-date information dealing with chromatographic techniques and methodologies for solving separation problems. It presents the fundamentals of problem-solving and materials identification, as well as real-world applications, in a comprehensive, easy-to-read format supplemented with an abundance of up-to-date key references.

The *Encyclopedia of Chromatography* has kept up with the development of new chromatographic technologies, which have been progressing by leaps and bounds over the past several years. Our goal of developing a living compendium of information for novices and seasoned chromatographers alike has been realized; the *Encyclopedia* is now the leading reference in its field.

This second edition has been completely updated. Articles from the first edition have been revised and all of the articles that were published online since the encyclopedia went digital have been added. The growth has been significant; in fact, the second edition is published in two volumes.

Like the first edition, this second edition is also online and will continue to track the developments of new techniques, instrumentation, and applications of chromatography. The Editor heartily thanks all contributors for a job well done. They have contributed their time and expertise to benefit the scientific community at large.

Jack Cazes,
Editor

ENCYCLOPEDIA OF
CHROMATOGRAPHY

Absorbance Detection in Capillary Electrophoresis

Robert Weinberger
CE Technologies, Inc., Chappaqua, New York, U.S.A.

INTRODUCTION

Most forms of detection in high-performance capillary electrophoresis (HPCE) employ on-capillary detection. Exceptions are techniques that use a sheath flow such as laser-induced fluorescence[1] and electrospray ionization mass spectrometry.[2]

In HPLC, postcolumn detection is generally used. This means that all solutes are traveling at the same velocity when they pass through the detector flow cell. In HPCE with on-capillary detection, the velocity of the solute determines the residence time in the flow cell. This means that slowly migrating solutes spend more time in the optical path and thus accumulate more area counts.[3]

Because peak areas are used for quantitative determinations, the areas must be normalized when quantitating without standards. Quantitation without standards is often used when determining impurity profiles in pharmaceuticals, chiral impurities, and certain DNA applications. The correction is made by normalizing (dividing) the raw peak area by the migration time. When a matching standard is used, it is unnecessary to perform this correction. If the migration times are not reproducible, the correction may help, but it is better to correct the situation causing this problem.

LIMITS OF DETECTION

The limit of detection (LOD) of a system can be defined in two ways: the concentration limit of detection (CLOD) and the mass limit of detection (MLOD). The CLOD of a typical peptide is about $1 \, \mu g/mL$ using absorbance detection at 200 nm. If 10 nL are injected, this translates to an MLOD of 10 pg at three times the baseline noise. The MLOD illustrates the measuring capability of the instrument. The more important parameter is the CLOD, which relates to the sample itself. The CLOD for HPCE is relatively poor, whereas the MLOD is quite good, especially when compared to HPLC. In HPLC, the injection size can be 1000 times greater compared to HPCE.

The CLOD can be calculated using Beer's Law:

$$\text{CLOD} = \frac{A}{ab} = \frac{5 \times 10^{-5}}{(5000)(5 \times 10)^{-3}}$$
$$= 2 \times 10^{-6} \, \text{M} \tag{1}$$

where A is the absorbance (AU), a is the molar absorptivity (AU/cm/M), b is the capillary diameter or optical path length (cm), and CLOD is the concentration (M). The noise of a good detector is typically 5×10^{-5} AU. A modest chromophore has a molar absorptivity of 5000. Then in a 50-μm-inner diameter (i.d.) capillary, a CLOD of 2×10^{-6} M is obtained at a signal-to-noise ratio of 1, assuming no other sources of band broadening.

DETECTOR LINEAR DYNAMIC RANGE

The noise level of the best detectors is about 5×10^{-5} AU. Using a 50-μm-i.d. capillary, the maximum signal that can be obtained while yielding reasonable peak shape is 5×10^{-1} AU. This provides a linear dynamic range of about 10^4. This can be improved somewhat through the use of an extended path-length flow cell. In any event, if the background absorbance of the electrolyte is high, the noise of the system will increase regardless of the flow cell utilized.

CLASSES OF ABSORBANCE DETECTORS

Ultraviolet/visible absorption detection is the most common technique found in HPCE. Several types of absorption detectors are available on commercial instrumentation, including the following:

1. Fixed-wavelength detector using mercury, zinc, or cadmium lamps with wavelength selection by filters
2. Variable-wavelength detector using a deuterium or tungsten lamp with wavelength selection by a monochromator
3. Filter photometer using a deuterium lamp with wavelength selection by filters

Encyclopedia of Chromatography DOI: 10.1081/E-ECHR-120039866

7. Filter photometer using a deuterium lamp with wavelength selection by filters
8. Scanning ultraviolet (UV) detector.
9. Photodiode array detector.

Each of these absorption detectors have certain attributes that are useful in HPCE. Multiwavelength detectors such as the photodiode array or scanning UV detector are valuable because spectral as well as electrophoretic information can be displayed. The filter photometer is invaluable for low-UV detection. The use of the 185-nm mercury line becomes practical in HPCE with phosphate buffers because the short optical path length minimizes the background absorption.

Photoacoustic, thermo-optical, or photothermal detectors have been reported in the literature.[4] These detectors measure the nonradiative return of the excited molecule to the ground state. Although these can be quite sensitive, it is unlikely that they will be used in commercial instrumentation.

OPTIMIZATION OF DETECTOR WAVELENGTH

Because of the short optical path length defined by the capillary, the optimal detection wavelength is frequently much lower into the UV compared to HPLC. In HPCE with a variable-wavelength absorption detector, the optimal signal-to-noise (S/N) ratio for peptides is found at 200 nm. To optimize the detector wavelength, it is best to plot the S/N ratio at various wavelengths. The optimal S/N is then easily selected.

EXTENDED PATH-LENGTH CAPILLARIES

Increasing the optical path length of the capillary window should increase S/N simply as a result of Beer's Law. This has been achieved using a z cell (LC Packings, San Francisco CA),[5] bubble cell (Agilent Technologies, Wilmington, DE), or a high-sensitivity cell (Agilent Technologies). Both the z cell and bubble cell are integral to the capillary. The high-sensitivity cell comes in three parts: an inlet capillary, an outlet capillary, and the cell body. Careful assembly permits the use of this cell without current leakage. The bubble cell provides approximately a threefold improvement in sensitivity using a 50-μm capillary, whereas the z cell or high-sensitivity cell improves things by an order of magnitude. This holds true only when the background electrolyte (BGE) has low absorbance at the monitoring wavelength.

INDIRECT ABSORBANCE DETECTION

To determine ions that do not absorb in the UV, indirect detection is often utilized.[6] In this technique, a UV-absorbing reagent of the same charge (a co-ion) as the solutes is added to the BGE. The reagent elevates the baseline, and when nonabsorbing solute ions are present, they displace the additive. As the separated ions migrate past the detector window, they are measured as negative peaks relative to the high baseline. For anions, additives such as trimellitic acid, phthalic acid, or chromate ions are used at 2–10 mM concentrations. For cations, creatinine, imidazole, or copper(II) are often used. Other buffer materials are either not used or added in only small amounts to avoid interfering with the detection process.

It is best to match the mobility of the reagent to the average mobilities of the solutes to minimize electrodispersion, which causes band broadening.[7] When anions are determined, a cationic surfactant is added to the BGE to slow or even reverse the electroosmotic flow (EOF). When the EOF is reversed, both electrophoresis and electro-osmosis move in the same direction. Anion separations are performed using reversed polarity.

Indirect detection is used to determine simple ions such as chloride, sulfate, sodium, and potassium. The technique is also applicable to aliphatic amines, aliphatic carboxylic acids, and simple sugars.[8]

REFERENCES

1. Cheng, Y.F.; Dovichi, N.J. Fluorescence detection in capillary electrophoresis. SPIE **1988**, *910*, 111.
2. Huang, E.C.; Wachs, T.; Conboy, J.J.; Henion, J.D. Atmospheric pressure chemical ionization: detection. Anal. Chem. **1990**, *62*, 713–724.
3. Huang, X.; Coleman, W.F.; Zare, R.N. Analysis of factors causing peak broadening in capillary zone electrophoresis. J. Chromatogr. **1989**, *480*, 95–100.
4. Saz, J.M.; Diez-Masa, J.C. J. Liq. Chromatogr. **1994**, *17*, 499.
5. Chervet, J.P.; van Soest, R.E.J.; Ursem, M. Z-shaped flow cell for UV detection in capillary electrophoresis. J. Chromatogr. **1991**, *543*, 439.
6. Jandik, P.; Jones, W.R.; Weston, A.; Brown, P.R. Violet diode laser for metal ion determination by capillary electrophoresis-laser induced fluorescence. LC-GC **1991**, *9*, 634.
7. Weinberger, R. Am. Lab. **1996**, *28*, 24.
8. Xu, X.; Kok, W.T.; Poppe, H. Capillary electrophoresis using air and helium as cooling fluids. J. Chromatogr. A. **1995**, *716*, 231.

Acids: Derivatization for GC Analysis

Igor G. Zenkevich

Chemical Research Institute, St. Petersburg State University, St. Petersburg, Russia

INTRODUCTION

The class "acids" includes various types of compounds with active hydrogen atoms usually having $pK_a < 7$. The most important group of organic acids is the compounds with carboxyl fragment –COOH. Some other compounds can be classified not only as O-acids [e.g., hydroxamic acids, $–CONHOH \rightleftharpoons –C(OH)=NOH$], but C–H acids [with the presence of structural fragments $–CH(NO_2)_2$, $–CH(CN)_2$, etc.]. Well-known substances of this class for GC analysis are semivolatile fatty acids of triglycerides and lipids, numerous nonvolatile polyfunctional biogenic compounds (including such phenol carboxylic acids like gallic, vanillic, and syringic acid), different acidic herbicides (for example, 2,4-D, 2,4,5-T, MCPB, MCPA, fenoprop, haloxyfop, etc.), and many other substances. Strong inorganic acids like volatile hydrogen halides (HHal) and nonvolatile H_2SO_4, H_3PO_4, etc. can be objects of GC analysis too.

The simplest monofunctional carboxylic acids have boiling points at atmospheric pressure without decomposition and, hence, can be analyzed directly by GC. However, owing to the high polarities of carboxyl compounds, a typical problem of their GC analysis with standard nonpolar phases is the nonlinear sorption isotherm. As a result, these compounds yield broad nonsymmetrical peaks, which leads to poor detection limits and unsatisfactory reproducibility of their retention indices. The recommended stationary phases for direct analysis of free carboxylic acids are polar polyethylene glycols (Carbowax 20M, DBWax, SP-1000, FFAP, etc.). However, these phases have lower thermal stability compared with polydimethyl siloxanes (ca. 225–250 vs. 300–350 °C). This means that the upper limit of GC columns with these polar phases in Retention index (RI) units is not more than 3000–3500 IU. High homologs even of monocarboxylic acids cannot be eluted within this RI window (this is confirmed by the absence of RI data for palmitic acid, $C_{15}H_{31}COOH$, on the mentioned types of polar phases).

Some dicarboxylic acids can also be distilled without decomposition under reduced pressures. This is at least the theoretical grounds for the possibility of their direct GC analysis. Few successive attempts have been described, but these analytes require "on-column" injection of samples and extremely high inertness of chromatographic systems. Many types of polyfunctional carboxylic acids (hydroxy-, mercapto-, amino-, etc.) cannot be analyzed in free, underivatized form owing to either nonvolatility and/or absence of thermal stability. These features are the principal reasons for the conversion of carboxylic acids before their GC analysis into less polar derivatives without active hydrogen atoms.

METHODS OF ACID DERIVATIZATION

The general method of carboxylic acid derivatization is their esterification with the formation of alkyl (arylalkyl, halogenated alkyl) or silyl esters:[1–4]

$$XCO_2H + RY \rightarrow XCO_2R + YH$$

$$XCO_2H + ZSi(CH_3)_3 \rightarrow XCO_2Si(CH_3)_3 + ZH$$

Some of the most widely used reagents for the synthesis of alkyl carboxylates are listed in Table 1. The general recommendations for the silylation of mono- and polyfunctional carboxylic acids [trimethylsilyl (TMS)[5,6] and more stable *tert*-butyldimethylsilyl (TBDMS) derivatives[7,8]] are the same as those for other hydroxy containing compounds.

In general, the simplest methyl esters of carboxylic acids are more stable than TMS-esters to hydrolysis and, hence, they are the preferable derivatives for their GC analysis.[9,10] The most available esterification reagents are the corresponding alcohols, ROH, themselves. Different esters have been used as the analytical derivatives of carboxylic acids: Me, Et, Pr, *iso*-Pr, isomeric Bu (excluding *tert*-Bu esters owing to their poorer synthetic yields), and so forth. This method requires the use of excess of dry alcohol and acid catalysis by BCl_3, BF_3, CH_3COCl, $SOCl_2$, etc. Otherwise, the alcohol used can be saturated by gaseous HCl, which must then be removed by heating the reaction mixtures after completion of the reaction.

Compound	pK_a	T_b,°C	$RI_{nonpolar}$	RI_{polar}
Acetic acid	4.75	118	638 ± 10	1428 ± 30
Palmitic acid	4.9	351.5	1966 ± 7	No data
Benzoic acid	4.2	250	1201 ± 24	2387 ± 5
Phenylacetic acid	4.2	266	1290 ± 44	No data

All RIs with standard deviations are randomized interlaboratory data.

Encyclopedia of Chromatography DOI: 10.1081/E-ECHR-120039943

Table 1 Physicochemical and GC properties of some alkylating derivatization reagents

Reagent (abbreviation)	MW	T_b, °C	$RI_{nonpolar}$	By-products ($RI_{nonpolar}$)
Methanol/BCl_3, BF_3, HCl, DCC, etc.	32	64.6	381 ± 15	—
Diazomethane (in ethyl ether solution)	42	−23	None (unstable)	—
Methyl iodide/DMFA, K_2CO_3	142	42.8	515 ± 7	CH_3OH (381 ± 15)
Dimethyl sulfate/*tertiary*-amines	126	188.5	853 ± 22	CH_3OH (381 ± 15)
1-Iodopropane/DMFA, K_2CO_3	170	102	711 ± 11	C_3H_7OH (552 ± 13), $(C_3H_7)_2O$ (680 ± 6)
2-Bromopropane/LiH, DMSO	122	59.4	571 ± 5	$(CH_3)_2CHOH$ (486 ± 9), (*iso*-Pr)$_2$O (598 ± 5)
Methyl chloroformate	94	71	582 ± 17	CH_3OH (381 ± 15)
Ethyl chloroformate	108	—	640 ± 12	C_2H_5OH (452 ± 18)
Butyl chloroformate	136	—	832 ± 10	C_4H_9OH (658 ± 12)
Pentafluorobenzyl bromide (PFB-Br)	260	174–175	991 ± 11^a	$C_6F_5CH_2OH$ (934 ± 16)^a
3,5-*bis*-(Trifluoromethyl)benzyl bromide (BTB-Br)	306	—	1103 ± 9^a	$(CF_3)_2C_6H_3CH_2OH$ (1046 ± 15)^a
Tetramethylammonium hydroxide (TMAH; in 25% aqueous solution)	74	—	Nonvolatile	$(CH_3)_3N$ (418 ± 9)
Trimethylanilinium hydroxide (TMPAH; in 0.2 M methanol solution)	136	—	Nonvolatile	$C_6H_5N(CH_3)_2$ (1065 ± 9)
3,5-*bis*-(Trifluoromethylbenzyl) dimethylanilinium fluoride (BTBDMA-F)	258	—	Nonvolatile	3,5-$(CF_3)_2C_6H_3CH_2N(CH_3)_2$ (no data), $C_6H_5N(CH_3)_2$ (1065 ± 9)
2-Bromoacetophenone (phenacyl bromide)	198	260	1321 ± 4	$C_6H_5COCH_2OH$ (1118)^b
Silylating reagents^c				

^{a}Estimated RI values.
^{b}Single experimental value.
^{c}See *Hydroxy Compounds: Derivatization for GC Analysis* in this volume.

The same procedures are used for the synthesis of 2-chloroethyl ($RCO_2CH_2CH_2Cl$), 2,2,2-trifluoroethyl ($RCO_2CH_2CF_3$), 2,2,2-trichloroethyl ($RCO_2CH_2CCl_3$), and hexafluoroisopropyl esters [$RCO_2CH(CF_3)_2$] for GC analysis with selective detection. Instead of acid catalysis of this reaction, some reagents for the coupling of formed water were recommended, namely 1,1′-carbonyldiimidazole (I) and 1,3-dicyclohexylcarbodiimide (DCC, II):

The application of any additive reagents usually leads to the appearance of additional peaks on the chromatograms (including the peaks of by-products, for example imidazole, $RI_{nonpolar}$ 1072 $\pm$ 17), which must be reliably identified and excluded from data interpretation. The by-product from compound (II) — 1,3-dicyclohexylurea—is nonvolatile for GC analysis.

Another class of esterification reagents are halogenated compounds (alkyl iodides, substituted benzyl[11]

and phenacyl bromides, etc.), which need basic media for their reaction [K_2CO_3 or DMFA (dimethyl formamide) is used usually for the neutralization of HBr or HCl as acid by-product]. For methylation of carboxylic acids, some tetra-substituted ammonium hydroxides or halides can be used, namely tetramethylammonium hydroxide (in aqueous solutions) or trimethylanilinium hydroxide (in methanol solution). The intermediate ammonium carboxylates are thermally unstable and can produce methylalkanoates during the following heating of reaction mixtures or even their introduction into the hot injector of the gas chromatograph (flash methylation):

$$RCO_2H + XNMe_3^+OH^-$$
$$\rightarrow [RCO_2^-NMe_3^+] \rightarrow RCO_2Me \quad (X = Me, Ph)$$

The possible by-products of these reactions are the corresponding amines (Me_3N or $PhNMe_2$). A similar method has been proposed for the butylation of organic acids.[12] If the appearance of any volatile by-products is undesirable, the methylation of carboxylic acids by diazomethane is recommended.

This reagent (*warning*: highly toxic) is synthesized by alkaline cleavage of *N*-methyl-*N*-nitrosourea (III) or *N*-methyl-*N*-nitrosotoluenesulfamide (IV) and owing to its low boiling point (−23 °C) can be used only in diethyl ether solutions prepared immediately before use.

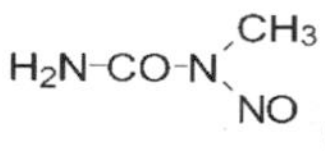

In the absence of acid catalysis, diazomethane reacts only with carboxylic acids (pK_a 4–5) and phenols (pK_a 9–10), but has no influence on aliphatic OH– groups. Besides CH_2N_2, some more complex diazocompounds (diazoethane, diazotoluene) have been recommended for the synthesis of other esters (ethyl and benzyl, respectively). For the synthesis of benzyl (or substituted benzyl) esters, some special reagents have also been proposed, for example, *N,N'*-dicyclohexyl-*O*-benzyl-urea (V) and 1-(4-methylphenyl)-3-benzyltriazene (VI):

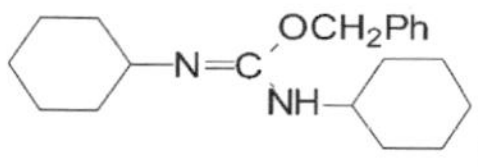

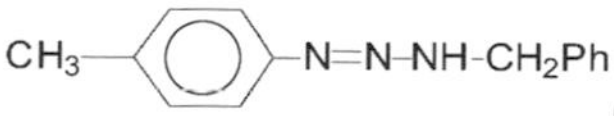

The esterification of carboxylic acids can also be accomplished using synthetic equivalents of acetals of alkanols $RCH(OR')_2$ (by acid catalysis), ortho-esters $RC(OR')_3$ (by acid catalysis), and dialkylcarbonates $CO(OR)_2$ (by base catalysis). The series of bifunctional reagents of this type—dimethylformamide dialkyl-acetals $(CH_3)_2N–CH(OR)_2$—is commercially available. Besides the esterification of carboxyl groups, these compounds react with primary amino groups used for GC analysis of amino acids:

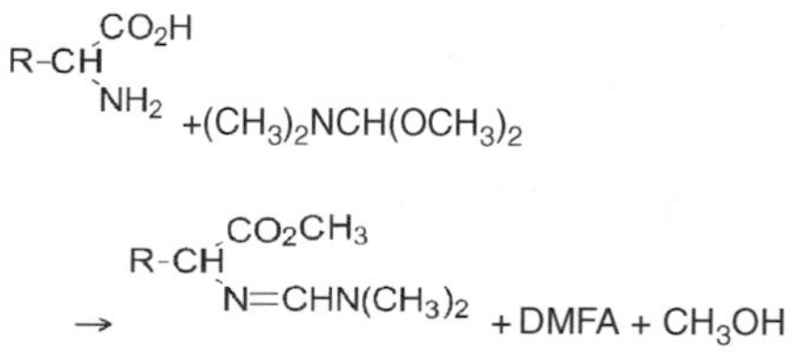

A "sandwich" technique (flash methylation) can also be used in this case. It implies the injection of the combined sample and reagent in the same syringe into the gas chromatograph, e.g., successively placed 1 mL of derivatization reagent, 1 μL of pyridine with internal standard, and 1 mL of the solution of analytes in the same solvent.

Alkyl chloroformates, $ClCO_2R$ (R = Me, Et, Bu), have been proposed as convenient alkylating reagents for carboxylic acids:[13]

$$RCO_2H + ClCO_2R' + B$$
$$\rightarrow RCO_2R' + CO_2 + BH^+Cl^-$$

Two-stage single-pot derivatization of carboxylic acids (with intermediate formation of chloroanhydrides with thionyl chloride followed by their conversion into amides) was recommended preferably for HPLC analysis, but the simplest dialkylamides and anilides[14] are volatile enough for GC analysis also (the mixture of Ph_3P and CCl_4 can be used in this reaction instead of $SOCl_2$). Moreover, the same procedure is used for the synthesis of diastereomeric derivatives of enantiomeric carboxylic acids (see below):

$$RCO_2H + SOCl_2 \rightarrow RCOCl + SO_2 + HCl$$
$$RCOCl + R'R''NH + B$$
$$\rightarrow RCONR'R'' + BH^+Cl^-$$

The reactivities of carboxy and hydroxy groups in the polyfunctional hydroxy- and phenol carboxylic acids are different. This indicates the possibility of an independent two-stage derivatization of these compounds, for example:

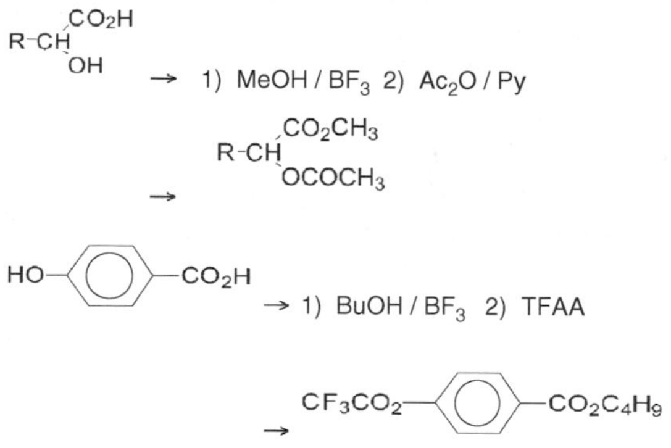

If these functional groups are located in vic (aliphatic series) or ortho positions (arenecarboxylic acids), methyl or butyl boronic acids are convenient reagents for their one-step derivatization with the formation of cyclic methyl(butyl) boronates:

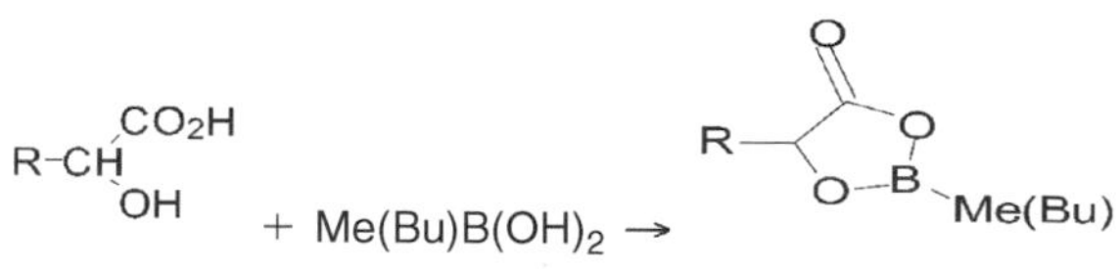

A similar method for simultaneous derivatization of two functional groups is the formation of cyclic silylene derivatives for the same types of compounds:[15]

A special type of carbonyl group derivatization is aimed at GC/MS determination of double bond (C=C) positions in unsaturated long-chain acids. The analytical derivatives for the solution of this problem are nitrogen-containing heterocycles. These compounds can be synthesized by high temperature condensation of carboxylic acids with 2-amino-2-methyl-1-propanol (2-substituted 4,4-dimethyloxazolines), 2-aminophenol (2-substituted benzoxazoles), and so forth.

Methyl esters of carboxylic acids form the same derivatives, but this also requires the heating of reaction mixtures up to 180 °C and, hence, seems inconvenient in analytical practice.[16]

GC separation of enantiomeric carboxylic acids on nonchiral phases is based on the formation of their esters or amides with optically active alcohols [for example, (–)-menthol, VII] or amines (α-methyl-benzenemethaneamine, VIII), usually through the intermediate chloroanhydrides. These diastereomeric products are not so volatile as other acid derivatives but, owing to the presence of two chiral carbon atoms (*) in the molecule, can be separated on nonchiral phases:

$R^*CO_2H \rightarrow [R^*COCl] +$ (VII) $\rightarrow$

$R^*CO_2H \rightarrow [R^*COCl] + C_6H_5C^*H(CH_3)NH_2$ (VIII)

$\rightarrow R^*CONHC^*H(CH_3)C_6H_5$

A problem closely related to the derivatization of free carboxylic acids is the determination of their composition in biogenic triglycerides, lipids, and so forth. The sample preparation includes the re-esterification

(preferably with formation of methyl esters) of these compounds in acid (MeOH/BF$_3$, MeOH/AcCl, etc.) or basic (MeONa, MeOH/KOH, etc.) media. Methyl esters of fatty acids are a group of compounds well characterized by both standard mass spectra and GC retention indices on standard phases. The combination of these analytical parameters provides their reliable identification.

The general method of GC analysis of anions of inorganic acids is their silylation. The values of retention indices on standard nonpolar phases (SE-30) are known for TMS derivatives of the most important among them:[3]

Anion	Volatile derivative for GC analysis	RI$_{nonpolar}$
Borate	B(OTMS)$_3$	1010
Carbonate	CO(OTMS)$_2$	1048
Phosphite	P(OTMS)$_3$	1115
Sulfate	SO$_2$(OTMS)$_2$	1148
Arsenite	As(OTMS)$_3$	1149
Phosphate	PO(OTMS)$_3$	1273
Vanadate	VO(OTMS)$_3$	1301
Arsenate	AsO(OTMS)$_3$	1353

CONCLUSIONS

Both strong inorganic and weak organic acids usually need derivatization prior to GC analysis. The existence of active hydrogen atoms in the molecules explains the significant contribution of ionic structures, which are responsible for the high polarity and low volatility of these substances.

Most universal methods of derivatization of acids are silylation (TMS and TBDMS) and alkylation (the simplest methyl esters are preferable). Other methods have an auxiliary predetermination and can be recommended for the solution of special analytical problems.

ARTICLES OF FURTHER INTEREST

Amines, Amino Acids, Amides, and Imides: Derivatization for GC Analysis, p. 57.
GC System Instrumentation, p. 682.
Hydroxy Compounds: Derivatization for GC Analysis, p. 809.

REFERENCES

1. Blau, K., King, G.S., Eds.; *Handbook of Derivatives for Chromatography*; Helden: London, 1977; 576.

2. Knapp, D.R. *Handbook of Analytical Derivatization Reactions*; John Wiley & Sons: New York, 1979; 741.

3. Drozd, J. *Chemical Derivatization in Gas Chromatography*; Journal of Chromatography Library; Elsevier: Amsterdam, 1981; Vol. 19, 232.

4. Blau, K., Halket, J.M., Eds. *Handbook of Derivatives for Chromatography;* 2nd Ed.; J. Wiley & Sons: New York, 1993; 369.

5. Wurth, C.; Kumps, A.; Mardens, Y. Urinary organic acids: retention indices on two capillary GC columns. J. Chromatogr. **1989**, *491*, 186–192.

6. Lefevere, M.F.; Verkaeghe, B.J.; Declerk, D.H.; Van Bocxlaer, J.F.; De Leenheer, A.P.; De Sagher, R.M. Metabolic profiling of urinary organic acids by single and multicolumn capillary gas chromatography. J. Chromatogr. Sci. **1989**, *27* (1), 23–29.

7. Rodriguez, I.; Quintana, J.B.; Carpinteiro, J.; Carro, A.M.; Lorenzo, R.A.; Cela, R. Determination of acidic drugs in sewage water by GC–MS as *tert.*-butyl dimethylsilyl derivatives. J. Chromatogr. A. **2003**, *985*, 265–274.

8. Crouholm, T.; Norsten, C. Gas chromatography–mass spectrometry of carboxylic acids in tissues as their *tert.*-butyl dimethylsilyl derivatives. J. Chromatogr. B. **1985**, *344*, 1–9.

9. Gonzalez, G.; Ventura, R.; Smith, A.K.; De la Torre, R.; Segura, J. Determination of non-steroidal anti-inflammatory drugs in equine plasma and urine by gas chromatography–mass spectrometry. J. Chromatogr. A. **1996**, *719*, 251–264.

10. Nilsson, T.; Baglio, D.; Galdo-Miquez, I.; Madsen, O.J.; Facchetti, S. Derivatization/ solid-phase microextraction followed by GC–MS for the analysis of phenoxy acid herbicides in aqueous samples. J. Chromatogr. A. **1998**, *826*, 211–216.

11. Gabelish, C.L.; Crisp, P.; Schneider, R.P. Simultaneous determination of chlorophenols, chlorobenzenes and chlorobenzoates in microbial solutions using pentafluorobenzyl bromide derivatization and analysis by GC with electron capture detection. J. Chromatogr. A. **1996**, *749*, 165–171.

12. Burke, D.G.; Halpern, B. Quaternary ammonium salts for butylation and mass spectral identification of volatile organic acids. Anal. Chem. **1983**, *55* (6), 822–826.

13. Butz, S.; Stan, H.-J. Determination of chlorophenoxy and other acidic herbicide residues in ground water by capillary GC of their alkyl esters formed by rapid derivatization using various chloroformates. J. Chromatogr. **1993**, *643*, 227–238.

14. Umeh, E.O. Separation and determination of low molecular weight straight chain C_1–C_8 carboxylic acids by gas chromatography of their anilide derivatives. J. Chromatogr. **1970**, *51*, 147–154.

15. Brooks, C.J.W.; Cole, W.T. Cyclic di-*tert.*-butylsilylene derivatives of substituted salicylic acids and related compounds. A study by gas chromatography–mass spectrometry. J. Chromatogr. **1988**, *441*, 13–29.

16. Fay, L.; Richli, U. Location of double bonds in polyunsaturated fatty acids by GC–MS after 4,4-dimethyloxazoline derivatization. J. Chromatogr. **1991**, *541*, 89–98.

A

Additives in Biopolymers: Analysis by Chromatographic Techniques

Roxana A. Ruseckaite
Research Institute of Material Science and Technology (INTEMA), University of Mar del Plata,
Mar del Plata, Argentina

Alfonso Jiménez
Department of Analytical Chemistry, University of Alicante, Alicante, Spain

INTRODUCTION

Biopolymers are naturally occurring polymers that are formed in nature during the growth cycles of all organisms; they are also referred to as natural polymers.[1] Their synthesis generally involves enzyme-catalyzed, chain growth polymerization reactions, typically performed within cells by metabolic processes.

Biodegradable polymers can be processed into useful plastic materials and used to supplement blends of the synthetic and microbial polymer.[2] Among the polysaccharides, cellulose and starch have been the most extensively used. Cellulose represents an appreciable fraction of the waste products. The main source of cellulose is wood, but it can also be obtained from agricultural resources. Cellulose is used worldwide in the paper industry, and as a raw material to prepare a large variety of cellulose derivatives. Among all the cellulose derivatives, esters and ethers are the most important, mainly cellulose acetate, which is the most abundantly produced cellulose ester. They are usually applied as films (packaging), fibers (textile fibers, cigarette filters), and plastic molding compounds. Citric esters (triethyl and acetyl triethyl acetate) were recently introduced as biodegradable plasticizers in order to improve the rheological response of cellulose acetate.[2]

Starch is an enormous source of biomass and most applications are based on this natural polymer. It has a semicrystalline structure in which their native granules are either destroyed or reorganized. Water and, recently, low-molecular-weight polyols,[2] are frequently used to produce thermoplastic starches. Starch can be directly used as a biodegradable plastic for film production because of the increasing prices and decreasing availability of conventional film-forming materials. Starch can be incorporated into plastics as thermoplastic starch or in its granular form. Recently, starch has been used in various formulations based on biodegradable synthetic polymers in order to obtain totally biodegradable materials. Thermoplastic and granular starch was blended with polycaprolactone (PCL),[3] polyvinyl alcohol and its co polymers, and polyhydroxyalcanoates (PHAs).[4] Many of these materials are commercially available, e.g., Ecostar (polyethylene/starch/unsaturated fatty acids), Mater Bi Z (PCL/starch/natural additives) and Mater Bi Y (polyvinylalchol-co-ethylene/starch/natural additives). Natural additives are mainly polyols.

The proteins, which have found many applications, are, for the most part, neither soluble nor fusible without degradation. Therefore, they are used in the form in which they are found in nature.[1] Gelatin, an animal protein, is a water-soluble and biodegradable polymer that is extensively used in industrial, pharmaceutical, and biomedical applications.[2] A method to develop flexible gelatin films is by adding polyglycerols. Quite recently, gelatin was blended with poly(vinyl alcohol) and sugar cane bagasse in order to obtain films that can undergo biodegradation in soil. The results demonstrated the potential use of such films as self-fertilizing mulches.[5]

Other kinds of natural polymers, which are produced by a wide variety of bacteria as intracellular reserve material, are receiving increasing scientific and industrial attention, for possible applications as melt processable polymers. The members of this family of thermoplastic biopolymers are the PHAs. Poly(3-hydroxy)butyrate (PHB), and poly(3-hydroxy) butyrate-hydroxyvalerate (PHBV) copolymers, which are microbial polyesters exhibiting useful mechanical properties, present the advantages of biodegradability and biocompatibility over other thermoplastics. Poly(3-hydroxy)butyrate has been blended with a variety of low- and high-cost polymers in order to apply PHB-based blends in packaging materials or agricultural foils. Blends with nonbiodegradable polymers, including PVAc, PVC, and PMMA, are reported in the literature.[4] Poly(3-hydroxy)butyrate has been also blended with synthetic biodegradable polyesters, such as poly(lactic acid) (PLA), poly(caprolactone), and

Encyclopedia of Chromatography DOI: 10.1081/E-ECHR-120018660

natural polymers including cellulose and starch.[2] Plasticizers are also included into the formulations in order to prevent degradation of the polymer during processing. Polyethylene glycol, oxypropylated glycerol, dibutylsebacate (DBS), dioctylsebacate (DOS), and polyisobutylene (PIB) are commonly used as PHB plasticizers.[6]

As was pointed out above, the processing and in-use biopolymer properties depend on the addition of other materials that provide a more convenient processing regime and stabilizing effects. Therefore the identification and further determination of these additives, as well as the separation from the biopolymer matrix, is necessary, and chromatographic techniques are a powerful tool to achieve this goal.

Many different compounds can be used as biopolymer additives, most of them are quite similar to those used in traditional polymer formulations. The use of various compounds as plasticizers, lubricants, and antioxidants has been recently reported.[7–9] Antioxidants are normally used to avoid, or at least minimize, oxidation reactions, which normally lead to degradation and general loss of desirable properties. Phenol derivatives are mostly used in polymers, but vitamin E and α-tocopherols are those most commonly found in biopolymer formulation.[10]

IDENTIFICATION AND DETERMINATION OF ADDITIVES IN BIOPOLYMERS

The modification and general improvement of properties caused by the addition of such compounds is a very interesting issue to be studied with a wide range of analytical techniques. Their identification and eventual determination is usually carried out by chromatographic techniques coupled to a variety of detection systems, most often MS. This powerful hyphenated technique, extensively used in many different analyses, combines the separation capabilities of chromatographic techniques with the potential use of MS to elucidate complicated structures and to identify many chemical compounds with low limits of detection and high sensitivities. The use of MS also permits the simultaneous detection and determination of several of those additives in a single analysis. This is especially valuable when only a small quantity of material is available, which is the usual case in some biopolymer formulations.

Some proposals have been recently reported to couple different chromatographic techniques with MS for the analysis of biopolymers and biocomposites, as well as additives used in such formulations. GC/MS was used in some particular determinations, but always with the need for complicated extraction procedures.

One example is the adaptation to biopolymers of a method for the simultaneous determination of diamines, polyamines, and aromatic amines in wines and other food samples.[11] While this method was successfully applied in such samples, it is not clear that its application to the determination of these additives in biopolymers will be easy, because of potential problems in the extraction of analytes prior to GC/MS. The proposed ion-pair extraction method is not always easily adaptable to solid samples. Therefore the potential application of this sensitive method to biopolymers is still under discussion.

SEC coupled to MS is the most successful chromatographic technique applied in the field of biopolymers. As is well known, SEC is a powerful analytical technique that allows separation of analytes based on their different molecular sizes. SEC is a common step in the separation and further purification of biopolymers, and the coupling with MS was firstly proposed for proteins and other biological samples.[12] One of the main drawbacks of traditional SEC, which was the limited range of molecular sizes to be measured, was recently overcome by the proposal of new columns with no limits in the molecular size of the species to be analyzed. This allows the possibility to separate and further analyze a large number of compounds, regardless of their chemical structures. The introduction of new packings and more stable columns allowed the development of high-performance size exclusion chromatography (HPSEC).

However, the on-line interfacing of HPSEC to MS for powerful detection is not as easy as in the case of conventional HPLC. A very promising possibility has been raised with the introduction of a new MS technique, which the authors named chemical reaction interface mass spectrometry (CRIMS).[13] This new approach permits the monitoring of any organic molecules, even the most complicated, after their derivatization and transformation to low-molecular-weight products, which are amenable to easy MS detection. By determination of some structural and compositional parameters, the CRIMS response is proportional to the amount of specific organic elements present in biopolymers. This method has been recently applied to the analysis of biopolymers of different chemical nature, such as polysaccharides and proteins;[14] its potential extension to other kinds of biopolymers is still under study.

SEC has been recently applied, with success, to the analysis of biopolymers derived from biomass, as it is used for the determination of molecular mass distributions of polymeric compounds in general, because of its short analysis time, high reproducibility, and accuracy.[15] This application of SEC has permitted the separation and further detection of polymeric and monomeric residues of biopolymers, as well as the estimation of the degree of polymerization and eventual uses of natural products as additives, not only in biocomposites, but in many industrial applications, e.g., food additives.

Another important development in the field of biopolymer analysis is the introduction of matrix-assisted laser desorption ionization (MALDI), which is a rather recent soft ionization technique that produces molecular ions of large organic molecules. In combination with time-of-flight (TOF) mass spectrometry, it was proposed as a valuable tool for the detection and characterization of biopolymers, such as proteins, peptides, and oligosaccharides, in many types of samples.[16] The use of these recently developed techniques has not decreased the use of chromatography in determinations of biopolymers. Some efforts on the adaptation of the separation abilities of HPLC to the high potential of MALDI-TOF for the sensitive determination of additives in biocomposites are currently being carried out.

In all these applications, the separation step is one of the most critical during the whole analytical process. Solid phase extraction (SPE) and capillary electrophoresis (CE) were also proposed for high-resolution and quantitative separations of analytes. Therefore it is likely that the use of chromatographic techniques in this area will be increased in the near future. The development of adequate interfaces for such hyphenated techniques is the most important problem to be solved by researchers in the field of biopolymer analysis.

A recent study of separation and determination of antioxidants in polymers showed the potential use of HPLC for the separation and isolation of tocopherols in polymers and biopolymers.[10] It was shown that although a large number of HPLC product peaks are formed, they corresponded to different stereoisomeric forms of only a small number of oxidative coupling products of tocopherol. The chromatographic parameters determined in this way, coupled to the study of spectral characteristics, allowed the complete identification of all antioxidants used in these polymers.

PYROLYSIS OF BIOPOLYMERS AND BIOCOMPOSITES

It is recognized that pyrolysis of biopolymers and biocomposites results in a large variety of primary and secondary products, such as carbon dioxide, methane, and other hydrocarbons. These low-molecular-mass products must be investigated to understand the behavior of biopolymers at high temperatures, under degradation conditions. All of these compounds are volatile and can be detected by GC[17] or HPLC[18]

analysis. In the first study, a special two-stage GC system was used for the analysis of flash-pyrolysis products. With this system, the pyrolysis was directly conducted in inert carrier gas. Two different columns coupled to an MS detector allowed the analysis of the resulting volatile products.

To obtain these results, it is usual to couple GC and MS. The pyrolysis products are first separated in the column and then immediately analyzed in the mass spectrometer. Therefore it is possible to obtain reliable and reproducible results in a single run with a relatively short time of analysis. Therefore high-resolution MS, in combination with pyrolysis and GC, is a unique approach to develop quantitative information in the analysis of biopolymers. Problems arising in high-resolution MS are the increased loss of sensitivity with increasing resolving power and, also, the decreased signal-to-noise ratio caused by the use of internal standards. In the case of biopolymers, it is usual to combine high-resolution MS with low-energy ionization modes, such as chemical ionization (CI) and field ionization (FI), in order to avoid high fragmentation, which could lead to information losses. Electron impact ionization (EI) at the normal ionizing voltage (70 eV) causes excessive fragmentation. Thus much information is lost by such MS detection, as many small additive fragments are not specific. Methods such as FI and CI are useful because of the difficulties arising from EI, such as the variation of fragmentation depending on instrumental conditions and the fact that only low-mass ions are observed. Soft ionization methods allow conservation of more information about structures and molecular identity. However, one problem with the soft ionization methods is the higher cost of instrumentation.

The identification of the degradation processes of additives in biopolymers was also studied by pyrolysis GC/MS (Pyr-GC/MS). However, direct additive analysis by flash-pyrolytic decomposition is usually not easy for this kind of sample. Therefore a prior separation of additives, or additive fragments contained in the polymer matrix, is usually necessary. A major advantage of Pyr-GC/MS is the nonrequirement of pretreatment of the sample. The fragments formed in this way are then separated in the gas chromatograph and detected with the mass spectrometer. Additive detection in biopolymers with Pyr-GC/MS is influenced by fragmentation, which is conditioned by the ionization mode, the concentration of the analyte, and the structures of the additive and biopolymer fragments. It is usual that polymer matrix fragments, at high concentrations, are superimposed on the additive fragments. Therefore it is necessary to filter additive fragments from the background of the biopolymer matrix to permit seeing a difference between them. The degree of fragmentation depends on the pyrolysis temperature. Thus Pyr-GC/MS is of limited use for additive analysis in thermally labile and low-volatility products, which give a high fragmentation. For the same reason, it is also necessary to perform pyrolysis at temperatures that are not too high.

The use of Pyr-GC/MS is still not common in the analysis of biopolymers and biocomposites because of the large quantity of parameters to be controlled for the development of a method. It is not easy, in a dynamic system, to transfer from a flow of inert gas (Pyr-GC) to vacuum conditions (MS). On the other hand, quantification is based on the fact that degradation is ion-specific, and that a given substance always produces the same fragments. This is not the case with biopolymer additives, especially in natural products, where fragmentation can proceed in several directions. This requires the use of internal standards and multiple measurements of each sample. Therefore a complete quantification requires considerable time and effort.

Despite all these drawbacks, the potential use of Pyr-GC/MS in biopolymer analysis is quite promising when considering the latest developments in instrumentation. There is a current tendency in analytical Pyr-GC/MS to preserve and detect higher-molecular-weight fragments. This led to developments in instrumentation, such as improvement of the direct transfer of high-molecular-weight and polar products to the ion source of the mass spectrometer, the measurement of these compounds over extended mass ranges, and the use of soft ionization conditions. In addition, the potential of Pyr-GC/MS has been greatly enhanced by the use of high-resolution capillary columns combined with computer-assisted techniques.

CONCLUSIONS

The application of a wide variety of chromatographic techniques to the analysis of additives in biopolymers is a current tendency in many research laboratories around the world. The increasing interest in the use of biopolymers in many technological applications will raise the research in this field in the future. Therefore, the potential of chromatography for separation, identification, and quantification will be very important for the development of reliable and reproducible analytical methods.

REFERENCES

1. Chandra, R.; Rustgi, R. Biodegradable polymers. Prog. Polym. Sci. **1998**, *23*, 1273–1335.

2. Amass, W.; Amass, A.; Tighe, B. A review of biodegradable polymers: uses, current developments in the synthesis and characterization of biodegradable polymers and recent advances in biodegradation studies. Polym. Int. **1998**, *47*, 89–144.

3. Ishiaku, U.S.; Pang, K.W.; Lee, W.S.; Mohd-Ishak, Z.A. Mechanical properties and enzymatic degradation of thermoplastic and granular sogo starch filled poly(epsilon-caprolactone). Eur. Polym. J. **2002**, *38*, 393–401.

4. Avella, M.; Matuscelli, E.; Raimo, M. Properties of blends and composites based on poly(3-hydroxy)butyrate (PHB) and poly-(3-hydroxybutyrate-hydroxyvalerate) (PHBV) copolymers. J. Mater. Sci. **2000**, *35*, 523–545.

5. Chiellini, E.; Cinelli, P.; Corti, A.; Kenawy, E.R. Composite films based on waste gelatin: thermal-mechanical properties and biodegradation testing. Polym. Degrad. Stab. **2001**, *73*, 549–555.

6. Savenkova, L.; Gercberga, Z.; Nikolaeva, V.; Dzene, A.; Bibers, I.; Kalnin, M. Mechanical properties and biodegradation characteristics of PHB-based films. Proc. Biochem. **2000**, *35*, 573–579.

7. Wang, F.C.Y. Polymer additive analysis by pyrolysis-gas chromatography I. Plasticizers. J. Chromatogr. A. **2000**, *883*, 199–210.

8. Wang, F.C.Y.; Buzanowski, W.C. Polymer additive analysis by pyrolysis-gas chromatography III. Lubricants. J. Chromatogr. A. **2000**, *891*, 313–324.

9. Wang, F.C.Y. Polymer additive analysis by pyrolysis-gas chromatography IV. Antioxidants. J. Chromatogr. A. **2000**, *891*, 325–336.

10. Al-Malaika, S.; Issenhuth, S.; Burdick, D. The antioxidant role of vitamin E in polymers. V. Separation of stereoisomers and characterization of other oxidation products of DL-α-tocopherol formed in polyolefins during melt processing. J. Anal. Appl. Pyrolysis **2001**, *73*, 491–503.

11. Fernandes, J.O.; Ferreira, M.A. Combined ion-pair extraction and gas chromatography-mass spectrometry for the simultaneous determination of diamines, polyamines and aromatic amines in Port wine and grape juice. J. Chromatogr. A. **2000**, *886*, 183–195.

12. Kriwacki, R.W.; Wu, J.; Tennant, L.; Wright, P.E.; Siuzdak, G. Probing protein structure using biochemical and biophysical methods—proteolysis, matrix-assisted laser desorption/ionization mass spectrometry, high-performance liquid chromatography and size-exclusion chromatography. J. Chromatogr. A. **1997**, *777*, 23–30.

A

13. Lecchi, P.; Abramson, F.P. Analysis of biopolymers by size exclusion chromatography–mass spectrometry. J. Chromatogr. A. **1998**, *828*, 509–513.

14. Lecchi, P.; Abramson, F.P. Size exclusion chromatography–chemical reaction interface mass spectrometry: a perfect match. Anal. Chem. **1999**, *71*, 2951–2955.

15. Papageorgiou, V.P.; Assimopoulou, A.N.; Kyriacou, G. Determination of naturally occurring hydroxynaphthoquinone polymers by size-exclusion chromatography. Chromatographia **2002**, *55*, 423–430.

16. Kaufmann, R. Matrix-assisted laser desorption ionization (MALDI) mass spectrometry: a novel analytical tool in molecular biology and biotechnology. J. Biotechnol. **1995**, *41*, 155–175.

17. Pouwels, A.D.; Eijkel, G.B.; Boon, J.J. Curie-point pyrolysis–capillary gas chromatography–high resolution mass spectrometry of microcrystalline cellulose. J. Anal. Appl. Pyrolysis **1989**, *14*, 237–280.

18. Radlein, A.G.; Grinshpun, A.; Piskorz, J.; Scott, D.S. On the presence of anhydro-oligosaccharides in the syrups from the fast pyrolysis of cellulose. J. Anal. Appl. Pyrolysis **1987**, *12*, 39–49.

Additives in Polymer Formulations: Analysis by Chromatographic Techniques

Roxana A. Ruseckaite
Research Institute of Material Science and Technology (INTEMA), and Chemistry Department, Sciences Faculty, University of Mar del Plata, Mar del Plata, Argentina

Alfonso Jiménez
Department of Analytical Chemistry, Nutrition and Food Sciences, University of Alicante, Alicante, Spain

INTRODUCTION

During the past 50 years, polymers have essentially changed human life, and the plastics industry has developed materials increasingly adapted to specific uses.[1] The processing, durability, and end-use response of plastic items result from an adequate and precise combination of the desired polymer and additives. Additives such as lubricants, plasticizers, antioxidants, light stabilizers, colorants and dyestuffs, antistatic agents, surfactants, and preservatives are all commonly encountered in various polymer formulations, including synthetic polymers, biopolymers, composites, and biocomposites.[2–9]

There are hundreds of chemical compounds that are currently used as additives in polymer formulations. Table 1 summarizes some of the most common among them. Of all additives used to modify polymer properties, plasticizers have gained industrial importance because they reduce melt viscosity, lower the viscosity modulus, and increase the flexibility and workability of the polymeric materials.[10] Lubricants increase the overall rate of processing or improve the release properties during processing and molding operations.[11] Antioxidants, in particular hindered phenols (Table 1), which can be natural, as in the case of α-tocopherol,[12] preserve polymer chemical and physical–mechanical properties both during processing and under use conditions.[13] Natural antioxidants are used to replace synthetic ones mainly in biopolymers and biocomposites to render them completely compatible with different biologically active environments (e.g., natural environments, human body). Light stabilizers have the ability to reduce photo-oxidation and protect polymers from UV damage. In general, they respond to the structure of hindered amines, and are known as hindered amine light stabilizers (HALSs).[5] One strategy to prevent bacterial attachment and proliferation on a polymer material is to kill the microorganisms in contact with the substratum, which has led to the investigation of polymeric materials that contain immobilized biocides.[14] Nowadays, and in response to the fact that polymers are currently exposed to outdoor environments, additives with the capacity of preventing or reducing biological attack, e.g., antifungals and biocides, are also included in polymer formulations.[15] In contrast, aromatic ketones are being used to modify stable polymers commonly used in the manufacture of packaging materials, such as polystyrene and poly (vinylchloride) (PVC), in an attempt to make them degradable in natural environments.[16] Flame retardants are a class of materials that are compounded into plastics to provide certain defined reactions during combustion.[17]

The identification and eventual determination of polymer additives is an important issue in many fields.[18] In the area of packaging materials, additive migration from materials in contact with food may have potential toxic effects in humans.[3] In biomedical applications, plasticizers present in polymers [e.g., diethylhexylphthalate (DEHP) in PVC] can readily leach into the liquids passing through them, particularly lipid-containing fluids, e.g., blood. There is great concern about the toxicity of DEHP, especially for risk groups such as patients on hemodialysis.[19] International regulations require that pharmaceutical, biomedical, cosmetic, and packaging materials should not interact physically or chemically with their contents or environments. Therefore, the possible release of polymer additives by plastic items should be monitored and minimized. Chromatographic techniques, either on their own or in conjunction with powerful analytical techniques, in particular mass spectrometry (MS), have shown great potential in the identification and eventual determination of many of these compounds. As pointed out above, it is evident that the development of new reliable and rapid analysis methods for polymer additives is a challenging task for several reasons: quality control; additive depletion/stability during processing and lifetime; litigation; migration studies; contamination; government regulations; and development of new materials.[19]

Encyclopedia of Chromatography DOI: 10.1081/E-ECHR-120040092

Table 1 Some common polymer additives

Additive	Chemical name	Acronym/commercial name
Lubricants	Stearic acid	SA
	Stearamide	STA
	N,N-9-Ethylenebisstearamide	EBS
Plasticizers	Di(2-ethylhexyl)phthalate	DEHP; DOP
	Dibutylphthalate	DBP
	Di(2-ethylhexyl)adipate	DOA
	Trioctyltrimellitate	TOTM
	Triethyl citrate	TEC
	Dipropylene glycol dibenzoate	Benzoflex 9-88®
Antioxidants	Di-t-butyl-p-cresol	Bisphenol A®
	Dioctadecyl-(3,5-di-t-butyl-4-hydroxybenzyl)phosphate	Irganox®1093
	2,2'-Methylene-bis-(4-methyl-6-t-butylphenol)	Cyanox 2246®
	Octadecyl-3-(3,5-di-t-butyl-4-hydroxyphenyl)propionate	Irganox®1076
	α-Tocopherol	Vitamin E
Light stabilizers	2-(2-Hydroxy-3,5-di-t-amylphenyl)-2H-benzotriazole	Tinuvin 328®
	2-Hydroxy-4-n-octoxybenzophenone	Chimassorb 81®
	Poly{6-[(1,1,3,3-tetramethylbutyl)-imino]-1,3,5-triazine-2,4-diyl-(2,2,6,6-tetramethylpiperidyl)-imino-hexamethylene-[4-(2,2,6,6-tetramethylpiperidyl)]-imino}	Chimassorb 944®
Organic flame retardants	Tetrabromobisphenol A	TBBA
	Hexabromocyclododecane	HBCD
	1,2,3,4,7,8,9,10,13,13,14,14-Dodecachloro-1,4,4α,5,6,6α,7,10,10α,11,12,12a-dodecahydro-1,4,7,10-dimethanodibenzo[α,ε]-cyclo-octene	Dechlorane Plus®

EXTRACTION AND SAMPLING METHODS

Techniques currently in use for the analysis of polymer additives have been extensively reviewed.[10–12,17,19–21] The most common analytical procedures concerning solid materials usually involve sample preparation and extraction procedures prior to the determination step, which is usually carried out by chromatographic techniques coupled to a variety of detection systems. For environmental reasons, direct examination of polymers, such as by spectroscopy and nondestructive methods, is preferred over solvent-consuming techniques.[20] Recently, Poleunis, Médard, and Bertrand[22] have reported the use of time-of-flight secondary ion mass spectrometry (ToF-SIMS) to analyze additive migration toward polymer surface in thin films of an amorphous polyester containing variable quantities of an antioxidant (Irgafos® 168) and a UV stabilizer (Hostavin N30®). The results obtained are promising, but the authors have stated that ToF-SIMS data can only be comparable quantitatively if the surfaces have undergone identical treatments.

In spite of these new environmentally friendly strategies, conventional extraction methods such as liquid/liquid extraction (LLE), liquid/solid extraction (LSE), in particular Soxhlet extraction, and sonication are being traditionally applied for releasing additives. More recently, novel analytical procedures, such as supercritical fluid extraction (SFE) and microwave-assisted extraction (MAE), have gained attention. Supercritical fluid extraction is commonly applied to analyses that require extraction from solid matrices and has found important applications in the extraction of aromatic amines, organotin stabilizers, light stabilizers, and antioxidants from different polymers.[23–26] In general, compared with Soxhlet extraction, SFE assures higher extraction efficiencies.[23,24] New and improved SFE-based methods are currently being developed. Supercritical fluid extraction coupled to supercritical fluid chromatography (SFE/SFC) is a promising method and is believed to have great potential in determining polymer additives.[26] In MAE, sample and solvent (or a mixture of solvents) are allocated in a vessel and microwave radiation is applied at adequate temperature. Samples are then recovered, filtered, and analyzed, generally by a chromatographic technique. This method has been successfully applied to extract commercial antioxidants (present in concentrations varying from 0.05 to 0.35% w/w) from pharmaceutical and cosmetic polyolefin-based packaging materials.[27] For insoluble or highly cross-linked polymers, these approaches are difficult to apply; then, thermoanalytical extraction techniques that liberate the additive by heating are needed. Some thermoanalytical techniques, such as thermogravimetric analysis (TGA) and temperature programmed pyrolysis (TPPy), take specific advantage of relatively slow heating, in particular in combination with appropriate detection modes, such as MS [e.g., TG/MS, TG/Fourier-transform infrared (TG/FTIR), TPPy/MS]. In such volatile removal techniques, the additives are usually detected at temperatures below the decomposition temperature of the polymer.[20,21] A lot of work is being devoted to improving thermoanalytical extraction techniques. As an example, temperature-programmed pyrolysis coupled with metastable mass spectrometry (TPPy/MAB-MS) has been used to analyze additives in polyurethane-based car paints.[9] By using a low-energy ionization source [MAB(N$_2$)], molecular ions of light stabilizers present at low concentration were identified, with low fragmentation patterns. This is not easy to achieve with medium or high-energy ionization techniques because fragmentation is greatly stimulated.

SEPARATION AND DETECTION

Among all chromatographic techniques that can be applied for the separation and identification of polymer additives after the extraction step, high-temperature capillary GC,[2,10–12,17,18,23,24,28] HPLC,[25] nano-LC,[29] SFC,[26] SEC,[30] capillary electrophoresis (CE), and TLC[31] are the most frequently used. In general, these techniques are coupled to different detection systems, such as MS, matrix-assisted laser desorption ionization (MALDI) in combination with time-of-flight (ToF) mass spectrometry (MALDI/ToF), and FTIR spectroscopy and microscopy.[31]

A recent study of separation and determination of antioxidants in polymers showed the potential use of LC for the separation and isolation of tocopherols in polymers and biopolymers.[13] It was shown that although a large number of LC peaks formed, they mainly corresponded to different stereoisomeric forms of only a small number of oxidative coupling products of tocopherol. The use of nano-LC[29] can offer advantages over classical LC and, among them, the higher efficiency and lower consumption of the mobile phase are very attractive conditions to increase sensitivity and reduce expenses for both solvents and their waste. On the other hand, this technique requires extraction procedures prior to analysis and, at the moment, the instruments and packed columns used are very expensive. An SFE commercial system with the addition of a single six-portion valve and coupled to a reversed-phase LC system to perform on-line SFE/LC analysis has been used for the quantitative determination of Irganox® 1010, Irganox® 1076, and Irgafos 168 in poly(methylmethacrylate) (PMMA).[25] This SFE/LC on-line method proved to be more accurate because the analytes could be extracted at relatively low temperatures in the absence of light and under anaerobic conditions.

GC combined with different extraction and detection techniques, such as Py/GC, Py/GC/MS, SFE/GC, and thermodesorption–cooled injection system (TDC/CIS)–GC/MS among others, has been widely applied to the analysis of most additives. The use of Py/GC and Py/GC/MS in the determination of lubricants, antioxidants, plasticizers, and flame retardants has been discussed in detail by Wang and coworkers. In the case of antioxidants, Py/GC seems to provide an approach that minimizes sample preparation followed by a one-step effective separation and identification.[12] Fast identification of low-molecular-weight antioxidants has been achieved using MS detection, which desorbs better before the polymer chain undergoes decomposition.[28] Lubricants, which can be classified into fatty acids, their esters and amide derivatives, and waxes with a high number of carbon atoms, can be identified by Py/GC.[11] The most important advantages of Py/GC in plasticizer analysis are the elimination of sample preparation and the fact that all information can be obtained in a single experiment. However, the critical step is the pre-separation of this kind of additive from the original polymer. Plasticizers can be resolved by their pyrolysate pattern. The analysis of organic flame retardants requires the use of MS detection due to the presence of halogen atoms in their structure. Py/GC/MS can effectively detect/identify halogenated flame retardants by the specific isotope ratios of chlorine and bromine.[17] Off-line SFE/GC/MS has been used to determine phenol, citrates, and phthalates in PVC with efficiencies of nearly 100%.[23,24]

SEC is usually applied to characterize complex mixtures of low-molecular-weight additives with molar masses between 150 and 1000 Da. This technique allows the classification of individual components in complex mixtures based on molecular mass or hydrodynamic volume, and not on boiling point (as in GC) or polarity (needed in HPLC). This technique provides reasonable separation for methylene bridged hindered phenols and alkylated diphenylamines.[30] In this way, a fingerprint chromatogram can be generated for mixtures of additives with a determined chemistry. Compared with HPLC, SEC shows high reproducibility and simplicity of the chromatogram, which makes the qualitative interpretation easier.

TLC remains one of the most widely used techniques for simple and rapid qualitative separation.[31] The combination of TLC with spectroscopic detection techniques, such as FTIR or nuclear magnetic resonance (NMR), is a very attractive approach in the analysis of polymer additives. Infrared microscopy is a powerful technique that combines the image capabilities of optical microscopy with the chemical analysis abilities of infrared spectroscopy. FTIR microscopy allows one to obtain infrared spectra from microsized samples. Off-line TLC/FTIR microscopy has been used to analyze a variety of commercial antioxidants and light stabilizers.[31] Transferring and identification by FTIR takes about 20 min. However, the main drawbacks of TLC/FTIR are that TLC is a time-consuming technique and usually needs solvent mixtures, which makes it environmentally unsound, analytes must be transferred for FTIR analysis, and TLC/FTIR cannot be used for quantifying purposes.

Microemulsion electrokinetic chromatography (MEEKC) is an electrophoretic method that can be classified as an extension of MEKC in which the micelles are substituted by oil droplets. The high solubilizing ability of MEEKC allows the analysis of highly hydrophobic and aromatic polymer additives, such as antioxidants. This technique has been applied to separate commercial antioxidants from polypropylene (PP).[32] Two different approaches were investigated. When an acidic buffer and a negative separation voltage were used, incomplete separation of the analytes resulted. However, by using an alkaline buffer joined to a positive separation voltage, the baseline separation of all analytes was achieved in less than 25 min. MEEKC is a superior separation method for these analytes with respect to separation time, selectivity, and efficiency.

Finally, it is important to mention that huge effort is being invested in the development of more powerful extraction/chromatographic technique/detection system combinations. In this sense, highly specific stationary phases are being designed by molecular imprinting (MI).[33] Molecular imprinting is known as a technique that allows the fabrication of artificial receptors with recognition properties. Compounds of similar structure, like a group of β-lactam antibiotics, can be separated based on the application of molecularly imprinted polymers (MIPs) as highly specific stationary phases in HPLC or capillary electrochromatography. Synthetic antioxidants, such as butylated hydroxyanisole (BHA) and butylated hydroxytoluene (BHT), have been selected as templates for generating an MIP (BHA-MIP and BHT-MIP), as well as a control polymer (CP), in the absence of any template. BHA-MIP was successfully applied as stationary phase in liquid chromatography and showed a higher affinity than its CP. This method is still under study for real applications.

The increase in the number of published works in this area during the last four years is a clear sign of the importance of the methods based on chromatography for the qualitative and quantitative analysis of polymer additives. This trend is expected to continue in the next years.

REFERENCES

1. Chiellini, E. *Biodegradable Polymers and Plastics*; Vert, M., Feijen, J., Albertsson, A.C., Scott, G., Chiellini, E., Eds.; Royal Society of Chemistry: Cambridge, UK, 1992.
2. Balafas, D.; Shaw, K.J.; Whitfield, F.B. Phthalate and adipate esters in Australian packaging materials. Food Chem. **1999**, *65*, 279–287.
3. Lau, O.W.; Wong, S.K. Contamination in food from packaging material. J. Chromatogr. A. **2000**, *882*, 225–270.
4. Jana, T.; Roy, B.C.; Maiti, S. Biodegradable film 7 modification of the biodegradable film for fire retardancy. Polym. Degrad. Stabil. **2000**, *69*, 79–82.
5. Peña, J.M.; Allen, N.S.; Edge, M.; Liauw, C.M.; Valange, B. Studies of synergism between carbon black and stabilizers in LDPE photodegradation. Polym. Degrad. Stabil. **2001**, *72*, 259–270.
6. Choi, J.S.; Park, W.H. Effect of biodegradable plasticizers on thermal and mechanical properties of poly(3-hydroxybutyrate). Polym. Test. **2003**, *23*, 455–460.
7. Lucena, M.C.C.; De Alencar, A.E.V.; Mazzeto, S.E.; Da Soares, S. The effect of additives on the thermal degradation of cellulose acetate. Polym. Degrad. Stabil. **2003**, *80*, 149–155.
8. Harper, D.; Wolcott, M. Interaction between coupling agent and lubricants in wood–polypropylene composites. Composites A. **2004**, *35*, 385–394.
9. Boutin, M.; Lesage, J.; Ostiguy, C.; Bertrand, M.J. Temperature-programmed pyrolysis hyphenated with metastable atom bombardment ionization mass spectrometry (TPPy/MAB–MS) for the identification of additives in polymers. J. Am. Soc. Mass Spectrom. **2004**, *15*, 1310–1314.
10. Wang, F.C.Y. Polymer additive analysis by pyrolysis–gas chromatography. I. Plasticizers. J. Chromatogr. A. **2000**, *883*, 199–210.
11. Wang, F.C.Y.; Buzanowski, W.C. Polymer additive analysis by pyrolysis–gas chromatography. III. Lubricants. J. Chromatogr. A. **2000**, *891*, 313–324.
12. Al-Malaika, S.; Issenhuth, S.; Burdick, D. The antioxidant role of vitamin E in polymers. V. Separation of stereoisomers and characterisation of other oxidation products of dl-α-tocopherol formed in polyolefins during melt processing. J. Anal. Appl. Pyrolysis **2001**, *73*, 491–503.

13. Wang, F.C.Y. Polymer additive analysis by pyrolysis–gas chromatography. IV. Antioxidants. J. Chromatogr. A. **2000**, *891*, 325–336.

14. Grapski, J.A.; Cooper, S.L. Synthesis and characterization of non-leaching biocidal polyurethanes. Biomaterials **2001**, *15*, 2239–2246.

15. US registration brings global status for Avecias's new plastic antimicrobial. Plastic Additive & Compounding, July/August **2003**, 8.

16. Kaczmarek, H.; Swicatek, M.; Kaminska, A. Modification of polystyrene and poly (vinyl chloride) for the purpose of obtaining packaging materials degradable in the natural environment. Polym. Degrad. Stabil. **2004**, *83*, 35–45.

17. Wang, F.C.Y. Polymer additive analysis by pyrolysis–gas chromatography. II. Flame retardants. J. Chromatogr. A. **2000**, *886*, 225–235.

18. Bart, J.C.J. Polymer/additive analysis at the limits. Polym. Degrad. Stabil. **2003**, *82*, 197–205.

19. Wahl, H.G.; Hoffmann, A.; Häring, H.U.; Liebich, H.M. Identification of plasticizers in medical products by a combined direct thermo-desorption–cooled injection and gas chromatography–mass spectrometry. J. Chromatogr. A. **1999**, *847*, 1–7.

20. Bart, J.C.J. Direct solid sampling methods for gas chromatographic analysis of polymer/additive formulations. Polym. Test. **2001**, *20*, 729–740.

21. Bart, J.C.J. Polymer additive analysis by flash pyrolysis techniques. J. Anal. Appl. Pyrolysis **2001**, *58–59*, 3–28.

22. Poleunis, C.; Médard, N.; Bertrand, P. Additive quantification on polymer thin films by ToF–SIMS: aging sample effects. Appl. Surface Sci. **2004**, *231–232*, 269–273.

23. Garrigós, M.C.; Reche, F.; Pernías, K.; Jiménez, A. Optimization of parameters for the analysis of aromatic amines in finger-paints. J. Chromatogr. A. **2000**, *896*, 291–298.

24. Guerra, R.M.; Marín, M.L.; Sánchez, A.; Jiménez, A. Analysis of citrates and benzoates used in poly (vinyl chloride) by supercritical fluid extraction and gas chromatography. J. Chromatogr. A. **2002**, *950*, 31–39.

25. Ashraf-Khorassani, M.; Nazem, N.; Taylor, L.T. Feasibility of supercritical fluid extraction with on-line coupling reverse-phase liquid chromatography for quantitative analysis of polymer additives. J. Chromatogr. A. **2003**, *995*, 227–232.

26. Zhou, L.Y.; Ashraf-Khorassani, M.; Taylor, L.T. Comparison of methods for quantitative analysis of additives in low-density polyethylene using supercritical fluid and enhanced solvent extraction. J. Chromatogr. A. **1999**, *858*, 209–218.

27. Marcato, B.; Guerra, S.; Vianello, M.; Scalia, S. Migration of antioxidant additives from various polyolefinic plastics into oleaginous vehicles. Int. J. Pharm. **2003**, *257*, 217–225.

28. Herrera, M.; Matuschek, G.; Kettrup, A. Fast identification of polymer additives by pyrolysis–gas chromatography/mass spectrometry. J. Anal. Appl. Pyrolysis **2003**, *70*, 35–42.

29. Fanali, S.; Camera, E.; Chankvetadze, B.; D'Orazio, G.; Quaglia, M.G. Separation of tocopherols by nano-liquid chromatography. J. Pharm. Biomed. Anal. **2004**, *35*, 331–337.

30. Greene, S.V.; Gatto, V.J. Size-exclusion chromatography method for characterizing low-molecular-mass antioxidant lubricant additives. J. Chromatogr. A. **1999**, *841*, 45–54.

31. He, W.; Shanks, R.; Amarasinghe, G. Analysis of additives in polymers by thin-layer chromatography coupled with Fourier transform-infrared microscopy. Vib. Spectrosc. **2002**, *30*, 147–156.

32. Hilder, E.F.; Klampfl, C.W.; Buchberger, W.; Haddad, P.R. Separation of hydrophobic polymer additives by microemulsion electrokinetic chromatography. J. Chromatogr. A. **2001**, *922*, 293–302.

33. Brüggemanna, O.; Visnjevski, A.; Burch, R.; Patel, P. Selective extraction of antioxidants with molecularly imprinted polymers. Anal. Chim. Acta **2004**, *504*, 81–88.

Adhesion of Colloids on Solid Surfaces by Field-Flow Fractionation

George Karaiskakis
University of Patras, Patras, Greece

INTRODUCTION

The adhesion of colloids on solid surfaces, which is of great significance in filtration, corrosion, detergency, coatings, and so forth, depends on the total potential energy of interaction between the colloidal particles and the solid surfaces. The latter, which is the sum of the attraction potential energy and that of repulsion, depends on particle size, the Hamaker constant, the surface potential, and the Debye–Huckel reciprocal distance, which is immediately related to the ionic strength of carrier solution. With the aid of the field-flow fractionation (FFF) technique, the adhesion and detachment processes of colloidal materials on and from solid surfaces can be studied. As model samples for the adhesion of colloids on solid surfaces (e.g., Hastelloy-C), hematite (a-Fe$_2$O$_3$) and titanium dioxide (TiO$_2$) submicron spherical particles, as well as hydroxyapatite [Ca$_5$(PO$_4$)$_3$OH] submicron irregular particles were used. The experimental conditions favoring the adhesion process were those decreasing the surface potential of the particles through the pH and ionic strength variation, as well as increasing the effective Hamaker constant between the particles and the solid surfaces through the surface-tension variation. On the other hand, the detachment of the same colloids from the solid surfaces can be favored under the experimental conditions decreasing the potential energy of attraction and increasing the repulsion potential energy.

METHODOLOGY

FFF technology is applicable to the characterization and separation of particulate species and macromolecules. Separations in FFF take place in an open flow channel over which a field is applied perpendicular to the flow. Among the various FFF subtechniques, depending on the kind of the applied external fields, sedimentation FFF (SdFFF) is the most versatile and accurate, as it is based on simple physical phenomena that can be accurately described mathematically. SdFFF, which uses a centrifugal gravitational force field, is a flow-modified equilibrium sedimentation-separation method. Solute layers that are poorly resolved under static equilibrium sedimentation become well separated as they are eluted by the laminar flow profile in the SdFFF channel. In normal SdFFF, where the colloidal particles under study do not interact with the channel wall, the potential energy of a spherical particle, $\varphi(x)$, is related to the particle radius, a, to the density difference, $\Delta\rho$, between the particle (ρ_s) and the liquid phase (ρ), and to the sedimentation field strength expressed in acceleration, G:

$$\varphi(x) = \frac{4}{3}\pi a^3 \Delta\rho G x \qquad (1)$$

where x is the coordinate position of the center of particle mass.

When the colloidal particles interact with the SdFFF channel wall, the total potential energy, φ_{tot}, of a spherical particle is given by

$$\begin{aligned}
\varphi_{\text{tot}} = {} & \frac{4}{3}\pi a^3 \Delta\rho G x \\
& + \frac{A_{132}}{6}\left[\ln\left(\frac{h+2a}{h}\right) - \frac{2a(h+a)}{h(h+a)}\right] \\
& + 16\varepsilon a\left(\frac{kT}{e}\right)^2 \tan h\left(\frac{e\psi_1}{4kT}\right)\tan h\left(\frac{e\psi_2}{4kT}\right)e^{-\kappa x}
\end{aligned}$$

$$(2)$$

where the second and third terms of Eq. (2) accounts for the contribution of the van der Waals attraction potential and of the double-layer repulsion potential between the particle and the wall, respectively, A_{132} is the effective Hamaker constant for media 1 and 2 interacing across medium 3, h is the separation distance between the sphere and the channel wall, e is the dielectric constant of the suspending medium, e is the electronic charge, ψ_1 and ψ_2 are the surface potentials of the particles and the solid wall, respectively, k is Boltzmann's constant, T is the absolute temperature, and κ is the Debye–Huckel reciprocal length, which

Encyclopedia of Chromatography DOI: 10.1081/E-ECHR-120039868

is immediately related to the ionic strength, I, of the medium.

Eq. (2) shows that the total potential energy of interaction between a colloidal particle and a solid substrate is a function of the particle radius and surface potential, the ionic strength and dielectric constant of the suspending medium, the value of the effective Hamaker constant, and the temperature. Adhesion of colloidal particles on solid surfaces is increased by a decrease in the particle radius, surface potential, the dielectric constant of the medium and by an increase in the effective Hamaker constant, the ionic strength of the dispersing liquid, or the temperature. For a given particle and a medium with a known dielectric constant, the adhesion and detachment processes depend on the following three parameters:

1. The surface potential of the particles, which can be varied experimentally by various quantities one of which is the suspension pH
2. The ionic strength of the solution, which can be varied by adding to the suspension various amounts of an indifferent electrolyte
3. The Hamaker constant, which can be easily varied by adding to the suspending medium various amounts of a detergent. The later results in a variation of the medium surface tension.

APPLICATIONS

The critical electrolyte (KNO_3) concentrations found by SdFFF for the adhesion of a-Fe_2O_3(I) (with nominal diameter 0.148 μm), a-Fe_2O_3(II) (with nominal diameter 0.248 μm), and TiO_2 (with nominal diameter 0.298 μm) monodisperse spherical particles on the Hastelloy-C channel wall were 8×10^{-2} M, 3×10^{-2} M, and 3×10^{-2} M, respectively. The values for the same sample (a-Fe_2O_3) depend on the particle size, in accordance with the theoretical predictions, whereas the same values are identical for various samples [a-Fe_2O_3(II) and TiO_2] having different particle diameters. The latter indicates that these values depend also, apart from the size, on the sample's physico-chemical properties, as is predicted by Eq. (2). The detachment of the whole number of particles of the above samples from the channel wall was succeeded by decreasing the ionic strength of the carrier solution.

The critical KNO_3 concentration for the detachment process was 3×10^{-2} M for the a-Fe_2O_3(I) sample and 1×10^{-2} M for the samples of a-Fe_2O_3(II) and TiO_2. Those obtained by SdFFF particle diameters after the detachment of the adherent particles [0.148 μm for a-Fe_2O_3(I), 0.245 μm for a-Fe_2O_3(II), and 0.302 μm for TiO_2] are in excellent agreement with the corresponding nominal particle diameters obtained by transmission electron microscopy. The desorption of all of the adherent particles was verified by the fact that no elution peak was obtained, even when the field strength was reduced to zero. A second indication for the desorption of all of the adherent material was that the sample peaks after adsorption and desorption emerge intact and without degradation.

In a second series of experiments, the adhesion and detachment processes of hydroxyapatite (HAP) poly-disperse particles with number average diameter of 0.261 μm on and from the Hastelloy-C channel wall were succeeded by the variation of the suspension pH, whereas the medium's ionic strength was kept constant (10^{-3} M KNO_3). At a suspension pH of 6.8, the whole number of injected HAP particles was adhered at the beginning of the SdFFF channel wall, which was totally released when the pH increased to 9.7, showing that, except for the ionic strength, the pH of the suspending medium is also a principal quantity influencing the interaction energy between colloidal particles and solid surfaces. The number-average diameter of the HAP particles found by SdFFF after the detachment of the adherent particles ($d_N = 0.262$ μm) was also in good agreement with that obtained when the particles were injected into the channel with a carrier solution in which no adhesion occurs ($d_N = 0.261$ μm).

The variation of the potential energy of interaction between colloidal particles and solid surfaces can be also succeeded by the addition of a detergent to the suspending medium, which leads to a decrease in the Hamaker constant and, consequently, in the potential energy of attraction.

In conclusion, FFF is a relatively simple technique for the study of adhesion and detachment of sub-micrometer or supramicrometer colloidal particles on and from solid surfaces.

FUTURE DEVELOPMENTS

Looking to the future, it is reasonable to expect more experimental and theoretical work in order to quantitatively investigate the adhesion/detachment phenomena of colloids on and from solid surfaces by measuring the corresponding rate constants with the aid of FFF.

SUGGESTED FURTHER READING

Athanasopoulou, A.; Karaiskakis, G. Potential barrier gravitational field-flow fractionation based on the variation of the pH solution for the analysis of colloidal materials. Chromatographia **1996**, *43*, 369.

Giddings, J.C.; Myers, M.N.; Caldwell, K.D.; Fisher, S.R. *Methods of Biochemical Analysis*; Glick, D., Ed.; John Wiley & Sons: New York, 1980; Vol. 26, 79.

Giddings, J.C.; Karaiskakis, G.; Caldwell, K.D.; Myers, M.N. J. Colloid Interf. Sci. **1983**, *92* (1), 66.

Hansen, M.E.; Giddings, J.C. Retention perturbations due to particle–wall interactions in sedimentation field-flow fractionation. Anal. Chem. **1989**, *61*, 811–819.

Hiemenz, P.C. *Principles of Colloid and Surface Chemistry*; Marcel Dekker, Inc.: New York, 1977.

Karaiskakis, G.; Cazes, J. J. Liq. Chromatogr. & Rel. Technol. **1997**, *20* (16 & 17).

Karaiskakis, G.; Athanasopoulou, A.; Koliadima, A. Adhesion studies of colloidal materials on solid surfaces by field-flow fractionation. J. Micro. Separ. **1997**, *9*, 275.

Koliadima, A.; Karaiskakis, G. Potential-barrier field-flow fractionation, a versatile new separation method. J. Chromatogr. **1990**, *517*, 345–359.

Ruckenstein, E.; Prieve, D.C. Adsorption and desorption of particles and their chromatographic separation. AIChE J. **1976**, *22* (2), 276.

Adsorption Chromatography

Robert J. Hurtubise
University of Wyoming, Laramie, Wyoming, U.S.A.

INTRODUCTION

In essence, the original chromatographic technique was adsorption chromatography. It is frequently referred to as liquid–solid chromatography. Tswett developed the technique around 1900 and demonstrated its use by separating plant pigments. Open-column chromatography is a classical form of this type of chromatography, and the open-bed version is called TLC.

Adsorption chromatography is one of the more popular modern HPLC techniques today. However, open-column chromatography and TLC are still widely used.[1] The adsorbents (stationary phases) used are silica, alumina, and carbon. Although some bonded phases have been considered to come under adsorption chromatography, these bonded phases will not be discussed. By far, silica and alumina are more widely used than carbon. The mobile phases employed are less polar than the stationary phases, and they usually consist of a signal or binary solvent system. However, ternary and quaternary solvent combinations have been used.

Adsorption chromatography has been employed to separate a very wide range of samples. Most organic samples are readily handled by this form of chromatography. However, very polar samples and ionic samples usually do not give very good separation results. Nevertheless, some highly polar multifunctional compounds can be separated by adsorption chromatography. Compounds and materials that are not very soluble in water or water–organic solvents are usually more effectively separated by adsorption chromatography compared to reversed-phase liquid chromatography.

When one has an interest in the separation of different types of compound, silica or alumina, with the appropriate mobile phase, can readily accomplish this. Also, isomer separation frequently can easily be accomplished with adsorption chromatography; for example, 5,6-benzoquinoline can be separated from 7,8-benzoquinoline with silica as the stationary phase and 2-propanol : hexane (1 : 99). This separation is difficult with reversed-phase liquid chromatography.[1]

STATIONARY PHASES

Silica is the most widely used stationary phase in adsorption chromatography.[2] However, the extensive

work of Snyder[3] involved investigations with both silica and alumina. Much of Snyder's earlier work was with alumina. Even though the surface structures of the two adsorbents have distinct differences, they are sufficiently similar. Thus, many of the fundamental principles developed for alumina are applicable to silica. The general elution order for these two adsorbents is as follows:[1] saturated hydrocarbons (small retention time), olefins, aromatic hydrocarbons, aromatic hydrocarbons ≈ organic halides, sulfides, ethers, nitrocompounds, esters ≈ aldehydes ≈ ketones, alcohols ≈ amines, sulfones, sulfoxide, amides, carboxylic acids (long retention time). There are several reasons why silica is more widely used than alumina. Some of these are that a higher sample loading is permitted, fewer unwanted reactions occur during separation, and a wider range of chromatographic forms of silica are available.

Chromatographic silicas are amorphous and porous and they can be prepared in a wide range of surface areas and average pore diameters. The hydroxyl groups in silica are attached to silicon, and the hydroxyl groups are mainly either free or hydrogen-bonded. To understand some of the details of the chromatographic processes with silica, it is necessary to have a good understanding of the different types of hydroxyl groups in the adsorbent.[1,3] Chromatographic alumina is usually γ-alumina. Three specific adsorption sites are found in alumina: (a) acidic, (b) basic, and (c) electronacceptor sites. It is difficult to state specifically the exact nature of the adsorption sites. However, it has been postulated that the adsorption sites are exposed aluminum atoms, strained Al–O bonds, or cationic sites.[4] Table 1 gives some of the properties of silica and alumina.

The adsorbent water content is particularly important in adsorption chromatography. Without the deactivation of strong adsorption sites with water, nonreproducible retention times will be obtained, or irreversible adsorption of solutes can occur. Prior to using an adsorbent for open-column chromatography, the adsorbent is dried, a specified amount of water is added to the adsorbent, and then the adsorbent is allowed to stand for 8–16 h to permit the equilibration of water.[3–4] If one is using a high-performance column, it is a good idea to consider adding water to the mobile phase to deactivate the stronger adsorption sites on the adsorbent. Some of the benefits are less

Encyclopedia of Chromatography DOI: 10.1081/E-ECHR-120039869

Table 1 Some adsorbents used in adsorption chromatography

Type	Name	Form	Average particle area size (μm)	Surface area (m^2/g)
Silica[a]	BioSil A	Bulk	2–10	400
	μPorasil	Column	10	400+
Silica[b]	Hypersil	Bulk	5–7	200
	Zobax Sil	Bulk or column	6	350
Alumina[a]	ICN Al-N	Bulk	3–7, 7–12	200
	MicroPak Al	Bulk or column	5, 10	79
Alumina[b]	Spherisorb AY	—	5, 10, 20	95

[a]Irregular
[b]Spherical
(Adapted from Ref.[1].)

variation in retention times, partial compensation for lot-to-lot differences in the adsorbent, and reduced band tailing.[1] However, there can be some problems in adding water to the mobile phase, such as how much water to add to the mobile phase for optimum performance. Snyder and Kirkland [1] have discussed several of these aspects in detail.

MOBILE PHASES

To vary sample retention, it is necessary to change the mobile-phase composition. Thus, the mobile phase plays a major role in adsorption chromatography. In fact, the mobile phase can give tremendous changes in sample retention characteristics. Solvent strength controls the capacity factor's values of all the sample bands. A solvent strength parameter (ε^0), which has been widely used over the years, can be employed quantitatively for silica and alumina. The solvent strength parameter is defined as the adsorption energy of the solvent on the adsorbent per unit area of solvent.[1,3] Table 2 gives the solvent strength values for selected solvents that have been used in adsorption chromatography. The smaller values of ε^0 indicate

Table 2 Selected solvents used in adsorption chromatography

	Solvent strength (ε_0)	
	Silica	**Alumina**
Solvent		
n-Hexane	0.01	0.01
1-Chlorobutane	0.20	0.26
Chloroform	0.26	0.40
Isopropyl ether	0.34	0.28
Ethyl acetate	0.38	0.58
Tetrahydrofuran	0.44	0.57
Acetonitrile	0.50	0.65

(Adapted from Ref.[1].)

weaker solvents, whereas the larger values of ε^0 indicate stronger solvents. The solvents listed in Table 2 are single solvents. Normally, solvents are selected by mixing two solvents with large differences in their ε^0 values, which would permit a continuous change in the solvent strength of the binary solvent mixture. Thus, some specific combination of the two solvents would provide the appropriate solvent strength. In adsorption chromatography, the solvent strength increases with solvent polarity, and the solvent strength is used to obtain the proper capacity factor values, usually in the range of 1–5 or 1–10. It should be realized that the solvent strength does not vary linearly over a wide range of solvent compositions, and several guidelines and equations that allow one to calculate the solvent strength of binary solvents have been developed for acquiring the correct solvent strength in adsorption chromatography.[1,3] However, it frequently happens that the solvent strength is such that all of the solutes are not separated in a sample. Thus, one needs to consider solvent selectivity, which is discussed below.

To change the solvent selectivity, the solvent strength is held constant and the composition of the mobile phase is varied. It should be realized that because the solvent strength is directly related to the polarity of the solvent and polarity is the total of the dispersion, dipole, hydrogen-bonding, and dielectric interactions of the sample and solvent, one would not expect that solvent strength alone could be used to fine-tune a separation. A trial-and-error approach can be employed by using different solvents of equal ε^0. However, there are some guidelines that have been developed that permit improved selectivity. These are the "B-concentration" rule and the "hydrogen-bonding" rule.[1] In general, with the B-concentration rule, the largest change in selectivity is obtained when a very dilute or a very concentrated solution of B (stronger solvent) in a weak solvent (A) is used. The hydrogen-bonding rule states that any change in the mobile phase that results in a change in hydrogen-bonding between sample and mobile-phase molecules usually results in a large change in selectivity. A more comprehensive means for improving selectivity is the solvent-selectivity triangle.[1,5] The solvent-selectivity triangle classifies solvents according to their relative dipole moments, basic properties, and acidic properties. For example, if an initial chromatographic experiment does not separate all the components with a binary mobile phase, then the solvent-selectivity triangle can be used to choose another solvent for the binary system that has properties that are very different than one of the solvents in the original solvent system. A useful publication that discusses the properties of numerous solvents and also considers many chromatographic applications is Ref.[6].

MECHANISTIC ASPECTS IN ADSORPTION CHROMATOGRAPHY

Models for the interactions of solutes in adsorption chromatography have been discussed extensively in the literature.[7–9] Only the interactions with silica and alumina will be considered here. However, various modifications to the models for the previous two adsorbents have been applied to modern high-performance columns (e.g., amino-silica and cyano-silica). The interactions in adsorption chromatography can be very complex. The model that has emerged which describes many of the interactions is the displacement model developed by Snyder.[1,3–4,7–8] Generally, retention is assumed to occur by a displacement process. For example, an adsorbing solute molecule X displaces n molecules of previously adsorbed mobile-phase molecules M:[8]

$$X_n + nM_a \rightleftharpoons X_a + nM_n$$

The subscripts n and a in the above equation represent a molecule in a nonsorbed and adsorbed phase, respectively. In other words, retention in adsorption chromatography involves a competition between sample and solvent molecules for sites on the adsorbent surface. A variety of interaction energies are involved, and the various energy terms have been described in the literature.[7–8] One fundamental equation that has been derived from the displacement model is

$$\log\left(\frac{k_1}{k_2}\right) = \alpha' A_S(\varepsilon_2 - \varepsilon_1)$$

where k_1 and k_2 are the capacity factors of a solute in two different mobile phases, α' is the surface activity of the adsorbent (relative to a standard adsorbent), A_S is the cross-sectional area of the solute on the adsorbent surface, ε_1 and ε_2 are the solvent strengths of the two different mobile phases. This equation is valid in situations where the solute and solvent molecules are considered nonlocalizing. This condition is fulfilled with nonpolar or moderately polar solutes and mobile phases. If one is dealing with multisolvent mobile phases, the solvent strength of those solvents can be related to the solvent strengths of the pure solvents in the solvent system. The equations for calculating solvents strengths for multisolvent mobile phases have been discussed in the literature.[8]

As the polarities of the solute and solvent molecules increase, the interactions of these molecules become much stronger with the adsorbent, and they adsorb with localization. The net result is that the fundamental equation for adsorption chromatography with relatively nonpolar solutes and solvents has to be modified. Several localization effects have been elucidated, and the modified equations that take these factors into consideration are rather complex.[7–8,10] Nevertheless, the equations provide a very important framework in understanding the complexities of adsorption chromatography and in selecting mobile phases and stationary phases for the separation of solutes.

APPLICATIONS

There have been thousands of articles published on the application of adsorption chromatography over the decades. Today, adsorption chromatography is used around the world in all areas of chemistry, environmental problem solving, medical research, and so forth. Only a few examples will be discussed in this section. Gogou et al.[11] developed methods for the determination of organic molecular markers in marine aerosols and sediment. They used a one-step flash chromatography compoundclass fractionation method to isolate compound-class fractions. Then, they employed GC/MS and/or GC/flame ionization detection analysis of the fractions. The key adsorption chromatographic step prior to the GC was the one-step flash chromatography. For example, an organic extract of marine aerosol or sediment was applied on the top of a 30 × 0.7-cm column containing 1.5 g of silica. The following solvent systems were used to elute the different compound classes: (a) 15 mL of n-hexane (aliphatics); (b) 15 mL toluene : n-hexane (5.6 : 9.4) (polycyclic aromatic hydrocarbons and nitropolycyclic aromatic hydrocarbons); (c) 15 mL n-hexane : methylene chloride (7.5 : 7.5) (carbonyl compounds); (d) 20 mL ethyl acetate : n-hexane (8 : 12) (n-alkanols and sterols); (e) 20 mL (4%, v/v) pure formic acid in methanol (free fatty acids). This example illustrates very well how adsorption chromatography can be used for compound-class separation.

Hanson and Unger[12] have discussed the application of nonporous silica particles in HPLC. Nonporous silica packings can be used for the rapid chromatographic analysis of biomolecules because the particles lack pore diffusion and have very effective mass-transfer capabilities. Several of the advantages of nonporous silica are maximum surface accessibility, controlled topography of ligands, better preservation of biological activity caused by shorter residence times on the column, fast column regeneration, less solvent consumption, and less susceptibility to compression during packing. The very low external surface area of the nonporous supports is a disadvantage because it gives considerably lower capacity compared with porous materials. This drawback is counterbalanced partially by the high packing density compared to porous silica. The smooth surface of the nonporous silica offers better biocompatibility relative to porous silica. Well-defined nonporous silicas are now commercially available.

REFERENCES

1. Snyder, L.R.; Kirkland, J.J. *Introduction to Modern Liquid Chromatography*, 2nd ed.; John Wiley & Sons: New York, 1979.
2. Knox, J.H., Ed.; *High-Performance Liquid Chromatography*; Edinburgh University Press: Edinburgh, 1980.
3. Snyder, L.R. *Principles of Adsorption Chromatography*; Marcel Dekker, Inc.: New York, 1968.
4. Snyder, L.R. *Chromatography: A Laboratory Handbook of Chromatographic and Electrophoretic Methods*, 3rd ed.; Heftmann, E., Ed.; Van Nostrand Reinhold: New York, 1975; 46–76.
5. Snyder, L.R.; Glajch, J.L.; Kirkland, J.J. *Practical HPLC Method Development*; John Wiley & Sons: New York, 1988; 36–39.
6. Sadek, P.C. *The HPLC Solvent Guide*; John Wiley & Sons: New York, 1996.
7. Snyder, L.R.; Poppe, H. Mechanism of solute retention in liquidsolid chromatography and the role of the mobile phase in affecting separation: competition versus "sorption". J. Chromatogr. **1980**, *184*, 363.
8. Snyder, L.R. *High-Performance Liquid Chromatography*; Horvath, C., Ed.; Academic Press: New York, 1983; Vol. 3, 157–223.
9. Scott, R.P.W.; Kucera, P. Liquid chromatography theory. J. Chromatogr. **1979**, *171*, 37.
10. Snyder, L.R.; Glajch, J.L. Solven strength of multicomponent mobile phases in liquidsolid chromatograph: further study of different mobile phases and silica as adsorbent. J. Chromatogr. **1982**, *248*, 165.
11. Gogou, A.I.; Apostolaki, M.; Stephanou, E.G. J. Adsorption chromatography. Chromatogr. A. **1998**, *799*, 215.
12. Hanson, M.; Unger, K.K. Adsorption chromatography. LC-GC **1997**, *15*, 364.

Advances in Chiral Pollutants: Analysis by Capillary Electrophoresis

Imran Ali
National Institute of Hydrology, Roorkee, India

V. K. Gupta
Department of Chemistry, Indian Institute of Technology, Roorkee, India

Hassan Y. Aboul-Enein
*Pharmaceutical Analysis Laboratory, Biological and Medical Research Department,
King Faisal Specialist Hospital and Research Center, Riyadh, Saudi Arabia*

INTRODUCTION

At present, about 60,000 organic substances are used by human beings and, presumably, some of these compounds are toxic and contaminate our environment. Some of the pesticides, phenols, plasticizers, and polynuclear aromatic hydrocarbons are chiral toxic pollutants. About 25% of agrochemicals are chiral and are sold as their mixtures. Recently, it has been observed that one of the two enantiomers of the chiral pollutant/xenobiotic may be more toxic than the other enantiomer.[1] This is an important information to the environmental chemist when performing environmental analysis, as the data of simple, direct analysis do not distinguish which enantiomeric structure of a certain pollutant is present and which is harmful. Biological transformation of the chiral pollutants can be stereoselective; thus uptake, metabolism, and excretion of enantiomers may be very different.[1] Therefore the enantiomeric composition of the chiral pollutants may be changed in these processes. Metabolites of the chiral pollutants are often chiral. Thus to obtain information on the toxicity and biotransformation of the chiral pollutants, it is essential to develop a suitable method for the analysis of the chiral pollutants. Therefore diverse groups of people, ranging from the regulators to the materials industries, clinicians and nutritional experts, agricultural scientists, and environmentalists are asking for data on the ratio of the enantiomers of the chiral pollutants. Chromatographic modalities, e.g., GC and HPLC, have been used for the chiral analysis of the pollutants. The high polarity, low vapor pressure, and the need for derivatization of some environmental pollutants make the GC method complicated. The inherent limited resolving power, complex procedures involved in the optimization of the chiral resolution of the pollutant, and the use of large amounts of solvents and sample volumes are the main drawbacks of HPLC. Conversely, capillary electrophoresis (CE), a versatile technique of high speed and sensitivity, is a major trend in analytical science; some publications on the chiral analysis of pollutants have appeared in recent years. The high efficiency of CE is due to the flat flow profile originated and to a homogeneous partition of the chiral selector in the electrolyte which, in turn, minimizes the mass transfer. Recently, Ali et al.[2] reviewed the chiral analysis of the environmental pollutants by CE. Therefore in this article, attempts have been made to explain the art of the enantiomeric resolution of the chiral environmental pollutants by CE.

CHIRAL SELECTORS

As in the case of chromatography, a chiral selector is also required in CE for enantiomeric resolution. Generally, suitable chiral compounds are used in the background electrolyte (BGE) as additives and hence are called chiral selectors or chiral BGE additives. There are only a few publications available that deal with the chiral resolution on a capillary coated with the chiral selector in CE.[3] The analysis of the chiral pollutants discussed in this chapter is restricted only to using chiral selectors in the BGE. The most commonly used chiral BGE additives are cyclodextrins, macrocyclic glycopeptide antibiotics, proteins, crown ethers, ligand exchangers, and alkaloids.[4,5] A list of these chiral BGE additives is presented in Table 1.

APPLICATIONS

CE has been used for the analysis of chiral pollutants, e.g., pesticides, polynuclear-aromatic hydrocarbons, amines, carbonyl compounds, surfactants, dyes, and other toxic compounds. Moreover, CE has also been

Encyclopedia of Chromatography DOI: 10.1081/E-ECHR-120027335

utilized to separate the structural isomers of various toxic pollutants such as phenols, polyaromatic hydrocarbons, etc. Sarac et al.[11] resolved the enantiomers of 2-hydrazino-2-methyl-3-(3,4-dihydroxyphenyl) propionic acid using cyclodextrin as the BGE additive. The cyclodextrins used were native, neutral, and ionic in nature with phosphate buffer as BGE. Weseloh et al.[12] investigated the CE method for the separation of biphenyls, using a phosphate buffer BGE with cyclodextrin as the chiral additive. Miura et al.[13] used CE for the chiral resolution of seven phenoxy acid herbicides using methylated cyclodextrins as the BGE additives. Furthermore, the same group[14] resolved MCPP, DCPP, 2,4-D, 2,4-CPPA, 2,4,5-T, 2,3-CPPA, 2,2-CPPA, 2-PPA, and silvex pesticides using cyclodextrins, with negatively charged sulfonyl groups, as the chiral BGE additives. Gomez-Gomar et al.[15] investigated the simultaneous enantioselective separation of (±)-cizolirtine and its impurities, (±)-N-desmethylcizolirtine, (±)-cizolirtine-N-oxide, and (±)-5-(-hydroxybenzyl)-1-methylpyrazole, by CE. Otsuka et al.[16] described the latest advancement by coupling capillary electrophoresis with mass spectrometry; this setup was used for the chiral analysis of phenoxy acid herbicides. The authors also described an electrospray ionization (ESI) method for the CE–MS interface. Generally, nonvolatile additives in sample solutions sometimes decrease the MS sensitivity and/or signal intensity. However, heptakis(2,3,6-tri-O-methyl)-β-cyclodextrin (TM-β-CD) was used as a chiral selector; it migrated directly into the ESI interface. Using the negative-ionization mode, along with a methanol–water–formic acid solution as a sheath liquid, and nitrogen as a sheath gas, stereoselective resolution and detection of three phenoxy acid herbicide enantiomers was successfully achieved with a 20-mM TM-β-CD in a 50-mM ammonium acetate buffer (pH 4.6).[17] Zerbinati et al.[18] resolved the four enantiomers of the herbicides mecoprop and dichlorprop using an ethylcarbonate derivative of β-CD with three substituents per molecule of hydroxypropyl-β-CD and native β-CD.

Table 1 Some of the most commonly used chiral selectors

Chiral selectors (chiral BGE additives)	Refs.
Cyclodextrins	[5,6]
Macrocyclic glycopeptide antibiotics	[6]
Proteins	[6,7]
Crown ethers	[6,8]
Alkaloids	[6]
Polysaccharides	[6,9]
Calixarenes	[6,9]
Imprinted polymers	[10]
Ligand exchangers	[10]

The performances of these chiral selectors have been quantified by means of two-level full factorial designs and the inclusion constants were calculated from CE migration time data. The analysis of the chiral pollutants by CE is summarized in Table 2. To show the nature of the electropherograms, the chiral separation of dichlorprop enantiomers is shown in Fig. 1 with different concentrations of α-cyclodextrin.[18]

OPTIMIZATION OF CE CONDITIONS

The analysis of the chiral pollutants by CE is very sensitive and hence is controlled by a number of experimental parameters. The optimization parameters may be categorized into two classes, i.e., the independent and dependent parameters. The independent parameters are under the direct control of the operator. These parameters include the choice of the buffer, pH of the buffer, ionic strength of the buffer, type of chiral selectors, voltage applied, temperature of the capillary, dimension of the capillary, BGE additives, and various other parameters. Conversely, the dependent parameters are those directly affected by the independent parameters and are not under the direct control of the operator. These types of parameters are field strength (V/m), EOF, Joule heating, BGE viscosity, sample diffusion, sample mobility, sample charge, sample size and shape, sample interaction with capillary and BGE, molar absorptivity, etc. Therefore the optimization of chiral resolution can be controlled by varying all of the parameters mentioned above. For detailed information on the optimization of chiral analysis, one should consult our review.[2] However, a protocol for the optimization of the chiral analysis is given in Scheme 1.

DETECTION

Normally, the chiral pollutants in the environment occur at low concentrations and therefore a sensitive detection method is essential and is required in chiral CE. The most commonly used detectors in the chiral CE are UV, electrochemical, fluorescence, and mass spectrometry. Mostly, the detection of the chiral resolution of drugs and pharmaceutical in CE has been achieved by a UV mode[13,27] and therefore the detection of the chiral pollutants may be achieved by the same method. The selection of the UV wavelength depends on the type of buffer, chiral selector, and the nature of the environmental pollutants. The concentration and sensitivity of UV detection are restricted insofar as the capillary diameter limits the optical path length. It has been observed that some pollutants,

A

Table 2 Chiral analysis of some pollutants by CE

Chiral pollutants	Sample matrix	Electrolytes	Detection	Refs.
Fenoprop, mecoprop, and dichlorprop	—	20 mM tributyl-β-CD in 50 mM, ammonium acetate, pH 4.6	MS	[16]
Dichlorprop and 2-(2,4-dichlorophenoxy) propionic acid		100 mM acetic acid–sodium acetate buffer (pH 5.0) containing α-, β-, and γ-CDs	UV 206 nm	[18]
2-Phenoxypropionic acid, dichloroprop, fenoxaprop, fluaziprop, haloxyfop, and diclofop enantiomers	—	75 mM Britton–Robinson buffer with 6 mM Vancomycin	—	[19]
Imazaquin isomer	—	50 mM sodium acetate, 10 mM dimethyl-β-CD, pH 4.6	—	[20]
Phenoxy acid herbicides	—	200 mM sodium phosphate, pH 6.5 with various concentrations of OG	—	[20]
Diclofop		50 mM sodium acetate, 10 mM trimethyl-β-CD, pH 3.6	—	[20]
Imazamethabenz isomers		50 mM sodium acetate, 10 mM dimethyl-β-CD, pH 4.6	—	[20]
2-(2-Methyl-4-chlorophenoxy) propionic acid	—	0.05 M lithium acetate containing α-cyclodextrins	UV 200 nm	[21]
2-(2-Methyl-4,6-dichlorophenoxy) propionic acid		0.05 M lithium acetate containing β-cyclodextrin	UV 200 nm	[21]
2-(2,4-Dichlorophenoxy) propionic acid		0.05 M lithium acetate containing heptakis-(2,6-di-O-methyl)-β-cyclodextrin	UV 200 nm	[21]
2-(2-Methyl-4-chlorophenoxy) propionic acid and 2-(2,4-dichlorophenoxy)-propionic acid		0.03 M lithium acetate containing heptakis-(2,6-di-O-methyl)-β-cyclodextrin	UV 200 nm	[22]
2(2-Methyl-chlorophenoxy) propionic acid	—	0.05 M NaOAc, pH 4.5, with α-CD	UV 230	[23]
Phenoxy acid	—	0.1 M phosphate buffer, pH 6 with		
		Vancomycin	—	[24]
		Ristocetin	—	[24,25]
		Teicoplanin	—	[24,26]
		0.1 M phosphate and acetate buffer containing OM	—	[27]
		OG	—	[28]
Phenoxy acid derivatives	—	β-CD and TM-β-CD	—	[23,29]
Silvex	—	0.4 M borate, pH 10 containing N,N-bis-(D-gluconamidopropyl)-cholamide (Big CHAP) and -deoxycholamide (Deoxy big CHAP)	—	[30]

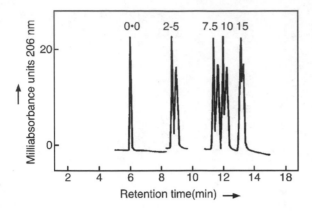

Fig. 1 Electropherograms of dichlorprop herbicide enantiomers with increasing concentrations of α-cyclodextrin. (From Ref.[18].)

especially organochloro pesticides, are UV transparent and therefore for such type of applications, electrochemical and mass spectrometry are the best detectors. Some of the chiral selectors, such as proteins and macrocyclic glycopeptide antibiotics, are UV-absorbing

in nature and hence the detection of enantiomers becomes poor.

Only a few reports are available in the literature dealing with the limits of the detection for the chiral resolution of environmental pollutants by CE, indicating mg/L to µg/L as the limits of the detection. Tsunoi et al.[14] carried out an extensive study on the determination of the limits of the detection for the chiral resolution of herbicides. The authors used a 230-nm wavelength for the detection and the minimum limit of the detection achieved was 4.7×10^{-3} M for 2,4-dichlrophenoxy acetic acid. On the other hand, Mechref and El Rassi[29] reported better detection limits, for herbicides, in the derivatized mode, in comparison to the underivatized mode. For example, the limit of the detection was enhanced by almost 1 order of magnitude from 1×10^{-4} M (10 pmol) to 3×10^{-5} M (0.36 pmol). In the same study, the authors reported 2.5×10^{-6} M and 1×10^{-9} M as the limits of detection for the herbicides by fluorescence and laser-induced fluorescence detectors, respectively.

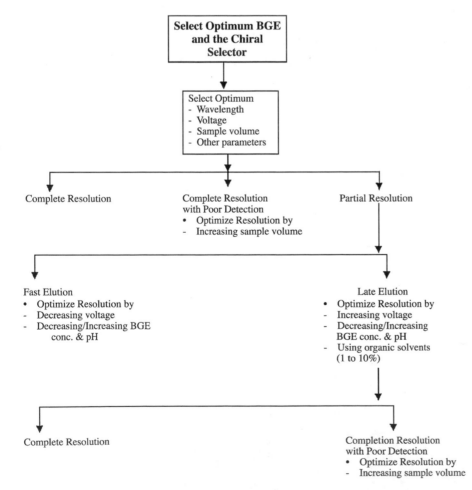

Scheme 1 The protocol for the development and optimization of CE conditions for the chiral resolution.

SAMPLE PREPARATION

Many of the impurities are present in samples of environmental or biological origin. Therefore sample pretreatment is very important and a necessary step for reproducible chiral resolution. Real samples often require the application of simple procedures, such as filtration, extraction, dilution, etc. A search of the literature conducted and discussed herein (Table 2) indicates that all of the chiral resolution of the environmental pollutants was carried out, by CE, in laboratory-synthesized samples only. Therefore no report is published on the sample pretreatment prior to the chiral resolution of the environmental pollutants by CE. Some reviews have been published, however, on the pretreatment and sample preparation methodologies for the achiral analysis of pollutants.[31,32] Therefore these approaches may be utilized for the preconcentration and sample preparation in the chiral CE of the environmental pollutants. Dabek-Zlotorzynska et al.[32] reviewed the sample pretreatment methodologies for environmental analysis before CE. Moreover, some reviews have also been published in the last few years on this issue.[33–35] Whang and Pawliszyn[36] designed an interface that enables the solid-phase microextraction (SPME) fiber hyphenation to CE. They prepared a semi-custom-made polyacrylate fiber to reach the SPME–CE interface. The authors tested the developed interface to analyze phenols in water and therefore the same may be used for the chiral resolution of the pollutants.

MECHANISMS OF THE CHIRAL SEPARATION

It is well known that a chiral environment is essential for the enantiomeric resolution of racemates. In CE, this situation is provided by the chiral compounds used in the BGE and is known as the chiral selector or chiral BGE additive. Basically, the chiral recognition mechanisms in CE are similar to those in chromatography using a chiral mobile-phase additive mode, except that the resolution occurred through different migration velocities of the diastereoisomeric complexes in CE. The chiral resolution occurred through diastereomeric complex formation between the enantiomers of the pollutants and the chiral selector. The formation of diastereomeric complexes depends on the type and nature of the chiral selectors used and the nature of the pollutants.

In the case of cyclodextrins, the inclusion complexes are formed and the formation of diastereomeric complexes is controlled by a number of interactions, such as π–π complexation, hydrogen bonding, dipole–dipole interaction, ionic binding, and steric effects. Zerbinati et al.[17] used ethylcarbonate-β-CD, hydroxypropyl-β-CD, and native α-CD for the chiral resolution of mecoprop and dichlorprop. The authors calculated the performance of these chiral selectors by means of a two-level full factorial design and calculated inclusion constants from CE migration time data. Furthermore, they have proposed the possible structure of inclusion complexes on the basis of molecular mechanics simulations. Recently, Chankvetadze et al.[37] explained the chiral recognition mechanisms in cyclodextrin-based CE using UV, NMR, and electrospray ionization mass spectrometric methods. Furthermore, the authors determined the structures of the diastereomeric complexes by an X-ray crystallographic method.

The macrocyclic antibiotics have some similarities and differences with the cyclodextrins. Most of the macrocyclic antibiotics contain ionizable groups and, consequently, their charge and possibly their three-dimensional conformation can vary with the pH of the BGE. The complex structures of the antibiotics containing different chiral centers, inclusion cavities, aromatic rings, sugar moieties, and several hydrogen donor and acceptor sites are responsible for their surprising chiral selectivities. This allows for an excellent potential to resolve a greater variety of racemates. The possible interactions involved in the formation of diastereomeric complexes are π–π complexation, hydrogen bonding, inclusion complexation, dipole interactions, steric interactions, and anionic and cationic binding. Similarly, the diastereomeric complexes are formed with other chiral selectors involving specific interactions. In this way, the diastereomeric complexes possessing different physical and chemical properties are separated on the capillary path (achiral phase). The different migration times of these formed diastereomeric complexes depend on their sizes, charges, and their interaction with the capillary wall and, as a result, these complexes are eluted at different time intervals.

CAPILLARY ELECTROPHORESIS VS. CHROMATOGRAPHY

Today, chromatographic modalities are used frequently for the analysis of chiral pollutants. The wide application of HPLC is due to the development of various chiral stationary phases and excellent reproducibility. However, HPLC suffers from certain drawbacks, as the chiral selectors are fixed on the stationary phase and hence no variation in the concentrations of the chiral selectors can be carried out. Moreover, a large amount of the costly solvent is consumed to establish the chiral resolution procedure. Additionally, the poor efficiency in HPLC is due to the profile of the laminar flow, mass transfer term, and possible additional interactions of enantiomers with the residual silanol groups of the stationary phase. GC also suffers from certain drawbacks as discussed in the "Introduction."

On the other hand, the chiral resolution in CE is achieved using the chiral selectors in the BGE. The chiral separation in CE is very fast and sensitive, involving the use of inexpensive buffers. In addition, the high efficiency of CE is due to the flat profile created and to a homogeneous partition of the chiral selector in the electrolyte which, in turn, minimizes the mass transfer. Generally, the theoretical plate number in CE is much higher in comparison to chromatography and thus a good resolution is achieved in CE. In addition, more than one chiral selector can be used simultaneously for optimizing the chiral analysis. However, reproducibility is the major problem in CE and therefore the technique is not popular for the routine chiral analysis. The other drawbacks of CE include the waste of the chiral selector as it is used in the BGE. In addition, chiroptical detectors, such as polarimetric and circular dichroism, cannot be used as detection devices because of the presence of the chiral selector in the BGE. Moreover, some of the well-known chiral selectors may not be soluble in the BGE and thus a stationary bed of a chiral selector may allow the transfer of the advantages of a stationary bed inherent in HPLC to electrically driven technique, i.e., CE. This will allow CE to be hyphenated with the mass spectrometer, polarimeter, circular dichroism, and UV detectors without any problem. Briefly, at present, CE is not a very popular technique as is chromatography for the chiral analysis of pollutants, but it will gain momentum in the near future.

CONCLUSIONS

Analysis of the chiral pollutants at trace levels is a very important and demanding field. In recent years, CE has been gaining importance in the direction of chiral analysis of various racemates. A search of the literature cited herein indicates a few reports on the chiral resolution of environmental pollutants by CE. It has not achieved a respectable place in the routine chiral analysis of these pollutants due to its poor reproducibility and to the limitations of detection. Therefore many scientists have suggested various modifications to make CE a method of choice. To achieve good reproducibility, the selection of the capillary wall chemistry, pH and ionic strength of the BGE, chiral selectors, detectors, and optimization of BGE have been described and suggested for the analysis of organic and inorganic pollutants.[38–43] In addition, some other aspects should also be addressed so that CE can be used as a routine method in this field. The most important

points related to this include the development of new and better chiral selectors, detector devices, and addition of a cooling device in the CE apparatus. In addition, chiral capillaries should be developed and the CE device should be hyphenated with mass spectrometer, polarimetric, and circular dichroism detectors, which may result in good reproducibility and improved limits of detection. The advancement of CE as a chiral analysis technique has not yet been fully explored and research in this direction is currently underway. In summary, there is much to be developed for the advancement of CE for the analysis of chiral pollutants. It is hoped that CE will be recognized as the technique of choice for chiral analysis of the environmental pollutants.

ABBREVIATIONS

BGE	Background electrolyte
CD	Cyclodextrin
CE	Capillary electrophoresis
2,2-CPPA	2-(2-Chlorophenoxy) propionic acid
2,3-CPPA	2-(3-Chlorophenoxy) propionic acid
2,4-CPPA	2-(4-Chlorophenoxy) propionic acid
2,4-D	(2,4-Dichlorophenoxy) acetic acid
DCPP	2-(2,4-Dichlorophenoxy) propionic acid
EOF	Electroosmotic flow
ESI	Electron spray ionization
GC	Gas chromatography
HPLC	High-performance liquid chromatography
MCPP	2-(4-Chlorophenoxy) propionic acid
MS	Mass spectrometer
NMR	Nuclear magnetic resonance
OG	n-Octyl-β-D-glucopyranoside
OM	n-Octyl-β-D-maltopyranoside
SPME	Solid-phase microextraction
SPME-CE	Solid-phase microextraction capillary electrophoresis
2,4,5-T	(2,4,5-Trichlorophenoxy) acetic acid
TM-β-CD	2,3,6-Tri-O-methyl-β-cyclodextrin
UV	Ultraviolet

REFERENCES

1. Ariens, E.J., van Rensen, J.J.S., Welling, W., Eds.; *Stereoselectivity of Pesticides, Biological and Chemical Problems: Chemicals in Agriculture*; Elsevier: Amsterdam, 1988; Vol. 1.
2. Ali, I.; Gupta, V.K.; Aboul-Enein, H.Y. Chiral resolution of the environmental pollutants by capillary electrophoresis. Electrophoresis **2003**, *24*, 1360–1374.

3. Jung, M.; Mayer, S.; Schurig, V. Enantiomer separations by GC, SFC and CE on immobilized polysiloxane bonded cyclodextrins. LC GC **1994**, *7*, 340–347.

4. Blaschke, G.; Chankvetadze, B. Enantiomer separation of drugs by capillary electromigration techniques. J. Chromatogr. A. **2000**, *875*, 3–25.

5. Zaugg, S.; Thormann, W. Enantioselective determination of drugs in body fluids by capillary electrophoresis. J. Chromatogr. A. **2000**, *875*, 27–41.

6. Chankvetadze, B. *Capillary Electrophoresis in Chiral Analysis*; John Wiley & Sons: New York, 1997.

7. Haginaka, J. Enantiomer separation of drugs by capillary electrophoresis using proteins as chiral selectors. J. Chromatogr. A. **2000**, *875*, 235–254.

8. Tanaka, Y.; Otsuka, K.; Terabe, S. Separation of enantiomers by capillary electrophoresis-mass spectrometry employing a partial filling technique with a chiral crown ether. J. Chromatogr. A. **2000**, *875*, 323–330.

9. Aboul-Enein, H.Y., Wainer, I.W., Eds.; *The Impact of Stereochemistry on Drug Development and Use*; John Wiley & Sons: New York, 1997; Vol. 142.

10. Gübitz, G.; Schmid, M.G. Chiral separation principles in capillary electrophoresis. J. Chromatogr. A. **1997**, *792*, 179–225.

11. Sarac, S.; Chankvetadze, B.; Blaschke, G. Enantioseparation of 3,4-dihydroxyphenylalanine and 2-hydrazino-2-methyl-3-(3,4-dihydroxyphenyl)propanoic acid by capillary electrophoresis using cyclodextrins. J. Chromatogr. A. **2000**, *875*, 379–387.

12. Welseloh, G.; Wolf, C.; König, W.A. New technique for the determination of interconversion processes based on capillary zone electrophoresis: studies with axially chiral biphenyls. Chirality **1996**, *8*, 441–445.

13. Miura, M.; Terashita, Y.; Funazo, K.; Tanaka, M. Separation of phenoxy acid herbicides and their enantiomers in the presence of selectively methylated cyclodextrin derivatives by capillary zone electrophoresis. J. Chromatogr. A. **1999**, *846*, 359–367.

14. Tsunoi, S.; Harino, H.; Miura, M.; Eguchi, M.; Tanaka, M. Separation of phenoxy acid herbicides by capillary electrophoresis using a mixture of hexakis(2,3-di-*O*-methyl)- and sulfopropyl-ether-α-cyclodextrins. Anal. Sci. **2000**, *16*, 991–993.

15. Gomez-Gomar, A.; Ortega, E.; Calvet, C.; Merce, R.; Frigola, J. Simultaneous separation of the enantiomers of cizolirtine and its degradation products by capillary electrophoresis. J. Chromatogr. **2002**, *950*, 257–270.

16. Otsuka, K.; Smith, J.S.; Grainger, J.; Barr, J.R.; Patterson, D.G., Jr.; Tanaka, N.; Terabe, S. Stereoselective separation and detection of phenoxy acid herbicide enantiomers by cyclodextrin-modified capillary zone electrophoresis-electrospray ionization mass spectrometry. J. Chromatogr. A. **1998**, *817*, 75–81.

17. Zerbinati, O.; Trotta, F.; Giovannoli, C. Optimization of the cyclodextrin-assisted capillary electrophoresis separation of the enantiomers of phenoxyacid herbicides. J. Chromatogr. A. **2000**, *875*, 423–430.

18. Zerbinati, O.; Trotta, F.; Giovannoli, C.; Baggiani, C.; Giraudi, G.; Vanni, A. New derivatives of cyclodextrins as chiral selectors for the capillary electrophoretic separation of dichlorprop enantiomers. J. Chromatogr. A. **1998**, *810*, 193–200.

19. Desiderio, C.; Polcaro, C.M.; Padiglioni, P.; Fanali, S. Enantiomeric separation of acidic herbicides by capillary electrophoresis using vancomycin as chiral selector. J. Chromatogr. A. **1997**, *781*, 503–513.

20. Penmetsa, K.V.; Leidy, R.B.; Shea, D. Enantiomeric and isomeric separation of herbicides using cyclodextrin-modified capillary zone electrophoresis. J. Chromatogr. A. **1997**, *790*, 225–234.

21. Nielen, M.W.F. (Enantio-)separation of phenoxy acid herbicides using capillary zone electrophoresis. J. Chromatogr. A. **1993**, *637*, 81–90.

22. Nielen, M.W.F. LIMS: A report on the 7th International LIMS Conference held at Egham, UK, 8–11 June, 1993. Trends Anal. Chem. **1993**, *12*, 345–356.

23. Garrison, A.W.; Schmitt, P.; Kettrup, A. Separation of phenoxy acid herbicides and their enantiomers by high-performance capillary electrophoresis. J. Chromatogr. A. **1994**, *688*, 317–327.

24. Gasper, M.P.; Berthod, A.; Nair, U.B.; Armstrong, D.W. Comparison and modeling of vancomycin, ristocetin A and teicoplanin for CE enantioseparations. Anal. Chem. **1996**, *68*, 2501–2514.

25. Armstrong, D.W.; Gasper, M.P.; Rundlet, K.L. Highly enantioselective capillary electrophoretic separations with dilute solutions of the macrocyclic antibiotic ristocetin A. J. Chromatogr. A. **1995**, *689*, 285–304.

26. Rundlet, K.L.; Gasper, M.P.; Zhou, E.Y.; Armstrong, D.W. Capillary electrophoretic enantiomeric separations using the glycopeptide antibiotic, teicoplanin. Chirality **1996**, *8*, 88–107.

27. Mechref, Y.; El Rassi, Z. Capillary electrophoresis of herbicides: III. Evaluation of octylmalto-

pyranoside chiral surfactant in the enantiomeric separation of phenoxy acid herbicides. Chirality **1996**, *8*, 518–524.

28. Mechref, Y.; El Rassi, Z. Capillary electrophoresis of herbicides: II. Evaluation of alkylglucoside chiral surfactants in the enantiomeric separation of phenoxy acid herbicides. J. Chromatogr. A. **1997**, *757*, 263–273.

29. Mechref, Y.; El Rassi, Z. Capillary electrophoresis of herbicides: 1. Pre-column derivatization of chiral and achiral phenoxy acid herbicides with a fluorescent tag for electrophoretic separation in the presence of cyclodextrins and micellar phases. Anal. Chem. **1996**, *68*, 1771–1777.

30. Mechref, Y.; El Rassi, Z. Micellar electrokinetic capillary chromatography with in-situ charged micelles: VI. Evaluation of novel chiral micelles consisting of steroidal–glycoside surfactant–borate complexes. J. Chromatogr. A. **1996**, *724*, 285–296.

31. Martinez, D.; Cugat, M.J.; Borrull, F.; Calull, M. Solid-phase extraction coupling to capillary electrophoresis with emphasis on environmental analysis. J. Chromatogr. A. **2000**, *902*, 65–89.

32. Dabek-Zlotorzynska, E.; Aranda-Rodriguez, R.; Keppel-Jones, K. Recent advances in capillary electrophoresis and capillary electrochromatography of pollutants. Electrophoresis **2001**, *22*, 4262–4280.

33. Haddad, P.R.; Doble, P.; Macka, M. Developments in sample preparation and separation techniques for the determination of inorganic ions by ion chromatography and capillary electrophoresis. J. Chromatogr. A. **1999**, *856*, 145–177.

34. Fritz, J.S.; Macka, M. Solid-phase trapping of solutes for further chromatographic or electrophoretic analysis. J. Chromatogr. A. **2000**, *902*, 137–166.

35. Pedersen-Bjegaard, S.; Rasmussen, K.E.; Halvorsen, T.G. Liquid-liquid extraction procedures for sample enrichment in capillary zone electrophoresis. J. Chromatogr. A. **2000**, *902*, 91–105.

36. Whang, C.; Pawliszyn, J. Solid phase microextraction coupled to capillary electrophoresis. Anal. Commun. **1998**, *35*, 353–356.

37. Chankvetadze, B.; Burjanadze, N.; Pintore, G.; Bergenthal, D.; Bergander, K.; Mühlenbrock, C.; Breitkreuz, J.; Blaschke, G. Separation of brompheniramine enantiomers by capillary electrophoresis and study of chiral recognition mechanisms of cyclodextrins using NMR spectroscopy, UV spectrometry, electrospray ionization mass spectrometry and X-ray crystallography. J. Chromatogr. A. **2000**, *875*, 471–484.

38. Pacakova, V.; Coufal, P.; Stulik, K. Capillary electrophoresis of inorganic cations. J. Chromatogr. A. **1999**, *834*, 257–275.

39. Liu, B.F.; Liu, B.L.; Cheng, J.K. Analysis of inorganic cations as their complexes by capillary electrophoresis. J. Chromatogr. A. **1999**, *834*, 277–308.

40. Valsecchi, S.M.; Polesello, S. Analysis of inorganic species in environmental samples by capillary electrophoresis. J. Chromatogr. A. **1999**, *834*, 363–385.

41. Timerbaev, A.R.; Buchberger, W. Prospects for detection and sensitivity enhancement of inorganic ions in capillary electrophoresis. J. Chromatogr. A. **1999**, *834*, 117–132.

42. Horvath, J.; Dolnike, V. Polymer wall coatings for capillary electrophoresis. Electrophoresis **2001**, *22*, 644–655.

43. Mayer, B.X. How to increase precision in capillary electrophoresis. J. Chromatogr. A. **2001**, *907*, 21–37.

Affinity Chromatography: An Overview

David S. Hage

Department of Chemistry, University of Nebraska, Lincoln, Nebraska, U.S.A.

INTRODUCTION

Affinity chromatography is a liquid chromatographic technique that uses a "biologically related" agent as a stationary phase for the purification or analysis of sample components. The retention of solutes in this method is based on the specific, reversible interactions that are found in biological systems, such as the binding of an enzyme with a substrate or an antibody with an antigen. These interactions are exploited in affinity chromatography by placing one of a pair of interacting molecules onto or within a solid support and using this as a stationary phase. This immobilized molecule is known as the *affinity ligand* and is what gives an affinity column the ability to bind to particular compounds in a sample.

Affinity chromatography is a valuable tool in areas such as biochemistry, pharmaceutical science, clinical chemistry, and environmental testing, where it has been used for both the purification and analysis of compounds in complex sample mixtures. The strong and relatively specific binding that characterizes many affinity ligands allows solutes to be quantitated or purified by these ligands with little or no interferences from other sample components. Often, the solute of interest can be isolated in one or two steps, with purification yields of one hundred to several thousand-fold being common. This review first examines the history of affinity chromatography and the general formats by which affinity chromatography is performed. The various components of an affinity chromatographic method are then discussed, including the types of affinity ligands, support materials, immobilization methods, and application or elution conditions that are employed in this technique. Several specific types of affinity chromatography are then considered.

HISTORY OF AFFINITY CHROMATOGRAPHY

The earliest known use of affinity chromatography was in 1910, when Emil Starkenstein examined the binding of insoluble starch to the enzyme α-amylase.[1] Similar work with starch and other insoluble ligands (acting as both binding agents and supports) was conducted in the 1920s through 1940s. These early studies mostly involved the use of affinity supports for the purification of enzymes. But work also began at this time on the selective purification of antibodies with biological ligands through immunoprecipitation.[2]

In the 1940s and 1950s, synthetic techniques became available for placing a broader range of ligands on insoluble materials. These efforts began by employing solids that contained a noncovalently adsorbed layer of ligand, followed later by methods for chemically bonding a ligand to the support. This was first used by Landsteiner and van der Sheer in 1936, when they adapted a diazo coupling technique used to prepare hapten conjugates for attaching a number of haptens to a solid material based on chicken erythrocyte stroma.[3] Another significant development in this area occurred in 1951, when Campbell and coworkers used an activated form of cellulose (*p*-aminobenzylcellulose) for immobilizing the protein serum albumin and isolating anti-albumin antibodies from rabbit serum.[4]

The modern era of affinity chromatography began in the late 1960s. This occurred following the creation of beaded agarose supports by Hjerten[5] and the discovery of the cyanogen bromide immobilization method by Axen, Porath, and Ernback.[6] These two methods were combined in 1968 by Cuatrecasas, Wilchek, and Anfinsen to create immobilized nuclease inhibitor columns. These columns were then used for purifying the enzymes staphylococcal nuclease, α-chymotrypsin, and carboxypeptidase A. It was also at this time that the term "affinity chromatography" was used to describe this separation technique.[7] Since that time, the number of articles containing the phrase "affinity chromatography" has grown to over 1000 per year.[2,8–11]

GENERAL FORMATS FOR AFFINITY CHROMATOGRAPHY

The most common scheme for performing affinity chromatography is by using a step gradient for elution, as shown in Fig. 1. This involves injecting a sample onto the affinity column in the presence of a mobile phase that has the right pH and solvent composition for solute–ligand binding. This solvent, which represents the weak mobile phase of the affinity column, is called the *application buffer*. During the application phase of the separation, compounds that are

Encyclopedia of Chromatography DOI: 10.1081/E-ECHR-120039873

A

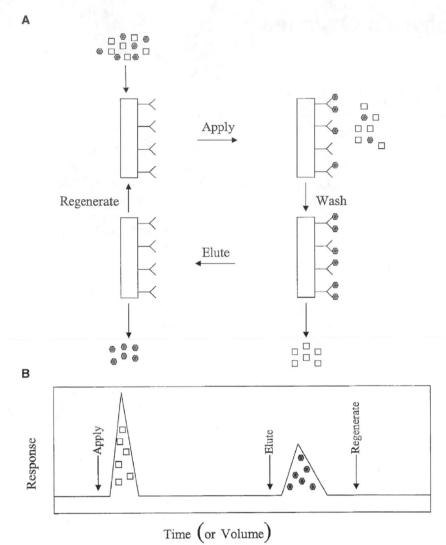

B

Fig. 1 (A) Typical separation scheme and (B) chromatogram for affinity chromatography. The filled circles represent the test analyte and the squares represent other, nonretained sample components.

complementary to the affinity ligand bind as the sample is carried through the column by the application buffer. However, due to the high selectivity of the solute–ligand interaction, the remainder of the sample components passes through the column unretained. After the unretained components have been completely washed from the column, the retained solutes can be eluted by applying a solvent that displaces them from the column or that promotes dissociation of the solute–ligand complex. This solvent represents the strong mobile phase for the column and is known as the *elution buffer*. As the solutes of interest elute from the column, they are either measured or collected for later use. The column is then regenerated by re-equilibration with the application buffer prior to injection of the next sample.

Even though the *step gradient*, or "*on/off*," *elution method* is the most common way of performing affinity chromatography, it is sometimes possible to use affinity methods under isocratic conditions. This

can be done if a solute's retention is sufficiently weak to allow elution on the minute-to-hour time scale and if the kinetics for its binding and dissociation are fast enough to allow a large number of solute–ligand interactions to occur as the analyte travels through the column. This approach is sometimes called *weak affinity chromatography* and is best performed if a solute binds to the ligand with an association constant that is less than or equal to about 10^4–$10^6\,M^{-1}$.[12,13]

TYPES OF AFFINITY LIGANDS

The most important factor in determining the success of any affinity separation is the type of ligand that is used within the column. A number of ligands that are commonly used in affinity chromatography are listed in Table 1. Most of these ligands are of biological origin, but a wide range of natural and synthetic molecules of nonbiological origin can also be used.

Table 1 Common ligands used in affinity chromatography

Type of ligand	Examples of retained compounds
High-specificity ligands	
Antibodies	Various agents (drugs, hormones, peptides, proteins, viruses, etc.)
Enzyme inhibitors and cofactors	Enzymes
Nucleic acids	Complementary nucleic acid strands and DNA/RNA-binding proteins
General ligands	
Lectins	Small sugars, polysaccharides, glycoproteins, and glycolipids
Protein A and protein G	Intact antibodies and Fc fragments
Boronates	Catechols and compounds that contain sugar residues, such as polysaccharides and glycoproteins
Synthetic dyes	Dehydrogenases, kinases, and other proteins
Metal chelates	Metal-binding amino acids, peptides, or proteins

Regardless of their origin, all of these ligands can be placed into one of two categories: high-specificity ligands or general ligands.[2,10,11]

The term *high-specificity ligand* refers to a compound that binds to only one or a few closely related molecules. This type of ligand is used in affinity systems where the goal is to analyze or purify a specific solute. Examples include antibodies (for binding antigens), substrates or inhibitors (for separating enzymes), and single-stranded nucleic acids (for the retention of a complementary sequence). As this list suggests, most high-specificity ligands tend to be of biological origin and often have large association constants for their particular analytes.

General, or *group-specific*, *ligands* are compounds that bind to a family or class of related molecules. These ligands are used when the goal is to isolate a class of structurally related compounds. General ligands can be of either biological or nonbiological origin and include compounds such as protein A or protein G, lectins, boronates, biomimetic dyes, and immobilized metal chelates. This class of ligands usually exhibits weaker binding for solutes than is seen with high-specificity ligands; however, some general ligands like protein A and protein G do have association constants that rival those of high-specificity ligands.

SUPPORT MATERIALS

Another important factor to consider in affinity chromatography is the material used to hold the ligand within the column. Ideally, this support should have low nonspecific binding for sample components, it should be easy to modify for ligand attachment, and it should be stable under the flow-rate, pressure, and solvent conditions that will be employed in the analysis or purification of samples. Depending on what type of support material is being used, affinity chromatogra-

phy can be characterized as being either a low- or high-performance technique.

In *low-performance* (or *column*) *affinity chromatography*, the support is usually a large-diameter, nonrigid material (e.g., a carbohydrate-based gel or one of several synthetic organic-based polymers). The low back-pressures that are produced by these supports mean that these materials can often be operated under gravity flow or with a peristaltic pump, making them relatively simple and inexpensive to use for affinity purification or sample pretreatment. Disadvantages of these materials include their slow mass transfer properties and their limited stabilities at high flow rates and pressures. These factors tend to limit the direct use of these supports in analytical applications, where both rapid and efficient separations are usually desired.[11,14]

High-performance affinity chromatography (HPAC) is characterized by a support that consists of small, rigid particles capable of withstanding high flow rates and/or pressures.[11,14,15] Examples of affinity supports that are suitable for work under these conditions include modified silica or glass, azalactone beads, and hydroxylated polystyrene media. The stability and efficiency of these supports allows them to be used with standard HPLC equipment. Although the need for HPLC instrumentation does make HPAC more expensive to perform than low-performance affinity chromatography, the better speed and precision of HPAC makes it the affinity method of choice for many analytical applications.

Although porous, particulate materials like agarose, polymethacrylate, and silica are used in most current applications of affinity chromatography, there are other types of supports that have recently become available. Materials that fall in this category include nonporous supports, membranes, flow-through beads, continuous beds, and expanded bed particles. The good mass transfer properties of nonporous beads with diameters of 1–3 μm make them appealing for fast analytical or micropreparative separations, as well as quantitative studies of affinity interactions. Similar properties are obtained using flow-through beds or continuous beds, such as monolithic supports. The flat geometry and shallow bed depth of affinity membranes allow their use at high flow rates, making them well suited for capturing proteins from dilute feed streams. Similarly, the presence of low back-pressures makes expanded bed particles attractive for use in isolating proteins from cell culture samples while allowing solid contaminants like cells and cell debris to pass through, thereby avoiding column clogging.[14]

IMMOBILIZATION METHODS

A third item to consider in affinity chromatography is the way in which the ligand is attached to the solid support, or the *immobilization method*. For a protein or peptide, this generally involves coupling the molecule through free amine, carboxylic acid, or sulfhydryl residues present in its structure. Immobilization of a ligand through other functional sites (e.g., aldehyde groups produced by carbohydrate oxidation) is also possible. All covalent immobilization methods involve at least two steps: 1) an *activation step*, in which the support is converted to a form that can be chemically attached to the ligand; and 2) a *coupling step*, in which the affinity ligand is attached to the activated support. Occasionally, a third step is necessary to remove remaining activated groups.

The method by which an affinity ligand is immobilized is important since it can affect the actual or apparent activity of the final affinity column. If the correct procedure is not used, a decrease in ligand activity can result from multisite attachment, improper orientation, and/or steric hindrance (see Fig. 2). *Multisite attachment* refers to the coupling of a ligand to the support through more than one functional group, which can lead to distortion of the ligand's active region and a loss of activity. This can be avoided by using a support with a limited number of activated sites or by using a method that couples through groups that occur only a few places in the structure of the ligand. *Improper orientation* can lead to a loss in activity by coupling the ligand to the support through its active region; this can be minimized by coupling through functional groups that are distant from this region. *Steric hindrance* refers to the loss of ligand activity due to the presence of a nearby support or neighboring ligand molecules that interfere with solute binding. This effect can be avoided through the use of a spacer arm or by using supports that contain a relatively low coverage of the ligand.[11,16]

Besides covalent immobilization, it is possible to place an affinity ligand within a column through other means. The simplest of these is through physical adsorption onto a surface through ionic or nonspecific interactions. In addition, a ligand can be held noncovalently on a column by means of a secondary ligand. A common example of this is the use of immobilized protein A or protein G to adsorb antibodies for use in immunoaffinity chromatography. It is also sometimes possible to entrap or encapsulate a ligand within a support if the size of the ligand is smaller than the pores in this material. Finally, there are situations in which the ligand itself can be used as both the support and stationary phases. This was illustrated in the first use of affinity chromatography, in which insoluble starch was used for isolation of the enzyme amylase.

APPLICATION AND ELUTION CONDITIONS

Two other items that must be considered in affinity chromatography are the application and elution buffers. Most application buffers in affinity chromatography are solvents that mimic the pH, ionic strength, and polarity experienced by the solute and ligand in their natural environment. This gives the solute its highest association constant for the ligand and, thus, its highest degree of retention on the column. The application buffer should also be chosen so that it minimizes nonspecific binding due to undesired sample components.

Elution conditions in affinity chromatography are usually chosen so that they promote either the fast or gentle removal of solute from the column. The two approaches used for this are *nonspecific elution* and *biospecific elution*, respectively.[1] Biospecific elution is based on the addition of a competing agent that gently displaces a solute from the column. This is done by adding an agent that either competes with the ligand for solute (i.e., *normal role elution*) or competes with the solute for ligand-binding sites (i.e., *reversed role elution*). Although it is a gentle method, biospecific elution does result in long elution times and broad solute peaks that are difficult to quantitate. Nonspecific elution uses a solvent that directly promotes weak solute-ligand binding. For instance, this is done by changing the pH, ionic strength, or polarity of the mobile phase or by adding a denaturing agent or chaotropic substance to the elution buffer. Nonspecific elution tends to be much faster than biospecific elution and results in sharper peaks with lower limits of detection. However, care must be exercised in nonspecific elution to avoid using a buffer that is too harsh and that causes solute denaturation or a loss of ligand activity.

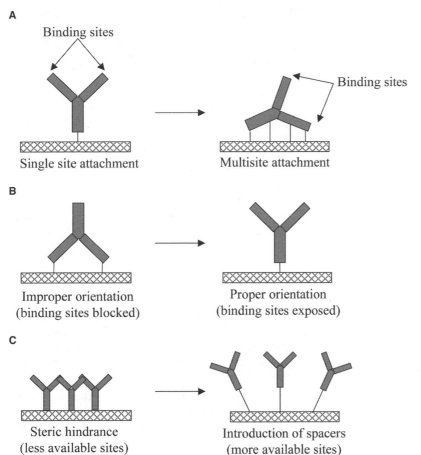

A

Binding sites

Single site attachment   Multisite attachment

B

Improper orientation
(binding sites blocked)

Proper orientation
(binding sites exposed)

C

Steric hindrance
(less available sites)

Introduction of spacers
(more available sites)

Fig. 2 Common immobilization effects during the coupling of affinity ligands to solid supports. (From Ref.[16].)

TYPES OF AFFINITY CHROMATOGRAPHY

There are many types of affinity chromatography that are in common use. *Bioaffinity chromatography* is probably the broadest category and includes any method that uses a biological molecule as the affinity ligand. *Immunoaffinity chromatography* (IAC) is a special category of bioaffinity chromatography in which the affinity ligand is an antibody or antibody-related agent.[17,18] This creates a highly specific method that is ideal for use in affinity purification or in analytical methods that involve complex samples. *Immunoextraction* is a subcategory of IAC in which an affinity column is used to isolate compounds from a sample prior to analysis by a second method. IAC can also be used to monitor the elution of analytes from other columns, giving rise to a scheme known as *postcolumn immunodetection*.[17,19]

Another common type of bioaffinity method is that which uses bacterial cell wall proteins like protein A or protein G for antibody purification. This is sometimes referred to as a subset of immunoaffinity chromatography but is more correctly viewed as a separate division of bioaffinity methods. In *lectin affinity chromatography*, immobilized lectins like concanavalin A or wheat germ agglutinin are used for binding to molecules that contain certain sugar residues. Additional types of bioaffinity chromatography are those that make use of ligands that are enzymes, inhibitors, cofactors, nucleic acids, hormones, or cell receptors.[8–11] Examples of these methods include *DNA affinity chromatography* and *receptor affinity chromatography*.

There are also many types of affinity chromatography that use ligands that are of nonbiological origin. For instance, the closely related methods of *dye–ligand affinity chromatography* and biomimetic affinity chromatography typically use an immobilized synthetic dye that mimics the active site of a protein or enzyme.[20] This is a popular tool for enzyme and protein purification. *Immobilized metal-ion affinity chromatography* (IMAC) is an affinity technique in which the ligand is a metal ion that is complexed with an immobilized chelating agent.[21,22] This is used to separate proteins and peptides that contain amino acids with electron donor groups. *Boronate affinity chromatography* employs boronic acid or a boronate as the affinity ligand. These ligands are useful in binding to compounds that contain *cis*-diol groups, such as catecholamines and glycoproteins.[23]

There are a number of other chromatographic methods that are closely related to traditional affinity chromatography. For instance, affinity chromatography can be adapted as a tool for studying solute–ligand interactions. This application is known as *analytical*, or *quantitative*, *affinity chromatography* and can be used to acquire information regarding the equilibrium and rate constants for biological interactions, as well as the number and type of binding sites that are involved in these interactions.[24,25]

Other methods that are related to affinity chromatography include *hydrophobic interaction chromatography* (HIC) and *thiophilic adsorption*. HIC is based on the interactions of proteins, peptides, and nucleic acids with short nonpolar chains, such as those that were originally used as spacer arms on affinity supports. Thiophilic adsorption, also known as *covalent* or *chemisorption chromatography*, makes use of immobilized thiol groups for solute retention. Applications of this method include the analysis of sulfhydryl-containing peptides or proteins and mercurated polynucleotides.

Many types of *chiral liquid chromatography* can be considered affinity methods since they are based on binding agents that are of biological origin. Examples include columns that use cyclodextrins or immobilized proteins for chiral separations. The growing use of molecularly imprinted polymers as stationary phases can also be considered to be a subset of affinity chromatography, since such materials are designed to mimic the multiple interactions and selectivity that are characteristic of many biological ligands.

CONCLUSIONS

In summary, affinity chromatography is a selective separation method that has numerous applications in the purification and analysis of compounds, especially those related to biological systems. The wide variety of ligands available for this technique makes it possible to design affinity separations for a large range of target compounds. Some of these ligands are of biological origin, while others are synthetic. It is also possible to perform affinity chromatography with support materials that allow its use either as a preparative tool or as an analytical method. Affinity chromatography can also be used as a means to study biological interactions. These features have made affinity methods valuable in a number of fields, including biotechnology, biochemistry, clinical analysis, pharmaceutical science, and environmental testing. The use of affinity chromatography in combination with other methods, such as mass spectrometry and capillary electrophoresis, is also an area that has seen considerable growth in recent years.

REFERENCES

1. Starkenstein, E. Ferment action and the influence upon it of neutral salts. Biochem. Z. **1910**, *24*, 210–218.
2. Hage, D.S.; Ruhn, P.F. An introduction to affinity chromatography. In *Handbook of Affinity Chromatography*; Hage, D.S., Ed.; Marcel Dekker: New York, 2005; (Chapter 1).
3. Landsteiner, K.; van der Scheer, J. Cross reactions of immune sera to azoproteins. J. Exp. Med. **1936**, *63*, 325–339.
4. Campbell, D.H.; Luescher, E.; Lerman, L.S. Immunologic adsorbents. I. Isolation of antibody by means of a cellulose-protein antigen. Proc. Natl. Acad. Sci. U.S.A. **1951**, *37*, 575–578.
5. Hjerten, S. The preparation of agarose spheres for chromatography of molecules and particles. Biochem. Biophys. Acta **1964**, *79*, 393–338.
6. Axen, R.; Porath, J.; Ernback, S. Chemical coupling of peptides and proteins to polysaccharides by means of cyanogen halides. Nature **1967**, *214*, 1302–1304.
7. Cuatrecasas, P.; Wilchek, M.; Anfinsen, C.B. Selective enzyme purification by affinity chromatography. Proc. Natl. Acad. Sci. U.S.A. **1968**, *68*, 636–643.
8. Turkova, J. *Affinity Chromatography*; Elsevier: Amsterdam, 1978.
9. Scouten, W.H. *Affinity Chromatography: Bioselective Adsorption on Inert Matrices*; John Wiley & Sons: New York, 1981.
10. Parikh, I.; Cuatrecasas, P. Affinity chromatography. Chem. Eng. News **1985**, *63*, 17–29.
11. Walters, R.R. Affinity chromatography. Anal. Chem. **1985**, *57*, 1099A–1114A.
12. Wang, W.T.; Kumlien, J.; Ohlson, S.; Lundblad, A.; Zopf, D. Analysis of a glucose-containing tetrasaccharide by high-performance liquid affinity chromatography. Anal. Biochem. **1989**, *182*, 48–53.
13. Standh, M.; Andersson, H.S.; Ohlson, S. Weak affinity chromatography. In *Affinity Chromatography*; Bailon, P., Ehrlich, G.K., Fung, W.J., Berthold, W., Eds.; Humana Press: Totowa, NJ, 2000 (Chapter 2).
14. Gustavsson, P.E.; Larson, P.O. Support materials for affinity chromatography. In *Handbook of Affinity Chromatography*; Hage, D.S., Ed.; Marcel Dekker: New York, 2005 (Chapter 2).
15. Ohlson, S.; Hansson, L.; Larsson, P.O.; Mosbach, K. High performance liquid affinity chromatography (HPLAC) and its application to the separation of enzymes and antigens. FEBS Lett. **1978**, *93*, 5–9.

16. Kim, H.S.; Hage, D.S. Immobilization methods for affinity chromatography. In *Handbook of Affinity Chromatography*; Hage, D.S., Ed.; Marcel Dekker: New York, 2005 (Chapter 4).

17. Hage, D.S. Survey of recent advances in analytical applications of immunoaffinity chromatography. J. Chromatogr. B. **1998**, *715*, 3–28.

18. Phillips, T.M. High performance immunoaffinity chromatography: an introduction. LC Mag. **1985**, *3*, 962–972.

19. Hage, D.S.; Nelson, M.A. Chromatographic immunoassays. Anal. Chem. **2001**, *73*, 198A–205A.

20. Labrou, N.E.; Clonis, Y.D. Immobilized synthetic dyes in affinity chromatography. In *Theory and Practice of Biochromatography*; Vijayalakshmi, M.A., Ed.; Taylor & Francis: London, 2002; 335–351.

21. Porath, J.; Carlsson, J.; Olsson, I.; Belfrage, B. Metal chelate affinity chromatography, a new approach to protein fraction. Nature **1975**, *258*, 598–599.

22. Chage, G.S. Twenty-five years of immobilized metal ion affinity chromatography: past, present and future. J. Biochem. Biophys. Meth. **2001**, *49*, 313–334.

23. Liu, X.C.; Scouten, W.H. Boronate affinity chromatography. In *Affinity Chromatography*; Bailon, P., Ehrlich, G.K., Fung, W.J., Berthold, W., Eds.; Humana Press: Totowa, NJ, 2000; 119–128.

24. Chaiken, I.M., Ed.; *Analytical Affinity Chromatography*; CRC Press: Boca Raton, FL, 1987.

25. Hage, D.S.; Tweed, S.A. Recent advances in chromatographic and electrophoretic methods for the study of drug-protein interactions. J. Chromatogr. B. **1997**, *699*, 499–525.

A

Aggregation of Colloids by Field-Flow Fractionation

Athanasia Koliadima
University of Patras, Patras, Greece

INTRODUCTION

The separation of the components of complex colloidal materials is one of the most difficult challenges in separation science. Most chromatographic methods fail in the colloidal size range or, if operable, they perform poorly in terms of resolution, recovery, and reproducibility. Therefore, it is desirable to examine alternate means that might solve important colloidal separation and characterization problems encountered in working with biological, industrial, environmental, and geological materials. One of the most important colloidal processes that is generally quite difficult to characterize is the aggregation of single particles to form complexes made up of multiples of the individual particles. Aggregation is a common phenomenon for both natural and industrial colloids. The high degree of stability, which is frequently observed in colloidal systems, is a kinetic phenomenon in that the rate of aggregation of such systems may be practically zero. Thus, in studies of the colloidal state, the kinetics of aggregation are of paramount importance. Although the kinetics of aggregation can be described easily by a bimolecular equation, it is not an easy thing to do experimentally.

One technique for doing this is to count the particles microscopically. In addition to particle size limitation, this is an extraordinarily tedious procedure. Light scattering can be also used for the kinetic study of aggregation, but experimental turbidities must be interpreted in terms of the number and size of the scattering particles.

In the present work, it is shown that the field-flow fractionation (FFF) technique can be used with success to study the aggregation phenomena of colloids.

The techniques of FFF appear to be well suited to colloid analysis. The special subtechnique of sedimentation FFF (SdFFF) is particularly effective in dealing with colloidal particles in the diameter range from 0.02 to 1 μm, using the normal or Brownian mode of operation (up to 100 μm using the steric-hyperlayer mode). As a model sample for the observation of aggregate particles by SdFFF, of poly(methyl methacrylate) (PMMA) was used, whereas for the kinetic study of aggregation by SdFFF, the hydroxyapatite (HAP) sample [$Ca_5(PO_4)_3OH$] consisting of submicron irregularly shaped particles was used.

The stability of HAP, which is of paramount importance in its applications, is dependent on the total potential energy of interaction between the HAP particles. The latter, which is the sum of the attraction potential energy and that of repulsion, depends on particle size, the Hamaker constant, the surface potential, and the Debye–Hückel reciprocal distance, which is immediately related to the ionic strength of carrier solution.

METHODOLOGY

FFF is a highly promising tool for the characterization of colloidal materials. It is a dynamic separation technique based on differential elution of the sample constituents by a laminar flow in a flat, ribbonlike channel according to their sensitivity to an external field applied in the perpendicular direction to that of the flow.

The total potential energy of interaction between two colloidal particles, U_{tot}, is given by the sum of the energy of interaction of the double layers, U_R, and the energy of interaction of the particles themselves due to van der Waals forces, U_A. Consequently,

$$U_{tot} = U_R + U_A \qquad (1)$$

For identical spherical particles U_R and U_A are defined as follows:

$$U_R = \frac{\varepsilon r \psi_0^2}{2} \ln[1 + \exp(-\kappa H)] \quad (\kappa r \gg 1) \qquad (2)$$

$$U_R = \frac{\varepsilon r \psi_0^2}{R} \exp(-\kappa H) \quad (\kappa r \ll 1) \qquad (3)$$

$$U_A = -\frac{A_{212} r}{12 H} \qquad (4)$$

where ε is the dielectric constant of the dispersing liquid, r is the radius of the particle, ψ_0 is the particle's surface potential, κ is the reciprocal double-layer thickness, R is the distance of the centers of the two particles, A_{212} is the effective Hamaker constant of two particles of type 2 separated by the medium of type 1, and H is the nearest distance between the surfaces of the particles.

Encyclopedia of Chromatography DOI: 10.1081/E-ECHR 120030874

Eqs. (2–4) show that the total potential energy of interaction between two colloidal spherical particles depends on the surface potential of the particles, the effective Hamaker constant, and the ionic strength of the suspending medium. It is known that the addition of an indifferent electrolyte can cause a colloid to undergo aggregation. Furthermore, for a particular salt, a fairly sharply defined concentration, called "critical aggregation concentration" (CAC), is needed to induce aggregation.

The following equation gives the rate of diffusion-controlled aggregation, u_r, of spherical particles in a disperse system as a result of collisions in the absence of any energy barrier to aggregate:

$$u_r = -k_r N_0^2 \qquad (5)$$

where k_r is the second-order rate constant for diffusion-controlled rapid aggregation and N_0 is the initial number of particles per unit volume.

In the presence of an energy barrier to aggregate, the rate of aggregation, u_s, is

$$u_s = -k_s N_0^2 \qquad (6)$$

where k_s is the rate constant of slow aggregation in the presence of an energy barrier.

The stability ratio, w, of a dispersion is defined as the ratio of the rate constants for aggregation in the absence, k_r and the presence, k_s, of an energy barrier, respectively:

$$w = \frac{k_r}{k_s} \qquad (7)$$

The aggregation process is described by the bimolecular kinetic equation

$$\frac{1}{N_i} = \frac{1}{N_0} + k_{app} t_i \qquad (8)$$

where N_i is the total number of particles per unit volume at time t_i and k_{app} is the apparent rate constant for the aggregation process. The measurement of the independent kinetic units per unit volume, N_i, at different times t_i can give the rate constant for the aggregation process.

Considering that d_{N_0} and d_{N_i} are the measured number-average diameters of the particles at times $t = 0$ and t_i, respectively, Eq. 8, for polydisperse samples, gives

$$d_{N_i}^3 = d_{N_0}^3 + d_{N_0}^3 N_0 k_{app} t_i \qquad (9)$$

Eq. (9) shows that from the slope of the linear plot of the $d_{N_i}^3$ versus t_i, the apparent rate constant k_{app} can be determined, as the N_0 values can be found from the ratio of the total volume of the injected sample to the volume of the particle, which can be determined from the diameter calculated from the intercept of the above plot.

APPLICATIONS

The observation of a series of peaks (Fig. 1) while analyzing samples of PMMA colloidal latex spheres by SdFFF suggests that part of the latex population has aggregated into doublets, triplets, and higher-order particle clusters. The particle diameter of the latex spheres was given as 0.207 μm. The aggregation hypothesis is confirmed by retention calculations and by electron microscopy. For this purpose, narrow fractions or cuts were collected from the first five peaks as shown in Fig. 1. A fraction was also collected for the peak which appeared after the field was turned off. The individual fractions were subjected to electron microscopy and as expected, cut No. 1 yielded singlets, cut No. 2 yielded doublets, cut No. 3 yielded triplets, cut No. 4 yielded quads, cut No. 5

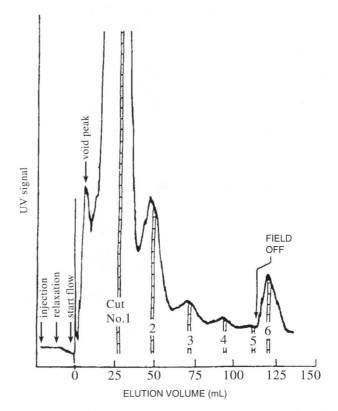

Fig. 1 SdFFF fractogram of 0.207-μm PMMA aggregate series from which six cuts were collected and analyzed by electron microscopy. Experimental conditions: field strength of 61.6 g and flow rate of 0.84 mL/min. (Adapted from Jone, H. K. et al. *J. Chromatogr.* **1988**, *455*, 1; Copyright Elsevier Science Publishers B.V.)

yielded quints, and the cut after the field was turned off yielded clusters from six individual particles.

Sedimentation field-flow fractionation was used also for the kinetic study of HAP particles' aggregation in the presence of various electrolytes to determine the rate constants for the bimolecular process of aggregation and to investigate the possible aggregation mechanisms describing the experimental data. The HAP sample contained polydisperse, irregular colloidal particles with number-average diameter $d_N = 0.262 \pm 0.046$ μm.

The number-average diameter, d_N, for the HAP particles increases with the electrolyte KNO_3 concentration until the critical aggregation concentration is reached, where the d_N value remains approximately constant. The starting point of the maximum d_N corresponds to the electrolyte concentration called CAC. The last value, which depends on the electrolyte used, was found to be 1.27×10^{-2} M for the electrolyte KNO_3.

According to Eq. 9, the plot of $d_{N_i}^3$ versus t_i at various electrolyte concentrations determines the apparent rate constant, k_{app}, of the HAP particles' aggregation. The found k_{app} value for the aggregation of the HAP particles in the presence of 1×10^{-3} M KNO_3 is 2.5×10^{-21} cm^3/s. It is possible to make a calculation which shows whether the value of k_{app} is determined by the rate at which two HAP particles can diffuse up to each other (diffusion control) or whether the rate of reaction is limited by other slower processes. The rate constant for the bimolecular collision (k_1) of the HAP particles, can be calculated by the Stokes–Einstein equation:

$$k_1 = \frac{8kT}{3n}\,\text{cm}^3/\text{s} \tag{10}$$

where n is the viscosity of the medium. The calculated value of $k_1 = 1.1 \times 10^{-11}$ cm^3/s is about 10 orders of magnitude greater than the value of k_{app} actually measured. So, the aggregation rates are slower than those expected if the process was simply diffusion controlled when electrostatic repulsion is absent. The latter indicates that the minimal mechanism for the aggregation process of the HAP particles would be

$$\text{Particle}_1 + \text{Particle}_2 \underset{k_{-1}}{\overset{k_1}{\rightleftharpoons}}$$

$$\text{Intermediate complex} \overset{k_2}{\rightleftharpoons} \text{Stable aggregate} \tag{11}$$

where k_{-1} is the rate constant for the dissociation of the intermediate aggregate and k_2 is the rate constant for the process representing the rate-determining step in the aggregation reaction. Because k_{app}, describing the overall process, is smaller than the calculated k_1

value, there must be rapid equilibration of the individual particles and their intermediate complexes followed by the slower step of irreversible aggregation. The stability factor, w, of HAP's particles found to be 4.4×10^9 is too high, indicating that the particles are very stable, even in the presence of significant quantity of the electrolyte KNO_3.

As a general conclusion, the FFF method can be used with success to study the aggregation process of colloidal materials.

FUTURE DEVELOPMENTS

Looking to the future, it is reasonable to expect continuous efforts to improve the theoretical predictions and more experimental work to investigate the aggregation phenomena of natural and industrial colloids.

SUGGESTED FURTHER READING

Athanasopoulou, A.; Karaiskakis, G.; Travlos, A. Colloidal interactions studied by sedimentation field-flow fractionation. J. Liq. Chromatogr. & Related Technol. **1997**, *20*, 2525–2541.

Athanasopoulou, A.; Gavril, D.; Koliadima, A.; Karaiskakis, G. Study of hydroxyapatite aggregation in the presence of potassium phosphate by centrifugal sedimentation field-flow fractionation. J. Chromatogr. A. **1999**, *845*, 293.

Caldwell, K.D.; Nguyen, T.T.; Giddings, J.C.; Mazzone, H.M. Field-flow fractionation of alkali-liberated polyhedrosis virus from Gypsy moth (*Lymantria dispar*, L.). J. Virol. Methods **1980**, *1*, 241–256.

Everett, D.H. *Basic Principles of Colloid Science*; Royal Society of Chemistry Paperbacks: London, 1988.

Family, F., Landan, D.P., Eds.; *Kinetic of Aggregation and Gelation*; North-Holland: Amsterdam, 1984.

Jones, H.K.; Barman, B.N.; Giddings, J.C. Resolution of colloidal latex aggregates by sedimentation field-flow fractionation. J. Chromatogr. **1988**, *455*, 1–15.

Koliadima, A. The kinetic study of aggregation of the sulphide $Cu_{0.2}Zn_{0.8}S$ particles by gravitational field-flow fractionation. J. Liq. Chromatogr. & Related Technol. **1999**, *22* (16), 2411.

Wittgren, B.; Borgström, J.; Piculell, L.; Wahlund, K.G. Conformational change and aggregation of kappa-carrageenan studied by flow field-flow fractionation and multiangle light scattering. Biopolymers **1998**, *45*, 85–96.

Alcoholic Beverages: Analysis by GC

Fernando M. Lanças
M. de Moraes
Laboratorio de CromatografiaInstituto de Quimica de São Carlos, Universidade de São Paulo, São Carlos, SP, Brazil

INTRODUCTION

Alcoholic beverages have been consumed by a significant range of worldwide population since the beginning of civilization until the present time. Therefore, there should be a great interest on consumption of beverages quality and, consequently, the usage of a suitable analytical technique to verify and control this desirable quality.

Alcoholic beverages are classified, in a general way, in fermented beverages (such as beer, wine, sake, etc.) and distilled ones (vodka, whisky, aguardente, tequila, cognac, liquors, etc.). The main volatile substances present in most alcoholic beverages belongs to the following chemical classes: alcohols (including ethanol, methanol, isobutanol, 3-methyl butan-2-ol, etc.), esters (such as ethyl acetate, methyl acetate, ethyl isobutyrate, isoamyl acetate, etc.), aldehydes (propanal, isobutanal, acetal, furfural, etc.), acids (acetic acid, propionic acid, butyric acid, etc.) and ketones (acetone, diacetyl, etc.).

Some of the substances are of greater concern than the others due to its relative quantities or to its flavored characteristic.[1] As an example, ethanol is the major compound in the group of alcohols being responsible for the formation of various other substances, such as acetaldehyde, resulting from ethanol oxidation, and it is the most abundant of the carbonylic compounds in distilled beverages. For the same reason, acetic acid is the major compound within its group, the carboxylic acids. Fusel alcohols (e.g., 1-propanol and 3-methyl butan-2-ol) as well as ethanol are also important substances in the alcohols group contributing to the flavor of distilled beverages because their odor is very distinctive and characteristic. There is an enormous variety of substances in small concentrations belonging to the esters group. Even so, ethyl acetate corresponds to more than 50% of the esters within this group.

These compounds are responsible to important characteristics as smell and taste, in which the large fraction of these substances originates from fermentation or during beverages storage.[1]

Because GC is an analytical technique in which separation and identification of volatile compounds occurs, it might be considered the best technique for this kind of sample.[2]

ANALYSIS BY GC

Analysis of alcoholic beverages by GC have as their main objectives to investigate flavoring compounds and contaminants which might be intentional or occasional. Whereas the former ones (the flavors) are analyzed to control their favorable characteristics to the beverage, the adulterants have to be controlled due to their deleterious contribution. Adulteration includes the addition of certain compounds to enhance a desirable flavor. Because these compounds are usually added as a racemic mixture, their presence can be verified using a suitable chiral column for enantiomer separation.[3] On the other hand, occasional contaminants are substances originating in small quantities during beverage production and might be carcinogenic. The main source might be raw materials like grape, sugar cane, and so forth contaminated with pesticides[4] or as a result of the fermentation process itself.[5–6]

Comments in each major part of the GC instrumentation as used for beverage analysis are presented.

SAMPLE INTRODUCTION

Distilled Beverages

Generally, samples are injected in the chromatographic system without any dilution or pretreatment step, using the split mode (i.e., with sample division), which is suitable for the analysis of the major compounds in beverages. When the objective of analysis is the determination of compounds present in small quantities (μg/L), some extraction and/or concentration step is necessary, followed by the sample injection in the splitless mode (without sample division). This last sample introduction mode is usually combined with extraction techniques such as liquid–liquid extraction (LLE), solid-phase extraction (SPE),[4] and solid-phase microextraction (SPME).[7–8]

Encyclopedia of Chromatography DOI: 10.1081/E-ECHR-120039878

Fermented Beverages

Sample introduction is basically the same compared to the distilled one. Nevertheless, in many complex samples (i.e., some kind of wines), it is not advisable to inject them into the chromatograph without a pretreatment step.

Recently, SPME has provided many improvements as the cleanup step of complex samples, particularly for the analysis of volatile compounds by *headspace* techniques.[8] SPME is a solventless extraction and concentration technique which has advantages as a simple and economic technique that reduces health hazards and environmental issues.

COLUMNS

Until 1960, all commercial chromatographic columns were packed in wide-bore tubing and separations had low resolution and low efficiency, taking a long time for a run to be completed. Since then, there have been significant improvements with the introduction of wall-coated open tubular (WCOT) columns, whose inner diameter was smaller than the packed ones are coated with a thin film of the liquid phase.[9]

When capillary fused-silica columns arose, a large number of separations of complex samples have obtained success as a result of the higher number of plates (about 30 times over the packed columns in average).

Despite the efficient separations, it has been noticed that some low-boiling compounds of alcoholic beverages coeluted because of the use of a polar stationary phase. This column separates mainly based on the boiling temperatures of chemical substances, but separation becomes very difficult if there are some compounds with similar boiling temperatures. A polar stationary phase like poly(ethylene glycol) (PEG) is a better choice for this sort of problem because these separations are based on compound structures. In all cases, cross-linked or immobilized phases are recommended because they are more thermolabile and also resistant to most solvents. This is particularly important when splitless injection is used in combination with PEG-type phases because otherwise a severe column bleeding might be observed after ~220°C.

Detection Systems

The flame ionization detector (FID) is one of the most commonly used detectors in beverage analysis by GC, as it is suitable to most groups of compounds investigated in alcoholic beverages.[9] This occurs because almost all compounds of interest in such samples are able to burn in the flame, forming ions and producing a potential difference measured by a collector electrode.

In trace analysis of contaminant substances, one can use specific detectors for certain compounds, such as a nitrogen–phosphorus detector (NPD), thus gaining detection ability for nitrogenated and phosphorylated compounds; the electron-capture detector (ECD) shows excellent performance for chlorinated substances and the flame photometric detector (FPD) is the most widely used for sulfur-containing compounds.

GC/MS combination has become one of the most important coupling in analytical chemistry used for the confirmation of results obtained by other detectors.[9] This technique is based on the fragmentation of the molecules that arrives into the detector. Ion formation occurs and they are separated by the mass/charge ratio (m/z) generally detected by a electron multiplier. Quantitative analysis can be realized through the single-ion monitoring (SIM) mode, where some characteristic ions are selected and monitored increasing the detection sensibility and selectivity. Another qualitative technique, GC–sniffing[10] is very much used for flavor analysis, despite discordance among researchers. In this case, the volatile substances from a extract are separated by GC, and as they leave the equipment through a specially designed orifice, a

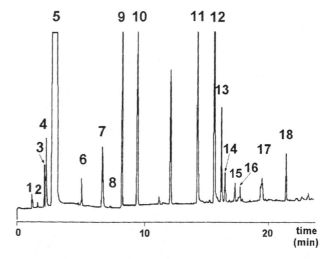

Fig. 1 GC analysis of white wine sample by fused-silica column poly(ethylene glycol) type (15.00 m × 0.53 mm × 1.00 µm). Chromatographic conditions: carrier gas: hydrogen (3.6 mL/min); column temperature: 40°C (4 min) → 8°C/min → 270°C; injector port temperature: 250°C; detector temperature: 300°C. Identity of the selected peaks: (1) acetaldehyde, (2) acetone, (3) ethyl acetate, (4) ethanol, (5) ethanol, (6) propanol, (7) isoamyl acetate, (8) *n*-butanol, (9) heptanone, (10) isoamyl alcohol, (11) acetic acid, (12) propionic acid, (13) isobutyric acid, (14) 2,3 butanediol, (15) 1,2 propanediol, (16) butyric acid, (17) isovaleric acid, (18) valeric acid.

trained analyst is able to smell some of the substances and tentatively identify them (Fig. 1).

More than 500 compounds have been found in concentrated flavor extracts in distilled beverages.[1] For this reason, there is an obvious necessity to find and optimize analytical techniques capable of investigating and to keep trading control of the alcoholic beverages. Among them, GC has been far from any other preferred tool for the analysis of alcoholic beverages.

ACKNOWLEDGMENTS

Professor Lanças wishes to express his acknowledgments to FAPESP (Fundação de Amparo à Pesquisa do Estado de São Paulo), CNPQ (Conselho Nacional de Desenvolvimento Científico e Tecnológico), and CAPES (Coordenação de Aperfeiçoamento e Pessoal de Nível Superior) for financial support to our laboratory and a fellowship to Marcelo de Moraes.

REFERENCES

1. Nykänen, L.; Nykänen, I. Distilled beverages. In *Volatiles Compounds in Foods and Beverages*; Maarse, H., Ed.; Marcel Dekker, Inc.: New York, 1991; 547–580.
2. Lanças, F.M.; Galhiane, M.S. Fast routine analysis of light components of alcoholic beverage using large bore open tubular fused silica column. Bol. Soc. Chil. Quim. **1993**, *38*, 177.
3. König, W.A. New developments in enantiomer separation by capillary gas chromatography. In *Analysis of Volatiles Methods, Applications*; Schreier, P., Ed.; Berlin, 1984; 77–91.
4. Kaufmann, A. Fully automated determination of pesticides in wine. J. AOAC Int. **1997**, *80* (6), 1302.
5. Lawrence, J.F.; Page, B.D.; Conacher, B.S. The formation and determination of ethyl carbamate in alcoholic beverages. Adv. Environ. Sci. Technol. **1990**, *23*, 457.
6. Shiomi, K. Determination of acetaldehyde, acetal and other volatile congeners in alcoholic beverages. J. High Resol. Chromatogr. **1991**, *14* (2), 136.
7. Pawliszyn, J. *Solid Phase Microextraction—Theory and Practice*; Wiley–VCH: New York, 1997; 1–247.
8. Garcia, D.; de la, C.; Reichenächer, M.; Danzer, K.; Hurlbeck, C.; Bartzsch, C.; Feller, K.H. Analysis of wine bouquet components using headspace solid phase micro extraction–capillary gas chromatography. J. High Resol. Chromatogr. **1998**, *21* (7), 373.
9. Smith, R.M. *Gas and Liquid Chromatography in Analytical Chemistry*; John Wiley & Sons: London, 1988; 1–401.
10. Abbott, N.; Etievant, P.; Langlois, D.; Lesschaeve, I.; Issanchou, S. Evaluation of the representativeness of the odor of beer extracts prior to analysis by GC eluate sniffing. J. Agric. Food Chem. **1993**, *41* (5), 777.

A

Alkaloids: Separation by Countercurrent Chromatography

Fuquan Yang
Yoichiro Ito
Laboratory of Biophysical Chemistry, National Heart, Lung, and Blood Institute,
National Institutes of Health, Bethesda, Maryland, U.S.A.

INTRODUCTION

Alkaloids are an important class of compounds that have pharmacological effects on various tissues and organs of humans and other animal species. More than 16,000 are known and most are derived from higher plants. Alkaloids have also been isolated from microorganisms, from marine organisms such as algae, dinoflagellates, and puffer fish, as well as from terrestrial animals, such as insects, salamanders, and toads.

Pelletier[1] defines an alkaloid as "a cyclic compound containing nitrogen in a negative oxidation state which is of limited distribution in living organisms." This definition includes both alkaloids with nitrogen as part of a heterocyclic system and the many exceptions with extracyclic bound nitrogen. From the viewpoint of analytical chemistry, the most important trait of alkaloids is their basicity, arising from a heterocyclic tertiary nitrogen atom. Many alkaloids are complex components which are biosynthetically derived from various amino acids, such as phenylalanine, tyrosine, tryptophan, ornithine, and lysine. The biogenesis of alkaloids is used for their classification, as this is directly linked with their molecular skeleton, e.g., the two largest groups are indole alkaloids and isoquinoline alkaloids. Other important groups are tropane alkaloids, pyridine, and pyrrolizidine alkaloids. In the past 10 years, there has been an increasing interest in the isolation and determination of alkaloids in plant materials, in pharmaceutical products, and in other samples of biological interest. Currently, much work is being carried out to discover new alkaloid molecules for different applications, such as new antiviral and antitumor treatments. So the separation and analysis of alkaloids are of great importance.[2,3] But the problem is that extensive tailing and irreversible adsorption may occur due to the interaction between the basic alkaloids and acidic silanol groups when alkaloids are separated by conventional column chromatography with silica gel or chemically bonded C_{18} on silica as the stationary phase.

OVERVIEW

Countercurrent chromatography (CCC), as a support-free liquid–liquid partition chromatographic separation technique, eliminates various complications such as irreversible adsorption loss, denaturation of sample, tailing of solute peaks, contamination, etc. that sometimes arise from the interaction between solute molecules and the solid support matrix present in most other chromatographic methods. Countercurrent chromatography utilizes an immiscible two-phase solvent system, one phase as the stationary phase and the other as the mobile phase. The chromatographic process in CCC is based on the partition of a solute between the stationary and mobile phases. The partition coefficient (K), being one of the most important parameter in CCC, is defined as the ratio of the concentration of a solute in the stationary phase (C_s) to that in the mobile phase (C_m). In the past, various CCC systems, such as droplet CCC, rotation locular CCC, and centrifugal partition chromatography, have been used for the final or partial separation of alkaloids from a crude sample. The high-speed CCC (HSCCC) technique developed in the early 1980s[4] improved in both the partition efficiency and separation speed, and has been successfully applied to the separation of alkaloids.

pH-Zone-refining CCC was developed for the large-scale separation of ionizable compounds, including alkaloids and organic acids in the mid-1990s by Ito.[5,6] It uses a retainer base (or acid) in the stationary phase to retain the solutes in the column and an eluter acid (or base) to elute the solutes according to their pK_a values and hydrophobicities, and produces a succession of highly concentrated rectangular solute peaks with minimum overlap while impurities are concentrated at the peak boundaries. This technique has been successfully used for large-scale separation of alkaloids.

SEPARATION OF ALKALOIDS BY STANDARD COUNTERCURRENT CHROMATOGRAPHY

The standard CCC separation is solely based on the difference in the partition coefficients of solutes between the two phases of a solvent system. Most alkaloids have basic properties with pK_a values ranging from 6 to 12, but usually 7 to 9. Although the free base

Encyclopedia of Chromatography DOI: 10.1081/E-ECHR-120016783

is soluble only in organic solvents, protonation of the nitrogen in the free base usually results in a water-soluble compound. This behavior serves as the basis for the selective extraction or isolation of alkaloids by liquid–liquid partitioning processes. On the other hand, quaternary alkaloids are poorly soluble in organic

Table 1 Separation of alkaloids by CCC

Alkaloids	Source	Two-phase solvent system	Instrument
Matrine and oxymatrine	*Sophora flavescens*	$CHCl_3$–0.07 M sodium phosphate (pH 6.4) (1:1)	HSCCC
Atropine, scopolamine, hyoscyamine	*Datura mete*	$CHCl_3$–0.07 M sodium phosphate (pH 6.5) (1:1)	HSCCC
Cephalotaxus alkaloids	*Cephalotaxus fortunei*	$CHCl_3$–0.07 M sodium phosphate–0.04 M citric acid (pH 5.0) (1:1)	HSCCC
Pyrrolizidine alkaloids	*Amsinckia tessellata*, etc.	$CHCl_3$–0.2 M potassium phosphate (pH 7.4, 6.0, 5.6, 5.0) (1:1)	HSCCC
Pentacyclic aromatic alkaloid	*Amphicarpa meridiana*	$CHCl_3$–MeOH–5% HCl (5:5:3)	HSCCC
Flavonoid alkaloids	*Buchenavia capitata*	$CHCl_3$–MeOH–0.5% HCl (5:5:3)	Sanki CPC
Isoquinoline alkaloids	*Ancistrocladus abbreviatus*	$CHCl_3$–MeOH–0.5% HBr (5:5:3)	HSCCC
Naphthyltetrahydroisoquinoline alkaloids	*Ancistrocladus korupensis*	$CHCl_3$–MeOH–0.5% HBr (5:5:3)	CPC
Aporphine alkaloids	*Dehaasia triandra*	$CHCl_3$–MeOH–0.5% HAc (5:5:3)	HSCCC
Naphthylisoquinoline alkaloids	*Ancistrocladus robertsoniorum*	$CHCl_3$–MeOH–0.1 M HCl (5:5:3)	HSCCC
Isoquinoline alkaloids	*Coptis chinensis*	$CHCl_3$–MeOH–0.2 M HCl (4:1.5:2)	HSCCC
Diterpenoid alkaloids	*Aconitum sinomontanum*	$CHCl_3$–MeOH–0.3/0.2 M HCl (4:1.5:2)	HSCCC
Pyrrolizidine alkaloids	*Symphytum officinale*	n-C_6H_{14}–EtOH–MeOH–0.05% TFA (5:5:5:5)	HSCCC
β-Carboline alkaloids	*Pachypellina* sp.	n-C_6H_{14}–MeCN–CH_2Cl_2 (10:7:3)	HSCCC
Vinca alkaloids	*Vinca rosea*	n-C_6H_{14}–EtOH–H_2O (6:5:5)	HSCCC
Isoquinoline alkaloids	*Stephania tetrandra*	n-C_6H_{14}–EtOAc–MeOH–H_2O (3:7:5:5), (1:1:1:1)	HSCCC
Diterpenoid alkaloids	*Consolida ambigua*	C_6H_6–$CHCl_3$–MeOH–H_2O (5:5:7:2)	CPC
Pyrroloquinoline alkaloids	*Bazella* sp.	n-C_6H_{14}–$CHCl_3$–MeOH–H_2O (4:7:4:3), (2:7:6:3); $CHCl_3$-i-Pr_2NH–MeOH–H_2O (7:1:6:4)	HSCCC
Pyrroloquinoline alkaloids	*Bazella* sp.	n-C_7H_{16}–$CHCl_3$–MeOH–H_2O (2:7:6:3), n-C_7H_{16}–EtOAc–$CHCl_3$–MeOH–H_2O (4:7:4:3)	HSCCC
Acridine alkaloids	*Dercitus* sp. and *Stellatta* sp.	CH_2Cl_2–MeOH–H_2O (5:5:3)	HSCCC
Quinoline alkaloids	*Camptotheca acuminata*	CCl_4–$CHCl_3$–MeOH–H_2O (2:2:3:1), CH_2Cl_2–$CHCl_3$–MeOH–H_2O (5:3:1)	HSCCC
Ergot alkaloids	*Stipa robusta*	$CHCl_3$–MeOH–H_2O (5:4:3)	HSCCC
Benzylisoquinoline alkaloids	*Anisocycla cymosa*	$CHCl_3$–MeOH–H_2O (10:10:1)	HSCCC
Imidazole alkaloids	*Discodermia polydiscus*	$CHCl_3$–MeOH–H_2O (5:10:6)	HSCCC
bis-Indole alkaloid	*Strychnos guianensis*	EtOAc–MeOH–H_2O (4:1:3)	HSCCC
Indole alkaloids	*Strychnos usambarensis*	n-BuOH–0.1 M NaCl (1:1)	HSCCC
Indole alkaloids	*Venezuelan curare*	n-BuOH–Me_2CO–H_2O (8:1:10)	HSCCC

solvents, but they are soluble in water, regardless of its pH.

Generally, a systematic search for the two-phase solvent systems for CCC is focused on the hydrophobicity of the solvent system for providing a proper range of partition coefficients of solutes. For the separation of ionizable compounds such as alkaloids, however, an additional adjustment is required with respect to the pH and ionic strength of the solvent system.

Halogen-containing organic solvents, such as chloroform and dichloromethane, have been widely used in alkaloid separation because of their relatively strong proton-donor character. The chloroform-containing, two-phase solvent systems (chloroform–aqueous phosphate or citrated buffer and chloroform–methanol–dilute

inorganic acid) have been widely used for the separation of alkaloids by CCC (Table 1).

Cai et al. first used a chloroform–0.07 M sodium phosphate buffer (pH 6.4–6.5) solvent system for the separation of matrine and oxymatrine from *Sophora flavescens*, atropine, scopolamine, and hyoscyamine from *Datura mete* L. by HSCCC.[7] Cooper et al. successfully used chloroform–0.2 M potassium phosphate buffer with an optimum pH value of each at 5.0, 5.6, 6.0, and 7.4 for the separation of pyrrolizidine alkaloids from various sources of *Senecio douglasii* var. *longilobus*, *Trichodesma incanum*, *Symphytum* spp. and *Amsinckia tessellata*, respectively.[8]

A two-phase solvent system composed of chloroform–methanol–dilute inorganic acid has been used for the separation of a variety of alkaloids including

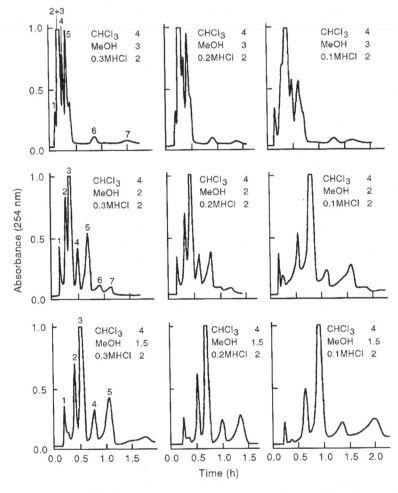

Fig. 1 Chromatograms of the crude alkaloids from *C. chinensis* Franch by analytical HSCCC. Nine chromatograms are arranged in such a way that the effects of methanol and HCl concentrations on the alkaloid separation are each clearly visualized. Experimental conditions: apparatus: analytical HSCCC instrument equipped with a multilayer coil of 0.85 mm I.D. and 30 mL capacity; sample: 2.5 mg of crude alkaloid extract of *C. chinensis* Franch; solvent system: shown above each chromatogram; mobile phase: lower organic phase; flow rate: 1 mL/min; revolution: 1500 rpm. Retention of the stationary phase was as follows: CHCl$_3$–MeOH–(0.1–0.3 M HCl) (4:3:2), 77%; CHCl$_3$–MeOH–(0.1–0.3 M HCl) (4:2:2), 80%; and CHCl$_3$–MeOH–(0.1–0.3 M HCl) (4:1.5:2), 77%.

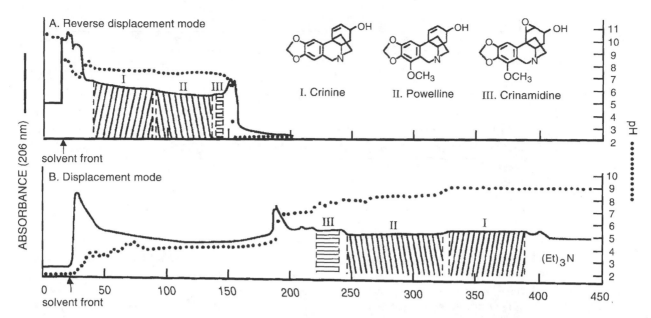

Fig. 2 Chromatograms of the crude alkaloid extract of *Crinum moorei* obtained by pH-zone-refining CCC. Experimental conditions were as follows: apparatus: high-speed CCC centrifuge equipped with a multiplayer coil of 1.6 mm I.D. and about 300 mL capacity; solvent system: methyl-*tert*-butyl ether–water; stationary phase: (A) upper phase (5 mM triethylamine) and (B) lower phase (10 mM HCl); mobile phase: (A) lower phase (5 mM HCl) and (B) upper phase (10 mM triethylamine); flow rate: 3.3 mL/min; sample: crude alkaloid extract of *C. moorei*, 3 g dissolved in 30 mL of each phase; revolution: (A) 800 rpm (600 rpm until 66 mL of mobile phase was eluted) and (B) 600 rpm throughout.

isoquinoline alkaloids, naphthyl-tetrahydroisoquinoline alkaloids, flavonoid alkaloids, pentacyclic aromatic alkaloids, diterpenoid alkaloids, aporphine alkaloids, etc. The following example illustrates a typical systematic solvent selection for the separation of palmatine, berberine, epiberberine, and coptisine from the crude alkaloids of *Coptis chinensis* Franch by analytical HSCCC.[9] In Fig. 1, nine chromatograms are arranged in such a way that the effects of the concentration of HCl (0.3–0.1 M) and the relative volumes of methanol (4:3:2–4:1.5:2, v/v) on the separation of alkaloids from *C. chinensis* Franch are each readily observed. As the concentration of HCl is reduced from 0.3 to 0.1 M in the solvent system, the retention time of alkaloids and their peak resolution are increased. A similar effect is also produced by decreasing the relative volumes of methanol in the solvent system from 4:3:2 to 4:1.5:2). Among those, the solvent system composed of CHCl3–MeOH–0.2 M HCl (4:1.5:2) produced the best separation of all four major alkaloids components. The optimum solvent system thus obtained led to the successful separation of alkaloids from *C. chinensis* Franch by preparative HSCCC. A similar method was also used for the separation of diterpenoid alkaloids, including lappaconitine, ranaconitine, *N*-deacetyllappaconitine, and *N*-deacetylranaconitine from *Aconitum sinomontanum*.[10]

More than 10 kinds of neutral two-phase solvent systems have been successfully used for the separation

of alkaloids by CCC.[11] Most of the alkaloids from marine sponge—including acridine, pyrroloquinoline, imidazole, β-carboline alkaloids, etc.—were separated with neutral two-phase solvent systems by HSCCC in the final or middle step of separation (Table 1).

SEPARATION OF ALKALOIDS BY H-ZONE-REFINING COUNTERCURRENT CHROMATOGRAPHY

Fig. 2 shows a typical separation of 3 g of three alkaloids from a crude extract of *Crinum moorei* using a binary solvent system composed of methyl-*tert*-butyl ether–water where triethylamine (5–10 mM) was added to the organic phase and HCl (5–10 mM) was added to the aqueous phase.[12] In Fig. 2A, where the aqueous phase was used as the mobile phase and HCl as the eluter, the alkaloids were eluted as HCl salts, while in Fig. 2B, where the organic phase is used as mobile phase and triethylamine as the eluter, the alkaloids were eluted as free bases. The same binary solvent system was also used for the separation of alkaloids from the root of *S. flavescens* Ait.[13] In order to increase the solubility of the alkaloids in methyl-*tert*-butyl ether–water solvent system, this solvent system was modified to methyl *tert*-butyl ether–tetrahydrofuran–water (2:2:3, v/v), while triethylamine (10 mM) was added to the upper organic stationary phase as a

Table 2 Separation of alkaloids by pH-zone-refining CCC

Alkaloids	Source	Solvent systems[a]	Retainer[b]	Eluter[c]
Amaryllis alkaloids	*Crinum moorei*	MtBE–H$_2$O (1;1)	TEA (5 mM/SP)	HCl (5 mM/MP)
		MtBE–H$_2$O (1:1)	HCl (10 mM/SP)	TEA (10 mM/MP)
Vinca alkaloids	*Vinca minor*	MtBE–H$_2$O (1:1)	TEA (5 mM/SP)	HCl (5 mM/MP)
Matrine, sophocarpine	*Sophora flavescens*	MtBE–H$_2$O (1:1)	TEA (10 mM/SP)	HCl (5–0 mM/MP)
Diterpenoid alkaloids	*Aconitum sinomontanum*	MtBE–THF–H$_2$O (2:2:73)	TEA (10 mM/SP)	HCl (10 mM/MP)
Isoquinoline alkaloids	*Hydrastis canadensis rhizomes*	CHCl$_3$–H$_2$O (1:1)	18–24 mM HCl	TEA (0.1%/MP)

[a]MtBE = methyl *tert*-butyl ether; THF = tetrahydrofuran.
[b]TEA = triethylamine; SP = stationary phase.
[c]MP = mobile phase.

retainer and hydrochloric acid (10 mM) to the aqueous mobile phase as an eluter. This solvent system was applied to the separation of diterpenoid alkaloids from a crude prepurified sample containing lappaconitine at about 90% purity. Up to 10.5 g of the sample loading yielded 9.0 g of lappaconitine at a high purity of over 99% as determined by HPLC.[14] As most alkaloids have a high solubility in chloroform, a binary solvent system composed of chloroform–water is suitable for a large-scale preparative separation when 18–24 mM HCl was added to the aqueous stationary phase as a retainer and 0.1% triethylamine (TEA) to the organic mobile phase as an eluter. This system was successfully used for the separation of berberine chloride, canadaline, canadine, β-hydrastine, and isocorypalmine from a methanolic extract of Goldenseal by pH-zone-refining CCC.[15] The separation of alkaloids by pH-zone-refining CCC is summarized in Table 2.

CONCLUSIONS

Countercurrent chromatography overcomes all of the problems caused by solid support matrix present in most other chromatographic methods. Countercurrent chromatography can purify alkaloids with high recovery and reproducibility.

pH-Zone-refining CCC extends the preparative capacity of HSCCC and provides many important advantages over the conventional CCC method, including an over 10-fold increase in sample loading capacity, high concentration of fractions with very high purity, concentration of minor impurities, etc.

REFERENCES

1. Pelletier, S.W., Ed.; *Alkaloids: Chemical and Biological Perspective*; John Wiley: New York, 1983; Vol. 1–6.

2. Muzquiz, M. Alkaloids/gas chromatography. In *Encyclopedia of Separation Science*; Wilson, I.D., Adlard, E.R., Cooke, M., Poole, C.F., Eds.; Academic Press: New York, 1938–1949; Vol. 5.

3. Verpoorte, R. Alkaloids/liquid chromatography. In *Encyclopedia of Separation Science*; Wilson, I.D., Adlard, E.R., Cooke, M., Poole, C.F., Eds.; Academic Press: New York, 1949–1956; Vol. 5.

4. Ito, Y. High-speed countercurrent chromatography. CRC Crit. Rev. Anal. Chem. **1986**, *17* (1), 65–143.

5. Ito, Y. pH-peak-focusing and pH-zone-refining countercurrent chromatography. In *High-Speed Countercurrent Chromatography*; Ito, Y., Conway, W.D., Eds.; John Wiley: New York, 1996; 121–175.

6. Ito, Y.; Ma, Y. pH-zone-refining countercurrent chromatography. J. Chromatogr. A. **1996**, *753* (1), 1–36.

7. Cai, D.G.; Gu, M.J.; Zhang, J.D.; Zhu, G.P.; Zhang, T.Y.; Li, N.; Ito, Y. Separation of alkaloids from *Datura mete* L. and *Sophora flavescens* Ait by high-speed countercurrent chromatography. J. Liq. Chromatogr. **1990**, *13* (12), 2399–2408.

8. Cooper, R.A.; Bowers, R.J.; Beckham, C.J.; Huxtable, R.J. Preparative separation of pyrrolizidine alkaloids by high-speed countercurrent chromatography. J. Chromatogr. A. **1996**, *732* (1), 43–50.

9. Yang, F.Q.; Zhang, T.Y.; Zhang, R.; Ito, Y. Application of analytical and preparative high-speed countercurrent chromatography for separation of alkaloids from *Coptis chinensis* Franch. J. Chromatogr. A. **1998**, *829* (1–2), 137–141.

10. Yang, F.Q.; Ito, Y. Preparative separation of lappaconitine, ranaconitine, *N*-deacetyllappaconitine and *N*-deacetylranaconitine from crude alkaloids of sample *Aconitum sinomontanum* Nakai by

high-speed countercurrent chromatography. J. Chromatogr. A. **2002**, *943* (2), 219–225.

11. Marston, A.; Hostettmann, K. Countercurrent chromatography as a preparative tool applications and perspectives. J. Chromatogr. A. **1994**, *658* (2), 315–341.

12. Ma, Y.; Ito, Y.; Sokoloski, E.; Fales, H.M. Separation of alkaloids by pH-zone-refining countercurrent chromatography. J. Chromatogr. A. **1994**, *685* (2), 259–262.

13. Yang, F.Q.; Quan, J.; Zhang, T.Y.; Ito, Y. Preparative separation of alkaloids from the root of *Sophora flavescens* Ait by pH-zone-refining

countercurrent chromatography. J. Chromatogr. A. **1998**, *822* (2), 316–320.

14. Yang, F.Q.; Ito, Y. pH-Zone-refining countercurrent chromatography of lappaconitine from *Aconitum sinomontanum* Nakai: I. Separation of prepurified extract. J. Chromatogr. A. **2001**, *923* (1–2), 281–285.

15. Chadwick, L.R.; Wu, C.D.; Kinghorn, A.D. Isolation of alkaloids from goldenseal (*Hydrastis canadensis rhizomes*) using pH-zone-refining countercurrent chromatography. J. Liq. Chromatogr. & Relat. Technol. **2001**, *24* (16), 2445–2453.

Alumina-Based Supports for LC

Esther Forgács
Tibor Cserháti
Institute of Chemistry, Chemical Research Center, Hungarian Academy of Sciences, Budapest, Hungary

INTRODUCTION

Various LC techniques offer a unique possibility for the separation and quantitative determination of a large variety of organic, metalloorganic, and inorganic compounds with highly similar molecular structures. These methods are indispensable in medical practice and research, pharmaceutical chemistry, food science and technology, environmental pollution control, legislation procedures, etc. The rapid development of the theory of the retention processes in chromatography have made it obvious that the efficient separation of various compounds (selection of the best separation method, support and mobile phase, and any other parameter influencing the efficacy) requires a profound knowledge of the impact of molecular characteristics of solutes, stationary and mobile phases, and their interplay at the molecular level on retention. The expert application of such knowledge will highly facilitate the rational design of optimal separation methods. As the chemistry and physicochemistry of the surface of the support determines the retention characteristics of stationary phases, physical methods such as nuclear magnetic resonance (NMR), Fourier transform infrared (FTIR), etc., have been frequently used to study the stationary phases in LC.

The overwhelming majority of LC separations are carried out in silica or in silica-based, reversed-phase (mainly octadecylsilica, ODS) stationary phases. Although the retention characteristics of silica and surface-modified silica supports are excellent, and they can be used for the successful separation of a wide variety of solutes, they also have some drawbacks. Thus, the acidic character of the free silanol groups exert a considerable impact on the retention behavior of both silica and silica-based supports; basic solutes are more strongly bonded onto the silica surface than neutral or acidic substances, resulting in unpredictable retention behavior. Moreover, silica and silica derivatives are not stable at alkaline pHs, making the separation of strongly basic compounds difficult. The objectives of the newest developments in LC are the development and practical application of more stable supports than silica and modified silicas, with different separation capacities (alumina, zirconia, titania, mixed oxides, and their modified derivatives, porous graphitized carbon, various polymer supports, etc.). Although these new supports show excellent separation characteristics, they are not well known, not frequently used, and the molecular basis of retention has not been elucidated in detail.

The objectives of this article are the enumeration and critical evaluation of the recent results obtained in the assessment of the relationship between the physicochemical characteristics and retention behavior of a wide variety of solutes on alumina stationary phases and the elucidation of the efficacy of various multivariate mathematical–statistical methods for the quantitative description of such relationships.

ALUMINA

Alumina offers another alternative to silica because of its inherent higher pH stability. However, in contrast to its extensive use as a medium in column chromatography for purification purposes, or for separations in the normal-phase mode, there are still relatively few reports concerning alumina-based materials in the reversed-phase mode.

The chromatographic aspects of the surface characteristics of alumina stationary phases have not been studied as profoundly as zirconia supports; however, the presence of hydroxyl and oxide ions on the surface has been reported.[1]

Crystalline alumina may exist in various forms; the τ-form is generally used in chromatography. Alumina strongly adsorbs water molecules, as depicted in Fig. 1. The two different hydroxyl groups show acidic or alkaline properties, resulting in amphoteric characteristics and ion exchange behavior of the alumina surface, as demonstrated in Fig. 2.[2] It has been further shown that the simultaneous interaction of pH, the composition of buffer and that of the mobile phase modifier governs the retention on alumina surfaces in ion exchange chromatography.[3–4] The good separation characteristics of alumina stationary phase were exploited, not only in the ion-exchange mode, but also in the adsorption (direct phase) chromatographic mode. Alumina supports have been frequently used

Encyclopedia of Chromatography DOI: 10.1081/E-ECHR-120014245

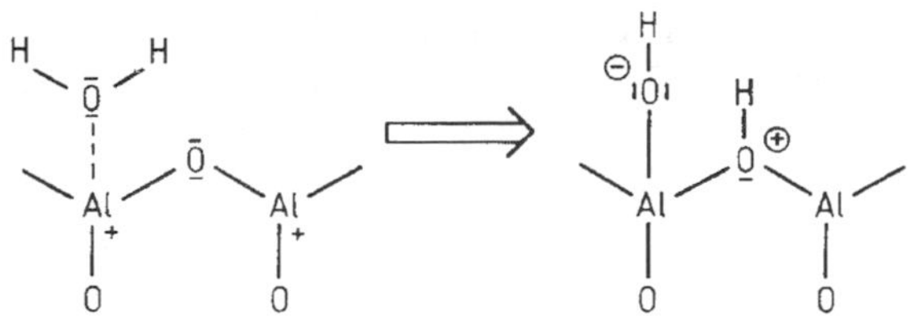

Fig. 1 Chemisorption of water on bare alumina. (From Ref.[2].)

in TLC, and the separation of inorganic and organometallic solutes on alumina layers was reviewed.[5] Nonionic surfactants (α-(1,1,3,3-tetramethylbutyl) phenyl ethylene oxide oligomers) were also separated on alumina layers using n-hexane mixed with ethyl acetate, dioxane, and tetrahydrofurane (THF), and the data were evaluated by spectral mapping technique. Surfactants were separated according to the number of ethylene oxide groups in the molecule; surfactants with longer ethylene oxide chain were eluted later. This finding indicates that the polar ethylene oxide units turn toward the stationary phase and that they are bonded to the adsorption sites on the alumina surface. Calculations proved that the solvent strength of THF was the highest, followed by dioxane and ethyl acetate. The selectivity of ethyl acetate differed considerably from the selectivities of THF and dioxane.[6] Other

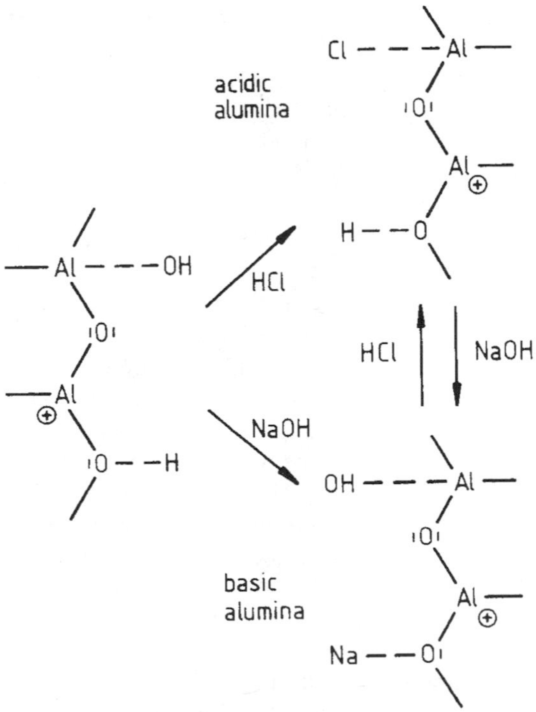

nonionic surfactants (nonylphenyl[7] and tetrabutylphenyl[8] ethyleneoxide oligomers) were successfully separated on an alumina HPLC column using n-hexane-ethyl acetate mixtures as mobile phases. Significant linear correlations were found between the molecular parameters of the solutes and their retention characteristics:

$$\log k_0' = 3.61 + (0.37 \pm 0.03)n_e \quad r = 0.9911 \tag{1}$$

$$b = 3.54 + (1.18 \pm 0.03)n_e \quad r = 0.9991 \tag{2}$$

where k_0' is the capacity factor of nonylphenyl ethyleneoxide oligomer surfactants, extrapolated to zero concentration of ethyl acetate in the mobile phase; b is the slope value of the linear relationship between the $\log k'$ values of surfactants and the concentration of ethyl acetate in the eluent ± standard deviation; and n_e is the number of ethylene oxide groups per molecule.

$$\log k_0' = 3.41 + (0.36 \pm 0.01)n_e + (0.13 \pm 0.03)\text{PI}$$
$$F = 217.65 \tag{3}$$

$$b = 3.56 + (0.78 \pm 0.07)n_e + (0.52 \pm 0.13)\text{PI}$$
$$F = 64.65 \tag{4}$$

where PI characterizes the position of the butyl substituents, other symbols are the same as in Eqs. (1) and (2).

The data entirely support the previous conclusions concerning the retention mechanism of surfactants on alumina. Moreover, Eqs. (3) and (4) indicate that alumina is an excellent support for the separation of tetrabutylphenyl ethylene oxide oligomers according to the number of ethylene oxide groups and the position of the alkyl substituents in one run. The efficiency of alumina support for the separation of positional isomers was also established with HPLC-mass spectrometry (HPLC-MS).[9]

Fig. 2 Surface behavior of alumina in basic and acidic media, respectively. (From Ref.[2].)

MODIFIED ALUMINA

Polymers have also been used for the coating of alumina. Thus, maleic acid adsorbed onto the alumina surface was, in situ, polymerized with 1-octadecene and cross-linked with 1,4-divinylbenzene.[10] It was assumed that the polymer forms a monolayer on the alumina, forming a reversed-phase surface. This assumption was substantiated by results showing that the retention order of model compounds was the same as on an ODS column. The lower separation capacity of the new stationary phase was tentatively explained by the lower surface porosity of alumina. Principal component analysis was employed for the elucidation of the relationship between the retention behavior of nonhomologous series of solutes on polybutadiene (PBD)-coated alumina and their physicochemical parameters.[11] Calculations revealed significant relationships between the capacity factor extrapolated to pure water (k_w') and the physicochemical parameters:

$$\log k_w' = 0.052\ 0.208(\log P) \tag{5}$$

$$n = 21;\ s = 0.279;\ r = 0.9711;\ F = 314$$

$$\log k_w' = 1.618\ 0.089\text{bonrefr}\ 2.505\text{delta} \tag{6}$$

$$n = 21;\ s = 0.500;\ r = 0.9090;\ F = 42.8$$

$$\log k_w' = 1.272 + 0.089\text{bonrefr}\ 2.648\text{delta}\ 0.598\text{ind} \tag{7}$$

$n = 21;\ s = 0.394;\ r = 0.9476;\ F = 49.8$ where $\log P$ is the hydrophobicity, "bondrefr" is the molecular refractivity, "delta" is the submolecular polarity parameter, "ind" indicator variable (0 for heterocyclics and 1 for benzene derivatives). Calculations indicated that PBD-coated alumina behaves as an RP stationary phase, the bulkiness and the polarity of the solute significantly influencing the retention. The separation efficiency of PBD-coated alumina was compared with those of other stationary phases for the analysis of *Catharanthus* alkaloids. It was established that the pH of the mobile phase, the concentration and type of the organic modifier, and the presence of salt simultaneously influence the retention. In this special case, the efficiency of PBD-coated alumina was inferior to that of ODS.[12] The retention characteristics of polyethylene-coated alumina (PE-Alu) have been studied in detail using various nonionic surfactants as model compounds.[13] It was found that PE-Alu behaves as an RP stationary phase and separates the surfactants according to the character of the hydrophobic moiety. The relationship between the physicochemical descriptors of 25 aromatic solutes and their retention on PE-coated silica (PE-Si) and PE-Alu was elucidated by stepwise regression analysis.[14]

$$\log k_{PEAlu}' = 0.144(\pm 0.092)$$
$$+ 0.9325(\pm 0.0505)\log k_{PESi}' \tag{8}$$

$$n = 25;\ r = 0.968;\ s = 0.333;\ F = 340$$

$$\log k_{PEAlu}' = 1.474(\pm 0.376) + 0.5976(\pm 0.2990)R_2$$
$$+ 0.9162(\pm 0.3025)\pi_2{}^* 0.8279(\pm 0.2624)$$
$$\times \sum \beta_2{}^H + 3.206(\pm 0.304)V_x \tag{9}$$

$$n = 24;\ r = 0.958;\ s = 0.397;\ F = 53$$

$$\log k_{PESi}' = 1.670(\pm 0.373)0.9167(\pm 0.2120)\pi_2{}^*$$
$$+ 3.842(\pm 0.267)V_x \tag{10}$$

$$n = 24;\ r = 0.956;\ s = 0.398$$

$$\log k_{PEAlu}' = 2.122(\pm 0.549)2.068(\pm 0.798)\delta_{max}$$
$$+ 0.3283(\pm 0.0897)\mu$$
$$+ 0.00940(\pm 0.00111)V_{aq} \tag{11}$$

$$n = 25;\ r = 0.926;\ s = 0.525;\ F = 42$$

$$\log k_{PESi}' = 2.401(\pm 0.456)2.424(\pm 0.662)\delta_{max}$$
$$+ 0.2821(\pm 0.0746)\mu$$
$$+ 0.01057(\pm 0.00092)V_{aq} \tag{12}$$

$n = 25;\ r = 0.953;\ s = 0.436;\ F = 69$ where μ is the total dipole moment, δ_{max} is the maximum electron excess charge (electron deficiency) on an atom in the solute molecule, V_{aq} is the solvent (water) accessible molecular volume. Calculations proved that both stationary phases behave as RP supports, with the retention strength of PE-Si being higher. The retention can be successfully related to the molecular parameters included in the calculations.

The differences observed may be attributed to free silanol groups on the silica surface not covered by the polyethylene coating.

Various synthetic methods were developed for the covalent binding of hydrophobic ligands to the surface of alumina.[15–16] These methods generally resulted in real RP stationary phases with higher pH stability than the silica-based stationary phase. The retention of 33 commercial pesticides was determined on OD alumina, ODS, and on alumina support.[17] Stepwise regression analysis proved that the retention of pesticides on OD alumina does not significantly depend on the lipophilicity of pesticides determined on a silica-based RP support. This discrepancy was tentatively explained by the different binding characteristics of the adsorption centers on the surfaces of silica and alumina not covered by the hydrophobic ligand. Using another set of solutes, the similarities of the retention order on both OD alumina and ODS was observed.[18]

The spectral mapping technique, combined with cluster analysis and nonlinear mapping, was used for the comparison of the retention behaviors of RP silica and RP alumina columns using tributylphenol ethylene oxide oligomers as model compounds.[19] The columns included in the experiments were C_1, C_2, C_6, C_8, and C_{18} silica, Pe-Si, alumina, and C_{18} alumina. The retention strengths of RP-HPLC columns showed considerable variations, the strongest and the weakest being C_{18} silica and PE-Si, respectively. The retention strength of alkyl-modified silicas depended linearly on the carbon load. The two-dimensional selectivity map and the cluster dendogram of the column selectivities are depicted in Figs. 3 and 4, respectively. From the results, it is clear that both the strength and the selectivity of retention show high variations. C_1, PE-Si, and PE-Alu exhibited similar retention selectivity.

This result may be attributed to the fact that polyethylene chains lie parallel to the surface of the support, with the end groups being anchored to the polar adsorption centers. According to this model, only the surface of the polyethylene coating pointing toward the mobile phase is available to the solutes in the mobile phase. This apolar polymeric layer is similar in thickness to the C_1 coating but differs markedly from the longer alkyl chains that are more or less immersed in the mobile phase and are providing more CH_2 groups for the binding of apolar solutes. The conclusions drawn from the two-dimensional non-linear mapping and cluster analysis are similar, suggesting that both methods can be used for the reduction of the dimensionality of a multidimensional spectral map.

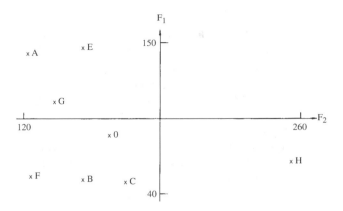

Fig. 4 Cluster dendogram of RP-HPLC columns. For symbols, see Fig. 3. (From Ref.[19].)

CONCLUSIONS

The development of new nonsilica-based stationary phases was mainly motivated by the poor stability of silica and modified silicas at extreme pH values. It was shown that alumina supports can be used successfully for the solution of a wide variety of analytical problems concerning the separation of natural products, pharmaceuticals, and xenobiotics at any mobile phase pH. Moreover, alumina shows different retention characteristics than silica (i.e., it shows higher separation capacity for positional isomers). The data may facilitate not only the solution of various practical separation problems in LC, but will also promote a better understanding of the underlying physico-chemical principles governing retention.

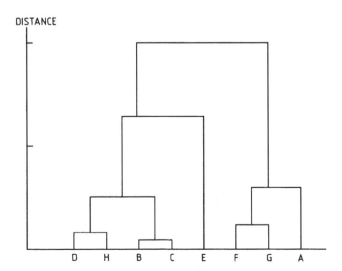

Fig. 3 Two-dimensional nonlinear selectivity map of reversed-phase HPLC columns. Number of iterations: 377. Maximum error: 2.1×10^{-3}. (A) C_1 silica; (B) C_2 silica; (C) C_6 silica; (D) C_8 silica; (E) C_{18} silica; (F) polyethylene coated silica; (G) polyethylene-coated alumina; (H) C_{18} alumina. (From Ref.[19].)

REFERENCES

1. Snyder, L.R. Separability of aromatic isomers on alumina: Mechanism of adsorption. J. Phys. Chem. **1962**, *72*, 489–492.
2. Laurent, C.; Billiet, H.A.H.; de Galan, L. The use of organic modifiers in ion exchange chromatography on alumina. The separation of basic drugs. Chromatographia **1983**, *17*, 253–258.
3. Laurent, C.; Billiet, H.A.H.; de Galan, L. On the use of alumina in HPLC aqueous mobile phases at extreme pH. Chromatographia **1983**, *17*, 394–398.
4. Laurent, C.; Billiet, H.A.H.; de Galan, L.; Byutenhuys, F.A.; van der Maeden, F.P.B. High-performance liquid chromatography of proteins on alumina. J. Chromatogr. **1984**, *287*, 45–49.
5. Ahmad, J. The use of alumina as stationary phase for thin layer chromatography of inorganic and

organometallic compounds. JPC, J. Planar Chromatogr. Mod. TLC **1996**, *9*, 236–239.

6. Szilagyi, A.; Forgács, E.; Cserháti, T. Separation of nonionic surfactants according to the length of the ethylene oxide chain on alumina layers. Toxicol. Environ. Chem. **1998**, *65*, 95–102.

7. Forgács, E.; Cserháti, T. Retention behavior of nonylphenyl ethyleneoxide oligomers on an alumina high-performance liquid chromatographic column. Fresenius J. Anal. Chem. **1995**, *351*, 688–689.

8. Forgács, E.; Cserháti, T. Retention behaviour of tributylphenol ethylene oxide oligomers on an alumina high performance liquid chromatographic column. J. Chromatogr. **1994**, *661*, 239–243.

9. Kòsa, A.; Dòbo, A.; Vèkey, K.; Forgács, E. Separation and identification of nonylphenylethylene oxide oligomers by high-performance liquid chromatography with UV and mass spectrometric detection. J. Chromatogr. A. **1998**, *819*, 297–302.

10. Mao, Y.; Fung, B.M. Use of alumina with anchored polymer coating as packing material for reversed-phase high performance liquid chromatography. J. Chromatogr. A. **1997**, *790*, 9–15.

11. Kaliszan, R.; Osmialowski, K. Correlation between chemical structure of non-congeneric solutes and their retention on polybutadiene-coated alumina. J. Chromatogr. **1990**, *506*, 3–16.

12. Theodoridis, G.; Papadoyannis, I.N.; Hermans-Lokkerbol, A.; Verpoorte, R. A study of the behaviour of some new column materials in the chromatographic analysis of *Catharanthus* alkaloids. Chromatographia **1997**, *45*, 52–57.

13. Forgács, E. Comparison of reversed-phase chromatographic systems with principal component and cluster analysis. Anal. Chim. Acta **1994**, *296*, 235.

14. Cserháti, T.; Forgács, E.; Payer, K.; Haber, P.; Kaliszan, R.; Nasal, A. Quantitative structure–retention relationships in separation mechanism studies on polyethylene-coated silica and alumina stationary phases. LC GC Int. **1998**, *11*, 240.

15. Pesek, J.J.; Lin, H.-D. Evaluation of synthetic procedures for the chemical modification of alumina for HPLC. Chromatographia **1989**, *28*, 565–574.

16. Pesek, J.J.; Sandoval, E., Jr.; Su, M. New alumina-based stationary phases for high-performance liquid chromatography. J. Chromatogr. **1993**, *630*, 95–103.

17. Cserháti, T.; Forgács, E. Separation of pesticides on an octadecyl-coated alumina column. Fresenius J. Anal. Chem. **1997**, *358*, 558–560.

18. Haky, J.E.; Vemulapalli, S.; Wieserman, L.F. Comparison of octadecyl-bonded alumina and silica for reversed-phase high performance liquid chromatography. J. Chromatogr. **1990**, *505*, 307–318.

19. Forgács, E.; Cserháti, T. Determination of retention behaviour of some non-ionic surfactants on reversed-phase high-performance liquid chromatography supports by spectral mapping in combination with cluster analysis or non-linear mapping. J. Chromatogr. A. **1996**, *722*, 281–287.

Amines, Amino Acids, Amides, and Imides: Derivatization for GC Analysis

Igor G. Zenkevich
Chemical Research Institute, St. Petersburg State University, St. Petersburg, Russia

INTRODUCTION

The amines are an extensive class of organic compounds of general formula RNH_2 (primary), $RR'NH$ (secondary), and $RR'R''N$ (tertiary). Their chemical and chromatographic properties are determined by the presence of a basic functional group and active hydrogen atoms in the molecule. The basicity is very different for aliphatic amines (pK_a 10.5 ± 0.8) and substituted anilines (pK_a 4.9 ± 0.3) owing to the p–π conjugation N–Ar. Amides are derivatives of carboxylic acids with structural fragments –CO–N<, >PO–N<, or –SO_2–N< (including cyclic structures), while imides have two acyl fragments connected through a nitrogen atom, e.g., –CO–NR–CO– or –CO–NR–SO_2– (for example, saccharine). The compounds with structural fragment >C=N– (so-called Schiff bases) also have the synonym "imines." As the last compounds have no active hydrogen atoms, they need no derivatization and, moreover, themselves are the derivatives of both amines and carbonyl compounds.

The simplest members of the amine class usually are volatile enough for their direct GC analysis. For example, the comparison of simple tertiary amines with the structurally analogous isoalkanes indicates that both the boiling points at atmospheric pressure and the values of retention indices (RIs) on standard nonpolar phases for the tertiary amines are lower than those for corresponding isoalkanes (!).

	T_b, °C	$RI_{nonpolar}$
tert-Amine R_3N		
Et_3N	89.4	677 ± 8
Pr_3N	156.5	933 ± 8
Hydrocarbon analog R_3CH		
Et_3CH	93.4	687 ± 3
Pr_3CH	157.5	936 ± 15

All RIs with standard deviations are randomized interlaboratory data.

Nevertheless, the principal reason for the derivatization of amines and related compounds is the high sensitivity of amino compounds to various chemical agents. Among the multitude of organic substances, only amines in acidic media form nonvolatile salts, which can make their GC analysis impossible. Of course, these salts can be reconverted to free bases by treatment with other (more basic) amines or a large excess of ammonia.[1] Besides that, these compounds are very sensitive to the action of various electrophilic reagents; their exhaustive alkylation gives nonvolatile quaternized ammonium salts, which cannot be restored to the initial analytes. Finally, amino compounds are easily oxidized.

Hence, owing to the abovementioned facts, the principal goal of derivatization of amines and related substances is to protect these compounds from chemical transformations prior to GC analysis by their conversion to more stable derivatives.

Another reason is in accordance with the general principles of derivatization. When other polar functional groups with active hydrogen atoms are present in the molecules, the derivatization of one or (better) all of them becomes necessary. A typical example of these compounds is amino acids, which exist in the form of inner-molecular salts $RCH(NH_3^+)CO_2^-$, in the solid state.

DERIVATIZATION OF AMINES

The principal directions of amino compound derivatization for GC analysis include the following types of chemical reactions:[2–5]

- Acylation: $RR'NH + R''COX + B$
  $$\rightarrow RR'NCOR'' + BH^+X^-$$
- Formation of Schiff bases (only for primary amines): $RNH_2 + R'R''CO$
  $$\rightarrow RN=CR'R'' + H_2O$$
- Alkylation: $RR'NH + R''X \rightarrow RR'NR'' + XH$
- Silylation: $RR'NH + XSi(CH_3)_3$
  $$\rightarrow RR'N–Si(CH_3)_3 + XH$$

The first group of reactions (acylation) includes the maximum number of examples. Numerous recommended reagents are listed in Table 1 and belong to

Encyclopedia of Chromatography DOI: 10.1081/E-ECHR-120039944

two classes of chemicals: anhydrides (X=OCOR'') and chloroanhydrides (X=Cl). The most widely used among them are acetic, trifluoroacetic (TFA), pentafluoropropanoic (PFP), and heptafluorobutanoic (HFB) anhydrides, as well as pentafluorobenzoyl chloride.[6] The by-products of acylation in all cases are acids; these reactions need basic media (additives of pyridine or tertiary-amines) to prevent the formation of nonvolatile salts from the analytes. The technique of derivatization is extremely simple: Sample mixtures are allowed to stand with acylating reagents for some minutes prior to analysis.

The anhydrides and chloroanhydrides of chlorinated acetic acids and pentafluorobenzoyl chloride (electrophoric reagents) are used for the synthesis of chlorinated amides for GC analysis with selective detectors. Diethylpyrocarbonate converts primary and secondary amines (including NH_3) into N-substituted carbamates:

$$RR'NH + O(CO_2C_2H_5)_2$$
$$\rightarrow RR'NCO_2C_2H_5 + CO_2 + C_2H_5OH$$

The next group of derivatization reactions is the formation of Schiff bases from primary amines with carbonyl compounds. Some recommended reagents are listed in Table 2. Aromatic aldehydes (including pentafluorobenzaldehyde as electrophoric reagent[7]) are much more reactive in this condensation compared with ketones and aliphatic compounds (of the latter, only low boiling acetone and cyclohexanone have been used in GC practice). All carbonyl compounds can be made to react with amines in the form of their different synthetic analogs, primarily acetals or ketals. So, dimethylformamide dialkylacetals, $(CH_3)_2N–CH(OCH_3)_2$, react with primary amines with formation of N-dimethylaminomethylene derivatives, $R–N=CH–N(CH_3)_2$. So long as these reagents simultaneously provide the esterification of carboxyl groups, they are recommended for single-stage derivatization of amino acids (discussed later).

Carbon disulfide as the thio-analog of carbonyl compounds reacts with primary amines with the resultant formation of alkyl isothiocyanates, which have lower RIs than other derivatives of primary

Table 1 Physicochemical and GC properties of some acylation reagents

Reagent (abbreviation)	MW	T_b, °C (P)	d_4^{20}	n_D^{20}	$RI_{nonpolar}$	By-products ($RI_{nonpolar}$)
Acetic anhydride	102	139.6	1.08	1.390	706 ± 9	CH_3CO_2H (638 ± 10)
Trifluoroacetic anhydride (TFA)	210	39–40	1.511	1.268	515 ± 6	CF_3CO_2H (744 ± 6)
Pentafluoropropionic anhydride (PFPA)	310	70–72	1.588	—	606 ± 6[a]	$C_2F_5CO_2H$ (781 ± 12)
Heptafluorobutyric anhydride (HFBA)	410	109–111	1.674	—	745 ± 4[a]	$C_3F_7CO_2H$ (863 ± 16)
bis-Trifluoroacetyl methylamine (MBTFA)	223	120–122	1.547	1.346	773 ± 16[a]	$CF_3CONHCH_3$ (540)
N-Trifluoroacetyl imidazole (TFAI)	164	137–138	1.442	1.424	830 ± 21[a]	Imidazole (1072 ± 17)
Chloroacetic anhydride	170	203	1.550	—	1116 ± 18[a]	$ClCH_2CO_2H$ (864 ± 3)
Dichloroacetic anhydride	238	214–216	—	—	1248 ± 14[a]	Cl_2CHCO_2H (1048 ± 23)
Trichloroacetic anhydride	306	222–224	1.691	1.484	1471 ± 27[a]	CCl_3CO_2H (1270)[a]
Benzoyl chloride	140	197–198	1.211	1.553	1046 ± 9	$C_6H_5CO_2H$ (1201 ± 24)
Pentafluorobenzoyl chloride	230	158–159	1.669	1.453	922 ± 14[a]	$C_6F_5CH_2OH$ (934 ± 16)[a]
Chlorodifluoroacetic anhydride	242	92–93	—	1.348	679 ± 8[a]	$ClCF_2CO_2H$ (793)[a]
Acetyl chloride	78	51.8	1.104	1.389	542 ± 7	CH_3CO_2H (638 ± 10)
Chloroacetyl chloride	112	106.1	1.419	1.454	622 ± 8	$ClCH_2CO_2H$ (864 ± 3)
Dichloroacetyl chloride	146	107–108	1.532	1.460	726 ± 19[a]	Cl_2CHCO_2H (1048 ± 23)
Trichloroacetyl chloride	180	118	1.629	1.470	778 ± 15	CCl_3CO_2H (1270)[a]
Pivaloyl anhydride	186	192–193	0.918	1.409	1053 ± 31[a]	$(CH_3)_3CCO_2H$ (804)
Diethylpyrocarbonate	162	93–94 (18)	1.101	1.398	—	C_2H_5OH (452 ± 18)

[a]Estimated RI values.

Table 2 Physicochemical and GC properties of some carbonyl reagents and their analogs for derivatization of amino compounds

Reagent (abbreviation)	MW	T_b, °C	d_4^{20}	n_D^{20}	$RI_{nonpolar}$
Acetone	58	56.2	0.791	1.359	472 ± 12
Pentafluorobenzaldehyde	196	166–168	1.588	1.450	943 ± 22[a]
Thiophen-2-carboxaldehyde	112	198	1.200	1.590	966 ± 9
Carbon disulfide	76	46.3	1.263	1.628	530 ± 9
Methyl isothiocyanate[b]	73	118	1.069	1.525	689 ± 16
Phenyl isothiocyanate[b]	135	219–221	1.130	1.652	1163 ± 7
Dimethylformamide dimethyl acetal (DMFDMA)[c]	119	106	0.897	1.397	726 ± 4
Dimethylformamide diethyl acetal (DMFDEA)[c]	147	134–136	0.859	1.400	826 ± 5[a]

[a]Estimated RI values.
[b]Only for derivatization of amino acids; with monofunctional amines, nonvolatile products can be formed.
[c]Bifunctional reagents; the by-products are MeOH (EtOH) and DMFA (749 ± 16).

amines, including *N*-trimethylsilylated amines. The sole by-product of this reaction is gaseous hydrogen sulfide:

$$RNH_2 + CS_2 \rightarrow RNCS + H_2S$$

	$RI_{nonpolar}$
R in RNHSi(CH$_3$)$_3$	
Me	689 ± 21
Et	756 ± 11
R in R–NCS	
Me	689 ± 16
Et	736 ± 5

All RIs with standard deviations are randomized interlaboratory data.

The alkylation of amines (including polyamines formed by the reduction of polypeptides) was a highly popular method of derivatization in peptide chemistry before the appearance of contemporary mass spectrometric techniques for the analysis of nonvolatile compounds [fast atom bombardment (FAB), matrix assisted laser desorption ionization (MALDI), etc.]. Direct alkylation of amines by alkyl halides (Hoffman reaction) can finally lead to nonvolatile ammonium salts and, hence, other "soft" reagents should be used. For example, exhaustive methylation without quaternization can be provided by the mixtures $CH_2O/NaBH_4/(H^+)$ or CH_2O/formic acid.

Silylation of amines is a well investigated,[8] but relatively rarely used, method for their derivatization at present owing to the facile hydrolysis of the resultant *N*-trimethylsilyl (TMS) compounds. It leads to the formation, in the reaction mixtures, of both *mono*-(RNHTMS) and *bis*-TMS [RN(TMS)$_2$] derivatives. This multiplicity of products from the same precursor creates some difficulties in data interpretation. The *N*-(*tert*-butyldimethylsilyl) (TBDMS) derivatives,

recently introduced into analytical practice, are more resistant to hydrolysis and their formation is unambiguous (only N-monosubstituted compounds are formed) due to steric reasons.

Sometimes, special derivatization methods are needed for amines for optimization of their determination in dilute water samples, especially in combination with preconcentration of analytes. This explains the choice of conversion of primary aromatic amines by iodine into the corresponding iodoarenes, $ArNH_2 + 1.5I_2 \rightarrow ArI + 2HI + 0.5N_2$.[9]

In the example of amines, it is interesting to note an unusual derivatization reaction, when a single reagent provides double functionalization of the protected group (primary NH_2) by different structural fragments. This reagent is trimethylsilylketene, $Me_3Si–CH=C=O$, which reacts with primary amines giving their TMS-acetyl derivatives:[10]

$$R–NH_2 + Me_3Si–CH=C=O$$
$$\rightarrow [R–N(COCH_3)–SiMe_3]$$
$$\rightarrow R–N=C(CH_3)–O–SiMe_3$$

This method underwent no further development owing to the absence of obvious advantages of these derivatives, but it illustrates the specific chemical properties of an interesting reagent.

Tertiary amines have no active hydrogen atoms, and their derivatization in the generally accepted sense is not required. If necessary (for GC analysis with selective detectors), the cleavage of N–CH$_3$ bonds by chloroformates can be used:

$$RR'N–CH_3 + CCl_3CH_2OCOCl$$
$$\rightarrow RR'N–CO_2CH_2CCl_3$$

DERIVATIZATION OF AMINO ACIDS

Mixtures of amino acids are some of the most important objects for chromatographic analysis. Some dozen methods have been proposed for derivatization of these compounds for their GC determination. The most widely used can be subdivided into two types:

1. Separate derivatization of functional groups $-CO_2H$ and $-NH_2$ by different reagents; and
2. Protection of both groups by only one reagent.

Typical derivatives of the first type are various esters (Me, Et, Pr, *iso*-Pr, Bu, *iso*-Bu, *sec*-Bu, Am, *iso*-Am, etc.) of *N*-acyl (acetyl, trifluoroacetyl, pentafluoropropionyl, heptafluorobutyryl, etc.) amino acids. The butyl esters of *N*-TFA amino acids even have a special abbreviation: TAB derivatives. The two-stage process includes the esterification of amino acids by an excess of the corresponding alcohol in the presence of HCl and, after the evaporation of volatile compounds, the treatment of the nonvolatile hydrochlorides of alkyl esters by acylating reagents:

$$\text{R-CH}\begin{smallmatrix}CO_2H\\NH_2\end{smallmatrix} \quad 1)\ +\ R'OH\,/\,H^+$$

$$\rightarrow \text{R-CH}\begin{smallmatrix}CO_2R'\\NH_3{}^+Cl^-\end{smallmatrix} \quad 2)\ +\ R''COX \rightarrow \text{R-CH}\begin{smallmatrix}CO_2R'\\NHCOR''\end{smallmatrix}$$

Some variations of this procedure are known. Instead of *N*-acylation, the treatment of intermediate esters by CS_2/Et_2N and CH_3OCOCl with the formation of 2-alkoxycarbonylisothiocyanates (I), or by carbonyl compounds, which leads to the Schiff bases (II), has been proposed. However, the analytical advantages of these derivatives are not obvious.

$$\text{R-CH}\begin{smallmatrix}CO_2R'\\NCS\end{smallmatrix} \quad (I) \qquad \text{R-CH}\begin{smallmatrix}CO_2R'\\N=CHR''\end{smallmatrix} \quad (II)$$

$$CF_3CH_2NH\!-\!CH_2\!-\!\underset{CH_2Ph}{\underset{|}{CH}}\!-\!NH\!-\!\underset{CH_2OSi(CH_3)_3}{\underset{|}{\overset{CH_2Ph}{\overset{|}{CH}}}} \quad (III)$$

The same types of *N*-acyl-*O*-alkyl derivatives can also be used for GC analysis of the simplest oligopeptides (at least dipeptides and tripeptides). Other, obsolete, recommendations on the GC analysis of these compounds include different sequences of their reduction by $LiAlH_4$ into polyaminoalcohols,

followed by *N*-acylation or permethylation and final silylation of OH-groups. For example, *N*-TFA dipeptide Phe-Phe, after reduction and silylation, gives a compound (III) with an RI of 2390 on semistandard phase Dexsil-300.

A typical example of single-stage derivatization of amino acids is their treatment by isopropyl bromide in the presence LiH with the formation of *N*-isopropylated isopropyl esters. Unfortunately, this reaction can take place only in high boiling bipolar aprotonic solvents like dimethyl sulfoxide (DMSO; T_b 189°C, $RI_{nonpolar}$ 790 ± 18), which is the significant restriction for its application in practice (the removal of such a high boiling solvent from reaction mixtures without loss of target analytes seems impossible). A more important method is based on the reaction of amino acids with methyl or phenyl isothiocyanates with the formation of 3-methyl (phenyl) hydantoins:

$$\text{R-CH}\begin{smallmatrix}CO_2H\\NH_2\end{smallmatrix} + Me(Ph)NCS\ (pH \approx 9)$$

$$\rightarrow \text{R-CH}\begin{smallmatrix}CO_2H\\NHCSNHMe(Ph)\end{smallmatrix} \quad (pH \approx 1) \rightarrow$$

The stable nonvolatile intermediate phenylthiocarbamoyl derivatives are formed in basic media and can be analyzed directly by RP HPLC. Their cyclization into hydantoins requires acid catalysis. This mode of derivatization is a very important supplement to the Edman's method of N-terminated sequencing of polypeptides. Before GC analysis, all hydantoins can be converted into *N*-trifluoroacetyl or enol-*O*-trimethylsilyl derivatives, which increases the selectivity of their determination in complex matrices.

N-Acylated amino acids in the presence of water coupling reagents [dicyclohexylcarbodiimide or an excess of trifluoroacetic anhydride (TFAA)] form other cyclic derivatives—azlactones (2,4-disubstituted oxazolin-5-ones):

$$\text{R-CH}\begin{smallmatrix}CO_2H\\NHCOR'\end{smallmatrix} - H_2O \rightarrow$$

One of the most popular methods of single-stage amino acid derivatization at present is their conversion into *N*,*O*(*S*)-*tert*-butyldimethylsilyl (TBDMS) derivatives [the reagents: *tert*-butyldimethylsilyl trifluoroacetamide (MTBSTFA) or its *N*-Me analog].[11–13]

Another way was proposed at the beginning of 1970s and is based on amino acid interaction with dimethylformamide dialkylacetals, $(CH_2)_2NCH(OR')_2$ (R = Me, Et, Pr, iso-Pr, Bu, Am), with the formation of N-dimethylaminomethylene derivatives of amino acids alkyl esters:[14]

$$R-CH \begin{array}{c} CO_2H \\ NH_2 \end{array} + (CH_3)_2NCH(OR')_2$$

$$\rightarrow R-CH \begin{array}{c} CO_2R' \\ N=CHN(CH_3)_2 \end{array} + (CH_3)_2NCHO + R'OH$$

The GC separation of enantiomeric D- and L-amino acids with nonchiral phases needs their conversion into diastereomeric derivatives. The second chiral center in the molecule (∗) arises after esterification by optically active alcohols [(R)- or (S)-2-BuOH, 2-AmOH, pinacolol, (–)-menthol, etc.] or NH_2-group acylation by chiral reagents, for example, α-methoxy-α-trifluoromethylphenylacetyl chloride [MTPAC, (IV)], N-trifluoroacetyl-L-prolyl chloride [N-TFA-L-Pro-Cl,(V)], N-trifluoroacetylthiazolidine-4-carbonyl chloride (VI), etc.

$$C_6H_5 \begin{array}{c} OCH_3 \\ | \\ -C-COCl \\ | \\ CF_3 \end{array} \quad (IV)$$

(V) structure with COCl, N-COCF₃
(VI) structure with S, COCl, N-COCF₃

DERIVATIZATION OF AMIDES

The selection of derivatization methods of amides and imides is not as great as for other classes of amino compounds. The active hydrogen atoms in the structural fragments –CO–NH– and SO_2–NH– are highly acidic and, hence, sometimes the recommended TMS, acetyl, or TFA derivatives of these compounds are unstable relative to hydrolysis. The best derivatization method is their exhaustive alkylation (preferably methylation), because permethylated amides and imides are volatile enough for GC analysis.

This general statement can be illustrated by the retention data of methylated derivatives of urea, $CO(NH_2)_2$, as the simplest amide: Both the initial compound and its mono- and two dimethyl homologs cannot be analyzed by GC owing to their nonvolatility. The RI of trimethyl urea on standard nonpolar polydimethylsiloxanes is 976 ± 28, whereas for tetramethyl urea it is 956 ± 5 (see the high interlaboratory reproducibility of the last RI value compared with the previous one).

The exhaustive methylation of amides can be realized with rather high yields by their reactions with dimethyl sulfate/EtN(iso-Pr)$_2$,[15] by CH_3J in acetone solution, with CH_3J in the presence of K_2CO_3 or LiH in DMSO, in heterophase "water–organic solvent" systems (together with the extraction of derivatives from matrices), and directly in the injector of gas chromatographs (so-called flash methylation) by $PhNMe_3^+OH^-$ (TMPAH). These modes of derivatization precede the GC determination of numerous diuretics (acetazolamide, ethacrinic acid, clopamide, etc.),[16] some barbiturates and their metabolites, xanthines (theophylline), different urea and carbamic pesticides (monuron, fenuron, linuron, and their metabolites), and so forth.

CONCLUSIONS

The main reason for the derivatization of amines and related compounds is their chemical lability. Compared with other classes of organic substances, only amines can reversibly form nonvolatile salts with acids. These compounds are very sensitive to the action of electrophilic reagents, which can irreversibly convert amines into nonvolatile ammonium salts.

Hence, one of the specific reasons for the derivatization of amines is to prevent chemical transformations of analytes prior to their GC analysis.

ARTICLE OF FURTHER INTEREST

GC System Instrumentation, p. 682.

REFERENCES

1. Nagase, M. Conversion of amines from their salts into free bases with ammonia. Bunseki Kagaku **1980**, *29* (5), 293–297 (in Japanese).
2. Blau, K., King, G.S., Eds.; *Handbook of Derivatives for Chromatography*; Heiden: London, 1977; 576.
3. Knapp, D.R. *Handbook of Analytical Derivatization Reactions*; John Wiley & Sons: New York, 1979; 741.
4. Drozd, J. *Chemical Derivatization in Gas Chromatography*; Journal of Chromatography Library; Elsevier: Amsterdam, 1981; Vol. 19, 232.
5. Blau, K., Halket, J.M., Eds.; *Handbook of Derivatives for Chromatography*; 2nd Ed.; John Wiley & Sons: New York, 1993; 369.
6. Jia, M.; Wu, W.W.; Yost, M.G.; Chadik, P.A.; Stacpoole, P.W.; Henderson, G.N. Simultaneous determination of trace levels of nine haloacetic acids in biological samples as their pentafluorobenzyl derivatives by gas chromatography/

tandem mass spectrometry in electron capture negative ion chemical ionization mode. Anal. Chem. **2003**, *75*, 4065–4080.

7. Avery, M.J.; Junk, G.A. Gas chromatographic/ mass spectrometric determination of water-soluble primary amines as their pentafluoro-benzaldehyde imines. Anal. Chem. **1985**, *57* (4), 790–792.

8. Iwase, H.; Takeuchi, Y.; Murai, A. Gas chromatography–mass spectrometry of TMS derivatives of amines. Chem. Pharm. Bull. **1979**, *27* (4), 1009–1014.

9. Schmidt, T.C.; Less, M.; Haas, R.; VanLow, E.; Steinbach, K.; Stork, G. Gas chromatographic determination of aromatic amines in water samples after solid phase extraction and derivatization with iodine. J. Chromatogr. A. **1998**, *810*, 161–172.

10. Coutts, R.T.; Jones, G.R.; Benderly, A.; Mac, A.L.C. A note on the synthesis and gas chromatographic mass-spectrometric properties of *N*-(trimethylsilyl)-acetates of amphetamine and analogs. J. Chromatogr. Sci. **1979**, *17* (6), 350–352.

11. Biermann, C.J.; Kinoshita, C.M.; Marlett, J.A.; Steele, R.D. Analysis of amino acids as *tert.*-butyldimethylsilyl derivatives by gas chromatography. J. Chromatogr. **1986**, *357*, 330–334.

12. Mawhinney, T.P.; Robinett, R.S.R.; Atalay, A.; Madson, M.A. Analysis of amino acids as their *tert.*-butyldimethylsilyl derivatives by gas–liquid chromatography and mass spectrometry. J. Chromatogr. **1986**, *358*, 231–242.

13. Chaves das Neves, H.T.; Vasconcelos, A.M.P. Capillary gas chromatography of amino acids, including asparagine and glutamine: sensitive gas chromatographic–mass spectrometric and selected ion monitoring gas chromatographic–mass spectrometric detection of the *N,O(S)-tert.*-butyldimethylsilyl derivatives. J. Chromatogr. **1987**, *392*, 249–258.

14. Horman, I.; Hesford, F.J. Amino acid mixture analysis by mass spectrometry in the form of their dimethylaminomethylene methyl esters. Biomed. Mass Spectrom. **1974**, *1* (2), 115–119; Horman, I.; Hesford, F.J. Nestle Res. News **1976/1977**, 100–103.

15. Nazareth, A.; Joppich, M.; Pauthani, A.; Fisher, D.; Giese, R.W. Alkylation with dialkyl sulphate and ethyl-diisopropyl amine. J. Chromatogr. **1985**, *319* (3), 382–386.

16. Carreras, D.; Imas, C.; Navajas, R.; Garcia, M.A.; Rodrigues, C.; Rodrigues, A.F.; Cortes, R. Comparison of derivatization procedures for the determination of diuretics in urine by GC–MS. J. Chromatogr. A. **1994**, *683*, 195–202.

Amino Acids: Analysis by HPLC: Advanced Techniques

Susana Maria Halpine
STArt! teaching Science Through Art, Playa del Rey, California, U.S.A.

INTRODUCTION

Amino acid analysis (AAA) is a classic analytical technique that characterizes proteins and peptides based on the composition of their constituent amino acids.[1,2] It provides qualitative identification and is essential for the accurate quantification of proteinaceous materials.

Amino acid analysis is widely applied in research, clinical facilities, and industry. It is a fundamental technique in biotechnology, used to determine the concentration of peptide solutions, to confirm protein binding in antibody conjugates, and for end-terminal analysis following enzymatic digestion. Clinical applications include diagnosing metabolic disorders in newborns. In industry, it is used for quality control of products ranging from animal feed to protein pharmaceuticals.

The analysis of a polypeptide typically involves four steps: hydrolysis (or deproteination with physiological samples), separation, derivatization, and detection. Hydrolysis breaks the peptide bonds and releases free amino acids, which are then separated by side-group using column chromatography. Derivatization with a chromogenic reagent enhances the separation and spectral properties of the amino acids, and is required for sensitive detection. A data processing system compares the resulting chromatogram, based on peak area or peak height, to a calibrated standard (Fig. 1A). The results, expressed as mole percent and micrograms of residue per sample, determine the percentage composition of each amino acid as well as the total amount of protein in the sample. Unknown proteins may be identified by comparing their amino acid composition with those in protein databases. Successful identification of unknown proteins may be achieved using internet search programs.[3]

Other techniques, such as capillary electrophoresis and matrix-assisted laser desorption ionization (MALDI) mass spectrometry, provide qualitative analyses–often with greater speed and sensitivity.[1] Nevertheless, AAA by HPLC complements other structural analysis techniques, such as peptide sequencing, and remains indispensable for quantifying the composition and absolute content of proteinaceous materials.[2]

PEPTIDE HYDROLYSIS

Acid hydrolysis of proteins and peptides yields 16 of the 20 DNA-coded amino acids; tryptophan is destroyed, cysteine recovery is unreliable, and asparagine and glutamine are converted to aspartic acid and glutamic acid, respectively.[1,2,4,5] Furthermore, some side-groups, such as the hydroxyl in serine, promote the breakdown of the residue, while aliphatic amino acids such as valine and leucine, protected by stearic hindrance, require longer hydrolysis time. This variation in yield can be overcome by hydrolyzing samples for 24, 48, and 72 hr and extrapolating the results to zero time point.

Conventional hydrolysis exposes the polypeptide to 6 M HCl under vacuum at 110°C for 20–24 hr. Protective agents, such as 0.1% phenol, are added to improve recovery. Gas-phase hydrolysis, in which the acid is delivered as a vapor, gives comparable results to liquid-phase hydrolysis. Additionally, the gas phase minimizes acid contaminants and allows parallel hydrolysis of standards and samples within the same chamber.

The reaction rate doubles with every 10°C increase, so that hydrolysis at 145°C for 4 hr gives results comparable to those from the conventional method. Microwave hydrolysis reduces analysis time to 30–45 min. Alternative hydrolysis agents include methane sulfonic acid, which often gives better recovery but is nonvolatile, and alkaline hydrolysis, used in the analysis of tryptophan, proteoglycans, and proteolipids.

Careful sample preparation and handling during the hydrolysis step are critical for maintaining accurate and reproducible results.[1,2,4–6] Salts, metal ions, and other buffer components remaining in a sample may accelerate hydrolysis, producing unreliable quantification. The Maillard reaction between amino acids and carbohydrates results in colored condensation products (humin) and decreased yield.[7] Routine method calibration with proteins and amino acid standards, use of an internal standard (1 nmol norleucine is used for sensitive analysis), and control blanks are strongly recommended, along with steps to minimize background contaminants (Fig. 1B). Attention to housekeeping details, such as cleaning glassware and baking in a muffle furnace, can minimize background contaminants. The practical limit for high sensitivity hydrolysis is 10–50 ng of sample; below this amount,

Encyclopedia of Chromatography DOI: 10.1081/E-ECHR-120039875

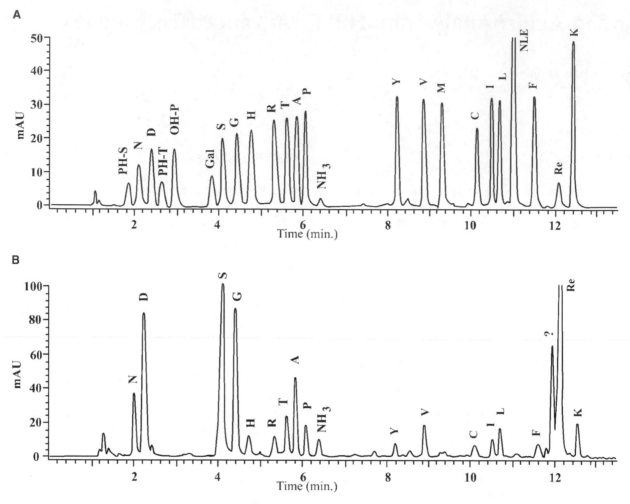

Fig. 1 A PTC-amino acid standard (200 pmol), including phosphoserine (PH-S), aspartate (N), glutamate (D), phosphothreonine (PH-T), hydroxyproline (OH-P), galactosamine (Gal), serine (S), glycine (G), histidine (H), arginine (R), threonine (T), alanine (A), proline (P), tyrosine (Y), valine (V), methionine (M), cysteine (C), isoleucine (I), leucine (L), norleucine (NLE, 1 nmol internal standard), phenylalanine (F), excess reagent (Re), and lysine (K). B Analysis of a human fingerprint, taken up from a watch glass using a mixture of water and ethanol. (Courtesy of the National Gallery of Art and the Andrew W. Mellon Foundation.)

background contaminants and losses during hydrolysis begin to play a larger role.

DERIVATIZING REAGENTS FOR ANALYSIS OF AMINO ACIDS BY HPLC

The first automated analyzer was developed by Moore, Stein, Spackman, and Hamilton in the 1950s. Hydrolysates were separated on an ion-exchange column, followed by postcolumn reaction with ninhydrin. Although this system remains the standard method, especially for physiological amino acids, its major drawback is low sensitivity, typically at the nanomole level. Several methods have since been developed offering high sensitivity and faster analyses without sacrificing reproducibility.[1–5,7–9] The choice of an optimal derivative technique depends on factors such as specific application requirements, sample size,

sample preparation time, and equipment maintenance.[10] Amino acid derivatives based on ultraviolet (UV) light detection provide accurate analysis at the picomole level and derivatives requiring fluorescence detection are accurate at the femtomole level.

Amino acids react with many reagents to form stable derivatives and strong chromophores (Table 1). Derivatization can precede (precolumn) or follow (in-line postcolumn) the chromatographic separation. Both pre- and postcolumn systems are currently employed: ninhydrin and phenylisothiocyanate (PITC) analyzers are widely used, while 6-aminoquinolyl-N-hydroxysuccinimidyl carbamate (AQC), ortho-phthalaldehyde (OPA), and OPA/9, fluorenylmethylchloro-formate (FMOC) systems provide the highest sensitivity.

Postcolumn systems typically use cation-exchange columns with either sodium citrate (for hydrolysates) or lithium citrate (for physiological samples) as the

Table 1 Amino acid derivatization reagents

Reagent	Chromophore	Detection limit	Separation time	Drawbacks	Advantages
AQC (6-aminoquinolyl-N-hyfroxysuccinimidyl carnamate)	Fluorescent (ex. 245 nm, em. 395 nm); UV 245 nm	160 fmol	35–50 min precolumn	Quaternary gradient elution required for complex, nonhydrolysate samples	Tolerates salts and detergents, rapid reaction, stable product, good reagent separation, high sensitivity and accurancy
Dansyl chloride(4-N, N-dimethylaminoazobenzene-4'-sulfonyl chloride)	Visible 436 nm	Low fmol	18–44 min precolumn	Multiple products, critical concentration	Stable product, good separation, high sensitivity
Dansyl chloride (5,N, N-dimethylaminonaphthalene-1-sulfonyl chloride)	Fluorescent (ex. 360–385 nm, em. 460–495 nm); UV 254 nm	Low pmol	60–90 min	Multiple products, critical concentration, difficult separation leads to long separation time	Stable product
Fluorescamine(4-phenyl-spiro[furan-2(3H), 1'-phthalan]-3,3'-dione)	Fluorescent (ex. 390 nm, em. 475 nm)	20-100 pmol	30–90 min postcolumn	Secondary amine pretreatment, critical concentraion, may give background interference	Rapid reaction, stable product, good reagent separation
FMOC (9-fluorenylmethylchloro formate)	Fluorescent (ex. 265 nm,1 pmol em, 320 nm); UV 265 nm	1 pmol	20–45 min precolumn	Multiple products, extraction of excess reagent	Stable product, used with OPA for detection of secondary amine
Ninhydrin (triketohydrindene hydrate)	Primary amine (440 nm), secondary amine (570 nm)	100 pmol	30 min postcolumn	Low sensitivity and resolution	Good reproducibility
OPA (ortho-phthalaldehyde)	Fluorescent (ex. 340 nm, em. 455 nm)	50 fmol	90 min postcolumn, 17–35 precolumn	Secondary amine pretreatment, slow reaction, unstable derivative, background interference	Good reagent separation, high sensitivity and reproducibility with automated system
PITC (phenylisothiocyanate)	UV 254 nm	1 pmol	15–27 min precolumn	Salt interference, requires refrigeration, excess reagent removed under vacuum	Ease of use, flexibility, good separation, reproducibility enhanced with automation

(From Refs.[1,2].)

mobile phase. Contaminating salts and detergents are better tolerated because the samples are "cleaned up" before reaction with the reagent. The additional pump for the reagent, however, may lead to sample dilution, peak broadening, baseline fluctuations, and longer analysis time (30–90 min). Fluorescent reagents are compatible with a wider range of buffers, but the buffers must be amine-free if used with precolumn methods.

Since the 1980s, precolumn derivatization methods have gained wider acceptance due to simpler preparation, faster analyses, and better resolution. The separation on reversed-phase C-8 or C-18 columns typically requires low-UV mobile phases, such as sodium phosphate or sodium acetate buffers, with acetonitrile or methanol as organic solvent. Separation times range from 15 to 50 min.

IMPROVED RECOVERY OF SENSITIVE AMINO ACIDS

Cysteine and tryptophan require special treatment for quantitative analysis.[1,11] Cystine/cysteine can be determined using three equally successful methods: oxidation, alkylation, and disulfide exchange. Oxidation to cysteic acid is commonly carried out by pretreatment with performic acid. Alkylation using pretreatment with 4-vinylpyridine or iodoacetate produces piridylethylcysteine (PEC) and carboxymethyl-cysteine (CMC), respectively. Disulfide exchange is achieved by adding reagents such as dithiodipropionic acid, dithiodiglycolic acid, or dimethylsulfoxide (DMSO) to the HCl during hydrolysis. The latter treatment offers ease of use as well as accurate yields.

The superior approach to tryptophan analysis involves the addition of dodecanethiol to HCl, especially when combined with automatic vapor-phase hydrolysis. Alternative hydrolysis agents such as methane sulfonic acid, mercaptoethanesulfonic acid, or thioglycolic acid can produce 90% or greater yields. Acid hydrolysis additives and alkaline hydrolysis using 4.2 M NaOH are also used with varying results.

Qualitative analysis of glycopeptides and phospho-amino acids is achieved through a separate, partial hydrolysis with 6 N HCl acid at 110°C for 1 and 1.5 hr, respectively.[1,12] Separation of cysteine, tryptophan, and amino sugars requires minimal chromatographic adjustments; phosphoamino acid separation is straightforward using reversed phase but cumbersome using ion exchange.

ANALYSIS OF FREE AND MODIFIED AMINO ACIDS

Blood, urine, and cerebrospinal and other physiological fluids contain a great number of post-translationally modified amino acids (approximately 170 have been studied to date) and in a wider range of concentrations than protein hydrolysates.[1,13,14] Additionally, plant sources produce about 500 nonprotein amino acids, and in geological samples, highly unusual amino acids may indicate extraterrestrial origin.[15,16]

Although the free amino acids in these samples do not require hydrolysis, blood plasma and cerebrospinal fluid must be deproteinated before analysis. Otherwise, proteins may bind irreversibly to ion-exchange columns, resulting in loss of resolution. Furthermore, any peptide hydrolases must be inactivated. For ion-exchange analysates, a protein precipitant is added before centrifugation. Sulfosalicylic acid, a common precipitating agent, is added in solid form to avoid sample dilution. For reversed-phase analysates, ultra-filtration, SEC, or organic solvent extraction is recommended. Samples with low protein and high amino acid concentrations, such as urine and amniotic fluid, need only to be diluted before analysis.

Precolumn derivatives are more tolerant of lipid-rich samples.[1] Changing the guard column routinely is recommended to avoid column buildup, especially for reversed-phase systems.

AMINO ACID RACEMIZATION ANALYSIS

L-Amino acid enantiomers are the most prevalent in nature. However, D-forms are increasingly found in living organisms, fossils, and extraterrestrial samples. D-Amino acids resulting from post-translational modifications now appear to be fundamental components of bacterial cell walls and microbial antibiotics.[2]

Racemization, the interconversion of amino acid enantiomers, occurs slowly in biological and geological systems. The rate increases with extreme pH values, high temperature, and high ionic strength. Rates also vary between amino acids: At 25°C, the racemization half-life of serine is about 400 yr, while that of isoleucine is 40,000 yr. Enantiomer analysis is used to confirm bioactivity of synthetic peptides and for geological dating.[1,2,7,16]

Hydrolysis itself accelerates racemization. Shorter acid exposure at higher temperatures, such as 160°C for 1 hr, decreases racemization by about 50% compared to conventional hydrolysis. Liquid-phase methanesulfonic acid, gas-phase microwave, conventional, and gas-phase microwave hydrolysis produce progressively higher rates of racemization. Additional phenol, however, significantly reduces racemization during microwave hydrolysis.[4]

The three general approaches to enantiomer separation entail a chiral stationary phase, a chiral mobile phase, or a chiral reagent. Tandem columns, with reversed and chiral stationary phases, were used to separate 18 D–L pairs of phenylthiocarbamyl (PTC)-amino acids in 150 min. OPA-amino acid enantiomers have been separated on both ion-exchange and reversed-phase columns using a sodium acetate buffer with an L-proline-cupric acetate additive. Chiral reagents, such as Marphey's reagent and OPA/IBLC (*N*-isobutyryl-L-cysteine), were successfully used for racemization analysis within 80 min.

CONCLUSIONS

Amino acid analysis continues to be an essential tool in protein chemistry. It has been described as "deceptively difficult": Accurate results require attention to sophisticated instrumentation, sample handling, and consideration of the chemistry of specific amino acids under investigation.[17] When the "art and practice" are carefully addressed, however, AAA provides a fundamental understanding of peptides and proteins unmatched by other techniques.[6]

ACKNOWLEDGMENT

The author would like to thank Drs. Steven Birken, Chun-Hsien Huang, Stacy C. Marsella, and Conceicao Minetti for their proofreading assistance.

REFERENCES

1. Cooper, C., Packer, N., Williams, K., Eds.; *Amino Acid Analysis Protocols*; Humana Press: Totowa, NJ, 2001.
2. Smith, B.J., Ed.; *Protein Sequencing Protocols*; Humana Press: Totowa, NJ, 2003; 111–194.
3. Schegg, K.M; Denslow, N.D.; Andersen, T.T.; Bao, Y.; Cohen, S.A.; Mahrenholz, A.M.; Mann, K. Quantitation and identification of proteins by amino acid analysis: ABRF-96AAA collaborative trial. In *Techniques in Protein Chemistry*; 8th Ed.; Marshak, D.R., Ed.; Academic Press: San Diego, CA, 1997; 207–216.
4. Fountoulakis, M.; Lahm, H.-W. Hydrolysis and amino acid composition analysis of proteins. J. Chromatogr. A. **1998**, *826*, 109–134.
5. Fini, C., Floridi, A., Finelli, V.N., Wittman-Liebold, B., Eds.; *Laboratory Methodology in Biochemistry, Amino Acid Analysis and Protein Sequencing*; CRC Press: Boca Raton, FL, 1990.
6. West, K.A.; Hulmes, J.D.; Crabb, J.W. Amino acid analysis tutorial: improving the art and practice of amino acid analysis. In *The Association of Biomolecular Resource Facilities (ABRF) Annual Meeting: Biomolecular Techniques*; San Francisco, CA March 30–April 2, 1996. www.abrf.org/ResearchGroups/AminoAcidAnalysis/EPosters/Archive/1c.html (accessed September 2004).
7. Barrett, G.C., Ed.; *Chemistry and Biochemistry of Amino Acids*; Chapman and Hall: London, 1985.
8. Tarr, G.E.; Paxton, R.J.; Pan, Y.C.E.; Ericsson, L.H.; Crabb, J.W. Amino acid analysis 1990: the third collaborative study from the Association of Biomolecular Resource Facilities (ABRF). In *Techniques in Protein Chemistry*; 2nd Ed.; Villafranca, J.J., Ed.; Academic Press: San Diego, CA, 1991; 139–150.
9. Hancock, W.S., Ed.; *CRC Handbook of HPLC for the Separation of Amino Acids, Peptides, and Proteins*; CRC Press: Boca Raton, FL, 1984; Vol. I.
10. Chin, D.; 2004 AAA Roundtable Summary. In *ABRF Amino Acid Analysis Research Group* www.abrf.org/ResearchGroups/AminoAcidAnalysis/EPosters/Chin_RT_Summary.pdf (accessed September 2004).
11. Strydom, D.J.; Andersen, T.T.; Apostol, I.; Fox, J.W.; Paxton, R.J.; Crabb, J.W. Cysteine and tryptophan amino acid analysis of ABRF92-AAA. In *Techniques in Protein Chemistry*; 4th Ed.; Hogue Angeletti, R., Ed.; Academic Press: San Diego, CA, 1993; 279–288.
12. Yuksel, K.U.; Andersen, T.T.; Apostol, I.; Fox, J.W.; Crabb, J.W.; Paxton, R.J.; Strydom, D.J. Amino acid analysis of phospho-peptides: ABRF-93AAA. In *Techniques in Protein Chemistry*; 5th Ed.; Crabb, J., Ed.; Academic Press: San Diego, CA, 1994; 231–240.
13. Haynes, P.A.; Sheumack, D.; Greig, L.G.; Kibby, J.; Redwood, J.W. Applications of automated amino acid analysis using 9-fluorenylmethyl chloroformate. J. Chromatogr. **1991**, *588*, 107–114.
14. Gibson, M. Amino acid analysis by HPLC/ninhydrin and tandem mass spectrometry detection. In *ABRF Amino Acid Analysis Research Group* http://www.abrf.org/ResearchGroups/AminoAcidAnalysis/EPosters/Gibson_MS.pdf (accessed September, 2004).
15. Rosenthal, G. *Plant Nonprotein Amino and Imino Acids: Biological, Biochemical, and Toxicological Properties*; Academic Press: New York, 1982.
16. Hare, P.E., Hoering, T.C., King, K., Eds.; *Biogeochemistry of Amino Acids*; John Wiley and Sons: New York, 1980.
17. Andersen, T.T. Practical amino acid analysis. In *ABRF Amino Acid Analysis Research Group*; 1995 www.abrf.org/ABRFNews/1994/September1994/sep94practicalaaa.html (accessed September 2004).

Amino Acids: Analysis by HPLC: An Introduction

Ioannis N. Papadoyannis
Georgios A. Theodoridis
Aristotle University of Thessaloniki, Thessaloniki, Greece

INTRODUCTION

Amino acids are small organic molecules that posses both an amino and a carboxyl group. Amino acids occur in nature in a multitude of biological forms, either free or conjugated to various types of compounds, or as the building blocks of proteins. The amino acids that occur in proteins are named α-amino acids and have the empirical formula $RCH(NH_2)COOH$. Only 20 amino acids are used in nature for the biosynthesis of the proteins, because only 20 amino acids are coded by the nucleic acids.

DISCUSSION

Amino acids show acid–base properties, which are strongly dependent on the varying R groups present in each molecule. The varying R groups of individual amino acids are responsible for specific properties: polarity, hydrophilicity–hydrophobicity.[1–2] Hence, the 20 α-amino acids could be categorized in the 4 distinct groups listed in Table 1.

Their dipolar (zwitterionic) behavior is a fundamental factor in any separation approach. At low pH, amino acids exist in their cationic form with both amino and carboxyl groups protonated. The ampholyte form appears at a pH of 6–7, whereas at higher values, amino acids are in their anionic form (carboxyl group dissociated). Another important parameter is that all α-amino acids (with the exception of Gly) are asymmetrical molecules exhibiting optical isomerization (L being the isomer found in nature). As can be seen in Table 1, amino acids are actually small [molecular weight (MW) ranging from 75 to 204] molecules, exhibiting pronounced differences in polarity and a few chromophoric moieties.

The determination of amino acids in various samples is a usual task in many research, industrial, quality control, and service laboratories. Hence, there is a substantial interest in the HPLC analysis of amino acids from many diverse areas like biochemistry, biotechnology, food quality control, diagnostic services, neuro-chemistry/biology, and so forth. As a result, the separation of amino acids is probably the most extensively studied and best developed chromatographic separation in biological sciences.

The most known system is the separation on a cation-exchange column and postcolumn derivatization with ninhydrin, which was described in 1951 by Moore and Stein. With this approach, a sulfonated polystyrene column achieved a separation of the 20 naturally occurring amino acids within approximately 6 h; modifications of the original protocol enhanced color stabilization of the derivatives and enabled the application of the method in various real samples. Since then, immense developments in instrumentation, column technologies, and automation established HPLC as the dominant separation technique in chemical analysis. Numerous published reports described the HPLC analysis of amino acids in a great variety of samples. To no surprise, a two-volume handbook is entirely devoted to HPLC for the separation of amino acids, peptides and proteins.[3] Many of the initial reports employed soft resins or ion exchangers such as polystyrene or cellulose as stationary phases. These materials show some disadvantages (e.g., compaction under pressure, reduced porosity, and wide particle size distribution). The last decades' developments in manufacturing silica-based materials resulted in the domination of reversed-phase (RP) silica-based packing in liquid chromatography. As a result, RP-HPLC is, at present, widely used for the separation of amino acids, because it offers high resolution, short analysis time, ease in handling combined with low cost, and environmental impact per analysis circle.

In ligand-exchange chromatography (LEC), the separation of analytes is due to the exchange of ligands from the mobile phase with other ligands coordinated to metal ions immobilized on a stationary phase. LEC has been used successfully for the resolution of free amino acids, amino acid derivatives, and for enantiomeric resolution of racemic mixtures.[3]

Apart from ninhydrin, many other derivatization reagents have been used; both precolumn and postcolumn derivatization modes have been extensively employed.[3–5] Derivatization procedures offer significant advantages in both separation and detection aspects and, thus, will be discussed in further detail. The rest of the entry will be divided into two sections: separation of underivatized amino acids, where the determination of free amino acids and postcolumn derivatization procedures are described; and

Encyclopedia of Chromatography DOI: 10.1081/E-ECHR-120040011

Table 1 Amino acids found in proteins

	Amino acid	Structure at pH 6-7	MW	pKa
Hydrophobic Aminoacids (Nonpolar R)	Alanine Ala	CH_3—CH—COO^- | NH_3^+	89	2.35, 9.69
	Leucine Leu	$CH(Me)_2CH_2$—CH—COO^- | NH_3^+	131	2.36, 9.60
	Isoleucine Ile	$C_2H_5CH(CH_3)$—CH—COO^- | NH_3^+	131	2.36, 9.68
	Valine Val	$CH(Me)_2$—CH—COO^- | NH_3^+	117	2.36, 9.68
	Proline Pro	COO$^-$ / NH_2^+ (ring)	115	1.99, 10.60
	Methionine Met	$CH_3SC_2H_5$—CH—COO^- | NH_3^+	149	2.28, 9.21
	Phenylanine Phe	(phenyl)—CH_2—CH—COO^- | NH_3^+	165	1.83, 9.13
	Tryptophan Trp	(indole)—CH_2—C—COO^- | NH_3^+	204	2.38, 9.39
Hydrophilic Aminoacids (Not Charged R)	Clycine Gly	H—CH—COO^- | NH_3^+	75	2.34, 9.6
	Serine Ser	CH_2OH—CH—COO^- | NH_3^+	105	2.21, 9.15
	Threonine Thr	CH_3CHOH—CH—COO^- | NH_3^+	119	2.63, 10.43
	Cysteine Cys	$HSCH_2$—CH—COO^- | NH_3^+	121	1.71, 10.78
	Tyrosine Tyr	HO—(phenyl)—CH_2—CH—COO^- | NH_3^+	181	2.20, 9.11
	Glutamine Gln	$NH_2COC_2H_5$—CH—COO^- | NH_3^+	146	2.17, 9.13
	Asparagine Asn	NH_2COCH_2—CH—COO^- | NH_3^+	132	2.02, 8.8
Acidic Aminoacids	Aspartic Acid Asp	$^-OOCCH_2$—CH—COO^- | NH_3^+	133	2.09, 3.86, 9.86
	Glutamic Acid Glu	$^-OOCC_2H_5$—CH—COO^- | NH_3^+	147	2.19, 4.25, 9.67

(*Continued*)

Table 1 Amino acids found in proteins *(Continued)*

	Amino acid	Structure at pH 6-7	MW	pKa
Basic Aminoacids	Lysine Lys	$^+H_3NC_4H_8$—CH—COO⁻ ; NH$_3^+$	133	2.18, 8.95, 10.53
	Arginine Arg	$H_2NCHNHC_3H_6$—CH—COO⁻ ; $^+NH_2$; NH$_3^+$	133	2.17, 9.04, 12.48
	Histidine His	(imidazole)—CH$_2$—CH—COO⁻ ; NH$_3^+$	155	1.82, 8.95, 10.53

separation of derivatized amino acids, where precolumn derivatization approaches are discussed.

SEPARATION OF UNDERIVATIZED AMINO ACIDS

The differing solubilities, polarities, and acid–base properties of free amino acids have been exploited in their separation by partition chromatography, ion-exchange chromatography, and electrophoresis. For example, the elution order obtained from a polystyrene ion-exchange resin with an acidic mobile phase corresponds to the amino acid classification depicted in Table 1: Acidic amino acids are eluted early, neutral between, and basic amino acids later. In this case, ionic interactions between the sample and the stationary phase are the driving force for the separation of the groups. However, hydrophobic van der Waals and π–π aromatic interactions are responsible for the separation of amino acids within the groups.[3]

The dominant stationary phase in HPLC is modified silica and, to be more specific, octadecyl silica (ODS). It should be pointed out that there could be great differences between various types of ODS materials or even between different batches of the same material. Carbon load, free silanol content, endcapping, type of silica, and coupling chemistry to the C_{18} moiety, not to mention the several physical characteristics of the packing material all involve the behavior of an ODS column. However, a rather safe generalization is that, in such material, hydrophobic interactions are a dominant mechanism of separation.[3–7]

In typical ODS materials, polar amino acids are very weakly retained on column; thus, they are insufficiently resolved. In contrast, nonpolar amino acids are stronger retained and adequately separated. To overcome the poor resolution of polar amino acids, two strategies are the most promising:

1. Derivatization (as discussed in the next section)
2. Modification of the mobile phase with the addition of ion-pairing reagents

Alkyl sulfates/sulfonates added to the mobile phase form a micellar layer interacting with both the stationary phase and the amino acids (which under these conditions are protonated). A mixed mechanism (ion-pairing and dynamic ion exchange) is observed. Furthermore, the ion-pairing reagent masks underivatized silanols of the ODS material, reducing nonspecific unwanted interactions. Sodium dodecyl sulfate (SDS) is the most often used ion-pairing reagent. Gradient elution is often required to achieve reasonable analysis time for nonpolar amino acids. Despite the above-mentioned advantages, ion-pairing shows some disadvantages, such as irreproducibility (especially in gradient runs), long equilibration times, and difficulties in ultraviolet (UV) detection.

Another possibility is the use of alternative stationary phases. A strong trend of the last decade is the employment of specialty phases in challenging and complex separations. Thus, newer C_8, NH_2, CN, mixedmode phases (materials incorporating both ion exchange and reversed-phase moieties), new polymeric phases, and zirconia-based materials offer attractive stationary-phase selectivities.

POSTCOLUMN DERIVATIZATION

The nonderivatized amino acids, following their chromatographic separation, can either be directly detected as free amino acids, on-line derivatized, or by postcolumn derivatization. Derivatization with ninhydrin, the classical amino acid analysis, was the first reported postcolumn derivatization method. Modern postcolumn derivatization protocols employ sophisticated instrumentation and achieve high resolution and sensitivity. In such configurations, derivatization occurs in a reaction coil placed between the analytical column and the detector. Additional pumps and valves are required; thus, such systems typically run fully automated and controlled by a computer. The major disadvantages of postcolumn derivatization are the need for sophisticated and complex instrumentation and the band broadening occuring in the reactor. *Ortho*-phthaldialdehyde (OPT) is the most common reagent in postcolumn derivatization. OPT reacts with primary amino acids under basic conditions, forming a fluorescent derivative (OPA derivative) that allows detection at femtomole levels. Disadvantages of OPA derivatization are the instability of the resultant derivatives and the fact that secondary amino acids are not detected.

If no derivatization takes place, detection is preferably accomplished by UV at a low wavelength (200–210 nm) in order to enhance detection sensitivity. However, detection selectivity is sacrificed at such low wavelengths. Electrochemical detection, when applied to the analysis of free amino acids, offers higher selectivity but suffers from a small linearity range. Furthermore, most amino acids (with the excepion of tryptophan, tyrosine, and cysteine) are not intrinsically electrochemically active within the current useful potential range.[5] Lately, the development of the evaporative light-scattering detector (ELSD) offers an attractive alternative for the determination of non-derivatized amino acids (Fig. 1).

SEPARATION OF DERIVATIZED AMINO ACIDS

Precolumn derivatization is the generally accepted approach for the determination of amino acids, because it offers significant advantages: increased detection sensitivity, enhanced selectivity, enhanced resolution, and limited needs for sophisticated instrumentation (in contrast with postcolumn derivatization techniques).

In modern instrument configurations, derivatization can take place in a conventional autosampler; the resultant derivatives are separated on the analytical column. Detection limits at the femtomole level are achieved, and the resolution of polar amino acids is greatly enhanced.

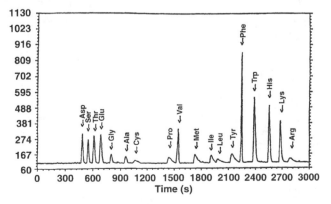

Fig. 1 HPLC analysis of 18 common amino acids. Conditions: stationary phase: CS-10 cation exchange; mobile phase: gradient of aqueous 0.01% TFA and ammonium acetate; detection at the ELSD. (From Ref.[11].)

The most common derivatization reagents are as follows:

- Dimethylamino azobenzene isothiocyanate (DABITC)
- 4-(Dimethylamino)azobenzene-4-sulfonyl chloride (dabsyl chloride or DABS-Cl) [dabsyl derivatives]
- 1-*N*-*N*-Dimethylaminonaphthalene-5-sulfonyl chloride [dansyl derivatives]
- Fluorodinitrobenzene (DNP derivatives)
- Fluorescamine
- 9-Fluorenylmethyl chloroformate (FMOC-Cl)
- 4-Chloro-7-nitro-2,1,3-benzoxadiazole (NBD-Cl)
- Phenylisothiocyanate (PITC) [phenylthiohydantoin (PTH) derivatives]
- Methylisothiocyanate (MITC) [methylthiohydantoin (MTH) derivatives]

Fig. 2 illustrates the structure of the product resulting from the derivatization of an amino acid with the above-mentioned reagents. Mixtures of derivatives with the most commonly used reagents (dansyl, DNP, PTH) are readily provided in kits, to be directly used as reference standards in HPLC analysis.

Phenylthiohydantoin derivatization offers a special value because it is actually performed during Edman degradation, the sequencing technique mostly used for the determination of the primary structure of proteins and peptides. PTH derivatives are separated in many different stationary phases, in either normal- or reversed-phase mode and are mostly detected at 254 nm.[8–9] Using radiolabeled proteins, sequencing of proteins down to the 1–100-pmol range can be achieved. The formed derivatives are basic and thus interact strongly with base silica materials. RP separations are mostly carried out in acidic conditions with the addition of appropriate buffers (sodium acetate mostly, but also phosphate, perchlorate, etc). Failings of PTH derivatization are the lengthy procedure and the higher detection limits obtained (compared to fluorescent derivatives). Potent advantages of the method are its robustness and reproducibility, and the extensive research literature that covers any possible requirement. An alternative to PTH is MTH derivatization, a method well suited for solid-phase sequencing.[3]

Dimethylamino azobenzene isothiocyanate microsequencing results in red–orange derivatives, which exhibit their absorbance maximum at 420 nm with $\varepsilon = 47.000$, in other words offering threefold higher sensitivity compared to PTH derivatives. DABITC derivatives are separated in C_8 or C_{18} columns in acidic environment, within 20 min.

Dabsyl chloride is an alternative to DABITC as a derivatization reagent to be used for manual sequencing. Dabsyl chloride reacts with primary and secondary amino acids forming red–orange derivatives that are

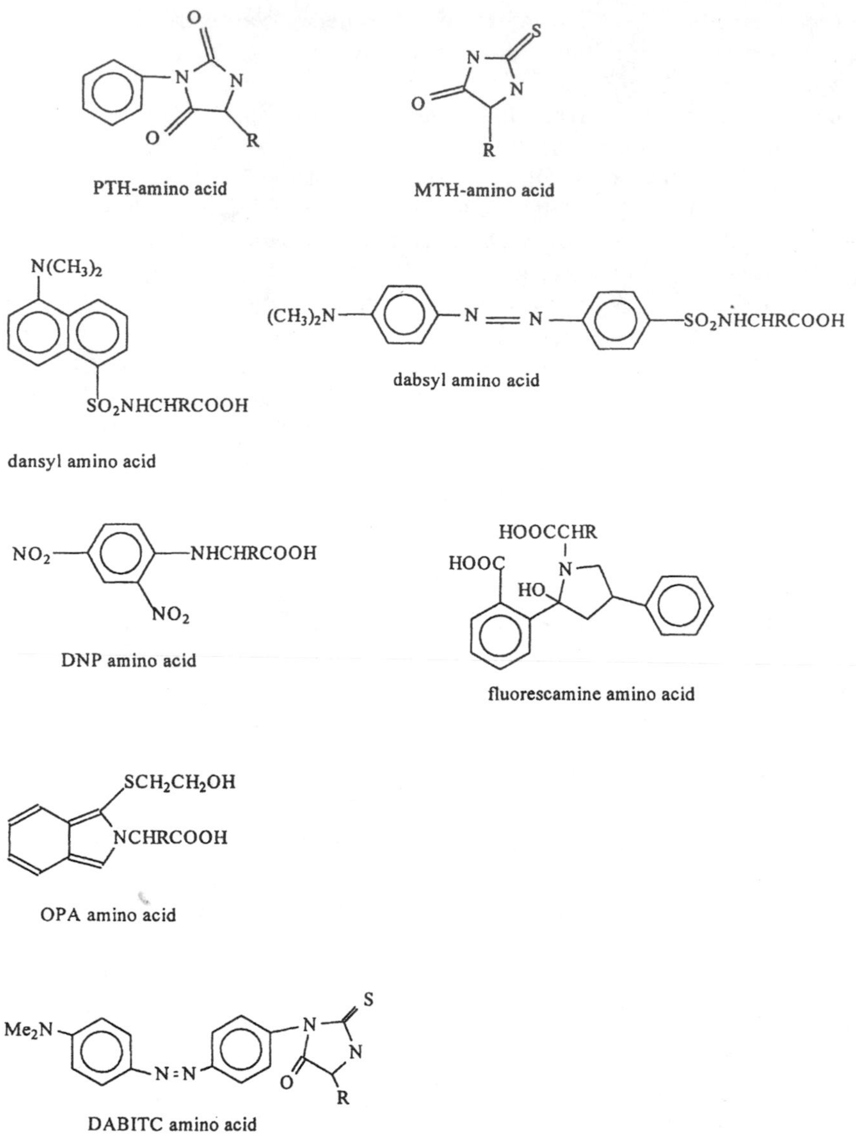

Fig. 2 Structures of the most common amino acid derivatives.

stable for months. The method offers excellent sensitivity, ease, and speed of preparation and high-resolution capabilities. However, it suffers from interferences with ammonia present in biological samples. Furthermore, it results in a relatively reduced column lifetime due to the utilization of excess of Dabsyl chloride.[9]

The dansyl derivatization has been extensively studied to label α- or ε-amino groups. DNS derivatives are formed within 2 min and are detected by either UV or fluorescence. A typical example of a separation of dansyl amino acids is illustrated in Fig. 3.

The FMOC derivatization offers high fluorescent detection sensitivity, but it requires an extraction step to remove unreacted FMOC and by-products. This step is a potential cause of analyte losses. Furthermore, it not suitable for Trp and Cys, because the corresponding derivatives exhibit a lower response due to intramolecular quenching of fluorescence.

The DNP derivatives are analyzed either in normal or reversed phase. Disadvantages of this method are the lower detection sensitivity (60 times less sensitive compared to dabsyl detection) and the lower separation resolution. However, this approach has proven useful for the determination of lysine in food materials.

Finally, the incorporation of an electroactive functionality into a chromatographic label is an attractive alternative for the HPLC of amino acids. Reagents like p–N and N-dimethylaminosothiocyanate have been used to facilitate amperometric detection of the derivatives.

CHIRAL SEPARATION OF AMINO ACIDS

The importance of chirality has rapidly evolved the last decade. Both analytical and preparative separations are needed for biochemical, pharmaceutical, and

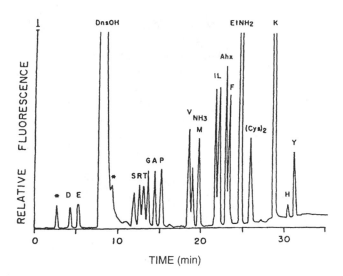

Fig. 3 RP–HPLC analysis of a mixture of dansyl amino acids. Conditions: stationary phase: 4 μm Nova Pak C_{18}; mobile phase: gradient of methanol and tetrahydrofuran versus aqueous phosphate buffer; detection in a fluorescent detector; excitation 338 nm, emission 455 nm. Amino acids are abbreviated by the one-letter system. (From Ref.[12].)

alimentary purposes. Amino acids are asymmetrical molecules. L is the form appearing in proteins; however the D form is also present in nature. Enantiomeric separation of amino acids has been achieved in various stationary phases, such as polystyrene and polyacrylamide to which chiral ligands were covalently bound. Metal ions, in conjunction with chiral ligands, have also been utilized in the mobile phase in the reversed-phase and ligand-exchange mode. Novel stationary chiral phases developed for enantiomeric analysis incorporate chiral ligands (e.g., cyclodextrins or even amino acids) immobilized on silica. Generally, L-amino acidbonded phases retain L-amino acids stronger than the D species.[3,5,10]

Recently, a strong trend in molecular recognition is the development of molecular imprinting polymers (MIP). MIPs have been used as synthetic antibodies in immunoassays and biosensors, but also as catalysts and separation media (employed both in analysis and extraction). One of the first applications of MIPs in separations was the enantiomeric separation of amino acids derivatives.

CONCLUSIONS

High-performance liquid chromatography, when compared to other instrumental methods [TLC, GC, automated amino acid analyzer], offers significant advantages in the analysis of amino acids: high resolution, high sensitivity, low cost, time saving (one third of the analysis time of an amino acid analyzer), and

a multivariate optimization scheme offering versatility and flexibility. Furthermore, optimization of HPLC determination enables the practitioner to overcome typical problems of other methods (e.g., the well-known interferences of ammonia in amino acid analyzer). An additional advantage of HPLC is its direct compatibility with mass spectrometry. The widespread use of LC–MS in proteomic analysis, which at present utilizes state of the art mass spectrometers (e.g., matrix-assisted laser desorption ionization–time-of-flight—mass spectrometry), is seen as a potent future trend.

The variety of instrumentation and experimental conditions (columns, buffers, organic modifiers, derivatization procedures, etc.) reported in the vast literature may hinder the novice from pinpointing the best method to use. The choice of the appropriate method depends on the specific needs of each analytical problem and the nature of the sample to be analyzed. Aspects such as specificity and speed of the derivatization reaction should always be considered. Furthermore, in such multivariate dynamic systems, precision, accuracy, and linearity of the chosen method is a very important factor. The practitioner should carefully follow the developed protocol; the use of automated systems, especially in derivatization procedures, could greatly enhance the reproducibility of the method.

REFERENCES

1. Silverman, L.M.; Christenson, R.H. Amino acids and proteins. In *Fundamentals of Clinical Chemistry*, 4th Ed.; Burtis, C.A., Ashwood, E.R., Eds.; Saunders: Philadelphia, 1996.
2. Matthews, C.K.; van Holde, K.E. *Biochemistry*, 2nd Ed.; Benjamin-Cummings: Menlo Park, CA, 1995; 129–214.
3. Hancock, W.S., Ed.; *CRC Handbook of HPLC for the Separation of Amino Acids, Peptides and Proteins*; CRC Press: Boca Raton, FL, 1984.
4. Hancock, W.S.; Sparrow, J.T. *HPLC of Biological Compounds*; Marcel Dekker, Inc.: New York, 1984; 187–207.
5. Papadoyannis, I.N. HPLC in the analysis of amino acids. In *HPLC in Clinical Chemistry*; Marcel Dekker, Inc.: New York, 1990; 97–154.
6. Kamp, R.M. High sensitivity amino acid analysis. In *Protein Structure Analysis*; Kamp, R.M., Choli-Papadopoulou, T., Wittman-Liebold, B., Eds.; Springer-Verlag: Berlin, 1997.
7. Lottspeich, F.; Hernschen, A. Amino acids, peptides, proteins. In *HPLC in Biochemistry*; Hernschen, A., Hupe, K.P., Lottspeich, F., Voelter, W., Eds.; VCH Weinheim, 1985.

8. Waterfield, M.D.; Scrace, G.; Totty, N. Analysis of phenylthiohydantoin amino acids. In *Practical Protein Chemistry—A Handbook*; Darbre, A., Ed.; John Wiley & Sons: Chichester, 1986.

9. Bergman, T.; Carlquist, M.; Jornvall, H. Amino acid analysis by high performance liquid chromatography of phenylthiocarbamyl derivatives, and amino acid analysis using DABS-Cl precolumn derivatization method, R. Knecht and J. Y. Chang. In *Advanced Methods in Protein Microsequence Analysis*; Wittmann-Liebold, B., Salnikow, J., Erdmann, V.A., Eds.; Springer-Verlag: Berlin, 1986.

10. Vollenbroich, D.; Krause, K. Quantitative analysis of D- and L-amino acids by HPLC. In *Protein Structure Analysis*; Kamp, R.M., Choli-Papadopoulou, T., Wittman-Liebold, B., Eds.; Springer-Verlag: Berlin, 1997.

11. Petterson, J.; Lorenz, L.J.; Risley, D.S.; Sanmann, B.J. Amino acid analysis of peptides using HPLC with evaporative light scattering detection. J. Liquid Chromatogr. & Related Technol. **1999**, *22*, 1009.

12. Martins, A.R.; Padovan, A.F. J. Liquid Chromatogr. & Related Technol. **1999**, *19*, 467.

Amino Acids and Derivatives: Analysis by TLC

L. Lepri
A. Cincinelli
University of Florence, Florence, Italy

INTRODUCTION

Amino acids are carboxylic acids in which a hydrogen atom in the side chain (usually on the α-carbon) has been replaced by an amino group. Hence, they are amphoteric. In weak acid solution, the carboxyl group of a neutral amino acid (with one amino group and one carboxyl group) is dissociated, and the amino group binds a proton to give a dipolar ion (zwitterion). The pH at which the concentration of the dipolar ion is maximum is called the isoelectric point (pI), which is calculated using the relation

$$pI = \frac{1}{2}(pK_1 + pK_2)$$

where pK_1 and pK_2 refer to the dissociation of the carboxyl group and the protonated amino group, respectively (most neutral aliphatic amino acids with a nonpolar side chain have $pI \cong 6.0$, which corresponds to $pK_{a_1} \cong 2.3$ and $pK_{a_2} \cong 9.7$).

Amino acids constitute the basic units of all proteins. The number of α-amino acids obtained from various proteins is about 40, but only 20 are present, in varying amounts, in all proteins. TLC is one of the most promising separation methods for these compounds, for which GC analysis is not suitable.

PREPARATION OF TEST SOLUTIONS

Amino acids should be as free from impurities as possible, since they exhibit a pronounced capacity for binding metal ions. The analysis of amino acids in natural fluids or extracts requires the removal of interfering compounds prior to chromatographic separation, in order to prevent tailing and deformation of the spots (e.g., high salt concentrations are found in urine samples and hydrolysates of proteins or peptides).

Salts can be conveniently removed by passing the sample through a cation-exchange resin column. Free amino acids from sanguine plasma can be obtained after centrifugation of the suspension resulting from the addition of a $Na_3(PW_{12}O_{40})$ solution to the samples for removing proteins.

Enrichment of amino acids in urine (10 ml) can be performed by extracting the lyophilized sample with 1 ml of a methanol–1 M HCl mixture (4 : 1 v/v) and applying an aliquot of supernatant liquid to the plate after centrifugation.

Multivitamin syrups and energy drinks are diluted with an appropriate aqueous–alcoholic mixture (80% ethanol), and the resulting solution is applied to plates for the determination of taurine and lysine.

CHROMATOGRAPHIC TECHNIQUES FOR AMINO ACID SEPARATION

Untreated Amino Acids

Standard solutions of amino acids have usually been prepared in water or in aqueous–alcoholic solvents (70% ethanol), with the addition of hydrochloric acid (0.1 M) for the dissolution of relatively insoluble amino acids (e.g., tyrosine and cystine). Detection is generally performed with ninhydrin reagent. After color development with the ninhydrin, treatment of the layer with a complex-forming cation (e.g., Cu^{II}, Cd^{II}, Ni^{II}) causes the color to change from blue to red and increases color fastness considerably. More specific coloration of amino acids can be achieved by adding bases such as collidine and cyclohexylamine to the detecting agent solution, or by using 4-hydroxybenzaldehyde–ninhydrin as spray reagent. For the location of tryptophan and its derivatives, a 1% solution of *p*-dimethylaminobenzaldehyde in ethanol–hydrochloric acid (1 : 1 v/v) can be used.

Amino acids have been separated on layers of a wide variety of inorganic and organic adsorbents, ion exchangers, and impregnated plates. The two most commonly used adsorbents are silica gel and cellulose.

Separation on Silica Gel and Cellulose Layers

It is interesting to note that by using neutral eluents such as ethanol or *n*-propanol–water, the acidic amino acids (e.g., Glu, Asp) travel much faster on silica gel than basic amino acids (e.g., Lys, Arg, His), which,

Encyclopedia of Chromatography DOI: 10.1081/E-ECHR-120039876

indeed, show very small R_f values. The difference is likely due to cation exchange between the protonated amino groups of basic amino acids and the acidic groups present on silica gel. The strong retention observed for these compounds when eluting with acidic solvents (Table 1) confirms this hypothesis. A similar phenomenon is also observed on cellulose plates and may be the cellulose carboxyl groups.

Furthermore, it is seen that the presence of a hydroxyl group in the molecule does not necessarily reduce the R_f value, as the chromatographic behavior of serine with respect to glycine on layers of silica gel and cellulose shows (Table 1). Some of the numerous eluents that have been used for the separation of amino acids on silica gel are acetone–water–acetic acid–formic acid (50:15:12:3), ethylacetate–pyridine–acetic acid–water (30:20:6:11), 96% ethanol–water–diethylamine (70:29:1), chloroform–formic acid (20:1), chloroform–methanol (9:1), isopropanol–5% ammonia (7:3), and phenol–0.06 M borate buffer pH 9.30 (9:1). On cellulose plates, ethylacetate–pyridine–acetic acid–water (5:5:1:3), n-butanol–acetic acid–water (4:1:1 and 10:3:9), n-butanol–acetone–acetic acid–water (35:35:7:23), n-butanol–acetone–ammonia–water (20:20:4:1), collidine–n-butanol–acetone–water (2:10:10:5), phenol–methanol–water (7:10:3), ethanol–acetic acid–water (2:1:2), and cyclohexanol–acetone–diethylamine–water (10:5:2:5) have also been used as eluents.

Recently, amino acids used in medical practice, as drugs for parenteral and per os feeding, in cattle breeding, and in poultry raising were separated and determined on silica plates.[1] Quantitative determination of serine, threonine, phenylalanine, tryptophan, lysine, ornithine, arginine, valine, and leucine was effected by videodensitometric scanning after selection of the optimum conditions for visualization of the spots on chromatograms by using the plate-immersion technique.

Separation efficiency can be increased by multiple developments or two-dimensional (2D) chromatography. Several solvent systems are suitable for 2D separation, and the combination of chloroform–methanol–17% ammonium hydroxide (40 + 40 + 20) and phenol–water (75 + 25) separates all protein amino acids, except leucine and isoleucine, on silica plates.

Table 1 hR_f values of the 20 common amino acids in different experimental conditions (ascending technique)

Amino acid and abbreviation	Silica gel G[a]	Microcrystalline cellulose[b]	Fixion 50-X8 (Na$^+$)[c]	Silanized silica gel + 4% HDBS[d]	pI
Glycine (Gly)	18	15	56	83	6.0
Alanine (Ala)	22	29	51	74	6.0
Serine (Ser)	18	16	67	85	5.7
Threonine (Thr)	20	21	67	83	6.5
Leucine (Leu)	44	64	22	26	6.0
Isoleucine (Ile)	43	60	28	31	6.1
Valine (Val)	32	48	43	54	6.0
Methionine (Met)	35	23	28	42	5.8
Cysteine (Cys)	7	3	56	—	5.0
Proline (Pro)	14	34	—	63	6.3
Phenylalanine (Phe)	43	55	14	21	5.5
Tyrosine (Tyr)	41	36	12	45	5.7
Tryptophan (Trp)	47	36	2	13	5.9
Aspartic acid (Asp)	17	15	72	86	3.0
Asparagine (Asn)	14	—	—	85	5.4
Glutamic acid (Glu)	24	27	35	83	3.2
Glutamine (Gln)	15	—	—	—	5.7
Arginine (Arg)	6	11	2	28	10.8
Histidine (His)	5	7	11	40	7.6
Lysine (Lys)	3	7	8	47	9.8

[a]Eluent: n-butanol–acetic acid–water (80 + 20 + 20 v/v/v).
[b]Eluent: 2-butanol–acetic acid–water (3:1:1 v/v/v).
[c]Eluent: 84 g citric acid + 16 g NaOH + 5.8 g NaCl + 54 g ethylene glycol + 4 ml concentrated HCl (pH = 3.3).
[d]Eluent: 0.5 M HCl + 1 M CH$_3$COOH in 30% methanol (pH = 0.7).

Separation on Ion Exchangers and Impregnated Plates

Cellulose ion exchangers (e.g., diethylaminoethyl cellulose) and ion-exchange resins have been widely used as stationary phases for TLC separation of untreated amino acids. Fixion 50-X8 commercial plates, which contain Dowex 50-X8 type resin, have been tested on both Na^+ and H^+ forms for 30 amino acids, and the results obtained for the 20 common protein amino acids are reported in Table 1. The isomer pair of leucine and isoleucine is well separated by this method. In addition, the hydroxyl group notably increases the R_f values owing to the hydrophobic properties of the resin, and the pairs serine/glycine and threonine/alanine can be resolved.

Many studies have recently focused on impregnated plates. The methods used for impregnation depend on whether the plates are homemade or commercially available. In the first case, the impregnation reagent is usually added to a slurry of the adsorbent, whereas ready-to-use plates are dipped in the solution of the reagent.

The resolution of amino acids has been effected by using different metal ions as impregnating agents at various concentrations. On silica gel impregnated with Ni^{II} salts, the results indicate a predominant role of the partitioning phenomenon when eluting with acidic aqueous and nonaqueous solutions (e.g., n-butanol–acetic acid–water and n-butanol–acetic acid–chloroform in the 3:1:1 v/v/v ratio). The impregnation of silanized silica gel with 4% dodecylbenzenesulfonic acid (HDBS) solution on both homemade and ready-to-use plates is particularly useful in resolving amino acids.[2] The parameters affecting the retention of amino acids on these layers are: type of adsorbent; concentration and properties of the impregnating agent; percentage and kind of organic modifier; pH; and ionic strength of the eluent.

The data in Table 1 show that complete resolution of basic amino acids (Arg, His, Lys) and of neutral amino acids that differ in their side-chain carbon atom number (i.e., Gly, Ala, Met, Val, Leu and Ile) is possible on home-made plates of silanized silica gel (C_2) impregnated with a 4% solution of HDBS in 95% ethanol. More compact spots can be obtained on RP-18 ready-to-use plates dipped in the same solution of the surfactant agent.

RESOLUTION OF AMINO ACID DERIVATIVES

The identification of N-terminal amino acids in peptides and proteins is of considerable practical importance because it constitutes an essential step in the process of sequential analysis of peptide structures. Many N-amino acid derivatives have been proposed for this purpose and the ones most commonly studied by TLC are 2,4-dinitrophenyl (DNP)- and 5-dimethylaminonaphthalene-1-sulfonyl (dansyl, Dns)-amino acids and 3-phenyl-2-thiohydantoins (PTH-amino acids). Recently, 4-(dimethylamino)azobenzene-4'-isothiocyanate (DABITC) derivatives of amino acids have also been investigated.

DNP-Amino Acids

The dinitrophenylation of amino acids, peptides, and proteins and their separation by 1D and 2D TLC have been reviewed by Rosmus and Deyl.[3]

DNP-amino acids are divided into those that are ether extractable and those that remain in the aqueous phase. Water-soluble α-DNP-Arg, α-DNP-His, ε-DNP-Lys, bis-DNP-His, O-DNP-Tyr, DNP-cysteic acid ($CySO_3H$), and DNP-cystine ($Cys)_2$ have been identified on silica gel plates in the n-propanol–34% ammonia (7:3 v/v) system. Although separation of DNP-Arg and ε-DNP-Lys is incomplete (R_f values 0.43 and 0.44, respectively), both of them can be detected because of the color difference produced in the ninhydrin reaction.

Ether-soluble DNP-amino acids have been investigated by 1D and 2D chromatography. The latter technique offers the possibility of almost complete separation of the two groups of derivatives.

The yellow color of DNP-amino acids deepens upon exposure to ammonia vapor, and it is sufficient by intense for visualizing even 0.1 μg. The detection limit is lower (about 0.02 μg) under UV light (360 nm with dried plates and 254 nm with wet ones), but it increases for 2D chromatography (about 0.5 μg). At present, the applications of DNP-amino acids are limited.

PTH-Amino Acids

The formation of PTH-amino acids by the Edman degradation[4] of peptides and proteins or by successive modifications of the method constitutes the most commonly used technique for the study of the structure of biologically active polypeptides today. Identification of PTH-amino acids in mixtures may be successfully achieved by TLC. Quantitative determination is based on UV adsorption (detection limit: 0.1 μg at 270 nm). An alternative is offered by the chlorine/toluidine test, which is very useful since the minimal amount required for detection is about 0.5 μg.

When 1D chromatography on alumina, polyamide, and silica gel is used, difficulties are encountered in resolving Leu/Ile and Glu/Asp pairs as well as other combinations of PTH-amino acids (e.g., Phe/Val/Met/Thr). The most common solvents used on polyamide plates are n-heptane–n-butanol–acetic acid (40:30:9), toluene–n-pentane–acetic acid (60:30:35), ethylene chloride–acetic acid (90:16), and ethylacetate–n-butanol–acetic acid (35:10:1), and those employed on silica gel are n-heptane–methylene chloride–propionic acid (45:25:30), xylene–methanol (80:10), chloroform–ethanol (98:2), chloroform–ethanol–methanol (89.25:0.75:10), chloroform–n-butylacetate (90:10), diisopropyl ether–ethanol (95:5), methylene chloride–ethanol–acetic acid (90:8:2), n-hexane–n-butanol (29:1); n-hexane–n-butylacetate (4:1), pyridine–benzene (2.5:20), methanol–carbon tetrachloride (1:20), acetone– methylene dichloride (0.3:8). The complete resolution of specific mixtures is possible with 2D chromatography by the use of certain of the solvent systems mentioned.

The characteristic colors of PTH-amino acids following ninhydrin spray and the colored spots observed under UV light on polyamide plates containing fluorescent additives are very useful in identifying those amino acids that have nearly identical R_f values.

TLC of PTH-amino acids has been reviewed by Rosmus and Deyl.[3]

Dns-Amino Acids

Dansylation in 0.2 M sodium bicarbonate solution is widely used for the identification of N-terminal amino acids in proteins, and it is the most sensitive method for the quantitative determination of amino acids, since dansyl derivatives are highly fluorescent under a UV lamp (254 nm). Much research has focused on silica gel and polyamide plates using both 1D and 2D chromatography.

No solvent system resolves all the Dns-amino acids by 1D chromatography. Also, 2D chromatography requires more than two runs for a complete resolution. The eluents most commonly used on polyamide layers are benzene–acetic acid (9:1), toluene–acetic acid (9:1), toluene–ethanol–acetic acid (17:1:2), water–formic acid (200:3), water–ethanol–ammonium hydroxide (17:2:1 and 14:15:1), ethylacetate–ethanol–ammonium hydroxide (20:5:1), n-heptane–n-butanol–acetic acid (3:3:1), chlorobenzene–acetic acid (9:1), and ethylacetate–acetic acid–methanol (20:1:1). On silica plates, acetone–isopropanol–25% aqueous ammonia (9:7:1), chloroform–benzyl alcohol–ethyl acetate–acetic acid (6:4:5:0.2), chloroform–ethyl acetate–acetic acid (38:4:2.8 or 24:4:5), and dichloromethane–methanol–propionic acid (21:3:2) are used.

A widely employed chromatographic system is one based on polyamide plates eluted with water–formic acid (200:3 v/v) in the first direction and benzene–acetic acid (9:1 v/v) in the second direction. A third run with 1 M ammonia–ethanol (1:1 v/v) or ethylacetate–acetic acid–methanol (20:1:1 v/v/v) in the direction of solvent 2 resolves especially basic Dns-amino acids or Glu/Asp and Thr/Ser pairs.

DABTH-Amino Acids

These derivatives are obtained in basic medium by the reaction of DABITC with the primary amino group of N-terminal amino acids in peptides. The color difference between DABITC (or its degradation products) and dimethylaminoazobenzenethiohydantoin (DABTH)-amino acids facilitates identification on TLC. These derivatives are colored compounds and, because of their stability and sensitivity, are usually used for qualitative and quantitative analyses of amino compounds such as amino acids and amines.

All DABTH-amino acids, except the Leu/Ile pair, can be separated by 2D chromatography on layers of polyamide, with water–acetic acid (2:1 v/v) and toluene–n-hexane–acetic acid (2:1:1 v/v/v) being solvents 1 and 2, respectively. Resolution of the Dns-Leu/Dns-Ile pair on polyamide is possible with formic acid–ethanol (10:9 v/v) and on silica plates using chloroform–ethanol (100:3 v/v) as eluent.

Cbo and BOC-Amino Acids

Carbobenzoxy (Cbo) and tert-butyloxycarbonyl (BOC) amino acids are very useful in the synthesis of peptides, and consequently their separation from each other and from unreacted components used in their preparation is very important. For this separation, various mixtures of n-butanol–acetic acid–5% ammonium hydroxide and of n-butanol–acetic acid–pyridine with or without the addition of water have been used on silica gel.

The BOC-amino acids give a negative ninhydrin test; however, if the plates are heated at 130°C for 25 min and immediately sprayed with a 0.25% solution of ninhydrin in butanol, a positive test is obtained.

RESOLUTION OF ENANTIOMERIC AMINO ACID AND THEIR DERIVATIVES

Amino acids are optically active and the separation of the enantiomeric pairs is an important aim. (The topic is discussed in the article *Enantiomeric Separations by TLC*.)

ARTICLE OF FURTHER INTEREST

Enantiomer Separations by TLC, p. 561.

REFERENCES

1. Krasikov, V.D.; Malakhova, I.I.; Degterev, E.V.; Tyaglov, B.V. Planar chromatography of free industrial amino acids. J. Planar Chromatogr. Mod. TLC **2004**, *17*, 113–122.
2. Lepri, L.; Desideri, P.G.; Heimler, D. Reversed-phase and soap-thin-layer chromatography of amino acids. J. Chromatogr. **1980**, *195*, 65–73; J. Chromatogr. **1981**, *209*, 312–315.
3. Rosmus, J.; Deyl, Z. The methods for identification of N-terminal amino acids in peptides and proteins. Part B. J. Chromatogr. **1970**, *70*, 221–339.
4. Edman, P. Method for determination of the amino acid sequence in peptides. Acta Chem. Scand. **1950**, *4*, 283–293.

SUGGESTED FURTHER READING

Bhushan, R.; Martens, J. Amino acids and their derivatives. In *Handbook of Thin-Layer Chromatography*; Sherma, J., Friend, B., Eds.; Marcel Dekker: New York, 1996; *71*, 389–425.

Amino Acids, Peptides, and Proteins: Analysis by Capillary Electrophoresis

Danilo Corradini
Institute of Chromatography, Rome, Italy

INTRODUCTION

Amino acids, peptides, and proteins are analyzed by a variety of modes of capillary electrophoresis (CE) which employ the same instrumentation, but are different in the mechanism of separation. A fundamental aspect of each mode of CE is the composition of the electrolyte solution. Depending on the specific mode of CE, the electrolyte solution can consist of either a continuous or a discontinuous system. In continuous systems, the composition of the electrolyte solution is constant along the capillary tube, whereas in discontinuous systems, it is varied along the migration path.

Capillary zone electrophoresis (CZE), micellar capillary electrokinetic chromatography (MECC), capillary gel electrophoresis (CGE), and affinity capillary electrophoresis (ACE) are CE modes using continuous electrolyte solution systems. In CZE, the velocity of migration is proportional to the electrophoretic mobilities of the analytes, which depends on their effective charge-to-hydrodynamic radius ratios. CZE appears to be the simplest and, probably, the most commonly employed mode of CE for the separation of amino acids, peptides, and proteins. Nevertheless, the molecular complexity of peptides and proteins and the multifunctional character of amino acids require particular attention in selecting the capillary tube and the composition of the electrolyte solution employed for the separations of these analytes by CZE.

DISCUSSION

The various functional groups of amino acids, peptides, and proteins can interact with a variety of active sites on the inner surface of fused-silica capillaries, giving rise to peak broadening and asymmetry, irreproducible migration times, low mass recovery, and, in some cases, irreversible adsorption. The detrimental effects of these undesirable interactions are usually more challenging in analyzing proteins than peptides or amino acids, owing to the generally more complex molecular structures of the larger polypeptides. One of the earliest, and still more adopted, strategy to preclude the interactions of peptides and proteins with the wall of bare fused-silica capillaries is the chemical coating of the inner surface of the capillary tube with neutral hydrophilic moieties.[1] The chemical coating has the effect of deactivating the silanol groups by either converting them to inert hydrophilic moieties or by shielding all the active interacting groups on the capillary wall. A variety of alkylsilane, carbohydrate, and neutral polymers can be covalently bonded to the silica capillary wall by silane derivatization.[2] Polyacrylamide (PA), poly(ethylene glycol) (PEG), poly(ethylene oxide) (PEO), and polyvinylpyrrolidone (PVP) can be successfully anchored onto the capillary surface treated with several different silanes, including 3-(methacryloxy)-propyltrimethoxysilane, 3-glycidoxy-propyltrimethoxysilane, trimethoxyallylsilane, and chlorodimethyloctylsilane. Alternatively, a polymer can be adsorbed onto the capillary wall and then cross-linked in situ. Other procedures are based on simultaneous coupling and cross-linking. Alternative materials to fused silica such as polytetrafluorethylene (Teflon) and poly(methyl methacrylate) (PMMA) hollow fibers has found limited application.

The deactivation of the silanol groups can also be achieved by the dynamic coating of the inner wall by flushing the capillary tube with a solution containing a coating agent. A number of neutral or charged polymers with the property of being strongly adsorbed at the interface between the capillary wall and the electrolyte solution are employed for the dynamic coating of bare fused-silica capillaries. Modified cellulose and other linear or branched neutral polymers may adsorb at the interface between the capillary wall and the electrolyte solution with the main consequence of increasing the local viscosity in the electric double layer and masking the silanol groups and other active sites on the capillary surface. This results in lowering or suppressing the electro-osmotic flow and in reducing the interactions with the capillary wall.

Polymeric polyamines are also strongly adsorbed in the compact region of the electric double layer as a combination of multisite electrostatic and hydrophobic interactions. The adsorption results in masking the silanol groups and the other adsorption active sites on the capillary wall and in altering the electroosmotic flow, which is lowered and, in most cases, reversed

Encyclopedia of Chromatography DOI: 10.1081/E-ECHR-120039877

from cathodic to anodic. One of the most widely employed polyamine coating agents is polybrene (or hexadimetrine bromide), a linear hydrophobic polyquaternary amine polymer of the ionene type.[3] Alternative choices are polydimethyldiallylammonium chloride, another linear polyquaternary amine polymer, and polyethylenimine (PEI). Very promising is the efficient dynamic coating obtained with ethylenediamine-derivatized spherical polystyrene nanoparticles of 50–100 nm diameter, which can be successively converted to a more hydrophilic diol coating by in situ derivatization of the free amino groups with 2,3-epoxy-1-propanol.[4]

In most cases, the electrolyte solution employed in CZE consists of a buffer in aqueous media. Although all buffers can maintain the pH of the electrolyte solution constant and can serve as background electrolytes, they are not equally meritorious in CZE. The chemical nature of the buffer system can be responsible for poor efficiency, asymmetric peaks, and other untoward phenomena arising from the interactions of its components with the sample. In addition, the composition of the electrolyte solution can strongly influence sample solubility and detection, native conformation, molecular aggregation, electrophoretic mobility, and electroosmotic flow. Consequently, selecting the proper composition of the electrolyte solution is of paramount importance in optimizing the separation of amino acids, peptides, and proteins in CZE. The proper selection of a buffer requires evaluating the physical–chemical properties of all components of the buffer system, including buffering capacity, conductivity, and compatibility with the detection system and with the sample.

Nonbuffering additives are currently incorporated into the electrolyte solution to enhance solubility, break aggregation, modulate selectivity, improve resolution, and allow detection, which is particularly challenging for amino acids and short peptides. In addition, a large number of amino compounds, including monovalent amines, amino sugars, diaminoalkanes, polyamines, and short-chain alkylammonio quaternary salts are successfully employed as additives for the electrolyte solution to aid in minimizing interactions of peptides and proteins with the capillary wall in bare fused-silica capillaries. Other additives effective at preventing the interactions of proteins, peptides, and amino acids with the capillary wall include neutral polymers, zwitterions, and a variety of ionic and nonionic surfactants.[5] Less effective at preventing these untoward interactions are strategies using electrolyte solutions at extreme pH values, whether acidic, to suppress the silanol dissociation, or alkaline, to have both the analytes and the capillary wall negatively charged.

Selectivity in CZE is based on differences in the electrophoretic mobilities of the analytes, which depends on their effective charge-to-hydrodynamic radius ratios. This implies that selectivity is strongly affected by the pH of the electrolyte solution and by any interaction of the analyte with the components of the electrolyte solution which may affect its charge and/or hydrodynamic radius.

Additives can improve selectivity by interacting specifically, or to different extents, with the components of the sample. Most of the additives employed in amino acid, peptide, and protein CZE are amino modifiers, zwitterions, anionic or cationic ion-pairing agents, inclusion complexants (only for amino acids and short peptides), organic solvents, and denaturing agents.

The capability of several compounds to ion-pair with amino acids, peptides, and proteins is the basis for their selection as effective additives for modulating the selectivities of these analytes in CZE.[5] Selective ion-pair formation is expected to enlarge differences in the effective charge-to-hydrodynamic radius ratio of these analytes, leading to enhanced differences in their electrophoretic mobilities, which determine improved selectivity. Several diaminoalkanes, including 1,4-diaminobutane (putrescine), 1,5-diaminopentane (cadaverine), 1,3-diaminopropane, and N,N,N',N'-tetramethyl-1,3-butanediamine (TMBD) can be successfully employed as additives for modulating the selectivity of peptides and proteins (Figs. 1 and 2). Moreover, several anions, such as phosphate, citrate, and borate, which are components of the buffer solutions employed as the background electrolyte, may also act as ion-pairing agents influencing the electrophoretic mobilities of amino acids, peptides, and proteins and, hence, selectivity and resolution. Other cationic ion-pairing agents include the ionic polymers polydimethyldiallylammonium chloride and polybrene, whereas myoinositol hexakis-(dihydrogen phosphate), commonly known as phytic acid, is an interesting example of a polyanionic ion-pairing agent.

Surfactants have been investigated extensively in CE for the separation of both charged and neutral molecules using a technique based on the partitioning of the analyte molecules between the hydrophobic micelles formed by the surfactant and the electrolyte solution, which is termed micellar electrokinetic capillary chromatography (MECC or MEKC). This technique is widely used for the analysis of a variety of peptides and amino acids,[6] but it is less popular for protein analysis.[7] The limited applications of MECC to protein analysis may be attributed to several factors, including the strong interactions between the hydrophobic moieties on the protein molecules and the micelles, the inability of large proteins to penetrate into the micelles, and the binding of the monomeric surfactant to the proteins. The result is that, even though the surfactant concentration in the electrolyte solution exceeds the critical micelle concentration, the

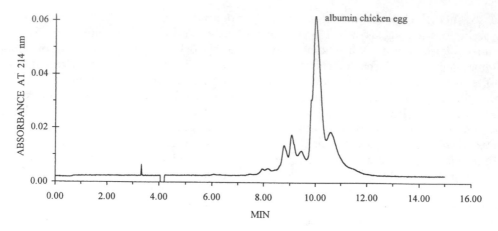

Fig. 1 Detection of microheterogeneity of albumin chicken egg by capillary zone electrophoresis. Capillary, bare fused-silica (50 μm × 37 cm, 30 cm to the detector); electrolyte solution, 25 mM Tris-glycine buffer containing 0.5% (v/v) Tween-20 and 2.0 mM putrescine; applied voltage, 20 kV; UV detection at the cathodic end.

protein–surfactant complexes are likely to be not subjected to partitioning in the micelles, as do amino acids, peptides, and other smaller molecules.

However, surfactants incorporated into the electrolyte solution at concentrations below their critical micelle concentration (CMC) may act as hydrophobic selectors to modulate the electrophoretic selectivity of hydrophobic peptides and proteins. The binding of ionic or zwitterionic surfactant molecules to peptides and proteins alters both the hydrodynamic (Stokes) radius and the effective charges of these analytes. This causes a variation in the electrophoretic mobility, which is directly proportional to the effective charge and inversely proportional to the Stokes radius. Variations of the charge-to-hydrodynamic radius ratios are also induced by the binding of nonionic surfactants to peptide or protein molecules. The binding of the surfactant molecules to peptides and proteins may vary

with the surfactant species and its concentration, and it is influenced by the experimental conditions such as pH, ionic strength, and temperature of the electrolyte solution. Surfactants may bind to samples, either to the same extent [e.g., protein–sodium dodecyl sulfate (SDS) complexes], or to a different degree, which can enlarge differences in the electrophoretic mobilities of the separands.

In CGE, the separation is based on a size-dependent mechanism similar to that operating in polyacrylamide gel electrophoresis (PAGE), employing as the sieving matrix either entangled polymer solutions or gel-filled capillaries.[8] This CE mode is particularly suitable for analyzing protein complexes with SDS. The separation mechanism is based on the assumption that fully denatured proteins hydrophobically bind a constant amount of SDS (1.4 g of SDS per 1 g of protein), resulting in complexes of approximately constant

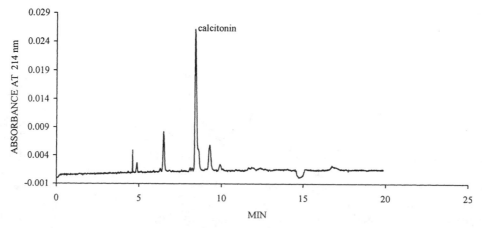

Fig. 2 Separation of impurities from a sample of synthetic human calcitonin for therapeutic use by CZE. Capillary, bare fused-silica (50 μm × 37 cm, 30 cm to the detector); electrolyte solution, 40 mM N,N,N',N'-tetramethyl-1,3-butanediamine (TMBD), titrated to pH 6.5 with phosphoric acid; applied voltage, 15 kV; UV detection at the cathodic end.

charge-to-mass ratios and, consequently, identical electrophoretic mobilities. Therefore, in a sieving medium, protein–SDS complexes migrate proportionally to their effective molecular radii and, thus, to the protein molecular weight. Consequently, SDS–CGE can be used to estimate the apparent molecular masses of proteins, using calibration procedures similar to those employed in SDS–PAGE.

Continuous electrolyte solution systems are also employed in ACE,[9] where the separation depends on the biospecific interaction between the analyte of interest and a specific selector or ligand. The molecules with bioaffinity for the analyte (the selector or ligand) can be incorporated into the electrolyte solution or can be immobilized, either to an insoluble polymer filled into the capillary or to a portion of the capillary wall. ACE is a useful and sensitive tool for measuring the binding constant of ligands to proteins and characterizing molecular properties of peptides and proteins by analyzing biospecific interactions. Examples of biospecific interactions currently investigated by ACE include molecular recognition between proteins or peptides and low-molecular-mass receptors, antigen–antibody complexes, lectin–sugar interactions, and enzyme–substrate complexes. ACE is also employed for the chiral separation of amino acids using a protein as the chiral selector.

Enantiomeric separations of amino acids and short peptides are performed using either a direct or the indirect approach [10]. The indirect approach employs chiral reagents for diasteromer formation and their subsequent separation by various modes of CE. The direct approach uses a variety of chiral selectors that are incorporated into the electrolyte solution. Chiral selectors are optically pure compounds bearing at least one functional group with a chiral center (usually represented by an asymmetric carbon atom) which allows sterically selective interactions with the two enantiomers. Among others, cyclodextrins (CDs) are the most widely chiral selectors used as additives in chiral CE. These are cyclic polysaccharides built up from D-(+)-glucopyranose units linked by α-(1,4) bonds, whose structure is similar to a truncated cone. Substitution of the hydroxyl groups of the CDs results in new chiral selectors which exhibit improved solubility in aqueous solutions and different chiral selectivity. Other chiral selectors include crown ethers, chiral dicarboxylic acids, macrocyclic antibiotics, chiral calixarenes, ligand-exchange complexes, and natural and semisynthetic linear polysaccharides. Chiral selectors are also commonly employed in combination with ionic and nonionic surfactants for enantiomeric separations of amino acids and peptides by MECC.

In discontinuous systems, the composition of the electrolyte solution is varied along the migration path with the purpose of changing one or more parameters responsible for the electrophoretic mobilities of the analytes. The discontinuous electrolyte solution systems employed in capillary isoelectric focusing (CIEF) have the function of generating a pH gradient inside the capillary tube in order to separate peptides and proteins according to their isoelectric points.[11] Each analyte migrates inside the capillary until it reaches the zone with the local pH value corresponding to its isoelectric point, where it stops moving as a result of the neutralized charge and consequent annihilated electrophoretic mobility. CIEF is successfully employed for the resolution of isoenzymes, to measure the isoelectric point (pI) of peptides and proteins, for the analysis of recombinant protein formulation, hemoglobins, human serum, and plasma proteins. Discontinuous electrolyte solution systems are also employed in capillary isotachophoresis (CITP), where the analytes migrate as discrete zones with an identical velocity between a leading and a terminating electrolyte solution having different electrophoretic mobilities. CITP finds large applications as an on-line preconcentration technique prior to CZE, MECC, and CGE. It is also employed for the analysis of serum and plasma proteins and amino acids.[12]

The majority of amino acids and short peptides have no, or only negligible, UV absorbance. Detection of these analytes often requires chemical derivatization using reagents bearing UV or fluorescence chromophores. High detection sensitivity, reaching the attomolar (10^{-21}) mass detection limit can be obtained using fluorescence labeling procedures in combination with laser-induced fluorescence detection.[13] A variety of fluorescence and UV labeling reagents are currently employed, including o-phthaldehyde (OPA), fluorescein isothiocyanate (FITC), 1-dimethylaminonaphthalene-5-sulfonyl chloride (dansyl chloride), 4-phenylspiro[furan-2(3H)-1' phthalene (fluorescamine), and naphthalene dicarboxaldehyde (NDA). However, derivatization may reduce the charge-to-hydrodynamic radius ratio differences between analytes, making separations difficult to achieve. In addition, precolumn derivatization is not suitable for large peptides and proteins, due to the formation of multilabeled products. These problems can be overcome using postcolumn derivatization procedures.

Another, very attractive alternative is indirect UV detection.[14] This indirect detection procedure makes use of a UV-absorbing compound (or "probe"), having the same charge as the analytes, that is incorporated into the electrolyte solution. Displacement of the probe by the migrating analyte generates a region where the concentration of the UV-absorbing species is less than that in the bulk electrolyte solution, causing a variation in the detector signal. In the indirect UV detection technique, the composition of the electrolyte solution is of critical importance, because it dictates separation performance and detection sensitivity.

Probes currently employed in the indirect UV detection of amino acids include *p*-aminosalicylic acid, benzoic acid, phthalic acid, sodium chromate, 4-(*N*,*N'*-dimethylamino)benzoic acid, 1,2,4,-benzenetricarboxylic acid (trimellitic acid), 1,2,4,5-benzenetetracarboxylic acid (pyromellitic acid), and quinine sulfate. Several of these probes are employed in combination with metal cations and cationic surfactants, which are incorporated into the electrolyte solution as modifiers of the electro-osmotic flow.

Coupling mass spectrometry (MS) to CE provides detection and identification of amino acids, peptides, and proteins based on the accurate determination of their molecular masses.[15] The most critical part of coupling MS to CE is the interface technique employed to transfer the sample components from the CE capillary column into the vacuum of the MS. Electrospray ionization (ESI) is the dominant interface which allows a direct coupling under atmospheric pressure conditions. Another distinguishing features of this "soft" ionization technique when applied to the analysis of peptides and proteins is the generation of a series of multiple charged, intact ions. These ions are represented in the mass spectrum as a sequence of peaks, the ion of each peak differing by one unit of charge from those of adjacent neighbors in the sequence. The molecular mass is obtained by computation of the measured mass-to-charge ratios for the protonated proteins using a "deconvolution algorithm" that transforms the multiplicity of mass-to-charge ratio signals into a single peak on a real mass scale. Obtaining multiple charged ions is actually advantageous, as it allows the analysis of proteins up to 100–150 KDa using mass spectrometers with an upper mass limit of 1500–4000 amu.

Concentration detection limits in CE/MS with the ESI interface are similar to those with UV detection. Sample sensitivity can be improved by using iontrapping or time-of-flight (TOF) mass spectrometers. MS analysis can also be performed off-line, after appropriate sample collection, using plasma desorption–mass spectrometry (PD–MS) or matrix-assisted laser desorption–mass spectrometry (MALDI–MS).

REFERENCES

1. Hjerten, S. Free zone electrophoresis. Chromatogr. Rev. **1967**, *9*, 122–219.

2. Rodriguez, I.; Li, S.F.Y. Surface deactivation in protein and peptide analysis by capillary electrophoresis. Anal. Chim. Acta **1999**, *383*, 1.

3. Wiktorowicz, J.E.; Colburn, J.C. Separation of cationic proteins via charge reversal in capillary elextrophoresis. Electrophoresis **1990**, *11*, 769–773.

4. Kleindiest, G.; Huber, C.G.; Gjerde, D.T.; Yengoyan, L.; Bonn, G.K. Capillary electrophoresis of peptides and proteins in fused silica capillaries coated with derivatized polystyrene nanoparticles. Electrophoresis **1998**, *19*, 262.

5. Corradini, D. Buffer additives other than the surfactant sodium dodecyl sulfate for protein separations by capillary electrophoresis. J. Chromatogr. B. **1997**, *699*, 221.

6. Muijselaar, P.G.; Otsuka, K.; Terabe, S. Micelles as pseudo-stationary phases in micellar electrokinetic chromatography. J. Chromatogr. A. **1998**, *780*, 41.

7. Strege, M.A.; Lagu, A.L. Micellar electrokinetic chromatography of proteins. J. Chromatogr. A. **1997**, *780*, 285.

8. Guttman, A. Capillary sodium dodecyl sulfate-gel electrophoresis of proteins. Electrophoresis **1996**, *17*, 1333.

9. Kajiwara, H. Affinity capillary electrophoresis of proteins and peptides. Anal. Chim. Acta **1999**, *383*, 61.

10. Wan, H.; Blomberg, L.G. Chiral separation of amino acids and peptides by capillary electrophoresis. J. Chromatogr. A. **2000**, *875*, 43.

11. Rodriguez-Diaz, R.; Wehr, T.; Zhu, M. Capillary isoelectric focusing. Electrophoresis **1997**, *18*, 2134.

12. Gebauer, P.; Bocek, P. Recent application and developments of capillary isotachophoresis. Electrophoresis **1997**, *18*, 2154.

13. Swinney, K.; Bornhop, D.J. Detection in capillary electrophoresis. Electrophoresis **2000**, *21*, 1239.

14. Hjerten, S.; Elenbring, K.; Kilar, F.; Liao, J.-L.; Chen, A.J.C.; Siebert, C.J.; Zhu, M.-D. Carrier-free zone electrophoresis, displacement electrophoresis and isoelectric focusing in a high-performance electrophoresis apparatus. J. Chromatogr. **1987**, *403*, 47.

15. Smith, R.D.; Loo, J.A.; Barinaga, C.J.; Edmonds, C.G.; Udseth, H.R. Capillary zone electrophoresis and isotachophoresis- mass spectrometry of polypeptides and proteins based upon an electrospray ionization interface. J. Chromatogr. **1989**, *480*, 211.

Analyte–Analyte Interactions: Effect on TLC Band Formation

Krzysztof Kaczmarski
Technical University of Rzeszów, Rzeszów, Poland

Mieczysław Sajewicz
Institute of Chemistry, Silesian University, Katowice, Poland

Wojciech Prus
*School of Technology and the Arts in Bielsko-Biała,
University of Bielsko-Biała, Bielsko-Biała, Poland*

Teresa Kowalska
Institute of Chemistry, Silesian University, Katowice, Poland

INTRODUCTION

Chromatographic separations are mainly used for analytical purposes and, as such, are termed analytical chromatography. Chromatography, however, is gaining increasing importance as a tool that enables isolation of preparative amounts of the desired substances. Such "preparative chromatography" is usually achieved with LC and HPLC, but also occasionally with TLC.

Each separation occurs because of the different interactions of each species with a sorbent. To describe the partitioning process, knowledge of the isotherm involved is needed. In analytical chromatography, the concentration of a species in an analyzed sample is very low, so description of the retention process typically requires knowledge of the slope of the isotherm when the concentration is zero.

When chromatography is used in the preparative mode, the entire dependence of the equilibrium on the concentrations of adsorbed and nonadsorbed solute must be established. The equilibrium isotherm is usually nonlinear and analysis of such isotherms is a necessary prerequisite to enable prediction of the retention mechanism.

OVERVIEW

Physicochemical description of retention processes in liquid chromatography (planar chromatography included) is far from complete and, therefore, new endeavors are regularly undertaken to improve existing retention models and/or to introduce the new ones. The excessive simplicity of already established retention models in planar chromatography is—among other reasons—because some types of intermolecular interaction in the chromatographic systems are disregarded. For example, none of the validated models focusing on prediction of solute retention takes into consideration so-called "lateral interactions," the term used to denote self-association of solute molecules.

The aim of this report is to give insight into the role of lateral interactions in TLC band formation.

THEORY OF CHROMATOGRAPHIC BAND FORMATION

Study of the mechanism of adsorption in TLC is more difficult than in column liquid chromatography. The nonlinear isotherm model in TLC can be designed in a qualitative way only, after investigation of chromatographic band shape and of the concentration distribution within this band; phenomena characteristic of TLC band formation can also have a major effect on the mechanism of retention.

Transfer Mechanism in TLC

In TLC, as most frequently practiced, transfer of mobile phase through the thin layer is induced by capillary flow. Solvents or solvent mixtures contained in the chromatographic chamber enter capillaries in the solid bed, attempting to reduce both their free surface area and their free energy. The free-energy gain ΔE_m of a solvent entering a capillary is given by the relationship:

$$\Delta E_m = -\frac{2\gamma V_n}{r} \tag{1}$$

Encyclopedia of Chromatography DOI: 10.1081/E-ECHR-120028840

where γ is the free surface tension, V_n denotes the molar volume of the solvent, and r is the capillary radius.

From Eq. 1, it follows that the capillary radius r has a very important effect on capillary flow; a smaller radius leads to more efficient flow. The methods used for preparation of commercial stationary phases and supports cannot ensure all pores are of equal, ideal diameter; this results in side effects that contribute to the broadening of chromatographic spots. Other mechanisms of spot broadening are described below.

Broadening of Chromatographic Bands as a Result of Eddy Diffusion and Resistance to Mass Transfer

The most characteristic feature of chromatographic bands is that the longer the development time and the greater the distance from the start, the greater become their surface areas. This phenomenon is not restricted to planar chromatography—it occurs in all chromatographic techniques. Band broadening arises as a result of eddy and molecular diffusion, the effects of mass transfer, and the mechanism of solute retention.

Eddy diffusion of solute molecules is induced by the uneven diameter of the stationary phase or support capillaries, which automatically results in uneven mobile phase flow rate through the solid bed. Some solute molecules are thus displaced more quickly than the average rate of displacement of the solute, whereas others are retarded.

Molecular diffusion is the regular diffusion in the mobile phase, the driving force of each dissolving process and, therefore, needs no further explanation.

The effects of mass transfer are different in the stationary and mobile phases. The resistance to mass transfer in the mobile phase varies with the reciprocals of mobile phase velocity and the diffusivity of the species. The resistance to mass transfer inside the stationary phase varies with the reciprocal of diffusivity and is proportional to the radius of the adsorbent granules attached to the chromatography plate, or the structural complexity of the internal pores in chromatographic paper. For both types of mass-transfer resistance, band stretching is proportional in each direction, as measured from the geometrical spot center, and increases in magnitude the greater the resistance.

All the aforementioned phenomena, which contribute jointly to spot broadening, are used to be described as the effective diffusion. Effective diffusion is a convenient notion, which, apart from being concise and informative, emphasizes that all the contributory phenomena occur simultaneously.

Broadening of Chromatographic Bands as a Result of the Mechanism of Retention

Mechanisms of solute retention, which are also responsible for spot broadening, differ from one chromatographic technique to another, and their role in this process is far less simple than that of diffusion and mass transfer. Use of densitometric detection has, however, furnished insight into concentration profiles across the chromatographic band, enabling estimation of the role of solute retention in peak broadening and prediction of the retention mechanism. Fig. 1 shows three examples of such concentration profiles in the absence of mass overload.

Numerous efforts have been made to describe the band broadening effect and the formation of the concentration profiles. The most interesting models are those that consider band broadening as a two-dimensional process.

Two models of two-dimensional band broadening were established by Belenky et al.[1,2] and by Mierzejewski.[3] In these models, nonlinearity of the adsorption isotherm was neglected so that elliptical spots only, with symmetrically distributed concentration (as shown in Fig. 1a), could be modeled. We will now focus our attention on the effect of the adsorption mechanism on the concentration profiles of chromatographic bands.

ADSORPTION EQUILIBRIUM ISOTHERMS

Isotherm models reflect interactions between active sites on the sorbent surface and the adsorbed species and, simultaneously, interactions occurring exclusively among the adsorbed species. The dependence of isotherm shapes on concentration profiles in TLC is fully analogous to relationships between HPLC peak profiles and the isotherm models, which have been discussed in depth by Guiochon et al.[4]

Let us briefly recall several chromatographic models and analyze the correspondence between concentration

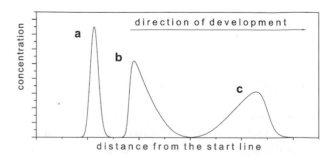

Fig. 1 Three examples of concentration profiles along the chromatographic stationary phase bed: (a) symmetrical without tailing, (b) skewed with tailing toward the mobile-phase front, and (c) skewed with tailing toward the origin.

profiles and types of isotherm. The simplest isotherm model is furnished by Henry's law.

$$q = HC \tag{2}$$

where q is the concentration of the adsorbed species, H is Henry's constant, and C is the concentration in the mobile phase. This isotherm is also called the linear isotherm, and concentration profiles obtained with its aid are similar to that shown in Fig. 1a. It should be stressed that, for the linear isotherm, peak broadening results from eddy diffusion and from resistance of the mass transfer only; it does not depend on Henry's constant. In practice, such concentration profiles are observed only for analyte concentrations that are low enough for the equilibrium isotherm to be regarded as linear.

One of the simplest nonlinear isotherm models is the Langmuir model.

$$q = \frac{q_s K C}{1 + KC} \tag{3}$$

where q_s is the saturation capacity and K the equilibrium constant. To make use of this isotherm, ideality of the liquid mixture and of the adsorbed phase must be assumed. Concentration profiles obtained with the aid of this isotherm are similar to that presented in Fig. 1c. The larger the equilibrium constant, the more stretched is the concentration tail (and the chromatographic band).

More complicated models take into account lateral interaction between the adsorbed molecules. One of these models was designed by Fowler and Guggenheim.[5] It assumes ideal adsorption on a set of the localized sites, with weak interactions among molecules adsorbed on neighboring sites. It also assumes that the energy of interactions between two adsorbed molecules is so small that the principle of random distribution of the adsorbed molecules on the sorbent surface is not significantly affected. For liquid–solid equilibria, the Fowler and Guggenheim isotherm is empirically extended and written in the form:

$$KC = \frac{\theta}{1 - \theta} e^{-\chi\theta} \tag{4}$$

where χ denotes the empirical interaction energy between two molecules adsorbed on nearest-neighbor sites, and θ is the degree of the surface coverage. For $\chi = 0$, the Fowler–Guggenheim isotherm simply becomes the Langmuir isotherm.

Another model, which takes into account lateral interaction and surface heterogeneity, is the Fowler–Guggenheim–Jovanovic isotherm.[6]

$$\theta = 1 - e^{-(aCe^{\chi\theta})^{\nu}} \tag{5}$$

where a is a constant and χ a heterogeneity term.

The next model, which assumes single-component localized monolayer adsorption with specific lateral interactions among all the adsorbed molecules, is the Kiselev model.[7–9] The final equation of this model is

$$\frac{\theta}{(1 - \theta)C} = \frac{K}{(1 - KK_a(1 - \theta)C)^2} \tag{6}$$

where $\theta = q/q_s$, K is the equilibrium constant for adsorption of analyte on active sites, and K_a is the association constant.

All these isotherms can generate the concentration profiles presented in Fig. 1b. The more pronounced the tailing, the stronger the lateral interactions. The concentration profiles presented in Fig. 1b could also be obtained if the adsorbed species formed multilayer structures.[10,11]

Multilayer isotherm models can be derived from the equations:

$$K_1 C(q_s - q_1 - q_2 - q_3 - \cdots - q_n) - q_1 = 0 \tag{7}$$

$$K_2 C q_1 - q_2 = 0 \tag{8}$$

$$K_3 C q_2 - q_3 = 0 \tag{9}$$

$$K_n C q_{n-1} - q_n = 0 \tag{10}$$

where the first equation describes the equilibrium between free active sites and adsorbed species, and subsequent equations depict equilibria between adjacent analyte layers. It is usually assumed that $K_2 = K_3 = \cdots = K_n = K_a$. This set of equations (i.e., Eqs. [7–10]) results in the isotherm:

$$q = q_s \frac{KC(1 + 2K_pC + 3(K_pC)^2 \cdots)}{1 + KC + KCK_pC + KC(K_pC)^2 \cdots} \tag{11}$$

The Retention Model

Qualitative modeling of the experimentally observed densitometric profiles for any given adsorption isotherm has been presented in Refs.[11,12] on the basis of the model:

$$\frac{\partial C}{\partial t} + w\frac{\partial C}{\partial x} + \Phi\frac{\partial q}{\partial t} = D_x\frac{\partial^2 C}{\partial x^2} + D_y\frac{\partial^2 C}{\partial y^2} \tag{12}$$

with the assumed boundary conditions:

$$\left.\frac{\partial C}{\partial x}\right|_{x=0,x=xl} = \left.\frac{\partial C}{\partial y}\right|_{y=0,y=yl} = 0 \tag{13}$$

where Eq. 12 represents the differential mass balance for the mobile phase and the solid state, w is the average mobile-phase flow rate, C and q are, respectively, the concentrations (mol dm^{-3}) of the analyte in the mobile phase and on the sorbent surface, D_x and D_y are, respectively, the effective diffusion coefficients lengthwise (x) and in the direction perpendicular to this direction (y), Φ is the so-called phase ratio, and $x1$ and $y1$ are the plate length and width, respectively. It was assumed that at time $t = 0$, analyte is concentrated in a rectangular spot at the start of the chromatogram.

The Role of Intermolecular Interactions: Multilayer Adsorption

When low-molecular–weight carboxylic acids are chromatographed on cellulose powder with a nonpolar mobile phase, the densitograms obtained are similar to those presented in Fig. 2. Carboxylic acids form associative multimers by hydrogen bonding because of the presence of the negatively polarized oxygen atom from the carbonyl group and the positively polarized hydrogen atom from the hydroxyl group. Direct contact of these cyclic acidic dimers with a sorbent results in forced cleavage of most of the dimeric rings (e.g., because of inevitable intermolecular interactions by hydrogen bonding with hydroxyl groups of the cellulose), thus considerably shifting the equilibrium of self-association toward linear associative multimers.

The tendency of carboxylic acid analytes to form associative multimers can also be viewed as multilayer adsorption. Analysis of the concentration profiles presented in Fig. 2 reveals that for low concentrations of the analyte, peaks a and b are similar to the band profiles simulated by use of the Langmuir isotherm, whereas peaks c–f resemble profiles obtained by use of the anti-Langmuir isotherm (tailing toward the

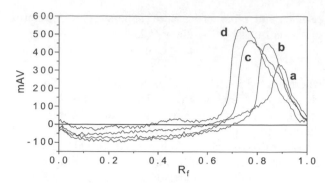

Fig. 3 Concentration profiles of 5-phenyl-1-pentanol obtained on Whatman No. 3 chromatography paper at ambient temperature with *n*-octane as mobile phase. Concentrations of the analyte solutions in 2-propanol were (a) 0.5, (b) 1.0, (c) 1.5, and (d) 2.0 M. The volumes of sample applied were 5 μL. (From Ref.[14].)

front of the chromatogram is more pronounced than tailing toward the start of the chromatogram.).

More spectacular results are obtained with some alcohols. Figs. 3 and 4 depict the densitometric profiles for 5-phenyl-1-pentanol chromatographed on Whatman No. 3 and Whatman No. 1 chromatography papers.

In this instance, very steep concentration profiles toward the start of the chromatogram are obtained; this is indisputably indicative of some kind of interaction among the adsorbed molecules. The concentration profiles presented in Figs. 2–4 can be obtained theoretically from the model given by Eqs. 12 and 13 combined with the isotherm (Eq. 11), assuming three-layer adsorption as a maximum.

As an example, qualitative reproduction of the experimental concentration profiles shown in Figs. 3 and 4 is given in Fig. 5. The Eq. 11 constants of the adsorption isotherm, the mobile phase velocity, and effective diffusion coefficients were chosen to

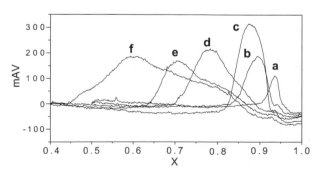

Fig. 2 Concentration profiles of 4-phenylbutyric acid on microcrystalline cellulose at 15°C with decalin as mobile phase. Concentrations of the analyte solutions in 2-propanol were (a) 0.1, (b) 0.2, (c) 0.3, (d) 0.4, (e) 0.5, and (f) 1.0 M. The volumes of sample applied were 3 μL. (From Ref.[13].)

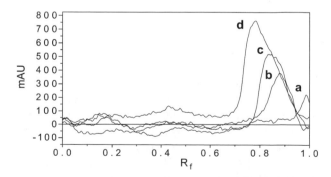

Fig. 4 Concentration profiles of 5-phenyl-1-pentanol obtained on Whatman No. 1 chromatography paper at ambient temperature with *n*-octane as mobile phase. Concentrations of the analyte solutions in 2-propanol were (a) 0.25, (b) 0.50, (c) 0.75, and (d) 1.0 M. The volumes of sample applied were 5 μL. (From Ref.[14].)

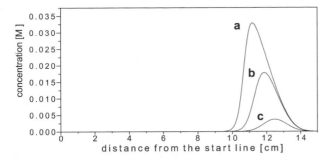

Fig. 5 The lengthwise cross section of the simulated chromatogram for a hypothetical alcohol or acid, according to the model given by Eqs. 12 and 13 in conjunction with the isotherm given by Eq. 11. Concentrations of the applied solutions were (a) 1.0, (b) 0.5, and (c) 0.1 M.

reproduce the shapes of the lengthwise cross sections of the chromatographic bands obtained in the experimental densitograms.

The calculations presented in graphical form in Fig. 5 were performed for $q_s = 1.5$, $K = 0.5$, $K_p = 5$, $w = 0.3\,cm\,min^{-1}$, $D_x = 0.007\,cm^2\,min^{-1}$, and an initial spot length of 0.06 cm. The phase ratio Φ was assumed to be 0.25.

From Fig. 5, it is apparent that the adsorption fronts are considerably less steep than the desorption fronts, and that the adsorption fronts simulated for different initial concentrations of the spots overlap. Similar behavior is apparent in the typical experimental densitograms, given in Figs. 3 and 4. In all these densitograms, the adsorption fronts for the different concentrations of acid also overlap.

CONCLUSIONS

Satisfactory qualitative agreement between experimental and theoretical concentration profiles for polar analytes suggests their retention is substantially affected by lateral interactions, which are probably even more complex than is assumed in this isotherm model. Overlapping of the adsorption fronts can be explained solely on the basis of the lateral interactions among the adsorbed molecules.

REFERENCES

1. Belenky, B.G.; Nesterov, V.V.; Gankina, E.S.; Smirnov, M.M. A dynamic theory of thin layer chromatography. J. Chromatogr. **1967**, *31*, 360–368.
2. Belenky, B.G.; Nesterov, V.V.; Smirnov, M.M. Theory of thin-layer chromatography. I. Differential equation of thin-layer chromatography and its solution (in Russian). Zh. Fiz. Khim. **1968**, *42*, 1484–1489.
3. Mierzejewski, J.M. The mechanism of spot formation in flat chromatographic systems. I. Model of fluctuation of substance concentration on spots in paper and thin layer chromatography. Chem. Anal. (Warsaw) **1975**, *20*, 77–89.
4. Guiochon, G.; Shirazi, S.G.; Katti, A.M. *Fundamentals of Preparative and Nonlinear Chromatography*; Academic Press: Boston, MA, 1994.
5. Fowler, R.H.; Guggenheim, E.A. *Statistical Thermodynamics*; Cambridge University Press: Cambridge, UK, 1960.
6. Quinones, I.; Guiochon, G. Extension of a Jovanovic–Freundlich isotherm model to multicomponent adsorption on heterogeneous surfaces. J. Chromatogr. A. **1998**, *796*, 15–40.
7. Berezin, G.I.; Kiselev, A.V. Adsorbate–adsorbate association on a homogenous surface of a nonspecific adsorbate. J. Colloid Interface Sci. **1972**, *38*, 227–233.
8. Berezin, G.I.; Kiselev, A.V.; Sagatelyan, R.T.; Sinitsyn, V.A. Thermodynamic evaluation of the state of the benzene and ethanol on a homogenous surface of a nonspecific adsorbent. J. Colloid Interface Sci. **1972**, *38*, 335–340.
9. Quinones, I.; Guiochon, G. Isotherm models for localized monolayers with lateral interactions. Application to single-component and competitive adsorption data obtained in RP/HPLC. Langmuir **1996**, *12*, 5433–5443.
10. Wang, C.-H.; Hwang, B.J. A general adsorption isotherm considering multi-layer adsorption and heterogeneity of adsorbent. Chem. Eng. Sci. **2000**, *55*, 4311–4321.
11. Kaczmarski, K.; Prus, W.; Dobosz, C.; Bojda, P.; Kowalska, T. The role of lateral analyte–analyte interactions in the process of TLC band formation. II. Dicarboxylic acids as the test analytes. J. Liq. Chromatogr. & Relat. Technol. **2002**, *25*, 1469–1482.
12. Prus, W.; Kaczmarski, K.; Tyrpień, K.; Borys, M.; Kowalska, T. The role of the lateral analyte–analyte interactions in the process of TLC band formation. J. Liq. Chromatogr. & Relat. Technol. **2001**, *24*, 1381–1396.
13. Kaczmarski, K.; Sajewicz, M.; Pieniak, A.; Piętka, R.; Kowalska, T. Densitometric acquisition of concentration profiles in planar chromatography and its possible shortcomings. Part 1. 4-Phenylbutyric acid as an analyte. Acta Chromatogr. **2004**, *14*, 5–15.
14. Sajewicz, M.; Pieniak, A.; Piętka, R.; Kaczmarski, K.; Kowalska, T. Densitometric comparison of the performance of Stahl-type and sandwich-type planar chromatographic chambers. J. Liquid Chromatogr. & Relat. Technol. **2004**.

Antibiotics: Analysis by TLC

Irena Choma
Department of Chromatographic Methods, M. Curie-Skłodowska University, Lublin, Poland

INTRODUCTION

Antibiotics are an extremely important class of human and veterinary drugs. Chemically, they constitute a widely diverse group with different functions and ways of operation. They can be derived from living organisms or obtained synthetically. Nowadays, the term "antibiotics" is often extended to so-called chemotherapeutics, such as the sulfonamides and quinolones. However, all of them exhibit antibacterial properties, i.e., either inhibit the growth of or kill bacteria.

Antibiotics are used both in human and veterinary medicine as well as in animal husbandry. They enable prevention and control of many bacterial diseases. However, there are many side effects connected with their use, such as: toxicity, allergies, or intestinal disorder. Additionally, overuse and misuse of these drugs can lead to the emergence of antibiotic resistant bacteria. Analysis of antibiotics embraces their determination in pharmaceuticals, body fluids, feed, and food. The most popular analytical methods are the chromatographic techniques. TLC is usually used as a screening method preceding HPLC analysis, but there are also many examples of quantitative TLC analysis. TLC is also applied in the purification of newly discovered antibiotics, analysis of antibiotic metabolites and impurities, search for new biologically active compounds, and studying interactions and retention behavior of antibiotics. Antibiotics can be also applied as stationary or mobile phase additives for chiral separations.

BACKGROUND INFORMATION

Penicillin, the first natural antibiotic, produced by genus *Penicillium*, discovered in 1928 by Fleming, and sulfonamides, the first chemotherapeutic agents, discovered in the 1930s, start a long list of presently known antibiotics. Beside β-lactams (penicillins and cephalosporins) and sulfonamides, the list includes aminoglycosides, macrolides, tetracyclines, quinolones, peptides, polyether ionophores, rifamycins, lincosamides, coumarins, nitroheterocycles, amphenicols, and others.

In principle, antibiotics should eradicate pathogenic bacteria in the host organism without causing significant damage to it. Nevertheless, most antibiotics are toxic, some of them even highly so. The toxicity of antibiotics for humans is not only due to medical treatment but also due to absorption of those drugs through contaminated food. In modern agricultural practice, antibiotics are administered to animals both for treatment of diseases and for prophylaxis as well as to promote growth as feed or water additives. All of this results in the appearance of unsafe antibiotic residues or their metabolites in edible products, e.g., milk, eggs, and meat. Some of them, like penicillins, can cause allergic reactions in sensitive individuals. Therefore, monitoring antibiotic residues should be an important task for government authorities.

There are many analytical methods for determining antibiotics in pharmaceuticals, body fluids, and food. They can be based on microbiological, immunochemical, and physicochemical principles. The most popular methods belonging to the latter group are chromatographic ones, mainly liquid chromatography, including HPLC and TLC.[1,2]

HPLC offers high sensitivity and separation efficiencies. However, it requires sophisticated equipment and is expensive. Usually, before HPLC analysis, tedious sample pretreatment is necessary, such as protein precipitation, ultrafiltration, partitioning, metal chelate affinity chromatography (MCAC), matrix solid-phase dispersion (MSPD), or solid-phase extraction (SPE). Generally, the sample clean-up procedures used before TLC separation are the same as for HPLC. Still, they can be strongly limited in the case of screening TLC or when plates with concentrating zones are applied.

TLC is cheaper and less complicated than HPLC, provides high sample throughput, and usually requires limited sample pretreatment. However, the method is generally less sensitive and selective and gives poor resolution. Some of these problems can be solved by high-performance TLC (HPTLC) or forced flow planar chromatography (FFPC), i.e., rotation planar chromatography (RPC), overpressured-layer chromatography (OPLC), and electro-planar chromatography (EPC). Lower detection limits can also be achieved using an autosampler for injection, applying special techniques of development and densitometry as a detection method, or/and spraying the plate after development with appropriate reagents. There is also a possibility of coupling TLC with autoradiography,

Encyclopedia of Chromatography DOI: 10.1081/E-ECHR-120039881

mass spectrometry (MS) or Fourier-transform infrared (FTIR). Then, TLC can reach selectivity, sensitivity, and resolution close to those of HPLC.

TLC stripped of the abovementioned attributes may still serve as a screening method, i.e., one that establishes the presence or absence of antibiotics above a defined level of concentration. Screening TLC methods show sensitivity similar to microbiological assays, which are the most popular screening methods, applied for controlling antibiotic residues in food in many countries. TLC/bioautography (TLC/B) is one of the TLC screening methods. The developed TLC plates are placed on or dipped in a bacterial growth medium seeded with an appropriate bacterial strain. The location of zones of growth inhibition gives information about antibiotic residues.[3,4]

In relation to the extremely diverse nature of antibiotics, a variety of different separation and detection modes are used in analytical practice. Characteristics in brief and some general rules of separation for the most popular classes of antibiotics are presented below.

PENICILLINS

The basic structure of penicillins is a thiazolidine ring linked to a β-lactam ring to form 6-aminopenicillanic acid, the so called "penicillin nucleus" (Fig. 1). This acid, obtained from *Penicillium chrysogenum* cultures, is a precursor for semisynthetic penicillins (ampicillin, amoxicillin, oxacillin, cloxacillin, dicloxacillin, and methicillin) produced by attaching different side chains to the "nucleus." Benzylpenicillin (penicillin G) and phenoxymethylpenicillin (penicillin V) are the naturally occurring penicillins.

The most widely used stationary phase for analysis of penicillins is silica gel, but silanized silica, cyanosilica, silica gel impregnated with tricaprylmethylammonium chloride, cellulose, and alumina plates are also employed. It is advantageous to add acetic acid to the mobile phase and/or spotting acetic acid before the sample injection in order to avoid the decomposition of β-lactams on silica gel. Mobile phases in reversed-phase (RP) systems usually contain pH 5–6

buffer and organic solvent(s).[5] The most popular detection methods are bioautography and UV densitometry, often coupled with spraying with proper reagents. A review paper on chromatographic analysis of penicillins in animal tissues, included TLC, was written by Boison.[6]

CEPHALOSPORINS

Cephalosporins are derived from natural cephalosporin C produced by *Cephalosporium acremonium.* Chemically, they are derivatives of 7-aminocephalosporanic acid (cephem nucleus) (Fig. 2). Cephalosporins are closely related to penicillins and exhibit the same mechanisms of action, i.e., they inhibit bacterial cell wall synthesis and are used mainly for treating staphylococcal and streptococcal infections in patients who cannot use penicillins. They are commonly divided into three classes differing in the spectrum and toxicity. Cephalosporins can be analyzed by both normal- (NP) and RP TLC; however, more efficient separation is obtained on silanized gel than on bore silica gel.[7,8] Silica gel is sometimes impregnated with Na_2EDTA, tricaprylmethylammonium chloride, transition metal ions, or hydrocarbon. Inorganic ion exchangers (e.g., stannic oxide) or silica gel mixed with an exchanger (e.g., with Mg/Al layered double hydroxide) can be also used as stationary phases for cephalosporin analysis. The mobile phases are polar and similar to those used for penicillins. Acetic acid or acetates are very often components of solvents for NP TLC, and ammonium acetate/acetic acid buffer for RP TLC. All cephalosporins can be detected at 254 nm. Applying reagents such as ninhydrin, iodoplatinate, chloroplatinic acid, or iodine vapor can lower the detection limit. An alternative to UV detection is bioautography with, for instance, *Neisseria catarrhalis.*

AMINOGLYCOSIDES

Aminoglycosides consist of two or more amino sugars joined via glycoside linkage to a hexose nucleus (Fig. 3).

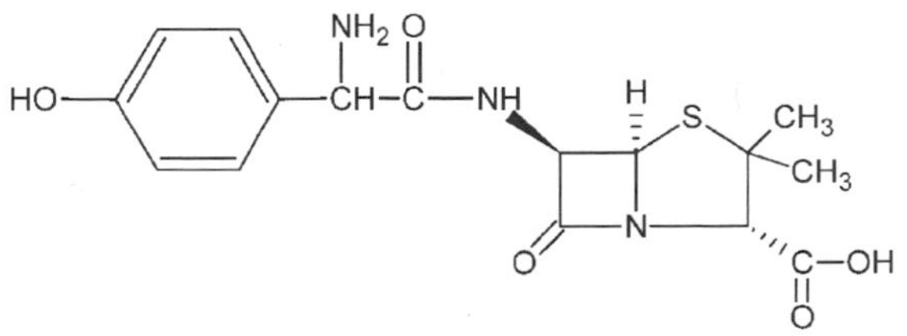

Fig. 1 Amoxicillin.

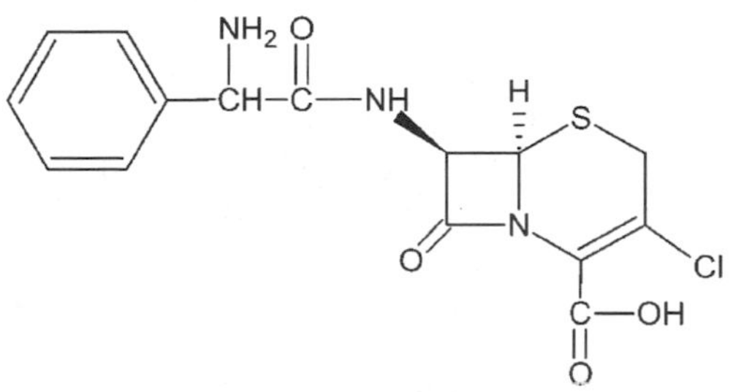

Fig. 2 Cefaclor.

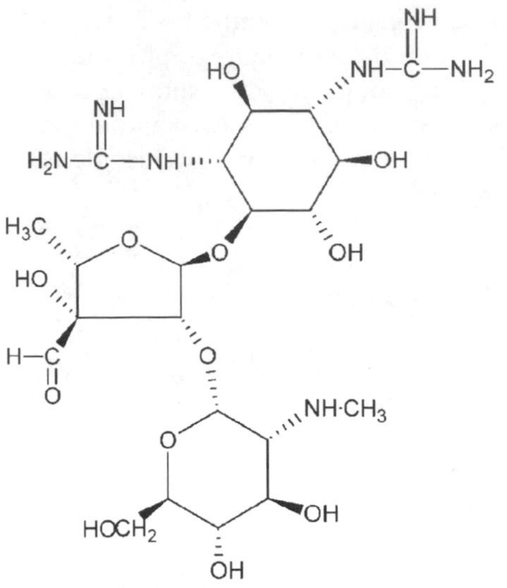

Fig. 3 Streptomycin.

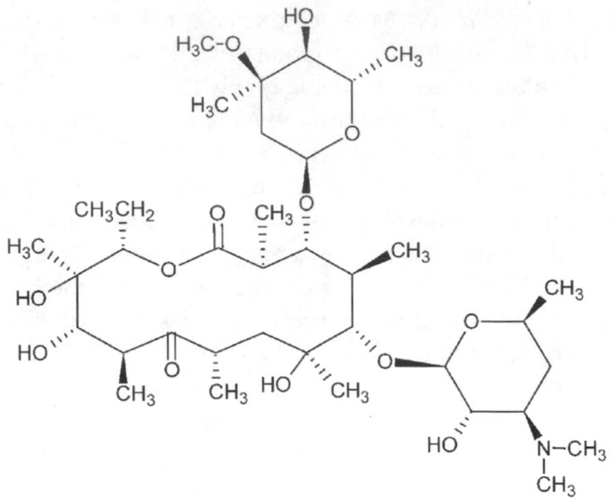

Fig. 4 Erythromycin.

Streptomycin was isolated in 1943 from *Streptomyces griseus*; then others were discovered in different *Streptomyces* strains. Aminoglycosides are particularly active against aerobic micro-organisms and against the tubercle bacillus, but because of their potential ototoxicity and nephrotoxicity, they should be carefully administered. Aminoglycosides, due to their extremely polar, hydrophilic character, are analyzed mostly on silica gel, but C-18 plates can also be used. Polar organic solvents (methanol, acetone, chloroform) mixed with 25% aqueous ammonia are the most popular mobile phases. Because the majority of aminoglycosides lack UV absorption, they must be derivatized by spraying or dipping after development with fluorescamine, vanillin, or ninhydrin solutions. They can be also detected by charring, treating with iodine vapor, or derivatization with 4-chloro-7-nitrobenzo-2-oxa-1,3-diazole (NBD-Cl) or with a mixture of diphenylboronic anhydride and salicylaldehyde. Bioautography with *Bacillus subtilis*, *Sarcina lutea*, and *Mycobacterium phlei* is also possible. Recently, a thorough review on aminoglycoside analysis appeared, embracing, among other methods, also TLC.[9]

MACROLIDES

Macrolides are bacteriostatic antibiotics composed of a macrocyclic lactone ring with one or more deoxy sugars attached to it (Fig. 4). The main representative of the class, erythromycin, was discovered in 1952 as a metabolic product of *Streptomyces erythreus*. Now, erythromycin experiences its renaissance because of its high activity against many dangerous bacteria such as *Campylobacter* and *Legionella*. The macrolide antibiotics group is still being expanded due to the

search for macrolides with pharmacokinetic properties better than those of erythromycin. The separation of macrolides is performed on silica gel, kieselguhr, cellulose, and RP layers.[10] Silica gel and polar mobile phases are extremely frequently applied, usually with the addition of methanol, ethanol, ammonia, sodium, or ammonium acetate. Because of the absence of chromophore groups, bioautography, derivatization, as well as charring are used, the last mainly by spraying with acid solutions (e.g., anisaldehyde/sulfuric acid/ethanol) and heating.

TETRACYCLINES

Tetracyclines, consisting of the octahydronaphthacene skeleton, are "broad-spectrum" antibiotics produced by *Streptomyces* or obtained semisynthetically (Fig. 5). They can be separated by both RP and NP TLC. Cellulose, kieselguhr, cyano-silica, or silica gel impregnated with EDTA or Na$_2$EDTA can be used. The last one is the most popular stationary phase in tetracycline analysis. Impregnation is necessary due to the very strong interaction of tetracyclines with hydroxyl groups

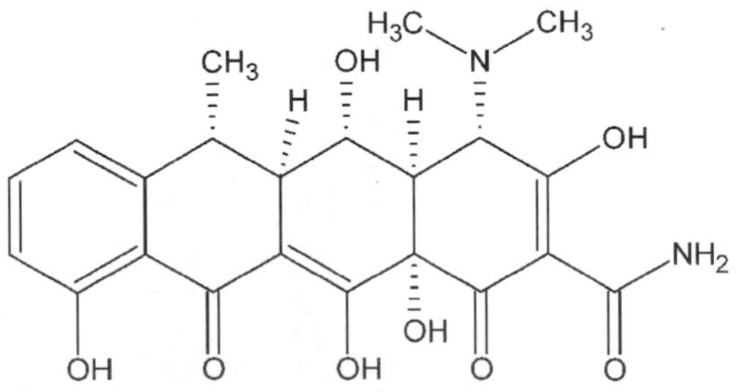

Fig. 5 Doxycycline.

and metal impurities. Also mobile phases, for both RP and NP TLC, should contain chelating agents such as Na₂EDTA, citric acid, or oxalic acid. Tetracyclines are amphoteric; thus, adjusting the pH of the mobile phases is very essential for their good separation. Tetracyclines give fluorescent spots, which can be detected by UV lamp, fixed at 366 nm or by densitometry. Spraying with reagents, for instance with Fast Violet B Salt solution, provides lower detection levels. Tetracyclines can also be detected by MS (TLC/FAB/MS, TLC/MALDI/MS) as well as by bioautography. Many TLC separation methods are described in the review on the analysis of tetracyclines in food.[11]

MACROCYCLIC ANTIBIOTICS

Peptides

Peptide antibiotics are composed of a peptide chain of amino acids D and L covalently linked to other moieties (Fig. 6). Most peptides are toxic and have poor pharmacokinetic properties. Peptide antibiotics are difficult to analyze in biological and food samples, as they are similar to matrix components. They can be separated on silica gel, amino silica gel, polyamide, modified cellulose, and silanized silica gel plates.[12] A variety of mobile phases are applied, from simple ones like chloroform/methanol to multicomponent ones like n-butanol/butyl acetate/methanol/acetic acid/water. Bioautographic detection can be employed with *Bacillus subtilis* and *Mycobacterium smegmatis* as well as fluorescence densitometry or densitometry after spraying the plate with reagents such as ninhydrin or Fluram®.

Rifamycins

Rifamycins (ansamycins) are structurally similar macrocyclic antibiotics, produced by *Streptomyces mediterranei*. Their characteristic "ansa" structure consists of aromatic rings spanned by an aliphatic bridge (Fig. 7). Rifamycins are active against Gram-positive bacteria and are mainly used in treating tuberculosis. They can be analyzed using silica gel, polyamide, diphenyl, or C-18 plates and various mobile-phase systems from neat organic solvents, through binary nonaqueous solvents, to binary aqueous–organic solvents.[12] Rifamycins are colored compounds and do not require special detection methods.

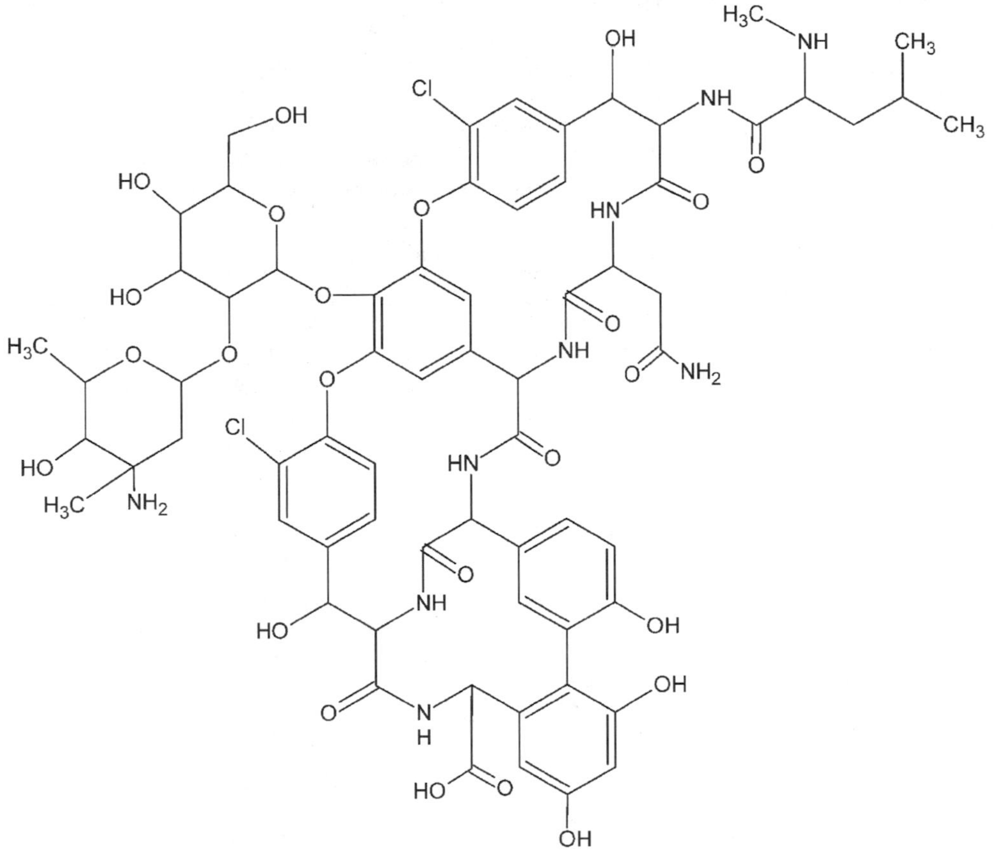

Fig. 6 Vancomycin.

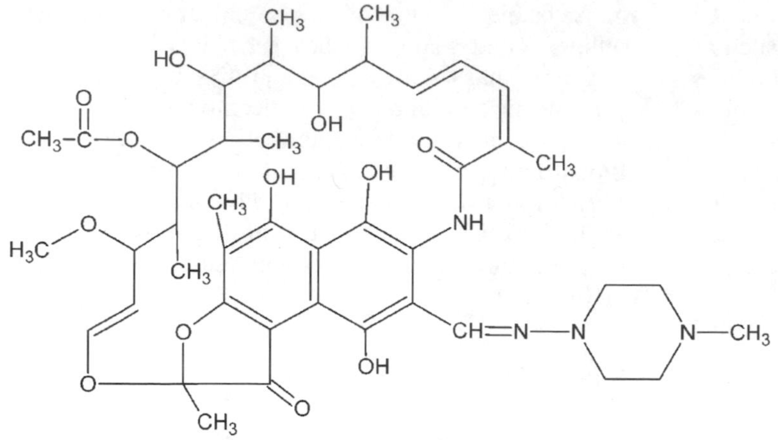

Fig. 7 Rifampicin.

POLYETHERS

Polyether or ionophore antibiotics, mainly produced by *Streptomyces* species, consist of cyclic ethers, a single carboxylic group, and several hydroxyl groups (Fig. 8). They are widely used anticoccidiosis agents for poultry. The main members of this class are salinomycin, monensin, narasin, and lasalocid. They can be analyzed on both silica gel and RP-18 plates with mixed organic phases. After derivatization with fluorescent pyrenacyl esters, they can be detected fluorodensitometrically at 360 nm. TLC/B with *Bacillus subtilis* can be used too.[13] There is also an example of coupling TLC with flame ionization detection.

AMPHENICOLS

Chloramphenicol is a highly effective broad-spectrum antibiotic originally isolated from *Streptomyces venezuelae* (Fig. 9). Nowadays, chloramphenicol is banned within the United States and the EC because it is believed to cause aplastic anemia. Other members of the amphenicol group are thiamphenicol and the recent one, florphenicol. These three antibiotics show strong UV absorption and can be determined directly, without any derivatization at 254 or 280 nm. Usually silica gel plates and simple organic or aqueous–organic solvents are used.[14]

SULFONAMIDES

Sulfonamide drugs are bacteriostatic synthetic compounds, the first chemotherapeutics used in human medicine. The progenitor of the class was a red azo dye, 2,4-diaminobenzene-4′-sulfonamide, called prontosil rubrum. The sulfonamides include sulfanilamide (4-amino-benzenesulfonamide) and numerous compounds related to it (Fig. 10).

Sulfonamide drugs are mainly used in veterinary practice and as growth promoters. They are used in the treatment of human infections to a lesser extent because they are toxic and some patients are hypersensitive to them.

Sulfonamides can be analyzed both by NP TLC (on silica gel, alumina, polyamide, and Florisil® layers) and by RP TLC (on silanized silica, RP-2, RP-8, and RP-18 layers).[15] Some sulfonamides have been separated by TLC on silica or polyamide impregnated with metal salts. Both aqueous and nonaqueous eluents are

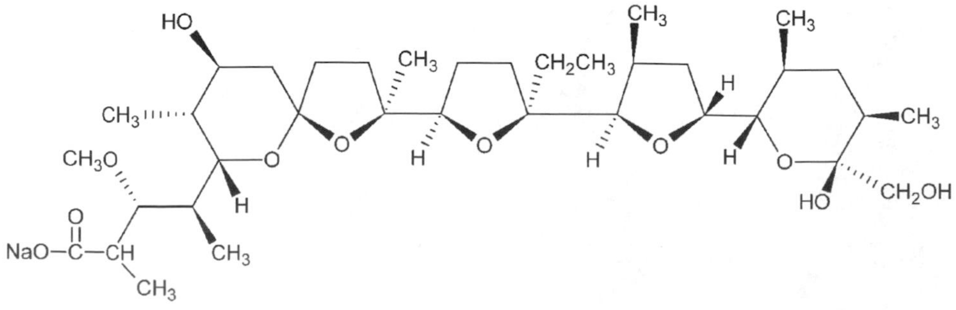

Fig. 8 Monensin.

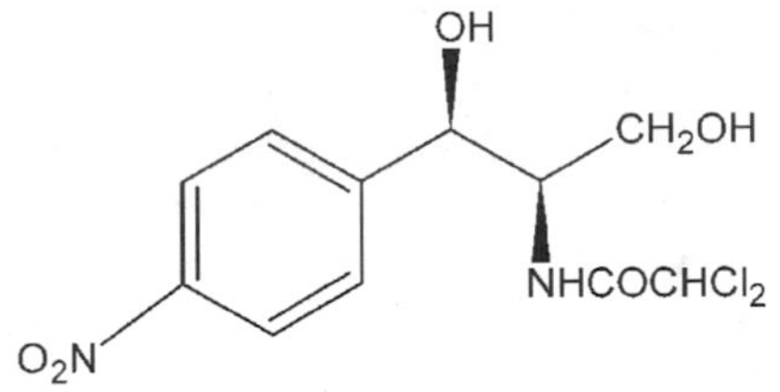

Fig. 9 Chloramphenicol.

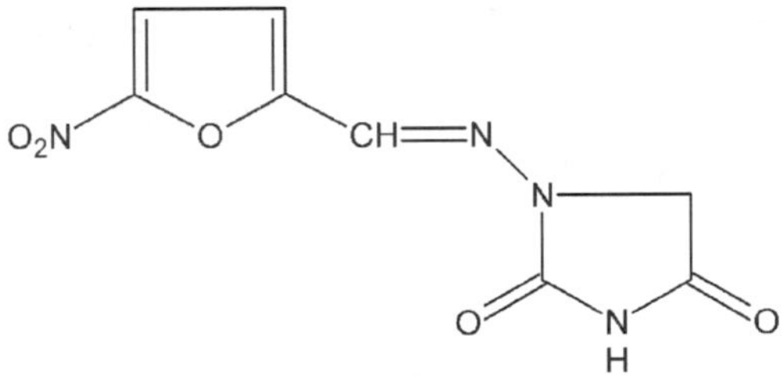

Fig. 11 Nitrofurantoin.

applied. Detection of sulfonamides can be performed on fluorescence layers at 254 nm and after derivatization with, for instance, fluorescamine solution at 366 nm.

NITROFURANS

Nitrofuran drugs are synthetic broad-spectrum chemotherapeutic agents, derivatives of nitrofuran (Fig. 11). Their application in human medicine is limited to some infections (e.g., nitrofurantoin is applied in treating urinary tract infections) or to external use. In veterinary practice, they are used as growth promoters and to prevent and treat diseases in poultry and swine.

Nitrofurans can be separated in NP systems on silica gel and can be detected as colored spots after spraying the plate with pyridine and illuminating with UV light at 366 nm.[16]

QUINOLONES

Nalidixic acid, discovered serendipitously in 1962, was the first member of this class, though of rather minor importance. In the 1980s, synthetic fluoroquinolones were developed and became valid antibiotics with broad spectrum and of good tolerance (Fig. 12). Quinolones are polar, mostly amphoteric compounds. They are usually analyzed on silica gel plates, preferably impregnated with Na_2EDTA or K_2HPO_4 to avoid strong adsorption. Multicomponent organic mobile phases are employed, usually with the addition of aqueous solutions of ammonia or acids to control pH. Micellar TLC with a cetyl trimethylammonium bromide/sodium dodecyl sulfate mixture as mobile

phase and polyamide as stationary phase can also be applied. Densitometry or fluorescence densitometry is a detection method of choice, sometimes preceded by postchromatographic derivatization. Bioautographic detection can also be applied.[4]

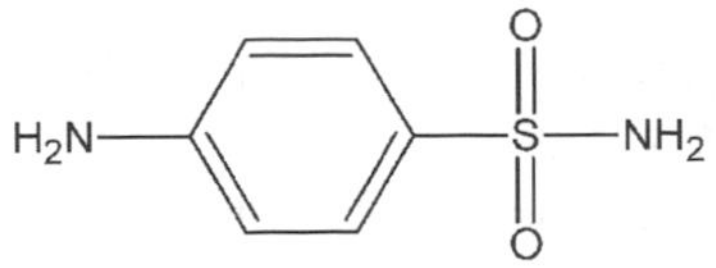

ANALYSIS OF ANTIBIOTICS BELONGING TO VARIOUS CLASSES

The analysis of antibiotics belonging to various classes is much more complicated than the analysis of members of one group only. Generally, it is necessary to divide the analyzed antibiotics preliminarily into subgroups. This can be achieved by developing the plate with different mobile phases or using gradient elution.[17] Different stationary phases are sometimes used for different antibiotic classes.[18] It is also possible to use one plate and one mobile phase but various modes of detection for different groups of antibiotics or to combine various modes of development with various modes of detection.[19] Bioautography is very often applied in the multiclass screening.[20] Scanning densitometry at different wavelengths on a hydrocarbon-impregnated silica gel HPTLC plate without solvent elution for direct quantification of many different classes of antibiotics is also described.[21]

OTHER APPLICATIONS

Beside typical antibiotic analysis, focused on the separation of antibiotics belonging to one or various

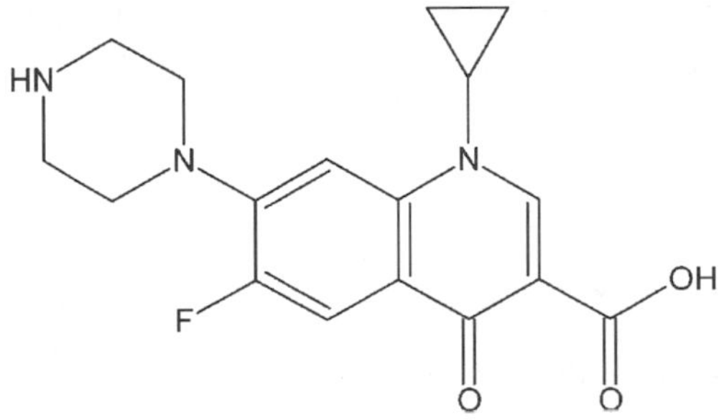

Fig. 10 Sulfanilamide.

Fig. 12 Ciprofloxacin.

classes, there are many examples of diverse TLC applications such as the following.

1. Purity control of antibiotics.
2. Examining the stability and breakdown products of antibiotics in solutions and dosage forms.[22]
3. Analysis of antibiotic metabolites.
4. Separation of antibiotic derivatives, obtained in the process of searching for new antibiotics.
5. Purification of newly discovered antibiotics before further testing. Antimicrobial substances are isolated from culture broths or plants and purified by preparative TLC on silica gel plates.
6. Chemical and biomolecular–chemical screening. Chemical screening is a systematic approach in the search for new biologically active compounds in extracts from micro-organisms or plants. Their chromatographic parameters calculated from the TLC plate as well as their chemical reactivity towards staining reagents allow one to obtain a picture of a microbial metabolite pattern (fingerprint). Biomolecular–chemical screening combines the chemical screening strategy with binding behavior towards DNA.[23]
7. Studying interactions of antibiotics with biological matrices. The interactions of the antibiotics with various compounds, cell membranes, and proteins present in biological matrices modify the biological efficacy and stability of the drugs. The interaction of 13 antibiotics with human serum albumin (HSA) was studied by charge-transfer RP TLC in neutral, acidic, basic, and ionic conditions and the relative strength of interaction was calculated.[24]
8. Applying some antibiotics as stationary or mobile phase additives for chiral separations. The macrocyclic antibiotics (ansamycins, glycopeptides, and polypeptides) can be used as chiral selectors in TLC. They can be used both as mobile phase additives (e.g., vancomycin) and for impregnation of TLC silica plates (e.g., erythromycin or vancomycin) for the separation of chiral compounds.[25]
9. Studying the retention behavior of antibiotics, e.g., determining the hydrophobicity parameters of antibiotics by RP TLC.

CONCLUSIONS

TLC is generally less sensitive and gives worse separation than HPLC. However, it predominates over HPLC in at least two aspects: It allows for the analysis of many samples at the same time, and it requires limited sample pretreatment. These features are very important in the analysis of antibiotics, which usually concerns controlling their level in many complicated matrices such as blood, urine, dietary products, and pharmaceuticals. Thus, TLC can be a very useful screening method preceding HPLC analysis. Nevertheless, there are also many examples of analytical applications of TLC, which can achieve selectivity and sensitivity comparable with those characteristic of HPLC. The future of the analytical option in antibiotic analysis is connected with progress in detection and the development of FFPC methods.

REFERENCES

1. Chromatography of antibiotics. J. Chromatogr. A. **1998**, 812 (1+2).
2. Choma, I. Antibiotics. In *Handbook of Thin-Layer Chromatography, Revised and Expanded*; 3rd Ed.; Sherma, J., Fried, B., Eds.; Marcel Dekker Inc.: New York, 2003; 417–444.
3. Botz, L.; Nagy, S.; Kocsis, B. Detection of microbiologically active compounds. In *Planar Chromatography*; 1st Ed.; Nyiredy, Sz., Ed.; Springer: Budapest, 2001; 489–516.
4. Choma, I.; Choma, A.; Komaniecka, I.; Pilorz, K.; Staszczuk, K. Semiquantitative estimation of enrofloxacin and ciprofloxacin by thin-layer chromatography–direct bioautography. J. Liquid Chromatogr. **2004**, *27* (13), 2071–2085.
5. Hendrickx, S.; Roets, E.; Hoogmartens, J.; Vanderhaeghe, H. Identification of penicillins by thin-layer chromatography. J. Chromatogr. **1984**, *291*, 211–218.
6. Boison, J.O. Chromatographic methods of analysis of penicillins in food–animal tissues and their significance in regulatory programs for residue reduction and avoidance. J. Chromatogr. **1992**, *624*, 171–194 (review).
7. Quintens, I.; Eykens, J.; Roets, E.; Hoogmartens, J. Identification of cephalosporins by thin layer chromatography and color reaction. J. Planar Chromatogr. Mod. TLC **1993**, *6*, 181–186.
8. Tuzimski, T. Two-dimensional thin layer chromatography of eight cephalosporins on silica gel layers. J. Planar Chromatogr. Mod. TLC **2004**, *17*, 46–50.
9. Stead, D.A. Current methodologies for the analysis of aminoglycosides. J. Chromatogr. B. **2000**, *747*, 69–93 (review).
10. Kanfer, I.; Skinner, M.F.; Walker, R.B. Analysis of macrolide antibiotics. J. Chromatogr. A. **1998**, *812*, 255–286 (review).

11. Oka, H.; Ito, Y.; Matsumoto, H. Chromatographic analysis of tetracycline antibiotics in foods. J. Chromatogr. A. **2000**, *882*, 109–133 (review).

12. Nowakowska, J.; Halkiewicz, J.; Lukasiak, J.W. TLC determination of selected macrocyclic antibiotics using normal and reversed phases. Chromatographia **2002**, *56*, 367–373.

13. VanderKop, P.A.; MacNeil, J.D. Separation and detection of monensin, lasalocid and salinomycin by thin-layer chromatography/bioautography. J. Chromatogr. **1990**, *508*, 386–390.

14. Freimüller, S.; Horsch, Ph.; Andris, D.; Zerbe, O.; Altorfer, H. Formation mechanism of solvent induced artefact arising from chromatographic purity testing of γ-irradiated chloramphenicol. Chromatographia **2001**, *53*, 323–325.

15. Bieganowska, M.L.; Doraczyńska-Szopa, A.D.; Petruczynik, A. The retention behavior of some sulfonamides on different thin layer plates J. Planar Chromatogr. Mod. TLC **1993**, *6*, 121–128.

16. Abjean, J.P. Qualitative screening for nitrofuran residues in food by planar chromatography. J. Planar Chromatogr. Mod. TLC **1993**, *6*, 319–320.

17. Krzek, J.; Kwiecień, A.; Starek, M.; Kierszniewska, A.; Rzeszutko, W. Identification and determination of oxytetracycline, tiamulin, lincomycin, and spectinomycin in veterinary preparations by thin-layer chromatography/densitometry. J. AOAC Int. **2000**, *83*, 1502–1506.

18. Vega, M.; Garcia, G.; Saelzer, R.; Villegas, R. HPTLC analysis of antibiotics in fish feed. J. Planar Chromatogr. Mod. TLC **1994**, *7*, 159–162.

19. Abjean, J.P. Planar chromatography for the multiclass, multiresidue screening of chloramphenicol, nitrofuran, and sulfonamide residues in pork and beef. J. AOAC Int. **1997**, *80*, 737–740.

20. Gafner, J.L. Identification and semiquantitative estimation of antibiotics added to complete feeds, premixes, and concentrates. J. AOAC Int. **1999**, *82*, 1–8.

21. Dhanesar, S.C.J. Quantitation of antibiotics by densitometry on a hydrocarbon-impregnated silica gel HPTLC plate. Part V: Quantitation and evaluation of several classes of antibiotics. J. Planar Chromatogr. Mod. TLC **1999**, *12*, 280–287.

22. Liang, Y.; Denton, M.B.; Bates, R.B. Stability studies of tetracycline in methanol solution. J. Chromatogr. A. **1998**, *827*, 45–55.

23. Maul, C.; Sattler, I.; Zerlin, M.; Hinze, C.; Koch, C.; Maier, A.; Grabley, S.; Thiericke, R. Biomolecular–chemical screening: a novel screening approach for the discovery of biologically active secondary metabolites—III. New DNA-binding metabolites. J. Antibiot. **1999**, *52*, 1124–1134.

24. Cserháti, T.; Forgács, E. Study of the binding of antibiotics to human serum albumin by charge-transfer chromatography. J. Chromatogr. A. **1997**, *776*, 31–36.

25. Ward, T.J.; Farris, A.B., III. Chiral separations using the macrocyclic antibiotics: a review. J. Chromatogr. A. **2001**, *906*, 73–89.

Antibiotics: Separation by Countercurrent Chromatography

M.-C. Rolet-Menet
Laboratoire de Chimie Analytique, UFR de Sciences Pharmaceutiques et Biologiques, Paris Cedex, France

INTRODUCTION

Antibiotics are chemical compounds made either by living organisms or by chemical synthesis. They have the property to inhibit, in small amounts, some vital processes of viruses, micro-organisms (such as bacteria and fungi), and certain cells of multicellular organisms (cancerous cells, parasitic cells, etc.). The development of antibiotics made by micro-organisms requires isolation and purification of the desired compound from a complicated matrix such as a fermentation broth. These bioactive microbial metabolites are often produced in very small quantities and have to be removed from other secondary metabolites and nonmetabolized media ingredients. Antibiotics are normally biosynthesized as mixtures of closely related congeners and many are labile molecules, thus requiring mild separation techniques with a high resolution capacity. Although recent advances in HPLC technology using sophisticated equipment and refined adsorbents greatly facilitate the isolation of antibiotics, some drawbacks remain, related to various complications arising from the use of a solid support, such as adsorptive loss, deactivation, and contamination. Moreover, HPLC purification always requires sample preparation, prepurification, concentration, etc. Liquid–liquid partition techniques and particularly countercurrent chromatography (CCC) are suitable for the separation of antibiotics because they utilize a separation column free of solid support matrix, made of Teflon® channels or tubes. Raw material can be injected into the column without any previous sample treatment, which simplifies the purification procedure.

ANTIBIOTICS

Antibiotics differ widely in their polarities because their chemical structures are very variable. They are synthesized by various living materials like bacterial strains (such as *Streptomyces*[1] and *Bacillus*) and marine sponges. Oka et al.[2] have gathered antibiotics purified by CCC from crude extract and fermentation broth. They have shown that CCC can be successfully applied to the separation of macrolides and of various other antibiotics, including various peptide antibiotics, which are strongly adsorbed to silanol groups on the silica gel used in the stationary phase in HPLC. Several CCC types are used, such as droplet CCC (DCCC)[3] and the more recent X-axis CCC, foam CCC, centrifugal partition chromatography (CPC), high-speed CCC (HSCCC), and Quattro CCC (QCCC). This discussion focuses on the separation of macrolides and polypeptide antibiotics by CCC.

Several separations of macrolide and polypeptide antibiotics by CCC are reported in the literature. Macrolides are heterosides in which the aglycone is a cyclic macrolactone with at least 14 atoms. They act by stopping protein synthesis. Polypeptide antibiotics are frequently cyclic molecules. They act by disorganizing the protein structure of the bacterial membrane. Figs.[1–5] show the structure of several molecules the purification of which is described subsequently. Sporaviridins[4] are produced by *Kutzneria viridogrisea*. They are very polar, water-soluble, basic glycoside antibiotics (Fig. 1). They consist of six components; each has a 34-membered lactone, seven monosaccharide units, a pentasaccharide (viridopentaose), and two monosaccharides. They are active against Gram-positive bacteria, acid-fast bacteria, and *Trichophyton*. WAP-8294A[5] complex (Fig. 2) is produced by *Lysobacter* sp. and consists of at least 19 closely related and very polar components. WAP-8294A2 is present as the major component and A1 and A4 are minor components. They show strong activity against Gram-positive bacteria, including methicillin-resistant *Staphylococcus aureus* (MRSA) and vancomycin-resistant enterococci. Ivermectins B1[6] are derived from avermectins B1, the natural fermentation products of *Streptomyces avermitilis*. The avermectins B1 have double bonds between carbon atoms 22 and 23, whereas the ivermectins B1 have single bonds in these positions (Fig. 3A). They have intermediate polarity. The ivermectins B1 are a mixture of two major homologs, ivermectins B1a (>80%) and ivermectins B1b (<20%), but a crude ivermectin complex also contains various minor components. Ivermectins B1 are broad-spectrum antiparasitic agents used against *Onchocerca volvulus* in human medicine and for food animals such as

Encyclopedia of Chromatography DOI: 10.1081/E-ECHR-120040132

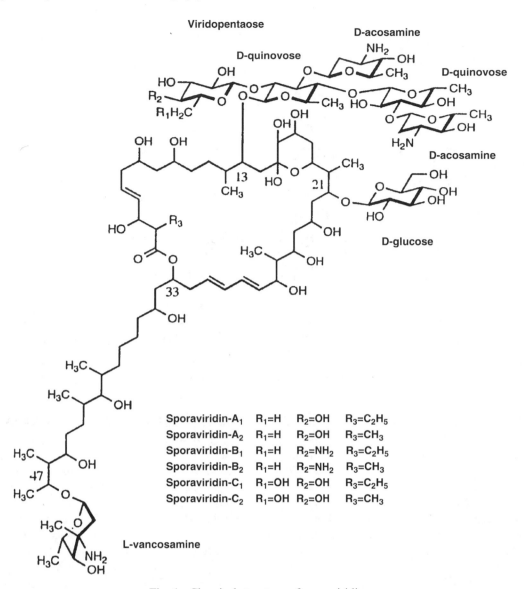

Viridopentaose
D-quinovose
D-acosamine
D-quinovose
D-acosamine
D-glucose
L-vancosamine

Sporaviridin-A$_1$	R$_1$=H	R$_2$=OH	R$_3$=C$_2$H$_5$
Sporaviridin-A$_2$	R$_1$=H	R$_2$=OH	R$_3$=CH$_3$
Sporaviridin-B$_1$	R$_1$=H	R$_2$=NH$_2$	R$_3$=C$_2$H$_5$
Sporaviridin-B$_2$	R$_1$=H	R$_2$=NH$_2$	R$_3$=CH$_3$
Sporaviridin-C$_1$	R$_1$=OH	R$_2$=OH	R$_3$=C$_2$H$_5$
Sporaviridin-C$_2$	R$_1$=OH	R$_2$=OH	R$_3$=CH$_3$

Fig. 1 Chemical structure of sporaviridins.

cattle, swine, and horse. The bryostatins have been isolated from the marine bryozoan *Bugula neritina*[7] (Fig. 3B). They are macrolides with intermediate polarity. They show significant activity against lymphocytic leukemia in vitro, with ED$_{50}$ values from 0.33 to 1.4 µg/ml, respectively. Ascomycin and related compounds[8] (Fig. 4) are macrolide antibiotics with intermediary polarity. They have been identified from *Streptomyces tsukubaensis* and *S. hygroscopicus* and are reported as immunosuppressants with higher potency than cyclosporin A.

Finally, bacitracins[9] are peptide antibiotics produced by *Bacillus subtilis* and *Bacillus licheniformis*. Over 20 components are contained in the bacitracin complex medium, among which the major active components are bacitracin A and F (Fig. 5). They exhibit inhibitory activity against Gram-positive

bacteria and are among the most commonly used antibiotics as animal feed additives.

SOLVENT SYSTEM

The polarity of the abovementioned molecules is very variable according to the saccharide unit number contained in the chemical formula. Several procedures to choose a solvent system are described in the literature. Usual solvent systems are biphasic and made of three solvents, two of which are nonmiscible. If the polarities of the solutes are known, the classification established by Y. Ito[2] can be taken as a first approach. He classified solvent systems into three groups according to their suitability for nonpolar molecules ("nonpolar" systems, based on *n*-hexane), intermediate polarity

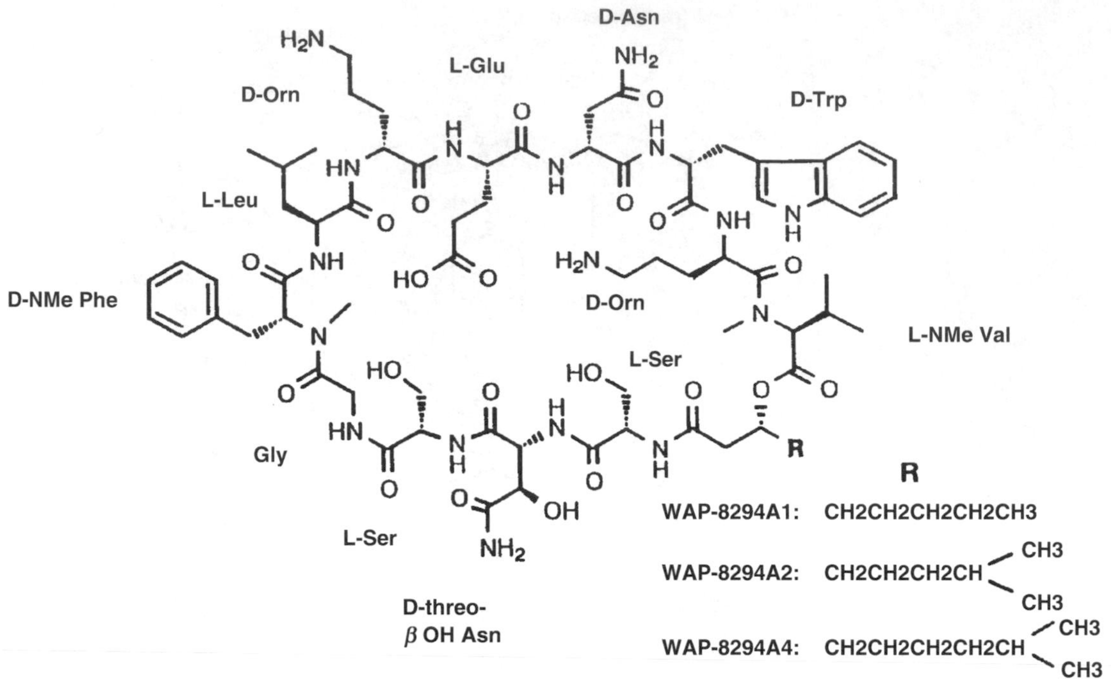

Fig. 2 Chemical structure of WAP-8294A complex.

molecules ("intermediary" system, based on chloroform), and polar molecules ("polar" system, based on *n*-butanol). The molecule must have a high solubility in one of the two nonmiscible solvents. The addition of a third solvent enables a better adjustment of the partition coefficients (K). Oka, Oka, and Ito[10] propose a choice of various solvent systems to purify antibiotics. They have to fulfill various criteria. The settling time of the solvent system should be shorter than 30 sec to ensure satisfactory retention of the stationary phase. The partition coefficient of the target compounds should be close to 1, and the separation factor (α) between the compounds should be larger than 1.5. Two series of solvent systems can provide an ideal range of K values for a variety of samples: *n*-hexane–ethylacetate–*n*-butanol–methanol–water and chloroform–methanol–water. These solvent series cover a wide range of hydrophobicity continuously from the nonpolar *n*-hexane–methanol–water system to the more polar *n*-butanol–water system. To select the solvent system, Wang-Fan et al.[8] measured the solubility of macrolides in a series of common solvents, where the polarities were ranked with dielectric constants. The partition coefficients of solutes were compared in various ternary solvent systems selected according to the solubility studies. A ternary solvent system was selected based on suitable partition coefficients of the antibiotics. Finally, in the further optimization of composition proportions, the quaternary solvent systems showed the best solvent selectivities by giving the most prominent differences of partition coefficient.

COUNTERCURRENT CHROMATOGRAPHY FOR PURIFICATION OF ANTIBIOTICS

Several CCC devices are commonly used to purify antibiotics, such as the rotating coil instruments particularly used in HSCCC and the cartridge instruments used in CPC. A chapter of this encyclopedia is entirely devoted to the various CCC devices, so that only some indications about performances of CCC as compared to preparative HPLC are given here.

Menet and Thiébaut[11] have compared the performances of CCC and preparative HPLC regarding the separation of two antibiotics X and Y. The CCC apparatus used was a centrifugal partition chromatograph (CPC, Sanki* LLN) of 250 ml internal volume. For the purpose, classical parameters of preparative scale chromatography were calculated: experimental duration, including the sample preparation and separation time; solvent consumption, including the volume of the mobile phase, the stationary phase, and the injection solvent; and purity of the purest fraction in Y. The parameter "purity in Y" was chosen because Y is the solute that is the most difficult to purify because of its physical properties (particularly hydrophobicity), which are close to those of the main impurities. The hourly yield (g/hr) is defined as the ratio of the recovered quantity to the experimental duration. The volumic yield (g/L) is defined as the ratio of the recovered quantity to the solvent consumption. Table 1 summarizes the results of separations of Y by CCC and preparative HPLC. The solvent volume

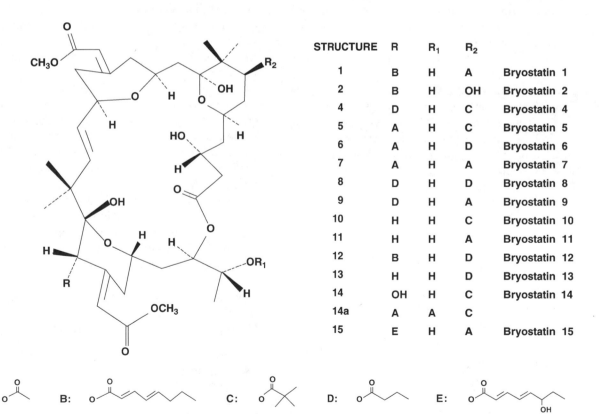

	R	-C22-X-C23-
Ivermectin B1a	C_2H_5	$-CH_2-CH_2-$
Ivermectin B1b	CH_3	$-CH_2-CH_2-$
Avermectin B1a	C_2H_5	$-CH=CH-$
Avermectin B1b	CH_3	$-CH=CH-$

STRUCTURE	R	R_1	R_2	
1	B	H	A	Bryostatin 1
2	B	H	OH	Bryostatin 2
4	D	H	C	Bryostatin 4
5	A	H	C	Bryostatin 5
6	A	H	D	Bryostatin 6
7	A	H	A	Bryostatin 7
8	D	H	D	Bryostatin 8
9	D	H	A	Bryostatin 9
10	H	H	C	Bryostatin 10
11	H	H	A	Bryostatin 11
12	B	H	D	Bryostatin 12
13	H	H	D	Bryostatin 13
14	OH	H	C	Bryostatin 14
14a	A	A	C	
15	E	H	A	Bryostatin 15

Fig. 3 Chemical structure of ivermectins (A) and bryostatins (B).

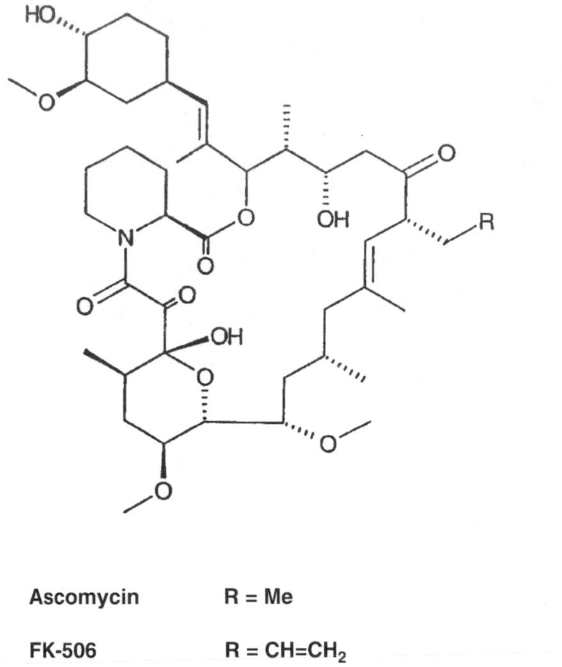

Ascomycin   R = Me

FK-506   R = CH=CH$_2$

Fig. 4 Chemical structures of ascomycin and derivatives.

Table 1 Comparison of CCC and HPLC performances

	CCC	HPLC
Crude extract purity in Y (%)	7	25
Injected quantity of Y (g)	0.28	1.59
Experiment duration (hr)	6.2	2.2[a]
Solvent volume consumption (L)	1.4	10.8
Purity of the purest fraction in Y	>95%	>95%
Hourly yield (g/hr)	0.035	0.72
Volumic yield (g/L)	0.20	0.15

[a]1 hr for column equilibration at 90 ml/min flow rate +1 hr for separation.

directly purifying crude extracts. Moreover, no preliminary purification of the extract is required, in contrast to preparative HPLC, which requires a 1 day enrichment of the crude extract from 7% to 25% in Y.

consumption is the volume of the stationary and mobile phases in CCC or the volume of the mobile phase used in HPLC and the samples. The injected sample in CCC was not prepurified to concentrate it in Y from 7% to 25%. So the injected quantity in Y in CCC is lower (0.28 g, as against 1.59 g in preparative HPLC). For similar volumic yields, i.e., 0.20 g/L in CCC and 0.15 g/L in preparative HPLC, the enrichment in Y is higher with CCC than with preparative HPLC. Indeed, starting from a crude extract at 7% in Y with CCC or from 25% in Y extract with preparative HPLC leads to the same 95% highest purity. These results demonstrate the advantage of CCC in

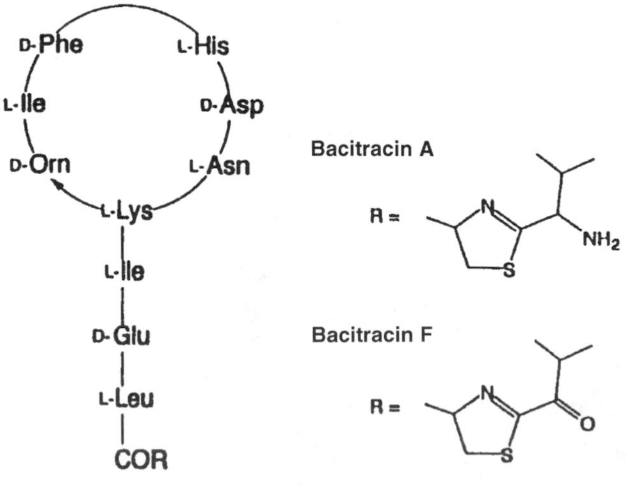

EXAMPLES OF PURIFICATION

Separation of Sporaviridins[4]

The chemical structures of sporaviridins are described in Fig. 1. They are only soluble in polar solvents such as water, methanol, and n-butanol. Therefore, a two-phase solvent system containing n-butanol was examined. A nonpolar solvent such as diethyl ether was added to the n-butanol–water system to decrease the solubility of molecules in n-butanol and to obtain partition coefficients close to 1. The partition coefficients K are defined as the ratio of the solute concentration in the upper phase (butanol rich) to its concentration in the lower one (water rich). A two-phase solvent system of n-butanol–diethylether–water (10:4:12, v/v/v) was selected because it allows one to obtain the almost equally dispersed partition coefficients among six components (C2, B2, A2, C1, B1, A1). The preparative separation of six components from sporaviridin complex by HSCCC was performed in 3.5 hr (500 ml elution volume). The six components were eluted in the order of their partition coefficients, yielding pure components A1 (1.4 mg), A2 (0.6 mg), B1 (0.7 mg), B2 (0.5 mg), C1 (1.1 mg), and C2 (1.4 mg) from 15 mg of the sporaviridin complex.

Separation of the Main Components of WAP-8294A Complex[5]

The chemical structures of WAP-8294A complex are described in Fig. 2. The structure of WAP-8294A2 has been elucidated as a cyclic depsipeptide with a molecular mass of 1561. High-speed CCC (type J apparatus, total capacity 300 ml) was applied to the

Fig. 5 Chemical structure of bacitracins.

separation of the main components of the WAP-8294A complex. Due to the high polarity of these compounds, a hydrophilic two-phase solvent system composed of n-butanol–ethylacetate–aqueous 0.005 M trifluoroacetic acid (1.25 : 3.75 : 5, v/v/v) was used, providing a suitable range of partition coefficient values. The preparative separation of six components from the WAP-8294A complex was performed in 13.3 hr, with the lower phase as mobile phase at 0.5 ml/min. Pure fractions (1–6 mg) were obtained from 25 mg of WAP-8294A complex.

Separation of Ivermectins[6]

These molecules have an intermediary polarity (Fig. 3A). A two-phase solvent system composed of n-hexane, ethyl acetate, methanol, and water was selected. In this case, the partition coefficients K are defined as the ratio of the solute concentration in the upper phase to its concentration in the lower one. A solvent mixture of n-hexane–ethylacetate–methanol–water (19 : 1 : 10 : 10, v/v/v/v) yielded the best K values; from 0 to 3.25 mg of crude ivermectin was separated in 4.0 hr. This separation yielded 18.7 mg of 99.0% pure ivermectin B1a, 1.0 mg of 96.0% pure ivermectin B1b, and 0.3 mg of 98.0% pure avermectin B1a.

Separation of Bryostatins[7]

An amount of 906.5 g of lymphocytic leukemia cell line active fraction was obtained by extraction from 1000 kg of *Bugula neritina*. Further purification was performed with HSCCC. Bryostatins have intermediary polarity, so that n-hexane–ethylacetate–methanol–water (3 : 7 : 5 : 5, v/v/v/v) was employed with the upper layer as mobile phase and lower layer as stationary phase. By this technique, from 300 mg to 3 mg of seven bryostatins have been isolated, including a new molecule, bryostatin 14 (Fig. 3B).

Separation of Ascomycin and Analogs[8,12]

Ascomycin and derivatives (Fig. 4) were purified by QCCC. The QCCC apparatus has four coils that are wound tightly on two separate bobbins on one rotor, each bobbin containing two concentrically wound coils. Optimization of solvent systems was based on solubility studies and measurements of partition coefficients for FK-506 and ascomycin. Hexane–*tert*-butylmethylether–methanol–water (1 : 3 : 6 : 5; v/v/v/v) showed the best solvent selectivity. Baseline separation of 25 mg of FK-506 and 50 mg of ascomycin was achieved in 6 hr.

Separation of Bacitracins[9]

Bacitracin complex (Fig. 5) was purified by foam CCC. The column design for foam CCC consists of a Teflon® tube. Simultaneous introduction of N_2 and the liquid phase through the respective flow tube produces a countercurrent between the gas and the liquid phase through the coil. The sample mixture injected through the middle portion of the column is separated according to the foaming capability: The foam active components travel through the coil with the gas phase and elute through the foam collection line, whereas the rest of components move with the liquid phase and elute through the liquid collection line.

After experiment, fractions from the foam and liquid outlets are collected and analyzed. The elution curve of bacitracin components from the foam outlet shows three major peaks, and the one from the liquid outlet, one peak. HPLC analysis of the fractions clearly indicates that the bacitracin components are separated in the order of hydrophobicity of the molecules in the foam fractions, and in increasing order of their hydrophilicity in the liquid fractions.

CONCLUSIONS

High-speed CCC successfully achieves preparative scale separations and purifications of numerous antibiotics from crude extracts. Moreover, the sample is directly analyzed without preliminary purification of the extract, as is required in preparative HPLC.

ARTICLES OF FURTHER INTEREST

Countercurrent Chromatography Solvent Systems,
  p. 401.
Countercurrent Chromatography/MS, p. 397.
Instrumentation of Countercurrent Chromatography,
  p. 847.

REFERENCES

1. Brill, G.M.; McAlpine, J.B.; Hochlowski, J.-E. Use of coil planet centrifuge in the isolation of antibiotics. J. Liq. Chromatogr. **1985**, *8* (12), 2259–2280.
2. Oka, H.; Harada, K.-I.; Ito, Y.; Ito, Y. Separation of antibiotics by counter current chromatography. J. Chromatogr. A. **1998**, *812*, 35–52.
3. Hostettmann, K.; Appolonia, C.; Domon, B.; Hostettmann, M. Droplet countercurrent chromatography—new applications in natural products chemistry. J. Liq. Chromatogr. **1984**, *7*, 231–242.

4. Harada, K.-I.; Kimura, I.; Yoshikawa, A.; Suzuki, M.; Nakazawa, H.; Hattori, S.; Ito, Y. Structural investigation of the antibiotic sporaviridin. XV. Preparative scale separation of sporaviridin components. J. Liq. Chromatogr. **1990**, *13*, 2373–2388.

5. Harada, K.-I.; Suzuki, M.; Kato, A.; Fujii, K.; Oka, H.; Ito, Y. Separation of WAP-8294A components, a novel anti-methicillin-resistant *Staphylococcus aureus* antibiotic, using high-speed counter-current chromatography. J. Chromatogr. A. **2001**, *932*, 75–81.

6. Oka, H.; Ikai, Y.; Hayakawa, J.; Harada, K.-I.; Suzuki, M.; Shimizu, A.; Hayashi, T.; Takeba, K.; Nakazawa, H.; Ito, Y. Separation of ivermectin components by high-speed counter-current chromatography. J. Chromatogr. A. **1996**, *723*, 61–68.

7. Pettit, G.R.; Gao, F.; Sengupta, D.; Coll, J.-C.; Herald, C.L.; Doubek, D.L.; Schmidt, J.M.; Van Camp, J.-R.; Rudloe, J.J.; Nieman, R.A. Isolation and structure of bryostatins 14 and 15. Tetrahedron **1991**, *47* (22), 3601–3610.

8. Wang-Fan, W.; Kusters, E.; Lohse, O.; Mak, C.P.; Wang, Y. Application of centrifugal counter-current chromatography to the separation of macrolide antibiotic analogues. I. Selection of solvent systems based on solubility and partition coefficient investigations. J. Chromatogr. A. **1999**, *864*, 69–76.

9. Oka, H.; Harada, K.-I.; Suzuki, M.; Nakazawa, N.; Ito, Y. Foam counter-current chromatography of bacitracin. I. Batch separation with nitrogen and water free of additives. J. Chromatogr. A. **1989**, *482*, 197–205.

10. Oka, F.; Oka, H.; Ito, Y. Systematic search for suitable two-phase solvent systems for high-speed counter-current chromatography. J. Chromatogr. A. **1991**, *538*, 99–108.

11. Menet, M.-C.; Thiébaut, D. Preparative purification of antibiotics for comparing hydrostatic and hydrodynamic mode counter-current chromatography and preparative high-performance liquid chromatography. J. Chromatogr. A. **1999**, *831*, 203–216.

12. Wang-Fan, W.; Kusters, E.; Mak, C.-P.; Wang, Y. Application of coil centrifugal counter-current chromatography to the separation of macrolide antibiotic analogues. III. Effects of flow-rate, mass load and rotation speed on the peak resolution. J. Chromatogr. A. **2001**, *925* (1–2), 139–149.

Antioxidant Activity: An Adaptation for Measurement by HPLC

Marino B. Arnao
Manuel Acosta
Antonio Cano
Department of Plant Biology (Plant Physiology), University of Murcia, Murcia, Spain

INTRODUCTION

The determination of antioxidant activity (capacity or potential) of diverse biological samples is generally based on the inhibition of a particular reaction in the presence of antioxidants. The most commonly used methods are those involving chromogenic compounds of a radical nature: the presence of antioxidant leads to the disappearance of these radical chromogens. They are either photometric or fluorimetric and can comprise kinetic or end-point measurements. Recently, there has been increasing interest in the adaptation of these methods for on-line determinations using LC. In this article, we present the adaptation to HPLC of our methods for the determination of the antioxidant activity in a range of samples. Advantages and disadvantages of these methods are discussed.

A biological antioxidant is a compound that protects biological systems against the potentially harmful effects of processes or reactions that cause excessive oxidation. Hydrophilic compounds, such as vitamin C, thiols, and flavonoids, as well as lipophilic compounds, such as vitamin E, vitamin A, carotenoids, and ubiquinols, are the best-known natural antioxidants. Many of these compounds are of special interest due to their ability to reduce the hazard caused by reactive oxygen and nitrogen species (ROS and RNS, some are free radicals), and have been associated with lowered risks of cardiovascular diseases and other illnesses related to oxidative stress.[1] Practically all the above mentioned compounds are obtained through the ingestion of plant products such as fruits and vegetables, nuts, flours, vegetable oils, drinks, and infusions, taken fresh or as processed foodstuffs.[2] A common property of these compounds is their antioxidant activity. The activity of an antioxidant is determined by:

1. Its chemical reactivity as an electron or hydrogen donor in reducing the free radical.
2. The fate of the resulting antioxidant-derived radical and its ability to stabilize and delocalize the unpaired electron.
3. Its reactivity with other antioxidants present.

Thus, antioxidant activity is a parameter that permits quantification of the capacity of a compound (natural or artificial) and/or a biological sample (from a wide range of sources) to scavenge free radicals in a specific reaction medium.[1,3–4]

METHODS TO MEASURE ANTIOXIDANT ACTIVITY

Antioxidant activity can be measured in a number of different ways. The most commonly used methods are those in which a chromogenic radical compound is used to simulate ROS and RNS; it is the presence of antioxidants that provokes the disappearance of these chromogenic radicals, as shown in the reaction model given in Scheme 1. In order for this method to be effective, it is necessary to obtain synthetic metastable radicals that can easily be detected by photometric or fluorimetric techniques. Nevertheless, different strategies for the quantification of antioxidant activity have been utilized: e.g., decoloration or inhibition assays. Details of these strategies and commonly used methods have been presented and reviewed elsewhere.[3–4]

When chromogenic radicals are used to determine antioxidant activity, the simplest method is to:

1. Dissolve the radical chromogen in the appropriate medium.
2. Add antioxidant.
3. Measure the loss of radical chromogen photometrically by observing the decrease in absorbance at a fixed time.
4. Correlate the decrease observed in a dose–response curve with a standard antioxidant (e.g., trolox, ascorbic acid), expressing the antioxidant activity as equivalents of standard antioxidant, a well-established parameter in this respect being Trolox Equivalent Antioxidant Capacity (TEAC).[3]

2,2'-Azino-bis-(3-ethylbenzthiazoline-6-sulfonic acid (ABTS) (Fig. 1) and α,α'-diphenyl-β-picrylhydrazyl

Encyclopedia of Chromatography DOI: 10.1081/E-ECHR-120013365

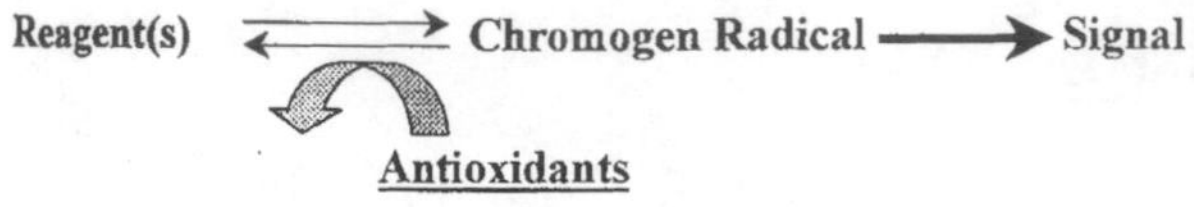

Reagent(s) ⇄ Chromogen Radical ⟶ Signal

Antioxidants

Scheme 1 Reaction model of antioxidant activity determination using chromogenic radicals.

radical (DPPH) are the two most commonly used synthetic compounds in antioxidant activity determinations. ABTS, when oxidized by the removal of one electron, generates a metastable radical. The ABTS radical cation (ABTS$^{\cdot+}$) has a characteristic absorption spectrum with maxima at 411, 414, 730, and 873 nm (Fig. 1), with extinction coefficients of 31 and 13 mM^{-1}cm^{-1} at 414 and 730 nm, respectively.[5] In the reaction between ABTS$^{\cdot+}$ and antioxidants, the radical is neutralized by the addition of one electron (see the reaction presented in Scheme 1. This leads to the disappearance of the ABTS$^{\cdot+}$, which can be estimated by the decrease in absorbance (virtually any wavelength between 400 and 900 nm can be selected to avoid exogenous absorption interferences). Generally, ABTS$^{\cdot+}$ is generated directly from its precursor in aqueous media by a chemical reaction (e.g., manganese dioxide, ABAP, potassium persulfate) or by an enzymatic reaction (e.g., peroxidase, hemoglobin, met-myoglobin) (see references in Ref.[5]).

Recently, we have developed a method based on ABTS$^{\cdot+}$ generated by horseradish peroxidase (HRP) that permits the evaluation of the antioxidant activity of pure compounds and plant-derived samples.[6] The method is easy, accurate, and fast to apply and presents numerous advantages because it avoids undesirable side reactions, does not require high temperatures to generate ABTS radicals, and allows for antioxidant activity to be studied over a wide range of pH values. This method is capable of determining both hydrophilic (in buffered media) antioxidant activity (HAA) and lipophilic (in organic media) antioxidant activity (LAA).[5] In the second case, ABTS$^{\cdot+}$ is generated directly in ethanolic medium by HRP, which is a powerful oxidizing biocatalyst that can act in nonaqueous media—a capacity that has been widely used in biotechnological applications. Thus, it is possible to estimate the antioxidant activity of both antioxidant types in the same sample (HAA and LAA). The antioxidant capacities of natural compounds, such as ascorbic acid, glutathione, cysteine, phenolic compounds (resveratrol, gallic acid, ferulic acid, quercetin, etc.), or synthetic antioxidants, such as BHT, BHA, or trolox (a structural analog of vitamin E), have been estimated, as well those of plant extracts or samples from other sources. Different applications of the method have determined antioxidant activity in a range of foodstuffs.[7] The ABTS$^{\cdot+}$ chromogen used in our method has been compared with another widely used radical chromogen, DPPH$^{\cdot}$; it was concluded that in the determination of the antioxidant potential of citrus and wine samples, the DPPH$^{\cdot}$ method could significantly underestimate TEAC by up to 36% compared to ABTS$^{\cdot+}$.[8] Also, we have applied the method to study the total antioxidant activity of different vegetable soups, obtaining relevant data on the relative contribution of hydrophilic (ascorbic acid and phenols) and lipophilic (carotenoids) components

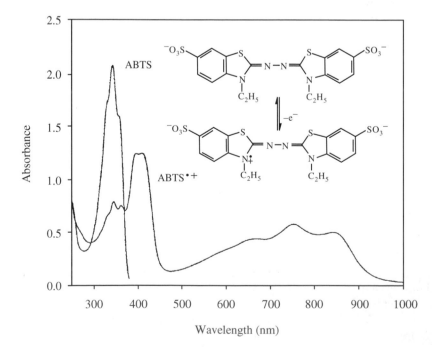

Fig. 1 Spectral characteristics of ABTS and its oxidation products, the ABTS radical (ABTS$^{\cdot+}$), showing absorbance of up to 1000 nm. The chemical structures show the nitrogen-centered radical cation of ABTS$^{\cdot+}$.

to their total antioxidant activity.[9] On the other hand, our methods have been used in animal physiological studies on changes in the plasma antioxidant status caused by the hormone melatonin in rats[10] and by other authors on the effect of "in vivo" oxidant stress in the rat aorta.[11]

Under our assay conditions, ABTS$^{\cdot+}$ generation progresses quickly and only 1–5 min is necessary to reach maximum absorbance. This is a decisive factor in the easy and rapid application of the assay with minimal reagent manipulation. In contrast, other assays that use ABTS$^{\cdot+}$ to measure the activity of lipophilic antioxidants have certain drawbacks, among which are: lengthy time (up to 16 hr) to chemically generate and stabilize ABTS$^{\cdot+}$ via potassium persulfate;[12] a previous filtration step when manganese dioxide is used; or, in the case of the assay that uses ABAP, high temperatures (45–60°C) that tend to affect ABTS$^{\cdot+}$ stability. The advantage of enzymatic ABTS$^{\cdot+}$ generation, as opposed to chemical generation, is that the reaction can be controlled by the amount of H_2O_2 added, while the exceptional qualities of HRP in ABTS$^{\cdot+}$ generation is an important feature in both the aqueous and the organic system.[5–6] The most significant limiting factor in this type of strategy is the fact that the ABTS$^{\cdot+}$ must be stable during the analysis; we were able to optimize the conditions to ensure >99% stability. During optimization, it was verified that the concentration ratio between radical (ABTS$^{\cdot+}$) and substrate (ABTS) is a determining factor for the stability of ABTS$^{\cdot+}$, although pH and temperature are also important elements. With respect to the sensitivity of these methods, the calibration using L-ascorbic acid presented a detection limit of 0.15 nmol and a quantification limit of 0.38 nmol. For lipophilic antioxidants, LOD of 0.08 and LOQ of 0.28 nmol of trolox were obtained. The LOD and LOQ of similar values were obtained for α-tocopherol and β-carotene.

ANTIOXIDANT ACTIVITY BY HPLC

The possibility of automating antioxidant activity determination and applying it to a large number of samples was an interesting objective. Previously, we have adapted our method as a microassay using a microplate reader to determine total antioxidant activity.[5] Recently, other authors have adapted radical chromogenic tests into methods that combine the advantages of rapid and sensitive chromogen radical assays with HPLC separation for the on-line determination of radical scavenging components in complex mixtures. Specifically, the DPPH$^{\cdot}$ method and the method of Rice-Evans, which uses ABTS$^{\cdot+}$ generated chemically with potassium persulfate, have been

adapted as such.[13–14] Nonetheless, the chemical generation of ABTS$^{\cdot+}$ via potassium persulfate required 16–17 hr to complete. Our methods resulted in faster and better controlled generation of stable ABTS radical because ABTS$^{\cdot+}$ was generated enzymatically in only 2–5 min, with perfect control over the amount of ABTS$^{\cdot+}$ formed and its stability (ABTS/ABTS$^{\cdot+}$ ratios).[6] The speedy generation of ABTS$^{\cdot+}$ permitted quick acquisition of the absorbance value desired in the detector by the addition of aliquots of H_2O_2 to the ABTS solution.

The adaptation of the ABTS$^{\cdot+}$ method as an on-line test required that the chromogen radical should be stable for sufficient time in different solvents to permit the utilization of isocratic or gradient elution programs. The on-line reaction time between ABTS$^{\cdot+}$ and potential antioxidants was an additional potential limiting factor.

For on-line measurement of the antioxidant activity of samples using LC, it is first necessary to consider the basic equipment required. Thus, the determination of antioxidant activities in separate components of samples by HPLC in a postcolumn reaction of analytes with preformed ABTS$^{\cdot+}$ requires at least:

1. Two pumps, one for the mobile-phase solutions and another for the preformed ABTS$^{\cdot+}$ solution. A pulse dampener is recommended to minimize pulse oscillations.
2. A sample injector.
3. The chromatography column.
4. A reaction coil of adequate length to give the desired reaction time.
5. A UV–visible (UV–VIS) detector.
6. An integration system (software) for data analysis.

Fig. 2 shows a schematic diagram of the equipment used in this study. In this case, because only one diode array detector was available, two injections of the sample were necessary: one to obtain the UV profile (at 250 nm) and another for the antioxidant activity profile at 600 nm (negative peaks). If two UV–VIS detectors had been used, only one injection would have been required to obtain the dual-HPLC profile but the chromatograms must be time-normalized.

In this type of analysis, a dual-HPLC profile was obtained. The UV profile (injection one) was of interest because all the main components of biological samples are absorbed in this wavelength range. The second injection detected absorbance changes at 600 nm or higher (see absorption spectrum of ABTS$^{\cdot+}$ in Fig. 1) to give the antioxidant activity profile. The photodiode array detector additionally recorded the absorption spectra of peaks and, consequently, could also provide data on the possible chemical nature of the analyzed

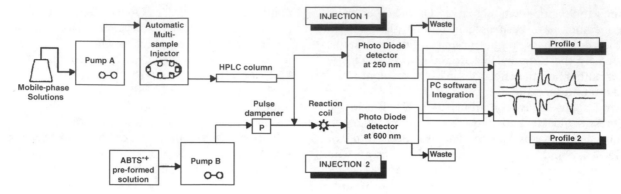

Fig. 2 Instrumental scheme for the determination of antioxidant activities by HPLC using ABTS$^{\cdot+}$ as chromogenic radical.

compounds. The HPLC-ABTS method can be used to characterize hydrophilic (ascorbic acid, phenolic compounds, organic acids, etc.) or lipophilic antioxidants (trolox, a synthetic standard antioxidant analog of vitamin E or carotenoids such as β-carotene, lycopene, xanthophylls, etc.). Using standard antioxidants, the dual-HPLC profile as shown in Fig. 3 could be obtained. In Fig. 3A, the upper chromatogram (trolox detected at 250 nm) and the lower chromatogram (the scavenging activity of trolox vs. ABTS$^{\cdot+}$ measured at 600 nm) were correlated. A calibration curve relating the antioxidant concentration and the signal (600 nm, as peak areas) was obtained and used as standard to

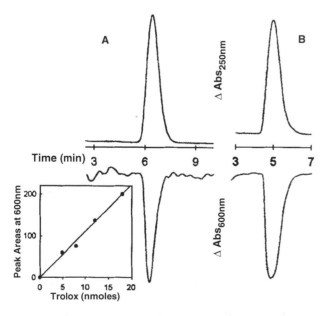

Fig. 3 Dual-HPLC plots of two antioxidants: trolox and resveratrol. Upper chromatograms show UV profiles registered at 250 nm and lower chromatograms ABTS$^{\cdot+}$ scavenging (antioxidant activity) profiles registered at 600 nm (negative peak). In (A), trolox was detected with a retention time of 6.2 min. Inset: Calibration curve of scavenging activity (peak areas at 600 nm) for different amounts of trolox. In (B), resveratrol was detected at 5.0 min.

express all data as TEAC. Generally, a known amount of trolox was injected into HPLC in any chromatographic conditions (to analyze hydrophilic or lipophilic compounds) to quantify its antioxidant activity and obtain the calibration curve. Thus, antioxidant activity was calculated from the sum of the peak areas of the chromatogram profile at 600 nm (negative peaks) and expressed as trolox equivalents (TEAC) using the previously mentioned calibration curves. An example of another important antioxidant (resveratrol) is shown in Fig. 3B.

Another significant aspect was the stability of the radical chromogen ABTS$^{\cdot+}$ in different solvents, in isocratic or gradient elution programs. We found that in the mobile phases used in our determinations (saline solutions and mixtures of organic solvents in different proportions), the observed fall was less than 0.01 expressed as $-\Delta Abs_{730\,nm}/min$.[15] This stability is high enough to obtain accurate data, approximately 10 times greater than the data reported in Ref.[14].

It was very important to guarantee at least 1 min of on-line reaction time between ABTS$^{\cdot+}$ and the antioxidants because fast antioxidants, such as trolox or ascorbic acid, reacted with ABTS$^{\cdot+}$ almost immediately, but other antioxidants required more time. In our case, trolox and ascorbic acid presented TEAC values of 1.0 and 0.99, respectively, using the HPLC-ABTS method (Table 1); similar values were obtained using the ABTS end-point method or the method of Rice-Evans.[3,6] In the method of Koleva et al.,[14] ascorbic acid presented time dependence: at 30 s, 60% of TEAC was expressed. In our system, and to guarantee sufficient on-line reaction time, a stainless steel reaction coil of 1 mL volume (2.5 m × 0.7 mm i.d.) coupled to a pump was connected to the chromatographic system (between the column and the diode detector) (Fig. 2). Thus, using a suitable elution program (0.5–0.7 mL/min of mobile phases) and introducing between 0.3 and 0.5 mL/min of the preformed ABTS$^{\cdot+}$ (0.2 mM), a total on-line reaction time of 1 min was obtained.

Table 1 Antioxidant activities of different compounds determined by the HPLC-ABTS and by the end-point method

Compound	Antioxidant activity (TEAC)	
	HPLC-ABTS method	End-point method[a]
L-Ascorbic acid	0.99	1.0
Trolox	1.0	1.0
Ferulic acid	0.87	1.94
Gallic acid	1.39	3.02
Resveratrol	1.32	2.34
Quercetin	2.83	4.30

[a](From Ref.[4].)

Under these conditions, a study of the antioxidant potential of pure compounds could be carried out. Table 1 shows the values of antioxidant activity (expressed as TEAC) of different compounds of interest, determined by the on-line method (HPLC-ABTS method) and compared with the values obtained by our conventional photometric end-point method.[6] As can be observed, the two most important standard antioxidants, trolox and ascorbic acid, presented similar TEAC using either method. Thus, either can be used as reference to express antioxidant activity, except that trolox has the advantage because it can be used in both hydrophilic and lipophilic assays. The TEAC values of phenolic compounds were underestimated by approximately half when the HPLC-ABTS method was used as compared to the end-point method. This was due to the different reactivities of antioxidants with ABTS[.+], and because, unfortunately, the time dependence of on-line scavenging activity determinations made it very difficult to obtain the total reaction for the slowest antioxidants resulting in a partial estimation of this activity. Nevertheless, the HPLC-ABTS method provided important additional information in the form of correlation between the different peaks of a sample and their antioxidant activities.

The HPLC-ABTS has been used in a study on the HAA and the LAA of fresh citrus and tomato juices.[15] The data obtained showed a good correlation between vitamin C content and HAA and slight underestimations of LAA. We are currently applying this method to different plant materials with the aim of finding out which compounds apport significant antioxidant properties to the foodstuffs studied.

CONCLUSIONS

Determinations of antioxidant activity are widely used in phytochemistry, nutrition, food chemistry, clinical chemistry, as well as in human, animal, and plant physiology, etc. Methods adapted to HPLC have appeared only recently but can be expected to have multiple applications in the future. ABTS[.+] is an excellent metastable chromogen for the detection and quantification of the HAA and LAA of biological samples. Thus, using a simple photometer (end-point method),[6] a microplate reader (multisample titration method),[5] or HPLC equipment, a broad range of possibilities are available for the characterization of diverse samples (animal- or plant-derived). Some applications of special interest could include:

1. Characterization of biological samples (e.g., plant extracts, foods).
2. Studies on the changes in the antioxidant activity of material during industrial or postharvest processing (e.g., thermal processes, Maillard reactions, and cold storage of foods, etc.).
3. The search for new natural antioxidants of vegetable or marine origin.
4. Clinical determinations.

ACKNOWLEDGMENTS

This work was supported by the Instituto Nacional de Investigación y Tecnología Agraria y Alimentaria (I.N.I.A., ministerio de Ciencia y Tecnología, Spain) project CAL00-062 and by the project PI-9/00759/FS/01 (Fundación Séneca, Murcia, Spain). A. Cano has a grant from the Fundación Séneca of the Comunidad Autónoma de Murcia (Spain). The authors wish to thank A.N.P. Hiner for checking the draft of this manuscript.

REFERENCES

1. Halliwell, B.; Gutteridge, J.M.C. *Free Radicals in Biology and Medicine*; 3rd Ed.; Halliwell, B., Gutteridge, J.M.C., Eds.; Oxford Univ. Press: New York, 2000.
2. Mackerras, D. Antioxidants and health. Fruits and vegetables or supplements? Food Aust. **1995**, *47*, S3–S23.
3. Rice-Evans, C.A.; Miller, N.J. Total antioxidant status in plasma and body fluids. Methods Enzymol. **1994**, *234*, 279–293.
4. Arnao, M.B.; Cano, A.; Acosta, M. Methods to measure the antioxidant activity in plant material. A comparative discussion. Free Radic. Res. **1999**, *31*, S89–S96.
5. Cano, A.; Acosta, M.; Arnao, M.B. Λ method to measure antioxidant activity in organic media:

Application to lipophilic vitamins. Red. Rep. **2000**, *5*, 365–370.

6. Cano, A.; Hernández-Ruiz, J.; García-Cánovas, F.; Acosta, M.; Arnao, M.B. An end-point method for estimation of the total antioxidant activity in plant material. Phytochem. Anal. **1998**, *9*, 196–202.

7. Arnao, M.B.; Cano, A.; Acosta, M. Total anti-oxidant activity in plant material and its interest in food technology. Rec. Res. Dev. Agric. Food Chem. **1998**, *2*, 893–905.

8. Arnao, M.B. Some methodological problems in the determination of antioxidant activity using chromogen radicals: A practical case. Trends Food Sci. Technol. **2000**, *11*, 419–421.

9. Arnao, M.B.; Cano, A.; Acosta, M. The hydrophilic and lipophilic contribution to total antioxidant activity. Food Chem. **2001**, *73*, 239–244.

10. Plaza, F.; Arnao, M.; Zamora, S.; Madrid, J.; Rol de Lama, M. Validación de un microensayo con ABTS·+ para cuantificar la contribución de la melatonina al estatus antioxidante total del plasma de rata. Nutr. Hosp. **2001**, *16*, 202.

11. Laight, D.W.; Gunnarsson, P.T.; Kaw, A.V.; Anggard, E.E.; Carrier, M.J. Physiological micro-assay of plasma total antioxidant status in a model of endothelial dysfunction in the rat following experimental oxidant stress in vivo. Environ. Toxicol. Pharmacol. **1999**, *7*, 27–31.

12. Re, R.; Pellegrini, N.; Proteggente, A.; Pannala, A.; Yang, M.; Rice-Evans, C.A. Antioxidant activity applying an improved ABTS radical cation deco-lorization assay. Free Radic. Biol. Med. **1999**, *26*, 1231–1237.

13. Koleva, I.I.; Niederländer, H.A.G.; van Beek, T.A. An on-line HPLC method for detection of radical scavenging compounds in complex mixtures. Anal. Chem. **2000**, *72*, 2323–2328.

14. Koleva, I.I.; Niederlander, H.A.G.; van Beek, T.A. Application of ABTS radical cation for selective on-line detection of radical scavengers in HPLC eluates. Anal. Chem. **2001**, *73*, 3373–3381.

15. Cano, A.; Alcaraz, O.; Acosta, M.; Arnao, M. On-line antioxidant activity determination: Comparison of hydrophilic and lipophilic antioxidant activity using the ABTS·+ assay. Red. Rep. **2002**, *7*, 103–109.

Application of Capillary Electrochromatography to Biopolymers and Pharmaceuticals

Ira S. Krull
Sarah Kazmi
Northeastern University, Boston, Massachusetts, U.S.A.

INTRODUCTION

Capillary electrochromatography (CEC) has grown considerably over the past few years, due to the developments in column technology and the appearance of several articles demonstrating the high efficiencies possible with this technique.[1] The literature has shown that there can be numerous applications for this technology, which was not possible with the earlier separation methods.

CEC Technique

Tsuda published an article that discussed the CEC technique in detail.[2] The technique itself is a derivative of high-performance capillary electrophoresis (HPCE) and HPLC, where the separations are performed using fused-silica tubes of 50–100 μm inner diameter (i.d.), that are packed with either a monolithic packing or small (3 μm or smaller) silica-based particles.[3–5] The packing is similar to the conventional HPLC; however, the mobile phase is driven by electro-osmosis, which results from the electric field applied across the capillary rather than by pressurized flow. The mobile phase is made up of aqueous buffers and organic modifiers [e.g., (ACN)]. An electro-osmotic flow (EOF) of up to 3 mm/s can be generated. It is a plug flow, where the linear velocity is independent of the channel width and there is no column back-pressure.[4] Partitioning or adsorption of the neutral analyte occurs in the same way as in HPLC. The analytes are separated while moving through the column with the EOF. Charged solutes have additional electrophoretic mobility in the applied electric field; therefore, the separation occurs by electrophoresis and partitioning. The selectivity in analysis of the charged analytes is increased by electromigration of the sample molecules. The flat flow profile results in a more efficient radial mass transport compared to the parabolic laminar flow in pressure-driven LC, and this results in a significant enhancement in separation performance and shorter analysis times.[1,5–6] The capillaries can be made shorter to offer the same plate count as HPLC, therefore reducing the back-pressure.

The packing material is smaller compared to HPLC, so with the high electric fields, the efficiency of this technique is very high (up to about 500,000 plates/m),[1] the peaks are sharp, the resolution is high, and the process is highly selective. An article by Angus et al.[7] demonstrated the separation efficiencies of 200,000–260,000 plates/m that were obtained by CEC and were reproducible from column to column for structurally related, polar neutral compounds of pharmaceutical relevance. The sample capacity in CEC is 10–100 times higher than that of capillary electrophoresis (CE), and this means that more sample volume can be placed on the CEC column to give better sensitivity. The high capacity comes from the high column load-ability that results from the stationary phase's retentive mechanism.[8] The absence of additives and predominantly organic mobile phases make CEC better suited for use in mass spectrometry (MS). In fact, nonaqueous CEC is already being practiced by analysts.[8] A recent article by Hansen and Helboe gives a detailed study of the possibility of using CEC to replace gradients or ion-pairing reagents. The group optimized the separation of six nucleotides using a background analyte consisting of 5 mM acetic acid, 3 mM triethylamine (TEA), and 98% acetonitrile and a C_{18} 3-μm column. This was accomplished in half the time taken for a similar separation in HPLC.[9]

A variation of gradient CEC is pressurized-flow CEC or PEC (pressurized flow electrochromatography). A pump forms the gradient and then allows part of this pressurized flow to pump the mobile phase through the packed bed. In this way, one can perform isocratic or gradient CEC with part of the mobile-phase driving force being pumped, part electrophoretic and part electroosmotic flows.[1,8,10]

Detection of Biomolecules and Pharmaceuticals in CEC

There are many different types of detectors used for pharmaceutical applications in CEC. They vary from indicating just the presence of a sample [fluorescence (FL)], to giving some qualitative information about a sample [photodiode array UV/vis detection (PDA)],

Encyclopedia of Chromatography DOI: 10.1081/E-ECHR-120039882

to absolute sample determination of the analyte (MS). The methods can be on-column, off-column, and end-column. With on-column, the solutes are detected while still on the capillary, in off-column, the solute is transported from the outlet of the capillary to the detector, and end-column is done with the detector placed right at the end of the capillary. Some modes of detection used in CEC are as follows:

1. UV/vis absorbance detection is widely used in CE Absorptivity depends on the chromophore (light-absorbing part) of the solute, the wavelength of the incident light, and the pH and composition of the run buffer. A photodetector measures light intensities and the detector electronics convert this into absorbance.[11]
2. Fluorescence detection is based on the fact that, when light energy strikes a molecule, some of that energy may be given off as heat and some as light. Depending on the electronic transitions within a molecule, the light given off may be fluorescent or phosphorescent.[12] Fluorescence occurs when an electron drops from an excited singlet to the ground state, as opposed to phosphorescence, which occurs when an electron's transition is from an excited triplet to the ground state.
3. LIF detection, such as argon ion, helium–cadmium,[5] and helium–argon lasers, can be used for this detection method. The criterion for choosing the laser is that the wavelength should be at or near the excitation maxima for the solute to be determined. The higher the power of the laser, the higher the intensity and the peak height and the laser's ability to focus the beam to a small spot.
4. MS detection[4] is the only detector that has high sensitivity and selectivity and can be used universally, thus, the increased interest in interfacing this technology with CEC compared to other detection methods. It can detect all solutes that have a molecular weight within the mass range of the MS. In the selected ion-monitoring mode, it detects only solutes of a given mass, and in the total ion chromatogram mode, it detects all the solutes within a given mass range.

Current applications of CEC use on-column UV or laser-induced fluorescence detection; however, for UV, the path length is quite short, which limits sensitivity, although bubble, Z-shaped, and high-sensitivity cells have helped to improve detection limits. However, UV and fluorescence are only good for samples that fluoresce and absorb light or are amenable to derivatization with fluorescing or absorbing chromophores. These detectors impose difficult cell volumes and sample size limits if high separation efficiencies are to be realized, and they are very expensive. All of these drawbacks are nonexistent for MS techniques, which are expensive but provide more structural information and high sensitivity and appear to have the greatest overall potential.[8]

According to an issue of LC/GC,[10] combining CEC with detection techniques such as MS, MS/MS and inductively coupled plasma (ICP)-MS are easier to accomplish, as the flow rates are at nanoliter per minute levels. Analysts must add makeup solvent after the capillary separation for certain ionization methods, and, because it is added later, users can select solvents that are more compatible with the detection technique. CEC mobile phases have a high organic solvent content that is more amenable to MS. Also, the low CEC flow rates means less maintenance and downtime for MS source cleaning.

In the references to the application of CEC to biopolymers, most of the work discusses CEC–electrospray ionization (ESI)/MS, much less to direct CEC-UV/FL methods. However, much of the work has evolved from the use of commercially available, prepacked capillaries, such as C_{18} or ion exchange or a mixed mode containing both ion exchange and reversed phase (RP). There are very few articles that have actually attempted to develop new phases specifically for biopolymers.

When using MS, the actual CEC conditions never really need to be fully optimized because the MS accomplishes the additional resolution and specific identification, as needed. The specific mobile-phase conditions in CEC/MS may be quite different than for CEC-UV/FL or HPLC, and thus optimization of CEC/MS conditions will be somewhat different than for CEC-UV/FL. This would include, just as for LC/MS, the use of volatile organic solvents and organic buffers, low flow rates, no void volumes, or loss of resolution in the CEC/MS interface and the usual interfacing requirements already developed and optimized for CE/MS.[10,13–23]

Few descriptions of quantitation have been reported so far. Most of the literature is qualitative by nature, simply demonstrating suitable, if not fully optimized, experimental conditions that provide evidence of the presence of certain biopolymers and their high resolution from other components in that particular sample. Absolute quantitation and validation needs to be developed and fully optimized for CEC, to become a more valuable and applicable separation mode for biopolymers.

SEPARATION OF PROTEINS

Capillary electrochromatography can accomplish high plate counts, as mentioned earlier; this means a high peak capacity (number of peaks that can be fitted into a typical separation time for a given length of column), therefore highly complex materials can be separated. The implication of better peak capacities is a better resolution of the peaks in a complex analyte. Because of the frequent overlap of peaks due to components in a complex sample, it is difficult to demonstrate peak purity with other methods. It is possible in CEC to quantitatively determine the presence of a particular analyte. CEC techniques have produced the separation of the enantiomers of amino acids.[24–27] This is done with limited use of solvents, buffer additives, salts, organics, chiral species, packing materials, and total time of analysis. Other groups have successfully utilized gradient elution to separate mixtures of dansylated amino acid mixtures on the ODS (octadecylsilane) stationary phase.[24] Also, microprocessor control of pressure flow and voltage, automated sample injection, automated data collection, automated capillary switching, and the ability to interface with a variety of detection instrumentation make CEC an appealing technique for protein separation and peptide mapping.

Proteins and peptides are water-soluble complex molecules that are composed of amino acids linked by peptidic and disulfide bonds. Proteins are really just larger peptides of higher molecular weight, and anti bodies are larger proteins of specific conformations, shape, size and immunogenicity, together with antigenic recognition properties.[28–31] The type, number, and sequence of amino acids in the chain determine the chemical characteristics of a peptide. The amino acid sequence determines the electrophoretic properties of the peptide. In addition to the amine terminus of the sequence, the amine and the guanidine residues of lysine and arginine are the main carriers of the positive charges, and the negative charge contribution is associated with the carboxylic acid terminus and the acidic groups of the aspartic and glutamic acids. The isoelectric and isoionic points of the peptide are their important characteristics in electrophoresis. These points are similar in peptides but not identical; the iso-electric point is determined by the given aqueous medium, whereas the isoionic point is related to the interactions with protons. The relation of the electrophoretic mobility of the peptides and their relative molecular weight is described by Offord's equation:

$$\mu_{rel} = \frac{Z}{[\sqrt[3]{(M^2)}]}$$

where μ_{rel} is the relative mobility, Z is the total net charge, and M is the molar mass in gram per mole. Calculation of the net charge cannot be done easily from the pK values of the acidic and basic groups for large peptides, but additional factors such

as conformational differences, primary sequence, chirality, and so forth need to be considered.

The popular methods of analysis of proteins currently are HPLC, HPCE, and MS. However, due to the complexity of proteins, LC approaches show a single, broad, ragged peak, which indicates that the method is unable to resolve the individual species.[1] In CEC, the success of the protein separation requires that the capillary packing material meet certain properties. Depending on the ionic characteristics of the biopolymers, pH-dependent "ideal" packings would be either RP or ion-exchange chromatography (IEC) or a combination of both.[1,32–46] There are several references that detail the possible application of SEC packings in CEC, but these have mainly been applied to synthetic organic polymers and much less to biopolymers.[47–50] Regardless of which packing is actually utilized, it should contain a stationary (bonded, not coated) phase that can successfully interact with the biopolymers, as in RP-HPLC, and prevent any unwanted silanol interactions with the underlying silica or ionic sites. It must also provide additional or programmable EOF, besides that from the uncoated fused-silica capillary walls. Perhaps an ideal packing would combine a cationic-exchange material (cationic-exchange chromatography (CIEC) together with RP, in order to allow separations based on RP (hydrophobicity) alone, a combination of RP and IEC, or just IEC alone, mobile phase (buffer) dependent. Also, that packing, be it single or mixed mode, should prevent unwanted biopolymer (e.g., peptide amino groups) interaction with the support, such as amine–silanol hydrogen-bonding in RP-HPLC for amine containing analytes (e.g., pharmaceuticals, peptides, and proteins). Additional articles on open tubular CEC (OT-CEC) applications, where a coating is applied on the inner surface of the capillary as in capillary GC have appeared.[51–54] There are articles on packed-bed CEC, where there is a real packing in the capillary.[55–56] Also, then there are methods that employ just CEC, without any additional, pressure-driven flow,[38–39] as well as pressurized CEC or PEC, with additional pressurized flow.[4,51–53] There is also electro-HPLC that utilizes gradient elution with an applied voltage, but it is mainly conventional HPLC with some voltage applied sporadically or continuously during the HPLC separation.[54]

CEC OF BIOPOLYMERS (PROTEINS, PEPTIDES, AND ANTIBODIES)

The majority of the applications of CEC for biopolymers have dealt with peptides, of varying sizes and complexity, utilizing different modes of CEC (OT–CEC, conventional isocratic CEC, gradient CEC,

PEC, and others). The following accounts are listed according to the work of different authors in the field.

Palm and Novotny applied the polymeric gel beds (monoliths) for peptide resolutions in CEC.[5] The peptide separation used the above packing beds with 29% as the ligand. Additional CEC conditions are indicated in Fig. 1, which depicts the separation of a series of tyrosine-containing peptides, with detection at 270–280 nm. In this particular study, peptide elution patterns were very sensitive to changes in pH and ACN concentrations. A gradient elution technique, not employed here, would have been more appropriate for such samples of peptides having small differences in their constitution. Attempts to elute protein samples were unsuccessful with these particular gel matrices, perhaps due to the high hydrophobicity of the packings.[5]

Euerby et al. reported the separation of an *N*-methylated, C- and N-protected tetrapeptide from its nonmethylated analog (Fig. 2).[38] These separations utilized a Spherisorb ODS-1, 3-μm packing material, without pressure-driven flow (true CEC using a commercially available CE instrument), and an ACN buffer. Using nonoptimized, nonpressurized CEC conditions (non-PEC, non-pressure-driven CEC), separation of the two tetrapeptides could be achieved in a run time of 21 min with efficiency values of 124,000 and 131,000 plates/m. In comparison, when

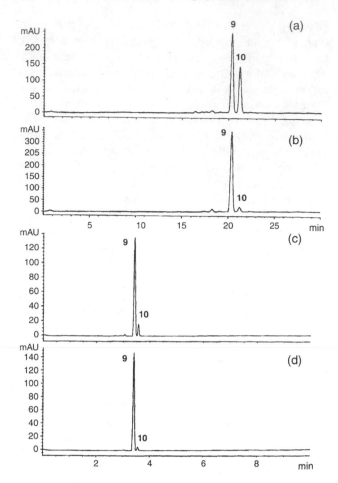

Fig. 2 Separation of synthetic, protected tetrapeptide intermediates: (9) *N*-methyl C- and N-protected tetrapeptide; (10) non-*N*-methyl C- and N-protected tetrapeptide. The structures of these compounds is proprietary information and consequently cannot be disclosed. Detection wavelength of 210 nm with a 10-nm bandwidth and a 1-s rise time. Electrochromatography was performed on a 250 mm × 50 μm i.d. spherisorb ODS-1 packed capillary using an ACN-Tris (50 mmol/L, pH 7.8) buffer 80:20 v/v mobile phase, capillary temperature of 15°C, and an electrokinetic injection of 5 kV/15 s. (a) Synthetic mixture of protected tetrapeptides 9 and 10. Efficiency values of 124,000 and 131,000 plates/m were obtained for analytes 9 and 10, respectively. (b) Chromatogram of synthetically prepared 9, the presence of residual nonmethylated tetrapeptide (10) can be seen. (c) Chromatogram of synthetically prepared 9, spiked with 10% of the nonmethylated tetrapeptide (10). Efficiency values of 83,000 and 101,000 plates/m were obtained for analytes 9 and 10, respectively. (d) Chromatogram of synthetically prepared 9, the presence of residual nonmethylated tetrapeptide (10) can be clearly seen at the 3% level. Additional conditions are indicated in Ref.[38]. (From Ref.[38]; with permission of the authors and John Wiley & Sons.)

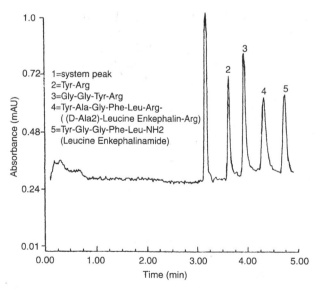

1=system peak
2=Tyr-Arg
3=Gly-Gly-Tyr-Arg
4=Tyr-Ala-Gly-Phe-Leu-Arg-
  ((D-Ala2)-Leucine Enkephalin-Arg)
5=Tyr-Gly-Gly-Phe-Leu-NH2
  (Leucine Enkephalinamide)

Fig. 1 Isocratic electrochromatography of peptides in a capillary filled with a macroporous polyacrylamide–polyethylene glycol matrix, derivatized with a C_{12} ligand (29%) and containing acrylic acid. Conditions: mobile phase, 47% acetonitrile in a buffer; voltage, 22.5 kV (900 V/cm), 7 μm; sample concentration, 4–10 mg/mL; detection, UV absorbance at 270 nm; other conditions are described in Ref.[5]. (From Ref.[5]; with permission of the authors and the American Chemical Society.)

a pressurized CE (buffer reservoirs and capillary were pressurized, with pressure-driven flow of buffer) system was used, separation of the components was

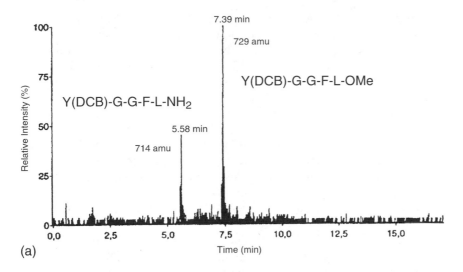

(a)

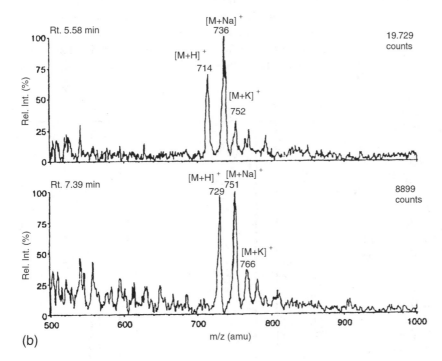

(b)

Fig. 3 Interfacing of pressure-driven CEC (PEC) for the separation of two simple peptides, enkephalin methyl ester (5.58 min) and enkephalin amide (7.39 min). (a) Extracted mass chromatogram of m/z 714 and 729 for the on-line peptide separation. Specific operating conditions are indicated in Ref.[4]. (b) Mass spectra taken from the chromatographic peaks in (a), illustrating true M_r and the presence of M + H, M + Na, and M + K cations at appropriate m/z (amu) values. (From Ref.[4]; with permission of the authors and the American Chemical Society.)

achieved within 3.5 min. According to Euerby et al. the separation of these two peptides using a pressure-driven HPLC gradient analysis took 30 min and gave comparable peak area results. Although this study illustrated the improved efficiency of both nonpressurized CEC and pressurized CEC over HPLC, no reasonable conclusions can be drawn from this work.

The attempts that have been made to utilize true chemometric optimization of operating conditions in CEC are unclear in most of the studies done utilizing CEC. This has been done for many years in GC and HPLC, as well as in CE, but there are no obvious articles that have appeared which have utilized true chemometric software approaches to optimization in CEC.[57–59] It is not clear that any true method optimization has been performed or what analytical figures of

merit were used to define an optimized set of conditions for biopolymer analysis by CEC. It is also unclear as to why a specific stationary phase (packing) was finally selected as the optimal support in these particular CEC applications for biopolymers. In the future, it is hoped that more sophisticated optimization routines, especially computerized chemometrics (expert systems, theoretical software, or simplex/optiplex routines) will be employed from start to finish.

The coupling of ESI and MS with a pressurized CEC system (PEC) has been shown to separate peptides.[4] This particular study of Schmeer et al. utilized a commercial RP silica gel, Gromsil ODS-2, 1.5-mm packing material, already utilized in capillary HPLC for peptide separations. It was never made perfectly clear why this particular packing material was selected or why PEC was selected for MS interfacing over conventional, isocratic CEC conditions. It is possible that the EOF alone with this packing material was in-sufficient to elute all peptides in a reasonable time frame and, thus, pressurized flow was introduced. No gradient elution PEC was demonstrated in this particular study. A mixture of enkephalin methyl ester and enkephalin amide was separated using the packed capillary column (Fig. 3a, 3b. The coupling of these two methods showed enhanced sensitivity and detectability. Like the Euerby study, this offered little insight into the capabilities of CEC to separate peptides; however, the study does provide a nice example of a peptide separation based on chromatographic and electrophoretic separation mechanisms, probably occurring simultaneously. This report also described the ability of easily interfacing CEC and PEC with ESI/MS.

The coupling of an MS with CEC or PEC provides several advantages. With the capillary columns of 100 mm inner-diameter (i.d.), flow rates of 1–2 L/min are obtained, which are ideal for electrospray MS.[4] No interface like a liquid sheath flow is required and the sintered silica gel frits allow direct coupling of the packed capillary columns without additional transfer capillaries. The spray is therefore formed directly at the outlet side of the column. Verheij et al. carried out the first coupling of a pseudoelectrochromatography system to a fast-atom bombardment (FAB)-MS in 1991.[6] However, this required transfer capillaries that caused a loss in efficiency, which was also a problem with other experimentations with this technique.

Lubman's group published several papers on the PEC/MS system.[51–54] RP open tubular columns (RP-OTC), which were prepared by a sol–gel process, were coated with an amine that enhanced the EOF in an acidic buffer solution and reduced the nonspecific adsorption between the peptides and the column wall. A six-peptide mixture was separated to baseline within 3 min using this system coupled to an on-line ion-trap storage–time-of-flight mass spectrometer (ITS–TOFMS). A full-range mass detection speed of 8 Hz was used in all these experiments, which was rapid to maintain the high efficiency and ultrafast separation. A high-quality total ion chromatogram could be obtained with only a couple of femtomoles of peptide samples, due to the high-duty cycle of the MS and the column path-length-independent and concentration-sensitive feature of the ESI process. The concentration limit of detection was also improved to about 1×10^{-6} M because of the preconcentration capability of the RP-CEC. A tryptic digest of horse myoglobin was successfully separated within 6 min on the gradient CEC system. The use of the MS increased the resolving power of this system by clearly identifying the coeluting components.

Another article by Wu et al.[52] dealt with a PEC coupled to an ion-trap storage/reflectron TOFMS (RTOFMS) for the analysis of peptide mixtures and protein digests. Taking advantage of the EOF, a high separation efficiency has been achieved in PEC due to a relatively flat flow profile and the use of smaller packing materials. With columns only 6 cm long, a tryptic digest of bovine cytochrome-c was successfully separated in about 14 min by properly tuning the applied voltage and the supplementary pressure. A relatively complex protein digest (tryptic digest) of chicken albumin gave 20 peaks (resolved) in the total ion current chromatogram in 17 min (Fig. 4). The sample concentrations were also on the order of about 1×10^{-5} M The detector increased the resolving power of PEC by unambiguously identifying coeluting components. The CEC was directly interfaced to the MS via an ESI, which provided the molecular-weight information of the protein digest products and structural information via MS/MS (Table 1). The device uses a quadrupole ion trap as a front-end storage

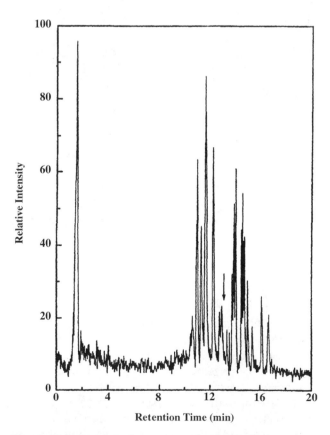

Fig. 4 The total ion chromatography (TIC) of the separation of a tryptic digest of chicken ovalbumin with a sample injection amount of 12 pmol corresponding to the original protein.[52] Column length, 6 cm. Conditions: 20 min, 0–40% acetonitrile gradient: 1000 V applied voltage with a 40-bar supplementary pressure. (From Ref.[52]; with permission of the authors and the American Chemical Society.)

device, which converts a continuous electrospray beam for TOF analysis. The storage property of the ion trap provides ion integration for low-intensity signals, whereas the nonscanning property of the TOFMS provides high sensitivity. A description of the MS is provided in an article by Wu et al.[54] According to an article by Verheij et al., problems that were encountered earlier, like formation of bubbles, have been overcome by using liquid junctions to apply the electric field over the column.[60]

A recent review article by the Lubman group points out that there are some serious advantages in using an open tubular column (OTC) for CEC as compared with packed-bed CEC.[54] OTCs with inner diameters around 10 μm have been found to have a smaller plate height when compared to packed columns. This is due to the lack of band-broadening effects associated with the presence of packing materials and end-column frits. OTC capillaries do not require end frits. High concentration sensitivity is another advantage of OTCs, as columns with very small dimensions are used. The small diameters of OTCs allow for the use of a higher voltage in CEC, without significant Joule heating. OTCs can also often provide more rapid separations than packed columns, by eliminating intraparticle diffusion, which is an important elimination for ultrafast separations in packed columns. However, there are some grave difficulties involved in using OTCs, perhaps because of the real difficulties with sample injection and detection. The injection volume of OTCs is in the low nanoliter or even picoliter range. The very small inner diameters of most OTCs make

Table 1 Comparison of calculated and measured tryptic peptides of chicken ovalbumin from PEC/MS

No.	Tryptic peptides	Calculated mass[a]	Determined mass[1–2][a,b]	Sequence
1	1–16	1709.0	1709.6	GSIGAASMEFCFDVFK
4, 5	47–55	1080.2	1079.7	DSTRTQINK
5	51–55	602.7	602.9	TQINK
6, 7	56–61	781.0	781.4	VVRFDK
7	59–61	408.5	408.4	FDK
10	105–110	779.8	780.1	IYAEER
11	111–122	1465.8	1466.3	YPILPEYLQCVK
12	123–126	579.7	579.7	ELYR
13	127–142	1687.8	1687.5	GGLEPINFQTAADQAR
16	182–186	631.7	631.6	GLWEK
16, 17	182–189	996.1	995.9	GLWEKAFK
17	187–189	364.4	364.5	AFK
18	190–199	1209.3	1209.0	DEDTQAMPFR
20	219–226	821.9	821.7	VASMASEK
21	227–228	277.4	277.6	MK
23	264–276	1581.7	1581.3	LTEWTSSNVMEER
24, 25	277–279	405.5	405.4	KIK
26	280–284	646.8	646.8	VYLPR
26, 27	280–286	924.2	924.4	VYLPRMK
27, 28	285–290	813.0	813.1	MKMEEK
28	287–290	535.6	535.5	MEEK
30	323–339	1773.9	1774.2	ISQAVHAAHAEINEAGR
31	340–359	2009.1	2008.5	EVVGSAEAGVDAASVSEEFR
32	360–369	1190.4	1190.2	ADHPFLFCIK
33	370–381	1345.6	1345.3	HIATNAVLFFGR
33, 34	370–385	1750.1	1749.5	HIATNAVLFFGRCVSP
34	382–385	404.5	404.4	CVSP

[a]Average masses.
[b]Average of all charge states observed.
From: Ref.[52]; reproduced with permission of the authors and the American Chemical Society.

optical detection difficult, but they are very compatible with a concentration-sensitive detection method, such as ESI–IT–TOFMS. With peptide mixtures, however, gradient elution CEC, with or without pressure-driven flow, is almost required over isocratic or step-gradient methods, because small changes in the mobile phase composition results in large changes in peptide retention times.

In a later study, Pesek et al. reported the separation of other proteins using a diol stationary phase.[61–64] The use of a diol stationary phase should result in a surface that is more hydrophilic than a typical alkyl-bonded moiety, like C_{18} or C_8. The overall results showed significant variations in retention times due to differences in solute-bonded phase interactions. Other factors, such as pH, could also influence this interaction, due to its influence on charge and protein conformations. Combining all these factors in the separation of peptides and proteins provides an experimentalist with many decisions to be made in the optimized experimental conditions to be used. Other chemical modifications of etched fused silica need to be studied in order to provide a better understanding of their interactions with proteins and peptides, as well as other classes of biopolymers.

CONCLUSIONS

At the present time, although there are several applications of CEC/PEC to biopolymer classes, these are to be considered only preliminary and not necessarily fully optimized in all possible parameters. At times, significant improvements in peak shape, plate counts, resolutions, efficiency, and the time of analysis can be realized. However, final optimizations of these separations have not been realized or possible. Some workers have utilized pressurized flow to solve the problems of obtaining reasonable EOF without silanol–analyte interaction; however, this does not solve the problem. It just forces the analyte to elute and approaches electro-HPLC, rather than true CEC. There are real differences between electro-HPLC, PEC and CEC that need to be recognized. There does not, in general, seem to have been any serious attempt to utilize any chemometric software approaches in CEC/PEC for biopolymer separation optimizations or rationale for doing so. At this time, packings are simply used because they were on the shelf in a laboratory or com-mercially available and not necessarily because they were really the best for protein–peptide separations in PEC/CEC. There remains a need for research-oriented column choices from commercial vendors to avoid the need to pack capillaries in-house with commercial HPLC supports.

REFERENCES

1. Krull, I.S.; Mistry, K.; Stevenson, R. CEC '98: the state-of-the-art. Am. Lab. August **1998**, *16A*.
2. Tsuda, T. Electrochromatography using high applied voltage. Anal. Chem. **1987**, *59*, 521.
3. Olivares, J.A.; Nguyen, N.T.; Yonker, C.R.; Smith, R.D. On-line mass spectrometric detection for capillary zone electrophoresis. Anal. Chem. **1987**, *59*, 1230.
4. Schmeer, K.; Behnke, B.; Bayer, E. Capillary electrochromatography-electrospray mass spectrometry: a microanalysis technique. Anal. Chem. **1995**, *67*, 3656.
5. Palm, A.; Novotny, M.V. Macroporous polyacrylamide/poly(ethylene glycol) matrixes as stationary phases in capillary electrochromatography. Anal. Chem. **1997**, *69*, 4499.
6. Verheij, E.R.; Tjaden, U.R.; Niessen, W.A.M.; van der Greef, J. Pseudo-electrochromatography-mass spectrometry: a new alternative. J. Chromatogr. **1991**, *554*, 339.
7. Angus, P.D.A.; Victorino, E.; Payne, K.M.; Demarest, C.W.; Catalano, T.; Stobaugh, J.F. Method development in pharmaceutical analysis employing capillary electrochromatography. Electrophoresis **1998**, *19*, 2073.
8. Majors, R.E. Perspectives on the present and future of capillary electrochromatography. LC–GC Magazine **1998**, *16* (2), 96.
9. Hansen, S.H.; Helboe, T. Separation of nucleosides using capillary electrochromatography. J. Chromatogr. A. **1999**, *836*, 315.
10. Landers, J.P., Ed.; *CRC Handbook of Capillary Electrophoresis—Principles, Methods, and Applications*; 2nd Ed.; CRC Press: Boca Raton, FL, 1997.
11. Baker, D.R. *Capillary Electrophoresis*; Techniques in Analytical Chemistry Series; John Wiley & Sons: New York, 1995.
12. Hercules, D.M. *Fluorescence and Phosphorescence Analysis: Principles and Applications*; Interscience: New York, 1966; 19.
13. Cunico, R.L.; Gooding, K.M.; Wehr, T. *Basic HPLC and CE of Biomolecules*; Bay Bioanalytical Laboratory: Richmond, CA, 1998.
14. Horvath, Cs., Nikelly, J.G., Eds.; *Analytical Biotechnology: Capillary Electrophoresis and Chromatography*; ACS Symposium Series; American Chemical Society: Washington, DC, 1990; Vol. 434.
15. Li, S.F.Y. *Capillary Electrophoresis: Principles, Practice and Applications*; Elsevier Science: Amsterdam, 1992.

16. Grossman, P.D., Colburn, J.C., Eds.; *Capillary Electrophoresis—Theory and Practice*; Academic Press: San Diego, CA, 1992.

17. Righetti, P.G., Ed.; *Capillary Electrophoresis in Analytical Biotechnology*; CRC Series in Analytical; CRC Press: Boca Raton, FL, 1996.

18. Camilleri, P., Ed.; *Capillary Electrophoresis: Theory and Practice*; CRC Press: Boca Raton, FL, 1993.

19. Mosher, R.A.; Thormann, W. *The Dynamics of Electrophoresis*; VCH: Weinhein, 1992, Chap. 7.

20. Altria, K.D.; Rogan, M.M. *Introduction to Quantitative Applications of Capillary Electrophoresis in Pharmaceutical Analysis, A Primer*; Beckman Instruments, Inc.: Fullerton, CA, 1995.

21. Weinberger, R.; Lombardi, R. *Method Development, Optimization and Troubleshooting for High Performance Capillary Electrophoresis*; Simon and Schuster Custom Publishing: Needham Heights, MA, 1997.

22. Karger, B.L., Hancock, Wm., Eds.; *High Resolution Separation and Analysis of Biological Macromolecules*; Methods in Enzymology Series, Part A, Fundamentals; Academic Press: San Diego, CA, 1996; Vol. 270.

23. Altria, K.D., Ed.; *Capillary Electrophoresis Guidebook, Principles, Operation and Applications*; Methods in Molecular Biology; Humana Press: Totowa, NJ, 1996.

24. Lurie, I.S.; Meyers, R.P.; Conver, T.S. Capillary Electrochromatography of cannabinoids. Anal. Chem. **1998**, *70*, 3255.

25. Sandra, P.; Dermaux, A.; Ferraz, V.; Dittman, M.M.; Rozing, G. J. Micro. Separ. **1997**, *9*, 409.

26. Dermaux, A.; Sandra, P.; Ksir, M.; Zarrouck, K.F.F. Analysis of the triglycerides and the free and derivatized fatty acids in fish oil by capillary electrochromatography. J. High Resolut. Chromatogr. **1998**, *21*, 545.

27. Li, D.; Knobel, H.H.; Kitagawa, S.; Tsuji, A.; Watanabe, H.; Nakshima, M.; Tsuda, T. J. Micro. Separ. **1997**, *9*, 347.

28. Mazzeo, J.R.; Krull, I.S. *Capillary Electrophoresis—Technology*; Guzman, N., Ed.; Marcel Dekker, Inc.: New York, 1993, Chap. 29.

29. Mazzeo, J.R.; Martineau, J.; Krull, I.S. *CRC Handbook of Capillary Electrophoresis: Principles, Methods, and Applications*; Landers, J.P., Ed.; CRC Press: Boca Raton, FL, 1994, Chap. 18.

30. Liu, X.; Sosic, Z.; Krull, I.S. Capillary isoelectric focusing as a tool in the examination of antibodies, peptides and proteins of pharmaceutical interest. J. Chromatogr. B. **1996**, *735*, 165.

31. Krull, I.S.; Dai, J.; Gendreau, C.; Li, G. HPCE methods for the identification and quantitation of antibodies, their conjugates and complexes. J. Pharm. Biomed. Anal. **1997**, *16*, 377.

32. Capillary Electrochromatography, Symposium, San Francisco, CA, organized by the California Separations Science Society, San Francisco, August 1997.

33. Royal Society of Chemistry Analytical Division, Northeast Region, Chromatography and Electrophoresis Group Symposium on New Developments and Applications in Electrochromatography, University of Bradford, Bradford, U.K., December, 3, 1997.

34. Tsuda, T., Ed.; *Electric Field Applications in Chromatography, Industrial and Chemical; Processes*; VCH: Weinheim, 1995.

35. Dittmann, M.M.; Weinand, K.; Bek, F.; Rozing, G.P. LC/GC Mag. **1995**, *13* (10), 800.

36. Dittmann, M.M.; Rozing, G.P. Capillary electrochromatography—a high-efficiency micro-separation technique. J. Chromatogr. A. **1996**, *744*, 63.

37. Ross, G.; Dittmann, M.; Bek, F.; Rozing, G. Capillary electrochromatography: enhancement of LC separation in packed capillary columns by means of electrically driven mobile phases. Am. Lab. March **1996**, 34.

38. Euerby, M.R.; Gilligan, D.; Johnson, C.M.; Roulin, S.C.P.; Myers, P.; Bartle, K.D. J. Micro. Separ. **1997**, *9*, 373.

39. Euerby, M.R.; Johnson, C.M.; Bartle, K.D.; Myers, P.; Roulin, S.C.P. Capillary electrochromatography in the pharmaceutical industry. Practical reality or fantasy? Anal. Commun. **1996**, *33*, 403.

40. Robson, M.M.; Cikalo, M.G.; Myers, P.; Euerby, M.R.; Bartle, K.D. J. Micro. Separ. **1997**, *9*, 357.

41. Miwaya, J.H.; Alesandro, M.S. LC/GC Mag. **1998**, *16* (1), 36.

42. Majors, R.E. LC/GC Mag. **1998**, *16* (1), 12.

43. Grant, I.H. *Capillary Electrochromatography*; Methods in Molecular Biology; Altria, K.D., Ed.; Humana Press: Totowa, NJ, 1996; Vol. 52, Chap. 15.

44. Euerby, M.R.; Johnson, C.M.; Bartle, K.D. LC/GC Int. January **1998**, 39.

45. Cikalo, M.G.; Bartle, K.D.; Robson, M.M.; Myers, P.; Euerby, M.R. Capillary electrochromatography. Analyst **1998**, *123*, 87R.

46. Wei, W.; Luo, G.; Yan, C. Am. Lab. January **1998**, *20C*.

47. Peters, E.C.; Lewandowsk, K.; Petro, M.; Frechet, J.M.J.; Svec, F. J.M.J. Anal. Commun. 1998, 35,83 a "Molded" Rigid Monolithic Capillary Column. In *HPLC 98*; St. Louis, MO, May 1998.

48. Peters, E.C.; Lewandowski, K.; Petro, M.; Svec, F.; Frechet, J.M.J. Chiral electrochromatography with a "molded" rigid monolithic capillary column. Anal. Commun. **1998**, *35*, 83.

49. Peters, E.C.; Petro, M.; Svec, F.; Frechet, J.M.J. Molded rigid polymer monoliths as separation media for capillary electrochromatography. Anal. Chem. **1997**, *69*, 3646.

50. Venema, E.; Kraak, J.C.; Poppe, H.; Tijssen, R. Electrically driven capillary size exclusion chromatography. Chromatographia **1998**, *48* (5/6), 347.

51. Wu, J.T.; Huang, P.; Li, M.X.; Qian, M.G.; Lubman, D.M. Open-tubular capillary electrochromatography with an on-line Ion trap storage/reflectron time-of-flight mass detector for ultrafast peptide mixture analysis. Anal. Chem. **1997**, *69*, 320.

52. Wu, J.T.; Huang, P.; Li, M.X.; Qian, M.G.; Lubman, D.M. Protein digest analysis by pressurized capillary electrochromatography using an ion trap storage/reflectron time-of-flight mass detector. Anal. Chem. **1997**, *69*, 2908.

53. Wu, J.T.; Huang, P.; Li, M.X.; Qian, M.G.; Lubman, D.M. Anal. Chem. **1997**, *69*, 2870.

54. Wu, J.T.; Huang, P.; Li, M.X.; Qian, M.G.; Lubman, D.M. On-line analysis by capillary separations interfaced to an ion trap storage/reflectron time-of-flight mass spectrometer. J. Chomatogr. A. **1998**, *794*, 377.

55. Yang, C.; El Rassi, Z. Capillary electrochromatography of derivatized mono- and oligosaccharides. Electrophoresis **1998**, *19*, 2061.

56. Zhang, M.; El Rassi, Z. Capillary electrochromatography with novel stationary phases. I. Preparation and characterization of octadecylsulfonated silica. Electrophoresis **1998**, *19*, 2068.

57. Bopp, R.J.; Wozniak, T.J.; Anliker, S.L.; Palmer, J. Pharmaceutical and biomedical applications of liquid chromatography. In *Progress in Pharmaceutical and Biomedical Analysis*; Riley, C.M., Lough, W.J., Wainer, I.W., Eds.; Pergamon/Elsevier Science: Amsterdam, 1994; Vol. 1, Chap. 10.

58. Dolan, J.W.; Snyder, L.R. Liquid chromatography expert systems: a modular approach. Am. Lab. May **1990**, *50*.

59. Snyder, L.R.; Kirkland, J.J.; Glajch, J.L. *Practical HPLC Method Development*, 2nd Ed.; John Wiley & Sons: New York, 1997, Chap. 10.

60. Verheij, E.R.; Tjaden, U.R.; Niessen, W.A.M.; van der Greef, J. Development of an instrumental configuration for pseudo-electrochromatography-electrospray mass spectrometry. J. Chromatogr. **1995**, *712*, 201.

61. Pesek, J.J.; Matyska, M.T.; Sandoval, J.E.; Williamsen, E.J. J. Liq. Chromatogr. & Related Technol. **1996**, *19* (17/18), 2843.

62. Pesek, J.J.; Matyska, M.T.; Mauskar, L. Separation of proteins and peptides by capillary electrochromatography in diol- and octadecyl-modified etched capillaries. J. Chromatogr. A. **1997**, *763*, 307.

63. Pesek, J.J.; Matyska, M.T. Electrochromatography in chemically modified etched fused-silica capillaries. J. Chromatogr. A. **1996**, *736*, 255.

64. Pesek, J.J.; Matyska, M.T. Separation of tetracyclines by high-performance capillary electrophoresis and capillary electrochromatography. J. Chromatogr. A. **1996**, *736*, 313.

Applied Voltage: Effect on Mobility, Selectivity, and Resolution in Capillary Electrophoresis

Jetse C. Reijenga
Eindhoven University of Technology, Eindhoven, The Netherlands

INTRODUCTION

Generally, migration times t_m in capillary electrophoresis (CE) are inversely proportional to the applied voltage.

DISCUSSION

In terms of analysis time, the voltage should, therefore, be as large as possible:

$$t_m \cong \frac{1}{V}$$

Under conditions optimized for limited power dissipation, effective mobilities and selectivities (defined as effective mobility ratios) are independent of the applied voltage.

Efficiency is also determined by the applied voltage, but in a much more complicated manner (see Band Broadening in Capillary Electrophoresis). If efficiency is limited by diffusion, a higher voltage also leads to a higher efficiency. Limitations are due to insulation properties and heat dissipation. Voltages larger than 30 kV should always be avoided because of danger of sparking and leaking currents, even more so in cases of significant atmospheric humidity. Excessive heat dissipation leads to an average temperature increase inside the capillary, which can be reduced by forced cooling. What cannot be reduced is the contribution of heat dissipation to band broadening. This can only be reduced by a lower conductivity, a lower current density, or a smaller inner diameter (see Band Broadening in Capillary Electrophoresis). In the case of diffusion-limited efficiency, the efficiency (as given by the plate number) is directly proportional to the applied voltage:

$$N \cong V$$

The ultimate criterion for quality of separation is the resolution R, given by the following relationship:

$$R = \frac{\Delta t_m}{4\sigma}$$

With the definition of plate number, it follows that $R \cong \sqrt{V}$.

Fig. 1 shows a computer simulation of the resolution and analysis time of a mixture of anions at 5, 10 and 25 kV. In order to better visualize the effect on resolution, a logarithmic x axis was chosen.

SUGGESTED FURTHER READING

Hjertén, S. Free zone electrophoresis. Chromatogr. Rev. **1967**, *9* (2), 122. Publication Types Review MeSH Terms Blood Protein Electrophoresis.

Jorgenson, J.W.; Lucaks, K.D. Capillary zone electrophoresis. Science **1983**, *222*, 266.

Li, S.F.Y. *Capillary Electrophoresis—Principles, Practice and Applications*; Elsevier: Amsterdam, 1992.

Reijenga, J.C.; Kenndler, E. Computational simulation of migration and dispersion in free capillary zone electrophoresis, part I, Description of the theoretical model. J. Chromatogr. A. **1994**, *659* (2), 403.

Reijenga, J.C.; Kenndler, E. Computational simulation of migration and dispersion in free capillary zone electrophoresis, part II, Results of simulation and comparison with measurements. J. Chromatogr. A. **1994**, *659* (2), 417.

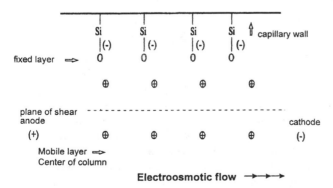

Fig. 1 Analysis of a mixture of weak anions at three different voltages. Suppressed EOF in a 400-mm capillary with negative inlet polarity. *Note*: The time axis is logarithmic.

Encyclopedia of Chromatography DOI: 10.1081/E-ECHR-120039884

Aqueous Two-Phase Solvent Systems for Countercurrent Chromatography

Jean-Michel Menet

Process Development Chemistry, Aventis Pharma, Vitry-sur-Seine, France

INTRODUCTION

Aqueous two-phase solvent (ATPS) systems are made of two aqueous liquid phases containing various polymers. Such systems are gentle toward biological materials and they can be used for the partition of biomolecules, membrane vesicles, cellular organites, and whole cells. They are characterized by a high content of water in each phase, by very close densities and refraction indices of the two phases, by a very low interfacial tension, and by high viscosities of the phases. As a result, settling times are particularly long and may last up to 1 h or longer.

The partition of a substance between the two phases depends on many factors. Theoretical studies have been carried out in order to better understand the reasons for the separation in two aqueous phases, thanks to the introduction of various polymers, and the role of various factors on the partition of the substances. However, no global theory is available to predict the observed behaviors. Hopefully, some empirical knowledge has been acquired which will help in the use of these unique solvent systems.

Various devices have been used for the partition of substances in ATPS systems. Countercurrent chromatography (CCC) has again revealed its unique features, because it has enabled the use of such very viscous systems at relatively high flow rates while obtaining a satisfactory efficiency and a good resolution for the separation. Many applications have been described in the literature for the use of ATPS systems with CCC devices.

ATPS SYSTEMS

For further information on ATPS systems and the par-titioning, we strongly recommend the books by Albertsson[1] and Walter et al.,[2] as reference books in this area.

Polymers Used for ATPS Systems

Many ATPS systems contain a polymer which is sugar based and a second one that is of hydrocarbon ether type. Sugar-based polymers include dextran (Dx), hydroxy propyl dextran (HPDx), Ficoll (Fi) (a polysaccharide), methyl cellulose (MC), or ethylhydroxyethyl cellulose (EHEC). Hydrocarbon ether-type polymers include poly(ethylene glycol) (PEG), poly(propylene glycol) (PPG), or the copolymer of PEG and PPG. Derivatized polymers can also be useful, such as PEG-fatty acids or di-ethylaminoethyl-dextran (Dx-DEAE).

Dextran, or α-1,6-glucose, is available in a mass range from 10,000 to 2,000,000. Dx T500 fractions, also called Dx 48 from Pharmacia (Uppsala, Sweden), are among the best known: their weight-average molecular weight (M_w) varies from 450,000 to 500,000. These white powders contain about 5–10% of water. PEG is a linear synthetic polymer which is available in many molecular weights, the most common being between 300 and 20,000.

Physical Properties of the ATPS Phases

Common characteristics of ATPS phases are their high content of water for both phases, typically 85–99% by weight and very close densities and refraction indices for the two phases. Moreover, both phases are viscous and the interfacial tension is low, from 0.1 to 0.001 dynes/cm. The settling times in the Earth's gravitational field range from 5 min to 1 h.

PRACTICAL USE OF ATPS

Because ATPS systems are particularly suited for protein separations, many research workers have worked in this area and have tried to model their behavior when varying the composition of these systems. However, there are still no theoretical models to calculate, a priori, the partition coefficient of a protein for a wide range of molecular weights of polymers and concentrations in salts and polymers. However, it remains possible to have qualitative explanations of the role of key factors on the partition of the substances.

Encyclopedia of Chromatography DOI: 10.1081/E-ECHR-120039885

CHOICE OF THE ATPS SYSTEM

The general principle for designing an ATPS system is to try various systems, either made from two phases containing polymers or from one phase containing a polymer and the other component a salt. The nature of the substances to be separated shall be taken into account: Fragile solutes may be denatured by a too high interfacial tension, as encountered in aqueous polymer–salt systems. Some substances may even aggregate in an irreversible way, or be altered in other ways by their contact with some polymers or salts. Moreover, some substances can require the specific use of given salts, or pH, or temperature. When all the previous considerations have been taken into account, the partition coefficients can then be determined in test-tube experiments. Afterward, the composition of the phases can be adjusted.

Systems Suited for the Separation of Molecules

A simple way consists in testing two ATPS systems, dextran 40/PEG-8000 and dextran 500/PEG-8000, which lead to a relatively quick settling and allow reproducible results. For charged macromolecules, the two key parameters are the pH with regard to the isoelectric point of the product and the nature and concentration of the chosen salt. The composition in polymer has a smaller influence, except for some neutral compounds.

Systems Suited for the Separation of Cells and Particles

The main parameter for such separations is the difference of concentrations of each polymer between the two phases. If the concentrations in polymer are quite high, particles tend to adsorb at the interface of the two phases, without any specificity. For instance, all human erythrocytes adsorb at the interface of the dextran 500/PEG-8000 systems with respective concentrations higher than 7.0% and 4.4% (w/w).

The goal is to find a system close to the critical point (in the phase diagram) to achieve the separation, as the partition coefficients all become close to 1. However, this requires increased attention to the experimental conditions in order to obtain reproducible results. If necessary, a change in the molecular weight of one of the polymers allows one to choose the aqueous phase in which the particle tends to accumulate. For instance, most mammalian cells partition between the interface and the upper phase rich in PEG in dextran 500/PEG-8000 systems,

whereas they partition between the interface and the lower phase rich in dextran in dextran 40/PEG-8000 systems.

ADJUSTMENT OF THE PARTITION COEFFICIENT

We note that the partition coefficient K is defined as the ratio of the concentration of the substance in the upper phase to its concentration in the lower phase. As cells and particles tend to partition between one phase and the interface, only molecules, such as proteins, are the subject of this section.

First, the partition coefficient of the substance should be determined in a standard system, such as dextran 500 (7.0%, w/w)/PEG-8000 (5.0%, w/w) with 5–10 mM of buffer added. Then, the following empirical laws can be used for the adjustment:

1. K is increased by diminishing the molecular weight of the polymer which is predominant in the upper phase (e.g., PEG) or by increasing the molecular weight of the polymer which is predominant in the lower phase (e.g., dextran). Reversing these changes decreases K.
2. K is significantly different than the unit value only if the concentrations of the polymers are high. K tends to the unit value when the concentrations of the polymers are decreased.
3. K can be adjusted by the addition of a salt, as long as the proteins are not close to their isoelectric points. For a negatively charged protein, K is decreased by following the series: phosphate < sulfate < acetate < chloride < thiocyanate < perchlorate and lithium < ammonium < sodium < potassium (for instance, lithium decreases K by a smaller amount than sodium). The influence of the nature of the salt may be amplified by an increase of the pH, which increases the negative net charge of the molecule. Positively charged proteins exhibit the opposite behavior. All these rules apply only for low concentrations of salts. Higher concentrations could, however, be used to favor the partition of the molecules in the upper phase.
4. Charged polymers can also be used; their influence is greater than that of the salts. The most common include charged polymers de-rived from PEG, such as PEG–TMA (trimethylamino) or PEG–S (sulfonate). Dextran can also be modified.
5. The derivatization by hydrophobic groups can also facilitate the extraction of molecules

containing hydrophobic sites. The most common polymer is PEG–P (PEG–palmitate).

K depends on the temperature, but in a complex way, so that its use is difficult for common cases.

Optimization of the Selectivity

Some general rules apply to proteins in PEG/dextran systems and are summarized as follows:

1. The concentration of the polymer is important: Decreasing the concentrations brings the system closer to the critical point (in the phase diagram), smoothing K values toward the unit value and finally decreasing the selectivity.
2. The nature of the salt is important. The most important effects are encountered for $NaClO_4$, which extracts positively charged molecules in the upper phase, and tetrabutyl ammonium phosphate, which extracts negatively charged proteins in the upper phase.

APPLICATIONS

These aqueous two-phase solvent systems are more difficult to handle than organic-based solvent systems, so that the number of applications in the literature is quite small as compared to the other systems. However, these applications are really specific, quite often striking in their separation power, and they truly reveal some unique features of CCC.

Former applications of ATPS systems on CCC devices were gathered by Sutherland et al.[3] For instance, both toroidal and type J [also called a high-speed countercurrent chromatograph (HSCCC)] CCC were successfully applied for the fractionation of subcellular particles. Using standard rat liver homogenates, plasma membranes, lysosomes, and endoplasmic reticulum were separated by a 3.3% (w/w) dextran T500, 5.4% PEG-6000, 10 mM sodium phosphate–phosphoric acid buffer (pH 7.4), 0.26 M sucrose, 0.05 mM Na_2EDTA, and 1 mM ethanol. Purification of torpedo electroplax membranes were also carried out, and the separation of various bacterial cells were also described, including the purification of different strains of *Escherichia coli* and the separation of *Salmonella typhirum* cells, using PEG–dextran ATPS systems. Moreover, these CCC devices were also applied to larger cells, such as the separation of various species of red blood cells.

In the same way, the separation of cytochrome-*c* and lysozyme was achieved in 1988 by Ito and Oka[4] using the type J (or HSCCC) device. The chosen ATPS

system consisted of 12.5% (w/w) PEG-1000 and 12.5% (w/w) dibasic potassium phosphate in water. The two peaks were resolved in 5 h using a 1-mL/min flow rate, but the retention of the stationary phase was as low as 26%. The limitation of this type of apparatus is definitely the low retention of the stationary phase for ATPS systems.

Several ATPS systems were also used with a centrifugal partition chromatograph (also called Sanki-type from the name of its unique manufacturer). Foucault and Nakanishi[5] tested PEG-1000/ammonium sulfate, PEG-8000/dextran, and other PEG-8000/hydoxypropylated starch on a test separation of crude albumin using a model LLN centrifugal partition chromatograph (CPC) containing six partition cartridges. They demonstrated that the systems could be used with the CPC apparatus, but the efficiency was particularly low (due to very poor mass transfer) and the flow rate was quickly limited by a strong decrease in the retention of the stationary phase (and not by the back-pressure). Afterward, CPC was then not considered as really suited for ATPS systems.

The third type, which is close in principle to the type J high-speed countercurrent chromatograph, was designed in the early 1980s and is named "cross-axis coil planet centrifuge." Such a new design has led to successful results with highly viscous polymer phase systems[6]. Indeed, it allows satisfactory retention of the stationary phase of ATPS systems, either in the polymer–salt form or the polymer–polymer form. Such an apparatus eliminates the main drawback of the previous CCC devices, as it allows one to maintain a good retention of the stationary phase with a sufficient flow rate to ensure an acceptable separation or purification time. Using such solvent systems, it has been applied to the separation and purification of various protein samples:

- Mixture of cytochrome-c, myoglobin, ovalbumin and hemoglobin[7]
- Histones and serum proteins[8]
- Recombinant uridine phosphorylase from *E. coli* lysate[9]
- Human lipoproteins from serum[10]
- Lactic acid dehydrogenase from a crude bovine heat extract[11]
- Profilin–actin complex from *Acanthamoeba* extract[12]
- Lyzozyme, ovalbumin, and ovotransferrin from chicken egg white[13]
- Acidic fibroblast growth factor from *E. coli* lysate[14]

The cross-axis coil planet centrifuge has, consequently, has been demonstrated to be particularly suited for the use of ATPS systems, leading to satisfactory retention of the stationary phase while keeping a sufficient flow rate of the mobile phase to limit the duration of the experiments.

REFERENCES

1. Albertson, P.-A. *Partition of Cell Particles and Macromolecules*, 3rd ed.; John Wiley & Sons: New York, 1986.

2. Walter, H.; Brooks, D.E.; Fisher, D. *Partitioning in Aqueous Two-Phase Systems*; Academic Press: New York, 1985.

3. Sutherland, I.A.; Heywood-Waddington, D.; Ito, Y. Counter-current chromatography: Applications to the separation of biopolymers, organelles and cells using either aqueous-organic or aqueous-aqueous phase systems. J. Chromatogr. **1987**, *384*, 197.

4. Ito, Y.; Oka, H. Horizontal flow-through coil planet centrifuge equipped with a set of multilayer coils around the column holder: Counter-current chromatography of proteins with a polymer-phase system. J. Chromatogr. **1988**, *457*, 393.

5. Foucault, A.P.; Nakanishi, K. Comparison of several aqueous two phase solvent systems (ATPS) for the fractionation of biopolymers by centrifugal partition chromatography (CPC). J. Liq. Chromatogr. **1990**, *13* (12), 2421.

6. Bhatnagar, M.; Oka, H.; Ito, Y. Improved cross-axis synchronous flow-through coil planet centrifuge for performing counter-current chromatography: II. Studies on retention of stationary phase in short coils and preparative separations in multilayer coils. J. Chromatogr. **1989**, *463*, 317.

7. Shibusawa, Y.; Ito, Y. Protein separation with aqueous-aqueous polymer systems by two types of counter-current chromatographs. J. Chromatogr. **1991**, *550*, 695.

8. Shibusawa, Y.; Ito, Y. Countercurrent chromatography of proteins with polyethylene glycol-dextran polymer phase systems using type-XLLL cross-axis coil planet centrifuge. J. Liq. Chromatogr. **1992**, *15*, 2787.

9. Lee, Y.W.; Shibusawa, Y.; Chen, F.T.; Myers, J.; Schooler, J.M.; Ito, Y. Purification of uridine phosphorylase from crude extracts of *Escherichia coli* employing high-speed countercurrent chromatography with an aqueous two-phase solvent system. J. Liq. Chromatogr. **1992**, *15*, 2831.

10. Shibusawa, Y.; Ito, Y.; Ikewaki, K.; Rader, D.J.; Bryan Brewer, J., Jr. Counter-current chromatography of lipoproteins with a polymer phase system using the cross-axis synchronous coil planet centrifuge. J. Chromatogr. **1992**, *596*, 118.

11. Shibusawa, Y.; Eriguchi, Y.; Ito, Y. Purification of lactic acid dehydrogenase from bovine heart crude extract by counter-current chromatography. J. Chromatogr. B. **1997**, *696*, 25.

12. Shibusawa, Y.; Ito, Y. Am. Biotechnol. Lab. **1997**, *15*, 8.

13. Shibusawa, Y.; Kihira, S.; Ito, Y. One-step purification of proteins from chicken egg white using counter-current chromatography. J. Chromatogr. B. **1998**, *709*, 301.

14. Menet, J.M. Thèse de Doctorat de l'Université Paris 6; 1995.

Argon Detector

Raymond P. W. Scott
Scientific Detectors Ltd., Banbury, Oxfordshire, U.K.

INTRODUCTION

The argon detector was the first of a family of detectors developed by Lovelock[1] in the late 1950s; its function is quite unique. The outer octet of electrons in the noble gases is complete and, as a consequence, collisions between argon atoms and electrons are perfectly elastic. Thus, if a high potential is set up between two electrodes in argon and ionization is initiated by a suitable radioactive source, electrons will be accelerated toward the anode and will not be impeded by energy absorbed from collisions with argon atoms. However, if the potential of the anode is high enough, the electrons will eventually develop sufficient kinetic energy that, on collision with an argon atom, energy can be absorbed and a *metastable* atom can be produced. A metastable atom carries *no* charge but adsorbs its energy from collision with a high-energy electron by the displacement of an electron to an outer orbit. This gives the metastable atom an energy of about 11.6 electron volts. Now 11.6 V is sufficient to ionize most organic molecules. Hence, collision between a metastable argon atom and an organic molecule will result in the outer electron of the metastable atom collapsing back to its original orbit, followed by the expulsion of an electron from the organic molecule. The electrons produced by this process are collected at the anode, generating a large increase in anode current. However, when an ion is produced by collision between a metastable atom and an organic molecule, the electron, simultaneously produced, is immediately accelerated toward the anode. This results in a further increase in metastable atoms and a consequent increase in the ionization of the organic molecules. This cascade effect, unless controlled, results in an exponential increase in ion current with solute concentration.

The relationship between the ionization current and the concentration of vapor was deduced by Lovelock [2-3] to be

$$I = \frac{CA(x + y) + Bx}{CA\{1 - a\exp[b(V - 1)]\} + B}$$

where A, B, a, and b are constants, V is the applied potential, x is the primary electron concentration, and y is the initial concentration of metastable atoms. The rapid increase in current with increasing vapor concentration, as predicted by the equation, is controlled by the use of a high impedance in series with detector power supply. As the current increases, more volts are dropped across the resistance, and less are applied to the detector electrodes.

THE SIMPLE OR MACRO ARGON DETECTOR SENSOR

A diagram of the macro argon detector sensor is shown in Fig. 1. The cylindrical body is usually made of stainless steel and the insulator made of PTFE or for high-temperature operation, a suitable ceramic. The very first argon detector sensors used a tractor sparking plug as the electrode, the ceramic seal being a very efficient insulator at high temperatures.

Inside the main cavity of the original sensor was a strontium-90 source contained in silver foil. The surface layer on the foil that contained the radioactive material had to be very thin or the β particles would not be able to leave the surface. This tenuous layer protecting the radioactive material is rather vulnerable to mechanical abrasion, which could result in radioactive contamination (strontium-90 has now been replaced by ^{63}Ni). The radioactive strength of the source was about 10 mCu which for strontium-90 can be considered a *hot* source. The source had to be inserted under properly protected conditions. The decay of strontium-90 occurs in two stages, each stage emitting a β particle producing the stable atom of zirconium-90:

$$^{90}\text{Sr} \xrightarrow[\substack{0.6 \\ \text{MeV}}]{\varepsilon} {}^{90}\text{Y} \xrightarrow[\substack{2.5 \\ \text{MeV}}]{\varepsilon} {}^{90}\text{Zr}$$

^{90}Sr half-life 25 year; ^{90}Y half-life 60 h; ^{90}Zr stable

The electrons produced by the radioactive source were accelerated under a potential that ranged from 800 to 2000 V, depending on the size of the sensor and the position of the electrodes. The signal is taken across a $1 \times 10^8 \Omega$ resistor, and as the standing current from the ionization of the argon is about 2×10^{-8} A, there is a standing voltage of 2 V across it that requires "backing off."

In a typical detector, the primary current is about 10^{11} electrons/s. Taking the charge on the electron as 1.6×10^{-19} C, this gives a current of 1.6×10^{-8} A. According to Lovelock,[1] if each of these electrons

Encyclopedia of Chromatography DOI: 10.1081/E-ECHR-120039886

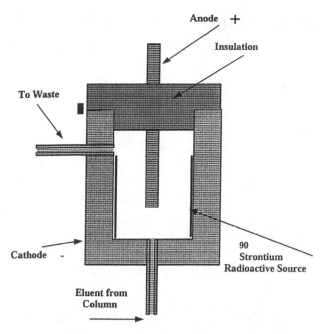

Fig. 1 The macro argon detector.

was not linear over more than two orders of magnitude of concentration ($0.98 < r > 1.02$) and its response was not predictable. In practice, nearly all organic vapors and most inorganic vapors have ionization potentials of less than 11.6 eV and thus are detected. The short list of substances that are not detected include and fluorocarbons. The compounds methane, ethane, acetonitrile, and propionitrile have ionization potentials well above 11.6 eV; nevertheless, they do provide a slight response between 1% and 10% of that for other compounds. The poor response to acetonitrile makes this substance a convenient solvent in which to dissolve the sample before injection on the column. It is also interesting to note that the inorganic gases H_2S, NO, NO_2, NH_3, PH_3, BF_3, and many others respond normally in the argon detector. As these are the type of substances that are important in environmental contamination, it is surprising that the argon detector, with its very high sensitivity for these substances, has not been reexamined for use in environmental analysis.

can generate 10,000 metastables on the way to the electrode, the steady-state concentration of metastables will be about 10^{10} per milliliter (this assumes a life span for the metastables of about 10^{-5} s at NTP). From the kinetic theory of gases, it can be calculated that the probability of collision between a metastable atom and an organic molecule will be about $1.6:1$. This would lead to a very high ionization efficiency and Lovelock claims that in the more advanced sensors ionization efficiencies of 10% have been achieved.

The minimum detectable concentration of a well-designed argon detector is about an order of magnitude higher than the FID (i.e., 4×10^{-13} g/mL). Although the argon detector is a very sensitive detector and can achieve ionization efficiencies of greater than 0.5%, the detector was not popular, largely because it

REFERENCES

1. Lovelock, J.E. *Gas Chromatography*; Scott, R.P.W., Ed.; Butterworths Scientific: London, 1960; 9.
2. Lovelock, J.E. A sensitive detector for gas chromatography. *J. Chromatogr.* **1958**, *1*, 35.
3. Lovelock, J.E. Nature (London) **1958**, *181*, 1460.

SUGGESTED FURTHER READING

Scott, R.P.W. *Chromatographic Detectors*; Marcel Dekker, Inc.: New York, 1996.

Scott, R.P.W. *Introduction to Gas Chromatography*; Marcel Dekker, Inc.: New York, 1998.

Aromatic Diamidines: Comparison of Electrophoresis and HPLC for Analysis

A. Negro
B. Rabanal
Analytical Chemistry Section, Faculty of Biological and Environmental Sciences, University of León, Leon, Spain

INTRODUCTION

HPLC can be considered to have been established by Ettre and Horvath. The popularity of HPLC may be explained by the versatility of this technique, which can be used to separate and quantify large or small; polar, nonpolar, or inorganic; and chiral or achiral molecules. In addition, its methods are easily automated, increasing the number of analyses that can be performed in a given time, improving accuracy and precision, as well as reducing costs. It was around 1960 that HPLC achieved its peak growth. This can be attributed, in large part, to its widespread acceptance by the pharmaceutical industry.

Electrophoresis is an analytical technique that was first introduced by Tiselius[2] in 1937. Thirty-five years ago, Hjertén[3] showed that it was possible to carry out electrophoretic separations in a 300.0-μm glass tube and to detect the separation of compounds by ultraviolet absorption. Capillary electrophoresis (CE) did not become popular until 1981, when Jorgenson and Lukacs[4] published work in which they demonstrated the simplicity of the instrumental setup required and the high resolving power of CE. The results shown were astonishing: sharp narrow peaks, 400,000 theoretical plates per meter (compared with 10,000 theoretical plates per meter of HPLC), and short analysis times. Galery introduced the first commercial instrument in 1988. There are excellent reviews of CE, among which should be mentioned are the ones done by Kuhr, Isaaq or Camilleri. These look at the increasing number of applications and future prospects.

CAPILLARY ELECTROPHORESIS

CE has had considerable success over the last 20 years. GC and HPLC[1] are still the dominant techniques. However, CE has several distinct advantages over other separation techniques.[5–7] One advantage CE has, relative to HPLC, is its simplicity and applicability for the separation of widely differing substances, such as organic molecules, inorganic ions, and so on, using the same instrument and, in most cases, the same capillary, while changing only the composition of the buffer used. This cannot be said with regard to any other separation techniques. In addition, CE offers the highest resolving power.

The aim of the work reported here was to study how changes in the principal parameters for each technique affect the separation processes when analyzing a series of aromatic diamidines, and, based on the results obtained, to establish comparisons between the two analytical techniques.

The aromatic diamidines are compounds of considerable pharmaceutical interest. This is, among others, for the following reasons: they have a strong antiprotozoan action and participate in the metabolism and transport of polyamines, inhibiting, for instance, S-adenosyl-L-methionine decarboxylase (SAMDC). Therefore, because this route is closely linked to cell proliferation processes, aromatic diamidines can slow down or prevent the growth of tumors.[8–10] The substances used in this work are as follows:[11,12]

1. Pentamidine: 4,4'-[1,5-pentanediyl *bis*(oxy)]*bis*-benzenecarboximidamide
2. Stilbamidine: 4,4'-(1,2-ethenediyl)*bis*-benzenecarboximidamide
3. DAPI: 4',6-diamidino-2-(4-amidinophenyl)indole dilactate
4. Propamidine: 4,4'-[1,3-propanediylbis(oxy)*bis*-benzenecarboximidamide
5. Hydroxystilbamidine: 4-[2-[4-(aminoiminomethyl)phenyl]ethenyl]-3-hydroxybenzenecarboximidamide
6. Phenamidine: 4,4'-diamidinodiphenylether
7. Diampron: 3,3'-diamidinocarbanilide
8. Berenil: 4,4'-diamidinodiazoamino benzene
9. Dibromopropamidine: 2',2''-dibromo-4',4''-diamidino-1,3-diphenoxypropane.

Encyclopedia of Chromatography DOI: 10.1081/E-ECHR-120038594

EXPERIMENTAL TECHNIQUES

Chemicals and Reagents

Pentamidine isethionate salt, berenil diaceturate salt, and DAPI dihydrochloride salt were obtained from Sigma-Aldrich Química SA (Madrid, Spain). Diampron isethionate salt, hydroxystilbamidine isethionate salt, propamidine isethionate salt, dibromopropamidine isethionate salt, phenamidine isethionate salt, and stilbamidine isethionate salt were generously donated by Rhône Poulenc Rorer (Dagenham, UK). The ion-pairing reagents, pentane sulphonate, hexane sulphonate, heptane sulphonate, octane sulphonate, and decane sulphonate sodium salts were supplied by Sigma-Aldrich Química SA. Methanol of HPLC grade and other chemicals of analytical grade were supplied by Merck (Darmstadt, Germany). The water used was purified with a Milli-Q system purchased from Millipore (Bedford, MA).

Chromatographic System

The HPLC system comprised a Beckman 116 programmable solvent pump with a Beckman 168 photodiode detector—this was checked and data were processed with the Gold Nouveau software system (Beckman Coulter, Palo Alto, CA) and a Beckman 507 automatic injector with a 100.0-μL loop and a heating chamber for the columns. Analyses were carried out with an Ultrasphere ODS column (5.00-μm particle size, 15.0 cm × 4.60 mm internal diameter) purchased from Beckman Coulter. A guard column (2.00 cm × 2.00 mm internal diameter), packed with Sperisorb RP-18 (30.0–40.0 μm pellicular), was supplied by Upchurch Scientific (Oak Harbor, WA).

Electrophoretic System

The CE system consisted of a P/ACE System 2100 high-performance CE apparatus (Beckman Coulter, Fullerton, CA). An untreated, fused silica capillary tube (Beckman Coulter) was used, with dimensions of 75.0-μm ID, $L_t = 57.0$ cm, and $L_d = 50.0$ cm, enclosed in a liquid-cooled cassette. Detection was performed with a UV-VIS detector ($\lambda = 200.0$ nm). Equipment was checked and data were processed with the Beckman P/ACE Station V 1.2 software (Beckman Coulter).

RESULTS AND DISCUSSION

To carry out a comparison of HPLC and CE, the effects of varying the parameters common to the two techniques with the greatest impact on the separation processes were evaluated. These were: pH of the mobile phase and electrolytes, buffer concentration, and temperature, with the gathered data compared in each case.

As these are separation techniques based on radically differing physical principles, it is evident that there are certain parameters, specific to a given technique, that have a strong influence over the separation process in only one of the two and are not comparable. In HPLC, there is the influence of concentration and chain length of the ion-pairing reagent and the methanol percentage; in CE, there is influence of the choice of electrolyte, length of the capillary, and voltage applied. Variations in these parameters were also taken into account because they provide extremely useful information for making an overall comparison of the two techniques.

Parameters Common to HPLC and CE

Influence of pH in the mobile phase

The pH value is the parameter with the greatest impact on the separation of ionizable molecules. To keep aromatic diamidines ionized, it is necessary to work at very low pH levels, in the range 3.00–4.50, as the diamidine groups twice present in each molecule confer on them a strongly basic character ($pK_a = 13.86$).[13] To determine the influence of pH in HPLC, five diamidines were analyzed using a mobile phase consisting of 25.0 mM citrate buffer, 45.0% methanol, and 4.00 mM octane sulphonate sodium salt, at a temperature (T) of 30.0°C and at pH values of 3.00, 3.25, and 3.70. The chromatograms obtained are shown in Fig. 1. It can be observed that, as pH increases, retention times are noticeably shortened for all the substances, with no significant variations being noted in resolution. The influence of pH in CE[14] was studied by using 25.0 mM citrate buffer at $T = 30.0$°C, 14.0 kV voltage, and pH levels of 3.50, 3.70, and 4.25. Fig. 1 shows that a drop in pH does not bring with it any large change in migration times, but it does produce a significant variation in resolution. It may be observed that a good separation of all nine diamidines is possible only at pH = 3.70. For all the substances, it can be noted that the times required for analyses using CE are around half those in HPLC and that efficiency is much greater in all cases with CE than it is with HPLC, with good resolution. The most appropriate pH levels for the analyses of these substances in aqueous solutions and in the serum and urine are very similar with the two techniques because both require the molecules to be strongly ionized.

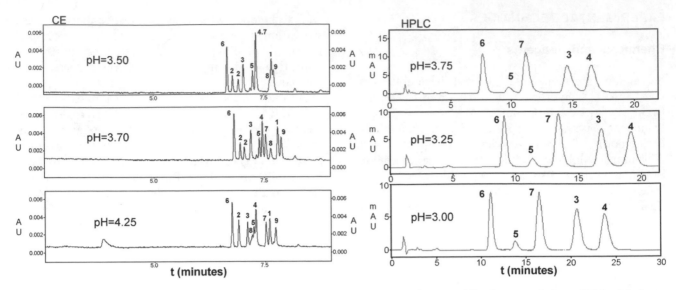

Fig. 1 Influence of pH in HPLC and CE. In HPLC, this effect was studied using a mobile phase consisting of 25.0 mM citrate buffer, 45.0% methanol, $T = 30.0°C$, 4.00 mM sodium octane sulphonate, and pH levels of 3.00, 3.25, and 3.75. In CE, a 25.0-mM citrate buffer electrolyte was used; pH values were 3.50, 3.70, and 4.25, and the voltage was 14.0 kV. *(View this art in color at www.dekker.com.)*

Influence of buffer concentration

For HPLC, it has been decided that the preferred buffer is citrate; it was necessary to establish the most suitable concentration. To study this influence, five diamidines were analyzed using a mobile phase consisting of 45.0% methanol, 4.00 mM octane sulphonate, and citrate buffer at various concentrations of 15.0, 25.0, and 35.0 mM, with $T = 30.0°C$ and pH = 3.25 in all cases. In Fig. 2, we see the results obtained. A change from 25.0 to 35.0 mM barely affects retention times for any of the substances, but a drop from 25.0 to 15.0 mM decreases retention times by almost 30.0% in every case.

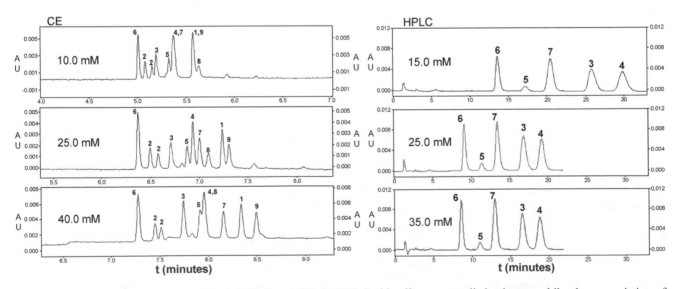

Fig. 2 Influence of buffer concentration in HPLC and CE. In HPLC, this effect was studied using a mobile phase consisting of citrate buffer at concentrations of 15.0, 25.0, and 35.0 mM, with 45.0% methanol, 4.00 mM sodium octane sulphonate, and pH = 3.25. In CE, a citrate buffer electrolyte was used at concentrations of 10.0, 25.0, and 40.0 mM, with pH = 3.70, voltage, 14.0 kV and $T = 30.0°C$. *(View this art in color at www.dekker.com.)*

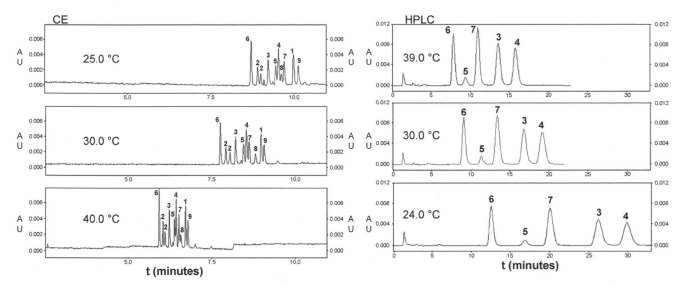

Fig. 3 Influence of temperature in HPLC and CE. In HPLC, this effect was studied using a mobile phase consisting of 25.0 mM citrate buffer, 45.0% methanol, 4.00 mM sodium octane sulphonate, pH = 3.25, and T = 24.0°C, 30.0°C, and 39.0°C. In CE, the electrolyte used was 25.0 mM citrate buffer, with pH = 3.70, 14.0 kV voltage, and T = 25.0°C, 30.0°C, and 40.0°C. (*View this art in color at www.dekker.com.*)

With CE, citrate buffer was also selected as the electrolyte for the study, and all nine diamidines were analyzed by using an electrolyte composed of citrate buffer at pH = 3.70, T = 30.0°C, and 14.0 kV voltage, at various concentrations (10.0, 25.0, and 40.0 mM), to determine which was the most appropriate. The results obtained are presented in Fig. 2. It can be seen that when concentrations go down to 10.0 mM, migration times are greatly reduced, but resolution also falls considerably; at 25.0 mM, resolution starts to be acceptable, and, at 40.0 mM, a good compromise between migration times and resolution is achieved. Comparison of the variations in buffer concentration in HPLC and CE allows one to conclude that, in both cases, a decrease in the concentration of the buffer reduces the time required for analyses. This reduction is much more striking in the case of CE and, in every instance, analysis times with CE are on the order of half of what they are with HPLC. Consequently, efficiency is much higher for all the compounds with CE than with HPLC, whereas resolution is good in both.

Influence of temperature

To study the effects of temperature in the analysis of these diamidines by means of HPLC, we used a mobile phase composed of 25.0 mM citrate buffer, with pH = 3.25, 45.0% methanol, and 4.00 mM octane sulphonate, at three different temperatures: 24.0°C, 30.0°C, and 39.0°C. It may be noted in Fig. 3 that increasing the temperature causes a notable drop in

retention times, whereas resolution remains at very good levels.

The temperature at which CE is carried out has to be selected carefully because this is one of the most

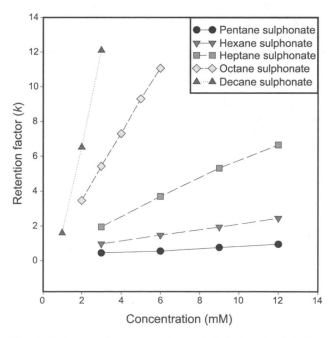

Fig. 4 Influence of concentration and chain length of the ion pair-forming agent in HPLC. This effect was studied using a mobile phase consisting of 25.0 mM citrate buffer, 45.0% methanol, and pH = 3.25, containing pentane sulphonate, hexane sulphonate, heptane sulphonate, octane sulphonate, and decane sulphonate sodium salts at concentrations of 0.00, 1.00, 2.00, 3.00, 4.00, 5.00, 6.00, 9.00, and 12.0 mM. (*View this art in color at www.dekker.com.*)

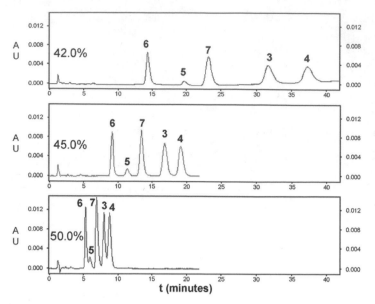

Fig. 5 Influence of methanol content in HPLC. This effect was studied using a mobile phase consisting of 25.0 mM citrate buffer, pH = 3.25, T = 30.0°C, 4.00 mM octane sulphonate, and 42.0%, 45.0%, and 50.0% methanol. *(View this art in color at www.dekker. com.)*

influential parameters in the CE process.[15] Precise temperature control during the CE process is of great importance in achieving good separation selectivity and, above all, good reproducibility.[16,17] To study temperature variation in CE, an electrolyte composed of 25.0 mM citrate buffer, with pH = 3.70 and 14.0 kV voltage, was used at temperatures of 25.0°C, 30.0°C, and 40.0°C. Fig. 3 shows that an increase in temperature causes a drastic reduction in migration times, but also reduces resolution excessively, causing serious problems for separation from 30.0°C onward.

If CE and HPLC at 30.0°C are compared, it will be noted, as in all previous cases, that CE has much shorter analysis times than HPLC and that efficiency

is much higher with CE than with HPLC, with good resolution being attained in both.

Parameters Exclusive to HPLC

Influence of concentration and chain length of the ion pair-forming agent

A technique often used in the analysis of ionic molecules is the formation of ion pairs[18] because this permits the separation of substances that are too ionized to separate by means of adsorption–partition methods, but are too insoluble in water to be analyzed through ion exchange techniques.[19] The pH

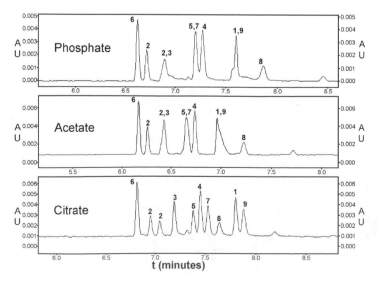

Fig. 6 Selection of electrolyte in CE. Electropherograms of nine diamidines using phosphate, acetate, and citrate electrolytes (25.0 mM), pH = 3.70, T = 30.0°C, and 14.0 kV voltage.

Table 1 Effect of capillary length on the volume of sample loaded and strength of the electric field

Length of capillary (cm)	Volume injected (nL)	Capillary occupied by the injection (mm)	Percentage of capillary occupied	Analyte loaded (ng)	Strength of electric field (V/cm)
77.0	21.77	4.92	0.70	8.80	181.0
57.0	29.41	6.65	1.33	147.0	245.0

Capillary, 75.0-μm ID; overall lengths, 77.0 and 57.0 cm (70.0 and 50.0 cm to the detector); electrolyte, 25.0 mM citrate buffer; pH = 3.70; voltage, 14.0 kV $T = 30.0°C$; injection under pressure for 5.00 sec.

of the mobile phase must be adjusted to ensure that the molecules are totally ionized and can combine with the ion pair-forming agent through the counter-ion. The substances most often used to form ion pairs are alkyl-sulphonate salts of varying chain length. In this work, several reagents of this type were evaluated, having chain lengths ranging from 5 to 10 carbons; the influence of the concentration of each was also investigated to determine which was the most suitable.

The effects of sulphonate salt concentration and chain length on the retention factor (k') were studied by measuring k', using only berenil as the diamidine, with a mobile phase consisting of 25.0 mM citrate buffer, pH = 3.25, 45.0% methanol, $T = 30.0°C$, with pentane sulphonate, hexane sulphonate, heptane sulphonate, octane sulphonate, and decane sulphonate sodium salts at concentrations of 0.00, 1.00, 2.00, 3.00, 4.00, 5.00, 6.00, 9.00, and 12.00 mM. The resultant data are shown in Fig. 4, where it may be observed that retention times and hence k' increase as the concentration of the ion pair-forming agent increases. This increase is much more pronounced when reagents with longer chain lengths are used.

Influence of methanol content

Two solvents were initially evaluated as organic modifiers for the mobile phase, these being acetonitrile and methanol. Methanol was finally selected, principally because of its greater solubility with respect to ion-forming reagents.

The effect of methanol percentage in the mobile phase on retention times was studied by using a mobile phase consisting of 25.0 mM citrate buffer, pH = 3.25, $T = 30.0°C$, and 4.00 mM octane sulphonate and methanol at 42.0%, 45.0%, and 50.0%. Fig. 5 shows that, with increasing percentages of methanol, retention times drop considerably and resolution decreases, but up to 50.0% methanol, this remains within acceptable limits.

Parameters Exclusive to CE

Selection of electrolyte

With a view to selecting the most suitable electrolyte for CE,[20,21] all the diamidines under study were analyzed by using various buffers (phosphate, acetate, and citrate), all at 25.0 mM, pH = 3.70, $T = 30.0°C$, and 14.0 kV voltage, as shown in Fig. 6. Only citrate gave useful values of resolution and efficiency for all the diamidines, together with migration times that were adequate for the kind of analysis they intended to optimize. Hence, citrate was also chosen for all the CE works.

Length of capillary

The length of the capillary is directly related to the electric field, efficiency, resolution,[22] and amount of

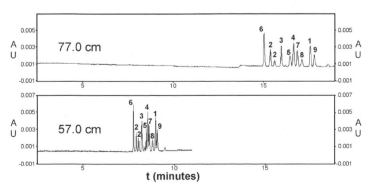

Fig. 7 Influence of capillary length in CE. Capillary, 75.0-μm ID; lengths, 50.0 and 70.0 cm to detector (57.0 and 77.0 cm overall); electrolyte, 25.0 mM citrate buffer; pH = 3.70; $T = 30.0°C$; and voltage, 14.0 kV.

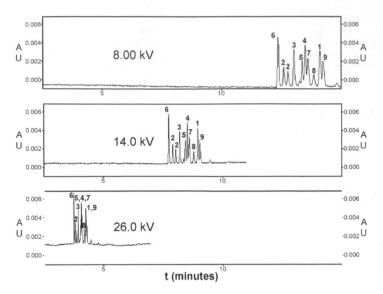

Fig. 8 Influence of variations in voltage in CE. Electrolyte, 25.0 mM citrate buffer; pH = 3.70; T = 30.0°C, and voltages, 8.00, 14.0, and 26.0 kV.

sample loaded. An increase from 50.0 to 70.0 cm, up to the detector (from 57.0 to 77.0 cm overall dimension) in the length of the capillary, causes the quantity of sample loaded to be reduced by approximately 26.0% and the strength of the field to be generated when applying the same potential drops by approximately the same amount (Table 1). To study the influence of the length of the capillary on migration times, the following conditions were used: capillary, 75.0-μm ID; lengths, 50.0 and 70.0 cm to the detector (57.0 and 77.0 cm overall length); electrolyte, 25.0 mM citrate buffer; pH = 3.70; T = 30.0°C; and voltage, 14.0 kV. Fig. 7 shows that as the length is increased from 50.0 to 70.0 cm, migration times are virtually doubled and a striking improvement in efficiency is achieved (i.e., an increase of between 10.0% and 15.0% in the number of theoretical plates), with good resolution.

Voltage applied

The voltage applied is one of the factors of greatest influence in a CE experiment because almost all the parameters governing separation are related to this voltage. The analysis time is inversely proportional to the applied voltage because of, among other things, the higher electrosomotic flow. An increase in the voltage also brings with it a growth in Joule heating[15,16] and, if this is not effectively eliminated, it may cause variations in resistance, pH, viscosity of the electrolyte, and so on, thus rendering the analysis impossible to reproduce. With a view to optimizing the voltage, the Ohm's law plot of intensity against voltage must be kept in mind. The maximum efficiency in electrophoretic separation is attained at the point where this plot begins to deviate from linearity.[23] In the work

reported here, this occurred from 24.0 kV upward, and, when this value was exceeded, a pronounced decrease in efficiency occurred.[24] In Fig. 8, the effects mentioned above can be readily seen; between the electropherogram at 8.00 kV and its counterpart at 14.0 kV, a clear increase in efficiency is observed, with resolution remaining at acceptable levels. On the other hand, in the electropherogram taken at 26.0 kV, outside the limits of linearity under Ohm's law, there is a complete loss of the improvements in both resolution and efficiency produced by higher voltage. This work was carried out using 25.0 mM citrate electrolyte, pH = 3.70, and T = 30.0°C, at voltages of 8.00, 14.0, and 26.0 kV.

CONCLUSIONS

A detailed study was undertaken of each of the parameters affecting the process of separation analysis in HPLC and CE for nine aromatic diamidines. The results obtained are noted and discussed; in the tables, comparative features of the two techniques that emerge from the data collected are recorded.

Performance of HPLC and CE in the Separation of Aromatic Diamidines

The data emerging from this work allow the selection of the optimum conditions for the analysis of each substance in aqueous solution, serum, and urine. For HPLC, they are: 25.0 mM citrate buffer, pH = 3.25, 45.0% methanol, column Ultrasphere ODS (5.00-μm particle size, 15.0 cm × 4.60 mm ID),

Table 2 Performance of HPLC and CE in separation of aromatic diamidines

	HPLC	CE
Detection limit (ng/mL)		
Pentamidine	20.00	300.0
Stilbamidine	10.00	300.0
DAPI	5.00	300.0
Propamidine	30.00	200.0
Hydroxystilbamidine	15.00	400.0
Phenamidine	20.00	150.0
Diampron	10.00	300.0
Berenil	60.00	500.0
Dibromopropamidine	40.00	600.0
Precision (% CV)		
Pentamidine	1.09	3.04
Stilbamidine	1.03	1.68
DAPI	1.16	2.88
Propamidine	1.53	2.37
Hydroxystilbamidine	0.70	3.73
Phenamidine	0.70	3.52
Diampron	0.97	2.43
Berenil	0.85	1.55
Dibromopropamidine	1.36	5.30
Efficiency (theoretical plates)		
Pentamidine	2.22×10^3	2.92×10^5
Stilbamidine	3.91×10^3	3.04×10^5
DAPI	4.60×10^3	2.93×10^5
Propamidine	3.19×10^3	2.92×10^5
Hydroxystilbamidine	3.87×10^3	2.79×10^5
Phenamidine	3.65×10^3	3.06×10^5
Diampron	4.57×10^3	2.78×10^5
Berenil	3.17×10^3	2.43×10^5
Dibromopropamidine	2.63×10^3	2.46×10^5

Table 4 Schematic table of the advantages of HPLC and CE

	HPLC	CE
Versatility	+++	++++
Speed of optimization of methods	++	++++
Stabilization time	++	++++
Analysis time	++	+++
Sensitivity	+++	++
Reproducibility of times	+++	++
Reproducibility of areas	+++	++
Precision	+++	++
Efficiency	++	++++
Amplitude of linear range	++++	++
Resolution capacity	++	++++
Interferences in complex samples	++	++++
Sample preparation	++	++++
Sample volume	++	++++
Application at pilot scale	++++	+
Automatization	+++	++++
Price of reagents and other consumables	++	++++

1.00 mL/min flow, and $T = 30.0°C$. The following features depend on the specific substance under analysis: pentamidine, 4.00 mM hexane sulphonate, $\lambda = 265.0$ nm; stilbamidine, 4.00 mM octane sulphonate, $\lambda = 330.0$ nm; DAPI, 8.00 mM heptane sulphonate, $\lambda = 350.0$ nm; propamidine, 6.00 mM heptane sulphonate, $\lambda = 265.0$ nm; hydroxystilbamidine, 4.00 mM octane sulphonate, $\lambda = 350.0$ nm; phenamidine, 4.00 mM octane sulphonate, $\lambda = 265.0$ nm; diampron, 4.00 mM octane sulphonate, $\lambda = 254.0$ nm; nm; berenil, 4.00 mM octane sulphonate, $\lambda = 370.0$ nm; and dibromopropamidine, 3.00 mM hexane sulphonate, $\lambda = 265.0$ nm.

Table 3 Operational differences between HPLC and CE

	HPLC	CE
Quantity of sample introduced into the system	10.0–1000.0 μL	1.00–50.0 nL
Size of the detector cell	8.00–12.0 mm^3	0.015 mm^3
Detection wavelength	Generally from 230.0 nm upwards	Possible to use wavelengths down to 185.0 nm
Interference	All components of the sample must pass through the detector	Possible to stop the analysis once the substance of interest has been detected
Flow	0.50–2.00 mL/min	Few microliters per minute
Equipment stabilization time	Requires balancing of the column with different timings before reliable results are obtained	Analysis can be carried out almost immediately after connection of equipment

For CE, the optimum values were: overall length of capillary, 57.0 cm (50.0 cm to the detector); 75.0-μm ID; electrolyte, 25.0 mM citrate; pH = 3.70, injection under pressure for 5.00 sec; voltage, 14.0 kV $T = 30.0°C$; and $\lambda = 200.0$ nm.

Analyses by means of HPLC and CE were carried out under these conditions for all the compounds, and comparative data for the two techniques are summarized in Table 2. The efficiency of CE is two orders of magnitude greater than HPLC for all the substances analyzed. The limits of detection for HPLC are much lower than in CE, there being some cases, such as DAPI, where the detection limit is 60 times lower with HPLC than with CE. Values for precision are significantly better with HPLC than with CE.

Operational Differences Between HPLC and CE

Table 3 shows some of the differences in working practices between HPLC and CE.

Advantages of HPLC and CE

To summarize the work reported here, there is a schematic presentation of views on the advantages and drawbacks of each technique in Table 4.

REFERENCES

1. Ettre, L.S.; Horvath, C. Foundations of modern liquid chromatography. Anal. Chem. **1975**, *47*, 422A.

2. Tiselius, A. A new apparatus for electrophoretic analysis of colloidal mixtures. Faraday Soc. **1937**, *33*, 524–531.

3. Hjertén, S. Free zone electrophoresis. Chromatogr. Rev. **1967**, *9*, 122–239.

4. Jorgenson, J.; Lukacs, K.D. Zone electrophoresis in open tubular glass capillaries. Anal. Chem. **1981**, *53*, 1298–1302.

5. Kuhr, W.G. Capillary electrophoresis. Anal. Chem. (Fund. Rev.) **1990**, *62*, 403R.

6. Issaq, H.J. Thirty-five years of capillary electrophoresis: advances and perspectives. J. Liq. Chromatogr. & Relat. Technol. **2002**, *25* (8), 1153–1170.

7. Camilleri, P. *Capillary Electrophoresis, Theories and Practice*, 2nd Ed.; Camolleri, P., Ed.; CRC Press: Boca Raton, FL, 1997; 1–22.

8. Grasilli, E.; Bettuzi, S.; Monti, D.; Ingletti, M.C.; Franceschi, C.; Corty, A. Studies on the relationship between cell proliferation and cell death: opposite patterns of SGP-2 and ornithine decarboxylase mRNA accumulation in PHA-stimulated human lymphocytes. Biochem. Biophys. Res. Commun. **1991**, *59*, 180.

9. Pegg, A.E. Recent advances in the biochemistry of polyamines in eukaryotes. Biochem. J. **1986**, *234*, 249.

10. Pegg, A.E. Polyamine metabolism and its importance in neoplastic growth and a target for chemotherapy. Cancer Res. **1988**, *48*, 759.

11. Rabanal, B.; Merino, G.; Negro, A. Determination by capillary zone electrophoresis of berenil, phenamidine, diampron and dibromopropamidine in serum and urine. J. Chromatogr. B. **2000**, *738*, 293–303.

12. Rabanal, B.; Negro, A. Study of nine aromatic diamidines designed to optimize their analysis by HPLC. J. Liq. Chromatogr. **2003**, *26* (20), 3499–3512.

13. Charton, M. The application of the Hammett equation to amidines. J. Org. Chem. **1965**, *30*, 969.

14. Bocek, P.; Deml, M.; Gebaner, P.; Dolnik, V. *Analytical Isotachophoresis*; VCH: Weinheim, 1988.

15. Rush, R.S.; Cohen, A.S.; Karger, B.L. Influence of column temperature on the electrophoretic behavior of myoglobin and α-lactalbumin in high-performance capillary electrophoresis. Anal. Chem. **1991**, *63*, 1346–1350.

16. Nelson, R.J.; Paulus, A.; Cohen, A.S.; Guttman, A.; Karger, B.L. Use of Peltier thermoelectric devices control column temperature in high performance capillary electrophoresis. J. Chromatogr. B. **1989**, *480*, 111–127.

17. Sepaniak, M.J.; Cole, R.O. Column efficiency in micellar electrokinetic chromatography. Anal. Chem. **1987**, *59*, 472–476.

18. Eksborg, S.; Lagerstom, P.; Modin, R.; Schill, G. Ion pair chromatography of organic compounds. J. Chromatogr. A. **1973**, *83*, 99–110.

19. Braithwaite, A.; Smith, F.J. *Chromatographic Methods*, 5th Ed.; Blackie Academic and Professional, 1996.

20. Issaq, H.J.; Atamna, I.Z.; Muschik, G.M.; Janini, G.M. The effect of electric field strength, buffer

type and concentration on separation parameters in capillary zone electrophoresis. Chromatographia **1991**, *32*, 155–161.

21. Nashabeh, W.; El Rassi, Z. Capillary zone electrophoresis of pyridylamino derivatives of maltooligosaccharides. J. Chromatogr. **1990**, *514*, 57–64.

22. Cohen, A.S.; Paulus, A.; Karger, B.L. High performance capillary electrophoresis using open tubes and gels. Chromatographia **1987**, *24*, 15–24.

23. Beckers, J.L.; Everaests, F.M. Isotachophoresis with two leading ions and migration behaviour in capillary zone electrophoresis: II. Migration behaviour in capillary zone electrophoresis. J. Chromatogr. A. **1990**, *508*, 19–26.

24. McLaughlin, G.M.; Nolau, J.A.; Lindahl, J.L.; Palmieri, R.H.; Anderson, K.N.; Morris, S.C.; Morrison, J.A.; Bronzert, T.J. Pharmaceutical drug separations by HPCE: practical guidelines. J. Chromatogr. **1992**, *15* (6&7), 961–1021.

A

Asymmetric Field-Flow Fractionation in Biotechnology

Thorsten Klein
Christine Hürzeler
Postnova Analytics, Munich, Germany

INTRODUCTION

The research and development in the fields of biochemistry, biotechnology, microbiology, and genetic engineering are fast-growing areas in science and industry. Chromatography, electrophoresis, and ultra-centrifugation are the most common separation methods used in these fields. However, even these efficient and widespread analytical methods cannot cover all applications. In this article, asymmetric flow field-flow fractionation (AF4) is introduced as a powerful analytical separation technique for the characterization of biopolymers and bioparticles. Asymmetric flow field-flow fractionation (FFF) can close the gap between analyzing small and medium-sized molecules/particles [analytical methods: HPLC, GFC, etc.] on the one hand and large particles (analytical methods: sedimentation, centrifugation) on the other hand,[1–2] whereas HPLC and GFC are overlapping with asymmetric field-flow fractionation in the lower separation ranges.

First publications about FFF by Giddings et al.[3] appeared in 1966. From this point, FFF was developed in different directions and, in the following years, various subtechniques of FFF emerged. Well-known FFF subtechniques are sedimentation FFF, thermal FFF, electric FFF, and flow FFF. Each method has its own advantages and gives a different point of view of the examined sample systems. Using sedimentation FFF shows new insights about the size and density of the analytes, thermal FFF gives new information about the chemical composition and the size of the polymers/particles, and electric FFF separates on the basis of different charges. Flow FFF, and especially asymmetric flow FFF (the most powerful version of flow FFF) is the most universal FFF method, because it separates strictly on the basis of the diffusion coefficient (size or molecular weight) 2, and it has the broadest separation range of all the FFF methods. It is usable for a large number of applications in the fields of biotechnology, pharmacology, and genetic engineering.

SEPARATION PRINCIPLE OF ASYMMETRIC FLOW FIELD-FLOW FRACTIONATION

All FFF methods work on the same principle and use a special, very flat separation channel without a stationary phase. The separation channel is used instead of the column, which is needed in chromatography. Inside the channel, a parabolic flow is generated, and perpendicular to this parabolic flow, another force is created. In principle, the FFF methods only differ in the nature of this perpendicular force.

The separation channel in AF4 is approximately 30 cm long, 2 cm wide, and between 100 and 500 μm thick. A carrier flow which forms a laminar flow profile streams through the channel. In contrast to the other FFF methods, there is no external force, but the carrier flow is split into two partial flows inside the channel. One partial flow is led to the channel outlet and, afterward, to the detection systems. The other partial flow, called the cross-flow, is pumped out of the channel through the bottom of the channel. In the AF4, the bottom of the separation channel is limited through a special membrane and the top is made of an impermeable plate (glass, stainless steel, etc.). The separation force, therefore, is generated internally, directly inside the channel, and not by an externally applied force.

Under the impact of the cross-flow, the biopolymers/particles are forced in the direction of the membrane. To ensure that the analytes do not pass through the membrane, different pore sizes can be used. In this way, the analytes can be selectively rejected and it is possible to remove low-molecular compounds before the separation. The analytes' diffusion back from this membrane is counteracted by the cross-flow, where, after a time, a dynamic equilibrium is established. The medium equilibrium height for smaller sized analytes is located higher in the channel than for the larger analytes. The smaller sized analytes are traveling in the faster velocity lines of the laminar channel flow and will be eluted first. As a result, fractograms, which show size separation of the fractions, are obtained as an analog to the chromatograms from HPLC or GFC.

APPLICATIONS OF AF4 ASYMMETRIC FLOW FIELD-FLOW FRACTIONATION IN BIOTECHNOLOGY

In addition to widespread applications in the field of polymer and material science or environmental research, AF4 can be used in bioanalytics, especially for the characterization of proteins, protein aggregates,

Encyclopedia of Chromatography DOI: 10.1081/E-ECHR-120039887

polymeric proteins, cells, cell organelles, viruses, lipo-somes, and various other bioparticles and biopolymers.

Cells and Viruses

The advantage of AF4, in contrast to chromatography, is the capability to separate bioparticles and bio-poylmers which usually stick onto chromatography columns. They are more or less filtered out (removed) by the stationary phase. Various applications using AF4 for the separation of shear-force sensitive bioparticles with high molecular weight and size have been reported in the literature. They deal with the efficient and fast separation of viruses[4–5] and bacteria.[5] Reference[4] discusses the investigation of a virus (STNV) with AF4 and the separation of the viral aggregates. In Ref.,[5] Litzen Wahlund report the separation of a virus (CPMV) together with different other proteins (BSA, Mab). They also present the characterization of bacillus streptococcus faecalis and its aggregates using AF4.

Proteins/Antibodies/DNA

The separation of proteins with AF4 has been demonstrated a number of times. For example, the fractionation of ferritin,[7] of HSA and BSA,[8] and of monoclonal antibodies,[8] including their various aggregates, were published. Asymmetric flow FFF is especially suitable for the separation and characterization of large and sensitive proteins and their aggregates because it is fast and gentle and aqueous solvents can be used that achieve maximum bioactivity of the isolated proteins and antibodies. Furthermore, even very large and sticky proteins can be analyzed because of the relatively low surface area and the separation in the absence a stationary phase. Nearly independent of the nature of the bioparticles, AF4 separates by size (diffusion coefficient). Therefore, DNA, RNA, and plasmids can be separated quickly and gently, together with proteins. Reference[6] deals with this issue and presents the AF4 separation of a mixture of cytochrome-c, BSA, ferritin, and plasmid DNA.

Artificial Polymeric Proteins

In addition to the characterization of well-known protein substances (serum proteins, aggregates, antibodies, etc.), AF4 is also a very promising separation/characterization technique for a new class of artificially made polymeric proteins from therapeutic/diagnostic applications, such as poly-streptavidin and polymeric

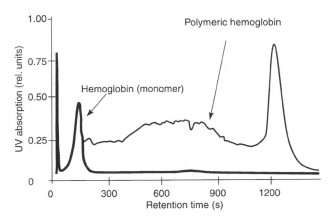

Fig. 1 Pig hemoglobin separated with AF4 and UV detection.

hemoglobin [personal communication of the authors]. These proteins usually have very high molecular weights and huge molecular sizes, and they are difficult to analyze by conventional GFC and related techniques. Very often, these proteins are also sticky and show adsorptive effects on the column material. Using AF4 without a stationary phase and without size-exclusion limit, these polymeric proteins can be readily separated and characterized. The application shown in Fig. 1 was done using an AF4 system (HRFFF 10,000 series, Postnova Analytics) and ultraviolet (UV) detection at 210 nm.

CONCLUSIONS

Asymmetric flow FFF is a new member in the FFF familiy of separation technologies; it is a powerful characterization technique, especially suited for the separation of large and complex biopolymers and bioparticles. Asymmetric flow FFF has many of the general benefits of FFF; it adds on several additional characteristics. In particular, these characteristics are as follows:

1. No sample preparation, or only limited sample preparation necessary.
2. Possibility of direct injection of unprepared samples.
3. Large accessible size molecular-weight range, no size-exclusion limit.
4. Very gentle separation conditions in the absence of a stationary phase.
5. Weak or no shear forces inside the flow channel.
6. Rapid analysis times, generally faster than GFC.
7. Fewer sample interactions during separation because of small surface area.

8. On-line sample concentration/large volume injection possible.
9. Gentle and flexible because it uses a wide range of eluents/buffers/detectors.
10. AF4 is a useful analytical tool, and when the limitations of the technology (e.g., sample interactions with membrane or the sample dilution during separation) are carefully observed, samples can be characterized where other analytical technologies fail or only yield limited information.

REFERENCES

1. Klein, T. Chemisch–physikalische Charakterisierung von schwermetallhaltigen Hydrokolloiden in natürlichen aquatischen Systemen mit Ultrafiltration und Flow-FFF. *Diploma thesis;* TU-Munich, 1995.
2. Klein, T. Entwicklung und Anwendung einer Asymmetrischen Fluß-Feldflußfraktionierung zur Charakterisierung von Hydrosolen. *Ph.D. thesis*; TU-Munich, 1998.
3. Giddings, J.C. A new separation concept based on a coupling of concentration and flow nonuniformities. Separ. Sci. **1966**, *1*, 123.
4. Litzen, A.; Wahlund, K.G. Zone broadening and dilution in rectangular and trapezoidal asymmetrical flow field-flow fractionation channels. Anal. Chem. **1991**, *63*, 1001.
5. Litzen, A.; Wahlund, K.G. Effects of temparature, carrier composition and sample load in asymmetrical flow field-flow fractionation. J. Chromatogr. **1991**, *548*, 393.
6. Kirkland, J.J.; Dilks, C.H.; Rementer, S.W.; Yau, W.W. Asymmetric-channel flow field-flow fractionation with exponential force-field programming. J. Chromatogr. **1992**, *593*, 339.
7. Tank, C.; Antonietti, M. Characterization of water-soluble polymers and aqueous colloids with asymmetrical flow field-flow fractionation. Macromol. Chem. Phys. **1996**, *197*, 2943.
8. Litzen, A.; Walter, J.K.; Krischollek, H.; Wahlund, K.G. Separation and quantitation of monoclonal antibody aggregates by asymmetrical flow field-flow fractionation and comparison to gel permeation chromatography. Anal. Biochem. **1993**, *212*, 169.

Automation and Robotics in Planar Chromatography

Wojciech Markowski
Department of Physical Chemistry, Medical University, Lublin, Poland

INTRODUCTION

Automation involves the use of systems (instruments) in which an element of nonhuman decision has been interpolated. It is defined as the use of combinations of mechanical and instrumental devices to replace, refine, extend, or supplement human actions and faculties in the performance of a given process, in which at least one major operation is controlled, without human intervention, by a feedback system. A feedback system is defined as an instrumental device combining sensing and commanding elements that can modify the performance of a given act.[1]

Three approaches to the automation process can be distinguished, taking into account the criterion of flexibility of the automation device.[2,3] The first, denoted as flexible, is characterized by the possibility of adaptation of the instruments to new and varying demands required from the laboratory; examples of these instruments are robots. The second approach, denoted as semiflexible, involves some restrictions for the tasks executed by the instrument; the tasks are controlled by software via a keyboard. As examples, autosamplers or robots of limited capacity can be cited. In the third approach, the instruments can execute one or two tasks, without the possibility of adapting to new requirements; examples include supercritical fluid extractors and diluters. Automation of an analytical laboratory gives several benefits: better reproducibility, increase in the number of samples that can be analyzed, and freedom of personnel to do more creative tasks (e.g., method development and interpretation of results). Harmful conditions in the laboratory or other workplace can be avoided, and the equipment of the laboratory can be more effectively utilized. Automation improves worker safety and product quality. It provides exact timing and uniform sample handling, which ensure precision and accuracy. It allows for the transfer of methods from one laboratory to another, since the methods are saved on any medium (diskette, CD), and they can be executed by instruments that are identical wherever they are implemented. Before illustration of the possibility of automation in TLC, let us consider the ideal, fully automated analytical laboratory. In such a laboratory, the sample is processed from its entry into the laboratory hopper, through the many operations until the final report is printed out and the sample is stored for future analysis. To illustrate the feasibility of automation in TLC,[4] the fundamental operations in an analytical laboratory, including the chromatographic process, are given in Fig. 1. The first and basic stage of the process, not limited to TLC, but also occurring in other chromatographic techniques, is the preparation of samples. This is the most tedious, time-consuming, and error-generating process in the whole analytical cycle and can be fully automated, or the automated stations may be complementary to operations or tasks executed individually. For instance, in a station, a volume of liquid is transferred from the first to a second container, an internal standard is added, and the solution is diluted and mixed. Further actions may be executed manually. Another, more advanced solution consists in automated execution of the tasks by the station, and the sample is transferred from one station to the other by a robot or another transport device. In a limited version, only the most critical stages are automated by the use of robots with limited, strictly defined movements; examples are automated processes of solid-phase extraction (SPE), heating, and mixing. The robots are controlled by software and the operator chooses the suitable values of the parameters from given ranges (e.g., autosampler).

At the end of the procedures, with chromatographic methods as the last step, the raw data collected during chromatographic analysis are critically reviewed by specific software designed for this purpose. If the results are not as expected, samples will be available for repeated analysis.

PREPARATION OF PLATES

In most laboratories using planar chromatography, precoated plates are used in everyday practice. Self-coating plates should be considered when special layers are required and when suitable precoated plates are not available on the market. Another reason for the application of the self-coating procedure is the cost of precoated plates following from the very high throughput—number of analyzed samples—and the limited budget of most laboratories. Special layers contain silver nitrate, buffer components, or a

Encylopedia of Chromatography DOI: 10.1081/E-ECHR-120039888

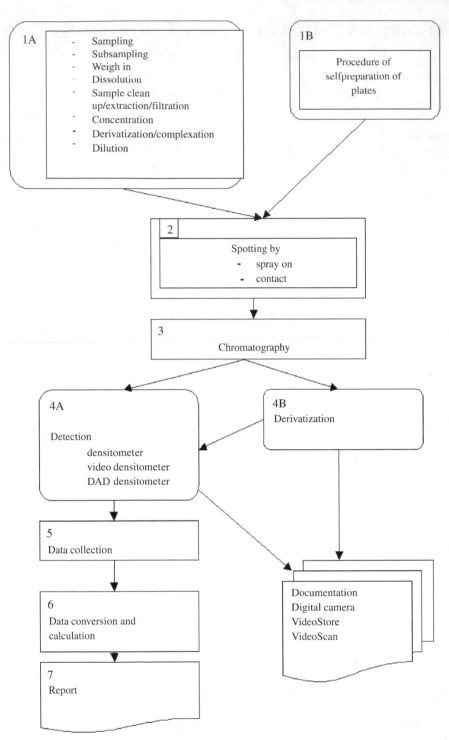

Fig. 1 Possible steps in an analytical laboratory for the process of automation.

mixture of adsorbents. A special case is when the binder used in a commercial precoated plate might interfere with detection. A self-prepared layer of good quality can be obtained by automatic coating. Fig. 2 presents an example of an automatic coating device.[5] The glass plates to be coated are conveyed underneath a hopper filled with the adsorbent suspension. The automatic TLC plate coater is supplied with a fixed gate for preset layers of 300 and 500 μm, an adjustable gate for layer thicknesses of 0–2 mm, and a plate holder for eight 20 × 20 cm plates. The plates are moved by a motorized conveying system at a uniform feed rate of 10 cm/sec to ensure a uniform layer.

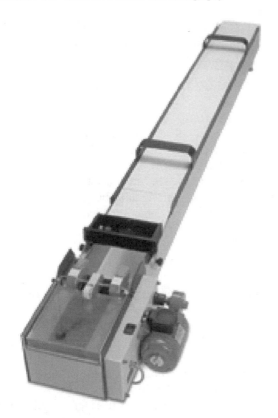

Fig. 2 Camag automatic TLC plate coater. (Courtesy of Camag.[5]) *(View this art in color at www.dekker.com.)*

SAMPLE APPLICATION

The selection of the sample application technique and the device to be used depends on sample volume, number of samples to be deposited, and precision and degree of automation required. During sample application, some minimum requirements must be fulfilled. The application of the sample onto the thin layer is a critical moment, owing to later localization by the scanner (densitometer) and the beginning of the chromatographic process at the moment of contact of the liquid sample with the chromatographic bed. There are two principal ways to deposit a sample onto the plate. They are contact spotting and spray-on application. Therefore, the applicator must guarantee the exact localization of the sample, particularly for quantitative analyses and uniform compact cross-section of the starting band. Typical values are from 1 to 4 mm for conventional layers (TLC), and for high-precision TLC (HPTLC) the upper limit is 1.5 mm. These demands limit the volume of sample applied to the layer. For TLC, the typical volume is from 0.5 to 5 μL. but for HPTLC, typical volumes spotted can be from 0.1 to 1.0 μL. The above limitation is valid when samples are applied as spots. In the case when the spray-on technique (narrow bands) is used, there is possibility of the application of larger volumes.

The simplest version of applicator is presented in Fig. 3 (Nanomat 4).[5] It serves for easy application of samples onto TLC and HPTLC plates or sheets, precisely positioned and without damage to the layer. The actual sample dosage is applied with disposable capillaries that are precisely guided by the universal capillary holder. Capillaries are loaded into the holder from dispenser magazines, then filled with sample solution and placed on the applicator head of the device. Capillary volumes of 0.5, 1, 2, and 5 μL. are available. The volume precision is $R = \pm 0.25\%$, $CV = \pm 0.6\%$. The next stage with a more complicated degree of automation represents semiautomatic applicators. Semiautomatic applicators presently available have the volume range of 20 nL–10 μL. (e.g., TLC—Spotter PS 01 Desaga). The sample is delivered from 0.5, 1.0, and 10 μL syringes. The piston stroke can be set in a continuous manner. To apply the sample, the piston is stopped and the solution is injected from the end of the capillary; the whole volume of the sample is displaced from the capillary. The position of the end of the capillary is adapted to the layer thickness; the spring-relieved syringe guide warrants that the capillary needle only lightly touches the adsorbent layer, avoiding damage to it. The change of position is automatic. The device permits application of spots or streaks at a distance of 5 mm from the edge of the plate; the syringe can be washed twice. The next step in automation is represented by Desaga TLC Applicator AS 30 or Camag Linomat 5. These automatic applicators can be operated in stand-alone mode or under the control of computer software. They are composed of an application module, interface, software, and an IBM PC-AT. The application module dispenses samples from a stainless-steel capillary that is connected to a dosage syringe operated by a stepping motor. Samples can be applied as spots or bands onto TLC plates or sheets up to 20 × 20 cm. Bandwise sample application uses the spray-on technique; for

Fig. 3 Simplest automatic applicator (Nanomat 4). (Courtesy of Camag.[5]). *(View this art in color at www. dekker.com.)*

Fig. 4 Full automatic applicator (ATS 4). (Courtesy of Camag.[5]) *(View this art in color at www.dekker.com.)*

spotwise application, either contact transfer or spraying may be selected. The samples are contained in vials, which may be sealed with regular septa. The vials are arranged in racks with 16 positions; two racks may be inserted per application program. The application pattern can be selected for normal development, for development from both sides, and for circular and anticircular chromatography. Camag Automatic TLC Sampler 4 (ATS 4) meets all the requirements for fully automatic sample application (Fig. 4).[5] The device allows for both methods of application of samples

and additionally for "overspotting," which can be used in prechromatographic derivatization, spiking, etc.

DEVELOPMENT OF THE CHROMATOGRAM

The next important stage is chromatogram development. Automatic developing chambers (Automatic Developing Chamber, Camag), DC-MAT (Byron), and TLC-MAT (Desaga) are automatically operating development systems. They increase the reproducibility of the chromatographic results because the development is carried out under controlled conditions. The progress of the solvent front is monitored by a sensors. The development process is terminated as soon as the mobile phase has traveled the programmed distance. Preconditioning, tank or sandwich configuration, solvent migration distance, and the drying conditions are selectable. All relevant parameters are entered via a keypad. The AMD2 system (Automated Multiple Development, Camag)[5] is a fully automated version of multiple development and stepwise technique with a free choice of mobile-phase gradient (Fig. 5). Because the chromatogram is developed repeatedly in the same direction and each individual run is somewhat farther than the last, a focusing of the separated substance zones takes place in the direction of development. The chromatography is reproducible because the mobile phase is removed from the separation chamber after each step and the layer is completely freed from

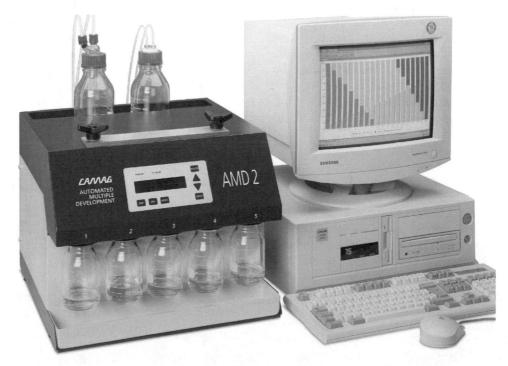

Fig. 5 Device for automated multiple development (AMD2). (Courtesy of Camag.[5]) *(View this art in color at www. dekker.com.)*

the mobile phase, in vacuum. Then, a fresh mobile phase is introduced for the next run. Provided all parameters, including solvent migration increments, are properly maintained, which is only possible with a fully automatic system, the densitogram of a chromatogram track can be superimposed with a matched-scale diagram of the gradient. Another device for automated development is the chamber constructed by Tyihak and Mincsovics,[6] in which the adsorbent layer is placed between two plates and the mobile phase flows under increased applied pressure. It can be operated in the linear or radial mode. Another automated device is the ultra-micro-rotation chromatograph (UMRC), where the eluent is delivered to the center of a rotating TLC plate.[7] A simple device was constructed by Delvordre, Reynault, and Postaire[8] in which the liquid is pumped out (by vacuum), which causes the flow of the mobile phase and decreases the vapor pressure.

DERIVATIZATION

Derivatization can be carried out both before and after development of the plate. In the latter case, it may be applied before detection or after scanning densitometry. Derivatization may be carried out using the device constructed by Kreuzig (Anton Paar KG), where the sprayer moves along a vertical guide while the plate moves horizontally.[9] Another method of derivatization consists in immersion of the plate into a suitable reagent solution. For this purpose, devices available from Camag or Desaga can be used (Camag Chromatogram Immersion Device III, or Desaga TLC Dip-Fix), in which a low-velocity motor causes the immersion and removal of the plate from the reagent solution (Fig. 6).[5]

EVALUATION

Thin-film chromatographic detection, in contrast to other chromatographic techniques, requires stopping of development, drying of the layer, and scanning with an appropriate detector. There are basically two alternatives for the evaluation of thin-layer chromatograms: elution of the separated substance from the layer followed by photometric determination (indirect determination), and in situ evaluation (scanning) directly on the TLC plate. The in situ evaluation of the chromatogram is carried out using a high-resolution chromatogram spectrophotometer (densitometer) (Fig. 7) [5] to scan each chromatogram track, from start to solvent front in the direction of development, by means of a slit. The measurements are carried out either in the visible-light range for colored or fluorescent substances or in the ultraviolet (UV) range

Fig. 6 Chromatogram immersion device. (Courtesy of Camag.[5]) *(View this art in color at www.dekker.com.)*

for UV-light-absorbing solutes. The wavelength of maximum absorption is generally selected as the measurement wavelength. The scanning process yields absorption or fluorescence scans (peaks), which are also used to assess the quality of chromatographic separation. TLC plates are generally scanned in the reflectance mode (diffuse reflectance), meaning that monochromatic light is directed by a mirror onto the layer surface at 90° and the diffuse reflectance is measured at 45° by means of a detector. The optical pathways used for absorption and fluorescence measurements are identical in commercially available scanners. The only difference is the light source: visible-light measurements are performed using tungsten lamps, whereas high-pressure mercury lamps are used for fluorescence measurements and deuterium lamps for absorption measurements in the UV range. In the case of fluorometric detection, it is also necessary to place a cutoff filter in front of the detector to prevent comeasurement of the short-wavelength excitation radiation. All functions of the scanner are controlled from a personal computer that is linked via an RS232 interface. The scanner transmits all measurement data, in digital form, to the computer for processing with the specific software. The final report is based on the following sequence: raw data acquisition, integration, calibration, and calculation. Integration is performed, post run, from the raw data gathered during scanning (i.e., after all tracks of a chromatogram plate

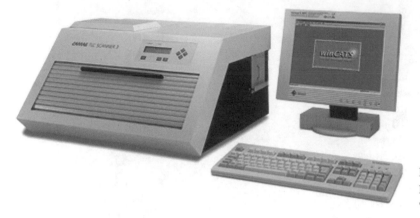

Fig. 7 Densitometer linked with personal computer. (Courtesy of Camag.[5]) *(View this art in color at www.dekker.com.)*

have been measured). Integration results can be influenced by selecting appropriate integration parameters. As all measured raw data remain stored on a disk, reintegration with other parameters is possible at any time. The system automatically defines and corrects the baseline and sets fraction limits. The operator can accept these or override the automatic process by video integration. All steps can be followed on the screen. The option to get a "visual impression" of the chromatogram is one of the inherent advantages of planar chromatography over all other chromatographic techniques. White light is required for imaging colored chromatogram zones; long-wave UV light reveals fluorescing substances. Ultraviolet-absorbing substances can be visualized under short-wave UV light, provided the layer contains a UV indicator. The well-established lighting unit Camag Reprostar (Fig. 8) [5] provides all three types of light. Combined with a

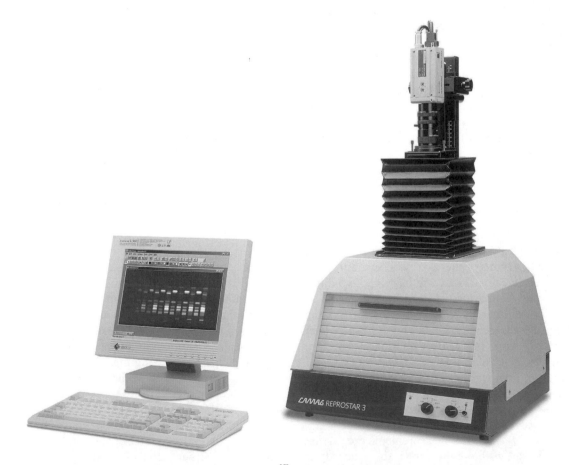

Fig. 8 Video scanner. (Courtesy of Camag.[5]) *(View this art in color at www.dekker.com.)*

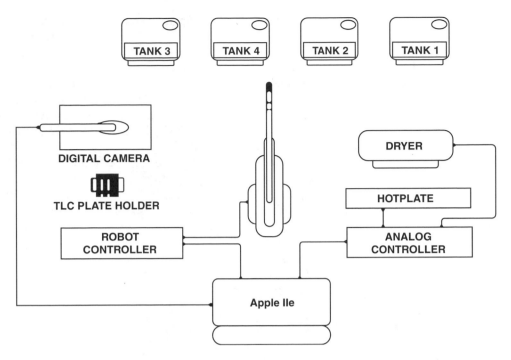

Fig. 9 Planar chromatography robot after Prosek. (From Ref.[4]. Copyright Research Institute for Medicinal Plants, Hungary, 2004.)

modern high-resolution digital camera, it forms an affordable documentation system for planar chromatograms and similar objects.

ROBOTS

Laboratory robots are adapted now for linking all of the steps between extraction and obtaining the analysis results. They are able to automate lengthy, routine, multistep analyses. They require electronic communication in real time to know the operating time and possible breakdowns exactly. A robotics system allows space saving and easier integration of equipment in the laboratory. Today, a conventional robot has a movable arm. The purpose of the arm is to extend the capabilities of the human arm. There are five basic parts to every robotic arm: controller, arm, drive, end effector, and sensor. There are also five basic functions: base, shoulder, elbow, pitch, and roll. Most modern robots belong to one of four categories: Cartesian, spherical, and cylindrical robots, and revolute arms.[10] In 1989, Prosek et al.[4] developed a planar chromatography robot (Fig. 9). Its arm, supported by a rotating base, executed movements with four degrees of freedom. Its work envelope comprised four tanks: the first for cleaning, the second for development, and the last two for derivatization by dipping. Also required were a hot plate, a drying system, and a digital camera for evaluation of the derivatized plate. The system was controlled by an Apple II e computer. The planar chromatography automaton was designed by Delvordre and Postaire with the objective of reducing the number of human movements required for the handling of precoated plates.[10] This device uses a conveyor-belt-like system to sustain all the chromatographic steps along with their own supply of reagents and tools. The procedure comprises six stages. Using this method, qualitative and quantitative data are obtained 50–150 min after starting the procedure.

CONCLUSIONS

Technological progress has enabled semi- and full automation of every stage of the planar chromatography technique. Coupling of automation of the sample preparation process and automation of TLC have strongly enhanced the advantages this separation method. Such improvements will now meet the requirements of the industrial sector (mainly pharmaceutical and plant industries) not only in terms of productivity, effectiveness, reduced cost, Good Laboratory Practice (GLP), and environmental quality, but also on the technical side (validation, flexibility, evolution). Complete chromatographic automation will bring planar chromatography to the same level as other chromatographic methods.

ARTICLES OF FURTHER INTEREST

Gradient Development in TLC, p. 702.
TLC Sorbents, p. 1645.

REFERENCES

1. *IUPAC Compendium of Analytical Literature*; Pergamon Press: Oxford, 1978; 22–23.
2. Majors, R.E.; Holden, B.D. Laboratory robotics and its role in sample preparation. LC–GC Int. **1993**, *6* (9), 530.
3. Luquede Castro, M.D.; Velasco-Arjona, A. Towards the most rational use of robotics within the overall analytical process. Anal. Chim. Acta. **1999**, *384*, 117–125.
4. Prosek, M.; Pukl, M.; Smidownik, A.; Medja, A. Automation of thin layer chromatography with a laboratory robot. J. Planar Chromatogr. **1989**, *2* (6), 244.
5. http://www.camag.com. **2004**.
6. Tyihak, E.; Mincsovics, E. Overpressured-layer chromatography (optimum performance laminar chromatography) (OPLC). In *Planar Chromatography. A Retrospective View for the Third Millennium*, 1st Ed.; Nyiredy, Sz., Ed.; Springer: Budapest, 2001; 137–176.
7. Nyiredy, Sz. Rotation planar chromatography. In *Planar Chromatography. A Retrospective View for the Third Millennium*; 1st Ed.; Nyiredy, Sz., Ed.; Springer: Budapest, 2001; 177–199.
8. Delvordre, P.; Reynault, C.; Postaire, E. J. Liq. Chromatogr. **1992**, *15*, 1673.
9. Kreuzig, F. Chromatographia. **1980**, *13*, 238.
10. Postaire, E.P.R.; Delvordre, P.; Sarbach, C. Automation and robotics in planar chromatography. In *Handbook of Thin-Layer Chromatography*. 2nd Ed.; Sherma, J., Fried, B., Eds.; Marcel Dekker, Inc.: New York; 373–385.

Band Broadening Correction Methods in GPC/SEC

Gregorio R. Meira
Jorge R. Vega
INTEC (Universidad Nacional del Litoral and CONICET), Santa Fe, Argentina

INTRODUCTION

In ideal SEC, fractionation is exclusively by hydro-dynamic volume. Unfortunately, perfect SEC fractionation is impossible due to the presence of secondary fractionations and band broadening (BB). Secondary fractionations result from physicochemical interactions between the polymer, the solvent, and the column packing[1] and are not discussed further. Band broadening is mainly due to axial dispersion in the columns, while other broadening sources include column end-fitting effects, finite injection volumes, finite detection cell volumes, and laminar flow profiles in the capillaries.[2,3] Mathematical models have been developed that describe the detailed fractionation processes in SEC. Their aim is to estimate the chromatograms from a priori knowledge of the molar mass distribution (MMD), the polymer–solvent–matrix interactions, the column characteristics, and the flow conditions.[4–7] Unfortunately, these complex models have not been applied so far for BB correction, and therefore they are not discussed further.

If a broad and smooth chromatogram is obtained with modern high-resolution columns, the BB effect is generally negligible, and no specific corrections are required. In contrast, corrections for BB may be important when analyzing: 1) narrow chromatograms of half-widths close to those of monodisperse samples appearing at similar elution volumes; and 2) broad but multimodal chromatograms, with sharp elbows and/or narrow peaks.

First, consider the simpler case of a mass-sensitive detector (typically, a differential refractometer, DR) in combination with a molar mass calibration (in turn, obtained from narrow standards of the analyzed polymer). Due to BB (and even in the simpler case of analyzing a linear homopolymer), a whole distribution of hydrodynamic volumes (and therefore of molar masses) is instantaneously present in the DR cell. This establishes that the mass chromatogram $w(V)$ (i.e., the instantaneous mass w vs. the elution time or elution volume V) is a broadened version of a hypothetically true (or corrected) mass chromatogram $w^c(V)$, as follows:[8]

$$w(V) = \int_0^\infty g(V, \overline{V}) w^c(\overline{V}) d\overline{V} \qquad (1)$$

where $g(V, \overline{V})$ is the (in general, nonuniform) BB (or spreading) function and $\overline{V}$ is a dummy integration variable that represents an average retention volume. At each $\overline{V}$, a different individual $g(V)$ function is defined. For any symmetrical $g(V)$ function, its $\overline{V}$ value is unambiguously assigned at its maximum (or mode). For skewed $g(V)$ functions, however, the average retention volume could be assigned at the mode, the mean, or any other measure of central tendency. This ambiguity in the origin of asymmetrical BB functions is still an unresolved question regarding the specification of $g(V, \overline{V})$. For uniform (or retention volume invariant) BB functions, Eq. (1) reduces to a simple convolution integral.

The molar mass calibration is normally expressed as $\log M(V)$. This calibration is obtained from narrow standards, by associating a set of average molar masses to a set of average retention volumes. If this association is carried out correctly, then the calibration is essentially unaffected by BB. When the MMD is estimated from a (broadened) mass chromatogram $w(V)$ and an unbiased (or "true") molar mass calibration $\log M(V)$, then the distribution is broader than real, the number-averaged molar mass $(\overline{M}_n)$ is underestimated, and the weight-averaged molar mass $(\overline{M}_w)$ is overestimated. The direct correction procedure for these biases is as follows: 1) From the knowledge of $w(V)$ and $g(V, \overline{V})$, calculate $w^c(V)$ by inversion of Eq. (1); and 2) from $w^c(V)$ and $\log M(V)$, obtain the unbiased MMD $w^c(\log M)$.

If the analyzed polymer is strictly monodisperse (both in hydrodynamic volume and in molar mass), then the corrected chromatogram $w^c(V)$ is an impulsive function, and the mass chromatogram is a direct measure of the BB function at the given $\overline{V}$. Thus, the global $g(V, \overline{V})$ could be obtained by interpolation, from a set of monodisperse (or uniform) standards. Unfortunately, uniform standards are only available for low molar masses (e.g., a pure solvent) and for some water-soluble biopolymers. "Almost" uniform standards have been produced by fractionating narrow (synthetic) standards through temperature-gradient interaction chromatography, and their chromatograms have been adequately fit with exponentially modified Gaussian (EMG) functions.[9] Inside the linear calibration range, these functions are quite uniform but skewed, with exponential decay (or tailing) toward the

Encylopedia of Chromatography DOI: 10.1081/E-ECHR-120039889

higher elution volumes. However, when approaching the limit of total exclusion, the BB function becomes narrower and more skewed, and cannot be well approximated by an EMG.[9] Even for a "linear" calibration, resolution in SEC falls exponentially with increasing molar mass,[10] while the BB function remains essentially uniform.[9] For this reason, the effects of BB are particularly serious at the higher molar masses.

Apart from the use of uniform (or almost uniform) standards, other methods for determining the BB function have been developed. For example, by assuming a uniform and Gaussian BB function with a linear molar mass calibration, it is possible to use the mass and molar mass chromatograms for simultaneously estimating the standard deviation of the BB function and the calibration coefficients.[11,12] Alternatively, if the shape of the MMD is known (e.g., it is a Poisson distribution on a linear molar mass axis), then the BB function can be estimated from the difference between the (mass or molar mass) chromatogram and its theoretical prediction in the absence of BB.[13] Finally, the BB function can be theoretically predicted from a representative fractionation model.[4,7] Unfortunately, however, this approach is so far unfeasible due to the difficulty in determining the associated physicochemical parameters.

Consider the BB problem when molar mass sensitive detectors are employed. First, let us analyze the ideal case of a chromatograph fit with perfect detectors and not exhibiting BB, secondary fractionations, or interdetector volumes. In this case, the instantaneous MMD is strictly uniform, and any molar mass detector type would provide the same result:[14]

$$M(V) = K_{LS} \frac{s_{LS}^c(V)}{w^c(V)}$$

$$= K_{IV} \left[\frac{s_{IV}^c(V)}{w^c(V)} \right]^{1/a} = K_{OS} \frac{w^c(V)}{s_{OS}^c(V)} \qquad (2)$$

where $s_{LS}^c(V)$, $s_{IV}^c(V)$, and $s_{OS}^c(V)$ are respectively the "true," "corrected," or unbroadened chromatograms obtained from a light-scattering (LS) detector, a specific viscosity (IV) detector, and a (still under development) colligative-property osmometer (OS) detector; a is the Mark–Houwink–Sakurada exponent at the given measuring conditions; and K_{LS}, K_{IV}, and K_{OS} are calibration constants. Eq. (2) provides an unbiased (or MMD-independent) molar mass calibration $\log M(V)$ that, in principle, is identical to that determined from uniform standards in a real chromatograph with BB.

The signal-to-noise ratios are generally poor at the chromatogram tails; for this reason, the signal ratios

of Eq. (2) are only precise in the mid-chromatogram region. Also, the molar mass sensitive sensor is normally connected in series with the DR, and this shows that the molar mass signal slightly leads the mass signal. To correct for this bias (and independent of BB), the LS signal must be adequately shifted toward higher retention volumes prior to calculating any quality variable [e.g., the molar masses of Eq. (2)].

The BB mainly occurs in the fractionation columns, and (to a first approximation) one can neglect the extra broadening introduced by the injector, the detector cells, and the interdetector capillaries. In this case, any generic chromatogram $s_k(V)$ is broadened by a common BB function $g(V, \overline{V})$ as follows:[15,16]

$$s_k(V) = K_k \int_{V_1^c}^{V_2^c} g(V, \overline{V}) s_k^c(\overline{V}) d\overline{V}$$

$$(k = DR, LS, IV, OS) \qquad (3)$$

Note that Eq. (3) reduces to Eq. (1) for $K_{DR} = 1$, $s_{DR} \equiv w$, and $s_{DR}^c \equiv w^c$. The instantaneous weight-, viscosity-, and number-averaged molar masses [$M_w(V)$, $M_v(V)$, and $M_n(V)$, respectively] are obtained from the signal ratios:[14,17,18]

$$M_w(V) = K_{LS} \frac{s_{LS}(V)}{w(V)}; \quad M_v(V) = K_{IV} \left[\frac{s_{IV}(V)}{w(V)} \right]^{1/a};$$

$$M_n(V) = K_{OS} \frac{w(V)}{s_{OS}(V)} \quad [M_w(V) \geq M_v(V) \geq M_n(V)]$$

$$(4)$$

Unlike $M(V)$ of Eq. (2), $M_w(V)$, $M_v(V)$, and $M_n(V)$ now depend on the analyzed MMD; therefore, $\log M_w(V)$, $\log M_v(V)$, and $\log M_n(V)$ can be thought of as "biased" or ad hoc molar mass calibrations. Even if $\log M_w(V)$, $\log M_v(V)$, and $\log M_n(V)$ were perfectly accurate, the MMDs directly derived from the (broadened) mass chromatogram and any of such calibrations are distorted with respect to the true $w^c(\log M)$. In spite of BB, if an instantaneous variable is accurately estimated, then its corresponding global variable is also exact. Thus, the MMD represented by $w(\log M_w)$ produces an exact global $\overline{M}_w$ but an overestimated global $\overline{M}_n$, while the MMD represented by $w(\log M_n)$ produces an exact global $\overline{M}_n$ but an underestimated $\overline{M}_w$. In both cases, the global polydispersity is underestimated.[19,20] The previous observation is generalized to any other global average obtained from multidetection SEC. For example, if an instantaneous copolymer composition is accurately calculated from a signals ratio, then the global composition will also be accurate, in spite of BB.[21]

The correction for BB in SEC is still a matter of active research, and a "state of the art" review has recently been published.[14] At present, the authors are participating in an IUPAC project entitled "Data Treatment in the Size Exclusion Chromatography of Polymers"; one of the project objectives is the evaluation and standardization of existing BB correction techniques. Thus, the present article can be considered a first contribution toward that aim.

CORRECTION FOR MASS CHROMATOGRAMS WITH INDEPENDENT CALIBRATIONS

Consider the direct inversion of Eq. (1), i.e., the calculation of $w^c(V)$ from the knowledge of $w(V)$ and $g(V, \overline{V})$ First, let us transform Eq. (1) into the following equivalent discrete expression:

$$\mathbf{w} = \mathbf{G}\mathbf{w}^c \qquad (5)$$

where $\mathbf{w}$ is an $(m \times 1)$-column vector containing the heights of $w(V)$ sampled at regular ΔV intervals in the elution volumes range $[V_1 - V_m]$; $\mathbf{w}^c$ is a $(p \times 1)$-column vector containing the heights of $w^c(V)$ calculated at the same elution volumes but in the narrower range $[\overline{V}_1 - \overline{V}_p]$; and $\mathbf{G}$ is an $(m \times p)$ rectangular matrix representing $g(V, \overline{V})$ in the range $[\overline{V}_1 - \overline{V}_p]$. A typical sampling interval is $\Delta V = 0.1 \, \text{ml}$.

Specification of Matrix G[22]

For a successful inversion of Eq. (5), it is important to adequately define matrix $\mathbf{G}$. First, it is recommendable to adjust $g(V, \overline{V})$ with a continuous analytical expression, and then to calculate the heights of the individual $g(V)$ functions from that expression. Many analytical functions (e.g., a Gaussian distribution) never strictly drop to zero, and this would produce "full" $\mathbf{G}$ matrixes with positive and nonzero elements. Instead, it is preferable to set to 0 all of the "almost-null" elements of $\mathbf{G}$ (e.g., those smaller than 1% of the maximum). Also, choose $\mathbf{G}$ of minimal dimensions, in the sense that: 1) its p columns must strictly cover the range of the corrected chromatogram $[\overline{V}_1 - \overline{V}_p]$; and 2) its m rows must strictly cover the range of the measured chromatogram $[V_1 - V_m]$.[22] Since, in general, the BB functions are skewed and nonuniform, it is convenient to specify each individual $g(V)$ to contain $(c + 1 + d)$ nonzero points, where c and d are the number of points before and after $\overline{V}$, respectively. Thus, the number of columns of $\mathbf{G}$ results: $p = m - c - d$, and the

matrix is defined as follows:

$$\mathbf{G} = \begin{bmatrix} g(V_1, \overline{V}_1) & \cdots & 0 & & 0 \\ \vdots & \ddots & 0 & & \vdots \\ g(V_{c+1}, \overline{V}_1) & & g(V_j, \overline{V}_j) & & \\ \vdots & \ddots & \vdots & \ddots & 0 \\ g(V_{c+1+d}, \overline{V}_1) & & g(V_{c+j}, \overline{V}_j) & & g(V_p, \overline{V}_p) \\ 0 & \ddots & \vdots & \ddots & \vdots \\ \vdots & & g(V_{c+j+d}, \overline{V}_j) & & g(V_{c+p}, \overline{V}_p) \\ & & 0 & \ddots & \vdots \\ 0 & \cdots & 0 & \cdots & g(V_m, \overline{V}_p) \end{bmatrix}$$

$$(m > p)$$

$$(6)$$

where each jth column of $\mathbf{G}$ contains $(c + 1 + d)$ nonzero heights of the discrete $g(V)$, with $\overline{V}_j = V_1 + (c + j - 1)\Delta V$. Note that by adopting $\overline{V}$ at the mode of $g(V)$, in each column, the largest element is c rows below the corresponding (j, j) "diagonal" element.

The direct inversion of Eq. (5) through, for example, the pseudoinverse $\hat{\mathbf{w}}^c = [\mathbf{G}^T\mathbf{G}]^{-1}\mathbf{G}^T\mathbf{w}$ (where "^" indicates estimated value) is not recommended because the square matrix $[\mathbf{G}^T\mathbf{G}]$ is generally ill-conditioned, and this produces highly oscillatory solutions with negative peaks. The propagation of errors is determined by: 1) the condition number of $\mathbf{G}^T\mathbf{G}$ (i.e., the ratio between the largest and the smallest eigenvalue of $\mathbf{G}^T\mathbf{G}$); and 2) the type and amplitude of the noise that contaminates $w(V)$.

In what follows, several BB correction techniques are presented and evaluated. To illustrate the effect of BB on the MMD, a molar mass calibration is adopted. The evaluated techniques are classified into two groups: 1) Methods I–III, which numerically invert Eq. (1) prior to calculating the MMD; and 2) Methods IV and V, which (avoiding the numerical inversion) calculate the corrected MMD in a single step. Methods I–III are more general, in the sense that they admit nonuniform and skewed BB functions. In contrast, Methods IV and V are strictly applicable to Gaussian chromatograms with Gaussian BB functions. Methods I–III have been developed to improve the (highly oscillatory) solution of the direct pseudoinverse, but at the cost of requiring an algorithm adjustment. The solutions of Methods I–III normally involve a trade-off between an excessively "rich" corrected chromatogram (with high-frequency oscillations and negative peaks) and an excessively smoothened solution (where some of the high-frequency components of the corrected chromatogram are lost).

Method I: Difference Function[23]

This iterative procedure was originally presented as Method 1 by Ishige, Lee, and Hamielec[23] and is based on the following recursive equation:

$$\Delta_i \mathbf{w} = \Delta_{i-1}\mathbf{w} - \mathbf{G}\,\Delta_{i-1}\mathbf{w}; \quad \text{with} \quad \Delta_0 \mathbf{w} = \mathbf{w} \qquad (7a)$$
$$(i = 1, 2, \ldots)$$

where i is the iteration step. After a few r iterations, $\Delta_i \mathbf{w}$ tends to almost zero, and at that point the corrected chromatogram is obtained from

$$\hat{\mathbf{w}}^c = \sum_{i=1}^{r} \Delta_i \mathbf{w} \qquad (7b)$$

Method II: Singular Value Decomposition[24]

The final expression of this least-squares estimation procedure is

$$\hat{\mathbf{w}}^c = \sum_{k=1}^{r} \frac{\mathbf{u}_k^{\mathrm{T}}\mathbf{w}}{\sigma_k}\mathbf{v}_k \quad (r \le p) \qquad (8)$$
$$(\sigma_1 \ge \sigma_2 \ge \cdots \ge \sigma_r \ge \cdots \ge \sigma_p \ge 0)$$

where $\mathbf{u}_k$ and $\mathbf{v}_k$ are the eigenvectors of $\mathbf{GG}^{\mathrm{T}}$ and $\mathbf{G}^{\mathrm{T}}\mathbf{G}$, respectively; σ_k are the singular values[24] of $\mathbf{G}$; and p is the (full) rank of $\mathbf{G}$. In Eq. (8), the number of "effective" summation terms is limited to r. The reason for discarding the lower σ_k's is to avoid amplifying the measurement noise. The lowest admissible σ_r is selected to slightly exceed the inverse of the lowest signal-to-noise ratio (normally encountered at the chromatogram tails).

Method III: Kalman Filter[25]

This fast and effective digital algorithm is based on a linear stochastic model that is equivalent to Eq. (1). The theoretical background is beyond the scope of the present article, and some knowledge on basic Kalman filtering theory is necessary for an adequate adjustment of the algorithm.[24–26] The adjustment involves estimating the variances of the measurement noise and of the expected solution.

Method IV: Rotation of the Linear Calibration[27]

Several (rather restrictive) conditions are here imposed: 1) The true mass chromatogram $w^c(V)$ is Gaussian (for example, because it corresponds to a Wesslau MMD and a linear calibration); 2) the BB function is uniform and Gaussian; and 3) the molar mass calibration is linear. Under these conditions, the ad hoc calibrations $\log M_w(V)$, $\log M_v(V)$, and $\log M_n(V)$ are all linear and rotated counterclockwise with respect to the unbiased linear calibration $\log M(V)$.[20,27,28] For a non-Gaussian chromatogram, the ad hoc calibrations are generally nonlinear but less steep than $\log M(V)$.[19,22,29] The method aims at recuperating unbiased estimates of the global averages $\overline{M}_n$ and $\overline{M}_w$ from an "effective" linear molar mass calibration defined by $M(V)|_{\mathrm{IV}} = D_1' \exp(-D_2'V)$, with[27]

$$D_1' = D_1 \exp\left\{\frac{D_2\sigma_g^2[D_2(\sigma_w^2 - \sigma_g^2) - 2\overline{V}]}{2\sigma_w^2}\right\} \qquad (9a)$$

$$D_2' = D_2 \exp\left(1 - \frac{\sigma_g^2}{\sigma_w^2}\right) \qquad (9b)$$

where σ_g^2 and σ_w^2 are the variances of $g(V)$ and $w(V)$, respectively; and $\overline{V}$ is the retention volume of the chromatogram peak. Even though the method is strictly applicable to Gaussian chromatograms, it will be tested here on a non-Gaussian chromatogram (but still satisfying the other requirements of a linear calibration and a uniform and Gaussian BB).

Method V: Approximate "Analytical" Solution[30]

This approach is again based on the following (rather strict) assumptions: 1) The BB function is Gaussian (but generally nonuniform), of variance $\sigma_g^2(V)$; and 2) at each retention volume, the integrand of Eq. (1) can be approximated by the product of the measured chromatogram $w(V)$ and a Gaussian "correction" function of variance $\sigma_0^2(V)$ and averages $\overline{V}(V)$. The molar mass calibration may be nonlinear, and is given by $M(V) = D_1(V) \exp[-D_2(V)V]$. The corrected chromatogram is obtained from

$$\hat{w}^c(V) = w(V)\left[\frac{\sigma_g(V)}{\sigma_0(V)}\right]\exp\left[-\frac{[V - \overline{V}(V)]^2}{2\sigma_0^2(V)}\right] \qquad (10a)$$

with

$$\overline{V}(V) = V + \frac{1}{D_2(V)}$$
$$\times \ln\left\{\frac{w[V + D_2(V)\sigma_g^2(V)]}{\sqrt{w[V - D_2(V)\sigma_g^2(V)]w[V + D_2(V)\sigma_g^2(V)]}}\right\} \qquad (10b)$$

$$\sigma_0^2(V) = \sigma^2(V) + \frac{1}{D_2^2(V)}$$
$$\times \ln\left\{\frac{w[V - D_2(V)\sigma_g^2(V)]w[V + D_2(V)\sigma_g^2(V)]}{w^2(V)}\right\} \qquad (10c)$$

Evaluation Example

Correction methods are normally evaluated on numerical examples. This is because their real (or true) solutions are a priori known, and therefore the quality of their recuperations is properly quantified. In contrast, in real measurements, the true corrected chromatograms (and/or the true MMDs) are never exactly known. Consider, in what follows, a synthetic example that has been previously attempted on several occasions.[25,31–33]

The raw data are the corrected chromatogram $w^c(V)$ of Fig. 1A, the (uniform) broadening function $g(V)$ of Fig. 1A, and the linear calibration $\log M(V)$ of Fig. 1B. By convolution of $w^c(V)$ and $g(V)$, a noise-free "measurement" was first obtained. Then, this noise-free function was rounded to the last integer;[33] this procedure is equivalent to adding a zero-mean random noise of uniform probability distribution in the range ± 0.5. The resulting "chromatogram" is $w(V)$ of Fig. 1A. Note that the multimodality of $w^c(V)$ is lost in $w(V)$. This example is particularly demanding because $w^c(V)$ is multipeaked, and the variance of $w^c(V)$ is similar to that of $g(V)$.

In previous publications,[25,31–33] only the ability of several inversion algorithms was evaluated, but not the effect of BB on the MMD. Here, the methods are compared on the basis of their performance in recuperating the true MMD. From $\log M(V)$ and $w^c(V)$, the true MMD $w^c(\log M)$ of Figs. 1C–F is obtained. The aim is to estimate $w^c(\log M)$ from $w(V)$, $g(V)$, and $\log M(V)$. Note that the selection of a uniform and Gaussian BB is not an impediment to adequately evaluating the (more comprehensive) Methods I–III.

In Figs. 1C–F, the MMDs recuperated through Methods I–V are compared with the real distribution; Table 1 presents the real and estimated average molar masses and polydispersities. In Method I, the best results were found after only 4 iterations (Fig. 1C). In Method II, the signal-to-noise ratio at the chromatogram tails suggested truncation of the summation of Eq. (8) at $r = 16$ (while the full rank of $\mathbf{G}$ is $p = 61$). The resulting solution exhibits a negative oscillation (Fig. 1D); and for comparison, the less "rich" solution with $r = 9$ is also presented. The Kalman filter of Method III was adjusted as follows: 1) The measurement noise variance was assumed time invariant, and estimated from the baseline noise; and 2) the solution variance was assumed time varying, and estimated by simply squaring the measured chromatogram heights (Fig. 1E). For Method IV, the "effective" linear calibration was calculated through Eqs. (9), and is presented in Fig. 1B. In Method V, it was verified that (for a linear calibration) the solution becomes almost independent of $D_2(V)$.[30] The solutions of Methods IV and V are shown in Fig. 1F.

In relation to Methods IV and V, the instantaneous MMDs were simulated with the aim of calculating the (noise-free) calibrations $\log M_n(V)$ and $\log M_w(V)$ (Fig. 1B). These functions were obtained from the (noise-free or nontruncated) mass chromatogram in order to illustrate their "true" shapes in the complete range of the measured chromatogram. The resulting ad hoc calibrations are nonlinear, generally less steep than $\log M(V)$, and close to the "effective" linear calibration $\log M(V)|_{IV}$ (Fig. 1B).

The following is observed. Only Method III (and to a lesser extent Method II with $r = 16$) was capable of recuperating the fine details of the true MMD, while all the other techniques yielded unimodal solutions. Methods III and II considerably improve the highly oscillatory direct pseudoinverse solution (not presented here for space reasons). The recuperated average molar masses of Table 1 are in all cases quite acceptable.

CORRECTION METHODS FOR MOLAR MASS SENSITIVE DETECTORS

Now, we wish to determine an unbiased MMD from measurements of mass and molar mass sensitive detectors. Rewrite Eq. (3) as follows:

$$\mathbf{s}_k = K_k \mathbf{G} \mathbf{s}_k^c \quad (k = DR, LS, IV, OS) \tag{11}$$

where $\mathbf{s}_k$ is an $(m \times 1)$-column vector containing the nonzero heights of $s_k(V)$, sampled at regular ΔV intervals; $\mathbf{s}_k^c$ is a $(p \times 1)$-column vector containing the nonzero heights of $s_k^c(V)$; and $\mathbf{G}$ is the $(m \times p)$ rectangular matrix of Eq. (6).

Inversion Methods

Two methods are described.[15,22] They both aim at correcting the raw chromatograms for BB prior to calculating the MMD, and are strictly applicable to linear homopolymers. Netopilík[15] proposed an iterative procedure for simultaneously estimating the MMD and the standard deviation of a uniform and Gaussian BB function (σ_g). The procedure is as follows: 1) Select a σ_g value; 2) estimate $w^c(V)$ and $s_k^c(V)$ by inversion of Eqs. (1) and (3), respectively; 3) use Eq. (2) for estimating a molar mass calibration $\log \widehat{M}(V)$; and 4) iterate until the slope of $\log \widehat{M}(V)$ coincides with that of an (independently determined) molar mass calibration. The method was theoretically tested on a narrow Schulz–Zimm MMD; while the original distribution was well recuperated, the standard deviation differed considerably from its original value.[15]

More recently, Vega and Meira[22] proposed a numerical method that does not impose any restriction

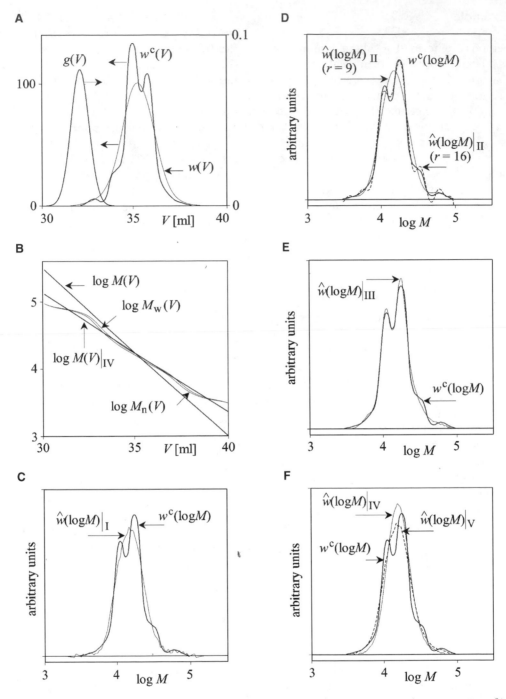

Fig. 1 Simulated example of DR detection and molar mass calibration. (A) "True" mass chromatogram $w^c(V)$; uniform BB function $g(V)$; and resulting "measured" chromatogram $w(V)$. (B) "True" molar mass calibration $\log M(V)$; rotated "effective" calibration according to Method IV[27] $\log M(V)|_{IV}$; and ad hoc calibrations assuming perfect molar mass sensors $\log M_n(V)$ and $\log M_w(V)$. (C)–(F) Comparison between the true MMD $w^c(\log M)$ and its estimates according to Methods I,[23] II,[24] III,[25] IV,[27] and V,[30] respectively.

on the shapes of the MMD or the BB function, and only implies a linear calibration in the range of the measured chromatograms. This last requirement is generally satisfied (especially if the MMD is narrow), and it is also easily verified from an independent calibration with narrow standards. The method is as

follows: 1) Estimate $w^c(V)$ and $s_k{}^c(V)$ by inversion of Eqs. (1) and (3), respectively; 2) calculate an unbiased molar mass calibration $\log \widehat{M}(V)$ through Eq. (2), and use its mid-chromatogram region to adjust a linear calibration $\log \widehat{M}_{lin.}(V)$; and 3) from $\hat{w}^c(V)$ and $\log \widehat{M}_{lin.}(V)$ estimate the unbiased MMD

Table 1 Simulated example that assumes a mass chromatogram, a linear calibration, and a Gaussian BB function: "true" and recuperated average molar masses

	"True" values	Without BB correction	I ($i = 4$)	II ($r = 9$)	($r = 16$)	III	IV	V
$\overline{M}_n$	13 975	13 464	14 029	13 993	14 033	14 084	14 871	14 038
$\overline{M}_w$	17 342	18 041	17 315	17 310	17 313	17 156	17 224	17 335
$\overline{M}_w/\overline{M}_n$	1.241	1.340	1.234	1.237	1.234	1.218	1.158	1.235

(Correction method no., with II spanning $(r = 9)$ and $(r = 16)$.)

$\hat{w}^c(\log \widehat{M}_{lin.})$.[22] Note the following: a) Any of the previously described Methods I–III can be used for solving step 1; b) by extrapolating a linear calibration toward the chromatogram tails, the technique also solves the problem of an oscillatory ad hoc calibration; and c) the resulting $\log \widehat{M}_{lin.}(V)$ can be verified with an independent calibration with narrow standards of the analyzed polymer.

Reconsider the numerical example of Vega and Meira.[22] The basic raw data are: 1) the "true" or corrected mass chromatogram $w^c(V)$ of Fig. 2A; 2) the molar mass calibration $\log M(V)$ of Fig. 2C; and 3) the nonuniform and skewed broadening function $g(V, \overline{V})$ of Fig. 2A. All of these functions are discrete, with their heights sampled every $\Delta V = 0.1$ ml. The true mass chromatogram contains $p = 70$ nonzero points in the range $[\overline{V}_1 - \overline{V}_p]$, and it is presented in Fig. 2A. Then, the "true" molar mass chromatogram was calculated by assuming the relationship: $s_{LS}^c(V) = 0.02[M(V)w^c(V)]$ (Fig. 2B).

From $w^c(V)$ and $\log M(V)$, the "true" MMD $w^c(\log M)$ of Fig. 2D was obtained, and its average molar masses are shown in the second row of Table 2. The $\log M$ values of Fig. 2D vertically correspond (through the linear calibration) with the V values of Figs. 2A–C.

The nonuniform $g(V, \overline{V})$ is represented by an EMG of constant skewness, variable standard deviation, and $\overline{V}$ averages adopted at the peaks of the individual $g(V)$ functions. Each $g(V)$ exhibits 39 nonzero points (with $c = 10$ points before the maximum and $d = 28$ points after the maximum). In Fig. 2A, only the two limiting and one intermediate $g(V)$ functions are presented. The (noise-free) mass and molar mass chromatograms were calculated through Eqs. (1) and (3). Then, the "measured" chromatograms $w(V)$ and $s_{LS}(V)$ of Figs. 2A,B were obtained by adding a zero-mean Gaussian noise to the noise-free chromatograms. The broadened chromatograms contain $m = 70 + 39 - 1$ nonzero points. From $w(V)$ and $\log M(V)$, the (broadened) MMD $w(\log M)$ of Fig. 2A was obtained; its average molar masses are given in the third row of Table 2. Both averages are underestimated, while the global polydispersity is overestimated. The underestimation of

the average molar masses is a result of having adopted the $\overline{V}$ averages at the maxima of skewed $g(V)$ functions.

At each retention volume of $w(V)$, the instantaneous MMDs in the detector cells were calculated by considering the contributions (toward that V) of all the hypothetical molecular species in the distribution, as determined by the discrete $w^c(V)$. From such instantaneous distributions, the ad hoc nonlinear calibrations $\log M_n(V)$ and $\log M_w(V)$ were calculated.[22] From $w(V)$, $\log M_n(V)$, and $\log M_w(V)$, the MMD estimates $w(\log M_n)$ and $w(\log M_w)$ were obtained; their averages are presented in Table 2. As expected, $w(\log M_n)$ accurately estimates $\overline{M}_n$ but underestimates $\overline{M}_w$, while $w(\log M_w)$ accurately estimates $\overline{M}_w$ but overestimates $\overline{M}_n$. Thus, the global polydispersity is underestimated in both cases.

In a standard data treatment without BB correction, $M_w(V)$ would have been directly estimated from $\hat{M}_w(V) = s_{LS}(V)/[0.02w(V)]$. Due to the measurement noise, the resulting $\log \hat{M}_w(V)$ of Fig. 2C is oscillatory at the chromatogram tails, and these oscillations make it impossible to recuperate an MMD.

The proposed procedure was applied to the noisy chromatograms $w(V)$ and $s_{LS}(V)$. The dimension of **G** is $(m \times p) = (108 \times 70)$; the inversions were carried out through the singular value decomposition expression of Eq. (8). The algorithm was adjusted with the criterion of producing smooth solutions with minimal negative peaks, yielding $r = 14$ for the mass chromatogram and $r = 12$ for the molar mass chromatogram. The final estimates were $\hat{w}^c(V)$ and $\hat{s}_{LS}^c(V)$ of Figs. 2A,B. These functions are smooth and almost coincident with the true $w^c(V)$ and $s_{LS}^c(V)$.

The resulting (unbiased) calibration of Fig. 2C, $\log \widehat{M}(V)$, almost overlaps the "true" $\log M(V)$ in the mid-chromatogram region, while it diverges at the tails. The linear calibration $\log \widehat{M}_{lin.}(V)$ was obtained from the points of $\log \widehat{M}(V)$ contained in the "adjustment range" of Fig. 2C. Finally, the unbiased distribution $\hat{w}^c(\log \widehat{M}_{lin.})$ of Fig. 2D was obtained from $\hat{w}^c(V)$ and $\log \widehat{M}_{lin.}(V)$. This distribution is smooth and close to the "true" $w_c(\log M)$. Accordingly, the estimated average molar masses are very close to the real values (Table 2).

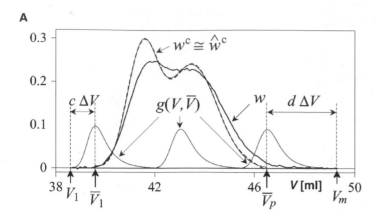

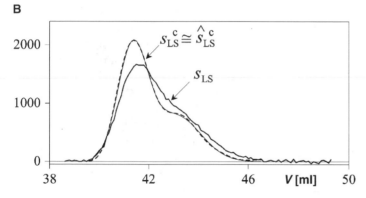

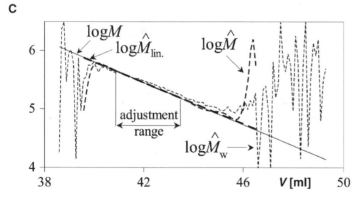

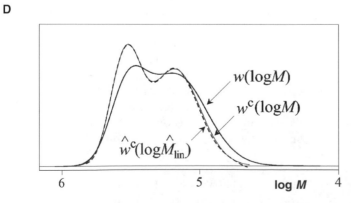

Fig. 2 Simulated example of DR/LS detection (after Ref.[22]). (A) "True" and "measured" mass chromatograms [$w^c(V)$, $w(V)$]; three samples of the BB function $g(V, \overline{V})$; and estimated corrected chromatogram [$\hat{w}^c(V)$]. (B) "True" and "measured" molar mass chromatograms [$s_{LS}{}^c(V), s_{LS}(V)$]; and estimated corrected molar mass chromatogram $\hat{s}_{LS}{}^c(V)$. (C) Unbiased linear calibration [$\log M(V)$]; estimated ad hoc calibration [$\log \hat{M}_w(V)$]; estimated unbiased calibration [$\log \hat{M}(V)$], and estimated linear unbiased calibration [$\log \hat{M}_{lin.}(V)$]. (D) "True" MMD [$w^c(\log M)$]; MMD estimate obtained from $w(V)$ and $\log M(V)$ [$w(\log M)$]; and MMD estimate obtained from $\hat{w}^c(V)$ and $\log \hat{M}_{lin.}(V)$ [$\hat{w}^c(\log \hat{M}_{lin.})$].

Table 2 Simulated example for molar mass sensitive detectors: "true" and recuperated average molar masses from several MMDs

MMD	$\overline{M}_n$	$\overline{M}_w$	$\overline{M}_w/\overline{M}_n$
$w^c(\log M)^a$	182 000	242 000	1.33
$w(\log M)^b$	160 000	224 000	1.40
$w(\log M_n)^c$	182 000	233 000	1.28
$w(\log M_w)^d$	191 000	242 000	1.27
$\hat{w}^c(\log \widehat{M}_{lin.})^e$	185 000	243 000	1.31

[a]"True" base distribution (Fig. 2D).
[b]Based on the linear calibration, without BB correction (Fig. 2D).
[C]Obtained from DR/OS detection, without BB correction.
[D]Obtained from DR/LS detection, without BB correction.
[E]Obtained from DR/LS detection, with BB correction (Fig. 2D).

Direct Calculation of the Corrected MMD

The interdetector volume compensation generally involves shifting the (leading) molar mass signal toward higher elution volumes. Independent of this compensation, a BB correction procedure has been proposed, which calculates the MMD in a single step by appropriately reducing the normal interdetector volume shift.[34] The procedure is equivalent to rotating the linear molar mass calibration counterclockwise.[28,35,36] Therefore, it is based on the following (rather strict) assumptions: 1) Both the (uniform) BB function and the measured mass chromatogram are Gaussian functions of (known) standard deviations σ_g and σ_w, respectively; 2) the calibration is linear and given by $M(V) = D_1 \exp(-D_2 V)$; and 3) the interdetector volume introduces a pure signal shift, but no additional BB. In the case of an LS/DR combination, the LS signal must suffer a (secondary) shift that involves a small reduction in the normal lag. This secondary lag reduction is given by

$$\Delta V_{LS} = \sigma_{w^c}(\sigma_w - \sigma_{w^c})D_2 \qquad (12a)$$

with

$$\sigma_{w^c} = (\sigma_w^2 - \sigma_g^2)^{1/2} \qquad (12b)$$

where σ_{w^c} is the standard deviation of the corrected chromatogram.

The commercially available molar mass sensitive detectors do include a correction for BB in their software. Unfortunately, the applied correction procedures are not fully disclosed, but they seem to involve an interdetector volume readjustment. For example, the Viscotek™ Model 200 detector combines a DR in parallel with a specific viscometer. First, the (mass and molar mass) chromatograms of several narrow standards must be measured to determine the interdetector volume and a (uniform EMG) BB function. Then, the MMD is corrected for BB in an unspecified manner. Similarly, Wyatt Corp. has recently introduced a patented BB correction procedure for their triple-detector system (multiangle LS, DR, and specific viscosity sensors).

CONCLUSIONS

Band broadening correction in SEC is still not a totally resolved issue, even when the MMD of a linear homopolymer is determined with a mass detector and a molar mass calibration. Fortunately, modern SEC columns are highly efficient, and the correction for BB is mainly limited to the case of narrowly distributed polymers. Even in the presence of BB, if an instantaneous quality variable is accurately measured, its corresponding global average will also be accurate.

Numerical inversion techniques aim at correcting the raw chromatograms prior to determining the MMD or any other polymer quality characteristics. Their main advantage is that they admit arbitrary shapes for the chromatograms or the BB function. Their disadvantage, however, is the ill-posedness of numerical inversions, which amplify the measurement noise. With molar mass sensitive detectors, two independent inversions are required prior to calculation of the molar masses. From the comparison of Methods I–III, Method II (a singular value decomposition technique) has shown a good compromise between a reasonably good solution and a relatively simple adjustment procedure. To improve the ill-conditioned nature of the numerical inversions, it is important to set to zero all the ultra-low elements of the BB matrix (normally placed at its upper-right and lower-left corners).

The techniques that avoid numerical inversions, correct the MMDs in a single step, and are based on either: 1) rotating the linear calibration (when only a mass chromatogram and a linear calibration are available); or 2) modifying the interdetector volume shift (when molar mass sensitive detectors are employed). Their main advantage is that they produce smooth and unique solutions. Their limitation, however, is that they produce only approximate solutions.

In general, the BB function seems to be moderately uniform in the linear calibration range, but definitely skewed toward the higher retention volumes. Its determination is only simple for low molar masses. In general, the BB function would be simpler to specify if the manufacturers of narrow standards provided the true MMDs of their samples.

A proper correction for BB and other sources of error seems essential for quantitative determinations in SEC. The following developments are expected in the future: (1) simpler techniques for determining the BB function; (2) a validation of several BB algorithms on real experimental data; and (3) a standardization of the "best" BB correction procedures (possibly, a trade-off between accuracy and simplicity).

ACKNOWLEDGMENTS

This work was carried out in the framework of Project 2003-023-2-G.Meira (IUPAC): "Data Treatment in the Size Exclusion Chromatography of Polymers," http://www.iupac.org/projects/2003/2003-023-2-400.html. Also, we are grateful for the financial support received from the following Argentine institutions: CONICET, Universidad Nacional del Litoral, and SECyT.

ARTICLE OF FURTHER INTEREST

Band Broadening in SEC, p. 163.

REFERENCES

1. Berek, D.; Marcinka, K. Gel chromatography. In *Separation Methods*; Deyl, Z., Ed.; Elsevier: Amsterdam, 1984; 271–299.
2. Hupe, K.; Jonker, R.; Rozing, G. Determination of band-spreading in high-performance liquid chromatographic instruments. J. Chromatogr. **1984**, *285*, 253–265.
3. Wyatt, P. Mean square radius of molecules and secondary instrumental broadening. J. Chromatogr. **1993**, *648*, 27–32.
4. Potschka, M. Mechanism of size-exclusion chromatography. I. Role of convection and obstructed diffusion in size-exclusion chromatography. J. Chromatogr. **1993**, *648*, 41–69.
5. Netopilík, M. Relations between the separation coefficient, longitudinal displacement and peak broadening in size exclusion chromatography of macromolecules. J. Chromatogr. A. **2002**, *978*, 109–117.
6. Dondi, F.; Cavazzini, A.; Remelli, M.; Felinger, A.; Martin, M. Stochastic theory of size exclusion chromatography by the characteristic function approach. J. Chromatogr. A. **2002**, *943*, 185–207.
7. Pasti, L.; Dondi, F.; van Hulst, M.; Schoenmakers, P.; Martin, M.; Felinger, A. Experimental validation of the stochastic theory of size exclusion chromatography: retention on single and coupled columns. Chromatographia **2003**, *57* (Suppl.), S171–S186.
8. Tung, L. Method of calculating molecular weight distribution function from gel permeation chromatograms. III. Application of the method. J. Appl. Polym. Sci. **1966**, *10*, 1271–1283.
9. Busnel, J.P.; Foucault, F.; Denis, L.; Lee, W.; Chang, T. Investigation and interpretation of band broadening in size exclusion chromatography. J. Chromatogr. A. **2001**, *930*, 61–71.
10. Belenkii, B.; Vilenchik, L. General theory of chromatography. In *Modern Liquid Chromatography of Macromolecules*; Journal of Chromatography Library; Elsevier: Amsterdam, 1983; Vol. 25, 1–67.
11. Lederer, K.; Imrich-Schwarz, G.; Dunky, M. Simultaneous calibration of separation and axial dispersion in size exclusion chromatography coupled with light scattering. J. Appl. Polym. Sci. **1986**, *32*, 4751–4760.
12. Billiani, J.; Rois, G.; Lederer, K. A new procedure for simultaneous calibration of separation and axial dispersion in SEC. Chromatographia **1988**, *26*, 372–376.
13. Schnöll-Bitai, I. The direct determination of axial dispersion in size exclusion chromatography based on Poissonian chain length distributions. Chromatographia **2003**, *58*, 375–380.
14. Baumgarten, J.; Busnel, J.; Meira, G. Band broadening in size exclusion chromatography of polymers. State of the art and some novel solutions. J. Liq. Chromatogr. & Relat. Technol. **2002**, *25* (13–15), 1967–2001.
15. Netopilík, M. Correction for axial dispersion in gel permeation chromatography with a detector of molar masses. Polym. Bull. **1982**, *7*, 575–582.
16. Hamielec, A. Correction for axial dispersion. In *Steric Exclusion Liquid Chromatography of Polymers*; Chromatographic Science; Janča, J., Ed.; Marcel Dekker, Inc.: New York, 1984; Vol. 25, 117–160.
17. Jackson, C.; Barth, H. Molecular weight sensitive detectors for size exclusion chromatography. In *Handbook of Size Exclusion Chromatography and Related Techniques*; 2nd Ed.; Chromatographic Science Series; Wu, Ch., Ed.; Marcel Dekker, Inc.: New York, 2004; Vol. 91, 99–138.
18. Lehmann, U.; Köhler, W.; Albrecht, W. SEC absolute molar mass detection by online membrane osmometry. Macromolecules **1996**, *29*, 3212–3215.
19. Prougenes, P.; Berek, D.; Meira, G. Size exclusion chromatography of polymers with molar mass detection. Computer simulation study

on instrumental broadening biases and proposed correction method. Polymer **1998**, *40*, 117–124.

20. Netopilík, M. Effect of local polydispersity in size exclusion chromatography with dual detection. J. Chromatogr. A. **1998**, *793*, 21–30.

21. Meira, G.; Vega, J. Characterization of copolymers by size exclusion chromatography. In *Handbook of Size Exclusion Chromatography and Related Techniques*; 2nd Ed.; Chromatographic Science Series; Wu, Ch., Ed.; Marcel Dekker, Inc.: New York, 2004; Vol. 91, 139–156.

22. Vega, J.; Meira, G. SEC of simple polymers with molar mass detection in presence of instrumental broadening. Computer simulation study on the calculation of unbiased molecular weight distribution. J. Liq. Chromatogr. & Relat. Technol. **2001**, *24* (7), 901–919.

23. Ishige, T.; Lee, S.; Hamielec, A. Solution of Tung's axial dispersion equation by numerical techniques. J. Appl. Polym. Sci. **1971**, *15*, 1607–1622.

24. Mendel, J. Least-squares estimation: singular-value decomposition. In *Lessons in Estimation Theory for Signal Processing, Communications, and Control*; Prentice Hall: New Jersey, 1995; 44–57.

25. Alba, D.; Meira, G. Inverse optimal filtering method for the instrumental broadening in SEC. J. Liq. Chromatogr. **1984**, *7*, 2833–2862.

26. Felinger, A. Signal enhancement. In *Data Analysis and Signal Processing in Chromatography*; Data Handling in Science and Technology; Elsevier: Amsterdam, 1998; Vol. 21, 143–181.

27. Jackson, C.; Yau, W. Computer simulation study of size exclusion chromatography with simultaneous viscometry and light scattering measurements. J. Chromatogr. **1993**, *645*, 209–217.

28. Netopilík, M. Combined effect of interdetector volume and peak spreading in size exclusion chromatography with dual detection. Polymer **1997**, *38*, 127–130.

29. Yau, W.; Stoklosa, H.; Bly, D. Calibration and molecular weight calculations in GPC using a new practical method for dispersion correction—GPCV2. J. Appl. Polym. Sci. **1977**, *21*, 1911–1920.

30. Hamielec, A.; Ederer, H.; Ebert, K. Size exclusion chromatography of complex polymers. Generalized analytical corrections for imperfect resolution. J. Liq. Chromatogr. **1981**, *4*, 1697–1707.

31. Chang, K.; Huang, R. A new method for calculating and correcting molecular weight distributions from permeation chromatography. J. Appl. Polym. Sci. **1969**, *13*, 1459–1471.

32. Gugliotta, L.; Alba, D.; Meira, G. Correction for instrumental broadening in SEC through a stochastic matrix approach based on Wiener filtering theory. In *Detection and Data Analysis in Size Exclusion Chromatography*; ACS Symposium Series No. 352; Provder, T., Ed.; American Chemical Society: Washington, 1987; 287–298.

33. Gugliotta, L.; Vega, J.; Meira, G. Instrumental broadening correction in size exclusion chromatography. Comparison of several deconvolution techniques. J. Liq. Chromatogr. **1990**, *13*, 1671–1708.

34. Jackson, C. Evaluation of the "effective volume shift" method for axial dispersion corrections in multi-detector size exclusion chromatography. Polymer **1999**, *40*, 3735–3742.

35. Cheung, P.; Lew, R.; Balke, S.; Mourey, T. SEC–viscometer detector systems. II. Resolution correction and determination of interdetector volume. J. Appl. Polym. Sci. **1993**, *47*, 1701–1706.

36. Netopilík, M. Effect of interdetector peak broadening and volume in size exclusion chromatography with dual viscometric-concentration detection. J. Chromatogr. A. **1998**, *809*, 1–11.

Band Broadening in Capillary Electrophoresis

Jetse C. Reijenga
Eindhoven University of Technology, Eindhoven, The Netherlands

INTRODUCTION

As in chromatography, band broadening in capillary electrophoresis (CE) is determined by a number of instrumental and sample parameters and has a negative effect on detectability, due to dilution. Also, as in chromatographic techniques, the user can minimize some, but not all, of the parameters contributing to band broadening. In CE, injection and detection are generally on-column, so that band broadening is limited to on-column effects. As will be shown, several effects are similar in chromatography; others are specific for CE and, in particular, for the potential gradient as a driving force. General equations for CE in open systems are given where the relative contribution of electro-osmosis is given by the electromigration factor f_{em}, given by

$$f_{em} = \frac{\mu_{eff}}{\mu_{eff} + \mu_{EOF}}$$

in which is the effective mobility and μ_{EOF} is the electro-osmotic flow mobility. This electromigration factor is unity for systems with suppressed EOF.

The band-broadening contributions can be described in the form of a plate-height equation, where one usually assumes, as in chromatography, mutual independence of terms.

INJECTION

Band broadening due to injection is naturally proportional to the injection volume, relative to the capillary volume, but, in contrast to chromatography, sample stacking or destacking may decrease or respectively increase the injection band broadening thus defined. Without stacking or destacking, the following plate-height term can be used:

$$H_{inj} = \frac{\delta_{inj}^2}{12L_d}$$

in which δ_{inj} is the length in the capillary of the sample plug and L_d is the length of the capillary to the detector. Naturally, the above relationship can be rewritten in terms of sample and capillary volume, which are in the order of 10 nL and 1 µL, respectively.

DIFFUSION

As in chromatography, the effect of diffusion on band broadening is generally pronounced. It is directly proportional to the diffusion coefficient and the residence time between injection and detection. This effect can, consequently, be reduced by increasing the voltage, or by increasing the electro-osmotic flow, in cases where cations are analyzed at positive inlet polarity, where it further shortens the analysis times. The effect is less at lower temperatures (as the diffusion coefficient decreases approximately 2.5% per degree Celsius of temperature drop), but most significantly decreases with increasing molecular mass of the sample component.

$$H_{diff} = \frac{2Dt_m}{L_d}$$

in which L_d is the capillary length to the detector, D is the diffusion coefficient, and t_m is the migration time. Substituting the diffusion coefficient, using the Nernst-Einstein relation, yields

$$H_{diff} = \frac{2RTf_{em}}{z_{eff}EF}$$

in which R is the gas constant, T is the temperature, z_{eff} is the overall effective charge of the sample ion, E is the electric field strength, and F is the Faraday constant. In this relationship, z_{eff} and E, by definition, have opposite signs for negative values of f_{em} only.

DETECTION

The detector time constant and detector cell volume are both involved. The slit width along the length of the capillary is proportional to the latter. A value of 200 mm for the slit width in the case of 10^5 plates in a 370-mm capillary has negligible contribution to band broadening:

$$H_{slit} = \frac{\delta_{det}^2}{12L_d}$$

where δ_{det} is the detector slit width along the capillary axis. In cases of diode array detection, larger slit widths

Encyclopedia of Chromatography DOI: 10.1081/E-ECHR-120039890

are usually applied; this reduces the noise level but may affect the peak shape at high plate numbers ($>10^5$).

The contribution of the detector time constant τ is modeled by the following relation:

$$H_\tau = L_d \left(\frac{\tau}{t_m}\right)^2$$

A detector time constant τ of 0.2 s is generally safe.

THERMAL EFFECTS

In cases of a relatively high current density, power dissipation in the capillary may result in significant radial temperature profiles. The plate-height contribution is given by

$$H_{\text{ther}} = \frac{f_T^2 \kappa^2 E^5 R_i^6 z_{\text{eff}} F f_{\text{em}}}{1536 RT \lambda_s^2}$$

where f_T is the temperature coefficient for conductivity, κ is the specific conductivity of the buffer, E is the electric field strength, R_i is the capillary inner diameter, and λ_s is the thermal conductivity of the solution.

As the effective mobility increases with the temperature at approximately 2.5% per degree, radial mobility differences may accumulate to significant band-broadening effects. The effect increases with increasing current density and capillary inner diameter. In a 75-μm-inner diameter capillary, a power dissipation of 1–2 W/m is generally safe. This value is calculated by multiplying the voltage and the current and dividing by the capillary length. Under these conditions, the radial temperature profile in the capillary is less than 0.5°C and the contribution to peak broadening negligible. In the case of higher conductivity buffers (e.g., a pH 3 phosphate buffer), the power dissipation and temperature profile can be 10 times higher and the effect on peak broadening significant. It should be emphasized that more effective cooling has no effect on thermal band broadening; the only effect is decreased averaged temperature inside the capillary.

ELECTRO-OSMOTIC EFFECTS

Electro-osmosis in open systems is generally considered not to contribute to peak broadening. In hydrodynamically closed systems with nonsuppressed electroosmosis, or in cases of axially different electro-osmotic regimes, however, a considerable contribution may result. The corresponding plate-height term is

$$H_{\text{EOF}} = \frac{R_i^2 \zeta^2 \varepsilon^2 z_{\text{eff}} EF}{24 RT \eta^2 \mu_{\text{eff}}^2}$$

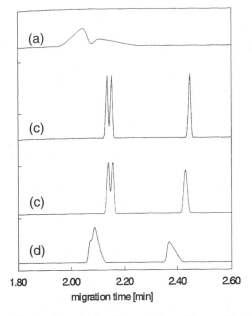

Fig. 1 Electrophoretic bandbroadening effects of benzoates as sample. Destacking trace a (1-mM sample in 1-mM buffer), stacking trace b (0.01-mM sample in 25-mM buffer), and trace c (1-mM sample in 25-mM buffer) and peak triangulation trace d (1-mM sample in 25-mM chloride buffer).

where ε is the dielectric constant and η is the local viscosity of the buffer at the plane of shear. This relationship shows that, in closed systems, the ζ-potential should be close to zero and that a viscosity increase near the capillary wall will be advantageous.

ELECTROPHORETIC EFFECTS

Peak broadening due to electrophoretic effects are generally proportional to the conductivity (and thus the ionic strength) of the sample solution, relative to that of the buffer. This effect is readily understood when considering that in the case of a high sample concentration, the electric field strength (and, consequently, the linear velocities) in the sample plug are much lower than in the adjacent buffer. Due to this, a dilution (destacking) of the sample occurs. This is illustrated in curve a in Fig. 1—the separation of a concentrated 1 mM solution benzenesulfonic, p-toluene sulfonic, and benzoic acid, dissolved in a buffer of 1 mM propionic acid/Tris to pH 8.

Alternatively, when injecting a low-conductivity (diluted, 0.01 mM) sample in a 25-mM buffer of same composition (curve b in Fig. 1–100 times amplified with respect to the others), the local field strength in the sample compartment is higher than in the adjacent buffer, resulting in a rapid focusing of ionic material at the sample-buffer interface (stacking), and resulting in very sharp sample injection plugs and high plate counts. This stacking takes place during the first second after switching on the high voltage. It may thus be advantageous to inject a larger volume of a more diluted sample for better efficiency. Choosing a higher-conductivity buffer also enhances the effect, where one has to consider that this may result in more pronounced band broadening due to other effects. Curve c in Fig. 1 shows that in such a high-conductivity buffer, even the 1-mM sample is separated to reasonable extent.

Peak symmetry is another important issue. Generally, capillary zone electrophoresis peaks are non-Gaussian and show nonsymmetry.This peak triangulation increases with increasing concentration overload. It is also proportional to the difference in effective mobility of the sample ion and the co-ion in the buffer. For instance, analyzing the same 1-mM sample mixture in a buffer consisting of, for example, 25 mM chloride/Tris to pH 8 will give triangular peaks (curve d in Fig. 1) because the effective mobility of benzoic acid is much lower than that of the buffer anion chloride: The buffer co-ion is not properly tuned to the sample component mobilities.

SUGGESTED FURTHER READING

Giddings, J.C. *Treatise on Analytical Chemistry*; Kolthoff, I.M., Elving, P.J., Eds.; John Wiley & Sons: New York, 1981; Part I, Vol. 5.

Hjertén, S. Chromatogr. Rev. **1967**, *9*, 122.

Jorgenson, J.W.; Lucaks, K.D. Science **1983**, *222*, 266.

Kenndler, E. J. Capillary Electrophoresis **1996**, *3* (4), 191.

Reijenga, J.C.; Kenndler, E. Computational simulation of migration and dispersion in free capillary zone electrophoresis, part I, Description of the theoretical model. J. Chromatogr. A. **1994**, *659*, 403.

Reijenga, J.C.; Kenndler, E. Computational simulation of migration and dispersion in free capillary zone electrophoresis, part II, Results of simulation and comparison with measurements. J. Chromatogr. A. **1994**, *659*, 417.

Virtanen, R. Acta Polytech. Scand. **1974**, *123*, 1.

Band Broadening in SEC

Jean-Pierre Busnel
Université Du Main, Le Mans, France

INTRODUCTION

In classical chromatography, band broadening (BB), which defines the shape of the chromatogram of a pure solute, is one of the factors limiting the resolution, but individual peaks are generally observable and the discussion of BB extent is direct. In SEC, the situation is more complex, as we observe, generally, only the envelope of a large number of individual peaks (Fig. 1). Imperfect resolution and its consequences on results cannot be directly observed. A few years after the pioneer publication on SEC by Moore,[1] Tung[2] presented the general mathematical problem of band-broadening correction (BBC). Until 1975, a number of simplified procedures have been proposed in order to compensate for the limited resolution of columns. After 1975, a spectacular increase in column resolution rendered the problem less important, but, recently, there is a growing interest in BBC as SEC users intend to obtain more and more detailed information on molecular-weight distributions (MWDs) and not only average MW values. For this reason, this discussion is separated into three parts:

- Experimental determination of extent of BB
- Interpretation of BB processes
- Correction methods for BB

Experimental Determination of the Extent of Broad-Banding

It is useful to choose a solute which is really eluted by a size-exclusion process, without adsorption or any additional interaction phenomena which might modify the shape of the peak. The most trivial method is to analyze the shape of a low-MW pure substance. This is usually used to determine the number of theoretical plates, $N = (V_r/\sigma)^2$ where V_r is the retention volume (volume at peak top) and σ is the standard deviation. σ can be computed from the weighing of each data point of the peak or can be estimated from the width at 10% maximum height ($\sigma = W_{0.1}/4.3$).

For this reason, when using THF as eluent and styrene/DVB gels, methanol or toluene are not good candidates; octadecane is preferred. For aqueous SEC, saccharose is the classical standard.

For polymers, a number of authors have claimed that the peak width increases as the MW increases, but to discuss band broadening for polymers, several precautions are required. First, it is necessary to be sure that the injected solution is sufficiently dilute to prevent any viscous effect. (Practically no viscous effect is observable, even for narrow standards when $[\eta]C < 0.1$; for flexible polymers, this corresponds roughly to a concentration $<1\,\mathrm{mg/mL}$ for MWs up to 500,000; for a higher MW, it is necessary to reduce the concentration.) Then, the real difficulty is to analyze very narrow standards for which polymolecularity is sufficiently low, so as not to participate in the peak width, or at least which polymolecularity is very precisely known.

Commercial indications on standards are in progress, but suppliers rarely guarantee the exact value for $I_p = M_w/M_n$. A tendency is to guarantee that I_p is lower than a given value, but that is not sufficient for precise BB study. Usual values for medium MW standards are $I_p < 1.03$ and $I_p < 1.05$ for high MW standards; the situation is worse for aqueous SEC with values around 1.1.

Recently, possibilities appeared with the results from thermal gradient interaction chromatography (TGIC).[3] That method has a much better resolution than SEC and allows a very precise determination of I_p. It is even possible to use it for a preparative scale to obtain extremely narrow standards, but, until now, it is only available for organic-soluble standards.

With all these precautions and when using modern columns, it clearly appears that peaks are not Gaussian, but systematically skewed. New computing facilities allow one to analyze more precisely such peaks by various functions. The best results seems to be obtained by exponentially modified gaussian (EMG) functions,[4] which are the convolution of a Gaussian dispersion and an exponential decay. In that case, two parameters define the shape of the peak s and t, which allow quantitative mapping of BB characteristics for further correction. Generally, σ and τ are constant or admit a limited increase with MW for samples eluted well after the void volume. However, a dramatic increase of σ and τ occurs near the total exclusion volume.[5]

Encyclopedia of Chromatography DOI: 10.1081/E-ECHR-120039891

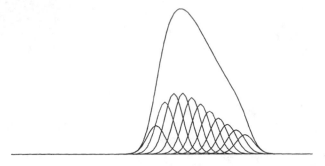

Fig. 1 Example of imperfect resolution: 10 peaks, $R = 0.25$ between neighbors.

BAND-BROADENING INTERPRETATION

As previously indicated, this discussion is organized for chromatograms from very narrow polymer standards for which we can consider that the effect of molecular weight distribution is negligible and for which the unique separation process is size exclusion. With these limitations, the contribution to band broadening is conveniently separated into extra column effects, eddy dispersion, static dispersion, and mass transfer. In the most classical chromatographic interpretation, extra-column effects are not discussed and the three other contributions are considered as Gaussian, so there is simply the addition of their variances. The number of theoretical plates is defined as $N = (V_e/\sigma)^2$ and the influence of v, the linear velocity of the eluent, is summarized by the so-called Van Deemter equation:

$$H = \frac{L}{N} = a + \frac{b}{v} + cv$$

This classical interpretation is not sufficient, as experimental results clearly indicate that there is peak skewing; for this reason, it is useful to study each contribution separately.

Static Dispersion

This classical contribution corresponds to the diffusion of the sample along the axis of the column by Brownian motion during the time t_0 spent in the interstitial volume. That spreading effect is Gaussian and its standard deviation σ is related to D, the diffusion coefficient of the solute: $\sigma = (2Dt_0)^{1/2}$. In classical operating conditions in modern LC, that contribution is generally a minor one, due to the use of relatively high flow rate. In SEC, the diffusion coefficients of polymers are very low and that contribution becomes negligible.

Extra-Column Effects

Generally, these effects are not discussed in detail, considering it is only necessary to select a chromatographic apparatus with a proper design to render these effects negligible. Recent results indicate that this situation tends to be different for macromolecular solutes. Using a chromatographic apparatus for which the column is replaced by tubings of various lengths, the end of elution is characterized by an exponential decay, the time of which is dependent not only on geometry but also strongly increases for high-MW solutes. The explanation is related to the more or less rapid averaging of radial positions of the solute in a cylindrical tube. In the case of high-molecular-weight solutes, the diffusion coefficient is small. Solute molecules which enter near the center of the tube stay in the high-velocity zone and those which enter near the walls stay in low velocity zones; this introduces skewing which is much more important than for low-molecular-weight solutes.

Eddy Dispersion

This contribution is related to the variety of channels available for any solute molecule throughout the elution process. These channels are defined by the interstitial volume between the beads of the column package, so they correspond to a variety of shapes and flow velocities. This produces a distribution in elution time which is classically considered as Gaussian and weakly depends on flow rate. As a rule of thumb, the theoretical plate height corresponding to this effect can be considered as being equal to the bead diameter of the packing for well-packed columns.

More detailed results take into account a "wall effect" to explain why elution profiles are skewed, even for nonretained solutes in modern columns. Detailed experimental results were recently presented by Farkas and Guiochon[6] on the radial distribution of flow velocity using local multichannel detection devices. On average, the flow velocity is very homogeneous in the center of the column, but, inevitably, it becomes lower near the walls. Similarly, the peak shape from a local microdetector situated near the wall is clearly distorted and skewed compared with that of a similar detector situated near the center of the column.

Mass Transfer

The simple model of a theoretical plate, which is simply the affirmation of the existence of N successive equilibrium steps, is not satisfactory, as it assumes a Gaussian spreading.

To obtain more realistic information, it is necessary to discuss the rate of exchange between the interstitial volume and pores.[7] Potschka[8] proposed taking into account the competition between diffusion and convection in the special situation of "perfusion chromatography," where very large pores exist inside the beads, which allow some distribution of the solute by convection.

Recently, a model has been proposed for which pores are simply long cylinders and time of residence corresponds to a one-dimension Brownian motion.[5] Exact mathematical expressions are available for describing that process[9] and the distribution of such time of residence is highly skewed. Additionally, for each solute molecule, the number of visited pores obeys a Poisson distribution, and when the average number of visits becomes small, the distribution of elution time becomes wider and more skewed. That explains, precisely, why strong peak distortion is observed for samples eluted near the total exclusion limit.

Band-Broadening Correction

As first stated by Tung, the general starting point is that the experimental chromatogram $H(V)$ (from a concentration detector) is the convolution of $g(V, V_r)$, the spreading function defining the elution of a single species with a peak apex position of V_r, and $w(V_r) \, dV_r$, the weight fraction of species which peak apex, is between V_r and $V_r + dV_r$:

$$H(V) = \int_{V_1}^{V_2} g(V, V_r) w(V_r) dV_r$$

In any case, a second step consists of converting $w(V_r)$ into $w(M)$ by defining a calibration curve which correlates M and V_r.

As there is only a finite number of data points and, with some instrumental noise, generally such an inversion problem is ill-defined, the stability of the values of $w(V_r)$ depends on the algorithm which is used. Among the huge number of articles treating such problems, a review by Meira and co-workers[10] gives useful information on the mathematical aspects and a detailed review by Hamielec[11] presents a variety of instrumental situations, from the simplest one (constant Gaussian spreading function and simple concentration detector) to the most complex (general spreading function, multidetection).

To take into account experimental evidence which clearly indicates systematic skewing, this discussion will no longer concern methods limited to Gaussian spreading functions.

SIMPLEST SITUATION: CONSTANT SPREADING FUNCTION, LINEAR CALIBRATION CURVE, SINGLE CONCENTRATION DETECTOR

This situation corresponds to a useful approximation in many cases, and it is almost strictly exact when the sample has a narrow MWD. From the chromatogram of an ideal isomolecular sample, considering that it is defined by a set of h_i values equidistant on the elution volume axis, classical summations give the uncorrected molecular-weight values:

$$M_{n_{\text{uncorrected}}} = \frac{\Sigma h_i}{\Sigma (h_i / M_i)} \quad \text{and} \quad M_{w_{\text{uncorrected}}} = \frac{\Sigma (h_i M_i)}{\Sigma h_i}$$

and the peak apex position gives the real molecular weight: M_{peak}.

Therefore, two correction factors exist for M_n and M_w:

$$K_n = \frac{M_{\text{peak}}}{M_{n_{\text{uncorrected}}}} \quad \text{and} \quad K_w = \frac{M_{\text{peak}}}{M_{w_{\text{uncorrected}}}}$$

For any other isomolecular sample analyzed on the same system, the change is simply a shift along the elution volume axis and each value M_i is multiplied by the same factor. Thus, the correction factors are unchanged. Finally, any broad MW sample analyzed on the same system is the addition of a set of isomolecular species; therefore, when summing, there is factorization of the correction factors and

$$M_{n_{\text{corrected}}} = K_n M_{n_{\text{uncorrected}}}$$
$$M_{w_{\text{corrected}}} = K_w M_{w_{\text{uncorrected}}}$$

This very simple BBC can always be used, at least to give a preliminary indication of the extent of BB. When using extremely narrow standards, as obtained by preparative TGIC, the result is accurate; more easily, it is possible to set the correction factors between two limits: lower one using data from a low-MW pure chemical and a higher limit using data from an imperfect narrow standard.

GENERAL SITUATION: SPREADING FUNCTION DEPENDS ON ELUTION VOLUME AND CALIBRATION CURVE IS NOT LINEAR, SINGLE CONCENTRATION DETECTOR

In such cases, as stated by Meira and co-workers,[10] the quality of results depends on computational refinements. It is necessary to add specific constraints related to the chromatographic problem: rejection of negative values or unrealistic fluctuations in the weight distribution. The normal way is to invert the large matrix defining the spreading function for any position on the elution volume scale. With modern computational facilities, that

B

becomes easy, but it is still not trivial to obtain stable results, and proper filtering processes are useful.

Good results can be obtained by using a more direct iterative method which can be briefly presented: n equidistant values are chosen on the elution volume scale; let us note these values as V_i (typically, n can be 200). Any sample is arbitrarily defined as the sum of n isomolecular species, whose positions at the peak apex are V_j. For each peak j, n values of heights h_{ij} for each V_i value are computed, normalizing the surface (the peak shape is defined by interpolation from experimental BB data).

The chromatogram corresponding to the sample is defined by a set of n H_i values at positions V_i.

To define the weight distribution of the sample, it is simply necessary to adjust a set of w_i values until the summation converges toward the experimental H_i values. For the first attempt, $w_i = H_i$. This gives, by addition, a chromatogram ($H1_i$). Then, $w_i = w_i * H_i / H1_i$; that gives $H2_i$ and so on until convergence.

The method is reasonably efficient. Stable convergence is observed except for very large samples for which the problem is too severely ill-conditioned. Applying it to narrow PS standards allows one to find, again, the true polymolecularity index.

MULTIDETECTION PROBLEM

The aim of multidetection, especially LS/DRI coupling, is to find a useful calibration curve directly from the sample data and without external information from standards. A crude calibration curve is obtained by plotting, on a semilogarithmic scale, the instantaneous weight average M_{wi} values for each data point at elution volume V_i. At this point, correction for BB is rarely used, as the accuracy of the calibration curve is generally poor and does not justify sophisticated corrections. Additionally, for complex polymers (blends, copolymers, branched polymers, etc.), a variety of molecular weights are eluted at the same position, even in the absence of band broadening,[12] so it is still very difficult to propose a general solution for the problem and results are available only for simplified situations. Normally, it would be necessary to find the exact weight distribution w_i as in the preceding paragraph, simply using the DRI signal; then, it becomes possible to adjust the calibration curve until the calculated M_{wi} values converge toward the experimental set of M_{wi} values.

CONCLUSIONS

Band broadening in SEC has several specific aspects, compared with other chromatographic processes. Solutes may have very low diffusion coefficients and that introduces additional tailing due the imperfect averaging of radial positions all along the tubing. Mass transfer can be described from Brownian motion properties, and for samples eluted near total exclusion volume, as the number of visited pores become small, this introduces a significant increase of skewing and a very important loss of resolution.

For correcting band broadening, the main difficulty is in obtaining precise mapping of the spreading function of the system. Normally, this needs very high quality standards and TGIC offers new possibilities in that area. Computational techniques are now sufficiently efficient to solve the general inversion problem associated with band-broadening correction, but it still needs some precautions to obtain stable results without artificial oscillations. Finally, as corrections become very important and unstable near total exclusion volume, it remains very imprudent to interpret data when part of the sample is totally excluded. It is better to first find a well-adapted column set, able to efficiently fractionate the whole sample.

REFERENCES

1. Moore, J.C. Gel permeation chromatography. I. A new method for molecular weight distribution of high polymers. J. Polym. Sci. A-2 **1964**, *835*.
2. Tung, L.H. Method of calculating molecular weight distribution function from gel. J. Appl. Polym. Sci. **1966**, *10*, 375.
3. Lee, W.; Lee, H.C.; Park, T.; Chang, T.; Chang, J.Y. Temperature gradient interaction chromatography of low molecular weight polystyrene. Polymer **1999**, *40*, 7227.
4. Jeansonne, M.S.; Foley, J.P. J. Chromatogr. Sci. **1991**, *29*, 258.
5. Busnel, J.-P.; Foucault, F.; Denis, L.; Lee, W.; Chang, T. J. Chromatogr. **2001**, *A 930*, 61 in press.
6. Farkas, T.; Guiochon, G. Contribution of the radial distribution of the flow velocity to band broadening in HPLC columns. Anal. Chem. **1997**, *69*, 4592.
7. Kim, D.H.; Johnson, A.F. *Size Exclusion Chromatography*; ACS Symposium Series; Provder, T., Ed.; American Chemical Society: Washington, DC, 1984; 245.
8. Potschka, M. Role of convection and obstructed diffusion in size-exclusion chromatography. J. Chromatogr. **1993**, *648*, 41.
9. Karatzas, I.; Shreve, S. *Brownian Motion and Stochastic Calculus*; Springer-Verlag: New York, 1991.
10. Gugliotta, L.M.; Vega, J.R.; Meira, G.R. J. Liq. Chromatogr. **1990**, *13*, 1671.
11. Hamielec, A.E. *Steric Exclusion Liquid Chromatography of Polymers*; Janca, J., Ed.; Marcel Dekker, Inc.: New York, 1984; 117–160.
12. Radke, W.; Simon, P.F.W.; Muller, A.H.E. Macromolecules **1996**, *29*, 4926.

Barbiturates: Analysis by Capillary Electrophoresis

Chenchen Li
Huwei Liu
Institute of Analytical Chemistry, College of Chemistry and Molecular Engineering, Peking University, Beijing, P.R. China

INTRODUCTION

Capillary electrophoresis (CE) is becoming a popular analytical tool for determining drugs because of its simplicity, high speed, and high efficiency. The present review studies different modes of CE used in the determination and chiral separation of barbiturates, as well as current developments in sample preparation for barbiturates in biological fluids. The comparison of different modes of CE with other separation approaches is also discussed.

BARBITURATES

Barbiturates, derivatives of barbituric acid, are found in a variety of pharmaceuticals, such as sedatives, hypnotics, and antiepileptics. However, their levels in body fluids have to be regulated within a narrow therapeutic window to avoid toxicity. Therefore determination of barbiturates in serum, plasma, and urine is important for investigation of intoxication, therapeutic drug monitoring, and pharmacokinetic and metabolic studies. Hence, many instrumental approaches for the analysis of barbiturates in body fluids, including immunoassays[1] and chromatographic methods, such as GC,[2] GC/MS,[3] and HPLC,[4–6] have been developed. Immunological techniques offer high performance, speed of analysis, and sensitivity. However, they are not specific enough to distinguish a single compound because barbiturates interfere with each other. Chromatographic methods have been applied for the determination of all common barbiturates, even while they require time-consuming sample pretreatment and are characterized by a low sample throughput.[7]

In recent years, CE has been successfully applied in the field of biochemical and analytical chemistry. It has been found to be attractive for pharmaceutical analysis because of its advantages related to excellent separation efficiency, high mass sensitivity, minimal use of samples and solvents, and the possibility of using different direct and indirect detection systems. This review focuses on analytical assays for barbiturates by CE.

SAMPLE PREPARATION FOR BARBITURATES IN BODY FLUIDS

Sample preparation is important here because matrices of biological fluids are so complicated that interfering signals are likely to appear in typical separation-based determinations. Among various sample preparation techniques, there are two major approaches combined with CE: liquid–liquid extraction (LLE) and solid-phase extraction (SPE) [or solid-phase microextraction (SPME)].

Liquid–Liquid Extraction

LLE with chloroform from acidified serum is widely executed[7–11] after the method of Shiu and Nemoto.[12] However, chloroform is of great toxicity and is not suitable for routine use. LLE with pentane at pH 6.4[7] was investigated for feasibility with seven barbiturates and was found to be specific for thiopental—an important barbiturate being used for anaesthetic medication and treatment of head trauma with severe brain injury. Comparing acetonitrile deproteinization with chloroform deproteinization, Shihabi demonstrated the feasibility of acetonitrile for extraction of pentobarbital. However, electropherograms obtained by extraction with chloroform are more sensitive and cleaner.[9] Wu et al.[10] used LLE with ether for extraction from serum, and chloroform for extraction from urine. They obtained recoveries of six barbiturates from 86.6% to 118%.

Solid Phase Extraction and Solid-Phase Microextraction

Although LLE is useful, it is being replaced by SPE or SPME because LLE is a time-consuming and laborious process that involves consumption of large volumes of organic solvents. The use of SPE for subsequent analysis is achieved in one step, and the results have shown that this is a reproducible, safe, convenient, and time-saving alternative to LLE. Most SPE applications use disposable cartridges or columns packed with C_{18}-bonded silica, which is the most classic packing for this technique.[7,10,11,13,15] Moreover, SMPE is based on

Encyclopedia of Chromatography DOI: 10.1081/E-ECHR-120040662

partition equilibrium of target compounds between the sample matrices and a polymeric stationary phase, such as plasticized poly (vinyl chloride) (PVC),[14] which is coated onto a fused silica fiber.

Among these techniques, the SMPE device is reported to be the easiest to construct and it performs very reliably. Thus alkaline or neutral compounds are expected not to be extracted or backextracted, and will not interfere with analysis of barbiturates. Barbiturate concentrations of 0.1–0.3 ppm in urine and about 1 ppm in serum can be determined.[14,16]

CAPILLARY ELECTROPHORETIC MODES OF OPERATION

Capillary zone electrophoresis (CZE) and MEKC are the most common CE modes used in determining barbiturates. We will discuss them separately.

Analysis by Capillary Zone Electrophoresis

In CZE, the separation mechanism is based on differences in the charge/mass ratios of ionic analytes. An electric field is applied to the capillary filled with a running buffer, and cations go to the cathode, whereas anions migrate to the anode. But because of the electroosmotic flow (EOF) of the buffer, which is the driving force of CE, all analytes will move in the direction of EOF (usually toward the negative electrode as the inner surface of fused silica capillary is negatively charged). Although barbiturates are negatively charged at high pH, their migration velocities are close to each other and their separations by CZE are more or less insufficient. A separation of some common barbiturates using a 500 mmol/L borate buffer (pH 8.5) and an applied voltage of 11 kV is

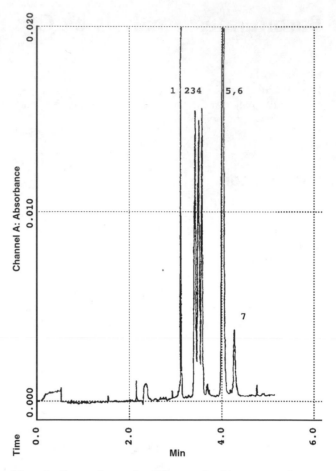

Fig. 1 Separation of different barbiturates using a 500-mmol/L borate buffer, pH 8.5. (1) Internal standard; (2) pentobarbital; (3) secobarbital; (4) amobarbital; (5) phenobarbital; (6) butabarbital; and (7) contamination.

illustrated in Fig. 1.[9] The phenobarbital peak was not resolved from butabarbital. Boone et al.[17] improved separation performance by using a 90 mmol/L borate buffer (pH 8.5) and an applied

Table 1 Running conditions of MEKC for determination of barbiturates

Analytes	Running buffer	Refs.
Barbital, allobarbital, phenobarbital, butalbital, thiopental, amobarbital, and pentobarbital	50 mM SDS, 9 mM $Na_2B_4O_7$, 15 mM $Na_2H_2PO_4$ (pH 7.8)	[7]
Barbital, phenobarbital, methyl phenobarbital, amobarbital, thiopental, pentobarbital, and secobarbital	100 mmol/L SDS:100 mmol/L $Na_2H_2PO_4$: MeOH : H_2O (70 : 15 : 5 : 10)	[10]
Seven barbiturates and 14 benzodiazepines	100 mM borate, 10 mM SDS, and 5 M urea (pH 8.5)	[11]
Heptabarbital, hexobarbital, pentobarbital, butalbital, and phenobarbital	30 mmol/L SDS, 30 mmol/L borate (pH 9.3), with 200 mL/L acetonitrile	[13]
Twenty-five barbiturates	50 mM SDS, 20 mM phosphate (pH 8.4)	[17]

Table 2 Comparison of CZE and MEKC for determining barbiturates

	CZE	MEKC
Total analysis time	No more than 10 min	About 15 min
Separation efficiencies	High	High
Analytical window	Relatively small window because of the resemblance of the chemical structures and the pK values of the barbiturates[17]	Provides a significantly increased analytical window
Resolution	Some barbiturates interfere with each other	Improved
Pretreatment process	Complex and time-consuming (unsatisfied resolution with LLE)	Both LLE and SPE produce good results

voltage of 30 kV. It was found that an applied voltage of 30 kV resulted in faster separations and a better resolution compared to lower voltages. In addition, the Tapso–Tris buffer has a higher buffering capacity but lower conductivity. Baseline separation of 10 barbiturates and eight benzoates was achieved in a 50-mM Tapso–Tris buffer system.[14] Recently, Delinsky et al.[15] completely resolved barbiturates from meconium in a 150-mM Tris running buffer after sample preparation by SPE.

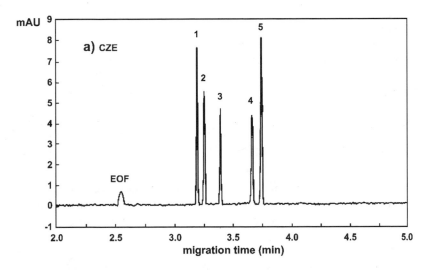

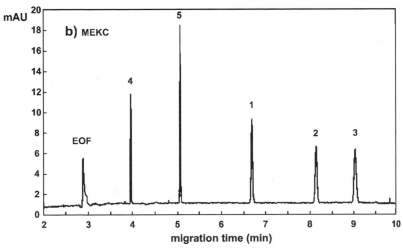

Fig. 2 Electropherograms of the separation of five barbiturate standards using CZE (a) and MEKC (b). Peaks: (1) hexobarbital; (2) methohexital; (3) secobarbital; (4) barbital; and (5) phenobarbital.

Analysis by MEKC

MEKC is another CE method based on the differences between interactions of analytes with micelles present in the separation buffer, which can easily separate both charged solutes and neutral solutes with either hydrophobic or hydrophilic properties. Micelles are formed by adding a surfactant [at a concentration above its critical micelle concentration (CMC)] to the separation buffer, and act as the so-called pseudo-stationary phase. The most striking observation is the resolution of MEKC, which is more likely to separate complex mixtures than CZE. Because of their similar migration velocities, barbiturates are most commonly investigated in the MEKC mode.[7,8,10–13,16–19] There are various buffers and modifiers that can be selected and combined in MEKC to optimize separation; some typical examples are summarized in Table 1.

COMPARISON OF CZE AND MEKC

A comparison of the CZE and MEKC modes for determining barbiturates is summarized in Table 2.

Fig. 2[17] shows electropherograms of a mixture of five barbiturate standards; it can be observed that the addition of micelles in MEKC clearly resulted in a different separation mechanism, reflected in various changes in elution order, compared to CZE. The migration behavior in MEKC depends largely on the hydrophobic interaction of the analytes with the micelles. Hydrophobic components are more solubilized in the micelles, resulting in a slower migration compared to less hydrophobic compounds.

OPTIMIZATION OF SEPARATION

Various types of coated capillaries have been applied to the CE separation of barbiturates for improvement of separation selectivity and efficiency.[11,20,21] Jinno et al. reported a series of investigations on PAA and AA-co-IPAAM-coated columns in both CZE and MEKC modes. By the use of polyacrylamide (PAA) coating for the capillary, EOF was sufficiently eliminated, leading to a shorter analysis time and better resolution.[11,20] In addition, the elution order of barbiturates partly changed because of the hydrophobic nature of PAA. It must be noted that a noncross-linked copolymer containing IPAAM had thermosensitive properties, and elution order was different at elevated temperatures when compared with that at ambient temperature,[21] as illustrated in Fig. 3.

In CE, the use of a reproducible identification parameter is very important because it influences identification power (IP). Boone et al.[17] used μ_{eff} and μ_{eff}^{c}

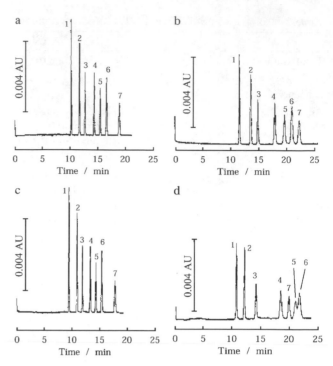

Fig. 3 Separation of barbital by the use of a capillary coated with (a and c) 10% T PAA and (b and d) 10% T poly(AA-co-IPAAM) containing 85% IPAAM. Experiments were carried out at (a and b) ambient temperature or (c and d) elevated temperature. Conditions: capillary column, 50 cm × 0.075 mm ID (25 cm effective length); buffer, 100 mM Tris–150 mM boric acid (pH 8.3); field strength, 300 V/cm; injection, electromigration for 5 sec at the side of cathode. Peak identification: (1) phenobarbital; (2) barbital; (3) mephobarbital; (4) amobarbital; (5) secobarbital; (6) pentobarbital; (7) metharbital.

instead of migration times to enhance reproducibility, to reduce the upward trend of RSD with increasing migration time, and to correct for outliers.

CHIRAL SEPARATION OF BARBITURATES

Chiral separation is one of the major outstanding advantages for CE compared to other separation techniques. As enantiomers have identical electrophoretic mobilities, some chiral complexing reagents, called chiral selectors, must be added to the separation buffer to form diasteromeric complexes in dynamic equilibrium. One of the most versatile techniques for achieving chiral recognition is the addition of cyclodextrins (CDs). It is reported that determination of (R)-secobabital and (S)-secobabital from serum was achieved by addition of hydroxypropyl-γ-cyclodextrin (HPGCD)[16] or hydroxypropyl-β-cyclodextrin (HPBCD)[18] in the running buffer. Fig. 4 illustrates a typical electropherogram of pentobarbital and

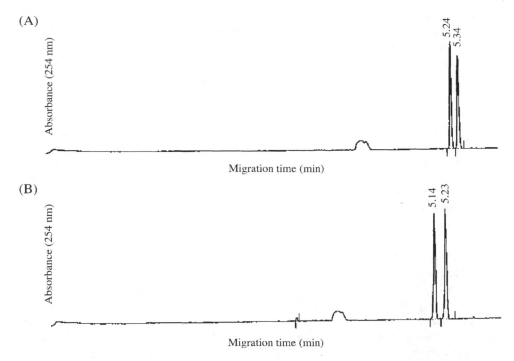

Fig. 4 Typical electropherograms of enantiomers of (A) pentobarbital and (B) secobarbital on a 57 cm × 75 μm ID fused silica capillary. The run buffer contained 40 mM HPBCD in 20 mM ammonium acetate buffer (pH 9.0) with detection at 254 nm. The capillary was thermostated at 25°C and the run voltage was set to 15 kV.

secobabital enantiomers, separated by HPBCD. Conradl and Vogt et al.[22] investigated the separation of (R)-thiopental and (S)-thiopental, and (R)-phenobarbital and (S)-phenobarbital by using α-CD and β-CD as chiral selectors, separately. They found out that the enantiomers of thiopental were not separated by β-CD, whereas cyclobarbital enantiomers were separated by both α-CD and β-CD. It is assumed that the size of the cavity of β-CD allows the side chain of thiopental, with the asymmetric carbon atom, to penetrate completely. Therefore the formation of stereoselective complexes is unlikely. In the case of

α-CD, the C_5 substituents of thiopental are optimal in size and structure to fit into the cavity whereas; for cyclobarbital, a partial inclusion of the cyclohexene ring or the methyl group at the C_5 of the heterocycle into the cavity is probable. However, the separation of cyclobarbital with low α-CD concentration was less efficient.

COMPARISON OF CE, GC, AND LC

LC, GC, and CE are three typical analytical separation approaches used to determine barbiturates. The advan-

Table 3 Comparative study of three techniques to determine barbiturates

Techniques	Advantages	Disadvantages
GC	High sensitivity and selectivity	Inadequate for polar, thermolabile, and low-volatility analytes
	Coupled with MS for the identification of unknowns	High consumption of expensive, high-purity gases
LC	Application to all organic solutes in spite of volatility or thermal stability	Insufficient separation selectivity and efficiency
	Both mobile and stationary phase compositions are variables	Related long analysis time
		Large amounts of expensive and toxic organic solvents used as mobile phase, such as acetonitrile[5] and methanol.[6]
CE	High separation efficiency	Inadequate detection limit
	Small consumption of expensive reagents and toxic solvents	Lack of highly selective detectors

tages and disadvantages of the three techniques are summarized in Table 3.

CONCLUSIONS

CE offers fast analysis, low consumable expenses, ease of operation, high separation efficiency, and selectivity. Now, CE has undoubtedly become an attractive technique for the determination and therapeutic monitoring of barbiturates and other drugs, and it is becoming more and more important in biological and toxicological analyses. It was also indicated that CE has a good potential for systematic toxicological analysis (STA).[17]

At present, the coupling of CE to MS is attractive because it facilitates the identification of analytes and improves detection sensitivity. Combining CE with fluorescence detection greatly improves the sensitivity and selectivity for determining proteins and other biological molecules. Future trends of CE will be used to develop more coupling methods to combine CE with highly selective detectors and to achieve miniaturization and automation for extremely fast, easy, and real-time determination in pharmaceutical and biological analyses.

REFERENCES

1. Colbert, D.L.; Smith, D.S.; Landon, J.; Sidki, A.M. Single-reagent polarization fluoroimmunoassay for barbiturates in urine. Clin. Chem. **1984**, *30*, 1765.
2. Berry, D.J. Gas chromatography analysis of the commonly prescribed barbiturates at therapeutic and overdoes levels in plasma and urine. J. Chromatogr. **1973**, *86*, 89.
3. Soo, V.A.; Bergert, R.J.; Deutsch, D.G. Screening and quantification of hypnotic sedatives in serum by capillary gas chromatography with a nitrogen-phosphorus detector, and confirmation by capillary gas chromatography–mass spectrometry. Clin. Chem. **1986**, *32*, 325.
4. Elisabeth, I.; Rene, S.; Dieter, J. Screening for drugs in clinical toxicology by high-performance liquid chromatography: Identification of barbiturates by post-column ionization and detection by a multiplex photodiode array spectrophotometer. J. Chromatogr. **1988**, *428*, 369.
5. Feng, C.L.; Liu, Y.T.; Luo, Y. HPLC-DAD analysis of thirteen soporific sedative drugs in human blood. Acta Pharm. Sin. **1995**, *30* (12), 914–919.
6. Coppa, G.; Testa, R.; Gambini, A.M.; Testa, I.; Tocchini, M.; Bonfigli, A.R. Fast, simple and cost-effective determination of thiopental in human plasma by a new HPLC technique. Clin. Chim. Acta **2001**, *305*, 41–45.
7. Thormann, W.; Meier, P.; Marcolli, C.; Binder, F. Analysis of barbiturates in human serum and urine by high-performance capillary electrophoresis-micellar electrokinetic capillary chromatography with on-column multi-wavelength detection. J. Chromatogr. **1991**, *545*, 445–460.
8. Meier, P.; Thormann, W. Determination of thiopental in human serum and plasma by high-performance capillary electrophoresis-micellar electrokinetic chromatography. J. Chromatogr. **1991**, *559*, 505–513.
9. Shihabi, Z.K. Serum pentobarbitutal assay by capillary electrophoresis. J. Liq. Chromatogr. **1993**, *16* (9/10), 2059–2068.
10. Wu, H.; Guan, F.; Luo, Y. Determination of barbiturates in human plasma and urine by high performance capillary electrophoresis. Yaowu Fenxi Zazhi **1996**, *16* (5), 316–321.
11. Jinno, K.; Han, Y.; Sawada, H.; Taniguchi, M. Capillary electrophoretic separation of toxic drugs using a polyacrylamide-coated capillary. Chromatographia **1997**, *46* (5/6), 309.
12. Shiu, G.K.; Nemoto, E.M. Simple, rapid and sensitive reversed-phase high-performance liquid chromatographic method for thiopental and pentobarbital determination in plasma and brain tissue. J. Chromatogr. **1982**, *227*, 207.
13. Evenson, M.A.; Wilktorowicz, J.E. Automated capillary electrophoresis applied to therapeutic drug monitoring. Clin. Chem. **1992**, *38* (9), 1847–1852.
14. Li, S.; Weber, S.G. Determination of barbiturates by solid-phase microextraction and capillary electrophoresis. Anal. Chem. **1997**, *69*, 1217–1222.
15. Delinsky, D.C.; Srinivasan, K.; Solomon, H.M.; Bartlett, M.G. Simultaneous capillary electrophoresis determination of barbiturates from meconium. J. Liq. Chromatogr. & Relat. Technol. **2002**, *25* (1), 113–123.
16. Srinivasan, K.; Zhang, W.; Bartlett, M.G. Rapid simultaneous capillary electrophoretic determination of (R)- and (S)-secobarbital from serum and prediction of hydroxypropyl-γ-cyclodextrin—secobarbital stereoselective interaction using molecular mechanics simulation. J. Chromatogr. Sci. **1998**, *36*, 85–90.
17. Boone, C.M.; Franke, J.-P.; de Zeeuw, R.A.; Ensing, K. Evaluation of capillary electrophoretic techniques towards systematic toxicological analysis. J. Chromatogr. A. **1999**, *838*, 259–272.
18. Srinivasan, K.; Bartlett, M.G. Comparison of cyclodextrin-barbiturate noncovalent complexes using electrospray ionization mass spectrometry

and capillary electrophoresis. Rapid Commun. Mass Spectrom. **2000**, *14*, 624–632.

19. Wu, H.; Guan, F.; Luo, Y. A universal strategy for systematic optimization of high performance capillary electrophoretic separation. Chin. J. Anal. Chem. **1996**, *24* (10), 1117–1122.

20. Jinno, K.; Han, Y.; Hirokazu, S. Analysis of toxic drugs by capillary electrophoresis using polyacrylamide-coated columns. Electrophoresis **1997**, *18* (2), 284–286.

21. Sawada, H.; Jinno, K. Capillary electrophoretic separation of structurally similar solutes in non-cross-linked poly(acrylamide-*co*-*N*-isopropylacrylamide) solution. Electrophoresis **1997**, *18* (11), 2030–2035.

22. Conradl, S.; Vogt, C. Separation of enantiomeric barbiturates by capillary electrophoresis using a cyclodextrin-containing run buffer. J. Chem. Educ. **1997**, *74* (9), 1122–1125.

B

β-Agonist Residues in Food: Analysis by LC

Nikolaos A. Botsoglou
Laboratory of Nutrition, Faculty of Veterinary Medicine, Aristotle University, Thessaloniki, Greece

INTRODUCTION

β-Agonists are synthetically produced compounds that, in addition to their regular therapeutic role in veterinary medicine as bronchodilatory and tocolytic agents, can promote live weight gain in food-producing animals. They are also referred to as repartitioning agents because their effect on carcass composition is to increase the deposition of protein while reducing fat accumulation. For use in lean-meat production, doses of 5 to 15 times greater than the recommended therapeutic dose would be required, together with a more prolonged period of in-feed administration, which is often quite near to slaughter to obviate the elimination problem. Such use would result in significant residue levels in edible tissues of treated animals, which might in turn exert adverse effects in the cardiovascular and central nervous systems of the consumers.[1]

There are a number of well-documented cases where consumption of liver and meat from animals that have been illegally treated with these compounds, particularly clenbuterol, has resulted in massive human intoxification.[1] In Spain, a foodborne clenbuterol poisoning outbreak occurred in 1989–1990, affecting 135 persons. Consumption of liver containing clenbuterol in the range 160–291 ppb was identified as the common point in the 43 families affected, while symptoms were observed in 97% of all family members who consumed liver. In 1992, another outbreak occurred in Spain, affecting this time 232 persons. Clinical signs of poisoning in more than half of the patients included muscle tremors and tachycardia, frequently accompanied by nervousness, headaches, and myalgia. Clenbuterol levels in the urine of the patients were found to range from 11 to 486 ppb. In addition, an incident of food poisoning by residues of clenbuterol in veal liver occurred in the fall of 1990 in the cities of Roanne and Clermont-Ferrand, France. Twenty-two persons from eight families were affected. Apart from the mentioned cases, two farmers in Ireland were also reported to have died while preparing clenbuterol for feeding to livestock.

Although, without exception, these incidents have all been caused by the toxicity of clenbuterol, the entire group of β-agonists are now treated with great suspicion by regulatory authorities, and use of all β-agonists in farm animals for growth-promoting purposes has been prohibited by regulatory agencies in Europe, Asia, and the Americas. Clenbuterol, in particular, has been banned by the FDA for any animal application in the United States, whereas it is highly likely to be banned even for therapeutic use in the United States in the near future. However, veterinary use of some β-agonists, such as clenbuterol, cimaterol, and ractopamine, is still licensed in several parts of the world for therapeutic purposes.

MONITORING

Monitoring programs have shown that β-agonists have been used illegally in parts of Europe and United States by some livestock producers.[1] In addition, newly developed analogues, often with modified structural properties, are continuously introduced in the illegal practice of application of growth-promoting β-agonists in cattle raising. As a result, specific knowledge of the target residues appropriate to surveillance is very limited for many of the β-agonists that have potential black market use.[2] Hence, continuous improvement of detection methods is necessary to keep pace with the rapid development of these new, heretofore unknown β-agonists. Both gas and LC methods can be used for the determination of β-agonist residues in biological samples. However, LC methods are receiving wider acceptance because GC methods are generally complicated by the necessity of derivatization of the polar hydroxyl and amino functional groups of β-agonists. In this article, an overview of the analytical methodology for the determination of β-agonist in food is provided.

ANALYSIS OF β-AGONISTS BY LC

Included in this group of drugs are certain synthetically produced phenethanolamines such as bambuterol, bromobuterol, carbuterol, cimaterol, clenbuterol, dobutamine, fenoterol, isoproterenol, mabuterol, mapenterol, metaproterenol, pirbuterol, ractopamine, reproterol, rimiterol, ritodrine, salbutamol, salmeterol, terbutaline, and tulobuterol. These drugs fall into two major

Encyclopedia of Chromatography DOI: 10.1081/E-ECHR-120028860 categories, i.e., substituted anilines, including clenbuterol, and substituted phenols, including salbutamol. This distinction is important because most methods for drugs in the former category depend on pH adjustment to partition the analytes between organic and aqueous phases. The pH dependence is not valid, however, for drugs within the latter category, because phenolic compounds are charged under all practical pH conditions.

EXTRACTION PROCEDURES

β-Agonists are relatively polar compounds that are soluble in methanol and ethanol, slightly soluble in chloroform, and almost insoluble in benzene. When analyzing liquid samples for residues of β-agonists, deconjugation of bound residues, using 2-glucuronidase/sulfatase enzyme hydrolysis prior to sample extraction, is often recommended.[3,4] Semisolid samples, such as liver and muscle, require usually more intensive sample pretreatment for tissue breakup. The most popular approach is sample homogenization in dilute acids such as hydrochloric or perchloric acid or aqueous buffer.[3–6] In general, dilute acids allow high extraction yields for all categories of β-agonists, because the aromatic moiety of these analytes is uncharged under acidic conditions, whereas their aliphatic amino group is positively ionized. Following centrifugation of the extract, the supernatant may be further treated with β-glucuronidase/sulfatase or subtilisin A to allow hydrolysis of the conjugated residues.

CLEANUP PROCEDURES

The primary sample extract is subsequently subjected to cleanup using several different approaches, including conventional liquid–liquid partitioning, diphasic dialysis, solid-phase extraction, and immunoaffinity chromatography cleanup. In some instances, more than one of these procedures is applied in combination to achieve better extract purification.

LIQUID–LIQUID PARTITION

Liquid–liquid partitioning cleanup is generally performed at alkaline conditions using ethyl acetate, ethyl acetate/tert-butanol mixture, diethyl ether, or tert-butylmethyl ether/n-butanol as extraction solvents.[5,7,8] The organic extracts are then either concentrated to dryness, or repartitioned with dilute acid to facilitate back extraction of the analytes into the acidic solution. A literature survey shows that liquid–liquid partitioning cleanup resulted in good recoveries of

substituted anilines such as clenbuterol,[7,8] but it was less effective for more polar compounds such as salbutamol.[5] Diphasic dialysis can also be used for purification of the primary sample extract. This procedure was only applied in the determination of clenbuterol residues in liver using tert-butylmethyl ether as the extraction solvent.[6]

SOLID-PHASE EXTRACTION

Solid-phase extraction is, generally, better suited to the multiresidue analysis of β-agonists. This procedure has become the method of choice for the determination of β-agonists in biological matrices because it is not labor and material intensive. It is particularly advantageous because it allows better extraction of the more hydrophilic β-agonists, including salbutamol. β-Agonists are better suited to reversed-phase (RP) solid-phase extraction due, in part, to their relatively non-polar aliphatic moiety, which can interact with the hydrophobic octadecyl- and octyl-based sorbents of the cartridge.[9–11] By adjusting the pH of the sample extracts at values greater than 10, optimum retention of the analytes can be achieved. Adsorption solid-phase extraction, using a neutral alumina sorbent, has also been recommended for improved cleanup of liver homogenates.[5] Ion-exchange solid-phase extraction is another cleanup procedure that has been successfully used in the purification of liver and tissue homogenates.[12] Because multiresidue solid-phase covering β-agonists of different types generally present analytical problems, mixed-phase solid-phase extraction sorbents, which contained a mixture of RP and ion-exchange material, were also used to improve the retention of the more polar compounds. Toward this goal, several different sorbents were designed, and procedures that utilized both interaction mechanisms have been described.[5,9,13]

IMMUNOAFFINITY CHROMATOGRAPHY

Owing to its high specificity and sample cleanup efficiency, immunoaffinity chromatography has also received widespread acceptance for the determination of β-agonists in biological matrices.[3,4,12,14] The potential of online immunoaffinity extraction for the multiresidue determination of β-agonists in bovine urine was recently demonstrated, using an automated column switching system.[14]

SEPARATION PROCEDURES

Following extraction and cleanup, β-agonist residues are analyzed by LC. GC separation of β-agonists is generally complicated by the necessity of derivatization of their polar hydroxyl and amino functional groups. LC RP columns are commonly used for the separation of the various β-agonist residues due to their hydrophobic interaction with the C_{18} sorbent. Efficient RP ion-pair separation of β-agonists has also been reported, using sodium dodecyl sulfate as the pairing counterion.[15]

DETECTION PROCEDURES

Following LC separation, detection is often performed in the ultraviolet region at wavelengths of 245 or 260 nm. However, poor sensitivity and interference from coextractives may appear at these low detection wavelengths unless sample extracts are extensively cleaned up and concentrated. This problem may be overcome by postcolumn derivatization of the aromatic amino group of the β-agonist molecules to the corresponding diazo dyes through a Bratton–Marshall reaction, and subsequent detection at 494 nm.[15] Although spectrophotometric detection is generally acceptable, electrochemical detection appears more appropriate for the analysis of β-agonists due to the presence on the aromatic part of their molecule of oxidizable hydroxyl and amino groups. This method of detection has been applied in the determination of clenbuterol residues in bovine retinal tissue with sufficient sensitivity for this tissue.[8]

CONFIRMATION PROCEDURES

Confirmatory analysis of suspected LC peaks can be accomplished by coupling LC with MS. Ion spray LC/MS/MS has been used to monitor five β-agonists in bovine urine,[14] whereas atmospheric-pressure chemical ionization LC/MS/MS has been used for the identification of ractopamine residues in bovine urine.[9]

CONCLUSIONS

This literature overview shows that a wide range of efficient extraction, cleanup, separation, and detection procedures is available for the determination of β-agonists in food. However, continuous improvement of detection methods is necessary to keep pace with the ongoing introduction of new unknown β-agonists that have potential black market use, in the illegal practice.

REFERENCES

1. Botsoglou, N.A.; Fletouris, D.J. *Drug Residues in Food. Pharmacology, Food Safety, and Analysis*; Marcel Dekker: New York, 2001.
2. Kuiper, H.A.; Noordam, M.Y.; Van Dooren-Flipsen, M.M.H.; Schilt, R.; Roos, A.H. Illegal use of beta-adrenergic agonists—European community. J. Anim. Sci. **1998**, *76*, 195–207.
3. Van Ginkel, L.A.; Stephany, R.W.; Van Rossum, H.J. Development and validation of a multiresidue method for beta-agonists in biological samples and animal feed. J. AOAC Int. **1992**, *75*, 554–560.
4. Visser, T.; Vredenbregt, M.J.; De Jong, A.P.J.M.; Van Ginkel, L.A.; Van Rossum, H.J.; Stephany, R.W. Cryotrapping gas-chromatography Fourier-transform infrared spectrometry—A new technique to confirm the presence of beta-agonists in animal material. Anal. Chim. Acta **1993**, *275*, 205–214.
5. Leyssens, L.; Driessen, C.; Jacobs, A.; Czech, J.; Raus, J. Determination of beta-2-receptor agonists in bovine urine and liver by gas-chromatography tandem mass-spectrometry. J. Chromatogr. **1991**, *564*, 515–527.
6. Gonzalez, P.; Fente, C.A.; Franco, C.; Vazquez, B.; Quinto, E.; Cepeda, A. Determination of residues of the beta-agonist clenbuterol in liver of medicated farm-animals by gas-chromatography mass-spectrometry using diphasic dialysis as an extraction procedure. J. Chromatogr. **1997**, *693*, 321–326.
7. Wilson, R.T.; Groneck, J.M.; Holland, K.P.; Henry, A.C. Determination of clenbuterol in cattle, sheep, and swine tissues by electron ionization gas-chromatography mass-spectrometry. J. AOAC Int. **1994**, *77*, 917–924.
8. Lin, L.A.; Tomlinson, J.A.; Satzger, R.D. Detection of clenbuterol in bovine retinal tissue by high performance liquid-chromatography with electrochemical detection. J. Chromatogr. **1997**, *762*, 275–280.
9. Elliott, C.T.; Thompson, C.S.; Arts, C.J.M.; Crooks, S.R.H.; Van Baak, M.J.; Verheij, E.R.; Baxter, G.A. Screening and confirmatory determination of ractopamine residues in calves treated with growth-promoting doses of the beta-agonist. Analyst **1998**, *123*, 1103–1107.
10. Van Rhijn, J.A.; Heskamp, H.H.; Essers, M.L.; Van de Wetering, H.J.; Kleijnen, H.C.H.; Roos, A.H. Possibilities for confirmatory analysis of some beta-agonists using 2 different derivatives simultaneously. J. Chromatogr. **1995**, *665*, 395–398.

11. Gaillard, Y.; Balland, A.; Doucet, F.; Pepin, G. Detection of illegal clenbuterol use in calves using hair analysis. J. Chromatogr. **1997**, *703*, 85–95.

12. Lawrence, J.F.; Menard, C. Determination of clenbuterol in beef-liver and muscle-tissue using immunoaffinity chromatographic cleanup and liquid-chromatography with ultraviolet absorbency detection. J. Chromatogr. **1997**, *696*, 291–297.

13. Ramos, F.; Santos, C.; Silva, A.; Da Silveira, M.I.N. Beta(2)-adrenergic agonist residues— Simultaneous methylboronic and butylboronic derivatization for confirmatory analysis by gas-chromatography mass-spectrometry. J. Chromatogr. **1998**, *716*, 366–370.

14. Cai, J.; Henion, J. Quantitative multi-residue determination of beta-agonists in bovine urine using online immunoaffinity extraction coupled-column packed capillary liquid-chromatography tandem mass-spectrometry. J. Chromatogr. **1997**, *691*, 357–370.

15. Courtheyn, D.; Desaever, C.; Verhe, R. High-performance liquid-chromatographic determination of clenbuterol and cimaterol using postcolumn derivatization. J. Chromatogr. **1991**, *564*, 537–549.

β-Lactam Antibiotics: Effect of Temperature and Mobile Phase Composition on Reversed-Phase HPLC Separation

J. Martín-Villacorta
R. Méndez
N. Montes
J. C. García-Glez
Physical Chemistry Department, University of León, Campus de Vegazana s/n, León, Spain

INTRODUCTION

In previous papers,[1,2] we reported the effect of column temperature on resolution in RP-HPLC to separate various β-lactams (penicillins and cephalosporins) from a single sample. In this work we describe the effect of column temperature and volume fraction of an organic solvent on resolution in the isocratic elution conditions of some β-lactam antibiotics.

Mobile phase composition and column temperature are two important experimental parameters that can be altered when a mixture of several compounds is to be separated.[3–5]

The relationships between capacity factor, k', and organic modifier concentration in the mobile phase, and the effect of the column temperature on k' for the antibiotics studied have been used to define k' as a function of T and V (volume fraction) on the basis of a small number of experimental measurements for a given combination of column, organic solvent, and type of antibiotic. From calculated values of k', resolution values, R_S, may be estimated for adjacent band-pairs under all conditions. The method developed enables the optimization of RP-HPLC separations of the β-lactam antibiotics in the absence of difficult theoretical calculations, using a small number of experimental data, including the influence of the organic solvent in the mobile phase (isopropanol) and the column temperature.

INSTRUMENTS

The HPLC system consisted of a Model 600E multisolvent delivery system equipped with a heated column compartment, a Model 484 variable-wavelength detector, and a Model 745B computing integrator, all from Waters Assoc., Inc., Milford, MA. The chromatograph was equipped with a Spherisorb ODS column (10 μm particle size, 25 cm × 4.6 mm I.D.).

CHROMATOGRAPHIC PROCEDURE

The mobile phases used to separate the compounds were acetate buffer (pH 5.00, 0.1 M)/isopropanol, 96.5/3.5, 95/5, 93/7, and 90/10 (v/v). A precolumn (3 cm × 4.6 mm I.D.), packed with the same packing materials, was used to guard the main column. The detector was set at 254 nm. The flow rate of the mobile phase was 1.0 mL/min. The column dead time, t_0, was measured by injecting methanol.

RESULTS AND DISCUSSION

The mixture of β-lactams was chromatographed at each of five column temperatures from 20°C to 60°C, and at four different volume fractions of isopropanol in mobile phase, from 0.035 to 0.1. Fig. 1A shows chromatograms with the volume fraction of isopropanol ranging from 0.035 to 0.1 at a constant column temperature (30°C). Fig. 1B shows the chromatograms obtained for each column temperature at 0.05 volume fraction of isopropanol. As one can see, a marked effect is produced for both parameters on the chromatographic behavior of each antibiotic.

Capacity Factor as a Function of Volume Fraction of Isopropanol in the Mobile Phase

The volume fraction (V) of organic solvent in the mobile phase is one of the most important parameters controlling capacity factor in RP-HPLC. Many reported studies[6–10] show that, for a given solute and separation temperature, T, the relationship between capacity factor (k') and the volume fraction (V), or the eluent concentration, can be expressed as follows:

$$k' = aV^{-b} \tag{1}$$

where a and b are constants.

Encyclopedia of Chromatography DOI: 10.1081/E-ECHR-120027336

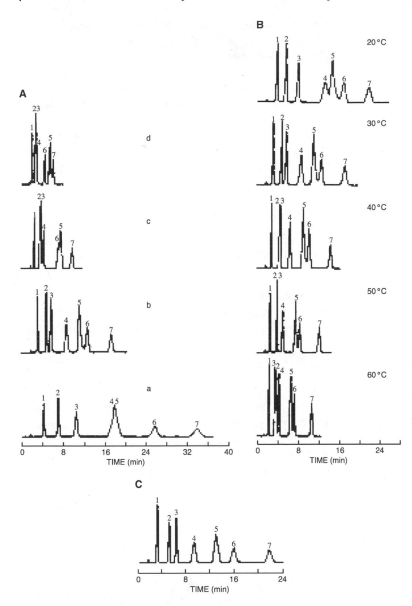

Fig. 1 (A) Effect of changing isopropanol volume fraction in the mobile phase on the elution profiles of cefonicid (1), cefaclor (2), cephazolin (3), cefodizime (4), cephaloridine (5), cephamandole (6), and cephalotin (7), using a mobile phase of 0.1 M acetate buffer (pH 5.00)/ isopropanol (v/v) $a = (96.5/3.5)$, $b = (95/5)$, $c = (93/7)$, and $d = (90/10)$ at a column temperature of 30°C. (B) Effect of column temperature on the elution profiles of cephalosporins studied [numbering as in (A)], using a mobile phase of 0.1 M acetate buffer (pH 5.00)/isopropanol (95/5) (v/v). (C) Isocratic elution profile under optimal conditions. Mobile phase 0.1 M acetate buffer (pH 5.00)/isopropanol (95.4/4.6) (v/v), column temperature, 32°C, numbering as in (A).

For all antibiotics, the plots of log k' vs. log V gave straight lines for all temperatures, with a correlation of 0.9992 or greater (Fig. 2A). As one can see, there are different slopes, which may probably be ascribed to different separation mechanisms.

Table 1 lists constants a and b, calculated from the intercepts and the slopes, respectively, by means of a least-squares method. The slopes for 1, 2, and 7 were found to be slightly affected by the column temperature, especially the slope for 1 (Cefonicid).

Capacity Factor as a Function of Column Temperature

Fig. 1B shows five chromatograms for the mixture of β-lactams, obtained at different temperatures. As one can see, the retention time of each antibiotic increased

strongly as the column temperature was increased from 20°C to 60°C.

The dependence of the capacity factor on temperature is given by the Van't Hoff equation:

$$\ln k' = -\Delta H°/RT + \Delta S°/R + \Phi \qquad (2)$$

where R is the gas constant, $\Delta H°$ and $\Delta S°$ are the enthalpy and entropy changes, respectively, associated with the solute retention process. The parameter Φ is the phase ratio and T is the absolute temperature.

Fig. 2B shows a Van't Hoff plot for each antibiotic. It can be seen that the lines generated from the ln k' of each compound at different temperatures are linear with a correlation coefficient of 0.995 or greater. The linearity of the plots supports the assumption that single sorption mechanisms are operative for each antibiotic.

A

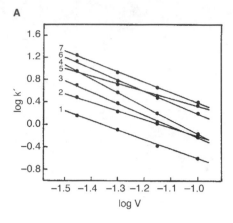

B

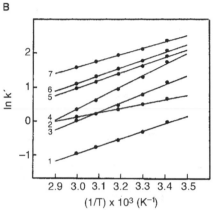

Fig. 2 (A) Effect of isopropanol volume fraction (V) in mobile phase on the capacity factor (k') at 30°C. Mobile phases as in Fig. 1B and numbering as in Fig. 1A. (B) Effect of column temperature on the capacity factor (k'). Mobile phase as in Fig. 1B and numbering as in Fig. 1A.

As for β-lactams and other compounds,[2,11,12] the slopes of these lines were not all the same, as one might expect if the effect of temperature was generalized.

For all antibiotics, the values of enthalpy change are negative, which indicates that the transfer of antibiotic from the mobile phase to sorption sites is favored.

Determination of Capacity Factor for Any Value of Volume Fraction of Isopropanol and Column Temperature

Following the methodology developed by Gant et al.,[3] it is possible to obtain the values for the capacity factor ($k_{T,V}'$) for any value of volume fraction of isopropanol and any column temperature. In this methodology, a standard state is defined by a temperature (T_S) and mobile phase composition (V_S). In the present study, $T_S = 20°C$ (293.3 K) and $V_S = 0.035$. The standard state value of k' for the solute in question is k_{T_S,V_S}'.

From Eq. (1), we can write:

$$\log k_{T_S,V}' = \log k_{T_S,V_S}' - b(\log V - \log V_S) \tag{3}$$

where $k_{T_S,V}'$ is the capacity factor for a value of V and the temperature T_S.

From Eq. (2), we can write at any value of T and V:

$$\log k_{T,V}' = \log k_{T_S,V}' - c(1/T_S - 1/T) \tag{4}$$

where $k_{T,V}'$ is the capacity factor for any value of T and V, and the parameter c varies with the β-lactam and with mobile phase composition.

According to Eqs. (3) and (4), the temperature coefficient c must be of the form:

$$c = d - e \log V \tag{5}$$

where d and e are constant for a given β-lactam and system.

Using Eqs. (3), (4), and (5), it is possible to calculate the capacity factor $k_{T,V}'$ for any value of T and V, after determining the values of b and c from these equations and the experimental data. The parameter c used for each volume fraction of isopropanol is determined

Table 1 Values of constants a and b in Eq. (1) at different column temperatures

| | Column temperature (°C) | | | | | | | | | |
| | 20 | | 30 | | 40 | | 50 | | 60 | |
Cephalosporin	b	$a \times 10^3$	b	$a \times 10^3$	b	$a \times 10^3$	b	$a \times 10^3$	b	$a \times 10^3$
1	1.43	13.9	1.67	5.07	1.82	2.48	2.18	0.665	2.24	0.472
2	1.41	29.5	1.49	20.0	1.55	14.1	1.52	13.4	1.65	7.86
3	2.01	7.86	2.03	5.23	1.99	4.23	2.20	1.71	2.14	1.60
4	2.50	3.22	2.46	2.29	2.48	1.54	2.71	0.559	2.70	0.410
5	1.24	163.7	1.35	92.6	1.36	68.7	1.33	60.3	1.42	37.6
6	1.96	22.8	2.03	14.1	2.03	10.4	2.04	8.25	2.20	4.30
7	1.74	59.5	1.86	33.0	1.87	25.9	2.02	13.7	1.99	12.6

Table 2 Experimental vs. calculated capacity factor (k') values for cephalosporins studied at different temperatures

Cephalosporin	Temperature (°C)									
	20		30		40		50		60	
	Experimental	Calculated	Experimental	Calculated	Experimental	Calculated	Experimental	Calculated	Experimental	Calculated
1	1.01	1.01	0.75	0.79	0.58	0.62	0.47	0.49	0.39	0.38
2	2.03	2.03	1.69	1.75	1.46	1.51	1.26	1.30	1.14	1.12
3	3.36	3.33	2.23	2.46	1.63	1.82	1.26	1.34	1.01	0.99
4	6.31	5.52	3.75	3.85	2.60	2.69	1.87	1.88	1.42	1.31
5	7.03	6.78	5.21	5.33	4.00	4.18	3.23	3.29	2.69	2.58
6	8.34	8.05	6.03	6.33	4.66	4.98	3.73	3.92	3.08	3.08
7	11.1	10.9	8.58	8.94	7.00	7.30	5.85	5.96	4.98	4.86

Volume fraction of isopropanol $V = 0.05$.
Numbering as in Table 1.

from Eq. (5), using the constants d and e previously determined by plotting experimental c values vs. log V. Table 2 compares the experimental and calculated values of the capacity factor, k', for the antibiotic studied, experimental k' values being predicted generally with good agreement.

EFFECT OF ELUTION CONDITIONS ON RESOLUTION

The conventional equation to evaluate the effect of elution conditions on resolution (R_S) is:

$$R_S = 1/4\sqrt{N}(k'/1 + k')(\alpha - 1/\alpha) \qquad (6)$$

Here, N is the plate number, α is the selectivity factor (defined as k_2'/k_1'), where k_1' and k_2' are the capacity factors for bands 1 and 2, and k' is the average capacity factor of k_1' and k_2'. The three factors, α, N, and k', control the resolution. It is assumed that the three terms of Eq. (6) are approximately independent, which allows their separate optimization. The resolution equation Eq. (6) must be applied to each of the adjacent band-pairs considered.

As one can see in Fig. 1B, the band broadening of any peak appearing at a fixed retention time decreases as the temperature increases; accordingly, the N value increases with increasing temperature. A linear relationship was obtained for the plot of N vs. T: $N = 24.7\,T - 6560$ (T has the dimension of absolute temperature) with a correlation coefficient of 0.991. The N value is practically independent of the volume fraction of isopropanol for the β-lactam studied.

In general, the contribution of $k'/1 + k'$ terms to R_S is approximately constant as long as k' is not small. As the selectivity factor (α) is of greater concern in Eq. (6), the separate optimization of the influence of the temperature and mobile phase composition in values is in many cases a good criterion for establishing the elution conditions of the separation.

Fig. 3 shows the influence of temperature and isopropanol volume fraction in the term: $(\alpha - 1/\alpha)$. There are examples of different situations: for pairs 5–6 and 4–5, the $(\alpha - 1/\alpha)$ term is markedly influenced by the isopropanol volume fraction; the column temperature also exerts an important influence on this term for pairs 2–3 and 4–5. It is interesting to note that both parameters (T and V) have important and similar effects on pairs 4–5 and 2–3 in this resolution term. For these cases, both parameters can be used to obtain a better resolution. Fig. 1C shows an elution profile under optimal elution conditions.

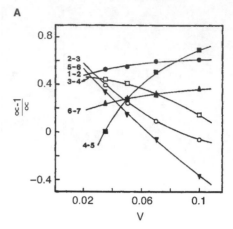

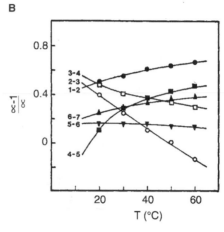

Fig. 3 (A) Effect of column temperature on the selectivity factor ($\alpha - 1/\alpha$) of the six pairs of sequentially resolved peaks. Mobile phase as in Fig. 1B and numbering as in Fig. 1A. (B) Effect of isopropanol volume fraction (V) in mobile phase on the selectivity factor ($\alpha - 1/\alpha$) of the six pairs of sequentially resolved peaks at 30°C. Mobile phases and numbering as in Fig. 1B and numbering as in Fig. 1A.

CONCLUSIONS

The present work demonstrates that there is linear relationship between the log k' and both the log V (volume fraction of organic solvent in the mobile phase) and the reciprocal of the absolute temperature. Therefore, it is possible with a small number of initial experimental data of the capacity factor (k') to predict k' for each cephalosporin as a function of T and V, which reveals the optimal elution conditions for the isocratic separation of a mixture of cephalosporins.

REFERENCES

1. Martín-Villacorta, J.; Méndez, R.; Negro, A. Effect of temperature on HPLC separation of penicillins. J. Liq. Chromatogr. **1988**, *11* (8), 1707.

2. Martín-Villacorta, J.; Méndez, R. Effect of temperature and mobile phase composition on RP-HPLC separation of cephalosporins. J. Liq. Chromatogr. **1990**, *13* (16), 3269.

3. Gant, J.R.; Dolan, J.W.; Snyder, L.R. Systematic approach to optimizing resolution in reversed-phase liquid chromatography, with emphasis on the role of temperature. J. Chromatogr. **1979**, *185*, 153.

4. Baba, Y.; Yoza, N.; Ohashi, S. Effect of column temperature on high-performance liquid chromatographic behaviour of inorganic polyphosphates: I. Isocratic ion-exchange chromatography. J. Chromatogr. **1985**, *348*, 27.

5. Atamna, I.; Gruska, E. Optimization by isochronal analysis: II. Changes in mobile phase velocity and temperature, and in mobile phase composition and temperature. J. Chromatogr. **1986**, *355*, 41.

6. Rothbart, H.L.; Weymouth, J.W.; Rieman, W., III. Separation of the oligophoshates. Talanta **1964**, *11*, 33.

7. Ohashi, S. Chromatography of phosphorous oxoacids. Pure Appl. Chem. **1975**, *44*, 415.

8. Ohashi, S.; Tsuji, N.; Ueno, Y.; Takeshita, M.; Muto, M. Elution peak positions of linear phosphates in gradient elution chromatography with an anion-exchange resin. J. Chromatogr. **1970**, *50*, 349.

9. Nakamura, T.; Kimura, M.; Waki, H.; Ohashi, S. The pH dependence of anion exchange chromatographic separation of tri- and tetraphosphate anions. Bull. Chem. Soc. Jpn. **1971**, *44*, 1302.

10. Schoenmakers, P.J.; Billiet, H.A.H.; Tijssen, R.; de Galan, L. Gradient selection in reversed-phase liquid chromatography. J. Chromatogr. **1978**, *149*, 519.

11. Chemielowiec, J.; Sawatzky, H. Entropy dominated high performance liquid chromatographic separations of polynuclear aromatic hydrocarbons. Temperature as a separation parameter. J. Chromatogr. Sci. **1979**, *17*, 245.

12. Diasio, R.B.; Wilburn, M.E. Effect of subambient column temperature on resolution of fluorouracyl metabolites in reversed-phase high performance liquid chromatography. J. Chromatogr. Sci. **1979**, *17*, 565.

B

Binding Constants: Determination by Affinity Chromatography

David S. Hage
Department of Chemistry, University of Nebraska–Lincoln, Lincoln, Nebraska, U.S.A.

INTRODUCTION

Numerous interactions within cells and the body are characterized by the specific binding that occurs between two or more molecules. Examples include the binding of hormones with hormone receptors, drugs with enzymes or receptors, antibodies with antigens, and small solutes with transport proteins. The study of these interactions is important in determining the role they play in biological systems. Because of this, there have been many methods developed to characterize such reactions. One of these approaches is affinity chromatography.

Affinity chromatography is a liquid chromatographic technique that makes use of an immobilized ligand, usually of biological origin, for the separation and analysis of analytes within a sample. However, it is also possible to use affinity chromatography as a tool for studying the interactions between the ligand and injected solutes. This application is known as *quantitative affinity chromatography, analytical affinity chromatography,* or *biointeraction chromatography.* Some attractive features of this approach include its relative simplicity, good precision and accuracy, and ability to use the same ligand for multiple studies. This article discusses various techniques employed for such studies. This includes methods for measuring both equilibrium constants and rate constants for biological processes, thus giving data on the thermodynamics and kinetics of these reactions.

One advantage of affinity chromatography is its ability to reuse the same ligand preparation for multiple experiments. This creates a situation in which only a relatively small amount of ligand is needed for a large number of studies. This helps give good precision by minimizing run-to-run variations. Other advantages include the ease with which affinity methods can be automated, especially when used in HPLC, and the relatively short periods of time required with such systems for solute binding studies (i.e., often 5–15 min per analysis). The fact that the immobilized ligand is continuously washed with an applied solvent is another advantage, since this minimizes the effects produced by soluble contaminants in the initial ligand preparation.

ZONAL ELUTION

The method of zonal elution is one of the most common techniques used in affinity chromatography to examine biological interactions.[1–5] An example of this type of experiment is shown in Fig. 1.[6] In its usual form, zonal elution involves the application of a small amount of analyte (in the absence or presence of a competing agent) to a column that contains an immobilized ligand. The retention of the analyte in this case depends on how strongly the analyte and competing agent bind to the ligand and on the amount of ligand that is in the column. This makes it possible to measure the equilibrium constants for these binding processes by examining the change in analyte retention as the competing agent's concentration is varied.

Zonal elution was first used for quantitative affinity chromatography in 1974 by Dunn and Chaiken, who examined the retention of staphylococcal nuclease on a low-performance column containing immobilized thymidine-5′-phosphate-3′-aminophenylphosphate.[1] By the late 1980s and early 1990s, reports also began to appear in which this approach was used with HPLC. It has since been used to examine numerous biological systems, including the binding of drugs with transport proteins, lectins with sugars, enzymes with inhibitors, and hormones with hormone-binding proteins.[2–5]

Eq. (1) represents one specific type of zonal elution study, in which the injected analyte and competing agent bind at a single common site on the immobilized ligand.

$$k = K_{a,A} m_L / \{V_M(1 + K_{a,I}[I])\} \tag{1}$$

Similar equations can be derived for other systems, such as those involving multiple binding sites or the presence of both soluble and immobilized forms of the ligand.[2–5] In Eq. (1), $K_{a,A}$ and $K_{a,I}$ are the association equilibrium constants for the binding of the ligand to the analyte (A) and competing agent (I) at their site of competition. The term [I] is the concentration of I applied to the column, m_L is the moles of common ligand sites for A and I, and V_M is the void volume of the column. The term k is the retention factor (or capacity factor) measured for A, as given by the relationship $k = (t_R/t_M) - 1$, where t_R is the retention

Encyclopedia of Chromatography DOI: 10.1081/E-ECHR-120039892

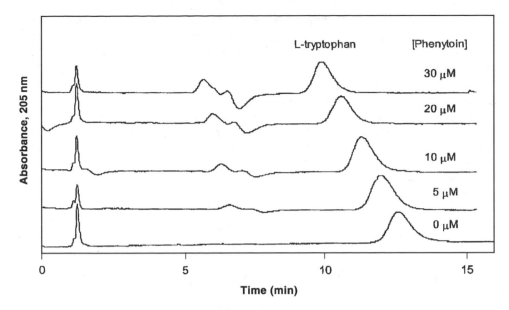

Fig. 1 Zonal elution studies for the injection of l-tryptophan onto an immobilized human serum albumin column in the presence of various concentrations of phenytoin as a mobile phase additive. (From Ref.[6].)

time for A and t_M is the void time. In this case, the values of the association constants $K_{a,A}$ and/or $K_{a,I}$ can be obtained by examining how the retention factor for A changes with [I].

There are a variety of ways in which zonal elution has been used to obtain information on the binding of solutes to a ligand. These include not only measurements of the degree and affinity of solute-protein binding but also studies examining changes in binding with variations in the mobile phase composition or temperature and experiments that consider how alterations in solute or protein structure affect these interactions. Each of these applications relies on the fact that the retention observed for an injected analyte is a direct measure of that analyte's interactions within the column. This is described by Eq. (2), which shows how the analyte's overall retention factor is related to the number of binding sites it has in the column and to the equilibrium constants for the analyte at these sites.[3]

$$k = (K_{a1}n_1 + \cdots + K_{an}n_n)m_L/V_M \qquad (2)$$

In this equation, the association equilibrium constants for the analyte are given by the terms K_{a1} through K_{an}, while the fraction of each type of site in the column is given by n_1 through n_n. From this equation, it can be seen that a change in the strength of binding, the number of binding sites, or the relative distribution of these sites can result in a shift in analyte retention.

One way zonal elution has been employed is as a means to measure the average extent of binding between a solute and immobilized ligand. This is based on the fact that the retention factor, when measured at true equilibrium, is equal to the fraction of an injected solute that is bound to the ligand (*b*) divided by the fraction of solute that remains free in the mobile phase (*f*), or $k = b/f$.[3] Another way in which the relative binding of two solutes can be compared is by taking the ratio of their retention factors on the same affinity column. According to Eq. (2), if both solutes have a single, common binding site on the ligand, the ratio of their retention factors should equal the ratio of their association constants at this site. However, caution must be exercised when using this approach with solutes that have multisite binding or different binding sites on a ligand, since these sites may have different susceptibilities to a loss of activity during immobilization.[7]

The most common use for zonal elution in quantitative affinity chromatography has been in competition and displacement studies. This is performed by injecting the analyte while a fixed concentration of a potential competing agent is passed through the column in the mobile phase. It is relatively easy from such work to determine qualitatively whether or not two compounds interact as they bind to the same immobilized ligand. But to obtain further information, such as the nature of this competition and the number of sites involved, it is necessary to compare the zonal elution data to the response expected for various models, such as that given for a system with 1:1 competition in Eq. (1).

A third way in which zonal elution can be used is to consider how changes in the reaction conditions affect solute–ligand binding. For instance, this can be examined by varying the pH, ionic strength, or general content of the mobile phase.[4] This is valuable in helping estimate the relative contributions of various forces to the formation and stabilization of a solute–ligand complex. As one example, changing the pH can affect the interactions between a ligand and solute by changing their conformations, net charges, and/or coulombic interactions. An increase in ionic strength tends to decrease coulombic interactions through a shielding effect, but at the same time may cause an increase in nonpolar solute adsorption. Adjusting the solvent's polarity by adding a small amount of organic modifier can alter solute–ligand binding by disrupting nonpolar interactions or by causing a change in structure.[4,8]

Temperature is another factor that can be varied during zonal elution studies. For instance, the following equation can be used for a system with 1 : 1 binding:

$$\ln k = -(\Delta H/RT) + \Delta S/R + \ln(m_L/V_M) \qquad (3)$$

where T is the absolute temperature at which the retention factor is measured, R is the ideal gas law constant, ΔH is the change in enthalpy for the reaction, ΔS is the change in entropy, and other terms are as defined previously. If it is known that there is no temperature dependence in the number of binding sites (m_L) for a ligand, the slope of a linear plot of $\ln k$ vs. $1/T$ can be used to determine the value of ΔH for a solute–ligand system.[9]

Yet another application of zonal elution in affinity chromatography has been its use in determining the locations and structures of binding regions on a ligand. For instance, if it is known where one agent interacts with a ligand, competition studies with this agent can be used to determine if other compounds bind at the same site. Another approach towards learning about binding sites is to study how changes in the structure of a solute or ligand affect their interactions. This is the principle behind the use of zonal elution to develop a *quantitative structure–retention relationship* (QSRR).

This involves measuring retention factors for a large set of structurally related compounds under constant temperature and mobile phase conditions. The resulting data are then compared to factors that describe various structural features of the solutes.[10,11] A complementary approach is to use zonal elution to investigate how solute retention changes as alterations are made to binding sites on a ligand, as has been performed in work with modified proteins and protein fragments.[4,12,13]

FRONTAL ANALYSIS

An alternative approach for equilibrium constant measurements is to use frontal analysis. In this technique, a solution containing a known concentration of the analyte is continuously applied to an affinity column at a fixed flow rate (Fig. 2). As the analyte binds to the immobilized ligand, the ligand becomes saturated and the amount of analyte eluting from the column gradually increases. This forms a characteristic breakthrough curve. The volume of the analyte solution required to reach the mean position of this curve is measured. If the association and the dissociation kinetics are fast, the mean position of the breakthrough curve is related to the concentration of the applied solute, the amount of ligand in the column, and the association equilibrium constants for solute–ligand binding. Frontal analysis experiments have been used to examine such systems as drug–protein binding, antibody–antigen interactions, and enzyme–inhibitor interactions.[3–5]

A simple example of a frontal analysis system is one where an applied analyte binds to a single type of immobilized ligand site. In this situation, the following equation can be used to relate the true number of

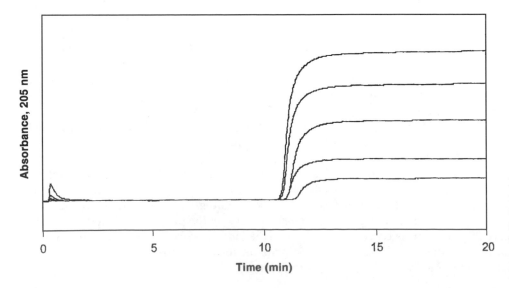

Fig. 2 Frontal analysis studies for the binding of phenytoin to immobilized human serum albumin at applied analyte concentrations (bottom-to-top) of 5, 10, 20, 30, and 40 µM. (From Ref.[6].)

active binding sites in the column (m_L) to the apparent moles of analyte ($m_{L,app}$) required to reach the mean position of the breakthrough curve.

$$1/m_{L,app} = 1/(K_{a,A}m_L[A]) + 1/m_L \qquad (4)$$

As defined earlier, $K_{a,A}$ is the association constant for the binding of A to L, and [A] is the molar concentration of analyte applied to the column. Eq. (4) predicts that a plot of $1/m_{L,app}$ vs. $1/[A]$ for a system with single-site binding will give a straight line with a slope of $1/(K_{a,A}m_L)$ and an intercept of $1/m_L$. The value of $K_{a,A}$ can be determined by calculating the ratio of the intercept to the slope, and $1/m_L$ is obtained from the inverse of the intercept. Similar relationships can be derived for cases in which there is more than one type of binding site or in which both a competing agent and solute are applied simultaneously to the column.

Frontal analysis with low-performance affinity columns was first employed for biological binding studies in the mid-to-late 1970s.[14–16] In the mid-1980s through early 1990s, this approach also began to see use with HPLC-based affinity columns. Like zonal elution, frontal analysis can provide a variety of information regarding a solute–ligand system, including the affinity and number of binding sites for a solute, the nature of this binding (e.g., single site or multisite), the effects of temperature or solvent on this binding, and the changes that occur in the presence of a competing agent.[3]

Most quantitative applications of frontal analysis have involved its use in providing data on the affinity and amount of ligand in a column. This is accomplished by measuring the breakthrough times for a solute at several concentrations and fitting the results to expressions like Eq. (4) based on a given reaction model.

The main advantage of frontal analysis over traditional zonal elution is that it can simultaneously provide information on both the association constant for a solute and its number of binding sites. This makes frontal analysis the method of choice when information is needed on the binding capacity. Frontal analysis is also preferred for accurate association constant measurements, since the values it provides for K_A can be determined independent of the binding capacity.

A second application of frontal analysis has been as a tool to examine the competition between solutes for an immobilized ligand. This is performed in a similar manner to that described for zonal elution, in which the change in analyte retention is measured as a function of the competing agent's concentration in the mobile phase. In frontal analysis, direct competition between the analyte and competing agent leads to a smaller breakthrough time for the analyte as the competing agent's level is increased. Positive or negative allosteric effects can also be observed, which lead to a shift to higher or lower breakthrough times, respectively, with an increase in the competing agent's concentration. The same technique can be used to examine how temperature, pH, ionic strength, or solvent polarity affect solute–ligand binding.[3] Like zonal elution, frontal analysis has been used to examine the binding of solutes to modified proteins to provide information on the nature of solute–ligand binding sites.[12,17]

One disadvantage of frontal analysis is that it requires a relatively large amount of analyte for study. However, frontal analysis does provide information on both the association constant for a solute and its total number of binding sites in a column. This feature makes frontal analysis the method of choice for high accuracy in equilibrium measurements, since the resulting association constants are essentially independent of the number of binding sites in the column.

BAND-BROADENING MEASUREMENTS

Another group of methods in quantitative affinity chromatography are those that examine the kinetics of biological interactions. Band-broadening measurements (also known as the *isocratic method*) represent one such approach. This is really a modification of the zonal elution method in which the widths of the eluting peaks are measured along with their retention times. Systems that have been studied with this method include the binding of lectins with sugars, the interactions of drugs and amino acids with serum albumin, and the kinetics of protein-based chiral stationary phases.[2,18,19]

This type of experiment involves injecting a small amount of an analyte onto an affinity column while carefully monitoring the retention time and width of the eluting peak. These injections are performed at several flow rates on both the affinity column and on a column of the same size that contains an identical support but with no immobilized ligand present. This control column is needed to correct for any band-broadening that occurs due to processes other than the binding and dissociation of analyte from the immobilized ligand. By comparing plots of the peak widths (or plate heights) for the affinity and control columns, it is possible to determine the value of the dissociation rate constant for the analyte–ligand interaction. An example of such a study is given in Fig. 3. A variation of this approach involves simultaneously determining the band-broadening for both a retained and a non-retained solute on an affinity column to examine the contributions of stationary phase mass transfer vs. other band-broadening processes.[20]

A

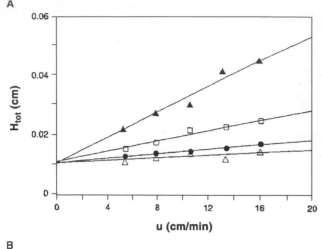

B

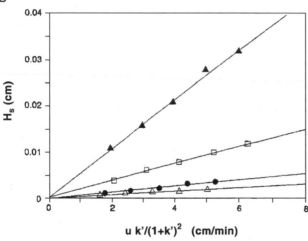

Fig. 3 Typical plots of (A) total plate height (H_{tot}) and (B) the plate height contribution due to stationary phase mass transfer (H_s) for injections of d-tryptophan at various flow rates onto an immobilized human serum albumin column. Symbols: u, linear velocity; k', retention factor. (From Ref.[19].)

SPLIT-PEAK EFFECT

Another way in which kinetic information can be obtained by affinity chromatography is the *split-peak method*.[2,21] This is based on an effect that occurs when the injection of a single solute gives rise to two peaks: the first representing a nonretained fraction and the second representing the retained solute. This effect can be observed even when only a small amount of analyte is injected and is the result of slow adsorption kinetics and/or slow mass transfer of analyte within the column (see Fig. 4). Such an effect can occur in any type of chromatography but is most common in affinity columns because of their smaller size, their lower amount of binding sites, and the slower association rates of affinity ligands compared to other types of stationary phases.

Split-peak measurements can be performed by injecting a small amount of analyte onto an affinity column at various flow rates. A plot of the inverse negative logarithm of the measured free fraction is then made vs. flow rate. The slope of this graph is related to the adsorption kinetics and mass transfer rates within the column. If the system is known to have adsorption-limited retention, or if the mass transfer rates are known, then the association rate constant for analyte binding can be determined. This approach has the advantages of being fast to perform and potentially has greater accuracy and precision than band-broadening measurements. Its disadvantages are that it requires fairly specialized operating conditions that may not be suitable for all analytes. Examples of biological systems that have been examined by the split-peak method include the binding of protein A and protein G to immunoglobulins and the binding of antibodies with antigens.[2,21–23]

PEAK-DECAY METHOD

The *peak decay method* is a third approach that can be used in affinity chromatography to examine the kinetics of an analyte–ligand interaction.[24] This technique is performed by first equilibrating and saturating a small affinity column with a solution that contains the analyte of interest or an easily detected analog of this analyte. The column is then quickly switched to a mobile phase in which the analyte is not present. The release of the bound analyte is then monitored over time, resulting in a decay curve. This decay is related to the dissociation rate of the analyte and the mass transfer kinetics within the column. If the mass transfer rate is known or is fast compared to analyte dissociation, then the decay curve can be used to provide the dissociation rate constant for the analyte from the immobilized ligand. Systems that have been studied with this approach include the dissociation of drugs from transport proteins and the dissociation of sugars from immobilized lectins.

FREE FRACTION ANALYSIS

Another recent method described for solute–ligand studies is chromatographic *free fraction analysis*. This uses small columns with antibodies that bind the solute of interest and are capable of extracting this solute in very short periods of time (i.e., 80–200 ms). With such a column, it is possible to isolate the nonbound fraction of a solute from a solution in which the most of this compound is bound to a soluble ligand, even when dissociation of the solute from this ligand occurs on the time scale of a few seconds. This approach has been used to examine the binding of *R*- and *S*-warfarin with

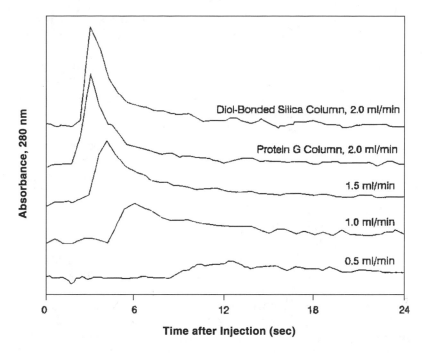

Fig. 4 Nonretained (or split-peak) fractions observed for injections of rabbit immunoglobulin G onto an immobilized protein G column at various flow rates. (From Ref.[22].)

human serum albumin (HSA) in solution, in which anti-warfarin antibodies were used to extract free warfarin fractions in less than 180 ms. The advantages of this approach are its speed and its ability to study the binding of solutes with ligands directly in solution.[25]

CONCLUSIONS

It has been shown that there are a variety of ways in which information on solute–ligand binding can be obtained by affinity chromatography. Zonal elution and frontal analysis are the approaches most commonly used for such studies. With these methods, information can be obtained on the extent of solute–ligand binding, binding affinity and stoichiometry, and the structure of binding sites. Kinetic information can also be obtained through a variety of methods, including band-broadening measurements, split-peak studies, peak decay analysis, and free fraction measurements.

ARTICLE OF FURTHER INTEREST

Affinity Chromatography: An Overview, p. 33.

REFERENCES

1. Dunn, B.M.; Chaiken, I.M. Quantitative affinity chromatography. Determination of binding constants by elution with competitive inhibitors. Proc. Natl. Acad. Sci. U.S.A. **1974**, *71*, 2382–2385.
2. Chaiken, I.M., Ed.; *Analytical Affinity Chromatography*; CRC Press: Boca Raton, FL, 1987.
3. Hage, D.S.; Tweed, S.A. Recent advances in chromatographic and electrophoretic methods for the study of drug–protein interactions. J. Chromatogr. B. **1997**, *699*, 499–525.
4. Hage, D.S.; Chen, J. Quantitative affinity chromatography: practical aspects. In *Handbook of Affinity Chromatography*; Hage, D.S., Ed.; Marcel Dekker: New York, 2005 (Chapter 22).
5. Winzor, D.J. Quantitative affinity chromatography: Recent theoretical developments. In *Handbook of Affinity Chromatography*; Hage, D.S., Ed.; Marcel Dekker: New York, 2005 (Chapter 23).
6. Chen, J.; Ohnmacht, C.; Hage, D.S. Studies of phenytoin binding to human serum albumin by high-performance affinity chromatography. J. Chromatogr. B. **2004**, *809*, 137–145.
7. Loun, B.; Hage, D.S. Characterization of thyroxine–albumin binding using high-performance affinity chromatography. I. Interactions at the warfarin and indole sites of albumin. J. Chromatogr. **1992**, *579*, 225–235.
8. Allenmark, S. *Chromatographic Enantioseparation: Methods and Applications*; 2nd Ed.; Ellis Horwood: New York, 1991 (Chapter 7).
9. Yang, J.; Hage, D.S. Role of binding capacity versus binding strength in the separation of chiral compounds on protein-based high-performance liquid chromatography columns. Interactions of D- and L-tryptophan with human

serum albumin. J. Chromatogr. A. **1996**, *725*, 273–285.

10. Kaliszan, R.; Noctor, T.A.G.; Wainer, I.W. Stereochemical aspects of benzodiazepine binding to human serum albumin. II. Quantitative relationships between structure and enantioselective retention in high performance liquid affinity chromatography. Mol. Pharmacol. **1992**, *42*, 512–517.

11. Wainer, I.W. Enantioselective high-performance liquid affinity chromatography as a probe of ligand–biopolymer interactions: an overview of a different use for high-performance liquid chromatographic chiral stationary phases. J. Chromatogr. A. **1994**, *666*, 221–234.

12. Chattopadhyay, A.; Tian, T.; Kortum, L.; Hage, D.S. Development of tryptophan-modified human serum albumin columns for site-specific studies of drug–protein interactions by high-performance affinity chromatography. J. Chromatogr. B. **1998**, *715*, 183–190.

13. Haginaka, J.; Kanasugi, N. Enantioselectivity of bovine serum albumin-bonded columns produced with isolated protein fragments. II. Characterization of protein fragments and chiral binding sites. J. Chromatogr. A. **1997**, *769*, 215–223.

14. Kasai, K.; Ishii, S. Affinity chromatography of trypsin and related enzymes. I. Preparation and characteristics of an affinity adsorbent containing tryptic peptides from protamine as ligands. J. Biochem. (Tokyo) **1975**, *78*, 653–662.

15. Nakano, N.I.; Oshio, T.; Fujimoto, Y.; Amiya, T. Study of drug–protein binding by affinity chromatography: interaction of bovine serum albumin and salicylic acid. J. Pharm. Sci. **1978**, *67*, 1005–1008.

16. Lagercrantz, C.; Larsson, T.; Karlsson, H. Binding of some fatty acids and drugs to immobilized bovine serum albumin studied by column affinity chromatography. Anal. Biochem. **1979**, *99*, 352–364.

17. Nakano, N.I.; Shimamori, Y.; Yamaguchi, S. Mutual displacement interactions in the binding of two drugs to human serum albumin by frontal affinity chromatography. J. Chromatogr. **1980**, *188*, 347–356.

18. Loun, B.; Hage, D.S. Chiral separation mechanisms in protein-based HPLC columns. 2. Kinetic studies of (*R*)- and (*S*)-warfarin binding to immobilized human serum albumin. Anal. Chem. **1996**, *68*, 1218–1225.

19. Yang, J.; Hage, D.S. Effect of mobile phase composition on the binding kinetics of chiral solutes on a protein-based high-performance liquid chromatography column: interactions of D- and L-tryptophan with immobilized human serum albumin. J. Chromatogr. A. **1997**, *766*, 15–25.

20. Talbert, A.M.; Tranter, G.E.; Holmes, E.; Francis, P.L. Determination of drug–plasma protein binding kinetics and equilibria by chromatographic profiling: exemplification of the method using L-tryptophan and albumin. Anal. Chem. **2002**, *74*, 446–452.

21. Hage, D.S.; Walters, R.R.; Hethcote, H.W. Split-peak affinity chromatographic studies of the immobilization-dependent adsorption kinetics of protein A. Anal. Chem. **1986**, *58*, 274–279.

22. Rollag, J.G.; Hage, D.S. Non-linear elution effects in split-peak chromatography. II. Role of ligand heterogeneity in solute binding to columns with adsorption-limited kinetics. J. Chromatogr. A. **1998**, *795*, 185–198.

23. Hage, D.S.; Thomas, D.S.; Beck, M.S. Theory of a sequential addition competitive binding immunoassay based on high-performance immunoaffinity chromatography. Anal. Chem. **1993**, *65*, 1622–1630.

24. Moore, R.M.; Walters, R.R. Peak-decay method for the measurement of dissociation rate constants by high-performance affinity chromatography. J. Chromatogr. **1987**, *384*, 91–103.

25. Clarke, W.; Chowdhuri, A.R.; Hage, D.S. Analysis of free drug fractions by ultrafast immunoaffinity chromatography. Anal. Chem. **2001**, *73*, 2157–2164.

Binding Molecules via –SH Groups

B

T. M. Phillips
Ultramicro Analytical Immunochemistry Resource, DBEPS, ORS, OD,
National Institutes of Health, Bethesda, Maryland, U.S.A.

INTRODUCTION

A prerequisite for producing a good affinity support is a firm, stable attachment of the ligand to the surface of the support. There are numerous linkage chemistries available for performing this task, and although the most popular approach is a reaction between the reactive side groups on the support and a primary amine on the ligand, there are a number of supports that can perform similar attachments through free thiol or sulfhydryl groups.

METHODOLOGY

Supports containing maleimide reactive side groups are specific for free sulfhydryl groups present in the ligand when the reaction is performed at pH 6.5–7.0. At pH 7.0, the interaction of maleimides with sulfhydryl groups is approximately a 1000-fold faster than with amine groups. The stable thioether linkage formed between the maleimide support and the sulfhydryl group on the ligand cannot be easily cleaved under physiological conditions, therefore ensuring a stable affinity matrix. Immobilization of sulfhydryl-containing molecules can also be achieved using either α-haloacetyl or pyridyl sulfide cross-linking agents. The α-haloacetyl cross-linkers [i.e., *N*-succinimidyl (4-iodoacetyl)aminobenzoate] contain an iodacetyl group that is able to react with sulfhydryl groups present in the ligand at physiological pH. During this reaction, the nucleophilic substitution of iodine with a thiol takes place, producing a stable thioether linkage. However, a shortcoming of this approach is that the α-haloacetyls interact with other amino acids, especially when a shortage or absence of free sulfhydryl groups exists. Linkage of pyridyl disulfides with aliphatic thiols at pH 4.0–5.0 produces a disulfide bond with the release of pyridine-2-thione as a by-product of the reaction. A disadvantage of this approach is the acidic pH of the reaction, which is essential for optimal linkage. The reaction can be performed at physiological pH, but under these conditions, the reaction is extremely slow.

Ligand immobilization through sulfhydryl groups can be advantageous due to its ability to be site-directed. Additionally, depending upon the linkage,

the ligand–support can be cleavable, allowing the same support to be reused. However, many useful affinity ligands do not possess free sulfhydryl groups and in such cases, free sulfhydryl groups can be engineered into the ligand via a series of commercially available reagents. Traut's reagent (2-iminothiolane) is the most common, although *N*-succinimidyl *S*-acetylthioacetate (SATA), dithio-*bis*-maleimidoethane (DTME), and *N*-succinimidyl-3-(2-pyridyldithio)-propionate (SPDP)[1] can also be used (Fig. 1). Traut's reagent reacts with primary amines present in the ligand introducing exposed sulfhydryl groups for further coupling reactions.

Chrisey, Lee, and O'Ferrall[2] describe an interesting use of sulfhydryl-mediated immobilization for binding thiol-modified DNA. A hetero-bifunctional cross-linker

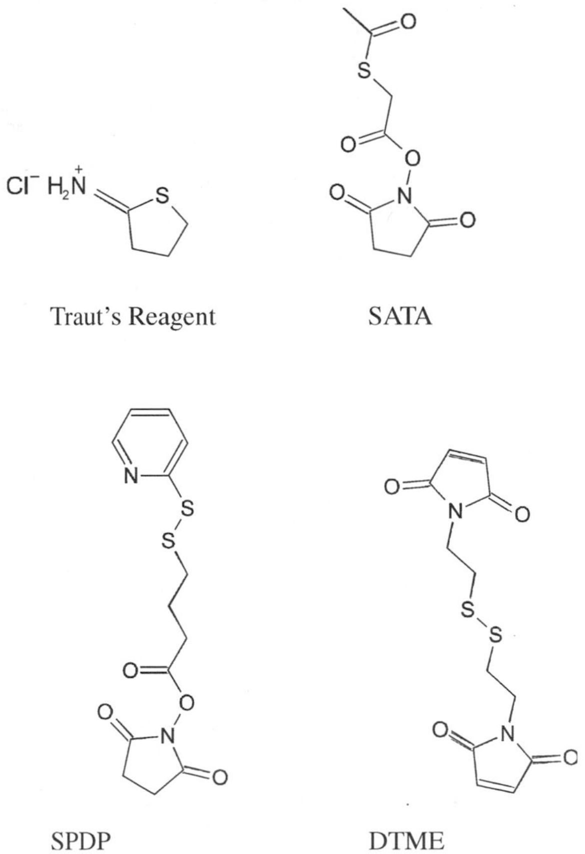

Fig. 1 Chemical structures of commercially available reagents for introducing sulfhydryl groups into molecules.

Encyclopedia of chromatography DOI: 10.1081/E-ECHR-120039893
Copyright © 2005 by Taylor & Francis. All rights reserved.

bearing both thiol and amino reactive groups was used to immobilize thiol-modified DNA oligomers to self-assembled monolayer silane films on fused silica and oxidized silicon substrates. The advantage of this approach was the use of site-directed immobilization to ensure the correct orientation of the DNA molecule.

ANTIBODY IMMOBILIZATION

Cleaving disulfide bonds already present in the ligand can also generate free sulfhydryl groups. The classic example of this approach is the digestion of the IgG antibody molecule to produce two monovalent, reactive Fab fragments each containing a free sulfhydryl group. In this case, reduction of the disulfide bridge (holding the two Fab arms together) is achieved using Cleland's reagent (dithiothreitol—DTT). Each Fab is then attached to free thiol groups present on the support by reforming a disulfide bond.[3–5] Free thiol groups can be condensed into silane-activated surfaces via sulfosuccinimidyl-4-(N-maleimidomethyl)-cyclohexane-1-carboxylate (sulfo-SMCC) or N-(β-maleimidopropyloxy) succinimide ester (BMPS). The advantage of this approach is that not only is a covalent linkage formed but also the linkage helps to orient the antigen receptor of the Fab away from the support matrix.

REFERENCES

1. Carlsson, J.; Drevin, H.; Axen, R. Protein thiolation and reversible protein–protein conjugation. N-Succinimidyl 3-(2-pyridyldithio)propionate, a new heterobifunctional reagent. Biochem. J. **1978**, *173*, 723.
2. Chrisey, L.A.; Lee, G.U.; O'Ferrall, C.E. Covalent attachment of synthetic DNA to self-assembled monolayer films. Nucleic Acids Res. **1996**, *24*, 3031.
3. Phillips, T.M. Determination of in situ tissue neuropeptides by capillary immunoelectrophores. Anal. Chim. Acta. **1998**, *372*, 209.
4. Karyakin, A.A.; Presnova, G.V.; Rubtsova, M.Y.; Egorov, A.M. Oriented immobilization of antibodies onto the gold surfaces via their native thiol groups. Anal. Chem. **2000**, *72*, 3805.
5. Phillips, T.M.; Smith, P. Analysis of intracellular regulatory proteins by immunoaffinity capillary electrophoresis coupled with laser-induced fluorescence detection. Biomed. Chromatogr. **2003**, *17*, 182.

SUGGESTED FURTHER READING

Hermanson, G.T.; Mallia, A.K.; Smith, P.K. *Immobilized Affinity Ligand Techniques*; Academic Press: New York, 1992.
Lundblad, R.L. *Techniques in Protein Modification*; CRC Press: Boca Raton, FL, 1995.
Phillips, T.M.; Dickens, B.F. *Affinity and Immunoaffinity Purification Techniques*; BioTechniques Books, Eaton Publishing: Natick, MA, 2000.
Wong, S.S. *Chemistry of Protein Conjugation and Cross-linking*; CRC Press: Boca Raton, FL, 1991.

Biopharmaceuticals: Analysis by Capillary Electrophoresis

Michel Girard
Centre for Biologics Research, Biologics and Genetic Therapies Directorate, Health Canada,
Sir F.G. Banting Research Centre, Ottawa, Ontario, Canada

INTRODUCTION

Capillary electrophoresis (CE) has rapidly established itself as one of the most versatile techniques for the analysis of biomolecules. In addition to providing exceptional separation efficiencies, it offers substantial advantages over conventional slab-gel electrophoretic techniques, namely, fast separation times, automation, reproducibility, and quantitative capabilities. Furthermore, owing to the different mechanisms by which products are separated in CE, data generated are generally complementary to those obtained by HPLC, thus allowing for more complete product characterization. Capillary electrophoresis methods can be successfully validated with respect to well-established analytical criteria (e.g., precision, accuracy, reproducibility, and linearity), making them a source of reliable information. These considerations are of key importance to the pharmaceutical industry in adopting CE as a frontline analytical technique for product characterization to meet the specific requirements associated with the manufacturing and testing of therapeutic substances.

Therapeutic biological compounds are collectively referred to as biopharmaceuticals and include recombinant proteins, monoclonal and polyclonal antibodies, antisense oligonucleotides, therapeutic genes, and recombinant and DNA vaccines. Although the majority of products on the market to date are proteins and antibodies, the first antisense oligonucleotide therapeutic, fomivirsen sodium, was approved in 1998 in the United States. Several other antisense and DNA-based products are being actively developed. Historically, biopharmaceuticals were obtained from biological sources, whether of human, animal, plant, or cellular origin, in the form of crude extracts or partially purified components of extracts. Because of the highly complex nature of these mixtures, only minimal physicochemical characterization could be carried out and product evaluation was generally based on biological response or surrogate bioassays. While a few traditional products remain in use today, newer production methods based on recombinant DNA or hybridoma technology are now being used for the large scale production of biopharmaceuticals. These developments have been paralleled by major advances in biomolecular separation and purification techniques and, consequently, have resulted in improvements in product development leading to the preparation of more consistent products with purity levels approaching those of conventional, small-molecule pharmaceuticals. A number of important therapeutics are produced in this manner. This is the case for somatropin (human growth hormone, hGH), insulin, erythropoietin (EPO), several interferons (IFN-α, -β, and -γ), and hepatitis B vaccine, to name just a few. These products are currently used for diseases not otherwise well treated with small-molecule pharmaceuticals.

Despite these advances, the characterization of biopharmaceuticals remains a challenge. The inherent structural complexity of the therapeutic substance and the potential presence of numerous process- and product-related impurities and contaminants in the final product complicate the analysis. In addition, biological activity determination is still often performed using costly and imprecise animal testing. Nevertheless, the regulatory approval of biopharmaceuticals is based in part upon a comprehensive chemistry and manufacturing submission with a strong emphasis on high resolution analytical methodologies. The determination of the quality of a product through the verification of product identity, strength, stability, and consistency of manufacture and the quantitative evaluation of impurities and contaminants are of prime importance. Therefore, the establishment of more precise and selective methods of analysis is highly relevant to the pharmaceutical and regulatory sectors.

COMMON CAPILLARY ELECTROPHORESIS SEPARATION MODES FOR BIOPHARMACEUTICALS

There are several CE modes, based on different separation mechanisms, that, alone or in combination, are commonly used for the analysis of biopharmaceuticals (Table 1). Capillary zone electrophoresis (CZE) is the simplest and most commonly used CE mode for the analysis of peptides and proteins, including glycoproteins and monoclonal antibodies (MAbs). In this mode, analytes migrate in a free solution according to their effective charge-to-size ratio. In capillary gel

Encyclopedia of Chromatography DOI: 10.1081/E-ECHR-120039894

Table 1 Capillary electrophoresis separation modes for the characterization of biopharmaceuticals

Mode	Separation mechanism	Applications
Capillary zone electrophoresis	Charge-to-size ratio	Proteins, peptides Glycoproteins Monoclonal antibodies Peptide mapping Monosaccharides, oligosaccharides
Capillary isoelectric focusing	Isoelectric point (pI)	Proteins, peptides Glycoproteins Monoclonal antibodies Isoelectric point determination Peptide mapping
Capillary gel electrophoresis	Size	Protein MW determination Aggregates Oligonucleotides, DNA fragments Polysaccharides
MEKC	Partition	Proteins, peptides Peptide mapping Monosaccharides, oligosaccharides

electrophoresis (CGE), analytes are separated according to their size through gel matrices by a molecular sieving mechanism analogous to that of polyacrylamide gel electrophoresis (PAGE). In capillary isoelectric focusing (CIEF), a stable pH gradient is formed inside the capillary and analytes migrate until they reach the pH equal to their isoelectric point (pI), at which time the net charge and mobility are zero and the analytes stop migrating. In MEKC, analytes interact with micelles formed by adding surfactants to the running buffer at concentrations above the critical micelle concentration. The separation mechanism involves partition of the analyte between the micelle and the electrolyte.

PRODUCT CHARACTERIZATION

The characterization of biopharmaceuticals involves carrying out tests to demonstrate that a given product meets established criteria and that it remains safe and efficacious. There is a wide range of physicochemical methods that are frequently used for comprehensive characterization. Methods used include, among others, electrophoresis, HPLC, mass spectrometry (MS), UV spectrophotometry, nuclear magnetic resonance, sequencing, and amino acid analysis. Capillary electrophoresis is particularly well suited for the characterization of complex biomolecules, especially with regards to product identity and for assessing product heterogeneity arising from translational and post-translational modifications, degradation, or genetic variation. These considerations are particularly important for biopharmaceuticals, since they are inherently labile substances,

especially when placed under nonphysiological conditions. Consequently, the formation of impurities may occur throughout the manufacturing process and during the shelf life of the product.

Proteins are particularly susceptible to modification and degradation through a variety of pathways, leading to the formation of several types of variants. Some of the more commonly encountered pathways include deamidation, oxidation, dimerization/aggregation, peptide bond cleavage through proteolysis or hydrolysis, N- or C-terminal truncation, and disulfide scrambling (Table 2). These transformations have a significant impact on the physicochemical properties of the molecule by altering its size, charge, mass, hydrophobicity, or conformation. For example, deamidation leads to the formation of a more acidic variant from the transformation of the amide side chain in asparagine or glutamine residues into a carboxylic acid functional group. Furthermore, degradation may occur at more than one residue, a situation that leads to the formation of complex mixtures. Consequently, the choice of a separation mode or detection system depends mainly on an in-depth evaluation of the physicochemical properties of the product under study, that is, hydrophobicity, isoelectric point, size, post-translational modifications, and susceptibility to degradation/aggregation or conformational stability, and on the type of information required.

In the following sections, an overview of the use of CE for the characterization of selected classes of biopharmaceuticals is presented. The emphasis is placed on the relationship between the test performed and its intended purpose, that is, in terms of its usefulness for product identity or product purity determination.

Table 2 Common protein modification and degradation pathways

Pathways	Reaction	Reaction site
Deamidation	$-CONH_2 \rightarrow -COOH$	Asn, Gln
Isomerization	Asn/Asp $\rightarrow$ isoAsp	Asn, Asp
Oxidation	$-SR \rightarrow -SOR, -SO_2R$	Met
	$-SH \rightarrow -SS-$	Cys
Dimerization/aggregation	Electrostatic, covalent, noncovalent bond formation	Various
Peptide bond proteolysis/hydrolysis	$-CONH- \rightarrow -COOH + NH_2-$	Various
N-, C-terminal truncation	Peptide bond cleavage	Various
Disulfide scrambling	R–SS–R' $\rightarrow$ R–SS–R''	Disulfide bridge (–SS–)

In-Process Monitoring

Capillary electrophoresis has widespread applications in the biopharmaceutical industry owing, in large part, to the rapidity with which methods can be developed and to its versatility by virtue of the wide range of separation modes available. Besides its use in the development, quality control, and batch release stages,[1] CE is also commonly applied for in-process monitoring.[2] Manufacturing processes using recombinant DNA technology involve, as a first step, the large scale production of the desired protein in a suitable expression system through fermentation or cell cultivation. Experimental conditions are critical at this stage for ensuring the production of the required substance. As such, the detection of the product or its precursor at this early stage of the manufacturing process clearly confers an economic benefit. Conventional gel electrophoretic techniques cannot be readily applied since they are labor intensive and require considerable periods of time (e.g., staining/destaining procedures). Capillary electrophoresis methods have a clear advantage in these situations by providing fast analysis, sometimes in less than 5 min. In addition, apart from enabling the confirmation of the identity of the substance, they can provide a quantitative assessment of the process. Methods based on CZE, CGE, and CIEF are commonly used for the characterization of the product in fermentation broth or cell cultivation and during the various purification steps. Optimization includes attaining maximal separation of the reference compounds, such as the intact product or the pro-product, because the presence of impurities lowers resolution and selectivity.

Product Identity

One of the critical aspects to be considered during the manufacturing of any drug is product identity. While in itself it does not fulfill all of the requirements for a safe and effective drug, product identity testing provides assurance that the product generated is that which is intended and offers a measure of the consistency of the manufacturing process. Capillary electrophoresis-based methods are applied to confirm the product identity of biopharmaceuticals. Approaches usually involve the comparison of the property of the substrate to that of a pre-established, well-characterized reference standard with demonstrated quality, efficacy, and safety. In some instances, primary reference standards are available from official organizations such as the World Health Organization, the European Pharmacopoeia (EP), or the United States Pharmacopeia (USP). These preparations are established through international collaborative studies and are generally intended for use in the characterization of in-house reference standards. Aside from performing a simple identity test involving comigration of the substrate with the reference standard, a number of methods have been devised to provide qualitative and quantitative information with respect to specific structural features of the molecule such as primary sequence, molecular weight or size, and carbohydrate or isoform profile and distribution.

Peptide mapping is one of the most powerful tools for the identification of proteins and has been successfully adapted to CE.[3] It involves the cleavage of the amino acid chain at specific sites, using proteases or chemical reagents, to generate a mixture of smaller peptides. Different proteins generate different peptides after digestion, and separation of these peptides leads to a characteristic map or "fingerprint" of that protein. The analysis of peptide digests is generally carried out by CZE, where products are separated based on differences in charge-to-mass ratios. Peptide mapping by CZE is generally considered an orthogonal technique to HPLC. Peptide mapping by CZE is usually faster than by HPLC and typically provides greater resolution of a larger number of peptides. For instance, highly hydrophilic peptides are often poorly resolved or elute in the column dead volume by reversed phase

HPLC. In CZE, these peptides migrate according to their charge-to-size ratio and are typically resolved. By contrast, peptides with similar net charges are not separated by CZE, and, in such cases, MEKC provides a useful alternative. The peptide map serves as a fingerprint of the substrate that, when compared to that of a reference standard, enables confirmation of the identity. Alternatively, it allows the detection and identification of amino acid and peptide modifications, which are indicative of the presence of product variants. In addition, it may be used to confirm the presence and position of disulfide bridges and glycosylation sites. Besides its application to simple proteins, peptide mapping by CZE can be particularly useful for the characterization of MAbs.[4]

Several important therapeutic proteins are glycoproteins [e.g., EPO and tissue plasminogen (tPA)] that exist as mixtures of closely related species that differ in their glycosylation patterns (glycoforms). These differences are often the result of both compositional and sequence variations of the glycan chains. Moreover, the biological activity of glycoproteins is frequently linked to the presence of these carbohydrates, and, consequently, the characterization of glycoprotein microheterogeneity represents one of the more challenging tasks in identity testing. Several CE approaches, based mostly on CZE and CIEF, have been devised.[5] For the frequently encountered sialoglycoproteins (i.e., sialic acid-containing glycoproteins), the analysis of the glycoform profile can be performed on the intact glycoprotein or on the glycopeptides obtained after the enzymatic digestion of the polypeptide chain. Alternatively, the analysis of the oligosaccharide profile can be performed following the chemical or enzymatic release of the glycan chains from the polypeptide backbone. In both cases, the profile obtained is an indication of the varying number of sialic acid residues on the oligosaccharide chains.

Carbohydrate analysis is essential to fully characterize a glycoprotein. Capillary electrophoresis methods are used for the analysis of the monosaccharide composition resulting from hydrolysis of the glycosylation chains. In those cases, monosaccharides must be derivatized with reagents such as 1-aminopyrene-3,6,8-trisulfonate (APTS) to provide both a readily ionizable group and a detectable chromophore. APTS has been used successfully for the derivatization of both oligosaccharides and monosaccharides and, coupled with CE/laser induced fluorescence (LIF), provides high sensitivity quantitative information with increased resolving power.

Capillary electrophoresis is a valuable tool for the confirmation of the structural integrity of biopharmaceuticals in final drug formulations.[6,7] Finished products generally contain low amounts of the active ingredient since the therapeutic effect can usually be achieved at low concentrations. In turn, formulations of low concentration proteins usually require the addition of large amounts of excipients to enhance product stability and to prevent nonspecific adsorption. Usual excipients include inorganic salts, amino acids, sugars, surfactants (e.g., polysorbate), and other proteins [e.g., human serum albumin (HSA)]. Typically, isotonic salt preparations are produced as most of these products are injectables and consequently high salt concentrations are present. Furthermore, many of these excipients may be present simultaneously, leading to complex mixtures. Such mixtures can interfere with traditional assay methodologies like HPLC or slab gel electrophoresis. Capillary zone electrophoresis is well suited for the direct analysis of products containing high salt concentrations, since conditions using highly concentrated buffer solutions increase the efficiency by contributing to the focusing effect. Fig. 1 shows the comparison of electropherograms obtained by reversed-polarity CZE[6] for a sample of formulated EPO-α (top trace) and a sample of its unformulated drug substance (bottom trace). Both traces show qualitatively similar glycoform profiles, a good indication of the structural integrity of the formulated product. These conditions also provide satisfactory results for assaying of the active ingredient and for quantitation of the glycoforms in the finished product.

In addition to other size-based analytical techniques such as size-exclusion HPLC or slab gel electrophoresis, CGE can be adequately used for the determination of a protein's apparent molecular weight.[8] Denaturing conditions are usually employed whereby sodium dodecyl sulfate (SDS)–protein complexes are formed with net overall negative charges that are proportional to their masses. These complexes migrate through the gel-filled capillary, acting as a sieving medium, in order of increasing molecular weight. The mobility of the substrate is used to estimate the molecular weight from a pre-established calibration plot of log molecular mass vs. mobility prepared from a series of protein standards of known molecular mass. Capillary gel electrophoresis separation of SDS–proteins has the advantage over SDS–PAGE of giving higher resolution and requiring shorter analysis time. Oligonucleotides and DNA size determination is also achieved by CGE.[9] In these cases, the sieving matrix is required to separate the individual components since they have nearly identical charge-to-size ratios. Similarly, CIEF is used to determine the pI of a protein by interpolation from a pI calibration plot generated from a series of protein standards with known isoelectric points.[10]

Several MAbs have been prepared for therapeutic purposes. They are among the most complex protein-based molecules, consisting of several light and heavy polypeptide chains, joined by multiple disulfide

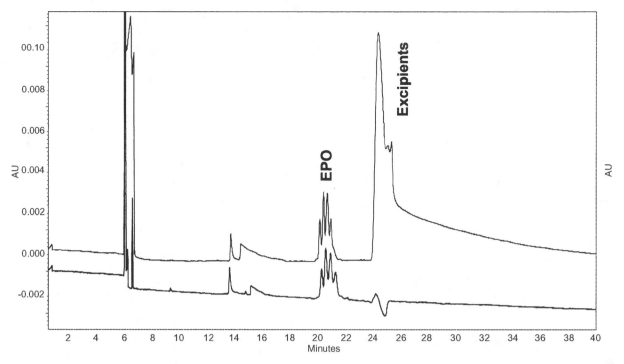

Fig. 1 Analysis of samples of EPO-α by reversed-polarity CZE: drug substance sample containing 8000 IU/ml (bottom trace) and drug product sample 10,000 IU/ml (top trace). Conditions were similar to those reported previously:[6] eCAP amine capillary (50 μm × 50 cm, 40 cm effective length); electrolyte: 200 mM sodium dihydrogen phosphate/1 mM nickel chloride, adjusted to pH 4.0 with acetic acid; 8 kV; UV detection at 200 nm. *(View this art in color at www.dekker.com.)*

bridges, and containing a number of glycosylation sites of varying sequences and arrangements. Typically, MAbs are very large molecules with molecular weights around 150,000 Da, a feature that, when combined with their structural complexity, makes high resolution chromatographic methods for the analysis of the intact molecule of little value. However, CE approaches have been highly successful for their characterization.[11] While all of the major CE separation modes have been applied, CIEF and CGE are particularly useful techniques. For instance, the high resolution achieved in CIEF allows monitoring of the profile of charge isoforms resulting from differential C-terminal processing (at lysine or arginine), a situation that frequently occurs in mammalian cell-derived products. Capillary gel electrophoresis analysis under denaturing conditions has been used to estimate MAb molecular weight as well as the presence of aggregates. When performed under denaturing and reducing conditions, CGE provides an effective way to monitor the light and heavy chains that make up the typical antibody structure.

With the recent progress made on mass spectrometric ionization modes, large biomolecules can now be readily analyzed. The most widely used ionization mode for proteins and peptides characterization is electrospray ionization (ESI).[12] Molecular weights can be readily determined for large proteins with accuracies in the range of ±0.01–0.05%. The ESI method is sensitive, presently requiring samples in the 100 fmol–10 pmol range for proteins. As a result, the coupling of CE to MS provides an extremely powerful tool for the unambiguous structural confirmation or identification of biomolecules. A widely used approach consists in the analysis of intact proteins, which enables the determination of exact molecular weights. However, little or no fragmentation results from ESI and, as a consequence, no information on the sequence is obtained. An alternative, and more commonly used, approach for protein identification by MS is peptide mapping. In this approach, the peptide fragments separated by CE are analyzed by MS to determine their respective molecular weight. The protein is then identified through a database search from which two or more of the separated peptides can be matched to the predicted ones. In some cases, it may be necessary to obtain sequence information for a given peptide in order to ascertain its identity. This can be accomplished by using CE/MS/MS.

Product Purity

Purity determination is an essential component of the assessment of the quality of any drug. However,

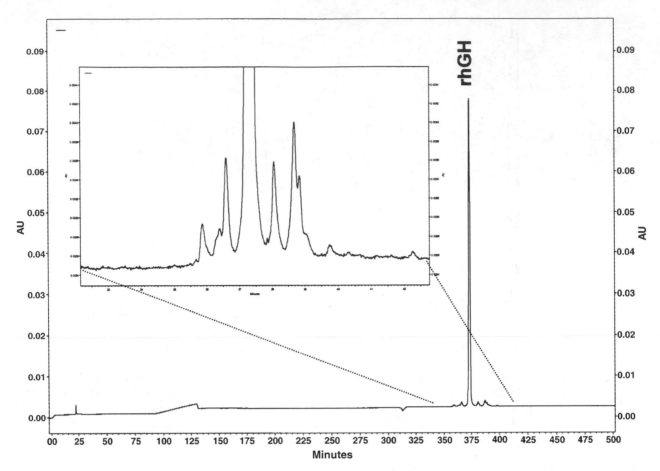

Fig. 2 Analysis of a sample of somatropin (recombinant hGH) by CZE. Inset: enlarged view showing the separation of impurities. Conditions were similar to those reported previously:[15] bare fused silica capillary (50 μm × 100 cm, 90 cm effective length); electrolyte: 0.1 M diammonium hydrogen phosphate, adjusted to pH 6.0 with phosphoric acid; 20 kV; UV detection at 200 nm.

the purity determination of biopharmaceuticals is not as straightforward as that of small-molecule pharmaceuticals, since biopharmaceuticals are structurally complex and have a wide range of potential impurities. Approaches usually involve the judicious choice of a combination of methods that enable the detection and quantitation of impurities from which an overall purity assessment can be derived. As mentioned previously, proteins undergo degradation or modification through a number of pathways. Most of the more common variants encountered in protein preparations can be detected by CE methods.[13] The high efficiency and quantitative properties of CGE can be used to detect nondissociable aggregates and clipped forms in proteins. This separation mode also provides an effective means to monitor the presence of deletion sequences in oligonucleotides.[14] The separation and detection of charge variants such as deamidation products and clipped forms resulting from the proteolytic cleavage of the polypeptide chain are readily amenable using

CE methods. In particular, CZE has proven to be highly effective for simple proteins having no carbohydrate-mediated heterogeneity present. The high efficiency of CZE, in some cases, allows the resolution of multiple charge variants, such as occur in hGH,[15] to be accomplished in a single run (Fig. 2). The coupling of CE to high sensitivity detection devices such as LIF detectors provides substantial enhancement of the detection limits of impurities.[16]

REGULATORY CONSIDERATIONS

Capillary electrophoresis is now recognized as a mature technique alongside HPLC and other modern analytical techniques. A harmonized general monograph[17] that presents both theoretical and practical considerations of the technique has been recently adopted by the Pharmacopoeial Discussion Group

(PDG) for implementation into the USP, the Japanese Pharmacopoeia (JP), and the EP. In addition, a CZE method is currently used as an identification test for EPO concentrated solution in the EP.[18] This represents the first example of the use of CE for the monitoring of a biopharmaceutical by an official method. A second example is in the final stage of implementation and involves the determination by CZE of charge variants in somatropin preparations.[19] This test, which was found to provide more precise quantitative data, will replace the current isoelectric focusing method.

Regulatory agencies around the world are seeing an increasing number of drug manufacturers that include CE methods in drug submissions. While most regulatory agencies recognize the applicability of CE for biopharmaceuticals, they require the methods to be validated based on well-established parameters that include accuracy, precision, specificity, limit of detection, limit of quantitation, linearity, and range. There is considerable guidance available for the validation of analytical methods. The International Conference on Harmonisation (ICH) has published guidelines that refer specifically to biopharmaceuticals: Q6B—Specifications: Test Procedures and Acceptance Criteria for Biotechnological/Biological Products; Q5C: Quality of Biotechnological Products: Stability Testing of Biotechnological/Biological Products; Q5E: Comparability of Biotechnological/Biological Products Subject to Changes in Their Manufacturing Process; Q2A: Text on Validation of Analytical Procedures; Q2B: Validation of Analytical Procedures: Methodology. Guidelines are available on the ICH website.[20]

CONCLUSIONS

The use of CE has become an integral part of the study of biopharmaceuticals, especially for the monitoring of product identity and purity. It is a powerful technique that, in many instances, is superior to the more conventional electrophoretic techniques and complementary to the widely used high resolution chromatographic techniques. It is particularly well suited to the study of complex mixtures of biomolecules such as glycoproteins and MAbs.

REFERENCES

1. Chen, A.B.; Canova-Davis, E. Capillary electrophoresis in the development of recombinant protein pharmaceuticals. Chromatographia 2001, 53 (Suppl.), S7–S17.

2. Klyushnichenko, V. Capillary electrophoresis in the analysis and monitoring of biotechnological processes. Meth. Mol. Biol. 2004, 276, 77–120.

3. Rickard, E.C.; Towns, J.K. The use of capillary electrophoresis for peptide mapping of proteins. In New Methods in Peptide Mapping for the Characterization of Proteins; Hancock, W.S., Ed.; CRC Press: New York, 1996; 97–118.

4. Liu, J.; Zhao, H.; Volk, K.J.; Klohr, S.E.; Kerns, E.H.; Lee, M.S. Analysis of monoclonal antibody and immunoconjugate digests by capillary electrophoresis and capillary liquid chromatography. J. Chromatogr. A. 1996, 735, 357–366.

5. Kakehi, K.; Honda, S. Analysis of glycoproteins, glycopeptides and glycoprotein-derived oligosaccharides by high performance capillary electrophoresis. J. Chromatogr. A. 1996, 220, 377–393.

6. Bietlot, H.P.; Girard, M. Analysis of recombinant human erythropoietin in drug formulations by high performance capillary electrophoresis. J. Chromatogr. A. 1997, 759, 177–184.

7. Park, S.S.; Cate, A.; Chang, B.S. Use of capillary electrophoresis to determine the dilute protein concentration in formulations containing interfering excipients. Chromatographia 2001, 53 (Suppl.), S34–S38.

8. Ma, S.; Nashabeh, W. Analysis of protein therapeutics by capillary electrophoresis. Chromatographia 2001, 53 (Suppl.), S75–S89.

9. Karger, B.L.; Foret, F.; Berka, J. Capillary electrophoresis with polymer matrices: DNA and protein separation and analysis. Meth. Enzymol. 1996, 271, 293–319.

10. Wehr, T.; Rodriguez-Diaz, R.; Zhu, M. Recent advances in capillary isoelectric focusing. Chromatographia 2001, 53 (Suppl.), S45–S58.

11. Krull, I.S.; Liu, X.; Dai, J.; Gendreau, C.; Li, G. HPCE methods for the identification and quantitation of antibodies, their conjugates and complexes. J. Pharm. Biomed. Anal. 1997, 16, 377–393.

12. Severs, J.C.; Smith, R.D. Capillary electrophoresis–electrospray ionization mass spectrometry. In Electrospray Ionization Mass Spectrometry; Cole, R.B., Ed.; John Wiley & Sons: New York, 1997; 343–382.

13. Teshima, G.; Wu, S.-L. Capillary electrophoresis analysis of recombinant proteins. Meth. Enzymol. 1996, 271, 264–293.

14. Srivatsa, G.S.; Pourmand, R.; Winters, S. Use of capillary electrophoresis for concentration analysis of phosphorothioate oligonucleotides. Meth. Mol. Biol. **2001**, *162*, 371–376.

15. Dupin, P.; Galinou, F.; Bayol, A. Analysis of recombinant human growth hormone and its related impurities by capillary electrophoresis. J. Chromatogr. A. **1995**, *707*, 396–400.

16. Lee, T.T.; Lillard, S.J.; Yeung, E.S. Screening and characterization of biopharmaceuticals by highperformance capillary electrophoresis with laser-induced native fluorescence detection. Electrophoresis **1993**, *14*, 429–438.

17. *European Pharmacopoeia*, 5th Ed.; 2005; 74–79 (Chapter 2.2.47).

18. *European Pharmacopoeia*, 5th Ed.; 2005; 1528–1532 (01/2005:1316).

19. Draft monograph for comment. Pharmeuropa **2004**, *16* (1), 72–73.

20. International Conference on Harmonisation. http://www.ich.org/(accessed on september 2004).

Biopolymer Separations by Chromatographic Techniques

Masayo Sakata
Chuichi Hirayama
*Department of Applied Chemistry and Biochemistry,
Kumamoto University, Kumamoto, Japan*

INTRODUCTION

Endotoxin (lipopolysaccharides; LPS) is an integral part of the outer cellar membrane of Gram-negative bacteria and is responsible for organization and stability. In the biotechnology industry, Gram-negative bacteria are widely used to produce recombinant DNA products such as peptides and proteins. Thus these products are always contaminated with LPS. Such contaminants have to be removed from drugs and fluids before use in injections, because their potent biological activities cause pyrogenic reactions.

To achieve selective removal of LPS from final biological products, such as proteins and protective antigens, it is necessary to consider not only the chemical and physical structures of LPS, but also those of the adsorbents and proteins, as well as the solution conditions. In physiological solutions, LPS aggregates form supramolecular assemblies (M_w: 4×10^5 to 1×10^6) with phosphate groups as the head group and exhibit a net-negative charge because of their phosphate groups. However, as proteins may release LPS monomers from the aggregates, we assume that LPS aggregates comprise a wide range of molecular sizes, with M_w from 2×10^4 to 1×10^6 in physiological solutions. On the other hand, the molecular weights of proteins are generally about 1×10^4 to 5×10^5. Therefore it is extremely difficult to separate LPS from protein solely by size-separation methods, such as SEC and ultrafiltration. Various procedures of LPS removal, such as ion exchange membrane, ultrafiltration, and extraction, have been developed for pharmaproteins. These procedures, however, are unsatisfactory with respect to selectivity, adsorption capacity, and protein recovery.

For the removal of LPS from final solutions of bioproducts, selective adsorption has proven to be the most effective technique. Therefore considerable effort is being put into the development of adsorbents capable of retaining high LPS selectivity under physiological conditions (ionic strength of $\mu = 0.05$–0.2, neutral pH). Recently, numerous cationic polymer adsorbents have been developed for removing LPS from protein solutions. This article will elucidate the chromatographic properties of various LPS adsorbents and will describe recent findings concerning methods for eliminating LPS from protein solutions using the adsorption technique.

CHROMATOGRAPHIC MATRICES WITH POLYCATIONIC LIGANDS

Lipopolysaccharide is an amphipathic substance[1–3] that has both an anionic region (the phosphoric acid groups) and a hydrophobic region (the lipophilic groups). From this point of view, an LPS-selective ligand should have, not only cationic properties, but also hydrophobic properties.[4–6] Fig. 1 shows structures of various cationic substances that are suitable as LPS-selective ligands. Through immobilization of polymyxin B on CNBr-activated Sepharose, Issekutz[7] created a polymyxin–Sepharose adsorbent for selective removal of LPS. This adsorbent is now commercially available. Although the polymyxin–Sepharose columns showed high LPS-adsorbing activity, protein losses during passage through the column have been noted (20% loss of BSA in Ref.[8]). This is due to the ionic interaction between the cationic region of the polymyxin B and the net-negatively charged proteins at low ionic strengths. Furthermore, polymyxin B is not suitable as a ligand for LPS removal from a solution for intravenous injection because it could escape from the column and would be physiologically active in solution. If any polymyxin is to be released into a solution, it would be physiologically active. Poly (ethyleneimine) (PEI)-immobilized cellulose fibers have been prepared by Morimoto et al.[9] and the PEI fibers showed significant LPS-adsorbing capacity under physiological conditions (neutral and ionic strength of $\mu = 0.1$–0.2). In a more recent publication, poly (ε-lysine) (PL) was covalently immobilized onto cellulose spherical particles and used for selective adsorption of LPS from protein solutions.[10] In addition, the PL (degree of polymerization: 35, pK_a: 7.6) (Chisso)[11] produced by *Streptomyces albulus*, which has become commercially available as a safe food preservative, is more suitable as a ligand than is polymyxin B. The high LPS adsorption of chromatographic matrices having polycationic ligands, such as

Encyclopedia of Chromatography DOI: 10.1081/E-ECHR-120038582

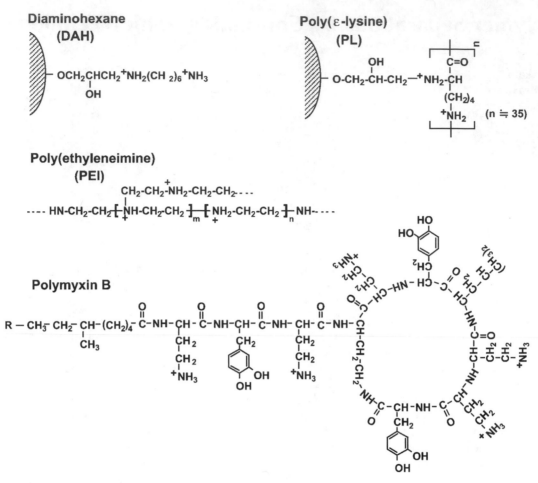

Fig. 1 Structure of LPS-selective ligands. Sepharose and cellulose particles are used as the matrix.

polymyxin B, PEI, or PL, is possibly due to the simultaneous effects of the cationic properties of the ligand and its hydrophobic properties.

EFFECTS OF VARIOUS FACTORS ON THE SEPARATION OF BIOPOLYMERS BY POLYCATIONIC ADSORBENTS

Effect of Pore Size of the Adsorbent on LPS Selectivity

To achieve the selective removal of LPS, it is important to determine the adsorbing activity of proteins. Table 1 shows the effect of the adsorbent pore size (molecular mass exclusion of polysaccharide, M_{lim})[12] on the adsorption of cellular products (biorelated polymers). The various PL-immobilized cellulose particles (PL cellulose) with pore sizes of M_{lim} 2×10^3 to $>2 \times 10^6$ were used as adsorbents. Lipopolysaccharide, DNA, and RNA, which are anionic biorelated polymers with phosphoric acid groups, were adsorbed very well by all the adsorbents. By contrast, the adsorption of protein

was more dependent on the M_{lim} of the adsorbent than its anion-exchange capacity (AEC). The adsorption of BSA (M_w 6.9×10^4), an acidic protein, increased from 5% to 68% with an increase in the M_{lim} from 2×10^3 to 1×10^4. The adsorption of γ-globulin (M_w 1.6×10^5), a hydrophobic protein, increased from 2% to 22% with an increase in the M_{lim} from 1×10^4 to $>2 \times 10^6$. Polymyxin–Sepharose with large pore size ($M_{lim} > 2 \times 10^6$) also adsorbed BSA (78%) and γ-globulin (26%), as shown in Table 1. Very little of the other neutral or basic proteins adsorbed onto the adsorbents under similar conditions. As a result, only when the PL cellulose (10^3), with a M_{lim} of 2×10^3 and AEC of 0.6 meq/g , was used as the adsorbent at pH 7.0 and ionic strength of $\mu = 0.05$ were LPS and DNA selectively well adsorbed.

The results reported in Table 1 show that the adsorption of protein was caused, mainly, by the entry of the protein into the pores of the adsorbent. This indicates that both BSA and γ-globulin can readily penetrate into a particle with an M_{lim} of $>2 \times 10^6$, but cannot penetrate into a particle with 2×10^3 (M_{lim}). On the other hand, it would also appear that

Table 1 Effect of adsorbent's pore size on adsorption of a bio-related polymer

Cellular product	(pl)	AEC^f (meq/g): Pore size $(M_{lim})^g$:	PL-cellulose $(10^3)^b$ 0.6 2×10^3	PL-cellulose $(10^4)^c$ 0.8 1×10^4	PL-cellulose $(10^6)^d$ 0.6 $>2 \times 10^6$	Polymyxin– sepharose^e 0.2 $>2 \times 10^6$
			Adsorption^a (%)			
Ovalbumin	4.6		2	65	85	75
BSA	4.9		5	68	82	78
Myoglobin	6.8		< 1	< 1	< 1	< 1
γ-Globulin	7.4		2	2	22	26
Lysozyme	11.0		< 1	< 1	< 1	< 1
DNA (salmon spermary)			99	99	99	99
RNA (yeast)			98	99	99	99
LPS (*E. coli* O111 : B4)			91	98	99	99
LPS (*E. coli* UKT-B)			89	96	99	99

^{a}The adsorption of a cellular product was determined using a batchwise method with 0.3 mL of wet adsorbent and 2 mL of a sample solution (100 μg/mL, pH 7.0, ionic strength of μ = 0.05).
^{b}Poly(ε-lysine)-immobilized Cellfine-GC-15.[10]
^{c}Poly(ε-lysine)-immobilized Cellfine-GC-700.[10]
^{d}Poly(ε-lysine)-immobilized Cellfine-CPC.[10]
^{e}Detixi-Gel.[7]
^{f}Anion-exchange capacity of the adsorbent.
^{g}Value deduced as molecular weight of polysaccharide.[12]

LPS aggregates are not able to enter the pores of the adsorbents with 2×10^3 (M_{lim}) because their molecular weights (4×10^5 to 1×10^6)[13] are significantly larger than the M_{lim} of the adsorbent. Many of the standard LPS molecules (*Escherichia coli* O111:B4 and UKT-B), however, were well adsorbed even by the adsorbent with an M_{lim} of 2×10^3, as shown in Table 1. We previously reported[13] that the LPS molecules were adsorbed by aminated poly (γ-methyl L-glutamate) particles not only into the pores of the particles but also on their surfaces. Poly (ε-lysine)-immobilized cellulose particles have similar characteristics.

Effect of Degree of Ligand Polymerization on Adsorption of Biopolymer

For selective removal of LPS from a protein solution, it is also necessary to select the ligand of the adsorbent. Fig. 2 shows the effects of a buffer's ionic strength and pH on the adsorption of LPS by diaminohexane- (DAH), PL-, or PEI-immobilized cellulose particles. Diaminohexane monomer (M_w: 116), PL (degree of polymerization: 35, M_w: 4.0×10^3), and PEI (degree of polymerization: 1600, M_w: 7×10^4) were used respectively as adsorbent ligands, and cellulose particles with M_{lim} 2×10^3 (Cellufine GC-15) were used as the matrix. As shown in Fig. 2a, the higher the ionic strength of the buffer the lower the LPS-adsorbing activity of the adsorbent. Both PEI cellulose and PL cellulose always showed a greater LPS-adsorbing

activity (99% to 82%) at a wide range of ionic strengths (μ = 0.05–0.8). The adsorbing activity of DAH cellulose decreased markedly when the ionic strength was increased to 0.2 or higher. Fig. 2b shows the effect of pH on the adsorption of LPS by various aminated cellulose adsorbents. The larger the molecular weight (polymerization degree) of an adsorbent's ligand the higher the LPS-adsorbing activity of the adsorbent. Over a wide pH range of 4.0–9.0 and at an ionic strength of μ = 0.05, PEI cellulose (AEC: 1.2 meq/g), with the largest polymerization degree of ligand, always showed the highest LPS-adsorbing activity (>98%). Poly(ε-lysine)-immobilized cellulose (AEC: 0.6 meq/g) also showed a high activity (>97%) over a pH range from 6.0 to 9.0, although it decreased from 99% to 75% as the pH decreased from 6.0 to 4.0. On the other hand, DAH cellulose (AEC: 0.1 meq/g) showed high adsorbing activity only at pH 7.0.

Fig. 3a and b shows the effect of a buffer's ionic strength on adsorption of BSA and γ-globulin, respectively, by aminated cellulose adsorbents. The stronger the ionic strength the lower the BSA-adsorbing activity of the adsorbent (Fig. 3a). By contrast, adsorption of BSA and γ-globulin is independent of ionic strength (Fig. 3b). Poly(ethyleneimine)-immobilized cellulose showed the highest adsorption of each protein among the three adsorbents.

From these results (Table 1, Figs. 2 and 3), we assumed that the adsorbing activity of aminated-cellulose adsorbents for biopolymers was induced by the simultaneous effects of their cationic properties and hydrophobic or other properties. The charge of

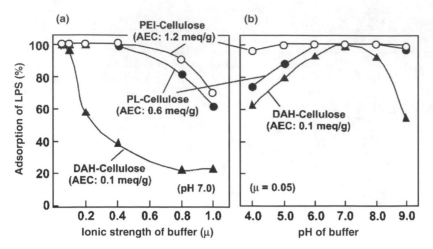

Fig. 2 Effects of a buffer's (a) ionic strength and (b) pH on the adsorption of LPS by various aminated cellulose adsorbents. The adsorption of LPS was determined using a batchwise method with 0.2 g of the wet adsorbent and 2 mL of a LPS (*E. coli* O111:B4, 1000 ng/mL) solution. Adsorbent and AEC of adsorbent: PEI-cellulose = poly(ethyleneimine)-immobilized Cellufine-GC15 (AEC: 1.2 meq/g); PL-cellulose = poly(ε-lysine)-immobilized Cellufine-GC15 (AEC: 0.6 meq/g); DAH-cellulose = diamino hexane-immobilized Cellufine-GC15 (AEC: 0.2 meq/g). M_{lim} of adsorbent: 2×10^3.

LPS is anionic at pH values greater than its pK_a (p$K_1 = 1.3$[14]). The charge of BSA is also anionic at pH values greater than 4.9 (its p*I*). The adsorption of LPS and BSA increased with increasing AEC content of the adsorbent (Figs. 2 and 3). It is also dependent on the ionic strength and pH values (Fig. 2a and b, respectively). These results suggest that aminated-cellulose adsorbents adsorb LPS and BSA mainly by ionic interaction. On the other hand, the ionic interaction of the adsorbent with γ-globulin (p*I*: 7.4) is not induced at pH 7.0, as the charge of the protein is cationic at a pH under its p*I* value. γ-Globulin is a weakly hydrophobic protein. Its adsorption was independent of ionic strength and increased with an increase in the hydrophobicity (ligand-polymerization degree) of the adsorbent (Fig. 3b). Hou and Zaniewski[15] also reported that a hydrophobic bond was formed between LPS and the polymeric affinity matrix. These findings suggest the participation of hydrophobic binding. Furthermore, as shown in Table 1, polycation-immobilized cellulose adsorbents bind more strongly with LPS than protein. This is because the pK_a of the phosphate residues of LPS is lower than the p*I* of protein (p*I*: 4.6–11.0), and, probably, because the LPS is adsorbed

by the adsorbent through its multipoint attachment onto the polycation chain of the adsorbent surface.

CHROMATOGRAPHIC RESULTS FOR SELECTIVE LPS REMOVAL

For selective LPS removal, it is necessary not only to select the ligand of the adsorbent, but also to control the conditions of the buffer (pH and ionic strength). The effect of ionic strength on the selective adsorption of LPS from a BSA-containing solution by various aminated cellulose adsorbents (M_{lim} 2×10^3) was examined (results are shown in Fig. 4a–c). A BSA solution, 500 µg/mL of BSA and 100 ng/mL of standard LPS, was used as a sample solution. The LPS-adsorbing activity of DAH cellulose decreased remarkably with an increase in the ionic strength. The DAH cellulose selectively adsorbed LPS only at $\mu = 0.05$ (Fig. 4a). Poly(ethyleneimine)-immobilized cellulose showed adsorbing activities for both LPS and BSA over a wide ionic strength, from $\mu = 0.05$ to 0.8 (Fig. 4c). It cannot therefore selectively adsorb

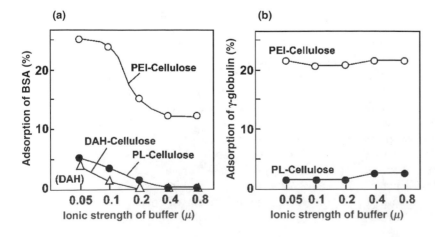

Fig. 3 Effect of a buffer's ionic strength on the adsorption of (a) BSA and (b) γ-globulin by various aminated cellulose adsorbents reported in Fig. 2. The adsorption of protein was determined using a batchwise method with 0.2 g of the wet adsorbent and 2 mL of a protein (100 µg/mL) solution.

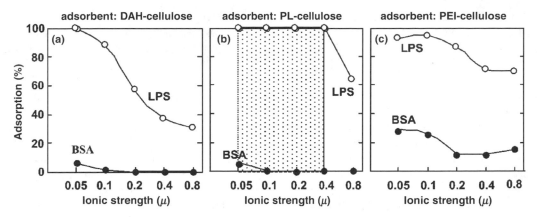

Fig. 4 Effects of ionic strength on selective adsorption of LPS from a BSA solution containing LPS by the various aminated cellulose adsorbents reported in Fig. 2. The selective adsorption of LPS was determined by a batchwise method with 0.2 g of the wet adsorbent and 2 mL of a sample solution [BSA: 500 μg/mL, LPS (*E. coli* O111:B4): 100 ng/mL, pH 7.0, and ionic strength of μ = 0.05–0.8].

LPS from the BSA solution at all ionic strengths. By contrast, PL cellulose selectively adsorbed LPS in the solution at ionic strengths of μ = 0.05 to 0.4 and pH 7.0, without adsorption of BSA (Fig. 4b). The residual concentrations of LPS after treatment were less than 100 pg/mL [1 endotoxin unit (EU)/mL] and each BSA recovery was over 95%.[10] The threshold level for intravenous application is set to 5 EU per kg body weight per hour by all pharmacopoeias.[2] The PL-cellulose adsorbent was able to remove LPS from a BSA solution to a level below 1 EU/mL.

As regards the adsorbing capacity of LPS, the cationic polymers with large pore sizes show a greater capacity, because of the entry of LPS molecules into the large pores. We have already reported[10] that PL cellulose with $M_{lim} > 2 \times 10^6$ can reduce the concentration of LPS to 0.1 EU/mL or under in a LPS solution, at a neutral pH and μ = 0.05–0.4. As shown in Table 1, Polymyxin B-Sepharose ($M_{lim} > 2 \times 10^6$) and PL cellulose (10^6) ($M_{lim} > 2 \times 10^6$) readily removed LPS from lysozyme and myoglobin solutions at pH 7 without a loss of protein. This is because the ionic interaction of the adsorbent with lysozyme (p*I* 11.0) and myoglobin (p*I* 6.8) is not induced at pH 7.0. Thus the cationic polymer particles having a large pore size are suitable as an adsorbent for removal of LPS from bioproducts (p*I*: 7.0–11.0) containing large quantities of LPS, such as a crude antigen solution originating from a Gram-negative bacterium.

CONCLUSIONS

The present results suggest that PL-cellulose spherical particles can reduce the concentrations of natural LPS to 1 EU/mL or lower in drugs and fluids used for intravenous injection, at a neutral pH and ionic strengths of μ = 0.05–0.4. These processes did not affect the recovery, even of acidic proteins such as BSA. The high LPS-adsorbing activity of the PL cellulose is possibly due to the cationic properties of the ligand and its suitable hydrophobic properties. The high LPS selectivity of the particles with small pore size is due to the size-exclusion effects on protein molecules. By contrast, that of the particles with large pore sizes is due to the decreases in ionic interactions for net-negative charged proteins, which arise when the buffer's ionic strength is adjusted to 0.2 or stronger.

For practical application, ease of regeneration is very important. The PL-cellulose spherical particles can be completely regenerated by frontal chromatography with 0.2 M sodium hydroxide followed by 2.0 M sodium chloride.[10] Their stable structures, even under extreme pH values, are due to their –CHNH– bonds. Of course, the development of even better adsorbents should be pursued, by continuing this search for materials. To achieve selective removal of LPS, it is important to not only select a suitable ligand, but also to adjust the pore size of the matrix.

REFERENCES

1. Vaara, M.; Nikaido, H. Outer Membrane Organization. In *Handbook of Endotoxin*; Rietschel, E.T., Ed.; Elsevier: Amsterdam, 1984; Vol. 1, 1–45.
2. Hirayama, C.; Sakata, M.; Ihara, H.; Ohkuma, K.; Iwatsuki, M. Effect of the pore size of an animated poly(γ-methyl L-glutamate) adsorbent on selective removal of endotoxin. Anal. Sci. **1992**, *8*, 805–810.

3. Li, L.; Luo, R.G. Protein concentration effect on protein-lipopolysaccharide (LPS) binding and endotoxin removal. Biotechnol. Lett. **1997**, *19*, 135–138.

4. Petsch, D.; Anspach, F.B. Endotoxin removal from protein solutions. J. Biotechnol. **2000**, *79*, 97–119.

5. Minobe, S.; Watanabe, T.; Sato, T.; Tosa, T.; Chibata, I. Preparation of adsorbents for pyrogen adsorption. J. Chromatogr. **1982**, *248*, 401–408.

6. Matsumae, H.; Minobe, S.; Kindan, K.; Watanabe, T.; Tosa, T. Specific removal of endotoxin from protein solutions by immobilized histidine. Biotechnol. Appl. Biochem. **1990**, *12*, 129–140.

7. Issekutz, A.C. Removal of Gram-negative endotoxin from solutions by affinity chromatography. J. Immunol. Methods **1983**, *61*, 275–281.

8. Anspach, F.B.; Kilbeck, O. Removal of endotoxins by affinity sorbents. J. Chromatogr. A. **1995**, *711*, 81–92.

9. Morimoto, S.; Sakata, M.; Iwata, T.; Esaki, A.; Hirayama, C. Preparations and applications of polyethyleneimine-immobilized cellulose fibers for endotoxin removal. Polym. J. **1995**, *27*, 831–839.

10. Todokoro, M.; Sakata, M.; Matama, S.; Kunitake, M.; Ohkuma, K.; Hirayama, C. Pore-size controlled and poly(ε-lysinc)-immobilized cellulose spherical particles for removal of lipopolysaccharides. J. Liq. Chromatogr. & Relat. Technol. **2002**, *25* (4), 601–614.

11. Shima, S.; Sakaki, H. Poly-L-lysine produced by *Streptomyces*: Part III. Chemical studies. Agric. Biol. Chem. **1981**, *45* (11), 2503–2508.

12. Hirayama, C.; Ihara, H.; Nagaoka, S.; Furusawa, H.; Tsuruta, S. Regulation of pore-size distribution of poly(γ-methyl L-glutamate) spheres as a gel permeation chromatography packings. Polym. J. **1990**, *22* (7), 614–619.

13. Hirayama, C.; Sakata, M.; Ihara, H.; Ohkuma, K.; Iwatsuki, M. Effect of pore size of an aminated poly(γ-methyl L-glutamate) adsorbent on the selective removal of endotoxin. Anal. Sci. **1992**, *8*, 805–810.

14. Hou, K.C.; Zaniewski, R. Depyrogenation by endotoxin removal with positively charged depth filter cartridge. J. Parenter. Sci. Technol. **1990**, *44*, 204–209.

15. Hou, K.C.; Zaniewski, R. The effect of hydrophobic interaction on endotoxin adsorption by polymeric affinity matrix. Biochem. Biophys. Acta **1991**, *1073*, 149–154.

Biopolymers: Analysis by Capillary Zone Electrophoresis

Feng Xu
Department of Medicinal Chemistry, Faculty of Pharmaceutical Sciences,
The University of Tokushima, CRESTJST, Shomachi, Tokushima and Analytical Instruments
Division, Shimadzu Corp., Kyoto, Japan

Yoshinobu Baba
Department of Medicinal Chemistry, Faculty of Pharmaceutical Sciences,
The University of Tokushima, CRESTJST, Shomachi, Tokushima, Japan

INTRODUCTION

Since Jorgenson and Lukacs[1] separated peptides in 1981 using free zone electrophoresis in glass capillaries, capillary electrophoresis (CE) has become a highly efficient technique for the separation of biopolymers such as proteins, peptides, carbohydrates, and DNA. Free solution CE, or capillary zone electrophoresis (CZE), is an electrophoresis in free homogeneous solution.[2,3] Charged solutes are simply separated on the basis of the solute charge-to-mass ratio, applied electric field, and the pH and ionic strength of the background electrolyte (BGE).

Separation of DNA is generally conducted by capillary gel electrophoresis (CGE), in which gel or polymer solutions act as sieving media, rather than by CZE, because each DNA has the same charge-to-mass ratio. The CZE separation of DNA can only be achieved by using some special modifications to break the charge-to-mass symmetry through either trapping analytes into sodium dodecyl sulfate (SDS) micelles or labeling analytes with large and weakly charged molecules at the DNA fragment ends (end-labeled free-solution electrophoresis).[4] Separation of DNA is not covered here. Interested readers are referred to our previous articles,[5,6] dealing with DNA size separation and sequencing, in the *Encyclopedia of Chromatography*. Here we concentrate on the CZE separations of three kinds of biopolymers (proteins, peptides, and carbohydrates), including the factors influencing the separations and the promising development of CZE in microchip and capillary array platforms.

PROTEINS AND PEPTIDES

Proteins are key participants in all biological activities. Peptides, generally shorter than proteins, have important biological functions, as hormones, neurotransmitters, etc. Owing to the similarity in structure, the general principles of the separations of proteins and peptides

are alike. A semiempirical relation exists between mobility, charge, and size of a peptide or protein:

$$\mu = A\frac{q}{M^p} + B$$

where μ is mobility, q is charge, M is molecular mass, A and B denote empirical constants, and p varies from 1/3 to 2/3 as a function of the pH and ionic strength of BGE, and shape of peptides and proteins, etc. For a protein, the most difficult part is estimation of the charge.[7]

Sample Preconcentration

The CZE of a low concentration protein or peptide requires preconcentration of the sample before analysis, either off-line or on-line. Preconcentration is commonly performed by using solid-phase packing material at the inlet end of the capillary and is based on chromatographic or electrophoretic principle. Solid-phase extraction (SPE) on C_{18} or C_8 cartridges can be on-line connected to CZE and the concentrated nonspecific analytes are released by an organic solvent (e.g., acetonitrile).[8] Alternatively, specific analytes can be concentrated by the use of antibody-containing cartridges and enzyme-immobilized microreactors. Desalting is important in protein processing. Microdialysis sampling can be used for the desalting of samples prior to introduction into electrospray ionization mass spectrometry (ESI/MS). Capillary isoelectric focusing (CIEF), with slow ramping of voltage, can achieve on-line desalting of proteins in cerebrospinal fluid. Head column stacking has a long history and is now a popular technique in which the injected sample is simply dissolved in water and is focused at the interface between the sample plug and separation buffer, as a result of different mobilities in the two solutions. Capillary isotachophoresis (CITP) can also be utilized to achieve on-column sample preconcentration.

Encyclopedia of Chromatography DOI: 10.1081/E-ECHR-120018663

The sample is introduced as a plug between a leading electrolyte, which has a higher mobility than the sample, and a trailing electrolyte, which has a lower mobility than the sample. After applying the voltage, the sample components are focused as a concentrated band. Very large sample zones can be accommodated by isotachophoresis to enhance the limit of detection.[9]

Detection

Peptide bonds enable proteins and peptides to be directly detected by ultraviolet (UV) radiation, at 200–220 nm, where the absorption is proportional to the number of peptide bonds. Sometimes, detection is performed around 254 or 280 nm, where proteins have modest absorbance in the presence of aromatic residues. However, the sensitivity of UV detection is relatively poor (to μM). Several commercially available Z-type cells and bubble-shaped cells can help, somewhat, in improving the sensitivity.

Laser-induced fluorescence (LIF) produces low limits of detection and a wide dynamic range. Native fluorescence of proteins is primarily associated with emission from tryptophan and tyrosine, between 300 and 400 nm. Proteins are easy to derivatize with a fluorescent reagent prior to electrophoresis. Postcolumn derivatization can be performed in sheath flow systems where the sheath flow cell acts as a postseparation labeling reactor. To avoid the formation of a mixture of multiply labeled products for originally single species, some special procedures utilizing Edman degradation chemistry or fluorescein isothiocyanate (FITC) labeling at lower than normal derivatization buffer pH have been developed. Noncovalent labeling is performed with dyes that interact with proteins, either by H bonding or through hydrophobic interactions. The sensitivity of LIF approaches the absolute limit of detection of a single molecule. A number of fluorescent dyes are widely used,[10] such as fluorescamine, 1-anilinonaphthalene-8-sulfonic acid, 4,4′-dianilino-1,1′-binaphthyl-5,5′-disulfonic acid, 2-(p-toluidino)naphthene-6-sulfonic acid, 3-(2-furoyl) quinoline-2-carboxaldehyde, naphthalene-2,3-dicarboxaldehyde, 4-chloro-7-nitrobenzofurazan, 4-fluro-7-nitrobenzofurazan, o-phthaldialdehyde-2-mercaptoethanol, and 3-(4-carboxybenzoyl)-2-quinolinecarboxaldehyde, etc.

Mass spectrometric detection has come into frequent use, as it provides significant information relating to the solute structure. Protein analysis using CZE/MS has been applied to abundant proteins in a single cell after the single intact cell was introduced into a separation capillary and after being lyzed. On the other hand, ESI/MS is a soft ionization technique that can form molecular ions. On-line ESI detection requires an external interface and volatile buffers to avoid contamination of the ionization chamber. Proteins can also be off-line identified by matrix-assisted laser desorption ionization/time-of-flight mass spectrometry (MALDI/TOF/MS), in which the matrix utilizes a low-molecular-weight organic acid that absorbs laser light and dissipates the energy in such a way that protein is evaporated, usually in a single protonated form. Time of flight has an advantage of having no upper mass-to-charge limit.

Coatings and Additives

The interaction of proteins with the negative charges of the ionized silanol groups on the inner capillary wall should be overcome prior to separation. The interaction results in peak tailing, poor resolution, unstable electroosmotic flow (EOF), and sample adsorption loss. A variety of approaches have been developed to overcome this problem, including the use of extreme buffer pH, coated capillaries, and additives in BGE.

The simplest way is to use extreme buffer pH, i.e., either a low pH at which the dissociation of silanol groups is suppressed so as to prevent their electrostatic interactions with positively charged polypeptides, or an alkaline pH that is at least two units above the isoelectric point of an analyzed polypeptide, which leads to electrostatic repulsion between the negatively charged polypeptide and the negatively charged silanol groups on the capillary wall. Unfortunately, even proteins that behave as anions still have a positively charged section on their surface that can interact with the capillary wall. In addition, more extreme pHs may denature proteins.

Static wall coatings are usually made by reacting the capillary wall with a small bifunctional reagent, which is then used to bind a polymer to the wall. The polymer is usually prepared in situ. γ-methacryloxypropyl-trimethoxysilane is a classical bifunctional reagent, while polyacrylamide is a frequently used polymer for coatings. A variety of polymer coatings can be anchored onto capillaries, such as polystyrene, polybrene, polyvinyl alcohol, polyethyleneimine, polyethylene glycerol, polyvinylpyrrolidone, polyethylene oxide, and nonionic surfactant coatings. Hydrophilic coating, epoxy-poly(dimethylacrylamide), can separate proteins over the range of pH 4–10 with the recovery of both cationic and anionic proteins over 90%.

Dynamic wall coatings are replaceable. There are two kinds of dynamic coatings: One is performed by introducing a neutral polymer to the BGE, and the other is performed by adding ions to the BGE. The neutral polymers not only shield capillary wall from interactions with solutes, but also increase the viscosity in the electric double layer and reduce EOF. Various cellulose derivatives and other hydrophilic polymers, such as polyvinyl alcohol, polyethyleneoxide, hydroxyethyl cellulose, copolymer of hydroxypropylcellulose and hydroxyethylmethacrylate, and polysaccharide guaran, have been used for this purpose. Ionic (mainly cationic) additives titrate the negative charge of the capillary wall so that the EOF is decreased, neutralized, or even reversed. Polyionic species reduce the pH dependence of the EOF if, for example, the coating contains sulfonic acid groups that are fully ionized over a wide pH range. One can also add, to the BGE,[11,12] oligoamines such as putrescine, spermine, and tetraethylene pentamine, cationic surfactants such as cetyltrimethylammonium bromide (CTAB) and didodecyldimethylammonium bromide (DDAB), high concentration of anionic surfactants such as SDS, zwitterionic phospholipids such as 1,2-dilauroyl-sn-phosphatidylcholine, high concentration of alkali salts and phytic acid, other compounds such as cyclodextrins, etc. A mixture of cationic and anionic fluorosurfactants produces efficient separations of acidic and basic proteins in a single run at neutral pH. Zwitterionic surfactants are typically superior to the nonionogenic ones. The protein interactions with the capillary wall are inversely proportional to the critical micellar concentration (CMC) of surfactants.

In addition to reducing the adsorption, the dynamic additives also play important roles in selectivity enhancement. pH is an important option for protein and peptide separation. An increase in the buffer ionic strength can increase the resolution. Any ion that displays a preferential affinity with the peptides has a potential modifying effect on the selectivity. The addition of organic modifiers, e.g., methanol, ethanol, and acetonitrile, in the BGE, can induce different solvation of the peptide chains and modify the migration order and selectivity of peptides. For separation of very hydrophobic proteins, e.g., lipoproteins, surfactants can act as buffer additives to improve solubilization.

Using amphoteric, isoelectric buffers at pH close to their isoelectric points (pI), at which the electrolytes have a net charge of zero, is an efficient way to decrease the BGE conductivity and apply extremely high field strength. Thus the separation time can be reduced to the order of a few minutes, and high resolution is achieved as a result of minimal diffusion-driven peak spreading. Several acidic isoelectric buffers, such as cysteic acid (pI 1.85), iminodiacetic acid (pI 2.23), aspartic acid (pI 2.77), and glutamic acid (pI 3.22), all at 50 mM concentration,[13] have been used in CZE separation of proteins and peptides. Fig. 1 shows a decrease of total running time from 70 to 12 min when using an isoelectric solution.

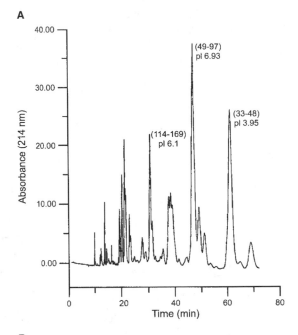

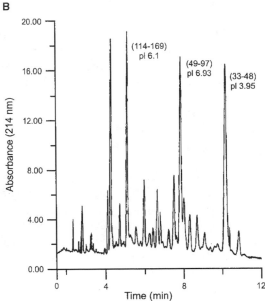

Fig. 1 CZE of tryptic digests of β-casein in 100 mm i.d. × 37- cm capillary. (A) BGE: 80 mM phosphate buffer, pH 2.0; injection: 0.5 p.s.i. for 3 sec; applied field strength: 110 V/ cm. The three major peaks are: 1) pI 6.1, fragment β-CN (114–169); 2) pI 6.93, fragment β-CN (49–97); and 3) pI 3.95, fragment β-CN (33–48). Note that the total running time is 70 min. (B) BGE: 50 mM isoelectric aspartic acid (pH = pI = 2.77) added with 0.5% hydroxyethyl cellulose (HEC) (M_n 27,000 Da) and 5% trifluoroethanol; applied field strength: 600 V/ cm. (From Ref.[13].)

CARBOHYDRATES

Carbohydrates are the third most important class of biopolymers, next to proteins and DNA. However, their analyses by CZE are still in infancy because of the complexity and diversity of their structures. In terms of ionization ability, carbohydrates may be divided into acidic and weakly ionizable classes. Acidic carbohydrates are negatively charged at neutral pH and can be conveniently separated. Weakly ionizable carbohydrates are neutral at mild pH, but deprotonate at extremely basic condition (e.g., pH >12).

Underivatized Carbohydrates

The borate buffer is the most effective buffer for the CZE separation of native (and derivatized) carbohydrates. Borate complexes with adjacent hydroxyl groups on carbohydrates to form a negatively charged complex, which has a 2- to 20-fold increased UV absorbance at 195 nm.[9] The stability of sugar–borate complexes increases with increasing pH and borate concentration, and depends on the number and configuration of the hydroxyl groups. The presence of a surfactant (e.g., tetrabutylammonium) in a borate buffer also enhances the solute selectivity by interacting with anionic borate–sugar complexes. Saccharides, e.g., sucrose, glucose, fructose, etc. can also be separated by chelating with Cu^{2+} present in the BGE. Elevated temperature up to 60°C facilitates the enhancement of resolution and efficiency.

Underivatized carbohydrates are mainly detected using low-wavelength UV (at 190–200 nm), indirect UV, or indirect LIF. The indirect method is based on the physical displacement of analytes with the added chromophoric or flurophoric compounds in the BGE. Indirect UV detection, using sorbic acid as both carrier electrolyte and chromophore, and employing high pH to achieve ionization of saccharides, permits the analysis of underivatized saccharides in low concentration. Fig. 2 shows the separation of underivatized acidic, neutral, and amino sugars and sugar alcohols by CZE-indirect UV.[14] Capillary electrophoresis-

pulsed amperometric detection (CE/PAD) in the analysis of oligosaccharides derived from glycopeptides provides structural information through simply modulating the detection potentials. Refractive index detection is a universal detection and is also useful for oligosaccharide analysis. However, the detection limit is rather poor.

Derivatized Carbohydrates

Derivatization provides the advantage of incorporating chromophoric or fluorophoric functions into carbohydrates to achieve highly sensitive detection. Electromigration can also be achieved by derivatization of neutral saccharides with reagents possessing ionizable functions. Derivatization of the reducing aldehyde and/or keto groups present in carbohydrates can be performed by reductive amination and condensation reaction.

The reaction of reductive amination is based on the reducing end of a saccharide reacting with the primary amino group of a chromophoric or fluorophoric reagent to form a Schiff base that is subsequently reduced to a stable secondary amine. For example, in the reductive amination of malto-oligosaccharides, 8-aminonaphthalene-1,3,6-trisulfonic acid (ANTS) introduces both electric charge and fluorescence to the saccharides.[15] The products are ionized at a pH as low as 2.5, allowing CZE of carbohydrates under conditions where both EOF and adsorption to the internal capillary wall are negligible, even in the absence of any coating. Aminobenzoic acid and related compounds have the merit to react with both ketoses and aldoses of oligosaccharides in less than 15 min at 90°C. Such fast reactions could be compatible with the CE separation speed. Other conventional fluorophores for carbohydrate labeling include 8-aminopyrene-1,3,6-trisulfonate (APTS),

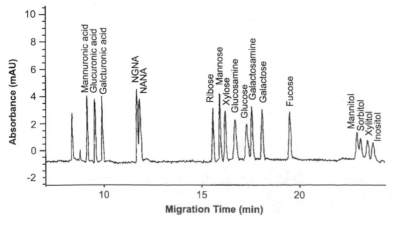

Fig. 2 Capillary zone electrophoresis of saccharides (1 mM each) standard mixture. Capillary: Fused silica, 50 μm i.d. × 80.5 cm (72-cm effective length); BGE: 20 mM 2,6-pyridinedicarboxylic acid (PDC) and 0.5 mM CTAB, pH 12.1; voltage: −25 kV. The signal wavelength was set at 350 nm with a reference at 275 nm. (From Ref.[14].)

7-amino-4-methylcoumarin, 3-(4-carboxybenzoyl)-2-quinolinecarboxaldehyde, 2-aminobenzamide, 4-aminobenzoate, 4-aminobenzonitrile, etc.

The derivatization of aldehydes in reducing carbohydrates can also be performed by a condensation reaction between the active hydrogens of 1-phenyl-3-methyl-5-pyrazolone (PMP) and the aldehyde functionality under slightly basic condition. The formed bis-PMP derivatives can be separated by CZE and detected by UV absorbance.[16] Oligosaccharides can also be separated as complexes with a variety of compounds, including acetate, molybdate, germanate, stannate, arsenite, wolframate, vanadate, and tellurate of various alkali and alkaline earth metal ions. In a BGE containing calcium, barium, or strontium acetate, a mixture of PMP-derivatized reducing carbohydrates such as arabinose, ribose, galactose, glucose, and mannose was fully resolved.[16]

Glycoprotein analysis requires both protein and glycan identification. CE/laser-induced fluorescence (CE/LIF) is widely used in the fingerprinting of fluorescently labeled glycans and in detection differences in maps of the oligosaccharides released from glycoproteins. The analysis of the complete composition of saccharides occurring in glycoproteins can be performed by separating the hydrolyzed sugars using CZE. Useful information about the glycoprotein structure can be obtained by combining CZE with MALDI-MS and ESI-MS. The techniques are well suited for the sensitive determination of the degree of heterogeneity, the site of glycosylation in a protein, and the composition and branching patterns of N- and O-linked glycans.

CONCLUSIONS

CZE, with its automation, simplicity, and rapid method development, is an attractive choice for biopolymer separation. Future methodological advances include novel capillary wall coatings, specific buffer additives, effective sample preconcentration, and highly sensitive detection. At the same time, CZE development will move toward two promising directions: High-throughput and ultrafast separations (in seconds not minutes). Previously, CZE was a sequential technique, which allowed analysis of one sample per analysis. One remedy to this problem is the use of the capillary array electrophoresis (CAE) technique. Array instruments use 100 or more capillaries in parallel, with the laser excited fluorescence signals from each channel simultaneously recorded, e.g., by a charge-coupled device (CCD) array. Another remedy for fast analysis is the use of microchip-based CE devices. All operations, e.g., sample manipulation, separation, and detection, are performed on the micro-

structures of the chips. The short sample injecting plug, essentially zero dead volume intersections, and high field strength result in extremely rapid and high efficient separation.[17] The advantage of microfabricated devices is also the potential of producing arrays of separation channels for high-throughput applications in a single run. After solving some technical problems, the microchip-based separations are expected to become a highly powerful tool in future separations of biopolymers.

REFERENCES

1. Jorgenson, J.W.; Lukacs, K.D. Zone electrophoresis in open-tubular glass capillaries. Anal. Chem. **1981**, *53* (8), 1298–1302.
2. Righetti, P.G., Ed.; *Capillary Electrophoresis in Analytical Biotechnology*; CRC Press: Boca Raton, FL, 1996.
3. Camilleri, P., Ed.; *Capillary Electrophoresis—Theory and Practice;* New Directions in Organic and Biological Chemistry Series; CRC Press: Boca Raton, FL, 1997.
4. Heller, C.; Slater, G.W.; Mayer, P.; Dovichi, N.; Pinto, D.; Viovy, J.-L.; Drouin, G. Free-solution electrophoresis of DNA. J. Chromatogr. A. **1998**, *806* (1), 113–221.
5. Baba, Y. DNA sequencing studies by CE. In *Encyclopedia of Chromatography*; Cazes, J., Ed.; Marcel Dekker, Inc.: New York, 2001; 259–261.
6. Kiba, Y.; Baba, Y. Nucleic acids, oligonucleotides, and DNA: capillary electrophoresis. In *Encyclopedia of Chromatography*; Cazes, J., Ed.; Marcel Dekker, Inc.: New York, 2001; 556–560.
7. Kašička, V. Recent advances in capillary electrophoresis of peptides. Electrophoresis **2001**, *22* (19), 4139–4162.
8. Dolník, V.; Hutterer, K.M. Capillary electrophoresis of proteins 1999–2001. Electrophoresis **2001**, *22* (19), 4163–4178.
9. Krylov, S.N.; Dovichi, N.J. Capillary electrophoresis for the analysis of biopolymers. Anal. Chem. **2000**, *72* (12), 111R–128R.
10. Bardelmeijer, H.A.; Waterval, J.C.M.; Lingeman, H.; van't Hof, R.; Bult, A.; Underberg, W.J.M. Pre-, on-, and post-column derivatization in capillary electrophoresis. Electrophoresis **1997**, *18* (12–13), 2214–2227.
11. Tabuchi, M.; Baba, Y. A separation carrier in high-speed proteome analysis by capillary electrophoresis. Electrophoresis **2001**, *22* (16), 3449–3457.
12. Cunliffe, J.M.; Baryla, N.E.; Lucy, C.A. Phospholipid bilayer coatings for the separation of

proteins in capillary electrophoresis. Anal. Chem. **2002**, *74* (4), 776–783.

13. Righetti, P.G.; Nembri, F. Capillary electrophoresis of peptides in isoelectric buffers. J. Chromatogr. A. **1997**, *772* (1–2), 203–211.

14. Soga, T.; Heiger, D.N. Simultaneous determination of monosaccharides in glycoproteins by capillary electrophoresis. Anal. Biochem. **1998**, *261* (1), 73–78.

15. El Rassi, Z. Recent developments in capillary electrophoresis and capillary electrochromatography of carbohydrate species. Electrophoresis **1999**, *20* (15–16), 3134–3144.

16. Honda, S.; Yamamoto, K.; Suzuki, S.; Ueda, M.; Kakehi, K. High-performance capillary zone electrophoresis of carbohydrates in the presence of alkaline earth metal ions. J. Chromatogr. **1991**, *588* (1–2), 327–333.

17. Effenhauser, C.S.; Bruin, G.J.M.; Paulus, A. Integrated chip-based capillary electrophoresis. Electrophoresis **1997**, *18* (12–13), 2203–2213.

Biotic Dicarboxylic Acids: CCC Separation with Polar Two-Phase Solvent Systems Using a Cross-Axis Coil Planet Centrifuge

Kazufusa Shinomiya
College of Pharmacy, Nihon University, Chiba, Japan

Yoichiro Ito
Laboratory of Biophysical Chemistry, National Heart, Lung, and Blood Institute, National Institutes of Health, Maryland, U.S.A.

INTRODUCTION

Among various types of countercurrent chromatographic instruments developed in the past, the cross-axis coil planet centrifuge (cross-axis CPC) is one of the most useful systems for separation of numerous kinds of natural and synthetic products.[1–3]

The cross-axis CPC produces a unique mode of planetary motion, such that the column holder rotates about its horizontal axis while revolving around the vertical axis of the centrifuge.[4,5] This motion provides satisfactory retention of the stationary phase for viscous, low-interfacial tension, two-phase solvent systems, such as aqueous–aqueous polymer phase systems. Our previous studies demonstrated that the cross-axis CPC equipped with a pair of multiplayer

coils or eccentric coil assemblies in the off-center position was very useful for the separation of proteins with polyethylene glycol-potassium phosphate solvent systems.[6–8] The apparatus is also useful for the separation of highly polar compounds such as

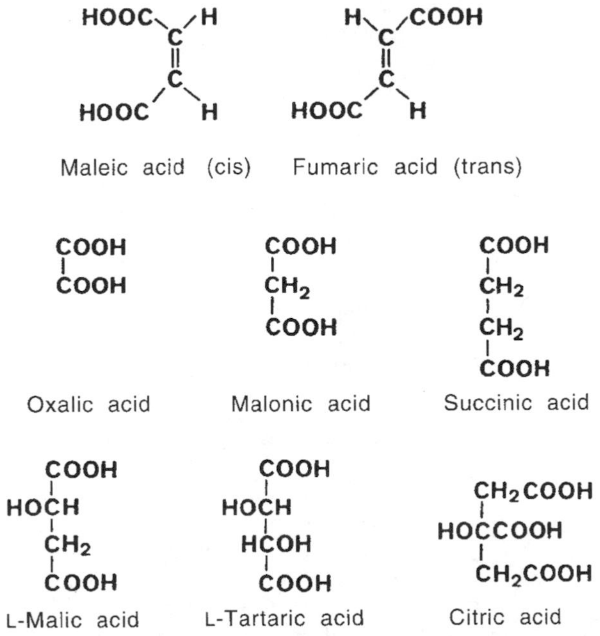

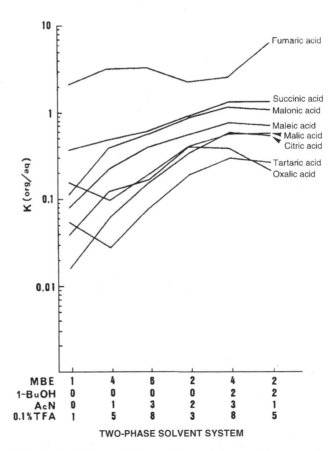

Fig. 2 Partition coefficients (K org/aq) of polar organic acids in methyl *t*-butyl ether/1-butanol/acetonitrile/aqueous 0.1% trifluoroacetic acid system. K is expressed by solute concentration in the organic phase divided by that in the aqueous phase. MBE = methyl *t*-butyl ether; 1-BuOH = 1-butanol; AcN = acetonitrile; TFA = trifluoroacetic acid.

Fig. 1 Chemical structures of polar organic acids.

Encyclopedia of Chromatography DOI: 10.1081/E-ECHR-120014224

sugars,[9] hippuric acid, and related compounds,[10] which require the use of polar two-phase solvent systems.

This article illustrates the separation of biotic dicarboxylic acids using the cross-axis CPC with eccentric coil assemblies.[11]

CCC APPARATUS AND SEPARATION COLUMNS

The cross-axis CPC produces a synchronous planetary motion of the column holder, which rotates about its horizontal axis and simultaneously revolves around the vertical axis of the apparatus at the same angular velocity. In the $X - 1.5L$ type of the apparatus, the column holder was mounted at an off-center position ($X = 10$ cm and $L = 15$ cm), which provides efficient mixing of the two-phase solvent systems and stable retention of the stationary phase in the coiled column.

The separation column was prepared using a pair of eccentric coil assemblies, which were made by winding a 1 mm-ID PTFE (polytetrafluoroethylene) tubing onto 7.6 cm long, 5 mm-OD nylon pipes forming a series of tight left-handed coils. A set of these coil units was symmetrically arranged around the holder hub of 7.6 cm diameter in such a way that the axis of each coil unit is parallel to the axis of the holder. A pair of eccentric coil assemblies was mounted on the rotary frame, one on each side, and serially connected with a flow tube.

CCC SEPARATION OF DICARBOXYLIC ACIDS

The chemical structures of eight typical biotic dicarboxylic acids are shown in Fig. 1. They are extremely hydrophilic and require a specific reaction for detection using the color-producing reagent such as 2-nitrophenylhydrazine hydrochloride.[12–13]

Fig. 2 illustrates the partition coefficient (K) values for these dicarboxylic acid samples in the polar two-phase solvent systems composed of methyl t-butyl ether (MBE)/1-butanol/acetonitrile (AcN)/

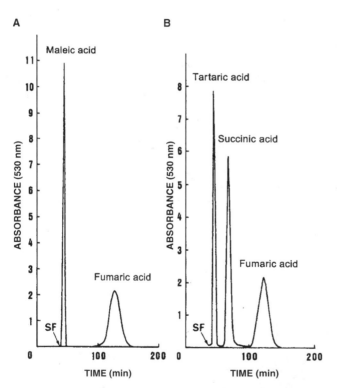

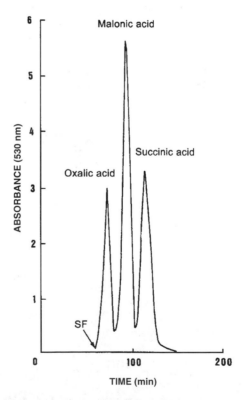

Fig. 3 CCC separation of polar organic acids by cross-axis CPC. Experimental conditions: apparatus, cross-axis CPC equipped with a pair of eccentric coil assemblies, 1 mm ID and 26.5 mL total column capacity; sample, (A) maleic acid (3 mg) and fumaric acid (3 mg); (B) tartaric acid (5 mg), succinic acid (5 mg) and fumaric acid (2.5 mg); solvent system, methyl t-butyl ether/1-butanol/acetonitrile/aqueous 0.1% trifluoroacetic acid (A) (1:0:0:1); (B) (2:0:2:3); mobile phase, lower phase; flow rate, 0.4 mL/min; revolution, 800 rpm. SF = solvent front.

Fig. 4 CCC separation of oxalic acid, malonic acid, and succinic acid by cross-axis CPC. Experimental conditions: sample, 3 mg each; solvent system, 1-butanol/water; flow rate, 0.25 mL/min. For other experimental conditions, see the Fig. 3 caption. SF = solvent front.

Biotic Dicarboxylic Acids: CCC Separation with Polar Two-Phase Solvent Systems Using a Cross-Axis Coil Planet Centrifuge   215

B

aqueous 0.1% trifluoroacetic acid (TFA), at various volume ratios. K values of the organic acids decreased as the hydrophobicity of the solvent system was increased, except for fumaric acid, which showed high K values regardless of the phase composition.

Fig. 3 illustrates the CCC chromatograms of dicarboxylic acids obtained with the above solvent system. In Fig. 3A, maleic acid and fumaric acid are separated at a volume ratio of $1:0:0:1$, which is used for the separation of aromatic acids such as hippuric acid.[10] Using a more polar solvent system, at a volume ratio of $2:0:2:3$, tartaric acid, succinic acid, and fumaric acid were well resolved by the lower aqueous mobile phase, as shown in Fig. 3B.

Fig. 4 illustrates a chromatogram of oxalic acid, malonic acid, and succinic acid obtained using the most polar binary solvent system, composed of 1-butanol/water. All components were well resolved from each other and eluted within 2.3 h, using the lower aqueous phase as the mobile phase.

As described above, various polar organic acids, such as dicarboxylic acids, can be separated with polar two-phase solvent systems, using the cross-axis CPC equipped with a pair of eccentric coil assemblies mounted in the off-center position.

REFERENCES

1. Mandava, N.B., Ito, Y., Eds.; *Countercurrent Chromatography: Theory and Practice*; Marcel Dekker: New York, 1988.
2. Conway, W.D. *Countercurrent Chromatography: Apparatus, Theory and Applications*; VCH: New York, 1990.
3. Ito, Y., Conway, W.D., Eds.; *High-peed Countercurrent Chromatography*; Wiley-Interscience: New York, 1996.
4. Ito, Y. Cross-axis synchronous flow-through coil planet centrifuge free of rotary seals for preparative countercurrent chromatography. Part I: Apparatus and analysis of acceleration. Sep. Sci. Technol. **1987**, *22*, 1971.
5. Ito, Y. Cross-axis synchronous flow-through coil planet centrifuge free of rotary seals for preparative countercurrent chromatography. Part II: Studies on phase distribution and partition efficiency in coaxial coils. Sep. Sci. Technol. **1987**, *22*, 1989.
6. Shinomiya, K.; Menet, J.-M.; Fales, H.M.; Ito, Y. Studies on a new cross-axis coil planet centrifuge for performing counter-current chromatography. I. Design of the apparatus, retention of the stationary phase, and efficiency in the separation of proteins. J. Chromatogr. **1993**, *644*, 215.
7. Shinomiya, K.; Inokuchi, N.; Gnabre, J.N.; Muto, M.; Kabasawa, Y.; Fales, H.M.; Ito, Y. Countercurrent chromatographic analysis of ovalalbumin obtained from various sources using the cross-axis coil planet centrifuge. J. Chromatogr. A. **1996**, *724*, 179.
8. Shinomiya, K.; Muto, M.; Kabasawa, Y.; Fales, H.M.; Ito, Y. Protein separation by improved cross-axis coil planet centrifuge with eccentric coil assemblies. J. Liq. Chromatogr. & Relat. Technol. **1996**, *19*, 415.
9. Shinomiya, K.; Kabasawa, Y.; Ito, Y. Countercurrent chromatographic separation of sugars and their p-nitrophenyl derivatives by cross-axis coil planet centrifuge. J. Liq. Chromatogr. & Relat. Technol. **1999**, *22*, 579.
10. Shinomiya, K.; Sasaki, Y.; Shibusawa, Y.; Kishinami, K.; Kabasawa, Y.; Ito, Y. Countercurrent chromatographic separation of hippuric acid and related compounds using cross-axis coil planet centrifuge with eccentric coil assemblies. J. Liq. Chromatogr. & Relat. Technol. **2000**, *23*, 1575.
11. Shinomiya, K.; Kabasawa, Y.; Ito, Y. Countercurrent chromatographic separation of biotic dicarboxylic acids with polar two-phase solvent systems using cross-axis coil planet centrifuge. J. Liq. Chromatogr. & Relat. Technol. **2001**, *24*, 2625.
12. Horikawa, R.; Tanimura, T. Spectrophotometric determination of carboxylic acids with 2-nitrophenylhydrazine in aqueous solution. Anal. Lett. **1982**, *15*, 1629.
13. Shinomiya, K.; Ochiai, H.; Suzuki, H.; Koshiishi, I.; Imanari, T. Simple method for determination of urinary mucopolysaccharides. Bunseki Kagaku **1986**, *35*, T29.

Bonded Phases in HPLC

Joseph J. Pesek
Maria T. Matyska
San Jose State University, San Jose, California, U.S.A.

INTRODUCTION

The development of chemically bonded stationary phases is one of the major factors that lead to the growth of HPLC and is responsible for its importance as a separation technique.

HISTORICAL BACKGROUND

In its earliest form, gravity flow moved the mobile phase through the column which was generally packed with a solid adsorbent such as silica or alumina. In a few instances, a high-molecular-weight liquid was coated on the solid particle to provide different types of selectivity. Under these circumstances, the column was similar to those used in GC, where a liquid stationary phase was held in place by physical forces alone. In GC, the requirement for the stationary phase to remain in place for a long time is low volatility. In LC, the requirement for durability is insolubility in the mobile phase. However, with the development of reliable high-pressure pumps that could produce stable flow rates for long periods of time, immiscibility with the mobile phase is not sufficient. At the pressures used to force solvents through most packed HPLC columns (from tens to a few hundred atmospheres), the shear forces developed at the interface between the stationary and the mobile phases are high enough to remove even insoluble liquids from the surface of the solid support. The stationary phase then is forced out of the column as an insoluble droplet. Removal of the stationary phase from a chromatography column is usually referred to as "column bleed." Therefore, it was necessary to develop a means of fixing the stationary phase on the solid support through a chemical bond. If the chemical bond between the surface of the solid support and the compound used as the stationary phase is stable under the experimental conditions of the HPLC experiment (temperature and mobile-phase composition), then column bleed will be avoided.

Fortunately, the most common support material used in LC experiments was silica. The chemistry of silica had been investigated for many years so a considerable amount of information was available about possible reactions on its surface. Silica can be considered as a polymer of silicic acid (H_2SiO_3). The terminal groups of the polymer located on the surface of the solid are hydroxide groups. These Si–OH functions are referred to as silanols. Because they come from an acid precursor, they are acidic themselves and generally have a pK_a near 5. This value is variable, depending on other constituents in the silica matrix such as metals. The structure of silica, including its major chemical features, are shown in Fig. 1. The polymeric unit consists of a series of siloxane bonds (–Si–O–Si–) that are slightly hydrophobic in nature. What is generally regarded as the most prominent feature on the surface is the silanol group, as indicated earlier. In a few cases, a single silicon atom will have two hydroxyl groups, which is called a geminal silanol. The silanols exist in two forms. First, they can be independent of other entities around them and are thus referred to as free or isolated silanols. If they are close enough to interact with a neighboring silanol, then these moieties become hydrogen-bonded or associated silanols. All forms of silanol are polar hydrophilic species. The relative number of free versus hydrogen-bonded silanols also has an influence on the pK_a value of the silica. Finally, because of the polar and hydrogen-bonding characteristics of the silanols, water is strongly adsorbed to the surface. This water is not easily removed, even at prolonged heating above 100°C. It is this complex matrix that must undergo a chemical reaction in order to attach a moiety to the surface as a stationary phase. According to the findings of early investigations on the reactivity of silica, it was determined that the silanol groups were the site of chemical modification on the surface.

The concept of attaching an organic moiety as a stationary phase to a silica surface was first applied in packed-column GC. The rationale for developing these materials was to prevent column bleed at the high temperatures required for some separations in GC. As long as the chemical bond was stable, the organic moiety would remain fixed to the surface. Some of the reactions utilized for the attachment of organic compounds in the synthesis of bonded stationary phases were originally developed for the modification of ordinary glass surfaces. Therefore, it was known that most of these modified surfaces were reasonably temperature stable and should be applicable to the bonding of organic groups

Encyclopedia of Chromatography DOI: 10.1081/E-ECHR-120039895

onto the porous silica particles used as supports in chromatography.

The first reaction used for the modification of porous silica in chromatography involves an alcohol as the organic species. This process is referred to as an esterification reaction. This may seem like incorrect nomenclature in order to describe the chemical process taking place between the silanol (Si–OH) and the organic compound (R–OH). However, the OH of the silanol is an acidic species, so the reaction taking place involves an acid and an alcohol, which, in typical organic chemistry nomenclature, is an esterification. The chemical reaction is illustrated in Fig. 2. The product of this reaction can be used as a stationary phase in GC because the material is thermally stable up to temperatures of approximately 300°C. However, the Si–O–C bond that exists between the surface and the bonded moiety is hydrolytically unstable in the presence of relatively small amounts of water. Therefore, these materials cannot be used for stationary phases in LC, where water comprises even a small fraction of the mobile phase.

The second reaction shown in Fig. 2 is the most common means used for the modification of silica surfaces. This method is referred to as organosilanization. Within this general reaction scheme, there are two possible approaches, as shown in Fig. 2. The first possibility involves the use of an organosilane reagent (RR'R'SiX) with only a single reactive group (X). The substituents on the silicon atom are as follows: X is a halide, most often Cl, methoxy or ethoxy; R is the organic moiety giving the surface the desired properties (i.e., hydrophobic, hydrophilic, ionic, etc.), and R' is a small organic group, typically methyl. This reaction leads to a single siloxane bond between the reagent and the surface. Because of the single point of attachment of the reagent, the resulting bonded material is referred to as a monomeric phase. The second approach to organosilanization involves a reagent with the general formula $RSiX_3$. The substituents on the silicon atom in this reagent are defined as above.

REACTION TYPE	REACTION	SURFACE LINKAGES
ESTERIFICATION	$Si - OH + R - OH \rightarrow Si - OR + H_2O$	Si - O - C
ORGANOSILANE	$Si - OH + X - SiR'_2R \rightarrow Si - O - SiR'_2R + HX$ 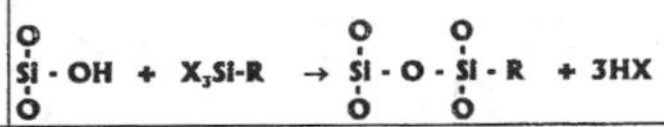	Si - O - Si - C
	$\overset{O}{\underset{O}{Si}} - OH + X_3Si-R \rightarrow \overset{O}{\underset{O}{Si}} - O - \overset{O}{\underset{O}{Si}} - R + 3HX$	
CHLORINATION FOLLOWED BY REACTION OF GRIGNARD REAGENTS OR ORGANOLITHIUM COMPOUNDS	$Si - OH + SOCl_2 \xrightarrow{Toluene} Si - Cl + SO_2 + HCl$ a). $Si - Cl + BrMgR \rightarrow Si - R + MgClBr$ or b). $Si - Cl + Li - R \rightarrow Si - R + LiCl$	Si - C
a). TES SILANIZATION b). HYDROSILATION	a). 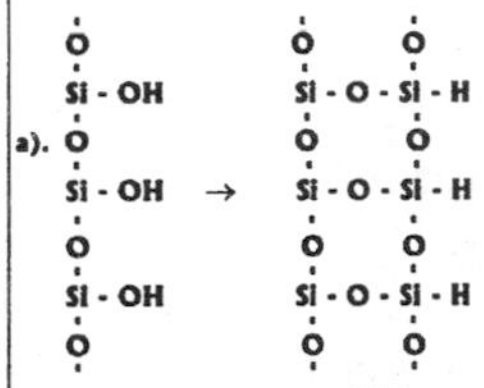 (Si - OH groups) $\rightarrow$ (Si - O - Si - H groups) b). $Si - H + CH_2 = CH - R \xrightarrow{Catalyst} Si - CH_2 - CH_2 - R$	a). Si - H monolayer b). Si - C

Fig. 2 Reactions for the modification of silica surfaces.

The basic difference between the approaches (as shown in Fig. 2) is that the reagent with three reactive groups results in bonding to the surface as well as cross-linking among adjacent bonded moieties and is referred to as a polymeric phase. This cross-linking effect provides extra stability to the bonded moiety but is less reproducible than the monomeric method. The one-step organosilanization procedure is relatively easy and the modification of the surface can be done by stirring the reagent continuously with the porous silica support. The reaction mixture is heated for about 1–2 h, then the reagent solution is removed, usually by centrifugation and/or filtration. The bonded phase is then washed with several solvents and dried under vacuum to remove as much of the rinse solutions as possible. Organosilanization accounts for virtually all of the commercially available chemically bonded stationary phases.

Another method that has been reported for the modification of silica supports is based on a chlorination/organometalation two-step reaction sequence. This process is also depicted in Fig. 2. In the first step, the silanols on the porous silica surface are converted to chlorides via a reaction with thionyl chloride. This step must be done under extremely dry conditions because the presence of any water results in the reversal of the reaction with hydroxyl replacing the chloride (Si–Cl), resulting in the regeneration of silanols (Si–OH). If the chlorinated material can be preserved (usually done in a closed vessel purged with a dry gas like nitrogen), then an organic group can be attached

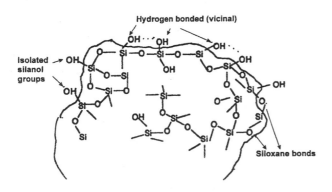

Fig. 1 Structure of silica showing the surface chemical features.

to the surface via a Grignard reaction or an organo-lithium reaction. The main advantage of this process is that it produces a very stable silicon–carbon linkage at the surface. However, the stringent reaction conditions for the first step and the possibility of forming salts that could affect chromatographic properties as by-products in the second reaction have resulted in relatively little commercial use of this process.

The final method shown in Fig. 2 involves, first, silanization of the silica surface, followed by attachment of the organic group through a hydrosilation reaction. In the first step, the use of triethoxysilane under controlled conditions results in a monolayer of the cross-linked reagent being deposited on the surface. This reaction results in the replacement of hydroxides by hydrides. In the second step, an organic moiety is attached to the surface via the hydride moiety by a hydrosilation reaction using a catalyst such as hexachloroplatinic acid (Speier's catalyst), but other transition metal complexes or a free-radical initiator have been reported as well. This process also results in a silicon–carbon bond at the surface, does not require dry conditions (water is required as a catalyst in the first step), and is applicable to a variety of unsaturated functional groups in the hydrosilation reaction, although terminal olefins are the most common. The silanization/hydrosilation method also has seen limited commercial utilization to date.

In all of the reactions described, the choice of the organic moiety on the reagent (R group) determines the properties of the material as a stationary phase. Therefore, selection of a hydrophobic moiety where R is typically an alkyl group leads to a stationary phase that selectively retains nonpolar analytes. These materials are typically used in reversed-phase chromatography. If the organic moiety contains a polar functional group such as amine, cyano, or diol, then the stationary phase selectively retains polar compounds. These materials are typically used in normal-phase chromatography.

The bonding of the organic group on the surface results in the replacement of silanols whose adsorptive properties are strong, especially for bases, and often nonreproducible. Although it is impossible to replace all silanols, the remaining Si–OH groups are often shielded from solutes by the steric hindrance of the bonded organic moiety. In many cases though, some silanols are accessible to typical solutes. In order to inhibit the interaction between analytes and residual silanols, the bonded phase is subjected to an additional reaction referred to as endcapping. In this case, a small organosilane, often trimethylchlorosilane, penetrates into the spaces between the bonded groups to react with the most accessible silanols. This process generally greatly reduces or eliminates solute interactions with silanols.

After the bonded phase is prepared, it must be packed into a column, usually a stainless-steel tube. In order for the material to form a uniform bed of high density that will not form voids after prolonged use, the packing process must be done under high pressure (>500 atm). The stationary phase is mixed with a solvent of approximately the same density as silica, so that a slurry is formed. This slurry is then forced into the column at high pressure with another solvent, usually methanol. After packing, most stationary phases require several hours of conditioning, with the mobile phase passing through the column at normal flow rates, before actual chromatographic analysis can begin.

SUGGESTED FURTHER READING

Iler, R.K. *The Chemistry of Silica*; John Wiley & Sons: New York, 1979.
Marciniec, B. *Comprehensive Handbook on Hydrosilylation*; Pergamon Press: Oxford, 1992.
Nawrocki, J. Chromatographia. **1991**, *31*, 177.
Nawrocki, J. Chromatographia. **1991**, *31*, 193.
Pesek, J.J.; Matyska, M.T. Interf. Sci. **1997**, *5*, 103.
Pesek, J.J.; Matyska, M.T.; Sandoval, J.E.; Williamsen, E.J. J. Liq. Chromatogr. & Related Technol. **1996**, *19*, 2843.
Unger, K.K. *Porous Silica*; Elsevier: Amsterdam, 1979.
Vansant, E.F.; Van Der Voort, P.; Vrancken, K.C. *Characterization and Chemical Modification of Silica*; Elsevier: Amsterdam, 1995.

Buffer Systems for Capillary Electrophoresis

Robert Weinberger
CE Technologies, Inc., Chappaqua, New York, U.S.A.

INTRODUCTION

The solution contained within the capillary in which the separation occurs is known as the background electrolyte (BGE), carrier electrolyte, or, simply, the buffer. The BGE always contains a buffer because pH control is the most important parameter in electrophoresis. The pH may affect the charge and thus the mobility of an ionizable solute. The electro-osmotic flow (EOF) is also affected by the buffer pH. Table 1 contains a list of buffers that may prove useful in high-performance capillary electrophoresis (HPCE). As will be seen later, only a few of these buffers are necessary for most separations.

Other reagents, known as additives, are often added to the BGE to adjust selectivity (secondary equilibrium), modify the EOF, maintain solubility, and reduce the adherence of the solute or sample matrix components to the capillary wall. Table 2 provides these applications, along with some of the commonly used reagents.

BUFFERS

The selection of the appropriate buffer is usually straightforward. For acids, start with a borate buffer (pH 9.3), and for bases, a phosphate buffer (pH 2.5). These two buffer systems, along with the appropriate additives will work well for most applications. Both buffers have good buffer capacity and the ultraviolet (UV) absorbance of each is low. If bases are not soluble in phosphate buffer, acetate buffer (pH 4) may be more effective. Higher pHs may be required for basic proteins to avoid solute adherence to the capillary wall. If pH 7 is desired, the phosphate buffer works well at that pH. If necessary, the buffer pH can be fine-tuned using a mobility plot.

Alternative buffer systems include zwitterions and dual-buffering reagents. Zwitterionic buffers such as bicine, tricine, CAPS, MES, and Tris may be useful for protein and peptide separations. An advantage of a zwitterionic buffer is low conductivity when the buffer pH is adjusted to its pI. There is little buffer capacity when the pK_a and pI are separated by more than 2 pH units. When the pI and pK_a are close together, the buffer is known as an isoelectric buffer.[1]

Selection of the appropriate counterion is also important. Lithium ion has the lowest mobility of the alkali earth metals. Its use provides for a low-conductivity buffer. Sodium salts are used more frequently due to purity and availability. It makes little sense to ever use a potassium salt. Dual-buffering systems with low-mobility ions and counterions (Tris-phosphate, Tris-borate, aminomethylpropanediol–cacodylic acid) are effective in minimizing buffer conductivity. These buffers are often used in the slab–gel, where low conductivity is particularly important.

The buffer concentration plays an important role in the separation. Typical buffer concentrations range from 20 to 150 mM. At the higher buffer concentrations, the production of heat may require the use of lower field strength or smaller-diameter capillaries (25 μm instead of 50 μm). An Ohm's law plot is used to select the appropriate voltage. The advantages of high-concentration buffers include improved peak shape, fewer wall effects, and increased sample stacking.

Low-concentration buffers (less that 20 mM) provide the fastest separations because solute mobility and EOF is inversely proportional to the square root of the buffer concentration. Because the conductivity of a dilute buffer is low, a high electric field strength can be used as well. Problems with low-concentration buffers are loading capacity, wall effects, and poor stacking. Sawtooth-shaped peaks from a process known as electrodispersion may occur whenever the solute concentration approaches the BGE concentration. It also becomes more likely that proteins will adhere to the capillary wall when the buffer concentration is low. Ionic-strength-mediated sample stacking relies on a high-conductivity BGE and a low-conductivity sample.[2] This important process is less effective at low buffer concentrations. When indirect detection is employed, the buffer (indirect detection reagent) concentration must be kept low to optimize sensitivity.[3] Sawtooth peaks are often observed when this technique is used.

It is important to refresh the BGE reservoirs frequently to avoid a process known as buffer depletion.[4] Electrolysis at the respective electrodes produces protons and hydroxide ions. This can cause pH changes in the buffer reservoirs.

High-pH buffers (>pH 11) are used for certain small ion separations and for the separation of

Encyclopedia of Chromatography DOI: 10.1081/E-ECHR-120039896

Table 1 Buffers for HPCE

Buffer	pK_a	Buffer	pK_a
Aspartate	1.99	DIPSO	7.5
Phosphate	2.14, 7.10, 13.3	HEPES	7.51
Citrate	3.12, 4.76, 6.40	TAPSO	7.58
β-Alanine	3.55	HEPPSO	7.9
Formate	3.75	EPPS	7.9
Lactate	3.85	POPSO	7.9
Acetate	4.76	DEB	7.91
Creatinine	4.89	Tricine	8.05
MES	6.13	GLYGLY	8.2
ACES	6.75	Bicine	8.25
MOPSO	6.79	TAPS	8.4
BES	7.16	Borate	9.14
MOPS	7.2	CHES	9.55
TES	7.45	CAPS	10.4

carbohydrates using indirect detection. Adsorption of carbon dioxide can cause the buffer pH to decline. It is best to use small containers filled to the top when storing these buffers.

BUFFER ADDITIVES

Secondary Equilibrium

If two solutes are inseparable based on pH alone, secondary equilibrium can be employed to effect a separation. The following equilibrium expressions can be written.[5]

$$A^+ + R \overset{K_A}{\rightleftharpoons} A^+R \tag{1}$$

$$B^+ + R \overset{K_B}{\rightleftharpoons} B^+R \tag{2}$$

If the equilibrium is pushed too far to the left, no separation can occur because A^+ and B^+ are inseparable. When the reagent interacts with the solute, the mobility decreases because the neutral reagent contributes mass without charge. However, if the equilibrium is pushed too far to the right, no separation occurs because A^+R and B^+R are inseparable. Separation only occurs when two conditions are met:

1. K_A does not equal K_B.
2. The equilibrium is not pushed to either extreme.

The next feature to consider is the charge of the reagent and solute. To separate charged solutes, the reagent can be charged or neutral. When the solute is neutral, the reagent must be charged.

Micelles and cyclodextrins are the most common reagents used for this technique. MECC or MEKC is generally used for the separation of small molecules.[6] Sodium dodecyl sulfate at concentrations from 20 to 150 mM in conjunction with 20 mM borate buffer (pH 9.3) or phosphate buffer (pH 7.0) represent the most common operating conditions. The mechanism of separation is related to reversed-phase LC, at least for neutral solutes. Organic solvents such as 5–20% methanol or acetonitrile are useful to modify selectivity when there is too much "retention" in the system. Alternative surfactants such as bile salts (sodium cholate), cationic surfactants (cetyltrimethylammonium bromide), nonionic surfactants (polyoxyethylene-23-lauryl ether), and alkyl glucosides can be used as well.

Cyclodextrins (CD) are frequently used for chiral recognition,[7] although they are quite useful for achiral applications as well. Many classes have been used including native, functionalized, sulfobutylether, and highly sulfated CDs. The latter two are generally most effective for chiral and structural isomer separations. The typical CD concentrations range from 1 to 20 mM in 20–50 mM of borate (pH 9.3) or phosphate buffer (pH 2.5). Other reagents useful for chiral recognition include macrocyclic antibiotics, bile salts, chiral surfactants, noncyclic oligosaccharides and polysaccharides, and crown ethers.

Additional reagents useful for secondary equilibrium include borate buffer for carbohydrates, chelating agents for transition metals, ion-pair reagents for acids and bases, transition metals for proteins and peptides, silver ion for alkenes, and Mg^{2+} for nucleosides.

Table 2 Buffer additives

Purpose	Reagent	Mechanism
Modify mobility	Borate	Complex with carbohydrates, diols
	Calixarenes	Inclusion complex
	Chelating agents	Complex formation with metals
	Crown ethers	Inclusion complex
	Cyclodextrins	Inclusion complex
	Dendrimers	Inclusion complex
	Macrocyclic antibiotics	Inclusion complex
	Organic solvents	Solvation
	Sulfonic acids	Ion-pair formation
	Surfactants	Micelle interaction
	Transition metals	Complex formation
	Quaternary amines	Ion-pair formation
Modify EOF	Cationic surfactant	Dynamic coating, EOF reversal
	Linear polymers	Dynamic coating
	Organic solvents	Affects viscosity
Reduce wall effects	Cationic surfactant	Dynamic coating, EOF reversal
	Linear polymers	Dynamic coating
Polyamines		Covers silanols
Maintain solubility	Organic solvents	Hydrophobicity
	Urea	"Iceberg effect"

Electro-Osmotic Flow Control

The control of EOF is critical to the migration time precision of the separation. Among the factors affecting the EOF are buffer pH, buffer concentration, buffer viscosity, temperature, organic modifiers, cationic surfactants or protonated amines, polymer additives, field strength, and the nature of the capillary surface.

At pH 2.5, the EOF is approximately 10^{-5} cm^2/V/s in 50 mM buffer. At pH 7, it is an order of magnitude higher. The EOF is inversely proportional to BGE viscosity and is proportional to temperature, up until the point where heat dissipation is inadequate. Organic modifiers such as methanol decrease the EOF because hydro-organic mixtures have higher viscosity compared to water alone. Acetonitrile does not strongly affect the EOF. Polymer additives such as methylcellulose derivatives increase viscosity as well as coat the capillary wall.

Cationic surfactants and protonated polyamines may reverse the direction of the EOF as they impart a positive charge on the capillary wall. This technique is used to prevent wall interactions with cationic proteins. Changing the direction of the EOF is important in anion analysis where comigration of anions and the EOF is required. Otherwise, highly mobile anions such as chloride migrate toward the anode, whereas lower mobility anions are swept by the EOF toward the cathode.

A new series of reagents (CElixir, MicroSOLV, Long Branch, NJ) have been shown to dramatically stabilize the EOF, resulting in highly reproducible run-to-run and capillary-to-capillary migration times.[8] First, a capillary is treated as usual with 0.1 N sodium hydroxide, followed by a rinse with a polycation solution. Then, a second layer consisting of a polyanion in a buffer at the desired pH is flushed through the capillary. Replicate runs are virtually superimposible, yielding reproducibility seldom found in HPCE. The reagents have been shown to work best for bases below at a pH below the pK_a.

Maintaining Solubility

All solutes and matrix components must remain in solution for an effective separation to occur. In aqueous systems, surfactants and urea are the most useful reagents. Organic solvents can be used as well, but this is less desirable because of evaporation. It can be difficult to separate solutes with widely different solubilities in a single run. In some cases, nonaqueous separations are necessary.

Reducing Wall Effects

Wall effects, or the adherence of material to the bare silica capillary wall, has been a difficult problem since the early days of HPCE, particularly for large molecules such as proteins. Small molecules can have, at most, one point of attachment to the wall and the kinetics of adsorption/desorption are rapid. Large molecules can have multiple points of attachment resulting in slow kinetics. Several solutions have been proposed, including the use of (a) extreme-pH buffers, (b) high-concentration buffers, (c) amine modifiers, (d) dynamically coated capillaries, and (e) treated or functionalized capillaries.

In the first case, it was recognized that if the buffer pH is greater that 2 units above the protein, the anionic protein would be repelled from the anionic capillary wall.[9] At a pH < 2, the capillary wall is neutral and does not attract the cationic protein. The problem with this approach is that a wide range of pHs are not available for use and separations of similar proteins may not occur. For high-pI proteins such as histones, a buffer pH of 13 is required. The conductivity and UV background of such an electrolyte is too high to be generally applicable.

The use of high-concentration buffers is effective in reducing wall effects. This includes electrolytes containing up to 250mM added salt. The problem with this approach is the high conductivity of the BGE. This requires a reduction in field strength resulting in lengthy separations. Zwitterionic buffers titrated to their pI can be used as well at concentrations approaching 1M. At that concentration, it is important to select a reagent with low UV absorption.

The latter three cases are most commonly employed to reduce wall effects. In the third case, amine modifiers such as polyamines are added to the BGE at concentrations ranging from 1 to 60mM.[10] These reagents coat the free silanols and reduce wall interactions. Now, any pH electrolyte can be employed. Diaminobutane, otherwise known as putrecein, is the preferred reagent because it is less volatile compared to diaminopropane. Monovalent amines such as triethanolamine are not as effective in this regard.

Dynamically coated capillaries (case d) are often used to reduce wall effects.[11] The mechanism of charge reversal is as follows. Ion-pair formation between the cationic head group of the surfactant and the anionic silanol group naturally occurs. The hydrophobic surfactant tail extending into the bulk solution is poorly solvated by water. The molecular need for solvation is satisfied by binding to the tail of another surfactant molecule. The cationic head group

of the second surfactant molecule now extends into the bulk solution. The capillary wall becomes positively charged and the EOF is directed toward the anode. Separations are performed using the reversed-polarity mode (inlet side negative). Following this approach, a buffer pH is selected that is below the pI of the target protein. The cationic protein is now repelled from the cationic wall.

When coated capillaries are employed (case e), conventional buffers without additives to reduce wall effects are used. Urea and/or organic solvents can be added to aid solubility. Reagents for secondary equilibrium can be used as well. It is best to operate at a pH below 8 to maximize the stability of the often labile coating material. Coated capillaries are also used simply to eliminate the EOF in some applications.

REFERENCES

1. Righetti, P.G.; Gelfi, C.; Perego, M.; Stoyanov, A.V.; Bossi, A. Capillary zone electrophoresis of oligonucleotides and peptides in isoelectric buffers: theory and methodology. Electrophoresis **1997**, *18*, 2145.
2. Burgi, D.; Chien, R.-L. Optimization in sample stacking for high-performance capillary electrophoresis. Anal. Chem. **1991**, *63*, 2042.
3. Jandik, P.; Jones, W.R.; Weston, A.; Brown, P.R. LCGC **1991**, *9*, 634.
4. Macka, M.; Andersson, P.; Haddad, P.R. Changes in electrolyte pH due to electrolysis during capillary zone electrophoresis. Anal. Chem. **1998**, *70*, 743.
5. Wren, S.A.C.; Rowe, R.C. Theoretical aspects of chiral separation in capillary electrophoresis: I. Initial evaluation of a model. J. Chromatogr. **1992**, *603*, 235.
6. Nishi, H.; Terabe, S. Micellar electrokinetic chromatography perspectives in drug analysis. J. Chromatogr. A. **1996**, *735*, 3.
7. Chankvetadze, B. *Capillary Electrophoresis in Chiral Analysis*; John Wiley & Sons: Chichester, 1997.
8. Weinberger, R. Am. Lab. **1999**, *31*, 59.
9. Lauer, H.H.; McManigill, D. Capillary zone electrophoresis of proteins in untreated fused silica tubing. Anal. Chem. **1986**, *58*, 166.
10. Bullock, J.A.; Yuan, L.-C. J. Microcol. Separ. **1991**, *3*, 241.
11. Wiktorowicz, J.E.; Colburn, J.C. Separation of cationic proteins via charge reversalin capillary electrophoresis. Electrophoresis **1990**, *11*, 769.

Buffer Type and Concentration: Effect on Mobility, Selectivity, and Resolution in Capillary Electrophoresis

Ernst Kenndler
Institute for Analytical Chemistry, University of Vienna, Vienna, Austria

INTRODUCTION

Resolution in capillary zone electrophoresis (CZE) is, as in elution chromatography, a quantity that describes the extent of the separation of two consecutively migrating compounds, i and j. It is the result of the counterplay of two effects: migration and zone dispersion. The different migration velocities of the two separands lead (at least potentially) to the separation of the sample zones. The simultaneous mixing of the samples with the background electrolyte (BGE), caused by a number of processes, results in zone broadening and counteracts separation. Both effects determine the overall degree of separation. A quantitative measure that describes this degree is the resolution a dimensionless number. It is expressed by the difference of the apex of the two peaks, on the one hand. It is of advantage not to measure this difference in an absolute scale (e.g., in seconds when the electropherogram is depicted in the time domain). In fact, a relative scale is taken, which is based on the widths of the peaks. We define the resolution as the difference in migration times, t, related to the peak width, taken, for example, by the mean standard deviation of the Gaussian peaks, as the scaling unit:

$$R_{ji} \equiv \frac{t_j - t_i}{2(\sigma_{t,i} + \sigma_{t,j})} \qquad (1)$$

Baseline separation is achieved for two peaks with the same area when the resolution is 1.5. For peak area ratios larger than unity, the resolution must be larger.

SELECTIVITY AND EFFICIENCY

This definitional equation (1) is not very operative and is, thus, transformed to an expression which more clearly visualizes the dependence of the resolution on sample properties and experimental variables. The migration times are substituted for by the mobilities of the separands, and the standard deviations by the plate height, H, or the plate number, N, respectively.

The resulting resolution is then

$$R_{ji} = \frac{1}{4} \frac{\mu_i - \mu_j}{\bar{\mu}} \sqrt{\frac{L}{\bar{H}}} = \frac{1}{4} \frac{\Delta\mu}{\bar{\mu}} \sqrt{\bar{N}} \qquad (2)$$

where $\bar{\mu}$, $\bar{H}$, and $\bar{N}$ are the average values; L is the migration distance.

The resolution consists of two terms, the selectivity term, $\Delta\mu/\bar{\mu}$, with the relative difference of the mobilities, and the efficiency term, the square root of the mean plate number. It must be pointed out that the plate height, $\bar{H}$, on which this plate number is based consists of all the contributions to peak broadening.

At this point, a differentiation should be made between two cases: the simple one, where migration is only caused by the electric force on the ionic separands, and the second, where an additional migration due to the occurrence of an electro-osmotic flow (EOF) takes place.

RESOLUTION IN ABSENCE OF EOF

Two main parameters determine the resolution: the effective mobility and the plate number. The effective mobility of a simple ion (e.g., the anion from a monovalent weak acid) is given by

$$\mu_{\text{eff}} = \frac{\mu_{\text{act}}}{1 + 10^{pK_a - pH}} \qquad (3)$$

We take, here, only protolysis into consideration and do not discuss such important other equilibria such as complexation or interactions with pseudo-stationary phases. It follows from Eq. 3 that the effective mobility depends on the actual mobility (that of the fully charged particle at the ionic strength of the experiment), on the pK_a value of the analyte, and on the pH of the BGE. It follows that all these properties determine the selectivity term in the resolution.

The actual mobility depends on the following:

1. *The solvent.* There is a more or less pronounced influence of the solvent viscosity, reflected by Walden's rule. However, this rule is obeyed

Encyclopedia of Chromatography DOI: 10.1081/E-ECHR-120039897

in rare cases; mainly in some mixed aqueous–organic solutions is an acceptable agreement found. On the other hand, in very viscous aqueous solutions of water-soluble polymers, such as poly(ethylene glycol), it was found that the actual mobility is independent of the viscosity.

2. *The size of the solvated ion.* Here, we must note that water is an excellent solvator for anions and cations as well, compared to most organic solvents. Only few exceptions for preferred solvation of the organic solvents are found (e.g., for Ag^+ and acetonitrile).

3. *The ionic strength of the BGE.* The dependence of the mobility on the ionic strength is expressed for simple systems (and simple ions) by the theory of Debye, Hückel, and Onsager. Without going into detail, we can state that the mobility decreases, in all cases, with increasing ionic strength of the BGE, and the decrease is more pronounced the higher the charge number of the ion.

4. *On the temperature.* In aqueous solutions, the mobility increases with temperature roughly by about 3% per degree. This is a strong effect as, for example, a temperature difference of only 5 K between the center and the wall of the separation capillary leads to a mobility difference of about 15%.

The pK_a value is also a function, mainly, of the solvent. Note that the pH scale is strongly dependent on the kind of solvent. Restricting the discussion to protolysis, it can be followed that the pH of the buffer has the most pronounced effect on the effective mobility, because the other effects change the mobility roughly in parallel for all separands. Again, it must be pointed out that other equilibria have an enormous potential to affect the effective mobility (cf. e.g., the use of cyclodextrins to introduce selectivity for the separation of enantiomers).

How is the efficiency influenced by the BGE? Peak broadening is the result of different processes in CZE occurring during migration [in addition, extracolumn effects contribute to peak width (e.g., that stemming from the width and shape of the injection zone, or the aperture of the detector cell)]. If the system behaves linearly, the individual peak variances (the second moments), are additive according to

$$\sigma_{tot}^2 = \Sigma\sigma_{ind}^2$$
$$= \sigma_{extr}^2 + \sigma_{dif}^2 + \sigma_{Joule}^2 + \sigma_{conc}^2 + \sigma_{ads}^2 \tag{4}$$

where the subscripts extr, dif, Joule, conc, and ads indicate the contributions from extracolumn dispersion,

longitudinal diffusion, Joule self-heating, concentration overload, and wall adsorption, respectively. All but one effect might be eliminated: The longitudinal diffusion is inevitable. Plate number expressing this contribution is dependent on the voltage, U, applied and on the charge number, z_i, of the analytes according to

$$N_i \approx 20z_iU \tag{5}$$

at 20°C. It is obvious that the charge number depends on the pH of the BGE, as it is related to the degree of ionization. It follows that the plate number is a function of the pH as well. Thus, the resolution is influenced by the pH of the BGE twofold: via the selectivity, on the one hand, and via the plate number, on the other hand.

In conclusion, it follows for the limiting case of longitudinal diffusion as the only peak-broadening effect, that the resolution depends on the following:

- Instrumental variables: U and T
- Analyte parameters: μ_{act} and pK_a
- Chemical conditions determining the degree of ionization, α, or charge number z.

RESOLUTION IN PRESENCE OF EOF

The EOF brings an additional, unspecific velocity vector to the electrophoretic migration of the separands. The total migration velocity of the analyte, i, is then

$$\nu_{i,tot} = (\mu_{i,eff} + \mu_{EOF})E \tag{6}$$

Note that the mobilities are taken as signed quantities. By convention, cations have positive electrophoretic mobilities and those of anions are negative. The mobility of the EOF when directed toward the cathode has positive sign, and vice versa.

The effect of the EOF on migration time and selectivity depends on the mutual signs of the mobilities of analytes and EOF, respectively. Concerning the change in separation selectivity, we refer to the expression of the selectivity term in the resolution equation. The difference between the mobilities of the two separands, i and j, will not be influenced by the EOF. However, the mean mobility is larger for the case of comigration. This means that the selectivity term in the expression for the resolution is always reduced in this case. In practice, selectivity is lost for cation separation when the EOF is directed, as is usual in uncoated fused-silica capillaries, toward the cathode. For this reason, cationic additives are applied in the BGE to reverse the EOF direction.

The effect of the EOF on separation selectivity (in comparison with the situation without EOF) can be quantified by the so-called electromigration factor, or reduced mobility, μ_i^*, defined as

$$\mu_i^* = \frac{\mu_{i,\text{eff}}}{\mu_{i,\text{eff}} + \mu_{\text{EOF}}} \tag{7}$$

The change of the selectivity term in the resolution is directly expressible by μ_j^*. Interestingly, the effect of the EOF on the dispersion effects, expressed by the plate height H, also depends directly on μ^*. For longitudinal diffusion, Joule self-heating, and concentration overload, the variation of the plate height in the presence of the EOF is directly dependent upon this reduced mobility according to

$$H^{\text{EOF}} = H^0 \mu^* \tag{8}$$

where the superscript 0 indicates the system without EOF. For wall adsorption, the corresponding effect is related to the reciprocal of μ^*.

An analysis of the effect of the EOF on the resolution brings the following result: For comigration of the analyte and EOF, the efficiency always increases and the selectivity term always decreases. As the decrease is directly proportional to μ^* but the gain in plate number is only increasing with the square root of μ^*, the resolution is always worse than without comigrating EOF.

For the case of countermigration, the situation is more complicated, because the overall effect depends on the magnitude of the mobility of analyte and that of the EOF. Roughly, it can be concluded that the resolution is increased for a given pair of analytes when the EOF is counterdirected, and it has a lower mobility than the analytes. Here, efficiency is lost, but selectivity is gained overproportionally. When the EOF mobility reaches a value that is twice as large as the analyte mobility (note that the signs of the mobilities are different), an analogous situation is found as without EOF. At mobilities of the EOF larger than twice the analyte mobility (conditions not impossible for high pH values of the BGE), resolution is worse here than without EOF, but the analysis time is shorter than in all other cases. It should be pointed out that all of these effects can be quantified by the reduced mobility defined in Eq. (7).

SUGGESTED FURTHER READING

Camillieri, P. *Capillary Electrophoresis, Theory and Practice*; CRC Press: Boca Raton, FL, 1998.

Giddings, J.C. Harnessing electrical forces for separation: CZE, IEF, FFF, SPLITT, and other techniques. J. Chromatogr. **1989**, *480*, 21.

Guzman, N.A. *Capillary Electrophoresis Technology*; Marcel Dekker, Inc.: New York, 1993.

Kenndler, E. J. Microcol. Separ. **1998**, *10*, 273.

Kenndler, E. *High Performance Capillary Electrophoresis, Theory, Techniques and Applications*; Khaledi, M.G., Ed.; John Wiley & Sons: New York, 1998; Vol. 146, 25–76.

Landers, J.P. *Handbook of Capillary Electrophoresis*, 2nd Ed.; CRC Press: Boca Raton, FL, 1997.

Reijenga, J.C.; Kenndler, E. Computational simulation of migration and dispersion in free capillary zone electrophoresis, part I, Description of the theoretical model. J. Chromatogr. A. **1994**, *659*, 403.

Calibration of GPC/SEC with Narrow Molecular-Weight Distribution Standards

Oscar Chiantore
Università degli Studi di Torino, Torino, Italy

INTRODUCTION

In SEC, polymer solutions are injected into one or more columns in series, packed with microparticulate porous packings. The packing pores have sizes in the range between ~5 and 10^5 nm, and during elution, the polymer molecules may or may not, depending on their size in the chromatographic eluent, penetrate into the pores. Therefore, smaller molecules have access to a larger fraction of pores compared to the larger ones, and the macromolecules elute in a decreasing order of molecular weights. For each type of polymer dissolved in the chromatographic eluent, and eluting through the given set of columns with a pure exclusion mechanism, a precise empirical correlation exists between molecular weights and elution volumes. This relationship constitutes the calibration of the SEC system, which allows the evaluation of average molecular weights (MWs) and molecular-weight distributions (MWDs) of the polymer under examination.

Direct column calibration for a given polymer requires the use of narrow MWD samples of that polymer, with molecular weights covering the whole range of interest. The polydispersity of the calibration standards must be less than 1.05, except for the very low and very high MWs ($<10^3$ and $>10^6$), for which polydispersity can reach 1.20. The chromatograms of such standards give narrow peaks and to each standard is associated the retention volume of the peak maximum.

There is a limited number of polymers for which narrow MWD standards are commercially available: polystyrene, poly(methyl methacrylate), poly(α-amethyl styrene), polyisoprene, polybutadiene, polyethylene, poly(dimethyl siloxane), polyethylen eoxide, pullulan, dextran, polystyrene sulfonate sodium salt, and globular proteins. In some cases, the standards available cover a limited molecular weight range, so it may be impossible to construct the calibration curve over the complete column pore volume.

Standard methods for calibration of SEC columns with narrow MWD samples have been published by the American Society for Testing and Materials (ASTM D2596-97) and the Deutsches Institut for Normung (DIN 55672-1).

PROCEDURE

Fresh solutions of the standards are prepared in the solvent used as chromatographic eluent. Calibration solutions should be as dilute as possible, in order to avoid any concentration dependence of sample retention volumes. The concentration effect causes an increase of retention volumes with increased sample concentration. As a rule of thumb, when high efficiency microparticulate packings are used, the concentration of narrow standards should be ≤0.025% (w/v) for MW over 10^6, ≤0.05% for MW between 10^6 and 2×10^5, and ≤0.1% for MW down to 10^4. With a lower MW and in the oligomer range, the sample concentration can be higher than the previously suggested values.

Two or more standards may be dissolved and injected together to determine several retention volumes with a single injection. In such a case, the MW difference between the samples in the mixture should be sufficient to give peaks with baseline resolution. A sufficient number of narrow MWD standards, with different MWs, are required for establishing the calibration of a SEC column system. At least two standards per MW decade should be injected, and a minimum of five calibration points should be obtained in the curve.

MW FRACTIONATION RANGE OF THE COLUMN SET

The maximum injection volume depends from column size and packing pore volumes, and for high-efficiency 300×8 mm columns, it is generally recommended not to exceed 100 μL per column.

The flow rate of the chromatographic apparatus must be extremely stable and reproducible: Flow rate fluctuations about the specified value should be lower than 3%, and long-term drift lower than 1%. Repeatability of flow rate setting is extremely important, as a 1% constant deviation of the actual flow rate from the required value may give 20% differences in calculated MW averages.

Encyclopedia of Chromatography DOI: 10.1081/E-ECHR-120039898

The systematic errors introduced by flow rate differences may be avoided by adding to the solutions a minimum amount of a low-molecular-weight internal standard (*o*-dichloro benzene, toluene, acetone, sulfur) which must not interfere with the polymer peaks. Flow rate is monitored in each chromatogram by measuring the retention time of the internal standard, and eventual variations may be corrected accordingly.

The peak retention times for the narrow polymer standards are measured from the chromatograms and transformed into retention volumes according to the real flow rate. For each standard, the logarithm of nominal molecular weight is plotted against its peak elution volume. Often, retention times are directly employed and plotted as the measured variable, and in this case, the condition of equal flow rate elutions for all the standards and for any subsequent sample analysis is achieved by means of the internal standard elution.

The molecular weight of the standards is supplied by the producers, either with a single value which should correspond to that of peak maximum, or with a complete characterization data sheet containing the values of M_n and M_w determined by osmometry and light scattering. In the latter case, the peak molecular weight to be inserted in the calibration plot is the mean value $(M_n M_w)^{1/2}$. A typical calibration curve for a threecolumn set, 300×7.5 mm, packed with a mixture of individual pore sizes is shown in Fig. 1. The calibration curve has a central part which is essentially linear and becomes curved at the two extremes: on the high-MW side when it approaches the retention value of totally excluded samples; on the low-MW side with a downward curvature until it reaches the retention time of total pore permeation.

The calibration curve, therefore, defines the extremes of retention times (or volumes) for the specific column system, the useful retention interval for sample analysis, and the related MW range. Columns packed with a balanced mixture of different pore sizes are capable of giving linear calibrations over the whole MW range of practical interest, from the oligomer region to more than 10^6.

The plot of $\log M$ versus peak retention volumes of narrow standards is represented in the more general form by a *n*th-order polynomial of the type

$$\log M = A + BV_r + CV_r^2 + DV_r^3 + \cdots$$

the coefficients of which are determined by regression on the experimental data. Most usually, when the linear plot is not sufficient to fit the points, a third-order polynomial will be adequate to represent the curve. Higher-order equations, although improving the fit, should be used with great care, as they can lead to unrealistic oscillations of the function.

The goodness of different equations fitted to the experimental data points is assessed by the results of statistical analysis or by simply considering the standard error of the estimate. It should be also considered that the adequacy of the calibration function for the determination of correct MW values is also dependent on the quality of the narrow MWD standards. Their nominal MWs are determined with independent absolute methods and are affected by experimental errors which may be different between samples with different MWs, or coming from different producers. A check of the quality of the narrow standards may be obtained by calculating the percent MW deviation of each standard from the calibration curve:

$$\Delta M(V_i)\% = \frac{M_{\text{pcak}}(V_i) - M_{\text{calc}}(V_i)}{M_{\text{pcak}}(V_i)} \times 100$$

A plot of $\Delta M(V_i)\%$ versus $\log M$ results in positive and negative values scattered around the MW axis, which allows one to visualize the limits of percent error into which the MW of standards are estimated by the calibration curve. If the MW error of some standard is found to be significantly larger than all the others, it is likely that its nominal MW is incorrect. The point of such sample should be removed from the calibration and the regression recalculated.

The calibration curve should always cover the MW of the samples that must be analyzed. Extrapolation of the calibration outside the range of injected polymer standards should be avoided in MW determinations.

From the calibration curve, the resolution power of the column set may also be evaluated. Resolution between two adjacent peaks, 1 and 2, is defined in

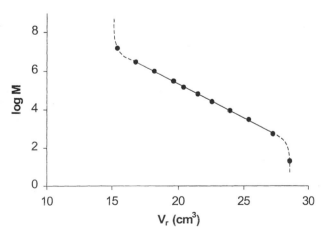

Fig. 1 Example of calibration curve with narrow MWD standards.

terms of their retention volumes, V_r, and peak widths, w:

$$R_S = \frac{2(V_{r2} - V_{r1})}{w_1 + w_2}$$

The calibration is often expressed in the form of $\ln M$ versus V_r, and assuming a linear function, it may be written as

$$\ln M = \ln D_1 - D_2 V_r$$

By solving for V_r and substituting into the relationship for R_S, we obtain

$$R_S = \frac{\ln(M_1/M_2)}{w D_2} = \frac{\ln(M_1/M_2)}{4\sigma D_2}$$

valid for peaks of similar width or standard deviation σ, where $w_1 \approx w_2 = w = 4\sigma$. The above equation shows that the MW fractionation of SEC columns is linked to both their useful pore volume (slope D_2 of the calibration curve) and to packing quality (column efficiency or number of plate heights, determining peak widths). Working with columns having linear calibration in their whole fractionation range guarantees equal resolution power over several MW decades.

SUGGESTED FURTHER READING

ASTM D5296-97. Standard Test Method for Molecular Weight Averages and Molecular Weight Distribution of Polystyrene by High Performance Size-Exclusion Chromatography (1997).

DIN 55672-1. Gelpermeationschromatographie Teil 1: Tetrahydrofuran als Elutionsmittel (1995–02) (1995).

Janca, J., Ed.; *Steric Exclusion Liquid Chromatography of Polymers*; Marcel Dekker, Inc.: New York, 1984.

Mori, S.; Barth, H. *Size Exclusion Chromatography*; Springer-Verlag: Berlin, 1999.

Yau, W.W.; Kirkland, J.J.; Bly, D.D. *Modern Size-Exclusion Liquid Chromatography*; John Wiley & Sons: New York, 1979.

Calibration of GPC/SEC with Universal Calibration Techniques

Oscar Chiantore
Università degli Studi di Torino, Torino, Italy

INTRODUCTION

Direct calibration of GPC/SEC columns requires well characterized polymer standards of the same type of polymer one has to analyze. However, narrow molecular-weight distribution (MWD) standards are available for a limited number of polymers only, and well-characterized broad MWD standards are not always accessible. The parameter controlling separation in GPC/SEC is the size of solute in the chromatographic eluent. Therefore, if different polymer solutes are eluted in the same chromatographic system with a pure exclusion mechanism, at the same retention volume, molecules with the same size will be found.

DISCUSSION

By plotting the logarithm of solute size versus retention volume, the points of all different polymers will be represented by a unique curve—a universal calibration curve. Thus, by application of the universal calibration, average molecular weights (MWs) and MWDs of any type of polymer may be evaluated from the SEC, provided that the relationship between molecular size and polymer molecular weight is known.

Several size parameters can be used to describe the dimensions of polymer molecules: radius of gyration, end-to-end distance, mean external length, and so forth. In the case of SEC analysis, it must be considered that the polymer molecular size is influenced by the interactions of chain segments with the solvent. As a consequence, polymer molecules in solution can be represented as equivalent hydrodynamic spheres,[1] to which the Einstein equation for viscosity may be applied:

$$\eta = \eta_0(1 + 2.5\phi_s) \tag{1}$$

η and η_0 are the viscosities of solution and solvent, respectively, and ϕ_s is the volume fraction of solute particles in the solution.

By expressing the solute concentration c in grams per cubic centimeter, the relationship holds:

$$\phi_s = \frac{cN_AV_h}{M} \tag{2}$$

where N_A is Avogadro's number and V_h and M are the hydrodynamic volume and the molecular weight of the solute, respectively. Substituting in Eq. (1) and taking into account that

$$[\eta] = \lim_{c \to 0}\left(\frac{(\eta - \eta_0)/\eta_0}{c}\right) \tag{3}$$

we obtain

$$[\eta]M = 2.5N_AV_h \tag{4}$$

Eq. (4) states that the hydrodynamic volume of a polymer molecule is proportional to the product of its intrinsic viscosity times the molecular weight.

The use of $[\eta]M$ as size parameter for GPC/SEC universal calibration was first proposed by Benoit and coworkers[2] and shown to be valid for homopolymers and copolymers with various chemical and geometrical structures. Their data are reported in the semilogarithmic plot of Fig. 1.

The hydrodynamic volume parameter $[\eta]M$ has been proven to be applicable also to the cases of rodlike polymers[3] and to separations in aqueous solvents[4] where, however, secondary nonexclusion mechanisms often superimpose and affect the sample elution behavior. In the latter situation, careful choice of eluent composition must be made in order to avoid any possible polymer-packing interaction.

The application of universal calibration requires a primary column calibration with elution of narrow MWD standards. For SEC in tetrahydrofuran, polystyrene (PS) standards are generally used. Intrinsic viscosities of the standards are either known or calculated from the proper Mark–Houwink equation, so that the plot of $\log[\eta]_{PS}M_{PS}$ values versus retention volumes V_r may be created. The universal calibration equation is obtained by polynomial regression, in the same way described for the calibration with narrow MWD standards.

Average molecular weights and MWDs of any polymer sample eluted on the same columns with pure exclusion mechanism may be calculated by considering that, at any retention volume, the following relationship holds:

$$[\eta]_iM_i = [\eta]_{PS,i}M_{PS,i} \tag{5}$$

Encyclopedia of Chromatography DOI: 10.1081/E-ECHR-120039899

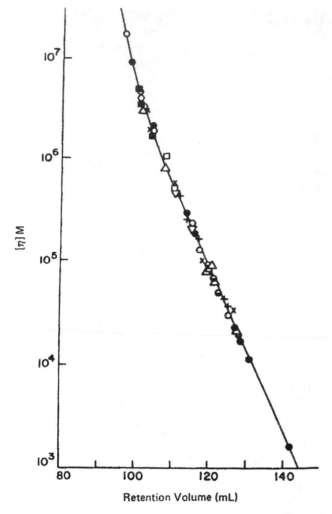

Fig. 1 Retention volume versus $[\eta]M$. (From Ref.[2].)

appropriate summations. The numerator in Eq. (7) is the value of the universal calibration at each retention volume.

The necessary conditions for application of the universal calibration method and for calculation of molecular weights through Eq. (6) is the knowledge of the $[\eta]_i$ values, which are obtained from the Mark–Houwink equations when the pertinent values of K and a constants are known. An alternative way is to make a continuous measurement of $[\eta]_i$ at the different elution volumes with an on-line viscometer detector coupled to the usual concentration detector system.

Methods for application of the universal calibration have been developed also for cases where K and a of the polymer of interest are not known and neither $[\eta]_i$ values are measured. Such methods are based on the availability of two broad MWD standards, having different molecular weights, of the polymer under examination.[5]

One important property of the universal calibration concept is that, in the SEC separation of complex polymers (i.e., polymers with different architectures or copolymers with nonconstant chemical composition), at each retention volume, $V_{r,i}$, molecules with same hydrodynamic volume but possibly different molecular weights will elute. It has been demonstrated that, in such a case, the application of the hydrodynamic volume parameter, $[\eta]M$ gives the number-average molecular weight, M_n, of the polymer.[6] In fact, at each retention volume, the intrinsic viscosity of the eluted fraction is given by the weight average over the n different molecular species present:

$$[\eta]_i = w_1[\eta]_1 + w_2[\eta]_2 + \cdots + w_n[\eta]_n \tag{8}$$

Eq. (8) may be written as

$$[\eta]_i = \frac{[\eta]_1 M_1 w_1}{M_1} + \frac{[\eta]_2 M_2 w_2}{M_2} + \cdots + \frac{[\eta]_n M_n w_n}{M_n} \tag{9}$$

As the condition holds, at each retention volume

$$[\eta]_1 M_1 = [\eta]_2 M_2 = \cdots = [\eta]_{PS} M_{PS} \tag{10}$$

Eq. (9) becomes

$$[\eta]_i = [\eta]_{PS} M_{PS} \sum \left(\frac{w_i}{M_i}\right) = \frac{[\eta]_{PS} M_{PS}}{M_{n,i}} \tag{11}$$

$$[\eta]_i M_{n,i} = [\eta]_{PS} M_{PS} \tag{12}$$

By considering all the fractions of the chromatogram, the M_n value of the whole sample may be then calculated.

from which

$$M_i = \frac{[\eta]_{PS,i} M_{PS,i}}{[\eta]_i} \tag{6}$$

To solve Eq. (6), the denominator must be known. Substituting into the denominator the Mark–Houwink expression $[\eta] = KM^a$ for the investigated polymer and rearranging, we obtain

$$M_i = \left(\frac{[\eta]_{PS,i} M_{PS,i}}{K}\right)^{1/1+a} \tag{7}$$

where K and a are the constants of the viscosimetric equation for that polymer, dissolved in the chromatographic eluent and at the temperature of analysis.

From Eq. (7), the molecular weight of each fraction in the chromatogram is obtained and average molecular weights may be calculated by application of the

Experimental aspects for the determination of molecular weight averages and MWD distributions by GPC/SEC using universal calibration are described in a standard ASTM method.[7] Detailed discussion on the validity and limitations of the method may be also found in Ref.[8].

REFERENCES

1. Flory, P.J. *Principles of Polymer Chemistry*; Cornell University Press: Ithaca, NY, 1953.

2. Grubisic, Z.; Rempp, P.; Benoit, H. A universal calibration for gel permeation chromatography. J. Polym. Sci. B. **1967**, *5*, 753.

3. Dawkins, J.V.; Hemming, M. Polymer **1975**, *16*, 554.

4. Dubin, P.L. *Aqueous Size Exclusion Chromatography*; Elsevier: Amsterdam, 1988.

5. Coll, H.; Gilding, D.K. J. Polym. Sci. A-2. **1970**, *8*, 89.

6. Hamielec, A.E.; Ouano, A.C.; Nebenzahl, L.L. J. Liquid Chromatogr. **1978**, *1*, 111.

7. ASTM D 3593-80, Standard Test Method for Molecular Weight Averages and Molecular Weight Distribution of Certain Polymers by Liquid Size-Exclusion Chromatography (Gel Permeation Chromatography—GPC) Using Universal Calibration; 1980.

8. Dawkins, J.V. *Steric Exclusion Liquid Chromatography of Polymers*; Janca, J., Ed.; Marcel Dekker, Inc.: New York, 1984.

C

Capacity

M. Caude
A. Jardy
ESPCI, Paris, France

INTRODUCTION

The capacity is closely related to the number of active sites of the stationary phase per volume or mass unit. Practically, there are two definitions corresponding to two different approaches to the problem. On the one hand, there is the linear capacity and, on the other, the maximum available capacity.

DISCUSSION

It is well known that when increasing the injected sample quantity, whether in volume or in concentration, peaks are distorted and/or shifted beyond a certain limit; the column is said to be overloaded. To quantify how much sample can be injected into a column without altering the resolution, it is convenient to define the column linear capacity. It is well known, for small injected quantities, that solute retention times and column efficiency are not affected by the sample size. However, above a critical sample size, a noticeable decrease in retention time and column efficiency are always observed.

Snyder has defined[1] the adsorbent linear capacity as the ratio (weight sample)/(weight stationary phase) giving a value of k' (or V_R) reduced by 10% relative to the constant k' values measured for smaller samples (Fig. 1). In Fig. 1, the adsorbent (Silica Davison) has a linear capacity close to 0.5 mg of dibenzyl per gram of silica. When the linear capacity of the column is exceeded, qualitative and quantitative analyses become much more complicated. Retention factors vary according to the injected solute quantity and the column efficiency can be tremendously decreased, entailing a degradation of resolution. Therefore, for analytical separations, it is always preferable to choose operating conditions corresponding to the linear capacity (k' and N values are constant whatever the injected sample sizes).

However, the practical interest of column linear capacity is very limited because its value varies according to various parameters: solute nature and retention and, even for the same quantity of injected solute, both the injected volume and the solute concentration of the injected solution. Thus, although widely accepted, the column linear capacity is misleading because it characterizes not only the thermodynamic nature of the chromatographic system but also the kinetic conditions (in term of column efficiency).

Consequently, it is preferable, according to Gareil et al.,[2] to define the concept of maximum available capacity C_D for a stationary phase: mass of solute entailing the saturation of the mass m of stationary phase contained in the column for given operating conditions:

$$C_D = \frac{Qs}{m} \tag{1}$$

with

$$k' = \frac{Qs}{Qm} = \frac{Qs}{V_m C_0} \tag{2}$$

The combination of Eqs. 1 and 2 gives

$$C_D = \frac{k' V_m C_0}{m}$$

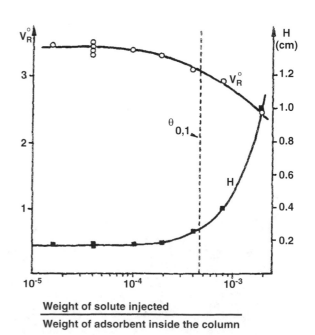

Fig. 1 Variation of the specific retention volume V_R^0 and of the height equivalent to a theoretical plate H as a function of the weight of injected solute (dibenzyl) related to the weight of adsorbent inside the column. (From Ref.[1].)

Encyclopedia of Chromatography DOI: 10.1081/E-ECHR-120039900

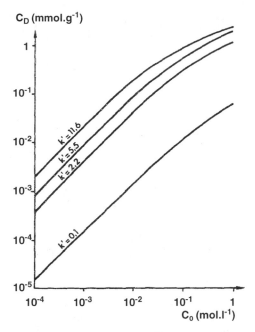

Fig. 2 Variation of the available capacity C_D as a function of the solute concentration in the module-phase C_0 (logarithm scales). In the case of RP chromatography, the stationary phase is n-octyl-bonded silica Lichroprep R.P.8 with 11.6% of carbon, the mobile phases are water-methanol mixtures, and the solute is phenol.

where k' is the solute retention factor measured for an analytical injection, V_M is the mobile-phase volume contained in the column, and C_0 is the solute concentratio-nin the mobile phase.

Fig. 2 shows, for various retention factors, the available capacity variation versus the solution concentration in the mobile phase in (RP) chromatography. These curves, called distribution isotherms, can be divided into two parts. In the first part, a linear variation of C_D versus C_0 is observed (bilogarithm scale); in the second part, a plateau is reached. In the first part and for the same retention (k' constant), the available capacity is independent of the solute nature.

The maximum available capacity is defined as the C_D limit value when both C_0 and k' are high ($C_0 \cong$ 1 mol/L, $k' \geq 10$). This value does not vary either with or k', or the solute nature (for the same family).

The maximum available capacity depends on the nature of the stationary phase:specific area for adsorption, the ion-exchange capacity for ion-exchange capacity, and the bonded rate for partition chromatography.

As a general rule, the maximum values of available capacity vary from 1.2 mmol/g (silica having a specific area close to 400 m^2/g) to 5 mEq/g for the cation exchanger (sulfonate groups).

REFERENCES

1. Snyder, L.R. Column efficiency in liquid-solid adsorption chromatography. H.E.T.P. [height equivalent to a theoretical plate] values as a function of separation conditions. Anal Chem. **1967**, *39*, 698.
2. Gareil, P.; Semerdjian, L.; Caude, M.; Rosset, R. J. High. Resolut. Chromatogr. Chromatogr. Commun. **1984**, *7*, 123.

SUGGESTED FURTHER READING

Knox, J.H., Ed.; *High Performance Liquid Chromatography*; Edinburgh University Press: Edinburgh, 1978; 27–28, 50.
Rosset, R.; Caude, M.; Jardy, A. *Chromatographies en phases liquide et supercritique*; Masson: Paris, 1991; 32–37.

Capillary Electrochromatography: An Introduction

Michael P. Henry
Chitra K. Ratnayake
Beckman Coulter, Inc., Fullerton, California, U.S.A.

INTRODUCTION

Capillary electrochromatography (CEC) is a technique in which a high direct voltage is applied, during analysis, across the ends of a capillary containing a solid stationary phase and a liquid mobile phase. It is thus a blend of HPLC and capillary electrophoresis (CE).[1]

Typically, fused silica capillaries are used with internal diameters of about 100 μm. Applied voltages of up to 30 kV are common. Mobile phases include low ionic strength aqueous buffer/organic solvent mixtures similar to those used in HPLC. Stationary phases generally consist of the same types of particulate materials as those employed in HPLC, although monolithic packings and surface-modified open tubular columns are gaining in popularity. Stationary phases of any kind usually extend only a portion of the way along the capillary, with particulate materials being held in place with small frits at either end of the packed region. Most CEC columns suffer from a degree of fragility due to the harm caused to the fused silica by the process of installing these two frits. Monolithic and wall-modified open tubular columns do not require frits and so offer an advantage here. Most instrumentation available for CEC is the same as that used for CE. Special capillary columns with a variety of dimensions and packing types are available. All sample classes that can be separated by HPLC can be separated by CEC and a great deal of work has been done to accumulate applications for this technique. There is still some controversy surrounding the immediate future of this technique, since certain technical challenges remain.

The objective of this article is a clear understanding of the basic concepts, history, instrumentation, applications, and future prospects for CEC.

TECHNIQUE OVERVIEW AND SEPARATION MECHANISMS

The technique and features of CEC are invariably compared to those of HPLC and CE. Krull et al.[1] have published a useful table comparing the three methods, and this is shown in Table 1 with some

changes. It can be seen that there are advantages of CEC over, for example, HPLC (efficiency, solvent consumption, and cost per run). On the other hand, HPLC has advantages over CEC (engineering, status of theory, frit integrity), not the least of which is the familiarity of the former and its integration into thousands of validated analyses.

Capillary electrochromatography is a technique in which a high direct voltage is applied across the ends of a capillary containing a solid stationary phase and a liquid mobile phase. The mobile phase is driven by the mechanism of electro-osmosis from one buffer vial through the stationary phase, through an unpacked region in the capillary, and finally into a second buffer vial (Fig. 1). A liquid sample is injected from a third vial (sample vial) onto the packed column by means of a brief application of either a lower voltage or pressure. The mobile phase in the buffer vial is again driven through the column at the elution voltage, bringing about the formation of flow-derived zones of separated sample components along the packed bed. Eventually the zones elute from the packed region of the column and pass by a window in the capillary that is adjacent to a detection/data system.

Elution voltages up to 30 kV are normal, and injection voltages from 2 to 5 kV for several seconds are used. Typically, the capillary is made of high electrical resistance fused silica coated with polyimide to improve flexibility. The capillary internal diameter (ID) may have values from 5 to 200 μm, and the total length 20–60 cm. The small capillary ID is important to minimize Joule heating effects, which depend in part upon the current generated within the mobile phase. The polyimide coating is UV-opaque, and so a short cylindrical piece is removed from the capillary forming a window that is transparent to radiation down to about 190 nm wavelength.

The interior surface of the unmodified capillary is typically negatively charged due to the presence of deprotonated acidic silanol groups. Mobile cations adjacent to this surface are drawn through an electric field towards the cathode and, in the process, drag the mobile phase against the forces of viscosity towards that electrode. This mechanism (e.o.f.) of mobilizing the solution through the capillary results

Encyclopedia of Chromatography DOI: 10.1081/E-ECHR-120040182

Table 1 Features of CEC compared with HPLC and CE

Feature	CEC	HPLC	CE
Sample types (without additives[a])	Nonpolar, polar, ionized	Nonpolar, polar, ionized	Ionized
Available theoretical plates	100,000–700,000	<50,000	200,000
Peak capacity	100+	About 50	About 100
Mechanical complexity	Simple	Complex	Simple
Compatibility with MS	Better than CE	Good	Limited
Mass detection limit	Slightly better than HPLC	Good	Often inadequate
Ease of operation	Difficult	Easy	Easy
Solvent consumption	<1% of HPLC	100 L/year	<1% of HPLC
pH operating range	2–11	2–14	2–11
Frits	An experimental problem only with particle-packed columns	Not a problem	Not applicable
Purchase price (US$)	30,000–50,000	20,000–50,000	30,000–50,000
Cost per run (US$, no labor)	0.02	2.00	0.04

[a]Such as surfactants, polymers.
(From Ref.[1], p. 28, with several changes.)

in the generation of an almost flat radial flow velocity profile across the interior, thereby substantially reducing a source of zone broadening.[1a] Sample component zones are therefore generally much narrower in CEC than in HPLC (where flow is generated hydrodynamically) and this leads directly to improved chromatographic resolution of mixture components.

As the mobile phase moves through the capillary containing the sorbent under the effect of this electroosmotic flow, sample components partition between the two phases in sorption and diffusive mechanisms characteristic of LC. Ions in the sample move both under the influence of e.o.f. and by their added attraction towards the oppositely charged electrode (electrophoresis). Uncharged components, on the other hand, move only under the influence of e.o.f. Thus, sample components in general separate by chromatographic and sometimes electrophoretic processes.

Typical mobile phases[1b] are aqueous buffer/organic mixtures, similar to those used in HPLC. Stationary phases[1c] are generally the same as those used in HPLC and are most often particulate in nature, although monolithic (single piece) bonded phases[1d] are also used. In addition, open tubular CEC is possible, in which an adsorptive layer is formed on, or deposited onto, the interior surface of the capillary.[1e] The first generally require end frits, whereas the last do not need frits to maintain column integrity. Actual packed lengths typically range from 10 to 20 cm or in some cases may fill the entire length of the capillary.

All sample classes that can be separated by HPLC can be separated by CEC (see section on "Applications"). Dipping the capillary inlet into a separate vial containing the sample in solution and applying an injection voltage for a brief time accomplish sampling. The capillary inlet is then moved back to the inlet buffer vial and elution is commenced at the (normally) higher voltage.

Samples are detected by the same techniques as those used in HPLC and CE. The uses of on-column detection with UV-vis spectrophotometers and lasers and postcolumn detection with mass spectrometers are well established.[1f] Instruments used for CEC are

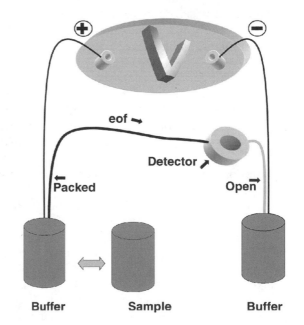

Fig. 1 Schematic illustration of a typical instrument for CEC.

generally the same as those used for CE and consequently have capabilities for full automation of analysis and data processing of the results.

Two books on CEC have been published in the third millennium.[1a–f,2]

HISTORY

Pretorius, Hopkins, and Schieke[3] were amongst the first investigators to carry out packed column LC in a tangential electric field (CEC) as a feasible alternative to using pressure. Their 1 mm ID quartz columns filled with 75–125 μm silica particles gave reduced plate heights of about 3 by CEC vs. the pressure mode values of about 8.

The advantages of electrochromatography, however, were not to be realized until the technology needed to create narrow capillaries (< 200 μm ID), stable frits, and sensitive detection systems had matured. Thus, in 1981, Jorgenson and Lukacs[4] carried out experiments in CEC using an instrument (see schematic in Fig. 1) whose basic design is still used today. Besides acting as a combined injector, separation medium, and flow cell, their column could also be used in the CE, CEC, and micro-LC modes. However, the fused silica capillaries (170 μm ID) drawn in their labs, packed with 10 μm particles and operated in neat acetonitrile, gave rather modest improvements in efficiency over standard liquid chromatographic techniques, with reduced plate heights of no less than 1.9. These disappointing results, coupled with an admission of the technical difficulty of using this technique, led these authors to conclude that CEC would only be useful in wider bore (several centimeters) preparative scale processes, a suggestion also made by Pretorius, Hopkins, and Schieke.[3]

Then, in 1991, Knox and Grant,[5] working carefully with 3 and 5 μm particles, showed that it was practical to achieve reduced plate heights of less than 1 in the CEC mode. These results confirmed their strongly optimistic view of the future of this technique, and a few years later, interest in CEC accelerated rapidly.

Over the past decade, CEC has become incorporated into the research and development programs of many universities and industrial organizations.[6] Progress has been made in the areas of column construction, column packing, general instrumentation, mobile phase optimization, applications, and method development.[1a–f] The elucidation of the physicochemical and electrochemical properties of CEC systems and the determination of their influence on the entire analysis is ongoing[7] and adds a fascinating dimension to the study of this technique.

OPERATIONAL LIMITS

Knox and Grant[5] have placed a general maximum limit of 200 μm on the ID of capillaries used in CEC in order to avoid problems with excessive internal heating, which harms column efficiency in aqueous/organic solutions. In principle, however, wider bore tubes can be used, provided the current and field strength are kept low, or the thermal conductivity of the system is kept high. Currents in general should be kept below 50 μA and field strengths held below 1000 V/cm. At higher field strengths and currents, outgassing of mobile phases can be a problem. This is often reduced in frequency if both ends of the capillary are pressurized. Most commercially available instruments have this capability. Bubbles are also a problem at higher temperatures; so cooling of the capillary is recommended.

Capillary electrochromatography in mobile phases containing organic solvents in any proportion is possible, provided the electrical double layer is formed with appropriate dissolved salts[3] in nonaqueous cases.

The ionic strength of most conventional buffers, such as phosphate, acetate, or borate, needs to be kept within the range 0.002–0.05 M, but care needs to be taken with the lower concentration to avoid buffer capacity depletion due to hydrolysis. Zwitterionic buffers such as morpholino ethane sulfonic acid (MES) (whose electrical conductivity is low) can be used in the range 0.01–0.1 M without undue heating problems, provided field strength and aqueous content are kept low and the capillary is cooled.

Since there are few or no pressure gradients generated within a CEC column, long packed capillaries containing very small particles are possible, but the longer columns mean slower chromatography and equilibration. Generally, columns in CEC are no longer than 60 cm. Times of analyses may be reduced by increasing flow rates. This is generally accomplished by increasing the applied voltage, while remaining within the limits mentioned above. Alternatively, one may increase the electro-osmotic flow rate by increasing pH or decreasing the amount of organic solvent (if any) in the mobile phase. All these changes (in flow rate, voltage, pH, and solvent), however, may cause simultaneous and undesirable shifts in resolution among the peaks of interest.

In some cases, hydrodynamic pressure may be exerted upon the inlet end of the capillary in order to increase the speed of analysis. This generally amounts to just 100 p.s.i., and so improvements are not often seen for conventionally packed columns. However, many types of monolithic and open tubular columns are sufficiently permeable to allow significant increases in flow rates when these low inlet pressures are used.[8] A disadvantage of using applied pressure in this fashion is that efficiency is reduced due to the intrusion of a parabolic radial flow profile into the capillary.

INSTRUMENTATION

Creative solutions to practical problems abound in the evolution of instruments designed to carry out CEC. The graphite electrodes, quartz tubing, glass wool frits, and on-column pressure injection of Pretorius, Hopkins, and Schieke[3] gave way to Jorgenson and Lukacs's[4] fused silica tubing, sintered frits, and electrokinetic injection. Commercially developed automated instruments designed for CE, whose appearance in 1988[9] followed these last authors' breakthrough research, have been used for most applications in CEC. Modern instruments therefore consist of the column, a cooling system, a detector, a voltage controller, an autosampler, and a data processor.

In-capillary optical focusing of UV, visible, and laser radiation has largely solved the problems of detection.[10] In-column (through the packing) detection of appropriate analytes by laser-induced fluorescence has improved general efficiency by avoiding the deterioration of peak shape that often occurs as the analyte zone passes through the outlet frit.[10] Mass spectrometers are also used for detection and identification in CEC since the low flow rates are compatible with the sampling kinetics required by these instruments.[1]

Columns for modern CEC have been prepared using standard HPLC particles, from 0.5 to 10 μm diameter, bearing C18, phenyl, C8, C6, C4, CN, amino, sulfo, and other functional groups and a variety of chiral polymers such as proteins and polysaccharides.[11] In situ sintered silica-based frits are most often used in these columns,[12] which are generally slurry-packed at high pressures. Several types of so-called monolithic (single piece) columns have been developed,[13,14] which dispense with frits, while generally maintaining high efficiency. Open tubular columns are becoming very popular, and many versions exist.[1]

Companies that have instruments suitable for CEC include Agilent Technologies, Beckman Coulter Inc., Prince Technologies, Unimicro Technologies, Micro-Tech Scientific, and Capital HPLC. More information on commercialized products for CEC can be found in Buyers' Guides to chromatographic instruments and accessories, such as the LC/GC North America publication, which is issued yearly in August.

APPLICATIONS

In 1998, Dadoo et al.[10] succeeded in achieving near baseline resolution of five polynuclear aromatic hydrocarbons over a 1 sec interval by CEC (Fig. 2). These high speeds were obtained from a combination of factors including a modest column length (6.5 cm), a high column plate number (13,000 plates) associated with 1.5 μm nonporous C18 particles, and a high voltage (28 kV). Although this separation is one of the fastest achieved in a liquid phase, higher column plate numbers have been obtained. Smith and Evans[15] report values of greater than 8 million plates per meter for the analysis of tricyclic antidepressants on 3 μm sulfopropyl bonded silica. These values are clearly due to a focusing effect within the column, and further work on this phenomenon has been done.[16] Dadoo et al.[10] have produced CEC columns that generate plate numbers of about 700,000 per meter when peaks were detected before they passed through the outlet column frit. These results illustrate how closely practical achievements in CEC have now approached predicted theoretical performance maxima.

Fu et al.[17] have prepared CEC columns from the highly efficient, strong cation exchange material Poly-SULFOETHYL A for the purpose of separating basic pharmaceuticals in human serum. This is a relatively rare example of the use of an ion exchanger to chromatograph positive organic ions. Fig. 3 shows the high quality of the capillary electrochromatogram achieved.

McKeown, Euerby, and Lomax[18] have made a detailed assessment of silica-based reversed-phase materials for the analysis of basic compounds. Walhagen et al.[19] have investigated the nature of the forces controlling the selectivity of peptides. Kang, Wistuba, and Schurig[20] have reviewed progress in chiral separations. A widely applicable cationic/hydrophobic monolithic column has been investigated

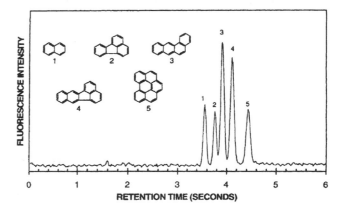

Fig. 2 Electrochromatogram of naphthalene (1), fluoranthene (2), benz[a]anthracene (3), benzo[k]fluoranthene (4) and benzo[ghi]perylene (5) using 1.5 μm nonporous ODS particles. Column dimensions: 100 μm ID × 6.5 cm packed length (10 cm total length). Mobile phase: 70% acetonitrile in a 2 mM Tris solution. Applied voltage for separation: 28 kV. Injection: electrokinetic at 5 kV for 2 sec. (From Ref.[10].)

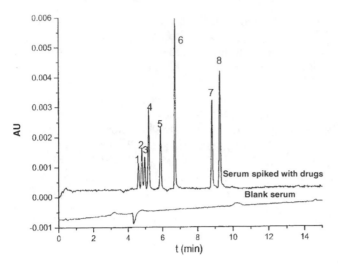

Fig. 3 Hydrophilic interaction CEC separation of basic drugs spiked in human serum. Column: fused-silica capillary, 27 cm (20 cm packed with 5 μm, 300 Å PolySULFOETHYL A particles). Mobile phase: 80% v/v acetonitrile in 100 mM TEAP buffer. TEAP buffer at pH 2.8. Applied voltage: 10 kV. Injection: 0.5 p.s.i. for 90 sec. Temperature: 25°C. Detection: UV at 214 nm. Concentration of drugs: 40 μg/ml. Solutes: (1) amobarbital; (2) phenobarbital; (3) barbital; (4) caffeine; (5) sulfanilamide; (6) theophylline; (7) 2,4-dimethylquinoline; (8) propranolol. (From Ref.[17].)

by Bedair and El Rassi[21] for the high-speed separation of proteins, pesticides, and amino acid derivatives.

Most chemical classes have been separated and analyzed by CEC.[11] These include many classes of pharmaceuticals, environmental chemicals, explosives, natural products, drugs of abuse, polypeptides, oligosaccharides, nucleosides and their bases, and polynucleotides. Applications of CEC, such as the review by Eeltink, Rozing, and Kok in *Electrophoresis*,[22] may be found in other journals such as the *Journal of Chromatography*, *Biomedical Chromatography*, *Chromatographia* and the *Journal of Separation Science*. Access to the literature of CEC in general is readily achieved through several user-friendly websites.[23–25]

PROBLEMS, ISSUES, AND FUTURE PROSPECTS

Fundamental theory of CEC that will provide a better understanding of mechanisms of separation is being developed. In particular, the work of Rathore and Horváth[7] in elucidating the electrical properties of packed columns is particularly interesting. Technological issues that remain to be addressed include the difficulty of dealing with bubble formation and the fragility of conventional columns due to the aggressive frit-forming methods currently used. Monolithic

columns[13,14] and open tubular columns[26] appear to have advantages in this regard.

Majors[6] has compiled the results of his *Perspectives* survey of 14 leading separation scientists with an interest in CEC. As expected, there is a wide divergence in the opinions of these leaders with regard to the current issues and future prospects for CEC. Few, however, underestimate the current technological difficulties in column manufacture, the reproducibility of chromatographic and electro-osmotic properties of the packed capillary, and the short-term problems of competing with HPLC or CE. But the majority of scientists interviewed believe that, as with any new technique, these problems will be overcome and that CEC will become a routine method of analysis in time.

On the other hand, the future of CEC may lie with the exciting developments in microfabrication described by Regnier[27] where capillaries here are open channels 1.5 μm wide and 4.5 cm long and can achieve efficiencies of 800,000 plates per meter.

CONCLUSIONS

This article has been a brief introduction to the technique of CEC. It is a technique in which a high voltage is applied across the ends of a capillary containing a solid stationary phase and a liquid mobile phase. Aspects of the method that were described included a brief overview of the technique, elements of the history of its development, the normal operational limits, aspects of modern instrumentation, a summary of selected applications, and future prospects for the development of CEC.

The future of the technique appears assured,[6] although the progress in moving it from the interesting stage to its use alongside the mature HPLC and CE has been quite slow. Fundamental theory of CEC continues to be developed. Several technological issues remain to be addressed, including the difficulty of dealing with bubble formation and the fragility of conventional columns. Monolithic and open tubular columns appear to have advantages in this regard.

REFERENCES

1. Krull, I.S.; Stevenson, R.L.; Mistry, K.; Swartz, M.E. *Capillary Electrochromatography and Pressurized Flow Capillary Electrochromatography: An Introduction*, 1st Ed.; HNB Publishing: New York, 2000; (a) 10–11; (b) 60–67; (c) 34–40; (d) 49–53; (e) 53–55; (f) 81–85.
2. Bartle, K.D., Myers, P. Eds.; *Capillary Electrochromatography*; 1st Ed.; RSC Chromatography

Monographs; Royal Society of Chemistry: Cambridge, 2001.

3. Pretorius, V.; Hopkins, B.J.; Schieke, J.D. Electro-osmosis: a new concept for high speed liquid chromatography. J. Chromatogr. **1974**, *99*, 23–30.

4. Jorgenson, J.W.; Lukacs, K.D. High resolution separations based on electrophoresis and electro-osmosis. J. Chromatogr. **1981**, *218*, 209–216.

5. Knox, J.H.; Grant, I. Electrochromatography in packed tubes using 1.5 to 50-micron silica gels and ODS bonded silica gels. Chromatographia **1991**, *32* (7–8), 317–327.

6. Majors, R.E. Perspectives on the present and future of capillary electrochromatography. LC/GC **1998**, *16*, 96–110.

7. Rathore, A.S.; Horváth, Cs. Axial non-uniformities and flow in columns for capillary electrochromatography. Anal. Chem. **1998**, *70*, 3069–3077.

8. Ratnayake, C.K.; Oh, C.S.; Henry, M.P. Characteristics of particle-loaded monolithic sol–gel columns for capillary electrochromatography. I. Structural, electrical and band broadening properties. J. Chromatogr. A. **2000**, *887*, 277–285.

9. Weinberger, R. The evolution of capillary electrophoresis: past, present, and future. Am. Lab. **2002**, *34* (10), 32–40.

10. Dadoo, R.; Zare, R.N.; Yan, C.; Anex, D.S. Advances in capillary electrochromatography: rapid and high-efficiency separations of PAHs. Anal. Chem. **1998**, *70*, 4787–4792.

11. Altria, K.D.; Smith, N.W.; Turnbull, C.H. A review of the current status of capillary electrochromatography technology and applications. Chromatographia **1997**, *46*, 664–674.

12. Dittman, M.M.; Rozing, G.R.; Ross, G.; Adam, T.; Unger, K.K. Advances in capillary electrochromatography. J. Capillary Electrophor. **1997**, *5*, 201–212.

13. Svec, F.; Peters, E.C.; Sykora, D.; Yu, C.; Fréchet, J.M.J. Monolithic stationary phases for capillary electrochromatography based on synthetic polymers: design and applications. J. High Resolution Chromatogr. **2000**, *23* (1), 3–18.

14. Ratnayake, C.K.; Oh, C.S.; Henry, M.P. Particle loaded monolithic sol–gel columns for capillary electrochromatography. J. High Resolution Chromatogr. **2000**, *23* (1), 81–88.

15. Smith, N.W.; Evans, M.B. The efficient analysis of neutral and highly polar pharmaceutical compounds using reversed-phase and ion-exchange electrochromatography. Chromatographia **1995**, *41*, 197–203.

16. Euerby, M.R.; Gilligan, D.; Johnson, C.M.; Roulin, S.C.P.; Myers, P.; Bartle, K.D. Applications of capillary electrochromatography in pharmaceutical analysis. J. Microcol. Sep. **1997**, *9*, 373–387.

17. Fu, H.; Jin, W.; Xiao, H.; Xie, C.; Guo, B.; Zou, H. Determination of basic pharmaceuticals in human serum by hydrophilic interaction capillary electrochromatography. Electrophoresis **2004**, *25*, 600–606.

18. McKeown, A.P.; Euerby, M.R.; Lomax, H. Assessment of silica-based reversed-phase materials for the analysis of a range of basic analytes by capillary electrochromatography. Electrophoresis **2002**, *25* (15–17), 1257–1268.

19. Walhagen, K.; Huber, M.I.; Hennessy, T.P.; Hearn, M.T.W. On the nature of the forces controlling selectivity in the high performance capillary electrochromatographic separation of peptides. Peptide Sci. **2003**, *71* (4), 429–453.

20. Kang, J.; Wistuba, D.; Schurig, V. Recent progress in enantiomeric separation by capillary electrochromatography. Electrophoresis **2002**, *23* (22–23), 4005–4021.

21. Bedair, M.; El Rassi, Z. Capillary electrochromatography with monolithic stationary phases III. Evaluation of the electrochromatographic retention of neutral and charged solutes on cationic stearyl-acrylate monolith and the separation of water soluble proteins and membrane proteins. J. Chromatogr. A. **2003**, *1013*, 47–56.

22. Eeltink, S.; Rozing, P.G.; Kok, W.Th. Recent applications in capillary electrochromatography. Electrophoresis **2003**, *24* (22–23), 3935–3961.

23. http://www3.interscience.wiley.com (Search CEC) (accessed October 2004).

24. http://www.separationsnow.com (Search CEC) (accessed October 2004).

25. http://www.ceexchange.com/nowjune2004.php (accessed October 2004).

26. Pesek, J.J.; Matyska, M.T.; Dawson, G.B.; Chen, J.I.-C.; Boysen, R.I.; Hearn, M.T.W Open-tubular electrochromatographic characterization of synthetic peptides. Electrophoresis **2004**, *25*, 1211–1218.

27. Regnier, F.E. Microfabricated monolith columns for liquid chromatography. Sculpting supports for liquid chromatography. J. High Resolution Chromatogr. **2000**, *23* (1), 19–26.

Capillary Electrophoresis in Nonaqueous Media

Ernst Kenndler
Institute for Analytical Chemistry, University of Vienna, Vienna, Austria

INTRODUCTION

Organic solvents are used in capillary electrophoresis (CE) for several reasons:

1. To increase the solubility of lipophilic analytes.
2. To affect the actual mobilities of the analytes (those of the fully charged species at the ionic strength of the solution).
3. To change the pK values of the analytes.
4. To influence the magnitude of the electro-osmotic flow.
5. To influence the equilibrium constant of association reactions between analytes and additives (e.g., for the adjustment of the degree of complexation; an important example is the separation of chiral compounds by the use of cyclodextrins).
6. In some rare cases, to allow homoconjunction or heteroconjugation of the analytes with other species present and, thus, enable separation. For such interactions, a low dielectric constant of the solvent is a prerequisite.

APPLICATION OF NONAQUEOUS SOLVENTS

The organic solvents are applied in many cases in order to enhance the separation selectivity by changing the effective mobilities of the analytes. They are either applied as pure solvents, or as nonaqueous mixtures, or as constituents of mixed aqueous–organic systems. Solvents used for CE, as described in the literature, are methanol, ethanol, propanol, acetonitrile, tetrahydrofuran, formamide, *N*-methylformamide, *N,N*-dimethylformamide, *N,N*-dimethylacetamide, dimethylsulfoxide, acetone, ethylacetate, and 2,2,2- trifluoroethanol.

Organic solvents have relevance in many fields of application: for the separation of inorganic ions, organic anions and cations, pharmaceuticals and drugs, amino acids, peptides, and proteins.

There are some practical restrictions for the use of organic solvents:

1. Many organic solvents have a significant ultraviolet (UV) absorbance in the range of wavelengths that are normally also used for the detection of the analytes. This property leads to a poor signal-to-noise ratio or limits the applicability to solutes with UV absorbances at a higher wavelength.
2. Many electrolytes cannot be used as buffers, due to their low solubilities in organic solvents.
3. The low dielectric constant of solvents suppresses ion dissociation and favors ion-pair formation.
4. Important physicochemical properties (e.g., ionization constants of weak acids and bases) are often not known, which leads to a more or less random experimental approach for the optimization of the resolution.
5. In this context, it should be mentioned that the clear determination of the pH scale in these solvents is not a straightforward task, which may introduce a certain inaccuracy for the description of the experimental conditions. As this aspect is not adequately considered in many articles on CE in nonaqueous solvents, it is discussed here in more detail.

ACIDITY SCALES IN ORGANIC SOLVENTS

When investigating the effect of organic solvents on the pK_a of an acid, the significance of the pH scale in this solvent must be questioned. We base such scales on the measurement of the activity of the solvated proton. We define the activity, a_i, of a particle, i, the proton in the case of interest, by the difference between the chemical potential, ω_i in the given and in a standard state (indicated by superscript 0)

$$\omega_i = \omega_i^0 + RT \ln a_i$$

In practice, we therefore differentiate a number of acidity scales: the standard, the conventional, the operational, and the absolute (thermodynamic) scale.

STANDARD ACIDITY SCALE

The standard state might be chosen in various ways (e.g., as the state at infinitely diluted solution). The resulting standard acidity scale is characterized by

Encyclopedia of Chromatography DOI: 10.1081/E-ECHR-120039903

the activity of the proton solvated by the given solvent, HS, according to

$$pH = -\log a_{SH_2^+} \qquad (1)$$

The range of this scale is defined by the ionic product of the solvent, pK_{HS}.

Measurements in the standard acidity scale are carried out in cells without liquid junctions (e.g., with the following setup: Pt/H$_2$/HCl in SH/AgCl/Ag). It is assumed, here, that the activities of the solvated proton and the counterion, chloride, are equal. In this case, the electromotive force (emf) of the cell can be expressed by

$$
\begin{aligned}
E &= E_S^0 - \frac{RT}{F}\ln a_{SH_2} - a_{Cl^-} \\
&= E_S^0 - \frac{2RT}{F}\ln(c_{HCl}\gamma_{HCl}) \qquad (2)
\end{aligned}
$$

where is c_{HCl} the concentration γ_{HCl} and is the mean activity coefficient of HCl. E_S^0 is the standard potential of the silver chloride electrode in the given solvent, S, after extrapolation of the measured emf to zero ionic strength. Rearrangement leads to the expression of the pH in the standard scale:

$$pH = -\frac{(E - E_S^0)F}{2.3RT} + \log c_{Cl^-} + \log \gamma_{Cl^-} \qquad (3)$$

CONVENTIONAL ACIDITY SCALE

The standard acidity scale, although well defined theoretically, has the limitation in practice that only the mean activity coefficient, but not the single-ion activity coefficient, is thermodynamically assessible. The single-ion coefficient depends on the composition of the solution as well. One way to circumvent this problem would be to have a defined value of the activity coefficient for one selected ion. Given that, all other activity coefficients could be obtained from the activity coefficients of the particular electrolytes and that special single-ion coefficient. The value of this selected coefficient could be used, then, as the base of the conventional acidity scale. This single-ion activity coefficient is derived for chloride by the Debye–Hückel theory. This choice is made by convention, initially proposed for aqueous solutions; it is accepted also for other amphiprotic, polar solvents. Note that the measurements of the proton activity are carried out in cells without liquid junction.

OPERATIONAL ACIDITY SCALE

Due to the disadvantage of working with cells without liquid junctions, in practice the operational scale uses buffer solutions with known conventional pH for the calibration of cells with liquid junction [e.g., the convenient glass electrode (with the calomel or silver electrode, respectively, as reference)]. After calibration of the measuring cell (with a buffer of known conventional pH), the acidities of unknown samples can be measured in the same solvent. It is clear that for the standard buffers used, the conventional and the operational pH are identical. However, we cannot assume such an identity for the unknown samples. This is because the activities and the mobilities of the different ionic species might change the potential on the boundary with all liquid junctions (even without taking effect of the nonelectrolytes into account).

ABSOLUTE (THERMODYNAMIC) SCALE AND MEDIUM EFFECT

This scale, in fact, would allow comparing the basicities of the different solvents in a general way. It is based on the question of the chemical potential of the proton (as a single-ion species) in water, W, and the organic solvent, S. Taking the hypothetical 1 M solution as the standard state, the chemical potential is given, according to Eq. (1), as

$$\omega_{H^+} = \omega_{H^+}^0 + RT\ln m_{H^+} + RT\ln \gamma_{H^+} \qquad (4)$$

where m is the molal concentration. The so-called medium effect on the proton is given by

$$
\begin{aligned}
\ln {}_W\gamma_{H^+} - \ln {}_S\gamma_{H^-} &= \ln\left(\frac{{}_W\gamma_{H^-}}{{}_S\gamma_{H^+}}\right) \\
&= \ln {}_m\gamma_{H^+} = \frac{{}_S\omega_{H^+}^0 {}_W\omega_{H^+}^0}{RT} \qquad (5)
\end{aligned}
$$

${}_m\gamma_{H^+}$ is named the transfer activity coefficient. The medium effect is proportional to the reversible work of transfer of 1 mol of protons in water to the solvent, S (in both solutions at infinite dilution). If the medium effect is negative, the proton is more stable in the solvent, S. It is, thus, an unequivocal measure of the basicity of the solvent, compared to water, as it allows us to establish a universal pH scale due to

$$-\log {}_W a_{H^+} = -\log {}_S a_{H^+} - -\log {}_m a_{H^+} \qquad (6)$$

It is a serious drawback that it is not possible to determine the transfer activity coefficient of the proton (or of any other single-ion species) directly by thermodynamic methods, because only the values for both the proton and its counterion are obtained. Therefore, approximation methods are used to separate the medium effect on the proton. One is based on the simple *sphere-in-continuum* model of

Born, calculating the electrostatic contribution of the Gibb's free energy of transfer. This approach is clearly too weak, because it does not consider solvation effects. Different extrathermodynamic approximation methods, unfortunately, lead not only to different values of the medium effect but also to different signs in some cases. Some examples are given in the following: $_m\gamma_{H^+}$ for methanol $+1.7$ (standard deviation 0.4); ethanol $+2.5$ (1.8), n-butanol $+2.3$ (2.0), dimethyl sulfoxide -3.6 (2.0), acetonitrile $+4.3$ (1.5), formic acid $+7.9$ (1.7), NH_3 -16. From these data, it can be seen that methanol has about the same basicity as water; the other alcohols are less basic, as is acetonitrile. Dimethyl sulfoxide, on the other hand, is more basic than water. However, the basicity of the solvent is not the only property that is important for the change of the pK values of weak acids in comparison to water. The stabilization of the other particles that are present in the acido-basic equilibrium is decisive as well.

SUGGESTED FURTHER READING

Bates, R.G. Medium effect and pH in nonaqueous and mixed solvents. In *Determination of pH, Theory and Practice*; John Wiley & Sons: New York, 1973; 211–253.

Covington, A.K.; Dickinson, T. Introduction and solvent properties. In *Physical Chemistry of Organic Solvents Systems*; Covington, A.K., Dickinson, T., Eds.; Plenum Press: London, 1973; 1–23.

Kolthoff, I.M.; Chantooni, M.K. General introduction to acid–base equilibria in nonaqueous organic solvents. In *Treatise on Analytical Chemistry, Part I, Theory and Practice*; Kolthoff, I.M., Elving, P.J., Eds.; John Wiley & Sons: New York, 1979; 239–301.

Popov, A.P.; Caruso, H. Amphiprotic solvents. In *Treatise on Analytical Chemistry, Part I, Theory and Practice*; Kolthoff, I.M., Elving, P.J., Eds.; John Wiley & Sons: New York, 1979; 303–347.

Sarmini, K.; Kenndler, E. Ionization constants of weak acids and bases in organic solvents. J. Biophys. Biochem. Methods **1999**, *38*, 123.

Sarmini, K.; Kenndler, E. Influence of organic solvents on the separation selectivity of capillary electrophoresis. J. Chromatogr. A. **1997**, *792*, 3.

Capillary Electrophoresis: Inductively Coupled Plasma-MS

Clayton B'Hymer
University of Cincinnati, Cincinnati, Ohio, U.S.A

INTRODUCTION

Capillary electrophoresis (CE) has many well-known advantages including low sample-volume require-ments, high plate number (i.e., peak efficiency), the ability to separate positive, neutral, and negatively charged species in a single run, and, when properly developed, relatively short analysis times. The ability of CE to separate ionic multispecies and to have low operational costs makes the technique superior in certain specific applications to conventional HPLC. The inductively coupled plasma-mass spectrometer (ICP-MS) has the advantages of possessing low detec-tion limits for the majority of the chemical elements. The ICP-MS detector has other additional positive attributes including linearity over a wide dynamic range, multielement detection capability, and the ability to perform isotopic analysis. Also, the ICP-MS is known to have minimal matrix-effect problems when compared to other detection systems. Sample matrix-effect problems are further reduced in CE-ICP-MS analysis owing to the small sample size and flow rates associated with CE. With all of these strong points, CE–ICP-MS is a rapidly growing hyphenated technique; the separation capability of CE is combined with the highly sensitive, element-specific detection system of ICP-MS.

HISTORICAL BACKGROUND AND USE

The first research papers describing CE–ICP-MS were written in 1995 by the Olesik, Lopez-Avila, and Barnes research groups.[1–3] The coupling of the ICP-MS detector with CE and HPLC has become the dominant analysis technique for elemental speciation analysis. Elemental speciation analysis is defined as the separa-tion, identification, and quantification of the different chemical forms (organometallic and inorganic) and oxidation states of specific elements in a given sample. Information on elemental speciation in clinical and environmental material is vital in the study of mechan-isms of element transport within living as well as envir-onmental systems, elemental bioavailability, metabolic pathways within living organisms, and toxicology.

THE FUNDAMENTALS OF CAPILLARY ELECTROPHORESIS–INDUCTIVELY COUPLED PLASMA-MS

The Inductively Coupled Plasma-MS Detector

MS has established itself as the detection system of choice for CE of trace metals and metalloids, as well as their chemical species. The ICP-MS has dominated CE analysis methods in recent years. The ICP-MS dif-fers from the more commonly used electrospray or ion-spray mass spectrometer method of ion generation. The electrospray MS can be described as using a "soft" ion source; that is, structural information can be obtained from molecular fragments. The ICP-MS is a "hard" ion source; that is, the plasma generally operates at an approximate temperature of 8000 K. Under these conditions, the ICP generates ions of ele-ments and a few polyatomic ions. The ICP-MS has been well documented since its early development by both the Houk[4] and Gray[5] research groups over 20 years ago. A diagram of a typical commercial ICP-MS detector is shown in Fig. 1. The inductively coupled plasma is formed from a flow of gas, typically argon, through a series of concentric tubes made of quartz called the torch. The ICP torch is surrounded by a copper load coil. The load coil is connected to a radio-frequency generator, which operates between 27 and 40 MHz at a power of 700–1500 W.[6] This induces an oscillating magnetic field near the exit of the torch. A plasma is formed while a spark is applied to the flowing gas stream to form gaseous ions. The free electrons created during this process are acceler-ated by the magnetic field and bombard other gas atoms; this causes further ionization and produces the plasma. Sample introduction into the plasma is via a carrier argon gas flow through the central tube of the ICP torch. Liquid samples are nebulized into an aerosol before being carried into the ICP torch, a function performed by a nebulizer and spray chamber. The nebulizer produces the aerosol, and the spray chamber separates and removes the large droplets from the aerosol to form a more uniform mist. Once the fine aerosol sample reaches the plasma, vaporization,

Encyclopedia of Chromatography DOI: 10.1081/E-ECHR-120028839

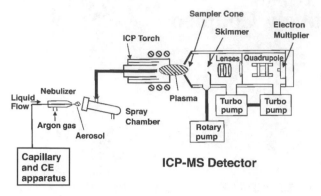

ICP-MS Detector

Fig. 1 The ICP-MS used as a detector for HPLC. The liquid sample passes through the capillary into a nebulizer where it is changed into an aerosol. The aerosol passes through a spray chamber and into the plasma. The analytes pass into the mass spectrometer. The CE interface is not in detail in this figure.

atomization, and ionization of the analyte to element ions occur almost simultaneously. Coolant and auxiliary gas are added to the ICP torch to keep the quartz from melting and to provide a tangential flow of gas, which serves to center and stabilize the plasma.

Beyond the ICP torch are the sampler and skimmer cones of a typical mass spectrometer (Fig. 1). Ions generated from the sample pass through the aperture of the cones into low-pressure chambers. Ion lenses, which are actually a series of electrodes, are used to "focus" the ion path, before reaching the quadrupole mass analyzer. Ions of only one mass-to-charge ratio are transmitted at a time and impacted onto an electron multiplier detector. The electron pulse is amplified and this signal is then recorded by the instrument's data system. The diagram in Fig. 1 displays a quadrupole mass analyzer, but other spectrometers have been used with CE including scanning instruments such as the double-focusing and sector-field mass detectors, and for fast separations of multielement mixtures of chemical species, the time-of-flight (TOF) MS.

Alternative plasmas have been occasionally used for elemental speciation analysis, including the microwave-induced plasma (MIP), which has been reviewed in Ref.[7] and the low-power helium plasma. Both of these plasma sources have the advantage of reduced gas and power consumption over the traditional ICP; however, the use of these plasmas with interfaces with CE has been very infrequent and does not warrant further discussion in this article. The MIP has been occasionally used with low flow rate liquid sample introduction. The low-power helium plasma has generally only been used with GC interfaces; their low-power levels are generally not capable of properly vaporizing and ionizing a liquid aerosol.

Interfacing Capillary Electrophoresis to the Inductively Coupled Plasma-MS

Overview of design considerations

The main design challenge of CE–ICP-MS is in the actual interface. In the typical practice of CE, a fused silica capillary filled with a buffer has both ends submerged or in physical contact with two buffer reservoirs. Electrodes placed in the buffer reservoir provide the application of a high electrical potential through the capillary. When attempting to interface CE to an ICP-MS, several problems need to be overcome. One is that CE has an extremely low flow rate (approximately $1\,\mu L\,min^{-1}$ or less). This requires the use of a low liquid flow nebulizer to be used in the interface. A low liquid flow rate nebulizer is required that maintains a high transport efficiency and delivers a large quantity of analyte to the plasma. The ICP-MS detector sensitivity is based on mass of the analytes, not concentration of the solution. Because CE injection volumes are low, high transport efficiency by the nebulizer is vital to reduce analyte loss to the MS detector. The second problem with CE–ICP-MS interfacing is that an electrical connection must be maintained to the end of the fused silica capillary, yet the capillary must still introduce the CE buffer flow into the nebulizer and produce a uniform aerosol for the analysis system. This problem has been solved by various designs, which usually involves the addition of a "make-up" buffer or sheath electrolyte added near the end of the fused silica capillary. Interfacing CE with ICP-MS has the advantage of requiring a low liquid flow rate, and therefore places a small demand on the desolvation and solvent load capacity of the inductively coupled plasma. This makes a more stable plasma less subject to long-term signal drift over the course of several CE runs. Two other design considerations of a CE–ICP-MS interface are countering or minimizing laminar flow through the capillary generated by the operation of the nebulizer[8] and minimizing band broadening for the separation of analytes. There are various strategies in reducing laminar flow through the electrophoretic capillary, and band broadening is minimized through geometry considerations in the design of the CE–ICP-MS interface. These points will be discussed in further detail in this article.

The nebulizer

A basic understanding of the nebulizer function and the types of nebulizers is necessary to successfully interface CE to the ICP-MS. Nebulization, as previously described, is the process to form an aerosol, i.e., to suspend a liquid sample into a gas in the form of a cloud of droplets. The quality of any nebulizer is based on many different parameters including mean droplet diameter, droplet size distribution, span of droplet size distribution, droplet number density, and droplet mean velocity. There are numerous nebulizers commercially available for the use with ICP-MS systems, and their detailed description can be found elsewhere.[9,10] Pneumatic designs, both concentric and cross flow, are the most popular for CE interfaces with the occasional use of the ultrasonic nebulizer (USN). Fig. 2 shows some typical nebulizers. The pneumatic nebulizer is either a concentric design (Fig. 2A), where both the gas stream and the liquid flow in the same direction or the cross-flow design (Fig. 2B), where the gas stream is at a right angle to liquid flow. Gas flowing past the tip of the liquid sample introduction tube generates the aerosol. The ultrasonic nebulizer (Fig. 2C) consists of a piezoelectric transducer and a liquid sample introduction tube. Liquid flow over the transducer plate forms a thin film and is nebulized by the high-frequency mechanical vibrations from the transducer.

There are several concentric-like pneumatic low liquid flow nebulizers commercially available that are often used in the construction of CE–ICP-MS interfaces. The Meinhard high-efficiency nebulizer (HEN)

(Meinhard Glass Products, Golden, Colorado) is a variation of the concentric nebulizer that has smaller internal dimensions and is specifically designed to operate at low liquid flow rates. A very similar nebulizer is also commercially available and is known as the MicroMist nebulizer (Glass Expansion Pty. Ltd., Victoria, Australia). Another low-flow commercial nebulizer, the Microconcentric Nebulizer (MCN) (CETAC Technologies, Inc., Omaha, Nebraska), has been used with liquid flow rates down to $10–30\,\mu L$ min^{-1}. The MCN is also concentric in nature, but it differs from both the Meinhard and MicroMist in having its outer body constructed of plastic instead of glass. Also, the MCN has its inner sample tube made of fused silica capillary tube, not drawn glass (Fig. 2D). All three of these commercial nebulizers have comparable analyte transport efficiencies.

Although there has been limited use with CE interfaces, the direct injection nebulizer (DIN) was first described by Shum et al.[11] and later used by Liu et al.[2] for CE (Fig. 2E). In this design, the nebulizer introduces the sample very near the plasma inside the ICP torch and eliminates the spray chamber assembly. Close to 100% analyte transport efficiency can theoretically be obtained with the DIN, but the nebulizer is

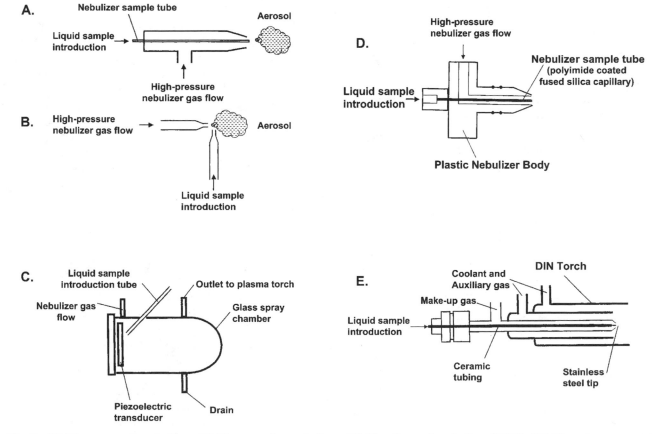

Fig. 2 (A) The concentric nebulizer. (B) The cross-flow nebulizer. (C) The ultrasonic nebulizer (USN). (D) The microconcentric nebulizer (MCN) by CETAC. The body of this nebulizer is made of plastic. (E) The direct injection nebulizer (DIN).

restricted to very low liquid flow rate and thus is well matched to CE interfacing. This design does induce local plasma cooling due the lack of desolvation and detection limits are only slightly improved over other nebulizer designs.[12]

Specific capillary electrophoresis interface designs

The CE–ICP-MS interface based on the sheath-flow (make-up buffer) and pneumatic concentric nebulizers described by Lu et al.[3] is the most widely used CE-ICP-MS interface, and it has also been applied to electrospray MS interfaces.[13] The sheath-flow or make-up buffer acts to complete electrical connection to the exit end of the electrophoretic capillary; grounding is achieved by having a metal tube or metal "tee" near the connection to the nebulizer (Fig. 3) or by coating the capillary with silver[1] (Fig. 4). The second function of the sheath flow is to compensate for the suction effect. Low pressure created near the tip of the pneumatic nebulizer by the flow gas of the operating nebulizer can induce laminar flow through the electrophoretic capillary. This can impair separation of analytes. Sheath flow can be introduced into the nebulizers by either self-aspiration with gravity siphoning control or by a pumping system. These strategies involve the precise addition of a sheath or make-up buffer to the nebulizer to prevent the degradation of the CE separation profile of the analytes. In the self-aspiration designs, the sheath flow is automatic, although adjustments to height of the sheath buffer reservoir (Fig. 3) can be used to optimize flow and separation to the nebulizer–CE interface. When a pumping system is used, the flow rate must be optimized to obtain the desired separation by reducing laminar flow through the capillary.

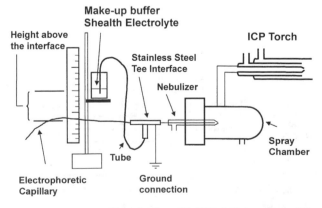

Fig. 3 A typical self-aspirating CE–ICP-MS interface. The make-up buffer/sheath electrolyte reservoir is positioned above the interface to provide the correct pressure and flow of buffer to the pneumatic nebulizer.

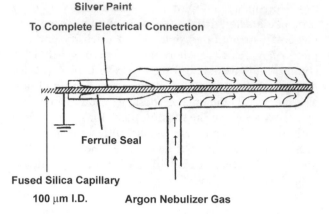

Fig. 4 Interface of an electrophoresis capillary and the concentric nebulizer. Silver paint was used to complete the electrical connection. (Reproduced from Ref.[1] with permission of the American Chemical Society.)

The use of controlled sheath-flow/make-up buffer rates to give equivalent CE/MS and CE-UV electopherograms was reported by Day et al.[14] It has also been reported that the sheath flow should be kept low and just compensate for laminar flow through the electrophoretic capillary.[8] This minimizes dead volume and band broadening of the CE separation. Precise pumping of a make-up buffer was demonstrated by Kinzer et al.[8] and later by Sutton et al.[15] The use of sol–gel frits near the exit tip of the electrophoretic capillary has been reported to reduce laminar flow effects and reduce the sheath flow.[16] Other important design and optimization considerations exit for the sheath-flow/make-up buffer interfaces. The positioning of the electrophoretic capillary into a concentric nebulizer is critical in these designs. Placement of the exit tip of the capillary close to the tip of the nebulizer can increase the "suction effect" and cause greater laminar flow to be generated through the capillary. Placement too far back from the tip of the nebulizer may induce greater band broadening from the extra dead volume. There has been some arguments about dilution effects of the sheath flow; that is, the sheath-flow lowers sensitivity of the MS detector from dilution of the analytes. The use of a sheath flow does not cause a decrease in sensitivity because of dilution, but actually because of a decrease in analyte transport efficiency through the spray chamber to the plasma. Large liquid flow rates generally have less-efficient analyte transport, droplet sizes, and distribution change so that the analyte loss is greater out the spray chamber drain. Generally, the sheath flow rate can be kept low and the low liquid flow nebulizers commercially available have high analyte transport efficiency, making loss of sensitivity minimal. Finally, if the sheath-flow/make-up buffer is different from the run buffer, changes to the separation will obviously occur.

Creation of a pH gradient across the electrophoretic capillary or isoelectric focusing from the use of a mismatch of the run and make-up buffer may cause undesirable results with the analyte separation.

Another less often used, but successful, technique to counter laminar flow in pneumatic-based nebulizer CE–ICP-MS interfaces is by the application of negative buffer reservoir pressure. This approach was first demonstrated by Lu et al.[3] and later by Taylor et al.[17] A matching mechanical counterbalance to the pneumatic nebulizer's suction was used in both of these interfaces. The theoretical advantage of non-sheath-flow system is that sensitivity of the CE–ICP-MS system is not reduced by dilution by the make-up buffer. Olesik et al.[1] originally described a sheathless pneumatic interface, but the main flaw in the design was the increased liquid flow through the electrophoretic capillary owing to the suction or Bernoulli effect of the operating nebulizer. The electro-osmotic flow of this CE system was measured at $0.05\,\mu L\ min^{-1}$, while the natural aspiration rate of the nebulizer vacuum was measured to be $2\,\mu L\ min^{-1}$. Some degradation of the separation of analytes was noted, but the high plate number of CE and the selective detection capability of MS allowed this to be a useful separation. Another problem encountered in some sheathless interfaces is the loss of electrical connectivity of the electrophoretic capillary.

Finally, other nebulizers have been used in CE–ICP-MS interfaces. In an interface developed by the Barnes research group[18] using the ultrasonic nebulizer (USN), the ground connection was provided by a make-up buffer/sheath-flow electrolyte. The separations obtained with the USN were demonstrated by Barnes' group to be superior to those obtained using a concentric pneumatic nebulizer in their study. Kirlew et al.[19] reported the comparison of a "home-made" ultrasonic nebulizer and a CETAC USN in CE–ICP-MS interfaces. Again, a make-up buffer was used, added to the system via a concentric capillary outside the electrophoretic capillary. In another work by Kirlew and Caruso,[20] an oscillating capillary nebulizer (OCN), which is a variation of the pneumatic concentric nebulizer built from flexible capillary tubes, was used in an interface. The OCN has had little application in CE interfaces, owing to its generally lower sensitivity performance when compared to other pneumatic nebulizers used with ICP-MS detection.[21] The direct injection nebulizer (DIN), previously described in "The Nebulizer," was used by Liu et al.[2] in a CE interface. The electrophoretic capillary was directly inserted through the central sample introduction capillary of the DIN. A platinum grounding electrode was positioned into a three-port connector. This connector contained the DIN sample introduction capillary as well as a make-up buffer flow. These alternative nebulizers have been successfully used in CE interfaces, but the pneumatic designs dominate the interface systems reported in the literature.

One last CE–ICP interface worthy of mention that is specific for the determination of elements capable of forming volatile compounds is by the use of a hydride generation system. Hydride generation followed by a gas–liquid separator in CE interfaces has been reviewed.[22] This technique has a drawback, because only arsenic, tin, lead, antimony, bismuth, germanium, selenium, and tellurium are capable of forming gaseous hydrides at room temperature. Hydride generation allows for the introduction of analyte species into the inductively coupled plasma nearly quantitatively; that is, the transport efficiency is nearly 100% percent less some loss by venting in the gas–liquid separator or other inefficiencies within the interface/sampling tube design to the plasma. In theory, detection limits are lower. In practice, hydride generation of the analytes may occur at different rates and the extra complexity and expense of the interface make these systems less useful as compared to the direct sample introduction systems previously described.

APPLICATIONS IN SAMPLE ANALYSIS

As mentioned in the Introduction, the main application of CE–ICP-MS is in the field generally known as elemental speciation analysis. There are a number of good reviews on elemental speciation CE advances including use MS detection; the two most recent were by Kannamkumarath et al.[23] and Timerbaev.[24] Work performed in this relatively young technique is very extensive, and only a few examples will be cited in this article.

Speciation analysis of arsenic and selenium is very active in the recent literature. The toxicity of arsenic varies widely and is dependent on the specific compound present. Arsenic in its various forms is also widely distributed in the environment and the food that human eats. Speciation of arsenic in human depends both on the form of the arsenic taken in and the metabolism within the body. Inorganic arsenic, in the forms of arsenite (As^{III}) and arsenate (As^{V}), is highly toxic. The common organic forms of arsenic have varying degrees of toxicity. Monomethylarsonic acid [MMA^{V}, $CH_3AsO(OH)_2$] and dimethylarsinic acid [DMA^{V}, $(CH_3)_2AsO(OH)$] exhibit a toxicity factor of 1 in 400 that of the inorganic forms. Arsenobetaine [$(CH_3)_3As^{+}CH_2COO^{-}$] and arsenocholine [$(CH_3)_3As^{+}CH_2CH_2O^{-}$] are commonly found in seafood and are relatively nontoxic. A variety of arsenic compounds are currently used as antifungal agents, herbicides, and pharmaceuticals; they are also used in semiconductor processing and there was an extensive use in pesticides before the invention of the more advanced organophophorus compounds. Thus the literature is filled with arsenic speciation analysis of food, the environment, and biological systems. Selenium has also been widely studied for many of the same reasons. Selenium intake in the human diet is essential, but excess intake can cause toxic reactions. A variety of selenium-containing species are present in the environment, natural foods, and food supplements.

In a previously mentioned work by Kirlew et al.,[19] electrophoretic separations of Se^{IV}, Se^{VI}, As^{III}, As^V, and dimethylarsinic acid were performed using various ultrasonic nebulizer (USN) interfaces. Using the optimized CE interface conditions and a borate run buffer at pH 8, a separation was accomplished within 10 min. Electrokinetic injections gave better sensitivities for the analytes as compared to hydrostatic sample injection. In the Kirlew study, arsenate and selenite ions had very similar migration times, but these analytes were easily resolved by the multielement capability of the ICP-MS detector. An electropherogram of this work is shown in Fig. 5. In an application to field samples, Van Holderbeke et al.[25] investigated arsenic speciation in three different sample matrices: drinking water, human urine, and soil leachate. All were run under basic conditions with 20 mM borate buffer (pH 9.40) and in the presence of cationic surfactant as the osmotic flow modifier (OFM) supplied by Waters Associates (Milford, Massachusetts). The separation of As^V, monomethylarsonic acid, dimethylarsinic acid, monomethylarsonic acid, arsenite As^{III}, arsenobetaine, and arsenocholine was obtained. Electropherograms from the Van Holderbeke study are shown in Fig. 6.

Capillary electrophoresis has been extensively used to separate biological molecules. Therefore it was only a natural extension that CE has been used for the analysis of metalloproteins and metal binding with other macromolecules. Also, the use of CE with the ICP-MS

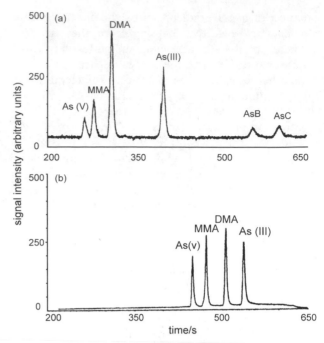

Fig. 6 (a) Electropherogram showing the separation of As^V, MMA, DMA, As^{III} AsB, and AsC, obtained with CE-ICP-MS before optimization. (b) Electropherogram of approximately 20 μg/L As^V, MMA, DMA, and As^{III} obtained after optimization of the CE-ICP-MS system. Conditions: 20 mM borate (pH 9.4), 2% OFM, 75 μm (id) capillary, total length 88 cm, 5 kPA for 40 sec plus 5 sec post injection, −25 kV. (Reproduced from Ref.[25] with permission of the Royal Society of Chemistry.)

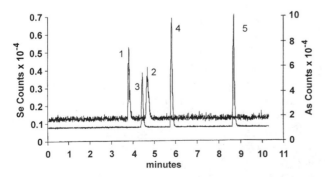

Fig. 5 Electropherogram of a mixed anion standard. Hydrodynamic injection (15 cm, 120 sec, 39.1 nL) with sodium borate buffer at pH 8. Peak 1: 3.2 ppm selenate; peak 2: 3.6 ppm selenite; peak 3: 1.9 ppm arenate; peak 4: 4.4 ppm DMA; peak 5: 4.7 ppm arsenite. (Reproduced from Ref.[19] with permission of Elsevier Science.)

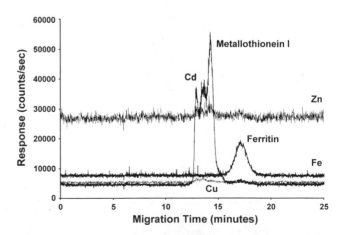

Fig. 7 Electropherogram showing separation of metallothionein I and ferritin. Zinc (mass 64) response at top and copper (mass 65) showed low signals for metallothionein. Cadmium (mass 114) showed a good response for metallothionein, and iron (mass 54) showed a good response for ferritin. This electropherogram was run using a run buffer of 15 mM Tris (hydroxymethyl) aminoethane (pH 6.8) at 15 kV, and the make-up buffer reservoir was positioned 2.5 cm above the MicroMist nebulizer. (Conditions are from Ref.[16].)

detector has been used in many studies reported in the literature. Metallothioneins are involved in metabolism and detoxification of several trace metals; thus the ability to monitor metallothioneins by CE–ICP-MS is of great importance. In a study of standard solutions of metallothionein I and ferritin, B'Hymer et al.[16] used various geometrical configurations of a microconcentric nebulizer to obtain different electropherograms. A run buffer of 15 mM Tris (hydroxymethyl) aminomethane adjusted to pH 6.8 by the addition of HCl was used at a potential of 15 kV. An example electropherogram is shown in Fig. 7. The MS detector has the advantage of being capable of simultaneous monitoring of various elements, thus resolving the relative quantities of cadmium, copper, and zinc bound to the metallothionein. Ferritin was clearly resolved from the iron signal.

CONCLUSIONS AND FURTHER READING

The advantages of CE coupled with those of the ICP-MS detector will certainly allow this relatively young hyphenated technique to grow in the areas of elemental speciation and the analysis of environmental and biological samples. It is doubtful that CE will replace HPLC; however, CE will certainly compliment the other more traditional separation techniques owing to its separation being based on a physical rather than chemical partitioning. The ability of CE to use an extensive variety of electrolyte/buffer solutions so that specific chemical species of interest can be maintained during the course of a separation is another advantage. The improvement in the mass spectrometer will of course lead to better detection limits and capabilities. Again, the low-flow pneumatic nebulizer will probably continue to lead the work performed in building CE–ICP-MS interfaces. Also, the current selection of commercial low-flow nebulizer will aid in the construction and use of interfaces. Finally, two books worthy of further reading are Akbar Montaser's *Inductively Coupled Plasma Mass Spectrometry* (Wiley-VCH, New York, 1998) and Joseph A. Caruso et al.'s *Elemental Speciation—New Approaches for Trace Elemental Analysis, Comprehensive Analytical Chemistry XXXIII* (Elsevier Science, Amsterdam, The Netherlands, 2000). Specific chapters were cited in this article, but both books comprise other chapters containing a wealth of information about capillary electrophoresis–inductively coupled plasma-mass spectrometry.

REFERENCES

1. Olesik, J.W.; Kinzer, J.A.; Olesik, S.V. Capillary electrophoresis inductively-coupled plasma spectrometry for rapid elemental speciation. Anal. Chem. **1995**, *67*, 1–12.
2. Liu, Y.; Lopez-Avila, V.; Zhu, J.J.; Weiderin, D.R.; Beckert, W.F. Capillary electrophoresis coupled online with inductively-coupled plasma-mass spectrometry for elemental speciation. Anal. Chem. **1995**, *67*, 2020–2025.
3. Lu, Q.; Bird, S.M.; Barnes, R.M. Interface for capillary electrophoresis and inductively-coupled plasma-mass spectrometry. Anal. Chem. **1995**, *67*, 2949–2956.
4. Houk, R.S.; Fassel, V.A.; Flesch, G.D.; Svec, H.L.; Gray, L.A.; Taylor, C.E. Inductively coupled argon plasma as an ion-source for mass-spectrometric determination of trace-elements. Anal. Chem. **1980**, *52*, 2283–2289.
5. Date, A.R.; Gray, A.L. Plasma source-mass spectrometry using an inductively coupled plasma and a high-resolution quadrupole mass filter. Analyst **1981**, *106*, 1255–1267.
6. Hill, S.J., Ed.; *Inductively Coupled Plasma Spectroscopy and Its Applications*; Sheffield Academic Press: Sheffield, England, 1999.
7. Olson, L.K.; Caruso, J.A. The helium microwave-induced plasma—An alternative ion-source for plasma-mass spectrometry. Spectrochim. Acta. Part B. **1994**, *49*, 7–30.
8. Kinzer, J.A.; Olesik, J.W.; Olesik, S.V. Effect of laminar flow in capillary electrophoresis: model and experimental results on controlling analysis time and resolution with inductively coupled plasma mass spectrometry detection. Anal. Chem. **1996**, *68*, 3250–3257.
9. Montaser, A.; Minich, M.G.; McLean, J.A.; Liu, H.; Caruso, J.A.; Mcleod, C.W. Sample introduction in ICPMS. In *Inductively Coupled Plasma Mass Spectrometry*; Montaser, A., Ed.; Wiley-VCH: New York, 1998; 1–47.
10. B'Hymer, C.; Caruso, J.A. Nebulizer sample introduction for elemental speciation. In *Elemental Speciation—New Approaches for Trace Elemental Analysis, Comprehensive Analytical Chemistry XXXIII*; Caruso, J.A., Sutton, K.L.M., Ackley, K.L., Eds.; Elsevier Science: Amsterdam, The Netherlands, 2000; 211–224.
11. Shum, S.C.K.; Neddersen, R.; Houk, R.S. Elemental speciation by liquid chromatography inductively coupled plasma mass-spectrometry with direct injection nebulization. Analyst **1992**, *117*, 577–582.
12. Shum, S.C.K.; Pang, H-M.; Houk, R.S. Speciation of mercury and lead compounds by microbore column liquid-chromatography inductively coupled plasma-mass spectrometry with direct injection nebulization. Anal. Chem. **1992**, *64*, 2444–2450.

13. Smith, R.D.; Barinaga, C.J.; Udseth, H.R. Improved electrospray ionization interface for capillary zone electrophoresis-mass spectrometry. Anal. Chem. **1988**, *60*, 1948–1952.

14. Day, J.A.; Sutton, K.L.; Soman, R.S.; Caruso, J.A. A comparison of capillary electrophoresis using indirect UV absorbance and ICP-MS detection with a self-aspirating nebulizer interface. Analyst **2000**, *125*, 819–823.

15. Sutton, K.L.; B'Hymer, C.; Caruso, J.A. UV absorbance and inductively coupled plasma spectrometric detection for capillary electrophoresis—A comparison of detection modes and interface designs. J. Anal. At. Spectrom. **1998**, *13*, 885–891.

16. B'Hymer; Day, J.A.; Caruso, J.A. Evaluation of a microconcentric nebulizer and its suction effect in a capillary electrophoresis interface with inductively coupled plasma-mass spectrometry. Appl. Spectrosc. **2000**, *54*, 1040–1046.

17. Taylor, K.A.; Sharp, B.L.; Lewis, D.J.; Crews, H.M. Design and characterisation of a microconcentric nebuliser interface for capillary electrophoresis—Inductively coupled plasma mass spectrometry. J. Anal. At. Spectrom. **1998**, *13*, 1095–1100.

18. Lu, Q.; Barnes, R.M. Evaluation of an ultrasonic nebulizer interface for capillary electrophoresis and inductively coupled plasma mass spectrometry. Microchem. J. **1996**, *54*, 129–143.

19. Kirlew, P.W.; Caruso, J.A.; Castillano, M.T.M. An evaluation of ultrasonic nebulizers as interfaces for capillary electrophoresis of inorganic anions and cations with inductively coupled plasma mass spectrometric detection. Spectrochim. Acta Part B. **1998**, *53*, 221–237.

20. Kirlew, P.W.; Caruso, J.A. Investigation of a modified oscillating capillary nebulizer design as an interface for CE–ICP-MS. Appl. Spectrosc. **1998**, *52*, 770–772.

21. B'Hymer, C.; Sutton, K.L.; Caruso, J.A. A comparison of four nebulizer/spray chamber interfaces for the high performance liquid chromatographic separation of arsenic compounds using ICP-MS detection. J. Anal. At. Spectrom. **1998**, *13*, 855–858.

22. Taylor, A.; Branch, S.; Fisher, A.; Halls, D.; White, M. Atomic spectrometry update. Clinical and biological materials, foods and beverages. J. Anal. At. Spectrom. **2001**, *16*, 421–446.

23. Kannamkumarath, S.; Wrobel, K.; Wrobel, K.; B'Hymer, C.; Caruso, J.A. Capillary electrophoresis-inductively coupled plasma-mass spectrometry: An attractive complementary technique for elemental speciation analysis. J. Chromatogr. A. **2002**, *975*, 245.

24. Timerbaev, A.R. Recent advances in capillary electrophoresis of inorganic ions. Electrophoresis **2002**, *23*, 3884–3906.

25. Van Hoderbeke, M.; Zhao, Y.; Vanhaecke, F.; Moens, L.; Dams, R.; Sndra, P. Speciation of six arsenic compounds using capillary electrophoresis inductively coupled plasma mass spectrometry. J. Anal. At. Spectrom. **1999**, *14*, 229–234.

Capillary Electrophoresis/MS: Large Molecule Applications

Ping Cao
Tularik, Inc., South San Francisco, California, U.S.A.

INTRODUCTION

Capillary electrophoresis (CE) is a modern analytical technique which permits rapid and efficient separation of charged components present in small-sample volumes. Separation occurs due to differences in electrophoretic mobilities of ions inside small capillaries. The impetus for CE method developments focused primarily on the separation of larger biopolymers such as polypeptides, proteins, oligonucleotides, DNA, RNA, and oligosaccharides.[1] MS has long been recognized as the most selective and broadly applicable detector for analytical separations. Currently, electrospray ionization (ESI) serves as the most common interface between CE and MS. Generation of multiply-charged species with an ESI extends the applicability of conventional mass analyzers of limited mass-to-charge (m/z) ranges to molecular mass and structure determination of larger biopolymers. CE/MS combines the advantages of CE and MS so that information on both high efficiency and molecular masses and/or fragmentation can be obtained in one analysis. This article focuses on larger-molecular analysis by on-line CE/MS interfaced via ESI sources.[2–3] However, CE/MS using continuous-flow fastatom bombardment (CF–FAB) sources employing either "liquid-junction" or "coaxial" interfaces and several off-line CE/MS combination should be noted.

When ESI–MS is employed as detector, the proper choice of a suitable electrolyte system is essential to both a successful CE separation and good quality ESI mass spectra. Even though a wide range of CE buffers were successfully electrosprayed when the liquid-junction and sheath flow CE/MS interfaces were employed since the low CE effluent flow is effectively diluted by a much large volume of sheath liquid;[4] the best detector response is produced by volatile electrolyte systems at the lowest practical concentration and ion strength and by minimizing other nonvolatile and charge-carrying components. Volatile reagents like ammonium acetate (pH 3.5–5.5) or formate (pH 2.5–5; both adjustable to high pH) and ammonium bicarbonate have been proven to be well suited for CE/ESI/MS.

Due to the inherent tendency to adsorb strongly to the inner walls of the fused-silica capillary, the analysis of proteins and peptides by CE has presented unique challenges to the analyst because this phenomenon gives rise to substantial peak broading and loss of separation efficiency. Successful separations of proteins and peptides by CE involve efficient suppression of adsorption to the fused-silica wall. Basically, there are two approaches to prevent protein adsorption: modification of the fused-silica surface by dynamic or static coating or by performing analysis under experimental conditions that minimize adsorption.[5] The static coating capillary is preferred under CE/ESI/MS /MS analysis of large molecules because the CE buffer composition is simplified. This article is meant only to provide the reader with a description of most common approaches taken to analyze large molecules, especially polypeptides and proteins, by CE/ESI/MS.

LARGE-MOLECULE ANALYSIS OF CE/MS BY NEUTRAL CAPILLARY

Because there is no ionizable groups of the coating in the neutral capillary, the interaction between charged molecules with ionic capillary surface is eliminated. Also, the electro-osmotic flow (EOF) of a neutral capillary is eliminated. However, a continuous and adequate flow of the buffer solution toward the CE capillary outlet is an important factor for routine and reproducible CE/ESI/MS analysis; in order to maintain a stable ESI operation, some low pressure applied to the CE capillary inlet is usually needed, especially when the sheathless interface is employed. The disadvantage of the pressure-assisted CE/ESI/MS is the loss of some resolution because the flat flow profile of the EOF is partially replaced by the laminar flow profile of the pressure-driven system. A typical neutral capillary is a LPA (linear polyacrylamide)-treated capillary. Karger and co-workers[6] used mixtures of model proteins, a coaxial sheath flow ESI interface, and a 75-μm-inner diameter (i.d.), 360-μm-outer diameter (o.d.), 50-cm-long LPA-coated capillary to evaluate CE/MS, CITP (capillary isotachophoresis)–MS, and the on-column combination of CITP/CE/MS. In the CE/MS experimental, 0.02 M 6-aminohexanoic acid + acetic acid (pH 4.4) was employed and a 18-kV constant voltage was applied during the experiment. Sevenmodel proteins were well resolved.

Encyclopedia of Chromatography DOI: 10.1081/E-ECHR-120039912

They showed that the sample concentration necessary to obtain a reliable full-scan spectrum was in the range of 10^{-5} M. However, by proper selection of the running buffers, they demonstrated that the on-column combination of both CITP and capillary zone electrophoresis (CZE) can improve the concentration detection limits for a full-scan CE/MS analysis to approximately 10^{-7} M.

LARGE-MOLECULE ANALYSIS OF CE/MS BY A POSITIVELY CHARGED CAPILLARY

To help overcome adsorption, positively charged coatings have been employed for the separation of positively charged solutes. In this approach, positively charged proteins are electrostatically repelled from the positively charged capillary inner wall. Two examples of such coatings are aminopropyltrimethoxysilane (APS) and polybrene, a cationic polymer. These coatings reverse the charge at the column–buffer interface and, thus, the direction of the EOF compared to uncoated capillaries.

The CE/MS analysis of the venom of the snake *Dendroaspis polylepis polylepis*, the black mamba, is reported by Tomer and co-workers.[7] A VG 12-250 quadrupole equipped with a Vestec ESI source (coaxial sheath flow interface) was employed for this experiment. The sheath fluid was a 50:50 methanol:3% aqueous acetic acid solution.The CE voltage was set at -30 kV during the analysis and the ESI needle was held at $+3$ kV. The CE running buffer used was 0.01 M acetic acid at pH 3.5. The APS column was flushed with buffer solution for 10 min prior to sample analysis. The snake venom was dissolved in water at a concentration of 1 mg/mL and 50 nL of

the analyte solution was injected into the column. They demonstrated the existence of at least 70 proteins from this venom.

One interesting example of intact protein analysis was described by Smith and co-workers.[8] They used the high sensitivity and mass accuracy of a Fourier transform ion cyclotron resonance (FTICR) MS detector to analyze hemoglobin α and β in a single human erythrocyte. Human erythrocytes were obtained from the plasma of a healthy adult male. A small drop of blood diluted with saline solution (pH 7.4) was placed on a microscope slide.With the help of a stereomicroscope and a micromanipulator, the etched terminus of the CE capillary was positioned within a few microns of the cell to be injected. Following electroosmotic injection of the cell, the end of the CE capillary wasplaced in a vial containing the CE running buffer (10 mM acetic acid, pH 3.4), and the cell membrane was lysed via osmotic shock from the running buffer and the cellular contents of the cell released for subsequent CE separation and mass analysis. A 1-m APS column and a sheathless interface employing a gold-coated capillary with -30 kV CE separation and $+3.8$ kV ESI voltage were used for this study. They demonstrated that adequate sensitivity needed to characterize the hemoglobin from a single human erythrocyte (~ 450 μmol) and mass spectra with average mass resolution in excess of 45,000 (full width at half-maximum) were obtained for both the α- and β-chain of hemoglobin. Fig. 1 shows the mass spectra obtained from this experiment.

In order to overcome the bubble formation associated with the sheathless CE/MS interface and quick degradation of the coated capillary, Moini et al.[9] introduced hydroquinone (HQ) as a buffer additive to suppress the bubbles formed due to the electrochemical oxidation of the CE buffer at the outlet

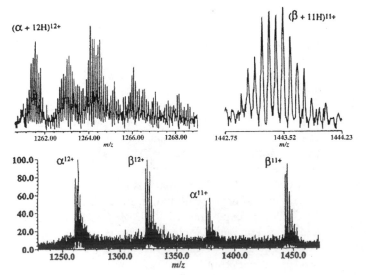

Fig. 1 Mass spectra obtained from CE/MS analysis of a single human erythrocyte using an FTICR mass analyzer. (Reproduced from Ref.[7], with permission.)

electrode. The oxidation of water ($2H_2O(l) \leftrightarrow O_2$ (g) + $4H^+$ $4e$) was replaced with that of more easily oxidized HQ (hydroquinone $\leftrightarrow$ p-benzoquinone + $2H^+$ + $2e$). Formation of p-benzoquinone, other than the formation of oxygen gas, effectively suppresses gas bubble formation. The APS-coated capillaries and 10 mM acetic acid CE running buffer containing 10 or 20 mM HQ were used for the experiments. The CE outlet/ESI electrode was maintained at +2 kV and the CE inlet electrode was held at −30 kV. Tryptic digest of cytochrome-c and hemoglobin were used as model proteins. They demonstrated that the combination of the in-capillary electrode sheathless interface using a platinum wire, HQ as a buffer additive, and pressure programming at the CE inlet provides a rugged high-efficiency setup for analysis of peptide mixtures.

Because the concentration limits of detection of CE are often inadequate for most practical applications (approximately 10^{-6} M), several analyte concentration techniques have been developed, including combining CITP with CE, transient isotachophoresis (tITP) in a single capillary, analyte stacking, and field amplification. Such electrophoretic techniques have extended the applicability of CE for the analysis of dilute analyte solutions. Chromatographic on-line sample concentration has been achieved by using an extraction cartridge which contains a bed of reversed-phase packing[10] or a membrane[11] with properties. Accumulated analyte on the cartridge can be prewashed to remove salts and buffers that are not suited for CE separation or ESI operation. Figeys and Aebersold[12] designed the solid-phase extraction (SPE)/CE/MS/MS system which consists of a small cartridge C_{18} of extraction material immobilized in a Teflon sleeve. Solutions of peptide mixtures typically derived by proteolysis of gel-separated proteins were forced through the capillary by applying positive pressure at the inlet and the peptides were concentrated on the SPE device. After equilibration with an electrophoresis buffer compatible with ESI, eluted peptides were separated by CE and analyzed by ESI/MS. A detection limit of 400 amol tryptic digest of bovine serum albumin (20 μL of solution at a concentration of 20 amol/μL was applied) was achieved in the ion trap mass spectrometer-based system. This method was successfully applied to the identification of yeast proteins separated by two-dimensional gel electrophoresis.

Applications of CE/MS to large molecules are progressing rapidly. As biology enters an era of large-scale systematic analysis of biological systems as a consequence of genome sequencing projects, rapid and sensitive identifications of large-scale (proteomewide) proteins that constitute a biological system is essential. CE/MS with its high separation efficiency, rapid

separation, and economy of sample size is complementary to microcolumn high-performance liquid chromatography (μHPLC)/MS. In addition, high-resolution, multiple-dimensional separations become increasingly attractive. HPLC/CE/MS, affinity CE/MS, capillary microreactor on line with CE/MS, and microchip based separations will be used in a broad range of future applications.

REFERENCES

1. Kuhr, W.G.; Monnig, C.A. Capillary electrophoresis. Anal. Chem. **1992**, *64*, 389.
2. Smith, R.D.; Udseth, H.R. *Pharmaceutical and Biomedical Applications of Capillary Electrophoresis*; Elsevier Science: New York, 1996; 229–276.
3. Banks, J.F. Recent advances in capillary electrophoresis/electrospray/mass spectrometry. Electrophoresis **1997**, *18*, 2255.
4. Smith, R.D.; Loo, J.A.; Edmonds, C.G.; Barinaga, C.J.; Udseth, H.R. New developments in biochemical mass spectrometry: electrospray ionization. Anal. Chem. **1992**, *62*, 882.
5. Thibault, P.; Dovichi, N.J. *Capillary Electrophoresis (Theory and Practice)*, 2nd ed.; CRC Press: Boca Raton, FL, 1998; 23–90.
6. Thompson, T.J.; Foret, F.; Vouros, P.; Karger, B.L. Capillary electrophoresis/electrospray ionization mass spectrometry: improvement of protein detection limits using on-column transient isotachophoretic sample reconcentration. Anal. Chem. **1993**, *65*, 900.
7. Perkins, J.R.; Parker, C.E.; Tomer, K.B. The characterization of snake venoms using capillary electrophoresis in conjunction with electrospray mass spectrometry: Black Mambas. Electrophoresis **1993**, *14*, 458.
8. Hofstadler, S.A.; Severs, J.C.; Smith, R.D. Rapid Commun. Mass Spectros. **1996**, *10*, 919.
9. Moini, M.; Cao, P.; Bard, A.J. Hydroquinone as a buffer additive for suppression of bubbles formed by electrochemical oxidation of the CE buffer at the outlet electrode in capillary electrophoresis/electrospray ionization-mass spectrometry. Anal. Chem. **1999**, *71*, 1658.
10. Figeys, D.; Aebersold, R. Electrophoresis **1998**, *19*, 885.
11. Tomlinson, A.J.; Benson, L.M.; Guzman, N.A.; Naylor, S. Preconcentration and microreaction technology on-line with capillary electrophoresis. J. Chromatogr. **1996**, *744*, 3.
12. Figeys, D.; Aebersold, R. Electrophoresis **1997**, *18*, 360.

Capillary Isoelectric Focusing: An Overview

Robert Weinberger
CE Technologies, Inc., Chappaqua, New York, U.S.A.

INTRODUCTION

Capillary isoelectric focusing (CIEF) employs a pH gradient developed within the capillary to separate zwitterions, usually proteins and peptides, based on each solute's pI. The technique is analogous to slab–gel CIEF[1] with several important differences: (a) Slab–gel IEF is a nonelution process. After running the electrophoretic step, the proteins are detected by staining. CIEF is usually an elution technique. The contents of the capillary are mobilized to pass through the detector region. (b) In the slab–gel, detection is by Commassie or silver staining. In CIEF, detection is by ultraviolet (UV) absorbance at 280 nm. (c) In the capillary format, no gel is required because mechanical stability is provided by the rigid capillary walls. (d) The field strength is at least an order of magnitude higher in the capillary format compared to the slab–gel.

The usual advantages of capillary electrophoresis apply equally to CIEF. Slab–gel IEF is extremely labor intensive and time-consuming. CIEF is simple to run, fully automated, and high speed and provides improved quantitative results compared to slab–gel IEF. This topic has been recently reviewed,[2–3] as is usually covered as a chapter in many high-performance capillary electrophoresis textbooks.

pH GRADIENT FORMATION

A solution containing 0.5–2.0% carrier ampholytes and 0.1–0.4% methylcellulose (1500 cP for a 2% solution) is filled into the capillary. A coated capillary is used to suppress the electro-osmotic flow in conjunction with the methylcellulose solution. The sample (protein) concentration in the ampholyte blend is usually between 50 and 200 µg/mL. The inlet reservoir (anolyte) is filled with 10 mM phosphoric acid in methylcellulose solution. The outlet reservoir (catholyte) contains 20 mM sodium hydroxide.

The condition of the capillary immediately upon activation of the voltage is illustrated at the top of Fig. 1. In this case, the capillary is filled with a pI 3–10 mixture of ampholytes. Assuming the pH of the solution is 7, charges have been assigned to the individual ampholytes. The ampholyte charge dictates the direction of migration. Should any ampholytes migrate

into a reservoir, the extreme pH conditions cause immediate charge reversal. Likewise, as the steady state is approached, should an ampholyte migrate into a more acidic or basic zone, charge reversal occurs as well. The result is the formation of a pH gradient as indicated at the bottom of Fig. 1. As each ampholyte approaches a pH equal to its individual pI, migration slows and then ceases. Because overlapping Gaussian zones of each ampholyte are formed, the gradient is smooth and linear.

The conventional pH range for CIEF is pH 3–10. Narrow-range gradients can be created by selecting custom ampholyte blends (e.g., pH 6–8). To avoid problems such as a step-gradient or the creation of water zones, the narrow-range ampholyte solution is usually supplemented with 20% pH 3–10 ampholytes. This ensures that there are sufficient ampholyte species to produce Gaussian overlaps.

The focusing step takes 2–5 min. at a field strength of 500–1000 V/cm. Overfocusing causes precipitation of proteins and can damage the capillary as well. Initially, the current is high as ampholytes and proteins are highly charged. As their pI's are approached, the current declines and reaches 1–4 µA when focusing is complete.

The 2 : 1 ratio of base : acid in the reservoirs is not coincidental. It is selected to minimize drift of the pH gradient. The pH of the anolyte must be lower than that of the most acidic ampholyte; likewise, the pH of the catholyte must be higher than the most basic ampholyte. Otherwise, ampholytes will migrate into the reservoirs and cause gradient drift. If the EOF is not reduced, a form of cathodic drift occurs as well. An exception to this is when one-step mobilization is employed.

Internal standards are always used to calibrate the pH gradient. This is important because the salt content of the sample can compress the gradient. The ideal internal standard absorbs at a wavelength other than 280 nm. It can then be added to the ampholytes and monitored without producing interference. In this case, photodiode array detection is used for multiwavelength monitoring. Methyl red, pI 3.8, is ideal in this regard, but other such markers have not been identified. Aminomethylphenyl dyes are often used with monitoring at 400 nm, but they have some absorbance at 280 nm.

Encyclopedia of Chromatography DOI: 10.1081/E-ECHR-120039904

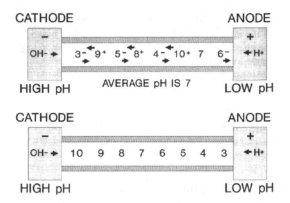

Fig. 1 Illustration of the process of pH gradient formation.

To prevent focused zones from reaching the blind side of the capillary past the detector, the ampholytes are either filled just prior to the detector or the reagent TEMED (N,N,N,N-tetramethylenediamine) is added to the blend at the appropriate concentration (0.5–2.0%). TEMED, a basic amine, then occupies that space past the detector window.

RESOLVING POWER

The resolving power, ΔpI of CIEF is described by

$$\Delta pI = 3\sqrt{\frac{D(dpH/dx)}{E(d\mu/dpH)}} \qquad (1)$$

where D is the diffusion coefficient, E is the field strength, μ is the mobility of the protein, and $d\mu/dpH$ describes the mobility–pH relationship. The term dpH/dx represents the change in the buffer pH per unit of capillary length. This adjustable parameter is controlled by selecting an appropriate ampholyte pH range as well as the capillary length. Under optimal conditions, a resolution of 0.02 pH units is possible.

MOBILIZATION

There are three ways of mobilizing the contents of the capillary: (a) chemical (salt) mobilization, (b) electro-osmotic (one-step) mobilization, and (c) hydrodynamic mobilization. Hydrodynamic mobilization is the simplest and most widely used method. Low pressure is used to evacuate the capillary with the voltage activated. Laminar band broadening is minimized by simultaneous focusing. In one-step mobilization, the EOF is reduced but not absent. If done correctly, focusing occurs prior to any proteins reaching the detector. This is the fastest method but has lower reso-

lution and linearity compared to hydrodynamic mobilization. Salt mobilization is infrequently used today, but it produces the highest resolution at the expense of run time and gradient linearity.

DETECTION

Because the ampholytes absorb below 250 nm, 280-nm detection is required. It is critical to run ampholyte blanks because the reagents are not checked by the manufacturers for UV absorption at 280 nm. The limit of detection is a few micrograms per milliliter of protein and this is usually limited by the UV reagent background. Proteins that are deficient in aromatic amino acids are poorly detected. Different lots of ampholytes from various manufacturers show variation in the UV background.

The combination of CIEF and mass spectrometry is analogous to two-dimensional electrophoresis.[4] In this case, the mass spectrometer provides the molecular-weight information instead of sodium dodecyl sulfate–polyacrylamide gel electrophoresis. This information can be obtained by on-line CIEF/MS[5] or by using CIEF as a micropreparative technique.[6]

For the on-line system, CIEF is performed conventionally in a 20-cm capillary mounted inside an electrospray probe. After focusing, the outlet reservoir (catholyte) is removed and the capillary tip set to 0.5 mm outside of the probe. A sheath liquid of 50% meth anol, 49% water, and 1% acetic acid (pH 2.6) pumped with a syringe pump at 3 μL/min produces a stable electrospray. Cathodic mobilization is produced by changing the anolyte to the sheath liquid. The ampholyte ions were observed up to m/z 800 and thus did not interfere with the protein signals.

ADDITIVES FOR HYDROPHOBIC PROTEINS

The tendency of hydrophobic proteins to aggregate and precipitate is a major problem in IEF whether in the slab–gel or capillary format. The focusing power of CIEF produces an increase in solute concentration by a factor of over 200.[7] Proteins also readily precipitate as the pI is approached because their charge and, thus, electrostatic repulsion approach zero.

Protein precipitation is indicated first by spikes in the electropherogram followed by clogging of the capillary. Additives are required to suppress the aggregation of hydrophobic proteins to keep them in solution. Two excellent review articles describe this in detail.[8–9]

Among the reagents used to prevent aggregation are urea, nonionic surfactants such as Brij-35, zwitterionic

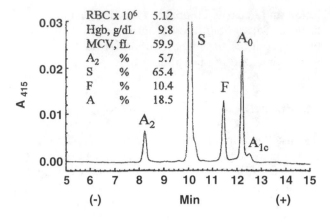

Fig. 2 Separation of hemoglobins by CIEF in blood from a patient suffering Hb S/β^+ from thalessemia. Capillary: 27 cm (20 cm to detector) × 50 µm i.d. DB-1 (J & W Scientific); ampholytes: 2% pH 6–8 : 10-3 (10 : 1) Pharmalytes (Pharmaceia Biotech) and 0.375% methylcellulose; catholyte: 20 mM sodium hydroxide; anolyte: 100 mM phosphoric acid in 0.375% methyl-cellulose; focusing: 5 min at 30 kV; mobilization: low pressure (0.5 psi) for 10 min at 30 kV; detection: UV, 415 nm. (Reprinted from *Electrophoresis*; Wiley–VCH; 1997; Vol. 18, 1785.)

detergents such as sulfobetains, polyols such as ethylene glycol, glycerol, amd sorbitol or nonreducing sugars. A strategy of mixing polyols and zwitterions is often successful in dealing with solubility problems.

APPLICATIONS

Capillary isoelectric focusing has been employed for separations of many proteins, recombinant proteins, monoclonal antibodies, and protein glycoforms. The most widely used method employing CIEF is the determination of hemoglobin variants for the screening of genetic disorders, including sickle cell anemia, thalessemisas, and other hemoglobinopathies. Fig. 2 illustrates the separation of hemoglobins in a patient with S/β^+ thalassemia.

REFERENCES

1. Righetti, P.G. *Isoelectric Focusing: Theory, Methodology and Applications*; Elsevier Biomedical Press: Amsterdam, 1983.
2. Righetti, P.G.; Bossi, A. Isoelectric focusing of proteins and peptides in gel slabs and in capillaries. Anal. Chim. Acta **1998**, *372*, 1.
3. Pritchett, T.J. Capillary isoelectric focusing of proteins. Electrophoresis **1996**, *17*, 1195.
4. Tang, Q.; Harrata, A.K.; Lee, C.S. Two-dimensional analysis of recombinant *E. coli* proteins using capillary isoelectric focusing electrospray Ionization mass spectrometry. Anal. Chem. **1997**, *69*, 3177.
5. Tang, Q.; Kamel Harrata, A.; Lee, C.S. Capillary isoelectric focusing-electrospray mass spectrometry for protein analysis. Anal. Chem. **1995**, *67*, 3515.
6. Foret, F.; Muller, O.; Thorne, J.; Gotzinger, W.; Karger, B.L. Analysis of protein fractions by micropreparative capillary isoelectric focusing and matrix-assisted laser desorption time-of-flight mass spectrometry. J. Chromatogr. A. **1995**, *716*, 157.
7. Yowell, G.G.; Fazio, S.D.; Vivilecchia, R.V. Analysis of a recombinant granulocyte macrophage colony stimulating factor dosage form by capillary electrophoresis, capillary isoelectric focusing and high-performance liquid chromatography. J. Chromatogr. **1993**, *652*, 215.
8. Rabilloud, T. Solubilization of proteins for electrophoretic analyses. Electrophoresis **1996**, *17*, 813.
9. Rodriguez-Diaz, R.; Wehr, T.; Zhu, N. Capillary isoelectric focusing. Electrophoresis **1997**, *18*, 2134.

Capillary Isotachophoresis

Ernst Kenndler
Institute for Analytical Chemistry, University of Vienna, Vienna, Austria

INTRODUCTION

Three analytical electrophoretic techniques can be distinguished. They differ in the kind of background electrolyte (BGE) and in its arrangements. Zone electrophoresis has a uniform BGE without a gradient; in isoelectric focusing, the separation is established by the aid of a BGE forming a continuous (linear) pH gradient. In contrast, isotachophoresis (ITP) is an electrophoretic method with a stepwise gradient of the background electrolyte.

ISOTACHOPHORESIS

In ITP, samples of only one charge type are separated in the same run (i.e., either anions or cations). The BGE in ITP is selected in the way that (the anion or cation of) the leading (L) and the terminating (T) electrolyte will have a higher and a lower mobility, μ, respectively, than the analytes of interest. Thus, the prerequisite for separation by ITP is that $\mu_L > \mu_{analytes} > \mu_T$.

Consider the case that the sample ions are anions (consisting of analytes A^- and B^-), and that the ions exhibit mobilities in the sequence $\mu_L > \mu_A > \mu_B > \mu_T$. The capillary is filled with and separated by a sharp boundary, and the sample is injected between the two electrolyte zones (for simplicity, it is assumed that the counterions, Q^+, are all the same). After application of an electric field, a certain field strength is established in the zones as depicted in Fig. 1 (it is assumed that the capillary has uniform diameter). Across the individual zone, the field strength is constant, but it increases due to the decreasing mobility (and increasing electrical resistance) from L to T. Across the zone of the mixed sample (A + B) it is constant as well, with strength E_{mix}. In this zone, the two analytes A and B are moving with different migration velocities, ν, according to $\nu_A = \mu_A E_{mix}$ and $\nu_B = \mu_B E_{mix}$. Due to the higher mobility of A, this analyte moves faster here than B: $\mu_A E_{mix} > \mu_B E_{mix}$. This effect leads to the migration of ions out of the mixed zone, forming a separate zone in front with pure A^- (which is always placed behind the zone of the leading electrolyte). For B^- the analogous situation occurs; it is moving slower in the mixed zone and forms a separate zone at the rear side (but in front of T).

The formation of zones of pure analyte ions (obviously with counterions, Q^+, the counterions of the leading electrolyte) is, therefore, observed. Five zones can be differentiated in this transient state: L^-, pure A^-, mixed $A^- + B^-$, pure B^-, and T^-. Due to the different mobilities, the field strength increases in this sequence. Separation takes place as long as the mixed zone exists. Finally, this zone disappears and the isotachophoretic condition is established: All zones and zone boundaries migrate with the same mean velocity ("isotachophoresis"); consequently,

$$\mu_L E_L = \mu_A E_A = \mu_B E_B = \mu_T E_T \tag{1}$$

This is the isotachophoretic condition. The conditions of electroneutrality must be fulfilled as well, which means, in the case considered, that in each zone the concentration of ions and counterions is equal. The third condition is Ohm's law, stating that (for given constant current) the product of electrical conductivity and field strength in each zone is equal. The combination of these conditions leads to the so-called regulation function (Kohlrausch), which reads in a simplified form (for monovalent strong electrolytes):

$$c_A = c_L \frac{\mu_A(\mu_L + \mu_Q)}{\mu_L(\mu_A + \mu_Q)} \tag{2}$$

It relates the concentration of a species in its own zone to the concentration of the species in the subsequent zone and, as a consequence, to the concentration of the first zone, the leading ion. It is also a function of the mobility of the ions involved: the analyte, the leading ion, and the counterion. This function allows two conclusions as follows.

As for given conditions, the mobilities in Eq. (2) are constant under ITP conditions; the concentration of the sample in its zone is constant as well. The concentration distribution is, therefore, given by a rectangular function. It follows that the temperature and the pH within the particular zone is constant, too, and changes stepwise at the boundary to the neighboring zone.

The concentration depends only on that of the leading ion; it is independent of the initial concentration in the sample. Therefore, ITP can act as an enrichment method, analogous to displacement chromatography and in contrast to zone electrophoresis and elution

Encyclopedia of Chromatography DOI: 10.1081/E-ECHR-120039906

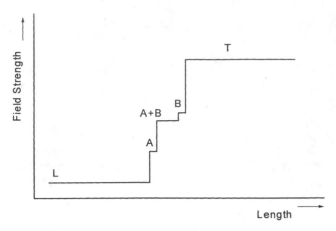

Fig. 1 Electrical field strength in the zones of the leading electrolyte, L, the sample consisting of A and B, and the termination electrolyte, T. For details, see the text.

chromatography. The concentration in the steady state is adjusted to the value given in Eq. (2). If the concentration of the analyte species is lower in the initial sample, the higher steady-state concentration is established. This concentration is independent of the migration distance: there is no dilution with a BGE as there is in capillary zone electrophoresis (CZE).

The adjustment of the steady-state concentration to a certain constant value has the consequence that the zone length of an analyte depends on the amount present in the sample. Increasing the amount results in an increase of the zone length under ITP conditions. The length is, therefore, the parameter for quantitative analysis.

The stepwise gradient of the electrical field at the zone boundary is the source of the "self-sharpening" effect in ITP. When, by diffusion, an ion migrates out of its own zone, into a neighboring zone (where the field strength is higher or lower, respectively), the condition given in Eq. (1) is not fulfilled any more. Therefore, the velocity of the considered ion is either accelerated or retarded, and the ion is pushed back into its initial zone. As a consequence, the boundary between the zones remains sharp and its shape is not dependent on the migration distance.

Isotachophoresis might have several advantages compared to zone electrophoresis. The adaptation to a considerably high concentration of the sample components in their own zones in the absence of further BGE favors the use of the electrical conductivity detector. The high concentration and the long sample zone have also some advantage in combining capillary electrophoresis with, for example, MS. Also, ITP can be used as an enrichment technique prior to zone electrophoretic separation, a phenomenon that is applied routinely in sodium dodecyl sulfate–polyacrylamide gel electrophoresis using a discontinuous buffer for sample introduction, and a technique called sample stacking in CZE. In fact, both methods rely on an isotachophoretic principle.

SUGGESTED FURTHER READING

Bocek, P.; Deml, M.; Gebauer, P.; Dolnik, V. *Analytical Isotachophoresis*; VCH: Weinheim, 1988.

Everaerts, F.M.; Beckers, J.L.; Verheggen, T.P.E.M. *Isotachophoresis: Theory, Instrumentation, and Applications*; Elsevier: Amsterdam, 1976.

Mosher, R.A.; Saville, D.A.; Thormann, W. *The Dynamics of Electrophoresis*; VCH: Weinheim, 1992.

Carbohydrates: Analysis by Capillary Electrophoresis

Oliver Schmitz
Division of Molecular Toxicology, German Cancer Research Center, Heidelberg, Germany

INTRODUCTION

Carbohydrates play an important role in many research and industrial domains. The huge number of stereoisomers, the immense combination possibilities of carbohydrate monomers in oligosaccharides, and the lack of chromophores are the major problems in the analysis of carbohydrates. Capillary electrophoresis (CE), in its various modes of operation, has been developed as a very useful tool in the analysis of carbohydrate species such as monosaccharides and oligosaccharides, glycoproteins, and glycopeptides.

DISCUSSION

Some simple sugar mixtures, such as monosaccharides and oligosaccharides, consisting of not more than about 15 monomer units, can be separated in free solution due to their mass-to-charge ratio (m/z). For an increase in selectivity or for analyzing neutral carbohydrates, MEKC can be used for analysis. In this case, charged amphiphilic molecules containing both hydrophilic and hydrophobic regions (e.g., sodium dodecyl sulfate) are used as buffer surfactants.

Higher oligosaccharides or polysaccharides possess unfavorable mass-to-charge ratios, preventing their effective resolution in open tubes. The separation of these carbohydrates is possible with capillary gel electrophoresis (CGE). The analytes are selectively retarded by a sieving network (gel or polymer matrix) due to differences in their sizes and structural conformations.

Complex carbohydrates (in particular, glycoproteins) play an important role in various biological processes and in biotechnological production of glycoproteinaceous pharmaceuticals. To elucidate the relationship between bioactivity and structures of complex carbohydrates, it is necessary to determine the sites of attachment of the oligosaccharide chains to the polypeptide backbone and to characterize the oligosaccharide class (N- or O-linked, high mannose, hybrid, etc.).

For this reason, glycoproteins must first be isolated from the biological matrix by dialysis, preparative chromatography, isoelectric focusing, and so forth or by a combination of several methods. For a structural determination, degradation steps such as a site-specific proteolysis (e.g., with trypsin), removal of oligosaccharides from the polypeptide (by an enzymatic hydrolysis or hydrazine treatment), or chemical hydrolysis, yielding a monosaccharide mixture may be applied. Then, the CE can function as a powerful end method in analytical and structural glycobiology. Due to the complexity of the carbohydrate-dependent microheterogeneity of glycoproteins, several electrophoretic techniques are usually needed, in concert, to characterize the various glycoforms of a given glycoprotein.

For detection of carbohydrates in principle, ultraviolet (UV), laser-induced fluorescence, refractive index, electrochemical, amperometric, and mass spectrometric detection can be used. MS, with its various ionization methods, has traditionally been one of the key techniques for the structural determination of proteins and carbohydrates. Fast-atom bombardment (FAB) and electrospray ionization (ESI) are the two on-line ionization methods used for carbohydrate analysis. The ESI principle has truly revolutionized the modern MS of biological molecules, due to its high sensitivity and ability to record largemolecule entities within a relatively small-mass scale.

The refractive index detection (RID), often used in high-performance liquid chromatography, is an interesting detection method in CE with a laser light source and a limit of detection (LOD) in the micromolar range. Electrochemical detection (ECD) and pulsed amperometric detection (PAD) of sugars are common and effective methods used in HPLC. Some recent communications show that the sensitivity of these detection methods in CE have an approximately 1000-fold better LOD than RID. Unfortunately, these detectors (RID, ECD, and PAD) are not commercially available for CE at the moment.

Indirect detection methods are a viable alternative for compounds lacking a chromophoric or a fluorophoric group. An electrolyte containing a chromophore or fluorophore allows the indirect detection of carbohydrates. This method is based on the displacement of the background electrolyte (BGE) by carbohydrates, which are dissociated in strongly alkaline electrolytes. The LOD with indirect LIF detection is in the nanometer range, but the lack of any specificity is a great disadvantage of this detection method,

Encyclopedia of Chromatography DOI: 10.1081/E-ECHR-120039909

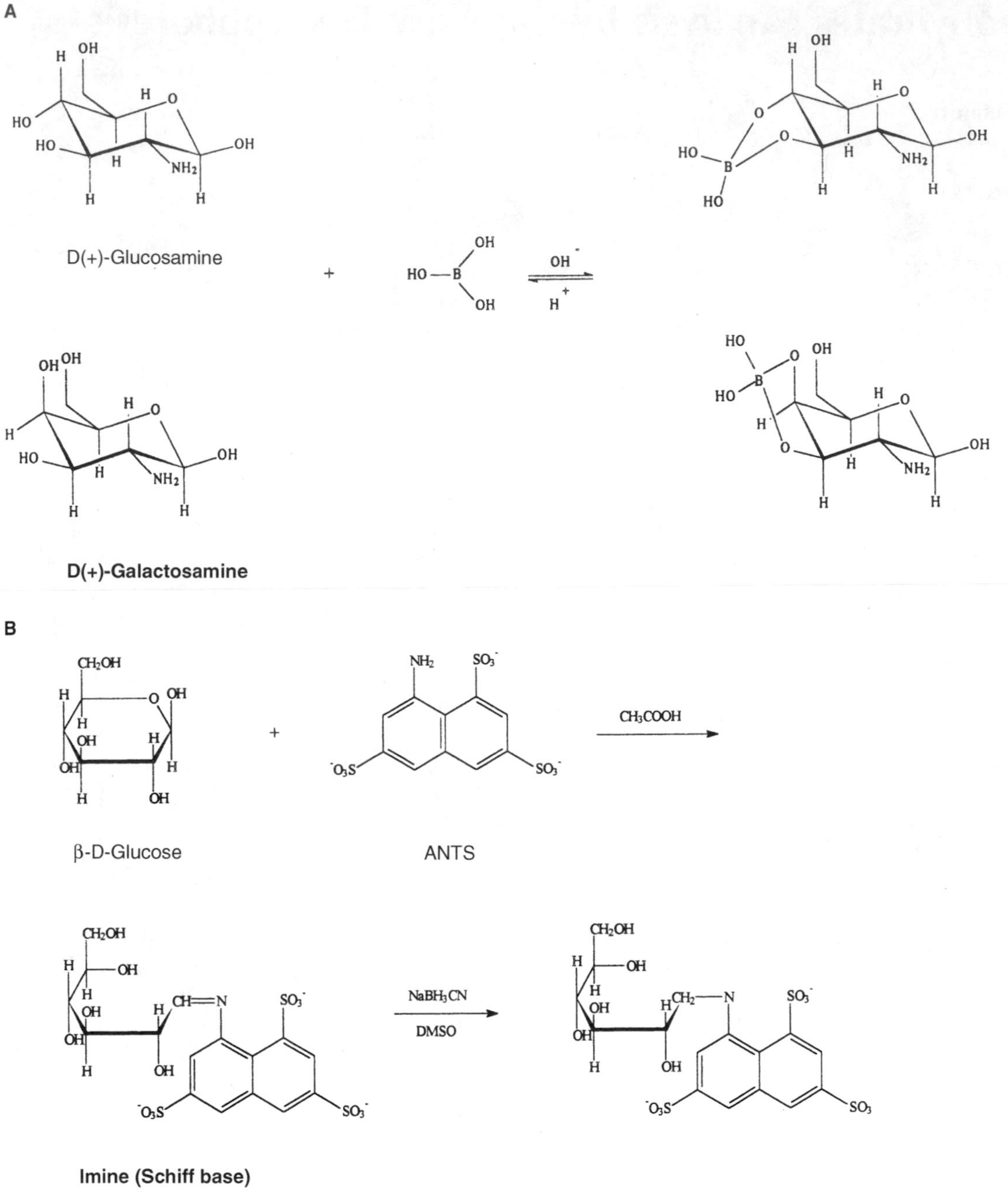

Fig. 1

because all sample compounds displace the BGE and the peak identification is only possible by the migration time.

Direct UV detection is the most versatile detection method in CE and is implemented in every commercial CE system. Unfortunately, its use for carbohydrates detection is restricted because of their lack of conjugated p-electron systems and, consequently, the relatively low extinction coefficients. Despite this fact, it is possible to detect carbohydrates with UV detection without any derivatization at 200 nm. Sensitivity and selectivity can be increased by the use of an alkaline borate buffer as the electrolyte by in situ complexation with the tetrahydroxyborate ion rather than the boric acid (Fig. 1A). The LOD is between micromolar and nanomolar. A further advantage of very high pH values (>10) is the negative charge of the carbohydrates, which are repelled by the negatively charged surface. Consequently, the surface problems in high-performance CE are much lower in carbohydrate analysis than in the analysis of proteins. Therefore, simple carbohydrates are often analyzed in uncoated fused-silica capillaries. Unlike the analysis of simple carbohydrates, for glycopeptides and glycoproteins, the use of coated capillaries such as hydroxypropyl-cellulose, hydroxyethylmethacrylate, polyether, or other commercially available coated fused-silica capillaries is necessary to achieve high resolution and reproducibility.

In complex matrices, the insufficient specificity at 200 nm and the low sensitivity of direct UV detection make the analysis of carbohydrates more difficult. Therefore, derivatization of carbohydrates with a suitable agents is still a preferred approach for the detection of monosaccharides and oligosaccharides. Derivatization agents like 2-aminopyridine, 8-amino-naphthalein-1,3,6-trisulfonate (ANTS) or 8-amino-pyren-1,3,6-trisulfonate (APTS) can be introduced mostly by reductive amination. This reaction is based on imine formation (Schiff base) by the condensation of the aldehyde group in a carbohydrate with the amino group in a primary amine (fluorescent tag), followed by reduction to an N-substituted glycamine with a reductant like sodium cyanoborohydride (see Fig. 1B). Selection of the suitable derivatization reagent is important, because the electrophoretic migration of the carbohydrates and, therefore, the separation power is influenced by the properties of the tags. Fluorescent dyes are better suitable than UV-active derivatization reagents, because CE analysis permits the use of laser-induced-fluorescence (LIF) detection with excellent sensitivity up to the femtomolar-level.

CONCLUSIONS

CE in carbohydrate analysis has advantages in both separation and detection over other techniques of electrophoresis, as well as chromatography. It allows high efficiency (up to a few million plate numbers) and very good sensitivities (up to femtomolar). In addition, CE permits analysis by a variety of separation modes simply by changing the electrolyte (capillary zone electrophoresis, MEKC, CGE).

SUGGESTED FURTHER READING

El assi, Z. Recent developments in capillary electrophoresis of carbohydrate species. Electrophoresis **1997**, *18*, 2400.

Grimshow, J. Analysis of glycosaminoglycans and their oligosaccharide fragments by capillary electrophoresis. Electrophoresis **1997**, *18*, 2408.

Linhardt, R.J.; Pervin, A. Separation of negatively charged carbohydrates by capillary electrophoresis. J. Chromatogr. **1996**, *720*, 323.

Novotny, M.V. Capillary electrophoresis of carbohydrates. In *High-Performance Capillary Electrophoresis*; Khaledi, M.G., Ed.; John Wiley & Sons: New York, 1998; 729–765.

Paulus, A.; Klockow, A. Detection of carbohydrates in capillary electrophoresis. J. Chromatogr. A. **1996**, *720*, 353.

Suzuki, S.; Honda, S. A tabulated review of capillary electrophoresis of carbohydrates. Electrophoresis **1998**, *19*, 2539.

Carbohydrates: Analysis by HPLC

Juan G. Alvarez
Beth Israel Deaconess Medical Center, Harvard Medical School, Boston, Massachusetts, U.S.A.

INTRODUCTION

Carbohydrates are widely distributed molecules in biological systems and pharmaceutical products, not only in free form but also in conjugated form. Because they are present in various forms and there are isomers and analogs, the separation of carbohydrates involves more difficult problems than those of proteins or nucleic acids. Difficulties are also found in detection, especially in biochemical and biomedical analyses due to their low abundance and the fact that photometric and fluorimetric methods cannot be applied directly because of the lack of chromophores and fluorophores.

Analysis of carbohydrates in body fluids by HPLC using anion-exchange columns was first reported in the 1970s.[1–5] This method has been greatly improved by the use of packing materials of fine, spherical particles and by the development of photometric and fluorimetric postcolumn labeling systems for sensitive detection. Honda et al. established rapid automated methods for microanalysis of aldoses,[6] uronic acids,[7] and sialic acids using a Hitachi 2633 anion-exchange resin and a photometric and fluorimetric postcolumn labeling system with 2-cyanoacetamide. Alditols[8] were fluorescence labeled by the use of sequential periodate oxidation and the Hantzsch reaction. Microanalysis of aminosugars was successful when their borate complexes were separated in the cation-exchange mode and detected by fluorescence generated either by the reaction with 2-cyanoacetamide[9] or by the Hantzsch reaction.[10] All these methods are suitable for routine analysis of clinical samples because of their speed of analysis and high sensitivity.

The United States Food and Drug Administration and the regulatory agencies in other countries require that pharmaceutical products be tested for composition to verify their identity, strength, quality, and purity. Recently, attention has been given to inactive ingredients as well as active ingredients. Some of these ingredients are nonchromophoric and cannot be detected by absorbance. Some nonchromophoric ingredients, such as carbohydrates, glycols, sugar, alcohols, amines, and sulfur-containing compounds, can be oxidized and, therefore, can be detected using amperometric detection. This detection method is specific for those analytes that can be oxidized at the selected potential, leaving all other nonoxidizable compounds transparent.[11] Amperometric detection is a powerful detection technique with a broad linear range and very low detection limits.

This review outlines current chromatographic methods utilized in the analysis of carbohydrates in biological systems and pharmaceutical products.

ANALYSIS OF CARBOHYDRATES BY PARTITION HPLC

Partition HPLC is an important type of chromatography for the analysis of monosaccharides and oligosaccharides. Analysis in this mode has the advantages that it requires a shorter analysis time and gives sharper peaks than anion-exchange chromatography of borate complexes, although it has the drawback of low sensitivity, as detection is usually performed by refractometry. Generally, silica gel whose silanol groups are substituted by alkyl or aminoalkyl groups is used as the stationary phase. HPLC separations using such a stationary phase has been applied successfully to separate oligosaccharides liberated from glycoproteins with hydrazine or borohydride in alkali, permitting quick separation within 60 min.[12–14] Previously, such oligosaccharides were separated and purified by tedious procedures involving gel permeation chromatography on Bio-Gel P-2 or P-4, paper chromatography, and paper electrophoresis. However, modified silica, especially amine-modified silica, has difficulties in durability, being unsuitable for routine analysis.

ANALYSIS OF CARBOHYDRATES BY ANION-EXCHANGE HPLC

The introduction in the 1980s by Honda and Suzuki[15] of the anion-exchange resin resulted in a significant improvement in the separation of carbohydrates by HPLC using the partition mode. Honda and Suzuki, using this mode, established analytical conditions common to aldoses, amino sugars, and sialic acids. Aldoses in the intact state, amino acids as their N-acetates, and

Encyclopedia of Chromatography DOI: 10.1081/E-ECHR-120039908

sialic acids as *N*-acylmannosamines were separated on a column of a proton-formed, sulfonated styrene–divinylbenzene copolymer and detected by measuring absorption at 280 nm after postcolumn labeling with 2-cyanoacetamide.

Postcolumn labeling is a characteristic feature of carbohydrate analysis in which no direct physical methods are available for sensitive detection. Many labeling methods have hitherto been developed. The methods with phenol in sulfuric acid,[16] orcinol in sulfuric acid,[17] anthrone in sulfuric acid,[18] tetrazolium blue in alkali,[19] copper(II)-2-2'-bicinchonitate,[20] and 2-cyanoacetamide[21] are used for photometric detection. The methods with 2-cyanoacetamide,[6] ethylenediamine,[22] ethanolamine,[23] taurine,[24] and arginine[25] are used for fluorimetric detection. Some labeling methods for electrochemical detection were reported by Honda and Suzuki in 1984.[26–27]

Quantification of Carbohydrates by Anion-Exchange HPLC and Amperometric Detection

Two main columns are used in the analysis of carbohydrates by amperometric detection: the CarboPac™ PA10 and the CarboPac MA1 anion-exchange columns. The CarboPac PA10 column packing consists of a nonporous, highly cross-linked polystyrene–divinylbenzene substrate agglomerated with 460-nm-diameter latex. The MicroBead™ latex is functionalized with quaternary ammonium ions, which create a thin surface-rich anion-exchange site. The packing is specifically designed to have a high selectivity for carbohydrates. The PA10 has an anion-exchange capacity of approximately 100 μEq/column.

The CarboPac MA1 resin is composed of a polystyrene–divinylbenzene polymeric core. The surface is grafted with quaternary ammonium anion-exchange functional groups. Its macroporous structure provides an extremely high anion-exchange capacity of 1450 μEq/column. The CarboPac MA1 column is designed specifically for sugar alcohol and glycol separations. The PA10 but not the MA1 is compatible with eluents containing organic solvents, which can be used to clean these columns.

The equipment used for the analysis of carbohydrates by anion exchange and amperometric detection include a Dionex DX-500 system consisting of a GP40 gradient pump, an ED40 electrochemical detector, a LC30 chromatography oven, and a PeakNet chromatography workstation. A gold electrode is used for both column applications. The flow rate used for the PA10 column is 1.5 mL/min and 0.4 mL/min for the MA1. Injection volumes are typically 10 μL

Table 1 Separation of carbohydrates, alditols, alcohols, and glycols using a CarboPac MA1 column and pulsed amperometry

Analyte	Retention time (min)
2,3-Butanediol	7.4
Ethanol	7.6
Methanol	7.8
Glycerol	9.0
Erythritol	10.1
Rhamnose	13.4
Arabitol	14.2
Sorbitol	16.3
Galactitol	18.0
Mannitol	19.5
Arabinose	21.8
Glucose	23.3
Galactose	27.4
Lactose	29.7
Ribose	32.0
Sucrose	46.5
Raffinose	52.9
Maltose	61.2

and the oven temperature 30°C. Eluent components include water and 200 mM sodium hydroxide for the PA10 column and water and 480 mM sodium hydroxide for the MA1 column. Eluent concentration for the PA10 column starts at 91% water/9% 200 mM sodium hydroxide for up to 11 min, 100% 200 mM sodium hydroxide from 11.1 to 17.6 min, and 91% water/9% 200 mM sodium hydroxide from 17.7 to 40.0 min. Eluent concentration for the MA1 column system starts at 52% water/48% 480 mM sodium hydroxide and is maintained for up to 60 min [28].

Table 1 shows the separation of alcohols (2,3-butanediol, ethanol, methanol), glycols (glycerol), alditols (erythritol, arabitol, sorbitol, galactitol, mannitol), and carbohydrates (rhamnose, arabinose, glucose, galactose, lactose, sucrose, raffinose, maltose) using a CarboPac MA1 column set with 480 mM sodium hydroxide eluent flowing at 0.4 mL/min. The alcohols, sugar alcohols (alditols), glycols, and carbohydrates are well resolved. Maltose elutes at about 60 min.

REFERENCES

1. Jolley, R.L.; Scott, C.D. Preliminary results from high-resolution analyses of ultraviolet- absorbing and carbohydrate constituents in several pathologic body fluids. Clin. Chem. **1970**, *16*, 687.

2. Butts, W.C.; Jolley, R.L. Gas-chromatographic identification of urinary carbohydrates isolated by anion-exchange chromatography. Clin. Chem. **1970**, *16*, 722.

3. Katz, S.; Dinsmore, S.R.; Pitt, W.W., Jr. A small, automated high-resolution analyzer for detemination of carbohydrates in body fluids. Clin. Chem. **1971**, *17*, 731.

4. Scott, C.D.; Chilcote, D.D.; Katz, S.; Pitt, W.W., Jr. Advances in the application of high resolution liquid chromatography to the separation of complex biological mixtures. J. Chromatogr. Sci. **1973**, *11*, 96.

5. Young, D.S.; Epley, J.A.; Goldman, P. Influence of a chemically defined diet on the composition of serum and urine. Clin. Chem. **1971**, *17*, 765.

6. Honda, S.; Takahashi, M.; Kakehi, K.; Ganno, S. Anal. Biochem. **1981**, *112*, 130.

7. Honda, S.; Suzuki, S.; Takahashi, M.; Kakehi, K.; Ganno, S. Automated analysis of uronic acids by high-performance liquid chromatography with photometric and fluorimetric postcolumn labeling using 2-cyanoacetamide. Anal. Biochem. **1983**, *134*, 34.

8. Honda, S.; Takahashi, M.; Shimada, S.; Kakehi, K.; Ganno, S. Automated analysis of alditols by anion-exchange chromatography with photometric and fluorimetric postcolumn derivatization. Anal. Biochem. **1983**, *128*, 429.

9. Honda, S.; Konishi, T.; Suzuki, S.; Takahashi, M.; Kakehi, K.; Ganno, S. Automated analysis of hexosamines by high-performance liquid chromatography with photometric and fluorimetric postcolumn labeling using 2-cyanoacetamide. Anal. Biochem. **1983**, *134*, 483.

10. Honda, S.; Konishi, T.; Suzuki, S.; Kakehi, K.; Ganno, S. Sensitive monitoring of hexosamines in high-performance liquid chromatography by fluorimetric postcolumn labelling using the 2,4-pentanedione-formaldehyde system. J. Chromatogr. **1983**, *281*, 340.

11. Rocklin, R.D. Detection in ion chromatography. J. Chromatogr. **1991**, *546*, 175.

12. Mellis, S.J.; Baenziger, J.U. Separation of neutral oligosaccharides by high-performance liquid chromatography. Anal. Biochem. **1981**, *114*, 276.

13. Dua, V.K.; Bush, C.A. Identification and fractionation of human milk oligosaccharides by proton-nuclear magnetic resonance spectroscopy and reverse-phase high-performance liquid chromatography. Anal. Biochem. **1983**, *133*, 1.

14. Dua, V.K.; Bush, C.A. Resolution of some glycopeptides of hen ovalbumin by reverse-phase high-pressure liquid chromatography. Anal. Biochem. **1984**, *137*, 33.

15. Honda, S.; Suzuki, S. Common conditions for high-performance liquid chromatographic microdetermination of aldoses, hexosamines, and sialic acids in glycoproteins. Anal. Biochem. **1984**, *142*, 167.

16. Simatupang, M.H. Hochdruckflssigkeitschromatographische Trennungen von Kohlenhydraten mit einem colorimetrischen Nachweisverfahren. J. Chromatogr. **1979**, *180*, 177.

17. Smith, D.F.; Zopf, D.A.; Ginsburg, V. Fractionation of sialyl oligosaccharides of human milk by ion-exchange chromatography. Anal. Biochem. **1978**, *85*, 602.

18. Kramer, K.J.; Speirs, R.D.; Childs, C.N. A method for separation of trehalose from insect hemolymph. Anal. Biochem. **1978**, *86*, 692.

19. Mopper, K.; Degens, E.T. A new chromatographic sugar autoanalyzer with a sensitivity of 10-10 moles. Anal. Biochem. **1972**, *45*, 147.

20. Mopper, K.; Gindler, E.M. A new noncorrosive dye reagent for automatic sugar chromatography. Anal. Biochem. **1973**, *56*, 440.

21. Honda, S.; Matsuda, Y.; Takahashi, M.; Kakehi, K.; Ganno, S. Fluorimetric determination of reducing carbohydrates with 2-cyanoacetamide and application to automated analysis of carbohydrates as borate complexes. Anal. Chem. **1980**, *55*, 1079.

22. Mopper, K.; Dawson, R.; Liebezeit, G.; Hansen, H.P. Borate complex ion exchange chromatography with fluorimetric detection for determination of saccharide ethylenediamine. Anal. Chem. **1980**, *52*, 2018.

23. Kato, T.; Kinoshita, T. Fluorometric detection and determination of carbohydrates by high-performance liquid chromatography using ethanolamine. Anal. Biochem. **1980**, *106*, 238.

24. Kato, T.; Kinoshita, T. Chem. Pharm. Bull. **1978**, *26*, 1291.

25. Mikami, H.; Ishida, Y. Post-column fluorometric detection of reducing sugars in high-performance liquid chromatography using arginine. Bunseki Kagaku **1983**, *32*, E207.

26. Rocklin, R.D.; Pohl, C.A. J. Liquid Chromatogr. **1983**, *6*, 1577.

27. Honda, S.; Konishi, T.; Suzuki, S. Electrochemical detection of reducing carbohydrates in high-performance liquid chromatography after post-column derivatization with 2-cyanoacetamide. J. Chromatogr. **1984**, *299*, 245.

Carbohydrates as Affinity Ligands

I. Bataille
D. Muller
Institut Galilee, Université Paris Nord, Villetaneuse, France

INTRODUCTION

Since its conception 30 years ago, affinity chromatography has been a powerful technique to separate or purify biological compounds, but also a method to study the interactions between living systems and molecules of biological or therapeutical interest. Among these molecules, oses, polysaccharides, or more complex molecules such as glycosaminoglycans constitute a family of potential ligands.

DISCUSSION

The majority of applications of affinity chromatography has been, for quite some time, in the field of protein purification. For example, some of the most rewarding applications have been in the area of purification of hormone and drug receptors. Among these drugs, carbohydrate-based structures are well known for their biological activity (e.g., their anticoagulant or antiproliferative properties). Although some of these biological properties are widely used in medical applications, the mechanisms of action at the molecular level is not accurately determined.

An example of the use of carbohydrate-based affinity chromatography is, thus, the separation of proteins that are responsible for the action of bioactive polysaccharides or oses. The strategy consists in the immobilization of these carbohydrates on classical low- or high-pressure affinity phases. We will distinguish two types of ligands: those based on osidic structures and those prepared from glycosamino glycans or polysaccharides.

OSIDIC LIGANDS

All osidic structures that can interact with proteins implicated in biological responses or phenomenons are, a priori, candidates as ligands in affinity chromatography. Here, we give some examples of oses which have been successfully immobilized to make selective chromatographic supports.

Sialic Acid (*N*-Acetylneuraminic Acid)

Purification of immunoglobulins G is of great interest in biological science. Among other separation methods, affinity chromatography has been used in different ways. Affinity supports were first prepared using protein A or protein G, whose affinity constants take different values according to the IgG subclasses (IgG1, 2, 3, or 4). Among interesting carbohydrates, sialic acid is known to specifically interact with IgGs. Indeed, immunoglobulins are able to react to the presence of tumoral cells. The antigens that are suspected to promote this reaction are gangliosids. Beyond a ceramide molecule, some oses are present in gangliosids, among which are included between one and three sialic acids.

The hypothesis has been made that sialic acid may lead to the formation of specific interactions with immunoglobulins G. Some workers have, thus, developed affinity supports bearing sialic acid in order to purify IgGs. Sialic acid can be extracted from swallow nests and coupled on activated coated silica. This support has been found to possess a good specificity for IgGs, in particular for the subclass IgG3.[1]

Supports prepared with sialic acid have also been used in the purification of insulin, as this sugar and *N*-acetylglucosamine have been found to take part in the interaction between insulin and its glycosylated receptor. The affinity between insulin and sialic acid-bearing supports has been found to be rather strong ($K_a \sim 10^9 M^{-1}$) and the system allows the separation of beef and pig insulins,[2] which differ by only two amino acids.

N-Acetylglucosamine

N-Acetylglucosamine also shows a specific interaction with insulin. Affinity chromatography experiments have evidenced results very close to those observed in the case of sialic acid. An interesting result is that the improvement of both affinity and capacity of supports bearing both sugars.[3] These supports are supposed to more accurately mimic the structure of insulin receptor.

Encyclopedia of Chromatography DOI: 10.1081/E-ECHR-120039907

265

Mannose and Derivatives

Mannose has been immobilized in order to separate mannose-binding proteins such as mannose-binding lectin. When grafted onto agarose, it constitutes a purification step of a specific lectin that does not bind to DEAE–cellulose or Affi-gel Blue gel. [4] Another way to obtain mannose-binding lectin is to perform expanded-bed affinity chromatography by immobilizing mannose on a DEAE Streamline support. [5] Affinity agarose supports, grafted with pentamannosyl phosphate, allowed the testing of the functionality (in terms of ligand binding) of truncated and glycosylation-deficient forms of the mannose 6-phosphate receptor from insect cells. [6] Phosphomannan affinity chromatography has been used to purify a human insulinlike growth factor II mannose 6-phosphate receptor. [7]

GLYCOSAMINOGLYCANS AND POLYSACCHARIDES

One of the most used glycosaminoglycans is heparin because of its anticoagulant properties. Other glycosaminoglycans or polysaccharides have also shown such properties; among them, sulfated dextran derivatives and naturally sulfated polysaccharides extracted from algae (fucans) will be discussed further.

Heparin

Anticoagulant properties are due to the formation of a complex between heparin and antithrombin (ATIII); heparin increases ATIII activity, inhibiting thrombin, which is responsible for the formation of the clot. [8] Although this complex is already characterized by a weak affinity, the exact mechanism of association between heparin and antithrombin is not exactly known. A multistep protocol of immobilization of heparin on silica beads permitted high-performance chromatographic phases to be obtained. Thus, it has been possible to evidence a slightly stronger affinity of heparin for antithrombin than for thrombin.

ATIII has been also used as a model protein to test a novel affinity chromatographic system: capillary affinity chromatography. [9] Separation quality has been found equivalent to that observed with classical affinity chromatography, whereas the necessary protein amount is strongly reduced to the nanogram level.

Heparin possesses an affinity for many molecules, among which is a phospholipid-binding protein, annexin V. Affinity chromatography evidenced the Ca^{2+} dependence of the binding mechanism; [10] von Willebrand factors with high and low molecular weights have been separated using their different affinities toward heparin. [11]

Heparin has been used in enzyme purification such as recombinant human mast cell tryptase. The purified enzyme is fully active. [12] Heparin-based affinity chromatography also permitted the isolation of growth factors such as basic fibroblast growth factor (bFGF). The affinity is lower when bFGF is complexed with acidic gelatin. [13] The elution of synthetic TFPI (tissue factor pathway inhibitor) peptidic fragments on immobilized heparin has allowed one to find the peptidic sequence responsible for the TFPI-heparin interaction. [14]

Heparin is able to inhibit smooth-muscle cell (SMC) proliferation in vitro. SMCs are present in blood vessel walls and may proliferate in the case of an internal injury. The antiproliferative action of heparin is due to its internalization in SMCs, which is probably mediated by membrane receptors. Heparin-based affinity chromatography of SMC membrane extracts allowed the separation of a few proteins, which could be implicated in the growth inhibition. [15]

The different actions taking place in the overall affinity mechanism of immobilized heparin for different biological compounds are not yet elucidated, but the influence of ionic strength demonstrates the important contribution of ionic interactions in this mechanism. However, the large spectrum of biological activities of heparin is also a limit for its specificity.

Other Glycosaminoglycans

Dermatan sulfate is known to specifically catalyze thrombin inhibition by the plasmatic inhibitor heparin cofactor II (HCII). Dermatan sulfate has been immobilized on a dextran- or agarose-coated silica matrix. These systems were tested as high-performance chromatographic supports for the purification of HCII from human plasma. The eluted HCII was obtained with no contamination of ATIII, the other main thrombin inhibitor. [16]

DEXTRAN DERIVATIVES

Phosphorylated dextran derivatives, called phosphodextrans, possess a strong affinity for K-vitamin-dependent coagulation factors, like factor II or prothrombin. This property was used to separate them by affinity chromatography on phosphodextrans, which interact in a similar way as phospholipids from the cell membrane. [17]

Heparinlike sulfated dextran derivatives, like carboxymethyldextran benzylamide sulfonates (CMDBS), have been immobilized on silica beads. By high-performance liquid affinity chromatography (HPLAC), they allow a good recovery of thrombin, with a yield of 80%. The affinity constant was estimated ($K_a \sim 10^5 M^{-1}$) and was found superior to the value obtained between thrombin and heparin. On the contrary, the affinity of dextran derivatives for ATIII is estimated at a lower value than that of heparin.[17]

FUCANS

Fucan is a sulfated polysaccharide, naturally present in algae such as *Fucus vesiculosus* or *Ascophyllum nodosum*. Fucan is a general name for a mixture of three polysaccharides; among them, fucoidan (or homofucan) can be theoretically considered as an homopolymer of α-1,2 L-fucose-4-sulfate and has been studied as a ligand in the same way as fucan itself. Their interaction with two proteins implicated in the coagulation process (thrombin and antithrombin) has been studied and is at least partially ionic. However, the dissociation of the complex fucan–antithrombin seems to include a slower step which could be attributed to a conformation change of the fucan.[18]

Fucan was also used as ligand for high-performance liquid affinity chromatography. In the same way as dextran derivatives, a good separation was obtained for thrombin, with a yield of 80%. The affinity constant was estimated in the same order as that obtained for CMDBS ($K_a \sim 10^5 M^{-1}$) and superior to the value obtained for heparin.[17]

CONCLUSIONS

In this article, we have described several uses of carbohydrate compounds as affinity ligands. These few examples clearly demonstrate the importance of such affinity supports in the separation of biological products. Affinity chromatography is also able to help in the determination of the interaction mechanisms of carbohydrate derivatives in biological reactions. This developing field of research will lead to improved quality and specificity of affinity-chromatographic phases.

REFERENCES

1. Serres, A.; Legendre, E.; Jozefonvicz, J.; Muller, D. Affinity of mouse immunoglobulin G subclasses for sialic acid derivatives immobilized on dextran-coated supports. J. Chromatogr. B. **1996**, *681*, 219.

2. Lakhiari, H.; Jozefonvicz, J.; Muller, D. Separation and purification of insulins on coated silica support functionalized with sialic acid by affinity chromatography. J. Liquid Chromatogr. & Related Technol. **1996**, *19*, 2423.

3. Lakhiari, H. Supports de silice enrobée à ligands biospécifiques: Synthèse, caractérisation et relations structure-propriétés de séparation. Application à la purification de l'insuline et des immunoglobulines G. In *Thesis*; University Paris, 1996; 13.

4. Ooi, L.S.M.; Ang, H.X.W.; Ng, T.B.; Ooi, V.E.C. Isolation and characterization of a mannose-binding lectin from leaves of the chinese daffodil Narcissus tazetta. Biochem. Cell Biol. **1998**, *76*, 601.

5. Bertrand, O.; Cochet, S.; Cartron, J.P. Expanded bed chromatography for one-step purification of mannose binding lectin from tulip bulbs using mannose immobilized on DEAE Streamline. J. Chromatogr. A. **1998**, *822*, 19.

6. Marron-Terada, P.G.; Bollinger, K.E.; Dahms, N.M. Characterization of truncated and glycosylation deficient forms of the cation-dependent mannose 6-phosphate receptor expressed in baculovirus-infected insect cells. Biochemistry **1998**, *37*, 17,223.

7. Costello, M.; Baxter, R.C.; Scott, C.D. Regulation of soluble insulin-like growth factor II mannose 6-phosphate receptor in human serum: Measurement by enzyme-linked immunosorbent assay. J. Clin. Endocrinol. Metab. **1999**, *84*, 611.

8. Björk, I.; Olson, S.T.; Shore, J.D. Molecular mechanisms of the accelerating effect of heparin on the reactions between antithrombin and clotting proteinases. In *Heparin, Chemical and Biological Properties, Clinical Applications*; Lane, D.A., Lindahl, U., Eds.; Edward Arnold: London, 1989; 229–255.

9. Wu, X.J.; Linhardt, R.J. Capillary affinity chromatography and affinity capillary electrophoresis of heparin binding proteins. Electrophoresis **1998**, *19*, 2650.

10. Capila, I.; Van der Noot, V.A.; Mealy, T.R.; Seaton, B.A.; et al. Interaction of heparin with annexin V. FEBS Lett. **1999**, *446*, 327.

11. Fischer, B.E.; Thomas, K.B.; Schlokat, U.; Dorner, F. Selectivity of von Willebrand factor triplet bands towards heparin binding supports structural model. Eur. J. Haematol. **1999**, *62*, 169.

12. Niles, A.L.; Maffit, M.; Haak-Frendscho, M.; Wheeless, C.J.; et al. Recombinant mast cell tryptase: Stable expression in Pichia pastoris and purification of fully active enzyme. Biotechnol. Appl. Biochem. **1998**, *28*, 125.

C

13. Muniruzzaman, M.; Tabata, Y.; Ikada, Y. Complexation of basic fibroblast growth factor with gelatin. J. Biomater. Sci. Polym. Ed. **1998**, *9*, 459.

14. Ye, Z.Y.; Takano, R.; Hayashi, K.; Ta, T.V.; et al. Structural requirements of human tissue factor pathway inhibitor (TFPI) and heparin for TFPI-heparin interaction. Throm. Res. **1998**, *89*, 263.

15. Clairbois, A.S.; Letourneur, D.; Muller, D.; Jozefonvicz, J. High-performance affinity chromatography for the purification of heparin-binding proteins from detergent-solubilized smooth muscle cell membranes. J. Chromatogr. B. **1998**, *706*, 55.

16. Sinninger, V.; Tapon-Bretaudière, J.; Zhou, F.L.; Bros, A.; et al. Immobilization of dermatan sulphate on a silica matrix and its possible use as an affinity chromatography support for heparin cofactor II purification. J. Chromatogr. **1991**, *539*, 289.

17. Zhou, F.L. Supports à base de silice enrobée par des polysaccharides pour chromatographie d'affinité haute performance: Préparation, caractérisation et application dans la purification de protéines. In *Thesis*; University Paris, 1990; 13.

18. Legendre, E. Etude par chromatographie d'affinité liquide haute performance des interactions entre des protéines de la coagulation et des polysaccharides sulfatés à activité anticoagulante immobilisés sur des supports de silice enrobée. In *Thesis*; University Paris, 1996; 13.

Carbohydrates: Derivatization for GC Analysis

Raymond P. W. Scott
Scientific Detectors Ltd., Banbury, Oxfordshire, U.K.

INTRODUCTION

As a result of the development of special bonded phases, carbohydrates or their derivatives are usually separated by LC. However, certain carbohydrate samples are still analyzed by GC due to the inherent high efficiencies obtainable from the technique and to the associated short elution times. In addition, GC/MS is a particularly powerful analytical technique for carbohydrates, especially for their identification. As a consequence, appropriate derivatives must be formed to render them sufficiently volatile but still easily recognizable from their mass spectra.

DISCUSSION

It is relatively easy to form the trimethylsilyl derivatives, employing standard silyl reagents such as trimethylchlorsilane or hexamethyltrisilazane, and the reactions normally can be made to proceed to completion. However, there is a major problem associated with the derivatization of natural sugars which arises from the formation of anomers and the pyranose–furanose interconversion. Reducing sugars in solution (e.g., glucose) exist as an equilibrium mixture of anomers. Consequently, each sugar produces five tautomeric forms—two pyranose, two furanose, and one open-chain form. In general, all the anomers can be separated by GC. This autoconversion can be minimized by mild and rapid derivatization. Equilibrium mixtures are to be expected from reducing sugars isolated from natural products.

Mixtures of hexamethyltrisilazane and trimethylchlorsilane are frequently used to derivatize sugars. Pure sugars (e.g., glucose, mannose, and xylose) can be readily derivatized using a mixture of hexamethyltrisilazane : trimethylchlorsilane : pyridine (2 : 1 : 10 v/v/v), giving single GC peaks; however, if the proportion of trimethylchlorsilane is doubled, small amounts of the anomeric forins are observed. One disadvantage of this procedure is the formation of an ammonium chloride precipitate, which, if injected directly onto the column, can cause column contamination and eventually column blockage. The formation of ammonium chloride can be avoided by extracting the derivative into hexane or by the use of an alternative derivatizing agent.

N,O-Bistrimethylsilyltrifluoroacetamide, usually combined with trimethylchlorosilane (10 : 1, v/v) is also a popular derivatizing agent for carbohydrates and the reaction mixture can be injected directly onto the column without fear of column contamination. The formation of anomers can also be avoided by preparing the alditols by treating with sodium borohydride. The alditols can then be separated after derivatizing with trimethylchlorosilane, a procedure that is considered preferable to the preparation of their acetates. The rapid preparation of trimethylsilylalidtols using trimethylsilylimidazole in pyridine mixtures at room temperature has the advantage over other methods, which require longer reaction times and higher temperatures. The use of silylaldolnitrile derivatives has been reported for the separation of aldoses. The sugars are reacted with hydroxylamine-O-sulfonic acid to form aldonitriles which are then silanated with N,O-bistrimethylsilyltrifluoroacetamide : pyridine (1 : 1 v/v). The silylaldonitrile derivatives are readily separated on open tubular columns.

Another silanization procedure for the derivatization of carbohydrates is the formation of trimethylsilyl oximes. The methyloxime is heated with hexarnethyldisilazane : trifluoroacetic anhydride (9 : 1 v/v) for 1 h at 100°C. Anthrone O-glucoside is an important ingredient in skin care cosmetics and can be fully silylated by reaction with N,O-bistrimethylsilylacetamide : acetonitrile mixture (1 : 1 v/v) for 1 h at 90°C, and subsequently separated by GC.

CONCLUSIONS

Acetylation and the use of trifluoroacetates, originally the more popular derivatives for the separation of carbohydrates by GC, are still used on occasion, but the various silanization methods are,

Encyclopedia of Chromatography DOI: 10.1081/E-ECHR-120039946

today, the most common and considered the most effective for GC carbohydrate analysis.

SUGGESTED FURTHER READING

Blau, K., Halket, J., Eds.; *Handbook of Derivatives for Chromatography*; John Wiley and Sons: New York, 1993.

Grant, D.W. *Capillary Gas Chromatography*; John Wiley and Sons: New York, 1996.

Scott, R.P.W. *Introduction to Analytical Gas Chromatography*; Marcel Dekker, Inc.: New York, 1998.

Scott, R.P.W. *Techniques of Chromatography*; Marcel Dekker, Inc.: New York, 1995.

Carbonyls: Derivatization for GC Analysis

Igor G. Zenkevich
Chemical Research Institute, St. Petersburg State University, St. Petersburg, Russia

INTRODUCTION

The carbonyl group in aldehydes (RCHO) and ketones (RCOR′) is one of the frequently encountered functional groups in the composition of organic compounds. This group has no active hydrogen atoms, excluding the cases of high content of enols for β-dicarbonyl compounds (β-diketones, esters of β-ketocarboxylic acids, etc.):

Hence, the derivatization at least of monofunctional carbonyl compounds is not an obligatory stage of sample preparation before their GC analysis. Nevertheless, the objective reasons for their derivatization are the following:

- The simplest aldehydes and ketones are the slightly polar low boiling substances with small retention indices (RIs) both on standard nonpolar and polar stationary phases, for example:

	$RI_{nonpolar}$	RI_{polar}
Aldehyde		
Ethanal	369 ± 7	701 ± 14
Propanal	479 ± 9	794 ± 9
Ketone		
Acetone	472 ± 12	820 ± 11
2-Butanone	578 ± 12	907 ± 14

All RIs with standard deviations are randomized interlaboratory data.

From these data, it is easy to conclude that RIs of both aldehydes and ketones exceed these parameters for *n*-alkanes with the same number of carbon atoms only at 176 ± 5 Retention index units (IU) on standard nonpolar phases. This small ΔRI value indicates that these compounds belong to the group of only slightly polar chemicals. At the same time, the average differences of RIs of carbonyl compounds on standard polar polyethylene glycols and nonpolar polydimethyl siloxanes is 293 ± 25 (aldehydes) and 282 ± 39 (ketones), which is only slightly more than, for example, the corresponding ΔRI value for alkyl substituted arenes (246 ± 30) (they are not classified as polar compounds).[1]

However, in the numerous GC analytical procedures, the first part of chromatograms very often can be overloaded by intense poorly resolved peaks of auxiliary compounds (solvents, by-products, etc.). The optimization of the determination of target carbonyl compounds usually needs the "replacement" of their analytical signals into less "populated" areas of chromatograms (the derivatives with retention parameters that exceed those for initial analytes).

- The simplest monofunctional compounds with active hydrogens in the functional groups –OH, NH_2, and so forth typically are volatile enough for their GC determinations. However, the presence in the molecules of any additional polar fragments (including C=O) makes the possibility of GC analysis of these compounds worse, and they usually need derivatization of one or (better) of both polar functional groups. Thus, in many cases, the necessity of derivatization of carbonyl fragment in polyfunctional compounds is a secondary procedure, which is caused by the presence of other functional groups.

- Both aliphatic and aromatic aldehydes are easily oxidized compounds (even by atmospheric oxygen). Hence, one of the aims of their derivatization is to prevent the oxidation of analytes with resultant formation of carboxylic acids.

METHODS OF DERIVATIZATION OF CARBONYL COMPOUNDS

Some methods of carbonyl compound derivatization have been known in "classical" organic chemistry since the 19th century. Their predetermination was simply the identification by comparison of melting points of purified solid derivatives with reference data. These derivatives are, for example, semicarbazones $RR′C=NNHCONH_2$, thiosemicarbazones $RR′C=NNHCSNH_2$, 2,4-dinitrophenyl hydrazones $RR′C=NNHC_6H_3(NO_2)_2$, and so forth. However, most of these derivatives with structural fragments =N–NH–C(=X)–NH_2 are not volatile for their direct GC analysis owing to the presence of three active hydrogen atoms in these fragments. The convenient derivatives for GC determinations should include not more then one of these atoms (monosubstituted

Encyclopedia of Chromatography DOI: 10.1081/E-ECHR-120039947

$$RR'CO + H_2NNHAr \rightarrow RR'C=NNHAr$$

$$RR'CO + H_2N-OR'' \rightarrow RR'C=N-OR''$$

$$RR'CO + H_2NOH \rightarrow RR'C=NOH \rightarrow RR'C=N-OSi(CH_3)_3$$

$$RR'CO + HXCH_2CH_2YH/(H^+)(X, Y=O, NH, S)-H_2O \rightarrow$$

$$RR'CH-CO-R'' + XSi(CH_3)_3 \rightarrow RR'C=CR''-OSi(CH_3)_3 + XH \qquad (1)$$

aryl hydrazones) or they need to be free of them (O-alkyl or O-benzyl ethers of oximes).[2–6] The oximes themselves can be synthesized by the reaction of carbonyl compounds with hydroxylamine, but they are usually used in GC practice in the form of O-trimethylsilyl (TMS) ethers.

Other types of carbonyl derivatives are acetals and/or ketals, including their thio- and aza-analogs, preferably cyclic 1,3-dioxolanes, 1,3-oxathiolanes, or thiazolidines. The last method of derivatization with 2-aminoethanethiol has been proposed for sampling and GC determination of volatile aliphatic aldehydes.[7] Some carbonyl compounds (especially steroids) can be analyzed as their enol-TMS derivatives [see (1) above].

All these processes (excluding the last one) can be classified as reactions of condensation. Hence, in all cases, the target derivatives are theoretically the sole components and no other volatile by-products excluding water (which has no influence on the results of GC analysis) have been formed. However, some features of these reactions should be considered:

• Both O-alkyl hydroxylamines and, especially, aryl hydrazines in the form of free bases are slightly oxidized compounds. Their salts with inorganic acids are more stable, but even in this case, the presence of any oxidizers in the reaction mixtures must be prevented. Nevertheless, in real practice, these mixtures very often contain some oxidation by-products (e.g., $ArNH_2$, $ArOH$, ArH, etc.). Usually, there are no problems in revealing their chromatographic peaks, because all of them have lower retention parameters than that for initial reagents and, moreover, all target derivatives.

• The condensation reaction of considered type can be characterized by statistically processed differences of RIs of products and initial substrates.[8] This additive scheme mode permits us to estimate these analytical parameters for any new derivatives on standard nonpolar polydimethyl siloxanes. For the simplest reaction scheme A + L → B + M, we have the following $\Delta RI = RI(B) - RI(A)$ values [$\Delta MW = MW(B) - MW(A)$]:

Scheme of reaction	ΔMW	<ΔRI>, IU
$RR'CO \rightarrow RR'=NNHCH_3$	28	229 ± 23
$ArRCO \rightarrow ArR=NNHCH_3$	28	320 ± 30
$RR'CO \rightarrow RR'C=NN(CH_3)_2$	42	302 ± 20
$RR'CO \rightarrow RR'C=NNHC_2H_5$	42	345 ± 16
$RR'CO \rightarrow RR'C=NNHC_6H_5$	90	858 ± 22
$ArCHO \rightarrow ArCH=NNHC_6H_5$	90	996 ± 50
$RR'CO \rightarrow RR'C(OCH_3)_2$	46	189 ± 17
$ROH \rightarrow ROSi(CH_3)_3$ (for comparison)	72	119 ± 18

This set of ΔRI values illustrates that the simplest alkyl hydrazones (methyl, ethyl, dimethyl, etc.) have appropriate GC retention parameters and theoretically can be recommended as the derivatives of carbonyl compounds. But in real analytical practice, these hydrazones are not used because of their low yields for the aliphatic ketones.

• All considered condensation reactions have some anomalies for the α,β-unsaturated carbonyl compounds. Most reagents with active hydrogen atoms can react not only with carbonyl groups, but with polarized conjugated double bonds C=C also. As a result of this regularity, three products instead of the estimated one target derivative are formed in the reactions of α,β-unsaturated carbonyl compounds with O-alkyl hydroxylamines [see (2) below].[9]

The relative ratio of these products depends on the excess of derivatization reagent and pH of the reaction mixtures. The chemical origin of products of the reaction of phenyl hydrazine with unsaturated carbonyl

$$>C=CR-COR + H_2NOR'' \rightarrow >C=CR-CR'=NOR''$$

$$\downarrow \qquad\qquad\qquad \downarrow$$

$$R''ONHC(<)-CHR-COR' \rightarrow R''ONHC(<)-CHR-CR'=NOR''(I) \qquad (2)$$

compounds depends on the same factors and, moreover, on the order of component mixing (it exerts influence on the current pH of the reaction media). 2-Pyrazolines have resulted in some cases as the sole reaction products instead of hydrazones.

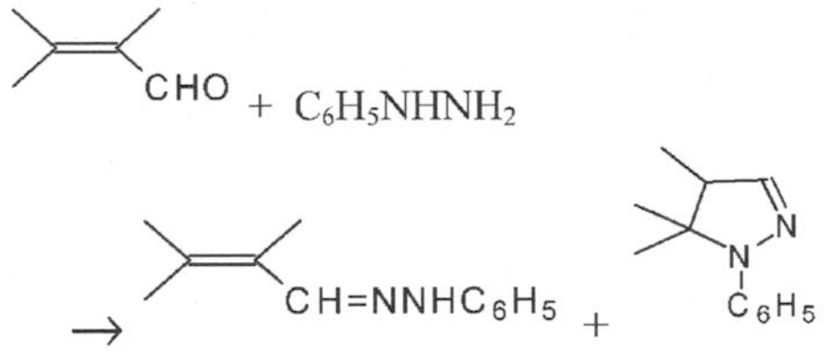

The existence of different products means that the quantitative yields of each of them are not enough, as is necessary for GC analysis of derivatives. It is quite probable that the same anomalies can take place in the reactions of unsaturated carbonyls with other reagents. For example, 2,4-dinitrophenyl hydrazones of unsaturated carbonyl compounds seem unstable at the high temperatures of injectors and GC columns. This fact explains the small number of published RI data for the derivatives of these compounds and the necessity to search for new types of derivatization reactions for them.

● Both hydrazones and O-alkyl oximes of asymmetrical carbonyl compounds exist in two isomeric structures with slightly different GC retention parameters:

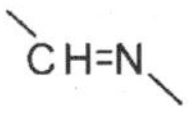

Hence, most of these derivatives give two analytical signals (anti-isomers typically are more stable and their peaks prevail over those of the syn-forms). If the assignment of these two peaks to the isomeric compounds is not obvious, in accordance with the usual chromatographic practice they can be marked by symbols #1 and #2 (in order of chromatographic elution). The same duplication of analytical signals takes place, for example, during RP HPLC analysis of 2,4-dinitrophenylhydrazones (2,4-DNPHs).

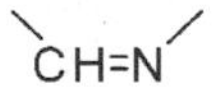

The condensation reaction of carbonyl compounds with halogen-containing derivatization chemicals (electrophortic reagents like pentafluorophenyl and 2,4,6-trichlorophenyl hydrazines) that has been considered can provide the possibility of their selective GC determination with electron capture detectors.[10] For GC analysis of monosaccharides (an important group of hydroxy carbonyl compounds), many combinations of derivatization reagents have been proposed, but most of them are based on the conversion of carbonyl groups into hydroximino ($>C=NOH$ with following silylation) and alkoxy(methoxy)imino ($>C=N–OR$).[11] An electrophortic reagent of the series of O-substituted hydroxylamines is O-(pentafluorobenzyl) hydroxylamine.[12]

For the polyfunctional organic compounds, the condensation reactions of carbonyls considered should be combined with different derivatization of other functional groups. So, the "standard" method of the derivatization of hydroxyketosteroids for GC analysis is their two-stage treatment by O-methyl hydroxylamine ($>C=O → >C=NOCH_3$) followed by silylation of OH– groups. The resulting trimethylsilylated methyl oxime derivatives (MO-TMS) of compounds of this class are well characterized by standard mass spectra and GC RIs on standard nonpolar phases for their identification. The use of most active silylating reagents permits us to exclude the stage of O-methyl oxime formation, so far as ketosteroids can form the enol-TMS ethers. For example, androstenedione (II) gives the bis-O-TMS derivative of bis-enol.[13] The enolization of the carbonyl group in the third position in similar structures can lead to the formation on a conjugate system of C=C double bonds not only in positions 3–5, but also in positions 2–4. These isomeric TMS-enol derivatives have different retention parameters [see (3) below].[14]

Because of the presence of p–π–(d) conjugated systems C=C–O–(Si), these derivatives indicate intense signals of the molecular ions in their mass spectra.

The formation of acetals from carbonyl compounds requires acid catalysis and (sometimes) the presence of water coupling reagents (for instance, anhydrous $CuSO_4$). The conversion of aliphatic aldehydes into dimethyl acetals slightly increases the retention parameters of analytes ($\Delta RI = 189 \pm 17$). Cyclic ethylene derivatives (1,3-dioxolanes, $\Delta RI = 212 \pm 7$; this value is valid only for acyclic carbonyl compounds) are more resistant to hydrolysis and used in GC practice preferably. Their important advantage for GC/MS

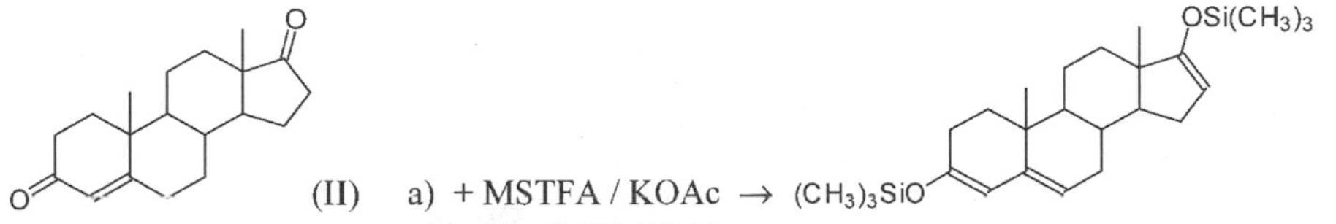

(II)   a) + MSTFA / KOAc →

        or b) + Me₃SiCH₂CO₂Et

(3)

$$+ \ H_2NNH\text{-}C^*H(CF_3)C_6H_5 \ (III) \ \rightarrow \quad \quad NNHC^*H(CF_3)C_6H_5$$

$$+ \ HO \quad OH \ (IV) \ \rightarrow \quad \quad \quad \quad \quad \quad \quad \quad \quad \quad \quad \quad \quad \quad (4)$$

analysis is the very specific fragmentation of molecular ions with the loss of substituents R, R′ in the second position of the cycle and the formation of daughter ions [M–R]$^+$ and [M–R′]$^+$, which gives useful information for the determination of the initial carbonyl compound structure. The use of 2-aminoethanethiol in this reaction instead of ethylene glycol leads to the oxathiolanes, which indicate intense signals of molecular ions.

The GC separation of chiral carbonyl compounds (at first natural terpenoids, for example camphor, menthone, carvone, etc.) on the nonchiral phases can be carried out in accordance with the same principles as those for enantiomers of other classes. Their conversion to diastereomers by reaction with chiral derivatization reagents is necessary. Examples of them are α-trifluoromethylbenzyl hydrazine (III) and optically active 2,3-butanediol (IV) [see (4) above].

In connection with the problem of carbonyl compound derivatization, it is expedient to touch upon a question of the application of new proposed reagents, which are not yet widely used in GC analytical practice, but seem to be promising ones in accordance with various criteria. One of them is low boiling trimethyltrifluoromethyl silane, $(CH_3)_3SiCF_3$, which converts the carbonyl groups into α-trifluoromethyl O-TMS carbinol fragments:

$$\begin{array}{c} R \\ R' \end{array}\!\!=\!O \ + \ (CH_3)_3SiCF_3 \ \rightarrow \ \begin{array}{c} R \\ R' \end{array}\!\!\!\!\begin{array}{c} CF_3 \\ OSi(CH_3)_3 \end{array}$$

The carbonyl groups are present in numerous derivatives of carboxylic acids, namely amides, imides, etc. There are no derivatization methods for these compounds at present (excluding reduction) that are based on the reactions of carbonyl group. Nevertheless, the amides react with $(CH_3)_3SiCF_3$ in a manner similar to other carbonyl compounds. The hydrolysis of intermediate O-TMS derivatives followed by dehydration leads to the low boiling α-trifluoromethyl enamines $RR'N\text{-}C(CF_3)\!=\!CR''R'''$. These structures are in mutually unambiguous accordance with the structures of initial amides and, hence, can be considered as new types of derivatives [see (5) below].

The RIs of members of this new class of derivatives are unknown at present (only the boiling points have been determined). Nevertheless, this information is enough for the estimation of RI values on the standard nonpolar phases with the following nonlinear equation:[15,16]

$$\log RI \ = \ a \log T_b \ + \ bMR_D \ + \ c,$$

$$(C_2H_5)_2NCOCH_3 + (CH_3)_3SiCF_3 \ \rightarrow \quad \begin{array}{c} (C_2H_5)_2N \quad CF_3 \\ CH_3 \quad OSi(CH_3)_3 \end{array} \ \rightarrow \ (C_2H_5)_2N\text{-}C(CF_3)\!=\!CH_2$$

T_b 185 °C $(RI_{nonpolar} \ 966 \pm 13)$ T_b 113.3 °C $(RI_{estd} \ 754 \pm 6)$

          + $(CH_3)_3SiCF_3$ $\rightarrow$ $\rightarrow$

no T_b and RI data T_b 152.1 °C $(RI_{estd} \ 897 \pm 9)$ (5)

Table 1 Physicochemical and GC properties of some reagents for derivatization of carbonyl compounds

Reagent (abbreviation)	MW	T_b (m.p.,°C)	$RI_{nonpolar}$
Phenyl hydrazine	108	244	1157 ± 11
Pentafluorophenyl hydrazine (PFPH)	198	(74–75)	1164 ± 16^a
2,4,6-Trichlorophenyl hydrazine (TCPH)	210	(141–143)	1654^b
2,4-Dinitrophenyl hydrazine	198	(200.5–201.5)	—
N-Aminopiperidine	100	146	859 ± 14
O-Methyl hydroxylamine (HCl)	47	47 (148)	c
O-Ethyl hydroxylamine (HCl)	61	68 (130–133)	c
O-Pentafluorobenzyl hydroxylamine (HCl) (PFBHA)	213	227 (subl.)	1068 ± 12^a
Hydroxylamine (HCl) (with following silylation of oximes)	33	(151)	c
Methanol/(H$^+$)	32	64.6	381 ± 15
Ethylene glycol/(H$^+$), CuSO$_4$	62	197.8	726 ± 28
Trimethyltrifluoromethyl silane	142	45	532 ± 3^a
Silylating reagents (for synthesis of TMS-ethers of enols)	See table in *Hydroxy Compounds: Derivatization for GC Analysis*		

^{a}Estimated RI values.
^{b}Single experimental data.
^{c}Nonvolatile salts.

where T_b is the boiling point, MR$_D$ the molar refraction (which can be estimated by additive schemes), and the coefficients of this equation are calculated by the least squares method with the data for compounds with known RI, T_b, and MR$_D$.

The results of calculations indicate that the RIs of α-trifluoromethyl enamines are lower than the RIs of initial amides on ca. 200 IU, which is a very important fact in the choice of an appropriate derivatization method for the more complex nonvolatile compounds of this class. Principal physicochemical and gas chromatographic properties of some reagents for derivatization of carbonyl compounds are summarized in Table 1.

CONCLUSIONS

Monofunctional carbonyl compounds belong to the group of slightly polar organic substances and typically needs no derivatization prior to GC analysis. However, the carbonyl functional group is one of the most encountered fragments in the composition of organic compounds. In accordance with the general principles of derivatization, all of the various groups present in molecules should be transformed into less polar fragments having no active hydrogen atoms. Following this concept, numerous methods have been proposed for the derivatization at first of polyfunctional carbonyl compounds.

Some special methods of derivatization of simplest low boiling aldehydes and ketones are used for the "displacement" of their analytical signals into less "populated" parts of chromatograms. This method increases the resulting selectivity of determination of target analytes.

ARTICLES OF FURTHER INTEREST

Derivatization of Analytes in Chromatography: General Aspects, p. 427.
Hydroxy Compounds: Derivatization for GC Analysis, p. 809.

REFERENCES

1. Zenkevich, I.G. Chemometrics characterization of differences of GC retention indices on standard polar and non-polar phases as the criterion of group identification of organic compounds. Zh. Anal. Khim. (Russ.) **2003**, *58* (2), 119–129.
2. *Handbook of Derivatives for Chromatography*; Blau, K., King, G.S., Eds.; Heiden: London, 1977; 576.
3. Knapp, D.R. *Handbook of Analytical Derivatization Reactions*; John Wiley & Sons: New York, 1979; 741.

4. Drozd, J. *Chemical Derivatization in Gas Chromatography*; Journal of Chromatography Library; Elsevier: Amsterdam, 1981; Vol. 19, 232.

5. *Handbook of Derivatives for Chromatography*, 2nd Ed.; Blau, K., Halket, J.M., Eds.; John Wiley & Sons: New York, 1993; 369.

6. Levine, S.R.; Harvey, T.M.; Waeghe, T.J.; Shapiro, R.H. *O*-Alkyloxime derivatives for GC–MS and GC determination of aldehydes. Anal. Chem. **1981**, *53* (6), 805–809.

7. Yasuhara, A.; Shibamoto, T. Determination of volatile aldehydes in the headspace of heated food oils by derivatization with 2-aminoethanethiol. J. Chromatogr. **1991**, *547*, 291–298.

8. Zenkevich, I.G. Chromatographic characterization of organic reactions by additivity of GC retention parameters of reagents and products. Zh. Org. Khim. (Russ.) **1992**, *29* (9), 1827–1840.

9. Zenkevich, I.G.; Artsybasheva, Ju.P.; Ioffe, B.V. Application of alkoxyamines for derivatization of carbonyl compounds in gas chromatography–mass spectrometry. Zh. Org. Khim. (Russ.) **1989**, *25* (3), 487–492.

10. Lehmpuhl, D.W.; Birks, J.W. New gas chromatographic electron capture detection method for the determination of atmospheric aldehydes and ketones based on cartridge sampling and derivatization with 2,4,6-trichlorophenyl hydrazine. J. Chromatogr. A. **1996**, *740*, 71–81.

11. Robards, K.; Whitelaw, M. Chromatography of monosaccharides and disaccharides. J. Chromatogr. **1986**, *373*, 81–110.

12. Biondi, P.A.; Mauca, F.; Negri, A.; Secchi, C.; Montana, N. Gas chromatographic analysis of neutral monosaccharides as their *O*-pentafluorobenzyloxime acetates. J. Chromatogr. **1987**, *411*, 275–284.

13. Gleispch, H. The use of different silylating agents for structure analyses of steroids. J. Chromatogr. **1974**, *91*, 407–412.

14. Iida, T.; Hikosaka, M.; Goto, J.; Nambara, T. Capillary gas chromatographic behaviour of *tert.*-hydroxylated steroids by trialkylsilylation. J. Chromatogr. A. **2001**, *937*, 97–105.

15. Zenkevich, I.G.; Kuznetsova, L.M. A new approach to the prediction of GC retention indices from physico-chemical constants. Collect. Czech Chem. Commun. **1991**, *56* (10), 2042–2054.

16. Zenkevich, I.G. Reciprocally unambiguous conformity between GC retention indices and boiling points within two- and multidimensional taxonomic groups of organic compounds. J. High Resolut. Chromatogr. **1998**, *21* (10), 565–568.

Catalyst Characterization by Reversed-Flow Gas Chromatography

Dimitrios Gavril
Physical Chemistry Laboratory, Department of Chemistry, University of Patras, Patras, Greece

INTRODUCTION

Reversed-flow gas chromatography (RF-GC) has been used to study the kinetics of surface-catalyzed reactions and the nature of the active sites. Reversed-flow gas chromatography is technically very simple, and it is combined with a mathematical analysis that gives the possibility for the estimation of various physicochemical parameters related to catalyst characterization, in a simple experiment, under conditions compatible with the operation of real catalysts. The experimental findings of RF-GC for the oxidation of CO over well-studied, silica-supported, platinum–rhodium bimetallic catalysts are in agreement with the results of other workers using different techniques, ascertaining that RF-GC methodologies can be used for the characterization of various solids with simplicity and accuracy.

OVERVIEW

During the last four decades, various chemical reactions concerning processes of technological and environmental interest have been related to the development of catalysts. Catalyst characterization is a necessary step, and it usually involves activity tests and investigation of the kinetics of the related reactions, as well as of the nature of the active sites.

Diffusion, adsorption, and surface reaction are closely interconnected in heterogeneous catalysis characterization studies. Chromatographic separation is a physicochemical process based also on diffusion, adsorption, as well as liquid dissolution. Based on the broadening factors embraced by the van Deemter equation, precise and accurate physicochemical measurements have been made by GC, using relatively low-cost instrumentation and a very simple experimental setup.

Reversed-flow gas chromatography is another gas chromatographic technique based on the perturbation of the carrier gas flow, which has been utilized for the measurement of physicochemical parameters. The fundamental difference of RF-GC from classical GC is the use of a T-form system of chromatographic columns (sampling and diffusion columns) placed perpendicularly, one in the middle of the other. The carrier gas flows continuously through the sampling column, while it is stagnant into the diffusion column. The solid catalyst under study is placed either near the injection point at the closed end of the diffusion column, for experiments under non-steady-state conditions, or at the middle of the sampling column, for experiments under steady-state conditions, as shown in Fig. 1. The fact that the solid material is under investigation also classifies RF-GC in inverse gas chromatographic (IGC) methodologies. The reversing of the carrier gas flow for short time intervals results in extra chromatographic peaks on the continuous concentration–time curve. Thus repeated sampling of the physicochemical phenomena occurring in the diffusion column is achieved, and, by using the appropriate mathematical analysis, the values of the relevant physicochemical quantities are determined.

EXPERIMENTAL APPROACH

The experimental setup of RF-GC for the study of catalytic processes has been presented elsewhere;[1–13] it is very simple. It comprises the following:

1. A conventional gas chromatograph equipped with the appropriate detector (e.g., flame ionization, thermal conductivity, etc.) depending on the reactant(s) and product(s). A separation column, L', may also be incorporated in the GC oven. The separation column can be filled with the appropriate material for the separation of the reactants and products, and it can be heated in the same or at a different temperature from the sampling cell.
2. The "sampling cell" is formed by the sampling column $l' + l$ and the diffusion column L, which is connected perpendicularly to the middle of the sampling column. The ends, D_1 and D_2, of the sampling column are connected through a four-port valve to the carrier gas inlet and the detector, as shown in Fig. 1.

Encyclopedia of Chromatography DOI: 10.1081/E-ECHR-120041124

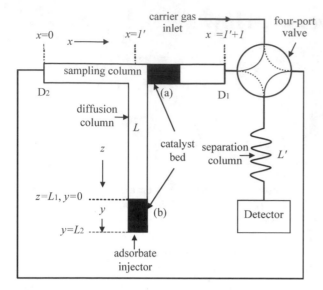

Fig. 1 Experimental setup used from RF-GC for the characterization of solid catalysts: (a) under steady-state conditions, with catalyst bed being put at a short length of sampling column l, near the junction of diffusion and sampling columns; (b) under non-steady-state conditions, with catalytic bed being put at the top of diffusion column L.

Performing flow perturbations, negative and positive abrupt fronts are made to appear in the chromatogram, forming the so-called sampling peaks (like those shown in Fig. 2 of Ref.[1]). The volumetric carrier gas flow rate does not affect the physicochemical phenomena occurring in the diffusion column, but only the speed of the sampling procedure.

THEORETICAL

The sampling peaks are predicted theoretically by the "chromatographic sampling equation," describing the concentration–time curve of the sampling peaks created by the flow reversals. The area or the height, H, of the sampling peaks is proportional to the concentration of the substance under study, at the junction, $x = l'$, of the sampling cell, at time t from the beginning of the experiment. If $\ln H$ is plotted against time, t, for each solute, the so-called diffusion band is obtained. (An example is shown in Fig. 3 of Ref.[1])

Under Steady-State Conditions

Having placed the catalyst bed at a short length of the sampling column l near the junction of the diffusion and sampling column (Fig. 1), the catalytic behavior under steady-state conditions can be studied. In that case, time-dependent fractional conversions X_t are

determined from the heights or the areas of the sampling peaks obtained after each flow reversal, and overall conversions X can be calculated from the total areas of the "diffusion bands" corresponding to reactants and products.[1]

Under Non-Steady-State Conditions

Having placed the catalyst bed at a short length from the entrance of the diffusion column L (Fig. 1), the catalytic behavior under non-steady-state conditions can be studied. In that case, not only conversions but also a large number of physicochemical parameters related to the interaction of the studied catalyst with the injected adsorbate can be determined. The whole treatment of experimental data is based on the fact that the heights of the "sampling peaks" are described by a clear function of time comprising the sum of two to four exponentials.

$$H^{1/M} = \sum_i A_i \exp(B_i t) \tag{1}$$

where H are the heights of the experimentally obtained chromatographic peaks, M the response factor of the detector, and t the time from the beginning of the experiment. Eq. (1) is not an a priori assumption but results from the solution of the mathematical model. The values of the pre-exponential factors A_i and the corresponding coefficients of time B_i are easily and accurately determined from the chromatogram by PC programs of nonlinear least-squares regression (c.f. Appendix of Ref.[9]).

The experimental pairs H and t are the variables of Eq. (1). By introducing them into the data lines of the GWBASIC program (such as given in the Appendix of Ref.[9]) together with other easily obtained quantities required by the input lines such as the geometric details of the diffusion column, mass and porosity of the catalyst bed, solute amount, as well as its diffusion coefficient in the carrier gas, and the carrier gas flow rate), the various physicochemical parameters related to the studied catalyst are calculated.

POTENTIAL OF THE METHODOLOGY AND INDICATIVE RESULTS

Reversed-flow gas chromatography methodologies have been utilized for the investigation of various catalytic processes, and a large number of physicochemical quantities related to the kinetics of the elementary steps (adsorption, desorption, surface reaction) and the nature of the active sites have been determined. These parameters are summarized as follows:

1. Time dependent, X_t, and overall, X, conversions, either under steady- or non-steady-state conditions.[1,2]
2. Adsorption, k_1, desorption, k_{-1}, and surface reaction, k_2, rate constants, and the respective activation energies, E_a.[3–5]
3. Local adsorption energies, ε, local adsorption isotherms, $\theta(p,T,\varepsilon)$, local monolayer capacities, c_{max}^*, and adsorption energy distribution functions, $f(\varepsilon)$, for the adsorption of gases on heterogeneous surfaces circumventing altogether the integral equation:[6,7]

$$\Theta(p,T) = \int_0^\infty \theta(\varepsilon,p,T)f(\varepsilon)\mathrm{d}\varepsilon \qquad (2)$$

where $\Theta(p,T)$ is the overall experimental adsorption isotherm.

4. The energy of the lateral molecular interactions on heterogeneous surfaces in a time-resolved procedure.[8]
5. Surface diffusion coefficients for physically adsorbed or chemisorbed species on heterogeneous surfaces in a time-resolved procedure.[9]
6. Standard free energy of adsorption and its probability density function over time, together with the geometrical mean of the London parts of the total surface free energy $(\gamma_1^L\gamma_2^L)^{1/2}$ of the adsorbed probe and the solid surface, accompanied by the relevant probability density functions over time.[10,11]
7. Investigation of the nature of the various groups of active sites of solid catalysts.[12]

The question naturally arising is how reliable are the physicochemical quantities determined by means of RF-GC. For this reason, the adsorption of CO, O_2, and CO_2, as well as the oxidation of CO, has been studied over well-studied, silica-supported, Pt–Rh bimetallic catalysts. The following are indicative conclusions extracted by using RF-GC, which are in agreement with the observations of other techniques:

1. The experimental data for carbon monoxide adsorption over the studied catalysts (in the absence of oxygen in the carrier gas), at temperatures higher than 300°C, suggest that the adsorption of CO is a dissociative process.[2,4]
2. There is a characteristic temperature of maximum catalytic activity, T_{max}, for every bimetallic catalyst. The temperatures found by

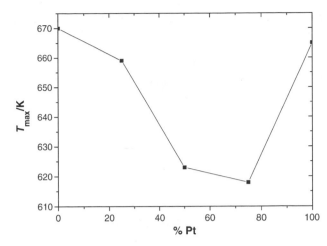

Fig. 2 Characteristic temperatures of maximum activity, T_{max} (K), for the oxidation of CO, over Pt–Rh alloy catalysts against the catalyst Pt content (% Pt). (From Ref.[3].)

RF-GC are very close to those found, for the same catalysts, by using different techniques.[1–4]

3. The bimetallic catalysts exhibit higher catalytic activity at lower temperatures in comparison with pure Pt and Rh ones, as shown in Fig. 2. Other workers have also observed this synergism for Pt–Rh bimetallic catalysts.[1–4]
4. The rate constants found by the RF-GC technique, such as those in Table 1, are very close to those determined experimentally by the frequency response method[2,4,5] for the adsorption of CO on Pt–SiO₂.
5. The values of the estimated activation energies for CO dissociative adsorption, given in Table 2, are low. Low activation energy values are indicative of corrugated surfaces. From the

Table 1 Rate constants for the adsorption (k_1), desorption (k_{-1}), and disproportionation reaction (k_2) of carbon monoxide over a Pt–SiO₂ catalyst, determined by RF-GC, at various temperatures

T (K)	$10^1 k_1$ (sec^{-1})	$10^4 k_{-1}$ (sec^{-1})	$10^4 k_2$ (sec^{-1})
555.0	1.33	6.09	2.80
573.6	1.41	6.48	2.91
595.7	1.59	6.76	3.62
615.6	1.63	6.85	4.04
633.6	1.86	7.34	4.13
643.2	1.84	7.63	3.78
657.7	1.83	7.82	3.95
673.2	1.98	8.23	4.00
692.3	2.13	8.74	4.15
707.5	2.27	9.29	4.58
723.7	2.68	9.62	4.53

(From Ref.[5].)

Table 2 Activation energies (kJ mol^{-1}) corresponding to the adsorption, E_{a1}, desorption, E_{a-1}, and disproportionation reaction, E_{a2}, of carbon monoxide over silica-supported Pt, Rh, and Pt$_{0.50}$–Rh$_{0.50}$ alloy catalysts

% Rh	E_{a1} (kJ mol^{-1})	E_{a-1} (kJ mol^{-1})	E_{a2} (kJ mol^{-1})
0	13.1 ± 1.2	9.4 ± 0.6	8.4 ± 1.3
50	18.2 ± 2.6	12.6 ± 1.7	—
100	19.4 ± 6.5	11.9 ± 2.2	3.8 ± 1.0

(From Ref.[2].)

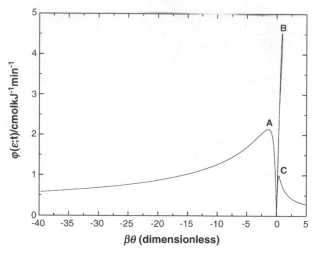

Fig. 4 Energy distribution function, $\varphi(\varepsilon;t)$ (cmol kJ^{-1} mol^{-1}), against the dimensionless product of the lateral interaction energy (β) and the local isotherm (θ)$\beta\theta$, for carbon monoxide adsorption over a bimetallic Pt$_{0.25}$–Rh$_{0.75}$ silica-supported catalyst, at 698 K. (From Ref.[12].)

difference in the found energy barriers, it is also concluded that CO adsorption is the rate-determining step for CO dissociative adsorption, followed by the dissociation step. These findings suggest a precursor-mediated mechanism for CO dissociative adsorption.[2]

6. The nature of the different groups of active sites, for the catalytic oxidation of CO, has also been investigated from the experimentally determined energy distribution functions. The existence of three groups of active sites is observed as shown in Figs. 3 and 4,[12] which is also expected from thermal desorption spectroscopy (TDS) for the adsorption of CO over group VIII noble metals.

Group A active sites correspond to the β states of TDS, arising from CO dissociative adsorption. The topography of these active sites is random, as the values of the lateral interaction energy, β, are negative. The B and C groups of active sites shown in our plots correspond to the α states of TDS. They are characterized by the positive values of β, which means that they

have a patchwise topography. Group C active sites correspond to higher β values, in comparison with B group active sites. They are indicative of CO island formation. The experimentally found results also explain the superior activity of Pt$_{0.25}$ + Rh$_{0.75}$ alloy catalyst (synergism) as a result not only of its capacity to adsorb a higher amount of carbon monoxide, at lower temperatures, but also because this catalyst is characterized by a more random topography in contrast with the other studied silica-supported pure Pt and Rh catalysts.

The utilization of RF/GC methodologies can be extended in the study of the surface properties of various solids and related processes. Thus, in a recent work, the effect of the presence of hydrogen in the adsorptive behavior of a Rh/SiO$_2$ catalyst was studied, as the selective oxidation of CO in a rich hydrogen atmosphere is a process of great technological and environmental interest, because it is related to the development of proton exchange membrane fuel cells.[13]

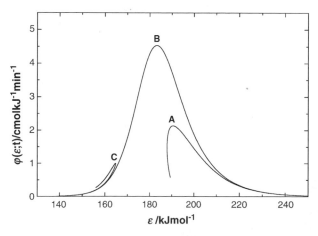

Fig. 3 Variation of the energy distribution function, $\varphi(\varepsilon;t)$ (cmol kJ^{-1} mol^{-1}), vs. the local adsorption energy, ε (kJ mol^{-1}), for the adsorption of carbon monoxide on a bimetallic silica-supported Pt$_{0.25}$–Rh$_{0.75}$ catalyst, at 698 K. (From Ref.[12].)

CONCLUSIONS

The usual inverse gas chromatography, in which the stationary phase is the main object of investigation, is a classical elution method that neglects the mass transfer phenomena; it does not take into account the sorption effect and it is also influenced by the carrier gas flow. In contrast to the integration method, the new methodology of reversed-flow gas chromatography (RF/GC), although being an inverse gas chromatographic technique, is a differential method not depending either on retention times and net retention volumes or on broadening factors and statistical moments of the elution bands.

The RF-GC methodology is technically very simple and it is combined with a mathematical analysis that gives the possibility for the estimation of various physicochemical parameters related to solid catalysts characterization in a simple experiment under conditions compatible with the operation of real catalysts. The experimentally determined kinetic quantities are not only consistent with the results of other techniques, but, moreover, they can give important information about the mechanism of the relevant processes, the nature of the active sites, and the topography of the heterogeneous surfaces.

The utilization of RF/GC methodologies can be extended in the study of the surface properties of various solids of technological and environmental interest.

REFERENCES

1. Gavril, D. Reversed flow gas chromatography: A tool for instantaneous monitoring of the concentrations of reactants and products in heterogeneous catalytic processes. J. Liq. Chromatogr. & Relat. Technol. **2002**, *25* (13–15), 2079–2099.
2. Gavril, D.; Loukopoulos, V.; Karaiskakis, G. Study of CO dissociative adsorption over Pt and Rh catalysts by inverse gas chromatography. Chromatographia **2004**, *59* (11), 721–729.
3. Gavril, D.; Katsanos, N.A.; Karaiskakis, G. Gas chromatographic kinetic study of carbon monoxide oxidation over platinum–rhodium catalysts. J. Chromatogr. A. **1999**, *852*, 507–523.
4. Gavril, D.; Koliadima, A.; Karaiskakis, G. Adsorption studies of gases on Pt–Rh bimetallic catalysts by reversed flow gas chromatography. Langmuir **1999**, *15*, 3798–3806.
5. Gavril, D.; Karaiskakis, G. Study of the sorption of carbon monoxide, oxygen and carbon dioxide on Pt-Rh alloy catalysts by a new gas chromatographic methodology. J. Chromatogr. A. **1999**, *845*, 67–83.
6. Katsanos, N.A.; Iliopoulou, E.; Roubani-Kalantzopoulou, F.; Kalogirou, E. Probability density function for adsorption energies over time on heterogeneous surfaces by inverse gas chromatography. J. Phys. Chem. B. **1999**, *103* (46), 10228–10233.
7. Gavril, D. An inverse gas chromatographic tool for the experimental measurement of local adsorption isotherms. Instrum. Sci. Technolog. **2002**, *30* (4), 397–413.
8. Katsanos, N.A.; Roubani-Kalantzopoulou, F.; Iliopoulou, E.; Vassiotis, I.; Siokos, V.; Vrahatis, M.N.; Plagianakos, V.P. Lateral molecular interaction on heterogeneous surfaces experimentally measured. Colloids Surf. A. **2002**, *201*, 173–180.
9. Katsanos, N.A.; Gavril, D.; Karaiskakis, G. Time-resolved determination of surface diffusion coefficients for physically adsorbed or chemisorbed species on heterogeneous surfaces, by inverse gas chromatography. J. Chromatogr. A. **2003**, *983* (1–2), 177–193.
10. Katsanos, N.A.; Gavril, D.; Kapolos, J.; Karaiskakis, G. Surface energy of solid catalysts measured by inverse gas chromatography. J. Colloid Interface Sci. **2003**, *270* (2), 455–461.
11. Margariti, S.; Katsanos, N.A.; Roubani-Kalantzopoulou, F. Time distribution of surface energy on heterogeneous surfaces by inverse gas chromatography. Colloids Surf. A. **2003**, *226*, 55–67.
12. Gavril, D.; Nieuwenhuys, B.E. Investigation of the surface heterogeneity of solids from reversed flow inverse gas chromatography. J. Chromatogr. A. **2004**, *1045* (1–2), 161–172.
13. Loukopoulos, V.; Gavril, D.; Karaiskakis, G. An inverse gas chromatographic instrumentation for the study of carbon monoxide's adsorption on Rh/SiO$_2$, under hydrogen-rich conditions. Instrum. Sci. Technolog. **2003**, *31* (2), 165–181.

Cell Sorting Using Sedimentation Field-Flow Fractionation: Methodologies, Problems, and Solutions—A "Cellulomics" Concept

Philippe J. P. Cardot
Yves Denizot
Serge Battu
*Laboratoire de Cellulomique Neurale and UMR CNRS 6101, Faculté de Médecine &
Pharmacie, Université de Limoges, Limoges, France*

INTRODUCTION

Sedimentation field flow fractionation (SdFFF) is an efficient, analytical/preparative-scale, cell sorting device, in particular, if coupled with sophisticated cellular characterization techniques, devices, or methods, including flow cytometry (FC). Cell population, by analogy with "polymers," appears very polydisperse and disperses in many dimensions, which may not be essentially biophysical (size, shape, density, and rigidity), but also functional (cell cycle, protein expression, and differentiation stage). Cell sample pretreatment has a unique goal, which is to provide a sterile and viable cell suspension at an appropriate concentration (1–10 million cells/mL). A particular elution mode, made possible in cell purification by exploiting complex hydrodynamic forces, is usually described as a "Hyperlayer." This elution mode exploits slight differences in physical characteristics of the cell (size, shape, density, rigidity, and surface characteristics); trends are opened to their association with complete functional cell characteristics analyses. It is shown that cells can be eluted, in some examples, according to functional parameters (cell cycle and differentiation stage), and that correlations may exist between functional parameters and the physical ones. If a given cell population is considered, considerable information provided by the separation and the complete physical and functional analyses allows one to define, characterize, and produce a new type of cell subpopulation. Such separation/characteri-characterization process for cells can be described as "cellulomics," by analogy with what is done for proteins of gene systems, keeping in mind the concept that the cell is the "fundamental and unique" localization and production factory of any gene and protein.

BACKGROUND

The last decades have shown considerable development of a large panel of cell sorting techniques.[1–5]

SdFFF, such as elutriation or centrifugation methods, belongs to the group of "physical methods," comparisons of which have already been assessed.[1] In the present report, we will focus on the specific features of cell elution in FFF and on the necessity of coupling such a physical separation/purification system with detectors of high functional specificity. Cell sorting with FFF emerged in the scientific literature in the early 1980s of the last century. A pioneering report was published in 1984 by Caldwell et al.[6] in which most of the separation rules and methodologies were described. At difference with all other species, cell sorting with FFF success was proved essential by using sedimentation subtechniques that required specific instrumentation and methodological setups. Systematic instrumentation development, allowing separation of living species in sterile conditions for further use, was initiated in the late 1990s in our laboratory.[7,8]

It must be noted here that the main objective of cell separation in FFF is not only analytical, but also preparative. The main goal is strongly linked to the possibility not only to characterize as completely as possible the cell subpopulation, but to provide or produce new living cell subpopulations for any use. It is essential to keep these cells in surviving condition. Therefore considerable attention must be given to characterize a potential subpopulation, and to define and control their survival. However, if nonstable subpopulations are recovered, cinematic studies of their properties must be developed in terms of "time or age"-dependent characteristic modifications. To be clear, cells are eluted at a given stage; they are characterized or used at that stage, in whatever way they can be cultivated, maintained in survival, and evolved to other critical stages (the must be characterized) where they can be functionally used. Such complex receiver-oriented characteristic (ROC) separation/characteriza-aracterization step can de described as "cellulomics."

In this report, specific instrumentation of SdFFF for cell sorting will be described. The interest and potential of the "Steric Hyperlayer"[6,9] elution mode

Encyclopedia of Chromatography DOI: 10.1081/E-ECHR-120040664

for cell sorting will be discussed, keeping in mind basic rules for physically or functionally oriented separation development. By analogy with polymers, experimentally driven definitions, descriptions, and characterizations of cell populations will be provided, leading to an information matrix described as "cell population multipolydispersity."[1]

A particular paragraph is devoted to some specific requirements for FFF cell sorting, such as sample preparation or separator poisoning. The battery of cell population or cell subpopulation characterization methods will be described with an FFF-dependent classification: physical methods compared to functional ones. Finally, experimental correlations between functional and physical cell characteristics lead to the isolation of very specific populations, thereby opening the field of "cellulomics." In this report, it is assumed that readers have a basic knowledge of separation sciences, in particular, in chromatography and FFF.

PRINCIPLE OF SdFFF AND INSTRUMENTATION

The FFF family encompasses a broad array of subtechniques; however, FFF techniques have in common the design of a channel, generally parallelepiped and often described as "ribbonlike," in which the critical dimensions are: 1) thickness, usually lower than $300\,\mu m$; 2) length, between 20 and 100 cm; and 3) breadth, from 0.5 to 2 cm; with 4) tapered channel ends. An external field acts on the great surface of the ribbon to achieve flow rate-dependent separations (Hyperlayer elution mode for micron-sized species). SdFFF techniques can be divided into two groups. The first uses simple gravity or gravity fractions [i.e., gravitational/subgravitational FFF (GrFFF/GFFF)]; the second uses the multigravity field created by centrifugation generically described as SdFFF. With common associated devices (injection, flow, and detection devices), their instrumental design and setup are completely different. The GFFF separator is very simple to set up and does not require more specific skills than a simple "exploration desire," whereas SdFFF is much more complex to set up and requires long-term know-how in the absence of commercially available devices devoted to cell sorting.

INSTRUMENTATION

Instrumentation for cell sorting can be very simple, using gravitational FFF devices as described by our group,[10–13] or by others.[14–16] A complete technical instrumentation has been also described.[17] The key parameter is the material used to construct the channel walls, which must be as "inert" as possible if considered against the cellular materials or their media (protein clotting). Media or sample-derived wall treatment may occur when changing, considerably, elution conditions. Such goal is empirically well assessed in the biological technology using materials of various "biocompatible" polymers. Therefore it is only necessary to choose as an appropriate material the one compatible with FFF instrumentation (low deformation under sealing pressure). It must be noticed here that some materials can be appropriate for some cell groups and be completely unusable for others, which requires a versatile instrumentation.

In SdFFF, the same strategy is employed with materials tested in GFFF, with the only price that sealing and deformability linked with centrifugation rate must be controlled. However, one particular point must be taken into account in SdFFF; long connection tubings are necessary and they must be chosen to be "biocompatible" and with appropriate inner diameter. This point is highly critical. Diameters that are too narrow may induce shear forces, destroying selective parts of the sample; too large diameters induce noncolumn band spreading, thereby limiting resolution. Channel dimensions and tubing choices must be chosen in the light of the sample and of the separation goals. Unfortunately, only a few laboratories exert an instrumental effort to define and construct biocompatible SdFFF systems.[18–20]

ELUTION MODE

Cells are in the 3–$40\,\mu m$ diameter range; channel thicknesses ranging from 70 to $250\,\mu m$ are commonly used, depending on elution selectivity requirements. Therefore cell size cannot be ignored in the light of the channel thickness leading to an experimentally developed elution mode described as "Steric Hyperlayer."[6,9] Again, in the light of the preceding paragraph, a "pure steric elution mode" must be avoided as possibly generating high particle–wall interactions, these being so far determined essentially on an experimental basis. Therefore elution conditions must meet the hypotheses developed in the Hyperlayer condition. As such, model cells are focused on different stream lines by the double and opposite actions of the external field and of forces generated by the cell in motion and described generically under the term of lift forces, driving the cell into an equilibrium position in the channel thickness. The lifting force characteristics are complex, were described for the first time by Ho and Leal,[21] and were developed on a theoretical basis for FFF by Martin and Williams.[22] Experimental proofs were given for cell sorting by Caldwell et al.[6] If little is known about the kinetics of this focusing process using retention ratio analysis, it is simplified in the determination of an average "equilibrium" position in the channel thickness of the cells eluted at a given retention ratio by a position(s) in the channel thickness given by the following equation:

$$R = \frac{3s}{w} \tag{1}$$

where R is the measured retention ratio, s is the distance between the average particle gravity center and the accumulation wall, and w is the channel thickness. The calculated s value does not describe the real distance at equilibrium, but can be considered an accurate evaluation of it, assuming that: 1) in identical experimental conditions, kinematics in the channel thickness of particles of analogous size, shape, and density is analogous; 2) flow injection reduces particle–wall interactions that are negligible; and 3) lifting forces are so intense that no particle–wall retardation effects occur. If s is greater than the cell greater radius, then elution is considered as a Hyperlayer. Therefore it is essential to obtain for cells an average position that is much larger than the cell radius; this ensures that cells have a low probability to interact with channel walls. Such considerations lead to a concept applied by Battu et al.[19,20] for cell sorting—the "safety Hyperlayer" elution mode in which cell recovery is maximum in the case of flow-established injections. Depending on the elution conditions (channel dimensions, external field intensity, and flow rate), it is, so far, possible to state that, for spherical particles of identical density, the elution order is size-dependent, with the larger ones being eluted first. It is also, so far, possible to assess that, at identical sizes, spherical particles are eluted according to their mass, with the least dense being eluted first. More complex situations arise if, as in the case of cells, the population presents independent size and density distributions.

There is a particular injection mode, described often as "stop flow" or primary relaxation step, in FFF. Suspensions are inserted in the channel under an external field, then the flow is stopped for a given time (stop flow time and primary relaxation time) and flow is reestablished for elution. This particular injection procedure is specific for FFF separations[6,23] and increases selectivity considerably, even in the case of the "Hyperlayer" elution mode. Historically, this was set up because of instrumental considerations, where the inlet tubing emerged at the depletion wall.[6,10,23] Simple instrumentation modification involving, in contrast, the accumulation wall made it possible to avoid such procedure driving the species into a geometric location in the channel thickness close to the accumulation wall.[7,8,11–13,18–20] If, experimentally, such modification was successful using a channel of reduced thickness, complex injection hydrodynamics occurs if the channels are thicker than 100 μm. Considerably

reducing the "stop flow time," or even avoiding it, is essential if cell sorting strategies are considered. Very little is known in terms of cell–channel interactions (even with biocompatible materials) and the stop flow procedure may be at the origin of (reversible or not) cell sticking, leading either to cell destruction (which may be selective) or cell differentiation by simple contact (stem cells). From injection to injection, the wall structure at the injection zone may be completely altered (modified), leading to analysis or separation biases. Therefore a price must be paid in terms of separation power to reduce these possible interactions. Limiting separation power by using systematic flow-established injections is essential if cell subpopulation integrity maintenance is required (cell differentiation process); in these conditions, the probability that any cell of the sample will interact with the walls is reduced—a process also limited by the kinetics (whose characteristics are not known) of the Hyperlayer focalization. As collateral consequences of flow injection, separation times are considerably reduced, recovery is enhanced, and channel poisoning is reduced. Fig. 1 shows the fractograms obtained in such conditions in the case of a mixture of ES cells and fibroblasts whose biological properties were analyzed.[7] Low-intensity spikes at the end of every fractogram signifies field stop and a low reversible release of trapped material, which may not necessarily be made of cells only.

Mouse E14 ES cells were routinely grown onto a monolayer of mitomycin-treated primary embryonic mouse fibroblasts as a source of cytokines and growth factors. The medium consisted of DMEM with 15% fetal calf serum (Gibco, Cergy Pontoise, France), 100 U/mL penicillin, 100 μg/mL streptomycin, 1 μM α-mercaptoethanol, and 10^3 U/mL leukemia-inhibitory factor (LIF; Gibco). ES cells were grown at 37°C in a humidified chamber with 10% CO_2. ES cells were recovered with trypsin treatment for SdFFF experiments or subcultures. The SdFFF

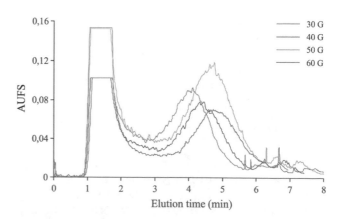

Fig. 1 Cell sorting with FFF. *(View this art in color at www.dekker.com.)*

separation device used in this study includes a separation channel, which consisted of two 870 × 30 × 2-mm polystyrene plates, separated by a Mylar spacer in which the channel (785 × 10 × 0.125 mm with two V-shaped ends of 70 mm) was carved. Inlet and outlet were 0.254 mm ID. Peek tubing (Upchurch Scientific, Oak Harbour, WA) was directly screwed into the accumulation wall. Then, polystyrene plates and Mylar spacers were sealed onto a centrifuge basket. The channel–rotor axis distance was measured at r (13.8 cm). Cleaning and decontamination procedures have been described in a previous report.[3] The elution signal was recorded at 254 nm.

MULTIPOLYDISPERSITY: PHYSICAL/FUNCTIONAL AND DETECTION REQUIREMENTS

A First-Order Definition of Multipolydispersity: Example of Mass Distribution

This concept has already been defined for cell sorting.[1] To precisely know the physical characteristics of cell populations, let us imagine a simple bidimensional one where cells are spherical and rigid, and show independent size and density distribution. According to the Steric Hyperlayer mode, with some subpopulations, coelution is possible, generating a need for size or mass detectors. Therefore granulometric or mass measurements appearing at the outlet (online/offline) are essential in terms of separation developments.

Fig. 2 shows a theoretical bidimentional polydispersity (size and density). The bidimensional Gaussian distribution is shown in Fig. 2A, where (S) is the size axis, (D) is the density, and C represents the number of particles. If such a sample is eluted according to the Hyperlayer elution mode, a broad fractogram is observed, as shown in Fig. 2B. From such a fractogram, every fraction corresponds to particles of different sizes and densities, as qualitatively shown in Fig. 2C. The front of the peak is associated with particles that are very different from the ones at the tail. It is possible to imagine granulometric detection all along the fractogram profile, as shown in Fig. 2D. If the density distribution coinvolved with the size distribution is considered, every fraction is associated with a particular particle size distribution (PSD). It is essential to note that, in the simulation considered here, density distribution is not disperse enough to invert or modify

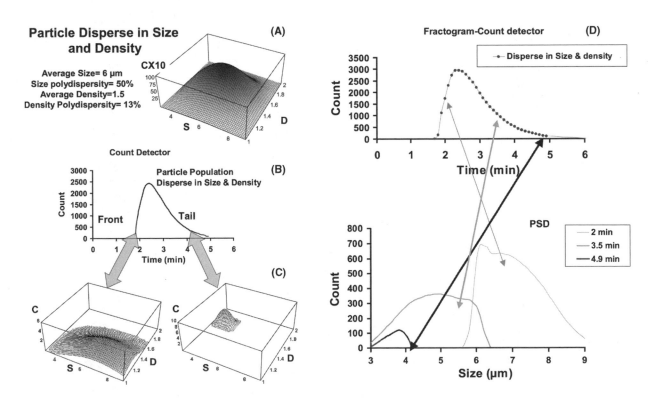

Fig. 2 The multipolydispersity concept: a bidimensional example (size and density distribution). Graph (A) represents the convolution if size and density distribution simulated by a bidimensional Gaussian. Fractogram (B) shows the elution profile obtained in the use of an ideal Steric Hyperlayer elution mode whose channel dimensions, external field, and carrier flow rate are chosen to produce discriminating separations on both size and density. Graph (C) shows bidimensional population pattern composing the fronting and tailing zones of the fractogram. The same fractogram is shown on graph (D), with simulation of fraction collection or granulometric detector time constant with three examples of PSD obtained.

the "size-dependent Hyperlayer elution order," where the biggest and most polydisperse particles elute first. It is now interesting to determine this PSD in the density distributions.

GRANULOMETRIC PATTERNS AND DENSITY DISTRIBUTION

It may be possible to discriminate density vs. size if a single particle mass detector were available for micron-sized species. In the absence of such a detector, only indirect information is affordable, exploring the consequences of the Hyperlayer elution mode. From the abovedescribed bidimentional population (size and density), the Hyperlayer mode allows us to draw isodensity and isosize retention patterns. Such information coinvolved with the PSD permits the determination of the density distribution envelope of the fractions considered, as shown in Fig. 3. Graph A shows the isoretention pattern of the population. It is observed that, at the peak front, a PSD ranging from 5 to 9 µm is obtained, with light small particles

(5 µm, 1.2 density) and heavy big ones (9 µm, 1.6 density). It is tempting to complicate such a chart in three dimensions, as shown in Fig. 3B, to draw an isosize retention chart as a function of density, whose bidimensional projection may be easier to read. Therefore it is observed that 3-µm particles (as well as 9-µm ones) are eluted at very different retention times, depending on their densities. It can be concluded that, in the light of what is known so far from the "Hyperlayer" elution mode, PSD obtained during elution can be linked to other dimensions (density, shape, and rigidity). Therefore it appears that large optimization processes are opened to understand the discriminating effects of channel dimensions, fields, and flow rates on a multi-polydisperse population.

HIGHER ORDERS: PHYSICAL AND FUNCTIONAL

If it is relatively easy to adapt offline/online granulometric detectors allowing PSD at every retention time; only very few cell-specific characterization devices are

Size and Density Elution Characteristics

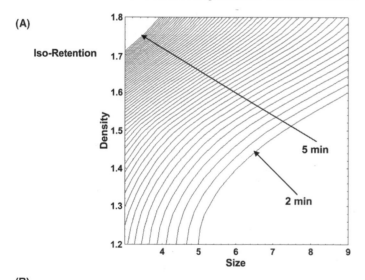

(A)

Iso-Retention

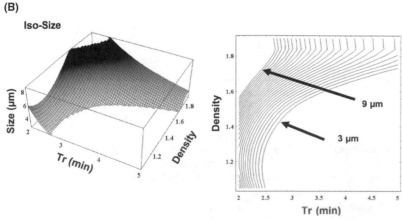

(B)

Iso-Size

Fig. 3 (A) Granulometric pattern and (B) density distribution. Particle characteristics are described in Fig. 2.

available, with the preferred one being FC.[24] FC's versatility is impressive, allowing particle counting as well as a series of particle characteristics and distribution measurements using light diffusion principles and/or fluorescence.[24] If used in its simplest mode, count histogram vs. diffused light leads to sophisticated information. When a single particle passes in front of a laser beam, the scattered (or emitted fluorescence) light generates a signal proportional to the studied parameter of that particle. If the collected light is at a large angle [90°, side scattering (SS)] from the incident laser beam, the signal is a complex function of the particle refractive index, composition, and surface characteristics.[24,25] If the diffused light is collected at low angles[24–26] [10°, forward scattering (FS)], diffraction predominates and the signal is a function of the particle volume described by the Mie law.[27] Therefore a multidimensional detection allowing population or subpopulation discrimination, whose pattern (fingerprint) can be associated with FFF retention properties, is made possible. Other detection methods can be used with specific functional properties (surface receptors and cell content), as already described by our group.[7,28,29] Every sample or collected fraction (during FFF elution) can be, as a consequence, described by a battery of information—some being possibly calibrated, and some others looked at only from a qualitative or comparative point of view.

SPECIFIC REQUIREMENTS

Suspension

It is essential to prepare single cell suspensions, which exist naturally in very few cases (cultured cells such as HeLa, Hel, and blood cells). In most critical cases (solid tumors, tissues, and neural cells), a preliminary step is necessary and classically requires transformation of the tissue in suspension. Two methods are used: the first is mechanical,[30] whereas the second requires enzymatic action.[30] It is essential to obtain a monocell suspension, whose characteristics have to be controlled by FC. Once this suspension is obtained, another critical parameter is its concentration; blood cells can be injected in a highly concentrated suspension, whereas neural ones must be inserted into the FFF system highly diluted, requiring large injection volumes. So far, no prediction rules have been established and the technique requires experience-based empirical knowledge.

FFF Device

Channel walls are assumed to be biocompatible, but so is the carrier phase, which can be added with specific surfactants to avoid aggregation during separations: so far, bovine serum albumin (BSA)[11–13] (and, recently, cholic acid) appeared to be more stable and at least as efficient.[31] Such precautions must be associated with systematic recovery measurements—a multidimensional concept associated with the detector pattern, and a comparison of the crude sample with every collected fraction.

Operating Conditions

A flow injection of appropriate cell number sometimes requires adapting the injection volume. In terms of elution, it is essential to observe or follow the safety Hyperlayer requirements that have been already defined. In this order, with the help of the granulometric detector, it is possible to determine the average sizes of cells eluting in the front and the tail of the fractogram, and to calculate, by means of the (S) position, their average position in the channel thickness; by experience, a useful recovery is found if the elution conditions (field and flow) drive to an average position in the channel thickness (s) at, at least, 0.75 times the diameter.

Poisoning/Cleaning

Cell suspensions are often introduced into the SdFFF separator in their cultivation media; these complex solutions encompass proteaceous hydrophilic compounds as well as hydrophobic ones. It has been also observed that cell recovery may reach 90% in optimized conditions with either a reversible trapping (released at field stopped) or an irreversible one, leading to the release of cellular material that may interact irreversibly with channel walls, thereby leading to surface modifications. Using a biocompatible carrier phase may generate local bacterial growth, either at the channel surface or within the channel; some workers use some toxic cellular killing compounds (e.g., sodium azide).[32] The experimenter must be warned about the toxicity, even at low concentration, of such compounds not only as far as the sample is concerned, but also the experimenter. With experience, it appears that channel poisoning is progressive, with constant recovery reduction, as well as baseline and noise evolution. Therefore there is a need to regularly clean the channel over the life of the separator. This procedure is relatively easy and must be performed in three steps: channel wall regeneration, which destroys sorbed proteins or biological compounds such as membranes or genetic materials; sterilization of the channel by means of an appropriate cocktail involving hypochlorite and ethanol; and rinsing of the separator with sterile mobile phase to completely wash the channel.

The effectiveness of instrument design and setup to reduce cell–accumulation wall interactions was demonstrated, first, by the recovery of cells in the corresponding elution peak (>70%). Second, it was shown by conservation of cell viability, which was, after SdFFF elution, similar to that of the control population. Finally, reduction of interactions was partially demonstrated by a very low cell release peak, which was observed at the end of the fractogram when channel rotation was stopped and the mean gravity was equal to zero (external field applied = $1g$; Fig. 4); such procedure is not possible in GFFF. This residual signal corresponded to reversible cell sticking due to weak interactions between cells and the accumulation wall.[7, 11–14] Cells or cellular materials can be released from the accumulation wall under the effect of the mobile phase flowing in the absence of channel rotation. However, irreversible cell trapping is due to strong cell–channel interaction and cannot be reversed under these conditions. This phenomenon cannot be observed on the fractogram and leads to channel poisoning, requiring routine channel wall regeneration procedures. For this purpose, cleaning cocktails used in FC or clinical instrumentation appear to be very effective.

These poisonings can be overcome by systematically performed cleaning and decontamination procedures.[19,20,28,29] Some experiments (not published) have shown that the absence of effective and systematic channel cleaning has led to many problems, in particular, an increase in apoptosis of separated cells, even though elution conditions were set up to enhance the "Hyperlayer" mode because the small portion of definitively trapped cell die released apoptotic signals into the separator, which could activate apoptosis in freshly separated cells.

The different steps and instrument setups used for cleaning and decontamination have been extensively described.[19,20,28,29] The cleaning procedure is based on the use of osmotic shock and injection of deproteinizing agent. The use of polystyrene plates and BSA-free mobile phase has simplified the previous steps;[10–13] it is now performed as an end-of-day cleaning–decontamination procedure. First, phosphate-buffered saline (PBS; pH 7.4) was replaced by flushing the entire system with sterile distilled water at a high flow rate. Second, the entire SdFFF device was flushed at 0.8 mL/min for 30 min with a protein cleaning agent (CLENZ; Beckman Coulter, Fullerton, CA). The system was rinsed with sterile distilled water for 1 hr at 0.8 mL/min. Then, the entire SdFFF device was flushed at 0.8 mL/min for 30 min with a 3–4°C sodium hypochlorite solution. The system was rinsed with sterile distilled water for 2 hr at 1 mL/min. The system was then ready to use by replacing the sterile water with sterile PBS. By implementing this cleaning–decontamination procedure, the same channel can be used for analysis of various cell populations without sample cross-contamination and microorganism proliferation.

FROM PHYSICAL/FUNCTIONAL TO BIOLOGICAL, AN EMPIRICAL EXPERIENCE: THE CONCEPT OF "CELLULOMICS"

ES mice embryonic stem cells are an important tool for the generation of transgenic and gene modified mice. We report the effectiveness of an SdFFF cell sorter to provide, from a crude ES cell preparation, a purified ES cell fraction with the highest in vivo developmental potential, to prepare mice chimeras having a high percentage of chimerism.

By taking advantage of biophysical properties (size, density, and shape), SdFFF sorts viable cells without labeling of any kind. SdFFF has a great potential with major biomedical applications, including hematology, neuroscience, cancer research, microorganism analysis, and molecular biology.[8,19,20,30] Embryonic stem cells are used in studies of gene disruption and transgenesis by using their ability to colonize the germline after introduction into blastocysts. Major limitations are the time required to obtain germline transmission, which sometimes required numerous experiments. ES cell suspensions are a mixture of cells at various stages of proliferation; heterogeneity within cells arises dur-

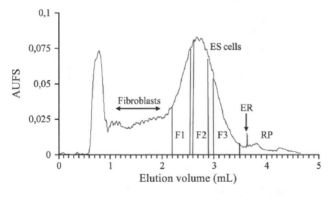

Fig. 4 Optimized SdFFF fractogram of ES Cells. Representative fractogram of ES cell suspensions after SdFFF elution. Elution conditions: Flow injection of 100 μL of ES suspension; flow rate, 0.6 mL/min (sterile PBS, pH 7.4); and external multigravitational field, 40 (0.1g; spectrophotometric detection at 254 nm). Fractions were collected as follows: PF1, 3 min 40 sec/4 min 15 sec; PF2, 4 min 20 sec/4 min 50 sec; PF3, 5 min 0 sec/5 min 50 sec. ER corresponds to the end of channel rotation. In this case, the mean externally applied field strength was equal to zero gravity; thus RP, a residual signal, corresponds to the release peak of reversible cell accumulation wall sticking *(View this art in color at www.dekker.com.)*

ing culture with respect to their in vivo developmental potential. We tested the effectiveness of SdFFF to provide, from crude ES cells, an enriched cell fraction with the highest in vivo developmental potential, to prepare chimeras having a high percentage of chimerism,[7] as shown in Fig. 5.

E14 ES cell cultures and the SdFFF device were used as previously described.[7,34] The optimal elution conditions (Hyperlayer mode[35]) were: flow injection 1.5×10^5 cells/0.1 mL, flow rate 0.6 mL/min, mobile phase PBS pH 7.4, external multigravitational field strength $40.0 \pm 0.1\,g$, and spectrophotometric detection $\lambda = 254$ nm. Three cell fractions were collected as shown in Fig. 4: F_1: 3 min 40 sec/4 min 15 sec; F_2: 4 min 20 sec/4 min 50 sec; F_3: 5 min/5 min 50 sec. Fractionated ES cells were stained with propidium iodide and analyzed for DNA cell status using an XL.2 flow cytometer (Beckman Coulter). The percentage of G_0/G_1 cells was higher in F_3 as compared with F_1 and F_2. The percentage of Ki67[+] cells (a protein expressed in all phases of the cell cycle, except G_0

and early G_1)[5] confirmed results with propidium iodide. Thus because F_2 and F_3 cells had the highest and lowest in vitro clonogenicity, respectively, 10 cells from these fractions were injected into C57 Bl/6 blastocysts to derive somatic chimeras. The frequency of chimeras obtained was similar (9 of 37 injected blastocysts). In contrast, a higher degree ($p < 0.02$, Mann–Whitney U-test) of color coat chimerism was obtained with F_3 ($87 \pm 5\%$) compared to F_2 ($63 \pm 9\%$). A germline transmission was obtained for all four F_3 males within 3 months as compared with one of four F_2 males after 6 months.

These results clearly show that SdFFF can sort, in a few minutes, the most convenient ES cell population to generate chimeras having the highest ability to colonize the germline. This result is, in particular, based on its newly described capacity to sort cells by their position in the cell cycle. In conclusion, SdFFF is of great interest to improve transgenesis and might also have further interesting applications for human gene therapy.

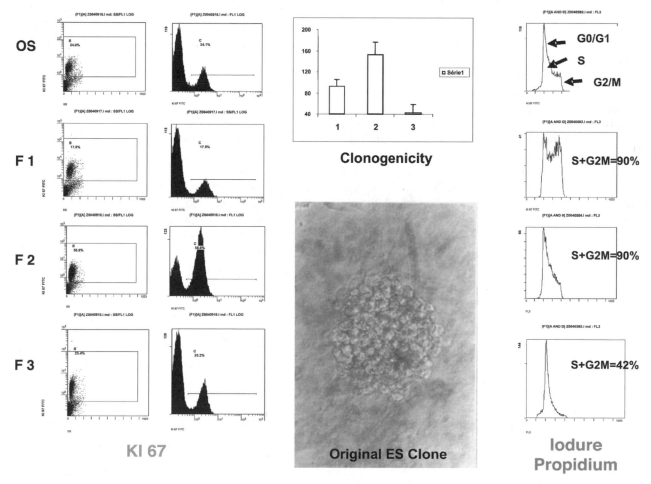

Fig. 5 ES cells, eluted as shown in Fig. 4, are characterized according to functional parameters described in the text: Bidimensional cytometric map of FS and fluorescence (Ki67). Right: Monodimensional propidium iodine fluorescence. Top center: Clonogenicity activities of the three collected fractions of Fig. 4. Bottom center: Microscopic image of ES cell clone.

CONCLUSIONS

The major difficulty in the development of FFF methodologies and instrumentations is linked to the complexity of correlating elution mode hypotheses (Hyperlayer) with experimental proofs, which depends essentially upon physical properties of the cellular material, such as size, density, shape, rigidity, and cellular viscosity, which are of greatest interest if the " physical point of view" is considered. This point of view is historical and is linked to the wide experience of FFF practitioners with latex or silica, micron-sized species, or starch granules, or others. There are some examples dealing with cellular materials (e.g., red blood cells or yeast) where elution is correlated with the abovedescribed physical properties; in these cases, "cellulomics" concepts are relatively simple.

More delicate are the situations where experimentally obtained functional properties, such as specific receptor presence or cell-dependent cycle elution, are the essential ROC and define the goal or the results of the separations. There is, as a consequence, a considerable task in understanding the very different origins of such links, where physics meets or generates biological results of interest, leading to the wide domain of "cellulomics."

The main general consideration that can be stated so far is: 1) the smoothness of the separations; 2) the absence of cell prelabeling; and 3) the development of biocompatible instrumentation, which allows subpopulation lineage to be produced, which are not only usable for fundamental studies such as differentiation pathways or apoptosis studies, but also for transplantation or genetic engineering. One must have in mind that the cell is definitively the place, home, and native localization of genes and proteins. The possibility of rapid, nondestructive separation, purification, and characterization of cells (cellulomics) opens fabulous dimensions for proteomics and genomics.

REFERENCES

1. Lucas, A.; Lepage, F.; Cardot, Ph. *Field Flow Fractionation Handbook*; Shimpf, M.E., Caldwell, K., Giddings, J.C., Eds.; Wiley-Interscience: New York, 2000; 471.
2. Lutz, M.P.; Geadick, G.; Hartmann, W. Anal. Biochem. **1992**, *2000*, 376.
3. Axen, R.; Porath, J.; Ernback, S. Chemical coupling of peptides and proteins to polysaccharides by means of cyanogen halides. Nature **1967**, *214*, 1302.
4. Hansen, E.; Hanning, K. Electrophoretic separation of lymphoid cells. Methods Enzymol. **1984**, *108*, 180.
5. Sharpe, P.T. *Laboratory Techniques in Biochemistry and Molecular Biology*; Burdon, R.H., Knippenberg, P.H., Eds.; Elsevier: Amsterdam, 1988; Vol. 18, 208.
6. Caldwell, K.D.; Cheng, Z.Q.; Hradecky, P.; Giddings, J.C. Separation of human and animal cells by steric field-flow fractionation. Cell Biophys. **1984**, *6*, 233.
7. Guglielmi, L.; Battu, S.; Le Bert, M.; Faucher, J.L.; Cardot, P.J.P.; Denizot, Y. Mouse Embryonic Stem Cell Sorting for the Generation of Transgenic Mice by Sedimentation Field-Flow Fractionation. Anal. Chem. **2004**, *76* (6), 1580.
8. Lautrette, C.; Cardot, P.J.P.; Vermot-Desroches, C.; Wijdenes, J.; Jauberteau, M.O.; Battu, S. Sedimentation field flow fractionation purification of immature neural cells from a human tumor neuroblastoma cell line. J. Chromatogr. B. **2003**, *791* (1–2), 149.
9. Williams, P.S.; Koch, T.; Giddings, J.C. Characterization of near-wall hydrodynamic lift forces using sedimentation field-flow fractionation. Chem. Eng. Commun. **1992**, *111*, 121.
10. Cardot, P.J.; Gerota, J.; Martin, M. Separation of living red blood cells by gravitational field-flow fractionation. J. Chromatogr. **1991**, *568* (1), 93.
11. Merino-Dugay, A.; Cardot, P.J.; Czok, M.; Guernet, M.; Andreux, J.P. Monitoring of an experimental red blood cell pathology with gravitational field-flow fractionation. J. Chromatogr. Aug 7, **1992**, *579* (1), 73–83.
12. Andreux, J.P.; Merino, A.; Renard, M.; Forestier, F.; Cardot, P. Exp. Hematol. **1993 Feb**, *21* (2), 326–330.
13. Bernard, A.; Paulet, B.; Colin, V.; Cardot, P.J.P. Red blood cell separations by gravitational field-flow fractionation: instrumentation and applications. TrAC, Trends Anal. Chem. **1995**, *14* (6), 266–273.
14. Sanz, R.; Galceran, M.T.; Puignou, L. Determination of viable yeast cells by gravitational field-flow fractionation with fluorescence detection. Biotechnol. Prog. **2004**, *20* (2), 613–618.
15. Urbankova, E.; Vacek, A.; Novakova, N.; Matulik, F.; Chmelik, J. Investigation of red blood cell fractionation by gravitational field-flow fractionation. J. Chromatogr. **1992** Nov 27, *583* (1), 27–34.
16. Urbankova, E.; Vacek, A.; Chmelik, J. Micropreparation of hemopoietic stem cells from the mouse bone marrow suspension by gravitational field-flow fractionation. J. Chromatogr. B. Biomed. Sci. Appl. Dec 13, **1996**, *687* (2), 449–452.
17. Cardot, Ph.; Chianea, T.; Battu, S. SdFFF of Living Cells in *Encyclopedia of Chromatography*;

Schimpf, M.E., Caldwell, K., Giddings, J.C., Eds.; Marcel Dekker, Inc.: New York, 2002; 742.

18. Metreau, J.M.; Gallet, S.; Cardot, P.J.; Le Maire, V.; Dumas, F.; Hernvann, A.; Loric, S. Sedimentation field-flow fractionation of cellular species. Anal. Biochem. 1997, 251 (2), 178.

19. Battu, S.; Delebasse, S.; Bosgiraud, C.; Cardot, P.J. Sedimentation field-flow fractionation device cleaning, decontamination and sterilization procedures for cellular analysis. J. Chromatogr. B. 2001, 751 (1), 131.

20. Battu, S.; Cook-Moreau, J.; Cardot, P.J.P. Sedimentation field-flow fractionation: methodological basis and applications for cell sorting. J. Liq. Chromatogr. & Relat. Technol. 2002, 25 (13–15), 2193–2210.

21. Ho, P.B.; Leal, L.G. Inertial migration of rigid spheres in twodimensional unidirectional flows. J. Fluid Mech. 1974, 65, 365.

22. Martin, M.; Williams, P.S. Theoretical Advances in Chromatography and Related Separation Techniques; NATO ASI Series; Dondi, F., Guiochon, G., Eds.; Kluwer Academic Publishers: London, 1991; Vol. 383, 513.

23. Lee, S.H.; Myers, M.N.; Giddings, J.C. Hydrodynamic relaxation using stopless flow injection in split inlet sedimentation field-flow fractionation. Anal. Chem. Nov. 1, 1989, 61 (21), 2439–2444.

24. Cram, L. Flow cytometry, an overview. Methods Cell Sci. 2002, 24 (1–3), 1

25. Papa, S.; Zamai, L.; Cecchini, T.; Del Grande, P.; Vitale, M. Cell cycle analysis in flow cytometry: Use of BrdU labelling and side scatter for the detection of the different cell cycle phases. Cytotechnology 1991, 5 (Suppl. 1), 103–106.

26. Petriz, J.; Tugues, D.; Garcia-Lopez, J. Relevance of forward scatter and side scatter in aneuploidy detection by flow cytometry. J. Eur. Soc. Anal. Cell. Pathol. May 1996, 10 (3).

27. Adams, J.M. Light extinction photometer for measurement of particle sizes in polydispersions. Rev. Sci. Instrum. 1968, 39 (11), 1748–1751.

28. Cardot, P.; Battu, S.; Simon, A.; Delage, C. Hyphenation of sedimentation field flow fractionation with flow cytometry. J. Chromatogr. B. 2002, 768 (2), 285–295.

29. Sanz, R.; Cardot, P.; Battu, S.; Galceran, M.T. Steric-hyperlayer sedimentation field flow fractionation and flow cytometry analysis applied to the study of Saccharomyces cerevisiae. Anal. Chem. Sep. 1, 2002, 74 (17), 4496–4504.

30. Battu, S.; Elyaman, W.; Hugon, J.; Cardot, P.J. Cortical cell elution by sedimentation field-flow fractionation. Biochim. Biophys. Acta 2001, 1528 (2–3), 89.

31. Reschiglian, P.; Zattoni, A.; Roda, B.; Cinque, L.; Melucci, D.; Min, B.R.; Moon, M.H. Hyperlayer hollow-fiber flow field-flow fractionation of cells. J. Chromatogr. A. 2003, 985, 519–529.

32. Hofstetter-Kuhn, S.; Rosler, T.; Ehrat, M.; Widmer, H.M. Characterization of yeast cultivations by steric sedimentation field-flow fractionation. Anal. Biochem. 1992, 206, 300–308.

33. Sanz, R.; Puignou, L.; Reschiglian, P.; Galceran, M.T. Gravitational field-flow fractionation for the characterisation of active dry wine yeast. J. Chromatogr. A. 2001, 919 (2), 339–347.

34. Giddings, J.C. Field-Flow Fractionation Handbook; Schimpf, M.E., Caldwell, K., Giddings, J.C., Eds.; Wiley-Interscience: New York, 2000.

35. Pinaud, E. Localization of the 3' IgH locus elements that effect long-distance regulation of class switch recombination. Immunity 2001, 15, 187–199.

Cells, Affinity Chromatography of

T. M. Phillips
*Ultramicro Analytical Immunochemistry Resource, DBEPS, ORS, OD,
National Institute of Health, Bethesda, Maryland, U.S.A.*

INTRODUCTION

Affinity cell separation techniques are based on similar principles to those described in procedures for the isolation of molecules and are used to quickly and efficiently isolate specific cell types from heterogeneous cellular suspensions. The procedure (Fig. 1) involves making a single cell suspension and passing it through a column packed with a support to which a selective molecule (ligand) has been immobilized. As the cells pass over the immobilized ligand-coated support (Fig. 1a), the ligand interacts with specific molecules on the cell surface, thus capturing the cell of interest (Fig. 1b). This cell is retained by the ligand-coated support, while nonreactive cells are washed through the column. Finally, the captured cell is released (Fig. 1c) by disrupting the bond between the ligand and its selected molecule, allowing a homogeneous population of cells to be harvested.

METHODOLOGY

Although affinity chromatography of cells is essentially performed in a manner similar to other affinity techniques, it is commonly used for both negative and positive selection. Negative selection removes specific cell types from the sample population, whereas positive selection isolates a single cell type from the sample. In the latter situation, the selected cells are recovered by elution from the immobilized ligand, thus yielding an enriched population. However, unlike molecules, cells are often quite delicate, and care must be exercised when choosing the chromatographic support and the method of retrieval. The support matrix must exhibit minimal nonspecific cell adhesion but be sufficiently porous to allow cells to pass through without physically trapping them or creating undue sheer forces likely to cause cell injury or death. Usually the support matrices of choice are loosely packed fibers, large pore cross-linked dextrans or agarose and large plastic or glass beads. Immunologists have long used the relatively nonspecific affinity of charged nylon wool to fractionate lymphocytes into different subpopulations. Such separations are achieved because certain subpopulations of lymphocytes express an affinity for the charged fibers while others do not. This negative selection process has been used to prepare pure suspensions of T-lymphocytes for many years, but has recently been replaced by the more selective antibody-mediated or immunoaffinity procedures. The use of immobilized ligands on magnetic beads has gained popularity, especially when employing immobilized lectins or antibodies as the capture ligand.[1,2]

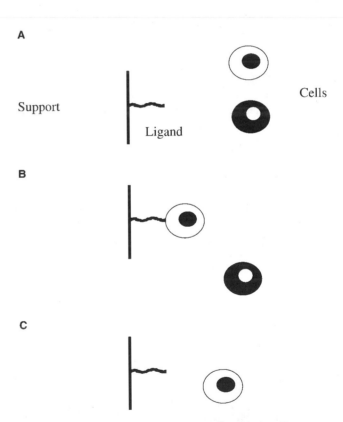

Fig. 1 Affinity isolation of specific cells. (A) A cell suspension containing the cell of interest (clear cytoplasm) and another cell type (dark cytoplasm) are passed over the support bearing a selective ligand immobilized to its surface. (B) The ligand interacts, with the surface molecules on the cell of interest, thus capturing it. The other cell type is not bound and passes through the column. (C) The bound cell is released by the addition of an elution agent to the running buffer of the column. This agent competes or disrupts the interaction between the ligand and the cell thus releasing the cell. The free cell is now washed through the column and harvested.

Encyclopedia of Chromatography DOI: 10.1081/E-ECHR-120039871

Table 1 Lectins and their reactive sugar moieties

Common name	Latin name	Reactive sugar residues
Castor bean RCA$_{120}$	*Ricinus communis*	β-D-Galactosyl
Fava bean	*Vicia faba*	D-Mannose
		D-Glucose
Gorse	*Ulex europaeus*	α-L-fucose
UEA I UEA II		*N,N'*-Diacetylchitobiose
Jacalin	*Artocarpus integrifolia*	α-D-galactosyl
		β-(1,3)-*n*-Acetyl galactosamine
Concanavalin A	*Canavalia ensiformis*	α-D-Mannosyl
		α-D-Glucosyl
Jequirity bean	*Abrus precatorius*	α-D-Galactose
Lentil	*Lens culinaris*	α-D-Mannosyl
		α-D-Glucosyl
Mistletoe	*Viscum album*	β-D-Galactosyl
Mung bean	*Vigna radiata*	α-D-Galactosyl
Osage orange	*Maclura pomifera*	α-D-Galactosyl
		N-Acetyl-D-Galactosaminyl
Pea	*Pisum sativum*	α-D-Glucosyl
		α-D-Mannosyl
Peanut	*Arachis hypogaea*	β-D-Galactosyl
Pokeweed	*Phytolacca americana*	N-Acetyl-β-D-glucosamine oligomers
Snowdrop	*Galanthus nivalis*	Non-reducing terminal end of α-D-mannosyl
Soybean	*Glycine max*	N-Acetyl-D-Galactosamine
Wheat germ	*Triticum vulgaris*	N-acetyl-β-D-glucosaminyl
		N-Acetyl-β-D-glucosamine oligomers

AFFINITY LIGANDS

Plant lectins are some of the most popular ligands for affinity cell separations. These molecules express selective affinities for certain sugar moieties often present on cell surfaces (Table 1), different lectins being used as selective agents for specific sugars. Whitehurst, Day, and Gengozian[3] found that the lectin *Pisum sativum* agglutinin could bind feline B-lymphocytes much more readily than T-lymphocytes and used lectin-coated supports to obtain pure subpopulations of T-lymphocytes by negative selection. Additionally, the retained cells were recovered by elution from the immobilized lectin with a suitable sugar. Lectins are efficient ligands for cell selection, but in many cases, their interaction with the selected cell surface molecule is highly stable and efficient, requiring mechanical agitation of the packing before recovery of the cells can be achieved. Pereira and Kabat[4] have reported the use of lectins immobilized to Sephadex or Sepharose beads for the isolation of erythrocytes.

Another useful ligand is protein A, which is a protein derived from the wall of certain *Staphylococcus* species of bacteria. This reagent binds selected classes of IgG immunoglobulin via their Fc or tail portion, making it an excellent ligand for binding immunoglobulins attached to cell surfaces, making it an ideal general-purpose reagent. Ghetie, Mota, and Sjoquist[5] demonstrated that protein A-coated Sepharose beads were useful for cell separation following initial incubation of the cells with IgG antibodies directed against specific cell surface markers. Surface IgG-bearing mouse spleen cells were pretreated with rabbit antibodies to mouse IgG prior to passage over the protein A-coated support. The cells of interest were then isolated by positive selection chromatography.

In addition to bacterial proteins, other binding proteins such as chicken egg white avidin have become popular reagents for affinity chromatography. These supports work on the principle that immobilized avidin binds biotin, which can be chemically attached to a variety of ligands including antibodies. Tassi et al.[6] used a column with an avidin-coated polyacrylamide support to bind and retain cells marked with biotinylated antibodies. Human bone marrow samples were incubated with monoclonal

Table 2 Elution buffers suitable for cell affinity chromatography

Acids
  0.33 M Citric acid
  0.15 M Formic acid

Chaotropic ions
  1–3.0 M Sodium and potassium thiocyanate
  0.9 M Sodium chloride
  0.5–1.5 M iodine

Competition agents
  Various sugars as competition agents for lectin binding (see Table 1)
  Antigens for competition binding in immunoaffinity matrices

mouse antibodies directed against the surface marker CD34, followed by a second incubation with biotinylated goat anti-mouse immunoglobulins. Binding of the biotin to the avidin support effectively isolated the antibody-coated cells.

A wide variety of immobilized antigens, chemicals, and receptor molecules have been used effectively for affinity cell chromatography. Sepharose beads coated with thyroglobulin have been used to separate thyroid follicular and parafollicular cells, while immobilized insulin on Sepharose beads has been used to isolate adipocytes by affinity chromatography. Dvorak, Gipps, and Kidson[7] reported the successful retrieval of a 95% pure fraction of chick embryonic neuronal cells using an affinity chromatography approach utilizing α-bungarotoxin immobilized to Sepharose beads.

Tlaskalova-Hogenova et al.[8] demonstrated the usefulness of affinity cell chromatography to isolate T- and B-lymphocytes from human tissues. These authors describe comparative studies on three popular approaches to the isolation of lymphocyte subpopulations, namely nylon wool columns, immunoaffinity cell panning (a batch technique using antibodies immobilized to the bottom of culture dishes), and immunoaffinity using anti-human immunoglobulins attached to Sephron (hydroxyethylmethacrylate) or Sepharose supports. These studies clearly indicate that the selectiveness of immobilized antibodies was superior for isolating defined subpopulations of cells.

Immobilized antibody ligands or immunoaffinity chromatography is now the approach of choice for cell separation procedures. Kondorosi, Nagy, and Denes[9] prepared columns packed with a support coated with nonimmune rat immunoglobulin and used these columns to isolate cells expressing surface Fc or immunoglobulin receptors, while van Overveld et al.[10] used anti-human IgE-coated Sepharose beads as an immunoaffinity chromatography step when fractionating human mast cells from lung tissue.

ELUTION TECHNIQUES

The elution agent used to recover affinity-selected cells must be carefully chosen. It must be able to either disrupt the binding of the ligand to the cell surface molecule or compete with the cell molecule for ligand binding. In many cases, such as lectin affinity chromatography, the elution agent is easy to select—it is usually a higher concentration of the sugar to which the ligand binds. Elution agents for other techniques such as immunoaffinity are harder to select. Harsh acid or alkaline conditions, although efficient at breaking antibody–antigen binding, are usually detrimental to cell membranes. Elution in these cases is often achieved using mild acids or mild chaotropic ion elution (Table 2).

REFERENCES

1. Putnam, D.D.; Namasivayam, V.; Burns, M.A. Cell affinity separations using magnetical stabilized fluidized beds: erythrocyte subpopulation fractionation utilization a lectin-magnetite support. Biotechnol. Bioeng **2003**, *81*, 650.
2. Vroemen, M.; Weidner, N. Purification of Schwann cells by selection of p75 low affinity nerve growth factor receptor expressing cells from adult peripheral nerve. J. Neurosci. Meth. **2003**, *124*, 135.
3. Whitehurst, C.E.; Day, N.K.; Gengozian, N. A method of purifying feline T lymphocytes from peripheral blood using the plant lectin from *Pisum sativum*. J. Immunol. Meth. **1994**, *175*, 189.
4. Pereira, M.E.; Kabat, E.A. A versatile immunoadsorbent capable of binding lectins of various specificities and its use for the separation of cell populations. J. Cell Biol. **1979**, *82*, 185.
5. Ghetie, V.; Mota, G.; Sjoquist, J. Separation of cell by affinity chromatography on SpA-sepharose 6MB. J. Immunol. Meth. **1978**, *21*, 133.
6. Tassi, C.; Fortuna, A.; Bontadini, A.; Lemoli, R.M.; Gobbi, M.; Tazzari, P.L. CD34 or S313 positive cells selection by avidin-biotin immunoadsorption. Haematologica **1991**, *76* (Suppl 1.), 41.
7. Dvorak, D.J.; Gipps, E.; Kidson, C. Isolation of specific neurones by affinity methods. Nature **1978**, *271*, 564.
8. Tlaskalova-Hogenova, H.; Vetvicka, V.; Pospisil, M.; Fornusek, L.; Prokesova, L.; Coupek, J.;

Frydrychova, A.; Kopecek, J.; Fiebig, H.; Brochier, J. Separation of human lymphoid cells by affinity chromatography and cell surface labelling by hydroxyethyl methacrylate particles using monoclonal antibodies. J. Chromatogr. **1986**, *376*, 401.

9. Kondorosi, E.; Nagy, J.; Denes, G. Optimal conditions for the separation of rat T lymphocytes on anti-immunoglobulin–immunoglobulin affinity columns. J. Immunol. Meth. **1977**, *16*, 1.

10. van Overveld, F.J.; Terpstra, G.K.; Bruijnzeel, P.L.; Raaijmakers, J.A.; Kreukniet, J. The isolation of human lung mast cells by affinity chromatography. Scand. J. Immunol. **1988**, *27*, 1.

SUGGESTED FURTHER READING

Phillips, T.M.; Dickens, B.F. *Affinity and Immunoaffinity Purification Techniques*; BioTechniques Books; Eaton Press: Natick, MA, 2000.

Sharma, S.K.; Mahendroo, P.P. Affinity chromatography of cells and cell membranes. J. Chromatogr. **1980**, *184*, 471.

C

Centrifugal Partition Chromatography

M.-C. Rolet-Menet
Laboratoire de Chimie Analytique, UFR de Sciences Pharmaceutiques et Biologiques, Paris, France

INTRODUCTION

Centrifugal partition chromatography (CPC) is a method based on countercurrent chromatography (CCC). Separation is based on the differences in partitioning behavior of components between two immiscible liquids. Like HPLC, the phase retained in the column is called the stationary phase, and the other one, the mobile phase. In CCC, there are two modes by which to equilibrate the two immiscible phases. They depend on the characteristics of the centrifugal force field, which permits retention of the stationary phase inside the column. Devices that equilibrate the phases according to the so-called "hydrodynamic mode" were developed by Mandava and Ito[1]. They use a centrifugal force variable in intensity and direction. Alternating zones of agitation and settling of both phases are present along the column. In contrast, CPC uses a so-called "hydrostatic mode," owing to a centrifugal force constant in intensity and direction. Therefore, the mobile phase penetrates the stationary phase either by forming droplets, or by jets stuck to the channel walls, broken jets, or atomization. The more or less vigorous agitation of both phases depends on the intensity of the centrifugal force, the flow rate of the mobile phase, and the physical properties of the solvent system. Chromatographic separations obtained in hydrostatic mode are less efficient than those in the hydrodynamic one. But the retention of the stationary phase is less sensitive to the physical properties of solvents systems, such as viscosity, density, and interfacial tension. This particularity justifies the wide application field of CPC.

APPARATUS

The CPC column is made of channels engraved in plates of an inert polymer (Fig. 1), and they are connected by narrow ducts. Several plates are put together to form a cartridge. The cartridges are placed in the rotor of a centrifuge and connected to form the chromatographic column. The mobile phase enters and leaves the column via rotary seals. Since two immiscible liquids are present in the channel, the denser liquid moves away from the axis because of the centrifugal force. The less dense liquid is pushed toward the axis. The mobile phase can be either the lighter or the denser phase. In the latter case, the mobile phase flows through the channels from the axis to the outside of the rotor. This is called the descending mode. The other case, where the mobile phase flows toward the axis, is called the ascending mode.

Hydrostatic apparatuses are manufactured by Sanki Engineering Limited (Kyoto, Japan). They include two types of devices: The first are designed for analytical or semipreparative scale applications, and the second for scale-up at industrial scale. Centrifugal partitioning chromatograph type LLN was introduced in 1984 but is no longer available since 1992. It could be thermostatted from 15 to 35°C in an ambient temperature of 25°C. Type HPCPC or Series 1000 supersedes type LLN. The HPCPC main frame is a 31 × 47 × 50 cm centrifuge operating in the range 0–2000 rpm; it cannot be thermoregulated. The rotor consists of two packs of six disks each, connected through a 1/16 in. tubing, and easily removable. Larger instruments have internal volumes from 1.4 to 30 L, can be used with flow rates ranging from 20 to 700 ml/min, and are custom designed for specific separation processes at a small industrial scale.

RETENTION OF STATIONARY PHASE

Before any use, the column is first filled with stationary phase and then rotated at the desired rotational speed. The mobile phase is then pumped into the cartridge at the desired flow rate and pushes out of the column a certain volume of stationary phase. Hydrostatic equilibrium is reached when the mobile phase is expelled at the column outlet. The retention of stationary phase, S_F, is defined as $S_F = V_s/V_t$, where V_s is the stationary phase volume in the column after equilibrium and V_t the total volume of the column.

The value of S_F depends on several parameters,[2,3] including the hydrodynamic properties of the channels, the centrifugal force (S_F increases to reach a maximum with the centrifugal force), the Coriolis force defined by the clockwise or counterclockwise column rotation[4] (higher retention of stationary phase is obtained with counterclockwise rotation), the mobile

Encyclopedia of Chromatography DOI: 10.1081/E-ECHR-120039913

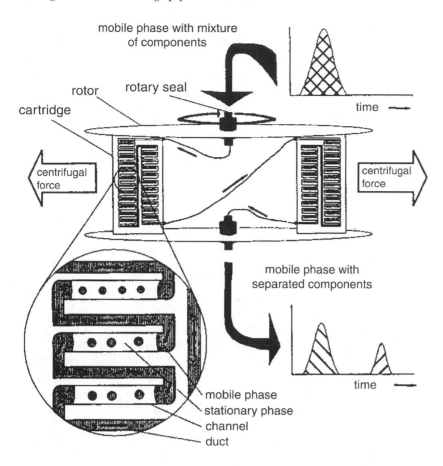

Fig. 1 Schematic representation of the CPC apparatus. (From Ref.[8].)

phase flow rate (S_F decreases linearly with mobile phase flow-rate), the physical properties of the solvent system (such as viscosity, density, interfacial tension), the sample volume, the sample concentration, the tensioactive properties of the solutes to be separated, etc. It is necessary to precisely monitor S_F because various chromatographic parameters depend on it, in particular the efficiency, the retention factor, and the resolution. Foucault[3] proposed an explanation for the variation of S_F with the various parameters described previously. He modeled the mobile phase in a channel as a droplet and applied the Stokes law, which relies on the density difference between the two phases, the viscosity of the stationary phase, and the centrifugal force. Then, he applied the Bond number derived from the capillary wavelength, which was formerly introduced for the hydrodynamic mode[5] and which relies on the density difference between the two phases, the interfacial tension, and the centrifugal force.

PRESSURE DROP

Van Buel, Van der Wielen, and Luyben[6] have proposed a model to explain the considerable pressure

drop arising in the column during CPC separation. The overall pressure drop is the sum of the hydrostatic pressure drop term and the hydrodynamic pressure drop terms over the individual parts of the system. The hydrostatic contribution is caused by the difference in density between the liquids in the ducts and in the channels ($\Delta P_{\text{stat}} = nl\Delta\rho\omega^2 R$, where n is the number of channels, l the height of stationary phase in the channel, $\Delta\rho$ the density difference between the phases, ω the rotational speed, and R the average rotational radius of the cartridge). The hydrodynamic contribution (ΔP_{hydr}) is caused by the friction of the mobile phase with the walls of the channels and ducts. This latter, in a channel and a duct, is proportional to the mobile phase density, the square of its linear velocity, the lengths of channel and duct, and the inverse of channel and duct diameter. Consequently, the overall pressure drop depends on the flow rate and rotational speed (input variables), the physical properties of the two-phase solvent system (variables), the geometry of the channels and ducts, the number of channel–duct combinations (apparatus variables), and the hold-up of stationary phase in the channel. The maximum pressure is limited by the rotary seals, which can support about 60 bars before leaking. Resolution and efficiency depend on the same variables as the pressure drop.

Therefore, it is important to determine which combinations of input variables and liquid two-phases systems can be applied, with respect to the maximum pressure that can be supported by the rotary seal.

EFFICIENCY

For a symmetrical peak, Efficiency (N) in CCC can be defined as in HPLC by

$$N = 16\left(\frac{V_r}{\omega}\right)^2$$

where V_r is retention volume of the solute and ω the peak base width expressed in volume units as V_r. For an asymmetrical peak, efficiency can be defined according to the Foley–Dorsey formula

$$N = 41.7\frac{(t_r/\omega_{0.1})^2}{(A/B) + 1.25}$$

where $\omega_{0.1}$ is the peak width at 10% of the peak height and A/B the asymmetry factor, with $A + B = \omega_{0.1}$.

Centrifugal partition chromatography apparatuses are still generally regarded as lacking efficiency (compensated by high, selectivity and S_F). The efficiency variation shows a minimum when the flow rate of the mobile phase is increased, which is the opposite of the usual HPLC Van Deemter plot. This observation has been modeled by Armstrong, Bertrand, and Berthod.[7] The mobile phase, when it comes out of the duct, flows very quickly to reach an intermediate emulsified layer and then settles in a third step before being transferred to another channel. In these conditions

$$\ln(1 - E) = \frac{A}{F} - BF^b$$

where

$$E = \frac{C_{m,t} - C_{m,0}}{C_{m,eq} - C_{m,0}}$$

$C_{m,t}$, $C_{m,0}$ and $C_{m,eq}$ are the solute concentrations in the mobile phase at a moment t, before equilibrium, and after equilibrium, respectively, and A depends on

S_F, B on the physical properties of the solvent system, and b on the solute and solvent systems. This variation is very interesting because it shows that a high mobile phase flow rate decreases the retention time without decreasing efficiency. However, it is observed that S_F decreases with the flow rate and the resolution Rs also decreases, as described in the following section. The flow rate of the mobile phase may be increased to lower the separation time but on condition that S_F remains adequate to maintain a sufficient Rs and consequently the quality of separation remains satisfactory.[2] Van Buel, Van der Wielen, and Luyben[8] first improved the understanding of the influence of flow patterns on the mass transfer between the two liquid phases and the chromatographic efficiency of CPC instruments. They directly visualized the mobile phase flow through the stationary phase (in a plane parallel to the rotation axis) as a function of the rotational speed and flow rate of the mobile phase. Four main types of flow states were observed: large droplets, jets stuck along the channel walls, broken jets, and atomization. Marchal et al.[9] visualized the mobile phase flow in a plane perpendicular to the rotation axis for different solvent systems [heptane/methanol (heptane/MeOH), chloroform/n-propanol/methanol/water (chloroform/n-ProOH/MeOH/W), heptane/chloroform/n-ProOH/MeOH/W, n-butanol/acetic acid/water (n-ButOH/acetic acid/W), aqueous two-phase systems]. They confirmed the observations of Van Buel et al. and observed deviations of jets or droplets from the radial direction caused by the Coriolis force. They correlated the chromatographic efficiency to the flow pattern observed: non-Gaussian peak corresponding to the jets stuck along the channel wall, and increase of efficiency when jets come unstuck from the walls. Increase of flow rate and rotation speed generally yielded better efficiencies.

RESOLUTION

Resolution (Rs) in CCC can be defined as in HPLC by

$$Rs = 2\frac{V_{r2} - V_{r1}}{\omega_1 + \omega_2} = 2V_s\frac{K_2 - K_1}{\omega_1 + \omega_2}$$

where V_{r1} and V_{r2}, ω_1 and ω_2, and K_1 and K_2 are, respectively, the retention volumes, the peak base widths expressed in volume units as V_r, and the partition coefficients of the first and second eluted solutes. Rs is directly proportional to volume V_s of the stationary phase and hence on the flow rate of the mobile phase and the centrifugal force.[2,10] Resolution in CCC, as in HPLC, is governed by the Purnell

Table 1 Applications in CPC

Species	Solvents system
Polyphenols and tannins[3]	CHCl$_3$/MeOH/W (7:13:8; v/v/v), CHCl$_3$/MeOH/ n-ProOH/W (9:12:2:8; v/v/v/v), n-BuOH/n-ProOH/W (4:1:5; v/v/v)
Triptolide and tripdiolide[3]	HEX/EtOAc/CH$_2$Cl$_2$/MeCN/MeOH/W (12:10:3:10:5:6; v/v/v/v/v/v)
Lanthanoids[3]	HEX containing bis(2-ethylhexyl)phosphoric/0.1 mol/L (H, Na)Cl$_2$CHCOO to an appropriate pH
Flavonoids[12]	CHCl$_3$/MeOH/W (5:6:4; v/v/v)
Polyphenols[12]	C$_6$H$_{12}$/EtOAc/MeOH/W (7:8:6:6; v/v/v/v)
Tannins[12]	n-BuOH/n-ProOH/W (2:1:3; v/v/v)
Naphthoquinones[12]	HEX/MeCN/MeOH (8:5:2; v/v/v)
Retinals[12]	C$_6$H$_6$/n-C$_5$H$_{12}$/MeCN/MeOH (500:200:200:11; v/v/v/v)
Chiral compounds[13]	Various systems containing chiral selectors

relation

$$Rs = \frac{\sqrt{N}}{4} \frac{k_2'}{1 + k_2'} \frac{1 - \alpha}{\alpha}$$

where k_2' is the retention factor of the second solute and α the separation factor. N is controlled by F, the centrifugal force, S_F, and the physical properties of solvent system, k' by the nature of the solvent system (through the partition coefficient, K, and S_F), and α mainly by the solvent system. This relation shows that it is essential in CCC to control technical parameters and to judiciously choose the solvent system to separate the products. Moreover, Ikehata et al.[4] showed that partition coefficients, efficiencies, and Rs were improved by rotating the column in the clockwise direction.

SOLVENT SYSTEMS

The choice of the solvent system is the key parameter to good separation. On one hand, its physical properties define S_F, N, and Rs, on the other hand, the relative polarities of its two phases define the partition coefficients of the solutes and, as a result, the selectivities and the retention factors. The usual solvent systems are biphasic and made of three solvents, two of which are immiscible. We only give guidelines for the choice of solvent system. If the polarities of the solutes are known, the classification established by Ito[1] can be taken as a first approach. He classified the solvent systems into three groups, according to their suitability for nonpolar molecules ("nonpolar" systems), intermediary polarity molecules ("intermediary" system), and polar molecules ("polar" system). The

molecule must have a high solubility in one of the two immiscible solvents. The addition of a third solvent enables better adjustment of the partition coefficients. When the polarity of the solutes is not known, the Oka[11] approach uses mixtures of n-hexane (HEX), ethyl acetate (EtOAc), n-BuOH, McOH, and water ranging from the HEX/MeOH/W (2:1:1; v/v/v) to the n-BuOH/W (1:1; v/v) systems and mixtures of chloroform, MeOH, and water. This solvent series covers a wide range of hydrophobicities, from the nonpolar HEX/MeOH/W system to the polar n-BuOH/W system. Moreover, all these solvent systems are volatile and yield a desirable two-phase volume ratio of about 1. The solvent system leading to partition coefficients close to 1 is selected.

APPLICATIONS

Numerous applications using CPC are described in reference books,[1,3] covering organic and mineral solutes (Table 1). We only give key examples extracted from the CPC literature.

Polyphenols and Tannins

Open column chromatography with silica gel and alumina is not applicable to the fractionation of tannins because of their strong binding to these adsorbents, which induces extensive loss of the compounds. Such losses do not occur with CCC, as it does not use a solid stationary phase. Such molecules are very polar, so that butanol-based solvent systems can be used. Centrifugal partition chromatography is more appropriate in this case compared to hydrodynamic CCC thanks to the good retention of the stationary phase of this solvent system. Okuda, Yoshida, and Hatano[14] separated castalagin from vescalagin by using the solvent system n-BuOH/n-ProOH/W (4:1:5; v/v/v). They are diastereoisomers that differ only in the configuration of the hydroxyl group of the central carbohydrate moiety. In the same way, these authors have separated oligomeric hydrolysable macrocyclic tannins oenothein B and woodfordins by using n-BuOH/n-ProOH/W (4:1:5; v/v/v). In spite of a small structural difference (presence or absence of a galloyl group), these dimers showed a considerable difference of partition coefficients in this solvent system (0.36 for woodfordin C and 0.19 for oenothein B).

Preparative Separation of Raw Materials

One of the major applications of CPC is the purification of natural products from vegetal extracts (flowers, roots, etc.) or crude extracts from fermentation broth without previous sample preparation. Hostettmann and coworkers[12] have described many examples of isolation of natural products by CPC. Some flavonoids are, for instance, purified by using solvent systems containing chloroform, some coumarins by using solvent systems containing HEX and EtOAc, and more polar products, such as tannins, by butanol-based systems. The main interest of this technique lies, however, in the possibility to overload its column so that all the applications of semipreparative chromatography are available. For instance, Menet and Thiébaut[15] have separated 140 mg of an antibiotic from a crude extract of a fermentation broth. Some fractions of up to 95% purity were collected, while the original extract contained only 7% of the molecule of interest. They have also compared the performances of CPC, preparative LC, and hydrodynamic mode CCC. They finally showed that the solvent consumption is the lowest for CPC, while the enrichment is the best.

The pH-zone refining mode was introduced by Ito.[16] It is a variant of displacement chromatography. It is devoted to the purification of compounds whose electric charge depends on pH. For example, a mixture of free acids is injected in the organic stationary phase along with an acid stronger than all the compounds to be separated. The compounds are moved along the column by pumping a basic aqueous mobile phase. Pure products are eluted from the column as salts, by contiguous rectangularly shaped peaks arranged according to the pK_a values and partition coefficients. This mode allowed a preparative isolation of indole alkaloids from *Catharanthus roseus*[17] The solvent system consisted of a mixture of methyl-*tert*-butyl ether, acetonitrile, and water. The upper organic phase was made basic with triethylamine and used as the mobile phase. The lower aqueous phase was acidified by hydrochloric acid.

log $P_{oct/water}$[18,19]

Octanol–water partition coefficients ($K_{o,w}$) have been established as the most relevant quantitative physical property correlated with biological activity. Centrifugal partition chromatography using octanol and water as the two phases is a useful alternative for providing octanol–water partition coefficients ($K_{o,w}$). It offers automation advantages compared to HPLC and the shake-flask method. Three approaches for determining $K_{o,w}$ by CPC have been described. The normal mode consists in equilibrating the CPC column according to a normal equilibrium and the overloading mode by artificially decreasing the volume of the stationary phase. $K_{o,w}$ is calculated according to the classical formula $K = k'(V_t - V_s)/V_s$ The second procedure is the dual-mode method,[20] which is based on the exchange of the role of the mobile and stationary phases during the experiment. Therefore, the determination range of partition coefficients can be extended. The third procedure, the cocurrent mode, relies on the simultaneous pumping of a mixture of a small flow of octanol and a larger flow of water to elute strongly retained compounds.

Elution Gradient[21,22]

Another way to extend the polarity range of analyzed compounds is the elution gradient. During the separation, the composition of the mobile phase is modified, while the composition of the stationary is kept constant. But in CPC, when the composition of the mobile phase is modified, the composition of the stationary phase changes. To prevent instability of the stationary phase during a gradient run, the change in stationary phase composition should be lower than 20% (v/v). Not all ternary two-phase systems are suitable for elution gradient. Foucault and Nakanishi[23] gave an overview of two-phase systems that are suitable for gradient elution.

CONCLUSIONS

Centrifugal partition chromatography is a method based on CCC. Devices equilibrate the phases according to a so-called "hydrostatic mode" owing to a centrifugal force constant in intensity and direction. So, the retention of the stationary phase is less sensitive to the physical properties of solvent systems compared to the "hydrodynamic mode." This particularity justifies the wide application field of CPC: use of *n*-But OH solvent systems, aqueous two-phase systems, and elution gradient. Moreover, the largest instruments have an internal volume from 1.4 to 30 L and are custom designed for specific separation processes at a small industrial scale. Finally, a better understanding of CPC (influence of Coriolis force) shows that the geometry of channel and duct are critical. A new CPC apparatus could take this into account.

REFERENCES

1. Mandava, B.N.; Ito, Y. Principles and instrumentation of counter current chromatography. In *Counter Current Chromatography. Theory and Practice*; Chromatographic Science Series; Mandava, B.N., Ito, Y., Eds.; Marcel Dekker, Inc.: New York, 1988; Vol. 44, 79–442.

2. Menet, J.-M.; Rolet, M.-C.; Thiébaut, D.; Rosset, R.; Ito, Y. Fundamental chromatographic parameters in counter-current chromatography: influence of the volume of stationary phase and the flow-rate. J. Liq. Chromatogr. **1992**, *15*, 2883–2908.

3. Foucault, A.P. Theory of centrifugal partition chromatography. In *Centrifugal Partition Chromatography*; Chromatographic Science Series; Foucault, A.P., Ed.; Marcel Dekker, Inc.: New York, 1995; Vol. 68, 25–50.

4. Ikehata, J.-I.; Shinomiya, K.; Kobayashi, K.; Ohshima, H.; Kitanaka, S.; Ito, Y. Effect of Coriolis force on counter-current chromatography separation by centrifugal partition chromatography. J. Chromatogr. A. **2004**, *1025*, 169–175.

5. Menet, J.-M.; Thiébaut, D.; Rosset, R.; Wesfreid, J.E.; Martin, M. Classification of countercurrent chromatography solvent systems on the basis of the capillary wavelength. Anal. Chem. **1994**, *66*, 168–176.

6. Van Buel, M.J.; Van der Wielen, L.A.; Luyben, K.Ch.A.M. Pressure drop in centrifugal partition chromatography. J. Chromatogr. **1997**, *773*, 1–12.

7. Armstrong, D.W.; Bertrand, G.L.; Berthod, A. Study of the origin and mechanism of band broadening and pressure drop in centrifugal partition chromatography. Anal. Chem. **1988**, *60*, 2513–2519.

8. Van Buel, M.J.; Van den Wielen, L.A.M.; Luyben, K.Ch.A.M. Pressure drop in centrifugal partition chromatography. In *Centrifugal Partition Chromatography*; Chromatographic Science Series; Foucault, A.P., Ed.; Marcel Dekker, Inc.: New York, 1995; Vol. 68, 51–69.

9. Marchal, L.; Foucault, A.; Patissier, G.; Rosant, J.M.; Legrand, J. Influence of flow patterns on chromatographic efficiency in centrifugal partition chromatography. J. Chromatogr. A. **2000**, *869*, 339–352.

10. Murayama, W.; Kobayashi, T.; Kosuge, Y.; Yano, H.; Nunogaki, Y.; Nunogaki, K. A new centrifugal counter-current chromatograph and its applications. J. Chromatogr. A. **1982**, *239*, 643–649.

11. Oka, H.; Harada, K.-I.; Ito, Y.; Ito, Y. Separation of antibiotics by counter-current chromatography. J. Chromatogr. A. **1998**, *812*, 35–52.

12. Maillard, M.; Marston, A.; Hostettmann, K. High speed counter current chromatography of natural products. In *High-Speed Counter Current Chromatography*; Chemical Analysis; Ito, Y., Conway, W.D., Eds.; John Wiley and Sons: New York, 1995; Vol. 132, 179–218.

13. Foucault, A. Enantioseparations in counter-current chromatography and centrifugal partition chromatography. J. Chromatogr. A. **2001**, *906*, 365–378.

14. Okuda, T.; Yoshida, T.; Hatano, T. Fractionation of plant polyphenols. In *Centrifugal Partition Chromatography*; Chromatographic Science Series; Foucault, A.P., Ed.; Marcel Dekker, Inc.: New York, 1995; Vol. 68, 99–132.

15. Menet, M.-C.; Thiébaut, D. Preparative purification of antibiotics for comparing hydrostatic and hydrodynamic mode counter-current chromatography and preparative high-performance liquid chromatography. J. Chromatogr. A. **1999**, *831*, 203–216.

16. Weisz, A.; Sher, A.L.; Shinomiya, K; Fales, H.M.; Ito, Y. A new preparative-scale purification technique: pH zone-refining countercurrent chromatography. J. Am. Chem. Soc. **1994**, *116*, 704–708.

17. Renault, J.H.; Nuzillard, J.-M.; le Crorérour, G.; Thépenier, P.; Zèches-Hanrot, M.; Le Men-Olivier, L. Isolation of indole alkaloids from *Catharanthus roseus* by centrifugal partition chromatography in the pH-zone refining mode. J. Chromatogr. **1999**, *849*, 421–431.

18. Berthod, A.; Talabardon, K. Operating parameters and partition coefficient determination. In *Counter Current Chromatography*; Chromatographic Science Series; Menet, J.M., Thiébaut, D., Eds.; Marcel Dekker, Inc.: New York, 1999; 121–148.

19. Wang-Fan, W.; Kusters, E.; Mak, C.-P.; Wang, Y. Application of centrifugal counter-current

chromatography to the separation of macrolide antibiotic analogues. II. Determination of partition coefficients in comparison with the shake-flask method. J. Chromatogr. A. **2000**, *23* (9), 1365–1376.

20. Bourdat-Deschamps, M.; Herrenknecht, C.; Akendengue, B.; Laurens, A.; Hocquemiller, R. Separation of protoberberine quaternary alkaloids from a crude extract of *Enantia chlorantha* by centrifugal partition chromatography. J. Chromatogr. A. **2004**, *1041* (1–2), 143–152.

21. Foucault, A.; Nakanishi, J.L.C. Gradient elution centrifugal partition chromatography comparison with HPLC gradients and use of ternary diagrams to build gradients. J. Liq. Chromatogr. **1990**, *13*, 3583–3602.

22. Van Buel, M.J.; Van der Wielen, L.A.M.; Luyben, K.Ch.A.M. Modelling gradient elution in centrifugal partition chromatography. J. Chromatogr. A. **1997**, *773*, 13–22.

23. Foucault, A.; Nakanishi, K. Gradient elution in centrifugal partition chromatography: use of ternary diagrams to predict stability of the stationary liquid phase and calculate the composition of initial and final phases. J. Liq. Chromatogr. **1989**, *12*, 2587–2600.

Centrifugal Precipitation Chromatography

Yoichiro Ito
*Center of Biochemistry and Biophysics, National Heart, Lung, and Blood Institute,
National Institutes of Health, Bethesda, Maryland, U.S.A.*

INTRODUCTION

For many years, proteins have been fractionated with ammonium sulfate (AS) by stepwise precipitation. In this conventional procedure, an increasing amount of AS is added to the protein solution and the precipitates are removed by centrifugation in each step. Recently, "centrifugal precipitation chromatography"[1,2] has been developed to replace the tedious manual procedure. This novel chromatographic system is capable of internally generating a concentration gradient of AS through a long separation channel under a centrifugal force field. Proteins introduced into the channel are exposed to a gradually increasing AS concentration and finally precipitated at different locations according to their solubility in the AS solution. Then, the AS concentration in the upper channel is gradually reduced so that the AS concentration gradient in the lower channel is proportionally decreased. This manipulation causes the precipitated proteins to be redissolved and eluted out by repeating precipitation and dissolution along the channel. As in LC, the effluent is continuously monitored with a UV monitor and fractionated in test tubes using a fraction collector.

PRINCIPLE AND DESIGN OF THE APPARATUS

The principle and unique design of the separation column is shown in Fig. 1A and B, respectively. In Fig. 1A, a pair of separation channels is partitioned by a dialysis membrane. A concentrated (C) AS solution is introduced through the upper channel at a high flow rate (V) from the right, whereas water is fed into the lower channel from the left at a much lower rate (v). This countercurrent flow of the two liquids through the channel results in AS transfer from the upper channel to the lower channel at every point, as shown by arrows across the membrane. Because the AS transfer rate through the membrane is proportional to the difference in AS concentration between the two channels, an exponential gradient of AS concentration (c) is formed through the lower channel. The system allows manipulation of the AS concentration in this gradient by modifying the AS concentration in the upper channel as described earlier. The separation column is fabricated from a pair of plastic disks (high

density polyethylene, 13.5 cm in diameter and 1.5 cm thick) equipped with mutually mirror-imaged spiral grooves (1.5 mm wide, 2 mm deep, and ca. 200 cm long), as shown in Fig. 1B. A regenerated cellulose membrane with desirable pore size (6000–8000 or 12,000–14,000) is sandwiched between these two disks, which are in turn tightly pressed between two metal plates with a number of screws. The capacity of each channel is about 5 ml. This column assembly is mounted on a seal-less continuous flow centrifuge, which allows continuous elution through the rotating column without the use of a conventional rotary seal device. Fig. 1C schematically illustrates the entire elution system of the present chromatographic system. Two sets of pumps are used, one (upper right) for pumping AS solution at a high rate, and the other (upper left) for eluting buffer solution and protein samples at a lower rate. The total of four flow tubes leading from these two pumps are bundled together and clamped at the top of the seal-less continuous flow centrifuge to reach the column assembly as illustrated. As mentioned above, these flow lines are twist free as the column rotates around the central axis of the centrifuge. Consequently, the system eliminates various complications such as leakage, clogging, and cross-contamination, which often arise from the use of the conventional rotary seal device for multiple flow lines.

BASIC STUDIES

A series of experiments was conducted to study the AS transfer rate through the dialysis membrane by pumping a concentrated AS solution into the upper channel at 1 ml/min and water into the lower channel at varied flow rates ranging from 1 to 0.1 ml/min without sample injection. In these experiments, the AS input concentration into the upper channel and the AS output concentration from the lower channel were compared. The rate of AS transfer rose, as expected, with decrease of flow rate through the water channel, and at a flow rate of 0.1 ml/min, the AS concentration collected through the lower channel reached nearly 100% that of the AS input in the upper channel. While AS diffuses from the upper channel toward the lower channel, water in the lower channel is absorbed into the upper channel. This water transfer rate is estimated by comparing

Encyclopedia of Chromatography DOI: 10.1081/E-ECHR-120039914

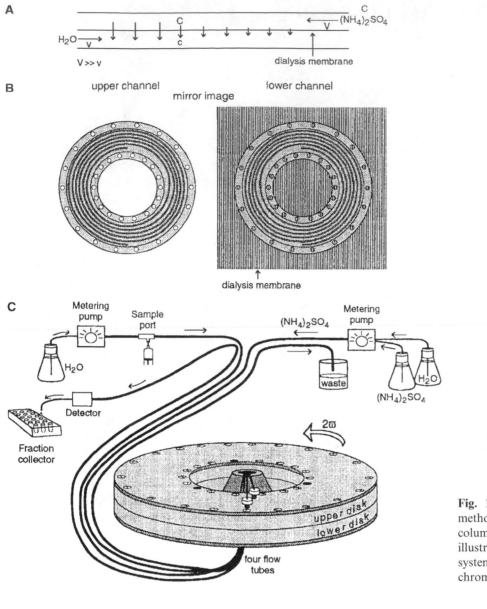

Fig. 1 (A) Principle of the present method; (B) design of the separation column assembly; and (C) schematic illustration of the entire elution system of centrifugal precipitation chromatography.

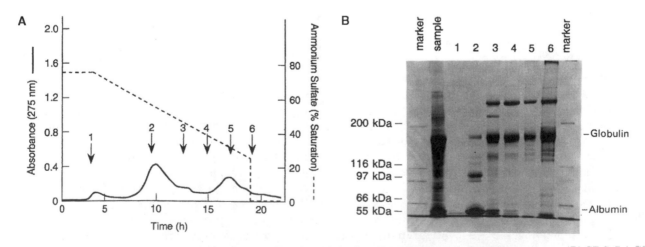

Fig. 2 Separation of human serum proteins by centrifugal precipitation chromatography. (A) Elution curve; (B) SDS–PAGE analysis of separated fractions.

A

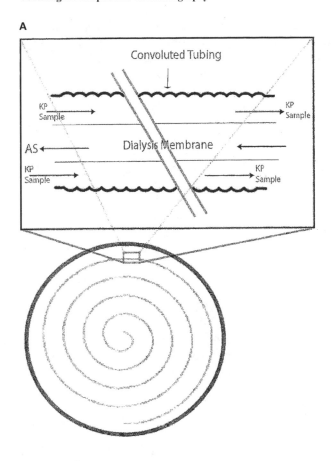

B

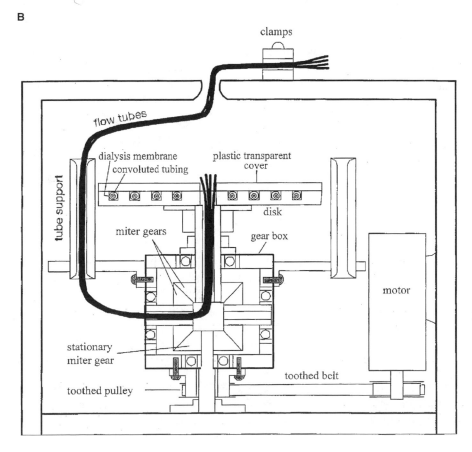

Fig. 3 Preparative centrifugal precipitation chromatograph. (A) Column design: a dialysis tubing (3 mm I.D.) is inserted into convoluted tubing (5.7 mm I.D. and 2.2 m length). (B) Cross-sectional view through the apparatus. The convoluted tubing is accommodated in a spiral groove of the centrifuge head.

input and output flow rates through the water channel. At an input rate of 0.1 ml/min, the outlet flow was decreased to one-fourth of the input rate, indicating that the separated fractions would be eluted in a highly concentrated state. Generating an AS concentration gradient and concentrating fractions are the two unique capabilities of the present system, which can be effectively utilized for the fractionation of proteins.

APPLICATIONS

Fig. 2 illustrates serum protein separation by centrifugal precipitation chromatography: the chromatographic tracing of the elution curve in Fig. 2A and SDS–PAGE analysis of separated fractions in Fig. 2B. In this example, 100 µl of normal human serum (pooled) was diluted to 1 ml and introduced into the separation channel. The experiment was initiated by filling both upper and lower channels with 75% AS solution followed by sample charge into the lower channel through the sample loop. After the separation column assembly was rotated at 2000 rpm, the upper channel was eluted with 75% AS solution at a flow rate of 1 ml/min, while the lower channel was eluted with 50 mM potassium phosphate at 0.06 ml/min. After 4 hr of elution, the AS concentration in the upper channel was linearly decreased down to 25%, as indicated in the chromatogram. The effluent from the lower channel was continuously monitored with a UV monitor (LKB Uvicord S) at 275 nm and fractionated into test tubes using a fraction collector (LKB Ultrorac), while the AS solution eluted from the upper channel was discarded. The chromatogram (Fig. 2A) produced two major peaks, one at AS saturation at 60–50% and the other at 35–30%. The SDS–PAGE analysis of peak fractions (Fig. 2B) revealed that the first peak represents albumin (MW 68,000) and the second peak, gamma-globulin (MW 160,000).

Centrifugal precipitation chromatography can produce highly purified protein fractions because the proteins are refined by repetitive precipitation and dissolution. The method has the following advantages over the conventional manual procedure: The method is programmed and automated; the fractions are almost free of small molecules, which are dialyzed through the membrane or otherwise quickly eluted out from the channel; noncharged macromolecules such as polysaccharides are washed out while charged biopolymers such as DNA and RNA may also be separated according to their solubility in AS solution; the method may be amenable for microscale to large-scale fractionation by design of the separation column in suitable dimensions.

The present system has been successfully applied to the fractionation of various protein samples, including serum proteins, monoclonal antibodies (IgM against mast cells) from hybridoma culture supernatant, minor protein components (less than 1% of total proteins) from a crude rabbit reticulocyte lysate containing a large amount of hemoglobin, and protein–polyethylene glycol conjugates. One important application of the present method is affinity separation using a ligand that can specifically bind to the target protein to substantially lower its solubility in the AS solution. This ligand–protein complex is then eluted out much later than most of other proteins. This affinity precipitation method has been demonstrated in the purification of recombinant proteins such as ketosteroid isomerase from a crude *Escherichia coli* lysate by adding an affinity ligand (17-estradiol–methylpolyethylene glycol-5000) to the sample solution. The application of the present method may be extended to fractionation of other biopolymers such as DNA and RNA using a pH gradient.

PREPARATIVE SCHEME

Recently, the sample loading capacity of the method was improved by a new column design, which uses dialysis membrane inserted into convoluted Teflon tubing (Fig. 3A). The column is snugly accommodated in a spiral groove of the rotary plate of the centrifuge (Fig. 3B). The AS solution is introduced through the dialysis tubing (membrane channel), while the phosphate buffer is eluted through the outside of the dialysis membrane (tubing channel) in the opposite direction. In this way, precipitated protein molecules can be retained more stably on the inner wall of the convolution tubing. A sample mixture containing 50 mg each of human serum albumin and gamma-globulin was separated by either stepwise or linear gradient elution of AS solution through the membrane channel.

ARTICLE OF FURTHER INTEREST

Instrumentation of Countercurrent Chromatography, p. 847.

REFERENCES

1. Ito, Y. Centrifugal precipitation chromatography applied to fractionation of proteins with ammonium sulfate. J. Liq. Chromatogr. & Rel. Technol. **1999**, *22* (18), 2825–2836.
2. Ito, Y. Centrifugal precipitation chromatography: principle, apparatus, and optimization of key parameters for protein fractionation by ammonium sulfate precipitation. Anal. Biochem. **2000**, *277*, 143–153.
3. Ng, V.; Yu, H.; Ito, Y. Preparative centrifugal precipitation chromatography using dialysis membrane inserted into convoluted tubing. J. Liq. Chromatogr. & Rel. Technol. *in press*.

Channeling and Column Voids

Eileen Kennedy
Novartis Crop Protection, Inc., Greensboro, North Carolina, U.S.A.

INTRODUCTION

Channeling can occur when voids that are created in the packing material of a column cause the mobile phase and accompanying solutes to move more rapidly than the average flow velocity. The most common result of channeling is band broadening and, occasionally, elution of peak doublets.

DISCUSSION

Column voids can develop in a poorly packed column from settling of the packing material or by erosion of the packed bed. In a properly packed column, voids can develop gradually over time or suddenly as the result of pressure surges. A void that forms in the inlet of a column may lead to poor peak shape, including severe band tailing, band fronting, or even peak doubling for every peak in the chromatogram. Filling the column inlet with the same or equivalent column packing can sometimes reduce voids. For this type of repair, the column should be held in a vertical position while the inlet frit is removed. The void will be evident as either settling of the packing material or as holes in the column surface. The new packing should be slurried with an appropriate mobile phase and packed into the column void with a flat spatula. Once the top of the new packing is level with the column end, a new inlet frit can be added and the end fitting reinstalled. The column should then be reconnected to the LC system and conditioned with the mobile phase at a fairly high flow rate to help settle the new column bed. After filling the void, the packing bed will generally be more stable if the repaired column is operated with the direction of flow reversed from the original direction. This repair procedure can be used to extend column life; however, it should be noted that the plate number of the repaired column would be, at best, only 80–90% of the original column. Columns that develop voids over time are often near the end of their useful life spans and in some cases it may be more cost efficient to discard such a column rather than to repair it.

SUGGESTED FURTHER READING

Dolan, J.W.; Snyder, L.R. *Troubleshooting LC Systems*; Humana Press: Totowa, NJ, 1989.

Majors, R.E. The care and feeding of modern HPLC columns. LC–GC. **1998**, *16*, 900.

Chelating Sorbents for Affinity Chromatography (IMAC)

Radovan Hynek
Anna Kozak
Jirí Sajdok
Jan Káš
Institute of Chemical Technology, Prague, Czech Republic

INTRODUCTION

Immobilized metal-ion affinity chromatography (IMAC) is a collective term that includes all kinds of affinity chromatography, where metal atoms or ions immobilized on polymer cause or dominate the interaction at the sorption site.

DISCUSSION

Metal-chelate affinity chromatography was introduced as a specific method for fractionation of proteins by Porath et al. in 1975.[1] The principle of this type of chromatographic method is that certain amino acid residues, such as histidine, cysteine, lysine, tryptophan, aspartic acid, glutamic acid, or phosphorylated amino acids, which are accessible on the protein surface, can interact through nonbonding lone-pair electron coordination with some metal ions. Metal cations Cu^{2+}, Ni^{2+}, Zn^{2+}, Co^{2+}, Fe^{3+}, Al^{3+}, and Cr^{3+}, and which have been chelated to ligands immobilized on support material, have already been used for such specific interactions.[2–3] The most widely used chelating ligands for the isolation of proteins are iminodiacetic acid (IDA) and its analogs, such as tricarboxyethylenediamine (TED). IDA is covalently coupled to an insoluble matrix (e.g., agarose or Sepharose) and forms stable coordinate compounds with a variety of divalent metal ions. These chelates create bases for the above-mentioned specific adsorption. Elution of adsorbed solutes from immobilized metal-ion affinity adsorbents can be provided by changing the pH of the elution buffer or by a specific competing solute, such as histidine, imidazole, or sodium phosphate, depending on the interaction types involved.

Ligands coupled to agarose gels were commonly used at the beginning of IMAC application; however, these sorbents were not suitable for (HPLC). Then, Small et al.[4] used a silica-based matrix and demonstrated that such IMA sorbents can be used in HPLC techniques. The metal-chelate adsorbent "TSK gel chelate-5PW,'' suitable for HPLC, which was prepared by coupling IDA to a hydrophilic resin-based matrix (TSK gel G 5000 PW), later became commercially available (Fig. 1).

Since the introduction of metal-ion affinity sorbents for the fractionation of proteins,[1] the method became popular for the purification of a wide variety of biomolecules. Metal-ion affinity sorbents are also widely used for the immobilization of enzymes. At present, IMAC is a powerful method for separation of phosphorylated macromolecules, particularly proteins and peptides. The significance of techniques for separation and characterization of phosphorylated biomolecules is now increasing, because phosphorylation modulates enzyme activities and mediates cell membrane permeability, molecular transport, and secretion. Phosphorylated peptides can be separated from a peptide mixture on IDA–Sepharose with Fe^{3+} ions (Fig. 2). The majority of peptides pass freely through an IMAC column, whereas acidic peptides, including phosphorylated ones, are retained and can be released by a pH gradient.

Acidic peptides are released in the pH range 5.5–6.2 and phosorylated peptides are eluted in the pH range 6.9–7.5.[5] Elution of retained peptides can also be performed with sodium phosphate. IMAC has been successfully used for the characterization of casein phosphopeptides in cheese extracts.[6] Phosphoproteins can be separated under very similar conditions as phosphopeptides. IMA sorbents were already used for fractionation of proteins according to the number of phosphate groups contained in their molecules.

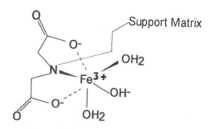

Fig. 1 Complex of water with IDA–Fe3+.

Encyclopedia of Chromatography DOI: 10.1081/E-ECHR-120039918

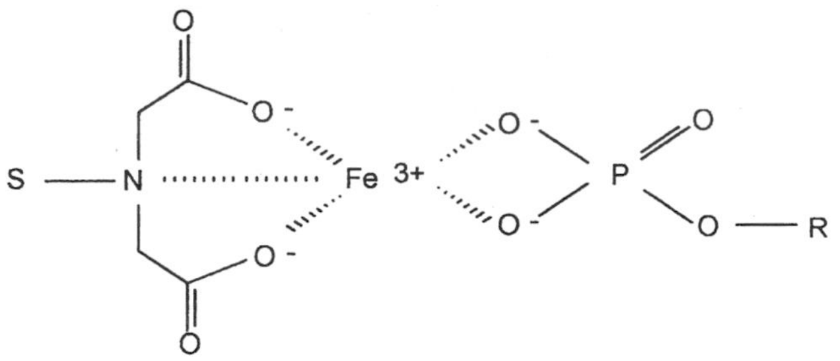

Fig. 2 Interaction of phosphate group with Fe^{3+} ion on IDA-Sepharose.

Separations of biomolecules on IMA sorbents achieved significant advances in the past few years, but detailed analyses of the mechanism of adsorption of molecules should be completed. Various factors, such as support matrix, chelating ligands, buffer composition, 0temperature, and so forth should be investigated in order to optimize analysis on metal-chelate ion affinity sorbents.

REFERENCES

1. Porath, J.; Carlsson, J.; Olson, I.; Belfrage, G. Nature **1975**, *258*, 598.
2. Sulkowski, E. Purification of protein by IMAC. Trends Biotechnol. **1985**, *3*, 1.
3. Hemdan, E.S.; Zhoa, Y.J.; Sulkowski, E.; Porath, J. Surface topography of histidine residues: A facile probe by immobilized metal ion affinity chromatography. Proc. Natl. Acad. Sci. USA **1989**, *86*, 1811–1815.
4. Small, D.A.P.; Atkinson, T.; Lowe, C.R. *Affinity Chromatography and Biological Recognition*; Chaiken, I.M., Wilchek, M., Parikh, I., Eds.; Academic Press: New York, 1983; 267.
5. Muszyńska, G.; Doborrowolska, G.; Medin, A.; Ekman, P.; Porath, J.O. Model studies on iron(III) ion affinity chromatography *1: II. Interaction of immobilized iron(III) ions with phosphorylated amino acids, peptides and proteins. J. Chromatogr. **1992**, *604*, 19–28.
6. Hynek, R.; Kozak, A.; Dráb, V.; Sajdok, J.; Káš, J. Adv. Food Sci. (CMTL) **1999**, *21* (5/6), 192.

Chemometrics in Chromatography

Tibor Cserháti
Esther Forgács
Department of Environmental Analysis, Institute of Chemistry, Budapest, Hungary

INTRODUCTION

During the last few decades, one of the major advances in chromatography has been the development and commercialization of automated chromatographic instruments. The output of retention data per unit of time has been considerably increased, and the evaluation of large data matrices, containing large amounts of chromatographic information (i.e., retention parameters of a homologous or nonhomologous series of solutes, measured on various stationary and mobile phases), is no longer possible without the application of high-speed computers and a wide variety of chemometric techniques. These methods allow the simultaneous evaluation of an almost unlimited amount of data, highly facilitating the clarification of both practical and theoretical problems. These chemometric procedures have been extensively employed in chromatography for the identification of the basic factors influencing retention and separation; for the comparison of various stationary and mobile phases; for the assessment of the relationship between molecular structure and retention behavior (quantitative structure–retention relationship, QSRR); for the elucidation of correlations between retention behavior and biological activity; etc. As each chemometric procedure generally highlights only one, or only a few features of the chromatographic problem under analysis, the concurrent application of more than one technique is rather a rule than an exception.

The objectives of this article are the enumeration, brief description, and critical evaluation of the recent results obtained in the application of various chemometric techniques in chromatography, and the comparison of the efficacy of various methods for the quantitative description of a wide variety of chromatographic processes. Fundamentals of chemometrics are discussed to an extent to facilitate the understanding of the principles at the application level.[1]

CHEMOMETRIC METHODS IN CHROMATOGRAPHY

Linear Regression Analyses

Linear and various multiple linear regression analysis techniques have been developed for the elucidation of the relationship between one dependent and one or more independent variables.[2] Because of their simplicity and excellent predictive power, they have been successfully applied in various fields of chromatography, such as gas–liquid chromatography (GLC), TLC, and HPLC.

Linear regression analysis with one independent variable

This simple technique can be employed in the case when the dependence of one parameter (dependent variable, Y) on another parameter (independent variable, X) has to be verified:

$$Y = a + bX \tag{1}$$

The result contains the intercept (a) values, an indicator of the amount of Y when X is equal to zero; slope (b) values measuring the change of Y at unit change of X; and the regression coefficient (r), an indicator of the extent of fit of equation to the experimental data, which serves for the determination of the significance level of the correlation and for the calculation of the variance of Y explained by X. Because of the restricted number of independent variables, the method found is only limited in applications in chromatography. It has been employed for the calculation of the dependence of the retention of one solute and the temperature of a column in GLC, and the concentration of one component in the mobile phase in TLC and HPLC. Furthermore, it can be used for the comparison of the separation characteristics of two (and no more) chromatographic systems. The log k'_o values of commercial pesticides, measured on alumina and on octadecyl-coated alumina columns, have been compared with this technique. They have been used for the calculation of the dependence of retention on the number of ethylenoxide groups of oligomeric nonionic surfactants on a porous graphitized carbon column; for the study of the effect of salt and pH on the hydrophobicity parameters of surfactant; for the assessment of the relationship between the retention and the hydrophobic surface area of nonylphenyl ethylene oxide oligomers on a polyethylene-coated zirconia HPLC column; for the determination of congenericity

Encyclopedia of Chromatography DOI: 10.1081/E-ECHR-120014246

of a set of 2,4-dihydroxythiobenzanilide derivatives by reversed-phase (RP) HPLC; etc.

Linear regression analysis with more than one independent variable (multiple linear regression analysis)

When the relationship between one dependent variable and more than one independent variables has to be calculated, Eq. (1) must be modified accordingly:

$$Y = a + b_1 X_1 \ldots + b_i X_i \ldots + b_k X_k \qquad (2)$$

In this instance, the r value is suitable only for the calculation of the variance of Y, explained by the X values; consequently, for the establishment of the significance level of the correlation, the F value has to be calculated and compared with the tabulated data. The path coefficients (normalized slope values) indicate the relative impact of the individual X values independent of their original dimensions. Because of the possibility to include more variables in the equation, the application field of multiple linear regression is more extended than that of simple linear regression analysis. Thus, it has been recently employed for the investigation of the molecular mechanism of separation, for the classification of modern stationary phases, for structure–retention relationship study in HPLC and in GLC, for the elucidation of the correlation between retention and biological activity, and for the study of the retention mechanism in adsorption and RP TLC. The method found further applications in the study of the effect of cyclodextrins and cyclo dextrin derivatives on the retention characteristics of a wide variety of bioactive compounds such as steroidal drugs, in the prediction of chromatographic properties of organophosphorous insecticides, and those of polychlorinated biphenyls in GLC.

Stepwise Regression Analysis

Stepwise regression analysis can be also used when the relationship between one dependent variable and more than one independent variables has to be assessed. In the common multiple linear regression analysis, the presence of independent variables exerting no significant influence on the change of dependent variable considerably decreases the significance level of the equation. Stepwise regression analysis automatically eliminates from the selected equation the dependent variables having no significant impact on the dependent variable, thereby increasing the reliability of calculation. The final form of the results of stepwise regression analysis is similar to Eqs. (1) or (2),

depending on the number of independent variables selected by the method. Because of its versatility and simplicity, the method has been frequently used in chromatography. It has found application in the study of the retention behavior of ethylene oxide surfactants and dansylated amino acids in adsorption and RP TLC, in the elucidation of the relative impact of various molecular parameters on the retention in adsorption and RP HPLC and GLC, in the determination of the molecular parameters significantly influencing the interaction of antibiotics with sodium dodecylsulfate measured by TLC, in the evaluation of the stability of pigments of paprika (*Capsicum annuum*) measured by HPLC, and in the determination of the relative impact of HPLC conditions on the retention behavior.

Partial Least Squares Regression (PLS)

When the independent variables are highly inter related, the application of traditional methods for the calculation of linear regressions may cause biased and unreliable results. PLS has been developed for the prevention of errors originating from such inter correlations. PLS has not been frequently employed in the analysis of chromatographic retention data; it has been only used in GLC for the study of the retention behavior of oxo compounds, in HPLC for the QSRR of chalcones, and in RP HPLC for the QSRR study of antimicrobial hydrazides.

Free–Wilson and Fujita–Ban Analysis

These special cases of multiple linear regression analysis have been developed for the determination of the impact of individual molecular substructures (independent variables) on one dependent variable. Both techniques are similar; yet, the Free–Wilson method considers the retention of the unsubstituted analyte as base, while Fujita–Ban analysis uses the less substituted molecule as reference. These procedures have not been frequently employed in chromatography; only their application in QSRR studies in RP TLC and HPLC have been reported.

Canonical Correlation Analysis (CCA)

CCA can be considered as a special case of multiple linear regression analysis, when the relationship between minimally more than one dependent variable (matrix I) and minimally more than one independent variable (matrix II) has to be elucidated. CCA acalculates the relationships between matrices I and II by extracting theoretical factors which explain the maximum of variance of the matrix with the lower number of variables. However, it can be employed only in the instances when the number of dependent variables is lower than that of independent variables. The maximal number of equations selected by CCA is equal to the number of columns in the smaller set of data. The results consist of the standard and weighted canonical coefficients [they are similar to the *b* values and path coefficients of Eq. (2)], of the *r* values related to the ratio of variance explained by the equations, and of the *X* (Greek Chi) value, indicating the fitness of equation to the experimental data. Despite its evident benefits, the technique has not been frequently employed for the analysis of chromatographic data. It has been applied for the elucidation of the relationship between the retention parameters of ring-substituted aniline derivatives determined on various HPLC columns (smaller matrix) and their calculated physicochemical parameters (larger matrix), for the study of the relationship between the physicochemical parameters of steroidal drugs and their retention characteristics in HPLC, and for the assessment of the correlation between the physicochemical parameters of tetrazolium salts and their retention behavior in various TLC systems.

Multivariate Mathematical–Statistical Methods

The prerequisite of the application of the regression analytical methods discussed above is that one or more chromatographic parameters have to be considered as being the dependent variables. However, when the simultaneous relationships among more retention parameters, or more retention parameters and more physicochemical parameters of a given set of analytes have to be elucidated, the linear regression methods cannot be employed. A considerable number of multivariate methods have been developed to overcome the disadvantages of regression analyses.[3] Various multivariate mathematical–statistical methods have been successfully employed for the elucidation of the relationship between the retention parameters and the structural descriptors of solutes for the comparison of more than two stationary phases, for the prediction of solute retention, for the assessment of the correlation between retention characteristics and biological activity, etc. As the information content of the mathematical–statistical methods considerably depends on the mode of calculation, the character of the problem to be elucidated limits, to some extent, the choice of the method.

Principal Component Analysis (PCA)

PCA can be used when the inherent relationships between the columns and rows of a data matrix have to be determined without one (stepwise regression analysis) or more (CCA) being the selected dependent variables. PCA is a versatile and easy-to-use multivariate mathematical–statistical method. It has been developed to contribute to the extraction of maximal information from large data matrices containing numerous columns and rows. PCA makes possible the elucidation of the relationship between the columns and rows of any data matrix without being one the dependent variable. PCA is a so-called projection method representing the original data in smaller dimensions. It calculates the correlations between the columns of the data matrix and classifies the variables according to the coefficients of correlation. The results of PCA generally contain the so-called eigenvalues which are related to the relative importance of the principal components calculated by PCA, the variance explained by the individual PCs, and the contributions (impacts) of the columns and rows of the original matrix to the principal component loadings and variables, respectively. Unfortunately, PCA does not define the principal components as concrete physical or physicochemical entities; it only indicates its mathematical possibility. Calculating linear regression between the principal component loadings and the chromatographic parameters and physicochemical characteristics may help the determination of the concrete constitution of principal components. Stepwise regression analysis is especially adequate to carry out such types of calculations.

Because of its simplicity, PCA has been frequently used in many fields of up-to-date chromatographic research. Thus, PCA has been employed for the evaluation of molecular lipophilicity, for QSRR studies, for the testing of the authenticity of edible oils, for the determination of the botanical origin of cinnamon, for the differentiation of Spanish white wines, for the characterization of RP supports, for the assessment of the relationship between molecular structure and retention behavior, etc. The method has found further applications in the classification of chili powders according to the distribution of pigments separated by TLC, in the determination of the molecular parameters of peptides and barbituric acid derivatives showing a significant impact on their retention on porous graphitized carbon column, in QSRR study of pesticide retention on polyethylene-coated silica column, in the study of the retention characteristics of titanium dioxide and polyethylene-coated titanium dioxide stationary phases, in the comparison of alumina stationary phases in TLC and HPLC, in the elucidation of the relationship between the retention of environmental pollutants on an alumina HPLC column and their physicochemical parameters, and in the study of the energy of interaction between commercial pesticides and a nonionic surfactant by GLC.

Spectral Mapping Technique (SPM)

The calculation methods discussed above classify the chromatographic systems (stationary and mobile phases) or solute molecules while simultaneously taking into consideration the retention strength and retention selectivity; thus, it cannot be applied when the separation of the strength and selectivity of the effect is required. SPM, another multivariate mathematical–statistical method, overcomes this difficulty.[4]

The SPM divides the information into two matrices using the logarithm of the data in the original matrix. The first one is a vector containing so-called potency values proportional to the overall effect; that is, it is a quantitative measure of the effect. The second matrix (selectivity map) contains the information related to the spectrum of activity, i.e., the qualitative characteristics of the effect. SPM first calculates the logarithm of the members of the original data matrix, facilitating the evaluation of the final plots in terms of log ratios. Subsequently, SPM subtracts the corresponding column-mean and row-mean from each logarithmic element of the matrix calculating potency values. The source of variation remaining in the centered data set can be evaluated graphically (selectivity map). This elegant and versatile calculation method has been used in chromatography for the characterization of stationary phases in TLC and HPLC, for the separation of the solvent strength and selectivity on a cyclodextrin-coated HPLC column using monoamine oxidase inhibitory drugs as solutes, for the investigation of the complex interaction between anticancer drugs and cyclodextrin derivatives, for the determination of the influence of storage conditions on pigments analyzed by HPLC, for the comparison of polymer-coated HPLC columns, and for the optimization of the microwave-assisted extraction of pigments for HPLC analysis.

Cluster Analysis (CA) and Nonlinear Mapping Technique (NLM)

Although both the PCA and the SPM techniques reduce the number of variables, the resulting matrices of PC loadings and variables and the spectral map are still multidimensional. The plot of PC loadings in the first vs. the second principal component has been frequently used for the evaluation of the similarities and differences among the observations. This method takes into consideration only the variance explained in the first two principal components and entirely ignores the impact of variances explained by the other principal components on the distribution of the matrix elements. The use of this approximation is only justified when the first two principal components explain the overwhelming majority of variance, which is not probable in the case of large original data matrices. As the evaluation of the distribution of data points in the multidimensional space is extremely difficult, calculation methods were developed for the reduction of the dimensionality of the matrices to one (CA) or to two (NLM). These methods can also be employed for the reduction of the dimensionality of the original data matrices before any other mathematical–statistical evaluation. CA has been employed for the elucidation of the retention behavior of anti-hypoxia drugs in adsorption TLC, that of barbituric acid derivatives and anti-inflammatory drugs in HPLC, for the classification of pharmaceutical substances according to their retention data, for the prediction of retention of phoshoramidate derivatives, etc.

However, both CA and NLM take into consideration the positive and negative sign of the coefficient of correlation and carry out the calculation accordingly. Therefore, the highly but negatively correlated points are far away on the maps and on the cluster dendograms in the same manner as the points that are not correlated. This procedure leads to correct assumptions in the case when the scientist is interested only in the positive correlations among variables and observations. To evaluate precisely the relationships between the points without taking into consideration the positive or negative character of the correlation, it is advisable to carry out the calculations with the absolute values of PC loadings and variables. The validity of this experimental approximation has been proven in the evaluation of the interaction of nonsteroidal anti-inflammatory drugs with a model protein studied by HPLC, and the parallel application of the original PC loadings and their absolute values in the data reduction techniques has been proposed. This procedure has been successfully used for the study of the effect of carboxymethyl-β-cyclodextrin on the hydrophobicity parameters of steroidal drugs measured by TLC, and for the assessment of the binding characteristics of environmental pollutants to the wheat protein, gliadin, investigated by HPLC.

The distances between the elements on the cluster dendograms and NL maps are a quantitative measure of similarity: Smaller distances indicate greater similarity. However, the fact that the differences among the elements are significant or not cannot be established on the traditional NL map or on the cluster dendogram. A graphical approximation has been developed for the inclusion of standard deviation in the NL maps and cluster dendograms. The data matrix for PCA has been composed from the main values of the matrix elements, the mean values minus twice their standard deviation, and the mean values plus twice their standard deviation. PCA has been carried out, and the cluster dendograms and NL maps have been calculated. A circle can be formed from the mean values and the mean value ± two standard deviations on the NL map, the center of the circle being the mean, and the radius of the circle being represented by the mean ± two standard deviations. It was assumed that the differences between the elements on the map are significant at the 95% significance level when the circles do not overlap. It was further assumed that the mean value and mean value ± two standard deviations of the matrix elements are close to each other (form a triad) on the cluster dendogram when they significantly differ from the others. The method has been employed for the classification of paprika (*C. annuum*) powders according to their pigment composition as determined by HPLC and for the comparison of HPLC and TLC systems.

Miscellaneous Multivariate Methods

The chemometric methods discussed above have found widespread applications in chromatography, and many theoretical and practical chromatographers have become familiar with these techniques and have applied them successfully. However, other less well-known methods have also found applicability in the analysis of chromatographic retention data. Thus, canonical variate analysis has been applied in pyrolysis GC/MS,[5] artificial neural network for the prediction of GLC retention indices, and factor analysis for the study of the retention behavior of *N*-benzylideneaniline derivatives.[6]

CONCLUSIONS

The examples enumerated above prove conclusively that chemometric techniques can be effectively employed for the elucidation of a large number of problems in chromatography, connected with the accurate and precise evaluation of large data matrices.

These methods allow not only the classification and clustering of any set of chromatographic systems but also exact determination of the relationship between the characteristics (physicochemical parameters or molecular substructures) of solutes and their retention behavior. It can be further concluded that chemometry considerably promotes a more profound understanding of the basic processes underlying chromatographic separations, increasing, in this manner, the efficiency (reliability, rapidity, etc.) of the methods.

REFERENCES

1. Cserháti, T.; Forgács, E. Use of Multivariate mathematical statistical methods for the evaluation of retention data matrices. In *Advances in Chromatography*; Brown, P.R., Grushka, E., Eds.; Marcel Dekker, Inc.: New York, 1996; Vol. 36, 1–63.
2. Mager, H. *Moderne Regressions analyse*; Salle, Sauerlander: Frankfurt am Main, Germany, 1982 .
3. Mardia, K.V.; Kent, J.T.; Bibby, J.M. Multivariate Analysis; Academic Press: London, 1979.
4. Levi, P.J. Spectral map analysis. Factorial analysis of contrast, especially from log ratios. Chemometr. Intell. Lab. Syst. **1989**, *5*, 105–116.
5. Kochanowski, B.K.; Morgan, S.L. Forensic discrimination of automotive paint samples using pyrolysis–gas chromatography–mass spectrometry with multivariate statistics. J. Chromatogr. Sci. **2000**, *38*, 100–108.
6. Ounnar, S.; Righezza, M.; Chretien, J.R. Factor analysis in normal phase liquid chromatography of *N*-benzylideneanilides. J. Liq. Chromatogr. & Relat. Technol. **1998**, *20*, 2017–2037.

SUGGESTED FURTHER READING

Acuna-Cueva, R.; Hueso-Urena, F.; Cabeza, N.A.J.; Jimenez-Pulido, S.B.; Moreno-Carretero, M.N.; Martos, J.M.M. Quantitative structure–capillary column gas chromatographic retention time relationships for natural sterols (trimethylsilyl esters) from olive oil. J. Am. Chem. Soc. **2000**, *77*, 627–630.

Al-Haj, M.A.; Kaliszan, R.; Nasal, A. Test analytes for studies of the molecular mechanism of chromatographic separations by quantitative structure–retention relationships. Anal. Chem. **1999**, *71*, 2976–2985.

Andrisano, V.; Bertucci, C.; Cavrini, V.; Recatini, M.; Cavalli, A.; Veroli, L.; Felix, G.; Wainer, I.W. Stereoselective binding of 2,3-substituted 3-hyroxy-propionic acids on an immobilized human serum albumin chiral stationary phase. Stereo-chemical characterisation and quantitative structure–retention relationship study. J. Chromatogr. A. **2000**, *876*, 75–86.

Dillon, W.R. *Multivariate Analysis*; John Wiley and Sons: New York, 1984; 213–254.

Geladi, P.; Kowlski, B.R. Partial least-squares regression: A tutorial. Anal. Chim. Acta **1986**, *185*, 1–17.

Gozalbes, R.; de Julián-Ortiz, J.; Antón-Fos, G.M.; Galvez-Alvarez, J.; Garcia-Domenech, R. Prediction of chromatographic properties of organophosphorous insecticides by molecular connectivity. Chromatographia **2000**, *51*, 331–337.

Hamoir, T.; Cuaste Sanchez, F.; Bourguignon, B.; Massart, D.L. Spectral mapping analysis: A method for the characterization of stationary phases. J. Chromatogr. Sci. **1994**, *32*, 488–498.

Heberger, K.; Gorgenyi, M. Principal component analysis of Kovats indices for carbonyl compounds in capillary gas chromatography. J. Chromatogr. A. **1999**, *845*, 21–31.

Ivaniuc, O.; Ivanciuc, T.; Cabrol-Bass, D.; Balaban, A.T.; Com, D.L. Spectral mapping analysis: A method for the comparison of weighting schemes for molecular graph descriptors. Application in quantitative structure–retention relationship models for alkylphenols in gas–liquid chromatography. J. Chem. Inf. Comput. Sci. **2000**, *40*, 732–743.

Jozwiak, K.; Szumilo, H.; Senczyna, B.; Niewiadomy, A. RP-HPLC as a tool for determining the congenericity of a set of 2,4-dihydroxythiobenzanilide derivatives. Chromatographia **2000**, *52*, 159–161.

Kaliszan, R.; van Straaten, M.A.; Markuszewski, M.; Cramers, C.A.; Claessens, H.A. Molecular mechanism of retention in reversed-phase high-performance liquid chromatography and classification of modern stationary phases by using quantitative structure–retention relationships. J. Chromatogr. A. **1999**, *855*, 455–480.

Monatana, M.P.; Pappano, N.B.; Debattista, N.B.; Raba, J.; Luco, J.M. High-performance liquid chromatography of chalcones. Quantitative structure–activity relationship using partial least squares (PLS) modeling. Chromatographia **2000**, *51*, 727–735.

Sammon, J.W., Jr. A nonlinear mapping for data structure analysis. IEEE Trans. Comput. **1969**, *C18*, 401–407.

Chiral Chromatography by Subcritical and Supercritical Fluid Chromatography

Gerald Terfloth
GlaxoSmithKline, King of Prussia, Pennsylvania, U.S.A.

INTRODUCTION

The intrinsic physical properties of supercritical fluids—increased diffusivity and reduced viscosity—when compared to "normal" liquid phases make sub-/supercritical fluid chromatography a very attractive technology when short cycle times are required. Chiral sub-/supercritical fluid chromatography typically is carried out using packed columns (pSFC) that frequently are identical in mechanical construction to the ones used in traditional HPLC. It should be noted, though, that capillary columns coated or packed with a chiral stationary phase (CSP) have been used for the separation of racemic mixtures. The direct separation of racemic mixtures by chromatographic means can be effected by using a CSP or chiral mobile phase additives. Both techniques have been used successfully in HPLC and pSFC. The use of chiral pSFC is not limited to analytical applications. The relative ease of solvent removal and recycling, typically carbon dioxide modified with a polar organic solvent such as methanol, makes pSFC a very attractive tool for preparative separations. Equipment for laboratory- and industrial-scale pSFC in traditional discontinuous batch-chromatography mode as well as continuous simulated moving bed (SMB) mode has been developed and is commercially available. pSFC can be used as an orthogonal method when techniques such as reversed-phase HPLC, capillary electrophoresis, or capillary electrochromatography provide insufficient or ambiguous results.

CHARACTERISTICS AND ADVANTAGES OF SUBCRITICAL AND SUPERCRITICAL FLUIDS

The advantages of using supercritical mobile phases in chromatography were recognized in the 1950s by Klesper et al., among others. Carbon dioxide is the most frequently used supercritical mobile phase due to its moderate critical temperature and pressure, almost complete chemical inertness, safety, and low cost. Virtually all chiral pSFC separations published have used carbon dioxide as the primary mobile phase component. Compared to most commonly used organic solvents, it is environmentally friendly. The reduced viscosity of carbon dioxide-based mobile phases, typically one order of magnitude less than that of water (0.93 cP at 20 °C), allows for efficient chromatography at higher flow rates. In addition, diffusion coefficients of compounds dissolved in supercritical mobile phases are about one order of magnitude larger than in traditional aqueous and organic mobile phases (D_M(naphthalene): 0.97×10^{-4} cm^2 s^{-1} in CO_2 at 25 °C, 171 bar, 0.90 g cm^{-3}). This directly translates to higher efficiency of the separation due to improved mass transfer.

The first chiral separation using pSFC was published by Caude and colleagues[3] in 1985. pSFC resembles HPLC. Selectivity in a chromatographic system stems from different interactions of the components of a mixture with mobile phase and stationary phase. Characteristics and choice of the stationary phase are described in the "Method Development" section. In pSFC, the composition of the mobile phase, especially for chiral separations, is almost always more important than its density for controlling retention and selectivity. Chiral separations are often carried out at $T < T_c$ using liquid modified carbon dioxide. However, high linear velocity and low pressure drop typically associated with supercritical fluids are retained with near critical liquids. Adjusting pressure and temperature can control the density of the sub-/supercritical mobile phase. Binary or ternary mobile phases are commonly used. Modifiers, such as alcohols, and additives, such as acids and bases, extend the polarity range available to the practitioner.

A typical pSFC instrument, at first glance, is designed like an HPLC system. The major differences are encountered at the pump, the column oven, and downstream of the column. pSFC is best carried out using pumps in a flow-control mode. A regulator mounted downstream of the column and ultraviolet (UV)/visible detector controls the pressure drop in the chromatographic system. Detection is not limited to UV. If pure carbon dioxide is used as the mobile phase, an easy-to-use, sensitive, and stable universal detector such as the FID can be employed. Other detection techniques are FT-IR and evaporative light scattering detection (ELSD), or hyphenated techniques

Encyclopedia of Chromatography DOI: 10.1081/E-ECHR-120039919

Table 1 Initial conditions for chiral method development using modified carbon dioxide as the mobile phase

Parameter	Unit	Value
Flow rate	ml/min	2.0
Pressure	bar	200
Temperature	°C	30
Methanol	%	5
Gradient	%/min	5
Gradient time	min	10
Injection volume	μl	5
Sample concentration	mg/ml	1
Detection	Diode array detector, 190–320 nm	

such as pSFC/MS and pSFC/NMR. Temperature control of mobile phase and column is achieved by a column oven allowing for operation under cryogenic conditions and/or from ambient temperature to 150°C. Capillary column supercritical fluid chromatography (cSFC), though, resembles GC at high pressures, with the pressure (density) programming taking the place of temperature programming in GC. Typical operating temperatures are up to 100°C.

METHOD DEVELOPMENT

Mechanistic considerations, e.g., the extensive work published on brush-type phases, or the practitioner's experience might help to select a CSP for initial work. Scouting for the best CSP/mobile phase combination can be automated by using automated solvent and column switching. More than 100 different CSPs have been reported in the literature to date. Stationary phases for chiral pSFC have been prepared from the chiral pool by modifying small molecules like amino acids or alkaloids, by the derivatization of polymers such as carbohydrates, or by bonding of macrocycles. Also, synthetic selectors such as the brush-type ("Pirkle") phases, helical poly(meth)acrylates, polysiloxanes and polysiloxane copolymers, and chiral selectors physically coated on graphite surfaces have been used as stationary phases.

Generally accepted starting conditions are summarized in Table 1. Typically, alkanol-modified carbon dioxide is used as the mobile phase. Depending on the nature of the analyte, acids or bases can be added to the modifier for controlling ionization of stationary phase and analyte. If partial selectivity is observed after the first injection, it is advisable to first adjust the modifier concentration. If the peak shape is not satisfactory, then the addition of 0.1% trifluoroacetic acid or acetic acid for acidic compounds or 0.1%

diethylamine or triethylamine for basic compounds to the modifier can bring about an improvement. In case the selectivity cannot be improved by the previous measures, decreasing the operating temperature can result in the desired separation. Although many chiral separations improve as the temperature is reduced, this does not occur in all cases. The temperature dependence of the selectivity does not necessarily follow the van't Hoff equation ($\ln \alpha \propto 1/T$), as one might expect based on experience with other chromatographic techniques. Stringham and Blackwell,[7] who have reported several examples of entropically driven separations, studied the effects of temperature in detail. In the temperature range between –10, 70 (T_{iso}), and 190°C, a reversal of elution order for the enantiomers of a chlorophenylamide was observed on an (S,S)-Whelk-O 1 CSP using 10% ethanol in carbon dioxide at a pressure of 300 bar. The potential for reversing the elution order can be valuable if just one enantiomer of the CSP affecting the separation is available. If all of the above adjustments should fail, a different CSP should be investigated. Due to the low viscosity of carbon dioxide-based mobile phases, multiple columns can be coupled. This provides the opportunity to increase chemical selectivity for the analysis of complex samples by coupling an initial achiral column with a chiral column. Also, the successful coupling of multiple different chiral columns has been reported.

CONCLUSIONS

Analytical applications of chiral pSFC in chemical and pharmaceutical research, development, and manufacturing comprise screening of combinatorial libraries, monitoring chemical and biological transformations from the laboratory to the process scale, following stereochemical preferences of drug metabolism and pharmacokinetics, and assessing toxicology and stability of drug substance and dosage form. Preparative applications are of considerable interest because of the relative ease with which the mobile phase can be removed and recycled. This is of particular interest in the pharmaceutical environment since a small amount of the desired product can be obtained almost free of solvent quite rapidly. Recent advances in automation and separation technology now allow for a predictable scale-up of the separation from a laboratory to a production scale.

REFERENCES

1. Klesper, E.; Corwin, A.H.; Turner, D.A. High pressure gas chromatography above critical temperatures. J. Org. Chem. **1960**, *27*, 700.

2. Gere, D.R. Supercritical fluid chromatography. Science **1983**, *222*, 253–259.

3. Mourier, P.A.; Eliot, E.; Caude, M.H.; Rosset, R.H. Supercritical and subcritical fluid chromatography on a chiral stationary phase for the resolution of phosphine oxide enantiomers. Anal. Chem. **1985**, *57*, 2819–2823.

4. Ruffing, F.J.; Lux, J.A.; Schomburg, G. Chiral stationary phases for Lc and SFC obtained by "polymer coating." Chromatographia **1988**, *26*, 19–28.

5. Anton, K.; Eppinger, J.; Fredriksen, L.; Francotte, E.; Berger, T.A.; Wilson, W.H. Chiral separations by packed-column super- and subcritical fluid chromatography. J. Chromatogr. **1994**, *666*, 395–401.

6. Terfloth, G. Enantioseparations in super- and subcritical fluid chromatography. J. Chromatogr. **2001**, *906*, 301–307.

7. Stringham, R.W.; Blackwell, J.A. Entropically driven chiral separations in supercritical fluid chromatography. Confirmation of isoelution temperature and reversal of elution order. Anal. Chem. **1996**, *68*, 2179–2185.

8. Phinney, K.W.; Sander, L.C.; Wise, S.A. Coupled achiral/chiral column techniques in subcritical fluid chromatography for the separation of chiral and nonchiral compounds. Anal. Chem. **1998**, *70*, 2331–2335.

9. Ying, L.; Lantz, A.W.; Armstrong, D.W. High efficiency liquid and super-/subcritical fluid based enantiomeric separations; an overview. J. Liq. Chromatogr. & Relat. Technol. **2004**, *27*, 7–9.

10. Chester, T.L.; Pinkston, J.D. Supercritical fluid and unified chromatography. Anal. Chem. **2004**, *76*, 4606–4613.

11. Depta, A.; Giese, T.; Johannsen, M.; Brunner, G. Separation of stereoisomers in a simulated moving bed–supercritical fluid chromatography plant. J. Chromatogr. **1999**, *865*, 175–186.

12. Gyllenhaal, O. Packed column supercritical fluid chromatography of a peroxysome proliferator-activating receptor agonist drug: achiral and chiral purity of substance, formulation assay and its enantiomeric purity. J. Chromatogr. **2004**, *1042*, 173–180.

SUGGESTED FURTHER READING

Anton, K.; Berger, C. *Supercritical Fluid Chromatography with Packed Columns*; Marcel Dekker, Inc.: New York, 1998.

Berger, T.A. *Packed Column SFC*; The Royal Society of Chemistry: Cambridge, 1995.

Chester, T.L.; Pinkston, J.D.; Raynie, D.E. Supercritical fluid chromatography and extraction. Anal. Chem. **1996**, *68*, 487–514.

Chiral Separations by GC

Raymond P. W. Scott
Scientific Detectors Ltd., Banbury, Oxfordshire, U.K.

INTRODUCTION

In GC, chiral selectivity is controlled solely by the choice of the stationary phase and the operating temperature. Thermodynamically, it is achieved by introducing an additional entropic component to the standard free energy of distribution. This is accomplished by employing a chiral stationary phase which will have unique spatially oriented groups or atoms that allow one enantiomer to interact more closely with the molecules of the stationary phase than the other. The enantiomer that can approach more closely to the stationary phase molecules will interact more strongly (the dispersive or polar charges being nearer) and, thus, the standard enthalpy of distribution of the two enantiomers will also differ. Consequently, the Van't Hoff curves will have different slopes and intersect at a particular temperature (see the entries Thermodynamics of Retention in GC and Van't Hoff Curves). At this temperature, the two enantiomers will co-elute and, hence, temperature is an important variable that must be used to control chiral selectivity. The farther the operating temperature of the column is away from the temperature of co-elution, the greater the separation ratio and the easier will be the separation (less theoretical plates, shorter column, faster analysis).

HISTORICAL BACKGROUND

The first effective chiral stationary phases for GC were the derivatized amino acids,[1] which, however, had very limited temperature stability. The first reliable GC stationary phase was introduced by Bayer and co-workers,[2] who synthesized a thermally stable, low-volatility polymer by attaching l-valine-tbutylamide to the carboxyl group of dimethylsiloxane or (2-carboxypropyl)-methylsiloxane with an amide linkage. This stationary phase was eventually made available commercially as Chirasil-Val and could be used over the temperature range of 30°C to 230°C. OV-225 (a well-established polar GC stationary phase) has also been used for the synthesis of chiral polysiloxanes, which, in this case, possess more polar characteristics than the (2-carboxypropyl)-methylsiloxane derivatives.

Although the polysiloxane phases carrying chiral peptides are still used in contemporary chiral GC, the presently popular phases are based on cyclodextrins. These materials are formed by the partial degradation of starch followed by the enzymatic coupling of the glucose units into crystalline, homogeneous, toroidal structures of different molecular sizes. The best known are the α-, β-, and γ-cyclodextrins which contain six (cyclohexamylose), seven (cycloheptamylose), and eight (cyclooctamylose) glucose units, respectively. The cyclodextrins are torus shaped macromolecules which incorporate the D(+)-glucose residues joined by α-(1-4)glycosidic linkages. The opening at the top of the torus-shaped cyclodextrin molecule has a larger circumference than that at the base. The primary hydroxyl groups are situated at the base of the torus, attached to the C_6 atoms. As they are free to rotate, they partly hinder the entrance to the base opening. The cavity size becomes larger as the number of glucose units increases. The secondary hydroxyl groups can also be derivatized to insert different interactive groups into the stationary phase. Due to the many chiral centers the cyclodextrins contain (e.g., β-cyclodextrin has 35 stereogenic centers), they exhibit high chiral selectivity and, as a consequence, are probably the most effective GC chiral stationary phases presently available.

DISCUSSION

The α-, β-, or γ-cyclodextrins that have been permethylated do not coat well onto the walls of quartz capillaries and must be dissolved in appropriate polysiloxane mixtures for stable films to be produced. In contrast, underivatized cyclodextrins can be coated directly onto the walls of the column with the usual techniques. The thermal stability of a mixed stationary phase can be improved by including some phenylpolysiloxane in the coating material. Phenylpolysiloxane also significantly inhibits any oxidation that might take place at elevated temperatures. However, unless some methylsiloxane is present the cyclodextrin may not be sufficiently soluble in the polymer matrix for successful coating.

The inherent chiral activity of the cyclodextrins can be strengthened by bonding other chirally active

Encyclopedia of Chromatography DOI: 10.1081/E-ECHR-120039922

groups to the secondary hydroxyl groups of the cyclodextrin. Certain derivatized cyclodextrins are susceptible to degradation, on contact with water or water vapor. Consequently, all carrier gases must be completely dry and all samples that are placed on the column must also be dry.

Derivatized cyclodextrins can interact with chiral substances in a number of different ways. If, the positions 2 and 6 are alkylated (pentylated), very dispersive (hydrophobic) centers are introduced that can strongly interact with any alkyl chains contained by the solutes. After pentylation of the 2 and 6 positions has been accomplished, the 3-position hydroxyl group can then be trifluoroacetylated. This stationary phase is widely used and it has been found that the derivatized γ-cyclodextrin is more chirally selective than the β material. It has been successfully used for the separation of both very small and very large chiral molecules. The cyclodextrin hydroxyl groups can also be made to react with pure "S" hydroxypropyl groups and then permethylated. As a result, the size selectivity of the stationary phase is reduced, but its interactive character is made more polar (hydrophilic). In general, the α or γ phases have less chiral selectivity than the β material. There are a considerable number of cyclodextrin based chiral stationary phases commercially available and, without doubt, there will be many more introduced in the future.

REFERENCES

1. Gil-Av, D.; Feibush, B.; Charles-Sigler, R. Separation of enantiomers by gas liquid chromatography with anoptically active stationary phase. Tetrahedron Lett. **1988**, 1009.
2. Frank, H.; Nicholson, G.J.; Bayer, E. J. Chromatogr. Sci. **1974**, *15*, 174.

SUGGESTED FURTHER READING

Beesley, T.E.; Scott, R.P.W. *Chiral Chromatography*; John Wiley & Sons: Chichester, 1998.
Scott, R.P.W. *Techniques of Chromatography*; Marcel Dekker, Inc.: New York, 1995.
Scott, R.P.W. *Introduction to Gas Chromatography*; Marcel Dekker, Inc.: New York, 1998.

Chiral Separations by HPLC

Nelu Grinberg
Richard Thompson
Merck Research Laboratories, Rahway, New Jersey, U.S.A.

INTRODUCTION

Chirality arises in many molecules from the presence of a tetrahedral carbon with four different substituents. However, the presence of such atoms in a molecule is not a necessary condition for chirality. An object is said to be chiral if it is not superposable with its mirror image and achiral when the object and its mirror image are superposable. A chiral pair can be distinguished through their interaction with other chiral molecules to form either long-lived or transient diastereomers. Diastereomers are molecules containing two or more stereogenic (chiral) centers and having the same chemical composition and bond connectivity. They differ in stereochemistry about one or more of the chiral centers.

LONG-LIVED DIASTEREOMERS

Long-lived diastereomers are generated by chemical derivatization of the enantiomers with a chiral reagent. They may be separated subsequently by achiral means. Their formation energies have no relevance to their chromatographic separation; it is, rather, due to the difference in their solvation energies. Differences in their shape, size, or polarity will affect the energy needed to displace solvent molecules from the stationary phase.[1]

There are several characteristics of diastereomeric chiral separations (also known as indirect enantiomeric separations) that are worth mentioning. Achiral phases that are cheaper, more rugged, and widely commercially available are used. The elution order can be controlled by choice of the chirality of the derivatizing agent. This feature is useful for the analysis of trace levels of enantiomers. The separation can be designed such that the minor enantiomer is eluted first, allowing for more accurate quantitation.

Derivatization requires that the species of interest must contain a functional group that can be chemically modified. There should be no enantioselectivity of the rate of the derivatization.[2] There are several disadvantages to an indirect chromatographic chiral separation. The derivatization procedure may be complex and time-consuming and there is always a possibility of racemization during the derivatization procedure. In the case of preparative chromatography of the diastereomeric species, they have to be chemically reversed to the initial enantiomers. Fig. 1 shows the main types of derivatives formed from amines, carboxylic acids, and alcohols in reaction with chiral reagents.[3]

There are several structural considerations to achieving a diastereomeric separation. The diastereomers should possess a degree of conformational rigidity in order to maximize their physical differences. Large size differences between the groups attached to the chiral center enhance the separation in most cases. The distance between the asymmetric centers should be minimal and ideally less than three bonds. The presence of polar or polarizable groups can enhance hydrogen-bonding, interactions with the stationary phase, resulting in increased resolution.

TRANSIENT DIASTEREOMERS

Objects that can distinguish between enantiomers are chiral receptors. Nature gives us plenty of examples of chiral receptors, such as enzymes and nucleic acids. There are also man-made chiral receptors such as chiral phases (CP) used in GC HPLC, supercritical fluid chromatography (SFC), and capillary electrophoresis (CE). The operation of a CP involves the formation of transient diastereomeric complexes between the enantiomer (selectand) and the CP (selector). They must be energetically nondegenerate in order to effect a separation. Because of their transient nature, it is usually not possible to isolate them.

There are specific criteria for the interaction between the selectand and the selector which leads to separation on a particular column 4:

1. Strong interactions, such as p–p interactions, coordinative bonds, and hydrogen bonds between the selector and selectand
2. Close proximity of the transient bonds to the respective asymmetric carbons
3. Inhibition of free rotation of the transient bonds
4. Minimal noncontributing associative forms that do not bring the respective asymmetric centers to proximity

Encyclopedia of Chromatography DOI: 10.1081/E-ECHR-120039923

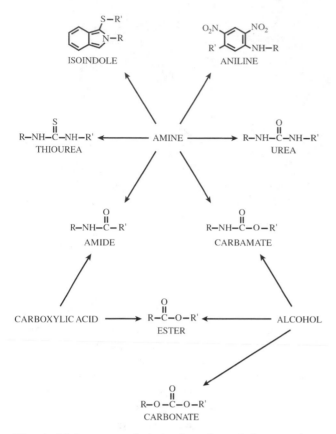

Fig. 1 Main types of derivatives formed from amines, carboxylic acids, and alcohols in reactions with chiral derivatizing reagents. (Adapted from Ref.[3].)

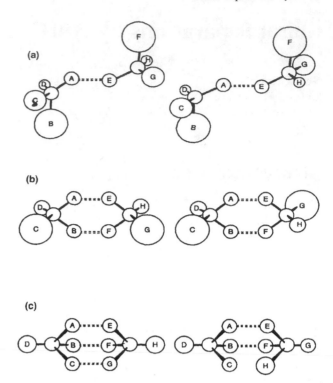

Fig. 2 Schematic representation of selectand–selector association. A dotted line represents a leading interaction between the two molecules. (a) The selectand forms a bond that involves only one substituent of its asymmetric carbon; (b) the selectand binds through two of its substituents; (c) the selectand binds through three substituents. (From Ref.[1].)

The diastereomeric associate between selectand and selector is formed through bonds between one or more substituents of the asymmetric carbon. These bonds are the leading selectand–selector interactions. Only when the leading bonds are formed and the asymmetric moieties of the two molecules are brought to close proximity do the secondary interactions (e.g., van der Waals, steric hindrance, dipole–dipole) become effectively involved (Fig. 2). The secondary interactions can affect the conformation and the formation energy of the diastereomeric associates. In Fig. 2a, the size, shape and polarity of the unbounded B, C, and D substituents of the selectand and their positions to the groups F, G, and H of the selector will determine the enantioselectivity of the system. One particular enantiomer of the selectands will interact more strongly with a particular selector. When the selective associate is formed through interactions A–E and B–F (Fig. 2b), enantioselectivity and elution order are determined by the effective size of unbounded groups C and D their relative positions, syn or anti, to groups G and H of the selector. In most of the cases that include hydrogenbonding or ligand–metal complexes, the

enantiomer with the larger nonbonded groups positioned syn to the selector's larger nonbonded group will elute last from the chiral column. When the selective association is formed through three leading interactions (Fig. 2c), the enantioselectivity is determined by the stereochemistry of the two enantiomers. One enantiomer in one configuration will establish three leading bonds (H bonds or a combination of H bonds and π–π interactions), whereas the other one will not.[1]

In chromatographic systems, the selectors are either added to the mobile phase [chiral mobile phases (CMP)] or are bonded to a stationary phase (e.g., silica gel) as chiral stationary phases (CSP).

CHIRAL MOBILE PHASES

In this mode of separation, active compounds that form ion pairs, metal complexes, inclusion complexes, or affinity complexes are added to the mobile phase to induce enantioselectivity to an achiral column. The addition of an active compound into the mobile phase

Table 1 Main classes of chiral additives and their applications

Mechanism	Additive	Application	Mode of separation	Refs.
Ion pair	(+)-10-camphorsulfonic acid	Aminoalcohols, alkaloids	HPLC	[5,6]
Ion pair	Quinines	Carboxylic acids	HPLC	[7,8]
Inclusion	Dimethyl β-cyclodextrin	Aminoalcohols, carboxylic acids	CE	[9,10]
Inclusion	Crown ether	Primary amines	CE	[11]
Ligand exchange	L-Proline/Cu^{2+}	Amino acids	HPLC	[12]
Proteins	α_1-Acid glycoprotein	Hexobarbitone	CE	[13]
Antibiotics	Rifamycin	Amino acids	CE	[14]

contributes to a specific secondary chemical equilibrium with the target analyte. This affects the overall distribution of the analyte between the stationary and the mobile phases, affecting its retention and separation at the same time. The chiral mobile phase approach utilizes achiral stationary phases for the separation. Table 1 lists several common chiral additives and applications.

CHIRAL STATIONARY PHASES

Compared to CMP, the mechanism of separation on a chiral stationary phase is easier to predict, due to a much simpler system. Because the ligand is immobilized to a matrix and is not constantly pumped through the system, the detection limits for the enantiomers are much lower. Depending on the ligand immobilized to the matrix, one can have different types of interactions between the selectand and selector: metal complexes, hydrogen-bonding, inclusion, π–π interactions, and dipole interactions, as well as a combination thereof.

CHIRAL SEPARATION WHERE THE LEADING INTERACTION IS ESTABLISHED THROUGH METAL COMPLEXES (LIGAND EXCHANGE)

Chiral separation using ligand-exchange chromatography involves the reversible complexation of metal ions and chiral complexing agents. The central ion, usually Cu^{2+} or Ni^{2+} forms a bis complex with bidentates ligands. If one of the chelating ligands is anchored to a support, the CSP can form diastereomeric adsorbates with the bidentate selectand. The metal ion is held by the stationary phase through coordination to the bound ligand. If the coordination sphere of the metal is unsaturated or is occupied by weakly bound solvent molecules, it can reversibly attach different solute ligands from the mobile phase. The solute ligands are then resolved according to differences in their binding constants. Ligand exchange is possible only in systems where the interaction of the mobile ligand with the sta-

tionary phase is reversible. The coordination bonds must be kinetically labile. If the chelating ligands are amino acids and the metal is copper (II), the amine and carboxylate groups of the ligands are arranged around the metal ion in a trans configuration, forming a square planar complex. A third interaction should take place to ensure enantioselectivity. The third interaction may arise through steric hindrance or attractive or repulsive interactions between the selector and the selectand.[15–16]

CHIRAL SEPARATION WHERE THE LEADING INTERACTION IS ESTABLISHED THROUGH HYDROGEN BONDING

A hydrogen bond is formed by the interaction between the partners R–X–H and :Y–R′ according to

$$R-X-H1:Y-R9 \rightarrow R-X-H^{...}Y-R'$$

R–X–H is the proton donor and :Y–R makes an electron pair available for the bridging bond. X and Y are atoms of higher electronegativity than hydrogen (e.g., C, N, P, O, S, F, Cl, Br, I). Hydrogen-bonding acceptors are the oxygen atoms in alcohols, ethers, and carbonyl compounds, as well as nitrogen atoms in amines and N heterocycles. Hydrogen-bonding donors are hydroxy, carboxyl, and amide protons. Interactions can be modified by changing the elution conditions. The more nonpolar the elution conditions, the stronger the H-bond interactions. Enantioselectivity is determined by the strength of the hydrogen bonds, which is, in turn, affected by secondary interactions such as steric hindrance or attractive or repulsive interactions between the selector and the selectand.

CHIRAL SEPARATION THROUGH CHARGE TRANSFER

Complexes formed by weak interactions of electron donors with electron-acceptor compounds are known as charge-transfer complexes. The necessary condition for the formation of a charge transfer complex is the presence of an occupied molecular orbital of sufficiently high energy in the electron-donor molecule, and the presence of a sufficiently low unoccupied orbital in the electron-acceptor molecule. Small unsaturated hydrocarbons are usually weak donors or weak acceptors. Polynuclear aromatic hydrocarbons are efficient π-donor molecules. Replacement of a hydrogen atom in the parent molecule with an electron-releasing substituent such as alkyl, alkoxy, or amino, increases the capability of the molecule to donate π-electrons. Aromatic molecules containing groups such as NO_2, Cl, C N are efficient electron acceptors. Carbonyl compounds are acceptors to aromatic hydrocarbons but are donors to bromine.

The overlapping and the orientation of the molecules in the crystal correspond to parallel planes if the bonding occurs only through π orbitals. π-donor–π-donor interactions do not occur in the same fashion because of repulsion between the π clouds. This repulsion leads to edge-to-face interactions, where weakly positive H atoms at the edge of the molecule point toward negatively charged C atoms on the faces of adjacent molecule. The dihedral ring planes are often close to perpendicular. Aromatic rings can act as hydrogenbond acceptors for the amidic proton.[17]

In general, the stability of a charge-transfer complex increases with the increase in the polarity of the solvent. To establish the enantiomeric separation under such conditions, secondary interactions must occur: namely the charge-transfer interactions have to be accompanied by hydrogen bonds and/or steric hindrance. Under these conditions, the mobile-phase conditions should be adjusted such that these interactions are achieved. Fig. 3 presents an example of a chiral stationary phase designed by Pirkle's group. This CSP allows for charge-transfer interaction with secondary interactions such as hydrogen-bonding and steric hindrance.[18]

CHIRAL SEPARATION THROUGH HOST–GUEST COMPLEXATION

Cyclodextrins and crown ethers are the main classes of compounds able to undergo host–guest complexes with

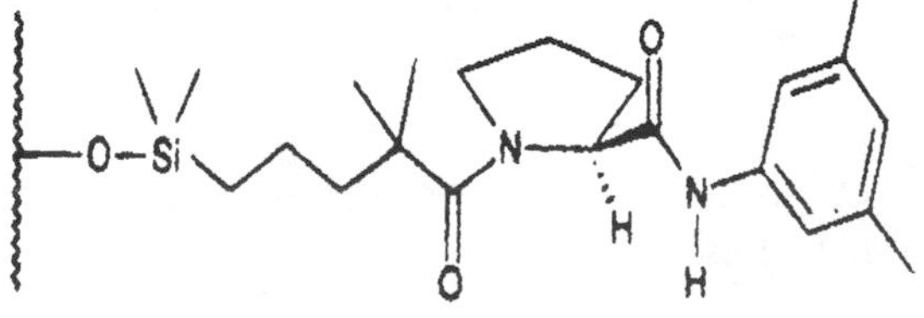

a particular pair of enantiomers. Cyclodextrins (CD) are natural macrocyclic polymers of glucose that contain 6–12 D-(+)-glucopyranose units which are bound through α-1,4-glucopyranose linkages. The number of glucose units per CD is denoted by a Greek letter: α for six, β for seven, and γ for eight (Fig. 4).[19] The inherent chirality of the CD renders them useful for chromatographic enantioseparations. In most cases, an inclusion complex is formed between the solute and the cyclodextrin cavity. The host–guest complexation is dependent on the polarity, hydrophobicity, size, and geometry of the guest, as well as the size of the internal cavity of the CD. Enantioselectivity is then determined by the fit in the cavity and by the interactions between substituents attached to or near the chiral center of the analyte and the unidirectional secondary hydroxyl groups at the mouth of the cavity. The temperature, pH, and the composition of the mobile phase influence the complexation.

Under reversed-phase conditions (RP), the presence of an organic modifier affects the binding of the guest molecule in the CD's cavity. The inclusion complex is usually strongest in water and decreases upon addition of organic modifiers. The modifier competes with the guest analyte for the cavity. Under normal-phase conditions, apolar solvents such as hexane and chloroform occupy the CD's cavity and cannot be easily displaced by the solute molecules. In these circumstances, the solute is usually restricted to interactions with the exterior of the CD. Chemical modifications of CD has opened new possibilities for enantiorecognition, widening the range of compounds that can be separated into enantiomers.[20]

Crown ethers, especially 18-crown-6 ethers, can complex not only inorganic cations but also alkylammonium compounds. The primary interactions occur between the hydrogens of the ammonium group and the oxygens of the crown ether. The introduction of bulky groups such as binaphtyl onto the exterior of the crown ether provides steric barriers and induces enantioselective interactions with the guest molecule.

The rigid binaphthyl units occupy planes that are perpendicular to the plane of the cyclic ether. One of the naphthalene rings forms a wall that extends along the sides and outward from the other face of the cyclic ether. The substituents attached at the 3-position of the naphthalene rings extend along the side or over the face of the cyclic ether. In the presence of a chiral primary amine, it forms a triple hydrogen bond with the primary ammonium cation. The same complex is formed whether the guest approaches from the top or from the bottom of the crown ether, as the crown ether has a C_2 axis of symmetry. In the complex, the large (L), medium (M), and small (S) groups attached to the asymmetric carbon of the guest must adjust themselves into two identical cavities. The L is placed in one

Fig. 3 The structure of the (S)-proline derivative chiral stationary phase. (From Ref.[18].)

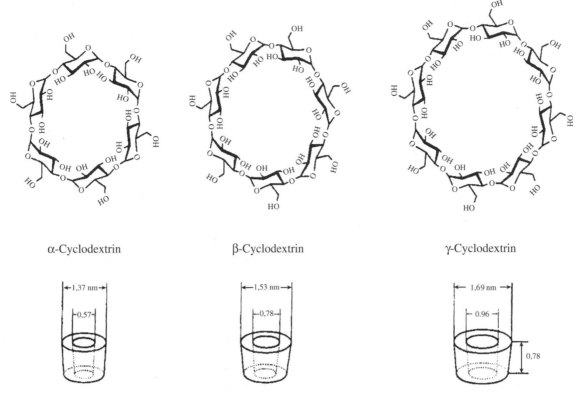

α-Cyclodextrin   β-Cyclodextrin   γ-Cyclodextrin

Fig. 4 Schematic representation of α-CD, β-CD, and γ-CD. (Adapted from Ref.[19].)

cavity and the M and S into the other cavity. M will reside in the pocket with S against the wall for the more stable diastereomeric complex (Fig. 5).[21]

CHIRAL SEPARATION THROUGH COMBINATION OF INTERACTIONS

Included in this category are stationary phases such as biopolymers (e.g., celluloses and cellulose derivatives, proteins),[22–23] as well as macrocyclic antibiotics.[24] These stationary phases exhibit interactions with a particular enantiomer through hydrogen-bonding, charge transfer, and inclusion interactions. They proved to be very effective in resolving a wide class of racemates encompassing a variety of structures. Describing the mechanism of such separation is very challenging due to the complexity of these stationary phases. Such stationary phases can be operated under RP conditions

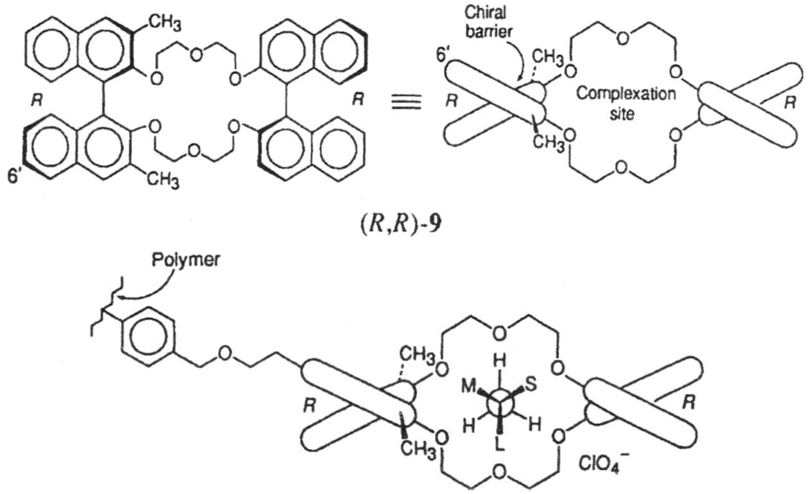

Fig. 5 Structure of the crown ether and the most stable complex. (From Ref.[21].)

(protein phases, cellulose phases, and macrocyclic antibiotics), as well as in the normal-phase conditions (cellulose phases and macrocyclic antibiotics). Conformational changes of biopolymers under the temperature and mobile-phase conditions can occur and they should be controlled such that the separation can be maximized.[25–26]

REFERENCES

1. Feisbush, B. Chirality **1998**, *10*, 382.
2. Lindner, W. *Chromatographic Chiral Separation*; Zieff, M., Crane, L.J., Eds.; Marcel Dekker, Inc.: New York, 1988; 91.
3. Ahnoff, M.; Einarsson, S. *Chiral Liquid Chromatography*; Lough, W.J., Ed.; Blackie and Son: Glasgow, 1989; 39.
4. Feibush, B.; Grinberg, N. *Chromatographic Chiral Separation*; Zieff, M., Crane, L.J., Eds.; Marcel Dekker, Inc.: New York, 1988; 1.
5. Pettersson, C.; Schill, G. Separation of enantiomeric amines by ion-pair chromatography. J. Chromatogr. **1981**, *204*, 179–183.
6. Pettersson, C.; Schill, G. Chiral separation of aminoalcohols by ion-pair chromatography. Chromatographia **1982**, *16*, 192.
7. Karlsson, A.; Pettersson, C. Separation of enantiomeric amines and acids using chiral ion-pair chromatography on porous graphitic carbon. Chirality **1992**, *4*, 323.
8. Pettersson, C.; No, K. Chromatographic separation of enantiomers of acids with quinine as chiral counter ion. J. Chromatogr. **1984**, *316*, 553–567.
9. Guttman, A. Novel separation scheme for capillary electrophoresis of enantiomers. Electrophoresis **1995**, *16*, 1900.
10. Guttman, A.; Cooke, N. Practical aspects in chiral separation of pharmaceuticals by capillary electrophoresis: II. Quantitative separation of naproxen enantiomers. J. Chromatogr. **1994**, *685*, 155.
11. Lin, J.-M.; Nakagama, T.; Hobo, T. Combined chiral crown ether and b-cyclodextrin for the separation of o-, m-, p-fluoro-D,L-phenylalanine by capillary gel electrophoresis. Chromatographia **1996**, *42*, 559.
12. Gil-Av, E.; Tishbee, S. Resolution of underivatized amino acids by reversed-phase chromatography. J. Am. Chem. Soc. **1980**, *102*, 5115.
13. Clar, B.; Mame, J. Resolution of chiral compounds by HPLC using mobile phase additives and a porous graphitic carbon stationary phase. J. Pharm. Biomed. Anal. **1989**, *7*, 1883.
14. Armstrong, D. Use of a macrocyclic antibiotic, rifamycin B, and indirect detection for the resolution of racemic amino alcohols by CE. Anal. Chem. **1994**, *66*, 1690.
15. Davankov, V.A. *Advances in Chromatography*; Giddings, J.C., Grushka, E., Cazes, J., Brown, P.R., Eds.; Marcel Dekker, Inc.: New York, 1980; Vol. 18, 139.
16. Davankov, V.A.; Kurganov, A.A.; Bochkov, A.S. *Advances in Chromatography*; Giddings, J.C., Grushka, E., Cazes, J., Brown, P.R., Eds.; Marcel Dekker, Inc.: New York, 1983; Vol. 22, 71.
17. Foster, R. *Organic Charge-Transfer Complexes*; Academic Press: London, 1969; 217.
18. Pirkle, W.H.; Selness, S.R. Chiral recognition studies: intra- and intermolecular 1H{1H}-nuclear overhauser effects as effective tools in the study of bimolecular complexes. J. Org. Chem. **1995**, *60*, 3252.
19. Konig, W.L. *Gas Chromatographic Enantiomer Separation with Modified Cyclodextrins*; Hütihig Buch Verlag: Heidelberg, 1992; 4.
20. Stalcup, A.M. *A Practical Approach to Chiral Separations by Liquid Chromatography*; VCH: Weinheim, 1994; 1994.
21. Cram, D.J.; Cram, J.M. *Container Molecules and Their Guests*; Royal Society of Chemistry: London, 1994; 56.
22. Okamoto, Y.; Kaida, Y. Resolution by high-performance liquid chromatography using polysaccharide carbamates and benzoates as chiral stationary phases. J. Chromatogr. **1994**, *666*, 403.
23. Allenmark, S.G.; Anderson, S. Proteins and peptides as chiral selectors in liquid chromatography. J. Chromatogr. **1994**, *666*, 167.
24. Ekborg-Ott, K.H.; Youbang, L.; Armstrong, D.W. Highly enantioselective HPLC separations using the covalently bonded macrocyclic antibiotic, ristocetin A, chiral stationary phase. Chirality **1998**, *10*, 434.
25. Waters, M.; Sidler, D.R.; Simon, A.J.; Middaugh, C.R.; Thompson, R.; August, L.J.; Bicker, G.; Perpall, H.J.; Grinberg, N. Chirality **1999**, *11*, 224.
26. O'Brien, T.; Crocker, L.; Thompson, R.; Thomson, K.; Toma, P.H.; Conlon, D.A.; Feibush, B.; Moeder, C.; Bocker, G.; Grinberg, N. Mechanistic aspects of chiral discrimination on modified cellulose. Anal. Chem. **1997**, *69*, 1999.

Chiral Separations by MEKC with Chiral Micelles

C

Koji Otsuka
Shigeru Terabe
Himeji Institute of Technology, Hyogo, Japan

INTRODUCTION

Since MEKC was first introduced in 1984, it has become one of major separation modes in capillary electrophoresis (CE), especially owing to its applicability to the separation of neutral compounds as well as charged ones. Chiral separation is one of the major objectives of CE, as well as MEKC, and a number of successful reports on enantiomer separations by CE and MEKC has been published. In chiral separations by MEKC, the following two modes are normally employed: (a) MEKC using chiral micelles and (b) cyclodextrin (CD)-modified MEKC (CD/MEKC).

MEKC USING CHIRAL MICELLES

An ionic chiral micelle is used as a pseudo-stationary phase; it works as a chiral selector. When a pair of enantiomers is injected to the MEKC system, each enantiomer is incorporated into the chiral micelle at a certain extent determined by the micellar solubilization equilibrium. The equilibrium constant for each enantiomer is expected to be different more or less among the enantiomeric pair; that is, the degree of solubilization of each enantiomer into the chiral micelle would be different for each. Thus, the difference in the retention factor would be obtained and different migration times would occur.

CD/MEKC

An ionic achiral micelle [e.g., sodium dodecyl sulfate (SDS)] and a neutral CD are typically used as a pseudo-stationary phase and a chiral selector, respectively. When a pair of enantiomers is injected into this system, two major distribution equilibria can be considered for the solutes or enantiomers: (a) the equilibrium between the aqueous phase and the micelle (i.e., micellar solubilization) and (b) the equilibrium between the aqueous phase and CD (i.e., inclusion complex formation). Each enantiomer may have a different equilibrium constant for the inclusion complex formation among the enantiomeric pairs due to the enantioselectivity of the CD. As a result, each

enantiomer exists in the aqueous phase at a different time among the enantiomeric pairs; hence, the time spent in the micelle would be varied.

In some cases, an ionic chiral micelle (e.g., a bile salt) is also used as a chiral pseudo-stationary phase with a CD. Moreover, cyclodextrin electrokinetic chromatography (CDEKC), where a CD derivative having an ionizable group is used as a chiral pseudo-stationary phase, has become popular recently since several commercially available ionic CD derivatives have appeared. Although the CDEKC technique is actually beyond the field of MEKC, it is an important method for enantiomer separation by CE.

In this section, chiral separation by MEKC with chiral micelles is mainly treated. The development of novel chiral surfactants adaptable to pseudo-stationary phases in MEKC for enantiomer separation is continuously progressing. It seems somewhat difficult for a researcher to find an appropriate mode of CE when one wants to achieve a specific enantioseparation. However, nowadays, various method development kits for chiral separation have been commercially available and some literature on the topic is also available, so that helpful information may be obtained without difficulty.

MEKC USING NATURAL CHIRAL SURFACTANTS

Bile Salts

Bile salts are natural and chiral anionic surfactants which form helical micelles of reversed micelle conformation. The first report on enantiomer separation by MEKC using bile salts was the enantioseparation of dansylated DL-amino acids (Dns-DL-AAs) and, since then, numerous papers have been available. Nonconjugated bile salts, such as sodium cholate (SC) and sodium deoxycholate (SDC), can be used at pH >5, whereas taurine-conjugated forms, such as sodium taurocholate (STC) and sodium taurodeoxycholate (STDC), can be used under more acidic conditions (i.e., pH >3). Several enantiomers, such as diltiazem hydrochloride and related compounds, carboline derivatives, trimetoquinol and related compounds, binaphthyl derivatives, Dns-DL-AAs, mephenytoin

Encyclopedia of Chromatography DOI: 10.1081/E-ECHR-120039924

and its metabolites, and 3-hydroxy-1,4-benzodiazepins have been successfully separated by MEKC with bile salts. In general, STDC is considered as the the most effective chiral selector among the bile salts used in MEKC.

The use of CDs with bile salt micelles has been also successful for enantiomer separations. For example, Dns-DL-AAs, baclofen and its analogs, mephenytoin and fenoldopam, naphthalene-2,3-dicarboxaldehyde derivatized DL-AAs (CBI-DL-AAs), diclofensine, ephedrine, nadolol, and other β-blockers, and binaphthyl-related compounds were enantioseparated by CD/ MEKC with bile salts.

Digitonin and Saponins

Digitonin, which is a glycoside of digitogenin and used for the determination of cholesterol, is a naturally occurring chiral surfactant. By using digitonin with ionic micelles, such as SDS or STDC as pseudo-stationary phases, some phenylthiohydantoin-DL-AAs (PTH-DL-AAs) were enantioseparated.

On the other hand, glycyrrhizic acid (GRA) and β-escin can be employed as chiral pseudo-stationary phases in MEKC. Chiral separations of some Dns-DL-AAs and PTH-DL-AAs were achieved.

MEKC USING SYNTHETIC CHIRAL SURFACTANTS

N-Alkanoyl-L-Amino Acids

Various N-alkanoyl-L-amino acids, such as sodium N-dodecanoyl-L-valinate (SDVal), sodium N-dodecanoyl-L-alaninate (SDAla), sodium N-dodecanoyl-L-glutamate (SDGlu), N-dodecanoyl-L-serine (DSer), N-dodecanoyl-L-aspartic acid (DAsp), sodium N-tetradecanoyl-L-glutamate (STGlu), and sodium N-dodecanoyl-L-threoninate (SDThr) have been employed as synthetic chiral micelles in MEKC; several enantiomers have been successfully separated (Fig. 1). In each case, the addition of SDS, urea, and organic modifiers such as methanol or 2-propanol were essential to obtain improved peak shapes and enhanced enantioselectivity.

N-Dodecoxycarbonyl-Amino Acids

Chiral surfactants of amino acid derivatives, such as (S)- and (R)-N-dodecoxycarbonylvaline (DDCV) and N-dodecoxycarbonylproline (DDCP) are available for enantiomer separation by MEKC: Several pharmaceutical amines, benzoylated amino acid methyl ester derivatives, piperidine-2,6-dione enantiomers, and aldose

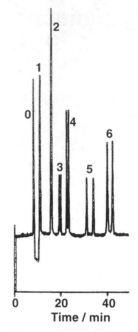

Fig. 1 Chiral separation of six PTH-DL-AAs by MEKC with SDVal. Corresponding AAs: (1) Ser, (2) Aba, (3) Nva, (4) Val, (5) Trp, (6) Nle; (0) acetonitrile. Micellar solution: 50 mM SDVal–30 mM SDS–0.5 M urea (pH 9.0) containing 10% (v/v) methanol; separation column: 50 μm inner diameter × 65 cm, 50 cm effective; applied voltage, 20 kV; current, 17 μA; detection wavelength, 260 nm; temperature, ambient. (Reprinted from K. Otsuka et al. J. Chromatogr. **1991**, *559*, 209, with permission.)

enantiomers were successfully resolved. Because both enantiomeric forms of DDCV or (S)- and (R)-forms are available, we can expect that the migration order of an enantiomeric pair would be reversed.

Alkylglucoside Chiral Surfactants

Anionic alkylglucoside chiral surfactants, such as dodecyl β-D-glucopyranoside monophosphate and monosulfate, and sodium hexadecyl D-glucopyranoside 6-hydrogen sulfate, were used as chiral pseudostationary phases in MEKC, where several enantiomers (e.g., PTH-DL-AAs and binaphthol) were resolved.

Several neutral alkylglucoside surfactants, such as heptyl-, octyl-, nonyl-, and decyl-β-D-glucopyranosides and octylmaltopyranoside, were also employed for the enantiomer separation of phenoxy acid herbicides, Dns-DL-AAs, 1,1'-bi-2-naphthyl-2,2'-diyl hydrogen phosphate (BNP), warfarin, bupivacaine, and so forth.

Tartaric Acid-Based Surfactants

A synthesized chiral surfactant based on (R,R)-tartaric acid was used for the enantiomer separation in MEKC, where enantiomers having fused polyaromatic rings were separated easier than those having only a single aryl group.

Some PTH-DL-AAs and drug enantiomers were successfully resolved by using tartaric acid-based chiral surfactants.

Steroidal Glucoside Surfactants

Neutral steroidal glucoside surfactants, such as *N,N*-bis-(3-D-gluconamidopropyl)-cholamide (Big CHAP) and *N,N*-bis-(3-D-gluconamidopropyl)-deoxycholamide (Deoxy Big CHAP), which contain a cholic or deoxycholic acid moiety, respectively, have been introduced for use as chiral pseudo-stationary phases in MEKC. By using a borate buffer under basic conditions, these surfactant micelles could be charged via borate complexation. Some binaphthyl enantiomers, Tröger's base, phenoxy acid herbicide, and Dns-DL-AAs were enantioseparated.

MEKC USING HIGH-MOLECULAR-MASS SURFACTANTS

The use of a high-molecular-mass surfactant (HMMS) or polymerized surfactant has been recently investigated as a pseudo-stationary phase in MEKC. Because a HMMS forms a micelle with one molecule, enhanced stability and rigidity of the micelle can be obtained. Also, it is expected that the micellar size is controlled easier than with a conventional low-molecular-mass surfactant (LMMS). The first report on enantiomer separation by MEKC using a chiral HMMS appeared in 1994, where poly(sodium *N*-undecylenyl-L-valinate) [poly(L-SUV)] was used as a chiral micelle and binaphthol and laudanosine were enantioseparated. The optical resolution of 3,5-dinitrobenzoylated amino acid isopropyl esters by MEKC with poly(sodium

(10-undecenoyl)-L-valinate) as well as with SDVal, SDAla, and SDThr was also reported.

As for the use of monomeric and polymeric chiral surfactants as pseudo-stationary phases for enantiomer separations in MEKC, a review article has been available.

The use of an achiral HMMS butyl acrylate/butyl methacrylate/methacrylic acid copolymer (BBMA) sodium salt was also investigated for enantiomer separations with CDs or as a CD/MEKC mode. A better enantiomeric resolution of Dns-DL-AAs was obtained by a β-CD/BBMA/MEKC system than an β-CD/SDS/MEKC system.

Polymerized dipeptide surfactants, which are derived from sodium *N*-undecylenyl-L-valine-L-leucine (L-SUVL), sodium *N*-undecylenyl-L-leucine-L-valine (L-SULV), sodium *N*-undecylenyl-L-leucine-L-leucine (L-SULL), and sodium *N*-undecylenyl-L-valine-L-valine (L-SUVV), were employed. Among these dipeptides, poly(L-SULV) showed the best enantioselectivity for the separation of 1,1'-bi-2-naphthol (BN).

SUGGESTED FURTHER READING

Camilleri, P. Chiral surfactants in micellar electrokinetic capillary chromatography. Electrophoresis **1997**, *18*, 2332.

Chankvetadze, B. *Capillary Electrophoresis in Chiral Analysis*; John Wiley & Sons: New York, 1997.

Otsuka, K.; Terabe, S. Enantiomer separation of drugs by micellar electrokinetic chromatography using chiral surfactants. J. Chromatogr. A. **2000**, *875*, 163.

Otsuka, K.; Terabe, S. Micellar electrokinetic chromatography. Bull. Chem. Soc. Jpn. **1998**, *71*, 2465.

Terabe, S.; Otsuka, K.; Nishi, H. Separation of enantiomers by capillary electrophoretic techniques. J. Chromatogr. A. **1994**, *666*, 295.

Chromatographic Methods Used to Identify and Quantify Organic Polymer Additives

Dennis Jenke
Technology Resources, Baxter Healthcare Corporation, Round Lake, Illinois, U.S.A.

INTRODUCTION

Plastic materials are widely used in numerous industries. The physiochemical nature of these materials provides a multitude of diverse products with their necessary, desirable performance characteristics. Commercial plastics are very complex materials. In addition to the various base polymers, commercially viable plastics contain a number of compounding ingredients (additives) whose purpose is to give the material its desired physical and/or chemical properties. Table 1 provides a brief summary of the types of additives typically encountered in commercial polymer systems.

Polymers and polymer systems are characterized for many reasons including the development of new materials or material sources, the end-use applications, the life test studies, the manufacturing control and troubleshooting, and the material or vendor identification. As it is typically the additive package that establishes the performance and processing properties of the commercial polymer, characterization of a polymer system for its additive package is essential in terms of material development, manufacturing, use, reuse, and, ultimately, disposal. A complete polymer characterization includes both the identities of the additives and their levels in the product. The identification and the quantification of additives in compounded polymers is generally a difficult task for the following reasons:

1. There is a wide variety of chemically diverse additive types. Literally, thousands of additives are commercially available, ranging from pure compounds, with molecular weights that vary from approximately 100 up to a few thousand mass units, to oligomers with up to 50 (or more) components.
2. Many additives are labile; thus, they are difficult to analyze without decomposition.
3. Complex mixtures of additives will normally be present in a commercial formulation.
4. Separation of the base polymer and the fillers from the organic additives are often required prior to additive analysis.

5. The levels of the organic additives in a commercial polymer may be quite low (and variable) compared to the base polymer and its associated fillers.

It is for these reasons that chromatographic methods of analysis have been widely employed in polymer characterization. In this paper, a review related to the chromatographic methods used to assess the identity and level of additives in polymer systems is provided.

DISCUSSION

Given the variety of additives used in commercial polymers, the task of characterizing such multicomponent systems for their additive packages can be daunting. While an analytical chemist has a multitude of chromatographic tools with which to perform an additive characterization, some guidance in terms of successfully applied strategies and methods can greatly facilitate the assessment. Thus this manuscript contains a general compilation of published chromatographic methods and strategies that have been successfully applied to the identification and the quantification of a large number of the more commonly encountered packaging material additives (Table 2 shows a listing of additives considered in this work's cited references). Examples are provided for each major separation strategy [e.g., HPLC, GC, TLC, supercritical fluid chromatography (SFC)] and for most commonly employed detection methods [e.g., ulraviolet (UV), MS, flame ionization detector (FID)]. While the compilations in Tables (3–10) are by no means exhaustive or comprehensive, they are sufficiently broad in scope to provide the investigator with a general overview of the ways in which chromatography has been applied to meet the objectives of a polymer's characterization.

Tables (3–10) provide general method details, such as separation medium, elution, and detection conditions, and other operating conditions. The level of detail associated with each citation reflects the level of detail provided by the citation's author(s). The materials investigated, as well as the specific additives examined, are also indicated. General comments are

Encyclopedia of Chromatography DOI: 10.1081/E-ECHR-120021147

Table 1 Common additives and fillers

Classification	Purpose	Examples
Antioxidants	Prevent thermal and/or oxidative degradation during processing, handling, and use. Typically are radical scavengers which interrupt the chain propagation steps of polymer autooxidation	Irganox 1010, Irgafos 168, BHT
Light stabilizers	Absorb UV light to prevent photooxidation	Tinuvin 327, 328, 384, 440, etc. (derivatives of 2-hydroxy-benzophenone)
Heat stabilizers	Protect polymers during thermal processing	Metallic salts, especially of weak fatty acids (e.g., zinc stearate)
Plasticizers	Increase the workability, flexibility or distensibility of polymer	Derivatives of organic acids such as adipic, azelic, citric, phosphoric, phthalic, trimellitic acids
Lubricants (slip agent)	Reduce polymer adhesion to metal surfaces during processing	Derivatives of fatty acids (esters, amides, metal salts, Erucamide, silicones)
Viscosity improvers	Control the flow and the sagging of prepolymers	Ethoxylated fatty acids
Accelerators, activators	Compounds that control the rate or the nature of cure of elastomers	Zinc oxide, stearic acid, 2-2'-dithiobis-benzothiazole, zinc dialkyldithio-carbamate
Mold release agents	Prevent adhesion between two surfaces (e.g., sticking of polymer and metal mold)	Derivatives of fatty acids, Montan wax, silicones, diethylene glycol monostearate, ethylene bis (stearamide)
Fillers (extenders)	Finely dissolved solids added to polymer systems to improve properties or to reduce cost	Calcium carbonate, kaolin, talc, alumina trihydrate
Flame retardants	Decrease flammability	Alumina trihydrate, mixtures of halogenated organics and antimony oxide
Antistatic agents	Dissipate electrostatic surface charge on polymer surfaces	Quaternary ammonium compounds, long-chain derivatives of glycols and polyhydric compounds, Atmos 150, N,N-bis (2-hydroxy-ethyl) alkyl-amine
Colorants	Improve the appearance of polymers, mask discoloration due to processing	Carbon black, titanium dioxide, azo-type dyes
Antimicrobial agents (biocides)	Reduce growth of microbes on polymer surfaces	Copper 8-hydroxyquinolate, n-(trichloromethylthio) phthalate
Cross-linking agents	Molecules that have two or more groups capable of reacting with the functional groups of polymer chains, where such a reaction connects or links the chains	2-Mercaptobenzothiazole, benzoyl peroxide, dicumyl peroxide, sulfur, toluene diisocyanate
Blowing agents	Gas-forming agents that facilitate the expansion of the polymer during processing	Nitrogen, toluenesufonyl semicarbazide, 1,1'-azobisforamide, phenyltetrazole

provided in terms of sample preparation. Given the number of methods cited, it is not possible to provide chromatographic profiles which are readily available in the cited references.

In generating this review, the author balanced two objectives. The first objective was to summarize the most current technologies that are utilized for the task of polymer characterization, thus providing the researcher with the most relevant and state-of-the-art tools for the task at hand. Thus emerging automated and hyphenated techniques, coupled with on-line sample preparation, high-efficiency separations, and selective and sensitive detection (e.g., supercritical fluid extraction (SFE)/HPLC/MS), have a prominent place in this review. However, this author also notes that more historically relevant methods, such as GC and

Table 2 Chemical names for the additives cited in this manuscript

Trade name	Chemical name	CAS RN
AcraWax C	*N,N'*-ethylenebisstearamide	110-30-5
Adkstab PEP-24G	Cyclinepentan tetrail bis (2,4-di-*tert*-butylphenyl) phosphite	29741-53-7
AM340		
ATBC	*o*-Acetyl-tri-*n*-butyl citrate (see Citroflex A-4)	—
Atmos 150	A mixture of glycerol mono- and distearate	11099-07-3
BAC-E	2,6-bis [(Azidophenyl) methylene]-4-ethylcyclohexanone	
BBP	Benzyl *n*-butyl phthalate	85-68-7
Benzoflex-2860	A mixture of 19% di (2-ethylhexyl) adipate, 57% diethyleneglycol dibenzoate, 24% triethyleneglycol dibenzoate	400609-45-2
BEHB	Butylated hydroxyethylbenzene	
Behenamide	Docosanoic acid amide	3061-75-4
BHA	2-*tert*-Butyl-4-hydroxyanisole	25013-16-5
BHT	2,6-Di-*tert*-butyl-*p*-cresol	128-37-0
BHET	bis (2-Hydroxyethyl) terephthalate	
Bis-A-bis azide	1,1'-(1-Methylethylidene)-bis [4-(4-axidophenoxybenzene]	
Bisphenol A	2,2'-bis (4-Hydroxyphenyl) propane	80-05-7
Brominated bisphenol A		
Brominated phenol		
Butyl oleate	9-Octadecanoic acid, butyl ester	142-77-8
Butyl palmitate	Hexadecanoic acid, butyl ester	111-06-8
Butyl stearate	Octadecanoic acid, butyl ester	123-95-5
Calcium stearate	Stearic acid, calcium salt	1592-23-0
Caprolactam	2-Oxohexamethyleneimine	105-60-2
Chimasorb 81	2-Hydroxy-4-*n*-octyloxybenzophenone	1843-05-6
Chimassorb 119 FL		
Chimassorb 944	Poly-[[6-[1,1,3,3,-tetramethylbutyl) amino]-1,3,5-triazine-2,4-diyl] [2.2.6.6-tetramethyl-4-piperidinyl) imino]-1,6-hexanediyl [2,2,6,6-tetramethly-4-piperidinyl) imino]]	71878-19-8
Citroflex A-4	2-Acetoxy-1,2,3-propanetricarboxylic acid tributyl ester	77-90-7
Cyanox 425	2,2'-Methylenebis (6-*tert*-butyl-4-ethylphenol)	88-24-4
Cyanox 1790	1,3,5-tris (4-*tert*-Butyl-3-hydroxy-2,6-dimethylbenzyl)-1,3,5-triazine-(1*H*,3*H*,5*H*) trione	40601-76-1
Cyanox 2246	See Irganox 2246	—
Cyasorb UV 9		
Cyasorb UV-24	2,2'-Dihydroxy-4-methoxybenzophenone	131-53-3
Cyasorb UV 531	2-Hydroxy-4-(octyloxy) benzophenone	1843-05-6
Cyasorb UV 1084	2,2'-Thiobis (4-*tert*-octylphenoxy) (nibutylamine) nickel	14516-71-3
Cyasorb UV 1164	2,4-bis (2,4-Dimethylphenyl)-6-(2-hydroxy-4-octyloxyphenyl)-1,3,5-triazine	2725-22-6
Cyasorb 2908	3,5-Di-*tert*-butyl-4-hydroxybenzoate	67845-93-6
Dechlorane Plus	1,2,3,4,7,8,9,10,13,13,14,14-Dodecachloro-1,4,4a,5,6,6a,7,10,10a,11,12,12a-dodecahydro-1,4,7,10-dimethanodibenzo [*a,e*] cyclooctene	13560-89-9
DEHA	Di-(2-ethylhexyl) adipate, dioctyl adipate	103-23-1
Di-Cup	Dicumyl peroxide	80-43-3

(Continued)

Table 2 Chemical names for the additives cited in this manuscript *(Continued)*

Trade name	Chemical name	CAS RN
Dibutyl sebacate		109-43-3
DEG	Diethylene glycol	111-46-6
DBP	Di-*n*-butyl phthalate	84-74-2
DEHP, DOP	Di (2-ethylhexyl) phthalate	117-81-7
Dinonyl phthalate	1,2-Benzenedicarboxylic acid, dinonyl ester	84-76-4
DMTDP	3,3'-Thiodipropionic acid di-*n*-tetradecyl ester	16545-54-3
DSTDP	Dioctadecyl 3,3'-thiodipropionate	693-36-7
DLTDP	Dilauryl 3,3'-thiodipropionate	123-28-4
Docosane		629-97-0
Eicosane		112-95-8
Epoxol 9.5	Epoxidized linseed oil	8016-11-3
EG	Ethylene glycol	107-21-1
Erucamide	*cis*-13-Docosenamide	112-84-5
Ethyl palmitate	Hexadecanoic acid ethyl ester	628-97-7
Ethyl linoleate	9,12-Octadecanoic acid ethyl ester	544-35-4
Ethyl oleate	9-Octadecanoic acid ethyl ester	111-62-6
Ethyl stearate	Octadecanoic acid ethyl ester	111-61-5
Ethanox 330	See Irganox 1330	—
Hexadecane		544-76-3
Hexacosane		630-01-3
Hostanox O3	bis [3,3-bis (4-Hydroxy-3-*tert*-butylphenyl) butanoic acid] ethylene glycol ester	32509-66-3
Hostavin TMN 20	2,2,4,4-Tetramethyl-21-oxo-7-oxa-3,20-diazadispiro [5.1.11.2] heneicosane	64338-16-5
Ionol 220	2,6-Di (*tert*-butyl)-4-methylphenol (see Topanol OC)	—
Ionox 100	4-Hydroxymethyl-2,6-di-*tert*-butylphenol	88-26-6
Ionox 129	2,2'-Ethylidenebis (4,6-di-*tert*-butylphenol)	35958-30-6
Ionox 220	4,4-Methylenebis (2,6-di-*tert*-butylphenol)	118-82-2
Irgafos 168	tris (2,4-Di-*tert*-butylphenyl) phosphite	31570-04-4
Irgafos P-EPQ	Tetrakis (2,4-di-*tert*-butylphenyl)-4,4'-biphenylene diphosphonite	38613-77-3
Irganox 245	Triethylene glycol bis-3-(3-*tert*-butyl-4-hydroxy-5-methyl) propionate	36443-68-2
Irganox 259	1,6-bis [3-(3,5-Di-*tert*-butyl-4-hydroxyphenyl) propionyloxy] hexane	35074-77-2
Irganox 565	2,4-bis (Octylthio)-6-(3,5-di-*tert*-butyl-4-hydroxyanilino)-1,3,5-triazine	991-84-4
Irganox 1010	Tetrakis-methylene-(3,5-di-*tert*-butyl-4-hydroxyhydrocinnamate)-methane	6683-19-8
Irganox 1035	2,2'-Thiodiethylene bis [3-(3,5-di-*tert*-butyl-4-hydroxyphenyl) propionate]	41484-35-9
Irganox 1076	Octadecyl-3-(3',5'-di(*tert*-butyl)-4'-hydroxyphenyl) propionate	2082-79-3
Irganox 1098	*N,N*-bis [3-(3,5-di-*tert*-butyl-4-hydroxyphenyl) propionyl] hexamethylene-diamine	23128-74-7
Irganox 1222	Diethyl (3,5-di-*tert*-butyl-4-hydroxybenzyl) phosphonate	976-56-7
Irganox 1330	1,3,5-Trimethyl-2,4,6-tris (3,5-di-*t*-butyl-4-hydroxy-benzyl)-benzene	1709-70-2

(Continued)

Table 2 Chemical names for the additives cited in this manuscript *(Continued)*

Trade name	Chemical name	CAS RN
Irganox 1425	Calcium bis (ethyl 3,5-di-*tert*-butyl-4-hydroxybenzylphosphonate)	65140-91-2
Irganox 2246	2,2'-Methylene-bis (4-methyl-6-*tert*-butylphenol)	119-47-1
Irganox 3052 FF	2,2'-Methylenebis (6-*tert*-butyl-4-methylphenol) monoacrylate	61167-58-6
Irganox 3114	1,3,5-tris (3,5-Di-*t*-butyl-4-hydroxybenzyl)-*s*-triazine-2,4,6-(1*H*,3*H*,5*H*) trione	27676-62-6
Irganox MD1024	3,5-bis (1,1-Dimethylethyl)-4-hydroxybenzenepropionic acid	32687-78-8
Irganox MD1025	*N*,*N*-bis [1-oxo-3(3,5-di-*tert*-butyl-4-hydroxyphenyl) propane] hydrazine	
Irganox PS800	Di-lauryl thio-dipropionate	123-28-4
Irganox PS802	Di-stearyl thio-dipropionate	693-36-7
Isonox 129	2,2'-Ethylidenebis [4,6-di-*tert*-butylphenol]	35958-30-6
Kemamide U	See Oleamide	—
Lauric acid	Dodecanoic acid	143-07-7
Lowinox 22M46	See Irganox 2246	—
MHET	Mono-(2-hydroxyethyl) terephthalate	155603-50-2
Myristic acid	Tetradecanoic acid	544-63-8
Naugard SP	*N*,*N*-bis (2-hydroxyethyl) alkyl-amine	94765-89-6
Naugard XL-1	2,2'-Oxamidobis [ethyl 3-(3,5-di-*tert*-butly-4-hydroxyphenyl) propionate]	70331-94-1
Naugawhite	2-2'-Methylenebis (4-methyl-6-nonylphenol)	7786-17-6
NC-4	1-(2,6-Dimethylphenylimino) imidazolidine	4859-06-7
Nonflex CBP	2,2'-Methylenebis (6-(1-methylcyclohexyl)-*p*-cresol)	77-62-3
Noclizer M-17	2,6-Di-*tert*-butyl-4-ethylphenol	4130-42-1
Octadecane		593-45-3
ODO	Octabromodiphenyl oxide	32536-52-0
Oleamide	9-Octadecenamide	301-02-0
Palmitic acid	Hexadecanoic acid	57-10-3
Palmitamide	Hexadecanoic acid amide	629-54-9
Permanax WSP	2,2'-Methylenebis [4-methyl-6-(1-methylcyclohexyl) phenol]	77-62-3
Sanol LS744		
Sanol LS770		
Santonox	See Yoshinox SR	—
Santowhite	4,4'-Butylidenebis (3-methyl-6-*tert*-butylphenol)	85-60-9
Seesorb 101	2-Hydroxy-4-methylbenzophenone	131-57-7
Seesorb 202	4-*tert*-Butylphenylsalicylate	87-18-3
Seenox DM	3,3'-Thio-dipropionic acid dimyristyl ester	16545-54-3
Sodium benzoate	Benzoic acid, sodium salt	532-32-1
Stearamide	Octadecanamide	124-26-5
Stearic acid	Octadecanoic acid	57-11-4
Synprolam	Quaternary ammonium compounds, di-C13—C15-alkylmethyl, chlorides	308074-61-5
Terephthalic acid		100-21-0
TETO	Glyceryl tri-epoxyoleate	

(Continued)

Table 2 Chemical names for the additives cited in this manuscript *(Continued)*

Trade name	Chemical name	CAS RN
Tetradecane		629-59-4
Tetracosane		646-31-1
Tinuvin 120	2′,4′-Di-*tert*-butylphenyl-3,5-Di-*tert*-butyl-4-hydroxybenzoate	4221-80-1
Tinuvin 144	2-*tert*-Butyl-2-(4-hydroxy-3,5-di-*tert*-butylbenzyl) [bis (methyl,2,2,6,6-tetramethyl-4-piperidinyl)] dipropionate	63843-89-0
Tinuvin 234	2[2′-Hydroxy-3,5-di91,1-dimethylbenzyl) phenyl]-2*H*-benzotriazole	70321-86-7
Tinuvin 292	bis (1-Methyl-2,2,6,6,tetramethylpiperidinyl) sebacate	41556-26-7
Tinuvin 312	*N*-(2-ethyoxyphenyl)-*N*′-(2-ethylphenyl)-ethanediamine	23949-66-8
Tinuvin 320	2-(2-Hydroxy-3,5-di-*tert*-butylphenyl)-2*H*-benzotriazole	3846-71-7
Tinuvin 326	2-(3-*tert*-Butyl-2-hydroxy-5-methylphenyl)-2*H*-5-chlorobenzotriazole	3896-11-5
Tinuvin 327	2-(2′-Hydroxy-3′,5′-di(tertbutyl) phenyl)-2*H*-5-chlorobenzotriazole	3864-99-1
Tinuvin 328	2-(2′-Hydroxy-3′,5′-di(tertbutyl) phenyl)-2*H*-chlorobenzotriazole	25973-55-1
Tinuvin 329	2-(2′-Hydroxy-5′-*tert*-octylphenyl) benzotriazole	3147-75-9
Tinuvin 350	2-(2*H*-benzotriazol-2-yl)-4-(1,1-dimethylethyl)-6-(2-methylpropyl) phenol	134440-54-3
Tinuvin 384	Octyl 3-[3-(2*H*-benzotriazol-2-yl)-5-*tert*-butyl-4-hydroxyphenyl] propionate	84268-23-5
Tinuvin 662		
Tinuvin 770	(2-(2-Hydroxy-3,5-bis (1-methyl-1-phenylether) phenyl) benzotriazole)	52829-07-9
Tinuvin 1130	α-[3-[2-(2*H*-benzotriazol-2-yl)-5-(1,1,-dimethylethyl)-4-hydroxyphenyl]-1-oxopropyl-hydroxy-poly (oxy-1,2-ethanediyl)	194810-48-2
Tinuvin P	2-(2-Hydroxy-5-methylphenyl)-2*H*-benzotriazole	2440-22-4
TNPP	tris (Nonylphenyl) phosphite	26523-78-4
Topanol CA	1,1,3-tri (3-*tert*-butyl-4-hydroxy-6-methylphenyl) butane	1843-03-4
Topanol OC	2,4,6-tri-*tert*-butylphenol	128-37-0
TPP	Triphenyl phosphate	115-86-6
Tributylacetylcitrate	See Citroflex A-4	—
Ultranox 626	bis (2,4-Di-*t*-butylphenyl)-pentaerythritol-diphosphite	26741-53-7
Uvitex OB	2,5-bis (5′-*tert*-Butyl-2′-benzoazolyl) thiophene	7128-64-5
Vulkanox CS		94766-18-4
Wingstay T		12674-05-4
Yoshinox 425	2,2′-Methylenebis (4-ethyl-6-*tert*-butylphenol)	88-24-4
Yoshinox BB	See Santowhite	—
Yoshinox SR	4,4′-Thiobis [3-methyl-6-*tert*-butylphenol]	96-69-5
Zinc stearate	Stearic acid, zinc salt	557-05-1

Table 3 Examples of HPLC methods (UV detection) used to identify and/or quantify packaging system additives

Material	Additive(s)	Sample preparation	Column	Mobile phase	λ (nm)	Other	Refs.
PE	Dicumyl peroxide, Santonox®	Extract with methanol, concentrate (evaporative)	Lichrosorb RP 18, 250 × 4.6 mm, 10 μm	Methanol/water, 80/20	254	1 mL/min	[1]
PVC	Di-(2-ethylhexyl) phthalate (DEHP), epoxidized linseed oil, tris (nonyphenyl) phosphite (TNPP)	Dissolve in tetrahydrofuran (THF), precipitate polymer with methanol, dry and dissolve residue	μPorasil C18	Carbon tetrachloride-dicloromethane (65/35)	280	1 mL/min	[2]
LDPE, PP	BHT, Cyasorb 531, Tinuvin 327, Irganox 1076	Reflux with CCl4 or THF, filter, concentrate	μBondapak C18, 60 × 0.39 cm	Methanol/water/THF, 63/7/30	254	2 mL/min	[3]
LDPE, HDPE	Tinuvin P, BHT, BEHB, Oleamide, Cyasorb UV 531, Isonox 129, AM340, Irganox 1010, Irganox 3114, Erucamide	Soxhlet, ultrasonic or microwave extraction with various solvent systems	Nova-Pak C18, 150 × 3.9 mm, 4 μm	A = water; B = acetonitrile; 3/2 initial, linear gradient to 100% B in 5 min	200	1.5 mL/min, T = 50°C	[4]
PP	Irgafos 168, Irganox 1076, Irganox 3114, Irganox 1010, Tinuvin P, Irgafos PEP-Q	Microwave extraction with 98/2 methylene chloride/2-propanol	Nova-Pak silica (150 × 3.9 mm, 4 μm) or resolve silica (150 × 3.9 mm, 5 μm)	70/30 n-butyl chloride/methylene chloride	225	1.5 mL/min, T = 30°C	[4]
Polyolefin	Irganox (245, 259, 565, 1010, 1035, 3114); Tinuvin (P, 234, 320, 326, 327, 328)	N/A[a]	Capcell Pak C18, 250 × 4.6 mm	Methanol/water mixtures (95/5, 90/10, 88/12, 85/15)	multiple	1 mL/min, T = 45°C, 20 μL	[5]
Polyolefin	BHT, BHA, Irganox 1010, Irganox 565, Tinuvin 327, Tinuvin P	N/A[a]	Licrosphere 100 RP 18, 250 × 4.6 mm, 5 μm	Various binary and tertiary mixtures of methanol, water, and acetonitrile	multiple	20 μL	[6]
General	BHT, Irganox antioxidants (245, 259, 565, 1010, 1035, 1076, 1098, 1222, 1330, etc.); UV absorbers (Tinuvin P, 312, 320, 327, 328, etc.)	N/A[a]	Ultrabase UB225, 250 × 4.6 mm, 5 μm; LiChrospher 100 RP 18e, 250 × 4.6 mm, 5 μm; Spheri 5-ODS, 250 × 4.6 mm, 5 μm	Quaternary gradient of THF, water, methanol, acetonitrile	Multiple wavelength UV and laser light scattering[b]	1 mL/min, 20 μL, ambient temperature	[7][c]
General	BHT, BHA	N/A[a]	Whatman Partisil PXS 10/25 ODS-2	0.05 M LiClO4 in 85% methanol	UV-electrochemical fluorescence	1 mL/min, 20 μL, T = 40°C	[8][d]
PP, ABS	Numerous[e]	Dissolve in MeCl2, precipitate polymer with methanol	Spherisorb S30DS2, 150 × 4.6 mm, 3 μm	A = Acetonitrile; B = water; Initial = 40% A, ramped to 100% A in 15 min, maintained at 100% for 17 min	210, 280	1 mL/min (2 mL/min after 22 min); 10 μL	[9]

C

	Additives	Sample preparation	Column	Mobile phase	Detection	Conditions	Ref.
PE	Topanol OC; Cyasorb UV 531; Irganox 1010, 1076, 1330	SFE contrasted to Soxhlet extraction with dichloromethane	Kaseisorb ODS-5, 550 × 0.53 mm	Methanol	254 nm	Column pressure = 100 atm	[54]
PP	Irganox 1010, Irgafos 168	SFE	Licrosorb RP-18, 200 × 4.6 mm, 5 μm	Gradient from methanol/water (95/5) to 100% methanol in 17 min	280 nm	1.5 mL/min, 20 μL	[11]
Polyolefin	Irganox PS802, 1010, 1076, 1425; calcium stearate; sodium benzoate	Microwave-assisted solvent extraction	Various; e.g., Microsper C$_{18}$, 250 × 4.6 mm, 5 μm	Water/acetonitrile/iso-propanol; Start = 12/88/0; 0.1 min = 5/65/30; 10 min = 0/65/30; 18 min = 0/65/30	273 nm, light scattering	2 mL/min, 10 μL, T = 50°C	[12]
PMMA	Irganox 1010, 1076; Irgafos 168	SFE	Nova-Pack C$_{18}$, 5 μm	Acetonitrile/water gradient; start at 80/20, change to 100/0 in 5 min	254 nm	50 μL	[13]
Dielectric resins	BAC-E, bis-A-bis-azide	Dissolution in THF	Zorbax RX C8, 100 × 4.5 mm, 5 μm[f]	A = 50/50 acetonitrile/water; B = 90/10 mixture. Gradient was start ramp form 0% A to 100% B in 5 min, hold at 100% B for 5 min	254 nm (380 nm for internal standard)	3 mL/min	[14]

ABS = acrylonitrile–butadiene–styrene; HDPE = high-density polyethylene; LDPE = low-density polyethylene; PC = polycarbonate; PE = polyethylene; PMMA = polymethylmethacrylate; PP = polypropylene; PS = polystyrene; PVC = polyvinyl chloride.

aThis reference examined the elution characteristics of the cited additives as a function of mobile phase and, thus, did not characterize actual polymers.

bThese authors report that the UV response for the analytes is typically 5 to 50 times greater than the light-scattering response.

cNumerous other examples of separations provided.

dFor BHA, the sensitivity was EC > UV (230 nm) > Fl. For BHT, the sensitivity was UV (280 nm) > EC > Fl.

eThis study examined the elution characteristics, sensitivity, and analytical recoveries of over 25 additives.

fPreanalytical column sample cleanup was achieved on-line by SEC. This preanalytical processing was accomplished with a 25 cm × 250 μm i.d. fused silica capillary column (PL-Gel 50 Å, 5 μm) and a THF mobile phase at 1.3 μL/min.

Table 4 Examples of HPLC methods [infrared (IR) detection] used to identify and/or quantify packaging system additives

Material	Additive(s)	Sample preparation	Column	Mobile phase	λ (nm)	Other	Refs.
Polyurethane	Irganox 1010	Extraction with n-hexane	μ-Porasil C$_{18}$	n-hexane/dichloromethane, 80/20	254[a]	1 mL/min	[15]
PVC, PP, PE	Ionol 220, Irganox 1076, Tinuvin 327, Tinuvin 328, Cyasorb UV 531	Extraction with acetonitrile or methanol, evaporative concentration	RP: Spherisorb ODS-2, 100 × 1.0 mm, 3 μm; SEC: PL-Gel, 250 × 4.6 mm, 5 μm, 500 Å	RP = methanol/water (95./5); SEC = Dicloromethane	280[b]	0.1–0.2 mL/min	[16]
PP	Irganox 1010, 3114; Tinuvin 326, 327	N/A[c]	Zorbax ODS, 250 × 4.6 mm		N/A[d]	0.5 mL/min, 20 μL	[17]
PP, PE	Irganox 245, 259, 1010, 1076, 1098, 3114; Irgafos 168; Tinuvin 234, 327, 328, 350; Sanowhite; Ethanox 330; Lowinox 22M46; Kemamide U; Naugard; BHT; Ultranox 626; Cyasorb 2908; Cyasorb UV 531	N/A[c]	Sperisorb ODS-2, 250 × 4.6 mm, 5 μm	Several cited[e]	280 nm and evaporative light scattering[f]	1 mL/min, 50 μL, ambient temperature	[18][g]

PE = polyethylene; PP = polypropylene; PVC = polyvinyl chloride.

[a]Peak collected on potassium bromide powder and an IR spectrum obtained off-line.

[b]On-line Fourier transform infrared spectroscopy (FTIR) with spray-jet interface, deposition on a moving zinc selenide substrate. UV for quantitation, IR for identification.

[c]This reference examined the detection characteristics of the cited additives and, thus, did not characterize actual polymers.

[d]Used surface-enhanced infrared spectroscopy with effluent deposition on an Ag metal film (BaF$_2$ substrate). Reported a detection limit of 10 ng.

[e]Several separations are cited in this manuscript. A gradient using methanol and water was used for the separation of nine additives. Initial composition, 94/6 methanol/water for 7 min, changed immediately to 100% methanol, hold for 14 min.

[f]Portion of column effluent deposited on a rotating germanium disk via drying nebulization.

[g]Detection limits by IR were generally near 0.2 μg, with quantities needed for accurate identification in the 0.5–1.0 μg range.

Table 5 Examples of HPLC methods (MS detection) used to identify and/or quantify packaging system additives

Material	Additive(s)	Sample preparation	Column	Mobile phase	Detection	Other	Refs.
General(PP)	BHT, Irganox 1010, Irganox 1076, Irganox 1330, Santowhite	N/A[a]	ODS, 250 × 2.1 mm, 5 µm	A = 75/25 acetonitrile/water; B = 50/50 THF/acetonitrile; 0 min = 100% A, 10 min = 60% A, 20 min = 100% B, 30 min = 100% B, 32 min = 100% A	UV, 280 nm. Moving belt MS interface, chemical ionization (CI) and electron input ionization (EI) spectra obtained	0.2 mL/min, 10 µL	[19]
PVC	Mono-alkyl esters and di-alkyl phthalates	N/A[a]	Symmetry C-8, 150 × 3.9 mm, 5 µm	A = 0.1 M ammonium acetate in 10/90 methanol/water; B = 0.1 M ammonium acetate in methanol 0% to 100% B over 25 min, hold for 5 min	UV, 277 nm. Thermospray MS, + ions	1 mL/min	[20]
PP	NC-4, Naugard-XL, Irganox 1076	Extract with acetonitrile at 60°C for 72 hr	Symmetry C8, 150 × 3.9 mm, 3 µm	A = acetonitrile; B = water; 0 = 30% A, 10 min = 100% A, 30 min = 30% A, 40 min = 30% A	Multiple λ UV. MS = EI and atmospheric pressure chemical ionization (APCI) [positive ion, single ion monitoring (SIM)]		[21]
PP	Irganox 245, BHA, BHT, Bispenol A, Topanol CA	N/A	Hypersil H5ODS, 100 × 4.6 mm, 5 µm	80/20 deuterated acetonitrile/water	UV, IR, MS, nuclear magnetic resonance (NMR)[b]	1 mL/min	[55]

PP = polypropylene; PVC = polyvinyl chloride.
[a]This reference examined the elution characteristics of the cited additives as a function of mobile phase and thus did not characterize actual polymers.
[b]The column effluent was split after the column (95/5) between the FTIR flowcell [attenuated total reflection (ATR) used] and the MS detector (single quad, positive ions). The effluent from the FTIR cell was directed through the UV and ultimately nmR detectors.

Table 6 Examples of SFC methods used to identify and/or quantify packaging system additives

Material	Additive(s)	Sample preparation	Column	Mobile phase	Detection	Other	Refs.
General (PE)	Numerous[a]	N/A[b]; SFE	C$_{18}$, 250 × 4.6 mm	CO$_2$ with methanol modifier; 2% methanol for 1 min, linear to 10% after 10 min, hold for 5 min	APCI-MS (+ and − ionization), UV	2 mL/min, 10 μL	[22][c]
PE	BHT, Tinuvin 326	SFE with various traps	Fused silica capillary (10 m × 0.1 mm i.d.) with octyl phase, 0.5-μm film	CO$_2$, pressure program (100 atm for 10 min, increased at 5 atm/min for 20 min)	FID	Column T = 90°C	[23]
Polyolefin (PP, PE)	BHT; Irganox PS800, PS802, 1010, 1076, 1330, 2246; Tinuvin P, 320, 326; Irgafos 168; Atmos 150	Soxhlet extraction with chloroform	Fused silica capillary (10 m × 0.05 mm i.d.) with 5% phenyl-polymethylsiloxane, 0.4-μm film	Various pressure and temperature programs with CO$_2$ were reported	MS, EI	1 μL injection	[5, 24][d,e]
PP	Numerous[f]	Soxhlet extraction with diethyl ether for 15 hr, precipitate polymer with ethanol	Fused silica capillary (15 m × 0.1 mm i.d.) with 5% phenyl-polymethylsiloxane, 0.5-μm film	CO$_2$ at 140°C; pressure = 150 atm for 12 min, ramp to 350 atm at 3 atm/min	FID, FTIR microscope	2 μL split injection, T = 150°C	[26]
PE	BHT, Irganox 1010, Irgafos 168	SFE (contrasted to Soxhlet extraction)	Fused silica capillary (15 m × 0.1 mm i.d.) with SB-Biphenyl-30, 0.5-μm film	CO$_2$ at 140°C; pressure = 100 atm, ramp to 200 atm at 3 atm/min, ramp at 10 atm/min to 400 atm	IR (flow through cell)	—	[27]
General	Tinuvin P, 326, 234, 770; Chimasorb 81, 770; Irganox 1010, 1076, 1330; Irgafos 168, Irgafos P-EPQ, N,N-bis-(2-hydroxyethyl) alkyl-amine	N/A[b]	Fused silica capillary (20 m × 0.1 mm i.d.) with DB-5, 0.4-μm film	CO$_2$; pressure = 10.6 MPa for 10 min, ramp to 15 MPa in 3.5 min, hold for 5 min; ramp at 0.5 MPa/min to 35 MPa, hold for 10 min	FID and MS (EI, 70 eV)	2 mL/min, T = 140°C	[10]

Polymer	Additives examined	Extraction	Column	Conditions	Detector	Temperature	Ref.
LDPE	BHT, BHEB, Isonox 129, Irganox 1010, 1076	On-line SFE	Deltabond cyano, 100 × 1.0 mm i.d., 5 μm	100 atm for 3 min, 100–330 atm for 7 min, 330–450 atm for 1.5 min	FID	Column T = 100°C	[28][g]
Polyurethane	BHT, Irganox 1010, Irganox 1076, dinonyl phthalate[h]	On-line SFE	30% biphenyl polysiloxane, 10 m × 50 μm i.d., 0.25-μm film	CO_2; hold at 100 atm for 5 min, ramp at 5 atm/min to 300 atm, ramp at 20 atm/min to 400 atm	MS (EI)	Column T = 100°C	[29]
Nylon, polystyrene	Caprolactam and oligomers, stearic acid, Irganox 1076	On-line SFE	Deltabond cyano, 100 cm × 1 mm, 5 μm	CO_2; hold at 100 atm for 2 min, ramp at 15 atm/min	FTIR	Column T = 100°C	[30]
PE, PP	BHT; Erucamide; Irgafos 168, Irganox 1010, 1076; Tinuvin 326, 770; Isonox 19; DLTDP	On-line SFE	Deltabond 300 Octyl, 250 cm × 1 mm	CO_2; hold at 1500 psi for 6 min, ramp at 200 psi/min to 6000 psi	FID	FID T = 350°C; Column T = 150°C	[31]

LDPE = low-density polyethylene; PE = polyethylene; PP = polypropylene.
[a] Additives examined included: BHT; Irganox 245, 1010, 1035, 1076, 1330, 1425, PS802; Irgafos 168; Tinuvin 327, 328, 384 440, 622, 770, 1130; Topanol CA; Cyasorb UV1164; Oleamide; Erucamide; Synprolam; Chimassorb 944.
[b] This reference examined the elution characteristics of the cited additives as a function of mobile phase and, thus, did not characterize actual polymers.
[c] This reference provides elution characteristics and relative intensities for specific positive and negative ions.
[d] The performance of the SFC method was compared to that of an isocratic reversed phase (RP)-HPLC method (UV detection).
[e] GC/MS was also used for compound identification.
[f] Additives included: Topanol OC; Tinuvin P, 292, 320, 326, 328, 770, 440, 144; Chimassorb 81; Erucamide; Irganox PS800, PS802, 245, 1010, 1035; MD1025, 1076, 1330, 3114.
[g] The SFE extraction was also coupled with HPLC separation and detection of the analytes.
[h] This method produced other additive peaks whose parent compound could not be identified.

Table 7 Examples of SEC/GPC methods used to identify and/or quantify packaging system additives

Material	Additive(s)	Sample preparation	Column	Mobile phase	Detection	Other	Refs.
PP	General[a]	Extraction with THF	Porogel A-1 (slurry-packed)	THF	Refractive index (RI) (collected fraction tested via IR for identification)	2 mL/min, column $T = 30°C$	[32]
PS, PVC	Tinuvin P, TNPP, TETO	Solvent extraction	500 Å, 100 Å and 50 Å PL-Gel (poly (styrene-divinylbenzene) 30 × 0.77 cm, 10 μm	THF	UV	2 mL/min, column $T = 30°C$	[33]
Polyolefin (PP, PE)	BHT; Cyasorb UV 9; Cyasorb UV 1084; Tinuvin 326, 327; Irganox 1076	General[b]	SEC (PL-Gel, 50 Å, 300 × 7.5 mm) coupled with normal phase (Nucleosil 100-7OH, 250 × 4.6 mm, 7 μm)	n-hexane/dichloromethane (73/27)	UV at 254 and 280 nm	0.9 mL/min, column $T = 35°C$	[34]
Polyolefin copolymers	Numerous[c]	Dissolution in THF	SEC (Uktrastyragel 10^4, 50 Å, 300 × 0.25 mm, 10 μm) coupled with GC (DB-1, 15 m × 0.32 mm i.d., 0.25-μm film)	SEC; THF mobile phase. GC; 100°C for 6 min, ramp at 16°C/min to 350°C	UV at 254 nm for SEC, MS (EI) for GC	For SEC: 3.0 μL/min, 0.2 mL injected	[35]

PE = polyethylene; PP = polypropylene; PS = polystyrene; PVC = polyvinyl chloride.
[a]This reference generated additive profiles for various test materials but did not specify the individual additives found.
[b]This reference examined the chromatographic characteristics of the cited additives as a function of mobile phase and, thus, did not characterize actual polymers.
[c]This reference documented chromatographic characteristics for many individual additives including plasticizers (phthalates, Citroflex A-4, TNPP); amides; Irganox and Irgafos antioxidants; UV absorbers (Tinuvin, Cyansorb); fatty acids (palmitic, stearic); Naugard XL-1; etc.

Table 8 Examples of TLC methods used to identify and/or quantify packaging system additives

Material	Additive(s)	Sample preparation	Plate	Mobile phase	Detection	Other	Refs.
Rubber	Numerous[a]	N/A[b]	Silica gel G, thickness of 250–300 μm	Numerous solvent systems examined vs. compound class	Various UV and visible developing reagents	Sample size: 3–5 μL	[36]
Polyolefin	Tinuvin 144, 770;Hostavin TMN 20	Extraction with chloroform, polymer precipitation, evaporative concentration	Alumina F254, 20 × 20 cm plate, 0.25-mm thickness	88/12, n-hexane/iso-propanol	Chlorination with chlorine gas, sprayed with potassium iodide starch solution	Sample size: 10 μL	[37]
Elastomers	BHT, Cyanox 2246, Cyanox 425, Permanax WSP, Naugawhite, Wingstay T, Naugard SP, Vulkanox CS	Soxhlet extraction with acetone, evaporative concentration	Merck 11845 silica gel	Benzene	Sprayed with 2.34% sodium tetraborate, 0.33% sodium hydroxide, and 0.1% methanolic solution of N-chlorodichloro-2,6-p-benzoquinone monoimine	Sample size: 20 μL	[38]
PP, PVC	Irganox 1010, Tinuvin 770, Chimassorb 119 FL, DEHP	N/A[b]	Silica gel 60 F254, 5 × 10 cm plate, 250-μm thickness	Toluene-diethyl ether (10/1) on first plate, acetone-formic acid (4/6) on second plate	UV for plate 1, plate 2 visualized by iodine vapor. Spots removed from plate 2 and analyzed by FTIR	Sample size: 2 μL	[39]

PP = polypropylene; PVC = polyvinyl chloride.
[a]This reference documents the chromatographic properties of over 100 rubber-related amine and phenolic antioxidants, antiozonants, guanidines, accelerators, and amine hydrochlorides.
[b]This reference examined the chromatographic characteristics of the cited additives and, thus, did not characterize actual polymers.

C

Table 9 Examples of GC methods used to identify and/or quantify packaging system additives

Material	Additives	Sample preparation	Column[a]	Oven program	Detection	Other	Refs.
PET	Diethylene glycol	High temperature and pressure extraction with water	Numerous[a]	Isothermal at 180°C	FID and thermal conductivity detector (TCD)	Injector T = 260°C; detector T = 270°C; He carrier at 50 mL/min	[40]
PET	BHET, EG, DEG, TA, MHET[b]	N/A;[c] samples TMS derivatized with BSFTA (80°C for 10 min)	3% OV101 on 80–100 mesh Chromosorb W, 6 ft, 0.125 in. o.d. 0.05 in. i.d.	80°C for 3 min, ramp at 15°C/min to 265°C	FID	Injector T = 270°C; detector T = 280°C; He carrier at 50 mL/min, 2 μL injection	[41]
Elastomers	Various antioxidants and additives	N/A[c]	3% SP2100 on 80–100 mesh Supelcoport	150°C for 2 min, ramp at 10°C/min to 250°C	MS	Injector T = 270°C; He carrier at 35 mL/min	[42]
Polyolefins	BHT; Irganox 1010, 1076, 2246; Irgafos 168, Santowhite	Refluxed in acetone for 2–3 hr, evaporated to dryness and dissolved in chloroform	PS264, 15 m × 0.32 mm i.d., 0.15-μm film	90°C to 280°C at 5°C/min	FID	Detector T = 300°C; He carrier at inlet pressure of 0.5 kg/cm²	[43]
PVC	Benzoflex 2860, ATBC, DEHA[d]	SFE	SPB-5 fused silica capillary, 30 m × 0.25 mm i.d., 0.25-μm film	110°C to 260°C at 10°C/min	FID	Detector, injector T = 300°C; He carrier at 50 cm/sec, 1 μL splitless	[44]
Polyolefins	Numerous[e]	Reflux with chloroform, evaporative concentration	HR-1701 (14% cyanopropylphenylmethyl-siloxane) fused silica capillary, 15 m × 0.53 mm i.d., 1.0-μm film	100°C for 2 min, 20°C/min to 210°C, 1.5°C/min to 222°C, 8°C/min to 350°C, hold for 10 min	FID	Injector and detector T = 350°C; He carrier at 5 mL/min, 1 μL splitless injection	[45]
Petroleum resin	Numerous[e]	Dissolve in n-hexane, cleanup with silica column	DB-1701 (14% cyanopropylphenylmethyl-siloxane) fused silica capillary, 15 m × 0.53 mm i.d., 1.0-μm film	150°C for 2 min, 20°C/min to 210°C, 1.5°C/min to 222°C, 8°C/min to 350°C, hold for 10 min	FID	Injector and detector T = 300°C; He carrier at 5 mL/min, 2 μL split injection (1:100 ratio)	[45]
PE	Numerous[f]	Dissolution	UA-1 HT (dimethylpolysiloxane) fused silica capillary, 30 m × 0.25 mm i.d., 0.1-μm film	Start at 50°C, ramp at 20°C/min to 300°C, hold for 10 min	MS, EI, 70 eV, 40–700 m/z	Injector T = 250°C; He carrier at 1 psi, 1 μL split injection (1/2 ratio)	[46]

PE = polyethylene; PET = polyethylene terephthalate; PVC = polyvinyl chloride.

[a]Columns used included: Carbowax 20M, Chromosorb W-HMDS, Aeropak Number 30, 10 ft × 1/8 in. × 0.055 in. i.d.

[b]BHET = bis (2-hydroxyethyl) terephthalate; EG = ethylene glycol; DEG = diethylene glycol; TA = terephthalic acid; MHET = mono-(2-hydroxyethyl)terephthalate.

[c]This reference examined the chromatographic characteristics of the cited additives and, thus, did not characterize actual polymers.

[d]ATBC = o-acetly-tri-n-butyl citrate; DEHA = di(ethylhexyl) adipate.

[e]Additives chromatographed included antioxidants (BHT; Irganox 1076, 1330; Yoshinow BB, SR, 2246 R; Irgafos 168; Ultranox 626; Topanol CA; DLTDP; DMTDP; DSTDP), light stabilizers (Sanol LS744, LS770; Tinuvin 120, 326, 327; UV 531), and slip agents (palmitic acid amide, oleic amide, stearic acid amide, erucic amide).

[f]This reference included the separation of 53 polymer additives including antioxidants, UV stabilizers, lubricants, and plasticizers. Performance details are provided.

Table 10 Examples of pyrolysis/GC methods used to identify and/or quantify packaging system additives

Material	Additives	Pyrolysis conditions	Column	Oven program	Detection	Other	Refs.
Polyolefin, PMMA	Hostanox O3; Hostavin N 20; Irganox 3052 FF; Irganox 3114; Tinuvin 320, 329, 350	100 μg sample at 550°C	RTX-5 fused silica capillary, 60 m × 0.32 mm i.d., 0.5-μm film	60° to 300°C at 7°C/min	MS (EI, 45–700 mass range)	Injector T = 260°C; detector T = 270°C; He carrier at 50 mL/min	[47]
PVC, Cellulose Copolymer^c, PS/PC blend, PU	DEHP, Dioctyl adipate, Dibutyl sebacate, tributyl acetylcitrate, TPP, Mixture of didecyl phthalate esters	0.5 mg sample at 700°C	DB-5 fused silica capillary, 30 m × 0.25 mm i.d., 1.0-μm film	40°C for 4 min, ramp at 10°C/min to 320°C, hold for 18 min	MS (EI, 15–650 mass range)	Injector T = 300°C; detector T = 300°C; 30/1 injection split	[48]^{a,b}
Epoxy resin^d, Poly-(diallyl-phthalate), ABS, PC-ABS blend	Brominated bisphenol A, Dechlorane Plus, Octabromodiphenyl oxide, Brominated phenol, triphenyl phosphate	0.5 mg sample at 950°C	DB-5 fused silica capillary, 30 m × 0.25 mm i.d., 1.0-μm film	40°C for 4 min, ramp at 10°C/min to 320°C, hold for 18 min	MS (EI, 15–650 mass range)	Injector T = 300°C; detector T = 300°C; 30/1 injection split	[49]^{a,e}
General	Various waxes, stearic acid, butyl stearate, zinc stearate, butyl oleate, butyl palmitate, stearamide, AcraWax C	0.5 mg sample at 950°C	DB-5 fused silica capillary, 30 m × 0.25 mm i.d., 1.0-μm film	40°C for 4 min, ramp at 10°C/min to 320°C, hold for 18 min	MS (EI, 15–650 mass range)	Injector T = 300°C; detector T = 300°C; 30/1 injection split	[50]^{a,f}
General	Irganox 565, 1010; MD1024, 1035, 1076, 1425, 3114; Irgafos 168	0.5 mg sample at 950°C	DB-5 fused silica capillary, 30 m × 0.25 mm i.d., 1.0-μm film	40°C for 4 min, ramp at 10°C/min to 320°C, hold for 18 min	MS (EI, 15–650 mass range)	Injector T = 300°C; detector T = 300°C; 30/1 injection split	[51]^{a,g}

ABS = acrylonitrile-butadiene-styerene; PC = polycarbonate; PMMA = polymethylmethacrylate; PS = polystyrene; PU = polyurethane; PVC = polyvinyl chloride.
[a] A more rapid oven program was also used to produce pyrograms via FID detection.
[b] This manuscript deals with plasticizers as a class of additives.
[c] Vinyl chloride–vinylidene chloride copolymer.
[d] Cross-linked epoxy resin (thermoset of bisphenol A diglycidyl ether).
[e] This manuscript deals with flame retardants as a class of additives.
[f] This manuscript deals with lubricants as a class of additives.
[g] This manuscript deals with antioxidants as a class of additives.

Table 11 Solubility of polymers

Polymer	Soluble in
Alkyd resin	Chlorinated hydrocarbons, lower alcohols, esters
Acrylonitrile-butadiene-styrene terepolymer	Methylene chloride
Polyacrylamide	Water
Polyamides	Phenols, *m*-cresol, concentrated mineral acids, formic acid
Polycarbonate	Ethanolamine, dioxane, chlorinated hydrocarbons, cyclohexanone
Polyethylene	Dichlorobenzene, pentachloroethylene, dichloroethylene, tetralin
Poly (ethylene terephthalate)	Cresol, concentrated sulfuric acid, chlorophenol, trichloroacetic acid
Polyformaldehyde	Dichlorobenzene, DMF, chlorophenol, benzyl alcohol
Polystyrene	Aromatic and chlorinated hydrocarbons, pyridine, ethyl acetate, dioxane, chloroform, acetone
Polytetrafluoroethylene	Fluorocarbon oil
Polyurethanes	Dioxane, THF, DMF, dimethylsulfoxide (DMSO), *m*-cresol, formic acid, 60% sulfuric acid
Poly (vinyl chloride)	Cyclohexanone, THF, dimethylformamide (DMF), ethylene dichloride
Poly (vinylidene chloride)	THF, ketones, DMF, chlorobenzene
Vinyl chloride–vinyl acetate copolymers	Methylene chloride, THF, cyclohexanone

HPLC with UV detection, remain capable of providing researchers with accurate, precise, and pertinent information. Thus this review attempts to maintain a historical perspective as well.

While it is not a chromatographic issue per se, pre-injection sample preparation, nevertheless, is an important consideration in the successful application of a complete analytical process. This fact is borne out in the observation that recent technological advances in polymer characterization focus not only on the analytical characterization of a test sample but also on the method of generation of that sample. Innovation in sample preparation is driven both by scientific considerations [the difficulty, but necessity, of isolating the analytes of interest (additives) from the sample matrix (bulk polymer) in such a way that the additives are not impacted by the process (e.g., Ref.[52])] and by practical considerations associated with all analytical chemistry (e.g., time and cost efficiency). While it is beyond the scope of this manuscript to exhaustively discuss sample preparation strategies for polymer characterization, this general information is provided to facilitate an important step of the overall analytical process. In terms of specific sample preparation methods, Nerin et al.[53] have recently provided a review of sample treatment techniques applicable to polymer extract analysis including headspace methods, supercritical fluid extraction, and solid-phase microextraction. In terms of general knowledge, Table 11 provides solubility information relevant to polymer dissolution.

REFERENCES

1. Duval, M.; Giguere, Y. Simultaneous determination of the antioxidant, the crosslinking-agent and decomposition products in polyethylene by reversed-phase HPLC. J. Liq. Chromatogr. **1982**, *5*, 1847–1854.
2. Sreenivasan, K. High-performance liquid chromatographic method for the simultaneous separation and determination of three additives in poly(vinyl chloride). J. Chromatogr. **1986**, *357*, 433–435.
3. Francis, V.C.; Sharma, Y.N.; Bhardwaj, I.S. Quantitative determination of antioxidants and ultraviolet stabilizers in polymer by high performance liquid chromatography. Angew. Makromol. Chem. **1983**, *113*, 219–225.
4. Nielson, R.C. Extraction and quantitation of polyolefin additives. J. Liq. Chromatogr. **1991**, *14*, 503–519.
5. Jinno, K.; Yokoyama, Y. Retention prediction for polymer additives in reversed-phase liquid chromatography. J. Chromatogr. **1991**, *550*, 325–334.
6. Lesellier, E.; Saint Martin, P.; Tchapla, A. Separation of six polymer additives using mobile-phase optimization software. LC GC Int. **1992**, *5*, 38–43.
7. Lesellier, E.; Tchapla, A. Sequential optimization of the separation of a complex mixture of plastic additives by HPLC with a quaternary gradient

C

and a dual detection system. Chromatographia **1993**, *36*, 135–143.

8. Masoud, A.N.; Cha, Y.N. Simultaneous use of fluorescence, ultraviolet, and electrochemical detectors in high performance liquid chromatography-separation and identification of phenolic antioxidants and related compounds. J. HRC & GC **1982**, *5*, 299–305.

9. Skelly, N.E.; Graham, J.D.; Iskandarani, Z.; Priddy, D. Reversed-phase liquid chromatographic separation of polymer additives combined with photodiode-array detection and spectral sort software. Polym. Mater. Sci. Eng. **1988**, *59*, 23–27.

10. Bucherl, T.; Gruner, A.; Palibroda, N. Rapid analysis of polymer homologues and additives with SFE/SFC-MS coupling. Packag. Technol. Sci. **1994**, *7*, 139–154.

11. Thilen, M.; Shishoo, R. Optimization of experimental parameters for the quantification of polymer additives using SFE/HPLC. J. Appl. Polym. Sci. **2000**, *76*, 938–946.

12. Marcato, M.; Vianello, M. Microwave-assisted extraction by fast sample preparation for the systematic analysis of additives in polyolefins by high-performance liquid chromatography. J. Chromatogr. A. **2000**, *869*, 285–300.

13. Nasim, N.; Taylor, L.T. Polymer-additive extraction via pressurized fluids and organic solvents of variously cross-linked poly(methylmethacrylates). J. Chromatogr. Sci. **2002**, *40*, 181–186.

14. Patrick, D.W.; Strand, D.A.; Cortes, H.J. Automation and optimization of multidimensional microcolumn size exclusion chromatography-liquid chromatography for the analysis of photocrosslinkers in Cyclotene 400 series advanced electronic resins. J. Sep. Sci. **2002**, *25*, 519–526.

15. Sreenivasan, K. A combined chromatographic and IR spectroscopic method to identify antioxidant in biomedical polyurethane. Chromatographia **1991**, *32*, 285–286.

16. Somsen, G.W.; Rozendom, E.J.E.; Gooijer, C.; Velthorst, N.H.; Brinkman, U.A.Th. Polymer analysis by column liquid chromatography coupled semi-on-line with Fourier transform infrared spectrometry. Analyst **1996**, *121*, 1069–1074.

17. Sudo, E.; Esaki, Y.; Sugiura, M. Analysis of additives in a polymer by LC/IR using surface-enhanced infrared absorption spectroscopy. Bunseki Kagaku **2001**, *50*, 703–707.

18. Jordan, S.L.; Taylor, L.T. HPLC separation with solvent elimination FTIR detection of polymer additives. J. Chromatogr. Sci. **1997**, *35*, 7–13.

19. Vargo, J.D.; Olson, K.L. Identification of antioxidant and ultraviolet light stabilizing additives in plastics by liquid chromatography/mass spectrometry. Anal. Chem. **1985**, *57*, 672–675.

20. Baker, J.K. Characterization of phthalate plasticizers by HPLC/thermospray mass spectrometry. J. Pharm. Biomed. Anal. **1996**, *15*, 145–148.

21. Yu, K.; Block, E.; Balogh, M. LC-MS analysis of polymer additives by electron and atmospheric-pressure ionization: identification and quantification. LC-GC **2000**, *18*, 162–178.

22. Carrott, M.J.; Jones, D.C.; Davidson, G. Identification and analysis of polymer additives using packed-column supercritical fluid chromatography with APCI mass spectrometric detection. Analyst **1998**, *123*, 1827–1833.

23. Daimon, H.; Hirata, Y. Directly coupled super-critical-fluid extraction/capillary supercritical-fluid chromatography of polymer additives. Chromatographia **1991**, *32*, 549–554.

24. Arpino, P.J.; Dilettato, D.; Nguyen, K.; Bruchet, A. Investigation of antioxidants and UV stabilizers from plastics: Part I. Comparison of HPLC and SFC; Preliminary SFC/MS study. J. High Resolut. Chromatogr. **1990**, *13*, 5–12.

25. Arpino, P.J.; Dilettato, D.; Nguyen, K.; Bruchet, A. Investigation of antioxidants and UV stabilizers from plastics: Part II. Application to polyolefin soxhlet extracts. J. High Resolut. Chromatogr. **1991**, *14*, 335–342.

26. Raynor, M.W.; Bartle, K.D.; Davies, I.L.; Williams, A.; Clifford, A.A.; Chalmers, J.M.; Cook, B.W. Polymer additive characterization by capillary supercritical fluid chromatography/Fourier transform infrared microspectrometry. Anal. Chem. **1988**, *60*, 427–433.

27. Wieboldt, R.C.; Kempfert, K.D.; Dalrymple, D.L. Analysis of antioxidants in polyethylene using supercritical fluid extraction/supercritical fluid chromatography and infrared detection. Appl. Spectrosc. **1990**, *44*, 1028–1034.

28. Zhou, L.Y.; Asharf-Khorassani, M.; Taylor, L.T. Comparison of methods for quantitative analysis of additives in low density polyethylene using supercritical fluid and enhanced solvent extraction. J. Chromatogr. A. **1999**, *858*, 209–218.

29. MacKay, G.A.; Smith, R.M. Supercritical fluid extraction-supercritical fluid chromatography-mass spectrometry for the analysis of additives in polyurethanes. J. Chromatogr. Sci. **1994**, *32*, 455–460.

30. Jordan, S.L.; Taylor, L.T.; Seemuth, P.D.; Miller, R.J. Analysis of additives and monomers in nylon and polystyrene. Text. Chem. Color. **1997**, *29*, 25–32.

31. Ryan, T.W.; Yocklovich, S.G.; Watkins, J.C.; Levy, E.J. Quantitative analysis of additives in polymers using coupled supercritical fluid extraction-supercritical fluid chromatography. J. Chromatogr. **1990**, *505*, 273–282.

32. Howard, J.M. Gel permeation chromatography and polymer additive systems. J. Chromatogr. **1971**, *55*, 15–24.

33. Shepherd, M.J.; Gilbert, J. Analysis of additives in plastics by high performance size exclusion chromatography. J. Chromatogr. **1981**, *218*, 703–713.

34. Nerfin, C.; Salafranca, J.; Cacho, J.; Rubio, C. Separation of polymer and on-line determination of several antioxidants and UV stabilizers by coupling size-exclusion and normal-phase high-performance liquid chromatography columns. J. Chromatogr. A. **1995**, *690*, 230–236.

35. Cortes, H.J.; Bell, B.M.; Pfeiffer, C.D.; Graham, J.D. Multidimensional chromatography using on-line coupled microcolumn size exclusion chromatography-capillary gas chromatography-mass spectrometry for determination of polymer additives. J. Microcolumn Sep. **1989**, *1*, 278–288.

36. Kreiner, J.G.; Warner, W.C. The identification of rubber compounding ingredients using thin-layer chromatography. J. Chromatogr. **1969**, *44*, 315–330.

37. Sevini, F.; Marcato, B. Chromatographic determination of some hindered amine light stabilizers in polyolefins. J. Chromatogr. **1983**, *260*, 507–512.

38. Airaudo, C.B.; Gayte-Sorbier, A.; Creusevau, R.; Dumont, R. Identification of phenolic antioxidants in elastomers for pharmaceutical and medical use. Pharm. Res. **1987**, *4*, 237–239.

39. He, W.; Shanks, R.; Amarasinghe, G. Analysis of additives in polymers by thin-layer chromatography coupled with Fourier transform-infrared microscopy. Vibr. Spectrosc. **2002**, *30*, 147–156.

40. Ponder, L.H. Gas chromatographic determination of diethylene glycol in poly(ethylene terephthalate). Anal. Chem. **1968**, *40*, 229–231.

41. Atkinson, E.R., Jr.; Calouche, S.L. Analysis of polyethylene terephthalate prepolymer by trimethylsilyation and gas chromatography. Anal. Chem. **1971**, *43*, 460–462.

42. Kiang, P.H. The application of gas chromatography/mass spectrometry to the analysis of pharmaceutical packaging materials. J. Parenter. Sci. Technol. **1981**, *35*, 152–161.

43. Pasquale, G.D.; Galli, M. Determination of additives in polyolefins by capillary gas chromatography. J. High Resolut. Chromatogr. Commun. **1984**, *7*, 484–486.

44. Guerra, R.M.; Marin, M.L.; Sanchez, A.; Jimenez, A. Analysis of citrates and benzoates used in poly(vinyl chloride) by supercritical fluid extraction and gas chromatography. J. Chromatogr. A. **2002**, *950*, 31–39.

45. Nagata, M.; Kishioka, Y. Determination of additives in polyolefins and petroleum resin by capillary GC. J. High Resolut. Chromtogr. **1991**, *14*, 639–642.

46. Kawamura, Y.; Watanabe, Z.; Sayama, K.; Takeda, Y.; Yamada, T. Simultaneous determination of polymer additives in polyethylene by GC/MS. Shokuhin Eiseigaku Zasshi **1997**, *38*, 307–318.

47. Meyer-Dulheuer, T.; Pasch, H.; Geissler, M. Direct analysis of additives in polymeric materials by pyrolysis-gas chromatography-mass spectrometry. KSG Kautsch. Gummi Kunstst. **2000**, *53*, 574–581.

48. Wang, F.C. Polymer additive analysis by pyrolysis-gas chromatography. I. Plasticizers. J. Chromatogr. A. **2000**, *883*, 199–210.

49. Wang, F.C. Polymer additive analysis by pyrolysis-gas chromatography: II. Flame retardants. J. Chromatogr. A. **2000**, *886*, 225–235.

50. Wang, F.C.; Buzanowski, W.C. Polymer additive analysis by pyrolysis-gas chromatography: III. Lubricants. J. Chromatogr. A. **2000**, *891*, 313–324.

51. Wang, F.C. Polymer additive analysis by pyrolysis-gas chromatography: IV. Antioxidants. J. Chromatogr. A. **2000**, *891*, 325–336.

52. Gasslander, U.; Jaegfeldt, H. Stability and extraction features in the determination of Irganox-1330 in a polyalkane copolymer. Anal. Chim. Acta **1984**, *166*, 243–251.

53. Nerin, C.; Rubio, C.; Salafranca, J.; Batlle, R. The simplest sample treatment techniques to assess the quality and safety of food packaging materials. Rev. Anal. Chem. **2000**, *19*, 435–465.

54. Hirata, Y.; Okamoto, Y. Supercritical fluid extraction combined with microcolumn liquid chromatography for the analysis of polymer additives. J. Microcolumn Sep. **1989**, *1*, 46–50.

55. Louden, D.; Handley, A.; Lenz, E.; Sinclair, I.; Taylor, S.; Wilson, I.D. Reversed-phase HPLC of polymer additives with multiple on-line spectroscopic analysis (UV, IR, 1H nmR, and MS). Anal. Bioanal. Chem. **2002**, *373*, 508–515.

Coil Planet Centrifuges

Yoichiro Ito
Laboratory of Biophysical Chemistry, National Heart, Lung, and Blood Institute,
National Institutes of Health, Bethesda, Maryland, U.S.A.

INTRODUCTION

We have defined "coil planet centrifuge" (CPC) as a term that designates all centrifuge devices in which the coiled separation column undergoes a planetary motion, i.e., the column rotates about its own axis while revolving around the central axis of the centrifuge. Except for the original CPC, all existing CPCs are equipped with a flow-through system so that the liquid can pass through the rotating coiled column. In most of these flow-through CPCs, the use of a conventional rotary seal device is eliminated. These sealless systems are classified into two categories according to their modes of planetary motion, i.e., synchronous and nonsynchronous [see instrumentation of countercurrent chromatography (CC), Fig. 3]. In the synchronous CPC, the coiled column rotates about its own axis during one revolution cycle, whereas in the nonsynchronous CPC, the rates of rotation and revolution of the coiled column are freely adjustable. Among several different types of CPCs, the following four instruments are described below in terms of their best applications: the original CPC, the type-J CPC, the cross-axis CPC, and the nonsynchronous CPC.

ORIGINAL CPC

This first CPC model was devised in an effort to improve the efficiency of lymphocyte separation which was conventionally performed in a short centrifuge tube. If a long tubing is wound into a coil and rotated in a centrifugal field, the particles present in the tube would travel through the tube from one end to the other at a rate depending on their size and density. This idea was implemented in the designs of the first device named the "CPC."[1]

Principle

Physical analysis of the motion of a particle in the rotating helical tube of a CPC has been carried out by means of a simple model is as follows.[1]

(A) We consider a tube filled with a fluid of density ρ_0 coiled into a helix of radius R, with its axis horizontal. If a spherical particle of radius a, density ρ, is placed in the tube, we can determine motion of the particle when the helix undergoes uniform rotation about its axis with angular velocity ω. As a first approximation, we neglect the lateral motion in the tube and assume that in a given turn of the coil, the particle moves on a vertical circle of radius R. The position of the particle can then be specified by the angle θ, as indicated in Fig. 1. With this approximation, the particle is acted on by only two forces, the Stokes drag

$$F_s = -6\pi a \eta R(d\theta/dt - \omega)$$

where η is the viscosity of the fluid, and the net gravitational force g tangent to the circular path

$$F_g = -(4\pi/3)a^3(\rho - \rho_0)g \sin \theta$$

The equation of motion is therefore

$$(4\pi/3)a^3\rho R(d^2\theta/dt^2) = F_s + F_g$$

and on introducing the total angle of rotation of the helix, $x = \omega t$, this can be written in the convenient form

$$d^2\theta/dx^2 + (1/\omega\tau)\{d\theta/dx - 1 + (\omega_e/\omega)\sin \theta\} = 0 \tag{1}$$

where τ is the relaxation time

$$\tau = 2\rho a^2/9\eta \tag{2}$$

and ω_e is the critical angular velocity

$$\omega_e = (2/9)\{(\rho - \rho_0)ga^2\}/\eta R = V_e/R \tag{3}$$

where V_e is the equilibrium Stokes velocity. We consider separately the motion for

$$\omega_e/\omega > 1 \quad \text{and} \quad \omega_e/\omega < 1$$

(B) $\omega_e/\omega > 1$. In this case, Eq. (1) has a singular point at $\omega_e/\omega \sin \theta = 1$, i.e., at $\theta = \theta_2 = \sin^{-1}$

Encyclopedia of Chromatography DOI: 10.1081/E-ECHR-120014242

349

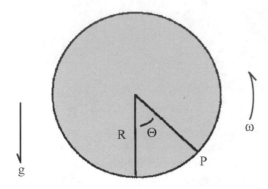

Fig. 1 The coil unit for mathematical analysis of the motion of a particle. (Adapted from Ref.[1].)

(ω/ω_e), $0 \leq \theta \leq \frac{\pi}{2}$, and at $\theta = \pi - \theta_e$. The terms θ_e and $\pi - \theta_e$ are the angles at which the drag of the liquid is just balanced by the gravitational force so that the particle is at equilibrium when at rest. It is readily seen from physical considerations that the equilibrium at θ_e is stable while that at $\pi - \theta_e$ is unstable. More precisely, an analysis in the neighborhood of the singular points shows that $\pi - \theta_e$ is an unstable saddle point, while θ_e is a stable node if $\omega\tau \leq 1/4 \tan \theta_e$ and a stable spiral point if $\omega\tau > 1/4 \tan \theta_e$. Consequently, after a time of the order of τ, the particle will remain fixed at θ_e. Its angular velocity relative to the tube will then be

$$\omega_{rel} = \omega\{d\theta/dx - 1\} = -\omega \qquad (4)$$

That is, it will spiral down the helix at a rate independent of its size and density.

(C) $\omega_e/\omega < 1$, $\omega\tau < 1$. When $\omega_e/\omega < 1$, Eq. (1) has no singular points. The character of the motion, however, depends to some extent on the size of $\omega\tau$, and we shall consider in detail only the physically interesting case $\omega\tau < 1$. It can readily be shown that within this limit, after a time of the order of τ, angular velocity of the particle adjusts itself in such a way that the inertial term, $\omega\tau \, d^2\theta/dx^2$, becomes negligible, so that Eq. (1) reduces to

$$d\theta/dx = 1 - \omega_e/\omega \, \sin \theta \qquad (5)$$

The particle then always rotates in the same direction as the tube, but more slowly than the tube when $0 < \theta < \pi$ and more rapidly when $\pi < \theta < 2\pi$. The two effects, however, do not quite cancel out, and the net effect is that, again, the particle spirals down the tube in a direction opposite to that of the tube rotation.

To determine the mean angular velocity of this spiraling motion, we first calculate the time required

for the particle to traverse one turn of the coil. We have from Eq. (5)

$$X(2\pi) - X(0) = \int_0^{2\pi} d\theta/\{1 - (\omega_e/\omega) \sin \theta\}$$
$$= 2\pi/\{1 - [(1 - \omega_e/\omega)^2]^{1/2}\}$$

The mean angular velocity of the particle relative to the tube is therefore

$$\omega_{rel} = \omega\{2\pi/[X(2\pi) - X(0)] - 1\}$$
$$= -\omega\{1 - [1 - (\omega_e/\omega)^2]^{1/2}\} \qquad (6)$$

which joins continuously on to the value at $\omega_e/\omega = 1$ of Eq. (4).

When ω is close to ω_e, ω_{rel} is rather insensitive to the size and density of the particle. When $\omega_e/\omega < 1$, however, Eq. (6) becomes

$$\omega_{rel} = -\omega_e^2/2\omega \qquad (7)$$

so that the rate of motion down the tube is proportional to a^4 and $(\rho - \rho_0)$. Thus, when $\omega_e < \omega < 1/\tau$, the method should be quite effective in segregating particles of different size and density.

In order to apply (A)–(C) to the CPC, the value of g should be replaced by that of the centrifugal force acting on the axis of the rotating helical tube.

Design of the Original CPC

Fig. 2 shows the first commercial model of the CPC manufactured by Sanki Engineering, Ltd. (Kyoto, Japan).[1,3] The main body of the apparatus consists of three parts, each capable to rotation as a unit: coil holder (6) and interchangeable gear (7) (part I); frame or a pair of arms (4) and discs (3) bridged with links (5) (part II); and central shaft (1) and gear 2 which interlocks to gear 7 of part I. Simultaneous rotation of parts II and III at different angular velocities results in revolution and rotation of part I as a planet. The rotation of part I is determined by the difference in angular velocity between parts II and III and by the gear ratio between 2 and 7.

The apparatus can produce $1000 \times g$ force where the ratio between the coil rotation and revolution can be adjustable to 1/100, 1/300, or 1/500 by interchanging the planetary gears (7) at the bottom of the column holder (6). Each coil holder is equipped with six straight grooves at its periphery to accommodate six coiled tubes that are covered by a transparent vinyl sheath protector tightly fitted to the outside of the

Fig. 2 The coil planet centrifuge fabricated by Sanki Engineering Ltd., Kyoto, Japan. (From Ref.[1].)

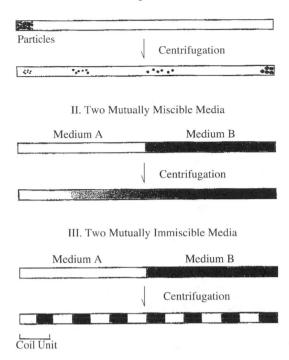

I. Single Medium

Particles

Centrifugation

II. Two Mutually Miscible Media

Medium A   Medium B

Centrifugation

III. Two Mutually Immiscible Media

Medium A   Medium B

Centrifugation

Coil Unit

Fig. 3 The principle of three different applications of the coil planet centrifuge. All tubes are shown uncoiled before and after centrifugation. (Adapted from Ref.[1].)

holder. Each coil is made by winding either polyethylene or Teflon tubing (typically 0.3-mm ID) onto a glass core (15 cm long and 6-mm diameter). After the liquid and sample are introduced into the tube, both ends of the coil are closed before loading onto the holder.

Three Different Modes of Operation

Preliminary experiments with the CPC revealed some interesting features of the apparatus. The results are summarized in Fig. 3.[1]

(I) Single medium: When the tube is filled with a single medium and a particle mixture is introduced at one end, centrifugation separates the particles according to Fig. 3. The difference in size and relative density. The method was effectively demonstrated by a model experiment using polyacrylic resin particles.[1]

(II) Two mutually miscible media (gradient method): When the tube is filled with two mutually miscible media, the heavier in one half and the lighter in the other half, centrifugation produces a density gradient. After centrifugation for some time, the gradient reaches a fairly stable state. In practice, such a stable gradient between water and isotonic saline solution can be introduced into the coil to measure osmotic fragility of erythrocytes. Erythrocytes introduced into the saline side of the coil are forced to travel through the gradient down to the point where hemolysis occurs, the distribution of hemoglobin thus formed, indicating the osmotic fragility of the sample.[2]

(III) Two mutually immiscible media CC. When two mutually immiscible media are used similarly, centrifugation forces these two media to undergo countercurrent motion, and, in the final stage, each turn of the coil is occupied by the two media nearly half and half as illustrated in Fig. 3-III. Consequently, a small amount of a sample injected beforehand at the interface of the two media, i.e., the middle portion of the tube, is distributed along the tube according to its partition coefficient. This CC method is applicable to microgram amounts of chemicals with a high efficiency that may be equivalent to near 1000 units of a Craig countercurrent distribution apparatus.

Applications to CC

Capability of the apparatus for performing microscale CC has been demonstrated using three different samples, i.e., a mixture of dyes, agal proteins, and mammalian erythrocytes.[3]

Separation of dyes

Fig. 4a shows CC separation of four basic dyes, i.e., methyl green (MG), methylene blue (MB), neutral red (NR), and basic fuchsin using a two-phase solvent system composed of isoamylalcohol–ethanol–acetic acid-distilled water (4 : 2 : 1 : 5, v/v). The first coil displays the separation of the mixture, and the other coils show distribution of individual dyes to demonstrate the reproducibility of the method. The separation was performed with 6 m of 0.35-mm-ID tubing (ca. 300 helical turns) at relative coil rotation of 0.25 rpm at 300 × g for 10 hr.

Separation of algal proteins

Phycoerythrin and phycocyanin were extracted from dried Asakusa-nori (*Porphyra tenera*) and subjected to partition with an aqueous polymer phase system (Table 1) using the above standard countercurrent method. Fig. 4b shows the results of separation, where two components are well resolved.

Erythrocyte separation

The separation of human and rabbit erythrocytes was performed with a modified method using a gradient between a pair of polymer phase systems A and B (Table 2), where the upper phase of A and the lower phase of B were used for separation. Before charging with sample, the coil was rotated at 1500 rpm at a relative rotation of 1/100 (15 rpm) for 30 min, which produced a gradient between the two phases along the coil. Then the sample cell mixture was loaded followed by centrifugation for 1 hr. Fig. 4c shows the partition of human and rabbit erythrocytes which were completely separated along the coil.

The results of the above studies using the original CPC led to the development of a series of new CPC devices equipped with a flow-through system that permits continuous elution through the column as in other chromatographic systems.

TYPE-J MULTILAYER CPC

Among all existing types of seal-free flow-through CPCs, the type-J CPC affords the most efficient and speedy separations or "high-speed CCC," and it has been extensively used for preparative separations of natural and synthetic products.

Mathematical Analysis of Planetary Motion

The type-J synchronous planetary motion of the coil holder is shown in Fig. 5a (see also instrumentation of CC, Fig. 3), where the holder rotates about its own axis and simultaneously revolves around the axis of the centrifuge at the same angular velocity but in the opposite direction. Simple mathematical analysis[4] is performed using a coordinate system shown in Fig. 5b, where the center of the revolution coincides with the center of the coordinate system (point O). For convenience of analysis, the center of rotation and an arbitrary point are initially located on the x-axis. After the lapse of time t, the holder circles around point O by $\theta = \omega t$, while the arbitrary point circles around the axis of rotation by 2θ to reach $P(x, y)$ where

$$x = R \cos \theta + r \cos 2\theta \tag{8}$$

$$y = R \sin \theta + r \sin 2\theta \tag{9}$$

The acceleration produced by the planetary motion is then obtained from the second derivatives of Eqs. (8) and (9),

$$d^2x/dt^2 = -R\omega^2(\cos \theta + 4\beta \cos 2\theta) \tag{10}$$

$$d^2y/dt^2 = -R\omega^2(\sin \theta + 4\beta \sin 2\theta) \tag{11}$$

where $\beta = r/R$.

From Eqs. (10) and (11), two centrifugal force components, i.e., Fr (radical component) and Ft (tangential component), are computed using the following formula:

$$Fr = R\omega^2(\cos \theta + 4\beta) \tag{12}$$

$$Ft = R\omega^2(\sin \theta) \tag{13}$$

Fig. 5c shows the distributions of force vectors computed from Eqs. (12) and (13) at various locations on the column holder. All vectors confined in a plane perpendicular to the holder axis. As the holder rotates, both the direction and the net strength of the force vector fluctuate in such a way that the vector becomes longest at the point remote from the centrifuge axis

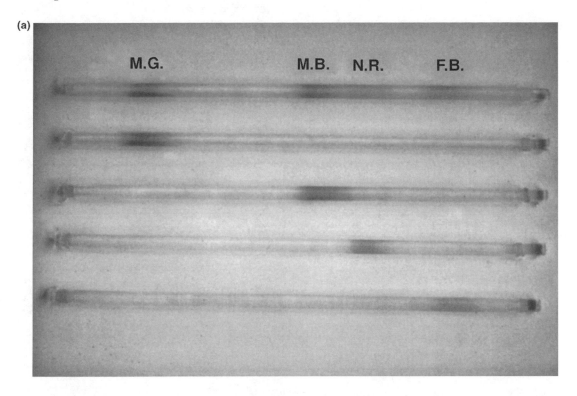

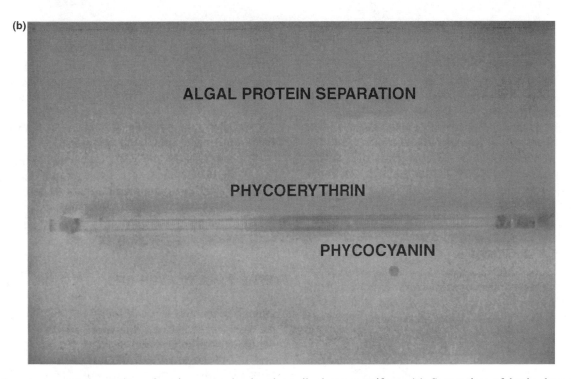

Fig. 4 Countercurrent separation of various samples by the coil planet centrifuge. (a) Separation of basic dyes with an organic/aqueous two-phase solvent system. M.G.: methyl green; M.B.: methylene blue; N.R.: neutral red; F.B.; basic fuchsin. (From Ref.[3].) Solvent system consisted of isoamyl alcohol/ethanol/acetic acid/water (4 : 2 : 1 : 5, v/v). *(Continued next page.)*

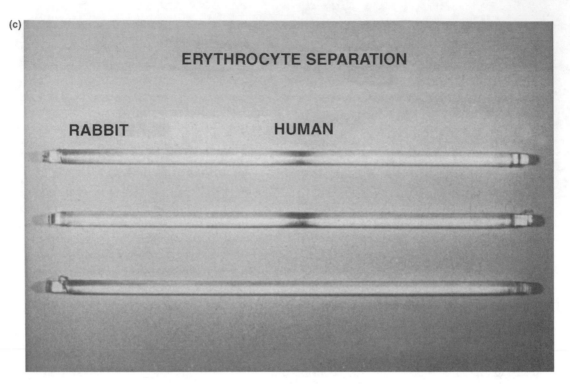

Fig. 4 (*Continued*)

and shortest at the point close to the central axis of the centrifuge. In most locations, the vectors are directed outwardly from the circle except for $\beta < 0.25$, where its direction is reversed at the vicinity of the center of revolution. This fluctuating centrifugal force field creates unique hydrodynamic effects on the two solvent phases in the coiled tube.

Stroboscopic Observation of Hydrodynamic Motion of Solvent Phases

Fig. 6 schematically illustrates motion of the two solvent phases in a spiral column, which is subjected to the type-J synchronous planetary motion. The upper diagram shows distribution of two solvent phases in the column observed under stroboscopic illumination. About one-fourth of the area near the center of revolu-

tion (point O) shows vigorous mixing of the two phases (mixing zone), whereas in the rest of the area, the two phases are separated by a strong centrifugal force in such a way that the lighter phase occupies the inner portion and the heavier phase occupies the outer portion of the tube. Four stretched tubes in the lower diagram illustrate the traveling pattern of the mixing zone through the spiral tube in one revolution cycle. In analogy to the motion of a wave advancing over water, the mixing zone travels one spiral turn for each revolution. This indicates that the solutes are subjected to an efficient partition process of repeating mixing and settling at a high rate of 13 times per second at 800 rpm of revolution. This accounts for the high partition efficiency of the present system, and we have called it "high-speed CCC."[4]

Design of the Apparatus

Fig. 7 shows a photograph of the most advanced model of the multilayer CPC, which holds a set of three multilayer coil separation columns symmetrically around the rotary frame.[5–6] All columns are connected in series with flow tubes which are supported by counterrotating pipes placed between the column holders. The type-J synchronous planetary motion of the holder is provided by engaging a planetary gear on the column holder with an identical stationary

Table 1 Polymer phase system for separation of algal proteins

20% (w/w) Dextran 500	35.0 g
30% (w/w) PEG 8000[a]	14.7 g
0.05 M KH_2PO_4	10.0 ml
0.05 M K_2HPO_4	10.0 ml
0.22 M KCl	10.0 ml
H_2O	20.3 ml

(From Ref.[3].)

Table 2 A pair of polymer phase systems for separation of erythrocytes

System A		System B	
20% (w/w) Dextran 500	25.0 g	20% (w/w) Dextran 500	25.0 g
30% (w/w) PEG 8000[a]	13.3 g	30% (w/w) PEG 8000[a]	13.3 g
0.55 M NaH$_2$PO$_4$	5.0 ml	0.55 M NaH$_2$PO$_4$	5.0 ml
0.55M Na$_2$HPO$_4$	5.0 ml	0.55M Na$_2$HPO$_4$	5.0 ml
25% HSA[b]	1.0 ml	25% HSA[b]	1.0 ml
H$_2$O	50.7 ml	1.5M NaCl	10.0 ml
		H$_2$O	40.7 ml

[a]PEG 8000 was labeled as PEG 6000 in early applications.
[b]HSA = human serum albumin.
(From Ref.[3].)

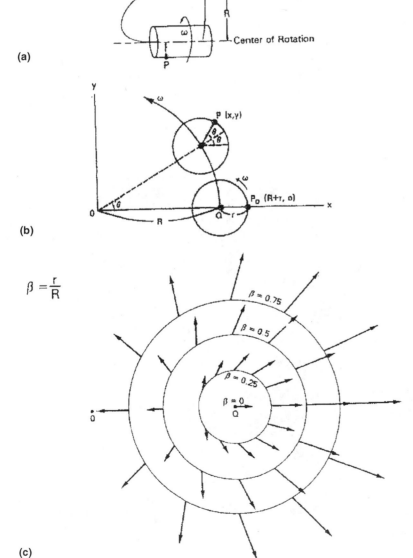

(a)

(b)

$$\beta = \frac{r}{R}$$

(c)

Fig. 5 Analysis of centrifugal force field for type-J planetary motion. (a) Planetary motion; (b) planetary motion in an *x–y* coordinate system for the analysis of centrifugal force; (c) distribution of the centrifugal vectors on the column holder. (From Ref.[12].)

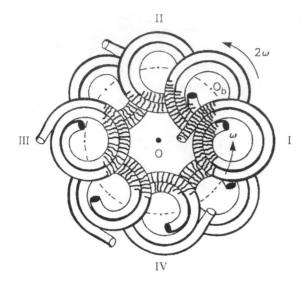

Fig. 6 Mixing and settling zones in the spiral column undergoing type-J planetary motion. (From Ref.[12].)

sun gear mounted around the central stationary shaft of the centrifuge. The counterrotation of the tube holder is effected by interlocking a pair of identical gears, one mounted on the holder and the other on the tube holder. Flow tubes from each end of the column assembly are passed through the central rotary shaft to exit the centrifuge on each side, where they are tightly affixed with a pair of clamps.

The multilayer coil separation column is prepared by winding a single piece of Teflon or Tefzel tubing around a spool-shaped column holder making multiple coiled layers between a pair of flanges. Currently, three different sets of multilayer coils are commercially available: the large preparative scale (2.6-mm ID, ca. 1000 ml total capacity); the standard preparative column (1.6-mm ID, ca. 320 ml total capacity); and the analytical scale (0.85-mm ID, ca. 120 ml capacity). The optimal revolution speed of the apparatus ranges from 800 to 1200 rpm.

Applications

Because of its rapid and high separation efficiency, the multilayer coil CPC has been extensively used for separation and purification of variety of compounds using suitable organic/aqueous solvent systems. The application also covers special CCC techniques such as peak-focusing CCC and pH-zone-refining CCC[7] (see pH-peak-focusing and pH-zone-refining CC, pp. 606–611); chiral and affinity CCC (see chiral

Fig. 7 Improved high-speed CCC centrifuge equipped with three column holders. (From Ref.[6].)

CC, pp. 160–163); foam CCC[8] (see foam CC, pp. 342–345), liquid–liquid dual CCC;[9] and CCC/MS (see CC/MS, pp. 208–211).

The method, however, fails to retain viscous, low interfacial tension polymer phase systems such as polyethylene glycol (PEG)–dextran systems[10] due to its intensive mixing effect which tends to produce emulsification, resulting in carryover of the stationary phase. This problem is largely eliminated by the cross-axis CPC described below.

CROSS-AXIS CPC

The cross-axis CPC has a specific feature in that it permits the universal use of two-phase solvent systems including aqueous–aqueous polymer phase systems which are useful for partitioning macromolecules and cell particles.[10]

Acceleration Field

The design of the cross-axis CPC is based on the hybrid between type-L and type-X planetary motions, which results in an extremely complex centrifugal force field with a three-dimensional fluctuation of force vectorsduring each revolution cycle of the holder. The pattern of this centrifugal force field produced by the cross-axis CPC somewhat resembles that produced by the type-J planetary motion (Fig. 5c), but it is superimposed by a force component acting in parallel to the axis of the coil holder. This additional force component acts to improve the retention of the stationary phase. This beneficial effect is greatest in type-L planetary motion and becomes smallest in the type-X planetary motion. A detailed mathematical analysis on this planetary motion is described elsewhere.[11]

Design of the Apparatus

Fig. 8 shows the general principle of various cross-axis CPCs. The geometrical parameter of the system is shown in Fig. 8a, where the orientation and planetary motion are indicated relative to the axis of the apparatus. The vertical axis of the apparatus and the horizontal axis of the coil are always kept perpendicular to each other at a fixed distance. The cylindrical column revolves at the central axis at the same rotational speed with which it rotates about its own axis. Three

(a)

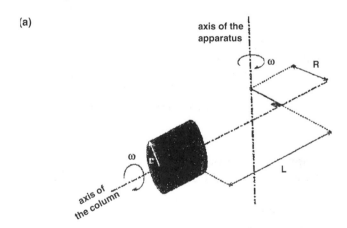

(b)

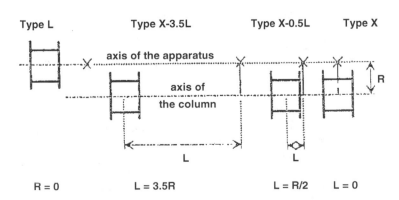

Fig. 8 General principle of various cross-axis CPCs. (a) Geometrical parameters; (b) some examples of prototypes built in NIH Machine Shop. X type in 1987; X-0.5L type in 1988; X-3.5L type in 1991; and L type in 1992. (From Ref.[12].)

parameters displayed in Fig. 8a explain various versions of the cross-axis prototypes: r is the radius of the column holder, R is the distance between the two axis, and L is the measure of the lateral shift of the column holder along its axis. The name of a cross-axis device is based on the ratio L/R, when $R \neq 0$. Types X and L represent the limits for the column positions; type X involves no shifting of the column holder, while type L corresponds to an infinite shifting. Some examples of prototypes fabricated at the machine shop in the National Institutes of Health, Bethesda, MD, are shown in Fig. 8b.

The two different column designs are schematically shown in Fig. 9: The first column (multilayer coil) is used for preparative scale separations, while the second column (eccentric coil) is for analytical scale separations. The multilayer coil is prepared by winding a large Teflon tubing (2.6-mm ID) directly onto the holder hub in such a way that after completing each coiled layer, the tubing is directly returned to the starting point to wind the second layer over this connecting tube segment, and so on. This results in multiple coiled layers of the same handedness that are connected in series with short-tube segments as shown in Fig. 9a. The eccentric coil is prepared by winding a piece of Teflon tubing (typically 0.85-mm ID) onto a set of multiple short cores (ca. 6-mm OD), which is arranged around the periphery of the holder hub as shown in Fig. 9b.

Fig. 10 shows a photograph of a recently designed cross-axis CPC equipped with a pair of multilayer coil separation columns.[12] The column can be mounted on

the rotary frame in two positions, X-1.5L (off-center position) and L (central position). The off-center position is used for both organic/aqueous and aqueous PEG—potassium phosphate systems, while the central position is used for viscous, low interfacial tension PEG–dextran systems.

Applications

Although cross-axis CPCs yield less efficient separations than the type-J multilayer CPC, they provide more stable retention of the stationary phase and are therefore useful for large-scale preparative separations with polar solvent systems. These are especially useful for the purification of proteins with aqueous–aqueous polymer phase systems composed of PEG and potassium phosphate. The cross-axis CPC has been used for the purification of various enzymes including choline esterase, ketosteroid isomerase, purine nucleoside phosphorylase, lactic acid dehydrogenase, uridine phosphorylase (see cross-axis coil planet centrifuge for the separation of proteins, pp. 212–213).

NONSYNCHRONOUS CPC

The nonsynchronous flow-through CPC is a particular type of planetary centrifuge which allows adjustment of the rotational rate of the coiled separation column at a given revolution speed. The first prototype was equipped with a dual rotary seal for continuous elution.[13] Later, an improved model[14] was designed to eliminate the rotary seal which had become a source of complications such as leakage, contamination, and clogging.

Design Principle of the Seal-Free Nonsynchronous CPC

Fig. 11a shows a cross-sectional view through the central axis of the apparatus. The rotor consists of two major rotary structures, i.e., frames I and II, which are coaxially bridged together with the center piece (dark shade).

Frame I consists of three plates rigidly linked together and directly driven by motor I. It holds three rotary elements, namely, the centerpiece (center), countershaft I (bottom), and countershaft II (top), all employing ball bearings. A pair of long arms extending symmetrically and perpendicularly from the middle plate forms the tube-supporting frame, which clears over frame II to reach the central shaft on the right side of frame II.

Frame II (light shade) consists of three pairs of arms linked together to rotate around the central shaft.

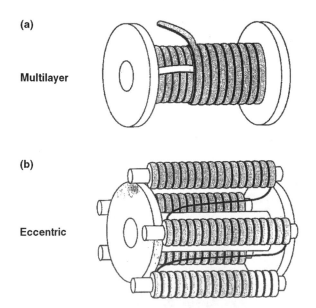

(a)

Multilayer

(b)

Eccentric

Fig. 9 Two different types of coiled columns for cross-axis CPC. Multilayer coil is for large-scale separations and eccentric coil assembly for small-scale separations.

Fig. 10 Photograph of X-1.5L and L prototypes. The rotary frame of the apparatus is driven by a motor (back of the apparatus) by coupling a pair of toothed pulleys one on the motor shaft and the other on the central shaft with a belt. Two cylindrical holders are mounted in the X-1.5L or off-center position. The inlet and the outlet Teflon tubes go through the upper plate of the apparatus inside the hollow part of the central vertical axis. A circular metallic plate around the rotary frame decreases the torque by an average of 30%. (From Ref.[12].)

It supports a pair of rotary shafts, one holding a coil holder assembly and the other the counterweight.

There are two motors, i.e., motors I and II, to drive the rotor. When motor I drives frame I, the stationary pulley 1 introduces counterrotation of pulley II through a toothed belt; therefore, countershaft I rotates at $-\omega_I$ with respect to rotating frame I. This motion is further conveyed to the centerpiece by 1 : 1 gearing between gears 1 and 2. Thus, the centerpiece rotates at $2\omega_I$ or at ω_I with respect to the rotating frame I. The motion of frame I also depends upon the motion of motor II.

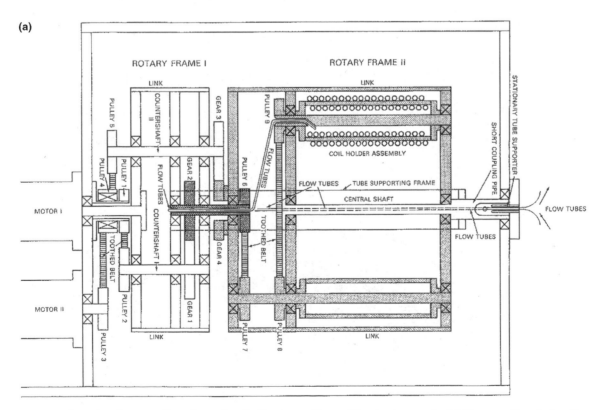

Fig. 11 Improved nonsynchronous flow-through coil planet centrifuge without rotary seals. (a) Cross-sectional view through the central axis of the apparatus; (b) photograph of the apparatus equipped with an eccentric coil assembly. (From Ref.[14].) *(Continued next page.)*

(b)

Fig. 11   *(Continued)*

If motor II is at rest, pulley 5 becomes stationary same with pulley 1 so that countershaft II counter-rotates at ω_I as does countershaft I. This motion is similarly conveyed to the rotary arms of frame II by 1 : 1 gearing between gears 3 and 4, resulting in rotation of frame II at $2\omega_I$ or the same angular velocity as that of the centerpiece. Consequently, coupling of pulleys 6 to 7 and 8 to 9 with toothed belts produces no additional motion to the rotary shaft, which simply revolves with frame II at $2\omega_I$ around the central axis of the apparatus.

When motor II rotates at ω_{II}, idler pulley 4 coupled to pulley 3 on the motor shaft rotates at the same rate, which, in turn, modifies the rotational rate of pulley 5 on countershaft II. Thus, countershaft II now counter-rotates at $\omega_I - \omega_{II}$ on frame I. This motion further alters the rotational rate of frame II through 1 : 1 gear coupling between gears 3 and 4. Subsequently, frame II rotates at $2\omega_I - \omega_{II}$ with respect to the earth or at $-\omega_{II}$ relative to the centerpiece which always rotates at $2\omega_I$. The difference in rotational rate between frame II and the center piece is conveyed to the rotary shafts through coupling of pulleys 6 to 7 and 8 to 9. Consequently, both rotary shafts rotates at ω_{II} about their own axes while revolving around the central axis of the apparatus at $2\omega_I - \omega_{II}$. This gives the rotation/revolution ratio of the rotary shaft

$$r/R \; = \; \omega_{II}/(2\omega_I - \omega_{II}) \qquad (14)$$

Therefore, any combination of revolutional and rotational speeds of the coil holder assembly can be achieved by selecting the proper values of ω_I and ω_{II}.

Coiled Column and Flow Tubes

Two different column configurations are used: a multilayer coil and an eccentric coil assembly. The multilayer coil column is prepared by winding a piece of Teflon tubing, typically 1.6-mm ID, directly onto the holder hub (ca. 2.5-cm OD), making multiple coiled layers as those for the type-J high-speed CCC. The eccentric coil assembly was made by winding a piece of Teflon tubing, typically 1-mm ID, onto a set of six units of 0.68-cm-OD stainless steel pipe in series. These coil units are arranged around the holder with their axis in parallel to the holder axis (see eccentric coil assembly in Fig. 9, lower diagram). A pair of flow tubes from each end of the coiled column is first led through the hole of the rotary shaft and then pass through the opening of the centerpiece to exit at the middle portion of frame I. The flow tubes are then led along from the tube support to clear frame II and then reach the side hole of the central shaft near the right wall of the centrifuge where they are tightly held by the stationary tube supporter.

Fig. 11b shows the overall photograph of the apparatus equipped with an eccentric coil assembly. The revolution speed of the coil holder assembly is continuously adjustable up to 1000 rpm combined with any given rotational rate between 0 and 50 rpm in either direction.

Applications

The nonsynchronous CPC is a most versatile centrifuge which can be applied to a variety of samples including cells, macromolecules, and small molecular weight compounds. Cell separations may be performed with a single phase such as physiological solution or culture medium[14–15] and also with PEG–dextran polymer two-phase systems.[16] DNA and RNA are partitioned with a PEG–dextran system by optimizing the pH.[14]

CONCLUSIONS

The CPC, which was originally developed for separating blood lymphocytes, has evolved into several useful instruments for separations of cells, macromolecules, and small molecular weight compounds. Among those, the type-J multilayer CPC is most extensively utilized for high-speed CCC separations of natural and synthetic products. The utility of the type-J CPC may be extended to the polymer phase separation of macromolecules and cell particles with a spiral disk assembly currently being developed in our laboratory.

ACKNOWLEDGMENT

The author is indebted to Dr. Henry M. Fales of Laboratory of Biophysical Chemistry, National Heart, Lung, and Blood Institute, National Institutes of Health, Bethesda, MD, for editing the manuscript.

REFERENCES

1. Ito, Y.; Weinstein, M.A.; Aoki, I.; Harada, R.; Kimura, E.; Nunogaki, K. The coil planet centrifuge. Nature **1966**, *212*, 985–987.
2. Harada, R.; Ito, Y.; Kimura, E. A new method of osmotic fragility test of erythrocytes with coil planet centrifuge. Jpn. J. Physiol. **1969**, *19*, 306–314.
3. Ito, Y.; Aoki, I.; Kimura, E.; Nunogaki, K.; Nunogaki, Y. New micro liquid–liquid partition techniques with the coil planet centrifuge. Anal. Chem. **1969**, *41*, 1579–1584.
4. Ito, Y. High-speed countercurrent chromatography. CRC Crit. Rev. Anal. Chem. **1986**, *17* (1), 65–143.
5. Ito, Y.; Oka, H.; Slemp, J.L. Improved high-speed counter-current chromatograph with three multilayer coils connected in series. I. Design of the apparatus and performance of semipreparative columns in DNP amino acid separation. J. Chromatogr. **1989**, *475*, 219–227.
6. Ito, Y.; Oka, H.; Lee, Y.-W. Improved high-speed counter-current chromatograph with three multilayer coils connected in series. II. Separation of various biological samples with a semi-preparative column. J. Chromatogr. **1990**, *498*, 169–178.
7. Ito, Y.; Ma, Y. pH-zone-refining countercurrent chromatography. J. Chromatogr. A. **1996**, *753*, 1–36.
8. Ito, Y. Foam countercurrent chromatography based on dual countercurrent system. J. Liq. Chromatogr. **1985**, *8*, 2131–2152.
9. Lee, Y.-W.; Fang, Q.-C.; Cook, C.E.; Ito, Y. The application of true countercurrent chromatography in the isolation of bio-active natural products. J. Nat. Prod. **1989**, *52* (4), 706–710.
10. Albertsson, P. *Partition of Cell Particles and Macromolecules*; Wiley Interscience: New York, 1986.
11. Ito, Y.; Oka, H.; Slemp, J.L. Improved cross-axis synchronous flow-through coil planet centrifuge for performing countercurrent chromatography. I. Design of the apparatus and analysis of acceleration. J. Chromatogr. **1989**, *463*, 305–316.
12. Ito, Y.; Menet, J.-M. Coil planet centrifuges for high-Speed countercurrent chromatography. In *Countercurrent Chromatography*; Menet, J.-M., Thiebaut, D., Eds.; Marcel Dekker: New York, 1999; 87–119.
13. Ito, Y.; Carmeci, P.; Sutherland, I.A. Nonsynchronous flow-through coil planet centrifuge applied to cell separation with physiological solution. Anal. Biochem. **1979**, *94*, 249–252.
14. Ito, Y.; Bramblett, G.T.; Bhatnagar, R.; Huberman, M.; Leive, L.; Cullinane, L.M.; Groves, W. Improved non-synchronous flow-through coil planet centrifuge without rotating seals. Principle and application. Sep. Sci. Technol. **1983**, *18* (1), 33–48.
15. Okada, T.; Metcalfe, D.D.; Ito, Y. Purification of mast cells with an improved nonsynchronous flow-through coil planet centrifuge. Int. Arch. Allergy Immunol. **1996**, *109*, 376–382.
16. Leive, L.; Cullinane, M.L.; Ito, Y.; Bramblett, G.T. Countercurrent chromatographic separation of bacteria with known difference in surface lipopolysaccharide. J. Liq. Chromatogr. **1984**, *7* (2), 403–418.

Concentration of Dilute Colloidal Samples by Field-Flow Fractionation

George Karaiskakis
University of Patras, Patras, Greece

INTRODUCTION

Many colloidal systems, such as those of natural water, are too dilute to be detected by the available detection systems. Thus, a simple and accurate method for the concentration and analysis of these dilute samples should be of great significance in analytical chemistry. In the present work, two methodologies of the field-flow fractionation (FFF) technique for the concentration and analysis of dilute colloidal samples are presented. Both the conventional and potential barrier methodologies of FFF are based on the "adhesion" of the samples at the beginning of the channel wall, followed by their total removal and analysis. In the conventional sedimentation FFF (SdFFF) concentration procedure, the apparent adhesion of a dilute sample is due to its strong retention, which can be achieved by applying high field strengths and low flow rates. In the potential barrier SdFFF (PBSdFFF) concentration procedure, the true adhesion of a dilute sample is due to its reverse adsorption at the beginning of the column, which can be achieved by the appropriate adjustment of various parameters influencing the interactions between the colloidal particles and the material of the channel wall. The total release of the adherent particles is accomplished either by reducing the field strength and increasing the solvent velocity (conventional SdFFF) or by varying the potential energy of interaction between the particles and the column material—for instance, by changing the ionic strength of the carrier solution (PBSdFFF).

METHODOLOGY

FFF is a one-phase chromatographic system in which an external field or gradient replaces the stationary phase. The applied field can be of any type that interacts with the sample components and causes them to move perpendicular to the flow direction in the open channel. The most highly developed of the various FFF subtechniques is sedimentation FFF (SdFFF), in which the separations of suspended particles are performed with a single, continuously flowing mobile phase in a very thin, open channel under the influence of an external centrifugal force field.

In the normal mode of the SdFFF operation, a balance is reached between the external centrifugal field, driving the particles toward the accumulation wall, and the molecular diffusion in the opposite direction. In that case, the retention volume increases with particle diameter until steric effects dominate, at which transition point there is a foldback in elution order.

PBSdFFF which has been developed recently in our laboratory, is based either on particle size differences or on Hamaker constant, surface potential, and Debye–Hückel reciprocal distance differences.

The retention volume of a component under study, in the normal SdFFF and the PBSdFFF methodologies is a function of the following parameters:

1. SdFFF:

$$V_r = f(d, \ G, \ \Delta\rho) \tag{1}$$

2. PBSdFFF:

$$V_r = f(d, \ G, \ \Delta\rho, \ \psi_1, \ \psi_2, \ A, \ I) \tag{2}$$

where d is the particle diameter, G is the field strength expressed in acceleration, $\Delta\rho$ is the density difference between solute and solvent, ψ_1 and ψ_2 are the surface potentials of the particle and of the wall, respectively, A is the Hamaker constant, and I is the ionic strength of the carrier solution.

The conventional concentration procedure in SdFFF consists of two steps: the feeding (or concentration) and the separation (or elution) step. In the feeding step, the diluted samples are fed into the column with a small flow velocity while the channel is rotated at a high field strength to ensure the "apparent adhesion" of the total number of the colloidal particles at the beginning of the channel wall as a consequence of the particles' strong retention.

Encyclopedia of Chromatography DOI: 10.1081/E-ECHR-120039927

In the separation step, the field is reduced and the flow rate is increased to ensure the total release and the consequence elution of the adherent dilute particles.

In PBSdFFF, the concentration step consists of feeding the column with the diluted samples at such experimental conditions, so as to decrease the repulsive component and increase the attractive component of the total potential energy of the particles under study. Because the stability of a colloid varies (increases or decreases) with a number of parameters (surface potential, Hamaker constant and ionic strength of the suspending medium), the proper adjustment of one or more of these parameters can lead not only to the adhesion of the dilute colloidal samples, which leads to their "concentration," but also to the total release of the adherent particles during the elution step.

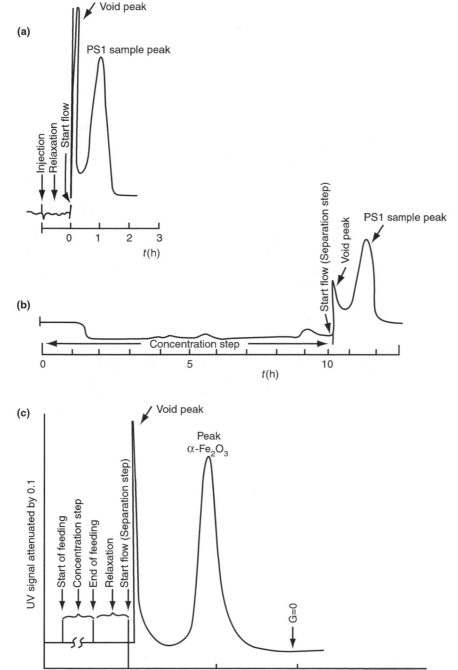

Fig. 1 Fractograms of the polystyrene latex beads of $0.357\,\mu m$ (PS1) obtained by the direct injection of $1\,\mu L$ of PS1 (a) and by the concentration procedure of the PS1 sample diluted in $10\,mL$ of the carrier solution (b) using the conventional SdFFF technique, as well as of the $\alpha\text{-}Fe_2O_3$ sample with nominal particle diameter of $0.271\,\mu m$ diluted in $6\,mL$ of the carrier solution obtained by the PBSdFFF concentration methodology (1c).

APPLICATIONS

Conventional SdFFF

As model samples for the verification of the conventional SdFFF as a concentration methodology monodisperse polystyrene latex beads (Dow Chemical Co.) with nominal diameters of 0.357 μm (PS1) and 0.481 μm (PS2) were used. They were either used as dispersions containing 10% solids or diluted with the carrier solution (triple-distilled water +0.1% (v/v) detergent FL-70 from Fisher Scientific Co. +0.02% (w/w) NaN_3) to study sample dilution effects. Diluted samples in which the amount of the polystyrene was held constant (1 μmL of the 10% solids) while the volume in which it was contained was varied over a 50,000-fold range (from 1 to 50 mL of carrier solution) were introduced into the SdFFF column. During the feeding step, the flow rate was 5.8 mL/h for the PS1 polystyrene, and 7.6 mL/h for the polystyrene PS2, and the channel was rotated at 1800 rpm for the PS1 sample and at 1400 rpm for the PS2 sample. In the separation (elution) step, the experimental conditions for the two samples were as follows:

> PS1: Field strength = 880 rpm,
>   flow rate = 12–53 mL/h
>
> PS2: Field strength = 500 rpm,
>   flow rate = 24–59 mL/h

Fig. 1 provides a comparison of fractograms for the 0.357 μm polystyrene injected as a narrow pulse (Fig. 1a) and injected at 10 mL dilution (Fig. 1b) by the conventional SdFFF concentration procedure described previously. Fig. 1b shows that the eluted peak from the diluted sample emerges intact and without serious degradation, compared to the peak of Fig. 1a, despite the fact that the sample volume (10 mL) is over twice the channel volume (4.5 mL). The same concentration procedure was also successfully applied to the separation of the two polystyrene samples initially mixed together in a volume of 10 mL, as well as to the concentration of the colloidal particles contained in natural water samples collected from the Colorado, Green, and Price rivers in eastern Utah (U.S.A.).

As a general conclusion, the on-column concentration procedure of the conventional SdFFF method works quite successfully in dealing with highly diluted samples. Optimization, particularly higher field strengths during the concentration step, would allow higher flow rates and increased analysis speed. However, experimental confirmation would be necessary to give assurance that the particle–wall adhesion is not irreversible at higher spin rates.

Potential Barrier SdFFF

As model samples to test the validity of the PBSdFFF as a concentration procedure of diluted samples the monodisperse colloidal particles of α-Fe_2O_3 with nominal diameters of 0.271 μm were used. Diluted samples of containing 2 μL of the 10% solid, in which the volume was varied over a 10,000-fold range (from 2 to 20 mL), were introduced into the column with a carrier solution containing 0.5% (v/v) detergent FL-70 + 3 × 10^{-2} M KNO_3 to ensure the total adhesion of the particles at the beginning of the SdFFF Hastelloy-C channel wall. In the separation step, the carrier solution was changed to one containing only 0.5% (v/v) detergent FL-70 (without electrolyte) to ensure the total detachment of the adherent particles. In that case, a sample peak appeared (cf. Fig. 1c) as a consequence of the desorption of the α-Fe_2O_3 particles. The mean diameter of the particles (0.280 mm) obtained by the proposed PBSdFFF methodology for the on_channel concentration procedure of the sample diluted in 8 mL of the carrier solution is very close to that found (0.271 μm) by the direct injection of the same particles into the channel, using a carrier in which no adsorption occurs.

As a general conclusion, one could say that the proposed PBSdFFF concentration procedure works quite successfully in dealing with highly dilute samples, separating them according to size, surface potential, and Hamaker constant. At the same time, as separation occurs, the particle sizes of the colloidal materials of the diluted mixture can be determined. The major advantage of the proposed concentration procedure is that the method can concentrate and analyze dilute mixtures of colloidal particles even of the same size but with different surface potentials and/or Hamaker constants. The method has considerable promise for the separation and characterization, in terms of particle size, of dilute complex colloidal materials, where particles are present in low concentration.

FUTURE DEVELOPMENTS

Looking to the future, we believe that the efforts of the researchers will be focused on the extension of the FFF concentration methodologies to the ranges of more dilute and complex colloidal samples, without lengthening the analysis time.

SUGGESTED FURTHER READING

Athanasopoulou, A.; Koliadima, A.; Karaiskakis, G. New methodologies of field-flow fractionation for the separation and characterization of dilute colloidal samples. Instrum. Sci. Technol. **1996**, *24* (2), 79.

Giddings, J.C.; Karaiskakis, G.; Caldwell, K.D. Separ. Sci. Technol. **1981**, *16* (6), 725.

Hiemenz, P.C. *Principles of Colloid and Surface Chemistry*; Marcel Dekker, Inc.: New York, 1977.

Karaiskakis, G.; Graff, K.A.; Caldwell, K.D.; Giddings, J.C. Sedimentation field-flow fractionation of colloidal particles in river water. Int. J. Environ. Anal. Chem. **1982**, *12*, 1.

Koliadima, A.; Karaiskakis, G. Sedimentation field-flow fractionation: a new methodology for the concentration and particle size analysis of dilute polydisperse colloidal samples. J. Liquid Chromatogr. **1988**, *11*, 2863.

Koliadima, A.; Karaiskakis, G. Potential-barrier field-flow fractionation, a versatile new separation method. J. Chromatogr. **1990**, *517*, 345.

Koliadima, A.; Karaiskakis, G. Concentration and characterization of dilute colloidal samples by potential-barrier field-flow fractionation. Chromatographia **1994**, *39*, 74.

C

Concentration Effects on Polymer Separation and Characterization by Thermal Field-Flow Fractionalization

Wenjie Cao
Mohan Gownder
Huntsman Polymers Corporation, Odessa, Texas, U.S.A.

INTRODUCTION

The understanding of the effects of sample concentration (sample mass) in field-flow fractionation (FFF) has being obtained gradually with the improvement of the sensitivity (detection limit) of HPLC detectors. Overloading, which was used in earlier publications, emphasizes that there is an upper limit of sample amount (or concentration) below which sample retention will not be dependent on sample mass injected into the FFF channels.[1] Recent studies show that such limits may not exist for thermal FFF (ThFFF) (may be true for all the FFF techniques in polymer separation), although some of the most sensitive detectors on the market were used.[2]

Experimental results indicate that the effects of sample mass include, but not exclusively, the following aspects.

Increased Polymer Retention

Fig. 1 shows the fractograms of ThFFF to show the concentration effects for poly(methyl methacrylate) (PMMA) in THF, where M_p is the peak average molecular weight, m is the sample mass in micrograms injected into the ThFFF channel, and t^0 is the retention time of a nonretained species such as toluene. When the molecular weight (MW) of a polymer is moderate or higher, say above 300×10^3 g/mol for PMMA in THF, a moderate increase in concentration will result in longer retention. As reported in Ref.,[2] the detector limits for the study was $0.09 \,\mu g$ of sample mass for 1000×10^3 g/mol polystyrene using an ultraviolet (UV) detector and $1 \,\mu g$ for 570×10^3 g/mol PMMA with an evaporative light-scattering detector. The retention was measured for sample masses ranging from these limits to more than $20 \,\mu g$ and was consistently found to increase with the increase in sample mass. The high limit, below which polymer retention is not dependent on concentration, was not found.

Broader Polymer Peaks

Increased concentration will increase band broadening in all chromatographic techniques, but it seems that the effect of concentration on band broadening is more serious in FFF, due to its concentration enhancement as shown by Fig. 1, and by Fig. 5 of Ref.[1]. More details will be discussed in the next section.

Distorted Peaks and Double-Topped Peaks (Ghost Peaks)

When the amount of sample mass injected into the ThFFF channel is moderate, say $1-10 \,\mu g$ for a typical channel, the peaks are pretty symmetrical and not much distortion may be observed for small polymers, as shown by Fig. 1. Increased retention may be observed for high-MW polymers.[2] When sample mass is further increased, say more than $20 \,\mu g$, distorted peaks, even double-topped peaks or ghost peaks, may be observed for high-MW polymers as shown by Fig. 1, Fig. 6 of Ref.,[1] and Fig. 3 of Ref.[2]. The detailed report and discussion of double-topped peaks can be found in both Refs.[1–2].

ENHANCED VISCOSITY IS BLAMED FOR THE SAMPLE CONCENTRATION EFFECTS

The viscosity of a polymer solution is highly dependent on concentration, temperature, and MW, as discussed below.

Concentration and Viscosity Enhancement in FFF

The amount of sample injected into the FFF channels can affect the retention time, primarily by influencing the viscosity of the solute–solvent mixture in the sample zone.[1–2] Unlike other chromatographic polymer separation methods [e.g., GPC–SEC and TREF, etc.], the viscosity of the fluid is not homogeneous at a given channel (column) cross section. In order for the samples to be retained by FFF, the concentration must be larger near the accumulation wall than that near the depletion wall.[3–4] In chromatography, sample

Encyclopedia of Chromatography DOI: 10.1081/E-ECHR-120039928

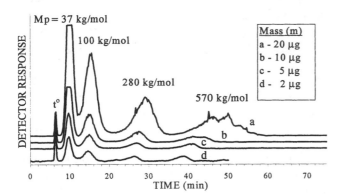

Fig. 1 Fractograms of PMMA in THF showing the effect of sample concentration on retention. Experimental conditions: cold-wall temperature, 25°C, T, 50°C; flow rate, 0.1 mL/min.

concentration changes only along one dimension (i.e., the flow axis) whereas in FFF, sample concentration varies along two dimensions, one being the flow axis and the other one is across the channel thickness, which is perpendicular to the flow axis. The concentration across the channel thickness varies due to the migration of the molecules under the influence of the temperature gradient across the ThFFF channel.[3] The concentration distribution is approximately exponential as given by

$$c(x) = c_0 \exp\left(\frac{-x}{\lambda w}\right) \tag{1}$$

where $c(x)$ is the concentration at distance x across the channel thickness measured from the accumulation wall, c_0 is the concentration at the accumulation wall, w is the channel thickness, and λ is the retention parameter or reduced mean thickness of the sample zone. Shortly after injection, the sample zone is assumed to broaden into a Gaussian distribution along the z axis, corresponding to the direction of flow down the channel. The two-dimensional concentration becomes[5]

$$c(x, z) = c_{00} \exp\left(\frac{-(z - Z)^2}{2\sigma^2}\right) \exp\left(\frac{-x}{\lambda w}\right) \tag{2}$$

where Z is the distance traveled by the center of the zone down the channel. The concentration at the accumulation wall at the center of the zone, c_{00}, is found from[6]

$$c_{00} \cong \frac{V_{inj} c_{inj} L}{(2\pi\sigma^2)^{1/2} V^0 \lambda} \tag{3}$$

where V_{inj} is the volume of sample injected, V^0 is the void volume (channel volume), c_{inj} is the concentration of the injected sample, L is the length of the channel, and σ is the sum of the variances contributing to the zone breadth.

A rough calculation using Eq. (3) indicates that the concentration of c_{00} can be as high as 20 times the concentration of the original polymer solution. The concentration of the sample zone, therefore, can be enhanced dramatically in field-flow fractionalization (FFF).

The relationship between viscosity and concentration of polymer solution is very complex. Several empirical equations are necessary to describe the viscosity behavior of a polymer solution's dependence upon concentration. As an example, the following equation can be used for dilute solutions:[7]

$$\eta = 0.54 + 1.3374C + 1.1593C^2 \tag{4}$$

where η is viscosity in centipoise and C is the concentration in grams per deciliter.

For the concentration where a microgel may be formed, the following equation is proposed:[8]

$$\eta = BM^3 C^{3.7} \tag{5}$$

where B is a constant and M is the polymer's molecular weight.

Although various empirical equations can be found in the literature, the common aspect is that the viscosity of a polymer solution is highly dependent on concentration and molecular weight.

Temperature Dependence of Viscosity

The effect of temperature on the viscosity of the carrier can be expressed as[5]

$$\frac{1}{\eta} = a_0 + a_1 T + a_2 T^2 + a_3 T^3 \tag{6}$$

where a_0, a_1, a_2, and a_3, are empirically obtained coefficients.

As Eq. (6) shows, the viscosity of a polymer solution is highly dependent on temperature. The sample zone of a high-MW polymer is pressed much closer to the cold wall in ThFFF. Its viscosity is more enhanced than with a low-MW polymer. The concentration effect, therefore, is more serious for high-MW polymers in ThFFF.

Molecular-Weight Dependence of Viscosity

If the temperature and concentration are kept the same, the viscosity of higher-MW polymer solution is higher, as shown by Eq. (2); thus, more distortion of its peak is expected, as shown by Fig. 1.

Unlike most of the elution separation methods, such as HPLC, GPC/SEC, GC, and so forth, where the concentration of the sample zone will never be higher than the stock solution before injection, FFF will concentrate the samples, that is to say that sample concentration will be enhanced near the accumulative wall of FFF and the cold wall in ThFFF. The higher the MW of the polymer, the more the concentration will be enhanced and the lower the temperature of the sample zone will be. All three factors, concentration, temperature, and MW, contribute simultaneously to enhance the viscosity of the sample zone of the polymers in ThFFF. The viscosity of the sample zone can reach such extension that there is a tendency for the carrier fluid to flow over the top of the zone, with increased velocity in the region above the sample zone. The moving fluid will go over the sample zone, thus resulting in a longer retention for the sample zone; this is like a sticky slump going slowly on the floor of a river. A longer retention will be observed even if the flow rate of the carrier is constant.

When the MW of the polymer is so large that its zone is compressed close to the cold wall, the temperature of the sample zone becomes, essentially, the temperature of the cold wall, 25°C in many experiments. The viscosity is enhanced so much that the flow velocity of the carrier fluid is further distorted, so that deformed or double-topped peaks will be produced.

For the double-topped peaks, pseudo-gel, formed near the cold wall, is also proposed due to the low temperature and high concentration of the sample zone in ThFFF.[2,9] The behavior of a pseudo-gel solution is quite different from the polymer solution from which it is formed. The diffusion coefficient of a pseudo-gel is much smaller than that of the original polymer, and the viscosity of the pseudo-gel solution will be much larger than that of the original polymer solution. The pseudogel, in theory, will be compressed closer to the cold wall and will elute out of the channel later than the parent molecules. However, as the size of the pseudo-gel cluster increases, hydrodynamic effects will result in an earlier emergence from the channel.[3] If either of these scenarios occurs in the ThFFF channel, double peaks might be observed for a sample of a single peak without "overloading."

CONCLUSIONS

Any attempts to obtain the parameters of the chromatograms and the physicochemical constants which are measurable in theory, by FFF, will be affected by the sample mass injected into the FFF channel. All of the concentration effects on the chromatograms discussed in the previous sections will be transferred, in turn, to those measured parameters and the physicochemical constants, such as the mass selectivity (S_m),

the common diffusion coefficient (D), the thermal diffusion coefficient (D_T) and so forth. The increased retention of large polymers will result in enhanced mass selectivity in ThFFF. For a long time, this enhanced selectivity, in turn, the enhanced ThFFF universal calibration constant n, has led to confusion concerning the accuracy and repeatability of FFF, because different research groups have reported different data for selectivity and physicochemical constants measured by FFF for a given polymer–solvent combination.[2,11] Recent studies show that the enhanced selectivity and the different values of the physicochemical constants reported by different laboratories, measured by ThFFF, may be caused by different concentrations (sample mass) used by different laboratories.

REFERENCES

1. Caldwell, K.D.; Brimhall, S.L.; Gao, Y.; Giddings, J.C. Sample overloading effects in polymer characterization by field-flow fractionation. J. Appl. Polym. Sci. **1988**, *36* (3), 703–719.
2. Cao, W.J.; Marcus, M.N.; Williams, P.S.; Giddings, J.C. Sample mass effects on thermal field-flow fractionation retention and universal calibration. Int. J. Polym. Anal. Charact. **1998**, *4*, 407.
3. Giddings, J.C. Field-flow fractionation: analysis of macromolecular, colloidal and particulate materials. Science **1993**, *260*, 1456.
4. Giddings, J.C. Universal calibration in size exclusion chromatography and thermal field-flow fractionation. Anal. Chem. **1994**, *66*, 2783–2787.
5. Giddings, J.C.; Yang, F.J.F.; Myers, M.N. Sedimentation field-flow fractionation. Anal. Chem. **1974**, *46*, 1917–1924.
6. Caldwell, K.D.; Brimhall, S.L.; Gao, Y.; Giddings, J.C. Sample overloading effects in polymer characterization by field-flow fractionation. J. Appl. Polym. Sci. **1988**, *36*, 703.
7. Tanford, C. *Physical Chemistry of Macromolecules*; John Wiley & Sons: New York, 1961; Chap. 6.
8. DeGennes, P.G. Dynamics of entangled polymer solutions. II. Inclusion of hydrodynamic interactions. Macromolecules **1976**, *9*, 594.
9. Tan, H.; Moet, A.; Hiltner, A.; Baer, E. Macromolecules **1983**, *16*, 28.
10. Hoyos, M.; Martin, M. Retention theory of sedimentation field-flow fractionation at finite concentration. Anal. Chem. **1994**, *66*, 1718–1730.
11. Sisson, R.M.; Giddings, J.C. Effects of solvent composition on polymer retention in TFFF. Anal. Chem. **1994**, *66*, 4043.

Conductivity Detection in Capillary Electrophoresis

Jetse C. Reijenga
Eindhoven University of Technology, Eindhoven, The Netherlands

INTRODUCTION

In contrast to component-specific detectors, such as ultraviolet (UV) absorbance and fluorescence, conductivity detection is a universal detection method. This means that a bulk property (conductivity) of the buffer solution is continuously measured. A migrating ionic component locally changes the conductivity and this change is monitored. As such, conductivity detection is universally sensitive because, in principle, all migrating ionic compounds show detector response, although not to the same extent.

TYPES OF CONDUCTIVITY DETECTION

Two kinds of conductivity detector are distinguished: contact detectors and contactless detectors. Both types were originally developed for isotachophoresis in 0.2–0.5-mm-inner diameter (i.d.) PTFE tubes. Contactless detectors are based on the measurement of high-frequency cell resistance and, as such, inversely proportional to the conductivity. The advantage is that electrodes do not make contact with the buffer solution and are, therefore, outside the electric field. As these types of detectors are difficult to miniaturize down to the usual 50–75-μm capillar inner diameter, their actual application in capillary electrophoresis (CE) is limited.

Contact detectors are somewhat easier to miniaturize. There are generally two subtypes: those with twin axially mounted electrodes and those with twin or quadruple radially mounted electrodes. The former can be operated in DC mode or AC mode. In the DC mode, the detector signal directly originates from the field strength between the electrodes and, given the current, is inversely proportional to the detector cell resistance. In the AC mode, both axially and radially mounted electrodes form part of a closed primary circuit of an isolation transformer, the output of which is also inversely proportional to the cell conductivity. Alternatively, the output can be linearized with respect to the conductivity.

CONDUCTIVITY DETECTOR RESPONSE

As mentioned, the detector continually measures the conductivity of the buffer solution in the capillary.

If an ionic component enters the detector cell, the local conductivity will change. At first glance, one would expect the conductivity to increase, because of additional ionic material. This is a simplified and incorrect approach, however. Suppose, in a buffer consisting of $0.01\,M$ potassium and $0.02\,M$ acetate (pH 4.7), a $10^{-4}\,M$ sodium solution is analyzed. Electroneutrality requires that with an increase of the sodium concentration from zero to, in this case, initially $10^{-4}\,M$, the potassium and/or charged acetate concentration cannot remain unchanged. This process is governed by the so-called Kohlrausch law. For strong ions, this equation reads

$$\Lambda = \sum_i \frac{c_i}{\mu_i}$$

in which Λ is the so-called Kohlrausch regulating function, c_i is the concentration of component i, and μ_i is the mobility of component i. Generally speaking, potassium will be partly displaced by sodium, whereas acetate will remain approximately (but not, by definition, exactly) constant. In the example given, the conductivity detector will give a negative response (see line A in Fig. 1), because potassium (with a high mobility and, hence, a higher contribution to conductivity) is, to some extent, replaced with sodium which has a ~30% lower mobility. From this example, it automatically follows that a potassium peak in a sodium acetate buffer, by contrast, will yield a positive amplitude. This makes interpretation of conductivity detector signals less straightforward.

SENSITIVITY OF CONDUCTIVITY DETECTION

A further example will illustrate aspects related to sensitivity. Suppose a 100 times more concentrated ($10\,mM$) solution of ammonium is coseparated in the potassium–acetate system mentioned earlier. Naturally, ammonium will displace potassium, but as the mobilities of potassium and ammonium are almost equal, the resulting change in conductivity is minor. Sensitivity in this example is, consequently, very low (line A in Fig. 1). On the other hand, $0.005\,mM$ lithium has a much lower conductivity

Encyclopedia of Chromatography DOI: 10.1081/E-ECHR-120039929

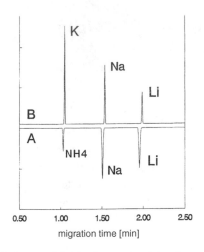

Fig. 1 Relative sensitivities in conductivity detection in CE. Trace A: sample of $10\,mM$ NH$_4$, $0.1\,mM$ Na, and $0.005\,mM$ Li in a $0.01\,M$ potassium–acetate buffer; trace B: sample of $0.1\,mM$ each of K, Na, and Li in a $10\,mM$ Tris–acetate buffer.

than sodium and, consequently, shows a higher specific response (line A in Fig. 1).

Generally, one cannot expect a high sensitivity anyhow, as the background signal (originating from the buffer) is generally much higher than the eventual change superimposed upon that background. One might argue that background conductivity can easily be decreased by diluting the buffer. Potential gain with this approach is very limited, because diluting the buffer below an ionic strength of $1\,mM$ will lead to unacceptable loss in buffering capacity and, moreover,

in severe sample overload. Another possibility to decrease the background conductivity is to use buffer components with lower mobility, such as GOOD buffers. This, however, will sooner lead to nonsymmetric peaks on sample overload (peak triangulation). Using low-mobility Tris as a buffer co-ion will lead to positive peaks for $0.1\,mM$ potassium, sodium, and lithium alike (line B in Fig. 1).

SUGGESTED FURTHER READING

Beckers, J.L. Isotachophoresis, some fundamental aspects. In *Thesis*; Eindhoven University of Technology, 1973.

Everaerts, F.M.; Beckers, J.L.; Verheggen, Th.P.E.M. *Isotachophoresis: Theory, Instrumentation and Applications*; Elsevier: Amsterdam, 1976.

Hjertén, S. Free zone electrophoresis. Chromatogr. Rev. **1967**, *9* (2), 122–219.

Kohlrausch, F. Ueber concentrations-verschiebungen durch electrolyse im innern von L—sungen und L—sungsgemischen. Ann. Phys. (Leipzig). **1897**, *62*, 209.

Li, S.F.Y. *Capillary Electrophoresis—Principles, Practice and Applications*; Elsevier: Amsterdam, 1992.

Reijenga, J.C.; Verheggen, Th.P.E.M.; Martens, J.H.P.A.; Everaerts, F.M. Buffer capacity, ionic strength and heat dissipation in capillary electrophoresis. J. Chromatogr. A. **1996**, *744*, 147.

Conductivity Detection in HPLC

Ioannis N. Papadoyannis
Victoria F. Samanidou
Aristotle University of Thessaloniki, Thessaloniki, Greece

INTRODUCTION

Conductivity detection is used to detect inorganic and organic ionic species in LC. As all ionic species are electrically conducting, conductometric detection is a universal detection technique, considered as the mainstay in high-pressure ion chromatography, in the same way as is ultraviolet (UV) detection in HPLC.

DISCUSSION

The principle of operation of a conductivity detector lies in differential measurement of mobile-phase conductivity prior to and during solute ion elution. The conductivity cell is either placed directly after an analytical column or after a suppression device required to reduce background conductivity, in order to increase the signal-to-noise ratio and, thus, sensitivity. In the first mode, known as *nonsuppressed* or *single-column ion chromatography*, aromatic acid eluents are used, with low-capacity fixed-site ion exchangers and dynamically or permanently coated reversed-phase columns. In the second mode, known as *eluent-suppressed ion chromatography*, the separated ions are detected by conductance after passing through a suppression column or a membrane, to convert the solute ions to higher conducting species (e.g., hydrochloric acid in the case of chloride ions and sodium hydroxide in the case of sodium ions). In the meantime, the eluent ions are converted to a low-residual-conductivity medium such as carbonic acid or water, thus reducing background noise.

Conductance G is the ability of electrolyte solutions in an electric field applied between two electrodes to transport current by ion migration. According to Ohm's law, ohmic resistance R is given by

$$R = \frac{U}{I} \tag{1}$$

where U is the voltage (V) and I is the current intensity (A). The reciprocal of ohmic resistance is the conductance G, where

$$G = \frac{1}{R} \tag{2}$$

expressed in Siemens in the International System of Units (SI), formerly reported in the literature as mho The measured conductance of a solution is related to the interelectrode distance d (cm) and the microscopic surface area (A) (geometric area × roughness factor) of each electrode (A is assumed identical for the two electrodes) as well as the ionic concentration, given by

$$G = \frac{kA}{d} \tag{3}$$

where k is the specific conductance or conductivity. The ratio d/A is a constant for a particular cell, referred as the cell constant K_c (cm^{-1}) and is determined by calibration. The usual measured variable in conductometry is conductivity k (S/cm)

$$k = GK_c \tag{4}$$

The conductance G (in μS) of a solution is given by

$$G = \frac{(\lambda^+ + \lambda^-)CI}{10^{-3}K_c} \tag{5}$$

where λ^+ and λ^- are limiting molar conductivities of the cation and anion, respectively, and C is the molarity and I the fraction of eluent that is ionized. If the eluent and solute are fully ionized, the conductance change accompanying solute elution is

$$\Delta G = \frac{(\lambda_s - \lambda_e)C_s}{10^{-3}K_c} \tag{6}$$

The specific conductance/conductivity k (S/cm) of salts measured by a conductivity detector is given by

$$k = \frac{(\lambda_{s+} + \lambda_{s-})C_s + (\lambda_{e+} + \lambda_{e-})C_e}{1000}$$
$$= \frac{\Lambda_s C_s + \Lambda_e C_e}{1000} \tag{7}$$

where C_s and C_e are the concentration (mol/L) of the solute and eluent ions, respectively, and Λ is the molar conductivity of the electrolyte.

The change in conductance when a sample solute band passes through the detector results from

Encyclopedia of Chromatography DOI: 10.1081/E-ECHR-120039930

replacement of some of the eluent ions by solute ions, although the total ion concentration C_{tot} remains constant:

$$C_{tot} = C_s + C_e \tag{8}$$

The background ion conductivity when $C_s = 0$ is

$$k_1 = \frac{\Lambda_e C_{tot}}{1000} \tag{9}$$

When a solute band is eluted, the ion conductivity k_2 is given by

$$k_2 = \frac{\Lambda_e C_{tot}}{1000} + \frac{(\Lambda_s - \Lambda_e)C_s}{1000} \tag{10}$$

The difference in conductivity is obtained after subtraction of the first equation from the second:

$$\Delta k = k_2 - k_1 = \frac{(\Lambda_s - \Lambda_e)C_s}{1000} \tag{11}$$

From Eq. (11), it is obvious that when a sample band is eluted, the observed difference in conductivity is proportional to the concentration of the sample solute C_s. However, the linear relation holds only for dilute solutions, as Λ is itself dependent on concentration, according to Kohlrausch's law:

$$\Lambda = \Lambda - A\sqrt{C} \tag{12}$$

where A is a constant and Λ° is the limiting molar conductivity in an infinitely dilute solution, given by the sum

$$\Lambda = \Lambda^+ + \Lambda^- \tag{13}$$

or

$$\Lambda = v_+\lambda_+ + v_-\lambda \tag{14}$$

where v_+ and v_- represent stoichiometric coefficients for the cation and anion, respectively, in the electrolyte.

Eq. (11) shows that the signal observed during solute ion elution is also proportional to the difference in limiting molar ionic conductivities between the eluent and the solute ions.

Values of limiting molar ionic conductivities for a few common ions are shown in Table 1. The data tabulated are referred to 25°C temperature. The term *limiting molar ionic conductivity* is used according to IUPAC recommendation, rather than the formerly used *limiting ionic equivalent conductivity*. The molar and equivalent values are interconvertible through stoichiometric coefficient z.

Table 1 Limiting molar ionic conductivities of some anions and cations at 25°C

Anions	λ^-	Cations	λ^+
OH$^-$	199.1	H$^+$	349.6
F$^-$	55.4	Li$^+$	38.7
Cl$^-$	76.4	Na$^+$	50.1
Br$^-$	78.1	K$^+$	73.5
I$^-$	76.8	NH$_4^+$	73.5
NO$_3^-$	71.46	Mg^{2+}	106
NO$_2^-$	71.8	Cu^{2+}	107.2
SO$_4^{2-}$	160.0	Ca^{2+}	120
Benzoate$^-$	32.4	Sr^{2+}	118.9
Phthalate^{2-}	76	Ba^{2+}	127.2
Citrate^{3-}	168	Ethylammonium	47.2
CO$_3^{2-}$	138.6	Diethylammonium	42.0
C$_2$O$_4^{2-}$	148.2	Triethylammonium	34.3
PO$_4^{3-}$	207	Tetraethylammonium	32.6
CH$_3$COO$^-$	40.9	Trimethylammonium	47.2
HCOO$^-$	54.6	Tetramethylammonium	44.9

Conductivity is measured by applying an alternating voltage to two electrodes of various geometric shapes in a flow-through cell, which results in anion migration, as negatively charged, toward the anode (positive electrode) and cation migration, as positively charged, toward the negative electrode (cathode). An AC potential (frequency 1000–5000 Hz) is required in order to avoid electrode polarization. The cell current is measured and the solution's resistance (or more strictly the impedance) is calculated by Ohm's law. Conductance is further corrected by the conductivity cell constant, thus giving conductivity.

The requirements for a typical conductivity detection cell are small volume (to eliminate dispersion effects), high sensitivity, wide linear range, rapid response, and acceptable stability. The cell generally consists of a small-volume chamber ($<5\,\mu$L) fitted with two or more electrodes constructed of platinum, stainless steel, or gold.

Most conductivity detectors function according to the Wheatstone Bridge principle. What is actually measured is resistance of the solution. Electronically, the electrodes are arranged in that way to constitute one arm of a Wheatstone Bridge. Eluting ions from chromatographic column subsequently enter the detector cell, leading to a change of electrical resistance and the out-of-balance signal is rectified with a precision rectifier. The DC signal is either digitized and sent to a computer data acquisition system or is passed to a potentiometric recorder, by means of a linearizing amplifier, which modifies the signal so that the output is linearly related to ion concentration. Sometimes, a variable resistance is situated in one of the other arms of the bridge and is used for zero adjustment to compensate for any signal from mobile-phase ions. As mentioned earlier, at constant voltage applied to the cell, the current will be proportional to the conductivity (Fig. 1).

The conductivity k is a characteristic property of the solution rather than a property of the cell used. It contains all the chemical information available from the measurement, such as concentration and mobilities of the ions present. Accordingly, the conductivity detector is a bulk property detector and, as such, it responds to all electrolytes present in the mobile phase as well as the solutes. Thus, the experimentally determined conductivity is the sum of the contributions from all ions present in the solution. The sensitivity of the conductivity detector depends on the difference between the limiting ionic conductivities of the solute and eluent ions.

The differential mode of detection is mostly effective, provided there is a significant difference in the values of the measured property between the eluent and solute ions. This difference may be either positive or negative. The former case refers to lower conductivity of the eluent ion, described as *direct detection method*, the latter to greater conductivity of the eluent ion, described as *indirect detection method* (Fig. 2).

The thermal stability of a conductivity detector is of great importance. Effective thermostating is highly required, as the temperature greatly affects the mobility of ions and, therefore, conductivity. A 0.5–3% increase of conductivity is usually expected per degree Celsius. Close temperature control is necessary to minimize background noise and maximize sensitivity;

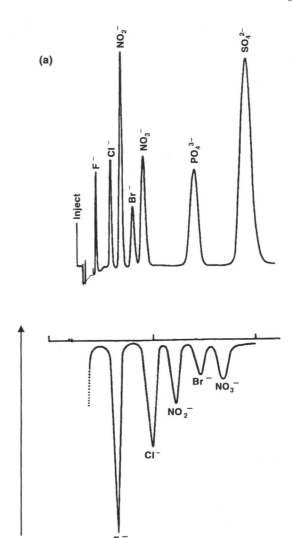

Fig. 2 Conductivity detection of anions in nonsuppressed (single column) ion chromatography using an eluent of (a) low background conductance (direct detection) and (b) high background conductance (indirect detection). The direction of the arrow indicates the increase of conductivity.

this is an especially important issue if nonsuppressed eluents are used.

Typical specifications for an electrical conductivity detector are as follows: sensitivity, 5×10^{-9} g/mL; linear dynamic range, 5×10^{-9} to 1×10^{-6} g/mL; response index, 0.97–1.03.

CONCLUSIONS

Conductivity detection in HPLC or, more precisely HPIC, can be applied to ionic species, including all anions and cations of strong acids and bases (e.g., chloride, sulfate, sodium, potassium, etc.). Ions of weaker acids and bases are detected provided that

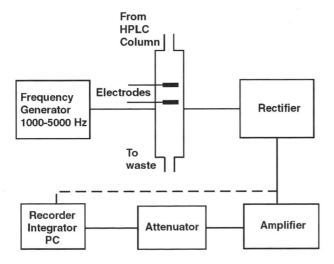

Fig. 1 Block diagram of electrical conductivity detector in HPLC.

the pH value of the eluent is chosen to maximize the analyte's ionization so as to increase sensitivity. The relatively simple design requirements, accuracy, and low cost contribute to its utility and popularity; thus, it is almost used in over 95% of analyses, where ion-exchange separation procedures are involved.

SUGGESTED FURTHER READING

Coury, L. Curr. Separ. **1999**, *18* (3), 91.

Papadoyannis, I.; Samanidou, V.; Zotou, A. J. Liquid Chromatogr. **1995**, *18* (7), 1383.

Parriott, D. *A Practical Guide to HPLC Detection*; Academic Press: San Diego, CA, 1993.

Schaefer, H.; Laubli, M.; Doerig, R. *Ion Chromatography*; Metrohm Monograph 50143; Metrohm AG: Herisau, 1996.

Scott, R. *Techniques and Practice of Chromatography*; Marcel Dekker, Inc.: New York, 1995.

Scott, R. *Chromatographic Detectors, Design, Function and Operation*; Marcel Dekker, Inc.: New York, 1996.

Tarter, J. *Ion Chromatography*; Chromatographic Science Series; Marcel Dekker, Inc.: New York, 1987; Vol. 37.

Congener-Specific PCB Analysis

George M. Frame, II
Wadsworth Laboratory, New York State Department of Health, Albany, New York, U.S.A.

INTRODUCTION

Polychlorinated biphenyls (PCBs) are complex mixtures of 209 possible chlorinated biphenyl molecules, referred to as congeners. There are from 3 to 46 isomers at each of the 10 possible levels of chlorination. Isomers of a given chlorination level are referred to as homologs. About 150 of these congeners appear at significant levels in commercial mixtures. These mixtures, trade named Aroclor (U.S.A.), Clophen (Germany), Kanechlor (Japan), and so forth, found use as electrical insulating fluids in transformers and capacitors and as binders for a wide variety of uncontained applications. Although their manufacture has been largely discontinued, their long-term stability, dispersion into the environment by prior uncontrolled releases, lipophilicity (resulting in biomagnification up food chains), and potential toxicity to humans and biota have sparked extensive research and the requirement for detailed analytical characterization of these mixtures.

DISCUSSION

This article does not discuss the extensive literature on sample preparation, cleanup, and proper instrumental operation. Adsorption column chromatography and HPLC procedures find application here, and the book by Erickson[1] provides exhaustive discussions of these and of the history of PCB use and analysis. Methods for measuring total PCB content or measuring and reporting the mixtures by their commercial designation are not detailed. Congener-specific PCB analysis demands separation and quantitation either of shortlists of priority PCB congeners or of the PCB content of all chromatographic PCB peaks that can be separated in particular system(s). This latter mode is referred to as comprehensive, quantitative, congener-specific analysis (CQCS). The methods of choice for CQCS PCB analysis employ high-resolution GC (HRGC) on capillary columns with sensitive and selective detection by electron-capture detectors (ECD), selected ion monitoring-MS (MS-SIM), or full-scan, ion-trap MS (ITMS). The most complete discussion of target congeners for specific research applications is in Ref.[2]. A descriptive overview of CQCS PCB

analysis appears in an Analytical Chemistry A-page article,[3] and extensive reviews[4–6] provide detailed discussions and large bibliographies.

Fig. 1 summarizes PCB congener structure, nomenclature, the Ballschmiter and Zell (BZ) congener numbering system, and the relative abundances in the commercial Aroclor mixtures as a function of single-ring chlorine-substitution patterns. The BZ numbers in the matrix correspond to the chlorine-substitution positions in each ring of the biphenyl structure, which are listed along the top and right sides of the figure matrix. The abbreviated nomenclature (e.g., 234–245 = PCB #138) defines each congener by the substitution pattern in each ring, with the dash representing the bond between the two phenyl rings. Rotation about this bond is possible except in congeners with three or four chlorines in the ortho (2 or 6) positions.

In the United States, commercial mixtures were manufactured until 1977 by Monsanto under the trade name Aroclor. In the four-digit Aroclor designations (e.g., Aroclor 1242), the 12 indicates a biphenyl nucleus and the 42 the weight percentage of chlorine in the mixture. Reference to the matrix in Fig. 1 reveals congeners in black cells which never exceed 0.1 wt% in the mixtures. These "non-Aroclor" congeners are absent due to unfavored or improbable formation in the electrophilic chlorination process employed in the manufacture of Aroclors.[7] Conversely, the ring chlorine-substitution patterns giving rise to the congeners in gray cells are especially favored.

Whereas three or more *ortho*-chlorines block rotation about the ring-connecting bond, congeners with no, or only one, *ortho*-chlorine can relatively easily achieve a planar configuration (colloquially referred to as "coplanars") and may behave as isosteres (compounds with similar shape, functionality, and polarity) to 2,3,7,8-tetrachlorodibenzodioxin (TCDD). These bind significantly to the "dioxin receptor," and measurement of their concentrations can be combined with their "dioxin-like toxic equivalency factors" (TEFs) to give an estimate of a PCB mixture's "dioxin-like toxic equivalency" (TEQ).[2] Thus, one "shortlist" analysis specified by the World Health Organization (WHO) is for 12 such congeners found in commercial mixtures, namely PCBs 77, 81, 105, 114, 118, 123, 126, 156, 157,

Encyclopedia of Chromatography DOI: 10.1081/E-ECHR-120039931

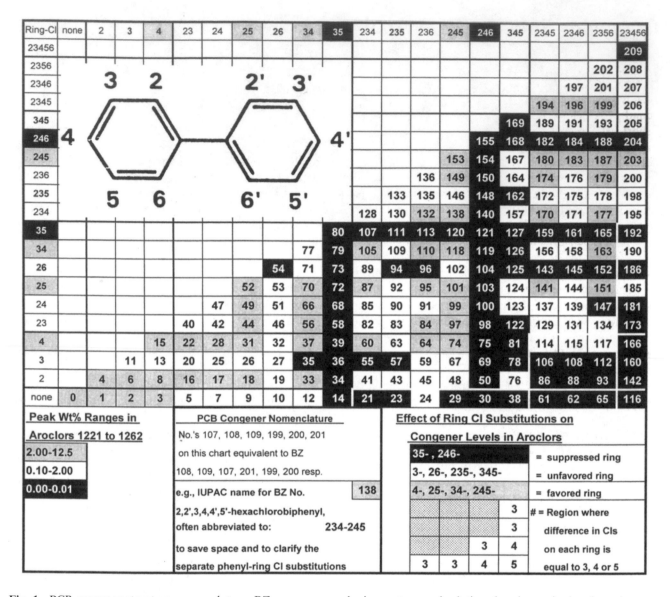

Fig. 1 PCB congener structure, nomenclature, BZ congener numbering system, and relative abundances in Aroclor mixtures.

167, 169, and 189. Although PCB 126 is generally a trace component, it has such a high TEF that it often dominates the TEQ calculation.

In the United States, the initial regulatory methods were the USEPA 8080 series. In 8080, packed column GC-ECD was recommended and calibration was against Aroclor mixtures and results were reported as Aroclor equivalents. Method 8081 encouraged use of higher-resolution capillary GC columns and MS-SIM detection, and the current version, 8082, extends this to suggest measuring some individual congeners against primary standards. An early version of CQCS analysis is EPA Method 680, which employs MS-SIM detection at the molecular ion cluster mass for each homolog level. It does not provide for actually identifying all congeners by determining their elution times and is, thus, not classified as congener-specific,

but rather homolog-class-specific. It is quantitatively calibrated against an average response at each level, resulting in less precise measurement of individual congeners, but it detects and measures all PCB-containing peaks whether Aroclor derived or not. It is thus superior to the 8080 series when a PCB mixture arising from a non-Aroclor source or a substantially altered Aroclor congener distribution is encountered. In Europe, the Community Bureau of Reference (BCR) specifies measurement of a shortlist of major persistent "indicator congeners," namely PCBs 28, 52, 101, 118, 138, 153, and 180. A number of other congener shortlists are detailed in Refs.[2,3]. A powerful but expensive and difficult-to-implement congener-specific PCB analysis method is USEPA Method 1668. The target list is the WHO list of coplanar PCBs with dioxin-like TEQs. The methodology is HRGC with >10,000 resolution high-resolution MS (HRMS) detection and ^{13}C-isotope dilution internal standards for all the analytes. The procedures are modeled on the well-established HRGC/HRMS USEPA Method 1613 for PCDD/Fs. The newer USEPA Method 1668 Revision A (December 1999) describes procedures for extending the analyte list to all 209 congeners that can be resolved on either an SPB-octyl capillary column or a DB-1 (100% methyl silicone) capillary. Primary standards for all 209 congeners are distributed among five calibration solutions, which avoid any isomer coelutions on the SPB-octyl column.

No single column, nor any pair of columns, can completely separate all 209 congeners, or even the 150 or so found in Aroclors. Analysts developing CQCS or even "shortlist" congener-specific PCB analyses must select GC stationary phases capable of resolving congeners in their target list. Many analysts have employed 5% phenyl-, 95% methyl-substituted silicone polymers (e.g., DB-5) since a very similar phase was the first one for which the relative retention times for all 209 PCB congeners were published.[3] Methyl silicone phases with 50% n-octyl or n-octadecyl substituents have PCB retention characteristics similar to those of hydrocarbon columns such as Apeizon L or Apolane, but much greater stability and higher temperature limits than the latter. They permit resolution of many pairs of lower homologs, which coelute on the more polar phases. This feature is valuable for characterizing the products of dechlorination by anaerobic bacteria.[7] Phases with arylene or carborane units substituted in the silicone backbone to decrease column bleed (e.g., DB-5MS, DB-XLB, HT-8) have been found to have particularly useful congener-separation capabilities.[3,7–9] Column manufacturers Restek and SGE have subsequently modified capillary columns of this type (Rtx-PCB and HT-8-PCB, respectively) to specifically support maximal resolution of PCBs in CQCS analysis.

A database of relative retention times and coelutions for all 209 congeners on 20 different stationary phases has been published.[8] For 12 of the most useful of these phases, the elution orders of 9 solutions of all 209 congeners are available from a standard supplier, which markets these solutions (AccuStandard, New Haven, Connecticut, U.S.A.). By surveying the database, one can determine the most suitable column(s) for a particular application and can quickly establish a method component table by nine injections of the standard mixtures. This greatly facilitates the development of new CQCS PCB analyses. Tables of the weight percentages of all congeners in each of the numbered Aroclor mixtures, from which the information condensed in the figure matrix was derived, are available.[7,8] These help

reduce the number of congeners that a CQCS method is required to separate when one anticipates analyzing only relatively unaltered Aroclor congener mixtures.

Prior to the availability of all 209 congeners in well-designed primary standard mixtures, much effort was expended to use structure–retention relationships on various phases to predict retention for congeners for which standards were not available.[5] In general, PCB retention times increase with chlorination level, and within chlorination levels, with less chlorine substitution in the ortho position (i.e., "coplanar PCBs" are more strongly retained). These relationships are of theoretical interest but are of less use now that accurate retention time assignments are possible with actual standards. The use of commercial mixtures such as Aroclors as quantitative secondary standards for CQCS PCB analysis is now to be discouraged,[4] as detailed studies of congener distributions show significantly different proportions among different lots.[7] In the case of Aroclor 1254, there are actually two different mixtures of radically different composition produced by totally different synthetic processes.[9]

The other major factor affecting the capability of CQCS PCB analyses is the selection of the GC detector. Initially, the ECD has been most useful for this application. It is selective for halogenated compounds, and its sensitivity is outstanding for the more chlorinated (Cl $\geq$ 4) congeners. Its drawbacks are twofold: It has a limited linear range, necessitating multilevel calibration, and the relative response factors vary widely from instrument to instrument and among congeners even at the same chlorination level.[3] For mono- and dichloro-substituted congeners, it is less sensitive than the corresponding MS detectors. Other halogenated compounds such as organochlorine pesticides produce ECD responsive peaks that may interfere by coelution with certain PCB congeners. For these reasons, CQCS PCB analyses with ECD detection often employ a procedure of splitting the injected sample to two columns (each with an ECD detector) of different polarity and PCB congener elution order.[3,7] To be reported, a congener must be measured on at least one column without coelution of PCB or another interfering compound. If separately measurable on each column, the quantities found must match within a preset limit to preclude the possibility of an unexpected coeluting contaminant on one of the columns. Given the large number of congeners that may need to be measured, the data reduction algorithm for such a procedure is complex and not easily automated.

Another approach to providing a second dimension to CQCS PCB analysis is to employ much more selective mass spectrometric detection.[3,6–8] In electrospray ionization-mass spectroscopy (EI-MS), the spectra consist of a molecular ion cluster of chlorine isotope MS peaks and similar fragment ion clusters resulting from the successive loss of chlorine atoms. Congeners differing by one chlorine substituent that coelute on the GC column may often be separately quantitated by MS detection, if the more chlorinated one is not in great excess. This is because the $[M-1Cl]^+$ fragment fragment that interferes with the lower congener's signal is from a ^{13}C isotope peak and typically has 0.5–12% the signal level of its M^+ peak.[9] In contrast to the ECD, the sensitivity of MS-SIM or full-scan ITMS is greater for the less chlorinated congeners, as their electron affinity is lower and the positive charge of the ions is distributed over a smaller number of fragments. The linearity of the MS detectors is better than that of ECDs, and the ions monitored are more specific for PCBs and less prone to interference from non-PCB compounds. Electron-capture detectors continue to hold the edge in absolute sensitivity (for the higher chlorinated congeners), and the dual-column/ECD detector systems are slightly less expensive than comparable bench-top, unit-mass-resolution, single-column GC/MS systems. Application to PCB analysis of more advanced (and expensive) MS detection systems, such as HRMS, MS/MS, and negative-ion MS, is described in several reviews.[4,6] Even higher throughput or more comprehensive CQCS PCB congener separations have recently been demonstrated using the latest capillary GC instrumental refinements. These are, respectively, fast GC with time-of-flight (TOF) MS detection,[10] and comprehensive 2D-GC with ECD detection.[11]

A final refinement of congener-specific PCB analysis arises from the fact that 19 of the congeners actually exist as stable enantiomeric pairs, either component of which can withstand racemization even at the elevated temperatures required to elute them from a capillary GC separation.[6] Some congeners containing either a 236- or a 2346-chlorine-substituted ring and three or more chlorines in the ortho position exist in two mirror-image forms by virtue of their inability to rotate around the bond between the two rings. These so-called atropisomers do not contain asymmetric carbon centers. They are PCB numbers 45, 84, 91, 95, 132, 135, 136, 149, 174, and 176 (containing the 236-ring), as well as PCB numbers 88, 131, 139, 144, 171, 175, 176, 183, 196, and 197 (containing the 2346-ring). They may be separated on chiral GC stationary phases, primarily those employing a family of modified cyclodextrins. A series of seven such columns have been found, which among them can achieve resolution of all 19 stable PCB atropisomers as well as separation of 11 of them from other possible coeluting PCBs if MS detection is employed.[12] Observation of PCB enantiomeric ratios significantly different from 1 is a certain indication of the action of an enzyme-mediated biological process operating on these congeners.

CONCLUSIONS

Comprehensive, quantitative, congener-specific PCB analysis requires use of high-resolution capillary GC separations, aided by selective ECD or MS detection. The availability of a range of well-documented stationary phases, complete sets of calibration and retention time standards for all 209 PCB congeners, and databases of retention data facilitates the efficient development of a CQCS assay procedure suitable for specific applications. The "holy grail" of a single system that can reliably and unambiguously identify and quantify any and all of the 209 congeners in a single run has not quite been achieved, but is being approached closely. Polychlorinated biphenyl-tailored stationary phases, fast TOF-MS detection, and comprehensive 2D-GC separations may well combine to achieve this goal.

REFERENCES

1. Erickson, M.D. *Analytical Chemistry of PCBs*; 2nd Ed.; Lewis Publishers: New York, 1997.

2. Hansen, L.G. *The Ortho Side of PCBs: Occurrence and Disposition*; Kluwer Academic: Boston, 1999.

3. Frame, G.M. Congener-specific PCB analysis. Anal. Chem. **1997**, *69*, 468A–475A.

4. Hess, P.; de Boer, J.; Cofino, W.P.; Leonards, P.E.G.; Wells, D.E. Critical review of the analysis of non- and mono-*ortho*-chlorobiphenyls. J. Chromatogr. A. **1995**, *703*, 417.

5. Larsen, B.R. HRGC separation of PCB congeners. J. High Resolut. Chromatogr. **1995**, *18*, 141.

6. Cochran, J.W.; Frame, G.M. Recent developments in the high resolution gas chromatography of polychlorinated biphenyls. J. Chromatogr. A. **1999**, *843*, 323.

7. Frame, G.M.; Cochran, J.W.; Bøwadt, S.S. Complete PCB congener distributions for 17 Aroclor mixtures determined by 3 HRGC systems optimized for comprehensive, quantitative, congener-specific analysis. J. High Resolut. Chromatogr. **1996**, *19*, 657–668.

8. Frame, G.M. A collaborative study of 209 PCB congeners and 6 Aroclors on 20 different HRGC columns: 1. Retention and coelution database, 2. Semi-quantitative Aroclor distributions. Fresenius J. Anal. Chem. **1997**, *357*, 701–722.

9. Frame, G.M. Improved procedure for single DB-XLB column GC–MS-SIM quantitation of PCB congener distributions and characterization of two different preparations sold as "Aroclor 1254." J. High Resolut. Chromatogr. **1999**, *22*, 533–540.

10. Cochran, J.W. Fast gas chromatography–time-of-flight mass spectrometry of polychlorinated biphenyls and other environmental contaminants. J. Chromatogr. Sci. **2002**, *40*, 254–268.

11. Korytár, P.; Danielsson, C.; Leonards, P.E.G.; Haglund, P.; de Boer, J.; Brinkman, U.Th. Separation of seventeen 2,3,7,8-substituted poly-chlorinated dibenzo-*p*-dioxins and 12 dioxin-like polychlorinated biphenyls by comprehensive two-dimensional gas chromatography with electron-capture detection. J. Chromatogr. A. **2004**, *1038*, 189–199.

12. Wong, C.S.; Garrison, A.W. Enantiomer separation of polychlorinated biphenyl atropisomers and polychlorinated biphenyl retention behavior on modified cyclodextrin capillary gas chromatography columns. J. Chromatogr. A. **2000**, *866*, 213.

Copolymer Composition by GPC/SEC

Sadao Mori
PAC Research Institute, Mie University, Nagoya, Japan

INTRODUCTION

Determination of the average chemical composition and polymer composition by SEC has been reported in the literature. Two different types of concentration detector or two different absorption wavelengths of an ultraviolet or an infrared detectors are employed; the composition at each retention volume is calculated by measuring peak responses at the identical retention points of the two chromatograms.

DISCUSSION

Synthetic copolymers have both molecular-weight and chemical composition distributions and copolymer molecules of the same molecular size, which are eluted at the same retention volume in SEC, may have different molecular weights in addition to different compositions. This is because separation in SEC is achieved according to the sizes of molecules in solution, not according to their molecular weights or chemical compositions.

Molecules that appear at the same retention volume may have different compositions, so that accurate information on chemical heterogeneity cannot be obtained by SEC alone. When the chemical heterogeneity of a copolymer, as a function of molecular weight, is observed, the copolymer is said to have a heterogeneous composition, but, even though it shows constant composition over the entire range of molecular weights, it cannot be concluded that it has a homogeneous composition.[1] Nevertheless, SEC is still extremely useful in copolymer analysis, due to its rapidity, simplicity, and wide applicability.

When one of the constituents, A or B of a copolymer A–B, has an ultraviolet (UV) absorption and the other does not, a UV detector–refractive index (RI) combined detector system can be used for the determination of chemical composition or heterogeneity of the copolymer. A point-to-point composition, with respect to retention volume, is calculated from two chromatograms and a variation of composition is plotted as a function of molecular weight. The response factors of the two components in the two detectors must first be calibrated.

Let A be a constituent that has UV absorption. K_A and K_B are defined as the response factors of an RI detector for the A and B constituents, and K_A' as the response of the UV detector for A. These response factors are calculated by injecting known amounts of homopolymers A and B into the SEC dual-detector system, calculating the areas of the corresponding chromatograms, and dividing the areas by the weights of homopolymers injected as

$$F_A = K_A G_A, \quad F_B = K_B G_B, \quad F_A' = K_A' G_A$$

where F_A, F_B, and F_A' are areas of homopolymers A and B in the RI detector and of homopolymer A in the UV detector, and G_A and G_B are the weights of homopolymers A and B injected into the SEC system.

The weight fraction $W_{A,I}$ of constituent A, at each retention volume I of the chromatogram for the copolymer, is given by

$$W_{A,I} = \frac{K_B}{R_I K_A' - K_A + K_B}$$

where $R_I = F_{RI,I}/F_{UV,I}$ for the copolymer at retention volume I. Retention volme I for the RI detector is not equal to the retention volume I for the UV detector. Usually, the UV detector is connected to the column outlet and is followed by an RI detector, and the dead volume between these two detectors must be corrected. The dead volume can normally be measured by injecting a polymer sample having a narrow molecular-weight distribution and by measuring the retention difference between the two peak maxima.

Because the additivity of the RI increments of homopolymers is valid for copolymers, the additivity of the response factors is also valid:

$$K_C = W_A K_A + W_B K_B$$

where K_C is the response factor for the copolymer in the RI detector. If the response factors of one or two homopolymers that comprise a copolymer cannot be measured because of insolubility of the homopolymer(s), then this equation is employed.

Alternatively, the extrapolation of the plot of RI response factors of copolymers of known compositions can be used. An example is that the RI response for

Encyclopedia of Chromatography DOI: 10.1081/E-ECHR-120039933

polystyrene was 2800 and that for polyacrylonitrile was 2250.

Although the values of these response factors are dependent on several parameters, the ratio of to K_A to K_B is almost constant in the same mobile phase.

An infrared detector can be used, at an appropriate wavelength, for detecting one component in copolymers or terpolymers and, thus, expand its range of applicability to copolymers analysis. Information on composition can be obtained by repeating runs, using different wavelengths to monitor different functional groups. A single-detector system is more advantageous than a dual-detector system, such as a combination of UV and RI detectors.

Instead of measuring chromatograms two or three times at different wavelengths for different functional groups, operation in a stop-and-go fashion was introduced for rapid determination of copolymer composition as a function of molecular weight.[2]

Pyrolysis gas chromatography has been widely used for copolymer analysis. This technique may offer many advantages over other detection techniques for copolymer analysis by SEC. One obvious advantage is the small sample size required. Another is the capability of application to copolymers which cannot utilize UV or IR detectors.[3]

Combination with other liquid chromatographic techniques is also reported by several workers. Orthogonal coupling of an SEC system to another HPLC system KB KA to achieve a desired cross-fractionation was proposed.[4] It was an SEC–SEC mode, using the same polystyrene column, but the mobile phase in the first system was chosen to accomplish only a hydrodynamic volume separation, and the mobile phase in the second system was chosen so as to be a thermodynamically poorer solvent for one of the monomer types in the copolymer, in order to fractionate by composition under adsorption or partition modes as well as size exclusion.

A combination of liquid adsorption chromatography with SEC has recently been developed by several workers. Poly(styrene–methyl methacrylate) copolymers were fractionated according to chemical composition by liquid adsorption chromatography and the molecular weight averages of each fraction were measured by SEC.[5–6]

REFERENCES

1. Mori, S. Comparison between size-exclusion chromatography and liquid adsorption chromatography in the determination of the chemical heterogeneity of copolymers. J. Chromatogr. **1987**, *411*, 355.
2. Mirabella, F.M., Jr.; Barrall, E.M., II; Johnson, J.F. A rapid technique for measuring copolymer composition as a function of molecular weight using gel permeation chromatography and infrared detection. J. Appl. Polym. Sci. **1975**, *19*, 2131.
3. Mori, S. Determination of the composition of copolymers as a function of molecular weight by pyrolysis gas chromatography-size-exclusion chromatography. J. Chromatogr. **1980**, *194*, 163.
4. Balke, S.T.; Patel, R.D. J. Polym. Sci. Polym. Lett. Ed. **1980**, *18*, 453.
5. Mori, S. Determination of chemical composition and molecular weight distributions of high-conversion styrene-methyl methacrylate copolymers by liquid adsorption and size exclusion chromatography. Anal. Chem. **1988**, *60*, 1125.
6. Mori, S. Trends Polym. Sci. **1994**, *2*, 208.
7. Mori, S.; Barth, H.G. *Size Exclusion Chromatography*; Chap. 12, 1999, Springer-Verlag: New York.
8. Mori, S. Copolymer analysis. In *Size Exclusion Chromatography*; Hunt, B.J., Hodling, S.R., Eds.; Blackie: Oxford, 1989.

Copolymer Molecular Weights by GPC/SEC

Sadao Mori
PAC Research Institute, Mie University, Nagoya, Japan

INTRODUCTION

It is well known that most copolymers have both molecular weight and composition distributions and that copolymer properties are affected by both distributions. Therefore, we must know average values of molecular weights and composition, and their distributions. These two distributions are inherently independent of each other. However, it is not easy to determine the molecular-weight distribution independently of the composition, even by modern techniques.

DISCUSSION

SEC is a rapid technique used to obtain the molecular-weight averages and the molecular-weight distributions of synthetic polymers. The objective of SEC for copolymer analysis must be not only the determination of molecular-weight averages and its distribution but also the measurement of average copolymer composition and its distribution. However, separation by SEC is achieved according to the sizes of molecules in the solution, not according to their molecular weights. Therefore, the retention volume of a copolymer molecule obtained by SEC reflects not the molecular weight, as in the case of a homopolymer, but simply the molecular size.

For example, the elution order of polystyrene (PS), poly(methyl methacrylate) (PMMA), and their copolymers [P(S–MMA)], both random and block, all having the same molecular weight are as follows: random copolymer of P(S–MMA), PS, block copolymer (MMA-S-MMA), and PMMA.[1] Copolymers having the same molecular weight but different composition are different in molecular size and elute at different retention volumes. Therefore, the accurate determination of the values of molecular-weight averages and the molecular-weight distribution for a copolymer by SEC might be limited to the case when the copolymer has the homogeneous composition across the whole range of molecular weights.

A calibration curve for a copolymer consisting of components A and B can be constructed from those for the two homopolymers A and B, if the relationships of the molecular weights and the molecular sizes of the two homopolymers are the same as their copolymer and if the size of the copolymer molecules in the solution is the sum of the sizes of the two homopolymers times the corresponding weight fractions. The molecular weight of the copolymer at retention volume I, $M_{C,I}$, is calculated using

$$\log M_{C,I} = W_{A,I} \log M_{A,I} + W_{B,I} \log M_{B,I}$$

where $M_{A,I}$ and $M_{B,I}$ are the molecular weights of homopolymers A and B, respectively, and $W_{A,I}$ and $W_{B,I}$ are the weight fractions of components A and B, respectively, in the copolymer at retention volume I. This equation was empirically postulated for block copolymers.[2]

The use of the so-called "universal calibration" is a theoretically reliable procedure for calibration. For ethylene–propylene (EP) copolymers, Mark–Houwink parameters in o-dichlorobenzene at 135°C are calculated as[3]

$$a_{EP} = (a_{PE}a_{PP})^{1/2}$$
$$K_{EP} = W_E K_{PE} + W_P K_{PP} - 2(K_{PE}K_{PP})^{1/2} W_E W_P$$

where W_E and W_P are the weight fractions of the ethylene and propylene units of the copolymer, respectively.

Calculated Mark–Houwink parameters for P(S–MMA) block and statistical copolymers at several compositions in tetrahydrofuran at 25°C are listed in Table 1.[4] The parameters for PS and PMMA used in the calculation are as follows:

$$PS : K = 0.682 \times 10^{-2} \text{mL/g}, a = 0.766$$
$$PMMA : K = 1.28 \times 10^{-2} \text{mL/g}, a = 0.69$$

If copolymer molecules and PS molecules are eluted at the same retention volume, then

$$[\mu]_C M_C = [\mu]_S M_S$$

Encyclopedia of Chromatography DOI: 10.1081/E-ECHR-120039934

Table 1 Calculated Mark–Houwink parameters for P(S–MMA) block and statistical copolymers at several compositions in teterahydrofuran at 25°C

Composition (styrene wt %)	Block copolymer		Statistical copolymer	
	$K \times 10^2$ (mL/g)	a	$K \times 10^2$ (mL/g)	a
20	1.124	0.705	1.044	0.718
30	1.054	0.714	0.953	0.731
40	0.989	0.721	0.879	0.742
50	0.929	0.729	0.821	0.750
60	0.872	0.736	0.779	0.756
70	0.820	0.744	0.747	0.760
80	0.771	0.751	0.722	0.763

where M_C and M_S are the molecular weights of the copolymer and PS, respectively, and $[\mu]_C$ and $[\mu]_S$ are the intrinsic viscosities of the copolymer and PS, respectively. A differential pressure viscometer can measure intrinsic viscosities for the fractions of the copolymer and PS continuously, followed by the determination of M_C of the copolymer fraction at retention volume I.

The application of a light-scattering detector in SEC does not require the construction of a calibration curve using narrow molecular-weight distribution polymers. However, this method is not generally applicable to copolymers because the intensity of light scattering is a function not only of molecular weight but also of the specific refractive index (the refractive index increment) of the copolymer in the mobile phase. The re-fractive index increment is also a function of composition. In the case of a styrene–butyl acrylate (30:70) emulsion copolymer, the apparent molecular weight of the copolymer in teterahydrofuran was only 7% lower than true one.[5] A recent study concluded that if refractive index increments of the corresponding homopolymers do not differ widely, SEC measurements combined with light scattering and concentration detectors yield good approximations to molecular weight and its distribution, even if the chemical composition distribution is very broad.[6]

REFERENCES

1. Dondos, A.; Rempp, P.; Benoit, H. Macromol. Chem. **1984**, *175*, 1659.
2. Runyon, J.R.; Barnes, D.E.; Rudd, J.F.; Tung, L.H. Multiple detectors for molecular weight and composition analysis of copolymers by gel permeation chromatography. J. Appl. Polym. Sci. **1969**, *13*, 2359.
3. Ogawa, T.; Inaba, T. Gel permeation chromatography of ethylene-propylene copolymerization products. J. Appl. Polym. Sci. **1988**, *21*, 2979.
4. Goldwasser, J.M.; Rudin, A. J. Liquid Chromatogr. **1983**, *6*, 2433.
5. Malihi, F.B.; Kuo, C.Y.; Provder, T. Determination of the absolute molecular weight of a styrene-butyl acrylate emulsion copolymer by low-angle laser light scattering (LALLS) and GPC/LALLS. J. Appl. Polym. Sci. **1984**, *29*, 925.
6. Kratochvil, P. International Symposium on Polymer Analysis and Characterization, 1995, Abstract L14.
7. Mori, S.; Barth, H.G. *Size Exclusion Chromatography*; Chap. 12, 1999, Springer-Verlag: New York.
8. Mori, S. Copolymer analysis. In *Size Exclusion Chromatography*; Hunt, B.J., Holding, S.R., Eds.; Blackie: Oxford, 1989.

Coriolis Force in Countercurrent Chromatography

Yoichiro Ito
*Center of Biochemistry and Biophysics, National Heart, Lung, and Blood Institute,
National Institutes of Health, Bethesda, Maryland, U.S.A.*

Kazufusa Shinomiya
College of Pharmacy, Nihon University, Chiba, Japan

INTRODUCTION

Coriolis force acts on a moving object on a rotating body such as the Earth or a centrifuge bowl. It was first analyzed by a French engineer and mathematician, Gaspard de Coriolis (1835).[1] The effect of the Coriolis force produced by the Earth's rotation is weak, whereas that on a rotating centrifuge is strong and easily detected. Fig. 1 illustrates the effect of Coriolis force on moving droplets in a rotating centrifuge, where the path of the sinking droplets shifts toward the direction opposite to the rotation (left); this effect is reversed for floating droplets (right).[2] Moving droplets in a rotating centrifuge have been photographed under stroboscopic illumination.[3,4] The effects of Coriolis force on countercurrent chromatography (CCC) have been demonstrated in the toroidal coil centrifuge, which uses a coiled tube mounted around the periphery of the centrifuge bowl.[2,5] When a protein mixture containing cytochrome *c*, myoglobin, and lysozyme was separated on an aqueous/aqueous polymer phase system composed of 12.5% (w/w) polyethylene glycol 1000 and 12.5% (w/w) dibasic potassium phosphate, the direction of elution through the toroidal coil had substantial effects on peak resolution, as shown in Fig. 2 and Table 1.[2,5] Since the toroidal coil separation column has a symmetrical orientation except for the handedness, the above effect is best explained on the basis of Coriolis force as follows: If the Coriolis force acts parallel to the effective coil segments (parallel orientation), the two phases form multiple droplets, which provide a broad interface area to enhance the mass transfer process, hence improving the partition efficiency (Fig. 3A). When the Coriolis force acts across the effecting coil segments, the two phases form a streaming flow, minimizing the interfacial area for mass transfer and resulting in lower partition efficiency (Fig. 3B). It is interesting to note that the above effects have not been observed during the separation of low molecular weight compounds such as dipeptides[2] and dinitrophenyl (DNP) amino acids[6] on conventional organic/aqueous two-phase solvent systems in the toroidal coil CCC centrifuge, except that at a relatively low revolution speed, the Coriolis force acting across the effective coil segments slightly improves the partition efficiency,

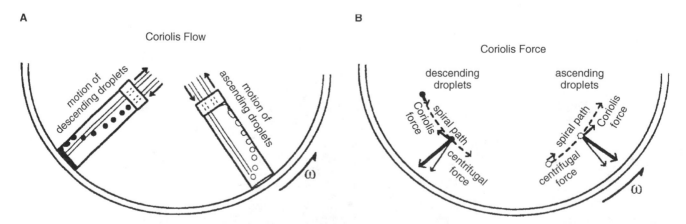

Fig. 1 Effects of Coriolis force on moving droplets in a rotating centrifuge. A. Motion of droplets in a flow through cell in a rotating centrifuge. B. Direction of Coriolis force acting on droplets on rotating centrifuge bowl.

Encyclopedia of Chromatography DOI: 10.1081/E-ECHR-120039935

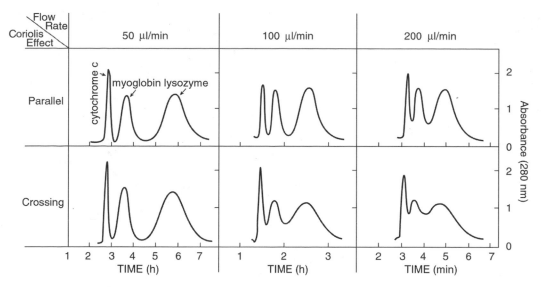

Fig. 2 Effects of Coriolis force on partition efficiency and retention of stationary phase in protein separation by toroidal coil centrifuge.

probably due to substantially higher retention of the stationary phase.

More recently, the effect of Coriolis force was demonstrated in the separation of organic acids with organic/aqueous two-phase solvent systems in a centrifugal partition chromatograph equipped with a separation column consisting of rectangular partition compartments connected in series.[6] As shown in Fig. 4, clockwise column rotation (CW) shows substantially better peak resolution than counterclockwise column rotation (CCW), especially in the separation of *p*-methyl hippuric acid and hippuric acid (middle column) with a two-phase solvent system composed of methyl *t*-butyl ether/aqueous 0.1% trifluoroacetic acid (1 : 1, v/v). Mathematical analysis is carried out to elucidate the effect of Coriolis force on the motion of the mobile-phase droplets.[6]

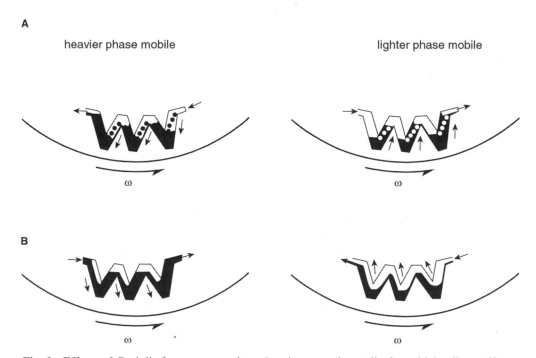

Fig. 3 Effects of Coriolis force on two-phase flow in separation coil of toroidal coil centrifuge.

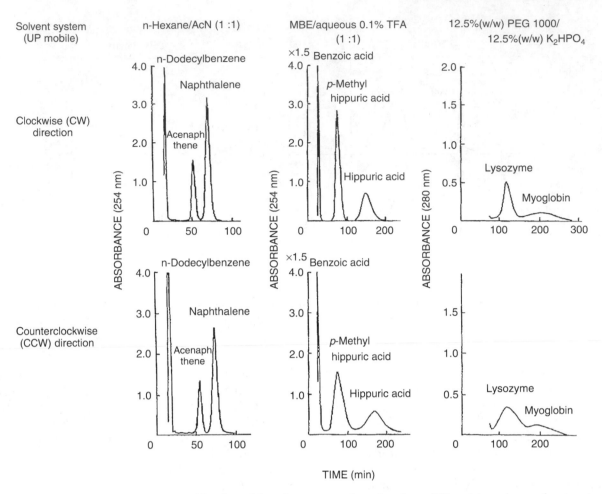

Fig. 4 Chromatograms obtained by centrifugal partition chromatography using three different two-phase solvent systems by eluting upper phase in ascending mode.

Table 1 Effects of Coriolis force on partition efficiencies of three stable proteins in toroidal coil CCC

Flow-rate (µl/min)	Analyte peak	TP (parallel/crossing)	R_s (parallel/crossing)	Retention (%) (parallel/crossing)
50	Cytochrome c	1860/1490		29.2/32.0
			1.62/1.39	
	Myoglobin	365/266		
			1.66/1.40	
	Lysozyme	156/104		
100	Cytochrome c	1760/821		30.0/30.3
			1.27/0.86	
	Myoglobin	433/172		
			1.39/0.84	
	Lysozyme	172/63		
200	Cytochrome c	1296/–		22.8/21.3
			0.84/–	
	Myoglobin	330/–		
			0.92/–	
	Lysozyme	123/–		

ARTICLE OF FURTHER INTEREST

Coil Planet Centrifuges, p. 349.

REFERENCES

1. *New Encyclopedia Britannica*; 1995; Vol. 3, 632.
2. Ito, Y.; Ma, Y. Effect of Coriolis force on counter-current chromatography. J. Liq. Chromatogr. **1998**, *21*, 1–17.
3. Marchal, L.; Foucault, A.; Patissier, G.; Rosant, J.M.; Legrand, J. J. Chromatogr. A., *in preparation*.
4. Morvan, A.; Foucault, A.; Patissier, G.; Rosant, J.M.; Legrand, J. J. Hydrodynamics, *in preparation*.
5. Ito, Y.; Matsuda, K.; Ma, Y.; Qi, L. Toroidal coil counter-current chromatography. Achievement of high resolution by optimizing flow-rate, rotation speed, sample volume and tube length. J. Chromatogr. A. **1998**, *808*, 95–104.
6. Ikehata, J.; Shinomiya, K.; Kobayashi, K.; Ohshima, H.; Kitanaka, S.; Ito, Y. Effect of Coriolis force on counter-current chromatographic separation by centrifugal partition chromatography. J. Chromatogr. A. **2004**, *1025*, 169–175.

C

Corrected Retention Time and Corrected Retention Volume

Raymond P. W. Scott
Scientific Detectors Ltd., Banbury, Oxfordshire, U.K.

INTRODUCTION

The corrected retention time of a solute is the elapsed time between the dead point and the peak maximum of the solute. The different properties of the chromatogram are shown in Fig. 1. The volume of mobile phase that passes through the column between the dead point and the peak maximum is called the corrected retention volume.

DISCUSSION

If the mobile phase is incompressible, as in liquid chromatography, the retention volume (as so far defined) will be the simple product of the exit flow-rate and the corrected retention time.

If the mobile phase is compressible, the simple product of the corrected retention time and flow rate will be incorrect, and the corrected retention volume must be taken as the product of the corrected retention time and the *mean* flow rate. The true corrected retention volume has been shown to be given

$$V_r{}' = V_{r'}{}'\left(\frac{\gamma^2 - 1}{\gamma^2 - 1}\right) = Q_0 t_r{}'\frac{2}{3}\left(\frac{\gamma^2 - 1}{\gamma^2 - 1}\right)$$

where the symbols have the meaning defined in Fig. 1, and $V_{r'}{}'$ is the corrected retention volume measured at the column exit and γ is the inlet/outlet pressure ratio.

The corrected retention volume, $V_r{}'$, will be the difference between the retention volume and the dead volume V_0, which, in turn, will include the actual dead volume V_m and the extra column volume V_E. Thus,

$$V_r{}' = V_r - (V_E + V_m)$$

The retention time can be taken as the product of the distance on the chart between the dead point and the peak maximum and the chart speed, using appropriate units. As in the case of the retention time, it can be more accurately measured with a stopwatch. Again, the most accurate method of measuring $V_r{}'$ for a non-compressible mobile phase, although considered antiquated, is to attach an accurate burette to the detector exit and measure the retention volume in volume units. This is an absolute method of measurement and does not depend on the accurate calibration of the pump, chart speed, or computer acquisition level and processing.

REFERENCE

1. Scott, R.P.W. *Introduction to Analytical Gas Chromatography*; Marcel Dekker, Inc.: New York, 1998; 77.

SUGGESTED FURTHER READING

Scott, R.P.W. *Techniques and Practice of Chromatography*; Marcel Dekker, Inc.: New York, 1996.
Scott, R.P.W. *Liquid Chromatography Column Theory*; John Wiley & Sons: Chichester, 1992; 19.

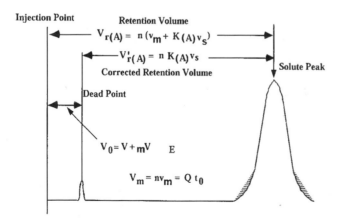

Fig. 1 Diagram depicting the retention volume, corrected retention volume, dead point, dead volume, and dead time of a chromatogram. V_0: total volume passed through the column between the point of injection and the peak maximum of a completely unretained peak; V_m: total volume of mobile phase in the column; $V_{r(A)}$: retention volume of solute A; $V_{r(A)}{}'$: corrected retention volume of solute A; V_E: extra column volume of mobile phase; v_m: volume of mobile phase, per theoretical plate; V_s: volume of stationary phase per theoretical plate; $K_{(A)}$: distribution coefficient of the solute between the two phases; n: number of theoretical plates in the column; Q: column flow rate measured at the exit.

Encyclopedia of Chromatography DOI: 10.1081/E-ECHR-120039936

Coumarins: Analysis by TLC

Kazimierz Glowniak
Jaroslaw Widelski
Department of Pharmacognosy with Medicinal Plant Laboratory,
Medical University of Lublin, Lublin, Poland

INTRODUCTION

Coumarins are natural compounds that contain the characteristic benzo[α]pyrone (2H-benzopyran-2-one) moiety. They are especially abundant in Umbelliferae, Rutaceae, Leguminosae, Compositae, and other plant families. Usually the substituents are at the positions C_5, C_6, C_7, and C_8 [e.g., umbelliferone (7-hydroxy-coumarin), hierniarin (7-methoxycoumarin), esculetin (6,7-dihydroxycoumarin), scopoletin (6-methoxy-7-hydroxycoumarin), and others].

In addition to simple coumarin derivatives, furano- and pyranocoumarins are also commonly encountered in the Umbelliferae and Rutaceae families. The essential chemical moiety of linear furanocoumarins consists of a 2H-furan[3.2-g]-benzo[b]pyran-2-one ring called psolaren (its derivatives include, e.g., bergapten, xanthotoxin, isopimpinelin, imperatorin, isoimpera-torin, oxypeucedanin, and others). The second type of furanocoumarins (the angular type of angelicin) has a 2H-furan[2.3-h]-benzo[b]pyran-2-one structure (isobergapten, pimpinelin, sphondin). There are also both linear and angular types of pyranocoumarins. In the linear type, which is named alloxanthiletin, the 2H,8H-pyran[3.2-h]-benzo[b]pyran-2-one ring is characteristic, whereas in the angular type called seselin, the 2H,8H-pyran[2.3-h]-benzo[b]pyran-2-one moiety is typical.

THIN LAYER CHROMATOGRAPHY

TLC is a very useful method for the separation of natural coumarins, furanocoumarins, and pyrano-coumarins. Natural coumarins exhibit fluorescence properties, which they display in ultraviolet (UV) light (365 nm). Their spots can be easily detected on paper and thin-layer chromatograms without the use of any chromogenic reagents. It is often possible to recognize the structural class of coumarin from the color it displays under UV detection (Table 1). Purple fluores-cence generally signifies 7-alkoxycoumarins, whereas 7-hydroxycoumarins and 5,7-dioxygenated coumarins tend to fluoresce blue. In general, furanocoumarins possess a dull yellow or ocher fluorescence, except for

psolaren, sphondin, and angelicin. Spot fluorescence is more intense or its color is changed after spraying the TLC chromatogram with ammonia (Table 1).[1]

Thin-layer chromatograms can also be detected by several nonspecific chromatogenic reactions:

1. 1% Aqueous solution of iron (III) chloride.
2. 1% Aqueous solution of potassium ferricyanide.
3. Diazotized sulfanilic acid and diazotized *p*-nitro-aniline.

None of these reagents is very specific for hydroxy-coumarins and their confirmation should be substan-tiated by other methods. Exposed groups present in many natural coumarins can be detected due to their susceptibility to cleavage by acids and applied over a phosphoric acid spot on a silica TLC plate.

The linear (psolarens) and angular (angelicins) furanocoumarins can be readily differentiated with the Emerson reagent. It is also used for detection of pyranocoumarins (selinidin, pteryxin) on TLC chro-matograms.

CONVENTIONAL TLC

Conventional TLC is a well-known technique, used for many years in systematic research on the coumarin content of numerous plant species, as well as for chemotaxonomic relationships between those species. Great progress in the optimization of the TLC separa-tion process was made by the design of modern, hori-zontal chambers for TLC. It is a universal design offering the possibility of developing chromatograms in the space saturated or nonsaturated with mobile-phase vapors. Moreover, it is possible to perform gradient elution, stepwise or continuous, or to accom-plish micropreparative separation of chemical com-pound composites (e.g., plant extracts).[2]

Gradient elution in TLC can be obtained in several ways:[1]

1. Multizonal development: the use of multi-component eluents that are partially separated during development (frontal chromatography),

Encyclopedia of Chromatography DOI: 10.1081/E-ECHR-120039937

Table 1 Chromatographic methods of coumarin identification: Fluorescence colors of coumarins under UV irradiation (365 nm)

Fluorescence color	Fluorescence color with ammonia	Coumarin or coumarin type
Blue	L. blue	7-Hydroxycoumarin
B. blue	V. blue	7-Hydroxycoumarins
Blue	B. blue	7-Hydroxy-6-alkoxycoumarins
Blue	Blue	5,7-Dialkoxycoumarins
B. blue	B. blue	6,7-Dialkoxycoumarins
Blue	Blue	6,7,8-Trialkoxycoumarins
W. blue	W. blue	5,6,7-Trialkoxycoumarins
W. blue	B. blue	7-Hydroxy-5,6-dialkoxycoumarins
Blue		Psolaren
Blue	B. blue	6-Methoxyangelicin
Blue	Green	7,8-Dihydroxycoumarin
Pink	Yellow	6-Hydroxy-7-glucosyloxycoumarin
Purple	Purple	8-Hydroxy-5-alkoxypsolarens
W. purple	Pink	6-Hydroxy-7-alkoxycoumarins
Purple	Green	Angelicin, coumestrol
Green		5-Methoxyangelicin
Green		8-Hydroxy-6,7-dimethoxycoumarin
Green	Yellow	7,8-Dihydroxy-6-methoxycoumarin
Yellow		3,4,5-Trimethoxypsolaren
Yellow		6-Hydroxy-5,7-dimethoxycoumarin
Yellow	Yellow	5-Hydroxy-6,7-dimethoxycoumarin
Yellow	Yellow	5-Hydroxypsolaren
Yellow	Yellow	5,6-Dimethoxyangelicin
Yellow	Yellow	8-Alkoxypsolarens
Yellow-green	Yellow-green	5-Alkoxypsolarens
B. yellow	B. yellow	5,8-Dialkoxypsolarens

B.= bright; V. = very bright; L. = light; W. = weak.
(From Ref.[1].)

forming an eluent strength gradient along the layer.

2. Development with a strong solvent (e.g., acetone) of an adsorbent layer exposed to vapors of a less polar solvent.
3. The use of mixed layers of varying surface area and activity (compare silica and Florisil®).
4. Delivery of an eluent whose composition is varied in a continuous or stepwise manner by introducing small volumes of more polar eluent fractions (e.g., 0.2 ml).

The possibility of zonal sample dosage in equilibrium conditions (after a front of mobile phase and continuous-chromatogram development, which is provided by a horizontal "sandwich" chamber) was utilized by Glowniak, Soczewinski, and Wawrzynowicz[3] in preparative chromatography of simple coumarins and furanocoumarins found in *Archangelica* fruits,

performed with a short-bed continuous development (SB-CD) technique.

The latter possibility was employed by Wawrzynowicz and Waksmundzka-Hajnos for micropreparative TLC isolation of furanocoumarins from *Archangelica*, *Pastinaca*, and *Heracleum* fruits on silica gel, silanized gel, and Florisil.

Superior coumarin compound separation with use of the described flat "sandwich" chambers is achieved with gradient chromatography on silica gel and stepwise variation of polar modifier concentrations in mobile phase, as less polar solvents (hexane, cyclohexane, toluene, or dichloromethane) and polar modifiers (acetonitrile, diisopropyl ether, ethyl acetate) are used.

Complex pyranocoumarin mixtures can be separated with the TLC technique by alternative use of two different polar adsorbents (silica gel, Florisil) and various binary and ternary eluents with different mechanisms of adsorption center effect on the molecules to be separated.[4] Improved separation can be achieved by high-performance TLC (HPTLC), which employs new, highly effective adsorbent of narrow particle size distribution, or with chemically modified surface. Because of its similarity, HPTLC is applied in designing optimal HPLC systems. Another gradient technique in coumarin compound separation is programmed multiple development (PMD), also called the "reversed gradient" technique, in which chromatograms are developed to increasing distances by a sequence of eluents with decreasing polarity, with eluent evaporation after each stage.

Two-dimensional TLC (2D-TLC) is particularly effective in the case of complex extracts when one-dimensional developing yields partial separation. Moreover, it offers the possibility of modifying separation procedures when the development direction is changed.

OVERPRESSURED LAYER CHROMATOGRAPHY

The term "overpressured layer chromatography" (OPLC) was originally introduced by Tyihak, Mincsovisc, and Kalasz[5] in the late 1970s. The crucial factor is pressurized mobile-phase flow through the planar medium. Short analysis time, low solvent consumption, high resolution, and availability of on-line and off-line modes are the main advantages of OPLC in comparison with the classical TLC techniques. Overpressured layer chromatography was proved effective in qualitative and quantitative analysis of furanocoumarins by densitometric on-line detection. Overpressured layer chromatography can also be performed in two-dimensional mode (2D-OPLC). This technique was used by Harmala et al.[8] for the separation of 16 closely related coumarins from *Angelica* genus.

Long-distance OPLC is a novel form of OPLC, in which chromatograms are developed over a long distance with optimal (empiric) mobile-phase flow. Used in combination with specialized equipment designs, it produces high performance (70,000–80,000 of theoretic plates) and excellent resolution. Botz, Nyiredy, and Sticher,[6] who initiated long-distance OPLC, proved its efficiency in the separation of eight furanocoumarin isomers and in the isolation of the furanocoumarin complex from *Peucedanum palustre* roots raw extract. This technique was used by Galand et al.[7] as well.

ROTATION PLANAR CHROMATOGRAPHY

Rotation planar chromatography (RPC), as with OPLC, is another thin-layer technique with forced eluent flow, employing a centrifugal force of a revolving rotor to move the mobile phase and separate chemical compounds. The RPC equipment can vary in chamber size, operative mode (analytical or preparative), separation type (circular, anticircular, or linear), and detection mode (off-line or on-line). The described technique was applied in analytical and micropreparative separation of coumarin compounds from plant extracts.

AUTOMATED MULTIPLE DEVELOPMENT

Automated multiple development (AMD), providing automatic chromatogram development and drying, is a novel form of the PMD technique. Automated multiple development as an instrumental technique can be used to perform normal-phase chromatography with solvent gradients on HPTLC plates. Most of the AMD applications use typical gradients: Starting with a very polar solvent, the polarity is varied by means of "base" solvent of medium polarity to a nonpolar solvent. Instrumentation for AMD was introduced by Camag (Switzerland) and provides a means for normal-phase gradient development in HPTLC. The developing distance increases while the solvent polarity decreases. Repeated development compresses the band on the plate, resulting in increased sensitivity and resolution.[9]

CONCLUSIONS

TLC is a suitable method of separation, characterization, and quantitative evaluation for any kind of coumarin compound. Modern TLC techniques such as AMD, HPTLC, and OPLC have been in use since many years in systematic research on the coumarin content of numerous plant species, as well as for chemotaxonomic relationships between those species.

REFERENCES

1. Murray, R.D.H. Coumarins. Nat. Prod. Rep. **1989**, 6, 591–624.
2. Soczewinski, E. Simple device for continuous thin-layer chromatography. J. Chromatogr. **1977**, *138*, 443–445.
3. Glowniak, K.; Soczewinski, E.; Wawrzynowicz, T. Optimization of chromatographic systems for the separations of components of the furocoumarin fraction of *Archangelica* fruits on a milligram scale. Chem. Anal. **1987**, *32*, 797–811.

4. Glowniak, K. Comparison of selectivity of silica and Florisil in the separation of natural pyranocoumarins. J. Chromatogr. **1991**, *552*, 453–461.

5. Tyihak, E.; Mincsovisc, E.; Kalasz, H. New planar liquid chromatographic technique: overpressured thin-layer chromatography. J. Chromatogr. **1979**, *174*, 75–81.

6. Botz, L.; Nyiredy, S.; Sticher, O. Applicability of long distance overpressured layer chromatography. J. Planar Chromatogr. **1991**, *4*, 115.

7. Galand, N.; Pothier, J.; Dollet, J.; Viel, C. OPLC and AMD, recent techniques of planar chromatography: their interest for separation and characterisation of extractive and synthetic compounds. Fitoterapia **2002**, *73*, 121–134.

8. Harmala, P.; Botz, L.; Sticher, O.; Hiltunen, R. Two-dimensional planar chromatographic separation of a complex mixture of closely related coumarins from the genus *Angelica*. J. Planar Chromatogr. **1990**, *3*, 515–520.

9. Gocan, S.; Cimpan, G.; Muresan, L. Automated multiple development thin layer chromatography of some plant extracts. J. Pharm. Biomed. Anal. **1995**, *14*, 1221–1227.

Countercurrent Chromatographic Separation of Vitamins by Cross-Axis Coil Planet Centrifuge

Kazufusa Shinomiya
College of Pharmacy, Nihon University, Narashinodai, Funabashi-shi, Chiba, Japan

Yoichiro Ito
Laboratory of Biophysical Chemistry, National Heart, Lung, and Blood Institute, National Institutes of Health, Bethesda, Maryland, U.S.A.

INTRODUCTION

Cross-axis coil planet centrifuges (cross-axis CPC), which have been widely used in the separation of natural and synthetic products, are some of the most useful models among various types of countercurrent chromatographic (CCC) apparatuses.[1–3] It produces a unique mode of planetary motion such that the column holder rotates about its horizontal axis while revolving around the vertical axis of the centrifuge.[4–5] The centrifugal force field produced by the planetary motion provides stable retention of the stationary phase for polar two-phase solvent systems such as aqueous–aqueous polymer phase systems with extremely low interfacial tension and high viscosity. Our previous studies demonstrated that cross-axis CPC, equipped with either a multilayer coil or eccentric coil assembly in the off-center position of the column holder, can be effectively applied for the separation of proteins[6–8] and sugars.[9]

This article illustrates the CCC separation of various vitamins by means of cross-axis CPC equipped with eccentric coil assemblies.[10–11]

APPARATUS AND SEPARATION COLUMNS

All existing cross-axis CPCs are classified according to their relative column position on the rotary frame, e.g., X,[4–5] XL,[12] XLL,[13] and XLLL,[14] where X is the distance from the column holder axis to the central axis of the centrifuge, and L is the distance between the holder and the middle point of the rotary shaft. Increasing the ratio L/X improves the retention of the stationary phase by moderating phase mixing. The separations of vitamins described below were performed using the $X-1.5L$ cross-axis CPC, which holds the separation column at $X = 10\,\mathrm{cm}$ and $L = 15\,\mathrm{cm}$.

The separation column was prepared by means of a pair of eccentric coil assemblies, which were made by winding a single piece of 1-mm-ID polytetrafluoroethylene (PTFE) tubing onto 7.6-cm-long, 5-mm-OD nylon pipes forming 20 units of serially connected left-handed coils. Then, a set of these coil units was arranged around the holder with their axes parallel to the holder axis. A pair of identical coil assemblies was connected in series to obtain a total column capacity of 26.5 mL.

CCC SEPARATION OF WATER-SOLUBLE VITAMINS

In CCC, the partition coefficient (K) is an important parameter which is used for selecting the optimal

Table 1 Partition coefficients of water-soluble vitamins in 1-butanol/aqueous 0.15 M monobasic potassium phosphate two-phase solvent systems

	1	4	8
1-Butanol	1	4	8
Ethanol	0	1	3
0.15 M KH$_2$PO$_4$	1	4	8
Thiamine nitrate (M.W. 327.36)	0.03	0.05	0.11
Thiamine hydrochloride (M.W. 327.27)	0.14	0.05	0.17
Riboflavin (M.W. 376.37)	0.54	0.83	1.08
Riboflavin sodium phosphate (M.W. 478.33)	0.11	0.18	0.34
Pyridoxine hydrochloride (M.W. 205.64)	0.23	0.43	0.59
Cyanocobalamin (M.W. 1355.38)	0.04	0.12	0.24
L-Ascorbic acid (M.W. 176.12)	0.07	0.11	0.15
Nicotinamide (M.W. 122.13)	1.84	1.41	1.34

Partition coefficients were calculated from the absorbance of the upper phase divided by that of lower phase at 260 nm.

Encyclopedia of Chromatography DOI: 10.1081/E-ECHR-120014225

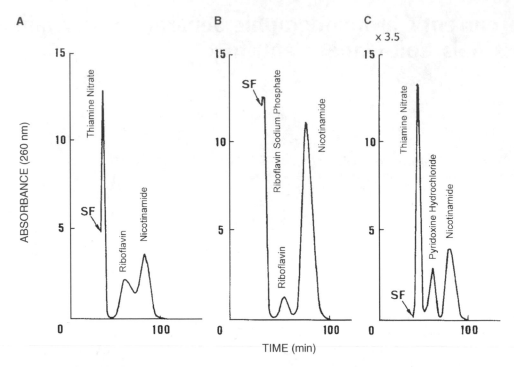

Fig. 1 CCC separation of water-soluble vitamins by cross-axis CPC. Experimental conditions: apparatus, cross-axis CPC equipped with a pair of eccentric coil assemblies, 1 mm ID and 26.5 mL capacity; sample. (A) Thiamine nitrate (2.5 mg) + ribo-riboflavin (1.5 mg) + nicotinamide (2.5 mg). (B) Riboflavin sodium phosphate (2.5 mg) + nicotinamide (2.5 mg). (C) Thiamine nitrate (2.8 mg) + pyridoxine hydrochloride (4.0 mg) + nicotinamide (3.0 mg); solvent system: (A) and (B) 1-butanol/ aqueous 0.15 M monobasic potassium phosphate (1 : 1) and (C) 1-butanol/ethanol/aqueous 0.15 M monobasic potassium phosphate (8 : 3 : 8); mobile phase: lower phase; flow rate: 0.4 mL/min; revolution: 800 rpm. SF = solvent front.

solvent system because it predicts the retention time of each component. Table 1 shows the K values of various water-soluble vitamins in the 1-butanol/aqueous 0.15 M monobasic potassium phosphate system. Most of the water-soluble vitamins were partitioned, almost unilaterally, into the aqueous phase in this solvent system, except that riboflavin and nicotinamide were distributed significantly into the organic phase

($K = 0.54–1.84$). Adding ethanol to the two-phase solvent system significantly increased the partition coefficients of riboflavin and pyridoxine hydrochloride.

Fig. 1A illustrates the CCC separation of thiamine nitrate, riboflavin, and nicotinamide by cross-axis CPC, with the 1-butanol/aqueous 0.15 M monobasic potassium phosphate (1 : 1) system. Riboflavin sodium phosphate is also resolved from riboflavin with the

Table 2 Effect of ion-pair reagents on partition coefficients of water-soluble vitamins in 1-butanol/aqueous 0.15 M KH$_2$PO$_4$ solvent systems

Ion-pair reagent concentration (%)	1-Butane sulfonic acid		1-Pentane sulfonic acid		1-Octane sulfonic acid
	1.5	2.5	1.5	2.5	1.5
Thiamine nitrate (M.W. 327.36)	0.18	0.29	0.76	1.01	2.45
Thiamine hydrochloride (M.W. 327.27)	0.17	0.25	0.73	0.97	2.42
Riboflavin (M.W. 376.37)	0.58	0.70	0.57	0.95	1.07
Riboflavin sodium phosphate (M.W. 478.33)	0.04	0.29	0.08	0.20	0.13
Pyridoxine hydrochloride (M.W. 205.64)	0.57	0.73	0.77	0.86	1.63
Cyanocobalamin (M.W. 1355.38)	0.08	0.44	0.37	0.35	0.47
L-Ascorbic acid (M.W. 176.12)	0.04	0.13	0.09	0.09	0.09
Nicotinamide (M.W. 122.13)	1.32	0.78	1.46	1.39	1.49

Partition coefficients were calculated from the absorbance of the upper phase divided by that of lower phase at 260 nm.

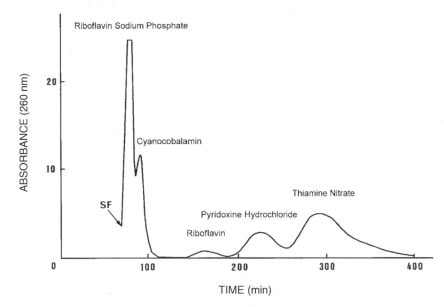

Fig. 2 CCC separation of water-soluble vitamins by cross-axis CPC. Experimental conditions: sample, riboflavin sodium phosphate (2.5 mg) + cyanocobalamin (2.5 mg) + pyridoxine hydrochloride (2.5 mg) + thiamine nitrate (2.5 mg); solvent system, 1-butanol and aqueous 0.15 M monobasic potassium phosphate containing 1.5% of 1-octanesulfonic acid sodium salt; mobile phase, lower phase; flow rate, 0.2 mL/ min. For other experimental conditions, see Fig. 1 caption. SF = solvent front.

same solvent system (Fig. 1B). When a more polar solvent system consisting of 1-butanol/ethanol/aqueous 0.15 M monobasic potassium phosphate (8:3:8) is used, thiamine nitrate, pyridoxine hydrochloride, and nicotinamide are well resolved from each other, as shown in Fig. 1C.

In order to improve the separation of each water-soluble vitamin by cross-axis CPC, three ion-pair reagents were added to the 1-butanol/aqueous 0.15 M monobasic potassium phosphate system. The K values of the vitamins in this solvent system are summarized in Table 2. Most of the K values increased by increasing the concentration and carbon number of ion-pairing reagents added in the two-phase solvent system. Among these ion-pairing reagents, 1-octanesulfonic acid sodium salt was found to be most suitable and was selected for the separation of water-soluble vitamins by cross-axis CPC.

Fig. 2 illustrates the CCC chromatogram of water-soluble vitamins obtained with the above solvent system, which comprise 1-butanol and aqueous 0.15 M monobasic potassium phosphate, including

1.5% of 1-octanesulfonic acid sodium salt. Riboflavin sodium phosphate, cyanocobalamin, riboflavin, pyridoxine hydrochloride, and thiamine nitrate were separated using the lower phase as the mobile phase. Riboflavin was found as an impurity contained in the riboflavin sodium phosphate sample.

CCC SEPARATION OF FAT-SOLUBLE VITAMINS

The K values of fat-soluble vitamins obtained with four different kinds of solvent systems are summarized in Table 3. The most suitable solvent system was found to be the 2,2,4-trimethyl pentane (isooctane)/methanol binary system.

Fig. 3A illustrates the CCC separation of fat-soluble vitamins using the cross-axis CPC equipped with eccentric coil assemblies. Calciferol, vitamin A acetate, and (±)-α-tocopherol acetate were well resolved from each other and eluted within 2.5 h. Vitamin K_3 and

Table 3 Partition coefficients of fat-soluble vitamins in four different two-phase solvent systems

	n-Hexane/aqueous 90% acetonitrile (1:1)	*n*-Hexane/ acetonitrile (1:1)	2,2,4-Trimethyl pentane/methanol (1:1)	*n*-Hexane/ethyl acetate/ methanol/water (1:1:1:1)
Vitamin A acetate	5.34	1.69	1.34	130
Calciferol (M.W. 396.66)	8.14	3.29	0.83	71.2
(±)-α-Tocopherol acetate (M.W. 472.25)	48.5	10.1	3.11	62.6
Vitamin K_1 (M.W. 450.71)	61.6	5.79	3.92	9.23
Vitamin K_3 (M.W. 172.19)	0.31	0.22	0.35	2.40

Partition coefficients were calculated from the absorbance of the upper phase divided by that of lower phase at 280 nm.

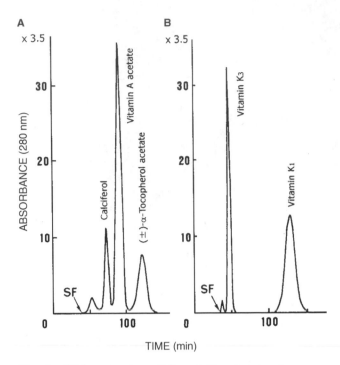

Fig. 3 CCC separation of fat-soluble vitamins by cross-axis CPC. Experimental conditions: sample, (A) calciferol (3 mg) + vitamin A acetate (30 mg) + (±)-α-tocopherol acetate (40 mg) and (B) vitamin K₃ (3 mg) + vitamin K₁ (10 mg); solvent system, 2,2,4-trimethyl pentane/methanol (1:1); mobile phase, lower phase. For other experimental conditions, see Fig. 1 caption. SF = solvent front.

Vitamin K₁ were also completely resolved with the same solvent system, as shown in Fig. 3B.

CONCLUSIONS

As described above, the cross-axis CPC, equipped with eccentric coil assemblies, can be used for the separation of both water-soluble and fat-soluble vitamins by selecting suitable two-phase solvent systems.

REFERENCES

1. Mandava, N.B., Ito, Y., Eds.; *Countercurrent Chromatography: Theory and Practice*; Marcel Dekker, Inc.: New York, 1988; 79.
2. Conway, W.D. *Countercurrent Chromatography: Apparatus, Theory and Applications*; VCH: New York, 1990; 37.
3. Ito, Y., Conway, W.D., Eds.; *High Speed Countercurrent Chromatography*; Wiley-Interscience: New York, 1996; 3.
4. Ito, Y. Cross-axis synchronous flow-through coil planet centrifuge free of rotary seals for preparative counter-current chromatography. Part I: Apparatus and analysis of acceleration. Sep. Sci. Technol. **1987**, *22*, 1971.
5. Ito, Y. Cross-axis synchronous flow-through coil planet centrifuge free of rotary seals for preparative counter-current chromatography. Part II: Studies on phase distribution and partition efficiency in coaxial coils. Sep. Sci. Technol. **1987**, *22*, 1989.
6. Shinomiya, K.; Menet, J.-M.; Fales, H.M.; Ito, Y. Studies on a new cross-axis coil planet centrifuge for performing counter-current chromatography. I. Design of the apparatus, retention of the stationary phase, and efficiency in the separation of proteins with polymer phase systems. J. Chromatogr. **1993**, *644*, 215.
7. Shinomiya, K.; Inokuchi, N.; Gnabre, J.N.; Muto, M.; Kabasawa, Y.; Fales, H.M.; Ito, Y. Countercurrent chromatographic analysis of ovalbumin obtained from various sources using the cross-axis coil planet centrifuge. J. Chromatogr. A. **1996**, *724*, 179.
8. Shinomiya, K.; Muto, M.; Kabasawa, Y.; Fales, H.M.; Ito, Y. Protein separation by improved cross-axis coil planet centrifuge with eccentric coil assemblies. J. Liq. Chromatogr. & Relat. Technol. **1996**, *19*, 415.
9. Shinomiya, K.; Kabasawa, Y.; Ito, Y. Countercurrent chromatographic separation of sugars and their p-nitro-phenyl derivatives by cross-axis coil planet centrifuge of stationary phase. J. Liq. Chromatogr. & Relat. Technol. **1999**, *22*, 579.
10. Shinomiya, K.; Komatsu, T.; Murata, T.; Kabasawa, Y.; Ito, Y. Countercurrent chromatographic separation of vitamins by cross-axis coil planet centrifuge with eccentric coil assemblies. J. Liq. Chromatogr. & Relat. Technol. **2000**, *23*, 1403.
11. Shinomiya, K.; Yoshida, K.; Kabasawa, Y.; Ito, Y. Countercurrent chromatographic separation of water-soluble vitamins by cross-axis coil planet centrifuge using an ion-pair reagent with polar two-phase solvent system. J. Liq. Chromatogr. & Relat. Technol. **2001**, *24*, 2615.
12. Ito, Y.; Oka, H.; Slemp, J. Improved cross-axis synchronous flow-through coil planet centrifuge for performing counter-current chromatography. I. Design of the apparatus and analysis of acceleration. J. Chromatogr. **1989**, *463*, 305.
13. Ito, Y.; Kitazume, E.; Bhatnagar, M.; Trimble, F. Cross-axis synchronous flow-through coil planet centrifuge (Type XLL). I. Design of the apparatus and studies on retention of stationary phase. J. Chromatogr. **1991**, *538*, 59.
14. Shibusawa, Y.; Ito, Y. Countercurrent chromatography of proteins with polyethylene glycol-dextran polymer phase systems using type-XLLL cross-axis coil planet centrifuge. J. Liq. Chromatogr. **1992**, *15*, 2787.

Countercurrent Chromatography/MS

Hisao Oka
Aichi Prefectural Institute of Public Health, Nagoya, Japan

Yoichiro Ito
*National Heart, Lung, and Blood Institute, National Institutes of Health,
Bethesda, Maryland, U.S.A.*

INTRODUCTION

Countercurrent chromatography (CCC) is a unique liquid–liquid partition technique which does not require the use of a solid support,[1–5] hence eliminating various complications associated with conventional LC, such as tailing of solute peaks, adsorptive sample loss and deactivation, contamination, and so forth. Since 1970, the CCC technology has advanced in various directions, including preparative and trace analysis, dual CCC, foam CCC, and, more recently, partition of macromolecules with polymerphase systems. However, most of these methods were only suitable for preparative applications due to relatively long separation times required. In order to fully explore the potential of CCC, efforts have been made to develop analytical high-speed CCC (HSCCC) by designing a miniature multilayer coil planet centrifuge; interfacing analytical HSCCC to a mass spectrometer (HSCCC/MS) began in the late 1980s.

Integration of the high-purity eluate of HSCCC with a low detection limit of MS has led to the identification of a number of natural products, as shown in Table 1.[6–10]

Various HSCCC/MS techniques and their applications are described herein.

INTERFACING HSCCC TO THERMOSPRAY MASS SPECTROMETRY

HSCCC/thermospray (TSP) MS was initiated using an analytical HSCCC apparatus of a 5-cm revolution radius, equipped with a 0.85-mm-inner diameter (i.d.) polytetrafluoroethylene (PTFE) column at 2000 rpm.[6–8] Directly interfacing HSCCC to the MS produced, however, a problem in that the high back-pressure generated by the TSP vaporizer often damaged the HSCCC column. To overcome this problem, an additional HPLC pump was inserted at the interface junction between HSCCC and MS to protect the column against high back-pressures. The effluent from the HSCCC column (0.8 mL/min) was introduced into the HPLC pump through a zero-dead-volume tee

fitted with a reservoir supplying extra solvent or venting excess solvent from the HSCCC system. The effluent from the HPLC pump, after being mixed with $0.3\,M$ ammonium acetate at a rate of $0.3\,mL/min$, was introduced into the TSP interface. This system has been successfully applied to the analyses of alkaloids,[6] triterpenoic acids,[7] and lignans[8] from plant natural products, thereby providing useful structural information. However, a large dead space in the pump at the interface junction adversely affected the resulting chromatogram, as evidenced by loss of a minor peak when HSCCC/UV and HSCCC/TSP–MS total ion current (TIC) chromatograms of plant alkaloids were compared.

In the subsequently developed techniques, the HSCCC effluent is directly introduced into the MS to preserve the peak resolution. Direct HSCCC/MS techniques have many advantages over the HSCCC/TSP method as follows:

1. High enrichment of sample in the ion source
2. High yield of sample reaching the MS
3. No peak broadening
4. High applicability to nonvolatile samples

Various types of HSCCC/MS have been developed using frit fast-atom bombardment (FAB) including continuous flow (CF) FAB, frit electron ionization (EI), frit chemical ionization (CI), TSP, atmospheric pressure chemical ionization (APCI), and electrospray ionization (ESI). Each interface has its specific features. Among those, frit MS and ESI are particularly suitable for directly interfacing to HSCCC, because they generate low back-pressures of approximately $2\,kg/cm^2$, which is only one-tenth of that produced by TSP.

INTERFACING OF HSCCC TO FRIT EI, CI, AND FAB–MS

In our laboratory, separations were conducted by newly developed HSCCC-4000 with a 2.5-cm revolution radius, equipped with a 0.3-mm or 0.55-mm-i.d.

Encyclopedia of Chromatography DOI: 10.1081/E-ECHR-120039938

Table 1 Summary of previously reported HSCCC/MS conditions

Sample	Column	Column capacity (mL)	Revolutional speed (rpm)	Solvent system	Mobile phase	Flow rate (mL/min)	Retention of stationary phase (%)	Ionization	Refs.
Alkaloids	0.85-mm PTFE tube	38	1500	n-Hexane–ethanol–water (6:5:5)	Lower phase	0.7	—	Thermospray	[6]
Triterpenoic acids	0.85-mm PTFE tube	38	1500	n-Hexane–ethanol–water (6:5:5)	Lower phase	0.7	—	Thermospray	[7]
Lignans	0.85-mm PTFE tube	38	1500	n-Hexane–ethanol–water (6:5:5)	Lower phase	0.7	—	Thermospray	[8]
Indole auxins	0.3-mm PTFE tube	7	4000	n-Hexane–ethyl acetate–methanol–water (1:1:1:1)	Lower phase	0.2	27.2	Frit-EI	[9]
Mycinamicins	0.3-mm PTFE tube7	7	4000	n-Hexane–ethyl acetate–methanol–8% ammonia (1:1:1:1)	Lower phase	0.1	40.4	Frit-CI	[9]
Colistins	0.55-mm PTFE tube	6	4000	n-Butanol–0.04M TFA (1:1)	Lower phase	0.16	34.3	Frit-FAB	[9]
Erythromycins	0.85-mm PTFE tube	17	1200	Ethyl acetate–methanol–water (4:7:4:3)	Lower phase	0.8	—	Electrospray	[10]
Didemnins	0.85-mm PTFE tube	17	1200	Ethyl acetate–methanol–water (1:4:1:4)	Lower phase	0.8	—	Electrospray	[10]

multilayer coil at a maximum revolution speed of 4000 rpm.[9] The system produced an excellent partition efficiency at a flow rate ranging between 0.1 and 0.2 mL/min, whereas the suitable flow rate for HSCCC/frit MS is between 1 and 5 μL/min. Therefore, the effluent of the HSCCC column was introduced into the MS through a splitting tee which was adjusted to a split ratio of 1:40 to meet the above requirement. A 0.06-mm-i.d. fused-silica tube was led to the MS and a 0.5-mm-i.d. stainless-steel tube was connected to the HSCCC column. The other side of the fused-silica tube extended deeply into the stainless-steel tube to receive a small portion of the effluent from the HSCCC column, and the rest of the effluent was discarded through a 0.1-mm-i.d. PTFE tube. The split ratio of the effluent depended on the flow rate of the effluent and the length of the 0.1-mm-i.d. tube. For adjusting the split ratio at 40:1, a 2-cm length of the 0.1-mm-i.d. tube was needed. Fig. 1 shows the HSCCC/MS system, including an HPLC pump, sample injection port, HSCCC/4000, the split tee, and mass spectrometer.[9]

In order to demonstrate the potential of HSCCC/frit MS, indole auxins, mycinamicins (macrolide antibiotics), and colistin complex (peptide antibiotics) were analyzed under HSCCC–frit, EI, CI, and FAB–MS conditions.

Three indole auxins, including indole-3-acetamide (IA, MW: 174), indole-3-acetic acid (IAA, MW: 175), and indole-3-butyric acid (IBA, MW: 203) were analyzed under frit EI–MS conditions. In comparison of a TIC with a UV chromatogram, both showed similar chromatographic resolution with excellent theoretical plate numbers ranging from 12,000 to 5500. The results indicate that MS interfacing does not adversely affect chromatographic resolution. In frit EI–MS, the mobile phase behaves like a reagent gas in CI–MS. Both molecular ions and protonated molecules appear in all mass spectra and these data are very useful for the estimation of the molecular weight. Common fragment ions originating from the indole nuclei are found at m/z 116 and 130.

A mixture of mycinamicins was analyzed under HSCCC/frit CI–MS conditions. Mycinamicins consist of six components, mycinamicins I to VI, and isolated mycinamicins IV (MN-IV, MW: 695) and V (MN-V, MW: 711) were used. The structural difference is derived from the hydroxyl group at C-14. These antibiotics were detected under CI conditions, but a reagent gas such as methane, isobutane, or ammonia was not introduced, because the mobile phase behaves like a reagent gas, as described earlier. Both UV and TIC chromatograms showed similar efficiencies, indicating that the MS interfacing does not affect peak resolution, as demonstrated in the analysis of indole auxins. An applicability of this HSCCC/MS system to less volatile compounds was examined under frit FAB–MS conditions.

A peptide antibiotic colistin complex consisting of two major components of colistins A (CL-A, MW: 1168) and B (CL-B, MW: 1154) is difficult to ionize by CI and EI–MS. For HSCCC analysis of these polar compounds, a wider column of 0.55 mm i.d. (instead of 0.3 mm i.d.) was used to achieve satisfactory retention of the stationary phase for a polar n-butanol–trifluoroacetic acid (TFA) solvent system. In addition, for obtaining FAB mass spectra, it is necessary to introduce a sample with an appropriate matrix such as glycerol, thioglycerol, and m-nitrobenzyl alcohol into the FAB–MS ion source. In the present study, glycerol was added as a matrix to the mobile phase at a concentration of 1%. Although a two-phase solvent system containing glycerol was the first trial for a HSCCC study, similarly satisfactory results were obtained in both retention and separation efficiency. Because of the use of a wider column with a viscous n-butanol–aqueous TFA solvent system, the separation was less efficient compared with those obtained from the above two experiments, but the peaks corresponding to CL-A and CL-B were clearly resolved. Mass chromatograms at individual protonated molecules showed symmetrical peaks without a significant loss of peak resolution due to MS interfacing. In all spectra, protonated

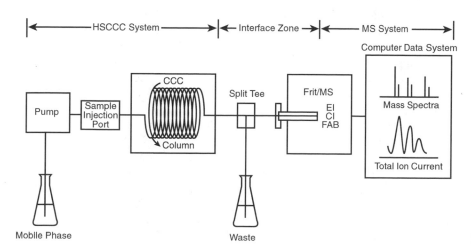

Fig. 1 HSCCC/frit MS system.

molecules appeared well above the chemical noise to indicate the molecular weights. These experiments demonstrated that the present HSCCC/frit MS system including EI, CI, and FAB is very potent and is applicable to various analytes having a broad range of polarity. For a nonvolatile, thermally labile and/or polar compound, HSCCC–frit FAB is most suitable, whereas both HSCCC–frit EI and CI can be effectively used for a relatively nonpolar compound.

INTERFACING HSCCC TO ESI–MS

The experiment was carried out using a small analytical coiled column (17 mL) at 1200 rpm. The effluent from the CCC column at 800 μL/min was split at a 1 : 7 ratio to introduce the smaller portion of the effluent into ESI–MS using a tee adaptor, as described earlier.

The performance of HSCCC–ESI–MS was evaluated by analyzing erythromycins and didemnins.[10] Because erythromycins (macrolide antibiotics) show weak UV absorbance and cannot be detected easily with a conventional UV detector, mass spectrometric detection is a very useful technique for analysis of these antibiotics. A mixture of erythromycin A (Er-A, MW: 733), erythromycin estolate (Er-E, MW: 789), and erythromycin ethyl succinate (Er-S, MW: 789) was analyzed using HSCCC–ESI–MS with a two-phase solvent system composed of n-hexane–ethyl acetate–methanol–water (4 : 7 : 4 : 3). TIC showed, clearly, four peaks corresponding to Er-A, Er-E, Er-S, and an unknown substance. The mass spectra of Er-E and Er-S gave $[M + H]^+$ at m/z 862 and 789 and $[M + H - H_2O]^+$ at m/z 844 and 772, respectively. In the mass spectrum of Er-A, $[M + H - H_2O]^+$ was observed at m/z 761; however, no $[M + H]$ was given. The mass spectrum of the unknown peak indicated that it consists of two components with molecular weights of 843 and 772, which correspond to dehydrated Er-S and Er-E, respectively.

Didemnin A (Did-A, MW: 942) is one of the main components of didemnins (cyclic depsipeptides) and is a precursor for synthesis of other didemnins which exhibit antiviral, antitumor, and immunosuppressive activities. Therefore, its purification is very important in the field of pharmaceutical science. However, large-scale purification of Did-A using conventional LC is difficult due to the presence of nordidemin A (Nordid-A), which contaminates the target fraction. HSCCC–ESI–MS has been successfully applied to the separation and detection of didemnins. Three peaks were observed on TIC corresponding to didemnins A and B and nordidemnin A. Their mass spectra gave only protonated molecules without fragmentation. The first eluted peak was didemnin B, which gave

$[M + H]^+$ and $[M + Na]^+$ at m/z 1112 and 1134, respectively. Did-A appeared as the second peak with $[M + H]^+$ at m/z 943 and $[M + Na]^+$ at m/z 965. The third peak was Nordid-A, showing $[M + H]^+$ at m/z 929 and $[M + Na]^+$ at m/z 951. The results indicated that Did-A can be isolated by HSCCC.

FUTURE PROSPECTS

HSCCC/MS has many desirable features for performing the separation and identification of natural and synthetic products, because it eliminates various complications arising from the use of solid support and offers a powerful identification capacity of MS with its low detection limit. We believe that the combination of these two methods, HSCCC–MS, has great a potential for screening, identification, and structural characterization of natural products and will contribute to a rapid advance in natural products chemistry.

REFERENCES

1. Mandava, N.B., Ito, Y., Eds.; *Countercurrent Chromatography: Theory and Practice*; Marcel Dekker, Inc.: New York, 1988.
2. Conway, W.D. *Countercurrent Chromatography: Apparatus, Theory and Applications*; VCH: New York, 1990.
3. Foucault, A., Ed.; *Centrifugal Partition Chromatography*; Marcel Dekker, Inc.: New York, 1995.
4. Ito, Y. CRC Crit. Rev. Anal. Chem. **1986**, *17*, 65.
5. Ito, Y., Conway, W.D., Eds.; *High-peed Countercurrent Chromatography*; Wiley–Interscience: New York, 1996.
6. Lee, Y.-W.; Voyksner, R.D.; Fang, Q.-C.; Cook, C.E.; Ito, Y. J. Liquid Chromatogr. **1988**, *11*, 153.
7. Lee, Y.-W.; Pack, T.W.; Voyksner, R.D.; Fang, Q.-C.; Ito, Y. J. Liquid Chromatogr. **1990**, *13*, 2389.
8. Lee, Y.-W.; Voyksner, R.D.; Pack, T.W.; Cook, E.; Fang, Q.-C.; Ito, Y. Application of countercurrent chromatography/thermospray mass spectrometry for the identification of bioactive lignans from plant natural products. Anal. Chem. **1990**, *62*, 244–248.
9. Oka, H.; Ikai, Y.; Kawamura, N.; Hayakawa, J.; Harada, K.-I.; Murata, H.; Suzuki, M. Direct interfacing of high-speed countercurrent chromatography to frit electron ionization, chemical ionization, and fast atom bombardment mass spectrometry. Anal. Chem. **1991**, *63*, 2861–2865.
10. Kong, Z.; Rinehart, K.L.; Milberg, R.M.; Conway, W.D. J. Liquid Chromatogr. **1998**, *21*, 65.

Countercurrent Chromatography Solvent Systems

T. Maryutina
Boris Ya. Spivakov
*Vernadsky Institute of Geochemistry and Analytical Chemistry, Russian Academy of Sciences,
Moscow, Russia*

INTRODUCTION

Countercurrent chromatography has been mainly developed and used for preparative and analytical separations of organic and bio-organic substances.[1] The studies of the last several years have shown that the technique can be applied to analytical and radio-chemical separation, preconcentration, and purifica-tion of inorganic substances in solutions on a laboratory scale by the use of various two-phase liquid systems.[2] Success in CCC separation depends on choosing a two-phase solvent system that provides the proper partition coefficient values for the com-pounds to be separated and satisfactory retention of the stationary phase. The number of potentially suita-ble CCC solvent systems can be so great that it may be difficult to select the most proper one.

DISCUSSION

Recent studies have made it possible to classify water-organic solvent systems in CCC for separation of organic substances on the basis of the liquid-phase density difference, the solvent polarity, and other para-meters from the point of view of stationary-phase retention in a CCC column.[1,3–9] Ito[1] classified some liquid systems as hydrophobic (such as heptane–water or chloroform–water), intermediate (chloroform–acetic acid–water and n-butanol–water) and hydrophi-lic (such as n-butanol–acetic acid–water) according to the hydrophobicity of the nonaqueous phase. Thirteen two-phase solvent systems were evaluated for relative polarity by using Reichardt's dye to measure solva-chromatic shifts and using the solubility of index compounds.[6]

However, the systems for inorganic separations are very different from those for organic separations, as, in most cases, they contain a complexing (extracting) reagent (ligand) in the organic phase and mineral salts and/or acids or bases in the aqueous phase. Thus, the complexation process, its rate, and the masstransfer rate can play a significant role in the separation pro-cess.[9] There are three important criteria for choosing a two-phase liquid system.

First, the systems must be composed of two immiscible phases. Each solvent mixture should be thoroughly equilibrated in a separatory funnel at room temperature and the two phases separated after the clear two phases have been formed. When the nature of the organic sample to be separated is known, one may find a suitable solvent system by searching the lit-erature for solvent systems that have been successfully applied to similar compounds.[1,3–8] In the case of organic-aqueous two-phase systems, the organic phase consists of one solvent or of a mixture of different sol-vents. Various nonaqueous–nonaqueous two-phase solvent systems have been used for separation of non-polar compounds and/or compounds that are unstable in aqueous solutions. Separation of macro-molecules and cell particles can be performed with a variety of aqueous–aqueous polymer-phase systems. Among the various polymer-phase systems available, the following two types are the most versatile for performing CCC.[1,8] Poly (ethylene glycol) (PEG)–potassium phosphate systems provide a convenient means of adjusting the partition coefficient of macro-molecules by changing the molecular weight of PEG and/or the pH of the phosphate buffer. The PEG 6000–Dextran 500 systems provide a physiological environment, suitable for separation of mammalian cells by optimizing osmolarity and pH with electro-lytes.

For preconcentration and separation of inorganic species, a stationary phase containing extracting reagents of different types (cation-exchange, anion-exchange, and neutral) in an organic solvent should be usually applied.[2,9–12] The mobile-phase compo-nents should not interfere with the subsequent analysis. Solutions of inorganic acids and their salts are most often used. The mobile phase may also contain specific complexing agents, which can bind one or several elements under separation.

Second, one of the phases (stationary one) must be retained in the rotating column to a required extent. The most important factor, which determines the separation efficiency and peak resolution for both organic and inorganic compounds, is the ratio of the stationary-phase volume retained in a column to the total column volume. The volume of the stationary

Encyclopedia of Chromatography DOI: 10.1081/E-ECHR-120039911

401

phase retained in the column depends on various factors, such as the physical properties of the two-phase solvent system, flow rate of the mobile phase, and applied centrifugal force field. In droplet CCC, where the separation is performed in a stationary column, a large density difference between the stationary solvent phases becomes the predominant factor for the retention of the stationary phase. In other CCC schemes, various types of two-phase solvent systems can be used under optimized experimental conditions. The influence of planetary centrifuge parameters and operation conditions on the stationary-phase retention have been well studied for some simple two-phase liquid systems consisting of water and one or two organic solvents.[1,3–8]

According to Ito's classifications,[1,3] hydrophobic organic phases are easily retained by all types of CCC apparatus. Intermediate solvent systems involve a more hydrophilic organic phase. Their tendency to evolve, after mixing, to a more stable emulsion than the hydrophobic systems decreases the retention of stationary phase. The hydrophilic two-phase systems containing a polar phase are even less retained in the column.

However, the addition of extracting reagents and mineral salts to a two-phase system (in case of inorganic separations) can strongly affect the physicochemical properties of liquid systems and, consequently, their hydrodynamic behavior and S_f value. Varying concentrations of the system constituents used for inorganic separation allows selective changing of a certain physicochemical parameter (interfacial tension γ, density difference between two liquid phases $\Delta\rho$ and viscosity of the organic stationary phase η_{org}). The type of the solvent may often have a great effect on the stationary-phase retention and, consequently, on the chromatographic process. The correlations between the physicochemical parameters of the complex liquid systems under investigation and their behavior in coiled columns are described in detail.[10] The composition and physicochemical properties of the organic phase in inorganic analysis were modified by adding an extracting reagent [e.g., di-2-ethylhexylphosphoric acid (D2EHPA), tri-n-butyl phosphate, trioctylamine].[2,10] The density and viscosity of the organic phase were varied by changing the amount of reagents in the stationary phase. For example, a small addition (5%) of D2EHPA in an organic solvent (n-decane, n-hexane, chloroform, and carbon tetrachloride) leads to a considerable increase in the factor in the organic solvent—$(NH_4)_2SO_4$—water systems (from 0 to 0.73 in the case of carbon tetrachloride).[10]

Third, the stationary phase should permit separate elution of the substances into the mobile phase and the selectivity toward samples of interest has to be sufficient to lead to separations with good resolution. The selectivity of solvent systems can be estimated by determination of the partition coefficients for each substance. The batch partition coefficients D^{bat} are calculated as the ratio of the component concentration in the organic phase to that in the aqueous phase. The dynamic partition coefficients of compounds are determined from an experimental elution curve.[7] Several solvent systems for organic separation were investigated.[4–8] The most efficient evolution usually occurs when the value of the partition coefficient is equal 1. However, in some CCC schemes, the best results are obtained with lower partition coefficient values of 0.3–0.5.[1,4]

CONCLUSIONS

In inorganic analysis with the use of CCC, the stationary phase should provide preconcentration of the elements to be determined, if necessary. It should be noted that the element elution depends on the operation conditions for the planetary centrifuge, which influence the quantity of the stationary phase in the column. A chromatographic peak shifts to left and narrows if the volume of the stationary phase lowers (all the other factors being the same).[2] The reagent concentration in the organic solvent also affects the elution curve shape and, therefore, the dynamic partition coefficient values. An increase of the reagent concentration in the organic phase leads to higher partition coefficients for the elements, and a better separation is achieved. However, a rather large volume of the mobile phase can be required for the elution of elements from the column.

The composition of the mobile phase also has an influence on the partition coefficients of inorganic substances and the separation efficiency. Concentrations of the mobile-phase constituents should provide partition coefficient values needed for the enrichment or separation of components under investigation. If a step-elution mode is used, partition coefficients higher than 10 and less than 0.1 are favorable for the enrichment of components into the stationary phase and their recovery into the mobile phase, respectively. Chemical kinetics factors may also play an important role in the separation of inorganic species by CCC.[9] It has been shown that the values of mass-transfer coefficients determine the type of elution (isocratic or step), which is necessary for the element separation. The data on batch extraction (mass-transfer coefficients and partition coefficients) and parameters of chromatographic peaks (half-widths) can be interrelated by some empirical expressions.[9] The application of CCC in inorganic analysis looks promising because various two_phase liquid systems, providing the separation of a variety of inorganic species, may be used for the separation of trace elements.

REFERENCES

1. Ito, Y. *Countercurrent Chromatography. Theory and Practice*; Mandava, N.B., Ito, Y., Eds.; Marcel Dekker, Inc.: New York, 1988.
2. Ya. Spivakov, B.; Maryutina, T.A.; Fedotov, P.S.; Ignatova, S.N. *Metal-Ion Separation and Preconcentration: Progress and Opportunities*; Bond, A.N., Dietz, M.L., Rodgers, R.D., Eds.; American Chemical Society: Washington, DC, 1999; 333–347.
3. Conway, W.D. *Countercurrent Chromatography. Apparatus, Theory and Applications*; VCH: New York, 1990.
4. Berthod, A.; Schmitt, N. Waterorganic solvent systems in countercurrent chromatography: Liquid stationary phase retention and solvent polarity. Talanta **1993**, *40*, 1489
5. Menet, J.-M.; Thiebaut, D.; Rosset, R.; Wesfreid, J.E.; Martin, M. Anal. Chem. **1994**, *66*, 168.
6. Abbott, T.P.; Kleiman, R. Solvent selection guide for counter-current chromatography. J. Chromatogr. **1991**, *538*, 109.
7. Drogue, S.; Rolet, M.-C.; Thiebaut, D.; Rosset, R. Separation of pristinamycins by high-speed counter-current chromatography I. Selection of solvent system and preliminary preparative studies. J. Chromatogr. **1992**, *593*, 363.
8. Foucault, A.P.; Chevolot, L. Counter-current chromatography: instrumentation, solvent selection and some recent applications to natural product purification. J. Chromatogr. A. **1998**, *808*, 3.
9. Fedotov, P.S.; Maryutina, T.A.; Pichugin, A.A.; Spivakov, B.Ya. Russ. J. Inorg. Chem. **1993**, *38*, 1878.
10. Matyutina, T.A.; Ignatova, S.N.; Fedotov, P.S.; Spivakov, B.Ya.; Thiebaut, D. J. Liquid Chromatogr. & Related Technol. **1998**, *21*, 19.
11. Kitazume, E.; Bhatnagar, M.; Ito, Y. Separation of rare earth elements by high-speed counter-current chromatography. J. Chromatogr. **1991**, *538*, 133.
12. Zolotov, Yu.A.; Spivakov, B.Ya.; Maryutina, T.A.; Bashlov, V.L.; Pavlenko, I.V. Fresenius Z. Anal. Chem. **1989**, *35*, 938.

Creatinine and Purine Derivatives: Analysis by HPLC

M. J. Arin
M. T. Diez
P. Garcia-del Moral
Departamento de Bioquímica y Biología Molecular, Facultad de Ciencias Biológicas y Ambientales, Universidad de León, León, Spain

J. A. Resines
Departamento de Física, Química y Expresión Gráfica, Facultad de Ciencias Biológicas y Ambientales, Universidad de León, León, Spain

INTRODUCTION

Several compounds, such as creatinine and purine derivatives (allantoin, uric acid, hypoxanthine, and xanthine) present in biological samples, are important analytes for diagnoses of certain types of metabolic diseases and can serve as markers for these processes. Analyses for such substances are crucial for the diagnosis and monitoring of renal diseases, metabolic disorders, and various types of tumorigenic activity.

OVERVIEW

Creatinine results from the irreversible, nonenzymatic dehydration and loss of phosphate from phospho-creatine (Fig. 1). Creatinine is used as an indicator of skeletal muscle mass because it is a by-product of the creatine kinase reaction and it is one of the most widely used clinical markers to assess renal function. Urine levels of creatinine are good indicators of the glomerular filtration rate of the kidneys (i.e., the amount of fluid filtered per unit time).

Allantoin is the catabolic end product of purines in most mammals. It is formed by the action of the enzyme uricase on urate. Humans and other primates lack uricase and excrete urate as the final product of purine metabolism. However, small amounts of allantoin are present in human serum. Some authors demonstrated that free radical attack on urate generates allantoin. Therefore small amounts of allantoin detected in human serum may provide a marker of free radical activity in vivo.

Uric acid is the major product of catabolism of purine nucleosides adenosine and guanosine. Hypoxanthine and xanthine are intermediates along this pathway (Fig. 2). Under normal conditions, they reflect the balance between the synthesis and breakdown of nucleotides. Levels of these compounds change in various situations (e.g., they decrease in experimental

tumors) when synthesis prevails over catabolism, and are enhanced during oxidative stress and hypoxia. Uric acid serves as a marker for tubular reabsorption of nephrons, in addition to glomerular filtration rate. An imbalance in uric acid level can lead to gout—the formation of urate crystals in joints. Uric acid and xanthines are also markers for metabolic disorders such as Lesch–Nyhan syndrome and xanthinuria, to indicate a few.

However, creatinine and purine derivatives are very important in the field of animal nutrition because measurements of urinary excretion of these compounds have been proposed as an internal marker for microbial protein synthesis.

SAMPLE PREPARATION

Determination of these compounds is carried out frequently in biological fluids. Analysis in urine requires a previous filtration to remove cells and other particulate matter; then, the samples are diluted and directly injected onto the column. With cerebrospinal fluid, the samples are obtained by lumbar puncture; each aliquot is centrifuged and decanted before analysis. Often in plasma or serum, some form of protein removal is needed because the presence of these compounds in injected samples can cause modifications of the column and bias in chromatographic results. Protein removal can be performed by various methods such as protein precipitation, ultrafiltration, centrifugation, liquid-phase or solid-phase extraction, and column-switching techniques.

ANALYSIS OF CREATININE

For determination of creatinine, the nonspecific Jaffé method, although subject to perturbation by many interfering substances of endogenous and exogenous

Encyclopedia of Chromatography DOI: 10.1081/E-ECHR-120041122

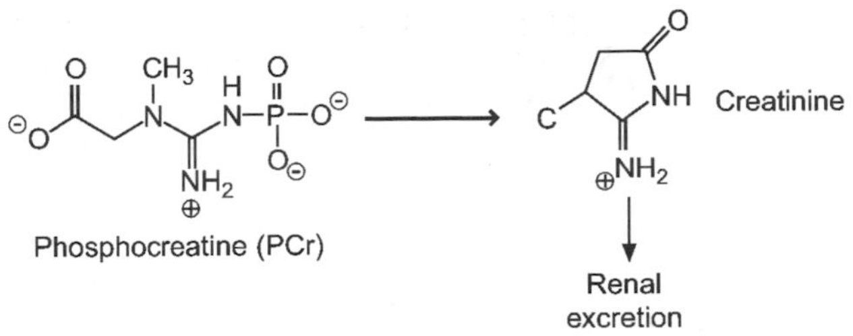

Fig. 1 Degradation of purine nucleotides and formation of purine derivatives.

origin, is the most widely used. However, a batchwise kinetic procedure and flow injection analysis have shown the possibility to determine creatinine in human urine samples by this reaction, free from any systematic error.[1] Enzymatic assays have higher specificities, but still suffer from interferences. To avoid these problems, new analytical methods were developed: potentiometry, capillary electrophoresis (CE), and isotope dilution–mass spectrometry (ID-GC/MS),[2] with the latter being proposed as a reference method.

Chromatographic techniques have been very useful for clinical analysis, with advantages of simultaneous measurements of different components and elimination of interfering species. Previous reviews have been realized for the determination of creatinine.[3] HPLC methods include ion exchange chromatography, reversed-phase (RP) chromatography, ion pair chromatography, and MEKC, and more complicated column-switching and tandem methods[4] have been described as well.

Urinary creatinine can be analyzed by HPLC using a variety of columns. Detection methods include absorption, fluorescence after postcolumn derivatization, MS,

and some other methods. Review of recent literature reveals that the method of choice for the measurement of creatinine has been RP-HPLC. C_8 and C_{18} columns, and UV or electrochemical detection (ED) with isocratic elution or gradient elution were the most commonly used.[5,6]

In most cases, RP ion pairing HPLC with UV photometric detection was used. The advantage of this technique is the broad range of parameters that may be conveniently adjusted to optimize the separation method; they include the concentration of organic modifier in the mobile phase, the type and concentration of buffer in the mobile phase, and the type and concentration of the counterion.[7] In addition to ion exchange methods, some authors have developed a procedure for the determination of creatinine and a wide range of amino acids that provides for fixed-site ion exchangers and eliminates the addition of ion-pairing agent to the mobile phase.[8] Creatinine has been analyzed in sera and tissues using HPLC and CE methods. Many of these determinations could also be applied to urinary creatinine analysis.

Various papers related to the simultaneous determination of creatinine and uric acid can be found in the literature. Several authors have developed capillary zone electrophoresis (CZE) methods for simultaneous analysis of these compounds in urine. The CE analysis of these renal markers offers some advantages when compared with chromatography, such as shortened separation time, reduced reagent consumption, and increased resolution. MEKC has been applied to the simultaneous separation of creatinine and uric acid in human plasma and urine. However, chromatographic techniques are widely accepted for the determination of these compounds. RP and ion pair HPLC methods have been applied for the simultaneous determination

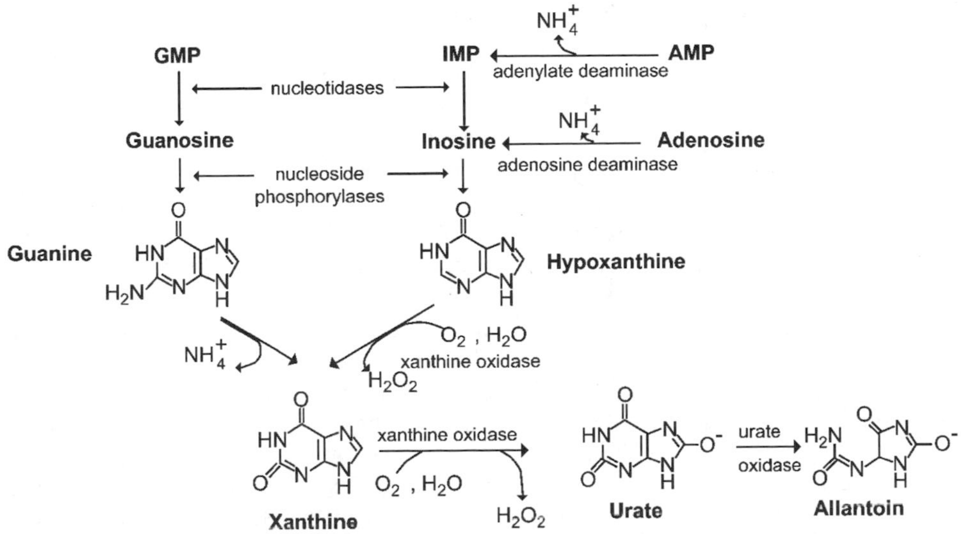

Fig. 2 Phosphocreatine metabolism.

of these compounds in sera.[9,10] These methods were consistent with the ID-GC-MS reference method.

PURINE DERIVATIVES: ALLANTOIN, URIC ACID, XANTHINE, AND HYPOXANTHINE

Traditionally, oxypurines allantoin, uric acid, and, in some cases, xanthines have been analyzed in biofluids by colorimetric methods.

The analysis of allantoin is based on the Rimini–Schriver reaction, in which allantoin is converted to glyoxylic acid by sequential hydrolysis under alkaline and acidic conditions, and then derivatizated with 2,4-dinitrophenylhydrazine to obtain the chromophore glyoxylate-2,4-dinitrophenylhydrazone.

The two predominant analyses of uric acid are the phosphotungstic acid (PTA) method and the uricase method. In the PTA method, urate reduces PTA to form a blue product. In the uricase analysis method, urate is oxidized by uricase oxidoreductase.

Xanthine and hypoxanthine have been often quantified colorimetrically or as uric acid following enzymatic conversion. Both approaches are problematic due to interference by compounds contained in biological fluids, whereas enzymatic conversion to uric acid has been often incomplete.

These photometric and enzymatic methods suffer from interferences by endogenous and exogenous compounds and can lead to inaccurate results. To avoid these problems, in recent years, CE, GC/MS, and HPLC methods have been developed. RP, normal phase, ion exchange, ion pair, column-switching, MEKC, and SEC were used for purine derivative determination.

Different methods, mostly colorimetric and chromatographic, for the determination of allantoin have been reviewed by Chen.[11] The chromatographic procedures are mainly based on separation by HPLC using C_{18} RP columns and monitoring at wavelengths around 200 nm. Many compounds present in plasma and urine also exhibit absorbance at these wavelengths; therefore, the detection of allantoin is not selective enough when the concentration is low (<60 μmol/L).[11] To avoid this problem, allantoin can be converted, prior to elution, to a derivative, which can be monitored at a more specific detection wavelength. However, allantoin is an extremely polar compound that has poor retention on C_{18} RP columns. To achieve a good separation from other polar compounds, the column length often needs to be extended (i.e., using long columns or two columns connected in series).[12] Another way to overcome the problem is the use of ion-pairing reagents to slow down the elution.[11]

For uric acid, ID-GC/MS[2] has been proposed as candidate reference method. RP methods have been

most widely employed. RP C_8 and C_{18} columns, ranging from 150 to 250 mm in length and usually with an internal diameter of 3.9–4.6 mm, were used. Both isocratic elution and gradient elution were applied. Variable wavelength UV or diode array detectors are the most commonly used, in the range 210–292 nm. Electrochemical and online combination of UV with ED detection improves the selectivity and sensitivity of analysis and decreases the probability that interfering substances are present in analyte peaks. Some methods make use of ion exchange HPLC for the determination of uric acid and other organic compounds. In most cases, RP ion pairing HPLC methods were used to determine uric acid in biological samples. One of these[13] has been proposed as a candidate reference method for the determination of uric acid. Data obtained by this method were compared with those from ID-GC/MS using [1,3-^{15}N$_2$] uric acid as internal standard.

The main advantage of these methods is that they allow direct determination of urine samples, whereas plasma samples only need deproteinization; in addition, they offer the possibility to simultaneously determine other purine derivatives.

Various methods have been proposed for the simultaneous determination of uric acid and allantoin in different biological matrices. For urine samples, GC/MS was applied for the determination of both compounds.[14] However, HPLC methods are still widely accepted.

Several HPLC methods have been reported for the determination of xanthine and hypoxanthine. Different RP-HPLC methods, using gradient elution and UV detection, have been described to determine these metabolites and other methylated purines.[15] To improve xanthine determination in urine samples, analyses were carried out with two columns—RP column and anion exchange column—connected by a column switch.[16] With this method, urinary hypoxanthine and xanthine can be measured without any sample preparation other than filtration.

Simultaneous determination of uric acid, hypoxanthine, and xanthine in different biological matrices such as urine, urinary calculi, cerebrospinal fluid, and plasma was carried out by RP-HPLC with isocratic elution and UV detection.[17,18]

Various analytical methods for the simultaneous determination of allantoin, uric acid, and xanthine in biological samples, such as urine, blood plasma, and serum, have been described. These procedures are mainly based on separation by HPLC using RP C_{18} columns, UV detection, and isocratic elution or gradient elution. In some cases, allantoin was converted to a derivative with a chomophoric group, but other authors avoid the disadvantages of the allantoin derivatization process.[19] The HPLC and CE methods have been compared for the determination of these compounds on plasma and atherosclerotic plaque. Comparison of results showed that CZE may have similar, or even superior, analytical performance to HPLC, especially for the determination of allantoin in biological samples.[20]

SIMULTANEOUS DETERMINATION OF CREATININE AND PURINE DERIVATIVES

Several papers can be found in the literature concerning the simultaneous determination of creatinine and uric acid or various purine metabolites; however, only a few have reported the simultaneous determination of creatinine and purine derivatives.

Kochansky and Strein[21] reviewed recent developments in chromatography and CE for the determination of creatinine, uric acid, and xanthine in biological fluids.

RP-HPLC procedures for the determination of creatinine and purine metabolites, such as allantoin, uric acid, xanthine, and hypoxanthine in ruminant urine, were described. Chromatography was achieved with a C_{18} column under isocratic conditions, and detection at 218 nm without allantoin derivatization. The chromatographic conditions were a compromise between the sensitivity and specificity of the measurement of each analyte, analysis time, and resolution of all analyte peaks from interfering compounds.[22,23] Uremic toxins creatine, creatinine, uric acid, and xanthine were simultaneously determined in human biofluids, simply after dilution, with UV detection at 200 nm. This method was compared, for creatinine and uric acid, with conventional routine methods and did not give significantly different results.[24]

An ion pair HPLC method for the determination of creatinine and purine derivatives (allantoin, uric acid, and hypoxanthine in sheep urine) using allopurinol as internal standard is described.[25] In this work, various variables were tested to optimize the simultaneous determination of these compounds, including alkyl chain length of the ion-pairing agent (C_6, C_8, and C_{10}), buffer concentration, pH, percentage of methanol of the mobile phase, and column temperature. The mobile phase composition was 10 mM phosphate buffer with 3 mM 1-octanesulfonic acid, sodium salt, pH 4, mixed with methanol:eluent A (5%) and eluent B (20%). The gradient program was 0–13 min, 0–100% B, flow rate 0.5 mL/min; 13–25 min, 100–0% B, flow rate 1.5 mL/min; reequilibration time at 0% B, 10 min. The injection volume was 20 μL. The column temperature was set at 30°C. The chromatographic conditions adopted represented a compromise between good separation and reasonable analysis time. Fig. 3 shows the chromatograms resulting from the injection

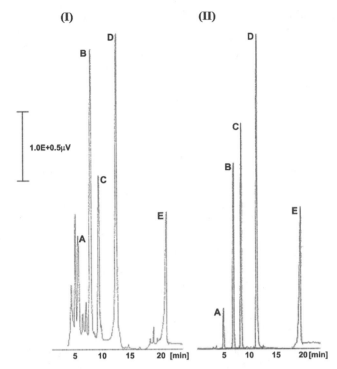

Fig. 3 (I) Chromatogram of 10-fold diluted sheep urine. (II) Chromatographic separation of standard solutions. Peaks: A = allantoin; B = uric acid; C = hypoxanthine; D = allopurinol (IS); and E = creatinine. (From Ref.[25].)

of pure standards and 10-fold diluted sheep urine sample under adopted chromatographic conditions. Under these conditions, there were no other endogenous urinary components that can interfere with the analyte peaks.

CONCLUSIONS

HPLC, when compared to other instrumental methods, presents significant advantages for the simultaneous analysis of creatinine and purine derivatives. The variety of instrumental and experimental conditions (columns, buffers, organic modifiers, detectors, etc.) of these methods reported in the literature offers versatility and flexibility. Chromatographic conditions for these analytes are not complicated when RP columns are employed. New stationary phases with high separation power provide short analysis times. The mobile phases used are also very simple ones (organic–water mixtures with controlled pH); both isocratic elution and gradient elution are recommended. Different sensitivity detectors (UV, electrochemical, fluorescence, and combined techniques such as HPLC/MS) are very valuable for the identification of all analyzed compounds. In some cases, only CE shows some advantages over HPLC, such as short analysis time, reduced reagent consumption, and increased resolution. However, detection limits are often inferior when using UV absorbance detectors.

REFERENCES

1. Campins, P.; Tortajada, L.A.; Meseger, S.; Blasco, F.; Sevillano, A.; Molins, C. Creatinine determination in urine simples by batchwise kinetic procedure and for injection analysis using the Jaffé reaction: Chemometric study. Talanta **2001**, *55*, 1079–1089.
2. Thienpont, L.M.; Van Nieuwenhove, B.; Reinuer, H.; De Leenheer, A.P. Determination of reference method values by isotope dilution–gas chromatography–mass spectrometry: A five years' experience of two European reference laboratories. Eur. J. Clin. Chem. Clin. Biochem. **1996**, *34*, 853–860.
3. Smith-Palmer, T. Separation methods applicable to urinary creatine and creatinine. J. Chromatogr. B. **2002**, *781*, 93–106.
4. Stokes, P.; O'Connor, G. Development of a liquid chromatography–mass spectrometry method for the high-accuracy determination of creatinine in serum. J. Chromatogr. B. **2003**, *794*, 125–136.
5. Hewavitharana, A.K.; Bruce, H.L. Simultaneous determination of creatinine and pseudouridine concentrations in bovine plasma by reversed-phase liquid chromatography with photodiode array detection. J. Chromatogr. B. **2003**, *784*, 275–281.
6. Mo, Y.; Dobberpuhl, D.; Dash, A.K. A simple HPLC method with pulsed EC detection for the analysis of creatine. J. Pharm. Biomed. Anal. **2003**, *32*, 125–132.
7. Resines, J.A.; Arín, M.J.; Díez, M.T.; García del Moral, P. Ion-pair reversed-phase HPLC determination of creatinine in urine. J. Liq. Chromatogr. & Relat. Technol. **1999**, *22* (16), 2503–2510.
8. Yokoyama, Y.; Horikoshi, S.; Takahashi, T.; Sato, H. Low-capacity cation-exchange chromatography of ultraviolet-absorbing urinary basic metabolites using a reversed-phase column coated with hexadecylsulfonate. J. Chromatogr. A. **2000**, *886*, 297–302.
9. Werner, G.; Schneider, V.; Emmert, J. Simultaneous determination of creatine, uric acid and creatinine by high-performance liquid chromatography with direct serum injection and multi-wavelength detection. J. Chromatogr. **1990**, *525*, 265–275.
10. Kock, R.; Seitz, S.; Delvoux, B.; Greiling, H. A method for the simultaneous determination of creatinine and uric acid in serum by high-performance-liquid-chromatography evaluated versus reference methods. Eur. J. Clin. Chem. Clin. Biochem. **1995**, *33*, 23–29.
11. Chen, X.B. Determination of allantoin in biological, cosmetic and pharmaceutical samples. J. AOAC Int. **1996**, *79* (3), 628–635.
12. Balcells, J.A.; Guada, J.M.; Peiro, J.A.; Parker, D.S. Simultaneous determination of allantoin and oxypurines in biological fluids by high-performance liquid chromatography. J. Chromatogr. **1992**, *575*, 153–157.
13. Kock, R.; Delvoux, B.; Tillmanns, U.; Greiling, H. A candidate reference method for the determination of uric acid in serum based on high-performance liquid chromatography, compared with an isotope dilution–gas chromatography–mass spectrometer method. J. Clin. Chem. Clin. Biochem. **1989**, *27*, 157–162.
14. Chen, X.B.; Calder, A.F.; Prasitkusol, P.; Kyle, D.L.; Jayasuriya, M.C.N. Determination of ^{15}N isotopic enrichment and concentrations of allantoin and uric acid in urine by gas chromatography/mass spectrometry. J. Mass Spectrom. **1998**, *33*, 130–137.
15. Di Pietro, M.C.; Vannoni, D.; Leoncini, R.; Liso, G.; Guerranti, R.; Marinello, E. Determination of urinary methylated purine pattern by high-performance liquid chromatography. J. Chromatogr. B. **2001**, *751*, 87–92.
16. Sumi, S.; Kidouchi, K.; Ohba, S.; Wada, Y. Automated determination of hypoxanthine and xanthine in urine by high-performance liquid chromatography with column switching. J. Chromatogr. B. **1995**, *670*, 376–378.
17. Safranow, K.; Zygmunt, M.; Ciechanowski, K. Analysis of purines in urinary calculi by high-performance liquid chromatography. Anal. Biochem. **2000**, *286*, 224–230.
18. Kuracka, L.; Kalnovicova, T.; Liska, B.; Turcani, P. HPLC method for measurement of purine nucleotide degradation products in cerebrospinal fluid. Clin. Chem. **1996**, *42* (5), 756–760.
19. Czauderna, M.; Kowalczyk, J. Quantification of allantoin, uric acid, xanthine and hypoxanthine in ovine urine by high-performance liquid chromatography and photodiode array detection. J. Chromatogr. B. **2000**, *744*, 129–138.
20. Terzuoli, L.; Porcelli, B.; Setacci, C.; Giubbolini, M.; Cinci, G.; Carlucci, F.; Pagani, R.; Marinello, E. Comparative determination of purine compounds in carotid plaque by capillary zone electrophoresis and high-performance liquid chromatography. J. Chromatogr. B. **1999**, *728*, 185–192.

21. Kochansky, C.J.; Strein, T.G. Determination of uremic toxins in biofluids: Creatinine, creatine, uric acid and xanthines. J. Chromatogr. B. **2000**, *747*, 217–227.

22. Resines, J.A.; Arín, M.J.; Díez, M.T. Determination of creatinine and purine derivatives in ruminant's urine by reversed-phase high-perfomance liquid chromatography. J. Chromatogr. **1992**, *607*, 199–202.

23. Shingfield, K.J.; Offer, N.W. Simultaneous determination of purine metabolites, creatinine and pseudouridine in ruminant urine by reversed-phase high-performance liquid chromatography. J. Chromatogr. B. **1999**, *723*, 81–94.

24. Samanidou, V.F.; Metaxa, A.S.; Papadoyannis, I.N. Direct simultaneous determination of uremic toxins: Creatine, creatinine, uric acid and xanthine in human biofluids by HPLC. J. Liq. Chromatogr. & Relat. Technol. **2002**, *25* (1), 43–57.

25. Del Moral, P.; Díez, M.T.; Resines, J.A.; Bravo, I.G.; Arín, M.J. Simultaneous measurements of creatinine and purine derivatives in ruminant's urine using ion-pair HPLC. J. Liq. Chromatogr. & Relat. Technol. **2003**, *26* (17), 2961–2968.

C

Cross-Axis Coil Planet Centrifuge for the Separation of Proteins

Yoichi Shibusawa
Division of Structural Biology and Analytical Science, School of Pharmacy, Tokyo University of Pharmacy and Life Science, Tokyo, Japan

Yoichiro Ito
Laboratory of Biophysical Chemistry, National Heart, Lung, and Blood Institute, National Institutes of Health, Bethesda, Maryland, U.S.A.

INTRODUCTION

Countercurrent chromatography (CCC) is a form of support-free liquid–liquid partition chromatography in which the stationary phase is retained in the column with the aid of the Earth's gravity or a centrifugal force.[1] Partition of biological samples such as proteins, nucleic acids, and cells has been carried out using various aqueous polymer-phase systems.[2] Among many existing polymer-phase systems, poly(ethylene glycol)–dextran (PEG–dextran) and PEG–phosphate systems have been most commonly used for the partition of biological samples. Whereas these polymer-phase systems provide an ideal environment for biopolymers and live cells, high viscosity and low interfacial tension between the two phases tend to cause a detrimental loss of stationary phase from the column in the standard high-speed CCC centrifuge system (known as type J).

The cross-axis coil planet centrifuge (CPC), with column holders at the off-center position on the rotary shaft, enables retention of the stationary phase of aqueous–aqueous polymer-phase systems such as PEG 1000–potassium phosphate and PEG 8000–dextran T500. Since the last decade, various types of cross-axis CPC (types XL, XLL, XLLL, and L) have been developed for performing CCC with highly viscous aqueous polymer-phase systems.[3,4] The separation and purification of protein samples, including lactic acid dehydrogenase (LDH) from bovine heart crude extract,[5] single stranded DNA binding protein (SSB) from *Escherichia coli* cell lysate,[6] glucosyltransferase from *Streptococcus mutans* cell lysate[7] and *S. sobrinus* cell culture medium,[8] and so on, were achieved using these cross-axis CPCs.[9]

APPARATUS

The cross-axis CPCs, which include types X and L and their hybrids, are mainly used for protein separations. These modified versions of the high-speed CCC centrifuge have a unique feature among the CPC systems in that the system provides reliable retention of the stationary phase for viscous polymer-phase systems. Fig. 1 presents a photograph of the latest type XL CPC unit and schematically illustrates the orientation and motion of the coil holder in the cross-axis CPC system, where R is the radius of revolution. There are five types of cross-axis CPC, in which the degree of lateral shift of the coil holder is conventionally expressed by L/R. This parameter for types X, XL, XLL, XLLL, and L CPCs is 0, 1, 2, 3.5, and infinity, respectively. Our studies have shown that the stationary-phase retention is enhanced by laterally shifting the position of the coil holder along the rotary shaft, apparently due to the enhancement of a laterally acting force field across the diameter of the tubing.

The polymer-phase system composed of PEG and potassium phosphate has a relatively large difference in density between the two phases, so that it can be retained well in both XL and XLL column positions, which provide efficient mixing of the two phases. On the other hand, the viscous PEG–dextran system, with an extremely low interfacial tension and a small density difference between the two phases, has a high tendency of emulsification under vigorous mixing. Therefore, the use of either the XLLL or L column position, which produces less violent mixing and an enhanced lateral force field, is required to achieve satisfactory phase retention of the PEG–dextran system.

The photograph of the XL cross-axis CPC ($L/R = 1.0$) equipped with a pair of multilayer coil separation columns is shown in Fig. 1. The multilayer coil separation column was prepared from a 2.6-mm inner diameter (i.d.) polytetrafluoroethylene (PTFE) tubing by winding it onto the coil holder hub, forming multiple layers.

POLYMER-PHASE SYSTEMS FOR PROTEIN SEPARATION

Countercurrent chromatography utilizes a pair of immiscible solvent phases that have been pre-equilibrated

Encyclopedia of Chromatography DOI: 10.1081/E-ECHR-120039939

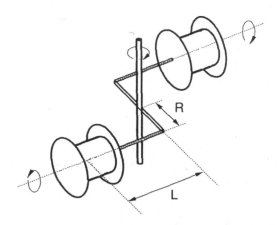

Fig. 1 Photograph of the latest type XL cross-axis coil planet centrifuge.

in a separating funnel: One phase is used as the stationary phase and the other as the mobile phase. A solvent system composed of 12.5% or 16.0% (w/w) PEG 1000 and 12.5% (w/w) potassium phosphate is usually used for the type XL and XLL cross-axis CPCs. These solutions form two layers: The upper layer is rich in PEG and the other layer is rich in potassium phosphate. The ratio of monobasic to dibasic potassium phosphates determines the pH of the solvent system. This effect can be used for optimizing the partition coefficient of proteins.

A solvent system composed of 4.4% (w/w) PEG 8000, 7.0% (w/w) dextran T500, and 10 mM potassium phosphate is used with the type XLLL and L cross-axis CPCs for the separation of proteins that are not soluble in the PEG–phosphate system. This two-phase solvent system consists of the PEG-rich upper phase and dextran-rich lower phase. The cross-axis CPC may be operated in four different elution modes: $P_I HO$, $P_{II} TO$, $P_I TI$, and $P_{II} HI$. The parameters P_I and P_{II} indicate the direction of planetary motion, where P_I indicates counterclockwise and P_{II} clockwise when observed from the top of the centrifuge. H and T indicate the head–tail elution mode, and O and I the inward–outward elution mode along the holder axis. In mode I (inward), the mobile phase is eluted against the laterally acting centrifugal force, and in mode O (outward), this flow direction is reversed. These three parameters yield a total of four combinations for the left-handed coils. Among these elution modes, the inward–outward elution mode plays the most important role in the stationary-phase retention for the polymer-phase systems. To obtain a satisfactory retention of the stationary phase, the lower phase should be eluted outwardly along the direction of the lateral force field ($P_I HO$ and $P_{II} TO$) or the upper phase in the opposite direction ($P_I TI$ or $P_{II} HI$).

APPLICATION OF CROSS-AXIS CPC FOR PROTEINS

Type XL Cross-Axis CPC

The performance of the XL cross-axis CPC, equipped with a pair of columns with a 165-ml capacity, was evaluated for purification of LDH from a crude bovine heart filtrate. Successful separation of the LDH fraction was achieved with 16% (w/w) PEG 1000–12.5% (w/w) potassium phosphate at pH 7.3. The separation was performed at 500 rpm at a flow rate of 1.0 ml/min using the potassium phosphate-rich lower phase as the mobile phase. The sodium dodecyl sulfate–polyacrylamide gel electrophoresis (SDS–PAGE) analysis of the LDH fractions showed no detectable contamination by other proteins. The enzymatic activity was also preserved in these fractions.

The purification of SSB from an *E. coli* lysate was also performed by CCC using the XL cross-axis CPC. About 5 ml of *E. coli* lysate was separated by CCC using a polymer-phase system composed of a 16% (w/w) PEG 1000 and 17% (w/w) ammonium sulfate aqueous polymer two-phase solvent system. The precipitation of proteins in the lysate took place in the CCC column, and the SSB protein was eluted in the fractions. Many other impurities were eluted immediately after the solvent front or precipitated in the column. The identities of the proteins in the fractions and the precipitate were confirmed by SDS–PAGE.

Type XLL Cross-Axis CPC

The XLL cross-axis CPC, with a 250-ml capacity column, was used for the purification of recombinant enzymes such as purine nucleoside phosphorylase (PNP) and uridine phosphorylase (UrdPase) from a crude *E. coli* lysate. The polymer-phase system used in this separation was 16% (w/w) PEG 1000–12.5% (w/w) potassium phosphate at pH 6.8. The separation was performed at 750 rpm at a flow rate of 0.5 ml/min using the upper phase as the mobile phase. About 1.0 ml of crude lysate, containing PNP in 10 ml of the above solvent system, was loaded into the multilayer coil. Purified PNP was harvested in 45-ml fractions. The SDS–PAGE analysis clearly demonstrated that PNP was highly purified in a one-step elution with the XLL cross-axis CPC.

The capability of the XLL cross-axis CPC was further examined in the purification of a recombinant UrdPase from a crude *E. coli* lysate under the same experimental conditions as described earlier. The majority of the protein mass was eluted immediately after the solvent front (between 105 and 165 ml elution volume), whereas the enzyme activity of UrdPase coincided with the fourth protein peak (between 225 and 265 ml elution volume). The result indicates that the recombinant UrdPase can be highly purified from a crude *E. coli* lysate within 10 hr using the XLL cross-axis CPC.

Type XLLL Cross-Axis CPC

Although PEG–phosphate systems yield a high-efficiency separation, some proteins show low solubility

due to a high salt concentration in the solvent system. In this case, the PEG–dextran polymer-phase system with a low salt concentration can be alternatively used for the separation of such proteins. Because the dextran–PEG system has high viscosity and extremely low interfacial tension, it tends to cause emulsification and loss of the stationary phase in the XLL or XL cross-axis CPCs. This problem is minimized using the XLLL cross-axis CPC, which provides a strong lateral centrifugal force to provide a more stable retention of the stationary phase.

Type L Cross-Axis CPC

This cross-axis CPC provides the universal application of protein samples with a dextran–PEG polymer-phase system. Using a prototype of the L cross-axis CPC with a 130-ml column capacity, a profilin–actin complex was purified directly from a crude extract of *Acanthamoeba* with the same solvent system as used for the serum protein separation earlier. The sample solution was prepared by adding the proper amounts of PEG 8000 and dextran T500 to 2.5 g of the *Acanthamoeba* crude extract to adjust the two-phase composition similar to that of the solvent system used for the separation. The experiment was performed by eluting the upper phase at 0.5 ml/min under a high revolution rate of 1000 rpm. The profilin–actin complex was eluted between 60 and 84 ml fractions and well separated from other compounds. The retention of the stationary phase was 69.0% of the total column capacity.

CONCLUSIONS

The overall results of our studies indicate that the retention of the stationary phase of polymer-phase systems in the cross-axis CPCs is increased by shifting the column holder laterally along the rotary shaft. Separation of proteins with high solubility in the PEG–phosphate system can be performed with the XL or XLL cross-axis CPC at a high partition efficiency. Proteins with low solubility in PEG–phosphate systems may be separated with a dextran–PEG system using the XLLL or L cross-axis CPC, which provides more stable retention of the stationary phase.

ARTICLE OF FURTHER INTEREST

Coil Planet Centrifuges, p. 349.

REFERENCES

1. Conway, W.D. *Countercurrent Chromatogaphy: Apparatus and Applications*; VCH: New York, 1990.

2. Albertsson, P.-A. *Partition of Cell Particles and Macromolecules*; 3rd Ed.; Wiley-Interscience: New York, 1986.

3. Shibusawa, Y.; Ito, Y. Protein separation with aqueous–aqueous polymer systems by two types of counter-current chromatographs. J. Chromatogr. **1991**, *559*, 695.

4. Shibusawa, Y.; Ito, Y. Countercurrent chromatography of proteins with polyethylene glycol–dextran polymer phase systems using type-XLLL cross axis coil planet centrifuge. J. Liq. Chromatogr. **1992**, *15* (15 & 16), 2787.

5. Shibusawa, Y.; Eriguchi, Y.; Ito, Y. Purification of lactic acid dehydrogenase from bovine heart crude extract by counter-current chromatography. J. Chromatogr. B. **1997**, *696*, 25.

6. Shibusawa, Y.; Ino, Y.; Kinebuchi, T.; Shimizu, M.; Shindo, H.; Ito, Y. Purification of single-strand DNA binding protein from an *Escherichia coli* lysate using counter-current chromatography, partition and precipitation. J. Chromatogr. B. **2003**, *793*, 275.

7. Yanagida, A.; Isozaki, M.; Shibusawa, Y.; Shindo, H.; Ito, Y. Purification of glucosyltransferase from cell-lysate of *Streptococcus mutans* by counter-current chromatography using aqueous polymer two-phase system. J. Chromatogr. B. **2004**, *805*, 155.

8. Shibusawa, Y.; Isozaki, M.; Yanagida, A.; Shindo, H.; Ito, Y. Purification of glucosyltransferase from *Streptococcus sobrinus* cell culture medium by combined use of batch extraction and counter-current chromatography with a polymer phase system. J. Liq. Chromatogr. & Rel. Technol. **2004**, *27* (14), 1.

9. Shibusawa, Y. *High-Speed Countercurrent Chromatography*; Chemical Analysis Series; Ito, Y., Conway, W.D., Eds.; Wiley-Interscience, 1996; Vol. 132, 121.

C

Cyanobacterial Hepatotoxin Microcystins: Purification by Affinity Chromatography

Fumio Kondo

Department of Toxicology, Aichi Prefectural Institute of Public Health, Tsuji-machi, Kita-ku, Nagoya, Japan

INTRODUCTION

Freshwater cyanobacteria *Microcystis*, *Oscillatoria*, *Anabaena*, and *Nostoc* produce several types of toxins, among which the most commonly detected are the hepatotoxic peptides microcystins. The general structure of the microcystins is cyclo-(D-Ala^1-X^2-D-MeAsp^3-Z^4-Adda^5-D-Glu^6-Mdha^7), in which X and Z represent variable L-amino acids, D-MeAsp^3 is D-*erythro*-β-methylaspartic acid, Mdha is *N*-methyldehydroalanine, and Adda is the unusual C_{20} amino acid, (2S, 3S, 8S, 9S)-3-amino-9-methoxy-2,6,8-trimethyl-10-phenyldeca-4(*E*),6(*E*)-dienoic acid (Fig. 1).[1] The structural differences in the microcystins mainly depend on the variability of the two L-amino acids (denoted X and Z), and secondarily on the methylation or demethylation of D-MeAsp and/or Mdha.[1] More than 60 microcystins have been isolated from bloom samples and isolated strains of cyanobacteria.[1]

Microcystins have caused the poisoning of wild and domestic animals worldwide, and in 1996, they caused the death of 76 people in Caruaru, Brazil, which was attributed to the use of microcystin-contaminated hemodialysis water.[2] Microcystins, like the well-documented tumor promoter, okadaic acid, strongly and specifically inhibit the protein phosphatases 1 and 2A and have a tumor-promoting activity in the rat liver.[3] In addition to acute hepatotoxicity, microcystins pose problems to human health—which could result from low-level, chronic exposure to microcystins in drinking water, as suggested by the high incidence of primary liver cancer in the Qidong and Haimen regions of China.[4] In 1998, the World Health Organization (WHO) proposed a provisional guideline level of 1.0 µg/L for microcystin-LR in drinking water.[4]

OVERVIEW

In order to achieve the rapid and precise determination of microcystins in complicated matrices, a systematic procedure is seriously required. This may include screening, sample purification, identification, and quantification processes.[5] Screening is intended to rapidly check for the presence of microcystins in a small amount of sample, through sensitive and simple methods. If a sample proves positive in the screening test, it will be necessary to follow through with sample purification, identification, and quantification analyses. The purpose of sample purification is to eliminate coexisting substances by a simple operation without the loss of any analyte and, considering that the microcystin concentration may be low, it also enables the enrichment of the analyte. Octadecyl silanized (ODS) silica gel has been extensively employed for this process because it retains microcystins and allows coexisting substances to pass through.[5] A method using ODS silica gel extraction followed by a purification on silica gel has been established to effectively eliminate the coexisting substances in lake water.[6] Although this method has been successfully applied to the analysis of microcystins in lake water samples, it had several problems. The method required a large water sample (5 L) to accumulate the low level of microcystins, resulting in a laborious and time-consuming extraction process. It was also shown that less hydrophobic coexisting substances still remained even after purification on silica gel,[6] indicating that a more effective purification method is required.

Recently, an immunoaffinity purification method using antimicrocystin-LR monoclonal antibodies (named M8H5) has been developed.[7] This purification method was found to be remarkably effective in the removal of coexisting substances and in the enrichment of microcystins in samples.[7–12] This work will focus on the immunoaffinity purification methods for microcystins in lake[10] and tap water samples,[11] and the analysis methods for microcystins and their metabolites in mouse and rat livers.[7–9] It will also cover the reuse of the immunoaffinity column.[12]

IMMUNOAFFINITY COLUMN

Preparation of Immunoaffinity Column

M8H5 antimicrocystin-LR monoclonal antibodies were produced by Nagata et al. as follows:[13,14] female

Encyclopedia of Chromatography DOI: 10.1081/E-ECHR-120016781

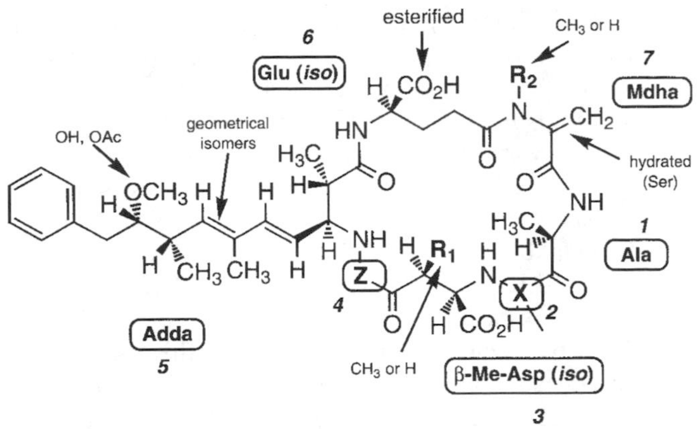

Fig. 1   Chemical structures of microcystins.

	X	Z	R₁	R₂	MW
Microcystin-LA	Leu	Ala	CH3	CH3	909
Microcystin-LR	Leu	Arg	CH3	CH3	994
Microcystin-YR	Tyr	Arg	CH3	CH3	1044
Microcystin-RR	Arg	Arg	CH3	CH3	1037
Microcystin-YM	Tyr	Met	CH3	CH3	1019
[D-Asp³]microcystin LR	Leu	Arg	H	CH3	980
[Dha⁷]microcystin LR	Leu	Arg	CH3	H	980

BALB/c mice were immunized with microcystin-LR conjugated proteins (e.g., bovine serum albumin [BSA]). The spleen cells of the immunized mice and SP2/O-Ag14 cells were fused with polyethylene glycol. From the serum-free cultured supernatants of the generated hybridomas, the M8H5 monoclonal antibodies were prepared by membrane ultrafiltration, ammonium sulfate, which were then finally purified using a protein G column (Pharmacia, Sweden). This monoclonal antibody shows extensive cross-reactivity to various microcystins and nodularin (pentapeptide sharing many common features with microcystins). Based on concentrations capable of inducing 50% inhibition of antibodies in a competitive enzyme-linked immunosorbent assay (ELISA), the cross-reactivities were 100% for microcystin-LR, 109% for microcystin-RR, 51% for [D-Asp³]microcystin-LR, 48% for [Dha⁷]microcystin-LR, 44% for microcystin-YR, 26% for microcystin-LA, and 20% for nodularin.

The antimicrocystin-LR monoclonal antibody-combined gel was prepared as follows:[7] Affi-Gel 10 (25 mL, Bio-Rad, Hercules, CA, U.S.A.) was washed with distilled ice water (250 mL) and mixed with an equal volume of 10 mg/mL of M8H5 antimicrocystin-LR monoclonal antibody in phosphate buffer saline (PBS) (pH 7.4). The gel mixture was incubated at 4°C for 24 h with gentle rocking. After the addition of 1 M ethanolamine hydrochloride (2.5 mL), the gel mixture was washed with distilled water (250 mL), followed by PBS (500 mL). The obtained gel cake was suspended in PBS containing 0.1% sodium azide (50 mL) and stored at 4°C. The gel mixture (0.5 mL) was transferred to a

polypropylene cartridge (Muromac column; Muromachi Kagaku Kogyo Kaisha, Tokyo, Japan) when used.

Protocol for Immunoaffinity Purification

The general protocol for immunoaffinity purification of the microcystins is as follows: sample extracts dissolved in PBS are loaded onto the immunoaffinity column and passed through the column. Gravity flow is usually used. After washing with PBS and distilled water, the microcystin fraction is eluted with 100% methanol.[7,10]

Direct immunoaffinity purification of the microcystins in complicated matrices is difficult because the coexisting substances in the sample occupy the column head, so that the process ends in failure.[10] Thus preliminary semipurification is indispensable when it comes to analyzing trace amounts of microcystins in samples containing substantial amounts of coexisting substances.

DETERMINATION OF MICROCYSTINS

Lake Water

A purification method for microcystins in lake water consists of solid-phase extraction on a Sep-Pak PS2® (styrene–divinylbenzene copolymer) or Excelpak SPE-GLF® (polymethacrylate) cartridge, followed by immunoaffinity purification.[10] A 1-L sample of lake water is filtered through a glass GF/C microfiber filter (Whatman, Maidstone, U.K.), and the filtrate is applied to a Sep-Pak PS-2 or Excelpak SPE-GLF cartridge at a flow rate of about 20 mL/min. The cartridge is washed with distilled water (10 mL), followed by 20% methanol–water (10 mL). Finally, the eluate from the cartridge with 100% methanol (10 mL) is collected and then evaporated to dryness under reduced pressure at 35°C. The resulting residue is dissolved in PBS containing 0.1% BSA (5 mL). Bovine serum albumin is used to prevent nonspecific binding of the microcystins with the immunoaffinity support Affi-Gel 10. The solution is subjected to an immunoaffinity column, which is preconditioned with PBS (5 mL), methanol (5 mL), distilled water (5 mL), and PBS containing 0.1% BSA (5 mL). After washing with PBS (5 mL) and distilled water (5 mL), the microcystin fraction is eluted with 100% methanol (10 mL). The eluate is then evaporated to dryness under reduced pressure at 35°C. The residue is dissolved in 0.5 mL of 30% methanol–water and then subjected to HPLC with ultraviolet (UV) detection and LC/MS analyses.

When an extract is prepared with a solid-phase extraction cartridge alone, the microcystins in lake water cannot be precisely detected because of the unstable baseline and many peaks on the chromatograms from coexisting substances (Fig. 2a). When this extract is further purified with the immunoaffinity column, the microcystin peaks are clearly detected (Fig. 2b) and effectively quantified because the coexisting substances are virtually eliminated. The recoveries from lake water (1 L) spiked with 100 ng each of microcystins-LR, -YR, and -RR are 92.2%, 89.2%, and 85.5%, respectively, with coefficients of variation of 3.3–7.6%. One of the advantages of this method is its speed; it took only 3 hr to complete the entire procedure, starting from the microcystin extraction, the immunoaffinity purification, and the quantification, whereas the previous procedures (tandem use of ODS silica gel and silica gel cartridges) took a day to complete. The detection limit for all of the three microcystins in lake water is 0.005 µg/L.

Fig. 2c shows a chromatogram of a sample taken from Lake Suwa, Japan, where water blooms occurred. Two peaks with retention times of 8.0 min (peak 1) and 17.5 min (peak 2) correspond to those of microcystins-RR and -LR, respectively, and they show the typical spectra of the respective microcystins with an

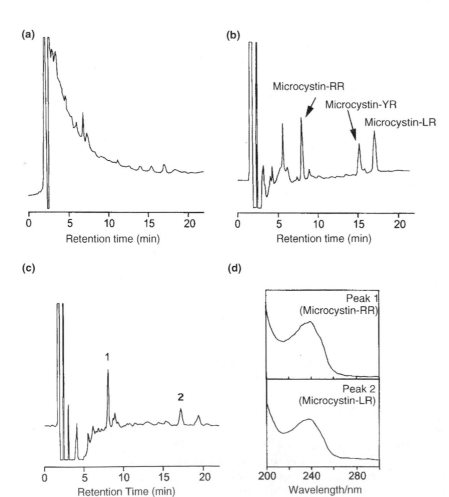

Fig. 2 HPLC analysis of lake water extracts. Microcystin-*free* lake water spiked with microcystins-LR, -YR, and -RR (100 ng each) was analyzed before (a) and after (b) purification with immunoaffinity column. (c) A water sample taken from Lake Suwa, Japan, in 1998 was analyzed after purification with the immunoaffinity column.

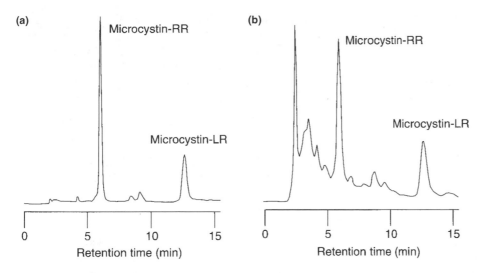

Fig. 3 HPLC analysis of cyanobacterial cells extracts. A cyanobacterial cell sample taken from Lake Suwa, Japan, in 1999 was analyzed after purification with (a) immunoaffinity column and (b) ODS silica gel cartridge.

absorption maximum at 238 nm (Fig. 2d), affording further confirmation.

Tap Water and Cyanobacterial Cells

Direct purification of the microcystins in tap water with immunoaffinity columns can be achieved because the tap water contains relatively smaller amounts of coexisting substances.[11] A tap water sample is filtered through glass GF/C microfiber filters, to which a 1/10 volume of 11 × PBS is added. The sample solution is loaded onto the immunoaffinity column and passed through the column. After washing with PBS (10 mL) and distilled water (10 mL), the microcystin fraction is eluted with 100% methanol (10 mL). The eluate is evaporated to dryness and the residue is dissolved in 0.05 mL of methanol for HPLC analysis. The chromatograms of the microcystin-added tap water with immunoaffinity purification show effective elimination of the coexisting substances compared to that with the ODS cartridge (data not shown). The mean recoveries of microcystin-LR, -YR, and -RR added to tap water are 91.8%, 86.4%, and 77.3%, respectively, in the range 2.5–100 µg/L.

Microcystins in cyanobacterial cells can also be directly purified via immunoaffinity columns. A 1-L sample of lake water containing cyanobacterial cells is filtered through a glass GF/C microfiber filter (Whatman), and the cells on the filter are suspended in distilled water (10 mL). The cell containing suspension is freeze-thawed three times and then filtered through a glass GF/C microfiber filter. After the addition of a 1/10 volume of 11 × PBS, the filtrate is loaded onto the immunoaffinity column and treated in a similar manner as described above. The HPLC chromatogram shows that the peaks of microcystins-LR and -RR are clearly detected and the peaks due

to the coexisting substances are almost negligible (Fig. 3a). On the other hand, the chromatogram of the same sample purified with the ODS silica gel cartridge shows that the removal of less hydrophobic coexisting substances is insufficient, especially in a shorter retention time area (Fig. 3b).

Mouse and Rat Livers

The in vivo tissue distribution, excretion, and hepatic metabolism of microcystins have been primarily investigated using variously radiolabeled ones.[4] The amounts of the injected microcystins were too small and the amounts of the contaminants in the tissues were too large to investigate the metabolites by instrumental analysis, such as HPLC with UV detection and

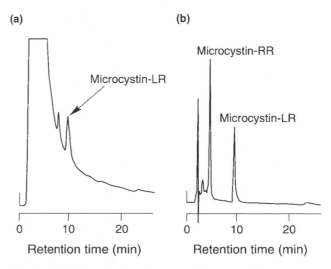

Fig. 4 HPLC analysis of a cytosolic extract from mouse liver spiked with 5 µg each of microcystins-RR and -LR. Before (a) and after (b) purification with immunoaffinity column.

LC/MS. Fig. 4 shows the HPLC profiles of a cytosolic extract from mouse liver spiked with 5 µg each of microcystins-LR and -RR. When the cytosolic extract is prepared by the method described by Robinson et al.,[15] which consists of heat-denaturation, pronase digestion, and ODS silica gel treatment (Fig. 4a), the two spiked microcystins cannot be precisely analyzed because of a substantial amount of coexisting substances. When the cytosolic extract is further purified with the immunoaffinity column, the coexisting substances are effectively eliminated and the peaks of the spiked microcystins-LR and -RR are clearly detected (Fig. 4b).

The method including heat-denaturation, pronase digestion, ODS silica gel treatment, and immunoaffinity purification has been applied to the analysis of microcystins and their metabolites in hepatic cytosols from mice and rats intraperitoneally administered microcystins. A purified cytosolic extract from mouse liver at 3 hr postinjection of microcystin-RR was analyzed by Frit-FAB LC/MS. The peaks of the microcystins and related compounds can be selectively detected by monitoring the mass chromatogram at m/z 135, which is the characteristic ion derived from Adda, a characteristic component of microcystins.[16] The mass chromatogram shows the relatively broad peak X (Fig. 5a), which is confirmed to contain at least four peaks by using slightly modified HPLC analysis conditions. The mass spectrum of the longer retention time area of peak X shows an ion at m/z 1345, whose molecular weight corresponds to that of the glutathione conjugate of microcystin-RR (Fig. 5b), the thiol of GSH having been nucleophilically added to the α,β-unsaturated carbonyl of the Mdha moiety in the microcystins. The glutathione conjugate was confirmed by comparison of the retention time with that of the

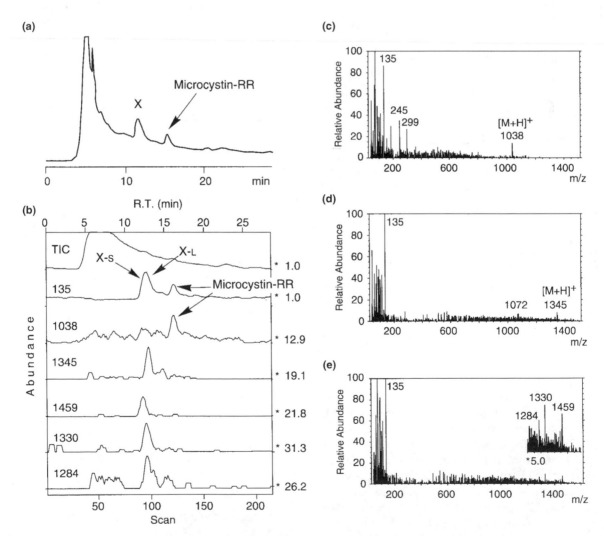

Fig. 5 Frit-FAB LC/MS analysis of a cytosolic extract from mouse liver at 3 hr postinjection of microcystin-RR. Shown are (a) simultaneously monitored UV chromatogram (238 nm), (b) total ion and mass chromatograms, (c) Frit-FAB LC/MS mass spectra of microcystin-RR, (d) longer retention time area of peak X (indicated by X-L), and (e) shorter retention time area of peak X (indicated by X-s).

chemically prepared one. The mass spectrum of the shorter retention time area of peak X shows ions at m/z 1284, 1330, and 1459, with relatively low intensities (Fig. 5c). This area was considered to contain a conjugate of the oxidized Adda diene. The cysteine conjugate was identified in a cytosolic extract from mouse liver at 24 hr postinjection of microcystin-RR. Both the glutathione and cysteine conjugates were also identified in hepatic cytosols from the rat-administered microcystin-LR.

The immunoaffinity purification method, followed by LC/MS analysis, has also been used in other toxicological studies. When aged mice (32 weeks) were orally administered microcystin-LR at 500 μg/kg, 62% of the aged mice showed hepatic injury, whereas such changes in the liver were not found in young mice (5 weeks).[8] Upon uptake of orally administered microcystin-LR at 500 μg/kg, the toxin into the liver was confirmed by Frit-FAB LC/MS after the immunoaffinity purification. When microcystin-LR was intraperitoneally injected 100 times at 20 μg/kg into male ICR mice (5 weeks old, Charles River Japan, Atsugi, Japan) for 28 weeks, multiple hyperplastic nodules up to 5 mm in diameter were observed in every liver.[9] Microcystin-LR and its cysteine conjugate were identified from the isolated mouse livers.

REUSABLE IMMUNOAFFINITY COLUMN

Reuse of the immunoaffinity column is apparently of great value because the immunoaffinity column requires a large amount of antibodies. The immunoaffinity column using Affi-Gel 10 as the immunoaffinity support could not be repeatedly used because the applied solutions tended to stick owing to the massive bubbles produced. In order to overcome this disadvantage, a new immunoaffinity column using the immunoaffinity support Formyl-Cellulofine® has been introduced.[12] Because of the spherical shape of the immunoaffinity support Formyl-Cellulofine, the applied solutions passed through the column smoothly even when they were used repeatedly.

The purification procedure using this immunoaffinity column has been optimized as follows: After extraction with a Sep-Pak PS2 cartridge, the extract is dissolved in 25 mM Tris–HCl buffer (pH 7.2) containing 1 mM EDTA, 0.15 M sodium chloride, and 0.1% sodium azide (Tris–HCl buffer A) with 0.1% BSA (Tris–HCl buffer B) (5 mL), and the solution is loaded onto an immunoaffinity column, which is preconditioned with Tris–HCl buffer B (10 mL). After washing with Tris–HCl buffer A (10 mL) and distilled water (10 mL), the microcystin fraction is eluted with 100% dimethylformamide (DMF) (2.5 mL). The eluate is then dried on a hot block (60°C) under a constant stream of nitrogen. The residue is dissolved in 30% methanol–water (0.5 mL) and then subjected to HPLC-photodiode array detection (HPLC-PDA) and LC/MS analysis. The immunoaffinity column is regenerated by washing with Tris–HCl buffer B (10 mL) before each reuse.

Recoveries of the spiked microcystins to lake water from the first use of the column are 87–88%, and 83–88% was recorded from the second and third uses. Recoveries gradually drop to 63–77% from the fourth to the fifth uses. These results indicate that the column can be repeatedly used up to three times.

CONCLUSIONS

The immunoaffinity column using antimicrocystin-LR monoclonal antibodies has proved to be an important purification tool for microcystin analysis in lake and tap water, and cyanobacterial cells and tissue samples from mice and rats. One of the advantages of the immunoaffinity purification is specificity. It enables operators to enrich the trace amounts of the microcystins and to eliminate large amounts of coexisting substances. As a result of the substantial elimination of these coexisting substances, the microcystin peaks were successfully and reliably identified with HPLC and LC/MS by excluding a few of the nonmicrocystin peaks that appear even after the immunoaffinity purification. On the other hand, one of the disadvantages of the analytical method using the immunoaffinity column is the requirement for large amounts of antibodies. Reuse of the immunoaffinity column is one approach to overcome this limitation. Although the immunoaffinity column for microcystins is not yet commercially available, it is believed that it will become more widely used within the next 2–3 years.

REFERENCES

1. Sivonen, K.; Jones, G. Cyanobacterial Toxins. In *Toxic Cyanobacteria in Water; a Guide to Their Public Health Consequences, Monitoring and Management*; Chorus, I., Bartram, J., Eds.; E & FN Spon: London, 1999; 41–111.
2. Carmichael, W.W.; Azevedo, S.M.F.O.; An, J.S.; Molica, R.J.R.; Jochimsen, E.M.; Lau, S.; Rinehart, K.L.; Shaw, G.R.; Eaglesham, G.K. Human fatalities from cyanobacteria: chemical and biological evidence for cyanotoxins. Environ. Health Perspect. **2001**, *109* (7), 663–668.
3. Falconer, I.R.; Bartram, J.; Chorus, I.; Kuiper-Goodman, T.; Utkilen, H.; Burch, M.; Codd, G.A. Safe levels and safe practices. In

Toxic Cyanobacteria in Water; A Guide to Their Public Health Consequences, Monitoring and Management; Chorus, I., Bartram, J., Eds.; E & FN Spon: London, 1999; 155–178.

4. Kuiper-Goodman, T.; Falconer, I.R.; Fitzgerald, J. Human Health Aspects. In *Toxic Cyanobacteria in Water; A Guide to Their Public Health Consequences, Monitoring and Management*; Chorus, I., Bartram, J., Eds.; E & FN Spon: London, 1999; 113–153.

5. Harada, K.-I.; Kondo, F.; Lawton, L. Laboratory analysis of cyanotoxins. In *Toxic Cyanobacteria in Water; A Guide to Their Public Health Consequences, Monitoring and Management*; Chorus, I., Bartram, J., Eds.; E & FN Spon: London, 1999; 999.

6. Tsuji, K.; Naito, S.; Kondo, F.; Watanabe, M.F.; Suzuki, S.; Nakazawa, H.; Suzuki, M.; Shimada, T.; Harada, K.-I. A clean-up method for analysis of trace amounts of microcystins in lake water. Toxicon **1994**, *32* (10), 1251–1259.

7. Kondo, F.; Matsumoto, H.; Yamada, S.; Ishikawa, N.; Ito, E.; Nagata, S.; Ueno, Y.; Suzuki, M.; Harada, K.-I. Detection and identification of metabolites of microcystins formed in vivo in mouse and rat livers. Chem. Res. Toxicol. **1996**, *9*, 1355–1359.

8. Ito, E.; Kondo, F.; Terao, K.; Harada, K.-I. Hepatic necrosis in aged mice by oral administration of microcystin-LR. Toxicon **1997**, *35*, 231–239.

9. Ito, E.; Kondo, F.; Harada, K.-I. Neoplastic nodular formation in mouse liver induced by repeated intraperitoneal injections of microcystin-LR. Toxicon **1997**, *35*, 1453–1457.

10. Kondo, F.; Matsumoto, H.; Yamada, S.; Tsuji, K.; Ueno, Y.; Harada, K.-I. Immunoaffinity purification method for detection and identification of microcystins in lake water. Toxicon **2000**, *38*, 813–823.

11. Tsutsumi, T.; Nagata, S.; Hasegawa, A.; Ueno, Y. Immunoaffinity column as clean-up tool for determination of trace amounts of microcystins in tap water. Food Chem. Toxicol. **2000**, *38*, 593–597.

12. Kondo, F.; Ito, Y.; Oka, H.; Yamada, S.; Tsuji, K.; Imokawa, M.; Niimi, Y.; Harada, K.-I.; Ueno, Y.; Miyazaki, Y. Determination of microcystins in lake water using reusable immunoaffinity column. Toxicon **2002**, *40*, 893–899.

13. Nagata, S.; Okamoto, Y.; Inoue, T.; Ueno, Y.; Kurata, T.; Chiba, J. Identification of epitopes associated with different biological activities on the glycoprotein of vesicular stomatitis virus by use of monoclonal antibodies. Arch. Virol. **1992**, *127*, 153–168.

14. Nagata, S.; Soutome, H.; Tsutsumi, T.; Hasegawa, A.; Sekijima, M.; Sugamata, M.; Harada, K.-I.; Suganuma, M.; Ueno, Y. Novel monoclonal antibodies against microcystin and their protective activity for hepatotoxicity. Nat. Toxins **1995**, *3*, 78–86.

15. Robinson, N.A.; Pace, J.G.; Matson, C.F.; Miura, G.A.; Lawrence, W.B. Tissue distribution, excretion and hepatic biotransformation of microcystin-LR in mice. J. Pharmacol. Exp. Ther. **1990**, *256*, 176–182.

16. Kondo, F.; Ikai, Y.; Oka, H.; Ishikawa, N.; Watanabe, M.F.; Watanabe, M.; Harada, K.; Suzuki, M. Separation and identification of microcystins in cyanobacteria by frit-fast atom bombardment liquid chromatography/mass spectrometry. Toxicon **1992**, *30*, 227–237.

Dead Point (Volume or Time)

Raymond P. W. Scott
Scientific Detectors Ltd., Banbury, Oxfordshire, U.K.

INRODUCTION

The *injection point* on a chromatogram is that position where the sample is injected. The *dead point* on a chromatogram is the position of the peak maximum of a completely unretained solute. The different attributes of the chromatogram are shown in Fig. 1. The *dead time* is the elapsed time between the injection point and the dead point. The volume that passes through the column between the injection point and the dead point is called the *dead volume*.

DISCUSSION

If the mobile phase is *incompressible*, as in LC, the *dead volume* (as so far defined) will be the simple product of the *exit* flow rate and the dead time. However, in LC, where the stationary phase is a porous matrix, the dead volume can be a very ambiguous column property and requires closer inspection and a tighter definition.

If the mobile phase is compressible, the simple product of dead time and flow rate will be incorrect, and the dead volume must be taken as the product of the dead time and the *mean* flow rate. The dead volume has been shown to be given by[1]

$$V_0 = V_0' \frac{3}{2}\left(\frac{\gamma^2 - 1}{\gamma^3 - 1}\right) = Q_0 t_0 \frac{3}{2}\left(\frac{\gamma^2 - 1}{\gamma^3 - 1}\right)$$

where the symbols have the meaning defined in Fig. 1, and V_0' is the dead volume measured at the column exit and γ is the inlet/outlet pressure ratio.

The dead volume will not simply be the total volume of mobile phase in the column system (V_m) but will include extra-column dead volumes (V_E) comprising volumes involved in the sample valve, connecting tubes, and detector. If these volumes are significant, then they must be taken into account when measuring the dead volume.

There are two types of dead volume (i.e., the *dynamic* dead volume and the *thermodynamic* dead

volume.[2]) The dynamic dead volume is the volume of the *moving phase* in the column and is used in kinetic studies to calculate mobile-phase velocities. In gas chromatography, both the dynamic dead volume and the thermodynamic dead volume can be taken as the difference between the dead volume and the extra-column volume. In LC, however, where a porous packing is involved, some of the mobile phase will be in pores (the *pore volume*) and some between the particles (the *interstitial volume*). In addition, some of the mobile phase in the interstitial volume which is close to the points of contact of the particles will also be stationary. The dynamic dead volume (i.e., the volume of the moving phase) is best taken as the retention volume of a relatively large inorganic salt such as potassiun nitroprusside. This salt will be excluded from the pores of the packing by ionic

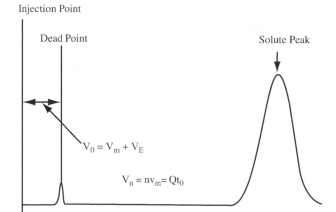

Fig. 1 Diagram depicting the dead point, dead volume, and dead time of a chromatogram. If the mobile phase is not compressible, then V_0 is the total volume passed through the column between the point of injection and the peak maximum of a completely unretained peak, V_m is the total volume of the mobile phase in the column, V_E is the extra column volume of the mobile phase, v_m is the volume of the mobile phase per theoretical plate, t_0 is the time elapsed between the time of injection and the retention time of a completely unretained peak, n is the number of theoretical plates in the column, and Q is the column flow rate measured at the exit.

Encyclopedia of Chromatography DOI: 10.1081/E-ECHR-120039940

exclusion and will only explore the moving volumes of the mobile phase.[2] The thermodynamic dead volume will include all the mobile phase that is available to the solute that is under thermodynamic examination. It is best measured as the retention volume of a solvent sample of very similar type to that of the mobile phase and of small molecular weight. If a binary solvent mixture is used (which is the more common situation), then one component of the mobile phase, in pure form, can be used to measure the thermodynamic dead volume. Careful consideration must be given to the measurement of the column dead volume when determining thermodynamic data by LC using columns packed with porous materials.

REFERENCES

1. Scott, R.P.W. *Introduction to Analytical Gas Chromatography*; Marcel Dekker, Inc.: New York, 1998; 77.
2. Alhedai, A.; Martire, D.E.; Scott, R.P.W. Analyst **1989**, *114*, 869.

SUGGESTED FURTHER READING

Scott, R.P.W. *Liquid Chromatography Column Theory*; John Wiley & Sons: Chichester, 1992; 19.
Scott, R.P.W. *Techniques and Practice of Chromatography*; Marcel Dekker, Inc.: New York, 1996.

Dendrimers and Hyperbranched Polymers: Analysis by GPC/SEC

Nikolay Vladimirov
Hercules, Inc., Wilmington, Delaware, U.S.A.

INTRODUCTION

Dendrimers and hyperbranched polymers are globular macromolecules having a highly branched structure, in which all bonds converge to a focal point or core, and a multiplicity of reactive chain ends. Because of the obvious similarity of their building blocks, many assume that the properties of these two families of dendritic macromolecules are almost identical and that the terms "dendrimer" and "hyperbranched polymer" can be used interchangeably. These assumptions are incorrect because only dendrimers have a precise end-group multiplicity and functionality. Furthermore, they exhibit properties totally unlike that of other families of macromolecules.

HISTORICAL BACKGROUND

Highly branched and generally irregular dendritic structures have been known for some time, being found, for example, in polysaccharides, such as amylopectin, glycogen, and some other biopolymers. In the area of synthetic structures, Flory discussed, as early as 1952, the theoretical growth of highly branched polymers obtained by the polycondensation of AB_x structures in which x is at least equal to 2. Such highly branched structures are now known as "hyperbranched polymers."

Today, regular dendrimers can only be prepared using rather tedious, multistep syntheses that require intermediate purifications. In contrast, hyperbranched polymers are easily obtained using a variety of one-pot procedures, some of which mimic, but do not truly achieve, regular dendritic growth.[1] The presence of such a large number of atoms within each dendritic or hyperbranched macromolecule permits an enormous variety of conformations with different shapes and sizes. The distribution of molecular weights focuses on the polydispersity index (M_w/M_n), and the requirements for gelation (or avoidance of gelation) when multimodal monomers are incorporated into the macromolecule. Each of these topics are discussed in Newcome's monograph.[2] Lists of reviews between 1986 and 1996 and Advances series are also given.

Buchard et al.[3] outlined some properties of hyperbranched chains. The dilute solution properties of branched macromolecules are governed by the higher segment density found with linear chains. The dimensions appear to be shrunk when compared with linear chains of the same molar mass and composition. It is shown that the apparent shrinking has an influence also on the intrinsic viscosity and the second virial coefficient. The broad molecular-weight distribution (MWD) has a strong influence on these shrinking factors, which can be defined and used for quantitative determination of the branching density (i.e., the number of branching points in a macromolecule). Here, the branching density can be determined only by SEC in on-line combination with light-scattering and viscosity detectors. The technique and possibilities are discussed in detail.

DISCUSSION

A dendritic structure generally gives rise to better solubility than the corresponding linear analog. For example, aromatic polyamide dendrimers and hyperbranched polymers are soluble in amide-type solvents and even in tetrahydrofuran. GPC was performed on a Jasco HPLC 880PU fitted with polystyrene–divinylbenzene columns (two Shodex KD806M and KD802) and a Shodex RI-71 refractive index detector in DMF containing 0.01 mol/L of lithium bromide as an eluent. Absolute molecular weights (M_w) of 74,600, 47,800 and 36,800 were determined by light scattering using a MiniDawn apparatus (Wyatt Technology Co.) and a Shimadzu RID-6A refractive index detector. A specific refractive index increment (dn/dc) of the polymer in DMF at 690 nm was measured to be 0.216 mL/g.[4]

Standards commonly employed[5] to calibrate SEC columns do not have a well-defined size. Carefully characterized spherical solutes in the appropriate size range are therefore of considerable interest. The chromatographic behavior of carboxylated starburst dendrimers—characterized by quasi-elastic light scattering and viscometry—on a Superose SEC column was explored. Carboxylated starburst dendrimers

Encyclopedia of Chromatography DOI: 10.1081/E-ECHR-120039942

appear to behave as noninteracting spheres during chromatography in the presence of an appropriate mobile phase. The dependence of the retention time on the solute size seems to coincide with data collected on the same column for Ficoll. Chromatography of the dendrimers yields to a remarkable correlation of the chromatographic partition coefficient with the generation number; this result is, in part, a consequence of the exponential relationship between the generation number and the molecular volume of these dendrimers. All measurements were made in a 9:1 mixture of pH = 5.5, 0.38 M, which has been previously known to minimize electrostatic interactions between a variety of proteins and this stationary phase.[4]

The SEC partition coefficient[6] (K_{SEC}) was measured on a Superose 6 column for three sets of well-characterized symmetrical solutes: the compact, densely branched nonionic polysaccharide, Ficoll; the flexible chain nonionic polysaccharide, pullulan; and compact, anionic synthetic polymers, carboxylated starburst dendrimers. All three solutes display a congruent dependence of K_{SEC} on solute radius, R. In accord with a simple geometric model for SEC, all of these data conform to the same linear plot of $K_{SEC}^{1/2}$ versus R. This plot reveals the behavior of noninteracting spheres on this column. The mobile phase for the first two solutes was 0.2 M NaH_2PO_4–Na_2HPO_4, pH 7.0. In order to ensure the suppression of electrostatic repulsive interactions between the dendrimer and the packing, the ionic strength was increased to 0.30 M for that solute.

The MWD[7] is derived for polymers generated by self-condensing vinyl polymerization (SCVP) of a monomer having a vinyl and an initiator group ("inimer") in the presence of multifunctional initiator. If the monomer is added slowly to the initiator solution (semibatch process), this leads to hyperbranched polymers with a multifunctional core. If monomer and initiator are mixed simultaneously (batch process), even at vinyl group conversions as high as 99%, the total MWD consists of polymers, which have grown via reactions between inimer molecules (i.e., the normal SCVP process) and those which have reacted with the initiator. Consequently, the weight distribution, $w(M)$, is bimodal. However, the z-distribution, $z(M)$, equivalent to the "GPC distribution," $w(\log M)$ versus $\log M$, is unimodal. Their theoretical studies showed that the hyperbranched polymers generated from an SCVP possess a very wide MWD $M_w/M_n \cong P_n$, where P_n is the number-average degree of polymerization. The evolution of the weight-distribution and z-distribution curves of the total resultant polymer during the SCVP in the presence of the core moiety with $f = 10$ is given. The weight distributions become less bimodal with increasing conversion. In contrast, all z-distributions are unimodal.

Striegel et al.[8] employed SEC with universal calibration to determine the molecular-weight averages, distributions, intrinsic viscosities, and structural parameters of Starburst dendrimers, dextrans, and the starch-degradation polysaccharides (maltodextrins). A comparison has been made in the dilute solution behavior of dendrimers and polysaccharides with equivalent weight-average molecular weights. Intrinsic viscosities decreased in the order $[\eta_{dexstran}] > [\eta_{dextrin}] > [\eta_{dendrimer}]$. A comparison between the molecular radii obtained from SEC data and the radii from molecular dynamics studies show that Starburst dendrimers behave as θ-stars with functionality between 1 and 4. Additionally, electrospray ionization MS was employed to determine M_w, M_n, and the PD of Astromol dendrimers.

SEC experiments were carried out on a Watters 150CV$^+$ instrument (Waters Associates, Milford, MA) equipped with both differential refractive index single-capillary viscometer detectors. The solvent/mobile phase was $H_2O/0.02\%$ NaN_3, at the flow rate of 1.0 mL/min. Pump, solvent, and detector compartments were maintained at 50°C. Separation occurred over a column bank consisting of three analytical columns preceded by a guard column: Shodex KB-G, KS-802, KS-803, and KB-804 (Phenomenex, Torrance, CA). Universal calibration was performed using a series of oligosaccharides (Sigma, St. Louis, MO), and Pullulan Standards (American Polymer Standards, Mentor, OH, and Polymer Laboratories, Amherst, MA).

The solution behavior of several generations of Starburst poly(amido amine) dendrimers, low-molecular-weight ($M_w < 60,000$) dextrans, and maltodextrins was also examined by SEC, using the universal calibration. For Starburst and Astramols, supplied M_w values are theoretical average molecular weights. Weight-average molecular weights for the dendrimers determined by SEC with universal calibration using oligosaccharide and polysaccharide narrow standards were slightly, albeit consistently lower than the theoretical averages. In general, the intrinsic viscosity of polymers tends to increase with increasing molecular weight (M), which accompanies an increase in the size of the macromolecule. Exceptions to this are the hyperbranched polymers, in which the Mark–Houwink double logarithmic $[\eta]$ versus M curve passes through a minimum in the low-molecular-weight region before steadily increasing. For solutions of the dendrimers studied in their experiments, it is evident that as M increases, $[\eta]$ decreases. This corresponds to the molecules growing faster in density than in radial growth. Fréchet has pointed out the special situation of this class of polymers, in which their volume increases cubically and their mass increases exponentially.[9] The exponent a in the Mark–Houwink equation for the dendrimers is −0.2 for convergent growth for the generation studied (located in the inverted region of the Mark–Houwink plot). This value for the Starburst dendrimers is comparable to the a value of −0.2 for convergent-growth dendrimers, generations 3–6, studied by Mourey et al.[9] When the results from SEC are combined with those from computer modeling by comparing the ratios of geometric to hydrodynamic radii for the trifunctional Starbursts to the ratios derived for the other molecular geometries, the dendrimers appear to resemble θ-stars.

SEC[9] with a coupled molecular-weight-sensitive detection is a simple convenient method for characterizing dendrimers for which limited sample quantities are available. The polyether dendrimers increase in hydrodynamic radius approximately linearly with generation and have a characteristic maximum in viscosity. These properties distinguish these dendrimers from completely collapsed, globular structures. The experimental data also indicate that these structures are extended to approximately two-thirds of the theoretical, fully extended length.

Puskas and Grasmüller characterized the synthesized star-branched and hyperbranched polyisobutylenes (PIBs) by SEC–light scattering in tetrahydrofuran (THF), with the dn/dc measured as 0.09 mL/g. The radius of gyration gave a slope of 0.3, demonstrating the formation of a star-branched polymer.[10]

Gitsov and Fréchet[11] reported the syntheses of novel linear-dendritic triblock copolymers achieved via anionic polymerization of styrene and final quenching with reactive dendrimers. For the characterization of the products in the reaction mixture, SEC with double detection was performed at 45°C on a chromatography line consisting of a 510 pump, a U6K universal injector, three Ultrastyragel columns with pore sizes 100 Å and 500 Å and Linear, a DRI detector M410, and a photodiode array detector M991 (all Millipore Co., Waters Chromatography Division). THF was used as the eluent at a flow rate of 1 mL/min. SEC with coupled PDA detection proves to be particularly useful in the separation and identification of all compounds in the reaction mixture. A detailed discussion can be found in Ref.[11]. SEC/VISC studies show that the ABA copolymers are not entangled and undergo a transition from an extended globular form to a statistical coil when the molecular weight of their linear central block exceeds 50,000.

The solution properties of hybrid–linear dendritic polyether copolymers are investigated by SEC with coupled viscometric detection from the same authors [12]. The results obtained show that the block copolymers are able to form monomolecular and multimolecular micelles depending on the dendrimer generation and the concentration in methanol–water (good solvent for the linear blocks).

Large macromolecular assemblies and agglomerates play an important role in living matter and its artificial reproduction. AB and ABA block copolymers are convenient tools used for modeling of these processes. Usually in a specific solvent–nonsolvent system, ABA triblocks form micelles with a core consisting of insoluble B blocks and a surrounding shell of A blocks that extend into the solvent phase. Two Waters/Shodex PROTEIN KW 802.5 and 804 columns were used for the aqueous SEC measurements. The columns were calibrated with 14 PEO and PEG standards. The radius of gyration was calculated from the intrinsic viscosity [η] and Unical 4.04 software (Viscotek). The calculated values for the Mark–Houwink–Sakurada constant a are 0.583 for PEG ($K = 9.616 \times 10^{-4}$) and 0.776 for PEO ($K = 2.042 \times 10^{-4}$). They are in close agreement with the data reported for the same polymer in other aqueous mixtures (compositions).

The significant decrease in the [η] of the copolymer solutions and the parallel decrease in R_g of the hybrid structures containing [G-4] blocks indicate that the block copolymers are undergoing intramolecular micellization. Unimolecular micelles consisting of a small, dense, dendritic core tightly surrounded by a PEO corona are formed. The influence of the size of the dendritic block was investigated with PEO7500. The solution behavior of ABA hybrid copolymers is documented. In general, materials containing more than 30 wt% of dendritic blocks are not soluble in methanol–water. However, it should be emphasized that the solubility of copolymers is also strongly influenced by the size of the dendritic block. Obviously, an optimal balance between the size of the dendrimer and the length of the linear block is required to enable the dissolution of the copolymer in the solvent composition.

Performing SEC with dual detection (DRI and viscometry) permmited application of the concept of universal calibration.

REFERENCES

1. Fréchet, J.M.J.; Hawker, C.J.; Gitsov, I.; Leon, J.W. Dendrimers and hyperbranched polymers: two families of three dimensional macromolecules with similar but clearly distinct properties. J.M.S.-Pure Appl. Chem. A **1996**, *33*, 1399.

2. Newcome, G.R.; Moorefield, C.N.; Vögtle, F. *Dendritic Molecules, Concepts, Syntheses, Perspectives*; VCH: Weinheim, 1996.

3. Buchard, W. Solution properties of branched macromolecules. Adv. Polym. Sci. **1999**, *143*, 113.

4. Yang, G.; Jikey, M.; Kakimoto, M. Synthesis and properties of hyperbranched aromatic polyamide. Macromolecules **1999**, *32*, 2215.

5. Dubin, P.L.; Eduards, S.L.; Kaplan, I.; Mehta, M.S.; Tomalia, D.; Xia, J. Carboxylated starburst dendrimers as calibration standards for aqueous size exclusion chromatography. Anal. Chem. **1992**, *64*, 2344.

6. Dubin, P.L.; Edwards, S.L.; Mehta, M.S.; Tomalia, D. Quantitation of non-ideal behavior in protein size-exclusion chromatography. J. Chromatogr. **1993**, *635*, 51.

7. Yan, D.; Zhou, Z.; Müller, A. Molecular weight distribution of hyperbranched polymers generated by self-condensing vinyl polymerization in the presence of a multifunctional initiator. Macromolecules **1999**, *32*, 245.

8. Strigel, A.M.; Plattner, R.D.; Willet, J.L. Dilute solution behavior of dendrimers and polysaccharides: SEC, ESI-MS, and computer modeling. Anal. Chem. **1999**, *71*, 978.

9. Mourey, T.H.; Turner, S.R.; Rubinstein, M.; Fréchet, J.M.J.; Hawker, C.J.; Wooley, K.L. The unusual behavior of dendritic macro-molecules: a study of the intrinsic viscosity, density and refractiveindex increment of poly-ether dendrimers. Macromolecules **1992**, *25*, 2401.

10. Puskas, J.E.; Grasmüller, M. Star-branched and hyperbranched polyisobutylenes. Macromol. Symp. **1998**, *132*, 117.

11. Gitsov, I.; Fréchet, J.M.J. Macromolecules **1994**, *27*, 7309.

12. Gitsov, I.; Fréchet, J.M.J. Solution and solid-state properties of hybrid linear-dendritic block copolymers. Macromolecules **1993**, *26*, 6536.

Derivatization of Analytes in Chromatography: General Aspects

Igor G. Zenkevich
Chemical Research Institute, St. Petersburg State University, St. Petersburg, Russia

INTRODUCTION

A priori information about analytes is available for most chromatographic analyses. Depending on the amount of information available, all determinations may be classified as: 1) preferably confirmatory (determined components are known); or 2) prospective (any propositions concerning their chemical nature are very approximate). Numerous differences in the design of analytical procedures in these two cases are manifested in the features of all stages of analysis—sampling, sample preparation, analysis itself, and interpretation of results. Preferably, for procedures classified as confirmatory, the stage of sample preparation should be supplemented by chemical treatment of the sample by different reagents for the optimization of subsequent chromatographic analysis. The most widely used kind of treatment is the synthesis of various chemical derivatives of target analytes, namely derivatization.

Derivatization is a special subgroup of organic reactions used in chromatography for compounds with selected types of functional groups. Not all known reactions can be applied as methods for derivatization, because these processes should be in accordance with some specific conditions. The consideration of numerous known recommendations[1–4] permits us to underline the following most important ones:

1. The experimental operations should be as simple as possible. The mixing of sample with reagent(s) at ambient temperature, without additional treatment of the mixture, is preferable. In chromatographic practice, the time needed for the completion of the derivatization reaction may be up to 24 hr (the so-called "stay overnight"). Instead of this long time, the heating of reaction mixtures in ampoules is also permitted. Some processes (including alkylation and silylation) may be realized by injection of the reaction mixtures into the hot injector of the GC equipment.
2. The number of stages of derivatization for any functional group in organic compounds should be minimal (one or two, but no more).

For multistage processes, the condition "single-pot synthesis" is necessary. Such operations as extraction or re-extraction are permitted only when the quantities of analytes are not very small or when it is necessary to isolate them from complex matrices. The large excess of derivatization reagent(s) and/or solvents should be easily removable, or have no influence on the results of the analysis. The use of high boiling solvents typically is not recommended. The possible by-products of reactions should have no influence on the results either.

3. The degree of transformation of initial compounds into products (yield, %) should be maximal and reproducible to provide for the quantitative determination of these compounds by analysis of their derivatives.
4. The chemical origin of the formed products should be strongly predictable. When a known derivatization reaction is put into practice for new compounds, this knowledge can be based on previously reported examples for the closest structural analogs of the target analytes.
5. Mutually unambiguous correspondence between the number of initial analytes and their derivatives should be assured. The optimal case for all compounds is $1 \rightarrow 1$, but numerous examples of type $1 \rightarrow 2$ are known (e.g., the derivatization of enantiomers by chiral reagents, which leads to the formation of a pair of diastereomers).[5] Similarly, the reaction of carbonyl compounds with *O*-alkoxyamines gives pairs of *syn-* and *anti-*isomers of oxime *O*-ethers etc. All processes that lead to further uncertainty (chemical multiplication of analytical signals), such as $1 \rightarrow N\,(N \geq 3)$, are not recommended for analytical practice. In connection with this, the number of reaction by-products should be minimal.
6. The feature of structural terminology of derivatives should be noted. If the number of newly added protecting groups in the molecules is unknown for derivatives of complex polyfunctional organic compounds, they can be classified in accordance with known derivatives.

Encyclopedia of Chromatography DOI: 10.1081/E-ECHR-120039945

For example, *N*-benzoyl glycine (PhCONHCH$_2$CO$_2$H) can form two trimethylsilyl (TMS) derivatives: *mono*-(PhCONHCH$_2$COOTMS) and *bis*-[PhC(OTMS)=NCH$_2$COOTMS]. If precise information on their chemical origin is unavailable, both of them can be named simply "benzoylglycine TMS," or "TMS #1" and "TMS #2" in the order of chromatographic elution.

7. The foregoing remarks follow from general features of derivatization reactions. It should be noted that this method is not used for completely unknown samples, because information about the presenting analytes is needed.

In accordance with the criteria mentioned, for example, *N,O*-trimethylsilyl derivatives of amino acids do not seem to be useful in analytical practice owing to the nonspecific *mono*- and *bis*-silylation of primary amino groups or postreaction hydrolysis of the resultant N–Si bonds. Even the simplest compounds of this class, H$_2$N–CHR–CO$_2$H, form three possible products, H$_2$N–CHR–CO$_2$TMS, TMSNH–CHR–CO$_2$TMS, and (TMS)$_2$N–CHR–CO$_2$TMS [TMS=Si(CH$_3$)$_3$], with different GC retention parameters. In the case of diamino monocarboxylic acids with nonequivalent amino groups [e.g., lysine, H$_2$N(CH$_2$)$_4$CH(NH$_2$)CO$_2$H], the number of similar semisilylated derivatives is theoretically increased up to nine.

MAIN FEATURES OF DERIVATIZATION REACTIONS

The greatest principal difference between organic reactions in general and those that can be considered chromatographic derivatization reactions is the de facto commonly accepted absence of necessity of product structure determination in the latter case. In "classical" organic chemistry, every synthesized compound must be isolated from the reaction mixture and characterized by physicochemical constants or spectral parameters for confirmation or estimation (for new objects) of its structure. Nevertheless, for the processes that have been classified as derivatization reactions, these operations are not necessary and generally not used in analytical practice. *The reaction itself is considered confirmation of the structure of the derivatives.* Of course, any exceptions to this important rule seem very dangerous and should be pronounced as special warnings for the application of any method of derivatization. Hence, in general cases, this method implies a risk in the ascribing of structures to the products formed in the chemical reactions. For new derivatives of complex organic compounds, the independent determination (or confirmation) of their structures seems desirable.

All organic reactions used for derivatization can proceed only in condensed phase, i.e., in solutions. None of these interactions are possible in gaseous media. Nevertheless, there is a special technique, flash derivatization, which involves joint (or more rarely, consecutive) injection of samples and reagents into GC equipment. It is noteworthy that in this case too, all reactions take place in the condensed phase, i.e., before evaporation of the samples. This method has been recommended for silylation, but is more often used for alkylation by a special group of chemicals—quaternized ammonium salts and hydroxides. For example, such reagents as 3,5-*bis*-(trifluoromethyl)-benzyl dimethylphenylammonium fluoride have been proposed for flash conversion of hydroxy compounds (preferably phenols and carboxylic acids) into their 3,5-*bis*-(trifluoromethyl)benzyl ethers.[6]

There are many examples where well-known derivatization reactions cannot be used in specific cases owing to the absence of mutually unambiguous correspondence between initial analytes and formed derivatives. For instance, methylation by diazomethane, CH$_2$N$_2$, is not recommended for barbiturates, because of the formation of mixtures of their *N*- and/or *O*-methyl derivatives. Another example is the interaction of dimethyl disulfide with conjugate dienes,[7] which gives complex mixtures of products and indicates the absence of regioselectivity. One of the frequently used derivatization methods for carbonyl compounds [RR′CO (including the important group of ketosteroids)] is their one-step treatment by *O*-alkyl hydroxylamines (R″ONH$_2$) with the expected formation of alkyl ethers of oximes (RR′C=NOR″). However, this reaction has an anomaly for compounds with C=C double bonds conjugated with carbonyl groups, that is, parallel addition of reagent with active hydrogen atoms to the polarized C=C bonds [see Eq. (1) below].[8]

This means that instead of one expected product with molecular weight (MW) = M_0 + 29 (with *O*-methyl hydroxylamine as reagent; M_0—molecular weight of initial carbonyl compound), reaction

$$R_2C=CR'-OR'' + CH_3ONH_2 \rightarrow R_2C=CR'-CR''=NOCH_3$$
$$\downarrow \qquad\qquad\qquad\qquad \downarrow$$
$$CH_3ONH-CR_2-CHR'-COR'' \rightarrow CH_3ONH-CR_2-CHR'-CR''=NOCH_3 \qquad (1)$$

mixtures may contain two additional products with MW $= M_0 + 47$ and $M_0 + 76$. This feature is negligible for the analysis of individual compounds, but when samples are mixtures of components of interest, the analysis becomes impossible because of the complexity of interpretation of the results.

In any case, the information about the estimated products of derivatization should be unambiguous. As an important example, it is interesting to note that for a long time, it was considered that various silylating agents could react only with compounds having active hydrogen atoms. However, only in 1999 was it shown[9] that typical reagents of this type (N-methyl trimethylsilyl acetamide, MSA) react with aromatic carbonyl compounds (preferably aldehydes) giving unusual products of acetamide addition to C=O bonds followed by their one- or two-step silylation. Keeping this fact in mind, it is better to name these products MSA adducts rather than TMS derivatives.

Only if the organic reaction is in accordance with all the abovementioned criteria may it be considered a method for derivatization. Finding new appropriate processes of this type is complex and often not obvious.

One of the main, but not sole, purposes of derivatization is the transformation of nonvolatile compounds into volatile derivatives. Each chromatographic method [GC, GC/MS, HPLC, capillary electrophoresis (CE), etc.] being supplemented by derivatization of analytes permits us to solve some specific problems. The principal among them are summarized briefly in Table 1; more detailed comments follow. Some of the derivatization methods mentioned can also be used in mass spectrometry, which includes no preliminary chromatographic separation of analytes,[10] but there are special derivatization techniques never used in chromatographic methods (for example, synthesis and analysis of isotopically labeled compounds).

Most monofunctional organic compounds [including alcohols (ROH), carboxylic acids (RCO$_2$H), amides (RCONH$_2$), etc.] are volatile enough for direct GC analysis. Exceptions are only those compounds with high melting points (sometimes with decomposition) because of strong intermolecular interactions

in their condensed phases {e.g., thiosemicarbazones (RR′C=N–NHCSNH$_2$), guanidines [RNH–C(=NH)–NH$_2$], etc.}. Ionic compounds [e.g., quaternary ammonium salts, $(R_4N)^+X^-$] are nonvolatile as well. If the compounds contain two or more functional groups with active hydrogen atoms [including the case of inner molecular ionic structures such as that seen in amino acids, $RCH(NH_3^+)CO_2^-$], their volatility decreases significantly. The purpose of derivatization of all these objects is to substitute active hydrogen atoms (better in all functional groups) by covalently bonded fragments that provide more volatile products. Direct GC analysis of highly reactive compounds {free halogens, hydrogen halides, strong inorganic acids like sulfonic acid (RSO$_2$OH), phosphonic acid [RPO(OH)$_2$], etc.} can be accompanied by their interaction with stationary phases of chromatographic columns and they also require derivatization.

If the initial compound A may be analyzed together with its derivative B, the comparison of their GC retention indices is a source of important information about the nature of these compounds. The average value of $\Delta RI = RI(B) - RI(A)$ may be used for the identification of both analytes and, if necessary, the reaction itself.[13] Selected ΔRI values for GC analysis on standard nonpolar polydimethyl siloxanes are presented in Table 2.

GC/MS analysis completely excludes the second item (see Table 1) from the possible aims of derivatization, insofar as the mass spectrometer itself is both a universal and selective GC detector. At the same time, two new important reasons for derivatization appear in this method. The intensities of molecular ion $(M^+ \cdot)$ signals are low for compounds having no structural fragments, which provides the effective delocalization of charge and unpaired electron in these ions. These fragments are conjugate bond and/or atom systems, or isolated heteroatoms with high polarizability (S, Se, I). In accordance with this regularity, O-TMS derivatives of alcohols in general show no $M^+ \cdot$ peaks in mass spectra, whereas for the TMS ethers of enols of carbonyl compounds (RCH$_2$COR′ → RCH=CR′–OSiMe$_3$) and TMS ethers of their oximes (RCH=N–OSiMe$_3$) (π–p–d conjugation systems), they are very intensive.

The determination of positions of C=C double bonds in the carbon skeletons of molecules is very often impossible owing to uncertain charge localization in molecular ions. The solution of this problem involves the conversion of unsaturated compounds into products whose molecular ions have sufficiently fixed charge localization. There are two methods by which to accomplish this localization: 1) addition of heteroatomic reagents directly to the C=C bond (the so-called "on-site" derivatization with the formation of TMS ethers of corresponding diols, adducts

Table 1 Principal applications of derivatization in different chromatographic techniques

Aims of derivatization	Typical examples
GC	
Transformation of nonvolatile, thermally unstable, and/or highly reactive compounds into stable volatile derivatives	$ROH \rightarrow ROSiMe_3$ $ArOH \rightarrow ArOCOCF_3$ $RR'CO \rightarrow RR'C{=}NOCH_3$
Synthesis of derivatives for element-specific GC detectors or conversion of nondetectable compounds into suitable products for minimization of detection limit	$ROH \rightarrow ROCOCCl_3$ (ECD) $RCO_2H \rightarrow RCO_2CH_2CCl_3$ (ECD) $HCO_2H \rightarrow HCO_2CH_2C_6H_5$ (FID)
Combination with stage of sampling (preferably in environment analyses when derivatization is used as method of chemisorption)	$RCHO \rightarrow 2,4\text{-}(NO_2)_2C_6H_3\text{--}NH\text{--}N{=}CHR$
Separation of enantiomeric compounds on nonchiral phases after their conversion into diastereomeric derivatives	$RR'C^*HNH_2 + C_6H_5C^*H(OMe)COCl$ $\rightarrow RR'C^*HNH\text{--}COC^*H(OMe)C_6H_5$
GC/mass spectrometry	
Determination of molecular weights of compounds with $W_M \approx 0$ at electron impact ionization (synthesis of derivatives with conjugated bond and/or atom systems)	$RR'CO \rightarrow RR'C{=}NNH\text{--}C_6F_5$ (π–p–π conjugation system) $RNH_2 \rightarrow RN{=}CH\text{--}NMe_2$ (p–π conjugation system)
Increase of specificity of molecular ion fragmentation for estimation of structure of analytes (e.g., determination of C=C double bond position in carbon skeleton of molecules)	"On-site" derivatization:[11] $RCH{=}CHR'$ $\rightarrow R\text{--}CH(SMe)\text{--}CH(SMe)\text{--}R'$ "Remote-site" derivatization:[12] $RCH{=}CH(CH_2)_nCO_2H$ $+H_2NCH_2CH_2OH \rightarrow$ 2-substituted oxazolines
HPLC with UV detection and CE	
Synthesis of chromogenic derivatives (with chromophores that provide adsorption within typical range of UV detection of 190–700 nm)	$C_6H_7O(OH)_5 \rightarrow C_6H_7O(OCOC_6H_5)_5$
Conversion of hydrophilic analytes into more hydrophobic derivatives	$RCH(NH_2)CO_2H \rightarrow RCH(NHCSNHC_6H_5)CO_2H$
Various chromatographic techniques	
Determination of number of functional groups with active hydrogen atoms using mixed derivatization reagents	$X(OH)_n + [(R_1CO)_2O + (R_2CO)_2O)] \rightarrow$ mixture of miscellaneous acyl derivatives

C^*—chiral carbon atoms.

with dimethyl disulfide,[11] etc.); and 2) introduction or formation of nitrogen-containing heterocycles rather far from the target C=C bond[12] ("remote-site" derivatization).

The formation of new chromophores for the optimization of UV detection of analytes in HPLC involves the synthesis of derivatives with conjugate systems in the molecule. Compared with GC, there are no restrictions on the volatilities of these derivatives for HPLC analysis. They may be synthesized before analysis (precolumn derivatization) or after chromatographic separation (postcolumn derivatization, or, in other words, reaction GC). The latter technique, as a method of identification of analytes, was highly popular until the 1970s. However, at present, this approach has practically lost its significance owing to the progress of GC/MS methods. Very few new GC applications of this method have been reported during the past dozens years or so

(see, for example, Ref.[14]), but it is still used in HPLC because it permits us to combine the measurement of retention parameters of initial analytes with detection of their derivatives.[15]

The range of most convenient hydrophobicity of organic compounds for reversed phase (RP) HPLC separation may be estimated approximately as $-1 \leq \log P \leq +5$ ($\log P$ is the logarithm of the partition coefficient of the compound being characterized in the standard solvent system 1-octanol/water). Highly hydrophilic substances with $\log P \leq -1$ need a special choice of analysis conditions, e.g., introduction of ion-pair additives into the eluents. Another approach is their conversion to more hydrophobic derivatives by the modification of functional groups with active hydrogen atoms.

The examples mentioned here for RP-HPLC analysis of monosaccharides in the form of their perbenzoates and amino acids as *N*-phenylthiocarbamoyl

Table 2 Average values of differences of GC retention indices (ΔRI) for some derivatization reactions

$\Delta MW = MW_B - MW_A$	Scheme of reaction A $\rightarrow$ B (for monofunctional compounds only)	$\Delta RI \pm s_{\Delta RI}$
14	$RCO_2H \rightarrow RCO_2Me$	-102 ± 28
14	$ArOH \rightarrow ArOMe$	-62 ± 16
28	$RCO_2H \rightarrow RCO_2Et$	-42 ± 7
28	$RR'CO \rightarrow RR'C=NNHMe$	229 ± 23
29	$ROH \rightarrow RONO$	-6 ± 27
42	$ROH \rightarrow ROCOMe$	142 ± 18
42	$ArOH \rightarrow ArOCOMe$	97 ± 20
42	$RNH_2 \rightarrow RNHCOMe$	437 ± 29
42	$ArNH_2 \rightarrow ArNHCOMe$	401 ± 7
42	$RR'CO \rightarrow RR'C=NNMe_2$	302 ± 20
72	$ROH \rightarrow ROSiMe_3$	119 ± 18
72	$RCO_2H \rightarrow RCO_2SiMe_3$	76 ± 16
96	$ROH \rightarrow ROCOCF_3$	-85 ± 24
114	$RCO_2H \rightarrow RCO_2SiMe_2\text{-}tert\text{-}Bu$	288 ± 29

derivatives (Table 1) satisfy both principal criteria: introducing the chromophores into molecules of analytes (C_6H_5CO- and $C_6H_5NH-CS-NH-$) and optimization of their retention parameters.

Sometimes, the generally prohibited multiplication of analytical signals of derivatives may be attained artificially for the solution of special problems. For example, the treatment of polyhydroxy compounds (phenols, phenol carboxylic acids, polyamines, etc.) $[X(OH)_n]$ by equimolar mixtures (1:1) of acylation reagents $[(R_1CO)_2O + (R_2CO)_2O]$ leads to the formation of $(n + 1)$ miscellaneous acyl derivatives $X(OCOR_1)_n, X(OCOR_1)_{n-1}(OCOR_2), \ldots, X(OCOR_2)_n$. The relative abundances of their chromatographic peaks should be close to the binomial coefficients, i.e., 1:1 (at $n = 1$), 1:2:1 ($n = 2$), 1:3:3:1 ($n = 3$), and so forth. Moreover, the differences of retention indices of all these mixed derivatives are close to each other. These two regularities permit us to determine the number of hydroxyl groups (n) in the molecules of analytes. This mode of derivatization can be realized in both HPLC[16,17] and GC[18] conditions.

Insofar as the derivatization can be considered one of the stages of sample preparation for chromatographic analysis, it can be combined with other procedures, for instance, the preconcentration of traces of analytes. For example, the yield of solid-phase extraction or microextraction of organic compounds from aqueous solutions with modified silica gels is better for more hydrophobic substances; the preliminary conversion of acidic compounds into suitable derivatives is recommended.[19]

CONCLUSIONS

Chemical derivatization as a stage of sample preparation is a widespread approach in various chromatographic and related techniques. It is used for the conversion of compounds that cannot be analyzed directly into suitable products (derivatives). The aims of this treatment of samples are very diverse and depend on the final goals of analyses as a whole. One of the most important problems to be solved using derivatization is the transformation of nonvolatile compounds into products volatile enough for GC analysis. Other aims of this method include the optimization of detection and structure evaluation of analytes.

A disadvantageous feature of derivatives of complex organic compounds can be the uncertainty in their structures. This requires their determination (or confirmation) by independent methods (mass spectrometry), or the exhaustive characterization of reactions classified as those of derivatization.

ARTICLES OF FURTHER INTEREST

Kovats' Retention Index System, p. 901.
Steroids: Derivatization for GC Analysis, p. 1600.

REFERENCES

1. Blau, K., King, G.S., Eds.; *Handbook of Derivatives for Chromatography*; Heiden: London, 1977; 576.
2. Knapp, D.R. *Handbook of Analytical Derivatization Reactions*; John Wiley & Sons: New York, 1979; 741.
3. Drozd, J. *Chemical Derivatization in Gas Chromatography*; Journal of Chromatography Library; Elsevier: Amsterdam, 1981; 19, 232.
4. Blau, K., Halket, J.M. Eds.; *Handbook of Derivatives for Chromatography*, 2nd Ed.; John Wiley & Sons: New York, 1993; 369.
5. Allenmark, S.G. *Chromatographic Enantioseparation: Methods and Applications*; Ellis Horwood Ltd.: New York, 1988; 268.
6. Amijee, M.; Cheung, J.; Wells, R.J. Development of 3,5-*bis*-(trifluoromethyl)benzyl-dimethylphenyl-ammonium fluoride, an efficient new on-column derivatization reagent. J. Chromatogr. A. **1996**, *738*, 57–72.
7. Vincentini, M.; Guglielmetti, G.; Gassani, G.; Tonini, C. Determination of double bond position in diunsaturated compounds by mass spectrometry of dimethyl disulfide derivatives. Anal. Chem. **1987**, *59* (7), 694–699.
8. Zenkevich, I.G.; Artsybasheva, Ju.P.; Ioffe, B.V. Application of alkoxyamines for derivatization of carbonyl compounds in gas chromatography–mass spectrometry. Zh. Org. Khim. (Russ.) **1989**, *25* (3), 487–492.
9. Little, J.L. Artifacts in trimethylsilyl derivatization reactions and ways to avoid them. J. Chromatogr. A. **1999**, *844*, 1–22.
10. Zaikin, V.G.; Mikaya, A.I. *Chemical Methods in Mass Spectromery of Organic Compounds* (in Russian); Nauka Publ. House: Moscow, 1987; 200.
11. Buser, H.-R.; Arn, H.A.; Guerin, P.; Rauscher, S. Determination of double bond position in mono-unsaturated acetates by mass spectrometry of dimethyl disulfide adducts. Anal. Chem. **1983**, *55* (6), 818–822.
12. Yu, Q.T.; Liu, B.N.; Zhang, J.Y.; Huang, Z.H. Location of double bonds in fatty acids of fish oil and rat testis lipids. GC–MS of the oxazoline derivatives. Lipids **1989**, *24* (1), 79–83.
13. Zenkevich, I.G. Chromatographic characterization of organic reactions by additivity of GC retention parameters of reagents and products. Zh. Org. Khim. (Russ.) **1992**, *29* (9), 1827–1840.
14. Mikaia, A.I.; Trusova, E.A.; Zaikin, V.G.; Zegelman, L.A.; Urin, A.B.; Volinsky, N.P. Reaction gas chromatography/mass spectrometry. IV. Postcolumn hydrodesulfurization in capillary GC/MS as an aid in structure elucidation of cyclic sulfides within mixtures. J. High Resolut. Chromatogr. Chromatogr. Commun. **1984**, *7* (11), 625–628.
15. Vassilakis, I.; Tsipi, D.; Scoullos, N. Determination of a variety of chemical classes of pesticides in surface and ground waters by off-line solid-phase extraction, GC with ECD and NP-detection and HPLC with post-column derivatization and fluorescence detection. J. Chromatogr. A. **1998**, *823*, 49–58.
16. Zenkevich, I.G. New applications of the retention index concept in gas and high performance liquid chromatography. Fresenius J. Anal. Chem. **1999**, *365* (4), 305–309.
17. Zenkevich, I.G. Determination of the number of functional groups with active hydrogen atoms in phenols and aromatic amines by HPLC. Zh. Phys. Khim. (Russ.) **1998**, *72* (6), 1131–1136.
18. Zenkevich, I.G.; Rodin, A.A. Gas chromatographic one-step determination of the number of hydroxyl groups in polyphenols using mixed derivatization reagents. Zh. Org. Khim. (Russ.) **2002**, *5* (7), 732–736.
19. Nilsson, T.; Baglio, D.; Galdo-Miques, I.; Madsen, O.J.; Facchetti, S. Derivatization/solid-phase microextraction followed by GC–MS for the analysis of phenoxy acid herbicides in aqueous samples. J. Chromatogr. A. **1998**, *826*, 211–216.

Detection in Countercurrent Chromatography

M.-C. Rolet-Menet
*Laboratoire de Chimie Analytique, UFR de Sciences Pharmaceutiques et Biologiques,
Paris Cedex, France*

INTRODUCTION

Detection of solutes is an essential link in the separation chain. It helps to reveal solute separation by detecting them in the column effluent, and in some cases, it could permit their characterization. These objectives are based on the various physical properties of the products.

Countercurrent chromatography (CCC) is a chromatographic method that separates solutes that are more or less retained in the column by a stationary phase (liquid in this case) and are eluted at the outlet of the column by a mobile phase. Two treatments of column effluent have been used up to now in CCC. Either the column outlet is directly connected to a detector commonly used in HPLC (on-line detection) or fractions of mobile phase are collected and analyzed by spectrophotometric, electrophoretic, or chromatographic methods etc (off-line detection).

The first one is more practical and rapid to carry out. It is commonly used in analytical applications of CCC and also in preparative CCC to analyze effluent continuously and to follow the steps of the separation.

The second one is often tedious because each fraction must be analyzed. However, it is of great interest in preparative applications of CCC, especially to measure the purity of fractions and the biological activity of separated compounds and also to recover a product from one fraction or some selected fractions to resolve its chemical structure.

ON-LINE DETECTION

This type of detection can be used as such in preparative CCC to monitor separations, before the fraction collector, if any, and in analytical CCC (for instance, during the determination of $\log P_{octanol/water}$).

Several detectors used in HPLC and in supercritical fluid chromatography (SFC) can be connected to the CCC column[1] to detect solutes and thus follow separation. They can be, for instance, fluorimeters (very sensitive and used without modifications in CCC), UV visible spectroscopes,[1] evaporative light scattering detectors,[1,2] atomic emission spectroscopes,[3] etc. Some detectors give more information

than the detection of the solute, such as structural information of separated components, as in infrared spectroscopy,[4] mass spectrometry,[5] or nuclear magnetic resonance.[6] These detectors are used either on-line with a collector fraction or in parallel if they are destructive.

UV Detection

The UV–visible detector is the universal detector used in analytical and preparative CCC. It does not destroy solutes. It is used to detect organic molecules with a chromophore moiety or mineral species after formation of a complex (the rare earth elements with Arsenazo III,[7] for instance). Several problems can occur in direct UV detection, as has already been described by Oka and Ito:[8] 1) carryover of the stationary phase due to improper choice of operating conditions, with appearance of stationary phase droplets in the effluent of the column; 2) overloading of the sample, vibrations, or fluctuations of the revolution speed; 3) turbidity of the mobile phase due to difference in temperature between the column and the detection cell; or 4) gas bubbling after reduction of effluent pressure. Some of these problems can be solved by optimization of the operating conditions, better control of the temperature of the mobile phase, and addition of some length of capillary tubing or a narrow-bore tube at the outlet of the column before the detector to stabilize the effluent flow and to prevent bubble formation. The problem of stationary phase carryover (especially encountered with hydrodynamic mode CCC devices) can be solved by the addition between the column outlet and UV detector of a solvent that is miscible with both stationary and mobile phases and that allows one to obtain a monophasic liquid in the cell of the detector[1] (a common example is isopropanol).

Evaporative Light Scattering Detection

Evaporative light scattering detection (ELSD) involves atomization of the column effluent into a gas stream via a Venturi nebulizer, evaporation of solvents by passing it through a heated tube to yield an aerosol of nonvolatile solutes, and finally measurement of the

Encyclopedia of Chromatography DOI: 10.1081/E-ECHR-120039950

intensity of light scattered by the aerosol. After a suitable evaporation step, in the worst case of segmented or emulsified mobile phase, the column effluent should always be an aerosol of the solutes before reaching the detection cell. It can be used without modifications. For molecules without chromophore or fluorophore groups or with mobile phases with a high UV cut-off (acetone, ethyl acetate, etc.), ELSD is useful.[1] But it cannot detect fragile or easily sublimable solutes because the nebulizate is heated. Moreover, this detection method does not preserve the solutes. To collect column effluent, a split must be installed at the outlet of the column to allow ELSD detection in a parallel direction to fraction collection with consequent loss of solutes.

Atomic Emission Spectrometry[3]

This detection mode can be used during ion separation. Kitazume et al. used a direct plasma atomic emission spectrometer (DCP, Spectra-Metrics Model Spectra-Span IIIB system with fixed-wavelength channels) for observation of the elution profile during the separation of nickel, cobalt, magnesium, and copper by CCC. For profile measurement of a single element, an analog recorder signal from the DCP was converted into a digital signal. The digital data were stored in a workstation and the elution profile was plotted. For simultaneous multielement measurement, the emission signal for each channel was integrated for 10 sec at intervals of 20 sec, and the integrated data were printed out.

Infrared Spectrometry[4]

Romanach and de Haseth have used a flow cell for liquid chromatography/Fourier transform-infrared

spectrometry (LC/FT-IRS) in CCC. The main difficulty is the absorbance of the liquid mobile phase. This problem is exacerbated in LC by low solute to solvent ratios in the eluates. In contrast, CCC leads to a high solute to solvent ratio so that it can be used with a very simple interface with a CCC column without any complex solvent removal procedures. High sample loadings are possible by using variable path length of the IR detector (from 0.025 to 1.0 mm).

Mass Spectrometry[5]

Several interfaces have been used in CCC/MS. The first employed is the thermospray (TSP). When the column is directly coupled with TSP MS, the CCC column often breaks due to the high backpressure generated by the TSP vaporizer. In contrast, other interfaces using methods such as fast atom bombardment (FAB), electron ionization (EI), and chemical ionization (CI) have been directly connected to a CCC column without generating high backpressure. Such interfaces can be applied to analytes with broad polarity. As it is suitable to introduce effluent from the column CCC into MS only at a flow rate between 1 and 5 µl/min, the effluent is usually introduced into MS through a splitting tee, which is adjusted to an adequate ratio.

Nuclear Magnetic Resonance

Nuclear magnetic resonance (NMR) gives maximum structural information and allows measurement of the relative concentrations of eluted compounds. Spraul et al.[6] experimented with coupling of pH zone refining centrifugal partition chromatography (CPC)

Table 1 Off-line detection

Molecules	Fraction analysis
Schisanhenol acetate 5 and 6 of *Schisandra rubriflora*[5]	TLC
	Purity control by HPLC
Bacitracin complex[5]	Absorbance measure at 234 nm
	Purity control by HPLC
Dye species[5]	Mass spectrometry
Thyroid hormone derivatives[5]	UV on line at 280 nm. Gamma radioactivity measure of fractions
	Purity control by TLC, HPLC, and UV spectra
Cerium chloride and erbium chloride[5]	Inductively coupled plasma–atomic emission spectroscopy (ICP–AES)
Recombinant uridine phosphorylase[9]	SDS–PAGE
	Enzymatic activity by Magni method
Torpedo electroplax membranes[9]	Percentage of cholinergic receptor

with NMR by using a biphasic system based on D_2O and an organic solvent.

On-line pH Monitoring

In pH zone refining, solutes are not eluted as separated peaks but as contiguous blocks of constant concentrations, so that it is highly difficult to monitor the separation by means of a UV detector. On-line pH monitoring is generally used, allowing the observation of transitions between solutes, since each zone has its own pH determined by the pK_a and the solute concentration. The experiment was carried out in stop–flow mode.

OFF-LINE DETECTION

The analysis of the mobile phase fractions collected at the outlet of the column is the oldest method used in CCC (droplet CCC and rotation locular CCC) to evaluate the quality of separation and to characterize solutes. With modern CCC methods such as CPC, CCC Type J, and cross axis, numerous applications have been described for preconcentration and preparative chromatography. Table 1 lists some applications described in reference books.[5,9] The type of detection used for each fraction depends on the isolated solute. They are TLC and HPLC on line with UV detector or mass spectrometer[10–12] (HPLC also enables an estimation of each fraction's purity,[13,14] a determination of fingerprints of medicinal plants,[15] etc.) for organic solutes, ICP–AES for mineral species, and polyacrylamide gel electrophoresis (PAGE) for biological molecules.[16,17] If the purity of the compound is satisfactory, a study by direct injection MS[18] and NMR[19] allows determination of its chemical structure. Biochemical tests are also available to verify the biological activity of biomolecules, which are often separated and collected in aqueous two-phase systems.[17]

CONCLUSIONS

High-speed CCC is mainly dedicated to preparative separations. Two types of detection are available: on-line and off-line detections. The first allows one to follow the quality of the separation. The second is suited to the analysis of fractions collected during preparative separation.

REFERENCES

1. Drogue, S.; Rolet, M.-C.; Thiébaut, D.; Rosset, R. Improvement of on-line detection in high-speed counter-current chromatography: UV absorptiometry and evaporative light-scattering detection. J. Chromatogr. A. **1991**, *538*, 91–97.

2. Bourdat-Deschamps, M.; Herrenknecht, C.; Akendengue, B.; Laurens, A.; Hocquemiller, R. Separation of protoberberine quaternary alkaloids from a crude extract of *Enantia chlorantha* by centrifugal partition chromatography. J. Chromatogr. A. **2004**, *1041* (1–2), 143–152.

3. Kitazume, E.; Sato, N.; Saito, Y.; Ito, Y. Separation of heavy metals by high-speed countercurrent chromatography. Anal. Chem. **1993**, *65*, 2225–2228.

4. Romanach, R.J.; de Haseth, J.A. Flow cell CCC/FT-IR spectrometry. J. Liq. Chromatogr. A. **1988**, *11* (1), 133–152.

5. Oka, H. High-speed counter current chromatography/mass spectrometry. In *High-Speed Counter Current Chromatography*; Ito, Y., Conway, W.D., Eds.; John Wiley & Sons: New York, 1995; 73–91.

6. Spraul, M.; Braumann, U.; Renault, J.-H.; Thépinier, P.; Nuzillard, J.-M. Nuclear magnetic resonance monitoring of centrifugal partition chromatography in pH-zone refining mode. J. Chromatogr. A. **1997**, *766*, 255–260.

7. Kitazume, E.; Bhatnagar, M.; Ito, Y. Separation of rare earth elements by high-speed counter-current chromatography. J. Chromatogr. A. **1991**, *538*, 133–140.

8. Oka, H.; Ito, Y. Improved method for continuous UV monitoring in high-speed counter-current chromatography. J. Chromatogr. A. **1989**, *475*, 229–235.

9. Lee, Y.W. Cross-axis counter current chromatography: a versatile technique for biotech purification. In *Counter Current Chromatography*; Menet, J.-M., Thiébaut, D., Eds.; Marcel Dekker Inc.: New York, 1999; 149–169.

10. Oka, H.; Harada, K.-I.; Suzuki, M.; Fuji, K.; Iwaya, M.; Ito, Y.; Goto, T.; Matsumoto, H.; Ito, Y. Purification of quinoline yellow components using high-speed counter-current chromatography by stepwise increasing the flow-rate of the mobile phase. J. Chromatogr. A. **2003**, *989* (2), 249–255.

11. Chen, L.-J.; Games, D.E.; Jones, J. Isolation and identification of four flavonoid constituents from the seeds of *Oroxylum indicum* by high-speed counter-current chromatography. J. Chromatogr. **2003**, *988* (1), 95–105.

12. Han, X.; Pathmasiri, W.; Bohlin, L.; Janson, J.-C. Isolation of high purity 1-[2′,4′-dihydroxy-3′,5′-di-(3″-methylbut-2″-enyl)-6′-methoxy]phenylethanone from *Acronychia pedunculata* by high-speed counter-current chromatography. J. Chromatogr. A. **2004**, *1022* (1–2), 213–216.

D

13. Chen, F.; Lu, H.-T.; Jiang, Y. Purification of paeoniflorin from Paeonia lactiflora by high-speed countercurrent chromatography. J. Chromatogr. A. **2004**, *1040* (2), 205–208.

14. Jiang, Y.; Lu, H.-T.; Chen, F. Preparative purification of glycyrrhizin extracted from the root of liquorice using high-speed counter-current chromatography. J. Chromatogr. A. **2004**, *1033* (1), 183–186.

15. Gu, M.; Ouyang, F.; Su, Z. Comparison of high-speed counter-current chromatography and high-performance liquid chromatography on fingerprinting of Chinese traditional medicine. J. Chromatogr. A. **2004**, *1022*, 139–144.

16. Yanagida, A.; Isozaki, M.; Shibusawa, Y.; Shindo, H.; Ito, Y. Purification of glycosyltransferase from cell-lysate of *Streptococcus mutans* by counter-current chromatography using aqueous polymer two-phase system. J. Chromatogr. B. Anal. Technol. Biomed. Life Sci. **2004**, *805* (1), 155–160.

17. Shibusawa, Y.; Fujiwara, T.; Shindo, H.; Ito, Y. Purification of alcohol dehydrogenase from bovine liver crude extract by dye–ligand affinity counter-current chromatography. J. Chromatogr. B. Anal. Technol. Biomed. Life Sci. **2004**, *799* (2), 239–244.

18. Wu, S.; Sun, C.; Cao, X.; Zhou, H.; Zhang, P.; Pan, Y. Preparative counter-current chromatography isolation of liensinine and its analogues from embryo of the seed of *Nelumbo nucifera* using upright coil planet centrifuge with four multiplayer coils connected in series. J. Chromatogr. A. **2004**, *1041* (1–2), 153–162.

19. Sannomiya, M.; Rodrigues, C.M.; Coelho, R.G.; dos Santos, L.C.; Hiruma-Lima, C.A.; Souza Brito, A.R.M.; Vilegas, W. Application of preparative high-speed counter-current chromatography for the separation of flavonoids from the leaves of *Byrsonima crassa* Niedenzu. J. Chromatogr. A. **2004**, *1035* (1), 47–51.

Detection Methods for Field-Flow Fractionation

Martin Hassellöv
Analytical and Marine Chemistry, Göteborg University, Göteborg, Sweden

Frank von der Kammer
Environmental Geoscience, Institute for Geoscience, Vienna University, Vienna, Austria

INTRODUCTION

The main purpose of the detector in a field-flow fractionation (FFF) system is to quantitatively determine particle number, volume, or mass concentrations in the FFF size-sorted fractions. Consequently, a number, volume, or mass dependent size distribution of the sample can be derived from detection systems applied to FFF [e.g., UV–VIS, fluorescence, refractive index, inductively coupled plasma ionization mass spectrometry (ICPMS)]. Further, on-line light scattering detectors can provide additional size and molecular weight distributions of the sample.

An analytical separation technique requires a detection method responding to some or all of the components eluting from the separation system. The choice of detector is determined by the demands of the sample and analysis. For FFF techniques, many of the detection systems have evolved from those used in LC techniques.

Detection can be carried out either with an on-line detector coupled to the eluent flow, or by collection and subsequent analysis of discrete fractions. For collected fractions, a range of analytical methods can be used, both quantitative (e.g., radioactive isotope labeling and metal analysis) and more qualitative (e.g., microscopic techniques). On-line detectors suitable for coupling to the FFF channels include both nondestructive flow through cell systems and destructive analysis systems. It is often desirable to use on-line detection if possible since the total analysis time is much less than for discrete fraction analysis. Regardless of detector type, the dead volumes and flows in the system between the FFF channel and detector or fraction collector must be accurately determined and corrected for.

If the signal from a detector is a factor of two properties, it is possible to use another detector on line to resolve the different properties, e.g., multiangle light scattering (MALS) in combination with differential refractive index (DRI), or continuous viscosimetry + DRI. Alternatively, the two detectors may respond to two different properties of interest: In either case, it is almost as simple to acquire multiple detector signals as a single one. Multiple detectors can be arranged either in series or in parallel. A parallel detector arrangement avoids the band broadening problem encountered in the serial arrangement, where the first detector may cause significant band broadening for the second, due to the dead volume in the flow cell. For a serial detector connection, it is best to have the one with the smallest dead volume first, as long as it is not a destructive detector. Some detectors, such as DRI detectors, have restrictions regarding the maximum allowable pressure in the flow cell and must be connected last in series. For parallel coupling, the outflow from the FFF channel needs to be split to two or more detectors, and it is then essential to have control of the individual flows, since changes can induce drift in sensitivity and shifts in the dead times between the channel and detector during a run.

When choosing detector and experimental conditions, one needs to consider analyte concentration, detector sensitivity, background level, and detection limits. The maximum amount of analytes that can be injected is usually limited by sample overloading in the FFF channel (interparticle interactions), which disturbs the separation. It is necessary to have an analyte detection limit well below the overloading sample concentration to be able to quantify the peaks without too noisy a background. When using multiple detectors on line, their sensitivity may be quite different either overall or as a function of size range. An example of this is the use of a MALS detector, which has much higher sensitivity for larger particles, together with a DRI detector, which has the opposite sensitivity properties, making the small and large particle ranges difficult to cover.

OPTICAL DETECTION SYSTEMS

UV–VIS spectrophotometers are the most commonly used detectors for FFF applications mainly due to their availability, simplicity, and low cost. The majority of FFF work to date has focused on separation

Encyclopedia of Chromatography DOI: 10.1081/E-ECHR-120039951

method development where the use of a UV–VIS detector showing the quality of the separations is sufficient. However, the quantification of the separated particles or macromolecules is not always straight forward, since the UV–VIS signal is actually a turbidimetric measure for solid particles that is fully based on light scattering principles and an absorption measure for light-absorbing macromolecules. The absorbance contribution is only dependent on concentration, but there is a more complicated relationship involved in the light scattering signal, and both principles may be applicable if the solid particles and macromolecules are of comparable. In case of turbidity, large particles scatter light much more effectively than smaller particles, and particles with varying composition and refractive indices give rise to further complications. The correction of the detector signal according to Mie scattering theory is complicated but can often be simplified with appropriate assumptions.[1] For particles larger than 1 µm, efforts have been put into development of an absolute or standard-free quantification method using UV–VIS detection for gravitational FFF.[2] For the scattering phenomenon of nanometer sized particles, a corrective method was developed recently by evaluation of the turbidity spectrum acquired with spectral resolved UV/Diode array detector (DAD) detection.[3] In principle, the light scattering signal is dependent on particle properties such as, e.g., concentration and size, but also on the observation angle and the wavelength of the incident light. In the case of UV/DAD and fluorescence detectors, where only one fixed observation angle is available, the light scattering may be evaluated as a function of the applied wavelength. This approach was successfully applied[3] for the correction of the turbidity signal of FFF fractionated latex beads, which were smaller in size than the applied wavelength ($\lambda = 254$ nm).

The DRI detector is very common in SEC and records any change in refractive index of the sample stream relative to a reference stream. It is a general detector with the advantage of responding to almost all solutes, and it is concentration selective. The sensitivity of a DRI detector is not always the best, but new detector models offer different lengths of optical path, so that the sensitivity can be adjusted to match sample concentration. The DRI detector is not sensitive to changes in flow rate, but is highly sensitive to temperature changes. It is probably the most used detector for FFF applications after the UV detector.

Flows through fluorescence detectors are very common in LC, mainly due to the high selectivity and good signal-to-noise ratio. Only a few papers on FFF with fluorescence detectors are published, but when the analytes have suitable fluorescence properties, this is an excellent choice.

For fluorescence detectors offering excellent stray light suppression, a special mode of operation is available that turns them into simple light scattering detectors. By setting the excitation wavelength equal to the emission wavelength, the light scattered by particles is observed at 90° to the incident light. In contrast to turbidimetry in UV–VIS detectors, this is termed nephelometry and not interfered by light-absorbing substances. The problem of the dependence of the light scattering signal on sample properties other than particle concentration, such as particle size and shape, is also present with this technique.

Photon correlation spectroscopy (PCS), also called dynamic light scattering or quasielastic light scattering, correlates the short-term fluctuations of the light scattering signal to the diffusion coefficients of the sample particles. Photon correlation spectroscopy is a valuable tool in validation of FFF separations. One prerequisite of PCS is that the sample itself be at rest and show nearly no motion other than the Brownian motion of the analyte. Hence, it is too slow to be of practical use as an on-line detector. Recent flow-through static light scattering detectors offer a PCS option. This can only provide the PCS detection on line and "in flow" for very small macromolecules and particles, which must be present at sufficiently large concentrations. Photon correlation spectroscopy has, for example, been used for verification of the average sizes obtained from FFF theory for discrete fractions of emulsion separated using sedimentation FFF.[4]

Improvements in light scattering theory and instrumentation have been going on for decades, but the development of the MALS instrument—from the earlier low angle light scattering (LALS) technology, now incorporating up to 18 detectors measuring the scattered light at individual angles—presents a breakthrough in particle sizing. Compared to LALS instruments, multiangle detection allows more physical properties of the particles to be derived from the results. Also, the MALS instrument has higher sensitivity and is less affected by dust particles in the sample. Light scattering techniques give average values of the properties of the particle population in the sample and do not describe the property distribution of the sample, but when coupled to a particle sizing technique, such as FFF, the distributions of the different properties are derived from each size fraction.

MALS theory has been thoroughly described in several papers by Wyatt[5] and will only be mentioned briefly here. For each size fraction or batch measurement, the following applies as long as the limits of the Rayleigh–Gans–Debye approximation are fulfilled. This means that the particles are smaller than the incident light's wavelength, the refractive index is similar to that of the solvent, and no light absorption occurs:

$$\frac{Kc}{R_\theta} = \frac{1}{M_w P(\theta)} + 2A_2 c + \cdots \qquad (1)$$

$$P(\theta) = 1 - a_1[2k\sin(\theta/2)]^2 + a_2[2k\sin(\theta/2)]^4 - \cdots \qquad (2)$$

K is a light scattering constant including refractive index increment and wavelength of the scattering light and A_2 is the second virial coefficient. If the sample is very dilute, the second term of Eq. (1) can be neglected and the excess Rayleigh ratio, R_θ (net light scattering contribution from each component at angle θ), becomes directly proportional to $M_w P(\theta)$. On plotting R_θ/Kc against $\sin^2(\theta/2)$, the intercept yields molecular weight (M_w) at the concentration c, and from the slope, the root mean square radius (RMS radius) can be derived. One great advantage with the MALS detector is that it does not demand calibration of the channel with reference materials. The absolute concentration of the analyte is necessary to determine the molar mass of the analyte since the signal includes a factor of concentration. The determination of the RMS radius is independent of concentration and can be achieved from the MALS signals alone. To acquire the sample concentration at each time slice, a concentration calibrated DRI detector is commonly used on line with the MALS detector. FFF/MALS/DRI is receiving much interest and attention and many applications have been developed in recent years, especially in synthetic polymer and biopolymer characterization. Thielking and Kulicke[6] have published papers on the coupling of FFF/MALS/DRI for analysis of both polystyrene particles and smaller polystyrene sulfonates (PSS). Fig. 1 shows their results of DRI derived concentration and molecular weight given from MALS data as a function of elution volume for seven PSS standards. However, for small molecules (<10 kDa), the sensitivity of the MALS detector is rather poor. The use of MALS together with FFF proved to be useful and complemented SEC/MALS techniques for polymers that could not be fractionated by SEC. After the first applications were documented,[6] several FFF subtechniques were coupled to MALS. The target analytes were mainly polysaccharides,[7] different kinds of polystyrene latex particles,[5,8] and starch polymers.[9] Nearly no characterization of natural particles using FFF/MALS is reported. Magnuson et al.[10] analyzed freshly precipitated iron oxides with FFF/MALS, and it was applied to analyze natural colloids extracted from soil.[11]

The technique complements FFF ideally, since FFF provides a prefractionation to overcome limitations of MALS on broad distributed bulk samples, and MALS delivers independent molar mass and RMS radius determination.

In an evaporative light scattering detector (ELSD), the sample is nebulized, and when the solvent in the resulting droplets is evaporated, their mass content is proportional to particle mass in the sample stream. The particles are detected with a laser light scattering detector and the signal is related to their size. The ELSD has not been extensively used in FFF. Oppenheimer and Mourey[12] showed that it can be a good complement to turbidimetric detection in sedimentation FFF for particles smaller than 0.2 μm. This detector is free from the problems associated with UV detectors when applied to a broad size range or samples with differing extinction coefficients over the size range. Further, it can be used for samples lacking absorbance characteristics. Compton, Myers, and Giddings[13] presented a single particle detector for steric FFF (1–70 μm) based on light scattering of single particles flowing through the laser light path. Today, there are several other commercial flow stream particle counters available.

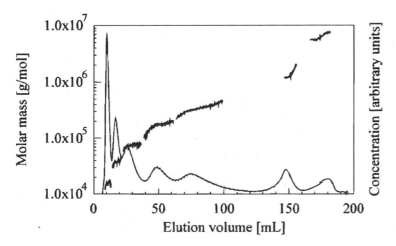

Fig. 1 Results from FlFFF/MALS/DRI analysis of seven PSS standards. Molecular weight derived from MALS data and concentration from the DRI detector. (From Ref.[6].)

The continuous viscosity detector has been shown to be a good detection tool for thermal FFF analysis of polymer solutions.[14] Due to the high sample dilution in FFF, the viscosity detector response above the solvent baseline, ΔS, is only dependent on the intrinsic viscosity of every sample point, $[\eta]$, multiplied by the concentration, c, at the corresponding points:

$$\Delta S = [\eta]c \qquad (3)$$

If a concentration-selective detector, such as a DRI detector, is connected on line with the viscosity detector, the ratio of the two signals yields the intrinsic viscosity distribution of the polymer sample. In polymer characterization, the intrinsic viscosity can be a property just as important as the molecular weight distribution. Furthermore, polymer intrinsic viscosity follows the Mark–Houwink relation to the molecular weight, M, where K and a are Mark–Houwink viscosity constants:

$$[\eta] = KM^a \qquad (4)$$

MASS SPECTROMETRIC DETECTION SYSTEMS

The MS detection methods covered here are mainly a selection of commonly used LC/MS methods, some of which have been optimized for FFF techniques or could potentially be good detection tools for FFF separations.

The issue in the coupling of a liquid-based separation method to a mass spectrometer is the ion source conversion of dissolved analytes to ions in the high vacuum mass analyzer, which for instance can be magnetic sectors, quadrupoles, ion traps, or time-of-flight (TOF) analyzers. Different ion sources give different information depending on the ionization mechanisms and will be discussed for each method below.

In most FFF separations, a moderate concentration of dispersion agent, electrolyte, or surfactant is used to improve the separations. A common feature for most MS instruments is that salt in the liquid entering the ion source leads to deterioration of the performance of the MS by lowering the signal-to-noise ratio and by condensing on surfaces inside the MS, thus continuously increasing the background level.

Today, the most frequently used LC/MS ion source is electrospray ionization (ESI), in which the sample stream ends in a narrow capillary, put on a high voltage (positive or negative). This potential, sometimes together with a sheath gas flow, gives rise to a spray of small charged droplets ($\sim 1\,\mu m$). When the solvent is evaporated from these droplets, electrostatic repulsion forces smaller droplets ($\sim 10\,nm$) to leave. Before entering the semivacuum region, free analytes with one or more net charges usually due to proton transfer or ion adducts (e.g., Li^+, Na^+, or NH_4^+) dominate.

Electrospray ionization is a mild ionization method, that is, almost no fragmentation of the ions occurs. It is applicable to all organic compounds involved in proton exchange or binding to ions in the gas phase, which includes almost all biomolecules and polymers. In ESI/MS, multiple charges occur with a charge distribution for all components. This charge envelope usually has maximum intensity at m/z about 1000 and rarely ranging beyond 2000 in m/z. This has the advantage that large molecules, such as peptides, DNA molecules, or polymers, can be analyzed by all common MS analyzers, but the drawback is that the resulting spectra can be complicated to interpret. For single mass molecules such as peptides, there are numerical models to deconvolute the single charge molecular weight from the ESI/MS m/z spectra, but for incompletely separated polymer components, the overlapping charge distributions for the individual polymer components make the interpretation complicated. ESI/MS sensitivity is dramatically reduced due to cluster formation in the presence of more than a few millimoles per liter of the salt, and surfactants can have a devastating effect on the ESI/MS spectra. Therefore, a volatile buffer should be used if possible (e.g., ammonium acetate, ammonium nitrate). ESI/MS has been used as a detector for FlFFF analysis of low molecular weight ethylene glycol polymers,[15] where the effect of different carriers on cluster formation was investigated. ESI/MS has been coupled to SEC in several applications for polymer analysis and other applications where FFF techniques can be successfully used, including proteins, neuropeptides, and DNA molecule segments. Modern ESI/MS has a broad range of flow rates from nanoliters up to a milliliter per minute.

Atmospheric pressure chemical ionization (APCI) is a method not yet applied to FFF, but could potentially be a good alternative to ESI for semivolatile analytes lacking a natural site for a charge. The analytes are evaporated and exposed to gas phase molecules ionized by a high voltage corona discharge electrode. The analytes are subsequently ionized by a charge transfer from the gas molecules. Atmospheric pressure chemical ionization has been shown to be less sensitive to buffer salts than ESI, and no fragmentation occurs in the ion source. Mainly singly charged ions are formed, making APCI less applicable to large molecules, depending on the upper range of the MS analyzer. Atmospheric pressure chemical ionization has a good flow rate compatibility ($0.3-1.5\,ml\ min^{-1}$) with FFF.

Matrix assisted laser desorption ionization (MALDI) is a frequently used ionization technique, but it is rarely used as an on-line detector. The sample stream is applied to a target plate, and it is allowed to cocrystallize with the matrix, which is subsequently desorbed, ionized with a laser, and analyzed in the MS. MALDI/TOF has been successfully used to determine molecular weight distributions of fractions collected after thermal FFF separation of polydisperse polymers.[16] MALDI is a good ion source, due to the soft ionization with high efficiency and simple mass spectra, even for heavier molecules, since the majority carry only a single charge.

ICPMS is an ion source for elemental analysis where the analyte stream is introduced into a high-energy plasma with efficient atomization and ionization, producing almost entirely singly charged elemental ions. It has multielement capability with excellent sensitivity and has good flow rate compatibility $(0.1–1.5 \, ml \, min^{-1})$ with FFF techniques. ICPMS has previously been applied to sedimentation FFF

for determination of major element composition in different size fractions of suspended riverine particles and soil particles in the size range 50–800 nm,[17] and recently flow FFF coupled on line to ICPMS has been used to determine elemental size distributions for over 55 elements in freshwater colloidal material (1–50 nm).[18,19] Fig. 2 shows a selection of elements and the signal from the UV detector, coupled on line before the ICPMS, from a river water sample. An interface between the FFF channel and the ICPMS was used to supply acid, to improve the performance of the nebulizer–spray chamber system, and internal standard. The interface also serves to dilute and split away about half of the salt content. The salt is necessary for the FFF separation, but harmful to the ICPMS.

DENSITY-BASED DETECTION

A continuous density detector works on the principle of a liquid flowing through an oscillating U-shaped glass tube, where the oscillating frequency is found relating the oscillations to the density of the flowing liquid. A densimetric detector has been evaluated for sedimentation FFF,[20] and the conclusions were that it is a universal concentration-selective detector without the need for signal correction or transformation. The sensitivity is however a limiting factor, since it is dependent on the density difference between the sample and the carrier liquid. A density difference of $0.2 \, g \, ml^{-1}$ is sometimes sufficient, but to achieve higher sensitivity a difference up to 1.0 is desirable, making densimetric detection suitable for inorganic particles, but less appropriate for lighter organic analytes. The densimeter detector is sensitive to temperature changes, but insensitive to flow changes, making it most suitable for flow programming applications.

CONCLUSIONS

A multidetector approach is often applied in FFF since all detection systems have some advantages and limitations over some size ranges, sample types, and detection limits. For example, the most common detector, the UV–VIS, has limitations that it is selective for both absorption and turbidity but it is still widely used. Elemental analysis of FFF fractions with ICPMS has been successfully developed during recent years, but other mass spectrometric hyphenations are still very few. On-line light scattering has proven to be a very valuable system to determine molecular weight and RMS radius distributions (MALS) as well as diffusion coefficients and hydrodynamic radius (DLS).

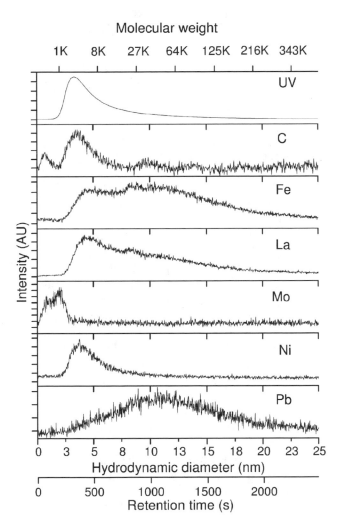

Fig. 2 Elemental size distributions of the colloidal material in a freshwater sample as given from an FlFFF coupled to ICPMS. A UV detector is placed on line prior to the ICPMS and the UV size distribution is included. The signals are plotted as a function of retention time, hydrodynamic diameter (from FFF theory), and molecular weight (from standardization with PSS standards). (From Ref.[18].)

REFERENCES

1. Yang, F.-S.; Caldwell, K.D.; Gidding, J.C. Colloid characterization by sedimentation field-flow fractionation. J. Colloid Interface Sci. **1983**, *92*, 81–91.
2. Reschiglian, P.; Melucci, D.; Zattoni, A.; Giancarlo, T. Quantitative approach to field-flow fractionation for the characterization of supermicron particles. J. Microcol. Sep. **1997**, *9*, 545–556.
3. Zattoni, A.; Loli Piccolomini, E.; Torsi, G.; Reschiglian, P. Turbidimetric detection method in flow-assisted separation of dispersed samples. Anal. Chem. **2003**, *75*, 6469–6477.
4. Caldwell, K.D.; Li, J. Emulsion characterization by the combined sedimentation field-flow fractionation–photon correlation spectroscopy methods. J. Colloid Interface Sci. **1989**, *132*, 256–268.
5. Wyatt, P.J. Submicrometer particle sizing by multiangle light scattering following fractionation. J. Colloid Interface Sci. **1998**, *197*, 9–20.
6. Thielking, H.; Kulicke, W.-M. On-line coupling of flow field-flow fractionation and multiangle laser light scattering for the characterization of macromolecules in aqueous solution as illustrated by sulfonated polystyrene samples. Anal. Chem. **1996**, *68*, 1169–1173.
7. Duval, C.; Le Cerf, D.; Picton, L.; Muller, G. Aggregation of amphiphilic pullulan derivatives evidenced by on-line flow field flow fractionation/multi-angle laser light scattering. J. Chromatogr. B. **2001**, *753*, 115–122.
8. Frankema, W.; van Bruijnsvoort, M.; Tijssen, R.; Kok, W.T. Characterisation of core-shell latexes by flow field-flow fractionation with multi-angle light scattering detection. J. Chromatogr. A. **2002**, *943*, 251–261.
9. Roger, P.; Baud, B.; Colonna, P. Characterization of starch polysaccharides by flow field-flow fractionation–multi-angle laser light scattering–differential refractometer index. J. Chromatogr. A. **2001**, *917*, 179–185.
10. Magnuson, M.L.; Lytle, D.A.; Frietch, C.M.; Kelty, C.A. Characterization of submicrometer aqueous iron III colloids formed in the presence of phosphate by sedimentation field-flow fractionation with multiangle laser light scattering detection. Anal. Chem. **2001**, *73*, 4815–4820.
11. Kammer, F.v.d.; Baborowski, M.; Friese, K. Field-flow fractionation coupled to multi-angle laser light scattering detectors: applicability and analytical benefits for the analysis of environmental colloids. Anal. Chim. Acta. Submitted for publication.
12. Oppenheimer, L.E.; Mourey, T.H. Use of an evaporative light-scattering mass detector in edimentation field-flow fractionation. J. Chromatogr. **1984**, *298*, 217–224.
13. Compton, B.J.; Myers, M.N.; Giddings, J.C. A single particle photometric detector for steric field-flow fractionation. Chem. Biomed. Environ. Instrum. **1983**, *12*, 299–317.
14. Kirkland, J.J.; Rementer, S.W.; Yau, W.W. Polymer characterization by thermal field flow fractionation with a continuous viscosity detector. J. Appl. Polym. Sci. **1989**, *38*, 1383–1395.
15. Hassellöv, H; Hulthe, G.; Lyvén, B.; Stenhagen, G. Electrospray mass spectrometry as online detector for low molecular weight polymer separations with flow field-flow fractionation. J. Liq. Chromatogr. & Rel. Technol. **1997**, *20*, 2843–2856.
16. Kassalainen, G.E.; Williams S.K.R., Coupling thermal field-flow fractionation with matrix-assisted laser desorption/ionization time-of-flight mass spectrometry for the analysis of synthetic polymers. Anal. Chem. **2003**, *75* (8), 1887–1894.
17. Taylor, H.E.; Garbarino, J.R.; Hotchin, D.M.; Beckett, R. Inductively coupled plasma-mass spectrometry as an element-specific detector for field-flow fractionation particle separation. Anal. Chem. **1992**, *64,* 5036.
18. Hassellöv, M.; Lyvén, B.; Haraldsson, C.; Sirinawin, W. Determination of continuous size and trace element distribution of colloidal material in natural water by on-line coupling of flow field-flow fractionation with ICMPS. Anal Chem. **1999**, *71*, 3497–3502.
19. Stolpe, B.; Hasseløv, M.; Andersson, K.; Turner, D.R. High resolution ICPMS as an on-line detector for flow field-flow fractionation; multi-element determination of colloidal size distributions in a natural water sample. Anal. Chim. Acta. **2005**, *in press.*
20. Kirkland, J.J.; Yau, W.W. Quantitative particle-size distributions by sedimentation field-flow fractionation with densimeter detector. J. Chromatogr. **1991**, *550*, 799–809.

Detection Principles

Kiyokatsu Jinno
*Department of Materials Science, Toyohashi University of Technology,
Tempaku-cho, Toyohashi, Japan*

INTRODUCTION

Various methods of detection are employed in chromatography. Each approach for the detection of solutes is based on their physical or chemical properties. Some of the more commonly used detectors are discussed here for LC, GC, and supercritical fluid chromatography (SFC).

LIQUID CHROMATOGRAPHY

The most commonly used detectors in LC are concentration-sensitive. The detector output signal is a function of the concentrations of the analytes passing through the detector cell. In order to use the information for quantitation, the detector must respond linearly to changes in concentration over a wide concentration range, which is called the linear dynamic range of the detector. Criteria for the evaluation of the quality or the suitability of the detector are as follows: the magnitude of the linear dynamic range, the noise level, the sensitivity, and the selectivity. The sensitivity is determined by the specific characteristics of the analytes and by the extent to which these differ from the characteristics of the sample matrix. The most important parameters are noise, drift, detection limit (sensitivity), selectivity, stability, and compatibility with various elution modes.

Noise is the high-frequency variation of the detector signal, which becomes visible when the baseline is recorded at the higher-sensitivity settings. To determine the noise level, parallel lines are drawn around the noise envelope and the distance between these lines; the actual noise level is expressed in detector signal units (for instance, AU, mV, or μA). This parameter is dependent upon the lamp, the amplifier, and the cell geometry, and is specified differently by many manufacturers. Static measurements usually provide better values for the noise level than those obtained under flow conditions. Noise levels are calculated using the mean of the baseline envelope. A measure for the sensitivity of a detector is the minimum detectable amount of a given compound (detection limit). Most LC detectors measure optical or spectroscopic characteristics of the analytes. Other detectors use electrical (conductivity) or electrochemical characteristics, such as oxidation or reduction (electrochemical detector, EcD) of the analytes. A detector is said to be more selective when it measures a more unique characteristic of an analyte.

Ultraviolet (UV) Detector

UV detection is the most popular, i.e., most commonly utilized, in the LC detection mode. Depending on instrumental design, three types of UV detectors are used today: single wavelength detectors, where a fixed wavelength is used for the absorbance monitoring of the analytes; variable wavelength detectors, with which one can choose the most appropriate wavelength for the analyte detection; and UV detectors which provide spectral information, such as fast-scanning UV detectors and diode array detectors.

UV detectors make use of the spectral absorption properties of the analytes in the UV and visible (Vis) wavelength range. The absorption measured at a given wavelength generally follows Beer's law and is transformed to a concentration-dependent signal. The change in absorption is proportional to the concentration when all parameters are kept constant. The detector cell volumes range from 5 and 10 μL; the light path typically ranges from 6 to 10 mm.

The fixed wavelength detector is the most widely used in LC. It is simple in design and, consequently, the least expensive although it is still the most sensitive. Its widespread use is historic in origin and is due to the fact that the strong emission line of the mercury lamp at 254 nm is well suited for absorption measurements of many organic compounds, provided that they possess an aromatic system. Some advantages of the fixed wavelength detector are as follows:

1. The simple design; the detector is relatively inexpensive.
2. The mercury spectral line is very strong and narrow.
3. The intensity of the light beam entering the system allows for a wide linear response range and high sensitivity.

Variable wavelength detectors use a continuous light source in combination with monochromators to

Encyclopedia of Chromatography DOI: 10.1081/E-ECHR-120013362

select the desired detection wavelength. The monochromator, in general, is a rotating diffraction grating which is positioned in the light path of the detector cell. Some instruments possess an additional variable bandwidth. In microprocessor-driven instruments, the desired wavelength and slit width can be selected and read on a display. Almost all instruments contain "classical" optics with spectral dispersion of the light occurring prior to passage through the flow cell. "Reversed optics" UV detectors have recently become popular. A holographic grating is placed between the flow cell and the photodiodes; it disperses the light beam through the detector cell. This design makes it possible to obtain spectral information at any point in time, which can then be further processed depending on the requirements of the analysis. Spectral information can be obtained from the diode array within 40–200 msec. Even for very narrow peaks, it is possible to scan spectra without stopping the flow. The diode array detector contains no moving parts; consequently, the spectra are of high quality in terms of resolution and reproducibility. Data can be acquired and evaluated, and relevant spectra can be stored and compared with spectral libraries via a computer. Of course, this type detector can be useful as a variable wavelength detector and, also, this provides information on peak identity and peak purity; it is routinely used in method development for separation of UV-active compounds.

Refractive Index (RI) Detectors

RI detection is the oldest in various LC detection modes and is commonly used in carbohydrate and polymer analysis. The RI (n) is a bulk property of the eluate. The RI detector is therefore a universal and rather nonspecific detector but offers relatively low sensitivity. In RI detection, the specific physical parameter is the RI increment, dn/dc, which detects the differential change in the RI (n) that is a dimensionless parameter, and dn/dc is therefore expressed in mL/g. For most compounds in common solvents, dn/dc lies between 0.8–0.15 mL/g. The actual parameter used in RI detection is the RI itself, whose minute changes are transformed into a detector signal. Three types of RI detectors are commercially available: Fresnel (reflection), deflection, and interference. Fresnel RI detectors measure the difference between the RI of a glass prism with reference to the eluate. At the glass/liquid phase boundary, part of the incident beam is completely reflected. The intensity of the transmitted light is then measured at a given angle. When the RI changes, the angle and the intensity of the beam hitting the photodiode changes accordingly.

In the deflection-type RI detector, the optical system is designed differently. The light beam actually passes through the detector cell twice. After passage through the detector and the reference cell, the light beam is reflected back through the detector cell. The beam is balanced by an optical zero control, divided into two beams of equal intensity by a prism, and focuses onto two photodiodes. One half of the detector cell is filled with pure mobile phase (reference cell), while the column eluate flows through the other half. The RI (n) is the same in each cell when only mobile phase passes through the cell. When an analyte passes through the flow cell, the RI (n) in the flow cell will change with respect to the reference cell. The light beam is deflected during forward and return passage through the cells. The resulting difference in light intensity is sensed by the photodiodes and the differential signal of the diodes is amplified and passed on to signal output devices.

In the interference-type RI detector, a monochromatic coherent light beam is split and the resulting two beams are directed through a reference and a sample cell, respectively. After passage through the cells, the difference in RI between the cells causes interference of two beams, which is measured by a photodiode.

Fluorescence (FL) Detector

Absorption of UV light by certain compounds triggers the emission of light with a longer wavelength. The spectral range of the emitted light depends on the excitation wavelength. Fluorescence emission can only be triggered at wavelengths at which the analytes absorb in the UV. Not all UV-absorbing compounds are also fluorescence emitters although some compounds possess native fluorescence. Nonfluorescent compounds can be converted into fluorescent compounds by derivatization with a suitable fluoresophore before (so-called pre-column derivatization) or after (post-column derivatization) LC separation. The most selective and most flexible design contains two diffraction gratings with variable wavelength monochromators at the excitation side and at the emission side of the system. These types of detectors contain an additional cut-off filter to control stray light/light scattering and noise. In the stopped-flow mode, excitation, as well as emission spectra, of labile compounds can be scanned, and the optimum wavelengths for routine measurements can be determined. Of course, photodiode array detection is also available for this detector type.

Electrochemical Detectors

EcDs are used for quantitation of compounds which can be easily oxidized or reduced by an applied potential. The standard reduction potential at the electrode is measured and transformed into a detector signal. The number of compounds which can be electrochemically detected is, however, considerably smaller than the number of optically detectable compounds by UV, RI, and FL. To become oxidized or reduced, a compound must possess electrochemically active groups. EcDs are mainly used in clinical, food, and environmental analysis.

Gas Chromatography

In contrast to LC detectors, GC detectors often require a specific gas, either as a reactant gas or as fuel (such as hydrogen gas as fuel for flame ionization). Most GC detectors work best when the total gas flow rate through the detector is 20–40 mL/min. Because packed columns deliver 20–40 mL/min of carrier gas, this requirement is easily met. Capillary columns deliver 0.5–10 mL/min; thus, the total flow rate of gas is too low for optimum detector performance. In order to overcome the problem when using capillary columns, an appropriate makeup gas should be supplied at the detector. Some detectors use the reactant gas as the makeup gas, thus eliminating the need for two gases. The type and flow rate of the detector gases are dependent on the detector and can be different even for the same type of detector from different manufacturers. It is often necessary to refer the specific instrument manuals for details to obtain the information on the proper selection of gases and flow rates. All detectors are heated, primarily to keep the analytes from condensing on the detector surfaces. Some detectors require high temperatures to function properly; some detectors are very sensitive to changes in temperature although others are only affected by very large changes in temperature. Some detectors are flow-sensitive; thus, their response changes or the baseline shifts according to the total gas flow rate through the detector.

Flame Ionization Detector (FID)

The FID is the most powerful and popular detector in GC, for which a basic structure is demonstrated in Fig. 1. Hydrogen and air are used to maintain a flame at the tip of the jet and into the flame where the organic components are burned. Ions are created in this combustion process; they are attracted to the charged collector electrode. This induced ion flow generates a current that can be measured and transformed to an output voltage by an electrometer. The amount of current generated should be dependent on the concentrations of the compounds introduced into the flame. The FID responds best to compounds containing a carbon–hydrogen bond. The lack of a carbon–hydrogen bond does not completely eliminate any response; however, the response is significantly depressed. Some notable compounds such as water, carbon monoxide, and carbon dioxide are nonresponding. Typical sensitivities for most organic compounds are 0.1–1 ng. The linear range is 5–6 orders magnitudes. Helium and nitrogen are typical as the makeup gases for capillary columns. Nitrogen is less costly and provides slightly better sensitivity. Near universal response, ease of use, wide linear range, and good sensitivity make the FID suitable for a wide variety of samples. Since FID is a simple detector, there is little routine maintenance required. The response of an FID is dependent on the hydrogen and gas flow rates; therefore, periodic measurement of these gas flow rates is necessary to maintain a stable performance. FIDs

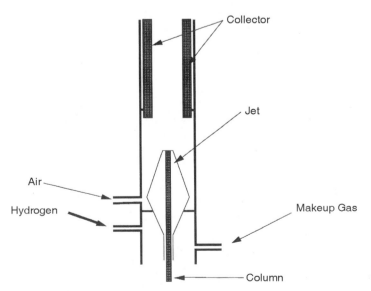

Fig. 1 Basic structure of FID for capillary GC.

are not very temperature-sensitive and temperature changes of 50°C or greater are needed before any performance alterations are observed. FIDs are not sensitive to changes in the carrier gas flow rates; thus, baseline shifts or drifting is rarely found by changing the experimental conditions.

Nitrogen–Phosphorus Detector (NPD)

An alkali metal bead, usually rubidium sulfate, is positioned above the jet in NPD, as seen in Fig. 2, and the current is applied to this bead, which causes it to achieve temperatures up to 800°C. The addition of hydrogen and air generates plasma around the bead, and carrier gas containing the solutes is delivered to the tip of the jet and the plasma. Specific ions are produced in the plasma from nitrogen- or phosphorus-containing compounds. These ions move to the charged collector. This movement of ions generates a current that is measured and transformed to an output voltage by an electrometer. The amount of the current is dependent on the amount of the compound introduced into the plasma. The NPD exhibits excellent selectivity and sensitivity for nitrogen- and phosphorus-containing compounds. The sensitivity is approximately 10,000 times greater for nitrogen and phosphorus compounds than for hydrocarbons. Typical NPD sensitivities are in the range of 0.5–1 pg with a linear range of 5–6 orders. Helium is the preferred makeup gas. Many compounds in the sample may not contain nitrogen or phosphorus, so a response is not observed for these compounds and, therefore, specific detection for nitrogen and phosphorus can be attained. Peak separation, identification, and measurement are made much easier because there are fewer peaks in the chromatogram. Also, less preparation of the sample prior to final determination is required because fewer of the potential interferences yield a response in the detector signal. Pesticides and pharmaceuticals analysis are generally the fields of application of the NPD.

Flame Photometric Detector (FPD)

Hydrogen and air are used to maintain a flame at the tip of the jet in the FPD. The carrier gas containing the analytes is delivered to the tip of the jet and into the flame. Some FPDs use a dual jet or burner design, but the overall process is not significantly different. The solutes are burned in the flame, forming S2 and HPO. Due to excitation in the flame, light at 392 nm for S2 and at 526 nm for HPO are emitted. A photomultiplier tube is used to measure the light intensity after the optical filter. A current is generated, which can be measured and transformed to an output voltage by an electrometer. The amount of light created at each wavelength is dependent on the amount of compound introduced into the flame. The selectivity of the FPD for sulfur and phosphorus compounds is 3–4 orders of magnitude. Typical sensitivities are 10–100 pg for

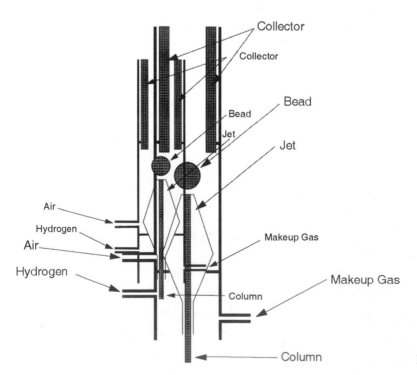

Fig. 2 Basic structure of NPD for capillary GC.

sulfur and 1–10 pg for phosphorus compounds. FPD responses are usually nonlinear for sulfur compounds. Nitrogen is the best makeup gas because its use results in the best sensitivity especially in the sulfur detection mode. FPDs are temperature- and flow-sensitive and, therefore, sensitivity changes are common when the temperature or carrier gas flow rates are changed. Increasing the detector temperature results in a decrease in sensitivity. FPD sensitivity for phosphorus compounds is comparable to NPD sensitivity. Due to some of the difficulties with NPDs, FPDs are frequently preferred for phosphorus-specific detection purposes. The increase of 500–1000 times in sensitivity of the FPD over the thermal conductivity detector (TCD) and FID often makes its nonlinear response behavior tolerable.

Electron-Capture Detector (ECD)

Carrier gas containing the solutes is delivered into the heated cell in the ECD. A ^{63}Ni source lining the cell acts as a source of electrons, and a moderating or auxiliary gas of nitrogen or argon/methane (95/5) is introduced into the cell to create thermal electrons which are attracted to the anode, creating a current. When an electronegative compound enters the cell, it captures thermal electrons and reduces the cell current. The amount of current reduction is measured and transformed to an output voltage by an electrometer. The size of the current loss is dependent on the amount of the compound entering the cell. The ECD primarily responds to compounds that contain a halogen, carbonyl, or nitrate group. Halogens have 100–100,000 times, nitrates have 100–1000 times, and carbonyls have 20–100 times better response than hydrocarbons. Polyhalogenated compounds, or compounds containing multiple nitrate or carbonyl groups, yield significantly increased detector responses. Also, the response for different halogens and types of carbonyls are varied, depending upon the structures of the analytes. Sensitivities approaching 1 pg for halogens, 10 pg for nitrates, and 50 pg for carbonyls are typical. The linear range is 2–3 orders and poor; therefore, multiple-point calibration curves are required for accurate quantitation. The auxiliary gas is often used as the makeup gas in situations where makeup gas is necessary. Nitrogen is less costly and provides slightly better sensitivity than others. Argon/methane provides a slightly better linear dynamic range than nitrogen. The primary application of ECDs is for the detection of halogenated compounds. The flow rates of the carrier gas and auxiliary gas have a pronounced effect on the sensitivity and linear range of an ECD. The ECD temperature also affects the sensitivity.

Thermal Conductivity Detector (TCD)

The TCD consists of two heated cells, one of which is a sample cell and the other is a reference cell. The carrier gas containing the separated compounds enters the sample cell, while the reference cell is supplied with the same type and flow rate of carrier gas that flows into the sample cell. There is a TCD design that utilizes a single cell and a switching valve to accomplish the same task. Current is applied to the filaments, which causes them to reach an elevated equilibrium temperature when the current and gas flows are constant. When a compound that has a thermal conductivity different from that of the carrier gas is eluted from the column, it induces a change in the filament temperature. Because the reference cell filament remains at a constant temperature, the temperature difference between the two filaments is compared. The difference is measured via a Wheatstone bridge, which produces an output voltage. It is dependent on the amount of compound entering the sample cell. The best TCD sensitivity is established when the difference in thermal conductivities between the carrier gas and the component is maximized. Helium or hydrogen is usually the carrier gas of choice because these gases have thermal conductivities 10–15 times greater than that of most organic molecules, whereas nitrogen is only seven times higher than most organic molecules. This means that the TCD is a universal detector in GC although the sensitivity is relatively low, 5–50 ng per component or 10–100 times less than that of the FID. The linear range is five orders magnitude. TCDs are flow- and temperature-dependent.

Other detectors

There are a number of other GC detectors commercially available. Photoionization detectors (PID) are primarily used for the selective, low-level detection of the compounds which have double or triple bonds or an aromatic moiety in their structures. Electrolytic conductivity detectors (ELCD) are used for the selective detection of chlorine-, nitrogen-, or sulfur-containing compounds at low levels. Chemiluminescence detectors are usually employed for the detection of sulfur compounds. The atomic emission detectors (AED) can be set up to respond only to selected atoms, or group of atoms, and they are very useful for element-specific detection and element-speciation work.

SFC

All detectors in GC and LC can be easily made useful for SFC. Basically, however, we can say that GC detectors are useful for capillary SFC, and LC detectors are useful for packed-column SFC. The most important and convenient features of SFC are that any detection systems available in chromatography can be useful and all work well.

SUGGESTED FURTHER READING

Gilbert, M.T. *High Performance Liquid Chromatography*; Wright Co.: Bristol, 1987.

Huber, L., George, S.A., Eds.; *Diode Array Detection in HPLC*; Marcel Dekker, Inc.: New York, 1993.

Jinno, K., Ed.; *Hyphenated Techniques in Supercritical Fluid Chromatography and Extraction*; Elsevier: Amsterdam, 1992.

Smith, R.M. *Gas and Liquid Chromatography in Analytical Chemistry*; John Wiley & Sons: Chichester, 1988.

Walker, J.Q., Ed.; *Chromatography Fundamentals, Applications, and Troubleshooting*; Preston Publications: Niles, IL, 1996.

Yeung, E.S., Ed.; *Detectors for Liquid Chromatography*; Wiley-Interscience: New York, 1986.

Detection (Visualization) of TLC Zones

Joseph Sherma
Department of Chemistry, Lafayette College, Easton, Pennsylvania, U.S.A.

INTRODUCTION

After development with the mobile phase, the TLC plate is dried in a fume hood and heated, if necessary, to completely evaporate the mobile phase. Separated compounds are detected on the layer by viewing their natural color, natural fluorescence, or quenching of fluorescence. These are physical methods of detection and are nondestructive. Substances that cannot be seen in visible or ultraviolet (UV) light can be visualized with suitable detection reagents to form colored, fluorescent, or UV absorbing compounds by means of derivatization reactions carried out pre- or post-chromatography. Although dependent upon the particular analyte, layer, and detection method chosen, sensitivity values are generally in the nanogram range for absorbance and picogram range for fluorescence.

Other detection methods include radioactivity for labeled compounds (a nondestructive physical method) and biological methods (e.g., immunochemical or enzymatic reactions). Coupled detection methods such as TLC/infrared spectrometry (TLC/IR) or TLC/mass spectrometry (TLC/MS) can be used for confirmation of zone identity as well as quantification in some cases.

DIRECT DETECTION

Compounds that are naturally colored (e.g., plant pigments, food colors, dyestuffs) are viewed directly on the layer in daylight, while compounds with native fluorescence (aflatoxins, polycyclic aromatic hydrocarbons, riboflavin, quinine) are viewed as bright zones on a dark background under longwave (366 nm) UV light. Compounds that absorb around 254 nm (shortwave), including most compounds with aromatic rings and conjugated double bonds and some unsaturated compounds, can be detected on an "F-layer" containing a phosphor or fluorescent indicator (often zinc silicate). When excited with 254 nm UV light, absorbing compounds diminish (quench) the uniform layer fluorescence and are detected as dark violet spots on a bright green background. Viewing cabinets or boxes (Fig. 1) incorporating 254 and 366 nm UV-emitting mercury lamps are available commercially for inspecting chromatograms in an undarkened room. Detection by natural color, fluorescence, or fluorescence quenching does not modify or destroy the compounds, and the methods are, therefore, suitable for preparative layer chromatography. Derivatization reactions modify or destroy the structure of the compounds detected, but they are often more sensitive than detection with UV radiation.

UNIVERSAL DETECTION REAGENTS

Postchromatographic universal reactions such as iodine absorption or spraying with sulfuric acid and heat treatment are quite unspecific and are valuable for completely characterizing an unknown sample. Absorption of iodine vapor from crystals in a closed chamber produces brown spots on a yellow background with almost all organic compounds except for some saturated alkanes. Iodine staining is nondestructive and reversible upon evaporation, while sulfuric acid charring is destructive. Besides sulfuric acid, 3% copper acetate in 8% phosphoric acid is a widely used charring reagent. The plate, which must contain a sorbent and binder that do not char, is typically heated at 120–130°C for 20–30 min to transform zones containing organic compounds into black to brown zones of carbon on a white background. Some charring reagents initially produce fluorescent zones at a lower temperature before the charring occurs at a higher temperature.

SELECTIVE DERIVATIZATION DETECTION

Selective derivatization reagents form colored or fluorescent compounds on a group- or substance-specific basis and aid in compound identification. They also allow the use of a TLC system with lower resolution, because interfering zones may not be detected. If the detection is to be the basis of a quantitative (densitometric) analysis, the reagent used should react with the analyte to produce the primary product in proportion to the quantity present in the zones and not produce any interfering secondary compounds.

Derivatization can be performed before or after development of the layer with the mobile phase. Prechromatographic derivatization is carried out either in solution prior to sample application or directly on the plate by applying the sample and reagents at the

Encyclopedia of Chromatography DOI: 10.1081/E-ECHR-120039952

449

Fig. 1 Darkroom viewing cabinet with two overhead ports that accept one or two 8 W combination shortwave/longwave portable UV lamps. (Photograph supplied by Analtech, Inc., Newark, Delaware.)

origin or in the preadsorbent (or concentration) area. Prechromatographic derivatization may enhance compound stability or chromatographic selectivity as well as serving for detection. Types of prechromatographic derivatization reactions that have been used include acid and alkaline hydrolysis, oxidation and reduction, halogenation, nitration and diazotization, hydrazone formation, esterification, and dansylation.

Derivatization reactions for detection are usually carried out postchromatography, and many hundreds of reagents have been reported in the literature. Selected examples are listed in Table 1.

APPLICATION OF DETECTION REAGENTS

Liquid chromogenic and fluorogenic detection reagents such as those in Table 1 can be applied by spraying or dipping the developed and dried layer. When a sequence of reagents is necessary, the layer is usually dried between each application.

Various types of aerosol sprayers that connect to air or nitrogen lines are available commercially for manual operation (Fig. 2), and this method is most widely used for reagent application. For safety purposes, spraying is carried out inside a laboratory fume hood or commercial TLC spray cabinet with a blower (fan) and exhaust hose, and protective eyeware and laboratory gloves are worn. The plate is placed on a sheet of paper or supported upright inside a cardboard spray box. The spray is applied from a distance of about 15 cm with a uniform up-and-down and side-to-side motion until the layer is completely covered. It is usually better to spray a layer two or three times lightly and evenly with intermediate drying rather than give a single, saturating application that might cause zones to become diffuse. Studies are required with each reagent to determine the optimum total amount of reagent that should be sprayed, but generally the layer is sprayed until it begins to become translucent. After

visualization, zones should be marked with a soft lead pencil because zones formed with some reagents may fade or change color with time. A cordless electropneumatic sprayer with separate spray heads for low and high viscosity reagents is manufactured by Camag (Fig. 3). The ChromaJet DS 20 (Fig. 4) is a PC-controlled automatic apparatus that reproducibly sprays derivatization reagents on selected tracks of the layer; minimal reagent volumes are used, and operation can be documented in conformity with good laboratory practice (GLP) standards.

As with proper application using a fine-mist manual sprayer or an automated sprayer, dipping can provide uniform reagent application that leads to sensitive, reliable detection and reproducible results in quantitative densitometric analysis. The simplest method is to manually dip for a short time (5–10 sec) in a glass or metal dip tank. More uniform dip application of reagents can be achieved by use of a battery operated automatic, mechanical chromatogram immersion instrument (Fig. 5), which provides selectable, consistent vertical immersion and withdrawal speeds between 30 and 50 mm sec^{-1} and immersion times between 1 and 8 sec for plates with 10 or 20 cm heights. The immersion device can also be used for impregnation of layers with detection reagents prior to initial zone application and development, and for postdevelopment impregnation of chromatograms containing fluorescent zones with a fluorescence enhancement and stabilization reagent such as paraffin. Dip application to the layer cannot be used when two or more aqueous reagents must be used in sequence without intermediate drying. Dip reagents must be prepared in a solvent that does not cause the layer to be removed from the plate or the zones to be dissolved from the layer or to become diffuse; dip reagents are usually the same concentration or less concentrated than corresponding spray reagents.

Detection reagents can also be applied to the layer as a vapor, as mentioned above for iodine. Other reagents delivered to the layer by vapor exposure include t-butyl hypochlorite and HCl, both of which form fluorescent derivatives with a variety of compounds. The Analtech Vapor Phase Fluorescence (VPF) Visualization Chamber provides detection of compounds such as sugars, lipids, steroids, flavonoids, and antibiotics by induced fluorescence after heating the sealed chamber, containing the plate and ammonium bicarbonate crystals, on a hotplate to a temperature that decomposes the salt to ammonia.

HEATING THE LAYER

Layers often require heating to eliminate residual mobile phase after development, and again after spray

Table 1 Reagents used for postchromatographic derivatization of different classes of compounds

Analyte(s)	Reagents and treatment	Result
Acidic or basic compounds	Solutions of pH indicators (e.g., bromocresol green, bromophenol blue)	Colored zones on pale background
Aldehydes	0.1 g 2,4-Dinitrophenylhydrazine in 100 ml methanol plus 1 ml conc. HCl	Orange-yellow or more colored zones on light-colored background
Alkaloids	0.85 g Basic bismuth nitrate, 40 ml water, and 10 ml glacial acetic acid mixed with 8 g potassium iodide in 20 ml water (Dragendorff reagent)	Yellow-brown zones
	0.3 g Hexachloroplatinic(IV) acid in 100 ml water mixed with 100 ml 6% potassium permanganate solution	Various colored and fluorescent zones; contrast of layer can be improved by heating
Amino acids	0.5% Ninhydrin in ethanol–glacial acetic acid (98 : 2); heat at 90–100 °C for 5–10 min	Blue to purple zones
Amino acid derivatives	0.1% p-Dimethylaminobenzaldehyde in ethanol mixed (1 : 1) with conc. HCl (Ehrlich's reagent); heat at 60 °C for 5 min	Different colored zones
Amphetamines	Spray with 0.5% aqueous Fast Black K salt, dry, spray with 0.5 M NaOH, spray again with Fast Black K solution	Orange-red and different violet colors
Antioxidants	Diazotized p-nitroaniline (800 mg p-nitroaniline in a mixture of 250 ml water and 20 ml HCl; 5 ml NaNO$_2$ added dropwise until solution is colorless)	Aromatic compounds give yellow to brown zones
Aromatic compounds	0.2 ml Formaldehyde (37%) in 10 ml conc. sulfuric acid (Marquis reagent); heat at 110 °C for 20 min	Colored zones on light pink background; some may be fluorescent
Ascorbic acid	1–50 mg Cacotheline in 50 ml water, heat at 110 °C for 20 min	Red-brown to violet zone on a yellow background
Carbohydrates	3% p-Anisidine hydrochloride in water-saturated butanol; heat at 110 °C for 10 min	Red to brown zones
	1.2 g Ammonium vanadate in 95 ml water and 5 ml conc. sulfuric acid (Mandelin reagent)	Blue zones on yellow background
Flavonoids	1% Aluminum chloride solution in ethanol	Fluorescent zones
Lipids	5 g Phosphomolybdic acid in 100 ml absolute ethanol; heat at 100–150 °C for 2–5 min	Blue zones on yellow layer background
Lipids and quinones	0.5–1.0 mg ml^{-1} Rhodamine 6G in ethanol	Fluorescent zones

(Continued)

Table 1 Reagents used for postchromatographic derivatization of different classes of compounds (*Continued*)

Analyte(s)	Reagents and treatment	Result
Metal cations	0.25% Ethanol solution of alizarin, then ammonia vapor	Red-violet zones on violet background
Organic acids	100 mg 2,6-Dichloroindophenol sodium salt in 100 ml ethanol (Tillman reagent); heat at 100°C for 5 min	Red-orange zones on violet background
Pesticides (carbamates), sulfonamides, and primary amino compounds	1 g Sodium nitrite in 20 ml water diluted to 100 ml with conc. HCl–ethanol (17 : 83); dry plate; then 1% *N*-(1-naphthyl)ethylenediamine dihydrochloride in 10 ml water and 90 ml ethanol (Bratton–Marshall reagent)	Pink to violet zones
Pesticides (organophosphate)	2% 2,6-Dibromoquinone-4-chlorimide in glacial acetic acid; heat at 110°C for 10 min	Pink, orange, and brown zones on pale yellow background
Phenols	2% 4-Aminoantipyrene in 80% ethanol, then 4% potassium hexacyanoferrate(III) in ethanol–water (1 : 1) (Emerson reagent)	Red zones on light yellow background
Phenols, amino compounds, aromatic hydrocarbons, and coumarins	1% 2,6-Dibromoquinone-4-chloroimide in methanol (Gibbs reagent); heat at 110°C for 2–5 min	Different colored zones
Steroids	0.5 ml Anisaldehyde dissolved in 8 ml conc. sulfuric acid and diluted with 85 ml methanol plus 10 ml glacial acetic acid; heat at 100°C for 5–10 min	Violet-blue zones on light pink or colorless background; sometimes fluorescent zones
Sugars	5 g α-Naphthol in 160 ml ethanol, 20 ml sulfuric acid, and 13 ml water; heat at 110°C for 5 min	Blue-purple zones
Sulfonamides	15% Fluorescamine in acetone	Fluorescent zones
Terpenoids	0.5 ml Anisaldehyde mixed with 8 ml conc. sulfuric acid and diluted with 85 ml methanol and 10 ml glacial acetic acid; heat at 100°C for 5–10 min	Violet-blue zones on light pink or colorless background
Vitamin B₁	10% Antimony(III) chloride in chloroform; heat at 120°C for 5–10 min	Variously colored zones
Vitamin E	10 mg Potassium hexacyanoferrate and 1 g NaOH in 7 ml water and 13 ml ethanol	Bluish fluorescent zones
	Mixture (1 : 1) of 0.1 g iron(III) chloride hexahydrate in 50 ml ethanol with 0.25 g 2,2′-bipyridine (α,α′-dipyridyl) in 50 ml ethanol (Emmerie–Engle reagent)	Red zone

D

Fig. 2 Glass TLC reagent sprayer for use with an air line or the rubber bulb shown. (Photograph supplied by Analtech.)

or dip application of the detection reagent in order to complete the reaction upon which detection is based and ensure optimum derivative formation. Typical conditions are 5–15 min at 100–110 °C. If a laboratory oven is used, the plate should be supported on a solid metal tray to help ensure uniform heat distribution. The plate heater shown in Fig. 6 usually provides more consistent heating conditions than an oven; it features a 20 × 20 cm flat, evenly heated ceramic surface; a grid to facilitate proper positioning of TLC or high performance TLC (HPTLC) plates; programmable temperature between 25 and 200 °C; and digital display of the programmed and actual temperatures. Prolonged heating time or excessive temperature can cause decomposition of the analytes and darkening of the layer background and should be avoided.

Some reagents can be impregnated into the layer before spotting of samples if the selectivity of the separation is not affected and the mobile phase does not strip the reagent during development. Detection takes place only upon heating after development. This method has been used for detection of lipids as blue spots on a

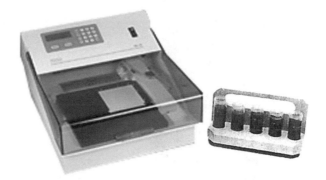

Fig. 4 ChromaJet DS 20 automated spray apparatus. (Photograph supplied by Desaga Sarstedt-Gruppe GmbH, Wiesloch, Germany.)

yellow background on silica gel layers preimpregnated with phosphomolybdic acid by dipping, spraying, or development. Analtech sells precoated silica gel plates impregnated with 5% ammonium sulfate; heating at 150–200 °C for 30–60 min in a closed container (the VPF Chamber, described above, can be used) generates sulfuric acid for charring detection of zones.

THERMAL DETECTION WITHOUT REAGENTS

Thermal derivatization (or thermochemical reaction) allows detection of zones without the use of reagents.

Fig. 3 TLC sprayer consisting of a charger and pump unit; homogeneous reagent aerosol particles in the 0.3–10 μm range are formed. (Photograph supplied by Camag, Wilmington, North Carolina.)

Fig. 5 Chromatogram immersion device set for 10 cm dipping depth with HPTLC plates. Vertical dipping and removal rates and the residence time in the reagent can be preselected. (Photograph supplied by Camag, Wilmington, North Carolina.)

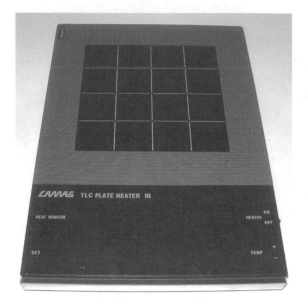

Fig. 6 TLC plate heater. (Photograph supplied by Camag.) *(View this art in color at www.dekker.com.)*

For example, simple heating of amino-modified silica layers causes conversion of sugars, oligosaccharides, creatine, catecholamines, steroid hormones, and other compound types to stable fluorescent compounds. Heating times of 3–45 min and temperatures of 140–200 °C have been used for different substances.

BIOLOGICAL DETECTION

Several detection methods are based on the biological activities of certain compounds. Cholinesterase inhibiting pesticides (e.g., organophosphates, carbamates) are detected sensitively by treating the layer with the enzyme and a suitable substrate, which react to produce a colored product over the entire layer except where colorless pesticide zones are located due to their inhibition of the enzyme–substrate reaction.

TLC/immunostaining has been used to detect solasodine glycosides by separation of the compounds on a silica gel layer, transfer to a polyvinylidene difluoride membrane, and treatment of the membrane with sodium periodate solution followed by bovine serum albumin (BSA), resulting in a solasodine–BSA conjugate. Individual zones were stained by monoclonal antibody against solamargine.

Bioluminescence has been used for specific detection of separated bioactive compounds on thin layers (BioTLC). After development and drying of the mobile phase by evaporation, the layer is coated with microorganisms by immersion of the plate. Single bioactive substances in multicomponent samples are located as zones of differing luminescence. The choice of the luminescent cells determines the specificity of the detection. A specific example is the use of the marine bacterium *Vibrio fischeri* with the BioTLC format. The bioluminescence of the bacterial cells is reduced by toxic substances, which are detected as dark zones on a fluorescent background with picogram level sensitivity.

BioTLC kits are commercially available from ChromaDex, Inc. (Santa Ana, California).

RADIOACTIVITY DETECTION

Radioactive zones can be detected on thin layers by film autoradiography, digital autoradiography with a multiwire proportional chamber, use of charged–coupled devices, or bioimaging/phosphor imaging techniques. These methods differ in terms of factors such as simplicity, speed, sensitivity, resolution, linear range, and accuracy and precision of quantification, and the method of choice depends on the available instrumentation, the type of experiment, and the information needed.

ZONE IDENTIFICATION AND CONFIRMATION

The identity of the detected TLC zones is obtained initially by comparison of characteristic R_f values between samples and reference standards chromatographed on the same plate, where R_f equals the migration distance of the center of the zone divided by the migration distance of the mobile phase front, both measured from the start (origin). Identity is more certain if a selective chromogenic detection reagent yields the same characteristic color for sample and standard zones. Because the chromatogram is stored on the layer, multiple compatible detection reagents can be applied in sequence to confirm the identity of unknown zones. As an example, almost all lipids are detected as light green fluorescent zones by use of 2,7-dichlorofluorescein reagent, while absorption of iodine vapor differentiates between saturated and unsaturated lipids or lipids containing nitrogen. The identity of zones is confirmed further by recording UV or visible absorption spectra directly on the layer using a densitometer (in situ spectra), or by direct or indirect (after scraping and elution) measurement of FT-infrared, Raman, or mass spectra.

CONCLUSIONS

Details of reagent preparation, application and heating procedures, results, and selectivity of many hundreds of reagents for detection of all classes of compounds and ions, including those in Table 1, are available in the literature references listed.

ARTICLES OF FURTHER INTEREST

TLC: Theory and Mechanism, p. 1700.
TLC Sorbents, p. 1645.

SUGGESTED FURTHER READING

Bauer, K.; Gros, L.; Sauer, W. *Thin Layer Chromatography—An Introduction*; EM Science: Darmstadt, Germany, 1991; 41–46.

Cimpan, G. Pre- and postchromatographic derivatization. In *Planar Chromatography*; Nyiredy, Sz., Ed.; Springer Scientific Publisher: Budapest, Hungary, 2001; 410–445.

Dyeing Reagents for Thin Layer Chromatography and Paper Chromatography; E. Merck: Darmstadt, Germany.

Fried, B.; Sherma, J. *Thin Layer Chromatography—Techniques and Applications*, 4th Ed.; Marcel Dekker, Inc.: New York, NY, 1999; 145–175 (detection and visualization), 249–267 (radiochemical techniques).

Hazai, I.; Klebovich, I. Thin-layer radiochromatography. In *Handbook of Thin Layer Chromatography*, 3rd Ed.; Sherma, J., Fried, B., Eds.; Marcel Dekker, Inc.: New York, NY, 2003; 339–360.

Jork, H.; Funk, W.; Fischer, W.; Wimmer, H. *Thin Layer Chromatography, Reagents and Detection Methods*; VCH Verlagsgesellschaft mbH: Weinheim, Germany, 1994; Vol. 1b.

Jork, H.; Funk, W.; Fischer, W.; Wimmer, H. *Thin Layer Chromatography, Physical and Chemical Detection Methods*; VCH Verlagsgesellschaft mbH: Weinheim, Germany, 1990; Vol. 1a.

Klebovich, I. Application of planar chromatography and digital autoradiography in metabolism research. In *Planar Chromatography*; Nyiredy, Sz., Ed.; Springer Scientific Publisher: Budapest, Hungary, 2001; 293–311.

Kreiss, W.; Eberz, G.; Weisemann, C. Bioluminescence detection for planar chromatography. Camag Bibliogr. Service (CBS) **2002**, *88*, 12–13.

Macherey-Nagel GmbH & Co. KG. TLC Catalog e2/5/0/1.98 PD; Dueren, Germany; A2–A81.

Maxwell, R.J. An efficient heating-detection chamber for vapor phase fluorescence TLC. J. Planar Chromatogr. Mod. TLC **1988**, *1*, 345–346.

Morlock, G.; Kovar, K.-A. Detection, identification, and documentation. In *Handbook of Thin Layer Chromatography*, 3rd Ed.; Sherma, J., Fried, B., Eds.; Marcel Dekker, Inc.: New York, NY, 2003; 207–238.

Stahl, E. *Thin Layer Chromatography—A Laboratory Handbook*; Academic Press: San Diego, CA, 1965; 485–502.

Tanaka, H.; Putalun, W.; Tsuzaki, C.; Shoyama, Y. A simple determination of steroidal alkaloid glycosides by thin layer chromatography immunostaining using monoclonal antibody against solamargine. FEBS Lett. **1997**, *404*, 279–282.

Touchstone, J.C. *Practice of Thin Layer Chromatography*, 3rd Ed.; Wiley-Interscience: New York, NY, 1992; 139–183.

Zweig, G.; Sherma, J. *Handbook of Chromatography*; CRC Press: Boca Raton, FL, 1972; Vol. 1, 103–189.

Detector Linear Dynamic Range

Raymond P. W. Scott
Scientific Detectors Ltd., Banbury, Oxfordshire, U.K.

INTRODUCTION

The linearity of most detectors deteriorates at high concentrations and, thus, the *linear dynamic range* of a detector will always be less than its dynamic range.

DISCUSSION

The symbol for the linear dynamic range is usually taken as (D_{LR}). As an example, the linear dynamic range of a flame ionization detector might be specified as

$$D_{LR} = 2 \times 10^5 \quad \text{for } 0.98 < r < 1.02$$

where r is the response index of the detector.

Alternatively, according to the ASTM E19 committee report on detector linearity, the linear range may also be defined as that concentration range over which the response of the detector is constant to within 5%, as determined from a linearity plot. This definition is significantly looser than that using the response index.

The lowest concentration in the linear dynamic range is usually taken as equal to the *minimum detectable concentration* or the *sensitivity* of the detector. The largest concentration in the linear dynamic range would be that where the response factor (r) falls outside the range specified, or the deviation from linearity exceeds 5% depending on how the linearity is defined. Unfortunately, many manufacturers do not differentiate between the dynamic range of the detector (D_R) and the linear dynamic range (D_{LR}) and do not quote a range for the response index (r). Some manufacturers do mark the least sensitive setting on a detector as N/L (nonlinear), which, in effect, accepts that there is a difference between the linear dynamic range and the dynamic range.

SUGGESTED FURTHER READING

Fowlis, I.A.; Scott, R.P.W. A vapour dilution system for detector calibration. J. Chromatogr. **1963**, *11*, 1.

Scott, R.P.W. *Chromatographic Detectors*; Marcel Dekker, Inc.: New York, 1996.

Encyclopedia of Chromatography DOI: 10.1081/E-ECHR-120039953

Detector Linearity and Response Index

Raymond P. W. Scott
Scientific Detectors Ltd., Banbury, Oxfordshire, U.K.

INTRODUCTION

It is essential that any detector that is to be used directly for quantitative analysis has a linear response. A detector is said to be truly linear if the detector output (*V*) can be described by the simple linear function

$$V = Ac$$

where *A* is a constant and *c* is the concentration of the solute in the mobile phase (carrier gas) passing through it.

DISCUSSION

As a result of the imperfections inherent in all electro-mechanical and electrical devices, true linearity is a hypothetical concept, and practical detectors can only approach this ideal response. Consequently, it is essential for the analyst to have some measure of detector linearity that can be given in numerical terms. Such a specification would allow quantitative comparison between detectors and indicate how close the response of the detector was to true linearity. Fowlis and Scott[1] proposed a simple method for measuring detector linearity. They assumed that for an approximately linear detector, the response can be described by the power function

$$V = Ac^r$$

where *r* is defined as the response index of the detector.

For a truly linear detector, $r = 1$, and the proximity of *r* to unity will indicate the extent to which the response of the detector deviates from true linearity. The response of some detectors having different values for *r* are shown as curves relating the detector output (*V*) to solute concentration (*c*) in Fig. 1. It is seen that the individual curves appear as straight lines but the errors that occur in assuming true linearity can be quite large. The errors actually involved are shown in the following, which is an analysis of a binary mixture employing detectors with different response indices:

Solute	$r = 0.94$	$r = 0.97$	$r = 1.00$	$r = 1.03$	$r = 1.05$
1	11.25%	10.60%	10.00%	09.42%	09.05%
2	88.75%	89.40%	90.00%	90.58%	90.95%

It is clear that the magnitude of the error for the lower-level components can be as great as 12.5% (1.25% absolute) for $r = 0.94$ and 9.5% (0.95% absolute) for $r = 1.05$. In general analytical work, if reasonable linearity is assumed, then $0.98 < r < 1.03$. The basic advantage of defining linearity in this way is that if the detector is not perfectly linear, but the value for *r* is known, then a correction can be applied to accommodate the nonlinearity.

There are alternative methods for defining linearity which, in the author's opinion, are somewhat less precise and less useful. The recommendations of the ASTM E19 committee on linearity measurement are as follows:

> The linear range of a detector is that concentration range of the test substance over which the response of the detector is constant to within 5% as determined from a linearity plot,—the linear range should be expressed as the ratio of the highest concentration on the linearity scale to the minimum detectable concentration.

This method for defining detector linearity is satisfactory up to a point and ensures a minimum linearity from the detector and, consequently, an acceptable quantitative accuracy. However, the specification is significantly "looser" than that given above, and it is not possible to correct for any nonlinearity that may exist, as there is no correction factor provided that is equivalent to the response index. It is strongly advised that the response index should be determined for any detector that is to be used for quantitative analysis. In most cases, *r* need only be measured once, unless the detector undergoes some catastrophic event that is liable to distort its response, in which case, *r* may need to be checked again.

There are two methods that can be used to measure the response index of a detector: the *incremental method* of measurement and the *logarithmic dilution method* of measurement.[2] The former requires no special apparatus, but the latter requires a log-dilution vessel, which, fortunately, is relatively easy to fabricate. The incremental method of measurement is the one recommended for general use.

The apparatus necessary is the detector itself with its associated electronics and recorder or computer system, a mobile-phase supply, pump, sample valve,

Encyclopedia of Chromatography DOI: 10.1081/E-ECHR-120039954

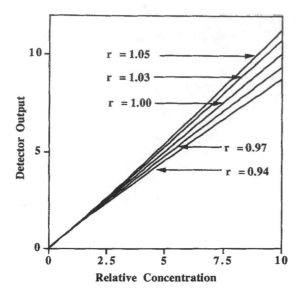

Fig. 1 Graph of detector output against solute concentration for detectors having different response indices.

and virtually any kind of column. In practice, the chromatograph to be used for the subsequent analyses is normally employed. The solute is chosen as typical of the type of substances that will be analyzed and a mobile phase is chosen that will elute the solute from the column in a reasonable time. Initial sample concentrations are chosen to be appropriate for the detector under examination.

Duplicate samples are placed on the column, the sample solution is diluted by a factor of 3 and duplicate samples are again placed on the column. This procedure is repeated, increasing the detector sensitivity setting where necessary until the height of the eluted peak is commensurate with the noise level. If the detector has no data acquisition and processing facilities, then the peaks from the chart recorder can be used. The width of each peak at 0.607 of the peak height is measured and the peak volume can be calculated from the chart speed and the mobile-phase flow rate. Now, the concentration at the peak maximum will be twice

the average peak concentration, which can be calculated from

$$c_p = \frac{ms}{wQ}$$

where c_p is the concentration of solute in the mobile phase at the peak height (g/mL), m is the mass of solute injected, w is the peak width at 0.6067 of the peak height, s is the chart speed of the recorder or printer, and Q is the flow rate (mL/min).

The logarithm of the peak height y (where y is the peak height in millivolts) is then plotted against the log of the solute concentration at the peak maximum (c_p). Now,

$$\log(V) = \log(A) + (r)\log(c_p)$$

Thus, the slope of the $\log(V)/\log(c_p)$ curve will give the value of the response index (r). If the detector is truly linear, $r = 1$ (i.e., the slope of the curve will be $\sin \pi/4 = 1$). Alternatively, if suitable software is available, the data can be curved fitted to a power function and the value of r extracted directly from the curve-fitting analysis. The same data can be employed to determine the linear range as defined by the ASTM E19 committee. In this case, however, a linear plot of detector output against solute concentration at the peak maximum should be used and the point where the line deviates from 45° by 5% determines the limit of the linear dynamic range.

REFERENCES

1. Fowlis, I.A.; Scott, R.P.W. A vapour dilution system for detector calibration. J. Chromatogr. **1963**, *11*, 1.
2. Scott, R.P.W. *Chromatographic Detectors*; Marcel Dekker, Inc.: New York, 1996.

Diffusion Coefficients from GC

George Karaiskakis
Department of Chemistry, University of Patras, Patras, Greece

INTRODUCTION

One of the most important physicochemical applications of GC is for the measurement of diffusion coefficients of gases into gases, liquids, and on solids. The gas chromatographic subtechniques used for the measurement of diffusivities are briefly reviewed, focusing on their accuracy and precision, as well as on the corresponding sources of errors responsible for the deviation of the experimental diffusion coefficients measured by GC from those determined by other techniques or calculated from known empirical equations.

The diffusion coefficients can be determined by various gas chromatographic techniques based either on the broadening of the elution peaks, or on the perturbation imposed on the carrier gas flow rate.

DIFFUSION IN GASES

The diffusion coefficient of a gas A into another gas B, D_{AB}, is a function of temperature, T, pressure, p, and composition, x, even for binary mixtures at low pressure D_{AB} is almost independent of the gas composition. Several empirical equations describing the dependence of D_{AB} on T and p are available, among which the most important are the following:[1,2]

1. The Stefan–Maxwell equation:

$$D_{AB} = \frac{a}{n\sigma_{AB}^2}\left[\frac{8RT}{\pi}\left(\frac{1}{M_A} + \frac{1}{M_B}\right)\right]^{1/2} \quad (1)$$

where a is a constant taking various values ($1/3\pi$, $1/8$, $1/2\pi$, and $3/32$), depending on the researcher, n is the number of gas phase molecules per cubic centimeter, σ_{AB} is the collision diameter between the gas molecules A and B, R is the gas constant, T is the absolute temperature, and M_A, M_B are the molecular masses of solute A and carrier gas B, respectively.

2. The Chapman–Enskog equation:

$$D_{AB} = \frac{0.00263T^{3/2}}{p\sigma_{AB}^2}\left(\frac{1/M_A + 1/M_B}{2}\right)^{1/2} \quad (2)$$

3. The Arnold equation:

$$D_{AB} = \frac{0.0083T^{3/2}(1/M_A + 1/M_B)^{1/2}}{p(V_A^{1/3} + V_B^{1/3})(1 + c_{AB}/T)} \quad (3)$$

where V_A and V_B are molar volumes in cubic centimeters, at the boiling points, while c_{AB} is Sutherland's constant, which can be estimated in various ways.[1,2] The above equation, which introduces a second temperature dependent term in the denominator to account for molecular "softness," shows a dependence varying from $T^{3/2}$ to $T^{5/2}$.

4. The Gilliland equation:

$$D_{AB} = \frac{0.0043T^{3/2}(1/M_A + 1/M_B)^{1/2}}{p(V_A^{1/3} + V_B^{1/3})} \quad (4)$$

5. The Hirschfelder–Bird–Spotz (HBS) equation:

$$D_{AB} = \frac{0.00186T^{3/2}(1/M_A + 1/M_B)^{1/2}}{p\sigma_{AB}^2\Omega_{AB}} \quad (5)$$

The term Ω_{AB} is the collision integral, depending in a complicated way on temperature and the interaction energy of the colliding molecules, ε_{AB}. Ω_{AB} values as a function of the reduced temperature $T^* = kT/\varepsilon_{AB}$, where k is the Boltzmann constant, have been tabulated.[3,4] The main disadvantage of the HBS equation is the difficulty encountered in evaluating σ_{AB} and Ω_{AB}.

6. Chen and Othmer provided the most explicit approximation of the HBS equation using the critical values of temperature, T_C, and volume, V_C:

$$D_{AB} = \frac{0.43\left(\frac{T}{100}\right)^{1.81}\left(\frac{1}{M_A} + \frac{1}{M_B}\right)^{1/2}}{p\left(\frac{T_{CA}T_{CB}}{10^4}\right)^{0.1405}\left[\left(\frac{V_{CA}}{100}\right)^{0.4} + \left(\frac{V_{CB}}{100}\right)^{0.4}\right]^2} \quad (6)$$

7. Fuller, Schettler, and Giddings[5] developed a successful equation in which atomic and

Encyclopedia of Chromatography DOI: 10.1081/E-ECHR-120041314

structural volume increments and other parameters were obtained by a least-squares fit to over 340 measurements. In the Fuller et al. (FSG) method, which provides the best practical combination of simplicity and accuracy,

$$D_{AB} = \frac{0.00143 T^{1.75}(1/M_A + 1/M_B)^{1/2}}{p\left[(\sum v)_A^{1/3} + (\sum v)_B^{1/3}\right]^2} \quad (7)$$

$\sum v$ is determined by summing the relative atomic contributions given in literature.[1]

The Broadening Techniques

The GC broadening techniques for the measurement of gaseous diffusivities are the *continuous elution method*, introduced by Giddings,[2,6,7] and its *arrested elution* modification invented by Knox and McLaren.[2,6,7]

The continuous elution method is conducted in an open tube with circular cross section. The carrier gas flow rate is chosen such that the plate height, H, depends mainly on only one of the van Deemter terms, namely the longitudinal diffusion term. For correction reasons, the use of two different length columns is necessary, and the equation for H is:[2]

$$H = (L_1 - L_2)\left[\frac{t_1^2 - t_2^2}{(t_1 - t_2)^2}\right] \quad (8)$$

where L_1 and L_2 are the lengths of the two columns, respectively, while $(t_1 - t_2)$ and $(t_1^2 - t_2^2)$ are the corresponding differences for the first and second moments of the time base. Replacement of Eq. (8) by the Golay equation, describing band broadening in open tube, gives the final equation from which the diffusion coefficient D_{AB} is obtained:[2]

$$D_{AB} = \frac{v}{4}\left[H \pm \left(H^2 - \frac{r^2}{3}\right)^{1/2}\right] \quad (9)$$

where v is the average carrier gas velocity and r is the radius of the tube. When the flow velocity is slow, the determination of D_{AB} is done from the positive root, while the negative root is used at higher flow velocities. One of the main advantages of the method is the high speed of collecting data with high precision.

In a typical experiment of the arrested elution method,[2,6,7] a solute sample is injected into the column and eluted in the normal way without arresting the gas, so that its outlet velocity, v, can be obtained. Then, during the elution of the solute band, about halfway along the particular column, the flow is switched to a dummy column of equal resistance. After a delay time, t, of 1–20 min, the flow is reconnected to the column and the peak is eluted. The spreading of the

band, during the delay time, can occur only by diffusion, and from the standard deviation δ (or the variance, σ^2) of the concentration profile, determined by the detector, the diffusion coefficient D_{AB} can be found from the following equation:[2]

$$\frac{d\sigma^2}{dt} = \frac{2D_{AB}}{v^2} \quad (10)$$

Eq. (10) shows that a plot of σ^2 against the delay time, t, should be a straight line with slope $2D_{AB}/v^2$, from which the D_{AB} value can be determined. The average reproducibility of the method, which is approximately ±2%, depends mainly on the accurate measurement of v, since v is to the second power in Eq. (10).

Although the arrested elution method has two drawbacks (the need of several runs to get a D_{AB} value with a precision of 2%, and of constant flow rates over long periods for runs at various arrested times), in comparison with the continuous elution method, the former has also the following advantages: 1) Effects of zone broadening other than axial molecular diffusion and nonuniform flow profile do not affect the measurement; 2) no assumptions are made about the precise form of the flow profile, the smoothness of the column wall, or the accuracy in the knowledge of the column diameter.

Diffusion coefficients, for various binary gas mixtures, at various temperatures and pressures with their accuracy and precision, measured by gas chromatographic broadening techniques (GC-BTs; continuous, as well as arrested elution methods), are given in Table 3 of Ref.[2] and in Table 1 of Ref.[6]. Representative data are collected here in Table 1.

The Flow Perturbation Techniques

The flow perturbation gas chromatographic methods used for the measurement of diffusion coefficients in gases are the *stopped-flow* and the *reversed-flow* techniques.

The stopped-flow technique[2,8] consists in stopping the carrier gas flow for short time intervals by using shut-off valves. Following each restoration of gas flow, a narrow peak (stop-peak) is recorded in the chromatographic trace, having the form of those of Fig. 2 in Ref.[2]. The problem to be solved here is to determine the area under the curve of each stop-peak as a function of the time of the corresponding stop in the flow of the carrier gas. Since the stop-peaks are fairly symmetrical and have a constant half-width, their height from the baseline, H, rather than their area, is used to plot $\ln(Ht^{3/2})$ vs. the inverse time, $1/t$, according

Table 1 Diffusion coefficients, D_{AB}, from GC-BTs[2,6]

Binary system A–B	T (K)	p (atm)	D_{AB} (cm^2 s^{-1})	Precision[a] (%)	Accuracy[b] (%)
CH$_4$–H$_2$	298.0	1	0.73	2.7	0
C$_2$H$_6$–H$_2$	298.0	1	0.54	1.9	1.9
C$_3$H$_8$–H$_2$	298.0	1	0.44	6.8	2.2
C$_4$H$_{10}$–H$_2$	298.0	1	0.40	3.8	—
n-C$_5$H$_{12}$–H$_2$	353.0	1	0.4895	1.5	—
	373.0	1	0.5324	3.9	—
	393.0	1	0.5830	4.3	—
	423.0	1	0.6300	0.06	—
	453.0	1	0.7425	0.47	—
n-C$_6$H$_{14}$–H$_2$	353.0	1	0.4990	0.94	—
	373.0	1	0.4740	1.5	10
	393.0	1	0.5310	0.38	—
	423.0	1	0.5923	0.30	—
	453.0	1	0.6520	0.00	—
CH$_4$–He	298.0	1	0.6776	0.22	—
	298.0	1	0.6735	0.12	—
	373.0	1	1.005	—	—
	373.0	1	1.007	—	—
	248.0	9.97	0.0501	—	—
	248.0	29.9	0.0169	—	—
	248.0	49.8	0.0103	—	—
	248.0	59.8	0.00872	—	—
	273.0	9.97	0.0588	—	—
	273.0	29.9	0.0198	—	—
	273.0	49.8	0.0119	—	—
	273.0	59.8	0.0101	—	—
	298.0	9.97	0.0681	—	—
	298.0	29.9	0.0229	—	—
	298.0	49.8	0.0139	—	—
	298.0	59.8	0.0117	—	—
	323.0	9.97	0.0781	—	—
	323.0	29.9	0.0265	—	—
	323.0	49.8	0.0159	—	—
	323.0	59.8	0.0134	—	—
n-C$_4$H$_{10}$–He	298.0	1	0.364	0.27	—
	372.6	1	0.477	2.1	—
	423.0	1	0.634	0.95	—
	473.0	1	0.797	0.75	—
n-C$_5$H$_{12}$–He	298.0	1	0.288	0.35	—
	372.6	1	0.422	0.71	—
	423.0	1	0.565	1.2	—
	473.0	1	0.695	17	—
n-C$_6$H$_{14}$–He	298.0	1	0.27	1.8	—
	372.6	1	0.390	1.5	—
	417.0	1	0.574	—	

(Continued)

Table 1 Diffusion coefficients, D_{AB}, from GC-BTs[2,6] *(Continued)*

Binary system A–B	T (K)	p (atm)	D_{AB} (cm^2 s^{-1})	Precision^a (%)	Accuracy^b (%)
	423.0	1	0.513	2.5	
	473.0	1	0.629	1.9	
C_2H_6–N_2	298.0	1	0.14	18	—
C_3H_8–N_2	298.0	1	0.11	—	—
n-C_4H_{10}–N_2	298.0	1	<0.07	—	—
	298.0	1	0.0954	—	0.63
	302.4	1	0.100	—	1.5
n-C_5H_{12}–N_2	353.0	1	0.136	—	—

^{a}Precision has been defined as: $100 \times$ Deviation/D_{AB}, where for two values, the high (D_{AB}^{high}) and the low (D_{AB}^{low}), the deviation is $(D_{AB}^{high} - D_{AB}^{low})/2$.

^{b}Accuracy has been defined as: $100 \times |D_{AB} - D_{lit}|/D_{AB}$, where D_{lit} are literature[2,6] values.

to the following equation:[8]

$$\ln(Ht^{3/2}) = \ln\left(\frac{mt_sL}{\pi^{1/2}D_{AB}^{1/2}}\right) - \frac{L^2}{4D_{AB}}\frac{1}{t} \qquad (11)$$

where m is the injected amount of solute in moles, t_s the stopped-flow interval in seconds, and L the length of the diffusion column (cf. Fig. 1 of Ref.[2]) in centimeters, which permits calculation of the diffusion coefficient D_{AB} from the slope $-L^2/4D_{AB}$.

Diffusion coefficients from the stopped-flow technique, which are very sensitive to the precision with which L is measured, since D_{AB} is proportional to L^2, are given in literature[2,8] for some binary gas mixtures.

The *reversed-flow GC (RF-GC) technique*, introduced in 1980,[9] is based on reversing the direction of flow of the carrier gas from time to time. Its experimental setup is very simple and consists of the following parts:[2,10,11]

1. A conventional gas chromatograph with any kind of detector capable of detecting the solute(s) contained in the carrier gas.
2. A so-called sampling column constructed from a glass or stainless steel chromatographic tube of any diameter, and having a total length 0.6–2.0 m, depending on the particular application. The sampling column, which is coiled and accommodated inside the chromatographic oven, should be completely empty of any solid material for the determination of the diffusion coefficient of gases into pure gases or mixtures of gases, or it can be filled with a common chromatographic material for the measurement of the diffusion coefficient of ternary gas mixtures into pure gases.

3. A diffusion column, which is constructed from the same material as the sampling column, is connected perpendicularly to it, usually at its midpoint. The other end of the diffusion column is closed with an injector septum and is used as the injection point of the solute under study. The diffusion column, which is empty of any solid material, is a straight or coiled, relatively short (30–80 cm) piece of empty tubing placed inside the chromatographic oven.
4. The sampling and the diffusion column form the sampling cell, and this cell must now be connected to the detector and to the carrier gas inlet in such a way that the carrier gas flow through the sampling column can be reversed in direction by a four- or six-port valve that is connected with the two ends D_1 and D_2 of the chromatographic column, as well as with the inlet of the carrier gas and the detector, as shown in Fig. 1.

If pure carrier gas passes through the sampling column, nothing happens on reversing the flow. But if a solute comes out of the diffusion column as a result of its diffusion into the carrier gas, filling the diffusion column and also running along the sampling column, the flow reversal records the concentration of the solute at the junction $x = l'$ (cf. Fig. 1) at the moment of the reversal. This concentration recording has the form of extra chromatographic peaks, called *sample peaks*, superimposed on the otherwise continuous detector signal. The loading of the carrier gas with other substance(s) is due to its (their) slow diffusion into the carrier gas passing through the sampling column. The enrichment of carrier gas in the gas(es) contained in the diffusion column depends on the rate with which gas(es) enters (enter) the sampling column at the junction $x = l'$ of the two columns. By reversing the

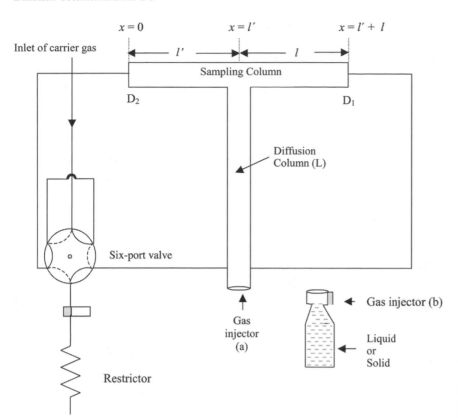

x = 0   x = l'   x = l' + l

Inlet of carrier gas

l'   l

Sampling Column

D_2 D_1

Diffusion Column (L)

Six-port valve

Gas injector (b)

Gas injector (a)

Liquid or Solid

Restrictor

Detector

Fig. 1 Schematic representation of the reversed-flow GC technique for measuring diffusion coefficients. (a) Injection point for gaseous diffusivities; (b) injection point for liquid and solid diffusivities.

flow now, one can perform a sampling of the concentration of the analyte gas at this junction, each sample peak measuring (by its height H) this concentration at the time of the flow reversal. Repeating this sampling procedure at various times, t, and using suitable mathematical analysis, the following equations describing the variation of H with t, when the slow process under study is the gaseous diffusion, were derived:[10–13]

$$\ln(H^{1/M}t^{3/2}) = \ln(gN_1) - \frac{L^2}{4D_{AB}}\frac{1}{t} \qquad (12)$$

$$\ln(H^{1/M}) = \ln(gN_2) - \frac{3D_{AB}}{L^2}t \qquad (13)$$

where

$$N_1 = \frac{mL}{\dot{V}(\pi D_{AB})^{1/2}} \qquad (14)$$

$$N_2 = \frac{\pi m D_{AB}}{\dot{V}L^2} \qquad (15)$$

M is the response factor of the detector [$M = 1$ for flame ionization detector (FID)], g a proportionality constant pertaining to the detector calibration, m the injected amount of solute in moles, and V the volumetric flow-rate in cubic centimeters per second.

Plotting the left-hand side of Eq. (12), $\ln(H^{1/M}t^{3/2})$ vs. $1/t$, one can obtain the diffusion coefficient value of D_{AB} from the slope $-L^2/4D_{AB}$ of the straight line and the known value of the diffusion column length (L).

Eq. (13) shows that a plot of $\ln H$ versus t (after the maximum of the diffusion band) is linear with the slope $-3D_{AB}/L^2$, from which the diffusion coefficient D_{AB} can be also determined.

The question that now naturally arises is: Which one of the two equations, namely Eqs. (12) and (13), is more accurate in the determination of the gaseous diffusion coefficients? The answer is not so simple and depends on both the gaseous system under study and the experimental conditions applied. Generally, Eq. (12) is used for short duration experiments and for long diffusion columns, while Eq. (13) is used for short column lengths (say 30–50 cm) and for experiments of long duration. The selection of the proper mathematical analysis to estimate the more accurate gaseous diffusion coefficient by RF-GC can also be based on the comparison of the two experimental values found from Eqs. (12) and (13) with those given in the literature or calculated from known empirical equations.

The diffusion coefficients of various gaseous hydrocarbons in carrier gases N_2, H_2, and He determined by the reversed-flow technique with the aid of Eq. (12) at various temperatures are compiled in Table 2. The values and their standard errors found by regression analysis using standard least-squares procedures are reduced to 1 atm after multiplication by the pressure of the experiment. This pressure is given in Table 2, so that one can find the actual values determined from the ratio D_{AB}/p. The precision of the method, defined as the relative standard deviation (%), can be judged from the data given for methane–helium. From the five values quoted, a precision of 0.9% is calculated. The experimental values of diffusion coefficient given in Table 2 are compared with those calculated theoretically with the equation of HBS [Eq. (5)]. The accuracy given in the last column

of Table 2 is a measure of the deviation of the values found by the RF-GC method from the calculated ones, defined as:

$$\text{Accuracy (\%)} = \frac{\left| D_{AB}^{\text{found}} - D_{AB}^{\text{calcd}} \right|}{D_{AB}^{\text{found}}} \times 100 \qquad (16)$$

All of the theoretical or semiempirical equations describing the dependence of the diffusion coefficient on temperature lead to the relationship[1,2,14]

$$D_{AB} = AT^n \qquad (17)$$

where A is a complex function including molar masses or volumes, critical volumes or temperatures, volume increments, pressures, etc. depending on the special equation used. Eq. (17) shows that the exponent n can be found from the slope of the linear plot of $\ln D_{AB}$ against $\ln T$. The various values of n calculated from these plots, which vary between 1.59 and 1.77,[13,14] are very close to the value 1.50 suggested by the

Table 2 Diffusion coefficients of various solutes into three carrier gases at ambient temperatures and reduced to 1 atm pressure, determined by reversed-flow GC

Carrier gas	Solute gas	T (K)	$\dot{V}$ (cm^{3}s^{-1})	p (atm)	D_{AB} (×10^3 cm^2 s^{-1})		Accuracy[b] (%)
					D_{AB}^{found}	$(D_{AB}^{\text{calcd}})^{\text{a}}$	
N_2	CH_4	296.0	0.260	1.96	272 ± 4	214	21.3
	C_2H_6	293.0	0.267	1.99	142 ± 0.03	144	1.4
	n-C_4H_{10}	295.5	0.300	2.15	98 ± 0.2	98.6	0.3
	C_2H_4	296.0	0.120	1.49	168 ± 2	156	7.1
		292.0	0.268	2.00	156 ± 0.4		0
		292.0	0.538	2.71	161 ± 0.4		3.1
	C_3H_6	298.0	0.260	1.96	124 ± 0.4	120	3.2
H_2	CH_4	293.0	0.287	1.70	699 ± 3	705	0.9
	C_2H_6	297.0	0.267	1.56	548 ± 5	556	1.5
	n-C_4H_{10}	296.0	0.273	1.60	386 ± 3	373	3.4
	C_2H_4	293.0	0.300	1.75	525 ± 5	559	6.5
	C_3H_6	296.0	0.273	1.60	485 ± 3	486	0.2
He	CH_4	295.7	0.250	1.78	527 ± 3	669	26.9
		295.0	0.283	2.03	520 ± 1		28.7
		296.0	0.283	2.03	522 ± 1		28.2
		296.0	0.283	2.03	514 ± 0.2		30.2
		296.7	0.283	2.03	522 ± 3		28.2
	C_2H_6	295.6	0.300	2.15	518 ± 3	507	2.1
	n-C_4H_{10}	290.0	0.283	2.03	333 ± 3	330	0.9
	C_2H_4	296.0	0.283	2.03	558 ± 4	544	2.5
	C_3H_6	291.0	0.283	2.03	412 ± 4	440	6.8

The actual values found at the pressure of the experiment, p, are simply D_{AB}/p. All errors given in this table are "standard errors" calculated by regressions analysis.
[b]Calculated by the HBS equation.
[a]This is defined by Eq. (16).

Stefan–Maxwell and the Gilliland and Arnold equations,[1] to the value 1.81 predicted by the Chen–Othmer equation,[1] and to the value 1.75 predicted by the FSG equation.[1,5]

The reversed-flow method for measurement of gas diffusion coefficients in binary mixtures can also be extended to simultaneous determination of effective diffusion coefficients for each substance in a multicomponent gas mixture.[15] This extension of the method is achieved by filling the column section *l* (Fig. 1) with a chromatographic material, which can affect the separation of all components of the gas mixture. Effective diffusion coefficients of various mixtures of gaseous hydrocarbons into the carrier gases N_2, H_2, and He, determined by RF-GC, with the aid of Eq. (12), can be found in the literature.[15]

Comparison of the Broadening with the Flow Perturbation Techniques

1. The accuracy of the RF-GC method, in comparison with the GC-BTs, is higher.[2] The mean percentage deviation of D_{AB} (for the binary mixtures C_2H_4–N_2 and C_2H_6–N_2 at various temperatures) determined by RF-GC[2] from the respective predicted values by means of the FSG equation is estimated at 3.4, while that of the GC-BTs is 5.7.[6]

2. The precision of the RF-GC method was calculated to be 0.9%. The precision of the continuous elution technique was about 1%, while that of the arrested elution method was about 2%.[2]

3. The time needed for the determination of D_{AB} values by the continuous elution method is very short ($\approx$5 min), while that by the arrested elution method is longer ($\approx$3 hr), due to the requirement of repeating the experiments at a number of delay times. The corresponding time for the RF-GC method is about 30–60 min, depending on the binary gas mixture and the length of the diffusion column used.

4. The length of the empty diffusion column in RF-GC is relatively short (30–80 cm), compared to that used in the continuous elution method, where much longer ($\sim$15 m) columns are used.

5. In the continuous elution method, in order to eliminate the effect of extra zone broadening factors, the use of two columns is necessary. Such a correction procedure is not necessary in the arrested elution technique, resulting in more reliable D_{AB} values through a more time consuming procedure. RF-GC, being a dynamic technique under steady-state conditions, has also the advantage that extra zone broadening factors are not implied in D_{AB} determinations.

DIFFUSION OF GASES IN LIQUIDS

Diffusion coefficients in liquids are of great significance in many theoretical and engineering calculations involving mass transfer, such as absorption, extraction, distillation, and chemical reactions. The measurement of accurate diffusion coefficients of gases in liquids is not an easy task. Different values are often obtained from different workers in different laboratories, even by using similar measuring techniques.

GC, being a dynamic technique, has been successfully used during the last three decades for the measurement of diffusion coefficients in a large number of gas–liquid systems, especially polymer–solvent systems.

From the 1980's, the *chromatographic broadening technique* has been "substituted" in the literature by *inverse gas chromatography* (IGC). The two main IGC methods for the measurement of diffusion coefficients in liquids are *packed column IGC*[7] and *capillary column IGC*.[7]

In packed column IGC, the preparation of the column is usually done by packing the column with a chromatographic material, running a solution of the solvent through it, and removing the solvent by evaporating with a stream of an inert gas. The presence of film irregularities due to the porous character of most chromatographic supports in packed columns is a problem that affects the accuracy of the measured diffusion coefficients.

An innovation in the use of IGC for the measurement of diffusion coefficients in liquids is the introduction of capillary columns, which solved the problem of the nonuniformity of the coating on the solid particles. A known concentration of a degassed solution is used to fill a capillary column, which is sealed at one end, and vacuum is applied to the other end. The evaporation of the solvent results in the deposition of a uniformly thin liquid layer on the capillary walls.

Data concerning diffusion coefficients of gases in liquids, determined by various IGC techniques, can be found in the literature.[2,6]

The Reversed-Flow Technique

The experimental setup of the RF-GC method for measuring diffusion coefficients of gases in liquids is similar to that of Fig. 1, the only difference being the end of the diffusion column, at which a vessel containing the liquid or the solid (when surface diffusion is studied) is placed.[16,17] Using suitable mathematical analysis, equations were derived by means of which diffusion coefficients of various gases into liquids, D_L, were determined.[16,17]

All calculations of D_L can be carried out simultaneously by a GW-BASIC personal computer program, written for nonlinear least-squares regression analysis[18] and based on the experimental pair values H and t in centimeters and seconds, respectively. The experimental D_L values are of the same order of magnitude and, in some cases, very close to those obtained by other techniques or calculated theoretically from empirical equations. The precision ($\sim$13%) and the accuracy (8–47%) of the RF-GC method, as determined from the D_L values of Refs.,[16,17] compared to those computed from the more accurate empirical equation of Wilke–Chang,[1] are relatively satisfactory, considering the difficulties in obtaining experimental D_L values, and the large dispersion of the predicted diffusion coefficients.[2]

SURFACE DIFFUSION

The only gas chromatographic method used for the measurement of diffusion coefficients of gases on solid surfaces is the RF-GC technique validating a recent mathematical analysis, also permitting the estimation of adsorption and desorption rate constants, local adsorbed concentrations, local isotherms, local monolayer capacities, and energy distribution functions.[19] The RF-GC technique has been successfully applied for the time-resolved determination of surface diffusion coefficients for physically adsorbed or chemisorbed species of O_2, CO, and CO_2 on heterogeneous surfaces of Pt/Rh catalysts supported on SiO_2.[19] All calculations for the surface diffusion coefficient, D_s, can be carried out simultaneously by the GW-BASIC personal computer program listed in Appendix A of Ref.[19]. The D_s values clearly show their dependence on time, and are in relative agreement with those found by nonchromatographic techniques.[19]

CONCLUSIONS

The gas chromatographic broadening and flow perturbation techniques have been proven to be useful tools for the accurate measurement of diffusion coefficients of gases in gases and liquids, and on solids.

ACKNOWLEDGMENTS

The author thanks Ms. Margarita Barkoula for her kind assistance.

REFERENCES

1. Giddings, J.C. *Dynamics of Chromatography*; Marcel Dekker, Inc.: New York, 1965; 237–241.
2. Karaiskakis, G.; Gavril, D. Determination of diffusion coefficients by gas chromatography. J. Chromatogr. A. **2004**, *1037*, 147–189.
3. Hirschfelder, J.O.; Curtis, C.F.; Bird, R.B. *Molecular Theory of Gases and Liquids*; John Wiley: New York, 1954; 1126–1127.
4. Bird, R.B.; Stewart, W.E.; Lightfoot, E.N. *Transport Phenomena*; John Wiley: New York, 1960; 744–746.
5. Fuller, E.N.; Schettler, P.D.; Giddings, J.C. A new method for prediction of binary gas-phase diffusion coefficients. Ind. Eng. Chem. **1966**, *58*, 19–27.
6. Maynard, V.R.; Grushka, E. Measurement of diffusion coefficients by gas-chromatography broadening techniques: a review. Adv. Chromatogr. **1975**, *12*, 99–140.
7. Yang, F.; Hawkes, S.; Lindstrom, F.T. Determination of precise and reliable gas diffusion coefficients by gas chromatography. J. Am. Chem. Soc. **1976**, *98*, 5101–5107.
8. Katsanos, N.A.; Karaiskakis, G.; Vattis, D. Diffusion coefficients from stopped-flow gas chromatography. Chromatographia **1981**, *14*, 695–698.
9. Katsanos, N.A.; Georgiadou, I. Reversed-flow gas chromatography for studying heterogeneous catalysis. J. Chem. Soc. Chem. Commun. **1980**, *5*, 242–243.
10. Katsanos, N.A.; Karaiskakis, G. Reversed-flow gas chromatography applied to physicochemical measurements. Adv. Chromatogr. **1984**, *24*, 125–180.
11. Katsanos, N.A. *Flow Perturbation Gas Chromatography*; Marcel Dekker, Inc.: New York, 1988; 113–161.
12. Katsanos, N.A.; Karaiskakis, G. Measurement of diffusion coefficients by reversed-flow gas chromatography instrumentation. J. Chromatogr. **1982**, *237*, 1–14.
13. Atta, K.A.; Gavril, D.; Karaiskakis, G. A new gas chromatographic methodology for the estimation of the composition of binary gas mixtures. J. Chromatogr. Sci. **2003**, *41*, 123–132.
14. Katsanos, N.A.; Karaiskakis, G. Temperature variation of gas diffusion coefficients measured by the reversed-flow sampling technique. J. Chromatogr. **1983**, *254*, 15–25.
15. Karaiskakis, G.; Katsanos, N.A.; Niotis, A. Measurement of diffusion coefficients in multicomponent gas mixtures by the reversed-flow technique. Chromatographia **1983**, *17*, 310–312.

16. Katsanos, N.A.; Kapolos, J. Diffusion coefficients of gases in liquids and partition coefficients in gas–liquid interphases by reversed-flow gas chromatography. Anal. Chem. **1989**, *61*, 2231–2237.

17. Atta, K.A.; Gavril, D.; Katsanos, N.A.; Karaiskakis, G. Flux of gases across the air–water interface studied by reversed-flow gas chromatography. J. Chromatogr. A. **2001**, *934*, 31–49.

18. Katsanos, N.A.; Arvanitopoulou, E.; Roubani-Kalantzopoulou, E.; Kalantzopoulos, A. Time distribution of adsorption energies, local monolayer capacities, and local isotherms on heterogeneous surfaces by inverse gas chromatography. J. Phys. Chem. B. **1999**, *103*, 1152–1157.

19. Katsanos, N.A.; Gavril, D.; Karaiskakis, G. Time-resolved determination of surface diffusion coefficients for physically adsorbed or chemisorbed species on heterogeneous surfaces by inverse gas chromatography. J. Chromatogr. A. **2003**, *983*, 177–193.

D

Displacement Chromatography

John C. Ford
Department of Chemistry, Indiana University of Pennsylvania, Indiana, Pennsylvania, U.S.A.

INTRODUCTION

One of the three basic modes of chromatographic operation, displacement chromatography is useful for preparative separations and trace enrichment. Most liquid chromatographic methods have been performed in displacement mode. Solutes purified by displacement chromatography include metal cations, small organic molecules, antibiotics, sugars, peptides, proteins, and nucleic acids.

Displacement operation involves the introduction of a volume of sample onto the column, which has been previously equilibrated with a weak mobile phase. Since this is a weak mobile phase, the individual sample components are significantly retained by the stationary phase. Displacement occurs with the introduction of a new mobile phase containing a species with a higher affinity for the stationary phase than that of any of the solutes. The solutes are displaced from the stationary phase by this higher-affinity displacer and move further down the column, readsorbing. The solute with the highest affinity for the stationary phase moves the least before readsorbing and that solute with the lowest affinity for the stationary phase moves the most. This process repeats until a series of separated, but adjacent, bands is formed. Each band moves at the same velocity as that of the displacer front. Following elution, the column must be regenerated and re-equilibrated with the initial mobile phase before any subsequent displacement separation. This re-equilibration step can be lengthy and is frequently considered a major limitation to efficient displacement operation.

DEFINITION

Displacement chromatography is one of the three basic modes of chromatographic operation, the other two being frontal analysis and elution chromatography. Displacement chromatography is rarely, if ever, used for analytical separations, but is useful for preparative separations. It has also been used for trace enrichment.

Many, if not most, retentive chromatographic methods have been performed in the displacement mode, including normal-phase, reversed-phase, ion exchange, and metal affinity chromatographies. Much of the recent work has focused on the use of ion exchange displacement chromatography for the preparative purification of biotechnological products.[1,2] Solutes purified by displacement chromatography include metal cations, small organic molecules, antibiotics, sugars, peptides, proteins, and nucleic acids.

Tswett recognized the difference between elution and displacement development, although Tiselius was the first to clearly define these differences. While displacement was popular in the 1940s, that popularity waned in the 1950s. In the 1980s, there was a resurgence of interest in displacement operation due to the efficient utilization of the stationary phase possible in that mode. Frenz and Horvath[3] have published a comprehensive review of the history and applications of displacement chromatography.

Displacement chromatography is characterized by the introduction of a discrete volume of sample into the chromatographic column that has been previously equilibrated with a weak mobile phase, termed the carrier. This carrier is chosen so that the individual components of the sample (the solutes) are significantly retained by the stationary phase. The displacement is accomplished by following the sample with a new mobile phase containing some concentration of the displacer, a molecule with a higher affinity for the stationary phase than that of any of the solutes. The solutes are displaced from the stationary phase by the higher-affinity displacer and move further down the column, readsorbing. That solute with the highest affinity for the stationary phase moves the least before readsorbing, and that solute with the lowest affinity for the stationary phase moves the most. This process is repeated as the displacer solution moves further down the column until a series of separated, but adjacent, bands is formed, termed the isotachic train. Each component of the train moves at the same velocity as the velocity of the displacer front. Following elution of the isotachic train and the displacer solution from the column, the column must be regenerated and re-equilibrated with the carrier before any subsequent displacement separation. This re-equilibration step can be lengthy and is frequently considered a major limitation to efficient displacement operation.

Encyclopedia of Chromatography DOI: 10.1081/E-ECHR-120039956

Displacement chromatography requires the competitive isotherms of the solutes and the displacer to be convex upward and to not intersect each other. The isotherm of the displacer must have a higher saturation capacity than any of the solutes. The width, not the height, of the solute band within the isotachic train varies as the amount of solute in the sample varies. The height of the solute band is determined by its isotherm. The concentration of the eluted solutes can be greater than their concentrations in the sample in displacement chromatography, unlike in isocratic elution chromatography, wherein dilution necessarily occurs.

The choice of displacer concentration is critical for successful displacement. If the displacer concentration is increased, then the solute concentrations in the isotachic train also increase. If the displacer concentration is decreased, then displacement does not occur and the solutes elute as overloaded peaks in the elution mode. Rhee and Amundson[5] have shown that there is a critical displacer concentration below which displacement cannot occur. This concentration depends primarily on the saturation capacities of the solutes and displacer.

When the solute isotherms cross one another, the situation becomes more complex. It then becomes possible to experience selectivity reversal, i.e., at one displacer concentration, the solutes elute in the order A first, then B, while at another displacer concentration, the order is B first, then A. In a study of this problem, Antia and Horvath[6] showed the existence of the separation gap. This is a region in the isotherm plane, the position of which depends on the ratio of the saturation capacities of the solutes in question. Outside the separation gap, displacement occurs in the normal fashion. However, within the separation gap, the displacement operation does not separate the displaced solutes, but results in the elution of a mixture of the solutes.

In addition to appropriate isotherm behavior and displacer concentration, other factors are important in determining the effectiveness of a displacement chromatographic method. Highly efficient columns and fast mass transfer kinetics are necessary to achieve sharp boundaries between the adjacent solute bands in the isotachic train. Diffuse boundaries mean significant regions of overlap between adjacent solute bands and thus low recovery of purified material.

Successful displacement requires the establishment of the isotachic train before elution of the solutes from the column. As might be expected, the column length is thus an important parameter in displacement chromatography. The column should be sufficiently long (or sufficiently efficient) to allow complete formation of the isotachic train, while lengths beyond that minimum do not improve the separation but increase the separation time. An inadequate length results in the elution of an incompletely formed isotachic train with inadequately resolved solute bands, again reducing the recovery yield.

Similarly, the sample size, column length, and displacer concentration jointly influence the establishment of the isotachic train and, thus, the effectiveness of the displacement separation. For a given displacer concentration and column length, increasing amounts of sample result in increasingly diffuse boundaries and, in sufficiently large samples, significant deterioration of the isotachic train. Likewise, for a given sample size and column length, increasingly high concentrations of displacer cause increasingly diffuse boundaries—termed overdisplacement.

Displacement chromatography has the attractive benefit of concentrating the solute. If the conditions are selected appropriately, large injection volumes of low concentration samples can result in isotachic trains having high solute concentrations, essentially identical to those obtained for narrow pulses of high concentration samples. This is one of the features that have caused the increased interest in displacement as a preparative mode. However, detailed comparisons of the production rates of displacement vs. overloaded elution operation are limited (see Ref.[4], pp. 641–648, and references therein). The limited experimental studies suggest that the displacement operation is superior, although regeneration time was not included in the production rate calculation. Alternately, extensive theoretical studies indicate that, for solutes having Langmuirian behavior, optimized overloaded elution chromatography is superior. Resolution of this issue currently awaits further studies.

ARTICLES OF FURTHER INTEREST

Distribution Coefficient, p. 476.
Elution Chromatography, p. 554.
Frontal Chromatography, p. 671.

REFERENCES

1. Antia, F.D.; Horvath, Cs. Displacement chromatography of peptides and proteins. In *HPLC of Peptides and Proteins: Separation, Analysis, and Conformation*; Mant, C., Hodges, R., Eds.; CRC Press: Boca Raton, 1990; 809–821.
2. Freitag, R. Displacement chromatography: application to downstream processing in

biotechnology. In *Bioseparation and Bioprocessing*; Subramanian, G., Ed.; Wiley-VCH Verlag GmbH: Weinheim, Germany, 1998; Vol. 1, 89–112.

3. Frenz, J.; Horvath, Cs. High-performance displacement chromatography. In *High-Performance Liquid Chromatography–Advances and Perspectives*; Horvath, Cs., Ed.; Academic Press: San Diego, 1988; 5, 211–314.

4. Guiochon, G.; Shirazi, S.G.; Katti, A.M. *Fundamentals of Preparative and Nonlinear Chromatography*; Academic Press: Boston, 1994.

5. Rhee, H.-K.; Amundson, N.R. Analysis of multicomponent separation by displacement development. AIChE J. **1982**, *28* (3), 423–433.

6. Antia, F.D.; Horvath, Cs. Analysis of isotachic patterns in displacement chromatography. J. Chromatogr. **1991**, *556* (1–2), 119–143.

Displacement TLC

Mária Báthori
Department of Pharmacognosy, University of Szeged, Szeged, Hungary

INTRODUCTION

Displacement takes place in a broad scale when a stronger species replaces a weaker one. Displacement of ligands on the receptor is a typical phenomenon in pharmacology which explains drug–drug interactions.

Displacement equilibrium is also known in chromatography, by which ligands compete for binding sites. This competition can be preferentially utilized for either analytical or preparative scale separations.

DISPLACEMENT CHROMATOGRAPHY WITH COLUMN HPLC

Chromatography may be performed as elution, frontal, or displacement. When the mode of development is not specified, a chromatographic separation is considered to be an elution.

The displacement phenomenon in chromatography was recognized from the beginning of separation processes in the present-day sense.[1] Classical displacement chromatography (DC) was used to separate biologically active compounds, such as amino acids, peptides, and fatty acids.[2] High-performance displacement chromatography (HPDC) was developed in Horváth's laboratory at Yale University. They realized displacement type of development on a microparticulate stationary phase; various applications were accomplished to demonstrate the power of HPDC in preparative scale separation of compounds of biological and medical interest.[3–6]

Experimental work of Kalász et al.[6] resulted in the statement of the characteristics and basic rules of DC. They conceived properties of the fully developed displacement train, factors affecting displacement development, efficacy of separation, analysis of displaced fractions, determination of displacement diagrams from Langmuirian isotherms, as well as selection of the column, carrier, and displacer for DC. Concentration of the sample is a particular feature of DC. However, the displacer in the carrier is also definitely concentrated through the development of the displacement train.

Furthermore, certain prerequisites of HPDC were stated, such as the limit of the mobile phase flow velocity, slight overlapping of peaks during fractionation, etc.

Displacement thin-layer chromatography (D-TLC) also stemmed from the activity of Horváth's group at Yale University. Experimental work with D-TLC has proven the validity of the rules of DC, found by using HPLC.[5,6] Kalász et al.[6–8] continued the research on D-TLC, mainly with the separation of steroids.

There are basic unique characteristics solely for DC. DC works with a mobile phase containing both the carrier and the displacer. A special front moves forward during the development; it is the front of the displacer. Of course, certain compounds are moving forward ahead of the displacer front; that is, the displacer (front) displaces the components from the binding sites of the stationary phase. This complex of the components is terminated by the displacer and the components, and also the displacer is moving forward in the form of clearly defined zone(s) instead of as the Gaussian peaks seen with elution chromatography. The concentration (zone height or peak height) of individual displaced zones is determined by the crossing point of the Langmuirian isotherm and the operating line, instead of the amount of sample components (Table 1).

DISPLACEMENT THIN-LAYER CHROMATOGRAPHY

D-TLC started with the experiments of Kalász and Horváth,[5–7] who stated the basic rules of D-TLC. A direct connection was found between the volumetric load of HPLC and the size (length) of sample in TLC. The size of displaced zones depends on the weight size of the load, but never on the volume of the injected sample with D-HPLC. Similarly, the dimensions of the displaced band are independent from that of the spotted sample. A surprisingly short distance of advancement totally developed the displacement train, e.g., a 20-mm development. It is even more surprising that 75 µg of the sample load could become part of a displacement train after having utilized about 250 mg of stationary phase.[6,7]

D-TLC has several analogs to column displacement HPLC, such as:

- A displacer is supplied after the samples have been loaded onto the stationary phase.

Encyclopedia of Chromatography DOI: 10.1081/E-ECHR-120027006

Table 1 Comparison of elution and displacement modes of development

Elution development	Displacement development
Linear and nonlinear (overloaded) elution.	
The components are diluted (relative to their concentrations in the sample).	The components are concentrated (relative to their concentrations in the sample).
There is one mobile phase, consisting of the elements of the eluent. Gradient elution can also be performed by the use of several mobile phases. Either stepwise gradient or continuous gradient elution is possible.	There are two consecutively supplied mobile phases, the carrier and the displacer. This latter consists of an adequate amount of displacer in the carrier.
The components travel with different migration speeds.	Isotachic migration of the component zones in the fully developed displacement train.
Gaussian (or quasi-Gaussian) peaks.	Adjacent square-wave-shaped zones.
The peak area is proportional to the amount of the component. In some approximations, the peak height is considered instead of peak area.	The zone height is determined by the crossing of the Langmuirian isotherm of the component with the operating line.
	The zone width is proportional to the amount of the component.
Separation means physical removal of the peaks from each other, the maximum of peaks have to be far away from each other, the distance should be at least one peak width measured at the baseline.	The adjacent zones of a totally developed displacement train touch each other, even in the optimal separation.

- A displacer front is generated in situ on the stationary phase.
- The sample components start to move because of the effect of displacer, and the displacer front forms a displacement train of adjacent spots.
- After having reached the state of a fully developed displacement train, all bands move with the same velocity, which is the velocity of the displacer front.

There are two conditions required for any component in the displacement train:

- The component has to show very moderate movement on the stationary phase from the effect of carrier.
- The displacer has to have stronger absorption to the displacement train than the component to be displaced.

There are four different migration types of any solute in the displacing system; for example,

1. The solute is displaced in the front of the displacer.
2. The solute migrates faster than the displacer front.
3. The solute migrates slower than the displacer front.
4. The solute remains at the start.

Cases 1–3 are given in Fig. 1. Case no. 1 generally represents the real displacement (bonafide

displacement).[9] Further proof of the bonafide displacing process can be given by the use of spacers, varying the front distance (and, thereby, the displacement front distance), and control of the load vs. spot length principle. This latter rule means that the higher the load, the longer will be the spot length. Bonafide displacement and quasi-displacement can also be differentiated on the basis of the chromatogram. Bonafide displacement gives a homogenous peak, which consists of both the displacer front and the displaced component. In the

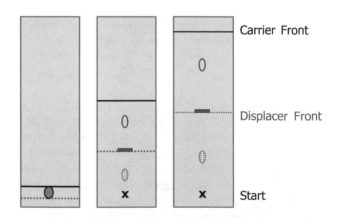

Fig. 1 Development of displacement chromatogram can take place on a plane. The faster-running front represents the carrier, and the slower-running front is the displacer in the carrier. Certain components can be eluted by the carrier (blue open circle), and another one is eluted with the displacer (green circle, shaded). If there is any component displaced, it is located just before the displacer front (red square, filled). (*View this art in color at www.dekker.com.*)

case of quasi-displacement, the two peaks are partially separated, and the peak of the displaced component is a little bit wider. A change of carrier generally solves this problem. Participation in the displacement train is the principal assignment.

Kalász et al.[10] showed that even a very short distance (e.g., 2 cm in the case of 5% displacer) may be sufficient for the formation of a fully developed displacement train. HPDC is devoted to preparative scale separation.[3] Fractions result in HPDC; the fractions are collected and analyzed "off-line." The appropriate fractions can be combined to yield the pure compounds. D-TLC was grown into an approach for scouting the optimum conditions for HPDC. Its goals are to find the proper carrier, the proper displacer, and the appropriate displacer concentration in the carrier. A series of experiments proved that the results of D-TLC can be transferred to HPDC.[5] Even the zone-concentrating effect of the increase of displacer concentration was directly proven using off-line quantitative determination of the fractions obtained by HPDC.[5] However, the major field of application of D-TLC remains qualitative analysis, such as identification of compounds such as metabolites. D-TLC is mainly performed for such analytical purposes. On plain silica stationary phase, the 20:1:4:2 mixture of n-butanol–hydrochloric acid–water–methanol properly displaced several morphine derivatives, including morphine, azido-dihydroisomorphine, 14-hydroxy-dihydromorphine, dihydromorphine, 14-hydroxy-azido-dihydroisocodeine, 14-hydroxy-azido-dihydroisomorphine, codeine, azido-dihydroisocodeine, and 14-hydroxy-dihydrocodeine. Nonetheless, the adequate choice of the displacement system can alter the order of components and even the participation of the components in the displacement train. Using the same stationary phase (plain silica), the chlorinated hydrocarbon carriers (chloroform, dichloromethane, and dichloroethane) make less migration possible, as compared with the n-butanol–methanol–water–hydrochloric acid system. When triethanolamine displacer in chloroform carrier was used, certain otherwise eluted components became part of the displacement train. The ECAM (O-ethyl-N-cyclopropyl-norazido-dihydroisomorphine), CAM (N-cyclopropyl-norazido-dihydroisomorphine), norazido-dihydroisomorphine, normorphine, and nalorphine changed their situation from eluted positions to displaced ones. The most surprising change happened in the case of 6-amino-dihydromorphine, which was eluted with the carrier well before the displacer front, but it was left behind the displacer front in chlorinated hydrocarbon carrier and triethanolamine displacer. This phenomenon can be explained by the reaction of amino group to the change of acidic to basic conditions caused by substituting triethylamine for the hydrochloric acid.

The change of the stationary phase from silica to alumina also altered the components, in the displacement train, and the lengths of the displaced zones were changed as well. Remarkably, certain components show faster mobility on alumina when chloroform carrier is used. In addition, the displacement zones are also generally longer.[11] The change of the stationary phase from plain silica to reversed-phase silica (TLC plate RP-18 $F_{254}s$, precoated, layer thickness 0.25 mm) turned the order of components upside-down (Fig. 2). Some of the components, however, remained in the displacement train, e.g., azido-dihydroisomorphine, ethylmorphine, and norazido-dihydroisomorphine. Preparative work can also be performed by scraping the spots from the plate. D-TLC separation of morphine analogs was performed using volatile mobile phases including the carrier and the displacer. The mixture of n-butanol–hydrochloric acid–water–methanol (20:1:4:2) can be preferentially used for preparative separation of semisynthetic morphine derivatives. The separation includes the removal of the intermediates of the synthesis as well as the purification of the member of metabolic pathway. Another important factor to be considered was the concentration of displacer in the carrier. The higher the displacer concentration, the larger will be the R_F of its front (i.e., R_D). The displacer front could reach the migration position of certain eluted components, which may be part of the displacement train.[9]

DC of substituted phenylisopropylamines can be easily accomplished. Both normal phase and octadecyl-substituted silica (reversed-phase) plates were used with the chloroform/triethanolamine and water–acetonitrile/tetrabutylammonium chloride carrier/displacer pairs, respectively. Only a few exceptionally behaving compounds moved faster or slower than the displacer front in the normal-phase and reversed-phase systems, respectively. The displacement chromatogram of

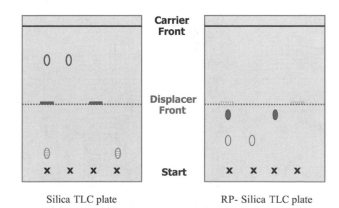

Fig. 2 The order of the spots is generally reversed when the stationary phase is changed from plain silica to reversed-phase silica. *(View this art in color at www.dekker.com.)*

HPDC can be characterized as series of sequentially increasing steps of the weight distribution or that of ultraviolet absorbance. This is the reason why off-line detection is used to characterize HPDC. D-TLC would have a similar profile, but specific detection may improve differentiation of the individual components. Further possibility is given by the application of spacers.

Kalász et al.[12] used spacer to improve separation of radiolabeled metabolites, and the method was called spacer D-TLC. They also constructed specific parameters of DC, especially for calculation of the resolution (R_D), yield (Y), loss (L), and efficiency (E). Displacement of radiolabeled compounds was readily visualized using X-ray film with contact autoradiography.[6,12,13] Two-dimensional chromatography can be simply arranged when a planar stationary phase is used. The stationary phase must be rotated 90° after the 1st dimensional development; then, the system is ready for the 2nd dimensional run. Elution type of development followed by displacement is an easy and useful means of two-dimensional thin-layer chromatography (2D-TLC). The 2D-TLC method has several increments to the elution–elution developments, such as the discrimination factors are different in the 1st and in the 2nd dimensional separations. The spots are concentrated through the 2nd dimensional development. Use of a spacer may further increase the separation of the spots that are located close to each other. Extensive work was devoted to the D-TLC using forced flow of the mobile phase (FFMP)[6,14] and to the comparison of the effect of forced flow of the mobile phase to that of the classical (capillary flow) TLC.[8]

The decrease of the mobile phase flow rate from 0.6 to 0.5 and then to 0.3 caused the appearance of the doubled fronts, such as alpha and beta fronts of the carrier and also α and β fronts of the displacer. This phenomenon can be explained by a particular characteristic of planar chromatography, as the mobile phase runs on a dry stationary phase. In addition to the chromatographic occurrences, there is an additional process; it is the wetting of the stationary phase with the components of the mobile phase. If the mobile phase supply is not adequate (this is the case of slow-traveling mobile phase), the wetting of the dry stationary phase is the rate-limiting step. Such cases are unknown in either HPDC or HPLC, when the stationary phase is presaturated and extensively washed by the mobile phase prior to loading a sample. Therefore development of HPDC is restrained by the high flow velocity of the mobile phase, and that of D-TLC is limited by using a slow velocity. 2D-TLC is the proper choice for separation of a single compound from a multicomponent mixture. When the 2nd dimensional run is carried out using D-TLC, one or several components can be well separated from all others. In addition, the displaced components are extensively concentrated.[14]

When varying the conditions of D-TLC, various components can be a part of the displacement train. An interesting application of DC is given by the identification of metabolically generated radiolabeled formaldehyde from the radiolabeled (−)-deprenyl [(−)-C^{14}-N-methyl-N-propynyl-phenyl-isopropylamine]. The analysis was carried out after reacting the metabolites in a urine sample with 2,4-dinitrophenylhydrazine. The essence of the identification included the use of a standard (2,4-dinitrophenylhydrazone of formaldehyde) and comparing the urine samples with, and without, reaction of 2,4-dinitrophenylhydrazine.[15,16] This method has been called reaction-displacement thin-layer chromatography.[16] Kalász et al.[9,17] reported front deformation when the stationary phase per load mass ratio was under 10, e.g., when 0.5 or 1 mg of solute was loaded onto 2.5 mg of the stationary phase.

The list of substances subjected to planar displacement chromatography includes a broad range of organic compounds. For instance, Kalász and Horváth[5] separated three corticosteroids, Reichstein's Q, Reichstein's S, and Reichstein's H compounds. Kalász[13] also separated phenylalkylamines, e.g., deprenyl, deprenyl metabolites, and related compounds. Báthori et al.[18–21] separated ecdysteroids; Kamano et al.[22] separated toad-poison bufadienolides, such as resibufogenin, cinobufagin, bufalin, bufotalin, cinobufotalin, telocinobufagin, and gamabufotalin. Kalász et al.[10] separated morphine and semisynthetic morphine derivatives from each other. Kalász et al.[15,16] also identified formaldehyde as an efferent metabolite of N-demethylation.

CONCLUSIONS

There are advantages which are offered by D-TLC. The planar stationary phase offers numerous advantages which cannot easily be realized using column (or capillary) arrangements. The entire displacement process can be readily visually followed. The separation can be directly evaluated. Sensitivity of the UV/visible scanners is well suited for the concentrations of substances in the displacement train. Radioactive spots on the planar stationary phase may be easily monitored by the use of X-ray film or digital autoradiography (DAR). Spacers can be inserted between the displaced bands. The plates are disposable; therefore the regeneration process is generally ignored. Two-dimensional developments (elution displacement) can be easily performed. The actual concentration of displacer in the carrier can be calculated on the basis of its retardation. A detailed general summary of D-TLC is also given in a recently published book on TLC.[23]

ACKNOWLEDGMENTS

This project was sponsored by the grant of OTKA T032185. The advice of Ms. Bogi Kalász is appreciated.

REFERENCES

1. Ettre, L.S. Evolution of Liquid Chromatography. In *High-Performance Liquid Chromatography. Advances and Perspectives*; Horváth, Cs., Ed.; Academic Press: New York, 1980; Vol. 1, 25.

2. Tiselius, A. Displacement development in adsorption analysis. Ark. Kemi, Mineral. Geol. **1943**, *16A*, 1–18.

3. Horváth, Cs.; Nahum, A.; Frenz, J. High-performance displacement chromatography. J. Chromatogr. **1981**, *218*, 365–393.

4. Kalász, H.; Horváth, Cs. Preparative scale separation of polymyxin B's by high performance displacement chromatography. J. Chromatogr. **1981**, *215*, 295–302.

5. Kalász, H.; Horváth, Cs. High-performance displacement chromatography of corticosteroids. Scouting for displacer and analysis of the effluent by thin-layer chromatography. J. Chromatogr. **1982**, *239*, 423–438.

6. Kalász, H.; Kerecsen, L.; Knoll, J.; Báthori, M. Displacement Chromatography of Steroids. In *Steroid Analysis*, Proceedings of the Symposium on the Analysis of Steroids, Sopron, Hungary, 1987; Görög, S., Ed.; Akadémiai Kiadó: Budapest, 1988, 405–410.

7. Kalász, H.; Horváth, Cs. Effects of Operating Conditions in Displacement Thin-Layer Chromatography. In *New Approaches in Liquid Chromatography*; Kalász, H., Ed.; Elsevier: Amsterdam, 1984; 57–67.

8. Kalász, H.; Báthori, M.; Kerecsen, L.; Tóth, L. Displacement thin-layer chromatography of some plant ecdysteroids. J. Planar Chromatogr. **1993**, *6*, 38–42.

9. Bariska, J.; Csermely, T.; Fürst, S.; Kalász, H.; Báthori, M. Displacement thin-layer chromatography. J. Liq. Chromatogr. & Relat. Technol. **2000**, *23*, 531–549.

10. Kalász, H.; Kerecsen, L.; Csermely, T.; Götz, H.; Friedmann, T.; Hosztafi, S. Displacement thin-layer chromatographic investigation of morphine and its semi-synthetic derivatives. J. Liq. Chromatogr. **1996**, *19*, 23–35.

11. Kalász, H.; Báthori, M.; Csermely, T. Planar versus microcolumn chromatography. Am. Lab. **2000**, *32* (9), 28–32.

12. Kalász, H.; Báthori, M.; Matkovics, B. Spacer and carrier spacer-displacement thin-layer chromatography. J. Chromatogr. **1990**, *520*, 287–293.

13. Kalász, H. Carrier displacement chromatography for identification of deprenyl and its metabolites. J. High Resol. Chromatogr. Chromatogr. Commun. **1983**, *6*, 49–50.

14. Kalász, H.; Báthori, M.; Ettre, L.S.; Polyák, B. Displacement thin-layer chromatography of some plant ecdysteroids with forced-flow development. J. Planar Chromatogr. **1993**, *6*, 481–486.

15. Kalász, H.; Szarvas, T.; Szarkáné-Bolehovszky, A.; Lengyel, J. TLC analysis of formaldehyde produced by metabolic *N*-demethylation. J. Liq. Chromatogr. & Relat. Technol. **2002**, *25*, 1589–1598.

16. Kalász, H.; Lengyel, J.; Szarvas, T.; Morovjan, Gy.; Klebovich, I. J. Planar Chromatogr. Submitted.

17. Kalász, H.; Báthori, M. Spacer displacement chromatography of steroids. Experiments, considerations and calculations. Invertebr. Reprod. Dev. **1990**, *18*, 119–120.

18. Kalász, H.; Kerecsen, L.; Nagy, J. Conditions dominating displacement thin-layer chromatography. J. Chromatogr. **1984**, *316*, 95–104.

19. Csermely, T.; Kalász, H.; Rischák, K.; Báthori, M.; Tarjányi, Zs.; Fürst, S. Planar chromatography of (−)-deprenyl and some structurally related compounds. J. Planar Chromatogr. **1998**, *11*, 247–253.

20. Lengyel, J.; Magyar, K.; Hollósi, I.; Bartók, T.; Báthori, M.; Kalász, H.; Fürst, S. Urinary excretion of deprenyl metabolites. J. Chromatogr. **1997**, *762*, 321–326.

21. Kalász, H.; Kerecsen, L.; Csermely, T.; Götz, H.; Friedmann, T.; Hosztafi, S. Thin-layer chromatographic investigation of some morphine derivatives. J. Planar Chromatogr. **1995**, *8*, 17–22.

22. Kamano, Y.; Kotake, A.; Nogawa, T.; Tozawa, M.; Pettit, G. Application of displacement thin-layer chromatography to toad-poison bufadienolides. J. Planar Chromatogr. **1999**, *12*, 120–123.

23. Kalász, H.; Báthori, M. Displacement Chromatography and Its Application Using a Planar Stationary Phase. In *Planar Chromatography, a Retrospective View for the Third Millenium*; Nyiredy, Sz., Ed.; Springer: Budapest, 2001; 220–233.

Distribution Coefficient

M. Caude
A. Jardy
ESPCI, Paris, France

INTRODUCTION

In HPLC, as in many chromatographic techniques, separations result from the great number of repetitions of the analyte distribution between the mobile and stationary phases that are linked. At each elementary step, the distribution is governed by the distribution equilibrium

$$X_M \rightleftharpoons X_S$$

where X stands for the solute and the subscripts M and S for the mobile phase and the stationary phase, respectively. Conventionally, this equilibrium is characterized by the distribution coefficient

$$K = \frac{C_S}{C_M} \tag{1}$$

where C_S and C_M are the molar concentrations of the solute X in the two phases. The distribution coefficient is also sometimes defined in terms of mole fractions of the solute in both phases. In elution analytical chromatography, concentrations are low enough to be comparable to the activities as an approximation. Although K should be dimensionless, as any thermodynamic constant, some authors use different units for the concentrations in both phases. For example, in LSC chromatography, concentrations are given in moles per square meter for the adsorbent. In such a case, K should be considered as an "apparent" constant, the dimension of which is m. However, in practice, the product in Eq. (2) must have the dimension of a volume.

K is interrelated with the retention factor k' by

$$k' = Kq$$

where q is the phase ratio, defined as the ratio of stationary phase volume V_S to the mobile phase volume V_M. Under linear conditions, K is linked to the solute retention volume (first-order moment of the elution peak) through

$$V_R = V_M + KV_S \tag{2}$$

The distribution coefficient is the reflection of the ternary interactions schematically represented by

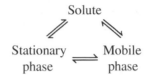

These interactions are as follows:

- Solute ↔ stationary phase for the retention

- Solute ↔ mobile phase for the solubilization

- Mobile phase ↔ stationary phase because of the competition between at least one constituent of the mobile phase (the strongest, also called the modifier) and the solute toward the active sites.

For an isobaric and isothermal process, the equilibrium constant K is given by

$$\ln K = -\frac{\Delta G}{RT} \tag{3}$$

where $\Delta G°$ is the difference in standard Gibbs free energy linked with the solute transfer from the mobile phase to the stationary phase:

$$\Delta G = -\frac{\Delta G}{RT} \tag{4}$$

where $H°$ and $S°$ are the enthalpy and entropy, respectively.

In many cases, the value of $H°$ is independent of temperature so that linear Van't Hoff plots ($\ln K$ versus $1/T$) are observed. However, irregular retention behaviors are often observed, in practice, due to the dependence of $H°$ on the temperature.

Relationship (3) shows, clearly, the obtaining of reproducible results needed to thermoregulate the chromatographic apparatus, especially the column and mobile phase. Similarly, Eq. (3) explains why temperature changes implemented in order to increase efficiency can affect, drastically, the selectivity.

Encyclopedia of Chromatography DOI: 10.1081/E-ECHR-120039957

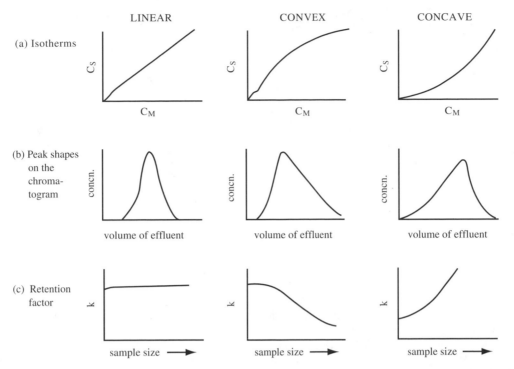

Fig. 1 Effect of isotherm shape on certain chromatographic properties. (a) Three different shapes of sorption isotherms encountered in chromatography; (b) peak shapes resulting from these isotherms; (c) dependence of the retention factor on the amount of solute injected. (Adapted from Karger, B.L.; Snyder, L.R.; Horvath, C. *An Introduction to Separation Science*; John Wiley & Sons, New York, 1973.)

From Eq. (1), *K* can be related to the sorption isotherm, as it corresponds to the chord slope at each point. Therefore, Eq. (2) is valid only if *K* is constant; that is, the sorption isotherm is linear or, at least, is in a region where it becomes linear (i.e., if the dilution is great enough). The effect of the isotherm shape is shown in Fig. 1.

For the low concentrations used in LC, the elution peak is Gaussian in shape and its retention factor is independent of the sample size. When isotherms are nonlinear (convex or concave, as illustrated in Fig. 1), an asymmetric elution peak is obtained and the retention factor measured at the peak apex is dependant on the sample size. This peak asymmetry is due to the dependence of the solute migration velocity versus the slope of the isotherm, which varies the solution concentration in the mobile phase.

SUGGESTED FURTHER READING

Karger, B.L.; Snyder, L.R.; Horvath, C. *An Introduction to Separation Science*; John Wiley & Sons: New York, 1973; 12–33.

Katz, E.; Eksteen, R.; Schoenmakers, P.; Miller, N. *Handbook of HPLC*; Marcel Dekker, Inc.: New York, 1998; Chap. 1.

Rosset, R.; Caude, M.; Jardy, A. *Chromatographies en phases liquide et supercritique*; Masson: Paris, 1991; 729–730.

DNA Sequencing Studies by Capillary Electrophoresis

Feng Xu

Department of Medicinal Chemistry, Faculty of Pharmaceutical Sciences, The 21st Century COE Program, The University of Tokushima, Tokushima, Core Research for Evolutional Science and Technology (CREST), Japan Science and Technology Agency (JST), and Analytical Instruments Division, Shimadzu Corporation, Kyoto, Japan

Yoshinobu Baba

Department of Medicinal Chemistry, Faculty of Pharmaceutical Sciences, The 21st Century COE Program, The University of Tokushima, Tokushima, and Single-Molecule Bioanalysis Laboratory, National Institute of Advanced Industrial Science and Technology, Takamatsu, Japan

INTRODUCTION

Cells store their hereditary information in the form of double-stranded DNA, formed of the same four monomers—adenine (A), thymine (T), cytosine (C), and guanine (G). Using capillary electrophoresis (CE), scientists can read out the complete sequence of monomers in DNA molecules and thereby decipher the hereditary information that each organism contains. Large-scale DNA sequencing projects need instruments that generate high throughput at low cost. One approach to increase the CE throughput is to run a large number of capillaries in parallel, so-called capillary array electrophoresis (CAE). The Human Genome Project was initiated in 1990 with the goal of sequencing the 3 billion nucleotides present within human chromosomes. In 2001, the "first draft" of the entire human genome was published,[1,2] which made it possible to see for the first time exactly how genes are arranged along human chromosomes. With the appearance of the first multicapillary instrument in 1990, traditional slab gel electrophoresis was gradually replaced by CAE due to higher DNA sequencing performance. The early experimental result of sequencing of 29–512 bases with >97% accuracy was obtained in 9.5 hr at a field strength of 50 V/cm by using a laser beam scanning across a capillary array.[3] Later, the appearance of 96-capillary array electrophoresis greatly speeded the DNA sequencing in the Human Genome Project. In the meantime, the new technique of microfabricated CAE (μCAE, or microchip electrophoresis) combines the advantages of CE (system automation, reproducibility, and accurate quantification) with those of microfabrication (high speed and multiplex analysis), and will find its position in DNA sequencing.

This article gives an overview of the fundamentals of DNA sequencing by CAE, μCAE, and four-color laser-induced fluorescence (LIF) detection, as well as some major factors (sieving matrix, sample preparation, electric field strength, etc.) influencing the sequencing accuracy and efficiency.

FOUR-COLOR LASER-INDUCED FLUORESCENCE DETECTION AND CAPILLARY ARRAY ELECTROPHORESIS

Laser-induced fluorescence is the standard CE detection method in DNA sequencing, due to its high sensitivity and the fact that the identity of the terminal base of each DNA can be encoded in the wavelength and intensity of the fluorescent emission. The DNA sequencing fragments are fluorescently labeled with four different dyes on each base, and are then detected by a four-color LIF detector.

Instrumentation design of detector systems for CAE and microfabricated devices has reached a mature stage. The two most successful LIF detector designs are the scanning confocal detector[4] and the multi-sheath flow detector.[5] A schematic of the scanning confocal detector (four-color planar fluorescence scanner) is shown in Fig. 1. The design adopts a scanning technology through capillaries that are illuminated by a single laser beam. The capillary bundle is placed on a planar translation stage, which moves at 1 cm/sec perpendicular to the direction of electrophoresis. The fluorescence, collected at right angles from the capillaries, is divided into four detection channels by dichroic beam splitters and band-pass filters, and then focused through a pinhole on four photomultiplier tubes and simultaneously recorded in four spectral channels. Using automated sample and gel-matrix loading, the total run time for sequencing more than 500 bases is <2 hr. The scanning confocal detector is adopted in the first commercial 96-capillary array MegaBACE 1000 system from Molecular Dynamics

Encyclopedia of Chromatography DOI: 10.1081/E-ECHR-120039958

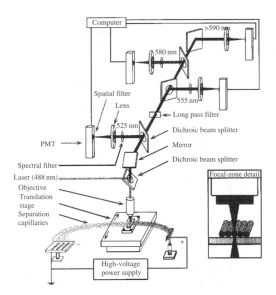

Fig. 1 Four-color planar confocal fluorescence CAE scanner using an excitation of 488 nm from argon ion laser. (From Ref.[33].)

(Amersham-Pharmacia Biotech)[6] for DNA sequencing by CAE. The system uses linear polyacrylamide (LPA) as the separation matrix, and has a turnaround time of 2 hr per run. The capillaries have a life of ca. 130 runs and can be run with both dye primer and dye terminator reactions. An average read length of >550 bases at 98.5% accuracy is feasible with the M13 standard template. Another high-throughput rotary-scanner detection device was designed (as shown in Fig. 2), in order to analyze over 1000 capillaries in parallel. Currently, the device accommodates 128 capillaries. A microscope objective and a mirror

assembly revolve inside a ring of capillaries, exciting fluorophores and collecting fluorescence from each capillary.

In the second detector design, multisheath flow detector, the capillaries are illuminated with a line-focused laser beam. Fluorescence is collected at right angles and imaged onto a CCD camera. The use of the CCD camera ensures that all capillaries are monitored simultaneously. To eliminate light scattering, the capillary array is inserted into a rectangular quartz cuvette that holds the capillaries like the teeth of a comb (Fig. 3). A simple siphon pumps the sheath fluid through the interstitial spaces between the capillaries, and draws a sample from each capillary as a thin stream in the open region below the capillaries. A single laser beam skims all flow streams. Since the laser beam only traverses the sheath fluid and DNA streams, the low-power beam can excite fluorescence from all samples simultaneously. The detector has no moving parts. This design has been incorporated in a second commercial system, ABI PRISM 3700 from PE Biosystems (Applied Biosystems).[7] The system uses four dyes, and simultaneously detects 96 capillaries with a turnaround time of 2 hr to obtain 550 bases with 98.5% accuracy using POP-6 (6% poly-N,N-dimethylacrylamide; pDMA) as the sieving matrix.

RIKEN Japan has produced an extra-high-throughput autosequencer (RISA sequencer) that consists of a

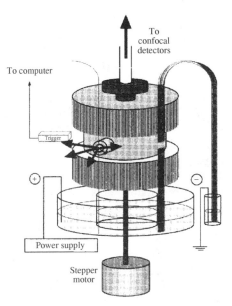

Fig. 2 Rotary confocal scanning detector. (From Ref.[34].)

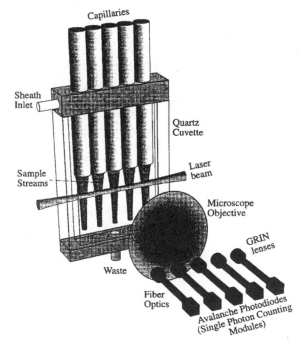

Fig. 3 The sheath-flow cuvette fluorescence detection chamber for an array of five capillaries. The chamber is tapered. A single laser beam is used to illuminate fluorescence from the five sample streams isolated by the sheath flow fluid. (From Ref.[35].)

384-capillary array. Cross-linked acrylamide is used as the sieving matrix and a scanning laser fluorescence as the detector.[8] The read length, with more than 99% accuracy, is 650 bases.

MICROCHIP ELECTROPHORESIS

Microchip electrophoresis is superior to CE in that it allows facile monolithic array construction, precise controlling of picoliter sample injection amount, quick electrophoretic separation speed, and potential for integration with sample pretreatment.

The first instance of DNA separation by microchip electrophoresis was in 1994.[9] A mixture of DNA oligomers from 10 to 15 bases was efficiently separated. Since then, microchip electrophoresis for DNA separation has developed very quickly. DNA sequencing of 200 bases by microchip takes only 10 min in cross-linked polyacrylamide gels.[10] By using 4% LPA in 7-cm-long coated microchannels, 600 bases were separated in 20 min at 160 V/cm.[11] Four-color LIF detection is feasible in micrototal analysis system (μTAS) use.

Long read lengths need long separation channels, because separation resolution scales with the square root of channel length. In order to adapt the radial chip design to modern wafer-scale fabrication for increasing the effective separation lengths, Mathies's group developed pinched turn geometries (hyperturns) in folding channels (Fig. 4).[12] A unique rotary design and a rotary confocal scanning system run 96 samples in a radial configuration at a time. DNA sequencing with an average read length of 430 bases at a rate of 1.7 kb/min was achieved in 96-lane microchip CAE. Ehrlich and coworkers[36] fabricated very long microchannels of 40 cm in large glass plates and obtained an average read length of 800 bases in 80 min with 98% accuracy (Fig. 5). Capillary array electrophoresis on a chip brings new potentials to high throughput DNA sequencing on miniaturized CE platforms.

FACTORS INFLUENCING SEQUENCING

A key parameter of an electrophoresis system is the read length per unit time. Since the read length for the present commercial instrument is limited to 500–600 bases with >99% accuracy in 2 hr, sequencing of long read lengths, e.g., >1000 bases, is very much required. Long read lengths in a short time reduce the number of sequencing reactions needed, because the number of primers required for directed sequencing strategy and the number of templates generated and sequenced in shotgun sequencing are inversely proportional to the read length. In addition, long read lengths minimize the computational effort required

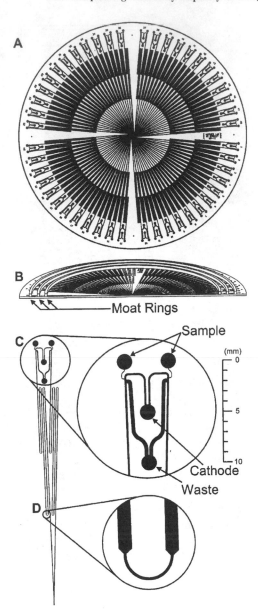

Fig. 4 Capillary array electrophoresis on a microchip with 96 channels. (A) Overall layout of the 96-lane DNA sequencing microchannel plate (MCP). (B) Vertical cut-away of the MCP. (C) Expanded view of the injector. Each doublet features two sample reservoirs and common cathode and waste reservoirs. (D) Expanded view of the hyperturn region. (From Ref.[12].)

to assemble shotgun-generated data into finished sequences. The read length is influenced by a number of factors, e.g., polymer matrix, capillary temperature, field strength, effective channel length, base-calling algorithms, etc., but the most important is resolution. Heller[13] summarized in detail how various factors affect resolution (Fig. 6). Diffusion is thought to be an ultimate limitation to the resolution of polymer matrix-based separation. Long separation channels, modest electric fields, high temperature, and medium

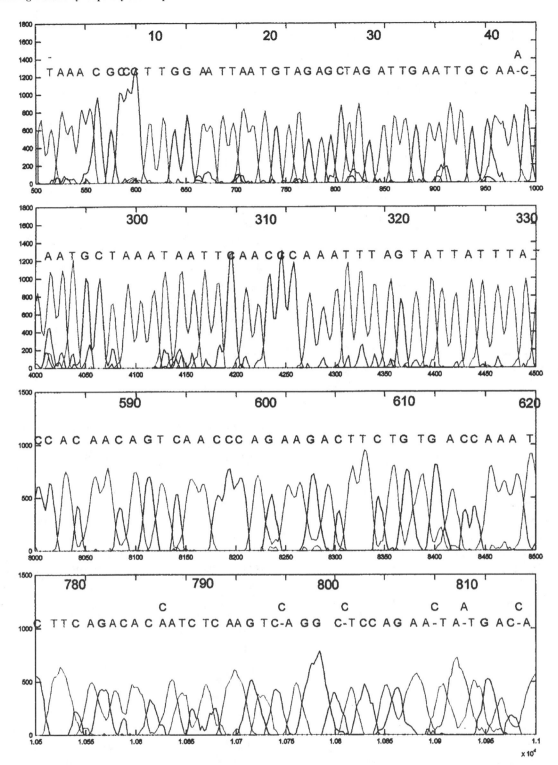

Fig. 5 Eight hundred base reads of a four-color DNA sequencing sample in a 40-cm-long microchannel. The four panels show the processed sequencing profiles and base calls at the beginning, the middle, and the end of the run. Conditions: 150 V/cm, 50°C, and 2% (w/v) LPA in 1 × TTE/7 M urea. (From Ref.[36].)

concentrations of polymer matrices are feasible for obtaining higher resolution and longer read length.

According to polymer theory, with a semidilute polymer solution above an overlap threshold concentration, c^*, fully entangled networks are formed by the interaction of polymer tangles, thereby forming dynamic pores in the polymer network for DNA separation. Depending on DNA size, the DNA

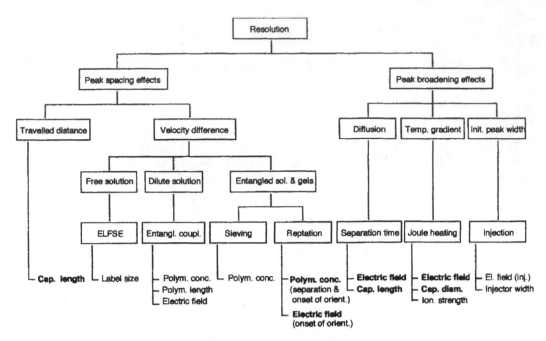

Fig. 6 Overview of the influence of different parameters on resolution during separation of DNA by CE. Only tunable parameters with strong influence are shown. (From Ref.[13].)

molecules can either be sieved through the polymer network (Ogston model) or reptate in a virtual tube (reptation model).[14] The Ogston model describes the separation of DNA molecules smaller than the polymer pore size. The reptation model can depict the electrophoresis behavior of DNA molecules larger than the polymer pore size. Larger DNA fluctuates in effective length during migration. The mobility of a DNA fragment is inversely proportional to its size (reptation without orientation). If the DNA molecule is very large, its mobility is size independent (reptation with orientation), in which case the resolution approaches zero. In DNA sequencing by electrophoresis, one can extend the reptation-without-orientation range by optimizing various parameters that influence resolution. Then the position of zero resolution is shifted to a higher base number.

Separation Matrix

Earlier DNA sequencing by CE used cross-linked polyacrylamide gels. The polyacrylamide capillary has a rather limited life. When the separation medium has degraded, the entire capillary must be replaced. The replacement is quite tedious because of alignment constraints of the optical system with the narrow-diameter capillaries. In contrast, low-viscosity polymers are attractive for DNA sequencing by CE. They can be pumped from the capillary and replaced with fresh matrix after each run without replacement of

the capillary or realignment of the optical system. A range of polymeric solutions have been tested. LPA, pDMA, and polyethylene oxide (PEO) are the most widely used ones. Nowadays, gel-filled capillaries have almost fully been substituted by polymer solution-filled capillaries.

Linear polyacrylamide represents the best replaceable sieving polymer in terms of read length and speed. It can generate DNA sequencing read lengths beyond 1000 bases within 1 hr with a run-to-run base calling accuracy of 99.2% for the first 800 bases and 98.1% for the first 900 bases.[15] At a high polymer concentration, both long and short chains give equally good resolution for single-stranded DNA. For separating a larger size range, it is better to use long polymer chains and lower polymer concentrations. This leads to a more uniform resolution in function of DNA size. Mixing two populations of polymers, each with a narrow but different range of molar mass, has become popular. It facilitates fine-tuning of the separation performance over a broad range of DNA sizes. With a novel LPA formulation comprising 0.5% w/w 270 kDa/2% w/w 17 MDa LPAs operated at 70°C and 125 V/cm, an ultralong read length of 1300 bases was reached in 2 hr with 98.5% accuracy.[16] The high performance is attributed to better thermostability of the LPA formulation, optimized temperature and electric field, adjustment of the sequencing reaction, and refinement of the base-calling software.

PEO, pDMA, and polyvinylpyrrolidone (PVP) have self-coating abilities, which allows DNA sequencing in bare fused-silica capillaries. A mixture of two molecular mass populations of PEO is possible to separate DNA sequencing fragments over 1000 bases, too, but the separation time exceeds 7 hr.[17] pDMA[18] and PVP[19] are low-viscosity sieving matrices (<100 cP). They have moderate separation efficiencies and can provide maximum read lengths of about 500–600 bases.

Copolymers of LPA and pDMA combine the excellent sieving performance of LPA with the good self-coating property of pDMA. On uncoated capillaries at room temperature, 700 bases can be obtained with resolution $R > 0.55$.[20]

Sieving matrices with "built-in thermal viscosity switches" are alternatives in DNA sieving. The 3–5% grafted copolymer solution, which has a hydrophilic LPA backbone and comblike poly(N-isopropylacrylamide) (pNIPA; M_w 650–1800 kDa) side chains, exhibits a viscosity lower than 300 cP at room temperature, and a high viscosity (10,000 cP) at 66°C, suitable for DNA sequencing with a read length of 800 bases ($R > 0.5$) in less than 1 hr.[21]

In order to avoid intramolecular base pairing of DNAs, sequencing has to be run under denaturing conditions. A denaturing agent, such as urea or formamide, is added during gel polymerization or to the buffer in the case of polymer solutions. Most researchers use urea at a concentration of 7–8 M. The addition of formamide up to 40% increases the denaturing capacity of the matrix. The denaturing power of these agents alone is not sufficient, and the separation has to be performed at an elevated temperature. Generally, a temperature of 50–60°C is used to keep the DNA fragments completely denatured. Furthermore, higher temperature operation increases sequencing rate and read length as well as resolutions.[22] For 600-base separation, the maximum separation efficiency was also found at 60°C.[23] Temperature stability is important for DNA sequencing by CE. Even millidegree temperature oscillations have a detrimental effect on DNA read lengths.[24] Beyond a point, the efficiency diminishes seriously.

Sample Preparation

Sample preparation is an important step in the sequencing protocol. The presence of impurities [salt, proteins, unincorporated deoxynucleotides (dNTPs), dideoxynucleotides (ddNTPs), etc.] in DNA samples has deleterious effects both on the sequencing run and on capillary lifespan. DNA sequencing samples are typically synthesized following the Sanger enzymatic method.[25] Four reactions are set up, each with a different A, T, C, or G modified by replacing an H atom from the OH group in the C3 position of the sugar. This ddNTP is incapable of forming the

next bond in the DNA chain; therefore, synthesis of that chain is terminated when a ddNTP is incorporated. The four reactions (with either labeled ddATP, ddTTP, ddCTP, or ddGTP) are performed, followed by ethanol precipitation to remove excess reagents and salts. The precipitated DNA is dissolved in formamide for electrophoretic injection and sequencing.

Since the beginning of DNA sequencing by CE, DNA labeled with four dyes (FAM, JOE, ROX, and TAMRA) has been standardized.[26] The four-color confocal fluorescence scanner utilizes excitation at either two laser wavelengths (e.g., 488 and 543 nm) or a single laser beam (e.g., 488 nm). New energy-transfer (ET) primers have higher molar absorbances. They contain a common donor dye at the 5' end and an acceptor dye about 8–10 nucleotides away. Using a single laser at 488 nm, the excitation is transferred by resonance ET to the acceptor dye; then higher fluorescence intensities are observed,[27] which results in longer read lengths, higher base-calling accuracies, and reduced template amount. Now the ET primers have gained wide acceptance and are widely used for DNA sequencing in CAE.[28]

Effect of Electric Field Strength

With decreasing electric field, the onset of reptation-with-orientation is shifted to larger DNA sizes, which extends the size range to be separated. This effect is confirmed by many experiments. An electric field strength of 600 V/cm produces an ultrafast analysis time of 3–4 min, but the read length degrades to 300 bases.[29] At the expense of an analytical time of 7 hr, a low field strength of 75 V/cm can extend the read length up to 1000 bases in PEO matrices.[17] Baba's group[30] proposed electric field step gradients, with an initial voltage ramp (up to 220 V/cm) for accelerating short fragments, followed by a voltage plateau, a voltage decrement, and a lower voltage constant of 90–130 V/cm for longer DNA fragments. A 20% extension of the read length was obtained, up to 800 bases at 60°C with high accuracy.

WALL COATING

Single-stranded DNA has a more hydrophobic character than double-stranded DNA and presents stronger interaction with the silica wall. Thus, capillary wall coating is greatly preferred. A few of polymers, such as pDMA, PEO, and PVP, as mentioned, have self-coating abilities. They can be used for DNA sequencing in bare capillaries or bare microchips. However, the life of the bare walls is limited because their performance deteriorates with repeated runs. Therefore, extensive and, sometimes, harsh rinsing is indispensable between runs.

Most polymeric matrices require an inner coating of separation channels to prevent both electro-osmotic flow (EOF) and DNA–channel wall interactions. The acrylamide coating procedure developed by Hjertén[31] is the most commonly used permanent covalent coating method. The disadvantage is that the coating cannot endure for long due to easy hydrolysis of the –Si–O–Si– bond. Hence, some more stable covalent coating procedures[32] have been developed to extend the wall coating life.

CONCLUSIONS

Capillary array electrophoresis has become a standard method to decipher genomes within a short time, and has played an important role in large-scale DNA sequencing. With further increase of the capillary number in the array, it will become difficult to manufacture and work with. The problem is being solved by a transition to µCAE systems. The quality of sequencing separations on microchips is rapidly approaching that obtained using conventional CAE. Hyphenation of sample automatic processing steps to µCAE will further lead to increase in efficiency and quality of DNA sequencing, reduction of overall cost, and automation of the whole process. It is plausible that high-throughput DNA sequencing by CAE and µCAE will afford a rapid means for human genetic counseling, disease diagnosis, and clinical therapy.

ACKNOWLEDGMENTS

The work was partially supported by the CREST program of the Japan Science and Technology Agency (JST); a grant from the New Energy and Industrial Technology Development Organization (NEDO) of the Ministry of Economy, Trade, and Industry, Japan; a Grant-in-Aid for Scientific Research from the Ministry of Health and Welfare, Japan; a Grant-in-Aid for Scientific Research from the Ministry of Education, Science, and Technology, Japan; a Grant-in-Aid of the 21st Century COE program, Human Nutritional Science on Stress Control, from the Ministry of Education, Science, and Technology, Japan; and a Grant-in-Aid from Shimadzu Corp., Japan.

REFERENCES

1. Lander, E.S.; Linton, L.M.; Birren, B.; Nusbaum, C.; Zody, M.C.; Baldwin, J.; et al. Initial sequencing and analysis of the human genome. Nature 2001, 409 (6822), 860–921.

2. Venter, J.C.; Adams, M.D.; Myers, E.W.; Li, P.W.; Mural, R.J.; Sutton, G.G.; et al. The sequence of the human genome. Science 2001, 291 (5507), 1304–1351.

3. Zagursky, R.J.; McCormick, R.M. DNA sequencing separations in capillary gels on a modified commercial DNA sequencing instrument. Biotechniques 1990, 9 (1), 74–79.

4. Mathies, R.A.; Huang, X.C. Capillary array electrophoresis: an approach to high-speed, high-throughput DNA sequencing. Nature 1992, 359 (6391), 167–169.

5. Kambara, H.; Takahashi, S. Multiple-sheathflow capillary array DNA analyzer. Nature 1993, 361 (6412), 565–566.

6. Bashkin, J.S.; Bartosievicz, M.; Roach, D.; Leong, J.; Barker, D.; Johnston, R. Implementation of a capillary array electrophoresis instrument. J. Capillary Electrophor. 1996, 3 (2), 61–68.

7. Swerdlow, H.; Zhang, J.Z.; Chen, D.Y.; Harke, H.R.; Grey, R.; Wu, S.L.; Dovichi, N.J.; Fuller, C. Three DNA sequencing methods using capillary gel electrophoresis and laser-induced fluorescence. Anal. Chem. 1991, 63 (24), 2835–2841.

8. Shibata, K.; Itoh, M.; Aizawa, K.; Nagaoka, S.; Sasaki, N.; Carninci, P.; et al. RIKEN integrated sequence analysis (RISA) system—384-format sequencing pipeline with 384 multicapillary sequencer. Genome Res. 2000, 10 (11), 1757–1771.

9. Effenhauser, C.S.; Paulus, A.; Manz, A.; Widmer, H.M. High-speed separation of antisense oligonucleotides on a micromachined capillary electrophoresis device. Anal. Chem. 1994, 66 (18), 2949–2953.

10. Woolley, A.T.; Mathies, R.A. Ultra-high-speed DNA sequencing using capillary electrophoresis chips. Anal. Chem. 1995, 67 (20), 3676–3680.

11. Liu, S.; Shi, Y.; Ja, W.W.; Mathies, R.A. Optimization of high-speed DNA sequencing on microfabricated capillary electrophoresis channels. Anal. Chem. 1999, 71 (3), 566–573.

12. Paegel, B.M.; Emrich, C.A.; Wedemayer, G.J.; Scherer, J.R.; Mathies, R.A. High throughput DNA sequencing with a microfabricated 96-lane capillary array electrophoresis bioprocessor. Proc. Natl. Acad. Sci. U.S.A. 2002, 99 (2), 574–579.

13. Heller, C. Principle of DNA separation with capillary electrophoresis. Electrophoresis 2001, 22 (4), 629–643.

14. Xu, F.; Baba, Y. Polymer solutions and entropic-based systems for double-stranded DNA capillary electrophoresis and microchip electrophoresis. Electrophoresis 2004, 25 (14), 2332–2345 (and references therein).

15. Salas-Solano, O.; Carrilho, E.; Kotler, L.; Miller, A.W.; Goetzinger, W.; Sosic, Z.; Karger, B.L. Routine DNA sequencing of 1000 bases in less than one hour by capillary electrophoresis with replaceable linear polyacrylamide solutions. Anal. Chem. **1998**, *70* (19), 3996–4003.

16. Zhou, H.; Miller, A.W.; Sosic, Z.; Buchholz, B.; Barron, A.E.; Kotler, L.; Karger, B.L. DNA sequencing up to 1300 bases in two hours by capillary electrophoresis with mixed replaceable linear polyacrylamide solutions. Anal. Chem. **2000**, *72* (5), 1045–1052.

17. Kim, Y.; Yeung, E.S. Separation of DNA sequencing fragments up to 1000 bases by using poly(ethylene oxide)-filled capillary electrophoresis. J. Chromatogr. A. **1997**, *781* (1–2), 315–325.

18. Madabhushi, R.S. Separation of 4-color DNA sequencing extension products in noncovalently coated capillaries using low viscosity polymer solutions. Electrophoresis **1998**, *19* (2), 224–230.

19. Gao, Q.F.; Yeung, E.S. A matrix for DNA separation: genotyping and sequencing using poly(vinylpyrrolidone) solution in uncoated capillaries. Anal. Chem. **1998**, *70* (7), 1382–1388.

20. Song, L.; Liang, D.; Kielscawa, J.; Liang, J.; Tjoe, E.; Fang, D.; Chu, B. DNA sequencing by capillary electrophoresis using copolymers of acrylamide and *N,N*-dimethylacrylamide. Electrophoresis **2001**, *22* (4), 729–736.

21. Sudor, J.; Barbier, V.; Thirot, S.; Godfrin, D.; Hourdet, D.; Millequant, M.; Blanchard, J.; Viovy, J.-L. New block–copolymer thermoassociating matrices for DNA sequencing: effect of molecular structure on rheology and resolution. Electrophoresis **2001**, *22* (4), 720–728.

22. Kleparnik, K.; Foret, F.; Berka, J.; Goetzinger, W.; Miller, A.W.; Karger, B.L. The use of elevated column temperature to extend DNA-sequencing read lengths in capillary electrophoresis with replaceable polymer matrices. Electrophoresis **1996**, *17* (12), 1860–1866.

23. Salas-Solano, O.; Ruiz-Martinez, M.C.; Carrilho, E.; Kotler, L.; Karger, B.L. A sample purification method for rugged and high-performance DNA sequencing by capillary electrophoresis using replaceable polymer solutions. B. Quantitative determination of the role of sample matrix components on sequencing analysis. Anal. Chem. **1998**, *70* (8), 1528–1535.

24. Voss, K.O.; Roos, H.P.; Dovichi, N.J. The effect of temperature oscillations on DNA sequencing by capillary electrophoresis. Anal. Chem. **2001**, *73* (6), 1345–1349.

25. Sanger, F.; Nicklen, S.; Coulson, A.R. DNA sequencing with chain-terminating inhibitors. Proc. Natl. Acad. Sci. U.S.A. **1977**, *74* (12), 5463–5467.

26. Carson, S.; Cohen, A.S.; Belenkii, A.; Ruiz-Martinez, M.C.; Berka, J.; Karger, B.L. DNA sequencing by capillary electrophoresis: use of a two-laser-two-window intensified diode array detection system. Anal. Chem. **1993**, *65* (22), 3219–3226.

27. Ju, J.; Glazer, A.N.; Mathies, R.A. Energy transfer primers: a new fluorescence labeling paradigm for DNA sequencing and analysis. Nat. Med. **1996**, *2* (2), 246–249.

28. Soper, S.A.; Legendre, B.L., Jr.; Willams, D.C. On-line fluorescence lifetime determinations in capillary electrophoresis. Anal. Chem. **1995**, *67* (23), 4358–4365.

29. Muller, O.; Minarik, M.; Foret, F. Ultrafast DNA analysis by capillary electrophoresis/laser-induced fluorescence detection. Electrophoresis **1998**, *19* (8–9), 1436–1444.

30. Endo, Y.; Yoshida, C.; Baba, Y. DNA sequencing by capillary array electrophoresis with an electric field strength gradient. J. Biochem. Biophys. Meth. **1999**, *41* (2–3), 133–141.

31. Hjertén, S. High-performance electrophoresis: elimination of electroendoosmosis and solute adsorption. J. Chromatogr. **1985**, *347*, 191–198.

32. Dolnik, V.; Xu, D.; Yadav, A.; Bashkin, J.; Marsh, M.; Tu, O.; Mansfield, E.; Vainer, M.; Madabhushi, R.; Barker, D.; Harris, D. Wall coating for DNA sequencing and fragment analysis by capillary electrophoresis. J. Microcol. Sep. **1998**, *10* (2), 175–184.

33. Kheterpal, I.; Scherer, J.R.; Clark, S.M.; Radhakrishnan, A.; Ju, J.; Ginther, C.L.; Sensabaugh, G.F.; Mathies, R.A. DNA-sequencing using a four-color confocal fluorescence capillary array scanner. Electrophoresis **1996**, *17* (12), 1852–1859.

34. Scherer, J.R.; Kheterpal, I.; Radhakrishnan, A.; Ja, W.W.; Mathies, R.A. Ultra-high throughput rotary capillary array electrophoresis scanner for fluorescent DNA sequencing and analysis. Electrophoresis **1999**, *20* (7), 1508–1517.

35. Zhang, J.; Voss, K.O.; Shaw, D.F.; Roos, K.P.; Lewis, D.F.; Yan, J.; Jiang, R.; Ren, H.; Hou, J.Y.; Fang, Y.; Puyang, X.; Ahmadzadeh, H.; Dovichi, N.J. A multiple-capillary electrophoresis system for small-scale DNA sequencing and analysis. Nucleic Acids Res. **1999**, *27* (24), e36.

36. Koutny, L.; Schmalzing, D.; Salas-Solano, O.; El-Difrawy, S.; Adourian, A.; Buonocore, S.; Abbey, K.; McEwan, P.; Matsudaira, P.; Ehrlich, D. Eight hundred-base sequencing in a micro-fabricated electrophoretic device. Anal. Chem. **2000**, *72* (14), 3388–3391.

D

Drug Residues in Food: Detection/Confirmation by LC/MS

Nikolaos A. Botsoglou
Laboratory of Nutrition, Faculty of Veterinary Medicine, Aristotle University, Thessaloniki, Greece

INTRODUCTION

Numerous detection systems based on almost all kinds of known analytical techniques have been developed for screening, identifying, and quantifying drug residues in food. Each detection system has its own advantages and drawbacks which must be carefully considered in the selection of the most convenient system for a particular analyte in a particular matrix. The problem of analyzing for drug residues in food is complicated by the fact that it is not known whether residues exist and, if they exist, the type and quantity are not known.

Microbiological or immunochemical detection systems offer the advantage to screen, rapidly and at low cost, a large number of food samples for potential residues, but cannot provide definitive information on the identity of violative residues found in suspected samples. For samples found positive by the screening assays, residues can be tentatively identified and quantified by means of the combined force of an efficient LC separation and a selective physicochemical detection system such as UV, fluorescence, or electrochemical detection. The potential of pre- or postcolumn derivatization can further enhance the selectivity and sensitivity of the analysis. Nevertheless, unequivocal identification by these methods is not possible unless a more efficient detection system is applied.

The possibility for unambiguous identification of the analytes is offered by liquid chromatography–mass spectrometry LC/MS. Mass spectrometry detection systems use the difference in mass-to-charge ratio (m/z) of ionized atoms or molecules to separate them from each other. Molecules have distinctive fragmentation patterns that provide structural information to identify structural components. The on-line coupling of LC with MS for the determination of drug residues in food has been under investigation for almost two decades.

LC/MS COUPLING

When LC is coupled with MS, three major problems generally arise. The first concerns the ionization of nonvolatile and/or thermolabile analytes. As MS operation is based on magnetic and electric fields that exert forces on charged ions in a vacuum, a compound must be charged or ionized in the source to be introduced in the gas phase into the vacuum system of the MS. This is easily attainable for thermally volatile samples, but thermally labile analytes may decompose upon heating. The second is due to mobile-phase incompatibility as a result of the frequent use of nonvolatile mobile-phase buffers and additives in LC. This is why routine or long-term use of nonvolatile mobile-phase constituents, such as phosphate buffers and ion-pairing agents, is prohibited by all current LC/MS methods. As far as the third problem is concerned, this is related to the apparent flow rate incompatibility as expressed in the need to introduce a mobile phase eluting from the column at a flow rate of around 1 mL/min into the high vacuum of the MS.

To eliminate these problems, several different interfaces that provide broad analytical coverage have been developed. The main limitation of LC/MS interfaces is the lack of fragmentation data provided for structure determination because most interfaces operate basically in a chemical ionization (CI) mode, providing mild ionization and making identification of unknowns difficult or impossible. Hence the choice of a suitable interface for a particular application always has to be related to the analytes considered, especially their polarity and molecular mass, and the specific analytical problem as well.[1]

Among the currently available interfaces for drug residue analysis, the most powerful and promising appear to be the particle-beam (PB) interface, the thermospray (TSP) interface that works well with substances of medium polarity, and the atmospheric pressure ionization (API) interfaces that have opened up important application areas of LC to LC/MS for ionizable compounds. Among the API interfaces, electrospray (ESP) and ionspray (ISP) appear to be the most versatile as they are suitable for substances ranging from polar to ionic and from low to high molecular mass. Ionspray, in particular, is compatible with the flow rates used with conventional LC columns. In addition, both ESP and ISP appear to be valuable in terms of analyte detectability.

Complementary to ESP and ISP interfaces, with respect to the analyte polarity, is the atmospheric

Encyclopedia of Chromatography DOI: 10.1081/E-ECHR-120038629

pressure chemical ionization (APCI) interface equipped with a heated nebulizer. This is a powerful interface for both structural confirmation and quantitative analysis.

Particle-Beam Interface

The PB interface is an analyte-enrichment interface in which the column effluent is pneumatically nebulized into a near atmospheric-pressure desolvation chamber connected to a momentum separator, where the high-mass analytes are preferentially directed to the MS ion source while the low-mass solvent molecules are efficiently pumped away. With this interface, mobile-phase flow rates within the range 0.1–1.0 mL/min can be applied. Particle-beam–mass spectrometry appears to have high potential as an identification method for residues of some antibiotics in foods as it generates library-searchable EI spectra and CI solvent-independent spectra. Limitations of the PB/MS interface, as compared with other LC/MS interfaces, include lower sensitivity, difficulty in quantification, and lower response with highly aqueous mobile phases. The low sensitivity can be attributed, in part, to chromatographic band broadening during the transmission of the sample through the interface and, in part, to nonlinearity effects that appear at low analyte concentrations.[2]

Liquid chromatography–particle beam–mass spectrometry has been investigated as a potential confirmatory method for the determination of malachite green in incurred catfish tissue,[3] and cephapirin, furosemide, and methylene blue in milk, kidney, and muscle tissue, respectively.[4] Liquid chromatography–particle beam–mass spectrometry has also been investigated for the analysis of ivermectin residues in bovine liver and milk.[5] The specificity required for regulatory confirmation was obtained by monitoring the molecular ion and characteristic fragment ions of the drug under negative-ion chemical ionization (NCI)–selective ion monitoring (SIM) conditions. Quantification and confirmation of tetracycline, oxytetracycline, and chlortetracycline residues in milk,[6] as well as chloramphenicol residues in calf muscle,[7] have also been carried out using LC/PB/NCI/MS.

Thermospray Interface

The thermospray (TSP) interface is widely used for the determination of drug residues in foods.[1] Thermospray is typically used with reversed-phase columns and volatile buffers. Aqueous mobile phases containing an electrolyte, such as ammonium acetate, are passed through a heated capillary prior to entering a heated ion source. As the end of the capillary lies opposite a vacuum line, nebulization takes place and

a jet of vapor containing a mist of electrically charged droplets is formed. As the droplets move through the hot source area, they continue to vaporize, and ions present in the eluent are ejected from the droplet and sampled through a conical exit aperture in the mass analyzer. The ionization of the analytes takes place by means of direct ion evaporization of the sample ion or by solvent-mediated CI reactions. With ionic analytes, the mechanism of ion evaporation is supposed to be primarily operative as ions are produced spontaneously from the mobile phase. Drawbacks of LC/TSP/MS are the requirements for volatile modifiers and the control of temperature, particularly for thermolabile compounds.[8] Also, ion evaporation often yields mass spectra with little structural information. Lack of structural information from LC/TSP/MS applications can be overcome by the use of LC/TSP/MS/MS. Use of this tandem MS approach provides enhanced selectivity, generally at the cost of loss of sensitivity as a consequence of decreased ion transmission.

Liquid chromatography–thermospray–mass spectrometry has been successfully applied for the detection/confirmation of nicarbazin residues in chicken tissues using negative-ion detection in SIM mode.[9] Liquid chromatography–thermospray–mass spectrometry in SIM mode has also been used for the quantification of residues of moxidectin in cattle tissues and fat,[10] and nitroxynil, rafoxanide, and levamisole in muscle.[11] Confirmatory methods based on LC/TSP/MS have been further reported for the determination of penicillin G,[12] cephapirin,[13] and various penicillin derivatives[14] in milk. Comparative evaluation of the confirmatory efficiency of LC/TSP/MS and LC/TSP/MS/MS in the assay of maduramycin in chicken fat showed the former approach to be marginally appropriate, whereas the latter is highly efficient.[15] Tandem LC/MS/MS has also been successfully applied in the analysis of residues of chloramphenicol in milk and fish.[16]

Electrospray Interface

The ESP interface, which is a widely applicable soft ionization technique, operates at the low microliter per minute flow rate, necessitating the use of either capillary columns or postcolumn splitting of the mobile phase. For ESP ionization, the analytes must be ionic or have an ionizable functional group, or be able to form an ionic adduct in solution; the analytes are commonly detected as deprotonated species or as cation adducts of a proton or an alkali metal ion. When using positive ion ESP ionization, the use of ammonium acetate as a mobile-phase modifier is generally unsuitable. Instead, organic modifiers, such as heptafluorobutyric or trifluoroacetic acid, usually at a concentration of 0.1%, are strongly recommended. For negative ion applications, the choice of the modifier is even more limited, triethylamine currently being the only suitable compound.

Liquid chromatography–electrospray–mass spectrometry has been successfully used for the multiresidue assay of penicillin G, ampicillin, amoxicillin, cloxacillin, and cephapirin in milk ultrafiltrate at the 100-ppb level after postcolumn splitting of the eluent and recording under SIM conditions in the positive-ion mode.[17] Significantly lower detection limits were reported by other workers who described an LC/ESP/MS confirmatory procedure for the simultaneous determination of five penicillins in milk and meat under SIM conditions in the negative-ion mode.[18] Electrospray has been further shown to be useful in the analysis of several classes of veterinary drugs, including sulfonamides and tetracyclines, which exhibit spectra with four common ions; however, this was not possible for the group of β-agonists because of their more diverse chemical structures.[19] Negative-ion ESP/MS has also been used for the detection/identification of a number of nonsteroidal anti-inflammatory drugs, including phenylbutazone, flunixin, oxyphenbutazone, and diclofenac, after their reversed-phase separation.[20] Liquid chromatography–electrospray–mass spectrometry has also been found suitable for the determination of four coccidiostats in poultry products.[21]

Ionspray Interface

The ISP interface is closely related to the ESP. However, unlike the ESP interface, ISP allows higher flow rates by virtue of pneumatically assisted vaporization. As both ESP and ISP produce quasi-molecular ions, more sophisticated techniques, such as LC/MS/MS, are required to obtain diagnostic fragment ions and thus analyte structure elucidation. Identification can often be achieved by using daughter ion MS/MS scans and collisionally induced dissociation (CID), most commonly on a triple quadrupole MS; in this way, dissociation of the quasi-molecular ion occurs and diagnostic structural information can be obtained.

Liquid chromatography–ionspray–mass spectrometry has been shown to be an attractive approach for the determination of semduramicin in chicken liver.[22] Tandem MS using the CID of the molecular ions further enhanced the specificity providing structure elucidation and selective detection down to 30 ppb. Liquid chromatography–ionspray–mass spectrometry has also been successfully applied for the assay of 21 sulfonamides in salmon flesh.[23] Coupling of LC with either ISP/MS or ISP/MS/MS has also been

investigated as an attractive alternative for the determination of erythromycin A and its metabolites in salmon tissue.[24] The combination of these methods permitted the identification of a number of degradation products and metabolites of erythromycin at the 10–50 ppb level. Tandem MS with CID has also been applied for the specific monitoring of danofloxacin and its metabolites in chicken and cattle tissues at levels down to 50 ppb.[25] Both ISP/MS in SIM mode and pulsed amperometric detection were found to be suitable for the determination of aminoglycoside antibiotics in bovine tissues.[26]

Atmospheric Pressure Chemical Ionization Interface

Complementary to ESP and ISP interfaces is the APCI interface equipped with a heated nebulizer. The nebulized liquid effluent is swept through the heated tube by an additional gas flow, which circumvents the nebulizer. The heated mixture of solvent and vapor is then introduced into the ionization source where a corona discharge electrode initiates APCI. The spectra and chromatograms from APCI are somewhat similar to those from TSP, but the technique is more robust, especially with gradient LC, and it is often more sensitive. Atmospheric pressure chemical ionization is particularly useful for heat labile compounds and for low-mass, as well as high-mass, compounds. In contrast to the TSP interface, no extensive temperature optimization is needed with APCI.

The applicability of the APCI interface is restricted to the analysis of compounds with lower polarity and lower molecular mass, compared with ESP and ISP. Applications include the LC-APCI-MS multiresidue determination of quinolone antibiotics,[27] the determination of tetracyclines in muscle at the 100-ppb level,[28] and the determination of fenbendazole, oxfendazole, and the sulfone metabolite in muscle at the 10-ppb level.[29]

CONCLUSIONS

The need to use drugs in animal husbandry will continue well into the future, and therefore monitoring of edible animal products for violative residues will remain an area of increasing concern and importance because of the potential impact on human health. The successful hyphenation of LC with MS has led to the development of highly flexible, computer-aided analytical methods that offer the required possibility for unambiguous identification of drug residues in food. Liquid chromatography–mass spectrometry is now in a mature state, but it still cannot be considered routine in the field of drug residue analysis. Possible reasons are the high initial cost, which is two to four times higher than that of GC/MS, and the poor detection limits, which are approximately 100 times higher than in GC/MS. Coupling of LC with tandem MS may be a solution for improving detection limits by reducing the background noise, but this combination is two or three times more expensive than its LC/MS analogue. Mass spectrometry has become a standard tool in every modern laboratory, but there will be a growing need for even more sophisticated couplings.

REFERENCES

1. Botsoglou, N.A.; Fletouris, D.J. *Drug Residues in Food. Pharmacology, Food Safety, and Analysis*; Marcel Dekker, Inc.: New York, 2001.

2. Tinke, A.P.; Van der Hoeven, R.A.M.; Niessen, W.M.A.; Tjaden, U.R.; Van der Greef, J. Some aspects of peak broadening in particle-beam liquid-chromatography mass-spectrometry. J. Chromatogr. **1991**, *554*, 119–124.

3. Turnipseed, S.B.; Roybal, J.E.; Rupp, H.S.; Hurlbut, J.A.; Long, A.R. Particle-beam liquid-chromatography mass-spectrometry of triphenyl-methane dyes—Application to confirmation of malachite green in incurred catfish tissue. J. Chromatogr. **1995**, *670*, 55–62.

4. Voyksner, R.D.; Smith, C.S.; Knox, P.C. Optimization and application of particle beam high-performance liquid-chromatography mass-spectrometry to compounds of pharmaceutical interest. Biomed. Environ. Mass Spectrom. **1990**, *19*, 523–534.

5. Heller, D.N.; Schenck, F.J. Particle beam liquid-chromatography mass-spectrometry with negative-ion chemical ionization for the confirmation of ivermectin residues in bovine milk and liver. Biol. Mass Spectrom. **1993**, *22*, 184–193.

6. Kijak, P.J.; Leadbetter, M.G.; Thomas, M.H.; Thompson, E.A. Confirmation of oxytetracycline, tetracycline and chlortetracycline residues in milk by particle beam liquid-chromatography mass-spectrometry. Biol. Mass Spectrom. **1991**, *20*, 789–795.

7. Delepine, B.; Sanders, P. Determination of chloramphenicol in muscle using a particle beam interface for combining liquid-chromatography with negative-ion chemical ionization mass-spectrometry. J. Chromatogr. **1992**, *582*, 113–121.

8. Niessen, W.M.A.; Van der Greef, J. *Liquid Chromatography–Mass Spectrometry, Principles and Application*; Marcel Dekker, Inc.: New York, 1992.

9. Lewis, J.L.; Macy, T.D.; Garteiz, D.A. Determination of nicarbazin in chicken tissues by liquid-chromatography and confirmation of identity by thermospray liquid-chromatography mass-spectrometry. J. Assoc. Off. Anal. Chem. **1989**, *72*, 577–581.

10. Khunachak, A.; Dakunha, A.R.; Stout, S.J. Liquid-chromatographic determination of moxidectin residues in cattle tissues and confirmation in cattle fat by liquid-chromatography mass-spectrometry. J. AOAC Int. **1993**, *76*, 1230–1235.

11. Cannavan, A.; Blanchflower, W.J.; Kennedy, D.G. Determination of levamisole in animal tissues using liquid-chromatography thermospray mass-spectrometry. Analyst **1997**, *120*, 331–333.

12. Boison, J.O.K.; Keng, L.J.-Y.; MacNeil, J.D. Analysis of penicillin-G in milk by liquid-chromatography. J. AOAC Int. **1994**, *77*, 565–570.

13. Tyczkowska, K.L.; Voyksner, R.D.; Aronson, A.L. Development of an analytical method for cephapirin and its metabolite in bovine milk and serum by liquid-chromatography with UV–Vis detection and confirmation by thermospray mass-spectrometry. J. Vet. Pharmacol. Ther. **1991**, *14*, 51–60.

14. Voyksner, R.D.; Tyczkowska, K.L.; Aronson, A.L. Development of analytical methods for some penicillins in bovine milk by ion-paired chromatography and confirmation by thermospray mass-spectrometry. J. Chromatogr. **1991**, *567*, 389–404.

15. Stout, S.J.; Wilson, L.A.; Kleiner, A.I.; Dacunha, A.R.; Francl, T.J. Mass-spectrometric approaches to the confirmation of maduramicin-alpha in chicken fat. Biomed. Environ. Mass Spectrom. **1989**, *18*, 57–63.

16. Ramsey, E.D.; Games, D.E.; Startin, J.R.; Crews, C.; Gilbert, J. Detection of residues of chloramphenicol in crude extracts of fish and milk by tandem mass-spectrometry. Biomed. Environ. Mass Spectrom. **1989**, *18*, 5–11.

17. Straub, R.F.; Voyksner, R.D. Determination of penicillin-G, ampicillin, amoxicillin, cloxacillin and cephapirin by high-performance liquid-chromatography electrospray mass-spectrometry. J. Chromatogr. **1993**, *647*, 167–181.

18. Tyczkowska, K.L.; Voyksner, R.D.; Straub, R.F.; Aronson, A.L. Simultaneous multiresidue analysis of beta-lactam antibiotics in bovine milk by liquid-chromatography with ultraviolet detection and confirmation by electrospray mass-spectrometry. J. AOAC Int. **1994**, *77*, 1122–1131.

19. Harris, J.; Wilkins, J. The application of HPLC-cone voltage assisted fragmentation electrospray mass spectrometry to the determination of veterinary drug residues. In *Residues of Veterinary Drugs in Food*, Proceedings of the Euroresidue

III Conference, Veldhoven, May, 6–8, 1996; Haagsma, N. Ruiter, A., Ed.; Fac. Vet. Med., Univ. Utrecht: The Netherlands, 1996.

20. Gowik, P.; Julicher, B. Behaviour of some selected NSAID's under electrospray LC/MS conditions. In *Residues of Veterinary Drugs in Food*, Proceedings of the Euroresidue III Conference, Veldhoven, May, 6–8, 1996; Haagsma, N. Ruiter, A., Ed.; Fac. Vet. Med., Univ. Utrecht: The Netherlands, 1996.

21. Blanchflower, W.J.; Kennedy, D.G. Determination of lasalocid in eggs using liquid chromatography-electrospray mass-spectrometry. Analyst **1995**, *120*, 1129–1132.

22. Schneider, R.P.; Lynch, M.J.; Ericson, J.F.; Fouda, H.G. Electrospray ionization mass-spectrometry of semduramicin and other polyether ionophores. Anal. Chem. **1991**, *63*, 1789–1794.

23. Pleasance, S.; Blay, P.; Quilliam, M.A.; O'Hara, G. Determination of sulphonamides by liquid-chromatography, ultraviolet diode-array detection and ion-spray tandem mass-spectrometry with application to cultured salmon flesh. J. Chromatogr. **1991**, *558*, 155–173.

24. Pleasance, S.; Kelly, J.; Leblanc, M.D.; Quilliam, M.A.; Boyd, R.K.; Kitts, D.D.; McErlane, K.; Bailey, M.R.; North, D.H. Determination of erythromycin-A in salmon tissue by liquid-chromatography with ionspray mass-spectrometry. Biol. Mass Spectrom. **1992**, *21*, 675–687.

25. McLaughlin, L.G.; Henion, J.D. Determination of aminoglycoside antibiotics by reversed-phase ion-pair high-performance liquid-chromatography coupled with pulsed amperometry and ion spray mass-spectrometry. J. Chromatogr. **1992**, *591*, 195–206.

26. Schneider, R.P.; Ericson, J.F.; Lynch, M.J.; Fouda, H.G. Confirmation of danofloxacin residues in chicken and cattle liver by microbore high-performance liquid-chromatography electrospray-ionization tandem mass-spectrometry. Biol. Mass Spectrom. **1993**, *22*, 595–599.

27. Doerge, D.R.; Bajic, S. Multiresidue determination of quinoloe antibiotics using liquid-chromatography coupled to atmospheric-pressure chemical-ionization mass-spectrometry and tandem mass-spectrometry. Rapid Commun. Mass Spectrom. **1993**, *9*, 1012–1016.

28. McCracken, R.J.; Blanchflower, W.J.; Haggan, S.A.; Kennedy, D.G. Simultaneous determination of oxytetracycline, tetracycline and chlortetracycline in animal tissues using liquid-chromatography, postcolumn derivatization with aluminum, and fluorescence detection. Analyst **1995**, *120*, 1761–1766.

29. Blanchflower, W.J.; Cannavan, A.; Kennedy, D.J. Determination of fenbendazole and oxfendazole in liver and muscle using liquid-chromatography mass-spectrometry. Analyst **1994**, *119*, 1325–1328.

Dry-Column Chromatography

Mark Moskovitz
Scientific Adsorbents, Inc., Atlanta, Georgia, U.S.A.

INTRODUCTION

Dry-column chromatography (DCC) is a modern chromatographic technique that allows easy and rapid transfer of the operating parameters of analytical TLC to preparative column chromatography (CC). The dry-column technique bridges the gap between preparative CC and analytical TLC.

DISCUSSION

TLC has become an important technique in laboratory work, because it permits the rapid determination of the composition of complex mixtures. TLC allows the isolation of substances in micro amounts. If, however, milligrams or even grams of substance are required, CC has to be applied, as TLC would involve a high cost and excessive time. In many cases, even the so-called thick layer or prep layer is but a poor choice because of time, cost, and sometimes inadequate transferability of the parameters of the analytical technique. In addition, the transfer from TLC to CC, however, often proves to be difficult because the CC adsorbent is not usually analogous to the TLC adsorbent.

It is imperative that when transferring conditions of TLC separations to preparative columns, the conditions responsible for the TLC separation be meticulously transferred. Both CC and TLC use the same principle of separation. For normal operating conditions, a TLC layer has a chromatographic activity of II–III of the Brockmann and Schodder scale. Therefore, the sorbent used for DCC has to be brought to the same grade of activity. TLC layers often contain a fluorescent indicator in which case the DCC sorbent has to contain the same phosphor.

In TLC, the silica or alumina layer is "dry" before it is used and contacts the solvent only after it has been placed into the developing chamber. This is why, in DCC, the dry column is charged with the sample. Contrary to the normal CC, DCC is a nonelution technique. Therefore, only a limited amount of eluent is used in DCC to merely fill the interstitial volume between the adsorbent particles.

Scientific Adsorbents, Inc. DCC adsorbents, which are commercially available from Scientific Adsorbents,

Inc. (Atlanta, GA, U.S.A.) are adjusted to meet the physical-chemical properties of TLC as closely as possible. These adjustments are made during the manufacturing cycle, and the material is packaged ready to use. With similar physical-chemical properties, the values obtained for the substances under investigation from TLC are practically identical to those obtained with DCC.

Using these especially adjusted adsorbents for DCC, one can use the same sorbent and the same solvent for the column work and can transfer the TLC results to a

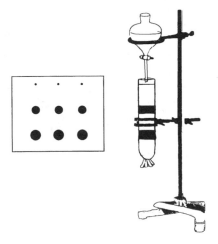

Fig. 1

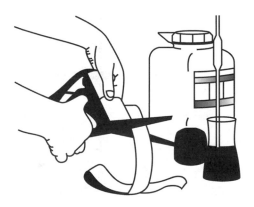

Fig. 2

Encyclopedia of Chromatography DOI: 10.1081/E-ECHR-120039959

Fig. 3

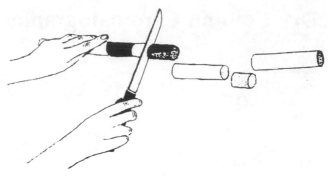

Fig. 6

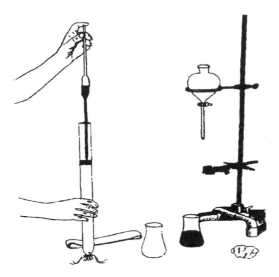

Fig. 4

Fig. 5

preparative scale column operation rapidly, saving time and money. DCC materials are available corresponding with the most common thin layers: silica DCC and alumina DCC.

These DCC sorbents have found wide use when it is necessary to scale up TLC separations in order to prepare sufficient quantities of compounds for further chemical reactions and/or analytical processes. DCC can be practically used for every separation achievable by TLC (Fig. 1).

SIMPLIFIED PROCEDURE

Preparation

1. Use the same solvent system that was developed on a TLC plate.
2. Cut a Nylon tube to the desired length. To isolate 1 g of material, use approximately 300 g of sorbent in a 1-m × 740-mm tube (Fig. 2).
3. Close the tube by rolling one end and securing it by a seal or a clip/staple.
4. Insert a small pad or wad of glass wool at the bottom of the column; pierce holes at the bottom with a needle.
5. Dry fill the column to three-fourths of its length (Fig. 3).
6. The sample to be separated should be combined with at least 10 times its weight of the same sorbent in a conical test tube.
7. Add an additional centimeter of sorbent on top of the sample, followed by a small pad of glass wool (Fig. 4).
8. Fasten the tube to a clamp on a stand.
9. Open the stopcock of the solvent reservoir and add solvent until it reaches the bottom of the column. Stop. Elapsed time: approximately 30 min. (Fig. 5).
10. Find the locations of the separated bands by visible, ultraviolet (UV), or UV quenching. Alternatively, cut a 1/16-in. vertical slice off

the tube. Spray the exposed area with an appropriate visualization reagent and align with the untreated column to identify (mark) the separated bands.

11. Mark the location of the bands on the Nylon tube.
12. Remove the column from the clamp.
13. Slice the column into the desired sections (Fig. 6).
14. Elute the pure compounds from the sliced sections with polar solvents.

SUGGESTED FURTHER READING

Love, B.; Goodman, M.M. Chem. Ind. (London) **1967**, 2026.

Love, B.; Snyder, K.M. Chem. Ind. (London) **1965**, 15.

Eddy Diffusion in LC

J. E. Haky
Florida Atlantic University, Boca Raton, Florida, U.S.A.

INTRODUCTION

Among the causes of widening of peaks corresponding to components of a mixture undergoing separation by LC is the phenomenon known as eddy diffusion. This results from molecules of a solute traversing a packed bed of a column through different pathways, in and around the stationary phase. Some molecules travel more rapidly through the column through more open, shorter pathways, whereas others will encounter longer, restricted areas and lag behind. The result is a solute band that passes through the column with a Gaussian distribution around its center.[1]

DISCUSSION

The degree of band broadening of any chromatographic peak may be described in terms of the height equivalent to a theoretical plate, H, given by

$$H = \frac{L}{N} \tag{1}$$

where L is the length of the column (usually measured in cm) and N is the number of theoretical plates, which can be calculated from Eq. (2), where t_R and W are the retention time and width of the peak of interest, respectively:

$$N = 16\left(\frac{t_R}{W}\right)^2 \tag{2}$$

Because higher values of N correspond to lower degrees of band broadening and narrower peaks, the opposite is true for H. Therefore, the goal of any chromatographic separation is to obtain the lowest possible values for H.

The contribution of eddy diffusion and other factors to band broadening in LC can be quantitatively described by the following equation, which relates the column plate height H to the linear velocity of the solute, μ:

$$H = A\mu^{0.33} + \frac{B}{\mu} + C\mu + D\mu \tag{3}$$

where A, B, C, and D are constants for a given column.[2] The linear velocity μ is related to the mobile-phase flow rate and is determined by

$$\mu = \frac{L}{t_0} \tag{4}$$

where t_0 (the so-called "dead time") is determined from the retention time of a solute which is known not to interact with the stationary phase of the column. The first term in Eq. (4), $A\mu^{0.33}$, includes the contribution of eddy diffusion to chromatographic band broadening. This term, which is dependent on the cube root of the linear velocity, is less dependent on mobile-phase flow rate than the other terms in the equation, which are either directly or inversely proportional to linear velocity.

Minimizing eddy diffusion in an LC column results in a lower $A\mu^{0.33}$ term in Eq. (3), which minimizes band spreading and gives narrower chromatographic peaks. The most common methods used for this purpose, in LC, are the following: (a) using a column of the smallest practical diameter; this obviously reduces the number of alternate pathways which a solute can take through the column; (b) using a stationary phase of smallest practical particle size; Giddings[3] and others have shown that the effects of eddy diffusion are directly proportional to the average diameter of stationary-phase particles; thus, smaller stationary phase particles give narrower peaks; (c) making sure the column is uniformly packed; again, this limits open space in the column, thus minimizing the number of pathways.

Those who prepare and/or manufacture LC columns must use the above methods to limit the effects of eddy diffusion on the chromatographic separations. However, there are practical limitations.

Encyclopedia of Chromatography DOI: 10.1081/E-ECHR-120039962

Column and stationary-phase particle diameters can only be reduced to points that are compatible with the pressure limitations of the pumps used in chromatographic instruments and the required sample capacities of the columns. The degree of training and experience of those who pack the columns may also limit the quality of the procedure used in packing the column. Nevertheless, most commercial manufacturers of LC columns have adopted column designs and packing procedures which generally reduce the effects of eddy diffusion on modern LC separations to an inconsequential level. Still, these effects may increase as a column ages, and

practicing chromatographers should be on the watch for them.

REFERENCES

1. Poole, C.F.; Poole, S.K. *Chromatography Today*; Elsevier: New York, 1991; Chap. 1.
2. Snyder, L.R.; Kirkland, J.J. *Introduction to Modern Liquid Chromatography,* 2nd Ed.; John Wiley & Sons: New York, 1979; 15–37.
3. Giddings, J.C. *Dynamics of Chromatography*; Marcel Dekker, Inc.: New York, 1965; 35–36.

E

Efficiency in Chromatography

Nelu Grinberg
Merck Research Laboratories, Rahway, New Jersey, U.S.A.

Rosario LoBrutto
Seton Hall University, South Orange and
Merck Research Laboratories, Rahway, New Jersey, U.S.A.

INTRODUCTION

One of the most important characteristics of a chromatographic system is the efficiency or the number of theoretical plates, N.

DISCUSSION

The number of theoretical plates can be defined from a chromatogram of a single band as

$$N = \left(\frac{t_R}{\sigma_t}\right)^2 = \frac{L^2}{\sigma_t^{\,2}} \tag{1}$$

where, for a Gaussian shaped peak, t_R is the time for elution of the band center, σ_t is the band variance in time units, and L is the column length.[1] N is a dimensionless quantity; it can also be expressed as a function of the band elution volume and variance in volume units:

$$N = \left(\frac{V_R}{\sigma_v}\right)^2 = 5.56\left(\frac{t_R}{W_{1/2}}\right)^2 \tag{2}$$

In a chromatographic system, it is desirable to have a high column plate number. The column plate number increases with several factors:[2]

- Well-packed column
- Longer columns
- Smaller column packing particles
- Lower mobile-phase viscosity and higher temperature
- Smaller sample molecules
- Minimum extracolumn effects.

In an open-bed system, N can be measured from the distance passed by a zone along the bed:

$$N = \left(\frac{d_R}{\sigma_d}\right)^2 \tag{3}$$

where d_R is the distance from the point of sample application to the point of the band center and σ_d is the variance of the band in distance units.[1]

In fact, the plate theory describes the movement of a particular zone through the chromatographic bed. As the zone is washed through the first several plates, a highly discontinuous concentration profile is obtained, with the solute being distributed in plates following the Poisson distribution.[3] At an intermediate stage (approximately 30–50 plates), much of the abrupt discontinuity disappears due to a similar concentration of the analyte in the neighboring plates. As the process continues (after 100 plates), the concentration profile is smooth and, even though the distribution is still Poisson, it can be approximated by a Gaussian curve. The standard deviation, , of the Gaussian curve, which is a direct measure of the zone spreading, is found to be

$$\sigma = \sqrt{HL} \tag{4}$$

where H is the plate height and L is the distance migrated by the center of the zone. In practice, the plate height is used to describe the zone spreading, including both nonequilibrium and longitudinal effects. In a uniform column, free from concentration and velocity gradients, the plate height is defined as

$$H = \frac{\sigma^2}{L} \tag{5}$$

In a nonuniform column, the zone spreading varies from point to point and its local value is

$$H = \frac{d\sigma^2}{dL} \tag{6}$$

which represents the increment of plate height in the variance σ^2 per unit length of migration. In practice, the smaller the value of H, the smaller the magnitude of band spreading per unit length of the column. The determination of H does not require the measurement

Encyclopedia of Chromatography DOI: 10.1081/E-ECHR-120039964

of , as long as N is known. Thus, combining Eqs. (1) and (6) yields[4]

$$H = L\left(\frac{\sigma}{L}\right)^2 = \frac{L}{N} \tag{7}$$

In practice, because the separation in a particular chromatographic column is linked to the time spent by the analyte in the stationary phase and the time spent by the analyte in the mobile phase is irrelevant for the separation, a new parameter is defined (i.e., *effective plate number*, N_{eff}). The effective plate number is related to the separation factor k' and N by

$$N_{eff} = N\left(\frac{k'}{1 + k'}\right)^2 \tag{8}$$

Similarly, an expression for H_{eff} can be written

$$H_{eff} = H\left(\frac{1 + k'}{k'}\right)^2 \tag{9}$$

The effective parameters are more meaningful when comparing different columns.[4]

There are several major contributions that will influence the band broadening and, consequently, H:[5] eddy diffusion, mobile-phase mass transfer, longitudinal diffusion, stagnant mobile-phase mass transfer, and stationary-phase mass transfer. The effect of each process on the band broadening and, consequently, on the plate height is related to all the experimental variables: mobile-phase velocity, u; particle diameter, d_p; sample diffusion coefficient in the mobile phase, D_m; the thickness of the stationary-phase layer, d_f; and the sample diffusion coefficient in the stationary phase, D_s. In general, H will vary with the velocity of the mobile phase, u, as it travels through the column. In a GC system, a plot of u versus H will lead to a curve which has a hyperbolic shape,[6] characterized by the equation

$$H = A + \frac{B}{u} + Cu \tag{10}$$

Eq. (10) is known as the van Deemter equation, and no correction was made for gas compressibility. Using the reduced parameters $h = Hd_p$ and $v = ud_pD_m$, Eq. (10) becomes

$$h = a + \frac{b}{v} + cv \tag{11}$$

where A, B, C, a, b, and c are constants for a particular sample compound and set of experimental conditions as the flow rate varies. The B term in Eq. (10) relates to band broadening occurring by diffusion in the gas phase in the longitudinal direction of the column. According to Einstein's equation for diffusion,

$$\sigma^2 = 2D_mt_0 = \frac{2D_mL}{u} \tag{12}$$

Because $H = \sigma^2L$, the B term becomes

$$B = 2\frac{D_m}{u} \tag{13}$$

The inverse velocity term in Eq. (13) becomes important at low velocities. Because the D_m in liquids is 10^5 times smaller than in gases, the longitudinal term plays no practical role in band broadening in LC. The A term in Eq. (10) describes the nonhomogeneous flow, also called eddy diffusion. In this case,

$$\frac{\sigma^2}{L} = 2\lambda d_p = A \tag{14}$$

where λ is a packing correction factor of $\sim$0.5. In classical GC, the A term is a constant, representing a lower limit on column efficiency, equivalent to $H = d_p$ or $h = 1$.

At velocities above H_{min}, the C term controls H and relates to nonequilibrium resulting from resistance to mass transfer in the stationary and mobile phases.[6]

In HPLC, the van Deemter equation still holds. However, Giddings [7] argued that the equation is too simplistic because it ignores the coupling that exists between the flow velocity and the radial diffusion in the void space of the packing around the particles. He suggested replacing the term A by a term $a(1 + bu^{-1})$ to account for the flow velocity, because both the eddy diffusion and the radial diffusion are responsible for the transfer of the molecules between the different flow paths of unequal velocity. To include the coupling between the laminar flow and the molecular diffusion in porous media, Horvath and Lin[8] introduced a new parameter, δ, which is the thickness of the stagnant film surrounding each stationary-phase particle. However, at high velocities required in HPLC, Horvath and Lin's model reduces to the Knox equation, which is a variation of the van Deemter equation:[9]

$$h = av^{0.33} + \frac{b}{v} + cv \tag{15}$$

where a, b, and c are empirical parameters related to the analyte and the experimental flow rate conditions.

REFERENCES

1. Karger, B.L.; Snyder, L.R.; Horvath, Cs. *An Introduction to Separation Science*; John Wiley & Sons: New York, 1973; 136.
2. Snyder, L.R.; Kirkland, J.J.; Glajch, J.L. *Practical HPLC Method Development*; John Wiley & Sons: New York, 1997; 42.
3. Giddings, J.C. *Dynamic of Chromatography, Part I, Principles and Theory*; Marcel Dekker, Inc.: New York, 1965; 23.
4. Horvath, Cs.; Melander, W.R. *Chromatography, Fundamentals and Applications of Chromatographic and Electrophoretic Methods, Part A: Fundamentals and Techniques*; Heftmann, E., Ed.; Elsevier Scientific: Amsterdam, 1983; A45.
5. Snyder, L.R.; Kirkland, J.J. *Introduction to Modern Liquid Chromatography*, 2nd Ed.; John Wiley & Sons: New York; 168.
6. Karger, B.L. *Modern Practice of Liquid Chromatography*; Kirkland, J.J., Ed.; Wiley–Interscience: New York, 1971; 23.
7. Giddings, J.C. *Dynamics of Chromatography, Part I, Principles and Theory*; Marcel Dekker, Inc.: New York, 1965; 61.
8. Horvath, Cs.; Lin, H.J. Band spreading in liquid chromatography: General plate height equation and a method for the evaluation of the individual plate height contributions. J. Chromatogr. **1978**, *149*, 43.
9. Guiochon, G.; Shirazi, S.G.; Katti, A.M. *Fundamentals of Preparative and Nonlinear Chromatography*; Academic Press: Boston, 1994; 201.

Efficiency of a TLC Plate

Wojciech Markowski
Department of Inorganic and Analytical Chemistry, Medical University,
Lublin, Poland

INTRODUCTION

Chromatography, by definition, is a separation metho-dology for a multicomponent sample mixture, which is based on differentiating movement zones of the sample. An essential feature of chromatographic separation is that the components of the sample are transported through the separation medium—in the case of TLC, through an open bed. Differences in interaction with the medium lead to a selective redistri-bution of the component zones, from overlapping zones at the start following injection, toward largely individual regions inside the separation medium. The appearance of individual component zones, after the development process, can be recorded with the aid of scanning densitometry, to convert the plate chromato-gram into realistic two-dimensional representation of the chromatographic process in a form suitable for evaluation of kinetic parameters. The underlying fun-damental processes responsible for chromatographic separations can be explained by thermodynamic and kinetic considerations. Thermodynamic relationships are responsible for retention and selectivity, and kinetic properties are responsible for band broadening. Thus, the position and separation of peaks in a chromatogram are thermodynamic properties, whereas the axial dimensions of the peaks are governed by kinetic considerations, and both phenomena must be considered to optimize resolution. As Giddings[1] emphases in his book, "separation is the art and science of maxi-mizing separative transport relative to the dispersive transport."

RESOLUTION

The most useful criterion for the estimation of the quality of a separation is the resolution. The resolution is given by:[2]

$$R_s = \frac{(z_f - z_o)(R_{f(2)} - R_{f(1)})}{0.5(w_1 + w_2)} \qquad (1)$$

where $R_{f(1)}$ and $R_{f(2)}$ are the R_f values of chromato-graphic spots 1 and 2, respectively; and w_1 and w_2 are widths of spots at the base. Eq. (1) clearly shows

the two competing aspects of a chromatographic separation: the separation distance achieved by the primary separation process (numerator) is opposed by the "blurring" action of the zone broadening (denomi-nator). Eq. (1) allows for direct calculation of R_s based on the parameters measured on the chromatogram. Eq. (1) can be transformed to the form:[3]

$$R_s = (1 - R_{f(1)})\left[1 - \frac{R_{f(1)}}{R_{f(2)}}\right]N^{0.5} \qquad (2)$$

Eq. (2) demonstrates that the plate resolution, as in other forms of chromatography, depends on the num-ber of theoretical plates N, the selectivity, and the retention coefficient of the solute for the particular layer concerned.

CONCEPT OF THEORETICAL PLATES AND THEIR MEASUREMENT

The measurement of plate efficiency is depicted in Fig. 1(a) and (b).[4] The number of "theoretical plates" is a measure of the quality or "efficiency" of a chroma-tographic layer. By analogy with the theoretical plates of a distillation column, the chromatographic separa-tions distance and the layer are divided into theoretical separation plates. For a given problem, sorbent, and solvent, a specific minimum number of theoretical plates is necessary to achieve the desired separation. For a capillary flow-controlled system, the mobile-phase velocity is not constant throughout the chroma-togram and, at any position within the chromatogram, its value depends on the system variables. The mobile-phase velocity is not under external control and its range cannot be varied independently to study the relationship between the layer plate height and the mobile-phase velocity. Because all zones do not migrate the same distance in TLC, individual zones experience only those theoretical plates through which they travel and the plate height is directly dependent on the migration distance. A further complicating factor is that the size of the starting zone applied to the layer is always a finite value with respect to the size of a developed zone. Therefore, it is not adequate to use the measured zone width as the starting point from which to determine the extent of zone broadening for

Encyclopedia of Chromatography DOI: 10.1081/E-ECHR-120022515

(a)

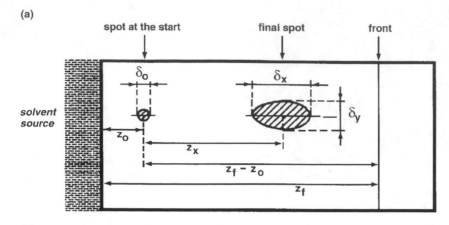

(b)

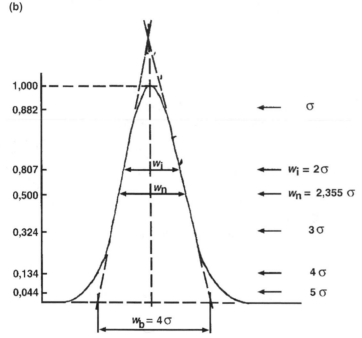

Fig. 1 (a) Basic symbols for TLC spot migration. z_x is the migration length of the spot, $z_f - z_o$ is the separation length, and z_f is the front migration length. (b) Width of a Gaussian band at different percentages of maximum height.

the layer. The plate number N can be experimentally determined via the "H value" (i.e., "height equivalent to a theoretical plate," or HETP).[5] The measured H value is an average over the separation length and the symbol H_{obs} or $\bar{H}$ is given and is obtained from the integration of the expression for the local plate height H_{loc} (quantity introduced by Giddings[1,4]):

$$N = \frac{z_f - z_o}{H} \tag{3}$$

$$H_{loc} = \frac{d\sigma_x^2}{dz} \tag{4}$$

and, in practical terms,

$$\bar{H} = H_{obs} = \frac{\int_{z_o}^{z_x} H_{loc} dz}{\int_{z_o}^{z_x} dz}$$

$$= \frac{\sigma_x^2}{z_x} = \frac{b_{0.5}^2}{5.54 z_x} \tag{5}$$

The average plate number N of the whole separation length is:

$$N \approx \bar{N} = \frac{z_f - z_o}{H_{obs}} = \frac{z_x}{H_{obs} R_f} \tag{6}$$

Hence, the average theoretical plate number can be calculated from the experimental data for $b_{0.5}$ and the lengths z_f and z_x in the chromatogram. To obtain reliable H values, it is recommended that the experiment comply with the following rules:[4]

1. The band maximum should be at least 10 times greater than the detection limit.
2. Asymmetrical or poorly resolved spots should not be used for determining H.
3. Only σ_{chrom}^2 should be used.

Failure to do this results in excessively high H values. The measured zone variance σ_x^2 in the direction of flow is composed of:

$$\sigma_x{}^2 = \sigma_{spotting}{}^2 + \sigma_{chrom}{}^2 + \sigma_{inst}{}^2 + \sigma_{other}{}^2 \qquad (7)$$

The contribution of the length of the starting zone to the length of the separated zone could be removed by considering their variances, such that:

$$\sigma_{chrom}{}^2 = \sigma_x{}^2 - \sigma_{spotting}{}^2 \qquad (8)$$

where $\sigma_x{}^2$ is the variance of the developed zone, $\sigma_{chrom}{}^2$ is the variance due to the zone expansion during the migration through the layer, and $\sigma_{spotting}{}^2$ is the variance associated with sample application; here, we are assuming that the contributions to zone broadening associated with the properties of the detection and recording devices $\sigma_{inst}{}^2$ are negligible. The form of the starting zone is immaterial—only its dimension and sample distribution along its axis parallel to the direction of development (first-order approximation) are significant. It is obvious that the characteristic dimension of the starting zone in the direction of migration is never infinitely small compared with the same characteristic dimension of the zone after normal development. The determination of peak variance is straightforward for developed zones, but presents some difficulty for the undeveloped starting zone. The starting zone is applied to the dry layer. At the start of the migration process, it is contacted by the advancing mobile phase, which is moving at its highest velocity and is probably not fully saturated at its leading edge. Several processes take place quickly, which can lead to changes in the dimensions of the starting zone at the moment the chromatogram begins. The solvent front contacts the bottom portion of the starting zone first, pushing it forward with a characteristic migration velocity (which depends on the solute R_f value) into the upper portion of the starting zone, which is fixed in position until it is contacted by the advancing mobile phase. This causes a reconcentration of the starting zone and a reduction of its characteristic dimension in the direction of development. In addition, because the flow of mobile phase is unsaturated, all the pores holding sample will not be filled simultaneously; adsorbed samples may not be displaced from the sorbent surface instantaneously, and localized solvent saturation may limit the rate of solute dissolution in the mobile phase. The dimensions of the starting zone in the direction of migration are too large to ignore.

It can be assumed that the variance of the starting zone is equivalent to the properties of the sample zone, after it has been transported a few millimeters from its point of application by the mobile phase. In this way, some account is taken of the capacity of the mobile phase to reshape the deposited sample zone at the beginning of the chromatogram.[6]

CAPILLARY FLOW

As normally practiced in TLC, capillary forces control the migration of the mobile phase through the layer. Under these conditions, the velocity at which the solvent front moves is a function of the distance of the front from the solvent entry position and declines as this distance increases.[4] There are two consequences of this effect:

1. The mobile-phase velocity is not constant throughout the chromatogram.
2. The mobile-phase velocity is set by the system variables and cannot be independently optimized unless forced flow development conditions are used.

If the migration distance is not excessively long, then the solvent front position as a function of time is adequately described by:

$$z_f{}^2 = \chi t \qquad (9)$$

where z_f is the distance of the solvent front position above the solvent entry position, χ is the mobile-phase velocity constant, and t is the elapsed time since the solvent commenced migration through the layer. At any position on the layer, the solvent front will be moving with a velocity given by:

$$u_f = \frac{\chi}{2z_f} \qquad (10)$$

There are two features of importance when using Eqs. (9) and (10). The velocity constant X depends on the identity of the solvent; layer characteristics such as average particle size, layer permeability, layer thickness, etc.; and the state of equilibrium between solvent vapors in contact with the layer and the bulk solvent moving through the layer. As the solvent permeates the layer, the channels of narrower diameter are filled first, leading to more rapid advancement of the mobile phase. Large pores below the solvent front fill more slowly, resulting in an increase in the thickness of the layer of mobile phase. The bulk mobile-phase velocity, representing saturated flow through the region occupied by the sample zones, is moving at a lower velocity than the solvent u_f front velocity. As a reasonable approximation, the bulk solvent velocity is usually taken to be $0.8\,u_f$.[6] The velocity constant χ is related to the experimental condition by Eq. (11):

$$\chi = 2k_0 d_p \frac{\gamma}{\eta} \cos \theta \qquad (11)$$

where κ_o is the layer permeability constant, d_p is the average particle diameter, γ is the surface tension, η is the viscosity of the mobile phase, and θ is the contact angle. The layer permeability constant is dimensionless and takes into account the effect of porosity on the permeability of the layer and the difference between the bulk liquid velocity and the solvent front velocity. A typical value of permeability is $1-2 \times 10^{-3}$,[6] virtually identical with typical column values. Assuming a narrow particle size distribution, Eq. (11) indicates that the velocity constant should increase linearly with average particle size. The solvent front velocity should be larger for coarse particle layers than for fine particle layers, in good agreement with experimental observations. In addition, from Eq. (11), we see that the velocity constant depends linearly on the ratio of the surface tension of the solvent to its viscosity, and the solvents that maximize this ratio are most useful for TLC. The contact angle for most mobile phases on polar adsorbent layers is generally close to zero and there does not exist the problem of wetting. In the case of reversed-phase layers containing bonded, long-chain, alkyl groups, it is not possible to apply the mobile phase with a content of water below 40%. The optimum mobile-phase velocity for a separation can be established by forced flow development[6] and it is considerably higher than the mobile-phase velocity obtained by the use of capillary flow under different experimental conditions. This is illustrated in Fig. 2(a) and (b).

BAND-BROADENING INTERPRETATION

The kinetic contributions to zone broadening are evaluated by fitting data for the column plate height, as a function of the mobile-phase velocity, to a mathematical model describing the relationship between the two parameters. Several models have been used in the above experiment, but those by de Ligny and Remijnsee[7] and Knox and Pryde,[11] and developed by Guiochon and Siouffi,[9] are most widely used and, at least for a first approximation, allow for comparison and determination of the differences between TLC and column chromatography:[7–11]

$$\bar{H} = \frac{3A_k d^{5/3}\theta^{1/3}}{2(2D_m)^{1/3}z_f^{1/3}} \frac{(z_f^{2/3} - z_0^{2/3})}{z_f - z_o}$$
$$+ \frac{B_k D_m}{\theta d_p}(z_f - z_o) + \frac{C_k \theta d_p^3}{2D_m(z_f - z_o)} \lg \frac{z_f}{z_o}$$

$$(12)$$

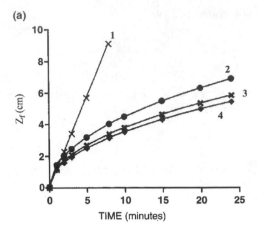

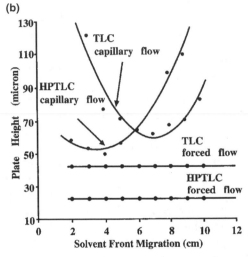

Fig. 2 (a) Plot of solvent front migration distance z_f for dichloromethane on a high-performance silica gel layer as a function of time under different experimental conditions. Identification: 1 = forced flow development at u_{opt}; 2 = capillary flow in a saturated developing chamber; 3 = capillary flow in a sandwich chamber; and 4 = capillary flow in an unsaturated developing chamber. (b) Variation of the observed plate height as a function of the solvent front migration distance for conventional TLC and high-performance thin-layer chromatography (HPTLC) silica layers under capillary flow and forced flow (u_{opt}) conditions.

where d_p is the average particle size, χ is the mobile-phase velocity constant, and D_m is the solute diffusion coefficient in the mobile phase. A_k, B_k, and C_k are dimensionless coefficients characterizing the packing quality (A_k), the diffusion in the mobile phase (B_k), and the resistance to mass transfer (C_k). Eq. (12) can be helpful in the interpretation of the influence of layer structure on plate height. Results of simulations of the relationship between plate height and different parameters are presented in Fig. 3 and parameters used in simulation are presented in Table 1.

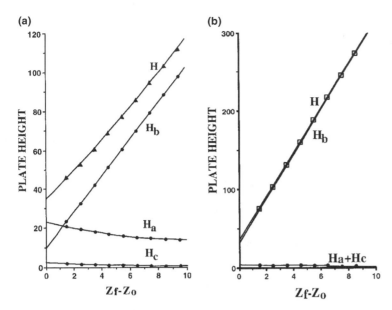

Fig. 3 (a) Simulation of the average plate height for a TLC layer by use (Eq. 12) and properties listed in Table 1. The contribution from flow anisotropy is represented by H_a, that from longitudinal diffusion by H_b, and that from resistance to mass transfer by H_c. (b) Simulation of the average plate height for an HPTLC layer by use (Eq. 12) and properties listed in Table 1. The contribution from flow anisotropy is represented by H_a, that from longitudinal diffusion by H_b, and that from resistance to mass transfer by H_c.

UNIDIMENSIONAL MULTIPLE DEVELOPMENT

Unidimensional multiple development provides a complementary approach to forced flow for minimizing zone broadening.[13] All unidimensional multiple development techniques employ successive repeated development of the layer in the same direction, with removal of the mobile phase between developments. Approaches differ in the changes made (e.g., mobile-phase composition and solvent front migration distance) between consecutive development steps; the total number of successive

development steps employed can also be varied. Capillary forces are responsible for migration of the mobile phase, but a zone-focusing mechanism is used to counteract the normal zone broadening that occurs in each successive development. Each time the solvent front traverses the stationary sample zone, the zone is compressed in the direction of development. The compression occurs because the mobile phase contacts the bottom edge of the zone first; here, the sample molecules start to move forward before those molecules are still ahead of the solvent front. When the solvent front has moved beyond

Table 1 Characteristic properties of precoated layers and HPLC columns

Property	HPTLC	TLC	HPLC
Porosity total	0.65–0.70	0.65–0.75	0.8–0.9
Interparticle	0.35–0.45	0.35–0.45	0.4–0.5
Intraparticle	0.28	0.28	0.4–0.5
Flow resistance parameter	875–1500	600–1200	500–1000
Apparent particle size (μm)	5–7	8–10	d_p
Minimum plate height (μm)	22–25	35–45	$2–3d_p$
Optimum velocity (cm sec^{-1})	0.03–0.05	0.02–0.05	0.2
Minimum reduced plate height	3.5–4.5	3.5–4.5	1.5–3
Optimum reduced velocity	0.7–1.0	0.6–1.2	3–5
Separation impedance 9,000–70,000	10,500–19,800	11,100–60,200	2,000–9,000
Mean pore diameter (Si 60) (nm)	5.9–7.0	6.1–7.0	
Knox coefficients			
A_k	0.75	2.83	0.5–1
B_k	1.56	1.18	1–4
C_k	1.42	0.84	0.05

(From Refs.[6,12].)

Table 2 Zone capacity calculated or predicted for different separation conditions in TLC[6]

Method	Dimensions	Zone capacity
(A) *Predictions from theory*		
Capillary flow	1	< 25
Forced flow	1	< 80 (up to 150, depending on pressure limit)
Capillary flow	2	< 400
Forced flow	2	Several thousands
(B) *Based on experimental observations*		
Capillary flow	1	12–14
Forced flow	1	30–40
Capillary flow (AMD)	1	30–40
Capillary flow	2	ca. 100
	2	
(C) *Predictions based on results in (B)*		
Forced flow	2	ca. 1500
Capillary flow (AMD)	2	ca. 1500

the zone, the focused zone migrates and is subject to the normal zone-broadening mechanisms. Experiment indicates that, beyond a minimum number of development steps, zone widths converge to a constant value that is roughly independent of migration distance.

SOLVENT GRADIENTS

A similar phenomenon, compression of chromatographic zones, occurs in gradient TLC when the concentration of mobile phase delivered to the layer is varied in a stepwise manner. In the case where the concentration front traverses the sample zone, the zone is compressed in the direction of development. The compression takes place on the length equal to the diameter of the spot. Application of multicomponent eluent for the development of the layer, when the components differ in polarity, causes the creation of a natural gradient in the mobile phase. The gradient appears as multiconcentration fronts.[14] As in step gradients, the compression takes place and the widths of the spots are much smaller. This should improve resolution.

MOVING PLATE

Changing the solvent entry position for each, or some, of the development steps enables the separation in each segment to be achieved in the shortest possible time under favorable capillary flow conditions. With as few as 10 developments, it is relatively easy to achieve 15,000–25,000 apparent theoretical plates for a zone migration distance of 6–11 cm.[6,15,16]

ZONE CAPACITY

The potential of a chromatographic system to achieve a particular separation can be estimated from its zone capacity, also referred to as the separation number (SN). It provides a method of comparison of different TLC systems and an indication of the possibility of separating a given mixture:[8]

$$SN = \frac{z_f}{b_0 + b_1} - 1 \qquad (13)$$

where b_0 is the extrapolated width of the starting spot at half-height of the concentration curve, and b_1 is the extrapolated width of the spot with $R_f = 1$. Some typical results for the zone capacity, either predicted from theory or by experiment, are summarized in Table 2.[6] Experimental observations are indicated for zone capacity of approximately 12–14 for a single development in a capillary flow, increasing to 30–40 if forced flow is used. Use of capillary flow and the zone-focusing mechanism of multiple development leads to a zone capacity similar to that for forced flow.

CONCLUSIONS

In capillary flow conditions, there is an inadequate range of mobile-phase velocities, which does not allow working at u_{opt} values; the role of the binder remains not completely clear. The zone-focusing mechanism causes an increase of separation performance of the system in the most simple way. Forced flow offers a modest increase in performance with a reduction in separation time.

REFERENCES

1. Giddings, J.C. *Unified Separation Science*; John Wiley & Sons, Inc.: New York, 1991; 10.
2. Kowalska, T. Theory and mechanism of thinlayer chromatography. In *Handbook of Thin Layer Chromatography*, 2nd Ed.; Sherma, J., Fried, B., Eds.; Marcel Dekker, Inc.: New York, 1996; Vol. 71, 49–80.
3. Cazes, J.; Scott, R.P.W. Thin layer chromatography. In *Chromatography Theory*; Marcel Dekker, Inc.: New York, 2002; 443–454.
4. Geiss, F. *Fundamentals of Thin Layer Chromatography*; Hüthig: Heidelberg, 1987; 9–82.
5. Van Deemter, J.J.; Zuiderweg, F.; Klinkenberg, A. Longitudinal diffusion and resistance to mass transfer as causes of nonideality in chromatography. Chem. Eng. Sci. **1956**, *5*, 271.
6. Poole, C.F. Kinetic theory of planar chromatography. In *Planar Chromatography. A Retrospective View for the Third Millenium*; Nyiredy, Sz., Ed.; Springer: Budapest, 2001; 13–32.
7. de Ligny, C.L.; Remijnsee, A.G. Peak broadening in paper chromatography and related techniques: III. Peak broadening in thin-layer chromatography on cellulose powder. J. Chromatogr. **1968**, *33*, 242–254; 257–268.
8. Zlatkis, A.; Kaiser, R.E. *HPTLC High Performance Thin Layer Chromatography*; Elsevier: Amsterdam, 1977; 15–38.
9. Guiochon, G.; Siouffi, A. Band broadening and plate height equation. III. Flow velocity. J. Chromatogr. Sci. **1978**, *16*, 470–481; 598–609.
10. Belenkii, B.B.; Nesterov, V.V.; Smirnov, V.V. Differential equation for thin layer chromatography and its solution. Russ. J. Phys. Chem. **1968**, *42*, 773–775; 1527–1530.
11. Belenkii, B.B.; Nesterov, V.V.; Smirnov, V.V. Comparison of theory with experimental results. Russ. J. Phys. Chem. **1968**, *42*, 773–775; 1527–1530.
12. Knox, J.H.; Pryde, A. Performance and selected applications of a new range of chemically bonded packing materials in high-performance liquid chromatography. J. Chromatogr. **1975**, *112*, 171–188.
13. Poole, C.K.; Poole, S.K. Instrumental thin-layer chromatography. Anal. Chem. **1994**, *66*, 27A–37A.
14. Poole, C.K.; Poole, S.K. *Chromatography Today*; Elsevier: Amsterdam, 1991.
15. Niederwieser, A.; Honegger, C.C. Gradient techniques in thin-layer chromatography. In *Advances in Chromatography*; Giddings, J.J., Keller, R.A., Eds.; Marcel Dekker, Inc.: New York, 1966; Vol. 2, 123.
16. Fernando, W.P.N.; Poole, C.F. Determination of kinetic parameters for precoated silica gel thin-layer chromatography plates by forced flow development. J. Planar Chromatogr. **1991**, *4*, 278–287.
17. Fernando, W.P.N.; Poole, C.F. Comparison of the kinetic properties of commercially available precoated silica gel plates. J. Planar Chromatogr. **1993**, *6*, 357–361.

SUGGESTED ADDITIONAL READING

Grinberg, N.; LoBrutto, R. Efficiency in chromatography. In *Encyclopedia of Chromatography*; Cazes, J., Ed.; Marcel Dekker, Inc.: New York, 2001; 274.

Grinberg, N., Ed.; *Modern Thin Layer Chromatography*; Marcel Dekker, Inc.: New York, 1990.

Tijssen, R. The mechanisms and importance of zone-spreading. In *Handbook of HPLC*; Katz, E., Eksteen, R., Schoenmakers, P., Miller, N., Eds.; Marcel Dekker, Inc.: New York, 1998; 55–142.

Electro-Osmotic Flow

Danilo Corradini
Institute of Chromatography, Rome, Italy

INTRODUCTION

Electro-osmosis refers to the movement of the liquid adjacent to a charged surface, in contact with a polar liquid, under the influence of an electric field applied parallel to the solid–liquid interface. The bulk fluid of liquid originated by this electrokinetic process is termed electro-osmotic flow (EOF). It may be produced both in open and in packed capillary tubes, as well as in planar electrophoretic systems employing a variety of supports, such as paper or hydrophilic polymers.

DISCUSSION

The formation of an electric double layer at the interfacial region between the charged surface and the surrounding liquid is of key importance in the generation of the EOF.[1–3] Most solid surfaces acquire a superficial charge when are brought into contact with a polar liquid. The acquired charge may result from dissociation of ionizable groups on the surface, adsorption of ions from solution, or by virtue of unequal dissolution of oppositely charged ions of which the surface is composed. This superficial charge causes a variation in the distribution of ions near the solid–liquid interface. Ions of opposite charge (counterions) are attracted toward the surface, whereas ions of the same charge (co-ions) are repulsed away from the surface. This, in combination with the mixing tendency of thermal motion, leads to the generation of an electric double layer formed by the charged surface and a neutralizing excess of counterion over co-ions distributed in a diffuse manner in the polar liquid. Part of the counterions are firmly held in the region of the double layer closer to the surface (the compact or Stern layer) and are believed to be less hydrated than those in the diffuse region of the double layer where ions are distributed according to the influence of electrical forces and random thermal motion. A plane (the Stern plane) located at about one ion radius from the surface separates these two regions of the electric double layer.

Certain counterions may be held in the compact region of the double layer by forces additional to those of purely electrostatic origin, resulting in their adsorption in the Stern layer. Specifically, adsorbed ions are attracted to the surface by electrostatic and/or van der Waals forces strongly enough to overcome the thermal agitation. Usually, the specific adsorption of counterions predominates over co-ion adsorption.

The variation of the electric potential in the electric double layer with the distance from the charged surface is depicted in Fig. 1. The potential at the surface (ψ_0) linearly decreases in the Stern layer with respect to the value of the zeta potential (ζ). This is the electric potential at the plane of shear between the Stern layer (plus that part of the double layer occupied by the molecules of solvent associated with the adsorbed ions) and the diffuse part of the double layer. The zeta potential decays exponentially from ζ to zero with the distance from the plane of shear between the Stern layer and the diffuse part of the double layer. The location of the plane of shear, a small distance further out from the surface than the Stern plane, renders the zeta potential marginally smaller in magnitude than the potential at the Stern plane (ψ_δ). However, in order to simplify the mathematical models describing the electric double layer, it is customary to assume identity of ψ_δ and ζ, and the bulk experimental evidence indicates that errors introduced through this approximation are usually small.

According to the Gouy–Chapman–Stern–Grahame (GCSG) model of the electric double layer,[4] the surface density of the charge in the Stern layer is related to the adsorption of the counterions, which is described by a Langmuir-type adsorption model, modified by the incorporation of a Boltzman factor. Considering only the adsorption of counterions, the surface change density σ_S of the Stern layer is related to the ion concentration C in the bulk solution by the following equation:

$$\sigma_S = zen_0 \frac{C}{V_m} \exp\left(\frac{ze\xi + \Phi}{kT}\right)$$
$$\times \left[1 + \frac{C}{V_m} \exp\left(\frac{ze\xi + \Phi}{kT}\right)\right]^{-1} \quad (1)$$

where e is the elementary charge, z is the valence of the ion, k is the Boltzman constant, T is the temperature, n_0 is the number of accessible sites, V_m is the molar volume of the solvent, and Φ is the specific adsorption potential of counterions.

Encyclopedia of Chromatography DOI: 10.1081/E-ECHR-120039969

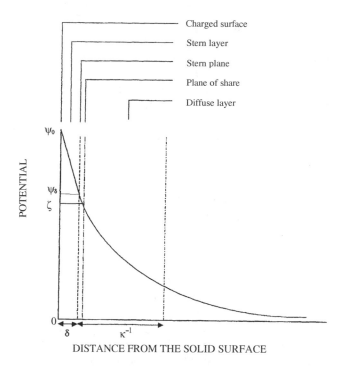

Fig. 1 Schematic representation of the electric double layer at a solid–liquid interface and variation of potential with the distance from the solid surface: ψ_0, surface potential; ψ_δ, potential at the Stern plane; ζ, potential at the plane of share (zeta potential); δ, distance of the Stern plane from the surface (thickness of the Stern layer); κ^{-1}, thickness of the diffuse region of the double layer.

The surface charge density of the diffuse part of the double layer is given by the Gouy–Chapman equation

$$\sigma_G = (8\varepsilon k T c_0) \sinh\left(\frac{ze\zeta}{2kT}f\right) \tag{2}$$

where ε is the permittivity of the electrolyte solution and c_0 is the bulk concentration of each ionic species in the electrolyte solution.

At low potentials, Eq. (2) reduces to

$$\sigma_G = \frac{\varepsilon\zeta}{\kappa^{-1}} \tag{3}$$

where κ^{-1} is the reciprocal Debye–Hückel parameter, which is defined as the "thickness" of the electric double layer. This quantity has the dimension of length and is given by

$$\kappa^{-1} = \left(\frac{\varepsilon kT}{2e^2 I}\right)^{1/2} \tag{4}$$

in which I is the ionic strength of the electrolyte solution.

Eq. (3) is identical to the equation that relates the charge density, voltage difference, and distance of separation of a parallel-plate capacitor. This result indicates that a diffuse double layer at low potentials behaves like a parallel capacitor, in which the separation distance between the plates is given by κ^{-1}. This explains why κ^{-1} is called the double-layer thickness.

Eq. (2) can be written in the form

$$\zeta = \frac{\sigma_G \kappa^{-1}}{\varepsilon} \tag{5}$$

which indicates that the zeta potential can change due to variations in the density of the electric charge, in the permittivity of the electrolyte solution, and in the thickness of the electric double layer, which depends, throughout the ionic strength [see Eq. (4)], on the concentration and valence of the ions in solution.

The dependence of the velocity of the electro-osmotic flow (v_{eo}) on the zeta potential is expressed by the Helmholtz–von Smoluchowski equation

$$v_{eo} = \frac{\varepsilon_0 \varepsilon \zeta}{\eta} E \tag{6}$$

where E is the applied electric field, ε_0 is the permittivity of vacuum, and ε and η are the dielectric constant and the viscosity of the electrolyte solution, respectively. This expression assumes that the dielectric constant and viscosity of the electrolyte solution are the same in the electric double layer as in the bulk solution.

The Helmholtz–von Smoluchowski equation indicates that under constant composition of the electrolyte solution, the EOF depends on the magnitude of the zeta potential, which is determined by the different factors influencing the formation of the electric double layer, as discussed earlier. Each of these factors depends on several variables, such as pH, specific adsorption of ionic species in the compact region of the double layer, ionic strength, and temperature.

The specific adsorption of counterions at the interface between the surface and the electrolyte solution results in a drastic variation of the charge density in the Stern layer, which reduces the zeta potential and, hence, the EOF. If the charge density of the adsorbed counterions exceeds the charge density on the surface, the zeta potential changes sign and the direction of the EOF is reversed.

The ratio of the velocity of the EOF to the applied electric field, which expresses the velocity per unit field, is defined as electro-osmotic coefficient or, more properly, electro-osmotic mobility (μ_{eo}).

$$\frac{v_{eo}}{E} = \mu_{eo} = \frac{\varepsilon_0 \varepsilon \zeta}{\eta} \tag{7}$$

Using SI units, the velocity of the electro-osmotic flow is expressed in meters per second (m/s) and the electric field in volts per meter (V/m). Consequently,

In analogy to the electrophoretic mobility, the electro-osmotic mobility has the dimension square meters per volt per second. Because electro-osmotic and electrophoretic mobilities are converse manifestations of the same underlying phenomenon, the Helmholtz–von Smoluchowski equation applies to electro-osmosis as well as to electrophoresis. In fact, when an electric field is applied to an ion, this moves relative to the electrolyte solution, whereas in the case of electro-osmosis, it is the mobile diffuse layer that moves under an applied electric field, carrying the electrolyte solution with it.

According to Eq. (6), the velocity of the EOF is directly proportional to the intensity of the applied electric field. However, in practice, the nonlinear dependence of the EOF on the applied electric field is obtained as a result of Joule heat production, which causes an increase of the electrolyte temperature with a consequent decrease of viscosity and variation of all other temperature-dependent parameters (protonic equilibrium, ion distribution in the double layer, etc.). The EOF can also be altered during a run by variations of the protonic and hydroxylic concentration in the anodic and cathodic electrolyte solutions as a result of electrolysis. This effect can be minimized by using electrolyte solutions with a high buffering capacity and electrolyte reservoirs of relatively large volume and by frequent replacement of the electrolyte in the electrode compartments with fresh solution.

The magnitude and direction of the EOF depend also on the composition, pH, and ionic strength of the electrolyte solution.[5–7] Both the pH and ionic strength influence the protonic equilibrium of fixed-charged groups on the surface and of ionogenic substances in the electrolyte solution which affect the charge density in the electric double layer and, consequently, the zeta potential. In addition, the ionic strength influences the thickness of the double layer (κ^{-1}). According to Eq. (4), increasing the ionic strength causes a decrease in κ^{-1}, which is currently referred to as the compression of the double layer that results in lowering the zeta potential. Consequently, increasing the ionic strength results in decreasing the EOF.

The charge density in the electric double layer and, hence, the EOF are also influenced by the adsorption of potential-determining ions in the Stern region of the electric double layer. A variety of additives can be incorporated into the electrolyte solution with the purpose of controlling the EOF by modifying the solid surface dynamically. These include simple and complex ionic compounds, ionic and zwitterionic surfactants, and neutral and charged polymers. The incorporation of these additives into the electrolyte solution may result either in increasing or in reducing the EOF, or even in reversing its direction. The impact of these additives on the EOF is generally concentration dependent. Such behavior is in accordance to the Langmuir-like adsorption model describing the variation of the charge density in the Stern layer on the concentration of adsorbing ions in the electrolyte solution [see Eq. (1)].

The proper control of the EOF can be also obtained by adding organic solvents to the electrolyte solution. The influence of organic solvents on the EOF may result from a multiplicity of mechanisms. Organic solvents are expected to influence both the dielectric constant and viscosity of the bulk electrolyte solution. Generally, this leads to the variation of the ratio of the dielectric constant to the viscosity of the electrolyte solution, to which the EOF depends according to Eq. (6). In addition, the local viscosity within the electric double layer[8] can be varied by the adsorption of the organic–solvent molecules in the Stern layer, which may also influence the adsorption of counterions, depending on the different solvation properties of the organic solvent. Organic solvents may also influence the zeta potential by affecting the ionization of potential-determining ions on the surface.

Different methods can be employed to measure the magnitude of the EOF.[9] One possibility involves measuring the velocity of the EOF by measuring the change in weight or in volume in one of the electrolyte solution reservoirs. The addition of an electrically neutral dye to one electrode reservoir and its detection in the other where the EOF is directed is another possible method. Other methods based on monitoring electric current while an electrolyte solution of different conductivity is drawn into the system by electro-osmosis or determining the zeta potential from streaming potential measurements are less popular and accurate. More common is the method of calculating the electro-osmotic velocity from the migration time of an electrically neutral marker substance incorporated into the sample solution. The selected compound must be soluble in the electrolyte solution, neutral in a wide pH range, and easily detectable. In addition, it should neither become partially charged by compellation with the components of the electrolyte solution nor interact with the capillary tube, the chromatographic stationary phase, or the slab gel employed in capillary electrophoresis (CE), capillary electrochromatography, and planar electrophoresis, respectively. This method has the advantage of simplicity and can be used to monitor the EOF during analysis in any of the above techniques, provided that the analytes and the EOF are directed toward the same electrode.

REFERENCES

1. Hiemenz, P.C. *Principles of Colloid and Surface Chemistry*, 2nd Ed.; Marcel Dekker, Inc.: New York, 1986; 677–735.

2. Adamson, A.W. *Physical Chemistry of Surfaces*, 5th Ed.; John Wiley & Sons: New York, 1990; 203–257.

3. Shaw, D.J. *Introduction to Colloid and Surface Chemistry*, 3th Ed.; Butterworths: London, 1980; 148–182.

4. Grahame, D.C. The electrical double layer and the theory of electro-capillarity. Chem. Rev. **1947**, *41*, 441–501.

5. Lukacs, K.D.; Jorgenson, J.W. J. High Resolut. Chromatogr. Chromatogr. Commun. **1985**, *8*, 407–411.

6. Altria, K.D.; Simpson, C.F. High voltage capillary zone electrophoresis: operating parameter effects upon electroendosmotic flows and electrophoretic mobilities. Chromatographia **1987**, *24*, 527–532.

7. Cikalo, M.G.; Bartle, K.D.; Myers, P. Attempt to define the role of the length of the packed section in capillary electrochromatography. J. Chromatogr. A. **1999**, *836*, 35–51.

8. Hjerten, S. Free zone electrophoresis. Chromatogr. Rev. **1967**, *9*, 122–219.

9. Van de Goor, A.A.A.M.; Wanders, B.J.; Everaerts, F.M. Modified methods for off- and on-line determination of electroosmosis in capillary electrophoretic separations. J. Chromatogr. **1989**, *470*, 95–104.

E

Electro-Osmotic Flow in Capillary Tubes

Danilo Corradini
Institute of Chromatography, Rome, Italy

INTRODUCTION

The electro-osmotic flow in open capillary tubes is generated by the effect of the applied electric field across the tube on the uneven distribution of ions in the electric double layer at the interface between the capillary wall and the electrolyte solution. In bare fused-silica capillaries, ionizable silanol groups are present at the surface of the capillary wall, which is exposed to the electrolyte solution. In this case, the electric double layer is the result of the excess of cations in the solution in contact with the capillary tube to balance the negative charges on the wall arising from the ionization of the silanol groups. Part of the excess cations are firmly held in the region of the double layer closer to the capillary wall (the compact or Stern layer) and are believed to be less hydrated than those in the diffuse region of the double layer.[1] When an electric field is applied across the capillary, the remaining excess cations in the diffuse part of the electric double layer move toward the cathode, dragging their hydration spheres with them. Because the molecules of water associated with the cations are in direct contact with the bulk solvent, all the electrolyte solution moves toward the cathode, producing a pluglike flow having a flat velocity distribution across the capillary diameter.[2]

DISCUSSION

The flow of liquid caused by electro-osmosis displays a pluglike profile because the driving force is uniformly distributed along the capillary tube. Consequently, a uniform flow velocity vector occurs across the capillary. The flow velocity approaches zero only in the region of the double layer very close to the capillary surface. Therefore, no peak broadening is caused by sample transport carried out by the electro-osmotic flow. This is in contrast to the laminar or parabolic flow profile generated in a pressure-driven system, where there is a strong pressure drop across the capillary caused by frictional forces at the liquid-solid boundary. A schematic representation of the flow profile due to electro-osmosis in comparison to that obtained in the same capillary column in a pressure-driven system, such as a capillary HPLC, is displayed in Fig. 1.

The dependence of the velocity of the electro-osmotic flow (v_{co}) on the applied electric field (E) is expressed by the Helmholtz-von Smoluchowski equation

$$v_{co} = -\frac{\varepsilon_0 \varepsilon_r \zeta}{\eta} E \tag{1}$$

where ζ is the zeta potential, ε_0 is the permittivity of vacuum, ε_r is the dielectric constant, and η is the viscosity of the electrolyte solution. This expression assumes that the dielectric constant and viscosity of the electrolyte solution are the same in the electric double layer as in the bulk solution. The term $-\varepsilon_0 \varepsilon_r \zeta / \eta$ is the defined electro-osmotic coefficient or, more properly, electro-osmotic mobility (μ_{co}) and expresses the velocity of the electro-osmotic flow per unit field. Accordingly, the Helmholtz–von Smoluchowski equation can be written

$$\mu_{co} = \frac{v_{co}}{E} \tag{2}$$

In a capillary tube, the applied electric field E is expressed by the ratio where V is the potential difference in volts across the capillary tube of length L_T (in meters). The velocity of the electro-osmotic flow, v_{co} (in meters per second), can be evaluated from the migration time t_{cot} (in seconds) of an electrically neutral marker substance and the distance L_D (in meters) from the end of the capillary where the samples are introduced to the detection windows (effective length of the capillary). This indicates that, experimentally, the electro-osmotic mobility can be easily calculated using the Helmholtz–von Smoluchowski equation in the following form:

$$\mu_{co} = \frac{L_T L_D}{V t_{co}} \, (\mathrm{m^2/V/s}) \tag{3}$$

which demonstrates that, by analogy to the electrophoretic mobility, the electro-osmotic mobility has the dimension of square meters per volt per second.

The electrically neutral marker substance employed to measure the velocity of the electro-osmotic flow has to fulfill the following requirements. The compound must be soluble in the electrolyte solution and neutral in a wide pH range and no interaction with the

Encyclopedia of Chromatography DOI: 10.1081/E-ECHR-120039970

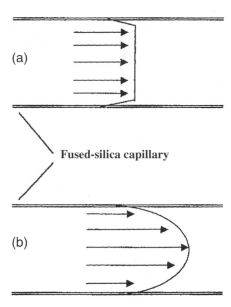

(a)

Fused-silica capillary

(b)

Fig. 1 Schematic representation of the flow profiles obtained with the same capillary column connected to an electric-driven system (a) and to a pressure-driven system (b). Arrows indicate flow velocity vectors.

capillary wall must occur. In addition, the electrically neutral marker substance should be easily detectable in order to allow a small amount to be injected. If the electrically neutral marker interacts with the capillary wall or becomes partially charged by complexation with the components of the electrolyte solution, the measured electro-osmotic velocity may appear slower or faster than the real flow. Some compounds that adequately serve as electrically neutral markers include benzyl alcohol, riboflavin, acetone, dimethylformamide, dimethyl sulfoxide, and mesityl oxide.

Alternatively, the velocity of the electro-osmotic flow can be measured by weighing the volume of the electrolyte solution displayed by electro-osmosis from the anodic to the cathodic reservoir. When detection is performed by ultraviolet (UV) absorbance, a "solvent dip" equal to the electro-osmotic flow appears in the electropherogram after any sample injection. In most cases, the sample solvent has a lower UV absorbance than the electrolyte solution, resulting in a negative UV signal. On the other hand, if the UV absorbance of the sample solvent is higher than that of the electrolyte solution, a positive system peak can be observed at the time corresponding to the velocity of the electro-osmotic flow. The time at which the "solvent dip" appears in the electropherogram can be used to measure the velocity of the electro-osmotic flow in a very simple but less accurate way than those using an electrically neutral marker substance or the weight of the displaced liquid.

The Helmholtz-von Smoluchowski equation indicates that under constant composition of the

electrolyte solution, the electro-osmotic flow depends on the magnitude of the zeta potential which is determined by many different factors, the most important being the dissociation of the silanol groups on the capillary wall, the charge density in the Stern layer, and the thickness of the diffuse layer. Each of these factors depends on several variables, such as pH, specific adsorption of ionic species in the compact region of the electric double layer, ionic strength, viscosity, and temperature.

Secondary equilibrium in solution, generation of Joule heat, and variation of protonic and hydroxylic concentration due to electrolysis may alter the hydrogen ion concentration in the capillary tube when electrolyte solutions having low buffering capacities are employed. A change in the protonic equilibrium directly influences the zeta potential through the variation of the charge density on the capillary wall resulting from the deprotonation of the surface silanol groups, which increases with increasing pH. The shape of a curve describing the dependence of the zeta potential on the electrolyte pH resembles a titration curve, the inflection point of which may be interpreted as the pK value of the surface silanol groups. At acidic pH, the ionization of the surface silanol groups is suppressed and the zeta potential approaches zero, determining the virtual annihilation of the electro-osmotic flow. Under alkaline conditions, the silanol groups are fully charged and the zeta potential reaches its maximum value, which corresponds to a plateau value of the electro-osmotic flow. Between these extreme conditions, the zeta potential rapidly increases with increasing pH up to the complete dissociation of the silanol groups, determining the well-known sigmoidal pH dependence of the electro-osmotic flow. The concentration and ionic strength of the electrolyte solution also have a strong impact on the electro-osmotic flow. The ionic strength influences the thickness of the diffuse part of the electric double layer to which the zeta potential is directly proportional. Because the thickness of the diffuse part of the electric double layer is inversely proportional to the square root of the ionic strength, the electro-osmotic flow decreases with the concentration of the electrolyte solution according to the following relationship:[3]

$$\mu_{co} \approx \frac{e}{3 \times 10^7 |z| \eta \sqrt{C}} \tag{4}$$

where, e, z, η, and C are the total charge per unit surface area, the electron valence of the electrolyte, the viscosity, and the concentration of the electrolyte in the bulk solution, respectively.

Another model that accounts for the decrease of the electro-osmotic flow with increasing the electrolyte concentration relates the electro-osmotic mobility to the concentration of a monovalent counterion, introduced with the buffer, according to the following relationship:[4]

$$\mu_{co} = \frac{Q_0}{\eta(1 + K_{wall}[M^+])}\left(d_0 + \frac{1}{K'\sqrt{[M^+]}}\right) \quad (5)$$

where η is the viscosity of the electrolyte solution and K' is a constant that, for a dilute aqueous solution at 25°C, is equal to $3 \times 10^9/m(mol/L)^{-1/2}$. The first term on the right side of Eq. (5) is related to the dependence of the surface charge on the concentration of the monovalent cation in the electrolyte solution. The model postulates that the initial charge per unit area at the surface of the silica capillary wall (Q_0) is reduced by the factor $1/1 + K_{wall}[M^+]$ upon incorporating a monovalent buffer of concentration $[M^+]$ into the electrolyte solution. This is a result of the neutralization of the free silanol groups on the capillary surface caused by the adsorption of the monovalent cations. The constant K_{wall} is defined as the equilibrium constant between the cations in the buffer solution and adsorption sites on the capillary wall. The second term describes the influence of the concentration of the monovalent cation on the thickness of the mobile region of the electric double layer. This is postulated to be composed of a fixed thickness (d_0) and the Debye-Hückel thickness $\delta = 1/K'[M^+]^{1/2}$, which is inversely proportional to the square root of the concentration of the monovalent ion. According to this model, increasing the concentration of the monovalent buffer cation in the bulk solution influences the electro-osmotic mobility by reducing the Debye-Hückel thickness of the diffuse double layer and by neutralizing the negative charges on the capillary wall resulting from the ionization of the silanol groups.

Certain counterions, such as polycationic species, cationic surfactants, and several amino compounds can be firmly held in the compact region of the electric double layer by forces additional to those of simple Coulombic origin. The specific adsorption of counterions at the interface between the capillary wall and the electrolyte solution results in a drastic variation of the positive charge density in the Stern layer, which reduces the zeta potential and, hence, the electro-osmotic flow. If the positive charge density of the adsorbed counterions exceeds the negative charge density on the capillary wall resulting from the ionization of silanol groups, the zeta potential becomes positive and the concomitant electro-osmotic flow is reversed from cathodic to anodic.

The dependence of the electro-osmotic flow on the specific adsorption of counterions in the electric double layer can be described by a model which correlates the electro-osmotic mobility to the charge density in the Stern part of the electric double layer (arising from the adsorption of counterions) and the charge density at the capillary wall (resulting from the ionization of silanol groups).[5] According to this model, the dependence of the electro-osmotic mobility on the concentration of the adsorbing ions (C) in the electrolyte solution is expressed as

$$\mu_{co} = \frac{\kappa^{-1}}{\eta}\left\{ zen_0 \frac{C}{V_m}\exp\left(\frac{ze\psi_d + \Phi}{kT}\right) \right.$$
$$\times \left[1 + \frac{C}{V_m}\exp\left(\frac{ze\psi_\delta + \Phi}{kT}\right)\right]^{-1}$$
$$\left. -\left(\frac{\gamma}{1 + [H^+]/K_a}\right)\right\} \quad (6)$$

where κ^{-1} is the Debye-Hückel thickness of the diffuse double layer, is the viscosity of the electrolyte solution, e is the elementary charge, z is the valence of the adsorbing ion, k is the Boltzman constant, T is the absolute temperature, is the number of accessible sites in the Stern layer, is the molar volume of the solvent, Φ is the specific adsorption potential of counterions, γ is the sum of the ionized and protonated surface silanol groups, $[H^+]$ is the bulk electrolyte hydrogen ion concentration, and is the silanol dissociation constant. According to this equation, at constant ionic strength, viscosity, and pH, the electro-osmotic mobility depends mainly on the surface density of the adsorbed counterions in the Stern region of the electric double layer, which follow a Langmuir-type adsorption model.

The reversal of the direction of the electro-osmotic flow by the adsorption onto the capillary wall of alky-lammonium surfactants and polymeric ion-pair agents incorporated into the electrolyte solution is widely employed in capillary zone electrophoresis (CZE) of organic acids, amino acids, and metal ions. The dependence of the electro-osmotic mobility on the concentration of these additives has been interpreted on the basis of the model proposed by Fuerstenau[6] to explain the adsorption of alkylammonium salts on quartz. According to this model, the adsorption in the Stern layer as individual ions of surfactant molecules in dilute solution results from the electrostatic attraction between the head groups of the surfactant and the ionized silanol groups at the surface of the capillary wall. As the concentration of the surfactant in the solution is increased, the concentration of the adsorbed alkylammonium ions increases too and reaches a critical concentration at which the van derWaals attraction forces between the hydrocarbon chains of adsorbed and freesurfactant molecules in solution cause their association into hemimicelles (i.e., pairs of surfactant molecules with one cationic group directed toward the capillary wall and the other directed out into the solution).

Lowering the velocity or reversing the direction of the electro-osmotic flow may have a beneficial effect of on the resolution of two adjacent peaks, as evidenced by the following expression for resolution in electrophoresis elaborated by Giddings:[7]

$$R_s = \frac{\sqrt{N}}{4}\left(\frac{\Delta\mu}{\mu_{av} + \mu_{co}}\right) \tag{7}$$

where N is the number of theoretical plates, $\Delta\mu$ and μ_{av} are the difference and the average value of the electrophoretic mobilities of two adjacent peaks, respectively, and μ_{co} is the electro-osmotic mobility. According to this equation, the highest resolution is obtained when the electro-osmotic mobility has the same value but opposite direction of the average electrophoretic mobility of the two adjacent peaks.

Neutral polymeric molecules, such as polysaccharides and synthetic polymers, may also adsorb onto the Stern layer, causing a variation of viscosity in the double layer with distance from the capillary wall, which affects the electro-osmotic mobility according to the following relationship:[2]

$$\mu_{co} = \frac{\varepsilon_r}{4\pi}\int_0^\zeta \frac{1}{\eta}d\psi \tag{8}$$

where ε_r is the dielectric constant, ζ is the zeta potential, η is the viscosity, and ψ is the electric potential. The value of the integral in this expression will approach zero when the viscosity in the double layer approaches infinity. Accordingly, the electro-osmotic flow is drastically reduced when the local viscosity of the double layer is increased as a result of the adsorption of a neutral polymer onto the Stern layer. It is worth noting that at constant value of the viscosity in the electric double layer, Eq. (8) is equivalent to the Helmholtz-von Smoluchowski expression for the electro-osmotic flow.

The incorporation of an organic solvent into the aqueous electrolyte solution also leads to a variation of the electro-osmotic flow.[8] The general trend is that the electro-osmotic flow decreases steadily with increasing concentration of the organic solvent in the hydro-organic electrolyte solution. This effect can be attributed, to some extent, to the increasing viscosity and decreasing dielectric constant of most hydro-organic electrolyte solutions with increasing concentration of organic solvent. However, in most cases, the decrease of the electro-osmotic flow is also observed at organic solvent concentrations greater than 50–60% (v/v), at which the ratio of the dielectric constant and the viscosity, ε_r/η is generally increasing. This indicates that the variation of the electro-osmotic flow caused by the incorporation of an organic solvent into the electrolyte solution cannot be solely related to the changes of the ratio ε_r/η.

Similar to the neutral polymers, organic solvents can adsorb at the interface between the capillary wall and the electrolyte solution, through hydrogen-bonding or dipole interaction, thus increasing the local viscosity within the electric double layer. Organic solvents may also influence the zeta potential by affecting the ionization of the silanol groups at the capillary surface, whose has been found to be shifted toward higher values with increasing the content of organic solvents in the electrolyte solution. The dependence of the zeta potential on the fraction of an organic solvent incorporated into the electrolyte solution may be also related to the variation of both the dielectric constant and the adsorption of counterions in the Stern layer. In practice, introducing a neutral polymer or an organic solvent into the electrolyte solution results in multiple changes, generally involving the viscosity and the dielectric constant of the bulk solution, the ionization of the silanol groups on the capillary wall, and the charge density in the Stern layer, as well as the local viscosity and the dielectric constant of the electric double layer.

REFERENCES

1. Hiemenz, P.C. *Principles of Colloid and Surface Chemistry*, 2nd Ed.; Marcel Dekker, Inc.: New York, 1986; 677–735.
2. Hjertén, S. Free zone electrophoresis. Chromatogr. Rev. **1967**, *9*, 122–219.
3. Tsuda, T.; Nomura, K.; Nakagawa, G. Open-tubular microcapillary liquid chromatography with electro-osmosis flow using a UV detector. J. Chromatogr. **1982**, *248*, 241–247.
4. Salomon, K.; Burgi, D.S.; Helmer, J.C. Valuation of fundamental properties of a silica capillary used for capillary electrophoresis. J. Chromatogr. **1991**, *559*, 69–80.
5. Corradini, D.; Rhomberg, A.; Corradini, C. Electrophoresis of proteins in uncoated capillaries with amines and amino sugars as electrolyte additives. J. Chromatogr. A. **1994**, *661*, 305–313.
6. Fuerstenau, D.W. J. Phys. Chem. **1956**, *60*, 981–985.
7. Giddings, J.C. Generation of variance, theoretical plates resolution and peak capacity in electrophoresis and sedimentation. Separ. Sci. **1969**, *4*, 181–189.
8. Schwer, C.; Kenndler, E. Electrophoresis in fused-silica capillaries: the influence of organic solvents on the electroosmotic velocity and the zeta potential. Anal. Chem. **1991**, *63*, 1801–1807.

Electro-Osmotic Flow Nonuniformity: Influence on Efficiency of Capillary Electrophoresis

Victor P. Andreev
Institute for Analytical Instrumentation, Russian Academy of Sciences, St. Petersburg, Russia

INTRODUCTION

There are two types of nonuniformities of electro-osmotic flow (EOF) that can contribute significantly to the solute peak broadening and are important for capillary electrophoresis (CE). The first is the transversal nonuniformity of the usual EOF in the capillary with the zeta potential of the walls and longitudinal electric field strength constant and independent of coordinates. The second one is the nonuniformity of EOF caused by the dependence of the zeta potential of the walls or electric field strength on coordinates.

DISCUSSION

The first type of EOF nonuniformity was described in the classical article by Rice and Whitehead[1] written much earlier than the first works on CE. The equation for the EOF velocity profile in the infinitely long tube with radius a was given by

$$V(r) = \frac{\zeta \varepsilon \varepsilon_0 E}{\eta}\left(1 - \frac{I_0(\kappa r)}{I_0(\kappa a)}\right) \tag{1}$$

where ζ is the zeta potential of the wall, ε and η are the dielectric constant and viscosity of the buffer, respectively, ε_0 is the permitivity of the free space, $\kappa^{-1} = (\varepsilon \varepsilon_0 \kappa_B T 2n e^2)^{1/2}$ is the Debye layer thickness, κ_B is the Boltzmann constant, T is the temperature, n is the number of ions per unit volume (proportional to the concentration of buffer C_0), e is the proton charge, and $I_0(x)$ is the modified Bessel function. It is evident from Eq. (1) that the nonuniformity of the EOF profile can be substantial only if the capillary radius and Debye length are commensurate. In fact, for the case of $\kappa a \approx 1$, the profile of EOF according to Eq. (1) is very close to parabolic. Luckily, it is not the case of usual capillaries for CE with $a \geq 25\,\mu m$ because, even for distilled water, $\kappa^{-1} \approx 0.1\,\mu m$. Another important result of Ref.[1] is the prediction of the EOF profile in the long capillary with closed ends. In such a capillary, liquid moves in one direction in the vicinity of the walls and in the opposite direction near the axis of the capillary, thus making the total flow through the cross section equal to zero. With this result, it is quite evident that CE must be realized in a capillary with open ends; otherwise, nonuniformity of EOF would ruin the separation.

Results of Ref.[1] were produced by employing a linear approximation of the exponential terms in the Poisson–Boltzmann equation for electrical potential and charge distribution. Strictly speaking, this linearization is valid only for $|\zeta| \ll kT/e \approx 0.03\,V$, whereas the range of the values of the zeta potential is $|\zeta| \leq 0.1\,V$. In Ref.,[2] the Poisson–Boltzmann equation and the Navier–Stokes equations for EOF velocity profile were solved numerically without linearization. The dependence of buffer viscosity on temperature and the existence of temperature gradients due to Joule heating were also taken into consideration. Calculated EOF profiles were compared with and predicted by Eq. (1), showing that the difference in flow profiles for $|\zeta| = 0.1\,V$ could be significant, especially for thin capillaries and low buffer concentrations ($ka \approx 10$). Calculated flow profiles were used to predict the stationary value of HETP by using the results of generalized dispersion theory. It was shown that for low buffer concentrations ($C_0 \leq 10^{-4}\,M$), the contribution of electro-osmotic flow nonuniformity to the HETP value could be larger than the contribution of the thermal effects and molecular diffusion.

A similar approach was used in Ref.[3], where the contributions of EOF nonuniformity (H_{co}) and molecular diffusion (H_{diff}) to HETP were compared for different values of solute diffusion coefficients. It was shown[3] that for the typical CE velocities of EOF (1–2 mm/s) and rather high buffer concentration ($C_0 = 10^{-2}\,M$), $H_{co}H_{diff} \geq 1$ for $D \leq 2 \times 10^{-12}/m^2/s$. For lower buffer concentrations, the influence of EOF nonuniformity is substantial for smaller molecules also ($H_{co} = 1.3 \times 10^{-8}\,m$, $H_{diff} = 1.6 \times 10^{-8}\,m$ for $D = 2.4 \times 10^{-11}\,m^2 s$, corresponding to α_2-macroglobulin).

The joint effect of EOF nonuniformity and particle–wall electrostatic interactions was studied in Ref.[4]. Two types of solute particles were examined: one with the charge of the same sign as the zeta potential of

Encyclopedia of Chromatography DOI: 10.1081/E-ECHR-120039971

the wall, and the other of the opposite sign. The particles of the first type are moving electrophoretically in the direction opposite to the direction of EOF and are electrostaticaly subtracted by the wall, whereas the particles of the second type are attracted by the wall and are moving electrophoretically in the same direction as EOF. Particles of the second type spend a large portion of time in the vicinity of the capillary wall and, so, EOF nonuniformity contributes significantly to peak broadening, whereas for the particles subtracted by the wall, the influence of EOF nonuniformity is negligible because their residence time in the vicinity of the wall is close to zero. For example, for the particles with $D = 5 \times 10^{-11} \, \text{m}^2\text{s}$, in the capillary with $a = 10 \, \mu\text{m}$, $\zeta = -0.1 \, \text{V}$, and $E = 40 \, \text{kV/m}$, filled with diluted buffer ($C_0 = 10^{-5} \, \text{M}$), one has HETP $\approx 10 \, \mu\text{m}$ for particles attracted by the wall, whereas for particles subtracted by the wall, HETP $\approx 0.1 \, \mu\text{m}$. For neutral particles not interacting with the walls, HETP $\approx 0.2 \, \mu\text{m}$ was predicted. The difference was much less dramatic for the case of the higher buffer concentrations and the lower zeta potential. For example, for $\zeta = -0.02 \, \text{V}$, $C_0 = 10^{-3} \, \text{M}$ and the rest of parameters being the same as described earlier, HETP is determined mainly by molecular diffusion and is close to $0.1 \, \mu\text{m}$.

The influence of EOF nonuniformity on efficiency of CE in the capillary with the zeta potential of the wall being the function $\zeta(x)$ of the longitudinal coordinate x was studied in Ref.[5]. To calculate the EOF velocity profile, an important approximation was justified by the fact that usually $\kappa a \gg$ in CE. Thus, the double-layer region was neglected and the following boundary condition was formulated:

$$V_x(x, a, t) = \frac{\varepsilon \varepsilon_0 E}{\eta} \zeta(x) \qquad (2)$$

With this boundary condition, the Navier–Stokes equations for longitudinal V_x and radial V_r components of EOF velocity were solved numerically, and the calculated EOF profiles were used to simulate the solute peak shapes. The situation where the part of the capillary length was modified to the zero value of the zeta potential and the part of capillary was not modified ($\zeta \neq 0$) was studied. It was shown that the radial component of the velocity is nonzero only in the rather short transition region between the uncovered and covered parts of the capillary. At a distance of a few capillary diameters from the transient region, the radial flows are negligible and the axial component of the velocity in the covered section of the capillary has an almost parabolic profile. Peak shapes and peak variances were studied, and the general conclusion of Ref.[5] was that the main contribution to the

peak width was from the parabolic velocity profiles, the contribution of the radial flow in the transient regions being less significant. Based on this result, the mathematical model of CE in the capillary made of several sections with various nonequal values of the zeta potentials and radii was developed.[6] For each of the sections, the total flow was considered to be the sum of EOF caused by electrical potential differences along the section and the Poiseuille flow, caused by the pressure drop along the section. The values of the pressure drops and potentials differences were determined by the solution of the set of $2N$ algebraic equations, where N is the number of sections in the capillary. These equations reflect the fact that the total flow of liquid and total current are constant along the capillary, and the sums of pressure drops and potential differences at the sections are equal to the total pressure drop and total potential difference at the whole capillary, respectively. The lengths of the sections were considered to be much larger than the capillary radius, so the results of the model are valid everywhere except the immediate vicinity of the points where the radius of capillary or the zeta potential of the wall change their values. When calculating the values of HETP in such a capillary, particle–wall electrostatic interactions were taken into consideration, and it was shown that HETP values are considerably larger for particles attracted by the wall. It was also shown that differences in the values of the zeta potential contributes to HETP much more significantly than the differences in radii values. The situation that might happen in the case of a bubble-cell detector was modeled and considerable growth of HETP was predicted.

In Ref.,[7] the case was studied in which the zeta potential of the wall was the linear function of the longitudinal coordinate. This situation may happen when the value of the zeta potential is controlled by the external electrical potential applied to the wall. Electrical potential value inside the capillary is naturally a linear function of the longitudinal coordinate x; therefore, if the electrical potential applied to the outer boundary of the capillary wall is constant, then the potential difference across the wall is a linear function of x. The theoretical approach used in Ref.[7] is similar to the one in Ref.[5]. Secondary parabolic flow was shown to be generated, leading to the increase of HETP. It was predicted theoretically, and verified experimentally, that a pressure profile superimposed on the capillary can, in some cases, compensate for the disturbed profile and reduce the HETP value.

In Ref.,[8] the mathematical model of CE in rectangular channels with nonequal values of the zeta potentials of the walls was developed. This model may be of interest for the case of CE on a microchip, where the microgroves are produced by wet chemical etching and, so, the walls of the groove can have different values of zeta potential than the cover plate that is not etched. Flow profiles for the channels with different aspect ratios and different combinations of the zeta potential values were examined. It was shown, for example, that a 10% difference in the values of the zeta potentials of upper and lower walls can cause a sixfold growth of the HETP value.

The above-mentioned examples show that EOF nonuniformities may occur in different situations and must be given considerable attention, as they can reduce the CE efficiency dramatically.

REFERENCES

1. Rice, C.L.; Whitehead, R. Electrokinetic flow in a narrow cylindrical capillary. J. Phys. Chem. **1965**, *69*, 4017.
2. Andreev, V.P.; Lisin, E.E. Investigation of the electroosmotic flow effect on the efficiency of capillary electrophoresis. Electrophoresis **1992**, *13*, 832.
3. Gas, B.; Stedry, M.; Kenndler, E. Contribution of the electroosmotic flow to peak broadening in capillary zone electrophoresis with uniform zeta potential. J. Chromatogr. A. **1995**, *709*, 63.
4. Andreev, V.P.; Lisin, E.E. On the mathematical model of capillary electrophoresis. Chromatographia **1993**, *37*, 202.
5. Potocek, B.; Gas, B.; Kenndler, E.; Stedry, M. Electroosmosis in capillary zone electrophoresis with non-uniform zeta potential. J. Chromatogr. A. **1995**, *709*, 51.
6. Andreev, V.P.; Shirokih, N.V. Electroosmotic flow profile in the capillary made of several sections, 20th Int. Symp. on Capillary Chromatography, Proceedings on CD, 1998 paper H 11.
7. Keely, C.A.; van de Goor, T.A.A.M.; McManigill, D. Modeling flow profiles and dispersion in capillary electrophoresis with nonuniform zeta potential. Anal. Chem. **1994**, *66*, 4236.
8. Andreev, V.P.; Dubrovsky, S.G.; Stepanov, Y.V. Mathematical modeling of capillary electrophoresis in rectangular channels. J. Microcol. Separ. **1997**, *9*, 443.

Electrochemical Detection

Peter T. Kissinger
Bioanalytical Systems, Inc. and Purdue University, West Lafayette, Indiana, U.S.A.

INTRODUCTION

With respect to chromatography, "electrochemical detection" means amperometric detection. Amperometry is the measurement of electrolysis current vs. time at a controlled electrode potential. It has a relationship to voltammetry similar to the relationship of a UV detector to spectroscopy. While conductometric detection is used in ion chromatography, potentiometric detection is never used in routine practice. Electrochemical detection has even been used in GC in a few unusual circumstances. It has even been attempted with TLC. Its practical success has only been with LC, and that will be the focus here.

Most chemists remember electrochemistry as a difficult subject they heard about in physical chemistry courses, and they regard it as having something to do with batteries. Both of these impressions arc true! What is important here is to understand that: 1) redox reactions can be made to occur at surfaces (electrodes); and 2) amazingly enough, such reactions are not just the fate of metals ($Fe^{+++} \rightarrow Fe^{++}$) but actually occur quite widely among organic compounds of interest, such as drugs, pesticides, explosives, food additives, neurotransmitters, DNA, etc. There are good references for the novice wishing to understand the analytical electrochemistry of organic substances.[1] I present a few common examples below for both oxidations (electrons are lost; the process is anodic) and reductions (electrons are gained by the analyte; the process is cathodic).

While we all remember (or try not to) the confusing math associated with electrochemistry and thermodynamics, all we need here is an appreciation of the fact that current, i, is proportional to the moles, N, of analyte reacted per unit time. The latter is proportional to concentration at a constant flow rate through a detector cell. The key equation is

$$i = \frac{dQ}{dt} = nF\frac{dN}{dt}$$

where Q is the amount of electricity (charge in coulombs), n is the number of electrons, and F is the Faraday constant. As one can see from the above examples, most organic analytes are involved in reactions where $n = 1$, 2, or 4. To use an electrochemical detector, it is very important to know that i is proportional to concentration and the amount injected, just as UV absorbance is proportional to concentration or the amount injected.

LC/electrochemistry (LCEC) is now over 30 years old.[1] In recent years, an emphasis has been placed on miniaturizing the technology to accommodate the study of smaller biological samples, often with a total available volume of only a few microliters. Both LC and electrochemistry are largely controlled by surface science. Considering this fact, both technologies benefit from reducing the distance from the bulk of the solution phase to the surface. In LC, this is accomplished by using smaller diameter stationary phase particles. In electrochemistry, it is accomplished by using packed bed or porous electrodes and/or thin-layer cells with greatly restricted diffusion pathways.

For analytical purposes, there is no loss in concentration detection limit by reducing the total surface area available in both methodologies. In LC, this reduction is accomplished by using smaller diameter columns and in EC, by using smaller electrodes. With LC column diameters of 0.1–1.0 mm and radial flow thin-layer cells with dead volumes of a few tens of nanoliters, it is possible to build analytical instruments suitable for routine use by neuroscientists, drug metabolism groups, and pharmacokinetics experts. LC/electrochemistry has been used for foods, industrial chemicals, and environmental work. Nevertheless, biomedical applications have dominated. It shows no potential for preparative chromatography and is

Encyclopedia of Chromatography DOI: 10.1081/E-ECHR-120039965

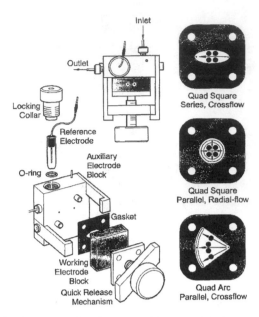

Fig. 1 One example of a sandwich type thin-layer LCEC detector with adjustable dead volume, flow pattern, and up to four channels. (From Curr. Sep. **2000**, *18*, 114.)

generally used when nanograms or picograms hold some appeal.

DETECTOR CELLS

A wide variety of detector cells have been used for LCEC.[1] The choice can be baffling to a non-expert. These all "work" to some degree. The key issues:

1. An electrode (the "working electrode") exposed to the mobile phase in a dead volume (small) appropriate to the column diameter chosen.
2. A place to locate at least one other electrode (a "counterelectrode") or preferentially two (an auxiliary electrode and a reference electrode).
3. The possibility for a choice of different working electrode materials (see following section).
4. The possibility of multiple channels in series or parallel.

Fig. 1 is representative of one choice that meets these criteria. Such a cell is normally described as a "thin-layer sandwich configuration." The working electrode(s) is (are) in the form of an interchangeable block. Electrodes of different sizes, shapes, or materials can be accommodated with a flow pattern established by a gasket shape and thickness. Such cells can easily be adapted for LC flow rates of 5–5000 μl/min. Different designs are used for capillary separation tools such as capillary electrophoresis (CE).

ELECTRODE MATERIALS

The most common electrode material used in LCEC is carbon, either as solid "glassy carbon" disks in thin-layer cells, or as a high surface area porous matrix through which the mobile phase can flow. Gold electrodes are useful to support a mercury film, and these are primarily used to determine thiols and disulfides, and also for carbohydrates using pulsed electrochemical detection (PED) with high pH mobile phases. Platinum electrodes are occasionally useful for specific analytes, but are most frequently employed to determine hydrogen peroxide following an oxidase immobilized enzyme reactor (IMER). More recently, copper electrodes have begun to attract serious interest for determination of carbohydrates in basic mobile phases. Glassy carbon is the overwhelming favorite choice due to its wide range of applicable potentials and its rugged convenience. Bulk glassy carbon is difficult to use in geometries other than disks and plates. There are a number of other geometries that have practical interest for multiple electrode detectors. One of the more valuable recent contributions to LCEC derives from the ability to deposit conducting vitreous carbon films on silicon or quartz substrates using lithography techniques. The lithography technology makes it possible to lay down a variety of electrode geometries, which could not possibly be manufactured in small sizes by traditional machining. While such electrodes are still at the research stage, they show considerable promise. Detector cells with 2, 4, or even 16 electrodes are commercially available. There are obvious parallels with diode array detection (DAD). When two electrodes are used in series, there are similarities to fluorescence or MS/MS in the way that selectivity is often enhanced.

PULSED ELECTROCHEMICAL DETECTION

There are many substances that would appear to be good candidates for LCEC from a thermodynamic point of view, but that do not behave well due to kinetic limitations. Johnson and co-workers at Iowa State University used some fundamental ideas about electrocatalysis to revolutionize the determination of carbohydrates, nearly intractable substances that do not readily lend themselves to ultraviolet absorption (LCUV), fluorescence (LCF), or traditional d.c. amperometry (LCEC).[2] At the time this work began, the LC of carbohydrates was more or less relegated to refractive index detection (LCRI) of microgram amounts. The importance of polysaccharides and glycoproteins, as well as traditional sugars, has focused a lot of attention on PED methodology. The detection limits are not competitive with d.c. amperometry of more easily oxidized substances such as phenols and aromatic amines; however, they are far superior to optical detection approaches.

POSTCOLUMN REACTIONS

Electrochemical detection is inherently a "chemical" rather than a "physical" technique (such as ultraviolet, infrared, fluorescence, or refractive index). It is therefore not surprising to find that many imaginative postcolumn reactions have been coupled to LCEC. These include photochemical reactions, enzymatic reactions, halogenation reactions, and Biuret reactions. In each case, the purpose is to enhance selectivity and therefore improve limits of detection. While simplicity is sacrificed with such schemes, there are many published methods that have been quite successful.

CAPILLARY ELECTROPHORESIS AND CAPILLARY ELECTROCHROMATOGRAPHY (CEC)

Since there is LCEC, it is only logical that there should be CEEC and CEC/EC. This area was pioneered by Andrew Ewing at the Pennsylvania State University. Richard Zare (Stanford University) and Susan Lunte (Kansas University) have explored this idea in a number of unique ways. The basic technology has been recently reviewed.[3] There are several fundamental problems that do not occur with LCEC. First, the capillaries must be of small diameter to properly dissipate resistive heating. Thus, the electrodes used in CEEC are normally carbon fibers or metallic wires placed in or at the capillary end. Second, the electrical current through the capillary that establishes the electro-osmotic pumping is much larger than the electrolysis current measured in determining analytes of interest. The ionic and electrolytic currents need to be "decoupled" in some way. A third concern is that the flow rate in CE or CEC is not independent of the choice of "mobile phase" or even the sample, whereas in LC it is easily predetermined and maintained by a volume displacement pump. In spite of these concerns, CE is very attractive because of its high resolution per unit time and the small sample volumes required. In the case of CEEC, the concentration detection limits are frequently superior to those of optical detectors for suitable analytes. This is because electrochemical detection is a surface (not volume) dependent technique. In the grand scheme of things, at this writing, CE and CEC are very rarely used vs. LC, and therefore CEEC and CEC/EC must be considered academic curiosities until this situation changes.

CONCLUSIONS

Electrochemical detection has matured considerably in recent years and is routinely used by many laboratories, often for a very specific biomedical application. The most popular applications include acetylcholine, serotonin, catecholamines, thiols and disulfides, phenols, aromatic amines, macrocyclic antibiotics, ascorbic acid, nitro compounds, hydroxylamines, and carbohydrates. As the last century concluded, it is fair to say that many applications for which LCEC would be an obvious choice are now pursued with LC/MS/MS. This only became practical in the 1990s and is clearly a more general method applicable to a wider variety of substances. In a similar fashion, LC/MS/MS has also largely supplanted LCF for new bioanalytical methods. Nevertheless, there remain a number of key applications for these more traditional detectors, known for their selectivity (and therefore excellent detection limits). Likewise, it will be quite a few years before LC/MS/MS is affordable worldwide. An interesting recent development is the combination of electrochemical detector flow cells with MS in various configurations including EC/LC/MS/MS and LC/EC/MS/MS.[4] This enables information-rich solution phase redox chemistry to be combined with gas phase ion chemistry, adding a new dimension for complex samples. It also provides new mechanistic information on the electrochemical detection process.

ARTICLE OF FURTHER INTEREST

Detection Principles, p. 443.

REFERENCES

1. Kissinger, P.T., Heineman, W.R., Eds.; *Laboratory Techniques in Electroanalytical Chemistry*, 2nd Ed; Marcel Dekker Inc.: New York, 1996.
2. LaCourse, W.R. *Pulsed Electrochemical Detection in High-Performance Liquid Chromatography*, John Wiley & Sons: New York, 1997.
3. Holland, L.A.; Lunte, S.M. Capillary electrophoresis coupled to electrochemical detection: a review of recent advances. Anal. Commun. **1998**, *35*, 1H–4H.
4. Bökman, C.F.; Zettersten, C.; Sjöberg, P.J.R.; Nyholm, L. A setup for the coupling of a thin-layer electrochemical flow cell to electrospray mass spectrometry. Anal. Chem. **2004**, *76* (7), 2017–2024.
5. Arakawa, R.; Yamaguchi, M.; Hotta, H.; Osakai, T.; Kimoto, T. Product analysis of caffeic acid oxidation by on-line electrochemistry/electrospray ionization mass spectrometry. J. Am. Soc. Mass. Spectrom. **2004**, *15*, 1228–1236.

Electrochemical Detection in Capillary Electrophoresis

Oliver Klett
Institute of Chemistry, Uppsala University, Uppsala, Sweden

INTRODUCTION

Capillary electrophoresis (CE) is a powerful separation tool which has its primary strength in the high separation efficiency and short analysis times.

DISCUSSION

By decreasing the internal diameter (i.d.) of the capillaries used, the situation can be further improved due to the possibility of using higher separation voltages. Such a miniaturization, however, often involves a challenge regarding how the detection is to be made in the narrow capillaries for sample volumes in the nanoliter to subpicoliter range. Electrochemical (EC) methods, usually based on the use of microelectrodes, are relatively inexpensive and are readily miniaturized and adapted to such low volumes and capillary sizes without loss of performance. Electrochemical detection is based on the monitoring of changes in an electrical signal due to a chemical system at an electrode surface, usually as a result of an imposed potential or current. The principles, advantages, and drawbacks of currently used EC methods will be discussed briefly below.

In a solution, the equilibrium concentrations of the reduced and oxidized forms of a redox couple are linked to the potential (E) via the Nernst equation

$$E = E^{0\prime} + \frac{RT}{nF}\ln\left(\frac{c(\mathrm{ox})}{c(\mathrm{red})}\right) \tag{1}$$

with $E^{0\prime}$ the standard potential and $c(\mathrm{ox})$ and $c(\mathrm{red})$ the concentration of the oxidized and reduced forms, respectively; the other symbols have their usual meaning.

In electrochemical detection, the potential of a working electrode can be measured versus a reference electrode, usually while no net current is flowing between the electrodes. This type of detection is referred to as "potentiometry." Alternatively, a potential is applied to the working electrode with respect to the reference electrode while the generated oxidation or reduction current is measured. This technique is referred to as "amperometry." When applying a negative potential to the working electrode, the energy of the electrons in the electrode is increased and,

eventually, an electron can be transferred to the lowest unoccupied level of a species in the nearby solution. This species is thus reduced; vice versa, species can be oxidized by applying a sufficiently high positive potential. In both cases, the generated current (i) can be expressed by

$$i = -aFDnc\delta_N^{-1} \tag{2}$$

with a being the electrode area, D the diffusion coefficient, δ_N the thickness of diffusion layer; the other symbols have their usual meaning. For each redox couple, there exists a potential, the standard potential E^0, for which the reduced and oxidized forms are present in equal concentrations. By applying a potential more positive than E^0, the concentration of the reduced form is forced to decrease at the electrode surface while the concentration of the oxidized form increases. This process is the cause of the current measured in amperometric techniques. By choosing the applied potential, it is also possible to discriminate between different analytes. The range of potentials that can be applied in amperometric detection is, however, generally limited by redox processes involving the solvent [e.g., the oxidative and reductive evolution of oxygen ($2H_2O \rightarrow O_2 + 4H^+ + 4e^-$) and hydrogen ($H_2O + e^- \rightarrow 0.5 H_2 + OH^-$), respectively, in water]. A wide range of physiologically and pharmacologically important substances, as well as many heavy metals, transition metals, and their complexes, exhibit standard potentials within this accessible potential range. In fact, many metabolic pathways involve redox processes taking place in aqueous systems. Neurotransmitters of the catechol type (o-dihydroxy benzene derivatives) were consequently among the first reported analytes for electrochemical detection in CE (CE/EC). Detection limits down to can be achieved in this way.

The choice of working electrode material is an important factor in amperometric detection. For catechols and similar substances, such as phenolic acids, electrodes made of glassy carbon have shown good performances. Other good detectable and biological important substances include thiols and disulfides (e.g., cysteine, glutathione, and their disulfides which are best detected on an Au/Hg amalgam electrode); amino acids and peptides, which can be detected using Cu electrodes, and carbohydrates,

Encyclopedia of Chromatography DOI: 10.1081/E-ECHR-120039966

glycopeptides, and nucleotides, detected on Au, Cu, or Ni electrodes.

Commonly in amperometric detection, a fiber or disk microelectrode is used where the electrode is positioned in or close to the outlet of the capillary (see Fig. 1). A complication when working with EC in CE is the need for careful alignment of the electrode(s) and capillary outlet. This alignment, which mostly is carried out with micromanipulators under a microscope, is essential to ensure both a good sensitivity and reproducibility and, hence, constitutes the main challenge while adapting EC for routine CE. Another complication in CEEC involves the interference of the high-voltage (HV) separation field on the EC detection. In the first combination of EC and CE, it was assumed that the HV field had to be totally removed from the detection area. This was done by various kinds of decouplers, which unfortunately also introduced additional band broadening and decreased sensitivities. Furthermore, the manufacturing of the decouplers requires considerable labor-intensive experience and skill.

A later approach is based on the utilization of small-inner-diameter ($<25\,\mu$m) capillaries to reduce the influence of the HV field.

In potentiometry, all ions present in the solution principally contribute to the potential of the working electrode. As the ratio between the analyte concentration and that of other species in the solution generally is rather low, the analyte contribution to the detector signal is often low, which results in relatively poor detection limits. To circumvent this problem, ion-selective membranes (ISM), which permit only some ions to pass through the membranes, are commonly employed. In this way, detection limits down to 10^{-7} mol/L can be achieved. The ISM also reduces the influence from matrix components, which allows measurements in complex matrices such as blood or serum without interferences. The long-term stability of these electrode may, however, be a problem, as the electrodes might have to be replaced after a few hours or days. Common analytes are inorganic anions and cations, especially alkali and alkaline earth metals ions. A further application is the indirect detection of amino acids, where the complexing of amino acids with Cu^+ ions selectively alters the potential of a copper electrode.

In conductometry, two working electrodes placed either in or at the end of the capillary, along the capillary axis, are commonly employed. A high-frequency AC potential is applied between the working electrodes and the conductance (L) of the solution is continually monitored. In this way, the passing of any zone deviating in its ion composition from the background is detected. As the ion mobilities contribute to the magnitude of the signal as seen in Eq. (3), slowly moving large molecules and low charged biomolecules are less straightforwardly detected, whereas detection limits down to some hundred parts per thousand have been reported for small inorganic and organic ions:

$$L = \frac{FA\Sigma i |z_i| u_i c_i}{l} \qquad (3)$$

with A as the cross-sectional area perpendicular to the AC field, l the electrode distance, and u the mobility. Applications have been described for ions up to a size of sulfate or Cd^+ and MES or benzylamine, respectively.

SUGGESTED FURTHER READING

The field of electrochemical detection in CE have been extensively reviewed in Refs.[1–3]. Instructive applications can be found for amperometry in Ref.,[4] for potentiometry in Ref.,[5] and for conductometry in Ref.[6]. An example of miniaturized on-chip EC/CE is given in Ref.[7]. The theoretical aspects of electrochemical detection have been discussed in detail in Ref.[8].

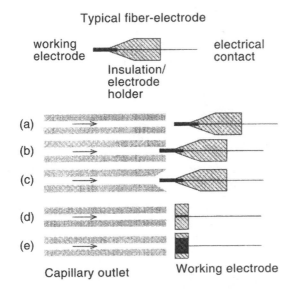

Typical fiber-electrode

working electrode — electrical contact

Insulation/ electrode holder

(a)
(b)
(c)
(d)
(e)

Capillary outlet   Working electrode

Fig. 1 Electrode setups for CE/EC with fiber microelectrodes: (a) end column, (b) on-column, (c) improved on-column, (d) wall tube, and (e) wall-jet detection.

REFERENCES

1. Voegel, P.D.; Baldwin, R.P. Electrophoresis **1997**, *18* (12–13), 2267–2278.

2. Lunte, S.M.; et al. Pharmaceut. Res. **1997**, *14* (4), 372–387.

3. Kappes, T.; Hauser, P.C. Electrochemical detection methods in capillary electrophoresis and applications to inorganic species. J. Chromatogr. A. **1999**, *834* (1–2), 89–101.

4. Holland, L.A.; Lunte, S.M. Postcolumn reaction detection with dual-electrode capillary electrophoresis-electrochemistry and electrogenerated bromine. Anal. Chem. **1999**, *71* (2), 407.

5. Kappes, T.; Hauser, P.C. Potentiometric detection in capillary electrophoresis with a metallic copper electrode. Anal. Chim. Acta **1997**, *354*, 129–134.

6. Haber, C.; et al. Conductivity detection in capillary electrophoresis–a powerful tool in ion analysis. J. Capillary Electrophoresis **1996**, *3* (1), 1–11.

7. Woolley, A.T.; Lao, K.; Glazer, A.N.; Mathies, R.A. Capillary electrophoresis chips with integrated electrochemical detection. Anal. Chem. **1998**, *70* (4), 684.

8. Brett, C.M.A.; Oliveira Brett, A.M. *Electrochemistry*; Oxford University Press: Oxford, 1993.

Electrokinetic Chromatography Including MEKC

Hassan Y. Aboul-Enein
Vince Serignese
Pharmaceutical Analysis Laboratory, King Faisal Specialist Hospital and Research Centre, Riyadh, Saudi Arabia

INTRODUCTION

Separation science technology has provided the analyst with numerous methods for quantitative determinations, the more established being HPLC and GC. With huge advancements in computer technology, extremely sensitive methods have been made available to the user such as tandem mass spectrometry (MS–MS), LC/MS, and GC/MS. However, an electrophoretic technique which has been developed through joint efforts from a number of scientific disciplines is rapidly generating interest for its wide applicability and highly sensitive assays. The field of capillary electrophoresis (CE) borrows principles from conventional electrophoresis, LC, and GC. High-performance capillary electrophoresis (HPCE) refers to all techniques that have been developed on the subject. In this article, the electrokinetic chromatographic (EKC) analysis method will be discussed, with emphasis on MEKC. Before doing so, some fundamental principles of CE will be discussed.

ELECTROPHORETIC AND ELECTRO-OSMOTIC MIGRATION

Capillary zone electrophoresis (CZE)[1] is a basic mode of HPCE and serves as a good starting point for laying the background information on EKC and MEKC. Only ionic or charged compounds are separated, based on their differential eletrophoretic mobilities. Fig. 1 illustrates the setup for a CE system. In summary, electrolyte buffer solutions and electrodes are present at both ends of the open-tube fused-silica capillary. A positive high-voltage power supply is the source of current. A sample is injected hydrostatically or electrokinetically at the positive end of the capillary (capillary head) and, in the presence of an applied voltage potential, moves toward the negative electrode, where it is detected by an ultraviolet (UV) absorbance instrument. The signal produced is recorded as a chromatogram where sample components are identified as chromatographic peaks according to their retention times, and peak areas or heights are calculated for quantitative purposes.

During analyte migration, electrophoresis and electro-osmosis are taking place. The velocity (v_s, cm/s) and mobility (μ_s, cm^2/V s) of the solute are defined by the following equations:

$$v_s = v_{co} + v_{cp} \tag{1}$$

$$\mu_s = \mu_{co} + \mu_{cp} \tag{2}$$

where v_{co} and μ_{co} are electro-osmotic velocity and mobility, respectively, v_{cp} and μ_{cp} and are electrophoretic velocity and mobility, respectively. The relationship between velocity and mobility is given by

$$v = \mu E \tag{3}$$

where E (V/cm) is the electric field strength. Because E is constant for all solutes in a separation analysis, the solute velocities are differentiated by their mobilities.

Electrophoresis is an electrokinetic phenomenon whereby charged compounds in an electric field move through a continuous medium and separate by preferentially obtaining different electrophoretic mobilities according to their charges and sizes. Cations move toward the negative electrode (cathode) and anions move toward the positive electrode (anode).

Electro-osmosis is created by the electric double-layer effect.[2] A fused-silica capillary at neutral pH attains fixed negative charges at the inner wall surface as its silanol groups undergo ionization. A layer of hydrated cations will form adjacent to the inner wall to counter the fixed negative charges. An applied voltage potential will cause the positively charged layer to migrate toward the cathode with a flat velocity profile, simultaneously dragging the bulk solution inside the capillary and transporting charged compounds at the electro-osmotic flow velocity. It is assumed that v_{co} is faster than v_{cp} and determines the direction of solute migration. Hence, a cation will have its sum total [Eq. (1)] greater than the individual component velocities ($v_s > v_{co}$) and the anion will migrate slower than the electro-osmotic flow ($v_s < v_{co}$).

Electro-osmosis need not be present in open-tube CE. Coatings exist that can be applied to the capillary surface to eliminate the electrical double layer. However, electro-osmotic flow can be used to reduce

Encyclopedia of Chromatography DOI: 10.1081/E-ECHR-120039967

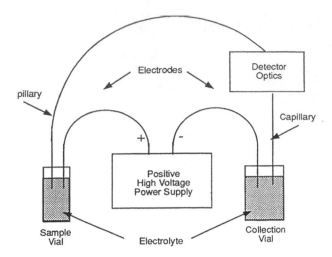

Fig. 1 Diagram of a CE system. [Reprinted from Waters Quanta 4000E CE System Operator's Manual, Waters Corp., 1993.]

solute retention times, which can be advantageous for certain analyses.

As previously mentioned, electrophoretic separations using open-tube capillaries are based on solute differential mobility, which is a function of charge and molecular size. A different approach is required for separating neutral or uncharged compounds. Because charge is absent, electrophoretic mobility is zero. Electro-osmotic flow would allow them to migrate, but their velocities would be equal. Separation would not be possible with the above method.

ELECTROKINETIC CHROMATOGRAPHY

Terabe[3] developed a method that separates neutral or uncharged compounds; he named it electrokinetic chromatography. The experimental design is that of CZE (Fig. 1). The difference lies in the separation principle. In LC, a solute freely distributes itself between two phases [i.e., a mobile phase usually made of a mixture of aqueous and organic solvents and a stationary phase (a solid material packed in a steel housing known as a chromatographic column)]. Under high pressure, the mobile phase is delivered by a liquid chromatographic pump and continuously solvates the stationary phase, thereby transporting nonvolatile compounds of interest that are introduced into the system via chromatographic injection. Separation is based on their phase-distribution profiles. EKC follows the above principle but uses electro-osmosis and electrophoresis to displace analytes and "chromatographic phases" in capillaries.

The electrolyte buffer solution is analogous to the mobile phase. A charged substance, referred to as the carrier, is dissolved in the electrolyte buffer. The neutral solute present in the separation medium will partition itself between the carrier (incorporated form) and the surrounding solution (free form). The carrier ("chromatographic phase") corresponds to the stationary phase in conventional chromatography with modifications, in that it is not a fixed support and exists homogeneously in solution. For this reason, the carrier is called the "pseudo-stationary phase." As discussed in the previous section, the charged carrier will transport the incorporated solute electrophoretically (here, is the carrier velocity) at a slower velocity than the free solute migrating with the electro-osmotic flow velocity in the opposite direction. The point to keep in mind is that the carrier migrates with a different velocity than the bulk solution. The variation of the ratio of the amount of incorporated solute to the amount of total solute between separands in a sample mixture will lead to sample component separation.

MICELLES IN ELECTROKINETIC CHROMATOGRAPHY

Different types of EKC have been developed. Cyclodextrins (CDEKC) have been used to form inclusion complexes with solutes to effect their separation. Other examples of EKC include microemulsion electrokinetic chromatography (MEEKC). The MEKC technique (for a detailed treatise, the reader is referred to Ref. [4]) utilizes the presence of micelles in the electrolyte buffer solution to influence the migration time of solutes. In this case, the separation carrier is the micelle.[5]

Surfactants produce micelles. Their amphophilic nature classifies them as detergents, surface-active agents that are composed of a hydrophilic group and a hydrophobic hydrocarbon chain. In addition to what is known as the critical micelle concentration (CMC), individual surfactant molecules (monomers) interact with each other to form aggregates or micelles, establishing a state of equilibrium between a constant monomer concentration and a rapidly increasing micelle concentration.

As shown in Fig. 2a, micelles are depicted as "roundlike" structures with their polar moieties exteriorly located in the vicinity of the aqueous medium and their hydrophobic tails oriented inward forming a cavity. The sizes and shapes of the structures formed when monomer units aggregate is affected by electrolyte concentration, pH, temperature, and hydrocarbon chain length. During aggregation, interactions occur not only between the aggregates but also among the monomer units within the aggregate structure. Monomer unit distribution (the number of surfactant molecules in an aggregate) is characteristic of the surfactant used.

E

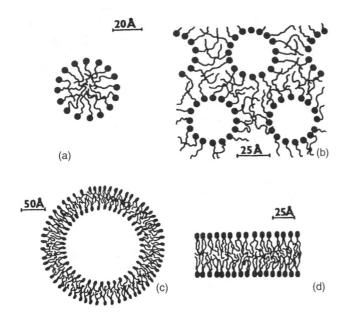

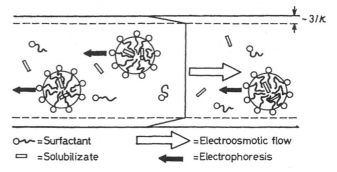

Fig. 3 Schematics of the separation principle of MEKC. (From Ref.[6].)

Fig. 2 Examples of aggregate structures of surfactants in solution: (a) micelle; (b) inverted micelles; (c) bilayer vesicle; (d) bilayer. (From Ref.[7].)

Fig. 2 displays several forms of aggregates that exist in solution and Table 1 gives examples of surfactants.

Basically, MEKC is an EKC application with the micelle as the designated carrier. A surfactant at a concentration above the CMC is added to the running buffer and initiates micelle formation. Because the separation principle has already been dealt with and the flow scheme in Fig. 3 is an illustrative summary notated for MEKC, it is clear that a neutral analyte residing in the hydrophobic interior of a micelle (depicted as a sphere in Fig. 3) will be transported with the micelle's velocity (v_{mc}). The free analyte will migrate with the electro-osmotic flow velocity (v_{co}).

The chromatographic aspect (solute partitioning) of the separation[6] can be explained in terms of a commonly used parameter in chromatography, the retention or capacity factor. We begin with the following equation:

$$k' = \frac{n_{\mathrm{mc}}}{n_{\mathrm{aq}}} \tag{4}$$

where n_{mc} and n_{aq} are the mole amounts of the analyte in the micellar and aqueous phases, respectively. The corresponding mole fractions are given by

$$\frac{n_{\mathrm{mc}}}{n_{\mathrm{mc}} + n_{\mathrm{aq}}} \quad \text{and} \quad \frac{n_{\mathrm{aq}}}{n_{\mathrm{mc}} + n_{\mathrm{aq}}}$$

where $n_{\mathrm{mc}} + n_{\mathrm{aq}}$ is the total amount of analyte present in the electrolyte buffer. The relationship in Eq. (4) and appropriate substitutions transform the above ratios into $k'/(1 + k')$ for the micelle analyte mole fraction and $1/(1 + k')$ for the aqueous analyte mole fraction. The total analyte velocity (v_s) takes the form

$$v_s = \frac{1}{1 + k'} v_{\mathrm{co}} + \frac{k'}{1 + k'} v_{\mathrm{mc}} \tag{5}$$

where $[1/(1 + k')]$ represents the velocity of the analyte mole fraction in the aqueous phase and

Table 1 Some common surface-active agents

Anionic	
Sodium stearate	$CH_3(CH_2)_{16}COO^-Na^+$
Sodium oleate	$CH_3(CH_2)_7CH{=}CH(CH_2)_7COO^-Na^+$
Sodium dodecyl sulfate	$CH_3(CH_2)_{11}SO_4^-Na^+$
Sodium dodecyl benzene sulfonate	$CH_3(CH_2)_{11}{\cdot}C_6H_4{\cdot}SO_3^-Na^+$
Cationic	
Laurylamine hydrochloride	$CH_3(CH_2)_{11}NH_3^+Cl^-$
Cetyltrimethylammonium bromide	$CH_3(CH_2)_{15}N(CH_3)_3^+Br^-$
Nonionic	
Polyethylene oxides	$CH_3(CH_2)_7{\cdot}C_6H_4{\cdot}(O{\cdot}CH_2{\cdot}CH_2)_8OH$

(From Ref.[9].)

(a)

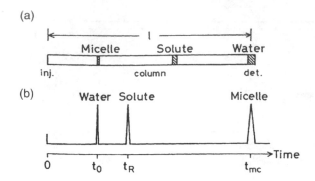

(b)

Fig. 4 (a) Representation of zone migration inside the capillary tube and (b) the corresponding chromatogram. (From Ref.[8].)

$[k'/(1 + k')]v_{mc}$ is the velocity of the analyte mole fraction in the micellar phase.

Because velocity is a function of length (the capillary length from the point of injection to the detector cell) over time ($v = l/t$), we can substitute and rearrange the terms in Eq. (5) to obtain a relationship between the migration time (t_R) of the analyte and k':

$$T_R = \frac{1 + k'}{1 + (t_0/t_{mc})k'}t_0 \qquad (6)$$

$$k' = \frac{t_R - t_0}{(1 - t_R/t_{mc})t_0}\left[\left(\frac{1 - t_R}{t_{mc}}\right)t_0\right]^{-1} \qquad (7)$$

Fig. 4a is a snapshot of the capillary tube following a sample injection at its positive end (inj.). The micelle, neutral solute, and aqueous solution (water) migrate toward the negative electrode (det.), establishing zones depicted by vertical bands, as they separate inside the capillary. The corresponding chromatogram in Fig. 4b shows the migration order where the t_R value for a neutral analyte is range bound between the migration times of the micelle (t_{mc}) and water (t_0). This limitation is reflected in the denominator of Eq. (6). When t_{mc} approaches infinity, the micelle is assumed to be stationary. The ratio t_0/t_{mc} will become zero and Eq. (6) turns into

$$t_R = 1 + k' \qquad (8)$$

In this case, a neutral solute completely solubilized within the micelle will have a t_R value approaching infinity as its k' value does the same. Solute elution is assumed not to occur. Similarly, the free neutral analyte is unretained and its k' is equal to zero. Therefore, the t_R value will be equal to t_0 [Eq. (8)]. This

explains why the neutral solute can migrate no slower than the micelle (t_{mc}) and no faster than the aqueous solution (t_0).

Although the above discussion focuses on neutral analyte separation, MEKC can be applied to ionic species which have their own electrophoretic mobilities and a broader migration time range.

CONCLUSIONS

The purpose of this article was to provide the reader with a basic understanding of CE and to describe how a technique such as MEKC uses basic principles of chromatography to perform separations which are not possible electrophoretically. As the applications for electrokinetic chromatography rapidly expand, the future direction will develop on two fronts:

1. The development of novel separation carriers that will broaden the species range of separable analytes. EKC is suitable for separating small molecules, considering the size of the cavities of the established carriers.
2. The scope for further partition mechanisms, as new separation carriers are discovered, is promising. The separation principle is basic chromatography and with research efforts introducing new carriers in the pipeline, modified versions of the separation mechanism are possible.

Its rapid analysis time, low sample and solvent volume requirements, high resolution, and selectivity will continue to attract researchers who are involved in separation analysis.

REFERENCES

1. Radola, B.J., Ed.; *Capillary Zone Electrophoresis*; VCH: Weinheim, 1993.
2. Karger, B.L.; Foret, F. Capillary electrophoresis: introduction and assessment. In *Capillary Electrophoresis Technology*; Guzman, N.A., Ed.; Marcel Dekker, Inc.: NewYork, 1993; 3–64.
3. Terabe, S. Electrokinetic chromatography: an interface between electrophoresis and chromatography. Trends Anal. Chem. **1989**, *8*, 129.
4. Muijselaar, P. Micellar electrokinetic chromatography: fundamentals and applications. In *Ph.D.*

thesis; Eindhoven University of Technology: Eindhoven, The Netherlands, 1996.

5. Foret, F.; Kivánková, L.; Boek, P. Principles of capillary electrophoretic techniques: micellar electrokinetic chromatography. In *Capillary Zone Electrophoresis*; Radola, B.J., Ed.; VCH: Weinheim, 1993; 67–74.

6. Terabe, S. Micellar electrokinetic chromatography. In *Capillary Electrophoresis Technology*; Guzman, N.A., Ed.; Marcel Dekker, Inc.: New York, 1993; 65–87.

7. Israelchvilli, J.N. Physics of amphiphiles: micelles, vesicles and microemulsions, Proceedings of the International School of Physics "Enrico Fermi," CourseXC, North-Hollsnd, Amsterdam, 1985; Degiorgio, V.Corti, M., Ed.; North-Holland, 24–37.

8. Terabe, S.; Otsuka, K.; Ando, T. Electrokinetic chromatography with micellar solution and open-tubular capillary. Anal. Chem. **1985**, *57*, 834

9. Shaw, D.J. *Introduction to Colloid and Surface Chemistry*; Butterworths: London, 1966; 57–72.

Electron-Capture Detector

Raymond P. W. Scott
Scientific Detectors Ltd., Banbury, Oxfordshire, U.K.

INTRODUCTION

The electron-capture detector (ECD) is probably the most sensitive GC detector presently available. However, like most high-sensitivity detectors, it is also very specific and will only sense those substances that are electron capturing (e.g., *halogenated* substances, particularly fluorinated materials).

DISCUSSION

The ECD detector was invented by Lovelock[1] and functions on an entirely different principle from that of the argon detector. A low-energy β-ray source is used in the sensor to produce electrons and ions. The first source to be used was tritium absorbed onto a silver foil, but, due to its relative instability at high temperatures, this was replaced by the far more thermally stable ^{63}Ni source. The detector can be made to function in two ways: either a constant potential is applied across the sensor electrodes (the DC mode) or a pulsed potential is used (the pulsed mode).

A diagram of the ECD is shown in Fig. 1. In the DC mode, a constant electrode potential (a few volts) is employed that is just sufficient to collect all the electrons that are produced and provide a small standing current. If an electron-capturing molecule (e.g., a molecule containing a halogen atom which has only seven electrons in its outer shell) enters the sensor, the electrons are captured by the molecules and the molecules become charged. The mobility of the captured electrons are much reduced compared with the free electrons and, furthermore, are more likely to be neutralized by collision with any positive ions that are also generated. As a consequence, the electrode current falls dramatically. In the pulsed mode of operation, which is usually the preferred mode, a mixture of methane in argon is usually employed as the carrier gas. Pure argon cannot be used very effectively as the carrier gas, as the diffusion rate of electrons in argon is 10 times less than that in a 10% methane–90% argon mixture. The period of the pulsed potential is adjusted such that relatively few of the slow negatively charged molecules reach the anode, but the faster moving electrons are all collected. During the "off-period," the electrons reestablish equilibrium with the gas. In general use, the pulse width is set at about 1 μs and the frequency of the pulses at about 1 kHz. This allows about 1 ms for the sensor to reestablish equilibrium in the cell before the next electron collection occurs. The peak potential of each pulse is usually about 30 V but will depend on the geometry of the sensor and the strength of the radioactive source. The average current resulting from the electrons collected at each pulse is about 1×10^{-8} A and usually has an associated noise level of about 5×10^{-12} A. Both the standing current and the noise will also vary with the strength of the radioactive source that is used.

The sensor consists of a small chamber, 1 or 2 mL in volume, with metal ends separated by a suitable insulator. The metal ends act both as electrodes and as fluid conduits for the carrier gas to enter and leave the cell. The cell contains the radioactive source, electrically connected to the conduit through which the carrier gas enters and to the negative side of the power supply. A gauze "diffuser" is connected to the exit of the cell and to the positive side of the power supply. In the pulsed mode, the sensor operates with oxygen-free nitrogen or argon–methane mixtures. The active source is ^{63}Ni, which is stable up to 450°C. The sensor is thermostatted in a separate oven which can be operated at temperatures ranging from 100°C to 350°C. The column is connected to the sensor at the base and makeup gas can be introduced into the base of the detector. If open tubular columns are employed, the columns are operated with hydrogen or helium as the carrier gas. The ECD is extremely sensitive (i.e., minimum detectable concentration $\sim 1 \times 10^{-13}$ g/mL) and is widely used in trace analysis of halogenated compounds—in particular, pesticides.

In the DC mode, the linear dynamic range is relatively small, perhaps two orders of magnitude, with the response index lying between 0.97 and 1.03. The pulsed mode has a much wider linear dynamic range and values up to five orders of magnitude have been reported. The linear dynamic range will also depend on the strength of the radioactive source and the detector geometry. The values reported will also rest on how the linearity is measured and defined. If a response index lying between 0.98 and 1.02 is assumed, then a linear dynamic range of at least three orders of magnitude should be obtainable from most pulsed-mode ECD.

Encyclopedia of Chromatography DOI: 10.1081/E-ECHR-120039968

ECD for Use with Constant Electrode Potential

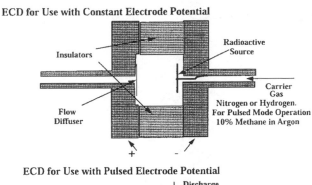

ECD for Use with Pulsed Electrode Potential

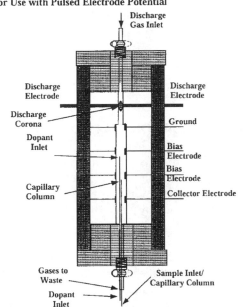

Fig. 1 The two types of electron capture detector. (Courtesy of Valco Instruments Company, Inc.)

PULSED-DISCHARGE ELECTRON-CAPTURE DETECTOR

The pulsed-discharge ECD is a variant of the pulsed ECD detector, a diagram of which is shown in the lower part of Fig. 1. The detector functions in exactly the same way as that of the traditional ECD but differs in the method of electron production. The sensor consists of two sections: the upper section, where the discharge takes place, has a small diameter and the lower section where the column eluent is sensed and the electron capturing occurs, has a wider diameter. The potential across the discharge electrodes is pulsed at about 3 kHz with a discharge pulse width of about 45 μs for optimum performance. The discharge produces electrons and high-energy photons (which can also produce electrons) and some metastable helium atoms. The helium doped with propane enters just below the second electrode, metastable atoms are removed, and electrons are generated both by the decay of the metastable atoms and by the photons. The electrons are collected by appropriate potentials applied to each electrode in the section between the third and fourth electrode and, finally, collected at the fourth electrode. The collector electrode potential (the potential between the third and fourth electrodes) is pulsed at about 3 kHz with a pulse width of about 23 μs and a pulse height of 30 V.

The device functions in the same way as the conventional ECD with a radioactive source. The column eluent enters just below the third electrode, any electron-capturing substance present removes some of the free electrons, and the current collected by the fourth electrode falls. The sensitivity claimed for the detector is 0.2–1.0 ng, but this is not very informative as its significance depends on the characteristics of the column used and on the k' of the solute peak on which the measurements were made. The sensitivity should be given as that solute *concentration* that produces a signal equivalent to twice the noise. Such data allow a rational comparison between detectors. The sensitivity or minimum detectable concentration of this detector is probably similar to the conventional pulsed an estimate from the published data. The modified form of the ECD, devoid of a radioactive source, is obviously an attractive alternative to the conventional device and appears to have similar, if not better, performance characteristics.

The high sensitivity of the ECD makes it very popular for use in forensic and environmental chemistry. It is very simple to use and is one of the less expensive, high-sensitivity selective detectors available.

REFERENCE

1. Lovelock, J.E.; Lipsky, S.R. J. Am. Chem. Soc. **1960**, *82*, 431.

SUGGESTED FURTHER READING

Scott, R.P.W. *Introduction to Gas Chromatography*; Marcel Dekker, Inc.: New York, 1998.
Scott, R.P.W. *Chromatographic Detectors*; Marcel Dekker, Inc.: New York, 1996.

Electrophoresis in Microfabricated Devices

Christa L. Colyer
Department of Chemistry, Wake Forest University, Winston-Salem, North Carolina, U.S.A.

Pertti J. Viskari
James P. Landers
Department of Chemistry, University of Virginia, Charlottesville, Virginia, U.S.A.

INTRODUCTION

It is no wonder that capillary electrophoresis (CE) has evolved into one of the premier separation techniques in use today, due to its extremely high efficiencies, fast analysis times, reduced sample and reagent consumption, and vast array of operating modes. The transposition of CE methods from conventional capillaries to channels on planar chip substrates is a more recent phenomenon and has been driven by several factors, including, but not limited to, the need for ever-more sensitive and selective assays, the need to manipulate increasingly smaller samples, and the desire to process many samples in parallel. Perhaps more important to the rapid development of this important field, however, is its amenability to the assimilation of multiple components of an assay—beyond simple separation of analytes—into a single, fully integrated device. The promise of the "lab-on-a-chip," although seemingly ambitious in concept, is clearly attainable, and microchip capillary electrophoresis (μ-chip CE) has quickly established itself as one of the most fundamental constituents of such systems.

The first published demonstration of CE on a chip appeared almost 15 years ago and although separation efficiencies and analysis times in this pioneering work did not represent significant improvements over those achievable by way of conventional CE, this work demonstrated the feasibility of miniaturizing a chemical analysis system involving electrokinetic phenomena for sample injection, separation, and solvent pumping. Within two years of the first publication in this field, analysis times on the order of seconds and even milliseconds had been demonstrated with similar μ-chip systems, and efficiencies in excess of 100,000 theoretical plates were routinely obtained. Subsequently, the integration of other functionalities, such as sample manipulations and chemical reactions, alongside the electrophoretic separation, has vaulted CE-on-a-chip to new heights. In order to provide a rudimentary introduction to this important new technique, this entry presents the fundamentals of microchip fabrication methods and materials, sample introduction and

detection methods, as well as selected "case studies" or notable applications of electrophoresis in microfabricated devices for various classes of analytes.

CHIP FABRICATION TECHNOLOGY

The evolution of microchip-based electrophoretic systems has been the direct result of the tremendous advances in semiconductor microfabrication technologies that have taken place over the past two decades. Although semiconducting substrates are not ideally suited to electrophoretic applications due to the high voltages applied for separation and fluid manipulation, many of the established semiconductor microfabrication techniques can be modified for the insulating glass or quartz substrates most commonly encountered in μ-chip electrophoresis. Here, the name μ-chip refers to the microscale channel dimensions as opposed to the actual substrate dimensions, which are commonly on the order of 1 mm thick and anywhere from 3 to 10 cm in length and width (or diameter for circular substrates). In many cases, standard photolithographic and wet etching techniques are employed in the manufacture of electrophoretic chips. To begin, the clean glass or quartz substrate is uniformly coated with sequential thin layers of chromium/gold and positive photoresist by sputter-coating and spin-coating methods, respectively. The design for the electrophoretic channel structure is then transferred to the substrate by exposure of the photoresist to UV light through a photomask of the channel pattern. After photoresist development, a series of wet etches are employed, first to remove the metal etch mask, and second to etch the channels into the substrate. Channels created in this fashion are trapezoidal in profile due to the isotropic etching of amorphous materials. Typical channel dimensions range from 5 to 40 μm deep, and 20–100 μm wide (at half the channel depth). Residual photoresist and metal film are stripped from the etched substrate prior to thermal bonding of a cover plate, thereby forming closed channels suitable for electrophoresis. Other common bonding methods include

Encyclopedia of Chromatography DOI: 10.1081/E-ECHR-120039901

UV glue where the glue is placed in extra bonding channels etched on the chip and cured with UV light, as well as adhesives, which are likewise placed between the plates that are then pressed together for bonding. Access to the channels is most commonly gained through holes drilled in the cover plate prior to bonding.

Electrophoretic chips so created are extremely rugged due to the monolithic nature of their structure, and they can withstand applied voltages in the same range (up to 30 kV) as those commonly encountered in conventional CE. In addition, they offer greater heat dissipation than conventional electrophoretic capillaries, thereby allowing for operation under conditions of higher power. Although quartz substrates have superior optical properties relative to their glass counterparts, both present an optically flat surface for detection schemes, which is a definite advantage over the curvature inherent to conventional capillary walls. Also, the void volumes associated with channel intersections on chip are virtually nonexistent. Despite their many advantages, these chips are still relatively time consuming and expensive to fabricate.

As such, alternative methods for CE chip fabrication are being developed, such as the creation of channels in polymeric materials by imprinting, molding, casting, and laser photoablation techniques.[1] Several different polymers or plastics have been utilized with these techniques, with the most frequently encountered materials being poly(methyl methacrylate) (PMMA), poly(dimethyl siloxane) (PDMS), polycarbonate (PC), and various polyesters.[2] These and other suitable polymers have to satisfy several important characteristics in order to be applicable for CE, where the analysis is commonly driven by electro-osmotic flow (EOF). As with any CE system, the chip materials need to be electric insulators so that the voltage drop can take place through the fluid in the microchannel. Also, since Joule heating is virtually unavoidable in electrophoresis in microscale channels, it is important to be able to remove the excess heat generated. Thus, the heat dissipation capability of the chip material is vital since excessive heat can remarkably reduce the efficiency of the separation and also cause deformation of the channel structure on the chip due to the low melting points of plastics. Finally, since the EOF on the chip is dependent upon the charges on the microchannel wall, the characteristics of the wall should be considered. Polymeric materials display a wide range of charges and charge densities; therefore, the EOF varies greatly between chips made of differing substrates. However, the wall chemistry can be controlled with the choice of polymer material, its fabrication method, and how the surface is chemically treated. A variety of methods have been exploited to modify the polymer surface. They include covalent binding, dynamic coating, and plasma oxidation.

Chip design and manufacture is not only limited to academic laboratories but is also available from several commercial sources. Agilent, Caliper, Aclara, Micralyne, and Nanogen are a few examples of companies that have established long records of microchip production and related CE instrument manufacturing.

The concept of a "lab-on-a-chip" or a "micro-Total Analysis System" (μ-TAS) has pushed the development of even more advanced design features on chips. Sophisticated channel features such as weirs, pressure valves, and posts (columns) can be found with increasing frequency on chips. Weirs can be used to outline a specific area in the channel where beads or sol–gels with immobilized enzymes can be packed. Liquid can flow through these physical obstacles but the beads are held in place due to their larger size. Elastomeric materials that function as diaphragm pumps or valves can be manufactured from PDMS sandwiched between the channel and cover plate of the chip. This way fluid movement in the chip can be manipulated by pressure as well as by electrokinetic means. A particularly intriguing channel feature is a collection of cylindrical or cubic posts micromachined on the chip surface. These greatly increase the surface area inside the channel and find applications comparable to capillary electrochromatography. They can also be coated with antibodies or reagents to tune them for specific applications. These posts, as well as other three-dimensional (3-D) structures, can be fabricated with an incredible accuracy of one-tenth of a micrometer,[3] as illustrated in Fig. 1.

SAMPLE INTRODUCTION ON CHIPS

Clever chip design permits the integration of the sample injector directly on the chip, thereby combining injection and separation functions by default on a single substrate. Most commonly and simply, the injector is fashioned as a simple cross or "double-t" arrangement of etched channels. One branch of this cross serves as the sample channel, and the other serves as the separation channel. Fluid flow is manipulated through this cross, just as with all other fluid manipulations on chip, by control of electrokinetic phenomena: electrophoresis and electro-osmosis. By first applying the appropriate voltage between the sample and sample waste reservoirs, sample solution crosses the separation channel, filling the channel intersection. Consequently, injection volumes are defined by the injector geometry. Typical sample plug volumes and lengths are on the order of about 10–100 pL and 50–200 μm, respectively. Provided the injection field strength and time are sufficient to ensure that the least

Fig. 1 Posts fabricated on PMMA substrate. The diameter of each post is 30 um and it is spaced 3 um apart. The scale bar in upper left corner is 1 mm in length. (From Ref.[3].)

mobile sample component has moved through the channel intersection, this method results in an unbiased injection, with all sample components represented in the intersection volume according to their original proportions. Additionally, matching the ionic strength and pH of the sample and buffer further ensures optimum reproducibility.[4] Having thus formed a sample plug, the voltage is switched so as to generate EOF along the separation channel, thus sweeping the sample plug out of the double-t injector and initiating separation along the second branch of the injector. The separation channel typically ranges from 1 to 10 cm in length. Further control of sample plug size and shape, and prevention of sample leakage, can be effected by carefully controlling the voltages applied to all four arms of the cross simultaneously during injection and separation phases—a procedure known as shaping or "pinching".[4,5]

As an alternative to such shaping or pinching of the injection plug by voltage control, optical gating methods have been adapted to microchip CE platforms. Optical gating permits injection and separation to occur without interruption of the applied electric field, thereby facilitating rapid serial sample introductions for ultrafast analyses. In this sample introduction method, a laser beam is split into a high power gating beam, which serves to photobleach fluorescently labeled analytes that are continuously moving through the separation channel, and a low power probe beam, which is used to detect a small fluorescent sample plug created by shuttering the probe beam for a short period of time. However, optically gated sample introduction does not remedy the problem of sample

carryover when changing samples or address the need for ready access from the macrosample world to the microchip world of separations.

Numerous recent advances[6] have been made in an effort to address these limitations and circumvent the largely manual nature of microchip sample introduction methods. Variations on flow injection or continuous flow schemes, for example, utilize differences in flow resistance between a wide sample introduction channel and a narrow separation channel to prevent pressure-driven sample flow from affecting the separation. This strategy essentially amounts to a split-flow interface fabricated directly on the chip, with numerous variations possible, such as a flow-through sampling reservoir with guided overflow or falling drop design, whereby samples can be changed by pumping from an access hole on the bottom side of the chip. Various microchip sample introduction schemes are illustrated in Fig. 2.

Furthermore, modifications made to the outlet end of the separation channel, such as its connection to a length of fused-silica capillary to facilitate chip interfacing with off-chip detection systems for example, can equally well be adopted for chip inlets and sample introduction. The use of conventional capillary to interface between macroscale sample reservoirs and microchip platforms has been successfully demonstrated,[7] and has even been commercialized in the form of "sipper chips," which are capable of sampling from up to 12 microtiter wells simultaneously via short lengths of conventional capillary interfaced to the bottom of the chip. Simpler still is the cutting away of chip substrate alongside the inlet end of the channel

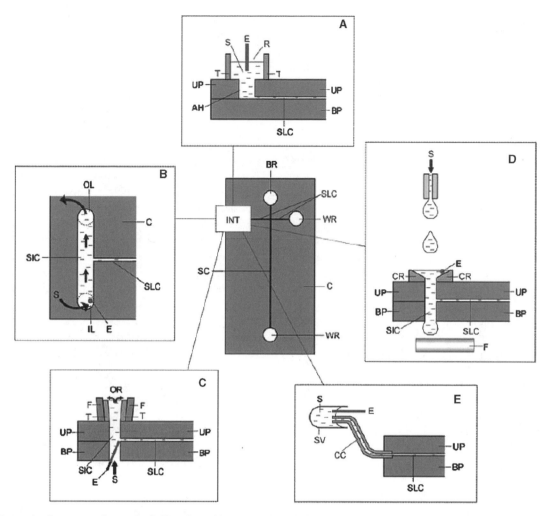

Fig. 2 Schematic diagrams of a typical CE microchip employing different sample introduction systems. Abbreviations are C, chip; BR, buffer reservoir; INT, interfacing; SLC, sample loading channel; WR, waste reservoir; SC, separation channel; UP, upper plate of the chip; BP, base plate; S, sample; AH, access hole; E, electrode; T, tube; R, reservoir; SIC, sample introduction channel; IL, inlet; OL, outlet; F, filter paper; OR, overflow reservoir; CR, conical reservoir; CC, connecting capillary sample probe, SV, sample vial. (A) Simple on-chip sample reservoir for electrokinetic cross or double-t type injector; (B) large flow channel for on-chip continuous sample introduction; (C) flow-through sampling reservoir with guided overflow design; (D) flow-through sampling reservoir with falling-drop modification; (E) fused-silica capillary "sampling probe" from sample vial to CE chip. (From Ref.[7] with permission. Copyright 2004, Springer.)

to produce a sharp inlet chip [analogous to sharp outlet chips used for electrospray ionization-mass spectrometry (ESI-MS) interfacing]. The sharp inlet chip can be inserted directly into regular sample reservoirs for electrokinetic injection with no sample carryover.[8]

DETECTION ON CHIPS

It is not surprising that the requirements for detection on chips are very stringent, especially given the extremely small sample sizes discussed earlier. This need for sensitivity, along with the optically flat chip surface, makes laser-induced fluorescence (LIF) detection a natural choice, and consequently, LIF detection on

chips is the most widespread of all chip detection types. However, since relatively few analytes are natively fluorescent, LIF detection often necessitates the development of selective and sensitive labeling strategies for each assay. One of the many benefits of μ-chip CE platforms, however, is the ability to integrate multiple functionalities onto a single platform; hence, fluorescent derivatization of the analyte to facilitate its detection by on-chip LIF can be performed directly on the chip without additional sample preparation. Specialized on-chip LIF detection schemes have rendered detection limits down to the hundreds-of-molecules' range or lower. For example, a confocal, epiluminescent system based on a 635 nm diode laser and cyanine-5 dye labeled antibody was able to detect 900 of 4560 injected molecules,[9] while a fluorescence burst counting technique with sample focusing achieved single-molecule counting of a DNA sample.[10] Other optical detection schemes aside from LIF, including UV/Vis absorbance, refractive index, and Raman spectroscopy, have been demonstrated but are not in widespread use with microchip substrates.

Electrochemical (EC) detection schemes also complement on-chip electrophoresis. This is because of the ability to incorporate detection electrodes directly onto the chip by standard microfabrication procedures. Decoupling the electric field used to drive the electrophoretic separation from the EC detection signal poses a challenge on chip substrates just as it does for conventional fused-silica CE protocols. Advances in decoupler fabrication, such as the recent development of a microfabricated palladium decoupler[11] shown to improve separation efficiency and sensitivity for dopamine and epinephrine samples, are increasing the likelihood of EC detectors finding more widespread use in chip-CE systems. Precise placement of the working electrode for detection relative to the end of the separation channel, however, can still have a significant impact on signal-to-noise ratios. The development of a sheath-flow EC detector design,[12] which relies on gravity-driven flow of a sheath buffer to carry the electroactive analytes to the detection electrodes, is able to reduce the dependence of the EC signal on electrode position (as shown in Fig. 3), further improving the potential of EC detection schemes for chip-based electrophoretic analyses.

Significant interest has been growing in the area of ESI-MS detection for electrophoretic chips because

of the potential for both analyte identification and quantitation by this method. Early work in this area utilized the chip simply as a sample delivery device for MS analysis,[13] but in 1999 three groups reported the first on-chip electrophoretic separations prior to ESI-MS detection.[14] Although the demonstrated ability to generate an electrospray directly from the edge of a CE chip is appealing in its simplicity, greater success has typically been met using carefully coupled short transfer capillaries to facilitate spray from the chip outlet to the MS inlet. Explorations of low-cost batch fabrication methods, such as the micromilling of ESI nozzles from polymer foil[15] or the addition of a thin hydrophobic membrane at the microchannel exit to control Taylor cone formation,[16] indicate widespread interest in the development of a routine and robust interfacing strategy. Furthermore, a CE microchip device capable of peptide and protein digest sample transfer from a standard microwell plate as well as electrophoretic separation and ESI interfacing[17] serves as a good example of the potential for automated, high-throughput CE/ESI-MS assays.

APPLICATIONS

Electrophoresis of DNA on Chips

Since the first DNA separations on a single-lane chip in the mid-1990s, improved fabrication and detection methods have allowed for a remarkable increase in the throughput and performance capabilities of these devices. The lane densities have soared from 48 to 96

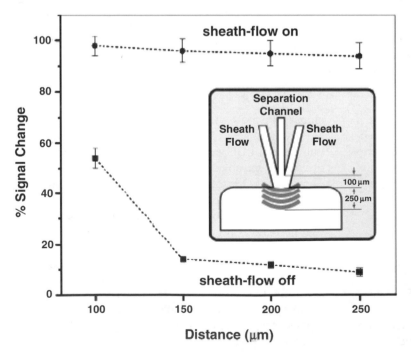

Fig. 3 Results of catechol separations presented as relative (%) signal change as a function of electrode distance calculated from triplicate measurements obtained in the (■) absence and (●) presence of sheath-flow support. Inset: Schematic representation of investigated EC detector configurations with working electrodes placed at a distance of 100, 150, 200, and 250 μm from the separation channel. (From Ref.[12] with permission. Copyright 2004, American Chemical Society.)

and currently to 384 separation lanes on a single chip. Mathies' group has pioneered the development of many high-throughput devices and recently introduced their 384-lane capillary array electrophoresis (CAE) bioanalyzer.[18] Their system was based on a round 200-mm diameter borosilicate glass wafer that had 384 separation lanes etched radially around the center of the chip, which also contained the central anode, common for all the lanes (see Fig. 4). In addition, each lane was served by two other electrodes located close to the edge of the chip, a cathode and sample injector, thus eliminating the usual fourth electrode at the sample waste arm. The separation length of each channel was 8.0 cm and they were 30 µm deep and 60 µm wide. They were coated to prevent DNA adsorption and EOF. As a sieving matrix, polydimethylacrylamide (PDMA) gel was pressure pumped into the channels through the central anode reservoir. Fluorescence detection was accomplished via a rotary confocal fluorescence scanner that consisted of an Ar-ion laser and a photomultiplier tube (PMT) as light source and detector, respectively. The utility of this system was demonstrated for hereditary hemochromatosis (HHC), not only one of the most common autosomal recessive diseases in the United States, but also one whose diagnosis remains troublesome. To validate the system for high-throughput analysis it was utilized in simultaneous genotyping of the DNA from 384 individuals for an HHC-related mutation in the human *HFE* gene. The genotyping success rate for the 384-lane chip was found to be 98.7% with a relative mobility deviation of <2.2%. Electrophoresis of the samples was completed in 325 sec. It was concluded that the microchip device outperformed the best commercially available 96-capillary instruments by a factor of 20 and that it could be ideal for gene mapping, pharmacogenomic screening, forensics, and proteomics.

In the last decade or so, forensic sciences have been transformed by vast improvements in DNA analysis techniques. Through development of the short tandem repeat (STR) method, DNA testing was revolutionized not only to serve the forensic field, but also to extend its usefulness to paternity testing, military purposes, historical investigations, and other biological

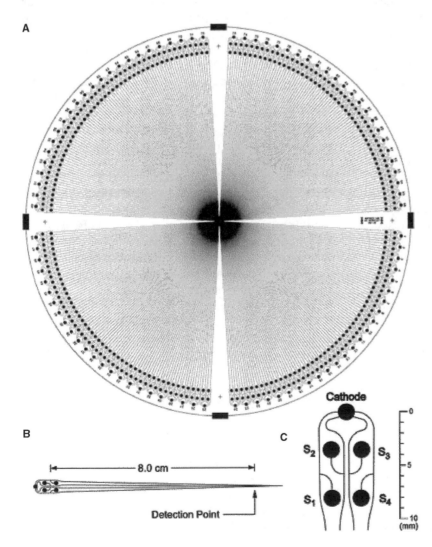

Fig. 4 (A) Layout of the 384-lane microcapillary array electrophoresis device on a 200-mm diameter wafer. Lanes are ~60 µm wide and 30 µm deep, and the effective separation length is 8.0 cm. (B) Expanded view of a single quartet of channels with their injectors. (C) All channels in a quartet share a common cathode reservoir located closest to the edge of the wafer. S_1–S_4 indicate individual sample reservoirs. (From Ref.[18] with permission. Copyright 2002, American Chemical Society.)

determinations. Short tandem repeats are short stretches of repetitive DNA sequences that are distributed throughout the genome of every individual. Each genetic locus consists of 7–20 repeats of a specific 2- to 7-base sequence. Through electrophoretic separation of these repeats, the DNA can be sequenced and genotyped. Because of the effectiveness of STR analysis, the FBI created a national criminal database—the Combined DNA Index System (CODIS)—in 1998. Due to the rigorous demands of the database, analysis techniques need to possess high reproducibility and throughput capability, and they should be fast, automated, and easily adapted by instrument operators. To address these requirements, Ehrlich's group developed "GeneTrack," a multiplexed DNA analysis instrument.[19] It consisted of two units, the support and the top unit, with the former containing the necessary power supplies and data acquisition system, and the latter containing the laser, optics, and the electrode board. The heart of the instrument, a microfabricated device, consisted of two glass plates thermally bonded together. Sixteen separation channels were wet etched on one of the plates. Each lane was 60 μm deep, 130 μm wide, and 20 cm long. Sample injection was achieved using a classic double-t injector. The inner channel surfaces were silanized by following a modified Hjerten method and finally filled with linear polyacrylamide (LPA) as the sieving media. Fluorescence detection was arranged with a multiline argon ion laser and four PMTs that allowed for four-color detection when coupled with a system of dichroic mirrors. The developed system was validated by analyzing the DNA template 9947a with the PowerPlex 16 PCR kit from Promega. All 16 lanes on the microchip were used for the analysis, which was repeated six times over 4 days. It was reported that 92 out of the 96 total lanes performed successfully. Also, after statistical testing, the instrument was shown to have 0.4–0.9 bp absolute signal accuracy for the full 13-locus STR DNA determination as required by the CODIS. For the specific TH01 DNA locus it was shown that the required single base pair resolution was reliably achieved. The authors concluded by stating that "The system is optimized for the forensics "casework" application and is expected to exceed current capillary systems in sample economy, critical temperature/environmental stability, and software features." However, the size of the chip device (>20 cm long) and the consequently long analysis times (up to 40 min) differ from the general trend toward smaller devices and faster analysis times.

To determine genomic alterations, DNA analysis by Southern blotting typically requires a large number of fragments for hybridization. It is also relatively time consuming, expensive, and tedious work. One genetic disease that is currently diagnosed by Southern blotting is Duchenne muscular dystrophy (DMD), which is caused by mutations in the X chromosome. The detection of duplicated or deleted exons in the dystropin gene helps to identify the affected male and female carriers. These extra or missing exons are typically found in "hot spot" areas of the gene and should, therefore, allow the diagnosis of DMD by the investigation of a limited number of PCR amplified DNA fragments. Due to the development of primers for a total of 18 known duplication/deletion loci, more than 97% of the Southern blot diagnosable cases can be detected rapidly and accurately by PCR. However, PCR fragment analysis most commonly uses slab gels that complicate quantitative detection. To improve the quantitative aspect and the speed of analysis, Ferrance, Snow, and Landers[20] were the first to evaluate the applicability of a commercial microchip electrophoresis instrument for diagnosis of DMD. They analyzed a total of 50 clinical samples from patients and compared the results with those from the conventional Southern blot method. The genomic DNA isolated from the blood of the patients was PCR amplified and analyzed in an Agilent 2100 BioAnalyzer using the commercially available DNA 500 LabChip kit. From the samples of 15 non-DMD patients, peak areas for each of the microchip electrophoresis detected exons were used to establish a range of reference values. The results from the DMD positive patients (diagnosed by Southern blot) were compared to the reference values to determine whether or not they were positive. Out of 35 patient samples that were known to be positive for DMD, all 35 were found to have exon concentrations that indicated the presence of the disease, thus confirming the diagnoses and highlighting the value of microchip analysis for DMD diagnosis. Also, the results were encouraging for determining the type of mutations present in the patients. By carefully optimizing the PCR conditions, it was speculated that the mutations could be correctly identified. It was shown that microchip electrophoresis using commercial instrumentation with standardized protocols and reagents is easily transferable to clinical laboratories, which would benefit from the high speed, lower labor and reagent costs, and the elimination of radioactive chemicals inherent to this new method.

Electrophoresis of Proteins on Chips

In the postgenomic era, proteomics pose a new and substantial challenge for analytical and biological chemists alike. According to the revised edition of the Merriam-Webster's *Medical Desk Dictionary* (2002), proteomics is *"a branch of biotechnology concerned with applying the techniques of molecular biology, biochemistry, and genetics to analyzing the structure, function, and interactions of the proteins produced by the genes of a particular cell, tissue, or organism, with organizing the information in databases, and with applications of the data (as in medicine or biology).* The most fundamental task in this study remains the separation and identification of proteins from a biological sample matrix. Traditionally, this task has been achieved by two-dimensional gel electrophoresis (2-D-GE) followed by spot excision, proteolytic digestion, and detection of the resulting fragments by mass spectrometry. Recently, Sluszny and Yeung[21] have miniaturized this process to achieve a much faster and highly sensitive proteomics tool. By successfully coupling isoelectric focusing (IEF) in one dimension with a 15 × 9 mm sodium dodecyl sulfate-poly(acrylamide) (SDS-PAGE) slab gel in the second dimension, Sluszny and Young were able to detect 200 protein spots from an *Escherichia coli* sample in about 2.5 hr. A detection limit of <0.25 μg of total protein was achieved by using a CCD camera to image native protein fluorescence induced by a simple Hg (Xe) lamp. This detection scheme provided superior sensitivity relative to traditional slab gel detection methods such as coomassie brilliant blue and silver staining. Furthermore, the native fluorescence detection scheme provided a dynamic range of more than two orders of magnitude, and did not require the time-consuming staining and destaining steps associated with the more traditional 2-D-GE detection methods.

Another variation of the traditional 2-D-GE system was achieved on an electrophoretic chip, which permitted orthogonal micellar electrokinetic chromatography (MEKC) and CE separations in microfluidic channels.[22] As shown in Fig. 5, this method was able to achieve a peak capacity of 4200 in <15 min for the 2-D separation of a bovine serum albumin (BSA) tryptic digest. This same system was able to identify peptides by standard addition and to distinguish between tryptic digests of human and bovine hemoglobin. Separation was achieved in the first dimension using an applied field strength of 200 V/cm with a 10 mM boric acid buffer (pH 8.4) with 20 mM SDS and 10% (V/V) 2-propanol, and in the second dimension using an applied field strength of 2400 V/cm with a 100 mM boric acid buffer (pH 8.4). The small quantity of SDS injected with effluent samples from the first to second dimension was rapidly diluted below the critical micelle concentration during the subsequent CE separation and so did not unduly influence the detection of peptides (prelabeled with 5-TAMRA dye) by LIF detection in the second dimension. Essential to this 2-D system was fast sampling of the effluent from the first 19.6 cm long serpentine MEKC channel, achieved by way of overlapping analysis times, to provide maximum throughput for the second, 1.3 cm long CE channel.

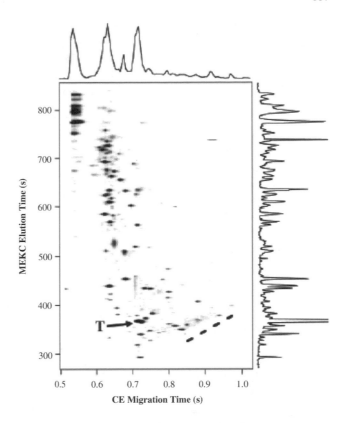

Fig. 5 Two-dimensional separation of a BSA tryptic digest. The projections of the 2-D data onto each axis are shown. Spots marked with a "T" represent unreacted 5-TAMRA dye. The dotted line illustrates correlation in the separation mechanisms. (From Ref.[22] with permission. Copyright 2003, American Chemical Society.)

An alternative to this inherently serial 2-D microchip design was presented by the simultaneous 2-D microchip design of Li et al.,[23] which consisted of an IEF microchannel orthogonal to an array of SDS-PEO gel microchannels on a plastic substrate as small as 2 cm × 3 cm. This design permitted the focused and concentrated protein fractions from the IEF channel to be simultaneously injected and separated (on the basis of protein size) in the multiple, parallel SDS-gel channels in the orthogonal array, yielding a comprehensive protein separation in <10 min with a peak capacity of about 1700. Furthermore, it was proposed that the peak capacity could be increased by simply raising the density of microchannels in the second-dimension array, which would have the effect of increasing the number of fractions from the IEF dimension that could be simultaneously analyzed in the size-based SDS-PEO separation dimension.

Beyond their rapidly developing role in 2-D protein separations, microchip electrophoresis platforms also offer considerable advantages in the area of protein separations as they pertain to immunoassay methods. Initial microchip-based immunoassay efforts involved offline mixing of immunoassay reagents, thus relegating the function of the microchip to the electrophoretic separation of antibody-bound and free antigen only. However, the ability to integrate multiple functionalities onto the chip was quickly exploited in order to extend the utility of chip-based immunoassays. For example, Cheng et al.[24] developed a six-channel microfluidic device with scanned fluorescence detection to conduct simultaneous, direct immunoassays for ovalbumin and for antiestradiol, with a limit of detection in the latter case of 4.3 nM. On-chip mixing and reaction of antibody and antigen reagents, followed by electrophoretic separation of free and bound species was achieved within 30 sec under optimized conditions, which included running a calibrant in several of the channels for simultaneous calibration and analysis.

Similarly, Roper et al.[25] more recently exploited on-chip mixing, reaction, separation, and detection to conduct a chip-based competitive immunoassay for insulin, with the additional capability of continuous live cell monitoring. In this work, anti-insulin antibody, fluorescein isothiocyanate (FITC) -labeled insulin, and insulin were electrophoretically sampled from separate chip reservoirs and allowed to mix as they traveled along a heated, 4 cm reaction channel. Subsequent injection from the reaction channel into a 1.5 cm separation channel allowed for the separation and detection of free FITC-insulin from FITC-insulin–antibody complex in 5 sec, with a limit of detection of 3 nM insulin achieved by this method. However, the true novelty of this chip-based assay lies in its ability to monitor insulin release from a single islet of Lagerhans—a micro-organ from the pancreas that consists of 2000–5000 endocrine cells, the majority of which are pancreatic β-cells that release insulin to regulate blood glucose levels. Although glucose is known to be the primary regulator of insulin secretion, the cause of dysfunctional insulin secretion is mostly unknown. Fluid in the immediate vicinity of a single islet maintained in a reservoir on the immunoassay chip could be monitored at 15-sec intervals to construct an insulin secretion profile. Hence, Roper et al.'s electrophoretic microchip provides an advanced tool not only for protein quantitation by immunoassay, but also for understanding the kinetics and mechanism of insulin secretion or any other cell or tissue release with high temporal resolution.

Electrophoresis of Cells on Chips

Understanding biological chemistry from both structural and functional perspectives necessitates tools capable of analysis on the cellular and molecular levels. Here, too, electrophoresis in microfabricated devices has contributed significantly. Extending the realm of the so-called "lab-on-a-chip" to the "lab-in-a-cell" concept requires three particular areas to be addressed, including cell sorting or trapping, cell treatment, and cell analysis. Most cells have a net negative charge near physiological pH and so an electric field can be used to transport them. Li and Harrison[26] first took advantage of this fact in their demonstration of electrophoretic and electro-osmotic pumping of three cell types—baker's yeast, E. coli and canine erythrocytes (red blood cells) —within a network of microfabricated capillary channels. Electro-osmotic flow in their uncoated glass chips exceeded the electrophoretic mobility of the cells, and so these researchers observed the net transport of cells toward the cathode at near-physiological pH values. Cell velocities of up to 0.5 mm/sec in channels with 15 × 50 μm cross-sections were observed. Care was taken to avoid cell lysis caused by the application of the separation voltage (less than 600 V/cm and typically about 100 V/cm), although high electric fields (1–10 kV/cm) have been shown to cause cell membrane permeation. Although yeast and E. coli cell adhesion to the channel walls was observed, reduction of cell counts significantly relieved this problem, as did wall treatment with a commercial trichlorohexadecylsilane agent, rendering the walls hydrophobic. This wall treatment, however, also substantially reduced EOF, so that the electrophoretic mobility of the negatively charged canine erythrocytes in isotonic solution (toward the anode) exceeded the EOF toward the cathode. Furthermore, Li and Harrison were able to demonstrate the second component of cell analysis—namely, cell treatment—by successfully using electrokinetic effects to mix streams of canine erythrocytes and SDS directly on the chip and subsequently observing the resulting cell lysis. Hence, this successful demonstration of cell sorting or directed cell transport and cell lysing by an electrokinetic, valveless control scheme within a system of four intersecting channels clearly established the utility of electrophoresis on microfabricated devices for cellular analysis.

Indeed, more sophisticated chip-based cellular assays followed this pioneering work, including several recent reports of microfluidic devices used to isolate and analyze individual cells and to perform high-throughput sequential separations of single cells.[27] These studies document the careful integration of microchip functionalities allowing the three key steps of cellular analysis to take place on a single substrate: cell sorting or selection; cell treatment (especially lysis by chemical, physical, or electrical means); and analysis of cellular components. Furthermore, each of these studies relies on high sensitivity LIF detection, and as such, fluorescent labeling is necessitated. Such labeling can occur off-chip (prior to lysis) or on-chip (subsequent to lysis). In the latter case, fluorescent labeling represents an additional cell treatment step and, therefore, an additional level of complexity that must be integrated onto the microchip substrate alongside the other functionalities. Wu, Wheeler, and Zare[27a] achieved this by designing and employing a poly(dimethylsiloxane) (PDMS) device with four sections: cell manipulation channels, reagent introduction channels, a reaction chamber (with a volume of 70 pL to avoid excessive sample dilution during lysing and derivatization), and a separation channel. The flow of cells and reagents for lysing and derivatization was controlled by a three-state valve, created by multilayered, soft lithographic fabrication. Individual Jurkat T cells were manipulated on the device, and after lysing by SDS, cell contents were derivatized by 2,3-dicarboxaldehyde (NDA). Separation by MEKC with an applied electric field of 250 V/cm in a 10 mM borate buffer with 1% SDS at pH 9.2 produced electropherograms with multiple peaks corresponding to derivatized amino acids from the lysed contents of single Jurkat T cells. The resulting "chemical cytometry" electropherogram had an unidentified large peak and suffered from poorer resolution than a comparable electropherogram obtained for the homogenized contents of a population of cells with prechip derivatization. The former difference was attributed to insoluble cell debris while the latter may have been caused by a number of factors, including lower analyte concentrations in the single-cell experiment, destacking due to concentration differences, and poorer performance of the reaction chamber relative to the standard double-t injector. This highly integrated system could be easily used to target other cell constituents by the use of alternative fluorescent probes in the on-chip reaction chamber prior to separation.

To enable higher cell throughput and to eliminate the possibility of cross-contamination from continuous sequential analysis schemes, Munce et al.[28] designed the first multichannel microchip platform capable of parallel injection and separation of single cells. In this work, optical tweezers were used for the selection and injection of single acute myloid leukemia cells, which were prelabeled with fluorescent dyes for detection purposes. Lysing was achieved by electromechanical shearing, whereby electrokinetic transport of the cell through a narrow taper in the separation channel resulted in cell rupture and the retention of large organelles such as the nucleus at the channel opening. This retention may prove advantageous, since it can prevent separation channel blockage resulting from the presence of large cellular components, and it creates the opportunity to harvest the nucleus, for example, for subsequent analysis. Although this parallel-channel system resulted in a throughput of only about 24 cells per hour (not including the prechip derivatization time), this could be increased by utilizing a faster cell selection system and many more than four parallel cell channels. Such modifications would provide the ability to examine, simultaneously, many individual cells from a given population in order to potentially identify analyte variability at the single-cell level or the potential existence of subpopulations within a collection of cells. Furthermore, it may provide an avenue to probe individual organelles at the molecular level.

There is, in fact, some precedence for the use of electrokinetic-based microchip separations to probe organelles. Lu et al.[29] reported on a microfabricated field flow fractionation device for the continuous separation of various subcellular structures by IEF. In their work, mitochondria from the lysate of HT-29 (human colon carcinoma) cells stained with MitoTracker Green migrated into a focused band (corresponding to a pI value of 4–5 for the organelles) in <6 min, as shown in Fig. 6. In addition, two different subpopulations of mitochondria could be separated from intact cells and nuclei by this method, and it was suggested that a quantitative assay to determine the extent of apoptosis could be developed by using this same device to measure the retention (or loss) of mitochondria membrane potential. It was suggested that by varying the device volume (2 µl was used with a population of about 2000 cells), this method could be used for analytical or preparatory purposes, and that it could be used as an individual component within a more complex, integrated microfluidic system for cell-based assays for systems biology.

Other Chip Functions and Applications

Despite the many examples of advanced chip functions and applications documented herein, efforts to expand functionality continue. For example, with the development of smaller and more complex microfluidic devices and drastic reductions in liquid and sample volumes, many groups have devoted considerable effort toward the development of sophisticated fluidic control features. Grover et al.[30] introduced monolithic elastomer membrane valves and diaphragm pumps that can be used for large-scale integration into high-throughput glass microfluidic devices. These valves and pumps were fabricated in three and four layers, each valve consisting of a glass manifold with a pneumatically actuated displacement chamber, a working PDMS membrane, and a glass fluidic channel wafer containing the channel to be valved. The three-layer design was assembled by applying a PDMS membrane over the wet-etched fluidic channel wafer and pressing the manifold wafer onto the PDMS membrane. This way, hybrid glass-PDMS fluidic channels were formed with valves in locations where a drilled or etched displacement chamber on the manifold was oriented

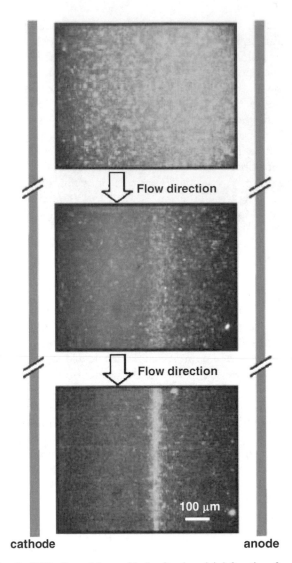

Fig. 6 IEF of crudely purified mitochondrial fraction from lysate of HT-29 cell stained with MitoTracker Green. Top views of three different sections of the channel at ~1, 6, and 18 mm from the inlet show that when mitochondria first flow into the channel, their distribution is relatively homogeneous (top). Further along the channel, some focusing is apparent (middle) and close to the exit of the device, the mitochondria form a narrow band (bottom). The electric field is applied laterally. The position of the electrodes is schematic, not to scale. A 3–6 pH buffer range was used; the mitochondria focus at a pI between 4 and 5 (assuming a linear pH profile). (From Ref.[29] with permission. Copyright 2004, American Chemical Society. *(View this art in color at www.dekker.com.)*

directly across the PDMS membrane from a valve seat. The four-layer design included an additional glass wafer with drilled holes to define all-glass fluidic channels with minimal fluid-PDMS contact for improved chemical and biochemical compatibility. The valves were actuated by applying a vacuum to the displacement chambers, which deflected the PDMS membrane

into the chamber and allowed fluid to flow across the gaps in the fluid channels. The fluid flow was stopped by applying a pressure that forced the PDMS layer back against the fluidic channel wafer. Different pump designs were tested on a 10-cm glass wafer that held 144 valves, thereby forming 48 different diaphragm pumps. It was found that the pumps had low dead volumes, could be fabricated in dense arrays, and could be addressed in parallel via an integrated manifold. It was concluded that these pumps offer precise and reliable control of nanoliter to microliter fluid volumes on glass microdevices, are easily designed and fabricated, and should be well suited for integration into bioassay devices.

There are many types of procedures, such as solvent extraction, which cannot be performed on standard electrophoretic microchip systems. If such procedures could be integrated onto microdevices, chip applications could be expanded even further. Kikutani et al.[31] introduced a novel chip design that showed the first example of chemical processing with continuous flow in a 3-D microchannel network. They used continuous flow chemical processing (CFCP) in glass microchips with pressure-driven laminar flow. The microchannels were constructed in three dimensions by laminating two 7.0×5.0 cm glass plates with etched channels to a third plate with holes drilled in it to connect the two channel structures. The chip had eight reagent inlets and four outlets, and it was designed to accommodate four analyses simultaneously. The system was used to analyze samples containing heavy metal cations Fe (II) and Co (II) and was shown to be able to perform 20 microscale operations for mixing, reaction, solvent extraction, and detection in a 3-D channel network on a single chip. The metal cations were first chelated before removing the interfering ions and finally, detection was by thermal layer microscopy (TLM) with YAG and He–Ne lasers. In this preliminary work, it was found that Fe and Co ions at micromolar levels could be analyzed with this system. Also, it was shown that the multiphase laminar flows were stable and that the microscale operations functioned properly without interferences. According to the authors, the integration offered by this kind of microdevice would be useful in proteomics, combinatorial chemistry, and pharmaceutical manufacturing, where high throughput of complex samples is required.

CONCLUSIONS AND FUTURE DIRECTIONS

The advantages typically associated with CE, such as reduced sample and reagent consumption, reduced analysis time, and increased separation efficiency, are augmented when the CE system is transposed to a chip substrate. More importantly, however, microchip electrophoresis offers the further advantage of integrating analytical processes beyond separation. Sample preparation, injection, reaction, and detection can be seamlessly tied to the electrophoretic separation stage of the analysis. The monolithic electrophoresis chips capable of separation coupled to some of these other analytical steps are precursors to the ultimate "lab-on-a-chip," which promises high-throughput, sensitive analyses with minimal user intervention. Applications of such a device in biochemical, clinical, forensic, and environmental analyses are seemingly unlimited. However, several challenges remain despite the great promise of these devices. In order to fully realize the advantages offered by the microfluidics regime of the chip, methods of physically addressing their small structures and volumes, and of interfacing to the macroscale world beyond the chip, must continue to be carefully managed. Although chip fabrication techniques are now well established, they are still not accessible to the majority of analysts. Fabrication processes must continue to be refined to allow routine chip construction at the hands of analysts in regular research labs, or ideally, the chips themselves must be made more readily available at low cost and in a variety of application designs for all potential users. Miniaturization, or careful arrangement of the apparatus accompanying the electrophoresis chip, including power supplies, detection systems, computer controllers, and other components, into compact and robust systems, must be considered in order to take full advantage of the chip's small size. Finally, true parallel processing facilities must be routinely developed on single chips in order to increase sample throughput and increase the applicability of these systems to large-scale analytical problems. These challenges are now being tackled by many research groups worldwide in order to successfully build labs-on-a-chip around the cornerstone of electrophoresis on a chip.

ACKNOWLEDGMENTS

C.L. Colyer gratefully acknowledges support of this work by the National Science Foundation under Grant No. 0138963. The authors thank Dr. Susan Barker for helpful comments during the preparation of this manuscript.

REFERENCES

1. Becker, H.; Locascio, L.E. Polymer microfluidic devices. Talanta **2002**, *56* (2), 267–287.
2. Bilitewski, U.; Genrich, M.; Kadow, S.; Mersal, G. Biochemical analysis with microfluidic systems. Anal. Bioanal. Chem. **2003**, *377* (3), 556–569.
3. HT Micro (2003), http://www.htmicro.com/htm/index.htm.
4. Schultz-Lockyear, L.L.; Colyer, C.L.; Fan, Z.H.; Roy, K.I.; Harrison, D.J. Effects of injector geometry and sample matrix on injection and sample loading in integrated capillary electrophoresis devices. Electrophoresis **1999**, *20* (3), 529–538.
5. Ermakov, S.V.; Jacobson, S.C.; Ramsey, J.M. Computer simulations of electrokinetic injection techniques in microfluidic devices. Anal. Chem. **2000**, *72* (15), 3512–3517.
6. Roddy, E.S.; Xu, H.; Ewing, A.G. Sample introduction techniques for microfabricated separation devices. Electrophoresis **2004**, *25* (2), 229–242.
7. Fang, Q. Sample introduction for microfluidic systems. Anal. Bioanal. Chem. **2004**, *378* (1), 49–51.
8. Chen, G.; Wang, J. Fast and simple sample introduction for capillary electrophoresis microsystems. Analyst **2004**, *129* (6), 507–511.
9. Jiang, G.F.; Attiya, S.; Ocvirk, G.; Lee, W.E.; Harrison, D.J. Red diode laser induced fluorescence detection with a confocal microscope on a microchip for capillary electrophoresis. Biosens. Bioelectron. **2000**, *14* (10–11), 861–869.
10. Haab, B.B.; Mathies, R.A. Single-molecule detection of DNA separations in microfabricated capillary electrophoresis chips employing focused molecular streams. Anal. Chem. **1999**, *71* (22), 5137–5145.
11. Lacher, N.A.; Lunte, S.M.; Martin, R.S. Development of a microfabricated palladium decoupler/electrochemical detector for microchip capillary electrophoresis using a hybrid glass/poly-(dimethylsiloxane) device. Anal. Chem. **2004**, *76* (9), 2482–2491.
12. Ertl, P.; Emrich, C.A.; Singhal, P.; Mathies, R.A. Capillary electrophoresis chips with a sheath-flow supported electrochemical detection system. Anal. Chem. **2004**, *76* (13), 3749–3755.
13. Figeys, D.; Ning, Y.B.; Aebersold, R. A microfabricated device for rapid protein identification by microelectrospray ion trap mass spectrometry. Anal. Chem. **1997**, *69* (16), 3153–3160.
14. (a) Li, J.; Thibault, P.; Bings, N.H.; Skinner, C.D.; Wang, C.; Colyer, C.L.; Harrison, D.J. Integration of microfabricated devices to capillary electrophoresis-electrospray mass spectrometry using a low dead volume connection: application to rapid analyses of proteolytic digests. Anal. Chem. **1999**, *71* (15), 3036–3045; (b) Zhang, B.; Liu, H.; Karger, B.L.; Foret, F. Microfabricated devices for capillary electrophoresis-electrospray mass spectrometry. Anal. Chem. **1999**, *71* (15), 3258–3264; (c) Lazar, I.; Ramsey, R.S.; Sundberg,

S.; Ramsey, J.M. Subattomole-sensitivity microchip nanoelectrospray source with time-of-flight mass spectrometry detection. Anal. Chem. **1999**, *71* (17), 3627–3631.

15. Schilling, M.; Nigge, W.; Rudzinski, A.; Neyer, A.; Hergenröder, R. A new on-chip ESI nozzle for coupling of MS with microfluidic devices. Lab. Chip **2004**, *4* (3), 220–224.

16. Wang, Y.-X.; Cooper, J.W.; Lee, C.S.; DeVoe, D.L. Efficient electrospray ionization from polymer microchannels using integrated hydrophobic membranes. Lab. Chip **2004**, *4* (4), 363–367.

17. Zhang, B.L.; Foret, F.; Karger, B.L. High-throughput microfabricated CE/ESI-MS: automated sampling from a microwell plate. Anal. Chem. **2001**, *73* (11), 2675–2681.

18. Emrich, C.A.; Tian, H.; Medintz, I.L.; Mathies, R.A. Microfabricated 384-lane capillary array electrophoresis bioanalyzer for ultrahigh-throughput genetic analysis. Anal. Chem. **2002**, *74* (19), 5076–5083.

19. Goedecke, N.; McKenna, B.; El-Difrawy, S.; Carey, L.; Matsudaira, P.; Ehrlich, D. A high-performance multilane microdevice system designed for the DNA forensics laboratory. Electrophoresis **2004**, *25* (10–11), 1678–1686.

20. Ferrance, J.; Snow, K.; Landers, J. Evaluation of microchip electrophoresis as a molecular diagnostic method for Duchenne muscular dystrophy. Clin. Chem. **2002**, *48* (2), 380–383.

21. Sluszny, C.; Yeung, E.S. One- and two-dimensional miniaturized electrophoresis of proteins with native fluorescence detection. Anal. Chem. **2004**, *76* (5), 1359–1365.

22. Ramsey, J.D.; Jacobson, S.C.; Culbertson, C.T.; Ramsey, J.M. High-efficiency, two-dimensional separations of protein digests on microfluidic devices. Anal. Chem. **2003**, *75* (15), 3758–3764.

23. Li, Y.; Buch, J.S.; Rosenberger, F.; DeVoe, D.L.; Lee, C.S. Integration of isoelectric focusing with parallel sodium dodecyl sulfate gel electrophoresis for multidimensional protein separations in a plastic microfluidic network. Anal. Chem. **2004**, *76* (3), 742–748.

24. Cheng, S.B.; Skinner, C.D.; Taylor, J.; Attiya, S.; Lee, W.E.; Picelli, G.; Harrison, D.J. Development of a multichannel microfluidic analysis system employing affinity capillary electrophoresis for immunoassay. Anal. Chem. **2001**, *73* (7), 1472–1479.

25. Roper, M.G.; Shackman, J.G.; Dahlgren, G.M.; Kennedy, R.T. Microfluidic chip for continuous monitoring of hormone secretion from live cells using an electrophoresis-based immunoassay. Anal. Chem. **2003**, *75* (18), 4711–4717.

26. Li, P.C.H.; Harrison, D.J. Transport, manipulation, and reaction of biological cells on-chip using electrokinetic effects. Anal. Chem. **1997**, *69* (8), 1564–1568.

27. (a) Wu, H.; Wheeler, A.; Zare, R.N. Chemical cytometry on a picoliter-scale integrated microfluidic chip. PNAS **2004**, *101* (35), 12809–12813; (b) Gao, J.; Yin, X.F.; Fang, Z.L. Integration of single cell injection, cell lysis, separation and detection of intracellular constituents on a microfluidic chip. Lab. Chip **2004**, *4* (1), 47–52; (c) McClain, M.A.; Culbertson, C.T.; Jacobson, S.C.; Allbritton, N.L.; Sims, C.E.; Ramsey, J.M. Microfluidic devices for the high-throughput chemical analysis of cells. Anal. Chem. **2003**, *75* (21), 5646–5655.

28. Munce, N.R.; Li, J.; Herman, P.R.; Lilge, L. Microfabricated system for parallel single-cell capillary electrophoresis. Anal. Chem. **2004**, *76* (17), 4983–4989.

29. Lu, H.; Gaudet, S.; Schmidt, M.A.; Jensen, K.F. A microfabricated device for subcellular organelle sorting. Anal. Chem. **2004**, *76* (19), 5705–5712.

30. Grover, W.H.; Skelley, A.M.; Liu, C.N.; Lagally, E.T.; Mathies, R.A. Monolithic membrane valves and diaphragm pumps for practical large-scale integration into glass microfluidic devices. Sens. Actuat. B. **2003**, *89* (3), 315–323.

31. Kikutani, Y.; Hisamoto, H.; Tokeshi, M.; Kitamori, T. Micro wet analysis system using multi-phase laminar flows in three-dimensional microchannel network. Lab. Chip **2004**, *4* (4), 328–332.

Electrospray Ionization Interface for Capillary Electrophoresis/MS

Joanne Severs
Bayer Pharmaceuticals, Berkeley, California, U.S.A.

INTRODUCTION

The development of the electrospray ionization (ESI) source for mass spectrometry provided an ideal means of detection for capillary electrophoretic (CE) separations. The ESI source is currently the preferred interface for CE/MS, due to the fact that it can produce ions directly from liquids at atmospheric pressure and with high sensitivity and selectivity for a wide range of analytes.

DISCUSSION

ESI is initiated by generating a high potential difference between the spray capillary tip and a counterelectrode.[1] This electric field leads to the production of micron-sized droplets with an uneven charge distribution, generally accepted to be due to an electrophoretic mechanism acting on electrolytes in the solvent.[2] This mechanism, combined with a shrinkage of the droplets due to solvent evaporation (aided by heat and an applied gas flow into the source), leads to electrostatic repulsion overcoming surface tension in the droplet. The "Rayleigh" limit is reached, a "Taylor cone" is formed, and smaller highly charged droplets are emitted, eventually leading to the production of gas-phase ions.[1–3] These ions are accelerated through a skimmer into successive vacuum stages of the mass analyzer. The ESI source has been demonstrated to act as an electrolytic cell, generating electrochemical oxidation and reduction.[2] The exact ionization mechanism will vary with experimental conditions and is still an area of continuing in-depth research and discussion.[3]

ESI is classified as a "soft" ionization technique. It produces molecular-weight information and very little, if any, fragmentation of the analyte ion, unless induced in the vacuum region of the mass analyzer. The number of charges accumulated by an analyte ion is proportional to its number of basic or acidic sites. The spray polarity and conditions, solution pH and nature, as well as solute concentration will all effect the charge state distribution observed in the mass spectrum. Multiple charging of an analyte ion encourages the release of very high-molecular-weight ions. It is mainly due to this fact that ESI has gained such enormous interest, especially among biochemists. Employing only small, relatively inexpensive mass analyzers, spectrometrists are able to obtain high-sensitivity information on analytes with molecular weights of up to 200 kDa. The multiple-charging phenomenon means that the mass-to-charge (m/z) range of the analyzer does not generally need to exceed 3000. A deconvolution algorithm,[4] generally built nowadays into the instrument software, can be applied to the series of multiply-charged, molecular-ion peaks, and a single peak, representing the molecular weight, is then displayed on a "true mass" scale. The m/z scale is calibrated with standards of known exact mass. Whereas ESI/MS (mass spectrometry) has made the largest impact on large biomolecule analysis, CE/ESI/MS has also been applied with great success to the analysis of many small-molecule applications.

The development of the first CE/MS was prompted by the early reports on electrospray ionization (ESI/MS) by Fenn and co-workers in the mid-1980s,[1] when it was recognized that CE would provide an optimal flow rate of polar and ionic species to the ESI source. In this initial CE/MS report, a metal coating on the tip of the CE capillary made contact with a metal sheath capillary to which the ESI voltage was applied.[5] In this way, the sheath capillary acted as both the CE cathode, closing the CE electrical circuit, and the ESI source (emitter). Ideally, the interface between CE and MS should maintain separation efficiency and resolution, be sensitive, precise, linear in response, maintain electrical continuity across the separation capillary so as to define the CE field gradient, be able to cope with all eluents presented by the CE separation step, and be able to provide efficient ionization from low flow rates for mass analysis.

Several research groups have presented work on the development of CE/ESI/MS interfaces. The interfaces developed can be categorized into three main groups: coaxial sheath flow, liquid junction, and sheathless interfaces. A schematic of the sheath-flow interface first developed for CE/ESI/MS by Smith et al.[6] is illustrated in Fig. 1a. A sheath liquid, with an

Encyclopedia of Chromatography DOI: 10.1081/E-ECHR-120039972

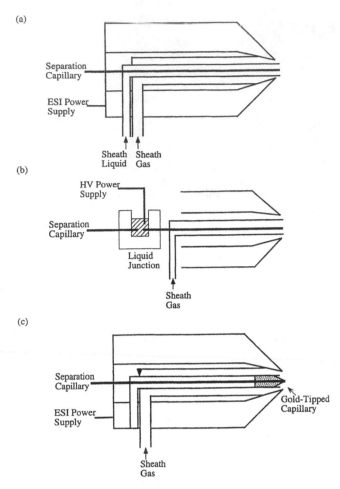

Fig. 1 Schematic illustration of CE/MS interfaces to an ESI source: (a) a coaxial sheath-flow interface; (b) a liquid-junction interface; (c) a sheathless interface.

electrolytic content, is infused into the ESI source at a constant rate, through the coaxial sheath capillary which surrounds the end of the separation capillary and terminates near the end of the separation capillary. This sheath liquid mixes with the separation buffer as it elutes from the tip of the CE capillary, thus providing the necessary electrical contact between the ESI needle and the CE buffer, and closing the CE circuit. Because the CE terminus and ESI source are at the same voltage, if the ESI source requires a high voltage (2–5 kV) (rather than ground potential), then the ESI voltage chosen also directly affects the potential difference across the separation capillary. To date, the sheath–liquid interface has been the most widely used and accepted system, being the simplest to construct, with numerous results published employing sheath liquids typically containing 60–80% organic solvent, modified with 1–3% acid in water and typically introduced at flow rates of 1–4 μL/min. The composition of the sheath liquid should be optimized for the specific systems under investigation. Recent reports have confirmed

that the relative dimensions and positioning of the separation and sheath capillaries also influence sensitivity and stability.

Although the additional flow of an organic-containing electrolyte into the ESI source moderately extends the range of CE buffer systems that can be used, the CE buffer composition still has a dramatic effect on the ESI signal, minimizing the buffer choice for best sensitivity to volatile solutions. Reports have also highlighted the need for a considered selection of sheath-liquid composition due to the possibility of formation of moving ionic boundaries inside the capillary.[7] The possibility of these effects occurring should be considered and minimized when transferring a CE method from an alternative detection system to MS. It should be noted, however, that these effects are minimized or eliminated when there is a sufficiently strong flow toward the CE terminus.

A "liquid-junction interface" has also been suggested and applied for CE/ESI/MS.[8] Electrical contact with this interface is established through the liquid reservoir which surrounds the junction of the separation capillary and a transfer capillary, as shown in Fig. 1b. The gap between the two capillaries is approximately 10–20 μm, allowing sufficient makeup liquid from the reservoir to be drawn into the transfer capillary while avoiding analyte loss. The flow of makeup liquid into the transfer capillary is induced by a combination of gravity and the Venturi effect of the nebulizing gas at the capillary tip.[8]

In comparisons of coaxial sheath-flow and liquid-junction interfaces, it has been noted that although both provide efficient coupling, the former is generally easier to operate. One of the major disadvantages in employing the liquid-junction interface is in establishing a reproducible connection inside the tee piece. Also, the use of a transfer capillary, which has no potential difference applied across it, can lead to peak broadening. Advantages of this interface, however, include the possibility of combining different outer-diameter capillaries through the junction and the extra mixing time provided for the makeup liquid and CE eluant.

The problem with both interfaces described so far is that they depend on the addition of excess electrolyte to the ESI source to maintain the circuit, generally leading to a decrease in analyte sensitivity. As previously mentioned, the first CE/MS interface reported made electrical connection between the separation buffer and the ESI needle via a metal coating on the tip of the CE capillary,[5] as represented in Fig. 1c. Although femtomole detection limits and separation efficiencies of up to half a million theoretical plates were achieved, problems included a high dependence on the buffer system used and the need to regularly replace the metal coating on the capillary tip.

The further development of interfaces which do not rely on an additional liquid flow are currently underway. Generally, they have employed metal deposition on the CE terminus that is tapered (by chemical etching or mechanical pulling) to provide an increased electric field at the capillary tip. These so-called "microspray" and "nanospray" approaches, with more effective ionization mechanisms, have been adopted recently by several groups for interfacing infusion systems, LC and CE to ESI/MS, and in all cases, significant gains in sensitivity and sample usage have been observed.[9] Attomole level detection limits from nanoliter sample volumes can now be attained, and the ability to form an electrospray from a purely aqueous solution is now possible. Alternative sheathless interfaces have also been briefly investigated.[10] Stability problems still need consideration in most cases. An interface which does not use an additional makeup flow can, as well as aiding sensitivity, also avoid such problems as charge state distribution shifts in the mass spectrum. In addition, the ability to electrospray purely aqueous systems is often advantageous for looking at fragile biological and noncovalently bound analytes. In some cases, however, a makeup liquid may be found necessary. For example, for certain separations, the EOF may need to be minimized or eliminated in the CE capillary, and thus flow rates into the source will not be sufficiently high as to maintain a stable electrospray. If a capillary needs to be coated to avoid analyte interaction with the capillary wall, then a cationic coating, which reverses the EOF rather than eliminating it, should preferably be chosen if a sheathless system is to be employed. Also, it may be found that a makeup liquid is necessary to increase the volatility of a specific CE electrolyte system.

Another disadvantage at present in using the sheathless interface is the time dispensed in preparing the tapered, coated tips. Although the coatings now employed are more stable than those initially used, the tips do not regularly survive more than a day or two of use. This can, however, be due to the tip "plugging" rather than the metal coating deteriorating. Filtering of electrolyte and analytes and rinsing of the capillary can, therefore, often prolong the capillary lifetime.

An instrumental attribute which aids the development and interfacing of CE to ESI/MS is the ability to pressurize the CE capillary, at low pressure for sample injection and higher pressures for capillary content elution. Balancing of the heights of the capillary termini is also an important consideration in order to avoid syphoning effects. In all cases of CE/ESI/MS application, safety, with respect to the electrical circuits, should be considered. It should be verified that all circuits have a common ground, and the addition of a resistor in the ESI power supply line when interfaced to CE is a wise precaution.

An incompatability that does need to be considered in CE/MS method development is the use of certain CE buffer systems and additives which are detrimental to the ESI process. For example, although sample concentration can be increased by the use of more conductive buffers, this approach is not advantageous for ESI/MS detection. These characteristics result in a significant demand upon ESI interface efficiency.[11] Ideally, the chosen CE buffer should be volatile, such as ammonium acetate or formate. The use of pure acids or bases rather than a true buffer has also been shown to be advantageous for certain molecules. Nonaqueous buffer systems are also being employed more widely.

Capillary electrokinetic chromatography (CEKC) with ESI/MS requires either the use of additives that do not significantly impact the ESI process or a method for their removal prior to the electrospray. Although this problem has not yet been completely solved, recent reports have suggested that considered choices of surfactant type and reduction of electroosmotic flow (EOF) and surfactant in the capillary can decrease problems. Because most analytes that benefit from the CEKC mode of operation can be effectively addressed by the interface of other separations methods with MS, more emphasis has until now been placed upon interfacing with other CE modes. For "small-molecule" CE analysis, in which micellar and inclusion complex systems are commonly used, atmospheric pressure chemical ionization (APCI) may provide a useful alternative to ESI, as it is not as greatly affected by involatile salts and additives.

The efficiency of the ESI detection process for CE/MS can be considered in terms of the simple model of Kebarle and Tang[3] and has been discussed in great detail by Smith and co-workers.[11] These considerations indicate that analyte sensitivity in CE/ESI/MS may be increased by reducing the mass flow rate of the background components. This decrease in background flow rates can be experimentally accomplished by decreasing the electric field or employing smaller-diameter capillaries, and this predicted increase in analyte sensitivity is now well supported by experimental studies.[11]

To reduce the elution speed of the analyte ions into the source, the electrophoretic voltage can be decreased just prior to elution of the first analyte of interest, minimizing the experimental analysis time while allowing more scans to be recorded without a significant loss in ion intensity.[11] Alternatively, the use of smaller-diameter capillaries than conventionally used for CE also increases sensitivity.[11] A capillary diameter should, ideally, be commercially available, amenable to alternative detection methods, provide the necessary detector sensitivity, and be free from clogging.

Capillary internal diameters of between 20 and 40 μm have been shown to be optimal and are compatible with "microspray" techniques.

The further development of microscale preconcentration and cleanup techniques and the resulting improvements in CE/MS concentration detection limits are likely to expand the use of this analytical technique. The more common use of small-diameter capillaries and even tiny etched microplate devices,[10] along with the improvements in ESI spray techniques are pushing research along. Further investigations into improving interface design, durability, reproducibility and sensitivity are still necessary. The availability of improved, less expensive, and smaller mass spectrometers will almost certainly lead to increased use of CE/MS. However, the sensitivity and selectivity already demonstrated by CE/MS systems, in combination with the minute analyte volumes sampled, already make this a highly powerful technique.

REFERENCES

1. Whitehouse, C.M.; Dreyer, R.N.; Yamashita, M.; Fenn, J.B. Electrospray interface for liquid chromatographs and mass spectrometers. Anal. Chem. **1985**, *57*, 675–679.
2. Kebarle, P.; Tang, L. From ions in solution to ions in the gas phase. Anal. Chem. **1993**, *65*, 972A.
3. Ikonomou, M.G.; Blades, A.T.; Kebarle, P. Electrospray-ion spray: a comparison of mechanisms and performance. Anal. Chem. **1991**, *63*, 1989–1998.
4. Mann, M.; Meng, C.K.; Fenn, J.B. Interpreting mass spectra of multiply charged ions. Anal. Chem. **1989**, *61*, 1702–1708.
5. Olivares, J.A.; Nguyen, N.T.; Yonker, C.R.; Smith, R.D. On-line mass spectrometric detection for capillary zone electrophoresis. Anal. Chem. **1987**, *59*, 1230–1232.
6. Smith, R.D.; Olivares, J.A.; Nguyen, N.T.; Udseth, H.R. Capillary zone electrophoresis-mass spectrometry using an electrospray ionization interface. Anal. Chem. **1988**, *60*, 436–441.
7. Foret, F.; Thompson, T.J.; Vouros, P.; Karger, B.L.; Gebauer, P.; Bocek, P. Liquid sheath effects on the separation of proteins in capillary electrophoresis/electrospray mass spectrometry. Anal. Chem. **1994**, *66*, 4450–4458.
8. Lee, E.D.; Mück, W.; Henion, J.D.; Covey, T.R. Liquid junction coupling for capillary zone electrophoresis/ion spray mass spectrometry. Biomed. Environ. Mass Spectrom. **1989**, *18*, 844–850.
9. Chowdhury, S.K.; Chait, B.T. Method for the electrospray ionization of highly conductive aqueous solutions. Anal. Chem. **1991**, *63*, 1660–1664, M. Wilm and M. Mann, Anal. Chem. 68: 1–8 (1996).
10. Figeys, D.; Aebersold, R. Electrophoresis **1998**, *19*, 885–892, and references therein.
11. Wahl, J.H.; Goodlett, D.R.; Udseth, H.R.; Smith, R.D. Electrophoresis **1993**, *14*, 448–457, J.P. Landers, Capillary electrophoresis–mass spectrometry, in *Handbook of Capillary Electrophoresis*, CRC Press, Boca Raton, FL, 1997, and references therein.

Eluotropic Series of Solvents for TLC

Simion Gocan
Department of Analytical Chemistry, "Babes-Bolyai" University, Cluj-Napoca, Romania

INTRODUCTION

The easiest way to vary the relative adsorption of the sample (and its R_f value, migration rate, respectively) is to change the solvent: strong (polar) eluents decrease adsorption, and weak (no polar) eluents increase it. When benzene, for instance, is used as solvent on silica gels or alumina layers, the ethers and esters are found on top of the chromatogram (with high R_f values), ketones and aldehydes are approximately in the center (medium R_f values), and the alcohols are below them (low R_f values), whereas the acids remain at the starting point ($R_f = 0$). Thus the separation sequence follows the polarities of the compounds.

The dielectric constant may be taken as an indication of solvent polarity, but the interfacial tension between solvents and polar adsorbents, approximated by the interfacial tension between the solvent and water, has been suggested as a fundamental basis for correlating solvent strength.

The physical factors that determine solvent strength in a given adsorption system have long been understood in general terms. Solvent strength can be interpreted in terms of the following basic contributions: 1) interactions between solvent molecules and a sample molecule in solution; 2) interactions between solvent molecules and a sample molecule in the adsorbed phase; and 3) interactions between an adsorbed solvent molecule and the adsorbent.

ELUOTROPIC SERIES

Normal-Phase Thin-Layer Chromatography (NPTLC)

A series of authors has defined the solvent relative strength for polar adsorbents in the form of eluotropic series, grouping them in order of their chromatographic elution strength, with both pure solvents and mixtures of solvents included. The most familiar of these is, of course, the one set up by Trappe,[1] and variations of this have been published from time to time. Trappe gave the following series (listed in order of increasing elution power): light petroleum, cyclohexane, carbon tetrachloride, trichloroethylene, toluene,

benzene, dichloromethane, chloroform, ether, ethyl acetate, acetone, *n*-propanol, ethanol, and methanol.

The eluotropic series of pure solvents is generally referred to a particular adsorbent. The magnitude of ε^o, the solvent strength parameter, can be defined as the adsorption energy of the solvent per unit of the standard activity surface. All these ε^o values are relative to the solvent pentane, for which ε^o is defined equal to zero. This parameter is defined as a measure of the degree of the adsorption interaction of the solvent with the stationary phase. As a function of this magnitude, the eluotropic series for various polar adsorbents is presented in Table 1.

It is well recognized that the eluotropic series of solvents, according to Snyder, is suited for monoactive site-type adsorbents, but in the case of multiactive site-type adsorbents (e.g., alumina), their imperfection becomes acute. Thus for multiactive-site-type adsorbents, the eluotropic series sequence and eluent strength values are highly dependent on the class of test solutes employed.[7]

A practical eluotropic series of solvents, based on the expended solubility parameter concept, was reported.[8] This series was defined based on partial specific solubility parameter (δ_s) that is equal to the sum of Keeson (δ_o) and acid–base ($2\delta_a\delta_b$), which represents the contribution to interaction forces introduced to characterize the solute, the mobile, and the stationary phase in liquid–solid chromatography. Exactly the same two interaction forces define ε^o and, consequently, there should exist a direct relation between ε^o and $\delta_s^2 = \delta_o^2 + 2\delta_a\delta_b$. Unfortunately, the general correlation for all the solvents on alumina is poor ($r^2 = 0.75$).

Snyder[2] has shown that there is a correlation between the ε^o values for a certain polar adsorbent (i.e., silica gel, Florisil, magnesia, and alumina). These values can be estimated from values for alumina using the following equations:

$$\varepsilon^o_{\text{silica gel}} = 0.77\varepsilon^o_{\text{alumina}}$$

$$\varepsilon^o_{\text{Florisil}} = 0.52\varepsilon^o_{\text{alumina}}$$

and

$$\varepsilon^o_{\text{magnesia}} = 0.58\varepsilon^o_{\text{alumina}}$$

Encyclopedia of Chromatography DOI: 10.1081/E-ECHR-120027340

Table 1 Relative strengths of different solvents on various adsorbents: eluotropic series

Solvent	ε^o				P_i' [3,4]	S_i [5]
	Al_2O_3 [2]	Silica gel [2]	Florisil [2]	MgO [2]		
n-Pentane	0.00	0.00	0.00	0.00		
n-Hexane	0.01				0.1 (−)[a] −0.14	
Cyclohexane	0.04				0.2 (VIa)	
Carbon tetrachloride	0.18	0.11	0.04	0.10	1.6 (−)[a] 1.56	
m,p-Xylene	0.25				2.7 (VII)	
Amyl chloride	0.26					
o-Xylene	0.27					
Isopropyl ether	0.28				2.4 (I)	
Isopropyl chloride	0.29					
Toluene	0.29				2.4 (VII)	
Benzene	0.32	0.25	0.17	0.22	2.7 (VII) 3.19	
Ethyl ether	0.38	0.38	0.30	0.21	2.8 (I) 3.15	
Chloroform	0.40	0.26	0.19	0.26	4.1 (VIII)[b] 4.31	
Methylene chloride	0.42	0.32	0.23	0.26	3.1 (V)	
Methyl isobutyl ketone	0.43					
1,2-Dichloroethane	0.43				3.1 (V) 4.29	
Acetone	0.56	0.47			5.1 (VIa) 5.10	3.4
Ethyl acetate	0.58	0.38			4.4 (VIa) 4.24	
Methyl acetate	0.60			0.28		
Amyl alcohol	0.61					
Dioxane	0.63	0.49			4.8 (VIa) 5.27	3.5
Pyridine	0.71				5.3 (III)	
Butyl-cellosolve (C_4H_9-O-C_2H_4OH)	0.74					
Acetonitrile	0.79	0.50			5.8 (VIb) 5.64	3.4
Tetrahydrofuran					4.0 (III) 4.28	4.4
Isopropanol	0.82				3.9 (II) 3.92	4.2
Methanol	0.95				5.1 (II) 5.10	2.9
Ethanol					4.3 (II)	3.6
Acetic acid					6.0 (IV) 6.13	
Water					10.2 (VIII) 10.2	0.0

I–VIII represent groups from Snyder's classification.
[a]Selectivity group irrelevant because of low P' values.
[b]Close to group and VIII.

The values calculated by means of these equations show the standard deviation of ε° values ± 0.004 units with respect to the experimental values.

The relationships between eluent strength and composition for binary (one to two) and ternary (one to three) solvents have been derived:[2]

$$\varepsilon_1 - \varepsilon_2 = \varepsilon_1 + \frac{\log[X_2 10^{\alpha n_2(\varepsilon_2 - \varepsilon_1)} + 1 - X_2]}{\alpha n_2} \quad (1)$$

$$\varepsilon_1 \rightarrow \varepsilon_3 = \varepsilon_2 - \frac{\log[X_3 10^{\alpha n_3(\varepsilon_3 - \varepsilon_2)} + X_2]}{\alpha n_3} \quad (2)$$

Eluent strength is assumed to increase in the order $\varepsilon_1 < \varepsilon_2 < \varepsilon_3$ for solvents A, B, and C; n_2 and n_3 are the effective molecular areas of adsorbed solvent molecules B and C, respectively (with the exception of certain very strong solvents); X_2 and X_3 are the mole fractions of B and C in the solvent mixtures; and α represents the adsorbent's surface activity. To use Eqs. (1) and (2), we must know the activity degree of the adsorbent. Consequently, the α (adsorbent surface activity function) values for a few adsorbents with respect to water content are presented in Table 2. Based on the data in Tables 1 and 2, and by using Eq. (1), an infinite number of such series can be established.

The principle of the variation of solvent composition while holding solvent strength constant was first developed by Neher[9] for separation of steroids. Fig. 1 is a representation of the eluotropic series of Neher for application in this fashion. Six solvents, which will act as solvent S_1 (100% concentration, on the left), are arranged vertically, whereas the same solvents, acting as solvent S_2 (100% concentration, on the right of the horizontal lines) in a binary solvent eluotropic series, are arranged horizontally. We can obtain a very large number of binary systems, of the

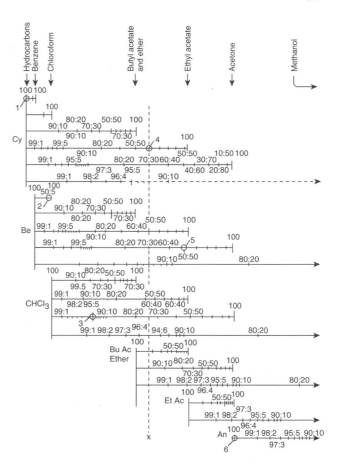

Fig. 1 Eluotropic series of Neher. Cy = cyclohexane; Be = benzene; BuAc = butyl acetate; EtAc = ethyl acetate; An = acetone.

same or different strength, by means of this nomogram. The dashed line X determines 12 compositions of binary systems of the same average eluotropic properties. These systems are called equieluotropic systems. The nomogram in Fig. 1 corresponds to adsorption on silica gel.

Saunders[10] obtained another nomogram by using six very common solvents. With the help of the nomogram presented in Fig. 2, we can achieve binary solvent mixtures of certain strength in the interval 0.0–0.75. In this graph, ε° is plotted across the top and in various binary solvent compositions in each of the horizontal lines below it. Each line corresponds to the range 0–100% by volume of binary solvent composition. Its manner of use is similar to that described for Neher's nomogram.

For NPTLC, the solvent strength weighting factor S_i is the same as the polarity index P' given in Table 1. The polarity index P' is given by the sum of the logarithms of the polar distribution constants for ethanol, dioxane, and nitromethane, and the selectivity parameters x_i is given as the ratio of polar distribution constant for solute I to the total solvent

Table 2 α Values for some common chromatographic adsorbents

H$_2$O[a] (%)	Alumina	Silica gel (wide pore) TLC	Florisil	Magnesia
0	1.00	0.83	1.61	1.00
0.5	0.90	0.79	1.18	1.00
1.0	0.84	0.75	1.00	1.00
2.0	0.75	0.71	0.90	0.98
4.0	0.63	0.70	0.81	0.93
7.0	0.59	0.69	0.79	0.86
10.0	0.59	0.69		
15.0	0.59			

(From Ref.[6].)

[a]Water added to activated adsorbent.

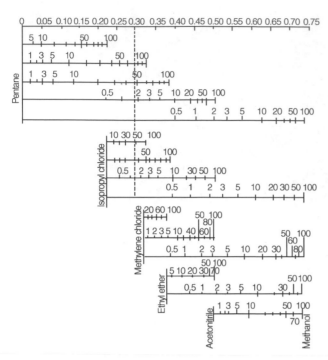

Fig. 2 Mixed solvent strengths on silica gel after Saunders. (From Ref.[10].)

polarity (P'). The sum of the three values for x_i will be normalized up to 1.0. Snyder was able to show that the many solvents available could be grouped into eight classes with distinctly different selectivities. Solvents within the same selectivity group exhibit similar separation properties (Table 1).[3] The polarity index P' of a mixed mobile phase is the arithmetic average of the solvent polarity index weighting factor adjusted according to the volume fraction of each solvent as is given by:[11]

$$P' = \Sigma_i P_i' \phi_i \qquad (3)$$

where P_i' is the polarity index weighting factor of solvent i (Table 1) and ϕ_i is the volume fraction of solvent i. For a binary solvent mixture containing 95% dichloromethane and 5% methanol, the polarity index of the mixed solvent is calculated as follows:

$$P' = (4.29)(0.95) + (5.10)(0.05) = 4.33$$

Reversed-Phase Thin-Layer Chromatography (RPTLC)

A general characteristic of RTLPC systems is that the stationary phase is nonpolar vs. the mobile phase that is polar. Thus a decrease in polarity of the mobile phase leads to a decrease in retention. This situation is the reverse of the general trends observed in NPTLC.

A decrease in polarity of the mobile phase can be realized by increasing the volume fraction of organic solvent in an aqueous organic mobile phase. The most common method for varying the chromatographic selectivity for neutral molecules is to change the type of organic modifier in the mobile phase. Eq. (4) can often be used as an acceptable approximation for the variation of the retention with the volume fraction of organic solvent in the mobile phase:[11]

$$\log k = \log k_w - S\phi \qquad (4)$$

where k is the solute capacity factor, k_w is the solute capacity factor with pure water as the mobile phase, ϕ is the volume fraction of organic solvent, and S is a solute-dependent factor related to the solvent strength of the organic solvent.

The literature data suggest that RPTLC solvent-strength varies as water (weakest) < methanol < acetonitrile < ethanol < tetrahydrofuran < propanol < (methylene chloride) (strongest). Thus solvent strength increases as solvent polarity decreases.

The eluotropic scale is a relative one; it is necessary to choose a reference. To obtain a positive value for solvent eluotropic strength, the reference solvent has to be water. The S values determined from the slope of Eq. (4) can be used as descriptor, in a semiquantitative way, of the solvent strength (S_i) of the organic solvent.[11] Some typical S_i values for common solvents are presented in Table 1. The solvent strength of a mixed mobile phase (S_T) is the arithmetic average of the solvent strength weighting factors adjusted according to the volume fraction ϕ of each solvent Eq. (5):

$$S_T = \Sigma_i S_i \phi_i \qquad (5)$$

where S_i is the solvent strength weighting factor of solvent i and ϕ_i is the volume fraction of solvent i. If we want to change solvent selectivity to adjust resolution, then the volume fraction of the new solvent required to obtain an isoeluotropic mixture could be calculated from Eq. (5). For example, for methanol–water (60:40), $S_T = 1.56$; using acetonitrile as an example and the same value of $S_T = 1.56$, we obtain $1.56 = 3.2\phi_a + 0\phi_w = 0.49$. Thus a mixture of acetonitrile–water (49:51) is similar in solvent strength to a mixture of methanol–water (60:40). In a similar manner, it is possible to calculate several eluotropic eluents.

TLC APPLICATIONS

In adsorption TLC, there exists a competition between the sample and eluent molecules for a place on the adsorbent surface. If we assume that one molecule of sample (X) replaces m solvent molecules (S) on adsorption, we may represent the adsorption process as:

$$X_1 + mS_a \Leftrightarrow X_a + mS_1 \qquad (6)$$

In this case, Snyder[2] found an important relationship for adsorption chromatography:

$$\log K^\circ = \log V_a + \alpha(S^\circ - A_s\varepsilon^\circ) \qquad (7)$$

where the sample adsorption distribution coefficient K° [mL/g] is defined as being equal to the ratio of sample concentrations in adsorbent and unadsorbed phases; V_a is adsorbent surface volume [mL]; S° is dimensionless free energy of adsorption of a sample compound on adsorbent of standard activity ($\alpha = 0$) from pentane as solvent; and A_s is molecular area of adsorbed sample. For adsorbents of the same type, Eq. (7) expresses K° as a function of some fundamental properties of the sample (S°, A_s), adsorbent (V_a, α), and solvent (ε°). The parameter ε° defines the effect of the solvent on the adsorption of a given sample; therefore it may be equated with solvent strength. The larger is the solvent strength parameter ε°, the smaller is the value of K° for a given sample and adsorbent. Eq. (7) predicts that sample separation order can vary with solvent for sample components of different sizes (A_s). Let K° and ε° for solvent 1 be K_1 and ε_1, and K_2 and ε_2 for solvent 2. Thus from Eq. (4), the ratio of K values for a sample component adsorbent from two solvents is given as:

$$\log(K_1/K_2) = \alpha A_s(\varepsilon_2 - \varepsilon_1) \qquad (8)$$

According to Eq. (8), the change in K on changing the solvent is predicted to be proportional to the difference in eluent strengths ($\varepsilon_2 - \varepsilon_1$) and to the sample molecule size A_s.

For the TLC or other bed, there were analogous relationships derived:[2]

$$R_{M'} = \log(V_a W/V^\circ) + \alpha(S^\circ - A_s\varepsilon^\circ) \qquad (9)$$

where $R_{M'} = \log[(1/\xi R_f) - 1]$, W is total weight of adsorbent in the bed, V° is bed void volume equal to the volume of solvent in a solvent wet bed, and $(R_f)_{true} = \xi(R_f)_{exp}$. Eq. (9) can be written for a given sample and adsorbent, for solvents 1 and 2,

respectively. Then, subtracting the second equation from the first gives:

$$(R_{M'})_1 - (R_{M'})_2 = \alpha A_s(\varepsilon_2 - \varepsilon_1) \qquad (10)$$

Eq. (10) can be useful in estimating the effect on sample R_f values of a change in solvent strength.

Eqs. (7) and (9) are a generally reliable relationship for adsorption systems with weak or moderately strong solvents. But, in their derivation, two major approximations were made:[2] 1) interactions between solvent and sample molecules are assumed unimportant; and 2) interactions between the adsorbent and various adsorbed molecules are assumed to be fundamentally similar in type. Both of these assumptions can be defended in the case of weak solvent systems, but they are poor approximations for strong solvents. These equations are concerned only with the primary effect of the solvent on sample adsorption. However, in the case of the majority of adsorption systems based on weak or moderately strong solvents, these approximate relationships are adequate for most purposes. For the strongest solvent systems (e.g., alcohols, acids, water, and their solutions), these equations can be corrected for secondary solvent effect by addition of a correction term Δ_{eas}:

$$R_{M'} = \log(V_a W/V^\circ) + \alpha(S^\circ - A_s\varepsilon^\circ) + \Delta_{eas} \qquad (11)$$

where Δ_{eas} represents the secondary adsorption effect and is a complex function of solvent elution strength, adsorbent activity, solute structure, and various possible interactions between the solute adsorbent, and solvent and adsorbent. These aspects are largely discussed in Ref.[2]. If $(\Delta_{eas})_1$ and $(\Delta_{eas})_2$ refer to the secondary solvent effects for a particular sample and solvents 1 and 2, respectively, then we can write:

$$(R_{M'})_1 - (R_{M'})_2 = \alpha A_s(\varepsilon_2 - \varepsilon_1) + (\Delta_{eas})_1 - (\Delta_{eas})_2 \qquad (12)$$

Eq. (12) gives the difference in $\Delta R_{M'}$ values for a particular sample (A_s) in two solvents 1 and 2, with ε_1 and ε_2, respectively, as a function of adsorbent activity α, and the secondary solvent terms for the particular solvent contribution. The largest Δ_{eas} values were found for samples with free hydroxyl groups, and between these samples, two free hydroxyl groups in the sample molecule give a larger value of Δ_{eas} than does a single free hydroxyl. Sample molecules with intramolecularly bonded hydroxyl groups give much smaller values of Δ_{eas}, but limited hydrogen bonding of these groups with basic solvents is still possible. For the same sample molecule, the Δ_{eas} value will be a function of the eluent composition.

Now let us write Eq. (11) for a particular eluent and sample molecules 1 and 2, respectively:

$$(R_{M'})_1 - (R_{M'})_2 = \alpha(S_1^o - S_2^o) + \alpha\varepsilon(A_{s2} - A_{s1}) + (\Delta_{eas})_1 - (\Delta_{eas})_2$$

(13)

In this equation, we find the different source and the selectivity between the two solutes, which can differ by the difference in energy of adsorption $(S_1^o - S_2^o)$, molecular size $(A_{s2} - A_{s1})$, and secondary adsorption effect $(\Delta_{eas})_1 - (\Delta_{eas})_2$. These considerations are valuable for a large number of compounds, but in the case of some isomers, the second terms of Eq. (13) can be considered practically equal to zero and only difference sources for selectivity remain in the difference in energy of adsorption and the secondary adsorption effect. The secondary adsorption effect plays a major role in the separation of the isomers. For instance, in adsorption on silica gel with benzene–pyridine (90 : 10, vol/vol) as eluent, the following Δ_{eas} values: -0.90 ± 0.17 and $+0.05 \pm 0.11$ for m-hydroxybenzaldehyde and o-hydroxybenzaldehyde, respectively, were obtained. These sample molecules can be very easily separated.

Adsorption TLC selection of the mobile phase is conditioned by sample and stationary-phase polarities. The following polarity scale is valid for various compound classes in NPTLC in decreasing order of K values: carboxylic acids > amides > amines > alcohols > aldehydes > ketones > esthers > nitro compounds > ethers > halogenated compounds > aromatics > olefins > saturated hydrocarbons > fluorocarbons. For example, retention on silica gel is controlled by the number and functional groups present in the sample and their spatial locations. Proton donor/acceptor functional groups show the greatest retention, followed by dipolar molecules, and, finally, nonpolar groups.

The activity degree is another important characteristic for adsorbents. As is well known, the adsorbent is in contact with large amounts of water in the thin-layer preparation process water that has to be removed by drying at 100–120°C for approximately 30–60 min. This process is known as activation. The activity of the layer is directly correlated with the water content of the adsorbent. The silanol groups show a great affinity for water, which is bound by hydrogen bonds.

Thus the activity degree can be controlled by the content of physisorbed water onto the adsorbent. Lower R_f values will be obtained on the adsorbent with a high degree of activity.

Generally speaking, the first problem with which the analyst is confronted concerns gathering information regarding the mixture to be separated, in terms of mixture polarity and the range of molecular masses. For example, if the mixtures that have to be separated are nonpolar, then we can select an active stationary phase and nonpolar mobile phase from the following scheme:

Sample to be separated:	nonpolar → medium polar → polar
Stationary phase:	active → medium active → inactive
Mobile phase:	nonpolar → medium polar → polar

Several mobile-phase optimization strategies in TLC are based on the use of isoeluotropic solvents (i.e., solvent mixtures of identical strengths but different selectivities). Selecting mobile phase will be achieved based on the eluotropic series (Figs. 1 and 2). These considerations are very general. The selection and optimization of one system of eluent is a more complex problem and must be discussed for each particular system. For example, the separation of 13 phenylurea and s-triazine herbicides was performed by overpressured layer chromatography (OPLC) with a binary mobile phase.[12] The optimization of the mobile-phase composition for the separation of these herbicides on silica gel was achieved by means of the "ELUO" method in which solvents are selected from an eluotropic series based on solvent power. The method is complementary to the "PRISMA" method for optimizing the compositions of binary, ternary, and quaternary mobile phase for OPLC and TLC.[13] The phenylurea herbicides could be separated with hexane–ethyl acetate (1 : 1, vol/vol), ethyl ether–benzene (9 : 1, vol/vol), or chloroform–ethyl acetate, and triazine herbicides could be separated with hexane–ethyl acetate (13 : 7, vol/vol), ethyl ether–benzene (9 : 1, vol/vol), or chloroform–ethyl acetate (3 : 1, vol/vol). The simultaneous separation of the herbicide classes could not be achieved and compounds that were chemically closely related were not well separated.

To maximize the differences in selectivity, solvents must be selected from different selectivity groups that are situated close to the Snyder's triangle apexes. For example, for NPTLC, a suitable selection could be solvents from several groups of Snyder's classifications (I, VII, and VIII), mixed with hexane to control solvent strength.[3]

REFERENCES

1. Trappe, W. Eluotropic series of solvents. Biochem. Z. **1940**, *305*, 150.
2. Snyder, L.R. *Principles of Adsorption Chromatography*; Marcel Dekker: New York, 1968.
3. Snyder, L.R. Classification of the solvent properties of common liquids. J. Chromatogr. Sci. **1978**, *16*, 223–234.
4. Poole, C.V.; Poole, S.K. *Chromatography Today*; Elsevier: Amsterdam, 1991.
5. Snyder, L.R.; Dolan, J.W.; Gant, J.R. Gradient elution in high-performance liquid chromatography: I. Theoretical basis for reversed-phase systems. J. Chromatogr. **1979**, *165*, 3–9.
6. Gocan, S. The mobile phase in thin-layer chromatography. In *Modern Thin-Layer Chromatography*; Grinberg, N., Ed.; Marcel Dekker: New York, 1990; 139, Chap. 3.
7. Kovalsca, T.; Klama, B. JPC, J. Planar Chromatogr. **1997**, *10*, 353–357.
8. Buchmann, M.L.; Kesselring, U.K. Pharm. Acta Helv. **1981**, *56*, 166–273.
9. Neher, R. *Thin Layer Chromatography*; Marini-Bettolo, G.B., Ed.; Elsevier: Amsterdam, 1964; 75–86.
10. Saunders, D.L. Solvent selection in adsorption liquid chromatography. Anal. Chem. **1974**, *46*, 470–473.
11. Poole, C.F.; Poole, S.K. *Chromatography Today*; Elsevier: Amsterdam, 1991.
12. Tekei, J. JPC, J. Planar Chromatogr. **1990**, *3*, 326–330.
13. Nyiredy, S.Z.; Meier, B.; Erdelmeier, C.A.J.; Sticher, O. PRISMA: Ageometrical design for solvent optimization in HPLC. HRC CC **1985**, *8*, 186–188.

E

Elution Chromatography

John C. Ford
Department of Chemistry, Indiana University of Pennsylvania, Indiana, Pennsylvania, U.S.A.

INTRODUCTION

By far the most common chromatographic mode of operation, elution chromatography is virtually the only mode used for analytical separations. Separation in elution chromatography occurs due to differences in migration velocities among the sample components. These differences are related to the affinities of the solutes for the mobile phase, of the solutes for the stationary phase, and of the mobile phase for the stationary phase, and to the properties of the stationary phase itself. Band broadening is typically caused by axial diffusion and mass transfer considerations. The plate number is a measure of the column efficiency, i.e., the ratio of separative to dispersive transport. The resolution is a measure of the overall quality of the separation of two solutes; resolution is a combination of the thermodynamic factors causing separative transport and the kinetic factors causing dispersive transport. Developing a useful elution separation requires more than obtaining the minimal resolution of the solutes of interest. A successful method should not only achieve the desired separation but should also do so in a cost-effective and robust manner. HPLC method development has its own, extensive literature, reflecting the importance of HPLC as an analytical technique. Some considerations for practical separations are also discussed.

DEFINITION

Elution chromatography is one of the three basic modes of chromatographic operation, the other two being frontal analysis and displacement chromatography. All three modes were known to Tswett in the early 1900s, although a systematic definition was not made until 1943. Elution chromatography is by far the most common chromatographic mode and is virtually the only mode used for analytical separations. Most theoretical work has been directed at the elution mode, although frequently the results are applicable to other modes as well.

The current IUPAC nomenclature for chromatography defines elution chromatography as "a procedure in which the mobile phase is continuously passed through or along the chromatographic bed and the sample is introduced into the system as a finite slug."[1] Typically, the volume of the sample is small compared to the volume of the column. The individual components of the sample (the solutes) move through the column at different average velocities, each less than the velocity of the mobile phase. The differences in velocities are caused by differences in the interactions of the solutes with the stationary and mobile phases. Assuming essentially equivalent interactions with the mobile phase, solutes that interact strongly with the stationary phase spend less time on average in the mobile phase and consequently have a lower average velocity than components that interact weakly with the stationary phase. If the difference between the average velocities of two solutes is sufficiently large, if the dispersive transport within the column is sufficiently small, and if the column is sufficiently long, the solute bands are resolved from one another by the time they exit the column.

Elution chromatography can be performed with a constant mobile phase composition (isocratic elution) or with a mobile phase composition that changes during the elution process (gradient elution). The following discussion focuses on isocratic operation. Further, each of the mechanistic categories of chromatography (ion exchange, reversed-phase, normal phase, etc.) can be performed in the elution mode and additional information on elution chromatography can be obtained by reference to the appropriate sections of this encyclopedia.

Elution chromatography is categorized as being linear or nonlinear, depending on the distribution isotherm, and as being ideal or nonideal, with ideal behavior requiring both infinite mass transfer kinetics and negligible axial dispersion. Although truly linear distribution isotherms are rare, at low solute concentrations or over small ranges of solute concentration, sufficient linearity may exist to approximate linear elution. Linear, ideal elution would result in band profiles that are identical to the injection profiles—an unrealistic situation. Under linear, nonideal elution conditions, thermodynamic factors control band retention and kinetic factors such as mass transfer resistances control the band shape. Guiochon, Shirazi, and Katti[2] have discussed the relationship between the isotherm and chromatographic behavior extensively.

Encyclopedia of Chromatography DOI: 10.1081/E-ECHR-120039973

The retention of a solute in elution chromatography is usually expressed as the retention factor, k (capacity factor or k'), given by $k = (t_R - t_M)/t_M$, where t_R is the retention time of the solute, and t_M is the hold up time (void time, dead time, or t_0). The hold up time is the time required to elute a component that is not retained at all by the stationary phase. One can relate k to the distribution coefficient, K, by $k = K\beta$, where ϕ is the phase ratio, the ratio of the stationary phase volume to the mobile phase volume. Rearranging the definition of retention factor, we find that $t_R = t_M(1 + k) = t_M(1 + K\beta)$. Since it is usually reasonable to assume that t_M and β are the same for different solutes, the retention time differences are due to distribution coefficient differences. Under appropriate conditions, the distribution coefficient can be related to the thermodynamic distribution constant and elution chromatographic measurements can be used for physicochemical determinations of thermodynamic parameters.

Differences in solute retention are usually expressed as the separation factor (selectivity coefficient or α), given by $\alpha = k_b/k_a$, where k_a and k_b are the retention factors of the two solutes in question. By convention, k_b is the more retained solute and $\alpha > 1$, although this is not always followed. Since again it is reasonable to assume that ϕ is the same for different solutes, $\alpha = K_b/K_a$, where K_a and K_b are the distribution coefficients of the two solutes, and again, retention time differences are due to distribution coefficient differences. If two solutes have the same distribution coefficient (i.e., $\alpha = 1$) in a particular combination of mobile and stationary phases, they cannot be separated by elution chromatography in that system. However, $\alpha \neq 1$ is a necessary, but not sufficient, condition for a successful separation.

As a solute moves through the column, it undergoes dispersive transport as well as separative transport. Under typical elution chromatographic conditions, the dispersive transport is caused by axial diffusion and mass transfer considerations, such as slow adsorption–desorption kinetics. This dispersive transport results in band spreading lowering column efficiency, which can prevent adequate separation of different solutes. The plate number (plate count, number of theoretical plates, theoretical plate number, or N), defined as $N = t_R^2/\sigma_t^2$, where σ_t^2 is the variance of the band in time units, is a measure of the column efficiency, i.e., the ratio of separative to dispersive transport. Several alternate forms of this equation are commonly used, usually based on the assumption of Gaussian peak shape. The effective plate number, N_{eff}, is a combination of the plate number and the capacity factor, i.e., $N_{eff} = N[k/(1 + k)]^2$, and is generally more useful than N for comparing the resolving power of different columns.

Another common measure of column efficiency is the plate height [height equivalent to a theoretical plate (HETP), H], defined by $H = L/N$, where L is the length of the column, usually in centimeters. This is frequently presented as the reduced plate height, h, the ratio of the plate height to the diameter of the packing material. A "good" column has a high plate count (a low plate height; $2 < h < 5$).

The overall quality of the separation of two solutes is measured by their resolution (R_s), a combination of the thermodynamic factors causing separative transport and the kinetic factors causing dispersive transport, and is an index of the effectiveness of the separation. Defined by $R_s = (t_{r,b} - t_{r,a})/\frac{1}{2}[(w_{t,b} + w_{t,a})]$, where a and b refer to the two solutes, $t_{r,x}$ is the retention time of solute x, and $w_{t,x}$ is the peak width at the base of solute x in units of time, it is frequently estimated by use of the fundamental resolution equation

$$R_s = \left(\frac{\sqrt{N}}{4}\right)\left(\frac{\alpha - 1}{\alpha}\right)\left(\frac{k_b}{1 + k_b}\right)$$

where k_b is the retention factor of the more retained solute, α is the separation factor of the solute pair under consideration, and N is the plate count. This equation assumes that the peak shapes are Gaussian and that the peak widths are equivalent.

Easy recognition of the two peaks over a wide range of relative concentrations is possible for $R_s = 1$, and this is essentially the practical minimum resolution desirable. It is usually stated that $R_s = 1$ corresponds to a peak purity of about 98%; however, this is correct only for equal concentrations of the two solutes. As the ratio of relative concentrations of the two solutes deviates from 1, the recovery of the lower concentration solute at a given level of purity becomes poorer.

Examination of the fundamental resolution equation shows that improvements in resolution can be obtained by: 1) increasing the column efficiency. The dependence of R_s on $\sqrt{N}$, rather than N, means that this method is most effective when the column efficiency is initially low. In other words, when using efficient columns to develop a separation, major improvements in R_s are not generally obtained by increasing N; 2) increasing α. If α is close to 1.0, the greatest increase in R_s can be obtained by changing those parameters that influence α, i.e., the mobile phase composition, the choice of stationary phase, the temperature, or, less frequently, the pressure. Increasing α from 1.1 to 1.2 increases R_s by more than 80%. However, as α increases, the amount of increase in R_s decreases, so that increasing α from 2.1 to 2.2 increases R_s by only about 4%; 3) increasing k. If k_b (and thus k_a) <1, R_s can be significantly increased by changing the mobile phase composition to increase k_b. As for α, the amount of increase decreases as k_b increases, so that while changing k_b from 0.5 to 1.5 improves R_s by about 80%, increasing k_b from 1.5 to 2.5 increases R_s by about 20%. Moreover, increasing k_b increases the analysis time, so that this approach is also of limited practicality.

To summarize, the most successful approach to obtaining adequate R_s is usually to increase α by varying the mobile phase composition—e.g., choice of solvent(s), pH, or temperature—or by varying the stationary phase. Increasing R_s by increasing N or k_b works in selected instances, but is not as generally applicable.

Developing an elution separation method to be used for the analysis of numerous samples requires more than obtaining the minimal resolution of the solutes of interest. A successful method should not only achieve the desired separation but should also do so in a cost-effective and robust manner. HPLC method development has its own, extensive literature, reflecting the importance of HPLC as an analytical technique.

Snyder, Kirkland, and Glajch state the goals of HPLC method development as: 1) precise and rugged quantitative analysis requires that R_s be greater than 1.5; 2) a separation time of <5–10 min; 3) ≤2% relative standard deviation (RSD) for quantitation in assays (≤5% for less-demanding analyses and ≤15% for trace analysis); 4) a pressure drop of <150 bar; 5) narrow peaks to give large signal/noise ratios; and 6) minimal mobile phase consumption per run. Additionally, thorough testing of the robustness of proposed methods is recommended.

ARTICLES OF FURTHER INTEREST

Dead Point (*Volume or Time*), p. 421.
Displacement Chromatography, p. 468.
Distribution Coefficient, p. 476.

Efficiency in Chromatography, p. 496.
Gradient Elution: Overview, p. 722.

REFERENCES

1. Section 9.2.1.2 of International Union of Pure and Applied Chemistry Compendium of Analytical Nomenclature; http://www.iupac.org/publications/analytical_compendium (accessed November 2004).
2. Guiochon, G.; Shirazi, S.G.; Katti, A.M. *Fundamentals of Preparative and Nonlinear Chromatography*; Academic Press: Boston, 1994.

SUGGESTED FURTHER READING

Bidlingmeyer, B.A. *Practical HPLC Methodology and Applications*; John Wiley & Sons: New York, 1992.

Giddings, J.C. *Unified Separation Science*; John Wiley & Sons: New York, 1991.

Karger, B.L.; Snyder, L.R.; Horvath, Cs. *An Introduction to Separation Science*; John Wiley & Sons: New York, 1973; 11–167.

Meyer, V.R. *Practical High-Performance Liquid Chromatography*, 3rd Ed.; John Wiley & Sons: New York, 1998.

Rizzi, A. Retention and selectivity. In *Handbook of HPLC*; Katz, E., Eksteen, R., Schoenmakers, P., Miller, N., Eds.; Marcel Dekker: New York, 1998; 1–54.

Snyder, L.R.; Kirkland, J.J.; Glajch, J.L. *Practical HPLC Method Development*, 2nd Ed.; John Wiley & Sons: New York, 1997.

Snyder, L.R.; Kirkland, J.J.; Glajch, J.L. *Introduction to Modern Liquid Chromatography*, 2nd Ed.; John Wiley & Sons: New York, 1979.

Elution Modes in Field-Flow Fractionation

Josef Chmelík
Institute of Analytical Chemistry, Academy of Sciences of the Czech Republic, Brno, Czech Republic

INTRODUCTION

Field-flow fractionation (FFF) is, in principle, based on the coupled action of a nonuniform flow velocity profile of a carrier liquid with a nonuniform transverse concentration profile of the analyte caused by an external field applied perpendicularly to the direction of the flow. Based on the magnitude of the acting field, on the properties of the analyte, and, in some cases, on the flow rate of the carrier liquid, different elution modes are observed. They basically differ in the type of the concentration profiles of the analyte. Three types of the concentration profile can be derived by the same procedure from the general transport equation. The differences among them arise from the course and magnitude of the resulting force acting on the analyte (in comparison to the effect of diffusion of the analyte). Based on these concentration profiles, three elution modes are described.

BACKGROUND INFORMATION

FFF represents a family of versatile elution techniques suited for the separation and characterization of macromolecules and particles. Separation results from the combination of a nonuniform flow velocity profile of a carrier liquid and a nonuniform transverse concentration profile of an analyte caused by the action of a force field. The field, oriented perpendicularly to the direction of the flow, forms a specific concentration distribution of the analyte inside the channel. Because of the flow velocity profile, different analytes are displaced along the channel with different mean velocities, and, thus, their separation is achieved.

According to the original concept,[1] the field drives the analytes to the accumulation wall of the channel. This concentrating effect is opposed by diffusion, driven by Brownian motion of the analytes, which causes a steady state when the convective flux is exactly balanced by the diffusive flux. The concentration profile is exponential and the corresponding elution mode is referred to as the normal mode. Recently, it has been called the Brownian elution mode.[2]

During last two decades, new elution modes were described[3–5] that were not suggested in the original concept.[1] Basically, they differ in the type of the analyte concentration profile.

In 1978, Giddings and Myers described another elution mode for large particles under conditions when diffusion effects can be neglected.[3] The particles form a layer on the channel wall and, under the influence of the carrier liquid flow, they roll on the channel bottom to the channel outlet. This elution mode is referred as the steric mode.[3]

The above-mentioned elution modes apply to the situation when the resulting force acting on the analytes does not change its orientation inside the channel. However, there exist conditions when the resulting force acting on analytes may change its orientation inside the channel [e.g., two counteracting forces, a gradient of a property (pH, density) of the carrier liquid, influence of hydrodynamic lift forces]. Under such conditions, the analytes form narrow zones at the positions where the resulting forces acting on them equal zero. The resulting force is changing its sign below and above this position in such a way that the analyte is focused into this equilibrium position. The concept of formation of narrow zones of analytes inside the FFF channel was first described in 1977 by Giddings.[6] In 1982, a technique utilizing sedimentation–flotation equilibrium and centrifugal field was suggested.[7] However, this technique has not been yet verified experimentally. Later, the general features of this elution mode were described and several techniques were implemented; for a review, see Ref.[8]. The mode was called either the hyperlayer[4] or focusing[5] elution mode.

Some other elution modes have been described. They are induced by various factors—cyclical field, secondary chemical equilibria, adhesion chromatography, asymmetrical electro-osmotic flow; for a review, see Ref.[2]. However, the number of their implementations is rather limited, and for this reason, these modes are not discussed here.

THEORY

FFF experiments are mainly performed in a thin ribbonlike channel with tapered inlet and outlet ends (Fig. 1). This simple geometry is advantageous

Encyclopedia of Chromatography DOI: 10.1081/E-ECHR-120039974

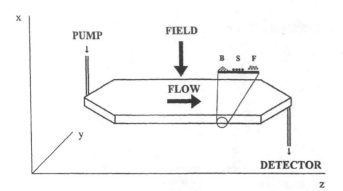

Fig. 1 The orientation of the field and flow in the given coordinate system. The zoomed inset shows the schematic representation of the zone shapes in particular elution modes (B for Brownian, S for steric, and F for focusing).

for the exact and simple calculation of separation characteristics in FFF. Theories of infinite parallel plates are often used to describe the behavior of analytes because the cross-sectional aspect ratio of the channel is usually large and, thus, the end effects can be neglected. This means that the flow velocity and concentration profiles are not dependent on the coordinate y. It has been shown that, under suitable conditions, the analytes move along the channel as steady-state zones. Then, equilibrium concentration profiles of analytes can be easily calculated.

Generally, the concentration profile of analytes in FFF can be obtained from the solution of the general transport equation. For the sake of simplicity, the concentration profile of the steady-state zone of the analyte along the axis of the applied field is calculated from the one-dimensional transport equation:

$$J_x = W_x c(x) - D\frac{\partial c}{\partial x} \qquad (1)$$

where J_x and W_x are the components of the flux density of the analyte and of the transport velocity of the analyte along the axis of the applied field, $c(x)$ is the analyte concentration distribution along the direction of the applied field, and D is the total effective diffusion coefficient. The term $W_x c(x)$ corresponds to the x component of the convective flux of the analyte and the term $D(\partial c/\partial x)$ corresponds to the x component of the diffusive flux of the analyte. W_x equals the sum of the x components of the transport velocity of the analyte induced by the external field applied U_x and the transport velocity of the analyte induced by the carrier liquid flow $v_x (W_x = U_x + v_x)$. Because of the direction of the carrier liquid flow inside the FFF channel, the component v_x equals zero (the x axis is perpendicular to the direction of the flow) and, thus, W_x equals U_x.

Following the treatment given by Giddings,[9] imposing for the condition of the steady-state zone of the analyte, which is characterized by the null flux density, and applying the equation of continuity, the general solution of the analyte concentration profile can be expressed in the form

$$c(x) = c_0 \exp\left[\int_0^x \left(\frac{U_x}{D}\right)dx\right] \qquad (2)$$

The integration limit $x = 0$ corresponds to the accumulation wall boundary. The particular solutions for the concentration profile are dependent on the course of the force field inducing the transport of the analyte, and the ratio of U_x and D.

The equation of the field-induced transport velocity was derived by Giddings:[10]

$$U_x = -ax^n \qquad (3)$$

where a is constant and n equals 0 or 1. If $n = 0$, then U_x is constant; if $n = 1$, then U_x is dependent on the position inside the channel.

DISCUSSION

Brownian Elution Mode

The field-induced velocity of the analyte in the separation channel is constant and comparable with its diffusive motion ($U_x = $ constant, $U_x t \approx \sqrt{2Dt}$, where t is time). The resulting concentration profile of the analyte is given by the exponential relationship[9]

$$c(x) = c_0 \exp\left(-\frac{|U_x|}{D}x\right) \qquad (4)$$

where c_0 is the maximum concentration at the accumulation channel wall. The elution mode with the exponential concentration profile is called Brownian.[2]

It is known that there are two main factors influencing the behavior of analytes in this elution mode: the properties of the analytes (characterized by the so-called analyte–field interaction parameter[11] and the diffusion coefficient) and the strength of the field applied.

In Brownian elution mode, the retention ratio R is indirectly dependent on both the applied force F and the thickness of the channel w, and independent on the flow rate.[9] It can be expressed in an approximate form:

$$R = \frac{6kT}{Fw} \qquad (5)$$

where k is the Boltzmann constant and T is the absolute temperature.

Steric Elution Mode

The velocity of transport induced by the force field in the separation channel is constant and much higher than the velocity caused by the diffusive motion of the analyte (U_x = constant, $U_x t \gg \sqrt{2Dt}$). In this case, the analyte forms a layer on the accumulation channel wall and its concentration in any other position inside the channel equals zero. The particle radius r_p describes the distance of the particle center from the accumulation wall:

$$c(r_p) = c_0 \quad \text{and} \quad c(x \neq r_p) = 0 \qquad (6)$$

The elution mode is called steric.[3] The retention ratio can be expressed in the form

$$R = \frac{6r_p}{w} \qquad (7)$$

This shows that R is independent of both the field applied and the flow rate, and it is dependent only on the particle radius and the channel thickness. In fact, the retention ratio values corresponding to the pure steric elution mode have been seldom observed experimentally.[12] The observed values often correspond to the focusing elution mode as a result of the action of some additional forces influencing retention behavior of analytes.

Focusing Elution Mode

In this elution mode, the velocity of analyte transport induced by a force field in the separation channel is dependent on the position across the channel ($U_x \neq$ constant). Based on Eq. (3), the nonconstant transport velocity can be, in the simplest case, described as

$$U_x = -a(x - s) \qquad (8)$$

where s is the distance of the center of the focused zone from the channel wall (i.e., the position where the resulting force acting on the analyte equals zero). Combining this equation with Eq. (2), we obtain a relation for the resulting concentration profile of the analyte:

$$c(x) = c_0 \exp\left(-\frac{a}{2D}(x - s)^2\right) \qquad (9)$$

where c_0 is the maximum concentration at the center of the focused zone at the position s. The concentration profile of the analyte across the channel thickness, in this simplest case, is Gaussian. In other cases, where other secondary effects act on the retention in the focusing elution mode, the observed concentration profile is more complex. However, even in these cases, the main feature remains the same; that is, the maximum concentration of the analyte is at the equilibrium position, where the resulting force acting on the analyte is zero, and not on the channel wall as in the case of the Brownian and steric elution modes. The elution mode is called focusing[5] or hyperlayer.[4] In the focusing elution mode, the retention ratio can be expressed in a form formally similar to the expression given for the steric elution mode [see Eq. (7)]:

$$R = \frac{6s}{w} \qquad (10)$$

At least two counteracting forces are necessary for the formation of the focused zone of the analyte. The center of the zone is located at the position s where the resulting force is zero. Changing of both forces can control the resulting position of the particle zone.

CONCLUSIONS

In the majority of FFF techniques, the retention ratio is dependent on the analyte size. This dependence for Brownian and steric mode is described by Eq. (11), derived by Giddings:[13]

$$R = 6(\alpha - \alpha^2)$$
$$+ 6\lambda(1 - 2\alpha)\left[\coth\left(\frac{1 - 2\alpha}{2\lambda}\right) - \frac{2\lambda}{1 - 2\alpha}\right] \qquad (11)$$

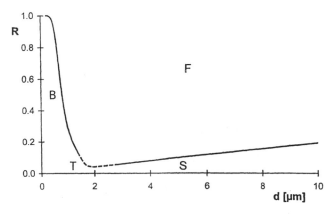

Fig. 2 Schematic representation of the dependence of the retention ratio R on the analyte size. Curve B corresponds to Brownian mode and the line S to the steric mode. The dashed part T denotes the transition between these two modes. The area F shows the range of applicability of the focusing mode.

where $\alpha = d/2w$, $\lambda = kT/Fw$ and d is the analyte diameter. The curve describing this dependence is shown in Fig. 2. The values of R for the focusing mode lie above the curve. This complex situation shows that determination of the elution mode is very important for evaluation of the measured retention data because different elution modes can act on particular analytes in the same experiment.

ACKNOWLEDGMENT

This work was supported by Grant No. A4031805 from the Grant Agency of Academy of Sciences of the Czech Republic.

REFERENCES

1. Giddings, J.C. A new separation concept based on a coupling of concentration and flow non-uniformities. Separ. Sci. **1966**, *1*, 123.
2. Martin, M. *Advances in Chromatography*; Brown, P.R., Grushka, E., Eds.; Marcel Dekker, Inc.: New York, 1998; Vol. 39, 1–138.
3. Giddings, J.C.; Myers, M.N. Separ. Sci. Technol. **1978**, *13*, 637.
4. Giddings, J.C. Separ. Sci. Technol. **1983**, *18*, 765.
5. Janća, J.; Chmelík, J. Focusing in field-flow fractionation. Anal. Chem. **1984**, *56*, 2481.
6. Giddings, J.C. Hyperlayer field-flow fractionation: state of development. Am. Lab. **1992**, *24*, 20D.
7. Janća, J. Sedimentation-flotation focusing field-flow fractionation. Makromol. Chem. Rapid Commun. **1982**, *3*, 887.
8. Janća, J.; Chmelík, J.; Jahnová, V.; Nováková, N.; Urbánková, E. Principle, theory and applications of focusing field-flow fractionation. Chem. Anal. **1991**, *36*, 657.
9. Giddings, J.C. *Unified Separation Science*; John Wiley & Sons: New York, 1991.
10. Giddings, J.C. Nonequilibrium theory of field-flow fractionation. J. Chem. Phys. **1968**, *49*, 81.
11. Giddings, J.C.; Caldwell, K.D. *Physical Methods of Chemistry*; Rossiter, B.W., Hamilton, J.F., Eds.; John Wiley & Sons: New York, 1989; Vol. 3B, 867.
12. Pazourek, J.; Wahlund, K.-G.; Chmelík, J. J. Microcol. Separ. **1996**, *8*, 331.
13. Giddings, J.C. Separ. Sci. Technol. **1978**, *13*, 241, 14: 869 (1979).

Enantiomer Separations by TLC

L. Lepri
A. Cincinelli
Department of Chemistry, University of Florence, Florence, Italy

INTRODUCTION

Enantiomers are compounds that have the same chemical structure but different conformations, whose molecular structures are not superimposable on their mirror images, and, because of their molecular asymmetry, these compounds are optically active. The most common cause of optical activity is the presence of one or more chiral centers, which are usually related to tetrahedral structures formed by four different groups around carbon, silicon, tin, nitrogen, phosphorous, or sulfur.

Many molecules are chiral, even in the absence of stereogenic centers; that is, molecules containing adjacent π systems, which cannot adopt a coplanar conformation because of rotational restrictions due to steric hindrance, can exist in two mirror forms (atropisomers). This is the case for some dienes or olefins, for some nonplanar amides, and for the biphenyl or binaphthyl types of compounds.

Optical isomers can be designated by the symbols D and L, which are used to indicate the relationship between configurations based on D(+)-glyceraldehyde as an arbitrary standard. If this relationship is unknown, the symbols (+) and (−) are used to indicate the direction of rotation of plane polarized light (i.e., dextrorotatory and levorotatory). In 1956, Cahn, Ingold, and Prelog[1] presented a new system, the *R* and *S* absolute configurations of compounds.

Many enantiomers show different physiological behaviors, and it is, therefore, desirable to have reliable methods for the resolution of racemates and the determination of enantiomeric purity. To this end, TLC is a simple, sensitive, economic, and fast method, which allows easy control of a synthetic process and can be used for preparative separations.

TLC SEPARATION OF ENANTIOMERS BY USE OF DIASTEREOMERIC DERIVATIVES

Because the stationary phases originally used in LC were achiral, much research was devoted to the separation of enantiomers as diastereomeric derivatives produced by reaction with an optically pure reagent (A_R). The resultant diastereomers could, because of

their different physicochemical properties, then be separated on conventional stationary phases:

$$\begin{matrix} E_R & & A_R E_R \\ & + \ A_R \ \Leftrightarrow & \\ E_S & & A_R E_S \\ \text{Enantiomers} & & \text{Diastereomers} \\ \text{(similar properties)} & & \text{(different properties)} \end{matrix}$$

In addition, a significant increase in the sensitivity of detection and the location, on the layers, of some compounds that are not otherwise identifiable can be achieved by this method. There are, however, some disadvantages: 1) It is necessary to use derivatization reagents with 100% optical purity; 2) quantitation is founded on the assumption that the reaction is complete and not associated with racemization; and 3) the two chiral centers should be as close as possible to each other in order to maximize the difference in chromatographic properties.

Many chiral derivatization reactions have been used, and the compounds examined are mostly amphetamines, β-blocking agents, amino acids, and anti-inflammatory drugs. Silica gel and, to a lesser extent, silanized silica have been used as stationary phases. The ΔR_f values obtained for the diastereomeric pairs were not usually very high (0.04–0.07), with the exception of amino alcohol and amino acid diastereomers obtained with Marfey's reagent, a derivative of L-alanine amide (0.06–0.22). This procedure has become more and more important owing to the occurrence of D-enantiomers of amino acids in tissues of various organisms. In addition, amino acid residues in dietary proteins have been reported to have been significantly racemized.

TLC SEPARATIONS OF ENANTIOMERS BY CHIRAL CHROMATOGRAPHY

In chiral chromatography, the two diastereomeric adducts $A_R E_R$ and $A_R E_S$ are formed during elution, rather than synthetically, prior to chromatography. The adducts differ in their stability with the use chiral stationary phases (CSP) or chiral coated phases (CCP) and/or in their interphase distribution ratio with

Encyclopedia of Chromatography DOI: 10.1081/E-ECHR-120039975

addition of a chiral selector to the mobile phase (CMP). The difference between the interactions of the chiral environment with the two enantiomers is called enantioselectivity.

According to Dalgliesh,[2] three active positions on the selector must interact simultaneously with the active positions of the enantiomer to reveal differences between optical antipodes. This is a sufficient condition for resolution to occur, but it is not necessary. Chiral discrimination may happen as a result of hydrogen bonding and steric interactions, making only one attractive force necessary in this type of chromatography. Moreover, the creation of specific chiral cavities in a polymer network (as in the "Molecular Imprinting Techniques" section) could make it possible to base enantiomeric separations entirely on steric fit.

Chiral Stationary Phases and Chiral Coated Phases

Few chiral phases are used in TLC; one of the main reasons for this is that stationary phases with very high ultraviolet (UV) background can be used only with fluorescent or colored solutes. For example, amino-modified ready-to-use layers bonded or coated with Pirkle-type selectors, such as N-(3,5-dinitrobenzoyl)-L-leucine or R(–)-α-phenylglycine, are pale yellow and strongly absorb UV radiation.

Another reason is the high price of most CSPs. In spite of this, Pirkle-type CSP, based on a combination of aromatic π–π bonding interactions, hydrogen bonding, and dipole interactions, allows the resolution of racemic mixtures of 2,2,2-trifluoro-1-(9-anthryl) ethanol, 1,1′-bi-2-naphthol, benzodiazepines, hexobarbital, and β-blocking agents derivatized with achiral 1-isocyanatonaphthalene. However, for ligand exchange chromatography (LEC), the most widely used CSPs or CCPs are polysaccharides and their derivatives (cellulose, cellulose triacetate, tribenzoate, and triphenylcarbamate) and silanized silica gel impregnated with an optically active copper (II) complex of (2S,4R,2′RS)-N-(2′-hydroxydodecyl)-4-hydroxyproline (ChiralPlate, Macherey-Nagel and HPTLC Chir, Merck, Germany). The chiral layer on the latter plates is combined with a so-called "concentrating zone." β-Cyclodextrin bonded to silica gel H has also been used for the resolution of some racemic drugs and binaphthalenes.

Ligand exchange chromatography is based on the copper (II) complex formation of a chiral selector and the respective optical antipodes. Differences in retention of the enantiomers are caused by dissimilar stabilities of their diastereomeric metal complexes. The requirement of sufficient stability of the ternary complex involves five-membered ring formation, and

compounds such as α-amino and α-hydroxy acids are the most suitable.

The resolution of optical antipodes on polysaccharides is mainly governed by the shape and size of the solutes (inclusion phenomena) and only to a minor extent by other interactions involving the functional groups of the molecules. In the case of microcrystalline cellulose triacetate (MCTA), the type and composition of the aqueous–organic eluent affect the separation because these result in different swelling of MCTA.

The use of silica gel impregnated with a chiral polar selector, such as D-galacturonic acid, (+)-tartaric acid, (–)-brucine, L-aspartic acid, erythromycin, vancomycin, or a complex of copper (II) with L-proline, should also be mentioned.

In CSP, owing to the nature of the polymer structure, the simultaneous participation of several chiral sites or several polymer chains is conceivable. In CCP, the chiral sites are distributed on the surface or in the network of the achiral matrix relatively far away from each other, and only bimolecular interaction is generally possible with the optical antipodes.

A survey of the optically active substance classes separated with ChiralPlate and HPTLC Chir layers and with MCTA plates is shown in Fig. 1.

Cellulose tribenzoate has recently been used for the separation of enantiomeric aromatic alcohols.[3]

Molecular Imprinting Techniques

This technique is based on the preparation of synthetic polymers with specific selectivity by using chiral imprinting molecules mixed with functional and cross-linking monomers (usually methacrylic acid and ethylene glycol dimethacrylate, respectively), capable of interacting with such molecules.

After polymerization, these molecules are removed by extraction, leaving cavities that correspond to those of the template. The resulting product is called a molecular imprinting polymer (MIP).

By using these stationary phases, many classes of optically active organic compounds have been separated (i.e., phenylethanolamine adrenergic drugs, β-blockers, and nonsteroidal anti-inflammatory drugs) other than the diastereomers of quinine alkaloids. Usually, the separation factor (α) is high, but R_s (resolution) is low because elongated spots are obtained.[4]

Chiral Mobile Phases

Chiral mobile phases enable the use of conventional stationary phases and show only minor detection problems compared to CSP or CCP. However, high cost chiral selectors (e.g., γ-cyclodextrin) are certainly not advisable for TLC. Enantiomer separations can be

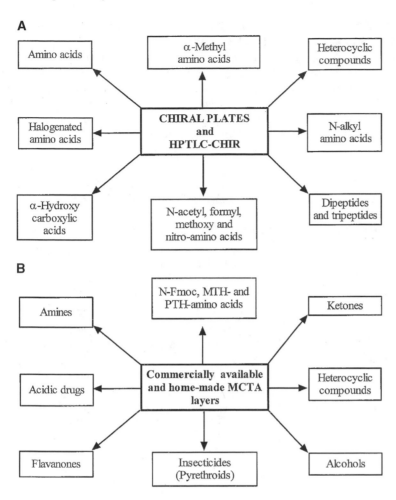

Fig. 1 Classes of chiral organic compounds resolved (A) by ligand exchange chromatography on ChiralPlate and HPTLC Chir plates and (B) on MCTA layers.

achieved using chiral mobile phases in both normal- and reversed-phase chromatography. The first technique uses silica gel and, mostly, diol F_{254} HPTLC plates (Merck) and, as chiral selectors, D-galacturonic acid for ephedrine, N-carbobenzoxy (CBZ)-L-amino acids or peptides and $1R(-)$-ammonium-10-camphorsulfonate for several drugs, and 2-O-[(R)-2-hydroxypropyl)]-β-cyclodextrin for underivatized amino acids. Extremely high ΔR_f values (0.05–0.25) have been observed for the various pairs of enantiomers, proving the strong enantioselectivity of this system.

Most separations have been obtained by reversed-phase chromatography on hydrophobic silica gel (RP-18W/UV_{254} and Sil C_{18}-50/UV_{254} from Macherey-Nagel, Germany; KC2F, KC18F, and chemically bonded diphenyl-F from Whatman, U.S.A., and RP-18W/F_{254} from Merck, Germany) as stationary phase and β-cyclodextrin and its derivatives, bovine serum albumin (BSA), and the macrocyclic antibiotic vancomycin as chiral agents. Enantiomers that interact selectively with β-cyclodextrin cavities are generally N-derivatized amino acids, whereas the use of BSA as chiral selector is able to resolve many

N-derivatized amino acids, tryptophan and its derivatives, derivatized lactic acid, and unusual optical antipodes such as binaphthols.

In particular, the resolution of dansyl-D- and L-amino acids by reversed-phase TLC using aqueous mobile phases containing methanol or acetonitrile and β-cyclodextrin as chiral selector is very useful for stereochemical analysis of amino acids from small peptides, since a significant number of naturally occurring peptides and peptide antibiotics isolated from plants and micro-organisms contain at least one amino acid in D-configuration.[5]

QUANTITATIVE ANALYSIS OF TLC-SEPARATED ENANTIOMERS

Although TLC/mass spectrometry (MS) has been shown to be technically feasible and applicable to a variety of problems, TLC is generally coupled with spectrophotometric methods for quantitative analysis of enantiomers. Optical quantitation can be achieved by in situ densitometry by measurement of UV/VIS absorption, fluorescence, or fluorescence quenching, or after extraction of solutes from the scraped layer. The evaluation of detection limits for separated enantiomers is essential because precise determinations of trace levels of a D- or L-enantiomer in an excess of the other become more and more important. Detection limits as low as 0.1% of an enantiomer in the other have been obtained.

REFERENCES

1. Cahn, R.S.; Ingold, C.K.; Prelog, V. Specification of asymmetric configuration in organic chemistry. Experientia **1956**, *12*, 81–94.
2. Dalgliesh, C.E. The optical resolution of aromatic amino acids on paper chromatograms. J. Chem. Soc. **1952**, *III*, 3940–3943.
3. Lepri, L.; Del Bubba, M.; Cincinelli, A.; Bracciali, M. Quantitative determination of enantiomeric alcohols by planar chromatography on tribenzoylcellulose. J. Planar Chromatogr. Mod. TLC **2002**, *153*, 220–222.
4. Suedee, R.; Srichana, T.; Saelim, J.; Thavonpibulbut, T. Thin layer chromatographic separation of chiral drugs on molecularly imprinted chiral stationary phases. J. Planar Chromatogr. Mod. TLC **2001**, *14*, 194–198.
5. Le Fevre, J.W.; Gublo, E.J.; Botting, C.; Wall, R.; Nigro, A.; Pham, M.L.T.; Ganci, G. Qualitative reversed-phase thin-layer chromatographic analysis of the stereochemistry of D- and L-α-amino acids in small peptides. J. Planar Chromatogr. Mod. TLC **2000**, *13*, 160–165.

SUGGESTED FURTHER READING

Gunther, R.; Möller, K. Enantiomer separations. In *Handbook of Thin Layer Chromatography*; Sherma, J., Fried, B., Eds.; Marcel Dekker, Inc.: New York, 1996; 621–682.

Lepri, L.; Del Bubba, M.; Cincinelli, A. Chiral separations by TLC. In *Planar Chromatography, A Retrospective View for the Third Millenium*; Nyiredy, Sz., Ed.; Springer Scientific Publisher: Hungary, 2001; 517–549.

Prosek, M.; Puki, M. Basic principles of optical quantitation in TLC. In *Handbook of Thin Layer Chromatography*; Sherma, J., Fried, B., Eds.; Marcel Dekker, Inc.: New York, 1996; 273–306.

Enantioseparation by Capillary Electrochromatography

Yulin Deng
University of Saskatchewan, Saskatoon, Saskatchewan, Canada

INTRODUCTION

Capillary electrochromatography (CEC) is considered to be a hybrid technique that combines the features of both capillary HPLC and capillary electrophoresis (CE). In CEC, a mobile phase is driven through a packed or an open tubular coating capillary column by electro-osmotic flow[1,2] and/or pressurized flow.[3] The first electrochromatographic experiments were done in early 1974 by Pretorius et al.,[4] who applied an electric field across a packed column. This allows the analyte to partition between the mobile and stationary phases. As a high voltage is applied, electrophoretic mobility should also contribute to the chromatographic separation for charged analyses. The ability of CEC to combine electrophoretic mobility with partitioning mechanisms is one of its strongest advantages. For electroosmotically driven capillary electrochromatography (ED-CEC), the resulting flow profile is almost pluglike; thus, a high column efficiency, comparable to that in CE, can be obtained. For pressure-driven capillary electrochromatography (PD-CEC), although dispersion caused by flow velocity differences causes zone broadening, plate numbers are higher than in capillary HPLC due to the contribution of the electric field to total flow rate. Unlike ED-CEC, the use of an HPLC pump provides stable flow conditions and, thus, offers improvements in retention reproducibility, in sample introduction (e.g., split injection), in suppression of bubble formation, and in gradient elution. More importantly, because the solvent can be mainly driven by pressurized flow, the change of the direction of electric field is no longer limited, and the separation of mixtures of cationic, anionic and neutral compounds becomes possible in a single run. Additionally, neutral molecules can be separated without micelles or other organic additives; this makes CEC more amenable to coupling with mass spectrometry.

Chiral separation in CE is usually achieved by the addition of chiral complexing agents to form *in situ* diastereometric complexes between the enantiomers and the chiral complexing agent. Many of the chiral selectors successfully used in HPLC[5] can also be applied in CE, and thus the experience from both HPLC and CE can be transferred to CEC. During the last few years, interest in CEC has increased due to the improvement in the preparation of capillary columns[6,7] and in the stability and efficiency of separations.[6–9] A limited but dramatically increasing number of chiral separations in CEC have been reported so far. This review will be mainly devoted to recent developments and applications. We are also interested in exploring the potential advantages offered by CEC and, in particular, its practical utility for enantioseparation.

ENANTIOSELECTIVITY IN CEC

CEC is a more complicated system than CE and HPLC due to the combination of both electrophoretic and chromatographic transport mechanisms. It is difficult to define an effective selectivity (separation factor) as in the case of general chromatography or general electrophoresis. To better illustrate the interactions that control selectivity, we defined a relative selectivity and postulated a model that illustrates the effect of separation parameters on the enantioselectivity.[10]

For enantioseparation chiral stationary phases (CSPs), an expression of the relative selectivity is obtained:

$$\alpha_r = \frac{\phi(K_{f2} - K_{f1})}{1 + \phi K_2 + \phi K_{f2}} \tag{1}$$

Interestingly, this equation indicates that the electrophoresis mechanism does not influence the enantioselectivity and the electric field only plays a role in driving the mobile phase.

For enantioseparation with chiral additives in CEC, we derived another expression:

$$\alpha_r = \frac{(K_{f1} - K_{f2})[\phi K v_c + (\mu_c - \mu_f)E][C]_m}{(1 + \phi K + K_{f2}[C]_m)(v_f + v_c K_{f1}[C]_m)} \tag{2}$$

where v_f and v_c are the apparent flow velocity of the free analyte and the complexed analyte, respectively. Both Eqs. (1) and (2) show that the enantioselectivity is not only dependent on the difference in formation constants (K_f) between a pair of enantiomers with

Encyclopedia of Chromatography DOI: 10.1081/E-ECHR-120039976

the chiral agents but also is influenced by some experimental factors. Substantially, chiral recognition of enantiomers is the direct result of the transient formation of diastereomeric complex between enantiomeric analytes and the chiral complexing agent (i.e., the difference in formation constants). However, the importance of experimental factors lies in the fact that they can convert the intrinsic difference into the apparent difference in migration velocity along the column. Therefore, the overall selectivity in chiral separation can be considered to be made up of two contributing factors: the intrinsic difference (intrinsic selectivity) in formation constants of a pair of enantiomers, and the conversion efficiency (exogenous selectivity) of the intrinsic difference into the apparent difference in the migration velocity. According to Eq. (2), these experimental factors may include the equilibrium concentration of a chiral selector, the electric field strength, and the properties of the stationary phase.

In CEC with chiral additives, Eq. (2) shows that there exists a maximum selectivity at the optimal concentration of chiral selector. The optimal concentration is not only dependent on the formation constants (K_{f1}, K_{f2}) but also on properties of the column (ϕ and K){i.e., $[C]_{opt} = \sqrt{(1 + \phi K)/K_{f1}K_{f2}}$}.

Unlike in the case of a chiral column, the selectivity in CEC with chiral additives is determined by both partition and electrophoresis, and the electric field either increases or decreases the selectivity. Table 1 summarizes the relationship between the direction of field strength and the electrophoretic mobility of the free and complexed analytes. For PD-CEC, the solvent is mainly driven by pressurized flow; thus, there is no limitation to change the direction of electric field.

For enantioseparation on CSPs in CEC, nonstereospecific interactions, expressed as ϕK, contribute only to the denominator as shown in Eq. (1), indicating that

any nonstereospecific interaction with the stationary phase is detrimental to the chiral separation. This conclusion is identical to that obtained from most theoretical models in HPLC. However, for separation with a chiral mobile phase, ϕK appears in both the numerator and denominator [Eq. (2)]. A suitable ϕK is advantageous to the improvement of enantioselectivity in this separation mode. It is interesting to compare the enantioselectivity in conventional CE with that in CEC. For the chiral separation of salsolinols using β-CyD as a chiral selector in conventional CE, a plate number of 178,464 is required for a resolution of 1.5. With CEC (i.e., $\phi K = 10$), the required plate number is only 5976 for the same resolution.[10] For PD-CEC, the column plate number is sacrificed due to the introduction of hydrodynamic flow, but the increased selectivity markedly reduces the requirement for the column efficiency.

CHIRAL SEPARATION IN CEC

There are different ways of performing chiral separation by CEC. Mayer and Schurig immobilized the chiral selectors by coating or chemically binding them to the wall of the capillary.[11,12] Permethylated β- or γ-CyD was attached via an octamethylene spacer to dimethylpolysiloxane (Chirasil-Dex) as the stationary phase. A high efficiency ($\sim 250,000/M$) was obtained for the separation of 1,1'-dinaphthyl-2,2'-diyl hydrogenphosphate. An alternative coating approach was developed by Sezeman and Ganzler. Linear acrylamide was coated on the capillary wall, and after polymerization, CyD derivatives were bound to the polymer.[13]

Chiral separation can also be performed with packed capillaries. β-CyD-bonded CSPs that are most frequently used in HPLC and CE were successfully applied in CEC. The separation of a variety of chiral compounds, such as some amino acid derivatives benzoin and hexobarbital was achieved by using CSPs bonded with different CyD derivatives.[14,15] Proteins are not ideal for use as buffer additives in CE because of their large detector response; however, CEC may be a good way to use this type of chiral selectors. Lloyd et al. have performed CEC enantioseparation by using commercially available protein CSPs, such as AGP and HAS.[16,17] The resolution obtained on protein CSPs was good; the efficiency, however, was rather poor. Another HPLC–CSP based on cellulose derivatives has been also reported for enantioseparation by CEC.[18] CSPs modified by covalent attachment of poly-N-acryloyl-L-phenylalanineethylester or by coating with cellulose tris(3,5-dimethylphenylcarbamate) can be performed in the reversed-phase mode.

Table 1 Relationship between the field strength and the electrophoretic mobility for getting high enantioselectivity in CEC

Direction of μ_{ep}		Size relationship (absolute value)	Direction of E
μ_f	μ_c		
+	+	$\mu_f < \mu_c$	+
+	+	$\mu_f > \mu_c$	−
+	−	a	−
−	−	$\mu_f < \mu_c$	−
−	−	$\mu_f > \mu_c$	+
−	+	a	+

aThe selection of direction of electric field is not influenced by size relationship in absolute values between the electrophoretic mobility of the free and complexed analytes.

Acetonitrile as organic modifier was found to be advantageous for this type of CSP. An anion-exchange-type CSP was recently developed for the separation of N-derivatized amino acids.[19] The new chiral sorbent was modified with a basic *tert*-butyl carbamoyl quinine. Enantioselectivity obtained in CEC was as high as in HPLC and efficiency was typically a factor of 2–3 higher than in HPLC. A recent innovative approach is the use of imprinted polymers as CSPs in CEC.[20,21] Imprinted polymers possess a permanent memory for the imprinted species, and, thus, their enantioselectivity is predetermined by the enantiomeric form of the templating ligand. The use of imprint-based CSPs in HPLC is hampered by their poor chromatographic performance. CEC, however, was found to greatly improve the efficiency of the imprint-based separation. The most successful approach is the use of capillary columns filled with a monolithic, superporous imprinted polymer obtained by an *in situ* photo-initiated polymerization process. This technique enables imprint-based column to be operational within 3 h from the start of preparation. Generally, the imprint-based CSPs show high enantioselectivity but somewhat low efficiency and are limited to the separation of very closely related compounds.

Enantioseparation can be achieved on a conventional achiral stationary phase by the inclusion of an appropriate chiral additive into the mobile phase. It is theoretically predicted that the enantioselectivity in CEC with a chiral additive may be higher than that using a chiral column with the same chiral selector.[10] Lelievre et al. compared an HP-β-CyD column and HP-β-CyD as an additive in the mobile phase with an achiral phase (ODS) to resolve chlortalidone by CEC.[22] It was demonstrated that resolution on ODS with the chiral additive was superior on the CSP; however, efficiency was low. With an increasing amount of acetonitrile, the peak shape was improved and the migration time was decreased. We achieved the separation of salsolinol by the use of CEC with β-CyD as a chiral additive in the mobile phase containing sodium 1-heptanesulfonate, as shown in Fig. 1. Salsolinol is a hydrophilic amine and is difficult to enantioseparate due to the small k' values on the reversed stationary phases. Sodium 1-heptanesulfonate was used as a counterion to improve the retention.

In conclusion, CEC has great potential in separation technology. Our theoretical model as well as many published practices in CEC show clearly that the benefit of combining electrophoresis and partitioning mechanisms in CEC is the increase in selectivity for the separation. The intrinsic difference in formation constants is critical, but the experimental factors, such as electric field or the stationary and mobile phases,

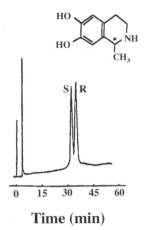

Time (min)

Fig. 1 Electrochromatogram of salsolinol enantiomers on a packed capillary column. Column: ODS-C18, 29 cm (23 cm effective length) × 75 μm ID; applied electric field strength: ~250 V/cm; mobile phase: 20 mM sodium phosphate buffer (pH 3.0) containing 12 mM β-cyclodextrin and 5 mM sodium 1-heptanesulfonate. The pump was set at the constant pressure of 100 kg/cm².

can also contribute to the improvement of the overall enantioselectivity via increasing the conversion efficiency. However, only when both electrophoretic and partitioning mechanisms act in the positive effects, can high overall enantioselectivity in CEC be obtained.

REFERENCES

1. Tsuda, T.; Nomura, K.; Nagakawa, G. Open-tubular microcapillary liquid chromatography with electro-osmosis flow using a UV detector. J. Chromatogr. **1982**, *248*, 241.
2. Jorgenson, J.W.; Lukacs, K.D. High-resolution separations based on electrophoresis and electro-osmosis. J. Chromatogr. **1981**, *218*, 209.
3. Tsuda, T. Direct chiral separations by capillary electrophoresis using capillaries packed with an .alpha.1-acid glycoprotein chiral stationary phase. LC–GC Int. **1992**, *5*, 26.
4. Pretorius, V.; Hopkins, B.J.; Schieke, J.D. J. Chromatogr. **1974**, *99*, 23.
5. Deng, Y.; Maruyama, W.; Kawai, M.; Dostert, P.; Naoi, M. *Progress in HPLC and HPCE*; VSP: Utrecht, 1997; Vol. 6, 301.
6. Boughtflower, R.J.; Underwood, T.; Paterson, C.J. Capillary electrochromatography: some important considerations in the preparation of packed capillaries and the choice of the mobile phase buffers. Chromatographia **1995**, *40*, 329.
7. Yan, C. U.S. Patent 54531631995.

8. Taloy, M.R.; Teale, P.; Westwood, S.A.; Perrett, D. Analysis of corticosteroids in biofluids by capillary electrochromatography with gradient elution. Anal. Chem. **1997**, *69*, 2554.

9. Eimer, T.; Unger, K.K.; Tsuda, T. Pressurized flow electrochromatography with reversed phase capillary columns. Fresenius J. Anal. Chem. **1995**, *352*, 649.

10. Deng, Y.; Zhang, J.; Tsuda, T.; Yu, P.H.; Boulton, A.A.; Cassidy, R.M. Modeling and optimization of enantioseparation by capillary electrochromatography. Anal. Chem. **1998**, *70*, 4586.

11. Mayer, S.; Schurig, V. Enantiomer separation by electrochromatography on capillaries coated with chirasil-dex. J. High Resolut. Chromatogr. **1992**, *15*, 129.

12. Mayer, S.; Schurig, V. Enantiomer separation by electrochromatography in open tubular columns coated with chirasil-dex. J. Liquid Chromatogr. **1993**, *16*, 915.

13. Sezemam, J.; Ganzler, K. Use of cyclodextrins and cyclodextrin derivatives in high-performance liquid chromatography and capillary electrophoresis. J. Chromatogr. A. **1994**, *668*, 509.

14. Li, S.; Lloyd, D.K. Packed-capillary electrochromatographic separation of the enantiomers of neutral and anionic compounds using β-cyclodextrin as a chiral selector: effect of operating parameters and comparison with free-solution capillary electrophoresis. J. Chromatogr. A. **1994**, *666*, 321.

15. Wistuba, D.; Czesla, H.; Roeder, M.; Schurig, V. Enantiomer separation by pressure-supported electrochromatography using capillaries packed with a permethyl β cyclodextrin stationary phase. J. Chromatogr. A. **1998**, *815*, 183.

16. Li, S.; Lloyd, D.K. Direct chiral separations by capillary electrophoresis using capillaries packed with an .alpha.1-acid glycoprotein chiral stationary phase. Anal. Chem. **1993**, *65*, 3684.

17. Lloyd, D.K.; Li, S.; Ryan, P. Protein chiral selectors in free-solution capillary electrophoresis and packed-capillary electrochromatography. J. Chromatogr. A. **1995**, *694*, 285.

18. Krause, K.; Girod, M.; Chankvetadze, B.; Blasehk, G. Enantioseparations in normal- and reversed-phase nano-high-performance liquid chromatography and capillary electrochromatography using polyacrylamide and polysaccharide derivatives as chiral stationary phases. J. Chromatogr. A. **1999**, *837*, 51.

19. Lammerhofer, M.; Lindner, W. High-efficiency chiral separations of N-derivatized amino acids by packed-capillary electrochromatography with a quinine based chiral anion exchange type stationary phase. J. Chromatogr. A. **1998**, *829*, 115.

20. Schweitz, L.; Andersson, L.I.; Nilsson, S. Capillary electrochromatography with predetermined selectivity obtained through molecular imprinting. Anal. Chem. **1997**, *69*, 1179.

21. Schweitz, L.; Andersson, L.I.; Nilsson, S. Molecular imprint-based stationary phases for capillary electrochromatography. J. Chromatogr. A. **1998**, *817*, 5.

22. Lelievre, F.; Yan, C.; Zare, R.N.; Gareil, P. Capillary electrochromatography: operation characteristics and enantiomeric separations. J. Chromatogr. A. **1996**, *723*, 145.

End Capping

Kiyokatsu Jinno
School of Materials Science, Toyohashi University of Technology, Toyohashi, Japan

INTRODUCTION

A typical stationary phase for chromatography, especially LC, is a chemically alkyl (C_{18})-bonded phase on silica gel particles. For the preparation of this type of bonded phase, alkylsilane is used to react with the silica gel surface by a silane-coupling reaction. In order to perform this synthesis, the silica gel to be bonded is treated to remove heavy metals and to prepare the surface for better bonding. Generally, only one of the functional groups bonds to form a Si–O–Si bond. Less often, two of the functional groups react to form adjacent Si–O–Si bonds. The remaining functional groups on each reagent molecule hydrolyze to form Si–O–H groups during workup, following the initial reaction. These groups, however, which form with the di- and tri- functional reagents, can cross-link with one another near the surface of the silica gel support. Thus, bonded phases made with any di- or tri- functional reagents are termed "polymeric" phases. A monofunctional silane reagent can only bond to the silanols and any excess is washed free as the ether resulting from hydrolysis of the reagent. Any packing made with a monofunctional silane reagent is referred to as a "monomeric" bonded phase. These schemes are summarized in Fig. 1(a)(i) and (ii). Other chemically bonded phases, such as cyano-, amino-, and shorter or longer alkyl phases are synthesized by similar bonding chemistries.

DISCUSSION

The products made by the above synthetic processes still have large numbers of residual silanols, which lead to poor peak shapes or irreversible adsorption, because chemically bonded groups on the silica gel surface have large, bulky molecular sizes and, after the bonding, the functionalized silane cannot react with the silanols around the bonded ligands. Because such alkyl-bonded phases are used for reversed-phase (RP) separations, especially for chromatography of polar molecules, any silanol groups that remain accessible to solutes after the bonding are likely to make an important contribution to the chromatography of such solutes; this is generally detrimental to the typical RP LC separations. It is a common fact that the residual silanols produce peak tailing for highly polar compounds which will interact with these silanol groups with deleterious effects. Therefore, the attempt to reduce the number of residual silanols on the silica gel is a common procedure in the preparation of chemically bonded stationary phases, where the surface of a RP material is ensured to be uniformly hydrophobic, for example, by blocking residual silanol groups with some functional groups. This process is the so-called "end capping." The end-capping process is possible with a smaller molecule than alkylchlorosilanes, such as a trimethyl-substituted silane (from trimethylchlorosilane or hexamethyldisilazane) as seen in Fig. 1(a)(iii). Because the molecular weights of these reagents are small, they do not add much to the total percent carbon, compared with the initial bonded phase. It must be known that all chemically bonded phases on silica gel cannot be end-capped by this process, because the above reagents can react with diol and amino phases, and not only with silanol groups on the surface. To block, end cap, and then unblock these phases would be very time-consuming and too expensive to be practical. If the final bonded phase is, in fact, a diol, this silane-bonding reagent is made from glycerol and has the structure $Si–O–CHOH–CH_2OH$. The cyano or amino phases are most often attached with a propyl group between the silicon atom and the CN or NH_2 group.

Often, when various bonded phases are studied for suitability for a particular separation, the question arises as to which is bonded most completely. This is a common question, because all phases, no matter how they are bonded, will have some residual silanols, even after an end-capping process. It is impossible for the bulkier bonding reagents to reach any but the most sterically accessible silanols. It is much easier for the smaller solutes to reach the silanols, however, and be affected by them. The final surface of the silica gel has three different structures, as demonstrated in Fig. 1(b)(i), (ii), and (iii), for a monomeric C_{18}, end capped by trimethylchlorosilane and residual silanols, respectively.

The presence of residual silanol groups can be detected most readily by using Methyl Red indicator,[1] which turns red in the presence of acidic silanol groups, but a more sensitive test is to chromatograph a polar solute on the RP material.

Encyclopedia of Chromatography DOI: 10.1081/E-ECHR-120039977

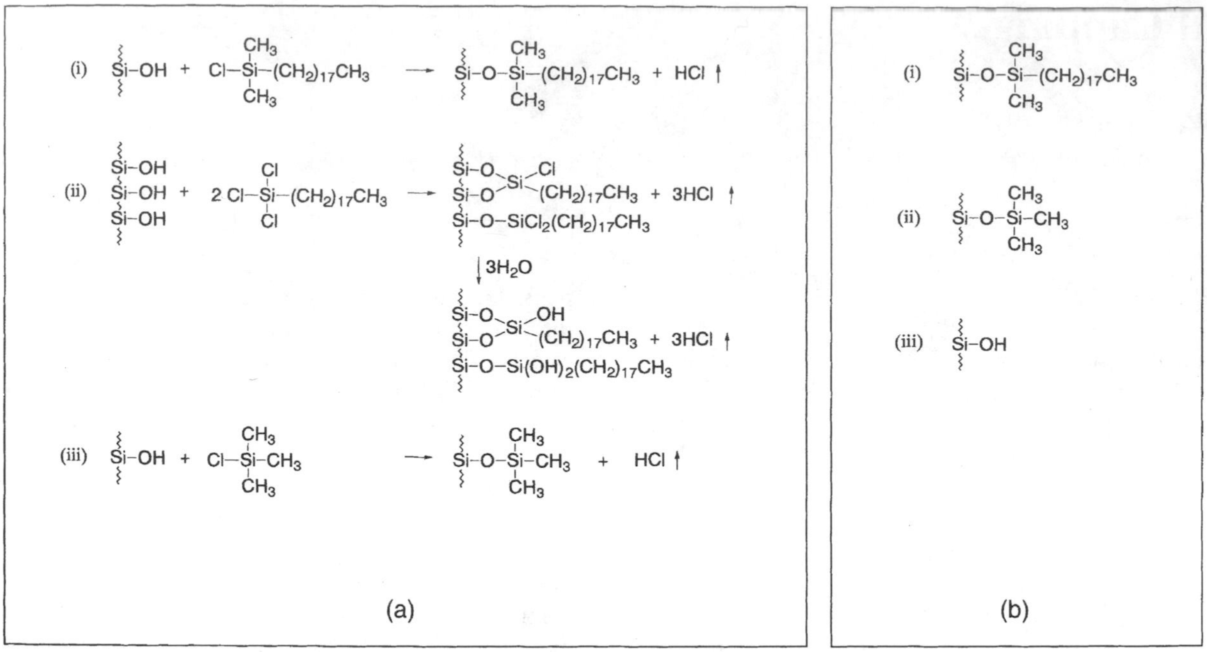

Fig. 1 (a) Scheme of bonding chemistry for chemically bonded C_{18} silica phase: (i) synthesis of monomeric C_{18}; (ii) synthesis of polymeric C_{18}; (iii) end-capping process. (b) Surface structure of a monomeric C_{18} phase: (i) monomeric C_{18} ligand; (ii) end-capped trimethyl ligand; (iii) residual silanol.

To test, chromatographically, any phase for residual silanols, the column has to be conditioned with heptane or hexane (which has been dried overnight with spherical 4A molecular sieves). The series of solvents to use if the column has been used with water or a water-organic mobile phase, such as water → ethanol → acetone → ethyl acetate → chloroform → heptane. Once activated, a sample of nitrobenzene or nitrotoluene is injected, eluted with heptane or hexane, and detected at 254 nm. The degree of retention is then a sensitive guide to the presence or absence of residual silanols; if the solute is essentially unretained, the absence of silanols may be assumed. The better the bonding, the faster the polar compound will be eluted from the column. A wel-bonded and end-capped phase will have a retention factor of between 0 and 1. Less well-covered silicas can have retention factors greater than 10. This is a comparative test, but it can also be useful for examining a phase to see if the end-capping reagent or primary phase has been cleaved by the mobile phase used over a period of time. Other methods to measure the silanol content of silica and bonded silica have been discussed by Unger.[2] Solid-state nuclear magnetic resonance spectrometry is the most powerful method to identify the species of residual silanol groups on the silica gel surface.[3]

In order to avoid the contribution of the residual silanols to solute retention, many packing materials that should not have silanols have been developed.[4] They are polymer-based materials and also polymer-coated silica phases. These polymer-based or polymer-coated phases can be recommended as very useful and stable stationary phases in LC separations of polar compounds; they also offer much better stability for use at higher pH alkaline conditions.

REFERENCES

1. Karch, K.; Sebestian, I.; Halasz, I. Preparation and properties of reversed phases. J. Chromatogr. **1976**, *122*, 3.
2. *Packings and Stationary Phases in Chromatographic Techniques*; Unger, K.K., Ed.; Marcel Dekker, Inc.: New York, 1990.
3. Pursch, M.; Sander, L.C.; Albert, K. Understanding reversed-phase LC with solid-state NMR. Anal. Chem. **1999**, *71*, 733A.
4. Unger, K.K. *A Guide to Practical HPLC*; GIT Verlag: Darmstadt, 1999.

Enoxacin: Analysis by Capillary Electrophoresis and HPLC

Hassan Y. Aboul-Enein
Imran Ali
Pharmaceutical Analysis Laboratory, Biological and Medical Research Department,
King Faisal Specialist Hospital and Research Center, Riyadh, Saudi Arabia

INTRODUCTION

Enoxacin, 1-ethyl-6-fluoro-1,4-dihydro-4-oxo-7-(1-piperazinyl)-1,8-naphthyridine-3-carboxylic acid (ENX; Fig. 1), is a new broad spectrum fluorinated 4-quinolone antibacterial agent.[1] It has a broad spectrum of antibacterial activity and is particularly potent against Gram-negative organisms and staphylococci.[2] The 4-quinolone antibiotics have been used in the treatment of many soft tissue infections including bacterial prostatitis.[3] ENX is excreted, mainly, in urine as the unchanged drug. It is metabolized by oxidation (to oxo-enoxacin), by conjugation with formic and acetic acid (ring opening), and by deamination of the piperazinyl ring. Its major metabolite, oxo-enoxacin, accounts for 10–15% of the administered dose and each of the other metabolites constitutes less than 1% of the dose.[2] It has also been reported that ENX has potent competitive inhibitory effects on theophylline metabolism, causing elevated plasma theophylline concentration and potential toxicity.[4] Because of these properties of ENX, it is very important to develop suitable analytical methods for this substance. HPLC is the most commonly employed method for the determination of ENX and its metabolites in plasma, urine, and tissues.[5–12] Capillary electrophoresis (CE) is becoming a reliable, preferable, and alternative method, especially for the analysis of drugs in biological matrices.[13–14] CE offers some advantages such as rapidity, short analysis time, and low cost.[15–17] Only one report on the analysis of ENX by CE is presented by Tuncel et al.[18] Further, the authors have also carried out the analysis of ENX by HPLC and compared this method with the CE method in pharmaceutical dosage forms and in biological fluids.

DETERMINATION OF ENX BY CE

Instruments

The experiments were conducted using a Spectrophoresis 100 system equipped with a modular injector and high-voltage power supply, and a model Spectra FOCUS scanning CE detector (Thermo Separation Products, California, U.S.A.) connected to a Model Etacomp 486 DX 4-100 computer which processed the data using PC 1000 (Version 2.6) running under the OS/2 Warp program (Version 3.0). The analysis was performed in a fused silica capillary which has a total length of 88 cm, an effective length of 58 cm, and an I.D. of 75 µm (Phenomenex, California, U.S.A.). The pHs of the solutions were measured with a Multiline P4 pH meter with SenTix glass electrode (WTW, Weilheim, Germany). All the solutions were filtered using a Phenex microfilter (25 mm, 0.45 µm) (Phenomenex) and were degassed using a model B-220 ultrasonic bath (Branson, Connecticut, U.S.A.).

Chemicals

Acetonitrile, methanol (HPLC grade), ethanol, propanol, hydrochloric acid, sodium hydroxide, borax, acetylpipemidic acid (internal standard (IS) for CE), and 3,4-dihydroxybenzylamine HBr (IS for HPLC) were from Merck (Darmstadt, Germany). Enoxacin was generously provided by Eczacibasi Ilac Sanayi ve Ticaret A.S. (Istanbul, Turkey). Blood samples were withdrawn from healthy volunteers after obtaining their consent. The serum samples were separated by centrifuging for 10 min at $5000 \times g$. Double-distilled water was used to prepare all the solutions. A stock solution of ENX (10 mg/25 mL of methanol) was prepared. Dilutions were made in the range of 2.5×10^{-5} to 1.2×10^{-4} M, each containing 0.25 µmol IS (acetylpipemidic acid) for CE and 0.11 µmol IS (3,4-dihydroxybenzylamine HBr) for HPLC. All the dilutions for CE were prepared in a background electrolyte. The background electrolyte was a 20 mM borate buffer at pH 8.6 for the CE experiments. The dilutions were analyzed by applying a +30-kV potential, injecting the sample 1s and detecting at 265 nm where ENX and acetylpipemidic acid (IS) absorb the monochromatic light equivalently.

Procedure for CE Analysis

The fused silica capillary tubing was filled with the background electrolyte (pH 8.6; 20 mM borate).

Encyclopedia of Chromatography DOI: 10.1081/E-ECHR-120013364

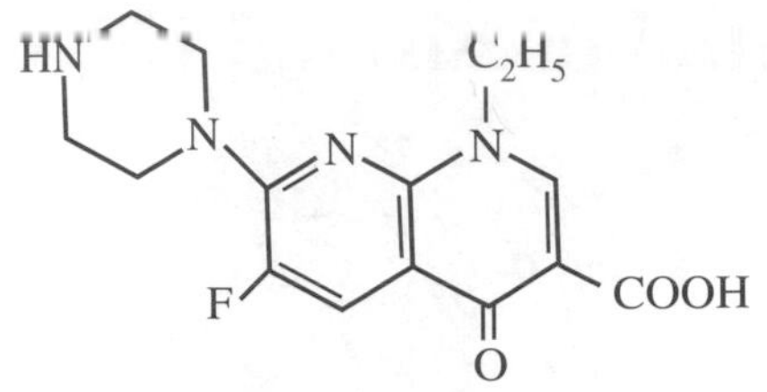

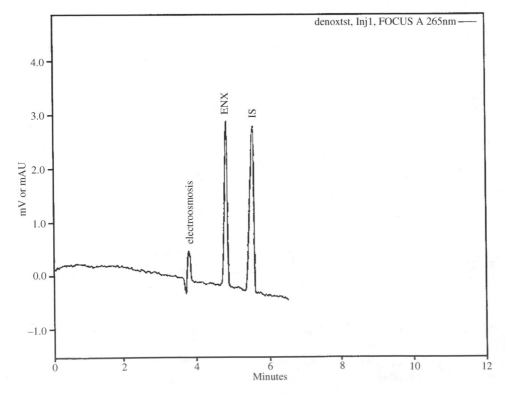

Fig. 1 The chemical structure of ENX.

Both ends of the tube were dipped into a reservoir (8 mL) and a vial (1.1 mL) filled with the background buffer. The end part where the sample (side of vial) was introduced was connected with a platinum electrode to the positive high-voltage side of the power supply. The reservoir side at the detector end was connected with a platinum electrode to the ground. Samples at a concentration of 7.7×10^{-5} M for the optimization of CE parameters were introduced by 1s of vacuum injection corresponding to almost 65 nL. Before each run, the capillary was purged for 2 min with 0.1 M sodium hydroxide solution, then for another 2 min with double-distilled water. It was then equilibrated by passing the background electrolyte for 5 min prior to operation.

A background electrolyte consisting of borax was preferred for conducting the initial CE experiments because ENX has a carboxylic group on its structure. Several pH values were tested in the range from 8.45 to 9.95 using the concentration of 20 mM borax buffer. It was observed that the ENX (1.26×10^{-4} M) peak appeared in all the studied pH values, but the migration time of ENX, as expected, increased with increasing pH. Phosphate and citrate buffers of the same pH and concentration (8.6, 20 mM) were used to compare the effect of the nature of the buffer components. The migration time (t_M) of ENX was not affected by the buffer components, but the repeatability of the peak areas decreased with the use of citrate and phosphate buffers. It is concluded that some optimization studies are required if these buffer systems are to be used.

The influence of borax buffer concentration was investigated in the range from 10 to 100 mM. The sharpest peaks were obtained in the use of 10–30 mM concentrations and the t_M of ENX was almost constant, and an increase was observed in the use of borax concentration above 30 mM, but peak deformation also occurred due to the heat production by the Joule effect. In order to achieve optimization of the proposed analytical procedure, low buffer concentration was considered to decrease the electrophoretic mobility that corresponds to short analysis time. Based on the above results, the most convenient buffer system was 20 mM borate buffer at pH 8.6. Since the separation depends on the conditioning of the capillary inner surface in the CE analysis, the t_M and peak integration values might be very similar to the HPLC techniques.

Fig. 2 Typical electropherogram of standard ENX (7.7×10^{-5} M) and IS (acetylpipemidic acid, 5.18×10^{-5} M). Conditions: 20 nM boratem pH 8.6; injection, hydrodynamically 1 s; applied voltage, +30 kV; capillary, uncoated fused silica, 75 μm I.D., 88 cm total, and 58 cm effective length; detection at 265 nm.

Table 1 Precision of peak areas (days = 3; n = 6)

Precision of peak areas (RSD%)	PN no IS	IS no PN	IS and PN
Repeatability	2.80	0.99	0.99
Intermediate precision	3.24	2.86	1.36

Table 3 Precision of the method (spiked placebos)

Concentration levels (%)	50	100	150
Repeatability (days = 3; n = 6, RSD%)	2.0	2.5	2.1
Intermediate precision (days = 3; n = 6, RSD%)	3.3	3.0	3.1

The electropherogram of ENX and acetylpipemidic acid (IS) in the background electrolyte is shown in Fig. 2. The signal of electroosmosis and the migration time of the peaks of ENX and IS appeared at 3.8, 4.8, and 5.5 min, respectively. From the integration data, the net mobility toward the cathode (electroosmosis) and the ENX and IS toward the anode (electrophoretic) were 7.56×10^{-4}, 5.96×10^{-4}, and 4.6×10^{-4} cm^2 V^{-1}s^{-1}, respectively. The capacity factors were 3.82 (ENX) and 4.53 (IS).

Certain evaluation methods were examined based on the quantification processes. These can be divided into three groups, employing only the values of peak normalization (PN). The effect of the use of IS and certain evaluation methods, such as correction of peak area (normalization), was calculated by dividing the related peak area into t_M on which the precision was examined. These can be divided into three groups: a) employing only the area values of PN (PN no IS); b) computing the ratio values (IS no PN); and c) using the area values of peak normalized IS and ENX (IS and PN). The precision of the peak areas was calculated as shown in Table 1.

The success of the CE experiments from an analytical point of view depends on the conditioning of the capillary surface. Therefore, cleaning and conditioning processes, as explained in the experimental part, must be repeated after each injection to provide optimum resolution and reproducibility. The precision of the peak area was also assessed by considering certain evaluations, such as the effects of correction of peak area (normalization), which were found by the division of the related peaks into the corresponding migration times; the use of an internal standard was studied.

As seen from Table 1, the lowest RSD% values were obtained from those of IS and PN. Thus, such evaluation was considered throughout the rest of the study. A series of standard ENX solutions in the concentration range of 2.5×10^{-5} through 1.2×10^{-4} M, with each solution containing 0.25 µmol at a fix concentration of IS, were prepared and injected (n = 3). Linear regression lines were obtained by plotting the ratios of normalized peak areas to those of the internal standard vs. the analyte concentration. The calibration equation was computed using a regression analysis program, considering the ratio values vs. the related concentrations. The results are presented in Table 2.

Method Validation and Accuracy

From the electropherogram in Fig. 2, no interference from the formulation excipients could be observed at the migration times of ENX and IS. The limit of the detection (LOD) was 3.85×10^{-7} M, while the limit of quantification (LOQ) was 1.16×10^{-6} M. The results indicate good precision. Method accuracy was determined by analyzing a placebo (mixture of excipients) spiked with ENX at three concentration levels (n = 6) covering the same range as that used for linearity. Mean recoveries with 95% confidence intervals are given in Table 3.

Determination of ENX by HPLC

HPLC experiments were carried out using a Model 510 Liquid Chromatograph equipped with a Model 481 UV detector (Waters Associates, Milford, Massachusetts, U.S.A.). The chromatograms were processed by means of a chromatographic workstation (Baseline 810). Separation was performed on a reversed-phase Supelcosil LC-18 column (250×4.6 mm I.D., 5 µm particle size) (Supelco, St. Louis, Missouri, U.S.A.).

Table 2 Linearity and accuracy of the method (spiked placebos)

Regression parameters			
Linearity	$r^2 = 0.9998$	intercept (mean ± SD) = $-0.02342 \pm 7.32 \times 10^{-3}$	slope (mean ± SD) = 9813.18 ± 102.31
Accuracy	50%	100%	150%
Mean recovery ±CI%	101.4 ± 2.12	101.1 ± 2.63	100.4 ± 2.25

The samples were injected into a 50-μL loop using a Rheodyne 7125 valve (Rheodyne, Cotati, California, U.S.A.).

Procedure for HPLC Analysis

The HPLC conditions were optimized by using different mobile and stationary phases. During HPLC experiments, a 10 mM phosphate buffer (pH 4.0)/acetonitrile (85:15, v/v) was used as a mobile phase. 3,4-Dihydroxybenzylamine-HBr (IS) was found to be a suitable internal standard for the HPLC experiments. The flow rate was 1.5 mL min^{-1} and detection was carried out at 260 nm. The ENX in tablet and the serum were identified by comparing the retention times of the pure ENX under the identical chromatographic conditions.

RESULTS AND DISCUSSION

Under the described chromatographic conditions, ENX has a retention time of 10.03 min, whereas IS eluted at 2 min. Peak area ratios were linearly proportional to ENX concentrations in the range 3.12×10^{-6} through 3.12×10^{-4} M, with a detection limit of 1.56×10^{-6} M. The calibration equation was found to be $[R = -0.51 + 2.1 \times 10^5 \ C$ (M); $r = 0.9992]$, where C (M) is the molar concentration of ENX. The results of the HPLC experiments were compared with those obtained by the CE experiments. As described earlier, various reports on enoxacin analysis by HPLC are available. The different stationary phases used are μ-Bondapak C_{18}, Spherisorb S5 ODS2, Nucleosil C_{18}, Hypersil ODS, etc. The mobile phases used for the analysis of ENX are different ratios of water–acetonitrile, buffers–acetonitrile, acetonitrile–salt solutions, methanol–salt solutions, etc.[5–12]

Analysis of ENX in Tablets by CE and HPLC Methods

Enoxacin tablets (containing 400 mg active material) were obtained from the local market. Ten ENX tablets were accurately weighed. The average weight of one tablet was calculated, and then the tablets were finely powdered in a mortar. A sufficient amount of tablet powder, equivalent to 10 mg of ENX, was accurately weighed, then transferred to a 25-mL volumetric flask, and methanol was added to dissolve the active material. It was magnetically stirred for 10 min and made up to the final volume with the related solvent. The solution was then centrifuged at $5000 \times g$ for 10 min. The supernatant and a fixed amount of IS solution were diluted with a background electrolyte or mobile phase to carry out either the CE or the HPLC assay. The electropherograms of ENX in tablets with IS are shown in Fig. 3.

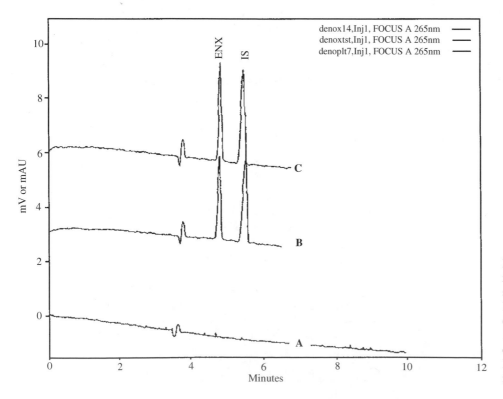

Fig. 3 Electropherograms of (a) inactive ingredients of a tablet solution of ENX; (b) standard ENX (7.7×10^{-5} M) and IS (acetylpipemidic acid, 5.18×10^{-5} M); and (c) enoxacin tablet solution containing IS. Conditions are the same as in Fig. 2.

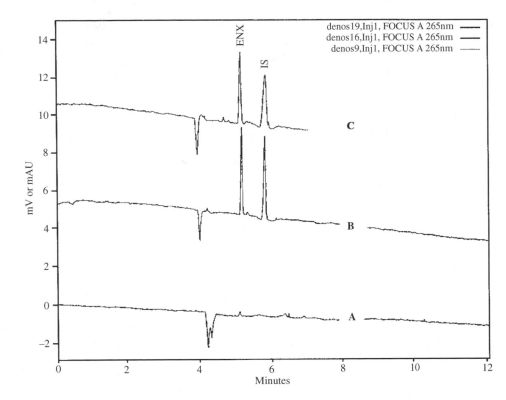

Fig. 4 Electropherogram of (a) blank serum deproteinized with ethanol; (b) serum spiked with the standard ENX solution (0.25 μmol) and IS (0.15 μmol); and (c) water spiked with the standard ENX solution (0.25 μmol). Conditions are the same as in Fig. 2.

Analysis of ENX in Serum by CE and HPLC Methods

For the CE analysis, 0.25 μmol of ENX (in 1 mL) was added to 1 mL of serum and was vigorously shaken. Then 3 mL of ethanol was added and mixed well using a shaker. The precipitated proteins were separated by centrifuging for 10 min at 5000× g. A specific amount of clear supernatant was transferred to a tube, IS solution was added, and the final solution was directly injected to the CE instrument under the same conditions. It was reported that some determinations have been carried out by directly injecting the supernatant of the homogenates and urine into the CE.[16] This kind of application shortens the total analysis time. For HPLC, the precipitation of proteins of 1 mL serum was achieved according to the methods described by Nangia et al.,[7] i.e., by adding 50 μL HClO$_4$ (60% w/v), centrifuging for 10 min at 5000× g, and directly injecting the supernatant into the column of the HPLC system under the conditions mentioned above. The electropherograms of ENX with IS in the serum are given in Fig. 4.

CONCLUSIONS

A typical electropherogram is shown in Fig. 2, which indicates no interferences from the tablet excipients. In order to examine the applicability and validity of

the CE method, ENX pharmaceutical tablets were analyzed by CE and HPLC methods. Results of the comparative studies are shown in Table 4. The results indicate that both methods, i.e., by CE and HPLC, show insignificant differences at the 95% probability level and the ENX tablet formulations satisfy the official requirements.[19] Certain experiments were conducted to elucidate the recovery of ENX and to validate the CE studies. Three sets of experiments with definite amounts of ENX were added to the serum and to the double-distilled water, and were analyzed.

Table 4 Comparative studies for the determination of ENX tablet

No. of experiment	Amount found (mg) using CE	Amount found (mg) using HPLC
1	419.7	410.1
2	428.5	417.4
3	422.3	414.9
4	417.1	419.9
5	419.4	417.4
Mean	421.4	415.9
RSD%	1.04	0.89
$t_{calculated}$	2.13	
t_{table}		2.78 ($p = 0.05$)

(Declared amount, 400 mg per tablet).

The same experiment was also performed without any ENX. The recovery was found to be 89.7 ± 0.63 (RSD%). The recovery experiments were also tested by HPLC and were found to be 78.8 ± 4.94 (RSD%). The difference between the methods could be due to the different precipitation procedures applied.

These results show that the proposed CE method is simple, rapid, and low cost, as compared to HPLC, especially for the quality-control analysis of ENX. It has also been observed that the amount of ENX found (Table 4) was always greater with CE than with HPLC. The presented CE method can be used for the analysis of ENX at trace levels in unknown matrices and also for routine quality control of ENX.

REFERENCES

1. Wolfson, J.S.; Hooper, C. *Quinolone Antimicrobial Agents*; American Society for Microbiology: Washington, DC, 1989.

2. Henwood, J.M.; Monk, J.P. Enoxacin: a review of its antibacterial activity, pharmacokinetic properties and therapeutic use. Drugs **1988**, *36*, 32–66.

3. Guimaraes, M.A.; Noone, P. The comparative in-vitro activity of norfloxacin, ciprofloxacin, enoxacin and nalidixic acid against 423 strains of gram-negative rods and staphylococci isolated from infected hospitalised patients. J. Antimicrob. Chemother. **1986**, *17*, 63–68.

4. Wijnands, W.J.; Vree, T.B.; Van Herwaarden, C.L. The influence of quinolone derivatives on theophylline clearance. Br. J. Clin. Pharmacol. **1986**, *22*, 677–683.

5. Vree, T.B.; Baars, A.M.; Wijnands, W.J.A. High performance liquid chromatography and preliminary pharmacokinetics of enoxacin and its 4-oxo metabolite in human plasma, urine and saliva. J. Chromatogr. Biomed. Appl. **1985**, *343*, 449–454.

6. Griggs, D.J.; Wise, R. A simple isocratic high pressure liquid chromatographic assay of quinolones in serum. J. Antimicrob. Chemother. **1989**, *24*, 437–445.

7. Nangia, A.; Lam, F.; Hung, C.T. Reversed phase ion-pair high performance liquid chromatographic determination of fluoroquinolones in human plasma. J. Pharm. Sci. **1990**, *79*, 988–991.

8. Goebel, K.J.; Stolz, H.; Ehret, I.; Nussbaum, W. A validated ion-pairing high performance liquid chromatographic method for the determination of enoxacin and its metabolite oxo-enoxacin in plasma and urine. J. Liq. Chromatogr. **1991**, *14*, 733–751.

9. Zhai, S.; Korrapati, M.R.; Wei, X.; Muppalla, S.; Vestal, R.E. Simultaneous determination of theophylline, enoxacin and ciprofloxacin in human plasma and saliva by high performance liquid chromatography. J. Chromatogr. Biomed. Appl. **1995**, *669*, 372–376.

10. Davis, J.D.; Aarons, L.; Houston, J.B. Simultaneous assay of fluoroquinolones and theophylline in plasma by high performance liquid chromatography. J. Chromatogr. Biomed. Appl. **1993**, *621*, 105–109.

11. Hamel, B.; Audran, M.; Costa, P.; Bressolle, F. Reversed phase high performance liquid chromatographic determination of enoxacin and 4-oxo-enoxacin in human plasma and prostatic tissue: Application to a pharmacokinetic study. J. Chromatogr. A. **1998**, *812*, 369–379.

12. Barbosa, J.; Berges, R.; Sanz-Nebot, V. Retention behaviour of quinolone derivatives in high performance liquid chromatography: Effect of pH and evaluation of ionization constants. J. Chromatogr. A. **1998**, *823*, 411–422.

13. Boone, C.M.; Douma, J.W.; Franke, J.P.; de Zeeuw, R.A.; Ensing, K. Screening for the presence of drugs in serum and urine using different separation modes of capillary electrophoresis. Forensic Sci. Int. **2001**, *121*, 89–96.

14. Lemos, N.P.; Bortolotti, F.; Manetto, G.; Anderson, R.A.; Cittadini, F.; Tagliaro, F. Capillary electrophoresis: a new tool in forensic medicine and science. Sci. Justice **2001**, *41*, 203–210.

15. Baker, D.R. *Capillary Electrophoresis*; J. Wiley & Sons, Inc.: New York, 1995.

16. Xu, Y. Capillary electrophoresis. Anal. Chem. **1995**, *67*, 463R–473R.

17. Altria, K.D. Overview of capillary electrophoresis and capillary electrochromatography. J. Chromatogr. **1999**, *856*, 443–463.

18. Tuncel, M.; Dogrukol-Ak, D.; Senturk, Z.; Ozkan, S.A.; Aboul-Enein, H.Y. Capillary electrophoretic behaviour and determination of enoxacin in pharmaceutical preparations and human serum. J. Liq. Chromatogr. & Relat. Technol. **2001**, *24*, 2455–2467.

19. United States Pharmacopoeia (USP) 22, NF-17; U.S. Pharmacopeial Convention: Rockville, Maryland, 1990.

Environmental Applications of Supercritical Fluid Chromatography

Yu Yang
East Carolina University, Greenville, North Carolina, U.S.A.

INTRODUCTION

Because supercritical fluids have liquid like solvating power and gaslike mass-transfer properties, supercritical fluid chromatography (SFC) is considered to be the bridge between GC and LC and possesses several advantages over GC and LC, as summarized in Table 1. For example, SFC can separate nonvolatile, thermally labile, and high-molecular-weight compounds in short analysis times. Another advantage of SFC is its compatibility with both GC and HPLC detectors. Because of these advantages of SFC, there is a large number of SFC applications in environmental analysis. However, only selected recent works are reviewed here. Although sample preparation is often required before SFC analysis to remove the analytes from environmental matrices and to enrich them, sample preparation is not intensively discussed in this review. To facilitate the discussion, the environmental pollutants are classified and reviewed separately in this article.

PESTICIDES AND HERBICIDES

The analysis of pesticides and herbicides has mainly been done either by GC with selective detectors or by HPLC with ultraviolet (UV) detection. As summarized in Table 1, GC is limited to thermally stable volatile compounds, whereas the HPLC with UV can only detect compounds with chromophores. These limitations of GC and HPLC led to the use of SFC in the analysis of pesticides and herbicides. Among the SFC works in environmental analysis, one-third of the works concerns the analysis of pesticides and herbicides.

Many detectors have been used to detect pesticides and herbicides in SFC. Among these detectors, the flame ionization detector (FID) is most commonly used for detection of a wide range of pesticides and herbicides, with a detection limit ranging from 1 ppm (for carbonfuran) to 80 ppm (for Karmex, Harmony, Glean, and Oust herbicides). The UV detector has frequently been used for the detection of compounds with chromophores. The detection limit was as low as 10 ppt when solid-phase extraction (SPE) was on-line

coupled to SFC. The mass spectrometric detector (MSD) has also been used in many applications as a universal detector. The MSD detection limit reached 10 ppb with on-line SFE (supercritical fluid extraction)–SFC. Selective detection of chlorinated pesticides and herbicides has been achieved by an electron-capture detector (ECD). The limit of detection for triazole fungicide metabolite was reported to be 35 ppb. Other detectors used for detection of pesticides and herbicides include thermoionic, infrared, photometric, and atomic emission detectors.

A variety of both packed and open tubular columns have been used for separation of pesticides and herbicides. The columns were either used separately or coupled in series to achieve better separations. Although environmental water samples were mostly analyzed by SFC, analyses of pesticides and herbicides from soil, foods, and other samples were also reported.

POLYCHLORINATED BIPHENYLS

Since 1929, polychlorinated biphenyls (PCBs) have been produced and used as heat-transfer, hydraulic, and dielectric fluids. Because of their chemical and physical stability, PCBs have been found in many environmental samples. Generally, PCBs have been analyzed by GC with electron-capture detection. There are many reports on subcritical and supercritical fluid extraction of PCBs, but only a few on supercritical fluid separation of PCBs.

Among the works of supercritical fluid separations of PCBs, UV has been the most popular detector. A Microbore C_{18} column was used to separate individual PCB congeners in Aroclor mixtures. Density and temperature programming was also utilized for separation of PCBs. Both packed (with phenyl and C_{18}) and capillary (Sphery-5 cyanopropyl) columns were used in this work. Carbon dioxide, nitrous oxide, and sulfur hexafluoride were tested as mobile phases for the separation of PCBs.

A HD and MSD were also used for detection of PCBs in SFC. Capillary columns packed with aminosilane-bonded silica and open tubular columns coated

Encyclopedia of Chromatography DOI: 10.1081/E-ECHR-120039978

Table 1 Comparison of characteristics of GC, SFC, and LC

	GC[a]	SFC	LC[b]
Suitability for polar and thermolabile compounds	Low	High	High
Size of analyte molecule	Small-Medium	Small-Large	Small-Large
Sample capacity	Low	High (packed column)	High
Possibility of introducing selectivity in the mobile phase	Low	High	Medium
Toxicity and disposal cost of the mobile phase	No	No (with pure CO_2) Low (with modifier)	High
Efficiency	High	Medium-High	Low
Use of gas-phase detectors	Yes	Yes	No
Analysis time	Medium	Medium	Long

[a]Only capillary GC is used for these evaluations. Fast GC and packed column GC are not included here.
[b]Capillary HPLC is not included.

with polysiloxane were employed for PCB separation in these works.

POLYCYCLIC AROMATIC HYDROCARBONS

Polycyclic aromatic hydrocarbons (PAHs) have routinely been analyzed by GC and LC. However, both techniques have limitations in terms of analyte molecular weight and analysis time. The greater molecular-weight range of SFC with respect to GC makes it better suited for determining a wide range of PAHs. SFC also has advantages over HPLC for the analysis of PAHs when the same kind of columns is used. Supercritical fluid has similar solvating power as a liquid does, and the solute diffusion coefficients are much greater than those found in liquids. Therefore, comparable efficiencies to HPLC can be obtained by SFC in shorter analysis time. Because of these characteristics of SFC, the separation of PAHs by SFC with different kinds of packed and capillary columns is a well-investigated and established method.

The most popular detector for PAHs is the UV detector. The detection limit was 0.2–2.5 ppb for 16 PAHs. A diode-array detector was also used for PAHs in SFC, and the detection limit was reported to be as low as 0.4 ppb. Other detections used for PAHs include mass spetrometric, thermoionic, infrared, photoionization, sulfur chemiluminescence, and fluorescence detectors.

Although has mainly been used as the mobile phase in SFC, modifiers have often been added to to increase the solvating power of the mobile phase. Although the most frequently used modifier has been methanol, many other modifiers were also tested. The modifier effect on retention is discussed separately in this encyclopedia. Because organic modifiers are incompatible with FID, FID was rarely used for PAHs in SFC.

Fast separations of 16 PAHs were achieved within 6–7 min using packed columns. A comparison study of the PAH molecular shape recognition properties of liquid-crystal-bonded phases in packed-column SFC and HPLC found that the selectivity was enhanced in SFC. The result of an interlaborotory round-robin evaluation of SFC for the determination of PAHs also shows that SFC possesses distinct advantages over GC/MS and nuclear magnetic resonance (NMR) including speed, cost, and applicability.

POLAR POLLUTANTS

Because carbon dioxide is nonpolar, the separation of polar compounds by supercritical carbon dioxide is difficult. Thus, polar modifiers are often used for the separation of phenols and amines. Derivatization has also been employed to obtain nonpolar analytes in some applications. The UV detector has mainly been used for the detection of polar compounds. Oxidative and reductive amperometric detection was also utilized with a detection limit of 250 pg for oxidative detection of 2,6-dimethylphenol. The detection of amines has generally been achieved by FID. Other detectors used for the detection of polar analytes include Fourier transform infrared (FTIR), photodiode array, and flame photometry.

It should be pointed out that separation of more than one class of organic compounds can be achieved by SFC. For example, Fig. 1 shows the chromatogram of 35 PAHs, herbicides, and phenols from a contaminated water sample. Solid-phase extraction was used for sample preparation. Five Hypersil silica columns were coupled in series for separation of these contaminants. The percentage of methanol (as modifier) was varied from 2% (5 min) to 10% (29 min) at 0.5%/min. A pressure program was also applied. A diode-array detector was used in this work.

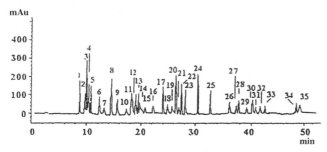

Fig. 1 Chromatogram of PAHs, herbicides, and phenols obtained by supercritical carbon dioxide modified with methanol. (Reprinted from L. Toribio, M.J. del Nozal, J.L. Bernal, J.J. Jimenez, and M.L. Serna, Packed-column supercritical fluid chromatography coupled with solid-phase extraction for the determination of organic microcontaminants in water, *J. Chromatogr. A 823*: 164, 1998. Copyright 1998, with permission from Elsevier Science.)

ORGANOTIN, MERCURY, AND OTHER INORGANIC POLLUTANTS

Organotin compounds are used extensively as biocides and in marine antifouling paints. These compounds accumulate in sediments, marine organisms, and water, as they are continuously released into the marine environment. Many of these organotin compounds are toxic to aquatic life. Most organotin separation techniques have been based on the GC resolution of volatile derivatives and coupled to elemental detection techniques that are often not sensitive enough to detect trace organotin compounds. However, the separation of organotin compounds was achieved by capillary columns (SB-Biphenyl-30 or SE-52) with pure CO_2 as the mobile phase. Inductively coupled plasma–mass spectrometry (ICP–MS) was used in most of the applications to improve the sensitivity for detecting trace organotin species. The reported detection limits range from 0.2 to 0.8 pg for tetrabultin chloride, tributyltin chloride, triphenyltin chloride, and tetraphenyltin. However, the detection limits obtained by FID are 15- to 45-fold higher than those obtained by ICP–MS for the above-mentioned organotin compounds. Flame photometric detector was also used to detect organotin species with a detection limit of 40 pg for tribultin chloride.

The separation of organomercury was conducted by using a SB-methyl-100 capillary column and pure CO_2 as the mobile phase. FID and atomic fluorescence were used for detection. The same column was also used for separation of mercury, arsenic, and antimony species using carbon dioxide as the mobile phase. A chelating reagent, bis(trifluoroethyl)dithiocarbamate, was used in this case to convert the metal ions to organometallic compounds before the separation. The detection limit of FID was 7 and 11 pg for arsenic and antimony, respectively.

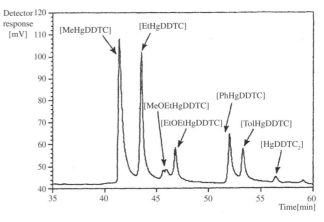

Fig. 2 Chromatogram of a standard mixture after complexation with sodium diethyldithiocarbamate. Composition of the standard: mercury dichloride, methylmercury chloride, ethylmercury chloride, methoxyethylmercury chloride, ethoxyethylmercury chloride, phenylmercury chloride, and tolymercury chloride. (Reprinted from A. Knochel and H. Potgeter, Interfacing supercritical fluid chromatography with atomic fluorescence spectrometry for the determination of organomercury compounds, *J. Chromatogr. A 786*: 192, 1997. Copyright 1997, with permission from Elsevier Science.)

Fig. 2 shows an example of separating organomercury using supercritical A 10-m 50-mm-inner CO_2. A 10-m × 50-μm-inner diameter SB-Methyl 100 column was used for the separation. Due to their poor solubility in supercritical carbon dioxide, monoorganomercury compounds were derivatized by diethyldithiocarbamate. An interface for a system consisting of SFC and atomic fluorescence spectrometry was developed for the detection of organomercurials.

In closing, supercritical fluid chromatography is a promising technique for the analysis of environmental pollutants. The analytes range from inorganic species to polar and nonpolar organic compounds. The sample matrices cover water, soil, sediments, sludge, and air particulate matters. The sample preparation has been done by solid-phase extraction, supercritical fluid extraction, or traditional solvent extraction. Modifiers are often used to enhance the solubility of analytes and to yield a better separation for polar and high-molecular-weight analytes. Packed columns are preferred for trace analysis because of their high sample capacity. Both gas-phase and liquid-phase detectors have been used in SFC to detect a wide range of environmental pollutants.

SUGGESTED FURTHER READING

Bayona, J.M.; Cai, Y. The role of supercritical fluid extraction and chromatography in organotin speciation studies. Trends Anal. Chem. **1994**, *13*, 327–332.

Berger, T.A. Separation of polar solutes by packed column supercritical fluid chromatography. J. Chromatogr. A. **1997**, *785*, 3–33.

Chester, T.L.; Pinkston, J.D.; Raynie, D.E. Supercritical fluid chromatography and extraction. Anal. Chem. **1998**, *70*, 301R–319R.

Dressman, S.F.; Simeone, A.M.; Michael, A.C. Supercritical fluid chromatography with electrochemical detection of phenols and polyaromatic hydrocarbons. Anal. Chem. **1996**, *68*, 3121–3127.

Juvancz, Z.; Payne, K.M.; Markides, K.E.; Lee, M.L. Multidimensional packed capillary coupled to open tubular column supercritical fluid chromatography using a valve-switching interface. Anal. Chem. **1990**, *62*, 1384–1388.

Knochel, A.; Potgeter, H. Optimisation of expression and purification of the recombinant Yol066 (Rib2) protein from Saccharomyces cerevisiae. J. Chromatogr. A. **1997**, *786*, 188–193.

Laintz, K.E.; Shieh, G.M.; Wai, C.M. Simultaneous determination of arsenic and antimony species in environmental samples using bis(trifluoroethyl)-dithiocarbamate chelation and supercritical fluid chromatography. J. Chromatogr. Sci. **1992**, *30*, 120–123.

Luffer, D.R.; Novotny, M. Element-selective detection after supercritical fluid chromatography by means of a Surfatron plasma in the near-infrared spectral region. J. Chromatogr. **1990**, *517*, 477–489.

Medvedovici, A.; Kot, A.; David, F.; Sandra, P. The use of supercritical fluids in environmental analysis. In *Supercritical Fluid Chromatography with Packed Columns*; Anton, K., Berger, C., Eds.; Marcel Dekker, Inc.: New York, 1998; 369–401.

Medvedovici, A.; David, F.; Desmet, G.; Sandra, P.J. Microcol. Separ. **1998**, *10*, 89–97.

Moyano, E.; McCullagh, M.; Galceran, M.T.; Games, D.E. Supercritical fluid chromatography-atmospheric pressure chemical ionisation mass spectrometry for the analysis of hydroxy polycyclic aromatic hydrocarbons. J. Chromatogr. A. **1997**, *777*, 167–176.

Mulcahey, L.J.; Rankin, C.L.; McNally, M.E.P. Environmental applications of supercritical fluid chromatography. In *Advances in Chromatography Vol. 34*; 1994; 251–308.

Shan, S.; Ashraf-Khorassani, M.; Taylor, L.T. Analysis of triazine and triazole herbicides by gradient-elution supercritical fluid chromatography. J. Chromatogr. **1990**, *505*, 293–298.

Smith, R.M.; Briggs, D.A. Separation of homologous aromatic alcohols and carboxylic acids by packed column supercritical fluid chromatography. J. Chromatogr. A. **1994**, *688*, 261–271.

Toribio, L.; del Nozal, M.J.; Bernal, J.L.; Jimenez, J.J.; Serna, M.L. Packed-column supercritical fluid chromatography coupled with solid-phase extraction for the determination of organic microcontaminants in water. J. Chromatogr. A. **1998**, *823*, 163–170.

Environmental Pollutants Analysis by Capillary Electrophoresis

Imran Ali
Hassan Y. Aboul-Enein
Pharmaceutical Analysis Laboratory, Biological and Medical Research Department,
King Faisal Specialist Hospital and Research Center, Riyadh, Saudi Arabia

INTRODUCTION

The quality of the environment is degrading continuously, due to the accumulation of various undesirable constituents. Water resources, the most important and useful components of the environment, are most affected by pollution. The ground and surface water at many places in the world are not suitable for drinking purposes, due to the presence of aesthetic and toxic pollutants. Therefore, the importance of water quality preservation and improvement is essential and continuously increasing.[1,2] The most important toxic pollutants are inorganic and organic chemicals. Therefore, determination of these water pollutants at trace levels is essential in environmental hydrology.

The analysis of these pollutants has been widely carried out using GC[3,4] and HPLC.[3,4] The high polarity, low vapor pressure, and required derivatization of some of the organic pollutants are the factors that complicate GC analysis. On the other hand, due to the inherently limited resolving power of conventional HPLC techniques, optimization of pollutant analysis often involves complex procedures or numerous experiments leading to the consumption of large amounts of solvents and sample volumes. Presently, capillary electrophoresis (CE), a versatile technique of high speed, high sensitivity, lower limit of detection, and reproducible results, is a major trend in analytical science and the number of publications on water pollutants analysis have increased exponentially in recent years.[3,5,11] Therefore, attempts have been made, here, to summarize the various CE methods for the analysis of different water pollutants. Critical comments are also presented to improve the application of CE in water pollutant analysis.

PRINCIPLE OF CAPILLARY ELECTROPHORESIS

The schematic representation of a CE apparatus is shown in Fig. 1. The mechanism of separation of water pollutants in CE is based on the electroosmotic flow (EOF) and electrophoretic mobilities of the pollutants. The EOF propels all pollutants (cationic, neutral, and anionic) toward the detector and, ultimately, separation occurs due to the differences in the electrophoretic migration of the individual pollutants. Under the CE conditions, the migration of the pollutant is controlled by the sum of the intrinsic electrophoretic mobility (μ_{ep}) and the electroosmotic mobility (μ_{eo}), due to the action of EOF. The observed mobility (μ_{obs}) of the pollutants is related to μ_{eo} and μ_{ep} by the following equation:

$$\mu_{obs} = \mu_{eo} + \mu_{ep} \tag{1}$$

The electrophoretic mobilities of cations (μ_{obs}) can be related to the limiting ionic equivalent conductivity, λ_{ekv}, by the following equation:

$$\mu_{obs} = \lambda_{ekv}/F = q_i/6\pi\eta r_i \tag{2}$$

where, F is the Faraday constant ($F = 9.6487 \times 10^4$ A sec mol^{-1}), λ_{ekv} (cm^2 mol^{-1} ohm^{-1}) is related, by the Stokes law, to the charge of the hydrated cation q_i, to the dynamic viscosity of the electrolyte, η (g cm^2 sec^{-1}), and to the radius of the hydrated cation r_i (cm). The μ_{ep} values can be calculated from the experimental data, the mobility of the cation (μ_{obs}), and the mobility of the EOF μ_{eo}, according to the following equation:

$$\mu_{ep} = \mu_{obsp} - \mu_{eo}$$
$$= [1/t_{m(ion)} - 1/t_{m(eo)}][l_T \cdot L_d/V] \tag{3}$$

where, $t_{m(ion)}$, $t_{m(eo)}$, l_T and L_d are the migration time of the cation (sec), migration time of the EOF (sec), the overall capillary length, and the length of the capillary to the detector (cm), respectively. Thus, EOF plays an important role in the determination of metal ions by CE. The determination of water pollutants can be carried out by several modes of CE. The various modes of CE include capillary zone electrophoresis (CZE), MECC, capillary isotachophoresis (CIEF), capillary gel electrophoresis (CGE), ion-exchange electrokinetic chromatography (IEEC), capillary isoelectric focusing (CIEF), affinity capillary electrophoresis (ACE),

Encyclopedia of Chromatography DOI: 10.1081/E-ECHR-120014229

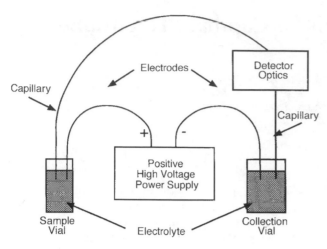

Fig. 1 Diagram of the CE system. (Reprinted from Waters Quanta 4000E Capillary Electrophoresis System Operator's Manual, Waters Corp., Milford, Massachusetts, U.S.A., 1993.)

capillary electrochromatography (CEC), separation on microchips (MC), and nonaqueous capillary electrophoresis (NACE).[8] However, most of the water pollutant analyses have been carried out in the CZE mode.

SAMPLE PRETREATMENT

The treatment of the samples from environmental matrices is an important issue in CE. Little attention has been given for the water sample treatment in CE analysis of environmental pollutants. Soil samples have been extracted by the usual methods. Besides, the sediment samples were digested using strong acids. The samples containing a highly ionic matrix may cause problems in CE. EOF in the capillary can be altered by the influence of the sample matrix, resulting in poor resolution. Additionally, the detector baseline is usually perturbed when the pH of the sample differs greatly from the pH of the background electrolyte (BGE). The samples containing UV absorbing materials are also problematic in the detection of the environmental pollutants. Due to all of these factors, some authors have suggested sample cleanup processes, solid/liquid phase extractions, and sample preparations prior to loading onto CE.[10–13] Real samples often require the application of simple procedures, such as filtration, extraction, dilution, etc. Electromigration of sample cleanup suffers severely from matrix dependence effects; even then, it has been used for pre-concentration in inorganic analysis. The sample treatment methods have been discussed in several reviews.[7–9,12] The use of ion-exchange and chelating resins to pre-concentrate the metal ion samples prior to CE application has been reported.[9,12] An on-line

dialysis sample cleanup method for CE analysis has also been presented. Besides, several reports have been published on dialysis and electrodialysis for sample cleanup prior to CE injection.[7–9,12]

DETECTION

Generally, UV detection is used for the determination of most environmental pollutants. However, the use of UV detectors in CE for metal ion and anion analysis is not suitable due to the poor absorbance of UV radiation by metal ions and anions. The most common method to solve this problem is indirect UV detection. The main advantage of the indirect UV detection method is its universal applicability. The complexation of metal ions with ligands also increases the sensitivity of their detection. The complexing agent is either added to the electrolyte (in situ, on-line complexation) or to the sample before the introduction into the capillary column (off-line complexation). The most commonly used ligands are azo dyes, quinoline dyes, porphyrin, dithiocarbamate, aminopolycarboxylic acids, 4-(2-pyridylazo) resorcinol, 8-hydroxyquinoline-5-sulfonic acid, ethylenediaminetetraacetic acid, cyanides, various hydroxy carboxylic acids, crown ethers, and other organic chelating agents. Besides, UV visualizing agents (probe) have also been used to increase the sensitivity of detection in the UV mode. The important probes include, e.g., Cu(II) salts, chromate, aromatic amines, and cyclic compounds, e.g., benzylamine, 4-methylbenzylamine, dimethylbenzylamine, imidazole, p-toludine, pyridine, creatinine, ephedrine, and anionic chromophores (benzoate and anisates).[7–9,12,14] Care must be taken to avoid the interaction of the cations and visualizing agent with the capillary wall. Besides, the visualizing agents should exhibit a mobility close to that of the cations, its UV absorbance should be as high as possible, and the detector noise as low as possible. Furthermore, sensitivity of the UV detection has been increased by using a double beam laser as the light source.

To overcome the problem of detection in CE, many workers have used inductively coupled plasma-mass spectrometry (ICP-MS) as the method of detection.[7–9,12–14] Electrochemical detection in CE includes conductivity, amperometry, and potentiometry detection. The detection limit of amperometric detectors has been reported to be up to 10^{-7} M. A special design of the conductivity cell has been described by many workers. The pulsed-amperometric and cyclic voltametry waveforms, as well as multi step wave forms, have been used as detection systems for various pollutants. Potentiometric detection in CE was first introduced in 1991 and was further developed by various workers.[7–9,12,14] 8-Hydroxyquinoline-5-sulfonic acid

Table 1 The applications of capillary electrophoresis for the determination of environmental pollutants

Pollutants	Sample matrix	Electrolytes	Detection	Detection limit
Metal Ions in Water, Sediment, and Soil[7–9]				
Arsenic and selenium	Drinking water	20 mM KHP, 20 mM Boric acid (pH 9.03) hydrodynamically modified EOF	Hydride generation ICP-MS	6–58 ng/L
		75 mM Dihydrogen phosphate, 25 mM tetraborate (pH 7.65)	Direct UV 195	12 µg/L
		Chromate, 0.5 mM TTAOH (pH 10.5)	Indirect UV 254 nm	10 µg/L
Mg, Ca, Na, and K	Well water	5 mM Imidazole, 6.5 mM HIBA 2 mM 18-crown-6 (pH 4.1)	Indirect UV 214 nm	—
Alkali and alkaline Earth metals	Tap and mineral waters	10 mM Imidazole (pH 4.5)	Indirect UV 214 nm	0.05 mg/L
Uranyl cation (UrO_2^{2+})	River water	10 mM Perchloric acid, 1 mM phosphate, 0.6 mM borate, 0.01–0.1 mM arsenazo III 650 nm, 50–150 mM NaCl, 10% MeOH	Direct UV–VIS	10 µg/L
Zn and other transition metals	Tap water	10 mM Borate buffer, 0.1 mM HQS (pH 9.2)	Direct UV 254 nm	3–225 µg/L
Al	River, reservoir, and spring waters	40 mM AcOH, 10 mM NH₄Ac (pH 4.0)	Fluorescence 419 and 576 nm	19 µg/L
Ca, Mg, Ba, Na, K, and Li	Mineral water	3–5 mM Imidazole, pH 4.5	Indirect 214 nm	0.05 ppb
Cu, Ni, Co, Hg, Mn, Fe, Pb, Pd, Zn, Cd, Mg, Sr, Ca, and Ba	River water	2 mM Na₂B₄O₇, 2 mM EDTA pH 4.4	Direct UV 200 and 214 nm	10 µM
Ca, Sr, Ba, Li, Na, K, Rb, Sc, and Mg	Tap, rain, and mineral waters	5 mM Benzimidazole, tartarate, pH 5.2, +0.1% HEC or methy-HEC, +40 mM 18C6	Indirect UV 254 nm	—
Chromate	Waste water	0.02 mM Phosphate buffer (pH 7)	Direct UV	—
Fe, Ni, Pd, Pt, and Cu(I) cyano complexes	Leaching solutions of automobile catalytic converters	20 mM Phosphate buffer, 100 Mm NaCl, 1.2 mM TBABr, 40 µM TTABr (pH 11)	Direct UV 208 nm	20 µg/L
Speciation of Metal Ions[7–9]				
Arsenic species	Drinking water	0.025 mM Phosphate buffer, pH 6.8	Direct UV 190 nm	<2 mg/L
	Water	50 mM CHES, 20 mM LiOH	Conductivity	0.4 mg/L
	Tin mining	15 mM Phosphate buffer, 1 mM	Conductivity	0.4 mg/L

(Continued)

E

Table 1 The applications of capillary electrophoresis for the determination of environmental pollutants (*Continued*)

Pollutants	Sample matrix	Electrolytes	Detection	Detection limit
As(III), As(V), and dimethyl arsenic acid	Process water	CTAB 50 mM CHES, 0.03% Triton X-100 20 mM LiOH, pH 9.4	ICP-MS	1 ppb
	—	60 mM Calcium chloride (pH 6.7) cetyltrimethyl ammonium bromide, pH 10.0		
Arsenic and selenium species	Drinking water	20 mM KHP, 20 mM Boric acid (pH 9.03) hydrodynamically modified EOF	Hydride generation ICP-MS	6 ng/L
	Tap and drinking waters	75 mM Dihydrogen phosphate 25 mM, tetraborate (pH 7.65)	Direct UV 12 µg/L 195 nm	—
	—	20 mM Na_2HPO_4, 5 mM DTPA, pH 8.0 or 8.5	Indirect UV 214 nm	10^{-6} M
Cr(IV) and Cr(VI)	Rinse water from chromium platings	1 mM CDTA, 10 mM Formate buffer (pH 3.8)	Direct UV 214 and 254 nm	10 µg/L
	Chromium plating water	10 mM Formate buffer, 1 mM CDTA (pH 3.0)	Direct UV 214 nm	10 ppb
	Electroplating water	10 mM Formate buffer (pH 3.0)	Indirect UV 214 nm	50 ppb
	Waste water	20 mM Na_2HPO_4, 0.05 mM TTAOH	Direct UV 214 and 254 nm	—
Fe(II) and Fe(III)	Electroplating waters	20 mM Phosphate buffer (pH 7.0)	Direct UV 214 nm	10^{-5} M
Hg(II), CH_3Hg^+, and $CH_3CH_2Hg^+$	—	25 mM $Na_2B_4O_7 \cdot 10\ H_2O$, pH 9.3	ICP-MS	81–275 ppb
Ir(II) and Ir(III)		4 mM H^+, 23 mM Cl^-, pH 2.4	Indirect UV 214 nm	—
Pb(II), triethyl lead (IV), trimethyl lead (IV), and diphenyl lead (IV)		Na_2HPO_4–$Na_2B_4O_7$, 2.5 mM TTHA, 2.0 mM SDS, pH 7.5	Indirect UV 220 nm	ppb level
Pt(II) and Pt(IV)	Electroplating bath	4 mM H^+, 23 mM Cl^-, pH 2.4	Indirect UV 214 nm	—
V(IV) and V(V)		20 mM Na_2HPO_4, 5 mM DTPA, pH 8.0 or 8.5	Indirect UV 214 nm	10^{-6} M
Anions Analysis[9][12]				
F^-	Rain water	1.13 mM PMA, 0.8 mM TEA, 2.13 mM HMOH, pH 7.7	Indirect UV 254 nm	0.6 µM
F^-, Cl^-, Br^-, SO_4^{-2}, NO_3^-, NO_2^-, PO_4^{-3}, and thiosulphate	Tap water	2.25 mM PMA, 6.5 mM NaOH, 0.75 mM HMOH, 1.6 mM TEA, pH 7.7	Indirect UV 250 nm	1–3 mg/L
$HClO_4^-$, Br^-, F^-, NO_3^-, and NO_2^-	Tap water	20 mM Sodium sulphate, pH 2.5	Ionselective microelectrode	5 µg/L

E

Analyte	Sample matrix	Buffer	Detection	Detection limit
Br⁻, BrO₄⁻, I⁻, IO₄⁻, NO₃⁻, NO₂⁻, and selenite organic and inorganic anions	River water	Phosphate buffer, pH 2.9	Direct UV 200 nm	—
	Water from dumping area	9 mM PDCA, 0.05 mM TTABr, pH 7.8	Indirect UV 254 nm	—
Br⁻, NO₂⁻, S₂O₃⁻, NO₃⁻, N₃⁻, Fe(CN)₆⁻⁴, MoO₄⁻², WO₄⁻², CrOₓ⁻³, and ReO₄⁻	—	10 mM Borate, 220 mM NaCl, pH adjusted to 8.5 by sodium hydroxide	UV 214 nm	—
Phenols and its Derivatives[3,10]				
Alkyl phenols		1.25 mM Na₂B₄O₇, 15 mM NaH₂PO₄ pH 11.0 with 0.001% HDB	UV 254 nm	—
Chlorophenols		50 mM Na₂HPO₄/NaH₂PO₄, pH 6.9	UV 214 nm	0.06 mg/L
Miscellaneous derivatives of phenols		10 mM Na₂B₄O₇/Na₃PO₄, pH 9.8 20 mM CHES, pH 10.1 15 mM Na₃BO₃, pH 9.9	UV 210 nm Amperometric Indirect fluorimetry	0.3 mg/L 0.03 mg/L 0.01 mg/L
Pentachlorophenols	Drinking water	40 mM Sodium borate, pH 10	—	ng level
Chloro- and nitrophenols	Tap water	20 mM Sodium borate	—	μg level
Pesticides[11]				
Hexazinone and its metabolite	Ground water	50 mM SDS, 12 mM Sodium phosphate, 10 mM Sodium borate, and 15% MeOH, pH 9.0	UV 220–247 nm	—
Primisulfuron and triasulfuron	Water and soil	25 mM NaH₂PO₄ + 50 mM LiDS buffer	UV at 214 nm	—
Triazines and chlorotriazines	Tap and river Water	30 mM Sodium borate, 30 mM SDS, pH 9.3	UV at 210 nm	—
Chlorinated acid herbicides and related compounds	Water	5 mM Ammonium acetate in isopropanol-water (40 : 60, v/v), pH 10	MS	—
Chlorphyrifos	Air and soil	—	—	—
Triazine pesticides		50 mM Ammonium acetate, 0.7 mM CTAB, pH 4.5	MS and UV (230 nm)	—
Polyaromatic Hydrocarbons[5]				
PAHs	Standard	8 mM Na₂B₄O₇, pH 9.0, 50 mM DOSS, 40% acetonitrile	UV 254 nm	—
	Standard	5 mM Resorcarene, pH 13.25, 6 M Urea, 50% acetonitrile	UV 260 nm	—
Amines[5]				

(Continued)

Table 1 The applications of capillary electrophoresis for the determination of environmental pollutants (*Continued*)

Pollutants	Sample matrix	Electrolytes	Detection	Detection limit
Methyl, dimethyl trimethyl, and ethyl amines	Atmospheric aerosols	5 mM DHBP, 6 mM glycine, 2 mM 18-crown-6 ether, pH 6.5	Indirect UV 280 nm	—
Substituted anilines	Tap and ground water, soil, and sediment	50 mM NaH_2PO_4, pH 2.35, 7 mM 1,3-diaminopropane	UV 280 nm	0.06 mg/L
Heterocyclic aromatic amines	Rain water	50 mM NaH_2PO_4–20 mM citric acid, 30 mM NaCl, 26% methanol	UV 190, 240, and 263 nm	0.05 mg/L
Carbonyls[5]				
Acetaldehyde, benzaldehyde, formaldehyde, and glyoxal	Rain water	5 mM Na_3PO_4–10 mM $Na_2B_4O_7$, pH 8.0, 20% acetonitrile	Laser-induced fluorescence 325 and 442 nm	—
Dyes[5,10]				
Synthetic cationic dyes	Standard	10 mM Citric acid, pH 3.0, 0.1% PVP	UV 214 nm	—
Anionic synthetic azo dyes	Standard	10 mM BTP-HCl, pH 6.5, 0.5% PEG, 0.05% PVP	UV 214 nm	—
Photoactive dyes	Coffee and beans	50 mM Boric acid/10 mM sodium borate, pH 8.5	—	0.08 µg L^{-1}
Chiral Separations[11,18]				
Fenoprop, mecoprop, and dichlorprop		20 mM Tributyl-β-CD in 50 mM, ammonium acetate, pH 4.6	MS	—
2-Phenoxypropionic acid, dichloroprop, fenoprop, fluaziprop, haloxyfop, and diclofop enantiomers	—	10^{-4} M 75 mM Britton-Robinson buffer with 6 mM Vancomycin	—	—

Analyte	Buffer	Detection							
Imazaquin isomer	50 mM Sodium acetate, 10 mM dimethyl-β-CD, pH 4.6	—	—	—	—	—	—	—	—
Phenoxy acid herbicides	200 mM Sodium phosphate, pH 6.5 with various concentrations of OG and NG	—	—	—	—	—	—	—	—
Diclofop	50 mM Sodium acetate, 10 mM trimethyl-β-CD, pH 3.6	—	—	—	—	—	—	—	—
Imazamethabenz isomers	50 mM Sodium acetate, 10 mM dimethyl-β-CD, pH 4.6	—	—	—	—	—	—	—	—
2-(2-methyl-4-chlorophenoxy) propionic acid,	0.05 M Lithium acetate containing α-cyclodextrins	UV 200 nm	—	—	—	—	—	—	—
2-(2-methyl-4,6-dichlorophe-noxy) propionic acid	0.05 M Lithium acetate containing β-cyclodextrin	UV 200 nm	—	—	—	—	—	—	—
2-(2,4-dichlorophenoxy) propionic acid	0.05 M Lithium acetate containing heptakis-(2,6-di-O-methyl)-β-cyclodextrin	UV 200 nm	—	—	—	—	—	—	—
propionic acid, and 2-(2-methyl-4-chlorophenoxy) propionic acid	0.03 M Lithium acetate containing heptakis-(2,6-di-O-methyl)-β-cyclodextrin	UV 200 nm	—	—	—	—	—	—	—
1,1'-Binaphthyl-2-2'-dicar-boxylic acid, 1,1'-binaphthyl-2,2'-dihydrogen phosphate, and 2,2'-dihydroxy-1-1'-binaphthyl-3,3'-dicarboxylic acid	0.04 M Carbonate, pH 9.0, with noncyclooligosaccharides	UV 215–235 nm	—	—	—	—	—	—	—

E

and lumogallion exhibit fluorescent properties and, hence, have been used for metal ion detection in CE by fluorescence detectors.[7–9,12,14] Overall, fluorescence detectors have not yet received wide acceptance in CE for metal ions analysis, although their gains in sensitivity and selectivity over photometric detectors are significant. Moreover, these detectors are also commercially available. Some other devices, such as chemiluminescence, atomic emission spectrometry (AES), refractive index, radioactivity, and X-ray diffraction, have also been used as detectors in CE for metal ions analysis,[14] but their use is still limited.

SEPARATION EFFICIENCY IN CAPILLARY ELECTROHPORESIS

From the literature available and discussed herein, it may be assumed that the selectivity of various environmental pollutants by CE is quite good. However, the detection sensitivity for metal ions and anions is poor. Therefore, many attempts have been made to solve this problem. Organic solvents have been added to the BGE to improve the selectivities of many of the pollutants. It has also been reported that the organic solvents ameliorate the solubility of hydrophobic complexes, reduce the adsorption onto the capillary wall, regulate the distribution of complexes between aqueous phase and micellar phase, adjust the viscosity of the separation medium and, accordingly, accomplish an improvement in detection. The pH of the electrolyte solution is very important from the selectivity point of view. The pH controls the behavior of EOF, acid/base dissociation equilibria of complexes, and the state of existing complexes. Therefore, the selectivity can be improved by adjusting the pH of the BGE. Besides, ion pairing can be used to improve the separation in CE. Six types of ion pairing agents have been developed and used. Ion pairing has also been used in MECC for the improvement of the separation of hydrophobic or weakly hydrophobic metal complexes. In addition, the selectivity of the separation of environmental pollutants has also been increased by varying the partition and ion-association (micellar interactions) mechanisms.

APPLICATIONS

During the last decade, CE has been increasingly used for the determination of environmental pollutants. Some of the methods of pre-treatment of waste environmental samples have been carried out prior to the injection into the CE system, as discussed above. CE has been applied for the determination of inorganic and organic pollutants. The major inorganic pollutants

include metal ions and anions. On the other hand, the most common toxic organic environmental pollutants analysed by CE are phenols, pesticides, polynuclear aromatic hydrocarbons, amines, carbonyl compounds, surfactants, dyes, and others. Recently, chiral separation of pollutants and xenobiotics has emerged as the most important issue for the environmental chemist. Therefore, CE has also been used for the analysis of the chiral pollutants. The application of CE for the determination of environmental pollutants is summarized in Table 1, which contains the type of pollutants, their sources, BGE used, detection method, and detection limit. As a specimen sample, a typical electropherogram of separated metal ions in a water sample by CE is shown in Fig. 2.[7]

Validation of Methods

There are only a few studies dealing with method validation of the determination of environmental pollutants by CE. However, some authors have demonstrated the application of their developed separation methods. The accuracy determination for Na, K, Ca, and Mg metal ions has been presented.[7–9] Similarly, the precision of migration times and peak areas for seven alkali and alkaline earth metals has been measured; RSD values were less than 0.4% for migration times and from 0.8% to 1.8% for peak

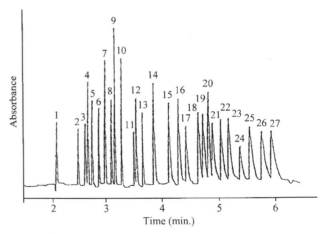

Fig. 2 The electropherogram of the separation of alkali, alkaline earth, transition metals, and lanthanoids on a fused silica capillary (60 cm × 75 μm) using 15 nM lactic acid, 8 mM 4-methylbenzylamine and 5% MeOH, pH 4.25, as running electrolyte with 30 kV as the separation voltage, 20°C temperature, and UV detection (214 nm).[7] Peaks: 1 = K; 2 = Ba; 3 = Sr; 4 = Na; 5 = Ca; 6 = Mg; 7 = Mn; 8 = Cd; 9 = Li; 10 = Co; 11 = Pb; 12 = Ni; 13 = Zn; 14 = La; 15 = Ce; 16 = Pr; 17 = Nd; 18 = Sm; 19 = Gd; 20 = Cu; 21 = Tb; 22 = Dy; 23 = Ho; 24 = Er; 25 = Tm; 26 = Yb; and 27 = Lu.

areas.[7–9] In one of the experiments, the reported %RSD varied from 2.79 to 3.38 for Zn, Cu, and Fe metal ions.[15] Several other studies have shown reliable results with recoveries close to 100%, or good agreement with the results obtained by other methods.[15] In spite of this, the precision of linearity, sensitivity, and reproducibility of CE methods for metal ions and anions analysis are not better than ion chromatography.

CONCLUSIONS

The determination of environmental pollutants at trace level is currently a very important and challenging issue. GC and HPLC have been used for the analysis of environmental pollutants but, in recent years, CE has also been used for the determination of environmental pollutants. A search of the literature indicates several reports of the analysis of environmental pollutants by CE, but CE could not have yet achieved a place in the routine analysis of these pollutants. The reason for this is the poor detection of metal ions and anions and the poor reproducibility of CE methods. Therefore, many workers have suggested various modifications and alternatives to make CE a method of choice. To obtain good sensitivity and reproducibility, the selection of the capillary wall chemistry, pH and ionic strength of the BGE, complexing and visualizing agents, detectors, and optimization of BGE have been described and suggested.[7–9,14–17]

Apart from the points discussed for improvement of CE applications for the determination of environmental pollutants, some other aspects should also be addressed so that CE can be used as the routine method of choice in this field. The important points relating to this include the development and wide use of fluorescent and radioactive complexing agents, since detection by fluorescent and radioactive detectors is more sensitive and reproducible with low limits of detection. To make CE application more reproducible, the BGE should be developed in such a way to ensure its physical and chemical properties remain unchanged during the experimental run. The nonreproducibility of the methods may be due to the heating of BGE during a long analysis. Therefore, to keep the temperature constant throughout the experiments, a cooling device should be included in the instrument. Besides, especially for the determination of anions, the CE instrument should be designed with the facility to reverse the electrodes. There are only a few reports dealing with method validation. To make the developed method more applicable, the validation of the methodology should be performed. All the capabilities and possibilities of CE have not yet been explored but they

are underway. However, eventually CE will be realized as a widely recognized method of choice for the determination of environmental pollutants.

In summary, there is much to be developed for the advancement of CE for the analysis of environmental pollutants. Definitely, CE will prove itself as the best technique for the determination of environmental pollutants within the next few years; it will achieve the status of the technique of routine analysis in most of the environmental laboratories.

LIST OF ABBREVIATIONS

AcOH:	Acetic acid
BTP:	Bis-trispropane
CD:	Cyclodextrin
CDTA:	Cyclohexane-1,2-diaminetetraacetic acid
CHES:	2-(N-Cyclohexylamino)-ethanesulphonic acid
CTAB:	Cetyltrimethylammonium bromide
DHBP:	1,1'-Di-n-heptyl-4,4'-bipyridinium hydroxide
DOSS:	Sodium dioctyl sulfosuccinate
DTPA:	Diethylenetriaminepenta acetic acid
EDTA:	Ethylenediaminetetraacetic acid
EOF:	Electroosmotic flow
HDB:	Hexadimethrine bromide
HEC:	Hydroxyethyl cellulose
HIBA:	α-Hydroxyisobutyric acid
HMOH:	Hexamethonium hydroxide
HQS:	8-Hydroxyquinoline-5-sulfonic acid
ICP-MS:	Inductively coupled plasma-mass spectrometer
KHP:	Potassium hydrogenphthalate
LiOH:	Lithium hydroxide
MeOH:	Methanol
MES:	Morpholinoethanesulfonic acid
NaAc:	Sodium acetate
NaCl:	Sodium chloride
NH_4Ac:	Ammonium acetate
NaOH:	Sodium hydroxide
NG:	Nonyl-β-D-glucopyranoside
OG:	Octyl-β-D-glucopyranoside
PAHs:	Polyaromatic hydrocarbons
PDCA:	Pyridine-2,6-dicarboxylic acid
PEG:	Polyethylene glycol
PMA:	Pyromellitic acid
PVP:	Polyvinylpyrrolidine
RSD:	Lower standard deviation
SDS:	Sodium dodecyl sulfate
TBABr:	Tetrabutylammonium bromide
TEA:	Triethanolamine
TTABr:	Tetradecyl-trimethylammonium bromide

TTAOH: Tetradecyl trimethylammonium
         hydroxide
TTHA: Triethylenetetraminehexa acetic acid
UV: Ultraviolet

REFERENCES

1. Franklin, L.B. *Wastewater Engineering: Treatment, Disposal and Reuse*; McGraw-Hill, Inc.: New York, 1991.

2. Droste, R.L. *Theory and Practice of Water and Wastewater Treatment*; John Wiley & Sons, Inc.: New York, 1997.

3. Crego, A.L.; Marina, M.L. Capillary zone electrophoresis versus micellar electrokinetic chromatography in the separation of phenols of environmental interest. J. Liq. Chromatogr. & Relat. Technol. **1997**, *20*, 1–20.

4. Kallenborn, R.; Huhnerfuss, H. *Chiral Environmental Pollutants: Trace Analysis and Ecotoxicology*; Springer-Verlag: Berlin, 2000.

5. Dabek-Zlotorzynska, E. Capillary electrophoresis in the determination of pollutants. Electrophoresis **1997**, *18*, 2453–2464.

6. Sovocool, G.W.; Brumley, W.C.; Donnelly, J.R. Capillary electrophoresis and capillary electrochromatography of organic pollutants. Electrophoresis **1999**, *20*, 3297–3310.

7. Pacakova, V.; Coufal, P.; Stulik, K. Capillary electrophoresis of inorganic cations. J. Chromatogr. A. **1999**, *834*, 257–275.

8. Liu, B.F.; Liu, B.L.; Cheng, J.K. Analysis of inorganic cations as their complexes by capillary electrophoresis. J. Chromatogr. A. **1999**, *834*, 277–308.

9. Valsecchi, S.M.; Polesello, S. Analysis of inorganic species in the environmental samples by capillary electrophoresis. J. Chromatogr. A. **1999**, *834*, 363–385.

10. Martinez, D.; Cugat, M.J.; Borrull, F.; Calull, M. Solid phase extraction coupling to capillary electrophoresis with emphasis on environmental analysis. J. Chromatogr. A. **2000**, *902*, 65–89.

11. Malik, A.K.; Faubel, W. A review of analysis of pesticides using capillary electrophoresis. Crit. Rev. Anal. Chem. **2001**, *31*, 223–279.

12. Haddad, P.R.; Doble, P.; Macka, M. Development in sample preparation and separation techniques for the determination of inorganic ions by ion chromatography and capillary electrophoresis. J. Chromatogr. A. **1999**, *856*, 145–177.

13. Dabek-Zlotorzynska, E.; Aranda-Rodriguez, R.; Keppel-Jones, K. Recent advances in capillary electrophoresis and capillary electro-chromatography of pollutants. Electrophoresis **2001**, *22*, 4262–4280.

14. Timerbaev, A.R.; Buchberger, W. Prospects for the detection and sensitivity enhancement of inorganic ions in capillary electrophoresis. J. Chromatogr. A. **1999**, *834*, 117–132.

15. Macka, M.; Haddad, P.R. Determination of metal ions by capillary electrophoresis. Electrophoresis **1997**, *18*, 2482–2501.

16. Horvath, J.; Dolnike, V. Polymer wall coating for capillary electrophoresis. Electrophoresis **2001**, *22*, 644–655.

17. Mayer, B.X. How to increase precision in capillary electrophoresis. J. Chromatogr. A. **2001**, *907*, 21–37.

18. Marina, M.L.; Crego, A.L. Capillary electrophoresis: a good alternative for the separation of chiral compounds of environmental interest. J. Liq. Chromatogr. & Relat. Technol. **1997**, *20*, 1337–1365.

Essential Oils: Analysis by GC

M. Soledad Prats
Alfonso Jiménez
Department of Analytical Chemistry, University of Alicante, Alicante, Spain

INTRODUCTION

Essential oils are highly concentrated substances extracted from flowers, leaves, stems, roots, seeds, barks, resins, or fruit rinds. The levels of essential oils found in plants can be anywhere from 0.01% to 10% of the total. This is why tons of plant materials are required for just a few hundred pounds of oil. These oils are often used for their flavor and their therapeutic or odoriferous properties, in a wide selection of products such as foods, medicines, and cosmetics.

Pure essential oils are mixtures of more than 200 components, normally mixtures of terpenes or phenylpropanic derivatives, in which the chemical and structural differences between compounds are minimal. They can be essentially classified into two groups: A volatile fraction, constituting 90–95% of the oil in weight, containing the monoterpene and sesquiterpene hydrocarbons, as well as their oxygenated derivatives along with aliphatic aldehydes, alcohols, and esters; and a nonvolatile residue that comprises 1–10% of the oil, containing hydrocarbons, fatty acids, sterols, carotenoids, waxes, and flavonoids.

OVERVIEW

Because of the enormous amount of raw products used to obtain a small amount of essential oil, many products on the market have been polluted with lower-quality commercial oils to reduce their cost, a fact not usually indicated on the label. This is why it is important to study the chemical composition of the volatile fraction once the essential oil is extracted. This fraction is characterized by the complexity in the separation of its components, which belong to various classes of compounds and which are present in a wide range of concentrations. Therefore it is complicated to establish a composition profile of essential oils.

The GC is almost exclusively used for the qualitative analysis of volatiles. The analysis of essential oils was developed in parallel with the technological developments in GC, such as stationary phases, detection devices, etc. However, advances in instrumentation were not the only important factor in the development of analytical methods for essential oils in plants. Sample extraction and concentration were also improved. The most outstanding improvements in the determination of the composition of essential oils came from the introduction of tandem techniques involving prior/further chromatography or spectroscopy. The great amount of information on the application of GC and hyphenated techniques to essential oils has led to much research in this field, and to the publication of recent reviews.[1–3]

EXTRACTION METHODS

Extraction of essential oils is one of the most time- and effort-consuming processes in the analysis of the constituents of plants. Various extraction methods were traditionally employed, depending on the material or the available devices. The most commonly used methods are steam distillation and distillation-solvent extraction. The introduction of innovative extraction methods, such as microwave-assisted extraction (MAE) and supercritical fluid extraction (SFE), has led to significant improvement, not only in the analytical performance, but also in the accuracy and reproducibility of methods.

Steam distillation has been traditionally used for isolation of essential oils, but some problems were recently reported; for example, degradation of certain monoterpenes can occur because of acid-catalyzed hydration.[4] An alternative method, useful for much smaller sample sizes, involves extraction with organic solvents, such as dichloromethane, followed by evaporation of solvent from the extract.[5] However, this approach is not very popular when the obtained extracts are to be used in the cosmetic or food industry, because of the possible toxic organic solvent residue.

Supercritical fluid extraction and microwave-assisted extraction have been recently applied to the extraction of essential oils. Both techniques are based on the application of high pressures and temperatures for the total extraction of analytes. Supercritical fluid extraction has shown much potential for the isolation of organic compounds from various samples by minimizing sample handling, providing relatively clean

Encyclopedia of Chromatography DOI: 10.1081/E-ECHR-120018661

extracts, expediting sample preparation, and reducing the use and disposal of environmentally aggressive solvents. Additionally, in many cases, SFE provides recoveries even better than those of conventional solvent extraction techniques.

One of the most important subjects of research in this field is the modeling of the essential oil extractions. Several kinds of models have been presented in recent literature.[6–8] Most of these models consider the natural matrix as a porous sphere and the extractable material as a single chemical species. The application to multicomponent systems permits the extension of the extraction model to simulate SFE in essential oils and the selective extraction of the mixed constituents.[7] Therefore the most important extraction parameters can be controlled for an optimized process with these complicated samples.

The use of SFE and MAE has been generalized to many essential oils and different samples. These techniques have improved recoveries in the determination of most organic additives, as well as permitted considerable reductions in solvent volume and extraction time. However, the comparison of extraction methods was usually reduced to relative recoveries of target analytes, ignoring important analytical parameters of the method. Selectivity is one of these, as the coextraction of other organics from the matrix usually requires a postextraction cleanup step before chromatographic analysis. There is still much effort to be carried out in this field in order to optimize the extraction of essential oils from different natural matrices. The selection of the best extraction method depends on the components to be extracted, and this is something to be carefully considered in each particular case.

CONCENTRATION OF ANALYTES

Another important aspect to take into account for a reproducible and accurate separation and determination of essential oils is the concentration of each component. In many cases, a preconcentration of the sample, prior to any other step in the analytical process, is necessary to assure a concentration range for an accurate determination. This is the way small amounts of each constituent in plants or complex matrices, such as pharmaceuticals, can be collected and concentrated using the headspace technique (HS-GC), which involves volatilization of the terpenoids and other substances in a closely confined space, followed with analysis of constituents in the gaseous phase. This process can be carried out as an equilibrium process (static headspace) or as a continuous process (dynamic headspace). Some essential oil applications of HS-GC were described using different enrichment and cryogenic techniques. Solid phase

microextraction (SPME) is one of the most promising ones.[9] A headspace-SPME system, coupled with GC/MS, has recently been reported as a powerful separation tool for essential oil analysis. Results were compared with those from steam-distilled samples and, in general, most of the monoterpene compounds were detected at higher levels by using HS-SPME with 30-sec extraction time. In addition, detailed information about terpenic compounds was obtained by using HS-SPME.[10]

GC ANALYSIS

The separation of essential oil components is usually carried out by GC with fused-silica capillary columns. The properties and conditions of columns used are variable, depending on the polarity of the components to be separated. The most used columns include stationary phases such as DB-1, Carbowax, OV-1, OV-101, PEG 20M, BP5, and DB-5, which cover a wide range of polarities. Column lengths normally range from 25 to 100 m, and stationary phase film thickness ranges from 0.2 to 0.7 μm. Elution of components is usually performed with a temperature gradient ranging from 50°C to 280°C.

New developments in stationary phases for use with essential oils have been recently reported.[2] These developments have led to the production of thermally and chemically stable phases, with greater selectivity and efficiency. It is advantageous to use a more selective phase for a given separation as the overlapping of peaks in the final chromatogram is often a significant drawback of chromatographic techniques in natural samples. The discovery of chiral phases (mostly based on cyclodextrin derivatives) allows the resolution of enantiomers of volatile components. These phases can give different elution sequences for a polarity range and provide a distinct advantage in identification because of large changes in solute relative retention times.

The information obtained from high-resolution GC analysis of the volatile fraction of essential oils must be sufficient to determine whether the product is genuine or not. If the product is adulterated, the kind and level of adulteration must be detected. Therefore a selective and accurate separation is absolutely necessary in the case of industrial analysis. On the other hand, GC sometimes permits the separation and further identification of some components of the nonvolatile residue as well.

An important drawback in the separation of essential oils is the time required for complete GC resolution of the components of interest, which sometimes can take hours. In fact, the analysis of essential oil samples is usually carried out with slow temperature programs, which take long times for the development of the whole chromatogram. There are several ways to reduce analysis time in GC. The most common approach is to use shorter capillary columns with reduced internal diameter and reduced film thickness. When using these columns, the optimum carrier gas velocity is higher, and it is possible to work with higher average linear velocity without the loss of efficiency. However, this increase in linear velocity must be linked with some specific conditions of measurement, such as fast oven heating, fast acquisition rate, high inlet pressures, and higher split ratios. Some work has been carried out in this area, using 10-m-long columns with 0.1-mm internal diameter and conventional instrumentation for the analysis of citrus essential oils by fast GC/flame ionization detection (GC/FID) and fast GC/MS.[11]

DETECTION AND CHARACTERIZATION OF CONSTITUENTS

The flame ionization detector (FID) is still widely applied for the detection and quantitation of some of the essential oil components, such as terpenoids. As usual in GC/FID, the primary criterion for the identification of peaks is the comparison of the standard retention times with the retention times of peaks in the sample's chromatogram. However, this procedure is sometimes not useful as the identification is quite difficult and overlapping of peaks makes determination not possible.

The easiest and most frequently used way to identify essential oil components when using GC/FID is comparison with Kovats retention indices (RI). The use of this type of retention data, derived from two GC columns of different polarities, allows highly reliable identification of large numbers of components in a particular sample.

By far, MS is the most popular detection technique for performing chromatographic studies of essential oils. The use of retention indices, in conjunction with GC/MS studies, is well established. Many laboratories use such procedures in their routine analyses to confirm the identities of unknown components. The identification of components is usually performed by comparing the mass spectra with an MS library. However, a feature of MS for essential oils is that mass spectra are not particularly unique in many cases because of the large numbers of isomers of the same molecular formula, but with different structures, that could exist. Therefore their mass spectra are similar and their identification is sometimes not so easy. The most common approach to solve this problem, as well as the presence of unknowns on whom very little other structural information is available, is the use of

algorithms and powerful MS databases, as has been recently proposed.[12] Two different MS databases are commonly used as references: National Institute of Standards and Technology (NIST)/Environment Protection Agency (EPA)/National Institutes of Health (NIH) and the Registry of Mass Spectral Data. The first one contains more than 62,000 mass spectra of different chemicals. The largest database is the Registry of Mass Spectral Data, called the Wiley database, containing more than 300,000 different spectra, resulting from the work of many researchers in the field of MS. One of the most used algorithms was proposed by Oprean et al.,[1] which considers two parameters as identification criteria for an unknown peak, i.e., the match index of the unknown mass spectrum with spectral libraries, and the relative retention indices computed from the retention times of the unknowns relative to a mixture of n-alkanes.

One of the most recently proposed methods to improve the analysis of complex mixtures, especially for deconvolution of overlapping mass spectra, is time-of-flight mass spectrometry (TOFMS). This technique allows assignment of a spectrum to each individual solute in significantly overlapping elution profiles. This is an important advantage that can be exploited when fast GC methods are applied for complex samples because each overlapping peak may be deconvoluted and the individual spectrum of each overlapping solute may be obtained. Although great efforts have been recently carried out in this field, it remains to be determined if TOFMS can be used on a routine basis.

The identification of compounds comprising more than 1% in the oils can be also carried out by ^{13}C-NMR and computer-aided analysis.[13] The chemical shift of each carbon in the experimental spectrum can be compared with those of the spectra of pure compounds. These spectra are listed in the laboratory spectral data bank, which contains approximately 350 spectra of mono-, sesqui- and diterpenes, as well as with literature data. Each compound can be unambiguously identified, taking into account the number of identified carbons, the number of overlapped signals, as well as the difference between the chemical shift of each resonance in the mixture and in the reference.

The combination of GC with olfactometry is another possibility for detection that has been used in essential oil analysis. Olfactometry adapters are commercially available and should include humidity of the GC effluent at the nose adapter and provide auxiliary gas flow. The correlation among eluted peaks with specific odors allows accurate retention indices or retention times to be established for the essential oil components. Some of them can be detected in such way after applying chemometric techniques, such as cluster analysis and principal component analysis, to the data from the sensors.

HYPHENATED OR MULTIDIMENSIONAL ANALYSIS OF ESSENTIAL OILS

Hyphenated or multidimensional techniques have been recently introduced for the analysis of essential oils. Various approaches were recently proposed to obtain better results in the identification and quantification of essential oil components. Thus it is possible to use systems that incorporate separations prior to GC, multi-column separations, and specific identification methods.

With respect to separations preceding GC analysis, some hyphenated techniques have been successfully used when there is a lack of resolution of the single capillary GC method. One of these is the combination of HPLC with GC. The prior HPLC step achieves the isolation of components of similar chemical composition, primarily based on polarity. Hence, this will separate saturated hydrocarbons from unsaturated or aromatic hydrocarbons, for example.[14] These systems are fully automated, but there is a problem with off-line sampling of HPLC fractions. The selection of the HPLC injection port will determine the particular method of separation. Therefore in this instrumental arrangement, each transferred fraction must be separately analyzed before the introduction of a subsequent fraction into the GC system. In general, the prior separation will be introduced to simplify the subsequent GC analysis, leading to improved resolution.

Multidimensional GC

The application of multidimensional GC (MDGC) to essential oil analysis is a great development in the determination of such complex samples. This is an appropriate approach when there are zones on the chromatogram where the peaks are not well resolved, which is a common situation in natural samples. The fractions corresponding to the zones with unresolved peaks are transferred to a second column containing a different stationary phase, where they are separated and completely resolved. Therefore MDGC permits the separation of poorly resolved peaks and increases resolution, with the final result of an improvement in both identification and quantification of components of essential oils.

However, this evidently improved approach in instrumentation is only available to relatively few regions of chromatographic analysis, as overlapping of peaks in the same area can be too complex for a complete resolution of each component, even by

applying multiple-column couplings. The use of conventional MDGC technology is not possible for the entire analysis because this would involve transferring all the components to the second column, with the inherent technical problems of loss of selectivity and sensitivity. This is why the MDGC analysis of essential oils is not focused on the increase of resolution for the whole sample, but only for specific components of interest in the quality control of the natural product. The use of chiral columns in one or both separation processes should be an additional improvement in the resolution of chromatograms, but this is an area requiring further attention.

A possible solution to the above problems would be the triple-dimensional analysis by using GC × GC coupled to TOFMS.[15] MS techniques improve component identification and sensitivity, especially for the limited spectral fragmentation produced by soft ionization methods, such as chemical ionization (CI) and field ionization (FI). The use of MS to provide a unique identity for overlapping components in the chromatogram makes identification much easier. Thus MS is the most recognized spectroscopic tool for identification of GC × GC-separated components. However, quadrupole conventional MS are unable to reach the resolution levels required for such separations. Only TOFMS possess the necessary speed of spectral acquisition to give more than 50 spectra/sec. This area of recent development is one of the most important and promising methods to improve the analysis of essential oil components.

CONCLUSIONS

The identification and determination of essential oils in many natural samples have improved greatly with the use of more powerful analytical techniques, such as fast extraction methods, better chromatographic detectors, and hyphenation. This improvement in analytical parameters open a great future for the development of analytical methods for essential oil determinations, even at low limits of detection.

REFERENCES

1. Oprean, R.; Tamas, M.; Sandulescu, R.; Roman, L. Essential oil analysis. I. Evaluation of essential oil composition using both GC and MS finger-prints. J. Pharm. Biomed. Anal. **1998**, *18*, 651–657.
2. Marriot, P.J.; Shellie, R.; Cornwell, C. Gas chromatographic techniques for the analysis of essential oils. J. Chromatogr. A. **2001**, *936*, 1–22.

3. Lockwood, G.B. Techniques for gas chromatography of volatile terpenoids from a range of matrices. J. Chromatogr. A. **2001**, *936*, 23–31.

4. Griffiths, D.W.; Robertson, G.W.; Birch, A.N.E.; Brennan, R.M. Evaluation of thermal desorption and solvent elution combined with polymer entrainment for the analysis of volatiles released by leaves from midge (*Dasineura tetensi*) resistant and susceptible blackcurrant (*Ribesnigrum* L.) cultivars. Phytochem. Anal. **1999**, *10*, 328–334.

5. Zhu, W.; Lockwood, G.B. Enhanced biotransformation of terpenes in plant cell suspensions using controlled release polymer. Biotechnol. Lett. **2000**, *22*, 659–662.

6. Spricigo, C.B.; Pinto, L.T.; Bolzan, A.; Novais, A.F. Extraction of essential oil and lipids from nutmeg by liquid carbon dioxide. J. Supercrit. Fluids **1999**, *15*, 253–259.

7. Tezel, A.; Hortaçsu, A.; Hortaçsu, O. Multicomponent models for seed and essential oil extraction. J. Supercrit. Fluids **2000**, *19*, 3–17.

8. Benyoussef, E.H.; Hasni, S.; Belabbes, R.; Bessiere, J.M. Modélisation du transfert de matiére lors de l'extraction de l'huile essentielle des fruits de coriandre. Chem. Eng. J. **2002**, *85*, 1–5.

9. Pawliszyn, J. *Applications of Solid Phase Microextraction*; Royal Society of Chemistry: Cambridge, 1999.

10. Rohloff, J. Essential oil composition of sachalinmint from Norway detected by solid phase microextraction and gas chromatography-mass spectrometry analysis. J. Agric. Food Chem. **2002**, *50*, 1543–1547.

11. Mondello, L.; Zappia, G.; Bonaccorsi, I.; Dugo, G.; McNair, H.M. Fast GC for the analysis of natural matrices. Preliminary note: The determination of fatty acid methyl esters in natural fats. J. Microcolumn September **2000**, *12*, 41–47.

12. Oprean, R.; Oprean, L.; Tamas, M.; Sandulescu, R.; Roman, L. Essential oils analysis. II. Mass spectra identification of terpene and phenylpropane derivatives. J. Pharm. Biomed. Anal. **2001**, *24*, 1163–1168.

13. Mundina, M.; Vila, R.; Tomi, F.; Gupta, M.P.; Adzet, T.; Casanova, J.; Cañigueral, S. Leaf essential oils of three Panamanian piper species. Phytochemistry **1998**, *47*, 1277–1282.

14. Mondello, L.; Dugo, G.; Bartle, K.D. On-line microbore high performance liquid chromatography capillary gas chromatography for food and water analyses. A review. J. Microcolumn September **1996**, *8*, 275–310.

15. Shellie, R.; Marriot, P.; Morrison, P. Concepts and preliminary observations on the triple-dimensional analysis of complex volatile samples by using GC × GC-TOFMS. Anal. Chem. **2001**, *73*, 1336–1344.

Evaporative Light-Scattering Detection: Applications

Juan G. Alvarez
*Beth Israel Deaconess Medical Center, Harvard Medical School,
Boston, Massachusetts, U.S.A.*

INTRODUCTION

HPLC is mainly carried out using light absorption
detectors as ultraviolet (UV) photometers and spectrophotometers (UVD) and, to a lesser extent, refractive
index detectors (RID). These detectors constitute the
main workhorses in the field.[1] The sensitive detection
of compounds having weak absorption bands in the
range 200–400 nm, such as sugars and lipids is, however, very difficult with absorption detectors. The use
of the more universal RID is also restricted in practice
because of its poor detection limit and its high sensitivity to small fluctuations of chromatographic
experimental conditions, such as flow rate, solvent
composition, and temperature.[2] Moreover, if the
separation of complex samples requires the use of gradient elution, the application of RID becomes almost
impossible. Although for some solutes the use of either
a reaction detector (RD) or a fluorescence detector
(FD) is possible, this is not a general solution. In this
regard, the analysis of complex mixtures of lipids or
sugars by HPLC remains difficult owing to the lack
of a suitable detector.

The miniaturization of detector cells is also extremely difficult and the technological problems have
not yet been solved because the detection limit should
also be decreased or, at least, kept constant.[2–6] Some
progress in the design of very small cells for UVD
and FD has been reported,[3–7] but the miniaturization
of RD and RID seems much more difficult in spite of
some suggestions.[8] Similarly, the development of open
tubular columns is plagued by the lack of a suitable
detector with a small contribution to band broadening.
A nonselective detector more sensitive than the RID
and easier to use with a small contribution to band
broadening is thus desirable in HPLC. The mass spectrometer would be a good solution if it were not so
complex[10] and expensive. The electron-capture detector (ECD)[11] and flame-based detectors have been suggested.[12] Both are very sensitive and could be made
with very small volumes. Unfortunately, the ECD
can be used only with volatile analytes and it is very
selective. Both ECD and flame-based detectors are
very sensitive to the solvent flow rate, and noisy signals
are often produced. The adaptability of these detectors
to packed columns is thus difficult. This probably
explains why the ECD has been all but abandoned.

The evaporative light-scattering analyzer,[13–14] on
the other hand, is an alternative solution which seems
very attractive for a number of reasons. As most analytes in HPLC have a very low vapor pressure at room
temperature and the solvents used as the mobile phase
have a significant vapor pressure, some kind of phase
separation is conceivable.

EVAPORATIVE LIGHT-SCATTERING DETECTOR

Principle of Operation

The unique detection principle of evaporative light-
scattering detectors involves nebulization of the
column effluent to form an aerosol, followed by
solvent vaporization in the drift tube to produce a
cloud of solute droplets (or particles), and then
detection of the solute droplets (or particles) in the
light-scattering cell.

Detector Components

Nebulizer: The nebulizer is connected directly to the
analytical column outlet. In the nebulizer, the column
effluent is mixed with a steady stream of nebulizing
gas, usually nitrogen, to form an aerosol. The aerosol
consists of a uniform dispersion of droplets. Two
nebulization properties can be adjusted to regulate
the droplet size of the analysis. These properties are
gas and mobile-phase flow rates. The lower the
mobile-phase flow rate, the less gas and heat are
needed to nebulize and evaporate it. Reduction of flow
rate by using a 2.1-mm-inner diameter column should
be considered when sensitivity is important. The gas
flow rate will also regulate the size of the droplets in
the aerosol. Larger droplets will scatter more light
and increase the sensitivity of the analysis. The lower
the gas flow rate, the larger the droplets. It is also
important to remember that the larger the droplet,
the more difficult it will be to vaporize in the drift
tube. An unvaporized mobile phase will increase the baseline
noise. There will be an optimum gas flow rate for each
method which will produce the highest signal-to-noise
ratio.

Encyclopedia of Chromatography DOI: 10.1081/E-ECHR-120039883

Drift tube: In the drift tube, volatile components of the aerosol are evaporated. The nonvolatile particles in the mobile phase are not evaporated and continue down the drift tube to the light-scattering cell to be detected. Nonvolatile impurities in the mobile phase or nebulizing gas will produce noise. Using the highest-quality gas, solvents, and volatile buffers, preferably a filter, will greatly reduce the baseline noise. Detector noise will also increase if the mobile phase is not completely evaporated. The sample may also be volatilized if the drift-tube temperature is too high or the sample is too volatile. The optimal temperature in the drift tube should be determined by observing the signal-to-noise ratio with respect to temperature.

Light-scattering cell: The nebulized column effluent enters the light-scattering cell. In the cell, the sample particles scatter the laser light, but the evaporated mobile phase does not. The scattered light is detected by a silicone photodiode located at a 90° from the laser. The photodiode produces a signal which is sent to the analog outputs for collection. A light trap is located 180° from the laser to collect any light not scattered by particles in the aerosol stream.

The signal is related to the solute concentration by the function $\Lambda = am^x$, where x is the slope of the response line, m is the mass of the solute injected in the column, and a is the response factor.

APPLICATIONS

Evaporative light-scattering detection finds wide applicability in the analysis of lipids and sugars. The analysis of lipids and sugars by HPLC has classicaly been hampered due to the lack of absorbing chromophores in these molecules. Accordingly, most analyses are carried out by GC, requiring derivatization in the case of the sugars or being especially difficult like the separation of the high-molecular-weight triglycerides, or even impossible for the important class of phospholipids, which cannot withstand high temperatures. Specific applications are as follows:

1. Use of evaporative light scattering detector in reversed-phase chromatography of oligomeric surfactants. Y. Mengerink, H. C. De Man, and S. J. Van Der Wal, J. Chromatogr. 552: 593 (1991).
2. A rapid method for phospholipid separation by HPLC using a light-scattering detector: W. S. Letter, J. Liq. Chromatogr. 15: 253 (1992).
3. Detection of HPLC separation of glycophospholipids: J. V. Amari, P. R. Brown, and J. G. Turcotte, Am. Lab. 23 (Feb. 1992).
4. Analysis of fatty acid methyl esters by using supercritical fluid chromatography with mass evaporative light-scattering detection: S. Cooks and R. Smith, Anal. Proc. 28, 11 (1991).
5. HPLC analysis of phospholipids by evaporative light-scattering detection: T. L. Mounts, S. L. Abidi, and K. A. Rennick, J. AOCS 69: 438 (1992).
6. Determination of cholesterol in milk fat by reversed-phase high-performance liquid chromatography and evaporative light-scattering detection: G. A. Spanos and S. J. Schwartz, LC–GC 10(10): 774 (19XX).
7. A qualitative method for triglyceride analysis by HPLC using ELSD: W. S. Letter, J. Liq. Chromatogr. 16: 225 (1993).
8. Detect anything your LC separates, P. A. As-mus, Res. Dev. 2: 96 (1986).
9. Rapid separation and quantification of lipid classes by HPLC and mass (light scattering) detection: W. H. Christie, J. Lipid Res. 26: 507 (1985).

REFERENCES

1. Scott, R.P.W. *Liquid Chromatography Detectors*; Elsevier: Amsterdam, 1977.
2. Colin, H.; Krstulovic, A.; Guiochon, G. Analysis **1983**, *11*, 155.
3. Scott, R.P.W.; Kucera, P. Mode of operation and performance characteristics of microbore columns for use in liquid chromatography. J. Chromatogr. **1979**, *169*, 51–72.
4. Knox, J.H.; Gilbert, M.T. Kinetic optimization of straight open-tubular liquid chromatography. J. Chromatogr. **1979**, *186*, 405–418.
5. Guiochon, G. Conventional packed columns vs. packed or open tubular microcolumns in liquid chromatography. Anal. Chem. **1981**, *53*, 1318–1325.
6. Guiochon, G. *Miniaturization of LC Equipment*; Kucera, P., Ed.; Elsevier: Amsterdam, 1983.
7. Kucera, P.; Umagat, H. Design of a post-column fluorescence derivatization system for use with microbore columns. J. Chromatogr. **1983**, *255*, 563–579.
8. Jorgenson, J.W.; Guthrie, E.J. Liquid chromatography in open-tubular columns: Theory of column optimization with limited pressure and analysis time, and fabrication of chemically

bonded reversed-phase columns on etched borosilicate glass capillaries. J. Chromatogr. **1983**, *255*, 335–348.

9. Jorgenson, J.W.; Smith, S.L.; Novotný, M. Light-scattering detection in liquid chromatography. J. Chromatogr. **1977**, *142*, 233–240.

10. Arpino, P.J.; Guiochon, G. Anal. Chem. **1979**, *51*, 682A.

11. Willmont, F.W.; Dolphin, R.J. A novel combination of liquid chromatography and electron capture detection in the analysis of pesticides. J. Chromatogr. Sci. **1974**, *12*, 695.

12. McGuffin, V.L.; Novotný, M. Micro-column high-performance liquid chromatography and flame-based detection principles. J. Chromatogr. **1981**, *218*, 179–187.

13. Charlesworth, J.M. Evaporative analyzer as a mass detector for liquid chromatography. Anal. Chem. **1978**, *50*, 1414–1420.

14. Macrae, R.; Dick, J.J. Analysis of carbohydrates using the mass detector. J. Chromatogr. **1981**, *210*, 138–145.

Evaporative Light-Scattering Detection for LC

Sarah Chen
Analytical Research Department, Merck Research Laboratories, Rahway, New Jersey, U.S.A.

INTRODUCTION

Evaporative light scattering detection (ELSD) is a powerful technique that can be applied in LC to all solutes having lower volatility than the mobile phase. It consists of a nebulizer that transforms the eluent from the HPLC into an aerosol, a drift tube to vaporize the solvent, and a light scattering cell (Fig. 1). When using ELSD in conjunction with LC, the eluent is nebulized immediately into a stream of warm gas. The solvent then vaporizes leaving a cloud of solute particles. The particles are subjected to a light source and scattering occurs in the scattering chamber. The amount of light scattered by the particles is proportional to the analyte concentration. These principles make ELSD a universal detector that can be used for analytes with low UV chromophores. Evaporative light scattering detection has been used for the detection of polymers in SEC.[1] It has also been used in the detection of small molecules in reversed phase and normal phase LC.[2–4]

THEORY AND INSTRUMENTATION

Light scattering detection cannot be used if the analyte particles or solvent vapors absorb at the wavelength range of the light source. When particles are hit with a beam of light, the light may be absorbed, refracted, reflected, or scattered. Reflection and refraction always occur together and they prevail when the wavelength of light approaches the particle's size.[5] The sum of the reflection and refraction intensities equals the intensity of the incident light when there is no absorbance. Scattering occurs when the particle diameter is close to one-tenth of the wavelength. There are two types of scattering: Mie scattering and Rayleigh scattering. Mie scattering occurs when the ratio of particle diameter to the wavelength of light is greater than 0.1.[6] Rayleigh scattering occurs when the ratio is less than 0.1.[7] These numbers are approximate and a transition region does exist. The property of scattering has been used in the evaporative light scattering detector (ELSD) for LC. In ELSD, the eluent for LC is nebulized, and the amount of light scattered by the nebulized eluent particles is proportional to the analyte concentration.

Evaporative light scattering detection involves three successive and interrelated processes: nebulization of the chromatographic eluent, evaporation of the volatile solvent (mobile phase), and scattering of light by residual analyte particles. The three major parts of the system are the nebulizer, drift tube, and light-scattering cell.

Nebulizer

The nebulizer is normally interfaced directly to the LC column. It combines the eluent with a stream of gas to produce an aerosol. Much of the theoretical and practical basis of nebulization comes from atomic spectroscopy. The average droplet diameter and uniformity of the aerosol are the most important factors for ELSD sensitivity and reproducibility. As larger solute particles scatter light more intensely, an aerosol with large droplets and a narrow droplet size distribution leads to the most precise and sensitive detection. A good nebulizer should produce a uniform aerosol of large droplets with narrow droplet size distribution. The droplets cannot be too large, however; otherwise, the solvent in a droplet will not be completely vaporized and errors in detection will occur. The nebulizer properties that can be adjusted to obtain the desired droplet properties are, primarily, the gas flow rate and the LC mobile phase flow rate.[8]

Drift Tube

Volatile components of the aerosol produced by the nebulizer are evaporated in the drift tube to produce nonvolatile particles in a dispersed mixture of carrier gas and solvent vapors. Ideally, the temperature in the drift tube should be high enough to ensure the complete evaporation of solvents, yet not so high as to be able to volatilize the analytes. If solvent removal is incomplete, detector noise will increase. When extremely large droplets reach the light scattering cell, they will be seen as spikes. If the drift tube temperature is

Encyclopedia of Chromatography DOI: 10.1081/E-ECHR-120038633

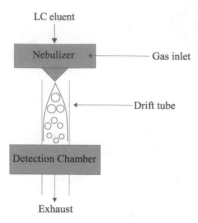

Fig. 1 Schematic diagram of evaporative light scattering detector. *(View this art in color at www.dekker.com.)*

too high, solute may be vaporized or partially vaporized, resulting in decreased sensitivity and accuracy. Droplet aggregation is another phenomenon that can occur in the drift tube. It can cause incomplete solvent removal and detector signal spiking. Overall, the drift tube should be wide enough, long enough, and hot enough to ensure complete and rapid solvent removal. Its outlet into the light-scattering cell should be shaped to send all of the particles past the detector window. There has been an increased need for a low temperature ELSD to address the detection of thermally labile compounds and volatile compounds. A low-temperature ELSD that can evaporate solvent at near-ambient temperature, 26–40°C, is now available from several vendors. These instruments are designed in a way that extremely large droplets are expelled to waste. Only droplets of optimum size can survive and the surface area of these droplets will allow maximum vaporization of solvents at moderate drift tube temperatures.

Light Scattering Cell

The particle cloud leaves the drift tube and enters into a light scattering cell. Laser light is passed into the cell through a window, scattered by the analyte, and detected at an angle to the incident light. Single wavelength light at 632, 650, or 670 nm has been generally used in various instrument designs. Other instruments use a polychromatic light source. It is believed that a polychromatic light source emits a distribution of wavelengths where specific absorbance effects are minimized and mass sensitivity predominates over structural sensitivity. The response increases monotonically with analyte mass because of the averaging effects that occur when polychromatic scattered light is collected. Thus a polychromatic light source

can be found more often than a single wavelength source in instruments. The detector is usually constructed in a way that material in the particle cloud will not stick to the window and fumes are properly vented. In addition, light traps are used to dissipate nonscattered light.

EXPERIMENTAL CONSIDERATIONS

The response factor of an ELSD largely depends on the size of analyte particles entering the detection chamber. After exiting from the HPLC column, the eluent stream is nebulized, and a scavenger gas stream carries the effluent cloud through a hot drift tube where the solvent vaporizes. The droplet shrinks to the volume of the nonvolatile material contained in the eluent. The average particle size in the cloud at a given time and the particle size distribution can be derived from the elution profile of the analyte and the droplet size distribution given by the nebulizer. The average diameter, D_o, of the particles formed in a concentric nebulizer is given by the Atkinson equation:[9]

$$D_o = \frac{A\sigma_1^{1/2}}{u\rho_1^{1/2}} + B\left(\frac{\eta_1}{(\sigma_1\rho_1)^{1/2}}\right)^{0.45}\left(\frac{1000Q_1}{Q_g}\right)^{1.5} \quad (1)$$

where A and B are constants, σ_1 is the surface tension of the mobile phase, ρ_1 is the density of the mobile phase, η_1 is the viscosity of the mobile phase, u is the relative velocity of the gas and liquid streams in the nebulizer (i.e., the cross-section average velocity of the gas stream between the gas and the liquid nozzles, minus the cross-sectional average velocity of the solvent in the liquid tube), Q_1 is the volume flow rate of the mobile phase, and Q_g is the volume flow rate of the scavenger gas.

Eq. (1) predicts that the average droplet size depends on the gas and solvent flow rate. It also predicts that the average droplet size will depend on the nature of the solvent, because of the dependency on the density, surface tension, and viscosity of the nebulized liquid. The initial droplet size formed in the nebulizer has little to do with the property of the analyte as it predominantly contains mobile phase. The final droplet size in the scattering chamber is dependent on the analyte concentration. When optimizing detector conditions, the experimental parameters that can be adjusted are nebulizer gas flow rate, mobile phase flow rate, and drift tube temperature.

Effect of Scavenger Gas Flow Rate

While keeping mobile phase constant, a plot of the response for a constant sample amount vs. the scavenger gas flow rate exhibits a maximum at an intermediate flow rate, and so does the plot of the signal-to-noise ratio vs. flow rate. At large flow rates, the decrease in response is due to the fact that the average particle size of the solute cloud decreases with increasing gas flow rate according to Eq. (1). The response decreases accordingly with the particle size. At low flow rate, the response factor decreases rapidly with decreasing flow rate, while the noise increases and spikes appear. This is related to the fact that the flow velocity of the scavenger gas in the concentric nebulizer should be in the sonic range in order for the nebulizer to function properly. A low gas flow rate results in very large droplets that vaporize too slowly; hence a spike appears. Precipitation or aggregation can occur if the nebulizer gas pressure is too low. Precipitation in the drift tube can also cause a decrease in sensitivity.

Effect of Mobile Phase Flow Rate

An increase in eluent flow rate will result in increased droplet size and high response factor. However, a flow rate that is too high will result in the incomplete vaporization of the mobile phase and high background noise. When ELSD is used in conjunction with LC, the effect of flow rate on the separation should also be considered.

Effect of Drift Tube Temperature

The solvent contained in the droplets formed in the nebulizer must be completely vaporized during the migration of these droplets down through the drift tube. Thus a compromise between the scavenger gas pressure, the mobile phase flow velocity, and drift tube temperature has to be chosen. The residence time of the droplets of solution in the drift tube should be large enough and the drift tube temperature should be high enough to ensure the complete vaporization of solvents. Meanwhile, the temperature must be low enough so that the analytes are not vaporized, as this would result either in a systematic error (small extent of analyte vaporization) or in a total loss of signal (total vaporization of analyte). Low temperatures avoid evaporation of semivolatile analytes and destruction of thermally labile compounds. Vaporization of the solvent is facile with organic mobile phases such as acetonitrile, hexane, or chloroform, but relatively difficult with aqueous mobile phases.

Other Considerations

The response of the evaporative light scattering detector is not linear. This is because the droplets scatter light with an intensity that increases much faster than the third power of their diameter.[10] The response of the detector is not linear but is given by

$$A = aC^b \qquad (2)$$

where A is the response of the detector, a and b are numerical coefficients, C is the concentration of solute. The data can be plotted as log A vs. log C to obtain a graph that has a large linear region with a slope b and an ordinate a. This region can be used for quantitation. Slope b tends to be similar for similar compounds and falls between 1 and 2. A slope of 2 is the limiting value for Rayleigh scattering.[11]

One advantage of ELSD is that a wide range of solvents can be used, including acetone and chloroform which are not useful with UV detection. One drawback is that the solvent must be significantly more volatile than the analytes; thus the use of non-volatile buffers should be strictly avoided. Only high-quality HPLC solvents with minimum particulates should be used.

If the solvents remain clean and totally volatilized, baseline drift should not be observed during gradient elution. However, sensitivity may change in solvent gradients, mainly due to the change in droplet size as a result of the change in eluent properties such as surface tension, viscosity, and density. In general, shallow gradients are preferable.

Outlet waste gas stream from the ELSD may contain organic solvent vapors. For safety reasons, it is essential to ensure that the outlet of ELSD is properly directed to a safe vented outlet (e.g., a fume hood). Waste ventilation should occur at atmospheric pressure. A vacuum or restriction may result in pressure changes within the optical detection chamber and cause detector baseline instability.

CONCLUSIONS

Evaporative light scattering detection can be used as a universal detector for LC. Its operation includes the nebulization of the eluent in the nebulizer, solvent evaporation in the drift tube, and scattered light detection at the light scattering chamber. Experimental conditions which can be adjusted in most ELSD systems to optimize the detector sensitivity are the nebulizer gas flow rate, mobile phase flow rate, and drift tube temperature. The detector response is nonlinear, but can be used in quantitative work if a calibration curve is obtained.

REFERENCES

1. Nagy, D.J. J. Appl. Polym. Sci. **1996**, *62* (5), 845.
2. Toussaint, B.; Duchateau, A.L.L.; van der Wal, Sj.; Albert, A.; Hubert, Ph.; Crommen, J. J. Chromatogr. A. **2000**, *890*, 239–249.
3. Risley, D.S.; Strege, M.A. Chiral separations of polar compounds by hydrophilic interaction chromatography with evaporative light scattering detection. Anal. Chem. **2000**, *72*, 1736–1739.
4. Chen, S.; Yuan, H.; Grinberg, N.; Dovletoglou, A.; Bicker, G. J. Liq. Chromatogr. & Relat. Technol. **2003**, *26* (3), 425–442.
5. Charlesworth, J. Enantiomeric separation of trans-2-aminocyclohexanol on a crown ether stationary phase using evaporative light scattering detection. Anal. Chem. **1978**, *50* (11), 1402–1414.
6. Righezza, M.; Guiochon, G. J. Liq. Chromatogr. **1988**, *11* (9,10), 1967–2004.
7. Mourey, T.; Oppenheimer, L. Principles of operation of an evaporative light-scattering detector for liquid chromatography. Anal. Chem. **1984**, *56*, 2427–2434.
8. Stolyhwo, A.; Colin, H.; Martin, M.; Guiochon, G. Study of the qualitative and quantitative properties of the light scattering detector. J. Chromatogr. **1984**, *288*, 253–275.
9. Nukiyama, S.; Tanasawa, Y. Trans. Soc. Mech. Eng. Tokyo **1938**, *4*, 86.
10. Guiochon, G.; Moysan, A.; Holley, C. Influence of various parameters on the response factors of the evaporative light scattering detector for a number of non-volatile compounds. J. Liq. Chromatogr. **1988**, *11* (12), 2547–2570.
11. Oppenheimer, L.; Mourey, T. Examination of the concentration response of evaporative light-scattering mass detectors. J. Chromatogr. **1985**, *323*, 297–304.

Extra-Column Dispersion

Raymond P. W. Scott
Scientific Detectors Ltd., Banbury, Oxfordshire, U.K.

E

INTRODUCTION

In addition to the dispersion that takes place during the normal function of the column, dispersion can also occur in connecting tubes, injection system, and detector sensing volume, and as a result of injecting a finite sample mass and sample volume onto the column.

DISCUSSION

The major sources of extra column dispersion are as follows:

1. Dispersion due to the sample volume (σ_S^2).
2. Dispersion occurring in valve-column and column-detector connecting tubes (σ_T^2).
3. Dispersion in the sensor volume from Newtonian flow (σ_{CF}^2).
4. Dispersion in the sensor volu.me from peak merging (σ_{CM}^2).
5. Dispersion from the sensor and electronics time constant (σ_t^2).

The sum of the variances will give the overall variance for the extra-column dispersion (σ_E^2). Thus,

$$\sigma_E^2 = \sigma_S^2 + \sigma_T^2 + \sigma_{CF}^2 + \sigma_{CM}^2 + \sigma_t^2 \qquad (1)$$

Eq. (1) shows how the various contributions to extra-column dispersion can be combined. According to Klinkenberg,[1] the total extra-column dispersion must not exceed 10% of the column variance if the resolution of the column is not to be seriously denigrated; that is,

$$\sigma_E^2 = \sigma_S^2 + \sigma_T^2 + \sigma_{CF}^2 + \sigma_{CM}^2 + \sigma_t^2 = 0.1\sigma_c^2$$

In practice, σ_T^2, σ_{CF}^2, σ_{CM}^2, and σ_t^2 are all kept to a minimum to allow the largest contribution to extra-column dispersion to come from σ_S^2. This will allow the largest possible sample to be placed on the column, if so desired, to aid in trace analysis. Each extra-column dispersion process can be examined theoretically and two examples will be the evaluation of σ_S^2 and σ_T^2.

MAXIMUM SAMPLE VOLUME

Consider the injection of a sample volume (V_i) that forms a rectangular distribution of solute at the front of the column. The variance of the final peak will be the sum of the variance of the sample volume plus the normal variance from a peak for a small sample. Now, the variance of a rectangular distribution of sample volume (V_1) is $V_i^2/12$, and assuming the peak width is increased by 5% due to the dispersing effect of the sample volume (a 5% increase in standard deviation is approximately equivalent to a 10% increase in peak variance), then by summing the variances,

$$\frac{V_i^2}{12} + [\sqrt{n}(v_m + Kv_s)]^2 = [1.05\sqrt{n}(v_m + Kv_s)]^2$$

where the dispersion due to the column alone is $[\sqrt{n}(v_m + Kv_s)]^2$ (see the entry Plate Theory). Simplifying and rearranging,

$$V_i^2 = n(v_m + Kv_s)^2(1.22)$$

Bearing in mind that

$$V_r = n(v_m + Kv_s)$$

then

$$V_i = \frac{1.1V_r}{\sqrt{n}}$$

Thus, the maximum sample volume that can be tolerated can be calculated from the retention volume of the solute concerned and the efficiency of the column. A knowledge of the maximum sample volume can be important when the column efficiency available is only just adequate, and the compounds of interest are minor components that are only partly resolved.

DISPERSION IN CONNECTING TUBES

The column variance is given by V_r^2/n, and for a peak eluted at the dead volume, the variance will be V_0^2/n (see the entry Plate Theory). Thus, for a connecting tube of radius r_t and length l_t, the dead volume (V_0)

Encyclopedia of Chromatography DOI: 10.1081/E-ECHR-120039980

(i.e., the volume of the tube) is

$$V_0 = \pi r_t^2 l_t$$

Thus,

$$\sigma_E^2 = \frac{0.1(\pi r_t^2 l_t)^2}{n}$$

Now, for the dead volume peak from an open tube, $n = 1/0.6 r_t$ (see the entry Golay Dispersion Equation for Open-Tubular Columns). Thus,

$$\sigma_E^2 = 0.06 \pi^2 r_t^5 l_t$$

However, when assessing the length of tube that can be tolerated, it must be remembered that the 10% increase in variance that can be tolerated before resolution is seriously denigrated involves *all* sources of extra-column dispersion, not just for a connecting tube. In practice, the connecting tube should be made as short as possible and the radius as small as possible commensurate with reasonable pressures and the possibility that if the radius is too small, the tube may become blocked. The different sources of extra-column dispersion have been examined in Refs.[2,3].

REFERENCES

1. Klinkenberg, A. *Gas Chromatography 1960*; Scott, R.P.W., Ed.; Butterworths: London, 1960; 194.
2. Scott, R.P.W. *Liquid Chromatography Column Theory*; John Wiley & Sons: New York, 1992; 19.
3. Scott, R.P.W. *Introduction to Gas Chromatography*; Marcel Dekker, Inc.: New York, 1998.

SUGGESTED FURTHER READING

Scott, R.P.W. *Chromatographic Detectors*; Marcel Dekker, Inc.: New York, 1998.

Extra-Column Volume

Kiyokatsu Jinno
*Department of Materials Science, Toyohashi University of Technology,
Tempaku-cho, Toyohashi, Japan*

INTRODUCTION

Three mechanisms produce dispersion of a band of solute in a chromatographic system as it passes through the separation column: 1) eddy diffusion; 2) longitudinal diffusion; and 3) mass transfer effects. These effects are discussed, in some detail, in this article.

EXTRA-COLUMN BAND BROADENING

Three mechanisms produce dispersion of a band of solute in a chromatographic system as it passes through the separation column.

a) Eddy diffusion and flow dispersion, which is the term for the dispersion produced because of the existence of different flow paths through which solutes can progress through the column. These differences of traveling distance arise because the stationary phase particles have different sizes and shapes, and because the packing of the column is imperfect, causing gaps or voids in the column bed. To reduce dispersion due to the multiple path effect, we need to pack the column with small particles, with as narrow a size distribution as possible.

b) Longitudinal diffusion, which also arises because of diffusion of solute in the longitudinal (axial) direction in the column. This is an important source of dispersion in GC, but less so in LC, because rates of diffusion are very much slower in liquids than they are in gases. This effect becomes more serious the longer the solute species spend in the column; so, unlike flow dispersion, using a rapid flow rate of mobile phase reduces this effect.

c) Mass transfer effects, which arise because the rate of the distribution process (sorption and desorption) of the solute species between mobile and stationary phases may be slow, compared to the rate at which the solute is moving in the mobile phase.

Except for the above general dispersions produced in the separation mechanisms, an unexpected, but important, dispersion can be produced outside the separation column by dead volumes in other parts of the chromatographic system, such as in the injector, the detector, the connecting tubing, and connectors. The combined effect of all of these parts is called "extra-column volume" and the dispersion produced by this volume is called "extra-column dispersion."[1–4] Fig. 1 demonstrates an example of this extra-column dispersion, in which different dead volumes are inserted between the column and the detector. One can see, from this figure, that the effect of extra-column volume can cause a serious loss in separation performance of the chromatographic system.

The variance (the square of the standard deviation) of the observed peak (σ^2) can be expressed as the sum of the peak variances caused only by the contribution of the column (σ_p^2) and all the contributions to the peak broadening due to the extra-column volume (σ_{ex}^2). This is expressed as

$$\sigma^2 = \sigma_p^2 + \sigma_{ex}^2 \tag{1}$$

Since the peak volume is four times the standard deviation (σ_s), Eq. 1 can be rewritten as

$$V_t^2 = V_p^2 + V_{ex}^2 \tag{2}$$

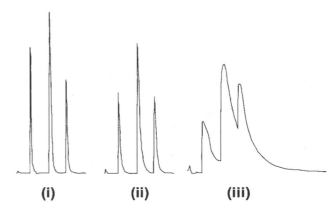

(i)   **(ii)**   **(iii)**

Fig. 1 Extra-column effects on the chromatogram. (i) Normal chromatogram for a test separation; (ii) chromatogram obtained by inserting 75 μL of extra volume between the column and the detector inlet tube; (iii) chromatogram obtained by inserting 2 mL of extra volume between the column and the detector inlet tube.

Encyclopedia of Chromatography DOI: 10.1081/E-ECHR-120013345

Table 1 Maximum extra-column peak volumes for peaks eluted by various columns[a]

| | Column type (length, mm; ID, mm) | | |
	Conventional (250; 4.6)	Semi-microcolumn (250; 1.5)	Microcolumn (250; 0.5)
Maximum extra column			
Peak volume (V_{ex}, μL)	53	5.5	0.6
Peak volume (V_p, μL)	116	12	1.4

[a]The above numbers have been estimated by assuming as follows: Column porosity = 0.7; retention factor = 0; and theoretical plate number = 10,000.

where V_p is the peak volume obtained only from column contribution and V_{ex} is the extra-column peak volume corresponding to the contributions of the injector, detector cell, and connecting tubing. Dividing Eq. 2 by V_p^2 produces

$$(V_t/V_p)^2 = 1 + (V_{ex}/V_p)^2 \qquad (3)$$

Therefore, if the observed peak is allowed to have a volume 10% greater than the column peak volume, the extra-column peak volume should be one-half (ca. 46%) of the column peak volume. Table 1 lists column peak volumes and maximum extra-column peak volumes for various types of columns for LC; because the largest contribution from this extra-column volume should be considered in liquid phase separations, the diffusion coefficients in liquids are very small.

From Table 1, it is very clear that, as the absolute volume of a microcolumn is relatively small, the extra-column volume will contribute significantly to disturb the separation performance of the chromatography system. Because the small extra-column volume is still a large portion of the total system volume which, in turn, is much smaller than the conventional column system, and it is hard to eliminate such small extra-column volume, even if attempts to reduce are applied, a serious problem would be produced. Most typical discussions on the applicability of microcolumn separations in LC are concerned with how to reduce the extra-column volume; this makes microcolumn LC techniques still unpopular, although many advantages are proven and acknowledged. In conclusion, the minimum column volume one can use will depend on the amount of extra-column dispersion and on what we consider to be an acceptable increase in peak width that is produced by the extra-column effects. In practice, this acceptable increase is assumed to be 10%, based on an unretained solute and, if we take 50 μL as a typical value for extra-column dispersion, then the minimum column diameter in LC works out to about 4.6 mm for a column 25 cm long, which is the most popular conventional LC separation column configuration that is commercially available.

REFERENCES

1. Knox, J.H. J. Chromatogr. Sci. **1977**, *15*, 352.
2. Golay, M.J.E.; Atwood, J.G. Early phases of the dispersion of a sample injected in poiseuille flow. J. Chromatogr. **1979**, *186*, 353.
3. Katz, E.D.; Scott, R.P.W. Low-dispersion connecting tubes for liquid chromatography systems. J. Chromatogr. **1983**, *268*, 169.
4. Hupe, K.P.; Jonker, R.J.; Rozing, G. Determination of band-spreading effects in high-performance liquid chromatographic instruments 1. J. Chromatogr. **1984**, *285*, 253.

Fast GC

Richard C. Striebich
University of Dayton Research Institute, Dayton, Ohio, U.S.A.

INTRODUCTION

The examination of ways to conduct fast GC has been a popular research topic since the 1960s, and even more so in the past 10 years. The need to analyze complex mixtures by GC is often a balance between the ability to separate adjacent peaks in a chromatogram (resolution) and analysis time. Especially with complex mixtures, analysts can use longer columns and much slower programming rates to increase resolution; however, there is a penalty to be paid in analysis time. Because petroleum samples are arguably the most complex samples known, a good deal of work has been performed to provide the greatest possible resolution without regard for the consideration of time. Some GC petroleum analyses have been reported, which take 2–4 hr and longer.[1] However, there is a definite application for faster analyses with less resolution.

The history, methods, and applications for conducting fast analyses by GC are delineated in several excellent reviews.[2–6] In these works, the authors discuss several ways to shorten analysis time, such as the following:

1. Decrease the column length
2. Increase the carrier gas flow rate
3. Use multichannel columns[6]
4. Provide rapid heating of the column with heating rates up to 1200°C/min and sometimes higher[2]

Fast GC in these instances is best described as conducting analyses as fast as is possible to provide just enough separation of the compounds of interest. Oftentimes, in the search for maximum resolution, compounds can be overseparated, which usually lengthens the time of analysis.

OVERVIEW

In petroleum analyses, and specifically for aviation fuels, there are a good many separations where complete resolution is not needed. GC fingerprinting of different types of fuels (diesel, gasoline, aviation fuels, kerosene, etc.) can be performed quickly to characterize the mixtures in useful ways. Simulated distillation[1]

is one good example of a chromatographic analysis that has low resolution, but can be conducted very quickly, (i.e., ≪5 min). Fortunately, excellent resolution is not usually necessary to obtain the critical information about distillation range,[7] and so this application is a good example of fast GC where limited resolution is acceptable. In this article, we introduce simple fast GC concepts that can speed up the low-resolution analysis of petroleum products.

EXPERIMENTAL

Short (3–7 m) microbore gas chromatographic columns (0.10 mm internal diameter, 0.17 μm film thickness) can be used to provide much faster analyses with acceptable resolution for at least two different types of useful analyses: *fuel GC fingerprinting* and *simulated distillation analysis*. The detector for the instrument used for these analyses is a hydrogen flame ionization detector (FID) capable of very fast sampling rates (adjustable up to 200 Hz), which is necessary because of the narrow peaks that are generated. Carrier gas is one of the parameters investigated; both helium and hydrogen carrier gases were used, with high-pressure hydrogen routinely providing the best and fastest separations (in agreement with previous work and theory). In this work, the programming rates were limited to that which was available using an Agilent 6890 instrument with a fast heating option (i.e., input rates to 120°C/min with trackable rates to approximately 75°C/min). No attempts were made to increase programming rate by resistively heating the column.

RESULTS AND DISCUSSION

Analysis of aviation fuels by capillary GC can be performed using a variable level of resolution (peak separation) by changing the conditions of the GC. For the purposes of this work, the approximate resolution required was that obtained with conventional methods. That is, an experiment that is completed in 20–30 min is typical because it is fast enough to be productive with regard to research and testing, and provides enough resolution to obtain the needed useful

Encyclopedia of Chromatography DOI: 10.1081/E-ECHR-120028084

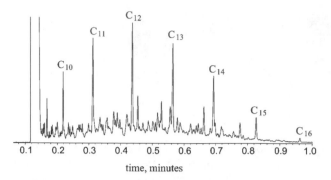

Fig. 1 Fast GC analysis of aviation turbine engine fuel using a typical laboratory GC instrument. Conditions: hydrogen carrier gas; temperature programming rate, 70–170°C at 120°C/min (actual 75°C/min); microbore column: 3 m × 0.10 mm ID.

information. In addition to this level of resolution, we briefly examined the output of a GC analysis conducted using a column of 50 μm internal diameter, half the diameter of typical fast GC columns. These conditions represent our laboratory's (present-day) limit of speed and resolving power (Fig. 1).

Effect of Column Dimensions

Table 1 shows a comparison of the general relationship that exists between GC efficiency as measured by the number of theoretical plates per meter and a particular column dimension. Because microbore columns (0.10 mm internal diameter) have more resolving power per meter, the length of these columns can be typically one third the length of standard bore columns (0.25 mm internal diameter) and still provide approximately the same resolution. Thus fast GC is really faster because of the use of shorter columns, which are more efficient because of their smaller inner diameter. Fig. 2a shows a typical high-resolution analysis using a conventional 30-m column (0.25 mm internal diameter) and a fast chromatogram with similar resolution, but with greatly

Table 1 Chromatographic column diameter vs. efficiency

Column internal diameter (ID) [mm]	Theoretical plates per meter
0.10	12,500
0.18	6,600
0.20	5,940
0.25[a]	4,750
0.32	3,710
0.45	2,640
0.53	2,240

[a]Most typically used in our laboratory.
(From Ref.[8].)

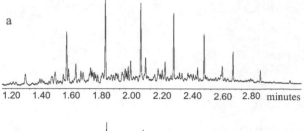

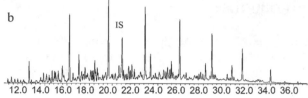

Fig. 2 Comparison of conventional 3 m × 0.25 mm ID (standard bore) column (b) with a 10 m × 0.10 mm ID column (a) with similar resolution.

reduced time. Changing column diameter and length is one of the easiest ways to decrease analysis times, but it is not without a price. Usually, this cost is in decreased column capacity (mass of solute chromatographed without overloading), or in the need to increase carrier head pressure.[4,6]

Effect of Carrier Gas

The widely accepted carrier gas for fast GC analyses is hydrogen, whereas in the United States, helium is the usual choice for conventional analyses. The use of hydrogen is typically better because faster optimal linear velocities are possible at the same generated resolution. Column efficiency is usually expressed in terms of H (i.e., the height equivalent of a theoretical plate, which, when minimized, expresses an optimal efficiency of separation between two components). By plotting the average velocity of the carrier gas vs. the H value, a van Deemter plot is generated. The optimal velocity of the carrier gas for the most efficient separation is higher for hydrogen carrier gas. Thus hydrogen can operate at higher carrier gas velocities without loss of resolution. Although helium carrier gas can be used above its optimal velocity, significant resolution decreases will occur at high flow rates. Because the van Deemter plot for hydrogen is flatter for higher velocities than it is for helium, less resolution is lost with higher velocities of hydrogen, compared to helium.

Effect of Temperature Programming

The ability to quickly ramp column temperature is an excellent way to increase analysis speed, given appropriate carrier gas flow rates and fast detector sampling rates. Along with column dimensions and operation above optimal velocities of hydrogen carrier gas, analyses can be performed extremely fast and with high resolution. Temperature programming for all of the experiments shown was at 70°C/min, which was approximately the fastest rate that the GC oven could reasonably track. Temperature programming rates of up to 120°C/min are possible to input into the GC, but the heaters cannot reliably heat the large oven at this rate. Clearly, faster temperature programs, using resistively heated columns and sheaths,[2] would lead to faster analyses. However, resolution is sacrificed if the column is heated so fast that large portions of the column act as a transfer line for GC solutes, whose boiling points have been exceeded too quickly. It is difficult to balance speed and accuracy with mixtures with wide boiling range.

APPLICATIONS

Fuel samples were examined using conditions similar to those used in this study, with 100-μm capillary columns and conventional GC instrumentation (Agilent 6890). Jet fuels are more of an analytical challenge because of their multicomponent nature; indeed, any analysis of jet fuel probably contains many unresolved solute zones. Enough chromatographic separation must be generated to conduct the particular task of the analysis, but must also be balanced with acceptable speed. The following applications show separations of various mixtures conducted with speed as the primary consideration.

Fuel GC Fingerprinting

Fig. 3 shows four examples of fuel "fingerprinting"; the normal alkane distribution helps to indicate the fuel type. Because of the high resolution in this example, it is possible to obtain useful information from the

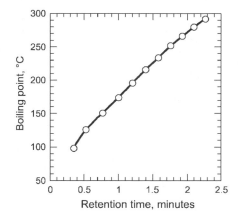

Fig. 4 Calibration curve for simulated distillation of JP-8, JP-TS, and JP-7.

chromatogram and to be able to compare this output tracing to those generated by other fuels. In many cases of fuel contamination, mixing of fuels, mislabeling of containers, and other commonly encountered problems, it is necessary to perform a "general pattern recognition" to identify or characterize the fuel.

Simulated Distillation

By conducting a calibration curve based on the boiling points of n-alkanes, it is possible to estimate the distillation temperatures according to the ASTM D2887 method.[9] Figs. 4 and 5 show the calibration and the simulated distillation curves, respectively, for JP-8, JP-TS, and JP-7, which were obtained from the fast

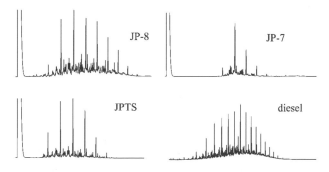

Fig. 3 Fast GC analysis of four fuels including (a) JP-8; (b) JP-7; (c) JP-TS; and (d) diesel fuel.

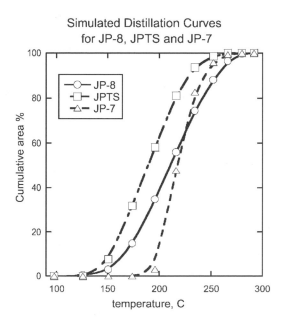

Fig. 5 Simulated distillation analysis of JP-8, JP-TS, and JP-7.

GC analyses. These analyses are directly comparable to fuel specification tests for distillation range. Even analyses faster than these are possible because simulated distillation is a technique where very low resolution is required and generated. Relatively high resolution was maintained for these runs because the chromatographic data were correlated to other specification properties, such as freeze point and flash point.[7]

CONCLUSIONS

Fast GC has great potential as a highly productive investigative tool in today's analytical laboratory. It is becoming more widely used as more advanced GC systems are introduced. We have shown applications of GC analyses with acceptable resolution for aviation fuels, analyzed in fewer than 5 min. The speed of analysis may eventually improve to the point where the GC could produce jet fuel analyses in much less than 1 min.

ACKNOWLEDGMENTS

This work was partially supported by the Fuels Branch of the Air Force Research Laboratory, Propulsion Sciences and Advanced Concepts Division, AFRL/PRSF under the program entitled "Advanced Integrated Fuel/Combustion System" (contract no. F33615-97-C-2719). Mr. Robert Morris was the technical monitor.

REFERENCES

1. Altgelt, K.H., Gouw, T.H., Eds.; *Chromatography in Petroleum Analysis*; Marcel Dekker, Inc.: New York, NY, 1979; 75–89.
2. McNair, H.M.; Reed, G.L. Fast gas chromatography: The effect of fast temperature programming. J. Microcolumn Sep. **2000**, *12* (6), 351–355.
3. Cramers, C.A.; Janssen, H.-G.; van Deursen, M.M.; Leclercq, P.A. High-speed gas chromatography: An overview of various concepts. J. Chromatogr. A. **1999**, *856*, 315–329.
4. Cramers, C.A.; Leclercq, P.A. Strategies for speed optimization in gas chromatography: An overview. J. Chromatogr. A. **1999**, *842*, 3–13.
5. David, F.; Gere, D.R.; Scanlan, F.; Sandra, P. Instrumentation and applications of fast high-resolution capillary gas chromatography. J. Chromatogr. A. **1999**, *842*, 309–319.
6. van Lieshout, M.; van Deursen, M.; Derks, R.; Janssen, H.-G.; Cramers, C. A practical comparison of two recent strategies for fast gas chromatography: Packed capillary columns and multicapillary columns. J. Microcolumn Sep. **1999**, *11* (2), 155–162.
7. Striebich, R.C. Fast gas chromatography for middle-distillate aviation turbine fuels. Assoc. Can. Stud. Pet. Chem. Prepr. **2002**, *47* (3), 219–222.
8. J&W Inc. GC Reference Guide; 1998; 13, Folsom, CA.
9. ASTM D2887-93 Boiling Range Distribution of Petroleum Fractions by Gas Chromatography. In *Section 5 Annual Book of ASTM Standards*; 1996; 192–201; Conshohocken, PA.

Field-Flow Fractionation Data Treatment

Josef Janča
Université de La Rochelle, La Rochelle, France

INTRODUCTION

Field-flow fractionation (FFF) methods are classified into two main categories:[1–3] *polarization* FFF and *focusing* FFF. Their basic characterization is given in the entry Field-Flow Fractionation Fundamentals. Whereas the polarization FFF methods allow to fractionate the samples on the basis of the differences in the extensive properties (such as the molar mass or particle size, etc.) of the individual species, the focusing FFF methods discriminate among the species, according to their intensive property differences (such as the charge or density, etc.). This article deals with the data treatment of the experimental results from polarization FFF, thus with the quantitative characterization of the extensive properties. However, a principally identical approach can be applied to the intensive properties data treatment of the results obtained from the focusing FFF experiments.

In general, the methodology of the data treatment, concerning the separations of the macromolecular or particulate samples, does not depend on the particular separation method or technique. The basics of this methodology were elaborated in parallel with the development of SEC[4] and of the techniques of particle size analysis,[5] but they originate at the very beginning[6,7] of LC of macromolecules and remain substantially unchanged until today.

Macromolecular or particulate samples fractionated by the FFF are usually not uniform but exhibit a distribution of the concerned extensive or intensive parameter[8] or, in other words, a polydispersity. Molar mass distribution (MMD), sometimes called molecular weight distribution (MWD), or particle size distribution (PSD) describes the relative proportion of each molar mass (molecular weight), M, or particle size (diameter), d_p, species composing the sample. This proportion can be expressed as a number of the macromolecules or particles of a given molar mass or diameter, respectively, relative to the number of all macro molecules or particles in the sample:

$$N(M) = \frac{n_i(M)}{\sum_{i=1}^{\infty} n_i}$$

$$N(d_p) = \frac{n_i(d_p)}{\sum_{i=1}^{\infty} n_i} \tag{1}$$

or as a mass (weight) of the macromolecules or particles of a given molar mass or diameter relative to the total mass of the sample:

$$W(M) = \frac{m_i(M)}{\sum_{i=1}^{\infty} m_i}$$

$$W(d_p) = \frac{m_i(d_p)}{\sum_{i=1}^{\infty} m_i} \tag{2}$$

Accordingly, the MMD (MWD) and PSD are called number or mass (weight) MMD or PSD, respectively. FFF provides a fractogram which has to be treated to obtain the required MMD or PSD. These distributions can be used to calculate various average molar masses or particle sizes and polydispersity indices.

AVERAGE MOLAR MASSES, PARTICLE SIZES, AND POLYDISPERSITIES

As mentioned, in addition to the MMD and PSD, various average molar masses, particle sizes, and polydispersity indexes can be calculated from the FFF fractograms. If the detector response, h, is proportional to the mass of the macromolecules or particles, the mass-average molar mass or mass average particle diameter can be calculated from

$$\overline{M_m} = \overline{M_w} = \frac{\sum_{i=1}^{\infty} M_i h_i}{\sum_{i=1}^{\infty} h_i}$$

$$\overline{d_m} = \overline{d_w} = \frac{\sum_{i=1}^{\infty} d_i h_i}{\sum_{i=1}^{\infty} h_i} \tag{3}$$

and the corresponding number average values are calculated from

$$\overline{M_n} = \frac{\sum_{i=1}^{\infty} M_i n_i}{\sum_{i=1}^{\infty} n_i} = \frac{\sum_{i=1}^{\infty} h_i}{\sum_{i=1}^{\infty} (h_i/M_i)}$$

$$\overline{d_n} = \frac{\sum_{i=1}^{\infty} d_i n_i}{\sum_{i=1}^{\infty} n_i} = \frac{\sum_{i=1}^{\infty} h_i}{\sum_{i=1}^{\infty} (h_i/d_i)} \tag{4}$$

Encyclopedia of Chromatography DOI: 10.1081/E-ECHR-120039981

The width of the MMD or PSD (polydispersity) can be characterized by the index of polydispersity:

$$I_{\text{MMD}} = \frac{\overline{M_m}}{\overline{M_n}}$$

$$I_{\text{PSD}} = \frac{\overline{d_m}}{\overline{d_n}} \tag{5}$$

PRACTICAL DATA TREATMENT

Provided that the correction for the zone broadening should not be applied, the first step in the data treatment is to convert the retention volumes (or the retention ratios R) into the corresponding molecular or particulate parameter, characterizing the fractionated species. Whenever the zone broadening correction procedure has to be applied, the data treatment protocol is modified, as described in the entry Zone Dispersion in Field-Flow Fractionation.

The dependences of the retention ratio R on the size of the fractionated species (molar mass for the macromolecules or particle diameter for the particulate matter) are presented for various polarization FFF methods in the entry Field-Flow Fractionation Fundamentals. The raw, digitized fractogram, which is a record of the detector response as a function of the retention volume, is represented by a differential distribution function $h(V)$. It can be processed to obtain a series of the height values h_i corresponding to the retention volumes V_i, as shown in Fig. 1. Subsequently, the retention volumes are converted into the retention ratios R_i:

$$R_i = \frac{V_0}{V_i} \tag{6}$$

The retention ratio R in polarization FFF is related to the retention parameter λ (see the entry Field-Flow Fractionation Fundamentals) by

$$R = 6\lambda \left[\coth\left(\frac{1}{2\lambda}\right) - 2\lambda \right] \tag{7}$$

or by an approximate relationship

$$(\lim R)_{\lambda \to 0} = 6\lambda \tag{8}$$

and the parameter λ is directly related to the molecular or particulate parameters by the general relationships

$$\lambda = f(M^{-n}) \quad \text{or} \quad \lambda = f(d_p^{-n}) \tag{9}$$

where the exponent $n = 1$, 2, or 3. As concerns the focusing FFF methods, similar relationships exist between the retention ratio R and the intensive properties of the fractionated species.

Having the V_i values converted into the R_i values by using Eq. (6), the corresponding molar mass M_i or the particle diameter d_i values are calculated by applying Eqs. (7–9). The difficulty is that Eq. (7) is a transcendental function $R = f(\lambda)$ for which the inversion function $\lambda = f'(R)$ does not exist. As a result, Eq. (8) can be used as a first approximation to estimate the λ_i values from the experimental R_i data, and by applying a rapidly converging iteration procedure, the accurate λ_i values can be calculated. The subsequent attribution of the corresponding M_i or d_i values to the calculated λ_i values, by using the appropriate relationship, Eq. (9), is not mathematically complicated.

In order to obtain an accurate result, the regular segmentation ΔV_i of the raw fractogram must be converted into the ΔR_i and, thereafter, into the appropriate increment of the molar mass ΔM_i or of the particle diameter Δd_i. The corresponding conversions of the raw experimental fractograms into the MMD or PSD can be carried out according to the following protocol. Eqs. (3) and (4) can be rewritten in integral form:

$$\overline{M_m} = \frac{\int_0^\infty W(M)M\,dM}{\int_0^\infty W(M)\,dM}$$

$$\overline{d_m} = \frac{\int_0^\infty d_p W(d_p)\,dd_p}{\int_0^\infty W(d_p)\,dd_p} \tag{10}$$

and

$$\overline{M_n} = \frac{\int_0^\infty N(M)M\,dM}{\int_0^\infty N(M)\,dM}$$

$$\overline{d_n} = \frac{\int_0^\infty d_p N(d_p)\,dd_p}{\int_0^\infty N(d_p)\,dd_p} \tag{11}$$

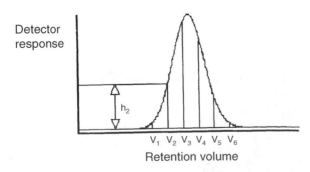

Fig. 1 Treatment of an experimental FFF fractogram of a polydisperse sample.

where it holds for the normalized MMD or PSD:

$$
\int_0^\infty W(M)dM = \int_0^\infty W(d_p)dd_p
$$
$$
= \int_0^\infty N(M)dM = \int_0^\infty N(d_p)dd_p = 1
$$

$$(12)$$

By considering all of the above-mentioned transformations, Eqs. (10–12) give

$$
\overline{M_m} = \int_0^\infty W(M)M\left(\frac{\partial M}{\partial \lambda}\right)\left(\frac{\partial \lambda}{\partial R}\right)\left(\frac{\partial R}{\partial V}\right)dV
$$
$$
\left(\int_0^\infty W(M)dV\right)^{-1}
$$

$$
\overline{d_m} = \int_0^\infty W(d_p)d_p\left(\frac{\partial d_p}{\partial \lambda}\right)\left(\frac{\partial \lambda}{\partial R}\right)\left(\frac{\partial R}{\partial V}\right)dV
$$
$$
\left(\int_0^\infty W(d_p)dV\right)^{-1}
$$

$$
\overline{M_n} = \int_0^\infty N(M)M\left(\frac{\partial M}{\partial \lambda}\right)\left(\frac{\partial \lambda}{\partial R}\right)\left(\frac{\partial R}{\partial V}\right)dV
$$
$$
\left(\int_0^\infty N(M)dV\right)^{-1}
$$

$$
\overline{d_n} = \int_0^\infty N(d_p)d_p\left(\frac{\partial d_p}{\partial \lambda}\right)\left(\frac{\partial \lambda}{\partial R}\right)\left(\frac{\partial R}{\partial V}\right)dV
$$
$$
\left(\int_0^\infty N(d_p)dV\right)^{-1}
$$

Any of Eqs. (13) can further be rewritten in a numerical form of Eqs. (3) and (4), which are convenient for the data treatment and calculations using the discrete M_i or d_i and h_i values. The acquisition of the experimental data and the treatment of the fractogram is easily performed by a computer connected on-line to the separation system.

REFERENCES

1. Janča, J. *Field-Flow Fractionation: Analysis of Macromolecules and Particles*; Marcel Dekker, Inc.: New York, 1988.
2. Janča, J. Isoperichoric focusing field-flow fractionation for characterization of particles and molecules. J. Liquid Chromatogr. & Related Technol. **1997**, *20*, 2555.
3. Cölfen, H.; Antonietti, M. "Field-flow fractionation techniques for polymer and colloid analysis" in "new developments in polymer analytics I." Adv. Polym. Sci. **2000**, *150*, 67.
4. Quivoron, C. *Steric Exclusion Chromatography of Polymers*; Janča, J., Ed.; Marcel Dekker, Inc.: New York, 1984.
5. Barth, H.G., Ed.; *Modern Methods of Particle Size Analysis*; John Wiley & Sons: New York, 1984.
6. Cazes, J. J. Chem. Educ. **1966**, *43*, A567.
7. Cazes, J. J. Chem. Educ. **1966**, *43*, A625.
8. Dawkins, J.V. *Comprehensive Polymer Science*; Booth, C., Price, C., Eds.; Pergamon Press: Oxford, 1989; Vol. 1.

Field-Flow Fractionation Fundamentals

Josef Janča

Department of chemistry, Université de La Rochelle, La Rochelle, France

INTRODUCTION

Field-flow fractionation (FFF) is a separation method suitable for the analysis and characterization of macromolecules and particles. The separation is based on the interaction of the effective physical or chemical forces (e.g., temperature gradient; electric, magnetic, gravitational, or centrifugal forces; chemical potential gradient; etc.) with the separated species. The field, acting across a separation channel, concentrates them at a given position inside the channel. The formed concentration gradient induces an opposite diffusion flux. This leads to a steady-state distribution of the sample components across the channel. The velocity of the longitudinal flow of the carrier liquid also varies across the channel. A flow velocity profile is established inside the channel. As a result, the components of the separated sample are transported in the longitudinal direction at different velocities depending on their transversal positions within the flow of the carrier liquid. This general principle of FFF is demonstrated in Fig. 1.

DISCUSSION

One of three mechanisms, *polarization*,[1] *steric*,[2] and *focusing*,[3] can lead to the formation of different concentration distributions across the fractionation channel. The components of the fractionated sample are either concentrated in the direction of the accumulation wall (polarization FFF), totally compressed at the accumulation wall of the channel (steric FFF), or focused at different positions (focusing FFF), as shown in Fig. 1. Steady state inside the channel is reached in a short time due to the small channel thickness. The strength of the field can be controlled within a wide range in order to manipulate the retention conveniently. Many operational variables in FFF can be manipulated during the experiment by suitable programming.

The *polarization* and *steric* FFF methods are classified according to the nature of the applied field, whereas the *focusing* FFF methods are classified by considering the combination of various gradients and fields emphasizing the focusing processes.

Polarization FFF methods make use of the formation of an exponential concentration distribution of each sample component across the channel with the maximum concentration at the accumulation wall, which is a consequence of the constant and position-independent velocity of transversal migration of the affected species due to the field forces. This concentration distribution is combined with the velocity profile formed in the flowing liquid. In steric FFF, the field strength is so high that all species interacting with the field are in contact with the accumulation wall. As a result, the proper size of the retained species determines their position in the flow velocity profile and, consequently, their elution velocity along the channel. Focusing FFF methods make use of transversal migration of each sample component under the effect of driving forces whose intensity varies across the channel. As a result, the sample components are focused at the positions where the intensity of the effective forces is zero and are transported longitudinally with different velocities according to the established flow velocity profile. The concentration distribution within a zone of a focused sample component can be described by Gaussian or similar distribution function.

PRINCIPLE AND THEORY

The carrier liquid flows in the direction of the channel's longitudinal axis, whereas the field forces act perpendicularly across the channel. The driving forces can be generated by a single field or by the coupled action of two or more different fields. Polarizing and focusing forces can operate simultaneously, resulting in a complex mechanism of separation. The field force F and, consequently, the velocity U are independent of position in the direction of the x-axis in polarization and steric FFF:

$$F \neq 0 \quad \text{and} \quad U \neq 0 \quad \text{for } 0 < x < w \qquad (1)$$

where w is the distance between the main channel walls in the direction of the x-axis, with $x = 0$ at the accumulation wall. On the other hand, the following hold

Encyclopedia of Chromatography DOI: 10.1081/E-ECHR-120039983

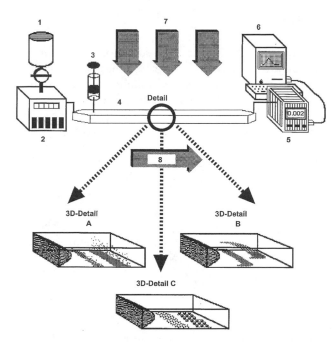

Fig. 1 Principle of field-flow fractionation. 1—Solvent reservoir, 2–carrier liquid pump, 3—injection of the sample, 4—separation channel, 5—detector, 6—computer for data acquisition, 7—transversal effective field forces, 8—longitudinal flow of the carrier liquid. A—Section of the channel demonstrating the principle of polarization FFF with two distinct zones compressed differently at the accumulation wall and the parabolic flow velocity profile. B—Section of the channel demonstrating the principle of focusing FFF with two distinct zones focused at different positions and the parabolic flow velocity profile. C—Section of the channel demonstrating the principle of steric FFF with two zones eluting at different velocities according to the distance of their centers from the accumulation wall.

for the x-axis-dependent direction of the field force in focusing FFF:

$$F = F(x) \quad \text{and} \quad U = U(x) \quad \text{within} \quad 0 < x < w \tag{2}$$

$$F(x) = 0 \quad \text{and} \quad U(x) = 0 \\ \text{for } x = x_{max} \text{ with } 0 < x_{max} < w \tag{3}$$

$$F(x) > 0 \quad \text{and} \quad U(x) > 0 \quad \text{for } x < x_{max} \tag{4}$$

$$F(x) < 0 \quad \text{and} \quad U(x) < 0 \quad \text{for } x > x_{max} \tag{5}$$

where the co-ordinate x_{max} corresponds to the position at which the concentration distribution of a sample component across the channel attains its maximal value.

POLARIZATION FFF

The equilibrium concentration distribution in the direction of the x-axis across the channel of a given component of the sample can be calculated from the continuity equation:

$$-D\frac{\partial c}{\partial x} - Uc = 0 \tag{6}$$

where D is the diffusion coefficient and c is the concentration. The solution of Eq. (6) gives the exponential concentration distribution of the sample component across the channel:[4]

$$c(x) = c(0)\exp(-xU/D) \tag{7}$$

By defining the mean layer thickness, $\ell = D/U$, Eq. (7) can be rewritten:

$$c(x) = c(0)\exp(-x/\ell) \tag{8}$$

The mean layer thickness is practically equal to the center of gravity of the concentration distribution.

STERIC FFF

The radius (or hydrodynamic equivalent of the radius) of the retained species determines the distance of the center of such a species from the accumulation wall. There is no distribution of concentration of the retained component across the channel. Consequently, the ratio α is a decisive parameter determining the retention in steric FFF:[2]

$$R = 6(\alpha - \alpha^2) \tag{9}$$

where $\alpha = r/w$, and r is the radius of the retained species.

FOCUSING FFF

It holds for a focused species at equilibrium that:

$$D\frac{\partial c}{\partial x} - U(x)c = 0 \tag{10}$$

The force $F(x)$, acting on one particle undergoing the focusing, can be written as:

$$F(x) = U(x)f \tag{11}$$

with the friction coefficient defined by:

$$f = kT/D \qquad (12)$$

where k is the Boltzmann constant and T is the absolute temperature. Then the following holds:

$$\frac{dc}{dx} = \frac{F(x)c}{kT} \qquad (13)$$

The focusing force can be approximated by:[5]

$$F(x) = -\left|\left(\frac{dF(x)}{dx}\right)_{x \approx x_{max}}\right|(x - x_{max}) \qquad (14)$$

where $[dF(x)/dx]_{x \approx x_{max}}$ is the gradient of the driving force. The solution is:

$$c(x) = c_{max} \exp\left[-\frac{1}{2kT}\left|\left(\frac{dF(x)}{dx}\right)_{x \approx x_{max}}\right|(x - x_{max})^2\right] \qquad (15)$$

which is the Gaussian concentration profile of a single focused component. A more accurate approach[5] is based on the real gradient of the focusing forces and results in a concentration distribution of the focused species that is not Gaussian.

FLOW VELOCITY PROFILES

The separation is usually carried out in a belt-shaped narrow channel of constant thickness. The cross-section of the channel is rectangular. The 2D velocity distribution in a plane parallel to the sidewalls in such a channel (provided that the flow is isoviscous) is parabolic:

$$v(x) = \frac{\Delta P x(w - x)}{2L\mu} \qquad (16)$$

where $v(x)$ is the longitudinal velocity at the x-coordinate, ΔP is the pressure drop along a channel of length L, and μ is the viscosity of the carrier liquid. The average velocity is:

$$\langle v(x) \rangle = \frac{\Delta P w^2}{12L\mu} \qquad (17)$$

Other shapes of flow velocity profiles can be formed in channels whose cross-section is not rectangular but, for example, trapezoidal. The use of such nonparabolic flow velocity profiles can be advantageous, especially in focusing FFF.

SEPARATION

Separation is due to the coupled action of the concentration and flow velocity distributions. The concentration distribution across the channel of each sample component is established and the sample components are eluted along the channel with different velocities depending on the distance of their centers of gravity from the accumulation wall. The average velocity of the zone of a retained sample component is:

$$\langle v \rangle = \langle c(x)v(x) \rangle / \langle c(x) \rangle \qquad (18)$$

The retention ratio R is defined as the average velocity of a retained sample component to the average velocity of the carrier liquid:

$$R = \frac{\int_0^w c(x)v(x)dx \int_0^w dx}{\int_0^w c(x)dx \int_0^w v(x)dx} \qquad (19)$$

where $v(x)$ and $c(x)$ are the local velocity and concentration, respectively, of the retained species. From the practical point of view, the retention ratio R can be expressed as the ratio of the experimental retention time t_0 or the retention volume V_0 of an unretained sample component to the retention time t_r or the retention volume V_r of the retained sample component. Provided that a relationship exists between the position of the center of gravity of the zone and the molecular parameters of the sample component, these parameters can be calculated from the retention data without calibration.

The retention ratio in polarization FFF is thus given by:[6]

$$R = 6\lambda\left[\coth\left(\frac{1}{2\lambda}\right) - 2\lambda\right] \qquad (20)$$

where $\lambda = l/w$. If λ is small, the following approximations hold:

$$\lim_{\lambda \to 0} R = 6(\lambda - 2\lambda^2) \quad \text{or} \quad \lim_{\lambda \to 0} R = 6\lambda \qquad (21)$$

The retention parameter λ relates the dispersive effect of thermal energy to the structuring effect of the field on the retained species:

$$\lambda = \frac{kT}{Fw} \qquad (22)$$

When the size of the separated species is commensurable with the thickness of the channel, the limit retention ratio in this mode of steric FFF is:[3]

$$\lim_{\alpha \to 0} R = 6\alpha \qquad (23)$$

where $\alpha = r/w$; r is the particle radius.

The retention ratio in focusing FFF carried out in a channel of rectangular cross-section is given by the approximate relationship:[7]

$$R = 6(\Gamma_{max} - \Gamma_{max}^2) \qquad (24)$$

where $\Gamma_{max} = x_{max}/w$ is the dimensionless co-ordinate of the maximal concentration of the focused zone.

METHODS AND APPLICATIONS

The retention is related to the size, charge, diffusion coefficient, thermal diffusion factor, and so forth of the separated species in polarization FFF, whereas it is exclusively the size that determines the retention in steric FFF. As concerns focusing FFF, the retention is usually related to the intensive properties of the fractionated species. Consequently, FFF can be used to characterize the properties related to retention. Only the polarization and steric FFF methods are described here.

The particular methods of polarization and steric FFF are denominated by the nature of the applied field. The most important of them are described in the following subsections.

Sedimentation FFF

Sedimentation FFF is shown schematically in Fig. 2a. The separation channel is situated inside a centrifuge rotor and the centrifugal forces are applied radially.[8] The method can be used for the analysis and characterization of various latexes, inorganic particles, emulsions, biological cells, etc. The retention parameter λ depends on the effective mass of the particles:

$$\lambda = 6kT/(\pi d^3 g w \Delta\rho) \qquad (25)$$

where g is the gravitational or centrifugal acceleration and $\Delta\rho$ is the density difference between the particles and the carrier liquid. The calculation of the particle size distribution is possible directly from the retention data.

Thermal FFF

Thermal FFF was the first experimentally implemented method.[9] It is used mostly for the fractionation of macromolecules. The temperature difference between two metallic bars, forming the channel walls with highly polished surfaces and separated by a spacer in which the channel proper is cut, produces the flux of the sample components, usually toward the cold wall. The channel for thermal FFF is shown in Fig. 2b.

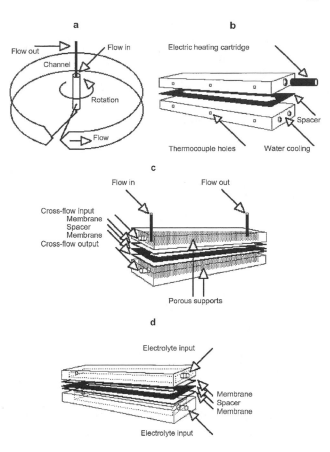

Fig. 2 Methods of polarization FFF. (a) Sedimentation FFF; (b) thermal FFF; (c) flow FFF; (d) electric FFF.

The relation between λ and the operational variables is given by:

$$\lambda = \frac{D}{wD_T(dT/dx)} \qquad (26)$$

where D_T is the coefficient of the thermal diffusion, which depends on the chemical composition and structure of the fractionated species but not on their size. On the other hand, the diffusion coefficient D depends on the size. As a result, the differences in thermal diffusion coefficients allow fractionation according to differences in chemical composition and structure, whereas different diffusion coefficients allow fractionation based on the size differences. The performances favor thermal FFF over its competing methods.

Flow FFF

Flow FFF is a universal method because the cross-flow field acts on all fractionated species in the same manner and the separation is due to the differences in diffusion coefficients.[10] The channel, schematically demonstrated in Fig. 2c, is formed between two parallel semipermeable membranes. The carrier liquid can permeate through the membranes but not the separated species. The retention parameter λ is related to the diameter d_p of the separated species:

$$\lambda = kTV_0/\left(3\pi\mu V_c w^2 d_p\right) \qquad (27)$$

where V_0 is the void volume of the channel, μ is the viscosity of the carrier liquid, and V_c is the volumetric velocity of the cross-flow. The separations of various kinds of particles such as proteins, biological cells, colloidal silica, polymer latexes, etc. as well as of soluble macromolecules have been described.

Electric FFF

Electric FFF uses the electric potential across the channel to generate the transversal flux of the charged species.[11] The walls of the channel can be formed by semipermeable membranes that allow the passage of small ions but not of the separated species. The channel is shown in Fig. 2d. The dependence of the retention parameter λ on the electrophoretic mobility μ_e, and on the diffusion coefficient of the charged particles, is given by:

$$\lambda = D/(\mu_e E w) \qquad (28)$$

where E is the electric field strength. As a result, the ratio of the diffusion coefficient to the electrophoretic mobility determines the retention. Species exhibiting only small differences in electrophoretic mobilities but significant differences in diffusion coefficients can be separated. Electric FFF is especially suited for the separations of biological cells as well as for charged polymer latexes and other colloidal particles and charged macromolecules.

Other Polarization FFF Methods

Other polarization FFF methods have recently been proposed theoretically and some of them implemented experimentally. Their use in current laboratory practice needs further development in methodology and instrumentation. One of the most recent review papers summarizes the state of the art of the polarization FFF methods.[12]

Steric FFF Methods

Any field force can be exploited to create conditions for effective action of the steric exclusion mechanism. The only condition is, as mentioned above, that the field strength be high enough to compress all retained species to the accumulation wall. In experimental practice, sedimentation FFF, flow FFF, and thermal FFF are the techniques actually applied in steric mode to separate effectively some particulate species.

ARTICLES OF FURTHER INTEREST

Field-Flow Fractionation Data Treatment, p. 611.
Focusing Field-Flow Fractionation of Particles and Macromolecules, p. 656.

REFERENCES

1. Giddings, J.C. A new separation concept based on a coupling of concentration and flow nonuniformities. Sep. Sci. **1966**, *1*, 123.
2. Giddings, J.C.; Myers, M.N. Steric field-flow fractionation: a new method for separating 1–100 μm particles. Sep. Sci. Technol. **1978**, *13*, 637.
3. Janča, J. Sedimentation–flotation focusing field-flow fractionation. Makromol. Chem. Rapid Commun. **1982**, *3*, 887.
4. Janča, J. *Field-Flow Fractionation: Analysis of Macromolecules and Particles*; Marcel Dekker, Inc.: New York, 1988.
5. Janča, J. *Chromatographic Characterization of Polymers, Hyphenated and Multidimensional Techniques*; Provder, T., Barth, H.G., Urban, M.W., Eds.; Advances in Chemistry Series 247; ACS: Washington, D.C., 1995.
6. Hovingh, M.E.; Thompson, G.H.; Giddings, J.C. Column parameters in thermal field-flow fractionation. Anal. Chem. **1970**, *42*, 195.
7. Janča, J.; Chmelik, J. Focusing in field-flow fractionation. Anal. Chem. **1984**, *56*, 2481.
8. Giddings, J.C.; Myers, M.N.; Moon, M.H.; Barman, B.N. In *Particle Size Distribution*; Provder, T., Ed.; ACS Symposium Series No. 472; ACS: Washington, D.C., 1991.
9. Jeon, S.J.; Schimpf, M.E. *Particle Size Distribution III: Assessment and Characterization*; Provder, T., Ed.; ACS: Washington, D.C., 1998.
10. Ratanathanawongs, S.K.; Giddings, J.C. *Chromatography of Polymers: Characterization by SEC and FFF*; Provder, T., Ed.; ACS Symposium Series 521; ACS: Washington, D.C., 1993.
11. Schimpf, M.E.; Caldwell, K.D. Electrical field-flow fractionation for colloid and particle analysis. Am. Lab. **1995**, *27*, 64.
12. Cölfen, H.; Antonietti, M. Field-flow fractionation techniques for polymer and colloid analysis. In new developments in polymer analytics I. Adv. Polym. Sci. **2000**, *150*, 67.

Field-Flow Fractionation with Electro-Osmotic Flow

Victor P. Andreev
Institute for Analytical Instrumentation, Russian Academy of Sciences, St. Petersburg, Russia

INTRODUCTION

It is well known that the essence of field-flow fractionation (FFF) is in the interaction between the distribution of the sample particles in the transversal field and the nonuniformity of the longitudinal flow profile. The classical FFF is realized in the channel with the flow driven by the pressure drop. The flow, in this case, is called Poiseuille flow and its profile is parabolic.

DISCUSSION

Electro-osmotic flow (EOF) is widely used for the propulsion of liquid in modern chromatographic methods, so it was natural to study the possibility of FFF with EOF, generated by applying an electric field, E, along a channel or a tube with charged (having the nonzero zeta-potentials) walls. The usual EOF is very close to uniform. For the cylindrical tube of radius a, the EOF velocity profile is described by

$$V(r) = \frac{\zeta \varepsilon \varepsilon_0 E}{\eta}\left(1 - \frac{I_0(\kappa r)}{I_0(\kappa a)}\right) \tag{1}$$

where ζ is the zeta-potential of the wall, ε and η are the dielectric constant and viscosity of the buffer, respectively, ε_0 is the permitivity of the free space, $\kappa^{-1} = (\varepsilon \varepsilon_0 k_B T/2ne^2)^{1/2}$ is the Debye layer thickness [k_B is the Boltzman constant, T is the temperature, n is the number of ions per unit volume (proportional to the concentration of buffer C_0), e is the proton charge], and $I_0(x)$ is the modified Bessel function. As can be seen from Eq. 1, the velocity profile of the EOF in the tube is very close to uniform everywhere except the Debye layer vicinity of the wall. Thus, it is hard to exploit such a profile for FFF unless the concentration of buffer is very low.

That is why it was proposed[1,2] to realize the asymmetrical FFF in the flat channel by making its walls of different materials or chemically modifying them. If the channel walls have nonequal values of the zeta-potentials, then the shape of the EOF profile can be quite different from uniform. The flow profiles that can be generated in the FFF channel with the applied electric field E and pressure drop Δp are presented in Fig. 1. These profiles can be described by

$$V(r) = \frac{\zeta_2 \varepsilon \varepsilon_0 E}{\eta} \cdot \left[(\zeta_R - 1)\frac{\sinh kY}{\sinh k} + (\zeta_R + 1)\right.$$
$$\left.\frac{\cosh kY}{\cosh k} + (1 - \zeta_R)Y - (1 + \zeta_R)\right] + V_0(1 - Y^2) \tag{2}$$

where $\zeta_R = \zeta_1/\zeta_2$ is the ratio of the zeta-potential of the accumulation wall to the zeta-potential of the depletion wall, $k = \kappa w/2$, w is the channel depth, $Y = 1 - 2y/w$, and $V_0 = \Delta p/2\eta L$.

For large values of k, the first two terms in the square brackets are substantially nonzero only in the Debye layer vicinity of the walls, whereas everywhere else the EOF profile is dominated by the last two linear terms in the square brackets. Therefore, the asymmetric EOF profile can be close to trapezoidal or close to triangular depending on the exact values of the zeta-potentials of the walls. If the signs of the zeta-potentials of the walls are different, then the liquid moves in one direction near one wall and in the opposite direction near another wall (this case can be interesting for the preseparation of the particles having different densities). The last term in Eq. 2 corresponds to the pressure-driven Poiseuille flow.

Having Eq. 2 for flow profile enables one to calculate the retention ratio R and χ coefficient describing the Taylor dispersion part of the theoretical plate height H for arbitrary flow profile, according to[3]

$$H = \frac{2D}{R\langle V \rangle} + \chi \frac{w^2 \langle V \rangle}{D} \tag{3}$$

where $\langle V \rangle$ is the average velocity of the flow and D is the diffusion coefficient of sample molecules. Usually, in FFF, the second term of Eq. 3 is much larger than the first one.

Comparison of R and χ values for the flow profiles presented in Fig. 1 for the case of the FFF parameter $\lambda \ll 1$ gives $R = 6\lambda$ and $\chi = 24\lambda^3$ for classical FFF with Poiseuille flow and $R = 2\lambda$ and $\chi = 8\lambda^3$ for FFF with a triangular EOF ($\zeta_1 = 0$). The most interesting result corresponds to the case of FFF with a combined triangular EOF and counterdirected Poiseuille flow (with $V_0 = \zeta_2 \varepsilon \varepsilon_0 E/4\eta$ leading to $dV/dY = 0$ for $Y = 0$). In this case, $R = 6\lambda^2$ and $\chi = 24\lambda^4$. Thus, the selectivity $S = d \ln R/d \ln \lambda = 2$ and is twice as large as in the case of classical

Encyclopedia of Chromatography DOI: 10.1081/E-ECHR-120039982

619

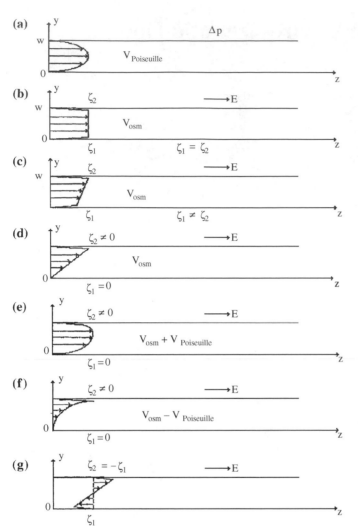

Fig. 1 Outline of the flow profiles in the FFF channels: (a) Poiseuille flow; (b) EOF (the equal zeta-potentials of the walls; (c) trapezoidal EOF (the nonequal zeta-potentials of the walls); (d) triangular EOF (the zero zeta-potential of the accumulating wall); (e) codirected triangular EOF and Poiseuille flow; (f) counterdirected triangular EOF and Poiseuille flow; (g) antisymmetric EOF (different signs of the zeta-potentials of the walls).

FFF and FFF with a triangular EOF. The χ coefficient is very small for $\lambda \ll 1$, so that the Taylor dispersion is very low and efficiency is high. The function $F = S/\sqrt{\chi}$ (fractionating power), characterizing the resolution for the given value of $\langle V \rangle$, is proportional to λ^{-2} in this case; in the rest of the cases, it is proportional to $\lambda^{-3/2}$. The situation with this kind of combined flow is very similar to the one described in Ref.[4] for the case of Poiseuille flow combined with the natural convection flow in the thermogravitational FFF channel.

High selectivity, efficiency, and fractionating power makes FFF with combined EOF and Poiseuille flows very interesting, as it can, at least theoretically, lead to finer separations for given values of λ. Experimental realization of FFF with asymmetrical EOF have not yet been reported due to some technical problems. However, considerable progress in this field is accomplished by a Finnish group (Riekkola, Vastamaki, and Jussila) working on the experimental realization of thermal FFF with asymmetrical EOF[5] and a Russian group (Andreev, Stepanov, and Tihomolov) working on gravitational FFF with asymmetrical EOF.[6]

The situation is more complicated, even theoretically, when the sample particles are charged. In this case, they are not only moving with the longitudinal flow (here, asymmetrical EOF) but are also forced by the longitudinal electric field to move along the channel electrophoretically. If the electrophoretic mobilities of the particles are different, then there are two types of separations combined: The FFF type due to the difference in λ values and the capillary zone electrophoresis (CZE) type due to the difference in electrophoretic mobilities. Agreat variety of variants of FFF and CZE combinations in the FFF channel with asymmetrical EOF could be imagined, depending on various factors such as the ratio of the zetapotentials of the channel walls, the sign and the value of the ratio of eletrophoretic and electro-osmotic velocities, and the type of the transversal field. Some of these combinations are examined in Ref.[6]. They could lead both to the new possibilities of the method and to some new complications in the interpretation of the experimental results.

Another possibility for realizing FFF with EOF is to reduce the concentration of the buffer, thus making the Debye length commensurate, if not with the depth of the channel, then with the thickness $l = \lambda w$ of the layer of sample particles compressed to the accumulating wall of FFF channel (for $C_0 = 10^{-5}M$ Debye length, $\kappa^{-1} = 0.1\,\mu m$). In this case, EOF will be nonuniform enough to realize FFF in a channel with equal zeta-potentials of the walls.

In Ref.[7] the mathematical model of CZE, taking into consideration EOF nonuniformity and particle–wall electrostatic interactions, was developed. It was shown that for the particles electrostatically attracted by the wall of the capillary, two mechanisms of separation exist. The first is the usual CZE mechanism and it dominates for the case of high buffer concentrations; the second is the FFF accompanying the CZE mechanism and it dominates for low buffer concentrations. As is usual in CZE, the total velocity of the particle is the sum of its electrophoretic velocity and electro-osmotic velocity of the flow. The larger the electrical charge of sample particles, the stronger they are attracted to the wall and the higher is their concentration in the Debye layer vicinity of the wall, where the EOF is substantially nonuniform. Thus, for the particles with a higher charge, the mean velocity of movement with EOF will be lower than for the particles with the lower charge. Especially interesting with this type of FFF is for the particles with equal electrophoretic mobilities but different charges. Such types of particles (e.g., DNA fragments) cannot be fractionated by usual CZE, but can be fractionated by FFF accompanying CZE, where the separation is due to the difference of electrical charges, not the difference of mobilities. Note that for this type of FFF, there is no need for any external transversal field, because the particles are attracted to the walls by the field of the electrical double layer. As is usual in FFF, there is the transition point from normal diffusional FFF to steric FFF mode, taking place when the size of the particle is commensurable with λw. In Ref.[8], it was theoretically predicted that steric FFF accompanying the CZE mode can be realized for the separation of DNA fragments in the range of 20–3000 bases with high resolution and speed. To realize this type of separation, one needs to develop a modified capillary with the positive value of the zeta-potential of the wall and without the irreversible sorption of DNA fragments on the walls. Such an attempt seems to be worthy because, unlike DNA separation by CZE in gel or polymer solution, in the case of FFF/CZE there is the possibility of on-line coupling with a mass spectrometer without the risk of gel particles going inside the spectrometer.

REFERENCES

1. Andreev, V.P.; Miller, M.E.; Giddings, J.C. Field-flow fractionation with asymmetrical electroosmotic flow. 5th Int. Symp on FFF, 1995.
2. Andreev, V.P.; Stepanov, Y.V.; Giddings, J.C. Field-flow fractionation with asymmetrical electroosmotic flow. I. Uncharged particles. J. Microcol. Separ. **1997**, *9*, 163.
3. Martin, M.; Giddings, J.C. Retention and nonequilibrium peak broadening for generalized flow profile in FFF. J. Phys. Chem. **1981**, *85*, 727.
4. Giddings, J.C.; Martin, M.; Myers, M.N. Thermogravitational FFF: an elution thermogravitational column. Separ. Sci. Technol. **1979**, *14*, 611.
5. Vastamaki, P.; Jussila, M.; Riekkola, M.-L. The effect of electrically nonconductive wall coating on retention in ThFFF. 7th Int. Symp. on FFF, 1998.
6. Andreev, V.P.; Stepanov, Y.V. Field-flow fractionation with asymmetrical electroosmotic flow. II. Charged particles. J. Liquid Chromatogr. & Related Technol. **1997**, *20*, 2873.
7. Andreev, V.P.; Lisin, E.E. On the mathematical model of capillary electrophoresis. Chromatographia **1993**, *37*, 202.
8. Andreev, V.P.; Stepanov, Y.V. Steric FFF accompanying capillary electrophoresis. 5th Int. Symp on FFF, 1995.

Flame Ionization Detector for GC

Raymond P. W. Scott
Scientific Detectors Ltd., Banbury, Oxfordshire, U.K.

INTRODUCTION

The flame ionization detector (FID) is, by far, the most commonly used detector in GC and is probably the most important. It is a little uncertain as to who was the first to invent the FID; some gave the credit to Harley and Pretorius,[1] others to McWilliams and Dewer.[2] In any event, it would appear that both contenders developed the device at about the same time, and independently of one another; the controversy had more patent significance than historical interest. The FID is an extension of the flame thermocouple detector and is physically very similar, the fundamentally important difference being that the ions produced in the flame are measured, as opposed to the heat generated.

DISCUSSION

The principle of detection is as follows. Hydrogen is mixed with the column eluent and burned at a small jet. Surrounding the flame is a cylindrical electrode and a relatively high voltage is applied between the jet and the electrode to collect the ions that are formed in the flame. The resulting current is amplified by a high-impedance amplifier and the output fed to a data acquisition system or a potentiometric recorder.

A detailed diagram of the FID sensor is shown in Fig. 1. The body and the cylindrical electrode is usually made of stainless steel and stainless-steel fittings connect the detector to the appropriate gas supplies. The jet and the electrodes are insulated from the main body of the sensor with appropriate high-temperature insulators. Some care must be taken in selecting appropriate insulators as many glasses (with the exception of fused quartz) and some ceramic materials become conducting at high temperatures (200–300°C).[3]

As a result of the relatively high voltages used in conjunction with the very small ionic currents being measured, all connections to the jet or electrode must be well insulated and electrically screened. In addition, the screening and insulating materials must be stable at the elevated temperature of the detector oven. In order to accommodate the high temperatures

that exist at the jet tip, the jet is usually constructed of a metal that is not easily oxidized, such as stainless steel, platinum, or platinum–rhodium. The detector electronics consist of a high-voltage power supply and a high-impedance amplifier. The jet and electrode can be connected to the power supply and amplifier in basically two configurations. The floating jet configuration is the most commonly used and in this arrangement, $+250$ to $+400\,V$ is applied to the cylindrical electrodes and the jet is connected to a ground by a very high resistance. The signal developed across the resistance is amplified, modified, and passed to a recorder of the data acquisition system. In the second alternative, the jet is grounded and the high-voltage power supply is electrically floated. Then, $+250$ to $+400\,V$ is applied to the cylindrical electrodes and the negative terminal of the power supply is connected to a ground by a very high resistance. The signal that is developed across the resistance is again amplified, modified, and passed to a recorder of the data acquisition system.

RESPONSE MECHANISM OF THE FID

The FID has a very wide dynamic range, has a high sensitivity, and, with the exception of about half a dozen low-molecular-weight compounds, will detect all substances that contain carbon. The response mechanism of the FID has been carefully investigated by a number workers. It was originally thought that the ionization mechanism in the FID flame is similar to the ionization process in a hydrocarbon flame, but it quickly became apparent that ionization in the hydrogen flame is many times higher than could be accounted for by thermal ionization alone. It would appear that the ionization potentials of organic materials become much lower when they enter the flame.

The generally accepted explanation of this effect is that the ions are not formed by thermal ionization but by thermal emission from small carbon particles that are formed during the combustion process. Consequently, the dominating factor in the ionization of organic material is not their ionization potential but the work function of the carbon that is transiently formed during their combustion. The flame plasma

Encyclopedia of Chromatography DOI: 10.1081/E-ECHR-120039984

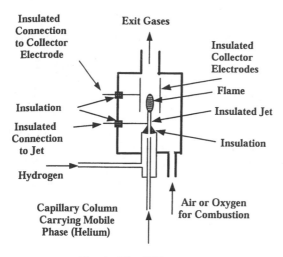

Fig. 1 The FID sensor.

contains both positive ions and electrons which are collected on either the jet or the plate, depending on the polarity of the applied voltage. Initially, the current increases with applied voltage, the magnitude of which depend on the electrode spacing. The current continues to increase with the applied voltage and eventually reaches a plateau at which the current remains sensibly constant. The voltage at which this plateau is reached also depends on the electrode distances.

As soon as the electron–ion pair is produced, recombination starts to take place. The longer the ions take to reach the electrode and be collected, the more the recombination takes place. Thus, the greater the distance between the electrodes and the lower the voltage, the greater the recombination. As a result, initially the current increases with the applied voltage and then eventually flattens out, and at this point, it would appear that all the ion–electron pairs were being collected. In practice, the applied voltage would be adjusted to suit the electrode geometry and ensure that the detector operates under conditions where all electrons and ions are collected.

It was also shown that the airflow should be at least six times that of the hydrogen flow for stable conditions and complete combustion. The base current from the hydrogen flow depends strongly on the purity of the hydrogen. Traces of hydrocarbons significantly increase the base current, as would be expected. Consequently, very pure hydrogen should be employed with the FID if maximum sensitivity is required. Employing purified hydrogen, Desty et al. reported a base current of 1.45×10^{-12} A for a hydrogen flow of 20 mL/min. This was equivalent to 1×10^{-7} C/mol. The sensitivity reported for n-heptane, assuming a noise level equivalent to the base current from hydrogen of $\sim 2 \times 10^{-14}$ A (a fairly generous assumption), was 5×10^{-12} g/mL at a flow rate of 20 mL/min. It

follows that although the sensitivity is amazingly high, the ionization efficiency is still very small ($\sim 0.0015\%$). The general response of the FID to substances of different type varies very significantly from one to another. For a given homologous series, the response appears to increase linearly with carbon number, but there is a large difference in response between different homologous series (e.g., hydrocarbons and alcohols).

The linear dynamic range of the FID covers at least four to five orders of magnitude for $0.98 < r < 1.02$. This is a remarkably wide range that also helps explain the popularity of the detector. Examination of the different commercially available detectors shows considerable difference in electrode geometry and operating electrode voltages, yet they all have very similar performance specifications.

OPERATION OF THE FID

The FID is one of the simplest and most reliable detectors to operate. Generally, the appropriate flow rates for the different gases are given in the detector manual. The hydrogen flow usually ranges between 20 and 30 mL/min and the airflow is about six times that of the hydrogen flow (e.g., 120–200 mL/min. The column flow that can be tolerated is usually about 20–25 mL/min, depending on the chosen hydrogen flow. However, if a capillary column is used, the flow rate may be less than 1 mL/min for very small-diameter columns. The mobile phase can be any inert gas—helium, nitrogen, argon, and so forth. To some extent, the detector is self-cleaning and rarely becomes fouled. However, this depends a little on the substances being analyzed. If silane derivatives are continuously injected on the column, then silica is deposited both on the jet and on the electrodes and may need to be regularly cleaned. In a similar way, the regular analysis of phosphate-containing compounds may eventually contaminate the electrode system. Electrode cleaning is best carried out by the qualified instrument service engineer.

Apparently, the sole disadvantage of the FID as a general detector is that it normally requires three separate gas supplies, together with their precision flow regulators. The need for three gas supplies is a decided inconvenience but is readily tolerated in order to take advantage of the many other attributes of the FID. The detector is normally thermostatted in a separate oven; this is not because the response of the FID is particularly temperature sensitive but to ensure that no solutes condense in the connecting tubes.

The FID has an extremely wide field of application and is used in the analysis of hydrocarbons, solvents, essential oils, flavors, drugs, and their metabolites—in fact, any mixture of volatile substances that contain carbon.

REFERENCES

1. Harley, J.; Nel, W.; Pretorius, V. Nature (London). **1958**, *181*, 177.
2. McWilliams, I.G.; Dewer, R.A. *Gas Chromatography 1958*; Desty, D.H., Ed.; Butterworths: London, 1957; 142.
3. Beres, S.A.; Halfmann, C.D.; Katz, E.D.; Scott, R.P.W. A new type of argon ionisation detector. Analyst. **1987**, *112*, 91.

SUGGESTED FURTHER READING

Scott, R.P.W. *Chromatographic Detectors*; Marcel Dekker, Inc.: New York, 1996.

Scott, R.P.W. *Introduction to Analytical Gas Chromatography*; Marcel Dekker, Inc.: New York, 1998.

Flavonoids: Analysis by Supercritical Fluid Chromatography

Xia Yang
Huwei Liu
Institute of Analytical Chemistry, College of Chemistry and Molecular Engineering,
Peking University, Beijing, P.R. China

INTRODUCTION

Several kinds of flavonoids are efficiently separated and analyzed using packed or capillary column supercritical fluid chromatography. The composition of mobile phase, stationary phase, temperature, and pressure all affect the resolution. This article mainly focuses on the separation of polymethoxylated flavones, polyhydroxyl flavonoids, and flavonol isomers.

FLAVONOIDS

As a result of the development of chromatography technology, supercritical fluid chromatography (SFC) has been used to separate more and more compounds, owing to the low viscosity and high diffusivity of its mobile phase compared to the liquid mobile phase in HPLC. Supercritical carbon dioxide is the most popular SFC mobile phase because it is nontoxic, nonflammable, and easy to obtain, and it has a near-ambient critical temperature (approximately 31°C at 74 bar). However, CO_2 has weak solvating power for polar compounds. Supercritical fluids that are substantially more polar than carbon dioxide generally tend to have extreme critical temperatures and pressures (e.g., water with a critical temperature near 400°C), which makes them difficult or dangerous to work with and raises questions about the effect of such conditions on labile solutes themselves.[1] So CO_2 is the best choice, but it is often modified by such polar organic solvents as methanol, ethanol, etc. However, this binary mobile phase cannot elute very polar compounds efficiently. To widen the applicability of SFC, a small amount ($< 1\%$) of additive is added to a modifier to form a ternary mixture with CO_2. Organic acids, such as trifluoroacetic acid and citric acid, were used as additives to cause polar solutes (e.g., hydroxybenzoic and polycarboxylic acids) to be eluted rapidly and efficiently from packed SFC columns.[2–4]

Flavonoids are a group of naturally occurring substances derived from flavone (phenyl-γ-benzopyrone) that are widely distributed in the plant kingdom and used in herbal medicines throughout the world. They are 15-carbon compounds consisting of two aromatic rings and, based on the oxidation level of another ring, are classified into several groups, i.e., chalcones, flavanones, flavones, isoflavones, and flavonols, which are collectively known as the "yellow pigments," and the colored "anthocyanin pigments." Flavonoid compounds may undergo further enzymatic hydroxylation, methylation, glycosylation, sulfonation, acylation, and/or prenylation reactions, resulting in the immense diversity of flavonoid structures. There are more than 5000 identified flavonoid compounds found in nature.

Traditionally, flavonoids have been separated and analyzed by HPLC[5,6] and GC.[7] However, recent developments of SFC may permit a more accurate and complete analysis of plant phenolic compounds. Supercritical fluid chromatography brings together the advantages of both HPLC and GC techniques because it may be readily employed in the analysis of nonvolatile and thermolabile compounds and provides facile coupling to detector technologies such as mass spectrometry and Fourier transform infrared (FT-IR) spectroscopy. In recent years, SFC has been used to separate flavonoid compounds, most of which are polymethoxylated flavones and polyhydroxylflavonoids.

SEPARATION OF FLAVONOIDS

Separation of Polyhydroxylflavonoids by Packed-Column SFC

Liu et al.[8] separated polyhydroxylflavonoids, quercetin, and risetin by packed-column SFC with a ternary mobile phase. They designed an SFC apparatus with two syringe pumps and a variable-wavelength UV detector. A manual back-pressure regulator was also used to control the flow rate. This experiment showed that there are several factors affecting the result.

Mobile Phase

Neither pure supercritical CO_2 nor ethanol-modified CO_2 eluted all the flavonoids tested in this experiment.

Encyclopedia of Chromatography DOI: 10.1081/E-ECHR-120041126

But when phosphoric acid and ethanol modifiers were added to the mobile phase together, the separation on a silica-based column was significantly improved, and quercetin and risetin were eluted rapidly and efficiently. With an increase of phosphoric acid concentration, the peak shapes were also improved. Because the phosphoric acid molecules could be adsorbed onto the active sites of the stationary phase, which could prevent solute molecules from being strongly adsorbed, the interaction between solutes and stationary phase was eliminated, making the solutes easily elute from the chromatographic system.

Stationary Phase

The polarity of the column packing is in the following order: cyanopropyl > phenyl > ODS C_{18}. It is understandable that the ODS column is not suitable for the separation because of its low polarity. Both the cyanoprophyl and phenyl columns could separate these solutes efficiently, but the latter exhibited shorter separation times.

Pressure and Temperature

The capacity factor for the separation of flavonoids is decreased with the increase of operating pressure, in addition to the effect of the decrease in modifier concentration, thus indicating that the pressure effect is a very important one.

The retention times of solutes slightly increased with increasing temperature in the range of 40.0–65.0°C. As the volatilities of the solutes were increased with an increase in the temperature (which is favorable to shorten the retention time), the density of the mobile phase decreases with the temperature, which is not favorable for eluting the solute. In the range of temperatures mentioned above, the second factor is dominant; thus a lower temperature is desirable for the separation of quercetin and risetin.

Separation of Polymethoxylated Flavones by Packed-Column SFC

Morin et al.[9] successfully separated polymethoxylated flavones (PMFs) by packed-column SFC, illustrating that the SFC procedure is considerably faster than HPLC, with good resolution and adequate accuracy for the quantitative analysis of the PMFs. The chromatographic system consisted of a bare silica column (250 × 4.6 mm I.D.) with a carbon dioxide–methanol mobile phase and UV detection (313 nm). The pressure was controlled by a manual back-pressure regulator connected in series after the detector and maintained at 40°C with a water bath. Six compounds, including tangeretin, heptamethoxyflavone, nobiletin, sinensetin, tetramethylisoscutellarein, and isosinensetin (Table 1), were separated in less than 12 min as shown in Fig. 1. The resolution between nobiletin and heptamethoxy-flavone is greater than 1.5, whereas reversed-phase HPLC required the use of water–tetrahydrofuran solvent to resolve these two compounds satisfactorily. If the carbon dioxide and methanol flow rates are increased from 3 and 0.3 mL/min to 9 and 0.9 mL/min, respectively, the polymethoxylated flavones (tangeretin, nobiletin, sinensetin, and tetramethylisoscutellatrein) are separated in 2 min without any significant loss of efficiency and resolution. Thus packed-column SFC appears to be useful for rapid analyses of the main polymethoxylated flavonones.

Separation of Flavonol Isomers by Packed-Column SFC

Flavonol isomers, which differ only in the position of hydroxyl group on their chemical structures, showed different chromatographic behaviors. Liu et al.[10] separated three flavonol isomers (3-hydroxyflavone, 6-hydroxyflavone, and 7-hydroxyflavone) by a lab-constructed packed column SFC system with carbon dioxide modified with ethanol containing 0.5% (V/V) phosphoric acid as the mobile phase. The effects of

Table 1 Structures of polymethoxylated flavones

PMF	Systematic name
Heptamethoxyflavone	3,5,6,7,8,3′,4′-Heptamethoxyflavone
Hexamethoxyflaxone	3,5,6,7,3′,4′-Hexamethoxyflavone
Nobiletin	5,6,7,8,3′,4′-Hexamethoxyflavone
Sinensetin	5,6,7,3′,4′-Pentamethoxyflavone
Tangeretin	5,6,7,8,4′-Pentamethoxyflaxone
Isosinensetin	5,7,8,3′,4′-Pentamethoxyflavone
Tetramethylisoscutellarein	5,8,7,4′-Tetramethoxyflavone
Tetramethylscutellarein	5,6,7,4′-Tetramethoxyflavone

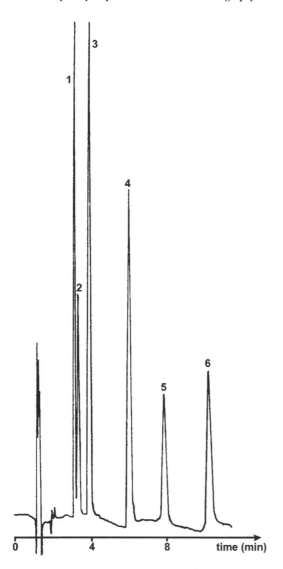

Fig. 1 SFC separation of synthetic mixture of poly-methoxylated flavones. Column, 250 mm × 4.6 mm I.D.; stationary phase, Zorbax (5 μm) silica; mobile phase, carbon dioxide modified with 10% methanol; inlet pressure, 220 atm; outlet pressure, 200 atm; column temperature, 40°C; carbon dioxide flow-rate, 3 mL/min; methanol flow-rate, 0.3 mL/min; UV detection at 313 nm. Peaks: 1 = tangeretin; tangeretin; 2 = heptamethoxyflavone; 3 = nobiletin; 4 = sinensetin; 5 = tetramethylisoscutellarein; 6 = isosinensetin.

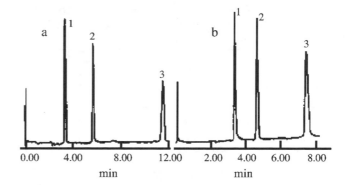

Fig. 2 Effect of stationary phase Pressure: 25 MPa, temperature: 50°C, mobile phase: carbon dioxide-ethanol with 0.5% phosphoric acid: 90 : 10, flowrate: 1.05 mL/min, stationary phase: a) cyano column, b) phenyl column.

temperature, pressure, composition of mobile phase, and packed-column type on the separation were studied. It was indicated that the addition of phosphoric acid to the mobile phase enabled flavonol isomers to be eluted from the column. It was also shown that a phenyl-bonded silica column was better and the ODS column was not as effective for the isomer separation. Increasing pressure shortened the retention time of each compound, with good resolution, and higher temperature led to longer retention times, and even the loss

of the bioactivities of these components. Under selected conditions, the separation of these isomers was very satisfactory, as illustrated in Fig. 2.

Separation of Polymethoxylated and Polyhydroxylated Flavones by Open-Tubular Capillary SFC

Solvent modifiers and additives can be used to adjust the retention and selectivity of separation in packed-column SFC. Similar effects have been reported with open-tubular capillary SFC.[11] The advantage of capillary column over packed column arises from the differences in permeability. Pressure ramps are much easier to use in capillary columns to modify the solvent strength (via density modification) as compared to packed columns. Therefore it should be entirely feasible, with capillary SFC, to combine the benefit of solvent density (pressure) programming with simultaneous modification of the solvent strength.[12,13]

Hadj-Mahammed et al.[11] analyzed a mixture of flavone, 5-methoxyflavone, and tangeretin by supercritical CO_2 SFC on capillary columns with two types of detectors: flame ionization (FID) and FT-IR. Peak identification was achieved with the help of the FT-IR fingerprint of each compound. However, the separation was satisfactory only by the use of supercritical CO_2 density programs, without the use of a phase modifier. The separations were accomplished using a Carlo Erba SFC system equipped with a Model SFC 300 pump and a Model SFC 3000 oven. The fused silica capillary columns were BP1 (12 m × 0.1 mm I.D.; 0.1-μm film of dimethylpolysiloxane) and DB5 (15 m × 0.1 mm I.D.; 0.4-μm film of 94% dimethyl-, 5% diphenyl-, and 1% vinylpolysiloxane). The two supercritical CO_2 density programs used in this work were P1 [from 0.127 g/mL (at a pressure of 73.3 bars) to 0.689 g/mL (324.2 bars) isothermally at 100°C] and

Table 2 Comparison of retention times and capacity factors of flavones analyzed using capillary columns DB5 and BP1 with supercritical CO_2 density program P2

Flavones	DB5		BP1	
	t_R (min)	k'	t_R (min)	k'
Flavone	17.93	2.40	9.52	1.36
5-Methoxyflavone	20.83	2.93	11.82	1.93
Tangeretin	25.84	3.90	16.53	3.10

P2 [from 0.111 g/mL (at a pressure of 79.0 bars) to 0.511 g/mL (318.5 bars) isothermally at 150°C].

On the DB5 capillary column, satisfactory separation of hydroxyl- and methoxyflavones could be obtained using either of the two gradient systems, but the retention times of the analytes for P2 were shorter than those for P1. This is due to the variation in the solubilities of flavones in the supercritical mobile phase when the density gradient was employed. The polarity effect of the stationary phase (BP1 phase is less polar than the DB5) is illustrated in Table 2. It can be seen that, on BP1, the flavones are less retained and the analysis time is decreased by nearly half, while conserving a satisfactory separation.

In summary, the use of a polar capillary column and an appropriate gradient of supercritical CO_2 density at a temperature of about 150°C permits the flavonoids to be separated rapidly and effectively.

CONCLUSIONS

Both packed-column and open-tubular capillary SFC can be used to separate flavonoids, and, in most cases, the separation is improved by changing the composition of mobile phase, stationary phase, temperature, and pressure.

Although HPLC has been used more often than SFC for the separation of flavonoids until now, SFC still has its particular merits and can be listed as the promising approach.

ACKNOWLEDGMENTS

This study is financially supported by the National Nature Science Foundation of China (NSFC), Grant Nos. 20275001 and 90209056.

REFERENCES

1. Berger, T.A.; Deye, J.F. Separation of benzene polycarboxylic acids by packed column supercritical fluid chromatography using methanol–carbon dioxide mixtures with very polar additives. J. Chromatogr. Sci. **1991**, *29*, 141.
2. Berger, T.A.; Deye, J.F. Separation of phenols by packed column supercritical fluid chromatography. J. Chromatogr. Sci. **1991**, *29*, 54–59.
3. Berger, T.A.; Deye, J.F. Separation of hydroxybenzoic acids by packed column supercritical fluid chromatography using modified fluids with very polar additives. J. Chromatogr. Sci. **1991**, *29*, 26–30.
4. Berger, T.A.; Deye, J.F. Separation of benzene polycarboxylic acids by packed column supercritical fluid chromatography using methane–carbon dioxide mixtures with very polar additives. J. Chromatogr. Sci. **1991**, *29*, 141–146.
5. Sendra, J.M.; Swift, J.L.; Izquierdo, L. C_{18} solid-phase isolation and high-performance liquid chromatography/ultraviolet diode array determination of fully methoxylated flavones in citrus juices. J. Chromatogr. Sci. **1988**, *26*, 443.
6. Hermburger, B.; Galensa, R.; Herrmann, K. High-performance liquid chromatography determination of polymethoxylated flavones in orange juice after solid-phase extraction. J. Chromatogr. **1988**, *439*, 481.
7. Drawert, F.; Leupold, G.; Pivernetz, H. Quantitative gaschromatographische bestimmung von Rutin, Hesperidin and Naringin in Orangensaft. Chem. Mikrobiol. Technol. Lebensm. **1980**, *20*, 111–114.
8. Liu, Z.; Zhao, S.; Wang, R.; Yang, G. Separation of polyhydroxylflavonoids by packed-column supercritical fluid chromatography. J. Chromatogr. Sci. **1999**, *37*, 155–158.
9. Morin, P.; Gallois, A.; Richard, H.; Gaydou, E. Fast separation of polymethoxylated flavones by carbon dioxide supercritical fluid chromatography. J. Chromatogr. **1991**, *586*, 171–176.
10. Liu, Z.; Zhao, S.; Wang, R.; Yang, G. Separation of flavonol isomers by packed column supercritical fluids chromatography. Chin. J. Chromatogr. **1997**, *15* (4), 288–291.
11. Hadj-Mahammed, M.; Badjah-Hadj-Ahmed, Y.; Meklati, B.Y. Behaviour of polymethoxylated and polyhydroxylated flavones by carbon dioxide supercritical fluid chromatography with flame ionization and fourier transform infrared detectors. Phytochem. Anal. **1993**, *4*, 275–278.
12. Schmitz, F.P.; Hilger, H.; Lorenschat, B.; Klesper, E. Separation of oligomers with UV-absorbing side groups by supercritical fluid chromatography using eluent gradients. J. Chromatogr. **1985**, *346*, 69.
13. Blilie, A.L.; Greibrokk, T. Gradient programming and combined gradient-pressure programming in supercritical fluid chromatography. J. Chromatogr. **1985**, *349*, 317–322.

Flavonoids: HPLC Analysis

Marina Stefova
Trajče Stafilov
Faculty of Science, Institute of Chemistry, Sts. Cyril and Methodius University,
Republic of Macedonia

Svetlana Kulevanova
Faculty of Pharmacy, Institute of Pharmacognosy, Sts. Cyril and Methodius University,
Republic of Macedonia

INTRODUCTION

Flavonoids are widely spread plant secondary metabolites called C_6–C_3–C_6 phenolics, which are classified in three groups, depending on the nature of the C_3 fragment and the type of the heterocyclic ring, as follows: 1) chromone derivatives (flavones, flavonols, flavanones, and flavanonols); 2) chromane derivatives (catechines and antocyanidines); and 3) flavonoids with open propane chain (chalcones) and with a furane ring (aurones). From all of these, flavones, flavonols, and flavanones are the most abundant in the plant kingdom and their skeleton is given in Fig. 1. Substitution in the positions 3, 5, 6, 7, 8, 2′, 3′, 4′, 5′, and 6′ gives all the compounds from these groups, with hydroxylation, methoxylation, and glycosylation being the most common substitution. Thousands of various flavonoids with various substitution patterns are recognized today as *free* flavones, flavonols, and flavanones, i.e., *aglycones*, and as *flavonoid glycosides*, which consist of flavonoid, nonsugar component *aglycone*, connected to the *sugar moiety* (mostly monosaccharides and disaccharides). Bonding to sugars makes flavonoids soluble in water and enables their easy transport within plants.

Flavonoids are a well-defined group of compounds with established physical and chemical characteristics. This especially counts for their absorption of ultraviolet (UV) radiation, which makes their UV spectra very characteristic and UV spectroscopy a method of choice for their characterization.[1] Two main absorption bands are observed: 1) band I (300–380 nm) due to absorption of ring B; and 2) band II (240–280 nm) due to absorption of ring A. The position of these bands gives information about the kind of the flavonoid and its substitution pattern; thus UV spectroscopy is used as a main method for the identification and the quantification of flavonoids for decades.

There are several published information regarding the isolation and the identification of flavonoids in plant material using different methods, mainly chromatographic and spectroscopic. Today, HPLC is

established as the most convenient method which enables separation and identification of flavonoids using various detection systems.[2–4] As for the quantitative analysis, much data have been published in the last few years confirming the suitability of this technique for simultaneous determination of flavonoid compounds in various samples, which gives an insight into the distribution of flavonoids in the studied material. HPLC methods are developed for qualitative and quantitative analyses of flavonoids in fruits and beverages, wine, honey, propolis, and, especially, in various plant materials[5–14] using different detection systems, from which UV diode array detectors are settled as the most suitable for these compounds and the most accessible as well.

EXPERIMENTATION

Stationary Phases

HPLC is the method of choice for the separation of complex mixtures containing nonvolatile compounds such as various flavonoids in extracts prepared from different samples. A survey of literatures revealed that most researchers have used C_{18}-reversed stationary phases, which proved to be superior to the normal phase technique. The reversed phases are suitable for separating flavonoids in a wide range of polarities, as Vande Casteele et al.[15] have demonstrated the separation of 141 flavonoids from polar triglycosides to relatively nonpolar polymetoxylated aglycones belonging to the classes of flavones, flavonols, flavanones, dihydroflavonols, chalcones, and dihydrochalcones.

The use of normal phase silica columns was also described but after acetylation of the flavonoids and then isocratic elution on silica gel.[16] Polystyrene divinylbenzene as a stationary phase was also found to give satisfactory separation and good peak shapes without using acidic mobile phases,[17] which are not so favorable for use with reversed phases.

Encyclopedia of Chromatography DOI: 10.1081/E-ECHR-120016287

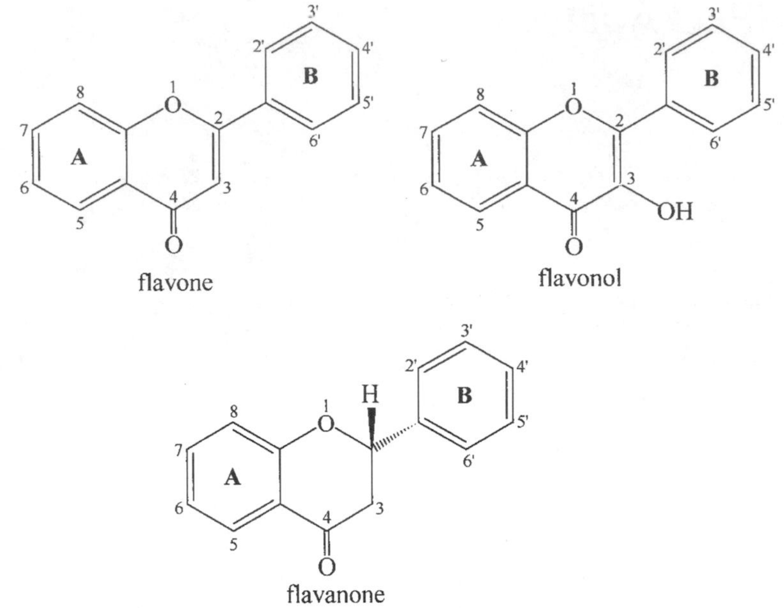

Fig. 1 Structure of flavone, flavonol, and flavanone.

The choice of a column involves matching the class of flavonoids to be separated to the characteristics of the stationary phase capable of providing satisfactory retention, selectivity, and peak shapes. As for the dimensions, the most popular are 150 and 250 mm long, with 5 μm particle size, although phases with 3 μm particles are becoming more popular because of better efficiency and less solvent consumption. The use of guard columns is recommended in order to extend their lifetime by protecting the analytical column from impurities in the samples prepared from various natural products.

Mobile Phases

The preferred solvent system used for the separation of flavonoids on reversed-phase stationary phases is methanol–water, followed closely by acetonitrile–water. Usually, acetic or formic acid is added (sometimes phosphoric acid, potassium dihydrogen phosphate, ammonium dihydrogen phosphate, and perchloric acid), which enables improved separation and prevention of peak tailing with respect to the phenolic character of the flavonoids. Ion pairing has also been used for improving the separation of neutral glycosides from flavonol sulfates.[4]

Often, gradient elution with a linear gradient, combined with isocratic steps, usually gives the best separation of flavonoids, differing in the degree of saturation, hydroxylation, methylation, and/or glycosylation. Optimization of the elution program is usually performed by changing the elution program until a satisfactory resolution and analysis time is achieved,

depending on the analysis purpose, i.e., qualitative or quantitative analysis.

Detection Systems

Identification of flavonoids separated by HPLC is commonly performed by comparing the obtained retention times with the ones of authentic samples as well as by analysis of their characteristics collected by the detector.

Detection of flavonoids separated by HPLC can be performed using several detection systems. Photodiode array detectors (DADs) are the most convenient for use because of their availability and easy maintenance, and mainly because of the valuable information for the identification of flavonoids contained in their UV [or ultraviolet–visible (UV–VIS)] spectra. The sophisticated software packages enable storage of all the spectra obtained during the elution process, which can later be analyzed and compared to the spectra from a library previously prepared from authentic samples of flavonoids. An experienced analyst working on flavonoids can recognize the peaks from flavonoids in the chromatogram from the spectra, which are, as previously mentioned, very characteristic, although they are not enough for identification. As for the quantitative analysis, scanning over a wide wavelength range enables measuring all the components at wavelengths of their absorption maxima, which provides maximum sensitivity.

Fluorescence detectors have also been employed for flavonoid determination, offering higher sensitivity and selectivity, as suggested by the results in the analysis of

3′,4′,5′-trimethoxyflavone, which has been determined by excitation at 330 nm and by detection at 440 nm.[18]

Overcoming the difficulty in introducing the liquid sample into the mass spectrometer in the last decade enabled their use for HPLC detection, even in routine practical applications. The atmospheric pressure chemical ionization, thermospray, and electrospray ionization systems have proved to be most convenient for flavonoid analysis by HPLC/MS.[18,19] Mass spectrometry is a more specific and extremely selective detection technique, although it is not enough for structure elucidation (it reveals the structural fragments that are present, but not the exact substitution pattern). It offers the molecular weight of the molecular ion, which helps the tentative identification of the flavonoid. The use of HPLC/MS/MS gives additional information about the characteristic fragmentation pattern—"fingerprint" of the substance, which is very useful for qualitative and quantitative analyses in complex matrices. An excellent illustration of the use of HPLC/MS for the identification of flavonoids is presented in the work of Huck et al.,[19] who have isolated and characterized polymethoxylated flavones from *Primulae veris flos* using HPLC/ESI/MS. The six detected flavonoid compounds were found to be mono-, di-, tri-, and pentametoxyflavones, but their exact substitution pattern could not be revealed, except for 3′,4′,5′-trimetoxyflavone, for which ^{13}C NMR spectral data were available.

As regards the sensitivity of the MS detection in HPLC analysis of flavonoids, this technique has proved to be the most sensitive as compared to UV and fluorescence detection. A very comprehensive comparison of the four detection systems—UV, fluorescence, and two MS systems [atmospheric pressure chemical ionization (APCI) and electrospray ionization (ESI)]—for the determination of the previously identified 3′,4′,5′-trimethoxyflavone[18] is presented in Table 1. Fluorescence detection is 10 times more sensitive than UV detection, whereas MS detection is 50 times more sensitive than UV detection and 5 times more sensitive than fluorescence detection.

Similar assay of the detection systems is performed by Stecher et al.,[9] who analyzed flavonols and stilbenes in wine and biological products and found that ESI/MS detection gave two, three, and nine times

lower limit of detection (LOD) for myricetin, quercetin, and kaempferol, respectively, as compared to UV absorbance detection at 377 nm (absorption maximum of flavonols).

A significant amount of literature regarding the antioxidant properties of flavonoids and other plant polyphenols is available. As the essence of redox chemistry involves electron transfer, it seems natural that electrochemical detection rivals spectrophotometric detection techniques for the compounds that are supposed to be antioxidants. With the improvements in electrochemical detector geometries and electronics over the last decade, coupled with a requirement for increased sensitivity, the use of electrochemical detectors offers significant additional advantages when combined with the traditional UV–VIS detection in the analysis of flavonoids and other plant polyphenols.[20]

APPLICATIONS

Several information regarding HPLC analysis of flavonoids, often together with other phenolic constituents, in various samples, such as fruits, vegetables, juices, wines, honey, propolis, and, especially, plant material, are published. The enormous interest in studying these compounds is a result of their potential importance to health and antioxidant defense mechanisms, which has imposed the need for developing methods for their identification and quantification in various natural products. Reversed-phase HPLC (RP HPLC) with combined isocratic and gradient elution with acidic mobile phases and UV diode array or mass spectrometer detector is the most often used system for flavonoid analysis. Depending on the nature of the sample, a variety of sample preparation techniques has been developed in order to achieve good recovery of the analyzed compounds in a simple sample preparation procedure. The procedure includes extraction of the flavonoids in a polar solvent; this extract is then used either for injection onto the column (after filtration) or for further fractionation and purification of the flavonoids to obtain a relatively "clean" sample for injection. The latter step can be carried out in several ways: by liquid–liquid extraction in suitable solvents; by chromatographic techniques (thin layer or column

Table 1 Regression equations for the calibration curves (logarithmic), regression coefficients, and detection limits for the determination of 3′,4′,5′-trimethoxyflavone

Detection method	Regression equation	Regression coefficient (R^2)	Detection limit
UV at 213 nm.	$y = 0.990x + 13.158$	0.9986	0.244 ng
Fluorescence at 330/440 nm	$y = 0.879x + 12.926$	0.9946	24.4 pg
ESI–MS	$y = 0.692x + 17.349$	0.9956	5.00 pg
APCI–MS	$y = 0.899x + 15.798$	0.9997	5.00 pg

(From Ref.[18].)

chromatography) on various stationary phases (silica, reversed phases, Amberlite, and Sephadex); or by solid-phase extraction techniques using different adsorbents, which in the last decade has been found very convenient for the isolation of flavonoids from complex matrices.

Flavonoids in Fruits, Juices, and Wine

During the past decades, extensive analytical research has been performed on the separation of phenolic constituents in various fresh fruits and fruit products. A very thorough examination was performed by Barberán et al.[11] on samples of several types of fresh nectarines, peaches, and plums using RP HPLC/DAD/ESI/MS and combined isocratic and gradient elution with a mobile phase of water–methanol with 5% formic acid. The samples were prepared by extraction of a homogenized frozen fruit material with water–methanol (2 : 8), by centrifugation, by filtration, and by injection into the column. This procedure recovers 85–92% of all phenolic compounds including hydroxycinnamates, procyanidins, flavonols, and anthocyanins. In the chromatograms obtained for nectarin and peach samples, three peaks from flavonols (quercetin 3-glucoside, quercetin 3-rutinoside, and, probably, quercetin 3-galactoside), several peaks from flavan-3-ols (catechin, epicatechin, and dimer procyanindines), and two anthocyanin pigments (cyanidine 3-glucoside and cyanidine 3-rutinoside) have been identified by comparing the retention data, the UV, and the mass spectra of the detected flavonoids with the ones obtained for authentic markers. Quantification was done using external standards: quercetin 3-rutinoside for flavonols at 340 nm, cyanidine 3-rutinoside for anthocyanins at 510 nm, and catechin for flavan-3-ols at 280 nm.

An HPLC/DAD method was developed for the separation and the determination of flavonoid and phenolic antioxidants in commercial and freshly prepared cranberry juice.[6] Two sample preparation procedures were used: with and without hydrolysis of the glycoside forms of flavonoids carried out by the addition of HCl in the step prior to solid-phase extraction (SPE). The flavonoid and phenolic compounds were then fractionated into neutral and acidic groups via a SPE method (Sep-Pak C_{18}), followed by a RP HPLC separation with gradient elution with water–methanol–acetic acid and a detection at 280 and 360 nm. A comparison of the chromatograms obtained for extracts prepared with and without hydrolysis showed that flavonoids and phenolic acids exist predominantly in combined forms such as glycosides and esters. In a freshly squeezed cranberry juice, for instance, 400 mg of total flavonoids and phenolics per

liter of sample was found, 56% of which were flavonoids. Quercetin was the main flavonoid in the hydrolyzed products, where it accounted for about 75% of the total flavonoids, while it was absent in the unhydrolyzed products.

The flavonoid constituents of *Citrus* have attracted attention in the last decade because of their biological activities (anticarcinogenic, anti-inflammatory effects, etc.) together with the chemotaxonomic importance of the specific flavanone glycosides found in this species. The presence of 12 flavonoids (9 flavanone glycosides, 2 flavanone aglycones, and 1 flavone glycoside) was detected in samples from leaves and fruits of *Citrus aurantium*.[7] The use of different extraction solvents was studied (methanol, dioxane–methanol 1 : 1, 0.1% NaOH aq., 0.01% KOH methanolic, pyridin, dimethylformamide, and dimethylsulfoxide), with dimethylsulfoxide representing the best results. Also, an exhaustive description of the optimization process by studying the quantitative chromatographic parameters—k', w, α, N, height equivalent to a theoretical plate (HETP), and R—is given with the discussion of the effects of the variation of the methanol (acetonitrile) content in the mobile phase, the degree of mobile phase acidity, together with the influence of the structural characteristics of the studied flavonoids on their behavior in the reversed-phase chromatographic system.

Another thorough procedure for flavonoid analysis in *Citrus* samples is presented in the work of Nogata et al.[8] They developed a method for separation of 9 flavanones, 10 flavones, and 6 flavonols using a RP HPLC system with gradient elution with 0.01 M phosphoric acid–methanol and for detection at 285 nm, which is presented in Fig. 2. Identification was carried out by comparing the retention data and the spectra of the sample components with the ones of the standards. For quantitative analysis, all flavonoids exhibited good linearity ($r = 0.988$–1.00) between concentration and peak area in the investigated concentration range (10–200 ppm) with detection limits from 0.5 to 2.5 ppm.

To investigate the probable health benefits of flavonoids and stilbenes in red wine, a RP HPLC method with enhanced separation efficiency, selectivity, sensitivity, and speed has been established for the determination of the flavonols quercetin, myricetin, and kaempferol and the stilbenes *cis*- and *trans*-resveratrol in a single run.[9] The sample preparation step for wines and grape juices included only filtration prior to injection. Identification was carried out by the comparison of the retention data, the UV, and the mass spectra of the flavonoids and the stilbenes with the ones obtained for standards. Quantitative analysis using external standards showed good linearity ($R^2 > 0.999$ for UV and $R^2 > 0.9878$ for MS detection), recoveries between 95% and 105%, and limits of detection in the nanogram range, with MS being more sensitive.

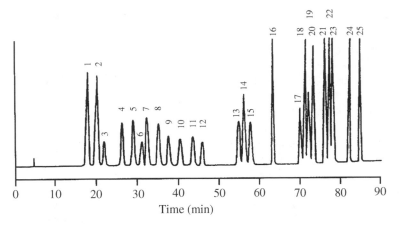

Fig. 2 Separation of 25 flavonoid standards (1—eriocitrin; 2—neoeriocitrin; 3—robinetin; 4—narirutin; 5—naringin; 6—rutin; 7—hesperidin; 8—neohesperidin; 9—isorhoifolin; 10—rhoifolin; 11—diosmin; 12—neodiosmin; 13—neoponcirin; 14—quercetin; 15—poncirin; 16—luteolin; 17—kaempferol; 18—apigenin; 19—isorhamnetin; 20—diosmetin; 21—rhamnetin; 22—isosakuranetin; 23—sinensetin; 24—acacetin; 25—tangeretin) using C_{18} Lichrospher 100, 250 × 4.0 mm, 5 μm, Merck; gradient elution with 0.01 M phosphoric acid–methanol with flow rate of 0.6 mL/min at 40°C and detection at 285 nm. (From Ref.[8].)

An interesting improvement in the sample preparation step has been suggested[10] using only filtering of the sample prior to injection into the system composed of two columns: clean-up column (50 × 2.1 mm packed with C_{18} stationary phase, 5 μm) and analytical column (250 × 2.1 mm, C_{18}, 5 μm). A column-switching procedure enables the sample introduced in the first column to be cleaned from the disturbing matrix compounds for 2 min (elution phosphate buffer, pH = 7, with 10% acetonitrile), and then by a gradient of the mobile phase (phosphate buffer, pH = 2.5, with acetonitrile); the retained analytes—flavonoids—are eluted to the analytical column, where they are separated and detected at 365 nm. The method was used for the determination of flavonoid profiles of berry wines containing the flavonols quercetin, myricetin, kaempferol, rutin, and isoquercitrin.

Flavonoids of Honey and Propolis

A very interesting application of flavonoid analysis by HPLC has been carried out by Barberán et al.[11] on honey samples. They used RP HPLC/DAD for the analysis of 20 flavonoid aglycones, which could be considered as markers for the floral origin of honey. Different solvent systems were applied to the analysis of flavonoids from citrus and rosemary honeys. The sample preparation included dilution with water and HCl (pH = 2–3) for hydrolysis of glycosides, followed by purification using chromatography on Amberlite XAD-2 and Sephadex LH-20. The flavonoid markers of the botanical origin, hesperetin and apigenin, were detected using methanol–water and acetonitrile–water mixtures with formic acid. Quercetin and kaempferol were separated using Prizma-optimized conditions,

whereas tectochrysin was eluted with methanol–water and acetonitrile–water mixtures by increasing the content of the organic solvent at the end of the elution programs. The use of diode array detection was found essential in studies of the floral origin of honey by flavonoid analysis.

Flavonoid analysis has also been carried out in propolis, another beehive product rich in flavonoids, which are partly responsible for its pharmacological activity.[12] Qualitative and quantitative analyses were performed by using a reversed-phase column and isocratic elution with water–methanol–acetic acid (60:75:5) and by monitoring at 275 and 320 nm. The propolis sample was cut into small pieces, extracted with boiling methanol, and the methanolic extract was then diluted with water and, subsequently, extracted with light petroleum and diethyl ether. The last extract contained the propolis flavonoids pinocembrin (21.4%), galangin (5%), chrysin (4.8%), quercetin (2.2%), and tectochrysin (1.1%).

Plant Material

Flavonoids play important roles in plant biochemistry and physiology; they are responsible for the biological effects of plants and their extracts as well as preparations on humans. HPLC is found very suitable for the detection and the determination of flavonoids present in various plants and plant products; the so-called HPLC fingerprint analysis is suggested for quality control and standardization because of the ability of good separation and resolution of complex mixtures as well as peak purity control. One of the limitations of the assays of flavonoids in plant material is the fact that many compounds, especially glycosides, are not commercially available. One of the possible solutions to this problem is the hydrolysis of flavonoid glycosides during the extraction procedure, which aids the identification on flavonoid aglycones. Such a procedure (the extraction in acetone with the addition of HCl for hydrolysis of glycosides) is proposed for the screening of flavonols and the determination of quercetin in medicinal plants using RP HPLC with gradient elution with water–acetonitrile–acetic acid and UV diode array detection.[13] Quercetin was found to be the most abundant, especially in *Hyperici herba* and *Pruni spinosae flos*, kaempferol in *Robiniae pseudoacaciae flos* and *P. spinosae flos*, whereas myricetin was detected only in *Betulae folium*, as can be seen in the chromatograms presented in Fig. 3. The content of quercetin ranged from 0.026% in *Bursae pastoris herba* to 0.552% in *Hyperici herba*.

Another quantitative RP HPLC method, based on the reduction of the complex flavonoid glycoside pattern by acid hydrolysis to one major aglycone (quercetin) and one C-glycoside (vitexin), was developed and employed for the characterization of Crataegus leaves and flowers.[14] A qualitative fingerprint method was also developed for the separation and the identification of all characteristic flavonoids (glycosides and aglycones). Samples for fingerprint analysis were prepared by extraction with 80% methanol and then filtered through Bond Elut C$_{18}$ cartridge prior to injection. Elution was performed by a mobile phase composed of tetrahydrofuran–acetonitrile–methanol and 0.5% orthophosphoric acid and was monitored at 370, 336, and 260 nm. For quantitative analysis, extraction was carried out with methanol in a Soxhlet apparatus, HCl was added to the methanolic extracts for the hydrolysis of glycosides, and, finally, they were filtered through Bond Elut C$_{18}$ before injection. Separation and quantification of vitexin and quercetin were performed for characterization and standardization of the plant material as well as its extracts and preparations.

Structure–RP HPLC Retention Relationships of Flavonoids

The ability to predict the chromatographic mobility of a compound under given conditions, based on its structure, offers many advantages in analysis. The elution sequence of individual flavonoids can be interpreted by assuming that the compounds are first adsorbed on the reversed stationary phase by "hydrophobic interaction," and then subsequently eluted with the mobile phase according to the extent of hydrogen bond formation. Therefore the hydrogen bond donating and/or accepting ability of a given substituent as well as its contribution to the hydrophobic interaction

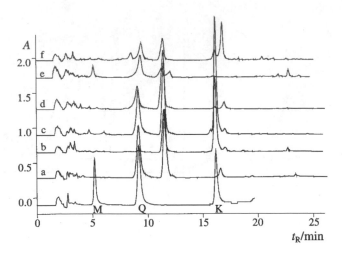

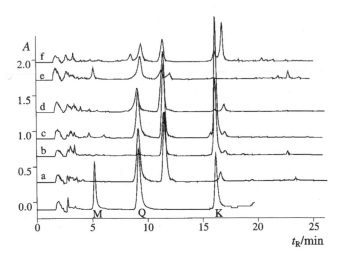

Fig. 3 Chromatograms obtained for extracts of the following: a) *H. herba*; b) *R. pseudoacaciae flos*; c) *P. spinosae flos*; d) *Sambuci flos*; e) *B. folium*; and f) *Primula flos*, and for mixture of authentic samples of myricetin (M), quercetin (Q), and kaempferol (K) using: C18 (250 × 4.6 mm, 5 μm, Varian); gradient elution with 5% acetic acid–acetonitrile with flow rate of 1.0 mL/min at 30°C and detection at 367 nm. (From Ref.[13].)

have to be considered. The retention data of 141 flavonoids[15] imply the balance between these two effects, resulting in almost identical retention of tricetin pentamethylether (pentamethoxy flavone) and unsubstituted flavone.

Hydroxylation in positions other than 3 and 5 decreases retention owing to increasing polarity (hydrogen bond formation ability). The presence of an OH group in positions 3 and 5 is specific because of the formation of an intramolecular hydrogen bond with the carbonyl group on C-4, which is the strongest hydrogen bond acceptor in flavones and isoflavones. This is the reason for the increase in retention, especially when OH is substituted in position 5, whereas another OH group in position C-3 only slightly lowers the retention. This produces poor separation of the so-called critical pairs flavone–flavonol differing only in the OH group in position 3.

Methylation of the OH groups more or less prevents their effect, which means that flavonoids and their partial methyl ethers are easily separated. On the other hand, the introduction of additional methoxy groups has little or no effect on retention—introducing another type of critical pair differing only in one methoxy group.

Glycosylation of an OH group means introducing a hydrophilic moiety together with shielding (by hydrogen bonding or just by steric hindrance) some hydrophilic substituents already present in the vicinity. The shielding effect plays a role when an OH group located *ortho* to another OH group is glycosylated (e.g., glycosylation of 7- or 4′-OH without an adjacent *ortho*-OH decreases the t_R values by 4.26–3.83 min, whereas, in the presence of an *ortho*-OH, the decrease is only 2.28–1.55 min[15]). The fact that the t_R value of luteolin-5-β-D-glucopyranoside is only 0.37 min smaller than that of the corresponding 7-β-D-glucopyranoside can also be explained by the shielding effect of sugar on the carbonyl group. The contributions of various types of sugars to the hydrophilic interaction decrease from hexoses through pentoses to methylpentoses.

Saturation of the C_3 ring, which means transformation of flavones to flavanones and of flavonols to dihydroflavonols, affects the retention in a very complex way. The saturation itself has a small effect; however, in the presence of OH groups, the retention is always decreased. This is explained by the interruption of the conjugation in the system, affecting the acidity and therefore the hydrogen bond accepting and donating abilities of the OH groups, especially the 3-OH groups which are phenolic in flavones and alcoholic in flavanones.

CONCLUSIONS

RP HPLC has proved to be the method of choice for the separation of a variety of flavonoids in different samples. The phenolic nature of these compounds requires the use of acidic mobile phases for satisfactory separation and peak shapes, whereas the detection is usually carried out with photodiode array detectors which are also very helpful for their identification of the characteristic absorption spectra of the flavonoids. In the last decade, mass spectrometers connected to HPLC systems introduced a greater selectivity and sensitivity in flavonoid analysis. Improving the characteristics of the stationary phases and developing more sophisticated instruments as well as devices for more

efficient and faster sample preparation are the challenges for all modern analysts. Discovering the beneficial health effects of flavonoids and the "going-back-to-nature" trend motivates the development of more efficient and fast procedures for their identification and quantification, with HPLC remaining the most powerful technique for their separation from the complex mixtures.

REFERENCES

1. Mabry, T.J.; Markham, K.R.; Thomas, M.B. *The Systematic Identification of Flavonoids*; Springer-Verlag: New York, 1970.
2. Harborne, J.B. *Phytochemical Methods (A Guide to Modern Techniques of Plant Analysis)*; Chapman and Hall: London, 1984.
3. Wollenweber, E.; Jay, M. Flavones and flavonols. In *The Flavonoids*; Harborn, J.B., Ed.; Chapman and Hall: London, 1988.
4. Daigle, D.J.; Conkerton, E.J. Analysis of flavonoids by HPLC: an update. J. Liq. Chromatogr. **1988**, *11* (2), 309–325.
5. Barberán, F.A.T.; Gil, M.I.; Cremin, P.; Waterhouse, A.L.; Hess-Pierce, B.; Kader, A.A. HPLC–DAD–ESIMS analysis of phenolic compounds in nectarines, peaches, and plums. J. Agric. Food Chem. **2001**, *49*, 4748–4760.
6. Chen, H.; Zuo, Y.; Deng, Y. Separation and determination of flavonoids and other phenolic compounds in cranberry juice by high-performance liquid chromatography. J. Chromatogr. A. **2001**, *913*, 387–395.
7. Castillo, J.; Benavente-García, O.; Del Rio, J.A. Study and optimization of *citrus* flavanone and flavones elucidation by reverse phase HPLC with several mobile phases: influence of the structural characteristics. J. Liq. Chromatogr. **1994**, *17* (7), 1497–1523.
8. Nogata, Y.; Ohta, H.; Yoza, K.I.; Berhow, M.; Hasegawa, S. High-performance liquid chromatographic determination of naturally occurring flavonoids in *citrus* with a photodiode-array detector. J. Chromatogr. A. **1994**, *667*, 59–66.
9. Stecher, G.; Huch, C.W.; Popp, M.; Bonn, G.K. Determination of flavonoids and stilbenes in red wine and related biological product by HPLC and HPLC–ESI–MS–MS. Fresenius' J. Anal. Chem. **2001**, *371*, 73–80.
10. Ollanketo, M.; Riekkola, M.L. Column-switching technique for selective determination of flavonoids in finnish berry wines by high-performance liquid chromatography with diode array detection. J. Liq. Chromatogr. & Relat. Technol. **2000**, *23* (9), 1339–1351.

11. Barberán, F.A.T.; Ferreres, F.; Blázquez, M.A.; García-Viguera, C.; Tomás-Lorente, F. High-performance liquid chromatography of honey flavonoids. J. Chromatogr. **1993**, *634*, 41–46.

12. Bankova, V.S.; Popov, S.S.; Marekov, N.L. High-performance liquid chromatographic analysis of flavonoids from propolis. J. Chromatogr. **1982**, *242*, 135–143.

13. Stefova, M.; Kulevanova, S.; Stafilov, T. Assay of flavonols and quantification of quercetin in medicinal plants by HPLC with UV-diode array detection. J. Liq. Chromatogr. & Relat. Technol. **2001**, *24* (15), 2283–2292.

14. Rehwald, A.; Meier, B.; Sticher, O. Qualitative and quantitative reversed-phase high-performance liquid chromatography of Crataegus leaves and flowers. J. Chromatogr. A. **1994**, *677*, 25–33.

15. Vande Casteele, K.; Geiger, H.; Van Sumere, C.F. Separation of flavonoids by reversed-phase high-performance liquid chromatography. J. Chromatogr. **1982**, *240*, 81–94.

16. Galensa, R.; Herrmann, K. Analysis of flavonoids by high-performance liquid chromatography. J. Chromatogr. **1980**, *189*, 217–224.

17. Jagota, N.K.; Cheathan, S.F. HPLC separation of flavonoids and flavonoid glycosides using a polystyrene/divinylbenzene column. J. Liq. Chromatogr. **1992**, *15* (4), 603–615.

18. Huck, C.W.; Bonn, G.K. Evaluation of detection methods for the reversed-phase HPLC determination of 3′,4′,5′-trimethoxyflavone in different phytopharmaceutical products and in human serum. Phytochem. Anal. **2001**, *12*, 104–109.

19. Huck, C.W.; Huber, C.G.; Ongania, K.H.; Bonn, G.K. Isolation and characterization of methoxylated flavones in the flowers of *Primula veris* by liquid chromatography and mass spectrometry. J. Chromatogr. A. **2000**, *870*, 453–462.

20. Milbury, P.E. Analysis of complex mixtures of flavonoids and polyphenols by high-performance liquid chromatography electrochemical detection methods. Methods Enzymol. **2001**, *335*, 15–26.

Flavonoids: Separation by Countercurrent Chromatography

L. M. Yuan

Department of Chemistry, Yunnan Normal University, Kunming, P.R. China

INTRODUCTION

Various countercurrent chromatography (CCC) methods employ two-phase solvent systems for separation of flavonoids. $CHCl_3 : CH_3OH : H_2O$ (4 : x : 2, for which x is between 2.5 and 4 for different samples) may be used for their prefractionation. Two-phase solvent systems can be divided into four types: $CHCl_3 : CH_3OH : H_2O$, $CHCl_3 : CH_3OH : n\text{-}BuOH : H_2O$, $EtOAc : PrOH : H_2O$ or $BuOH : HOAc : H_2O$, and $n\text{-}C_6H_{14} : EtOAc : CH_3OH : H_2O$, respectively. CCC fractionation permits the use of either stepwise elution or gradient elution, and analytical-scale CCC can also be carried out for separation of flavonoids.

FLAVONOIDS

Flavonoids are polyphenolic compounds and have a wide polarity range. Successful CCC depends on the appropriate choice of a solvent system. The partition of solutes between two immiscible solvent phases is an ideal method for the separation of flavonoids because the phenomenon of irreversible adsorption, tailing, sample loss, and denaturation, which plague analysts using other forms of LC, are avoided.

SEPARATION OF FLAVONOIDS BY CCC

Flavonoids are a specific class of polyphenols. It is generally believed that flavonoids include a wide variety of phenolic compounds, such as flavones, flavonols, flavanones, flavanonols, anthocyanidins, flavan-3, 4-diols, xanthones, flavan-3-ols, isoflavones, isoflavanones, chalcones, dihydrochalcones, aurones, and homoisoflavones. Their separation poses special problems because there is often irreversible adsorption and even hydrolysis on solid supports.

The development of CCC began in the mid-1960s. Among various types of CCC modes for the separation of flavonoids, the main techniques are droplet countercurrent chromatography (DCCC), rotary locular countercurrent chromatography (RLCCC), and centrifuge partition chromatography (CPC). Today, the cartridge and multilayer coil CPC methods are the leading techniques.

The selection of a solvent system is the most important step in performing CCC. Selecting a solvent system for CCC means *simultaneously* choosing the column and the eluent. The chromatographic literature contains numerous examples of solvent systems used in various CCC systems for separation of flavonoids,[1–6] and consultation of these references may give some leads as to possible systems that would be useful for a particular separation. The selection of appropriate biphasic solvent systems also may be aided by TLC or HPLC experiments.

One quite remarkable comparative separation of flavonoids by DCCC, RLCC, and cartridge and multilayer coil CPC methods, with the same solvent system (chloroform–methanol–water, 33 : 40 : 27), is shown in Fig. 1.[7,8]

Elutions of the three peaks were similar and according to the order of increasing polarity:

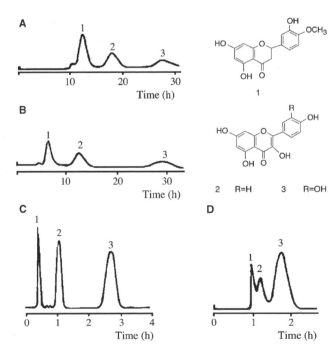

Fig. 1 Chromatograms of hesperetin (1), kaempferol (2), and quercetin (3) by different countercurrent chromatographic methods. Solvent system: chloroform–methanol–water (33 : 40 : 27). (A) DCCC; (B) RLCC; (C) multilayer coil CPC system; and (D) cartridge CPC system. (Reprinted with permission from Elsevier Science.)

Encyclopedia of Chromatography DOI: 10.1081/E-ECHR-120041127

Table 1 Separations of flavonoids by CCC

Sample	Solvent system
Flavonoids	$CHCl_3 : CH_3OH : H_2O$, 4:3:2
	$CHCl_3 : CH_3OH : H_2O$, 7:13:8
	$CHCl_3 : CH_3OH : H_2O$, 33:40:27
	$CHCl_3 : CH_3OH : H_2O$, 5:6:4
Flavonol glycosides	$CHCl_3 : CH_3OH : H_2O$, 7:13:8
Isoflavonol glycosides	$CHCl_3 : CH_3OH : H_2O$, 7:13:8
Baccharis trimera	$CHCl_3 : CH_3OH : H_2O$, 13:7:4
Orthosiphon spicatus	$CHCl_3 : CH_3OH : H_2O$, 13:7:4
Epilobium parviflorum	$CHCl_3 : CH_3OH : H_2O$, 7:13:8
Licorice	$CHCl_3 : CH_3OH : H_2O$, 7:13:8
Sea buckthorn	$CHCl_3 : CH_3OH : H_2O$, 4:3:2
Daphne genkwa sieb et zucc	$CHCl_3 : CH_3OH : H_2O$, 4:3:2
Strychnos variabilis	$CHCl_3 : CH_3OH : H_2O$, 5:6:4
Pericarpium citri reticulata	$CHCl_3 : CH_3OH : H_2O$, 4:4:2
Radix puerariae	$CHCl_3 : CH_3OH : H_2O$, 4:2.9:2
Radix glycyrrhizae	$CHCl_3 : CH_3OH : H_2O$, 4:4:2
Radix scutellariae	$CHCl_3 : CH_3OH : H_2O$, 4:3.8:2
Flos genkwa	$CHCl_3 : CH_3OH : H_2O$, 4:3.6:2
Flavonoids	$CHCl_3 : CH_3OH$: 5% HCl, 5:5:3
Flavonoid glycosides	$CHCl_3 : CH_3OH$: *n*-BuOH : H_2O, 10:10:1:6
Flavonol glycosides	$CHCl_3 : CH_3OH$: *n*-BuOH : H_2O, 7:6:3:4
	$CHCl_3 : CH_3OH$: *n*-BuOH : H_2O, 10:10:1:6
Galipea trifoliata	$CHCl_3 : CH_3OH$: *n*-BuOH : H_2O, 10:10:2:6
Arnica species	$CHCl_3 : CH_3OH$: *n*-BuOH : H_2O, 10:10:1:6
Flavonol glycosides	$CHCl_3 : CH_3OH$: *i*-PrOH : H_2O, 5:6:1:4
Bidens pilosa	$CHCl_3 : CH_3OH$: *i*-PrOH : H_2O, 5:6:1:4
Picea abies	$CHCl_3 : CH_3OH$: *i*-PrOH : H_2O, 5:6:1:4
Alangium premnifolium	$CHCl_3 : CH_3OH$: *i*-PrOH : H_2O, 9:12:2:8
G. biloba	$CHCl_3 : CH_3OH$: PrOH : H_2O, 5:6:1:4
Vaccinium uliginosum	$CHCl_3 : CH_3OH$: PrOH : 5% HOAc, 31.2:37.5:6.25:25
Oxytropis ochrocephala	$CHCl_3 : CH_3OH$: EtOH : H_2O, 2:5:8:5
Tephrosia vogelii	$CHCl_3 : CH_3OH$: EtOH : H_2O, 7:3:3:4
Flavonoid glycosides	EtOA : PrOH : H_2O, 4:2:7
Arnica montana	EtOA : PrOH : H_2O, 4:2:7
Stryphnodendron adstringens	EtOA : PrOH : H_2O, 35:2:2
Anthocyanidins	EtOA : PrOH : H_2O, 140:8:80
Flavonoid glycosides	EtOA : BuOH : H_2O, 2:1:2
Crossopteryx febrifuga	EtOA : BuOH : H_2O, 2:1:2
	EtOA : EtOH : H_2O, 2:1:2
Tephrosia vogelii	EtOA : 94% EtOH : H_2O, 2:1:2
Flavonoid glycosides	BuOH : HOAc : H_2O, 4:1:5
Anthocyanidins glycosides	BuOH : HOAc : H_2O, 4:1:5
Sambucus nigra	BuOH : HOAc : H_2O, 4:1:5
Pavetta owariensis	BuOH : PrOH : H_2O, 4:1:5

(Continued)

Table 1 Separations of flavonoids by CCC *(Continued)*

Sample	Solvent system
Arnica species	CH_2Cl_2 : PrOH : H_2O, 7 : 13 : 8
Biflavonoids	n-C_6H_{14} : EtOAc : CH_3OH : H_2O, 2 : 8 : 5 : 5
Brackenridgea zanguebarica anthocyanidins	n-C_6H_{14} : EtOAc : CH_3OH : H_2O, 8 : 8 : 6 : 6
	n-C_6H_{14} : EtOAc : CH_3OH : H_2O, 8 : 6 : 7 : 10
	n-C_6H_{14} : EtOAc : BuOH : HAc : 1% HCl, 2 : 1 : 3 : 1 : 5
Gradient elution	
G. biloba	EtOAc→EtOAc-i-BuOH, 6 : 4
Esenbeckia pumila	Et_2O→EtOAc-PrOH-H_2O (10 : 1 : 2)→EtOAc-PrOH-H_2O (4 : 1 : 2)

hesperetin, kaempferol, quercetin. The above peak effect can be seen in the various CCC methods, which have similar two-phase solvent systems, for the separation of flavonoids, but DCCC and RLCC require much more time than CPC.

A versatile two-phase solvent system for flavonoid prefractionation by high-speed CCC was introduced.[9] The flavonoid compounds encompass a wide polarity range and exhibit good solubility in methanol. The two-phase solvent system composed of $CHCl_3$: CH_3OH : H_2O = 4 : x : 2, in which x was between 2.5 and 4 for different samples, was selected for the separation of crude flavonoid extracts because this solvent system contains methanol and provides nearly equal volumes of the upper and lower phases with reasonably short settling times. Changing the ratio of methanol in the solvent system permitted changing, simultaneously, the selectivity of the upper and lower phases, as methanol can dissolve in chloroform and water and may change the polarity of the two phases.

The previous applications of CCC are summarized in Table 1. One may use this table to search for suitable solvent systems that have been previously used for similar compounds.

In Table 1, the various two-phase solvent systems are divided into four types. It has been shown that the solvent system chloroform–methanol–water has been used for the greatest number of applications. $CHCl_3$: CH_3OH : H_2O is a versatile CCC solvent system for flavonoid separations as well.

The second most used two-phase solvent system is $CHCl_3$: CH_3OH : n-BuOH (i-PrOH or PrOH) : H_2O. It is similar to $CHCl_3$: CH_3OH : H_2O, but CH_3OH is partially replaced by BuOH, i-PrOH, or PrOH. The desirable K value may be obtained by varying the ratios.

The third most used type of two-phase solvent system is EtOAc : PrOH (BuOH or EtOH) : H_2O or BuOH : HOAc (or PrOH) : H_2O. It is a more polar solvent system and is useful for the separation of flavonoid glycosides.

The final type of solvent system is n-C_6H_{14} : EtOAc : CH_3OH : H_2O. Sometimes CH_3OH also may be substituted by other solvents such as BuOH. It is a slightly more hydrophobic solvent system than $CHCl_3$: CH_3OH : H_2O.

Analogous to HPLC, CCC fractionation permits the use of either stepwise elution or gradient elution, provided that some precautions are taken. For example, in the isolation of extracts of *Ginkgo biloba*, one starts with water as a stationary phase, eluting with ethyl acetate with increasing amounts of i-butanol and finally reaching the 6 : 4 proportion of ethyl acetate : i-butanol at the end of the elution.[10]

Samples ranging in size from microgram to gram quantities can be separated with the range of available instruments. Analytical-scale CCC also was carried out for the separation of flavonoids.[11]

CONCLUSIONS

The application of CCC to the separation of flavonoids has been proven to be very successful. Chloroform–methanol–water can be chosen as starting point and, by modifying the relative proportions of methanol or by replacing methanol with other solvents, it is possible finally to obtain the required distribution of sample components between the two phases. EtOA : PrOH : H_2O and n-C_6H_{14} : EtOAc : CH_3OH : H_2O also are very useful solvent systems. The technique is versatile and can be employed for the initial fractionation of crude extracts for the separation of closely related flavonoids and/or the isolation of pure products.

ACKNOWLEDGMENT

This work is supported by the National Natural Science Foundation, Yunnan Province Natural Science Foundation, and TRAPOYT of China.

REFERENCES

1. Ito, Y. High-speed countercurrent chromatography. Crit. Rev. Anal. Chem. **1986**, *17*, 65–143.
2. Marston, A.; Slacanin, I.; Hostettmann, K. Centrifugal partition chromatography in the separation of natural products. Phytochem. Anal. **1990**, *1*, 3–17.
3. Marston, A.; Hostettmann, K. Counter-current chromatography as a preparative tool—applications and perspectives. J. Chromatogr. A. **1994**, *658*, 315–341.
4. Yuan, L.M.; Fu, R.N.; Zhang, T.Y. Separation of bioactive components from medicinal plants by high-speed countercurrent chromatography. Chin. J. Med. Anal. **1998**, *18* (1), 60–64.
5. Hostettmann, K.; Hostettmann, M.; Marston, A. Counter-current chromatography. In *Preparative Chromatography Techniques*; Springer-Verlag: Berlin, 1986; 80–126.
6. Ito, Y.; Conway, W.D. High-speed counter-current chromatography of natural products. In *High-Speed Countercurrent Chromatography*; Wiley-Interscience: New York, 1996; 189–251.
7. Slacanin, I.; Marston, A.; Hostettmann, K. Modifications to a high-speed counter-current chromatograph for improved separation capability. J. Chromatogr. **1989**, *482*, 234–239.
8. Marston, A.; Borel, C.; Hostettmann, K. Separation of natural products by centrifugal partition chromatography. J. Chromatogr. **1988**, *450*, 91–99.
9. Yuan, L.M.; Ai, P.; Chen, X.X.; Zi, M.; Wu, P.; Li, Z.Y.; Chen, Y.G. Versatile two-phase solvent system for flavonoid prefractionation by high-speed countercurrent chromatography. J. Liq. Chromatogr. & Relat. Technol. **2002**, *25* (5), 10.
10. Vonhaelen, M.; Vanhaelen-Fastre, R. Counter-current chromatography for isolation of flavonol glycosides from *Ginkgo biloba* leaves. J. Liq. Chromatogr. **1988**, *11*, 2969–2975.
11. Zhang, T.Y.; Xiao, R.; Xiao, Z.Y.; Pannell, L.K.; Ito, Y. Rapid separation of flavonoids by analytical high-speed counter-current chromatography. J. Chromatogr. **1988**, *445*, 199–206.

Flow Field-Flow Fractionation: Introduction

Myeong Hee Moon
Pusan National University, Pusan, Korea

INTRODUCTION

Flow field-flow fractionation (flow FFF or FlFFF) is one of the FFF subtechniques in which particles and macromolecules are separated in a thin channel by aqueous flow under a field force generated by a secondary flow. As with other FFF techniques, separation in FlFFF is based on the applied force directed across the axis of separation flow. In FlFFF, this force is generated by cross-flow of liquid delivered across the channel walls. In order to maintain the uniformity of cross-flow moving in a typical rectangular channel, two ceramic permeable frits are used as channel walls and the flow stream enters and exits through these walls. The force applied in FlFFF is a Stokes force that depends only on the sizes of sample components.

PRINCIPLES

In FlFFF, particles or macromolecules entering the channel are driven toward an accumulation wall by the cross-flow. Normally, a sheet of semipermeable membrane is placed at the accumulation wall in order to keep sample materials from being lost by the wall. While sample components are being transported close to the accumulation wall, they are projected against the wall by Brownian diffusion. The diffusive transport against the wall leads the sample components to be differentially distributed against the wall, according to their sizes: The larger particles, having a small diffusion coefficient, are placed at an equilibrium position closer to the vicinity of accumulation wall than the smaller ones. Thus, small particles, which are located further from the wall, will be exposed to the fast streamline of a parabolic flow profile, and they will be eluted earlier than the larger ones. This is the typical elution profile that can be observed in the normal operating mode of FFF (denoted as Fl/Nl FFF). Retention time in Fl/Nl FFF is inversely proportional to the diffusion coefficient of the sample; it is represented as

$$t_r = \frac{w^2}{6D} \frac{\dot{V}_c}{\dot{V}} \left(\text{where } D = \frac{kT}{3\pi\eta d_s} \right) \tag{1}$$

where w is the channel thickness, D is the diffusion coefficient, $\dot{V}_c$ is the cross-flow rate, and $\dot{V}$ is the channel flow rate. Because the diffusion coefficient D is

inversely proportional to the viscosity of carrier solution η and hydrodynamic radius d_s, the retention time can be simply predicted provided the particle diameter or the diffusion coefficient is known. Conversely, the particle diameter of an unknown sample can be calculated from experimental retention time by rearranging Eq. 1.

As the particle size becomes large at or above 1 μm, the diffusional process of particles becomes less dominant in FFF. In this regime, a particle's retention is largely governed by the particle size itself, in which the center of large particles is located at a higher position than small ones. Thus, large particles meet the faster streamlines and they elute earlier than the small ones; the elution order is reversed. However, it is known, from experimental results, that particles migrate at certain positions elevated from the wall due to the existence of hydrodynamic lift forces that act in the opposite direction to the field. This is described as the steric/hyperlayer operating mode of separation in flow FFF and is denoted by Fl/Hy FFF. Whereas the theoretical expectation of particle retention in Fl/Nl FFF is clearly understood, retention in Fl/Hy FFF is not predictable because the hydrodynamic lift forces are not yet completely understood. Therefore, the particle size calculation in Fl/Hy FFF relies on the calibration process in which a set of standard latex particles of known diameter is run beforehand as

$$\log t_r = -S_d \log d_s + \log t_{r1} \tag{2}$$

where S_d is the diameter-based selectivity and t_{r1} is the interpolated intercept representing the retention time of a unit diameter. The S_d values found experimentally are about 1.5 in Fl/Hy FFF. By using Eq. 2, the particle diameters of unknown samples can be calculated once the calibration parameters S_d and t_{r1} are provided.

TYPES OF CHANNEL IN FlFFF

There are two main categories of flow FFF channel systems, depending on the use of frit wall. The above-described flow FFF system has a frit on both walls; this is classified as a symmetrical channel, as shown in Fig. 1a. An asymmetrical channel system is being widely studied in which only one permeable frit

Encyclopedia of Chromatography DOI: 10.1081/E-ECHR-120039985

(a) symmetrical, rectangular

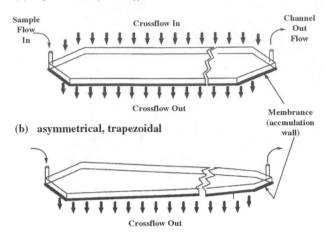

(b) asymmetrical, trapezoidal

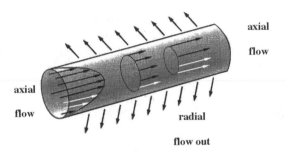

(c) hollow fiber

Fig. 1 Types of channel in FFF.

wall is used, at the accumulation wall, and the depletion wall is replaced with a glass plate (Fig. 1b). In an asymmetrical channel, part of the flow entering the channel is lost by the accumulation wall and this acts as a field force to retain the sample components in the channel, as does the cross-flow in a symmetrical channel.

The separation efficiency of an asymmetrical flow FFF system has been known to be higher than that of a conventional symmetrical channel. Because an asymmetrical channel utilizes only one frit, nonuniformity of flow that could arise from the imperfection of frits can be reduced. In addition, the initial sample band can be kept narrower in an asymmetrical channel, due to the focusing–relaxation procedure, which is an essential process in an asymmetrical channel. The relaxation processes, which provide an equilibrium status for sample components, are necessary in both symmetrical and asymmetrical channels for a period of time prior to the separation. For a symmetrical channel, this is normally achieved by stopping channel flow immediately after sample injection, while the cross-flow is applied.

During the relaxation process, sample components seek their equilibrium positions where the drag of the cross-flow is counterbalanced with diffusive transports (or lift forces) against the walls. After relaxation, flow is resumed and separation begins. However, in an asymmetrical channel, the relaxation process is achieved by two convergent focusing flow streams originating at the channel inlet and outlet (focusing–relaxation). Thus, injected sample can be focused at a certain position near the inlet end and the broadening of the initial sample band can be better minimized. This will lead to a decrease in band broadening of an eluted peak in an asymmetrical channel.

In asymmetrical flow FFF, two channel designs are utilized: rectangular and trapezoidal. Because flow velocity decreases along the axis of migration, a trapezoidal channel in which the channel breadth decreases toward the outlet is known to be more efficient in eluting low-retaining materials such as high-molecular-weight proteins. Retention in an asymmetrical flow FFF system follows the basic FFF principle and the retention time is calculated as

$$t_r = \frac{w^2}{6D} \ln\left(1 + \frac{\dot{V}_c}{\dot{V}_{\text{out}}}\right) \tag{3}$$

where $\dot{V}_c$ is the cross-flow rate and $\dot{V}_{\text{out}}$ is the outlet flow rate.

In addition to the rectangular channels in FlFFF described thus far, a cylindrical channel system has been developed with the use of hollow fibers in which the fiber wall is made of a porous membrane, as shown in Fig. 1c. It also requires a focusing–relaxation process, as does an asymmetrical channel. Retention in hollow-fiber flow FFF (HF-FlFFF) is controlled by the radial flow, which effectively acts as the cross-flow of a conventional flow FFF system, and the retention in a hollow fiber resembles that of an asymmetrical channel system.

However, the retention ratio in HF-FlFFF is approximately 4λ for a sufficiently retained component, which is somewhat different from that of a conventional channel system ($R \cong 6\lambda$). The retention time in a hollow fiber is calculated as

$$t_r = \frac{r_f^2}{8D} \ln\left(1 + \frac{\dot{V}_{\text{rad}}}{\dot{V}_{\text{out}}}\right) \tag{4}$$

where r_f^2 is the radius of the fiber and $\dot{V}_{\text{rad}}$ is the radial flow rate. Although a number of experiments have indicated a great potential of hollow fibers as an alternative for a flow FFF channel, a great deal of study related to their performance and optimization is needed.

SUGGESTED FURTHER READING

Giddings, J.C. Field flow fractionation: separation and characterization of macro molecular, colloidal, and particulate materials. Science **1993**, *260*, 1456.

Giddings, J.C.; Yang, F.J.; Myers, M.N. Theoretical and experimental characterization of flow field-flow fractionation. Anal. Chem. **1976**, *48*, 1126.

Jönsson, J.A.; Carlshaf, A. Flow field flow fractionation in hollow cylindrical fibers. Anal. Chem. **1989**, *61*, 11.

Litzén, A.; Wahlund, K.-G. Improved separation speed and efficiency for proteins, nucleic acids and viruses in asymmetrical flow field flow fractionation. J. Chromatogr. **1989**, *476*, 413.

Litzén, A.; Wahlund, K.-G. Zone broadening and dilution in rectangular and trapezoidal asymmetrical flow field-flow fractionation channels. Anal. Chem. **1991**, *63*, 1001.

Moon, M.H.; Kim, Y.H.; Park, I. Size characterization of liposomes by flow field-flow fractionation and photon correlation spectroscopy: effect of ionic strength and pH of carrier solutions. J. Chromatogr. **1998**, *813*, 91.

Ratanathanawongs, S.K.; Giddings, J.C. *Chromatography of Polymers: Characterization by SEC and FFF*; ACS Symposium Series; Provder, T., Ed.; American Chemical Society: Washington, D.C., 1993; Vol. 521, 13–29.

Fluorescence Detection in Capillary Electrophoresis

Robert Weinberger
CE Technologies, Inc., Chappaqua, New York, U.S.A.

INTRODUCTION

One cannot overestimate the importance of fluorescence detection in high-performance-capillary electrophoresis (HPCE).[1] The success of the human genome project along with the forthcoming revolutions in forensic testing and genetic analysis might not have occurred without the sensitivity and selectivity of laserinduced fluorescence (LIF) detection.

BASIC CONCEPTS

The stunning sensitivity of fluorescence detection arises from two areas: (a) detection is performed against a very dark background and (b) the use of the laser as an excitation source provides a high photon flux. The combination of the two can yield single-molecule detection in exceptional circumstances, although picomolar ($10^{-12}M$) is typically obtained. Under conditions that are easy to replicate, LIF detection is often times more sensitive compared to ultraviolet (UV) absorption detection.

Most molecules absorb light in the ultraviolet or visible portion of the spectrum, but only few produce significant fluorescence. This provides for the extreme selectivity of the technique. Molecular fluorescence is usually quenched through vibronic or collosional events resulting in a radiationless decay of excited singlet-state energy to the ground state. In aromatic structurally rigid molecules, quenching is less significant and the quantum yield increases.

The selectivity of fluorescence is to itself a problem because the technique is applicable to fewer separations. Sophisticated derivatization schemes have been developed for these applications to take advantage of the attributes contributed by fluorescence detection. Because there are two instrumental parameters to adjust, the excitation and emission wavelengths, the inherent selectivity of the method is further enhanced.

The fundamental equation governing fluorescence is

$$I_f = \Phi_f I_0 abc E_x E_c E_m E_{\text{pmt}}$$

where I_f is the measured fluorescence intensity, Φ_f is the quantum yield (photons emitted/photons absorbed), I_0 is the excitation power of the light source,

a, b, and c are the Beer's Law terms, and the E terms are the efficiencies of the excitation monochromator or filter, the optical portion of the capillary, the emission monochromator or filter, and the detector (photomultiplier or charge-coupled device), respectively. It is no wonder why optimization of fluorescence detection is difficult for the uninitiated.

EXCITATION SOURCES

The optimal excitation wavelength is usually a combination of the power of the light source and the molar absorptivity of the solute at the selected wavelength. The argon-ion laser is used for most DNA applications since the primers, intercalators, and dye terminators have been optimized for 488-nm excitation. For other applications, particularly for small molecules, where native fluorescence is measured, a tunable light source is desirable. The deuterium lamp is useful for low-UV excitation and the xenon arc is superior in the near-UV to visible region. With a 75-W xenon arc, the limit of detection (LOD) is 2 ng/mL ($6 \times 10^{-9}M$) for fluorescein using fiber-optic collection of the fluorescence emission.[2] This is a 100-fold improvement compared to absorption detection. By using a microscope objective to focus the light along with a sheath-flow cuvette (to reduce scattering, see below) and lens to collect the light, the LOD is reduced to 8×10^{-11}.[3] Nevertheless, the LOD using conventional tunable sources will never be superior to that found with the laser.

It is possible to select lasers other than the argonion laser for LIF detection. A 625-nm diode laser is available on a commercial unit (Beckman P/ACE and MDQ). Tunable dye lasers would be desirable but cost and reliability has precluded widespread use. The KrF laser is particularly useful because it emits in the UV at 248 nm. If fiber optics are employed to direct the laser light, then a UV transparent fiber optic must be used. A table of lasers and their wavelengths of emission is given in Table 1.

Lower-power lasers are often used in HPCE. Because scattered light is the factor that often limits detectibility, raising the power levels is ineffective. At high laser power, photobleaching becomes more likely to occur as well.

Encyclopedia of Chromatography DOI: 10.1081/E-ECHR-120039986

Table 1 Laser light sources for LIF detection

Laser	Available wavelengths
Ar ion (air-cooled)	457, 472, 476, 488, 496, 501, 514
Ar ion (full frame)	275, 300, 305, 333, 351, 364, 385, 457, 472, 476, 488, 496, 501, 514
Ar ion (full frame, frequency doubled)	229, 238, 244, 248, 257
ArKr	350–360, 457, 472, 476, 488, 496, 501, 514, 521, 514, 521, 531, 568, 647, 752
HeNe	543, 594, 604, 612, 633
Excimer	
XeCl (pulsed)	308
KrF (pulsed)	248
Nitrogen (pulsed)	337
Nitrogen-pumped dye (tunable)	360–950
Solid state	
YAG (frequency doubled)	532
YAG (frequency quadrupled)	266
Diode lasers	
Frequency doubled (LiNbo$_3$)	415
Frequency doubled (KTP)	424
Frequency tripled (Nd-doped YLiF)	349

(From Ref.[13]).

METHODS FOR COLLECTING FLUORESCENT EMISSION

The goal here is to minimize the collection of scattered radiation and optimize the collection of emitted fluorescence. Scattered radiation comes from two sources: Rayleigh scattering and Raman scattering. Rayleigh scattering occurs at the wavelength of excitation. To optimize the LOD, virtually all of this radiation must be excluded from detection. Raman scattering is observed at longer wavelengths than Rayleigh scattering and it is times less intense. Despite the weakness of Raman scattering, this effect can significantly elevate the background if left unchecked. Bandpass and/or cutoff filters are often used to reduce the impact of scattering. It is important to ascertain that the selected filter does not fluoresce as well.

Fiber optics held at right angles to the capillary can be employed to route emitted light toward the photomultiplier tube (PMT).[2] The Beckman LIF detector employs a collecting mirror to increase the

amount of collected emission. One problem with both of these approaches is the failure to prevent small amounts of scattered light from reaching the PMT. Cutoff and/or bandpass filters are not 100% efficient in this regard. This is particularly important when lasers are used because of the intense scattering of light.

The sheath-flow design is an important advance in reducing scattering because detection occurs after the solutes have exited the capillary.[4] Scattering occurs whenever a refractive index (RI) change occurs in the optical path. These RI changes include the air–capillary interface and the buffer–capillary interface. Eliminating the capillary from the optical path effectively removes four scattering surfaces. This becomes most important in multiple-capillary systems such as the DNA sequencer because many surfaces are now involved. The sheath-flow device patented in 1998[5] is illustrated in Fig. 1 for a five-capillary system. In actual practice, 96 capillaries are employed. The laser beam is sufficiently strong that attenuation is not significant or at least can be compensated for in the software. Fluorescence from each capillary is then imaged onto a charge-coupled device (CCD) camera.

For single-capillary systems, a conventional PMT is used for detection. Light is routed to that PMT either with fiber optics, a collecting mirror, epi-illumination microscopy, or a microscope objective. For multiple-capillary systems, the system must be scanned[6] or the light imaged onto a CCD camera.

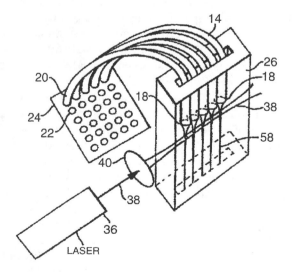

Fig. 1 Multiple-capillary instrument employing the sheath-flow technique. Key: 14, capillary; 18, capillary outlet; 20, capillary inlet; 22, buffer well; 24, microtiter plate; 26, quartz chamber; 36, laser; 38, laser beam; 40, lens; 58, fluidic stream. The electrodes are not shown nor is the device for delivering the sheath fluid. (Reprinted in part from U.S. Patent No. 5,741,412.)

DERIVATIZATION

Derivatization is important in capillary electrophoresis to enhance the detectibility of solutes that are non-fluorescent.[7] The chemistry can occur precapillary, on capillary, postcapillary. Typically, the solutes are amino acids, catecholamines, peptides, or proteins, all of which contain primary or secondary amine groups.

Reagents such as *ortho*-phthaldehyde (OPA), naphthalenedialdehyde (NDA), 3-(4-carboxy-benzoyl)-2-quinoline carboxaldehyde (CBQCA), fluorescein, and fluorenylmethyl chloroformate (FMOC) are all useful for precapillary derivatization, the most common of the three techniques. For carbohydrates, reagents such as aminopyrene naphthalene sulfonate (APTS) are used for precapillary derivatization. For chiral recognition, prederivatization with optically pure fluorenylethyl chloroformate (FLEC) provides for both enantioseparation by micellar electrokinetic capillary chromatography (MECC or MEKC) and a tag that absorbs at 260 nm and emits above 305 nm. Reagents for derivatizing carbonyl, hydroxyl, and other functional groups are also available.

For on-capillary and postcapillary derivatization, the reagent must not fluoresce until reacted with the solute. For these purposes, NDA and OPA are the best choices. With on-capillary derivatization, it is possible to use a reagent that fluoresces, but its removal prior to solute detection can be difficult.

The advantage of precapillary and on-capillary derivatization is the lack of the need for additional instrumentation beyond the basic HPCE instrumentation. The disadvantage of precapillary derivatization is the need for extra sample-handling steps. For postcapillary derivatization, the need for additional miniaturized instrumentation is the principle disadvantage. This problem may be overcome when dedicated microfabricated systems become available.

Important non-DNA application areas for precapillary derivatiation with LIF detection include the determination of amino acids and amines in cerebrospinal fluid to distinguish disease states such as Alzheimer's disease and leukemia from the normal population. In vivo monitoring of microdialysates from the brain of living animals has been employed for the determination neuropeptides, amphetamine, neurotransmitters, and amino acids. The contents of single neurons and red blood cells have been studied as well.

A variant of postcapillary derivatization is chemiluminescence (CL) detection.[8] In this case, the chemical reaction replaces the light source for excitation. The detector is a PMT run at high voltage. Solutes can be tagged with CL reagents such as luminol or directly excited via the peroxyoxalate reaction. The latter works best for aminoaromatic hydrocarbons such as dansylated amines. The LODs using CL detection approach laser levels because of the low background. However, the need for specialized apparatus has limited the applicability of CL detection.

FLUORESCENCE DETECTION FOR MICROFABRICATED SYSTEMS

The so-called micro-total analytical systems (mTAS) can integrate sample handling, separation, and detection on a single chip.[9] Postcapillary reaction detectors can be incorporated as well.[10] Fluorescence detection is the most common method employed for these chip-based systems. A commercial instrument (Agilent 2100 Bioanalyzer) is available for DNA and RNA separations on disposable chips using a diode laser for LIF detection. In research laboratories, polymerase chain reaction (PCR) has been integrated into a chip that provides size separation and LIF detection.[11]

INDIRECT FLUORESCENCE DETECTION

When detecting solutes that neither absorb nor fluoresce, indirect detection can be employed. With this technique, a reagent is added to the background electrolyte that absorbs or fluoresces and is of the same charge for the solute being separated. This reagent elevates the baseline. When solute ions are present, they displace the additive as required by the principle of electroneutrality. As the separated ions migrate past the detector window, they are measured as negative peaks relative to the high baseline. The advantage of indirect fluorescence compared to indirect absorption is an improved LOD.

The sensitivity of indirect detection is given by the following equation:[12]

$$C_{LOD} = \frac{C_R}{(DR)(TR)}$$

where the CLOD is the concentration limit of detection, is the concentration of the reagent, D_R is the dynamic reserve, and TR is the transfer ratio. Thus, the lowest CLOD occurs when the reagent concentration is minimized.

With 100 μm fluorescein, a mass limit of detection of 20 μM was measured for lactate and pyruvate in single red blood cells. Fluorescein is a good reagent because it absorbs at 488 nm and thus matches the argon-ion laser emission wavelength. In indirect absorption detection, the additive concentration is usually 5–10 mM. Band broadening due to electrodispersion is less unimportant in indirect fluorescence detection because the solute concentration is so low.

At higher solute concentrations, the system will be less useful because of electrodispersion. The concentration of the indirect reagent could be increased, but then indirect absorption detection becomes applicable.

With the advent of microfabricated systems that employ LIF detection, it is expected that indirect fluorescence will gain importance as a general-purpose detection scheme.

REFERENCES

1. MacTaylor, C.E.; Ewing, A.G. Critical review of recent developments in fluorescence detection for CE. Electrophoresis **1997**, *18*, 2279.

2. Albin, M.; Weinberger, R.; Sapp, E.; Moring, S. Fluorescence detection in capillary electrophoresis: evaluation of derivatizing reagents and techniques. Anal. Chem. **1991**, *63*, 417.

3. Arriaga, E.; Chen, D.Y.; Cheng, X.L.; Dovichi, N.J. High-efficiency filter fluorometer for capillary electrophoresis and its application to fluorescein thiocarbamyl amino acids. J. Chromatogr. **1993**, *652*, 347.

4. Cheng, Y.F.; Dovichi, N.J. SPIE **1988**, *910*, 111.

5. Dovichi, N.J.; Zhang, J.Z. U.S. Patent 5,741,412 (April 21, 1998).

6. Huang, X.C.; Quesada, M.A.; Mathies, R.A. Capillary array electrophoresis using laser-excited confocal fluorescence detection. Anal. Chem. **1992**, *64*, 967.

7. Bardelmeijer, H.A.; et al. Pre-on and postcolumn-derivatization in capillary electrophoresis. Electrophoresis **1997**, *18*, 2214.

8. Staller, T.D.; Sepaniak, M.J. Chemiluminescence detection in capillary electrophoresis. Electrophoresis **1997**, *18*, 2291.

9. Manz, A.; et al. Capillary electrophoresis integrated onto a planar microstructure (review). Analusis **1994**, *22*, M25.

10. Jacobson, S.C.; Koutny, L.B.; Hergenroeder, R.; Moore, A.W., Jr. Anal. Chem. **1994**, *66*, 4372.

11. Waters, L.C.; et al. Microchip device for cell lysis, multiplex PCR amplification, and electrophoretic sizing. Anal. Chem. **1998**, *70*, 158.

12. Yeung, E.S.; Kuhr, W.G. Anal. Chem. **1991**, *63*, 275.

13. Schwartz, H.E.; Ulfelder, K.J.; Chen, F.-T.A.; Pentoney, J. J. Capillary Electrophoresis **1994**, *1*, 36.

F

Fluorescence Detection in HPLC

Ioannis N. Papadoyannis
Anastasia Zotou
Aristotle University of Thessaloniki, Thessaloniki, Greece

INTRODUCTION

Detection based on analyte fluorescence can be extremely sensitive and selective, making it ideal for trace analysis and complex matrices. Fluorescence has allowed LC to expand into a high-performance technique. HPLC procedures with fluorescence detection are used in routine analysis for assays in the low nanogram per milliliter range and concentrations as low as picogram per milliliter often can be measured. The linearity range for these detectors is similar to that of ultraviolet (UV) detectors (i.e., 10^3–10^4).

DISCUSSION

One major advantage of fluorescence detection is the possibility of obtaining three orders of magnitude increased sensitivity over absorbance detection and its ability to discriminate analyte from interference or background peaks. Contrary to absorbance, fluorescence is a "low-background" technique. In an absorbance detector, the signal measured is related to the difference in light intensity in the presence of the sample versus the signal in the absence of the sample. For traces of analyte, this difference becomes extremely small and the noise level of the detector increases significantly. In a fluorescence detector, however, the light emitted from the analyte is measured against a very low-light (dark) background and, thus, against a very low noise level. The result is a much lower detection limit, which is limited by the electronic noise of the instrument and the dark current of the photomultiplier tube.

Another major advantage of fluorescence detection is selectivity. The increased selectivity of fluorescence versus absorbance is mainly due to the following reasons: (a) Most organic molecules will absorb UV/visible light but not all will fluoresce. (b) Fluorescence makes use of two different wavelengths (excitation and emission) as opposed to one in absorbance, thus decreasing the chance of detecting interfering chromatographic peaks.

Quantitative analysis can be performed with fluorescence detection even when poor column resolution occurs, provided there is enough detection selectivity to resolve the peaks.

One of the weak points of fluorescence is that relatively few compounds fluoresce in a practical range of wavelengths. However, chemical derivatization allows many nonfluorescent molecules containing derivatizable functional groups to be detected, thus expanding the number of applications. Fluorescence derivatization can be accomplished either via precolumn or postcolumn methods.

THEORETICAL BACKGROUND OF FLUORESCENCE DETECTION

Fluorescence is a specific type of luminescence. When a molecule is excited by absorbing electromagnetic radiation (a photon) supplied by an external source (i.e., an incandescent lamp or a laser), an excited electronic singlet state is created. Eventually, the molecule will attempt to lower its energy state, either by reemitting energy (heat or light) by internal rearrangement or by transferring the energy to another molecule through a molecular collision. This process distinguishes fluorescence from chemiluminescence, in which the excited state is created by a chemical reaction. If the release of electromagnetic energy is immediate or stops upon the removal of the excitation source, the substance is said to be fluorescent.

In fluorescence, the excited state exists for a finite time (1–10 ns). If, however, the release of energy is delayed or persists after the removal of the exciting radiation, then the substance is said to be phosphorescent.

Once a photon of energy $h\nu_{ecx}$ excites an electron to a higher singlet (absorbance) state (1 fs), emission of the photon $h\nu_{em}$ occurs at longer wavelengths. This is due to the competing nonradiative processes (such as heat or bond breakage) occurring during energy deactivation. The difference in energy or wavelength represented by $h\nu_{em} - h\nu_{exc}$ is called the *Stokes shift*.

The fluorescence signal, I_f, is given by

$$I_f = \varphi I_0\left(1 - e^{-kcl}\right)$$

where φ is the quantum yield (the ratio of the number of photons emitted to the number of photons

Encyclopedia of Chromatography DOI: 10.1081/E-ECHR-120039987

absorbed), I_0 is the intensity of the incident light, c is the concentration of the analyte, k is the molar absorbance, and l is the path length of the cell.

With few exceptions, the *fluorescence excitation spectrum of a single fluorophore in dilute solution is identical to its absorption spectrum*. Under the same conditions, the *fluorescence emission spectrum is independent of the excitation wavelength*, due to the partial dissipation of the excitation energy during the excited lifetime. The emission intensity is proportional to the amplitude of the fluorescence excitation spectrum at the excitation wavelength.

DEACTIVATION PATHWAYS IN FLUORESCENCE

The excited state exists for a finite time (1–10 ns) during which the fluorophore undergoes conformational changes and is also subject to several interactions with its molecular environment. The processes which deactivate the excited state may be radiational or nonradiational (see Fig. 1) and are the following.

Internal Conversion

A transition from a higher (S_3, S_2) to the first singlet excited energy state occurs (S_1) through an internal conversion (in 1 ps). Internal conversion is increased with increasing solvent polarity.

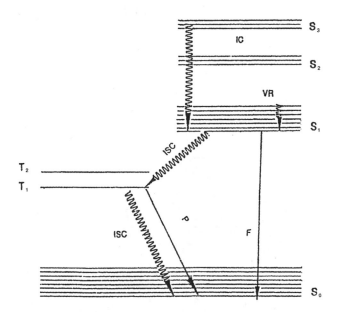

Fig. 1 Deactivation pathways in fluorescence; S_1, S_2, and S_3 are singlet excited states; S_0 is the ground state; T_1 and T_2 are triplet excited states; VR is vibrational relaxation, IC is internal conversion, ISC is intersystem crossing, P is phosphorescence, and F is fluorescence.

External Conversion (Quenching)

This is a chemical or matrix effect and can be defined as a bimolecular process that reduce the fluorescence quantum yield without changing the emission spectrum. Fluorescence radiation is transferred to foreign molecules after collisions.

Vibrational Relaxation

The energy of the first excited singlet state is partially dissipated through vibrations, yielding a relaxed singlet excited state. *Increased vibrations lower the fluorescence intensity*, due to the fact that they occur much faster (1 ps) than the fluorescence event. The molecular structure itself will determine the amount of vibrations. Rigid and planar molecules usually do not favor vibrations and they are prone to fluoresce.

Intersystem Crossing (Photobleaching)

This is a nonradiational process under high-intensity illumination conditions and in the same timescale as fluorescence (1–10 ns). It is defined as a transition from the first excited singlet (S_1) to the excited triplet (T_1) state. This is a "forbidden" transfer and necessitates the change of electron spin. *The quantum yield of fluorescence is reduced and phosphorescence also occurs.*

Phosphorescence

This event occurs due to a radiational relaxation to the ground singlet (S_1) state and in the 0.1 ms to 10 s time frame. Therefore, the emission is at even longer wavelengths than in fluorescence. Energy addition to the molecule in the form of heat or collisions of two triplet-state molecules can cause delayed fluorescence.

FACTORS AFFECTING FLUORESCENCE

Molecular structure and environmental factors such as acidity, solvent polarity, and temperature variations exert significant influence on fluorescence intensity. Also, variations in mobile-phase composition will cause excitation and emission-wavelength changes in the fluorophore.

Molecular Structure

Common fluorophores possess aromaticity and electron-donating substituents on the ring. Only compounds with a high degree of conjugation will fluoresce. The possible molecular transitions resulting in fluorescence are $\sigma \rightarrow \sigma^*$, occurring only on alkanes in the vacuum UV region, and $\pi \rightarrow \pi^*$ with very high extinction coefficients, occurring in alkenes, carbonyls, alkynes, and azo compounds. The majority of strong fluorophores undergo this transition and the excited state is more polar than the ground state.

Solvent Polarity

Polar solvents affect the excited state differently in $\pi \rightarrow \pi^*$ and $n \rightarrow \pi^*$ transitions. The excited state in $\pi \rightarrow \pi^*$ transition is stabilized. A reduction in the energy gap will occur and the emission will be shifted to a longer wavelength (red shift). Therefore, the difference between excitation and emission wavelengths will be greater in polar solvents.

Temperature

A rise in temperature increases the rate of vibrations and collisions, resulting in increased intersystem crossing, internal and external conversion. Consequently, the fluorescence intensity is inversely proportional to the temperature increase. Additionally, an increased temperature causes a red shift of the emission wavelength.

Acidity

Acidity can drastically affect the fluorescence intensity. The pK_a of concern is the pK_a of the excited state. Because protonation is faster than fluorescence, the pK_a can be quite different than it is for the molecule in the ground state. Therefore, a pH optimization versus fluorescence intensity is needed for molecules that are particularly prone to pH changes.

FLUORESCENCE DETECTOR INSTRUMENTATION

Fluorescence detectors for HPLC use come in many designs from the manufacturer. Differences in detector design can lead to markedly different results during interlaboratory comparisons.

Fluorescence detectors are based either on the straight-path design (similar to UV photometers) or on the more often encountered right-angle design. The common excitation source lamps used are continuous deuterium, xenon, xenon–mercury, and pulsed xenon. Recently, the use of high-power light sources for excitation, such as laser sources, allows the development of much smaller volume flow cells with less scatter (noise), resulting in improved efficiency. Photomultiplier

tubes are commonly used as the photodetectors (photocells) versus photodiodes in UV detectors. They convert a light signal to an electronic signal.

Detector flow cells are the link between the chromatographic system and the detector system. The cell cuvettes are made of quartz, with either cylindrical or square shapes and volumes between 5 and 20 μL. The sensitivity is directly proportional to the volume. However, resolution decreases with increasing volume. Fluorescence is normally measured at an angle perpendicular to the incident light. An angle of 90° has the lowest scatter of incident light. However, fluorescence from the flow cell is isotropic and can be collected from the entire 360°.

With the straight-path design, a standard UV cell can be used, but the filters must be selected so as to prevent stray light from reaching the photodetector. The right-angle design often uses a cylindrical cell. This design is less efficient than the straight-path cell because light-scattering problems result in a lower light intensity reaching the photodetector. However, this design is less susceptible to interference from stray light from the lamp, because the photodetector is not in line with the lamp.

With respect to monochromator type, three general detector designs are available: filter–filter, grating–filter, and grating–grating, where either a filter or monochromator grating is used to select the correct excitation and emission wavelengths. Gratings allow a choice of any desired wavelength, whereas filters are limited to a single wavelength.

Fluorescence detectors that use *filters* to select excitation and emission wavelengths are called *filter fluorometers*. This type of detector is the most sensitive,

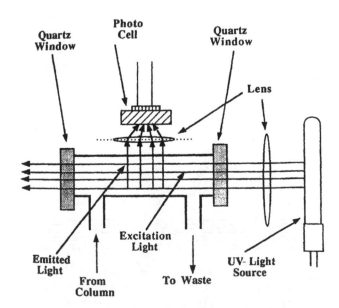

Fig. 2 Schematic of a single-wavelength Fluorescence detector.

yet the simplest and least expensive. A diagram of this simple form of fluorescence detector is shown in Fig. 2. Usually, in order to enhance the fluorescence collected from the flow cell, lenses are employed along with filters. The lenses are positioned before the excitation filter and after the flow cell to focus and collect the light.

The ultimate in fluorescence detection is a detector that uses a diffraction grating to select the excitation wavelength and a second grating to select the wavelength of the fluorescent light. These dual monochromatic *grating–grating* fluorescence detectors are called *spectrofluorometers*. If the gratings are used in the scanning mode, the detector is a *scanning spectrofluorometer*. A fluorescence or excitation spectrum can be provided by arresting the flow (stop-flow technique) of the mobile phase when the solute resides in the detecting cell or by scanning the excitation or fluorescent light, respectively. In this way, it is possible to obtain excitation spectra at any chosen fluorescent wavelength or fluorescence spectra at any chosen excitation wavelength.

The *grating–filter* detector is a hybrid between the filter–filter and the grating–grating types. Both high sensitivity and intermediate selectivity are achieved. The use of a filter in combination with gratings is ideal for lowering the background.

Grating–grating fluorometers are convenient for method development, because they permit selection of any excitation or emission wavelength. Filter–filter instruments, on the other hand, are simpler, easier in use, less expensive, more sensitive, and better suited for transferring an HPLC method between laboratories.

With the vast development of technology, fluorescence detectors have become programmable. Optimization of wavelength-pair maxima for each analyte can be time programmed during the chromatographic run.

The proper use of fluorescence detectors necessitates knowledge and understanding of noise sources. Dual monochromatic detectors have stray light leakage. When the wavelength pair is close, the background noise can significantly limit the detection limit. The *stray light*, along with *reflection* and *scattering*, increases the blank signal, resulting in reduced signal-to-noise ratio. Reflection occurs at interfaces that have a difference in the refractive index. Scattering can be of *Rayleigh* or *Raman* type.

In Rayleigh scatter, the wavelength of the absorbed and emitted photons are the same. Ultraviolet wavelengths scatter more than visible. Rayleigh scatter can be a significant problem when the wavelength pair overlaps (less than 50 nm) and instruments do not have filter accommodations and adjustable slits.

Raman scatter can also be troublesome. Depending on the wavelength pair of the sample, Raman scatter

from the mobile phase can overlap the fluorescence signal and, thus, can be misdiagnosed as the fluorescence signal itself. This problem arises during increasing instrument sensitivity. However, satisfactory separation can be achieved by changing the excitation wavelength because emission is independent of the excitation wavelength.

To summarize, in terms of instrumental operation, the following practices should be followed: proper zeroing of the blank and nontampering with the gain during serial dilutions. Increased sensitivity should be accomplished by varying the full-scale range.

The basic sequence in instrumental adjustments is to select the minimum gain necessary to allow a full-scale deflection, at the least sensitive scale. When linear curves are prepared, the gain need not be adjusted. Amplification should always be done using the range control. Any small changes in the gain during calibration will cause nonlinearity. Once the gain has been set, the zero can be set. To ensure reproducibility, zeroing the detector from time to time during the day is recommended, because the dark current can change during the day.

SUGGESTED FURTHER READING

Dolan, J.W.; Snyder, L.R. *Troubleshooting LC Systems*; Humana Press: Clifton, NJ, 1989; 337–339.

Gilbert, M.T. *High Performance Liquid Chromatography*; IOP Publishing: Bristol, U.K., 1987; 34–35.

Hancock, W.S.; Sparrow, J.T. *HPLC Analysis of Biological Compounds, A Laboratory Guide*; Marcel Dekker, Inc.: New York, 1984; 166–169.

Haugland, R.P. *Handbook of Fluorescent Probes and Research Chemicals*, 6th Ed.; Molecular Probes, Inc.: Eugene, OR, 1996; 1–4.

O'Flaherty, B. Fluorescence detection. In *A Practical Guide to HPLC Detection*; Parriott, D., Ed.; Academic Press: San Diego, CA, 1993; 111–139.

Papadoyannis, I.N. *HPLC in Clinical Chemistry*; Marcel Dekker, Inc.: New York, 1990; 74–75.

Scott, R.P.W. *Techniques and Practice of Chromatography*; Marcel Dekker, Inc.: New York, 1995; 288–292.

Scott, R.P.W. *Chromatographic Detectors, Design, Function and Operation*; Marcel Dekker, Inc.: New York, 1996; 199–211.

Snyder, L.R.; Kirkland, J.J. *Introduction to Modern Liquid Chromatography*, 2nd Ed.; John Wiley & Sons: New York, 1979; 145–147.

Snyder, R.L.; Kirkland, J.J.; Glajch, J.L. *Practical HPLC Method Development*; John Wiley & Sons: New York, 1997; 81–84.

Foam Countercurrent Chromatography

Hisao Oka
Aichi Prefectural Institute of Public Health, Nagoya, Japan

Yoichiro Ito
*Laboratory of Biophysical Chemistry, National Heart, Lung and Blood Institute,
National Institutes of Health, Bethesda, Maryland, U.S.A.*

INTRODUCTION

When a foam moves through a liquid, it carries particles caught at its interface, resulting in accumulation of these particles at the surface. For many years, this phenomenon has been utilized for the separation of minerals and metal ions. Since the method only employs inert gas and aqueous solution, it should have great potential for the separation of biological samples. This idea has been materialized using the high-speed countercurrent chromatographic (CCC) system. In this foam CCC method, foam and liquid undergo rapid countercurrent movement through a long, fine Teflon tube (2.6 mm I.D. × 10 m) under a centrifugal force field. This foam CCC technology has been applied to the separation of a variety of samples.

APPARATUS OF FOAM CCC

Fig. 1A illustrates a cross-sectional view of the foam CCC apparatus. The rotary frame holds a coiled separation column and a counterweight symmetrically at a distance of 20 cm from the central axis of the centrifuge. When the motor drives the rotary frame, a set of gears and pulleys produces synchronous planetary motion of the coiled column in such a manner that the column revolves around the central axis of the centrifuge while it rotates about its own axis at the same angular velocity in the same direction. The rotating force field resulted from this planetary motion induces countercurrent movement between the foam and its mother liquid through a long, narrow, coiled tube. Introduction of a sample mixture into the coil results in the separation of sample components. The foam active components are quickly carried with the foaming stream and are collected from one end of the coil, while the rest moves with the liquid stream in the opposite direction and is collected from the other end of the coil.

Fig. 1B illustrates the column design for foam CCC. The coiled column consists of a 10 m long, 2.6 mm I.D. Teflon tube of 50 ml capacity. The column is equipped with five flow channels. The liquid is fed from the liquid feed line at the tail and collected from the liquid collection line at the head. Nitrogen gas is fed from the gas feed line at the head and discharged through the foam collection line at the tail, while the sample solution is introduced through the sample feed line in the middle portion of the coil. The head–tail relationship of the rotating coil is conventionally defined by an Archimedean screw force, where all objects of different density are driven toward the head. Liquid feed rate and sample injection rate are each separately regulated with a needle valve, while the foam collection line is left open to the air.

APPLICATION

Foam CCC can be applied to two types of samples with: 1) affinity to the foam producing carrier; and 2) direct affinity to the gas–liquid interface.

Foam Separation Using Surfactants

This technique was demonstrated for the separation of methylene blue and dinitrophenyl (DNP)-leucine having affinity to the foam producing carrier. Sodium dodecyl sulfate (SDS) and cetyl pyridinium chloride (CPC) were used as carriers to study the effects of their electric charges on the foam affinity of various compounds. When the sample mixture was introduced with the anionic SDS surfactant, the positively charged methylene blue was adsorbed onto the foam and quickly eluted through the foam collection line while the negatively charged DNP-leucine was carried with the liquid stream in the opposite direction and eluted through the liquid collection line. Similarly, when the same sample mixture was introduced with the cationic CPC surfactant, the negatively charged DNP-leucine was totally eluted through the foam collection line and the positively charged methylene blue through the liquid collection line.

Encyclopedia of Chromatography DOI: 10.1081/E-ECHR-120039988

F

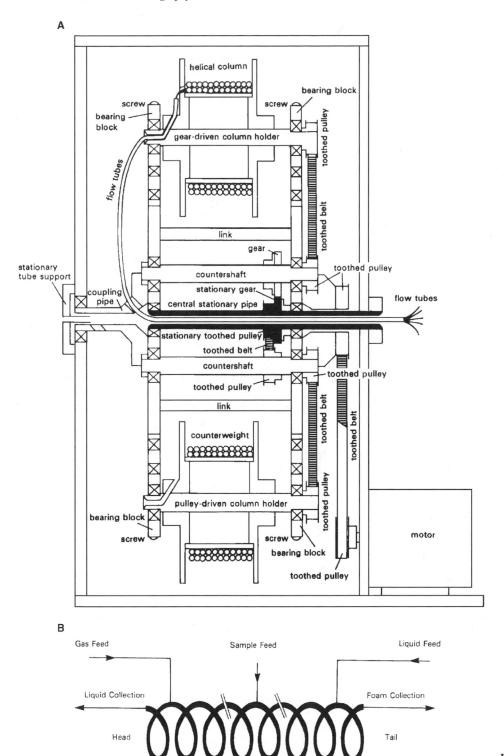

Fig. 1 A Foam CCC apparatus; B column design for foam CCC.

Foam Separation Without Surfactant

Many natural products have foaming capacity so that foam CCC may be performed without surfactant. This possibility was demonstrated using bacitracin complex (BC) as a test sample because of its strong foaming capacity. Bacitracin complex is a basic cyclic peptide antibiotic consisting of more than 20 components, but except for the major components BCs-A and -F, the chemical structures of the other components are still unknown.

The foam CCC experiment for the separation and enrichment of BC components was conducted using nitrogen gas and distilled water entirely free of surfactant or other additives.

Batch sample loading

Foam CCC of BC components was initiated by simultaneously introducing distilled water through the liquid feed line at the tail and nitrogen gas through the gas feed line at the head into the rotating column, while the needle valve in the liquid collection line was fully open. After steady state hydrodynamic equilibrium was reached, the pump was stopped and the sample solution was injected into the sample feed line at the middle of the column. After the lapse of a predetermined standing time, the needle valve opening was adjusted to the desired level and pumping was resumed. Effluents were collected at 15 sec intervals. The bacitracin components were separated in the order of hydrophobicity of the molecule in the foam fractions, with the most hydrophobic compounds being eluted first. This method can also be applied to continuous sample feeding as described below.

Continuous sample feeding

The experiment was initiated by introducing nitrogen gas into the gas feed line at the head of the rotating column. Then, a 2.5 L volume of the BC solution was continuously introduced into the coil through the sample feed line at 1.5 ml/min. The hydrophobic components produced a thick foam that was carried with the gas stream and collected from the foam collection line at the tail, while the other components stayed in the liquid stream and eluted from the liquid collection line at the head. HPLC analysis of the foam fraction revealed that the degree of enrichment increased with the hydrophobicity of the components. These results clearly indicate that the present method is quite effective for the detection and isolation of small amounts of natural products present in a large volume of aqueous solution.

Recycling sample injection

In this system, the effluent from the liquid outlet is directly returned into the column through the sample feed line so that the sample solution is continuously recycled for repetitive foam fractionation (Fig. 2). The utility of this system was demonstrated in the separations of microcystin extract and bacitracin complex from large volumes of sample solution. Microcystins were separated and enriched in decreasing order of hydrophobicity. Bacitracin A, a hydrophobic major component, in the bacitracin complex was highly enriched in the foam fraction and almost completely isolated from other components. This recycling foam CCC method may be effectively applied for the separation and enrichment of various foam-active components from crude natural products.

The general procedure for foam CCC using this recycling sample injection is as follows: 1) Clamp the liquid feed inlet (no liquid feed is employed); 2) rotate the column at 500 rpm; 3) fully open the needle valve; 4) introduce nitrogen gas at 80 psi through the gas feed line; 5) introduce the sample at a flow rate of 9.0 ml/min through the sample feed line for 20 min; 6) stop the pumping; 7) close the needle valve; 8) resume the pumping at a flow rate of 1.0 ml/min; 9a) when foam emerges, fractionate effluents from foam outlet at 2.5 min intervals; 10) increase the flow rate to 1.5 ml/min; 9b) when failing to elute at step 8, increase the flow rate to 1.5 ml/min or until the foam is eluted.

Foaming parameters

For application of foam CCC to various natural products, it is desirable to establish a set of physicochemical parameters that reliably indicate their applicability to foam CCC. Two parameters were selected for this purpose, i.e., "foaming power" and "foam stability," which can be simultaneously determined by the following simple procedure. In each test, the sample solution (20 ml) is delivered into a 100 ml graduated cylinder with a ground stopper and the cylinder vigorously shaken for 10 sec. The foaming power is expressed as the volume ratio of the resulting foam to the remaining solution, and the foam stability by the duration of the foam.

In order to correlate the foaming parameters to the foam productivity in foam CCC, the following five samples were selected because of their strong foaming capacities: bacitracin, gardenia yellow, rose bengal, phloxine B, and senega methanol extract. The results of our studies indicated that a sample having foaming power greater than 1.0 and foam stability of over 250 min could be effectively enriched by foam CCC.

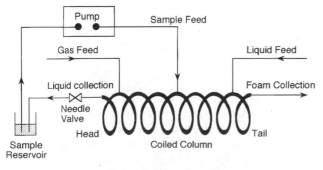

Fig. 2 Foam CCC system with recycle sample injection.

These minimum requirements of foaming parameters derived from the bacitracin experiment were found to be well applicable to four other samples.

CONCLUSIONS

Foam CCC can be successfully applied to a variety of samples having foam affinity with or without surfactants. The method offers significant advantages over conventional foam separation methods by allowing the efficient chromatographic separation of sample in both batch loading and continuous feeding. We believe that the foam CCC technique has great potential in the enrichment, stripping, and isolation of foam active components from various natural and synthetic products in both research laboratories and industrial plants.

ARTICLES OF FURTHER INTEREST

Countercurrent Chromatography Solvent Systems, p. 401.
Coil Planet Centrifuges, p. 349.

SUGGESTED FURTHER READING

Bhatnagar, M.; Ito, Y. Foam countercurrent chromatography on various test samples and the effects of additives on foam affinity. J. Liq. Chromatogr. **1988**, *11*, 21.

Ito, Y. Foam countercurrent chromatography: new foam separation technique with flow-through coil planet centrifuge. Sep. Sci. **1976**, *11*, 201.

Ito, Y. Foam countercurrent chromatography based on dual countercurrent system. J. Liq. Chromatogr. **1985**, *8*, 2131.

Ito, Y. Foam countercurrent chromatography with the cross-axis synchronous flow-through coil planet centrifuge. J. Chromatogr. **1987**, *403*, 77.

Oka, H.; Harada, K.-I.; Suzuki, M.; Nakazawa, H.; Ito, Y. Foam countercurrent chromatography of bacitracin with nitrogen and additive-free water. Anal. Chem. **1989**, *61*, 1998.

Oka, H.; Harada, K.-I.; Suzuki, M.; Nakazawa, H.; Ito, Y. Foam countercurrent chromatography of bacitracin I. Batch separation with nitrogen and water free of additives. J. Chromatogr. **1989**, *482*, 197.

Oka, H.; Harada, K.-I.; Suzuki, M.; Nakazawa, H.; Ito, Y. Foam countercurrent chromatography of bacitracin II. Continuous removal and concentration of hydrophobic components with nitrogen gas and distilled water free of surfactants or other additives. J. Chromatogr. **1991**, *538*, 213.

Oka, H. Foam countercurrent chromatography of bacitracin complex. In *High-Speed Countercurrent Chromatography*; Ito, Y., Conway, W.D., Eds.; John Wiley & Sons, Inc.: New York, 1996; 107–120 (Chapter 5).

Oka, H.; Iwaya, M.; Harada, K.-I.; Muarata, H.; Suzuki, M.; Ikai, Y.; Hayakawa, J.; Ito, Y. Effect of foaming power and foam stability on continuous concentration with foam countercurrentchromatography. J. Chromatogr. A. **1997**, *791*, 53.

Oka, H.; Iwaya, M.; Harada, K.-I.; Suzuki, M.; Ito, Y. Recycling foam countercurrent chromatography. Anal. Chem. **2000**, *72*, 1490.

Focusing Field-Flow Fractionation of Particles and Macromolecules

Josef Janča
Université de La Rochelle, La Rochelle, France

INTRODUCTION

The original idea of focusing field-flow fractionation (focusing FFF)[1] was introduced in 1982. Giddings[2] proposed the same principle in 1983 under the name hyperlayer FFF. A more detailed methodology of focusing FFF was developed later by the exploitation of various separation mechanisms.[3] The emerging discipline of isoperichoric focusing FFF represents a generalization of the original concept.[4]

The principle of focusing FFF is different from that of the polarization FFF. The crucial difference between the focusing and polarization mechanisms is that the intensity and direction of the driving field force must be dependent on the position across the channel and converging in focusing FFF, whereas it is position independent in polarization FFF. The sample components are focused at different altitudes across the channel and, consequently, eluted at different velocities corresponding to their positions within the flow velocity profile in focusing FFF, as shown in Fig. 1A. Although focusing FFF is as yet in a stage of fundamental investigation, some applications concerning the fractionation of macromolecular and particulate species have been published.

METHODS AND TECHNIQUES

Focusing can take place only if a gradient of the effective forces exists and the magnitude of these converging forces is position dependent and is zero at the focusing point. Various combinations of fields and gradients determining the focusing FFF methods and techniques can be exploited, as demonstrated in the following subsections.

Effective Property Gradient of the Carrier Liquid Combined with a Field Action

The gradient of an effective property of the carrier liquid combined with the action of a field can lead to the focusing of macromolecules or particles. For example, a density gradient combined with a gravitational or centrifugal field leads to focusing of the species at their isopycnic positions, amphoteric species focus at their isoelectric points in a pH gradient combined with an electrical field, and so forth. All these phenomena are known under the general term *isoperichoric focusing*, introduced by Kolin.[5]

Usually, the same primary field forces as those that produce the effective property gradient, are used to generate the focusing. However, the use of secondary field forces of a different nature to generate the focusing phenomenon within the corresponding gradient established by the primary field is possible; for example, isopycnic focusing of large-sized uncharged particles due to a weak gravitational field force was found effective under dynamic focusing FFF conditions, where the density gradient was generated by an electrical field acting on small charged colloidal particles suspended in the carrier liquid.[6] The construction of the fractionation channel is extremely simple, as shown in Fig. 1B. This principle, applied under static or dynamic FFF conditions, is promising for high-performance analytical and micropreparative separations.

Although focusing under static conditions, without the action of perpendicularly (with respect to the focusing axis) applied bulk flow, can lead to good separation of the focused species, theoretical calculations, as well as experimental tests, have shown increase in resolution under the dynamic conditions of focusing FFF.

Cross-Flow Velocity Gradient Combined with a Field Action

The velocity gradient of the carrier liquid across the fractionation channel, generated by transversal flow through semipermeable walls, which opposes the action of an external field, can produce the focusing phenomenon. A longitudinal flow is applied simultaneously. This method is called elutriation focusing FFF and has been used to separate model mixtures of polystyrene latex particles and silica particles in a trapezoidal cross-section channel.[7] The principle of this fractionation channel is shown in Fig. 1C. A similar focusing FFF principle can be utilized in a rectangular cross-section channel with two opposite semipermeable walls if the flow rates through the walls are different.[8]

Encyclopedia of Chromatography DOI: 10.1081/E-ECHR-120039989

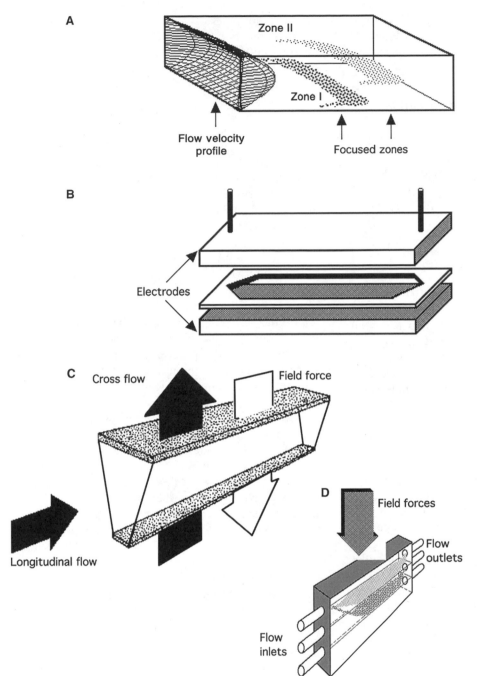

Fig. 1 Principle of focusing field-flow fractionation. (A) Section of the channel demonstrating the principle of focusing FFF with two distinct zones focused at different positions and the parabolic flow velocity profile. (B) Design of the channel for dynamic focusing FFF in coupled electrical and gravitational fields. (C) Schematic representation of the trapezoidal cross-section channel for elutriation focusing FFF. (D) Separation channel for continuous preparative focusing FFF operating in natural gravitational field with three inlet capillaries allowing one to preform the step density gradient by pumping three liquids of different densities and with three outlet capillaries to collect the separated fractions.

Lift Forces Combined with a Field Action

The hydrodynamic lift forces appearing at high flow rates of the carrier liquid, combined with field forces, are able to concentrate suspended particles into focused layers. While the field forces in polarization and steric FFF concentrate the retained species at the accumulation wall, the lift forces, becoming operational at high flow rates, pull the particles away from this wall. As a result, the transition from polarization or steric to focusing FFF appears first, followed by the proper focusing effect. In most published works,

the appearance of the lift forces generating the focusing effect has been observed in sedimentation and flow FFF. For example, Wahlund and Litzen[9] observed the interference of the lift forces in polarization flow FFF carried out in an asymmetrical channel with one semipermeable wall. Such interference perturbs the separation. However, in some cases, the lift forces have been exploited under optimized operational conditions and, as a result, high-speed and high-performance separations have been achieved. Two examples of such spectacular separations have been published for sedimentation FFF[10] and very recently for microthermal focusing FFF.[11] Although this focusing FFF method is just beginning to be used in real applications, its main advantages, high speed and high performance, seem to make it very promising.

Shear Stress Combined with a Field Action

A high shear gradient can lead to the deformation of macromolecular coils. The entropy gradient thus generated produces the driving forces that displace the macromolecules into a low-shear zone. The reversed elution order of high-molecular weight polystyrenes in thermal FFF at high flow rates could be attributed to this phenomenon,[12] but another possibility to explain the reversed elution order cannot be neglected.[13]

Gradient of a Nonhomogeneous Field Action

The use of a high-gradient magnetic field has been proposed to separate paramagnetic and diamagnetic species by a mechanism of focusing FFF.[14] Various aspects of focusing FFF carried out under these conditions have been discussed, but no experimental results have been published until now.

Preparative Fractionation

No principal difference distinguishes the analytical and preparative uses of focusing FFF. Both types of fractionation can be carried out under conditions of continuous operation,[15] which represents the high-performance experimental arrangement for preparative FFF. The fractionation channel, equipped with several outlet capillaries at various positions (and occasionally with several inlets to preform a stepwise gradient in the direction of the focusing), allows one to fractionate the sample, which is introduced continuously into the channel, and to collect the focused layers eluted at the individual outlets. A schematic representation of such a fractionation channel is shown in Fig. 1.

The experimental implementation of this technique has been demonstrated by the fractionation of various samples of silica particles by applying natural gravitation and a counteracting cross-flow gradient. The silica particles were separated according to size. Isopycnic or isoelectric focusing FFF, already performed on an analytical scale, can easily be transformed into such continuous large-scale separation.

APPLICATIONS

Focusing FFF represents an important contribution to the science and technology of separation and analysis of macromolecules of synthetic or natural origin. The range of molar masses and sizes of particles in submicron and micron ranges, supramolecular structures, organized biological species such as the cells and microorganisms, and so forth that can be fractionated by focusing FFF is very large.

Molecules that do not interact sufficiently with the imposed fields, such as low-molar-mass species, and that, consequently, do not exhibit the focusing effect can still be separated. The condition is that an equilibrium between them and the effectively focused species should be established. As a result, species that originally do not undergo the separation processes can be transported and thus fractionated with the "carrier" focused species.

The most important field of potential applications of focusing FFF is in research and technologies related to the life sciences and macromolecular chemistry. Problems related to trace analysis, which have enormous importance in the protection of the environment and many other scientific and technological activities, have already stimulated the development of new analytical separation methods. Focusing FFF is one of them, representing an alternative choice whenever macromolecular or particulate species are concerned.

The newest achievements in focusing FFF clearly indicate that most of the experimental implementations have been with model systems. Practical applications for daily laboratory use, elaborated to the minutest details, have rarely been described. However, the most significant advantages of these methods, already mentioned earlier, are evident. Some of these advantages are inherently related to the separation principle of focusing FFF, such as the absence of a large surface area inside the separation channel, which is of crucial importance for sensitive biological materials, which can be denatured by contact with active surfaces.

The operational variables, such as the strength of the field, the flow rate, and so forth, can be manipulated continuously in a very wide range. Another advantage is that although specific FFF apparatuses are already in production, the instruments commercially available for LC can easily be adapted for use with focusing FFF methodology. All specific components of the complete focusing FFF apparatuses are identical to those for LC, except the separation channel, which, in most cases, is not difficult to construct in the laboratory. Certainly, focusing FFF represents a large field of challenges, soliciting creativity and inventiveness in theory, methodology, and practical applications.

ARTICLE OF FURTHER INTEREST

Field-Flow Fractionation Fundamentals, p. 614.

REFERENCES

1. Janča, J. Sedimentation–flotation focusing field-flow fractionation. Makromol. Chem. Rapid Commun. **1982**, *3*, 887.
2. Giddings, J.C. Hyperlayer field-flow fractionation. Sep. Sci. Technol. **1983**, *18*, 765.
3. Janča, J. *Field-Flow Fractionation: Analysis of Macromolecules and Particles*; Marcel Dekker, Inc.: New York, 1988.
4. Janča, J. Isoperichoric focusing field-flow fractionation for characterization of particles and molecules. J. Liq. Chromatogr. & Rel. Technol. **1997**, *20*, 2555.
5. Kolin, A. *Electrofocusing and Isotachophoresis*; Radola, B.J., Graesslin, D., Eds.; de Gruyter: Berlin, 1977.
6. Janča, J.; Audebert, R. New concept in focusing field-flow fractionation and thin layer isopycnic focusing: coupling of primary electric field with secondary gravitational force. Mikrochim. Acta **1993**, *111*, 163.
7. Urbankova, E.; Janča, J. An attempt at experimental elutriation focusing field-flow fractionation. J. Liq. Chromatogr. **1990**, *13*, 1877.
8. Janča, J. Elutriation focusing field-flow fractionation. Makromol. Chem. Rapid Commun. **1987**, *8*, 233.
9. Wahlund, K.G.; Litzen, A. Application of an asymmetric flow field-flow fractionation channel to the separation and characterization of proteins, plasmids, plasmid fragments, polysaccharides, and unicellular algae. J. Chromatogr. **1989**, *461*, 73.
10. Koch, T.; Giddings, J.C. High speed separation of large ($>1\,\mu m$) particles by steric field-flow fractionation. Anal. Chem. **1986**, *58*, 994.
11. Janča, J.; Ananieva, I.A.; Menshikova, A.Yu.; Evseeva, T.G. Micro-thermal focusing field-flow fractionation. J. Chromatogr. B. **2004**, *800*, 33.
12. Giddings, J.C.; Li, S.; Williams, P.S.; Schimpf, M.E. High-speed separation of ultra-high molecular weight polymers by thermal/hyperlayer field-flow fractionation. Makromol. Chem. Rapid Commun. **1988**, *9*, 817.
13. Janča, J.; Martin, M. Influence of operational parameters on retention of ultra-high molecular weight polystyrenes in thermal field-flow fractionation. Chromatographia **1992**, *34*, 125.
14. Semyonov, S.N.; Kuznetsov, A.A.; Zolotaryov, P.P. Theoretical examination of focusing field-flow fractionation. J. Chromatogr. **1986**, *364*, 389.
15. Janča, J.; Chmelik, J. Focusing in field-flow fractionation. Anal. Chem. **1984**, *56*, 2481.

F

Food Colors: Analysis by TLC/Scanning Densitometry

Hisao Oka
Yuko Ito
Tomomi Goto
Aichi Prefectural Institute of Public Health, Tsuji-machi, Kita-ku, Nagoya, Japan

INTRODUCTION

Many synthetic and natural colors are used in foods all over the world. In Japan, 12 synthetic and 66 natural colors are generally permitted for use in foods. The Japanese government requires labeling on the package concerning kinds of colors that have been used in the contained foods. However, nonpermitted colors are also frequently detected in food, and also unlabeled foods are found in the market. Thus, the inspection of colors in foods has been performed by a public health agency.

The analyses of colors in foods have been mainly achieved by TLC, because TLC is a simple and effective technique for the separation of components in a mixture. However, the only useful information obtained from a TLC plate to identify a component is the Rf value; the identification of the separated components is difficult unless an appropriate spectrometric method, such as ultraviolet-visible absorption spectrometry, is used. A stepwise operation, including individual component separation by TLC and measurement of the spectrum, is laborious and time-consuming, because it requires extra steps such as extraction of the desired compound from the TLC plate and elimination of adsorbents.

TLC/scanning densitometry is a useful tool for the identification of the target compounds on a TLC plate, because the combined methods can separate and then directly measure ultraviolet-visible absorption spectra of the compounds without the laborious and time-consuming procedures described above. In this paper, we deal with the identification of synthetic and natural colors in foods using TLC/scanning densitometry.

SYNTHETIC COLORS

A simple and rapid identification method for synthetic colors in foods has been established, using TLC/scanning densitometry.[1] Forty-five synthetic colors were able to be completely separated on a C18 TLC plate by complementary use of the following four solvent systems: 1) acetonitrile–methanol–5% sodium sulfate (3:3:10); 2) methyl ethyl ketone–methanol–5% sodium sulfate (1:1:1); 3) acetonitrile–methanol–5% sodium sulfate (1:1:1); and 4) acetonitrile–dichloromethane–5% sodium sulfate (10:1:5). We measured the visible absorption spectra of the synthetic colors on the developed C18 TLC plates by scanning densitometry to identify them. The spectra of the colors purified from foods were in close agreement with those of the standard colors, and the reliability of identification was established.

Next, we successfully applied this technique to the identification of an unknown synthetic color in a pickled vegetable.[2] This color was suspected to be orange II (Or-II), which is not permitted for use in foods in Japan. It was difficult to identify Or-II by conventional analytical methods for food colors, including TLC and HPLC, because there are actually three isomers in total: orange I (Or-I), orange RN (Or-RN), and Or-II, due to differences in the positions of hydroxyl groups in the molecules. Under conventional TLC or HPLC conditions, it is hard to separate these isomers from each other. As shown in Fig. 1 (left), the unknown color showed the same Rf value as those of Or-II and Or-RN, although it showed a different Rf value from Or-I. In order to identify the unknown color, we measured the visible absorption spectrum of the color using TLC/scanning densitometry and compared it with those of Or-II and Or-RN. Both spectra of the unknown color and Or-II gave maximum absorption at only 485 nm; however, that of Or-RN showed maximum absorption at 485 and 400 nm. Therefore, we identified the unknown color in the pickled vegetable to be Or-II.

Thus, TLC/scanning densitometry is shown to be effective for the identification of an unknown synthetic color in foods.

NATURAL COLORS

Lac Color and Cochineal Color

Lac color is a natural food additive extracted from a stick lac, which is a secretion of the insect *Coccus*

Encyclopedia of Chromatography DOI: 10.1081/E-ECHR-120014243

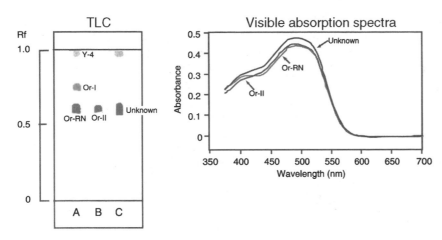

Fig. 1 TLC and visible absorption spectra of synthetic colors extracted from a pickled vegetable under TLC/scanning densitometry. (A) Standards of tartrazine (Y-4), orange I (Or-I), and orange RN (Or-RN). (B) Standard of orange II (Or-II). (C) Extract of the sample. TLC/scanning densitometric conditions. Plate: RP-18 (E. Merck). Solvent system: methyl ethyl ketone-methanol-5% sodium sulfate (1:1:1). Apparatus: Shimadzu CS-9000. Wavelength scanning range: 370–700 nm. Slit size: 0.4 × 0.4 mm. Measuring mode: reflecting absorption.

laccae (*Laccifer lacca* Kerr), and is widely used for coloring food. It is known that the red color is derived from a water-soluble pigment including laccaic acids A, B, C, and E. Cochineal color extracted from the dried female bodies of the scale insect (*Coccus cacti* L.) is water-soluble and has a reddish color. The main coloring component is carmic acid.

Because these colors are frequently used in juice, jam, candy, jelly, etc., it is required to establish a simple and rapid analysis method using TLC. However, as described in the "Introduction," the only useful information obtained from a TLC plate to identify a component is the Rf value. Therefore, we applied TLC/scanning densitometry to the identification of lac and cochineal colors in foods.[3]

TLC conditions

After various experiments, the best results were obtained using methanol–0.5 mol/L oxalic acid (5:4.5) as the solvent system, with a C18 TLC plate. As shown in Fig. 2 (left), the lac color standard was separated into two spots at Rf values of 0.60 and 0.29, and cochineal color standard gave a spot at Rf value of 0.52. Anthraquinone compounds, such as lac and cochineal colors, showed extreme tailing on the C18 TLC plate using conventional TLC conditions. We have previously found that the use of a solvent system containing oxalic acid is effective for controlling the tailing of anthraquinone compounds. Therefore, we decided to use a solvent system containing

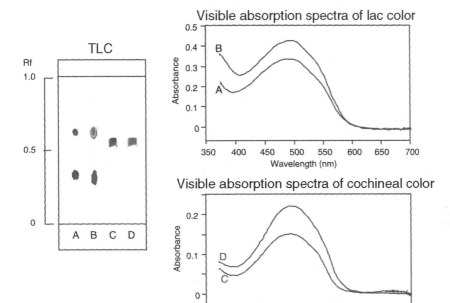

Fig. 2 TLC and visible absorption spectra of lac color and cochineal color extracted from commercial foods under TLC/scanning densitometric conditions. (A) Lac color standard. (B) Extract of jelly. (C) Cochineal color standard. (D) Extract of spagetti sauce. Plate: RP-18 (E. Merck). Solvent system: methanol-0.5 mol/L oxalic acid (5.5:4.5). Other conditions: see Fig. 1.

oxalic acid and tried various TLC conditions. Finally, we found the best conditions described above.

Measurement of visible absorption spectrum by scanning densitometry

Reflection spectra of the spots of lac color standard at Rf values of 0.60 and 0.29 and cochineal color standard at Rf value of 0.52 on the TLC plate were taken under the conditions described above. The obtained spectra showed good agreement with the spectra obtained from methanol solutions. Therefore, we considered that TLC/scanning densitometry is effective for the identification of these natural colors.

Application to commercial food

Reproducibility of the Rf Value by Reversed Phase TLC (RP-TLC).
In order to examine the effects of the contaminants contained in the sample on the Rf value, 122 commercial foods (41 foods for lac color and 81 foods for cochineal color) were analyzed by C18 TLC as described above. The obtained Rf values of the spots were then compared. The difference between the Rf value of the standard color and the Rf value of the color in the sample was expressed as the ratio between the Rf value of the color in the sample (Ra) and the Rf value of the standard color (Rs); the reproducibility was evaluated according to the coefficient of variation of this ratio.[4] With respect to lac color, the average Ra/Rs values were 0.99 with a coefficient of variation of 8.1% and 1.00 with 4.6% for spots at Rf values of 0.29 and 0.60, respectively. Cochineal color gave an average Ra/Rs value of 0.99 with a coefficient of variation of 5.9%. These results suggest that the spots extracted from the samples appear nearly at the same positions as those of the lac color and the cochineal color standard without being affected by contaminants in the sample, and that the identification of the color is reliable and reproducible.

Identification by TLC/Scanning Densitometry.
The visible absorption spectra of the spots of the lac and cochineal colors on the C18 TLC plates, for which the reproducibility of the Rf value had been evaluated, were measured using a scanning densitometer. Fig. 2 shows the typically obtained TLC chromatograms and spectra obtained from the spots. The spectra of the colors purified from foods were in good agreement with those of the standard colors; thus, the reliability of identification was then established.

Paprika Color

Paprika color is obtained by extraction from the fruit of red peppers (*Capsicum annuum*) and contains capsanthin and its esters, formed from acids, such as lauric acid, myristic acid, and palmitic acid, in large amounts as its color components. Commercially available paprika colors are known to have different compositions of these color components, depending on the material from which the paprika color is extracted; this makes the identification of paprika color, based on the analysis of the color components, impossible, causing difficulty in developing a simple, rapid, and reliable identification method for the paprika color in foods. Therefore, we investigated a TLC/scanning densitometric method for the identification of paprika color using capsanthin, which is a main product of saponification, as an indicator.[5]

TLC conditions

When a paprika color standard, before saponification, was subjected to C18 TLC, a number of overlapping spots were observed, and a satisfactory separation could not be obtained. This was probably due to the paprika color containing a large number of esters. Paprika color is known to be hydrolyzed into a carotenoid and a fatty acid when saponified under mild conditions. Thus, a paprika color standard, after saponification, was subjected to TLC using a solvent system of acetonitrile–acetone–*n*-hexane (11 : 7 : 2) on a C18 plate. It was found that the paprika color standard, after saponification, was satisfactorily separated into a main spot having an Rf value of 0.50 and two subspots having Rf values of 0.60 and 0.75 (Fig. 3A). The main spot was identical with the spot of the capsanthin standard in terms of its Rf value, color, and shape (Fig. 3C).

As described above, it was suggested that the paprika color is hydrolyzed into a carotenoid and a fatty acid by saponification under mild conditions. Next, the saponification conditions were investigated; based on various experimental results, the following saponification conditions were selected: reaction time, 24 hr; amount of 5% sodium hydroxide–methanol solution, 2 mL.

Measurement of visible absorption spectrum by scanning densitometry

The separated spots, obtained by subjecting a paprika color standard, after saponification, to C18 TLC under the conditions described above, were then subjected to scanning densitometry. The visible absorption spectra were scanned in the wavelength range of 370–700 nm, and excellent visible absorption spectra were obtained (Fig. 3A). The spectrum of the main spot (Rf = 0.50) of the paprika color, after saponification, showed its maximum absorption wavelength at 480 nm, which identically matches the spectrum of the capsanthin standard (Fig. 3C).

TLC

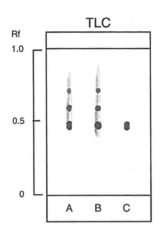

Visible absorption spectra

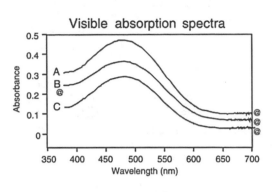

Fig. 3 TLC and visible absorption spectra of hydrolyzed paprika colors extracted from commercial foods under TLC/scanning densitometric conditions. (A) Hydrolyzed paprika color standard. (B) Hydrolyzed extract of rice-cracker. (C) Capsanthin. Plate: RP-18 (E. Merck). Solvent system: acetonitrile-acetone-n-hexane (11:7:2). Other conditions: see Fig. 1.

Application to commercial foods

Reproducibility of the Rf Value by RP-TLC. The paprika color in 42 samples from commercially available foods, that had a label stating the use of paprika color, were analyzed by C18 TLC to examine the influence of the coexisting substances from the sample on the Rf value. The obtained Rf values of the main spot (Rf = 0.50) of saponified paprika color were then compared, and Ra/Rs value was computed. The average Ra/Rs value was 1.01 with a coefficient of variation of 2.6%, suggesting that the spot extracted from the samples appear nearly at the same position as that of the paprika color standard without being affected by contaminants in the sample and that the identification of the color is reliable and reproducible.

Identification by RP-TLC/Scanning Densitometry. The visible absorption spectra of the main spot of the saponified paprika color on the C18 TLC plates, for which the reproducibility of the Rf value had been evaluated, were measured using a scanning densitometer. Fig. 3 shows the typically obtained TLC chromatograms and spectra obtained from the spots. The spectra of the colors purified from foods were in good agreement with those of the standard colors, and the identification reliability was then destablished.

Gardenia Yellow

Gardenia yellow is a yellow color obtained by extracting or hydrolyzing the fruit of the *Gardenia augusta* MERR. var. *gardiflora* HORT. with water or ethanol and is widely used for the coloring of noodles, candies, and candied chestnuts. The yellow color is derived from the carotinoids crocin and crocetin. Crocetin is the hydrolysis product of crocin. Gardenia yellow has been conventionally analyzed by a method based on reversed-phase chromatography/scanning densitometry using crocin as the indicator. However, when this method was applied to samples containing caramel or anthocyanins, their spots overlapped with that of crocin, which made it difficult to identify the gardenia yellow. Therefore, we evaluated an analytical method for gardenia yellow based on C18 TLC/scanning densitometry using crocetin as the indicator by hydrolyzing crocin, extracted from food samples, into crocetin.[6]

Hydrolysis and TLC conditions

In order to examine the optimal hydrolysis conditions of crocin, a standard crocin solution was hydrolyzed by varying the pH of the solution, temperature, and incubation time; the degree of hydrolysis was followed by C18 TLC as described below. Samples of crocin were completely hydrolyzed to crocetin by adjusting the pH to 11 or above with 0.1 mol/L sodium hydroxide and incubating them at 50°C for 30 min. Therefore, we applied these conditions to hydrolyze crocin to crocetin in the subsequent work.

Next, we investigated the optimal TLC conditions for the separation of crocin and crocetin and found that the combined use of a C18 TLC plate and solvent system of acetonitrile–tetrahydrofuran–0.1 mol/L oxalic acid (7:8:7) gave a satisfactory separation. Under these TLC conditions, crocin gives three spots at Rf values of 0.74, 0.79, and 0.83, and crocetin gives one spot at an Rf value of 0.51 (Fig. 4, left).

Measurement of visible absorption spectrum by scanning densitometry

Reflection spectra of the spots on the TLC plates separated under the conditions described above were measured at scanning wavelengths of 370–700 nm. Fig. 4 (right) shows the visible absorption spectra obtained; the maximum absorption wavelengths were 435 and 460 nm, being in complete agreement with the visible absorption spectrum for the standard preparation of crocetin.

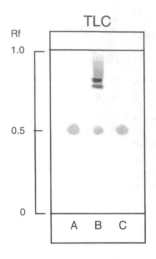

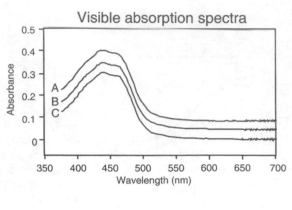

Fig. 4 TLC and visible absorption spectra of the hydrolyzed gardenia yellow extracted from commercial foods under TLC/scanning densitometry. (A) Hydrolyized gardenia yellow standard. (B) Candy containing gardenia yellow and anthocyanin. (C) Crocetin. Plate: RP-18 (E. Merck). Solvent system: acetonitrile-tetrahydrofuran-0.1 mol/L oxalic acid (7:8:7). Other conditions: see Fig. 1.

Application to commercial foods

As described above, foods that contained caramel or anthocyanins, for which the identification of gardenia yellow was impossible by the analytical method using crocin as an indicator due to the appearance of interfering spots at the same positions as the spots of crocin on the C18 TLC plates, were analyzed by the present method. As shown in Fig. 4 (left), crocetin appeared as a clear spot on the plate, and the shape and Rf value of the spot were in close agreement with those of the standard preparation. Hence, gardenia yellow can be identified using crocetin as the indicator.

Reproducibility of the Rf Value by RP-TLC. To examine the influence of the contaminants contained in the sample on the Rf value, 37 commercial foods were analyzed by C18 TLC. The obtained Rf values of the spots were then compared. The mean value of Ra/Rs was 0.99, and the coefficient of variation was 2.5%. These results suggest that the spots of crocetin generated by hydrolysis appear nearly at the same positions as those of the standard color, without being affected by contaminants in the sample, and that the identification of the color is reliable and reproducible.

Identification by RP-TLC/Scanning Densitometry. The visible absorption spectra of the crocetin spots on the reversed-phase TLC plates, for which the reproducibility of the Rf value had been evaluated, were measured using a scanning densitometer. Fig. 4 shows the typically obtained TLC chromatograms and spectra obtained. The spectra of the colors purified from foods were in close agreement with that of the standard dye, and the identification reliability was then established.

CONCLUSIONS

We introduced the identification of food colors in foods using TLC/scanning densitometry and consider the method to be sufficiently applicable to routine analyses at facilities such as the Centers of Public Health and the Food Inspection Office. Also, we consider that TLC/scanning densitometry is applicable to the identification of various food additives, drugs, and pesticides in foods. However, TLC/scanning densitometry has a limitation: It can be applied only to samples which have chromophores in the molecules. Recently, applications of the TLC/matrix-assisted laser desorption ionization time-of-flight mass spectrometry (TOF MS),[7] TLC/fast atom bombardment MS,[7] and TLC/multiphoton ionization TOF MS[8] have been reported. Combined uses of these techniques and TLC/scanning densitometry can develop further applications of TLC.

REFERENCES

1. Ohno, T.; Ito, Y.; Mikami, E.; Ikai, Y.; Oka, H.; Hayakawa, J.; Nakagawa, T. Identification of coal tar dyes in cosmetics and foods using reversed phase TLC/scanning densitometry. Jpn. J. Toxicol. Environ. Health **1996**, *42*, 53–59.
2. Ueno, E.; Ohno, T.; Oshima, H.; Saito, I.; Ito, Y.; Oka, H.; Kagami, T.; Kijima, H.; Okazaki, K. Identification of small amount of coal tar dyes in foods by reversed phase TLC/scanning densitometry with sample concentration techniques. J. Food Hyg. Soc. Jpn. **1998**, *39*, 286–291.
3. Itakura, Y.; Ueno, E.I.; Ito, Y.; Oka, H.; Ozeki, N.; Hayashi, T.; Yamada, S.; Kagami, T.; Miyazaki, Y.; Otsuji, Y.; Hatano, R.; Yamada, E.; Suzuki, R. Analysis of lac and cochineal colors in foods

using reversed phase TLC/scanning densitometry. J. Food Hyg. Soc. Jpn. **1999**, *40*, 183–188.

4. Ozeki, N.; Oka, H.; Ikai, Y.; Ohno, T.I.; Hayakawa, J.; Sato, T.; Ito, M.; Ito, Y.; Hayashi, T.; Yamada, S.; Kagami, T.; Miyazaki, Y.; Otsuji, Y.; Hatano, R.; Yamada, E.; Suzuki, R.; Suzuki, R. Application of reversed phase TLC to the analysis of coal tar dyes in foods. J. Food Hyg. Soc. Jpn. **1993**, *34*, 542–545.

5. Hayashi, T.; Ueno, E.; Ito, Y.; Oka, H.; Ozeki, N.; Itakura, Y.; Yamada, S.; Kagami, T.; Miyazaki, Y. Analysis of β-carotene and paprika color in foods using reversed phase TLC/scanning densitometry. J. Food Hyg. Soc. Jpn. **1999**, *40*, 356–362.

6. Ozeki, N.; Oka, H.; Ito, Y.; Ueno, E.; Goto, T.; Hayashi, T.; Itakura, Y.; Ito, T.; Maruyama, T.; Tsuruta, M.; Miyazawa, T.; Matsumoto, H. A reversed-phase thin-layer chromatography/scanning densitometric method for the analysis of gardenia yellow in food using crocetin as an indicator. J. Liq. Chromatogr. **2001**, *24*, 2849–2860.

7. Wilson, I.D. The state-of-the-art in thin-layer chromatography–mass spectrometry: A critical appraisal. J. Chromatogr. A. **1999**, *856*, 429–442.

8. Krutchinsky, A.N.; Dolgin, A.I.; Utsal, O.G.; Khodorkovski, A.M. Thin-layer chromatography–laser desorption of peptides followed by multiphoton ionization time-of-flight mass spectrometry. J. Mass Spectrom. **1995**, *30*, 375–379.

Forskolin Purification Using an Immunoaffinity Column Combined with an Anti-Forskolin Monoclonal Antibody

Hiroyuki Tanaka
Yukihiro Shoyama
Kyushu University, Fukuoka, Japan

INTRODUCTION

Forskolin, a labdane diterpenoid, was isolated from the tuberous roots of *Coleus forskohlii* Briq. (Lamiaceae).[1] *C. forskohlii* has been used as an important folk medicine in India. Forskolin was found to be an activator of adenylate cyclase,[2] leading to an increase of c-AMP, and now a medicine in India, Germany, and Japan. The production of forskolin is completely dependent on the commercial collection of wild and cultivated plants in India. We have already set up the production of monoclonal antibodies (MAbs) against forskolin.[3] The practical application of enzyme-linked immunosorbent assay (ELISA) for the distribution of forskolin contained in clonally propagated plant organs and the quantitative fluctuation of forskolin depend on the age of *C. forskohlii*.[4,5] As an extension of this approach, we present the production of the immunoaffinity column using anti-forskolin MAb and its application.[6]

MATERIALS AND METHODS

Chemicals

Bovine serum albumin (BSA) was provided by Pierce (Rockford, U.S.A.). Forskolin and 7-deacetyl forskolin were isolated from the tuberous root of *C. forskohlii*, as previously reported.[1] 1-Deoxyforskolin, 1,9-dideoxyforskolin, and 6-acetyl-7-deacetylforskolin were purchased from Sigma Chemical Company (St. Louis, MO, U.S.A.). The mixture (approximately 20 mg) of forskolin and 7-deacetyl forskolin, purified by the immunoaffinity column, was acetylated with pyridine and acetic anhydride mixture (each 100 ml) at 4°C for 2 h to give pure forskolin.

Preparation of Immunoaffinity Column Using Anti-Forskolin Monoclonal Antibody[6]

Purified IgG (10 mg) in PBS was added to a slurry of CNBr-activated Sepharose 4B (600 mg; Pharmacia Biotech) in coupling buffer (0.1 M NaHCO$_3$ containing 0.5 M NaCl). The slurry was stirred for 2 h at room temperature and then treated with 0.2 M glycine at pH 8.0 for blocking of activated groups. The affinity gel was washed four times with 0.1 M NaHCO$_3$ containing 0.5 M NaCl and 0.1 M acetate buffer (pH 4.0). Finally, the affinity gel was centrifuged and the supernatant was removed. The immunoaffinity gel was washed with phosphate buffer solution (PBS) and packed into a plastic mini-column in volumes of 2.5 ml. Columns were washed until the absorption at 280 nm was equal to the background absorption. The columns were stored at 4°C in PBS containing 0.01% sodium azide.

Direct Isolation of Forskolin from Crude Extractives of Tuberous Roots and Callus Culture of *C. forskohlii* by Immunoaffinity Column

The dried powder (10 mg dry weight) of tuberous root was extracted five times with diethyl ether (5 ml). After evaporation of the solvent, the residue was redissolved in MeOH and diluted with PBS (1 : 16), and then filtered by Millex-HV filter (0.45-µm filter unit; Millipore Products, Bedford, MA, U.S.A.) to remove insoluble portions. The filtrate was loaded onto the immunoaffinity column and allowed to stand for 90 min at 4°C. The column was washed with the washing buffer solution (10 ml). After forskolin disappeared, the column was eluted with PBSM (45%) at a flow rate of 0.1 ml/min. The fraction containing forskolin was lyophilized and extracted with diethyl ether. Forskolin was determined by TLC developed with C$_6$H$_6$–EtOAc (85 : 15) [Rf; forskolin (0.21), 7-deacetyl forskolin (0.16)] and ELISA.

RESULTS AND DISCUSSION

We established a simple and reproducible purification method for forskolin using an immunoaffinity column chromatography method. Because forskolin

Encyclopedia of Chromatography DOI: 10.1081/E-ECHR-120039991

is almost insoluble in water, various buffer solutions were tested for the solubilization of forskolin. It became evident that 6% MeOH in PBS was necessary for the solubilization of forskolin.[3,5] Next, the elution system for the immunoaffinity column was investigated by using various elution buffers based on PBS. Only 9% of bound forskolin can be recovered by the PBS supplemented with 10% of MeOH. The forskolin concentrations eluted increased rapidly from 20% of MeOH, and reached the optimum at 45% of MeOH.

To assess the capacity and the recovery of forskolin from the affinity column, 30 µg of forskolin was added and passed through the column (2.5 ml of gel), and the forskolin content was analyzed by ELISA. After washing with 5 column volumes of PBST, 22.5 µg of forskolin remained bound and was then completely eluted with the PBS containing 45% of MeOH. Therefore, the capacity of affinity column chromatography was determined to be 9.4 µg/mL.

The crude diethyl ether extracts of the tuberous root of *C. forskohlii* were loaded onto the immunoaffinity column chromatography system, washed five times with PBS containing 6% of MeOH, and eluted with the PBS containing 45% of MeOH. Fig. 1 shows a chromatogram detected by ELISA. Fractions 2–8

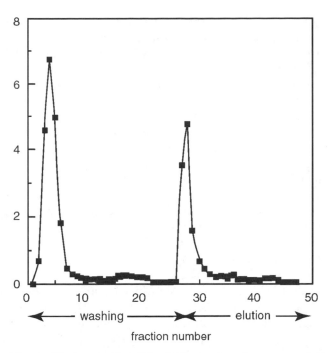

Fig. 1 Elution profile of forskolin in the tuberous root of *C. forskohlii* by purification on immunoaffinity column chromatography. The column was washed with PBSM, then eluted by PBS containing 45% of methanol after the forskolin disappeared. Individual fractions were assayed by ELISA.

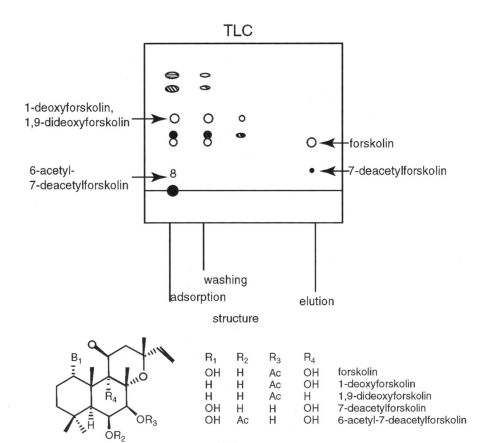

R_1	R_2	R_3	R_4	
OH	H	Ac	OH	forskolin
H	H	Ac	OH	1-deoxyforskolin
H	H	Ac	H	1,9-dideoxyforskolin
OH	H	H	OH	7-deacetylforskolin
OH	Ac	H	OH	6-acetyl-7-deacetylforskolin

Fig. 2 TLC of adsorption, washing, and elution solutions, and structures of forskolin and the related compounds.

contained 45 µg of forskolin that were over the column capacity, together with the related compounds 1-deoxyforskolin, 1,9-dideoxyforskolin, 7-deacetylforskolin and 6-acetyl-7-deacetylforskolin, and other unknown compounds which were detected by TLC, as indicated in Fig. 2. The peak of fractions 26–30 shows the elution of forskolin (21 µg) eluted with the PBS containing 45% of MeOH. Forskolin eluted by washing solution (fractions 2–8) was repeatedly loaded and finally isolated. However, forskolin purified by the immunoaffinity column chromatography was still contaminated with a small amount of 7-deacetyl forskolin (Fig. 2) because this compound has a 5.5% cross-reactivity against Mab, as previously indicated.[3] Therefore, the mixture was treated with pyridine and acetic anhydride at 4°C for 2 h to give pure forskolin. In our case, the stability of antibody against PBS containing 45% MeOH is also quite high, because the immunoaffinity column has been used over 10 times, under the same conditions, without any substantial loss of capacity. Therefore, we concluded that the PBS supplemented with 45% MeOH can be routinely used as an elution buffer solution.

REFERENCES

1. Bhat, S.V.; Bajwa, B.S.; Dornauer, H.; de Sousa, N.J.; Fehlhaber, H.W. Structures and stereochemistry of new labdane diterpiniods from coleus forskohlii briq. Tetrahedron Lett. **1977**, 1669–1672.
2. Metzger, H.; Lindner, E. The positive inotropic-acting forskolin, a potent adenylatecyclase activator. Drug Res. **1981**, *31*, 1248–1250.
3. Sakata, R.; Shoyama, Y.; Murakami, H. Cytotechnology **1994**, *16*, 101–108.
4. Yanagihara, H.; Sakata, R.; Shoyama, Y.; Murakami, H. Relationship between the content of forskolin and growth environments in clonally propagated *Coleus forskohlii* Brig. Biotronics **1995**, *24*, 1–6.
5. Yanagihara, H.; Sakata, R.; Shoyama, Y.; Murakami, H. Rapid analysis of small samples containing forskolin using monoclonal antibodies. Planta Med. **1996**, *62*, 169–172.
6. Yanagihara, H.; Minami, H.; Tanaka, H.; Shoyama, Y.; Murakami, H. Immunoaffinity column chromatography against forskolin using an anti-forskolin monoclonal antibody and its application. Anal. Chim. Acta **1996**, *335*, 63–70.

Frit-Inlet Asymmetrical Flow Field-Flow Fractionation

Myeong HeeMoon
Pusan National University, Pusan, South Korea

INTRODUCTION

Frit-inlet asymmetrical flow field-flow fractionation (FIA-FlFFF)[1–3] utilizes the frit-inlet injection technique, with an asymmetrical flow FFF channel which has one porous wall at the bottom and an upper wall that is replaced by a glass plate. In an asymmetrical flow FFF channel, channel flow is divided into two parts: axial flow for driving sample components toward a detector, and the cross-flow, which penetrates through the bottom of the channel wall.[4–5] Thus, the field (driving force of separation) is created by the movement of cross-flow, which is constantly lost through the porous wall of the channel bottom. FIA-FlFFF has been developed to utilize the stopless sample injection technique with the conventional asymmetrical channel by implementing an inlet frit nearby the channel inlet end and to reduce possible flow imperfections caused by the porous walls.

DISCUSSION

The asymmetrical channel design in flow FFF has been shown to offer high-speed and more efficient separation for proteins and macromolecues than the conventional symmetrical channel. However, an asymmetrical channel requires a focusing-relaxation procedure for sample components to reach their equilibrium states before the separation begins. The focusing-relaxation procedure is achieved by two counterdirecting flow streams from both the channel inlet and outlet to a certain point slightly apart from the channel inlet end for a period of time. This is a necessary step equivalent to the stop-flow procedure as is normally used in a conventional symmetrical channel system. Although the stop-flow and the focusing-relaxation procedures are essential in each technique (symmetrical and asymmetrical channels, respectively), they are basically cumbersome in system operation due to the stoppage of flow with valve operations. In addition, they often cause baseline shifts during the conversion of flow. For these reasons, the frit-inlet injection technique, which can be an alternative to bypass those flow-halting processes, is adapted to an asymmetrical flow FFF channel in order to take advantage of hydrodynamic relaxation of sample components.

The frit-inlet injection device was originally applied to the conventional symmetrical channel in order to bypass the stop-flow procedure.[6] However, the lowest axial flow rate that can be manipulated in a frit-inlet symmetrical system is limited, because the total axial flow rate becomes the sum of the injection flow rate and frit flow rate, and the incoming cross-flow penetrates through the bottom wall at the same rate. The relatively high axial flow rate in a symmetrical system needs a very high cross-flow rate in order to separate relatively low-retaining materials, such as proteins or low-molecular-weight components.

Compared to the limited choice in the selection of flow rate conditions, application of the frit-inlet injection technique to an asymmetrical flow FFF channel can be more flexible in allowing the selection of a low axial flow rate condition which is suitable for low-retaining materials without the need of using a very high cross-flow rates and for the reduction of injection amount resulting from the concentration effect.

In FIA-FlFFF, sample materials entering the channel are quickly driven toward the accumulation wall and are transported to their equilibrium positions by the compressing action of a rapidly flowing frit flow entering through the inlet frit. The schematic view of an FIA-FlFFF channel is shown in Fig. 1. In the relaxation segment of a FIA-FlFFF channel, the frit flow stream and the sample stream of relatively low speed will merge smoothly. During this process, sample materials are expected to be pushed below the inlet splitting plane formed by the compressing effect of frit flow, as illustrated in Fig. 1b. Thus, sample relaxation is achieved hydrodynamically in the relaxation segment (under the inlet frit region), and the sample components are continuously carried to the separation segment where the separation of sample components takes place. System operation requires only a simple one-step injection procedure, with no need for valve switching or interruption of flow. This is far simpler and more convenient than the operation of the conventional relaxation techniques, such as stop-flow and focusing-relaxation procedures.

In the first experimental work on FIA-FlFF,[1] the system efficiency was studied by examining the effect of the ratio of injection flow rate to frit flow rate on hydrodynamic relaxation; the initial tests showed a

Encyclopedia of Chromatography DOI: 10.1081/E-ECHR-120039993

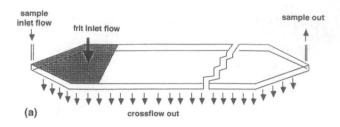

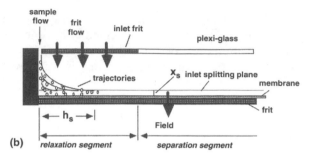

Fig. 1 Schematic view of an FIA-FlFFF channel.

possibility f using hydrodynamic relaxation in asymmetrical flow FFF with a number of polystyrene latex standards, in both normal and steric/hyperlayer modes of FFF. Normally, relaxational band broadening under hydrodynamic relaxation arises from a broadened starting band. The length of an initial sample band during hydrodynamic relaxation is dependent on flow rates as

$$h_s = \frac{\dot{V}_s}{\dot{V}_f} \frac{\dot{V}}{\dot{V}_c} L$$

where L is the channel length; $\dot{V}_s$, $\dot{V}_f$, $\dot{V}$, and $\dot{V}_c$ represent the flow rates of the sample stream, frit stream, effective channel flow, and cross-flow, respectively. Eq. (1) suggests that a small ratio of sample flow rate to frit flow rate, with a combined high cross-flow rate, is preferable in reducing h_s, leading to minimized relaxational band broadening.

Experimentally, the optimum ratio of $\dot{V}_s/\dot{V}_f$ has been found to be about 0.03–0.05 for the separation of latex beads and for proteins.

Retention in the separation segment of the FIA-FlFFF channel is expected to be equivalent to that observed in a conventional asymmetrical channel system, if complete hydrodynamic relaxation can be obtained. It will follow basic principles, as shown by the retention ratio, R, given by

$$R = \frac{t^0}{t_r} = 6\lambda \left[\coth\left(\frac{1}{2\lambda} - 2\lambda\right) \right] \left(\text{where } \lambda = \frac{D}{w^2} \frac{V^0}{\dot{V}_c} \right)$$

where t^0 is the void time, t_r is the retention time, λ is the retention parameter, D is the diffusion coefficient, w is the channel thickness, and V^0 is the channel void volume. The void time in an FIA-FlFFF channel system is complicated to calculate, because sample flow and frit flow enter the channel simultaneously, and part of the merged flow exits through the accumulation wall. For this reason, channel flow velocity varies along the axial direction of channel. By considering these, the determination of void time can be represented as

$$t^0 = \frac{V^0 A_f/A_c}{\dot{V}_f - \dot{V}_c A_f/A_c} \ln\left(\frac{\dot{V}_s + \dot{V}_f - \dot{V}_c A_f/A_c}{\dot{V}_s}\right)$$
$$+ \frac{\dot{V}^0}{\dot{V}_c} \ln\left(\frac{\dot{V}_s + \dot{V}_f - \dot{V}_c A_f/A_c}{\dot{V}_{out}}\right)$$

where $\dot{V}_{out}$ is the channel outflow rate and A_f and A_c are the area of the inlet frit and the accumulation wall, respectively. Eq. (3) represents the void time calculation in terms of volumetric flow rate and channel dimensions only; it is valid for any channel geometry, such as rectangular, trapezoidal, and even exponential design. Retention in FIA-FlFFF has been shown to follow the general principles of FFF with the confirmation of experimental work. It has also been found that the trapezoidal channel design provides a better resolving power for the separation of protein mixtures than a rectangular channel in FIA-FlFFF.

REFERENCES

1. Moon, M.H.; Kwon, H.S.; Park, I. Stopless flow injection in asymmetrical flow field-flow fractionation using a frit inlet. Anal. Chem. **1997**, *69*, 1436.
2. Moon, M.H.; Kwon, H.S.; Park, I. J. Liquid Chromatogr. & Related Technol. **1997**, *20*, 2803.
3. Moon, M.H.; Stephen Williams, P.; Kwon, H.S. Retention and efficiency in frit-inlet asymmetrical flow field-flow fractionation. Anal. Chem. **1999**, *71*, 2657.
4. Litzén, A.; Wahlund, K.-G. Zone broadening and dilution in rectangular and trapezoidal asymmetrical flow field-flow fractionation channels. Anal. Chem. **1991**, *63*, 1001.
5. Litzén, A. Separation speed, retention, and dispersion in asymmetrical flow field-flow fractionation as functions of channel dimensions and flow rates. Anal. Chem. **1993**, *65*, 461.
6. Giddings, J.C. Optimized field-flow fractionation system based on dual stream splitters. Anal. Chem. **1985**, *57*, 945.

Frontal Chromatography

Peter Sajonz
Merck Research Laboratories, Rahway, New Jersey, U.S.A.

INTRODUCTION

Frontal chromatography is a mode of chromatography in which the sample is introduced continuously into the column. The sample components migrate through the column at different velocities and eventually break through as a series of fronts. Only the least retained component exits the column in pure form and can, therefore, be isolated; all other sample components exit the column as mixed zones. The resulting chromatogram of a frontal chromatography experiment is generally referred to as a breakthrough curve, although the expression *frontalgram* has also been used in the literature.[1]

The exact shape of a breakthrough curve is mainly determined by the functional form of the underlying equilibrium isotherms of the sample components, but secondary factors such as diffusion and mass-transfer kinetics also have influence. The capacity of the column is an important parameter in frontal chromatography, because it determines when the column is saturated with the sample components and, therefore, is no longer able to adsorb more sample. The mixture then flows through the column with its original composition.

THE USE OF FRONTAL CHROMATOGRAPHY

Frontal chromatography can also be called *adsorptive filtration* because it can be used for the purpose of filtration. The purification of gases and solvents are two classical applications of frontal chromatography. Another important use is the purification of proteins, where a frontal chromatography step is used in the initialpurification procedure.[2,3]

One of the most important applications of frontal chromatography is the determination of equilibrium adsorption isotherms. It was introduced for this purpose by Shay and Szekely and by James and Phillips.[4,5] The simplicity as well as the accuracy and precision of this method are reasons why the method is so popular today and why it is often preferred over other chromatographic methods {e.g., elution by characteristic points (ECP) or frontal analysis by characteristic points (FACP).[6,7] Frontal chromatography as a tool for the determination of single-component adsorption isotherms will be discussed in the following section.

FRONTAL CHROMATOGRAPHY FOR THE DETERMINATION OF ISOTHERMS

Theory

First, the column is filled only with sample at concentration C_n; then, a step injection is performed (i.e., sample with the concentration C_{n+1} is introduced into the column). This results in a breakthrough curve, as shown in Fig. 1. The amount adsorbed at the stationary phase Q_{n+1} can be calculated by

$$Q_{n+1} = q_{n+1}Vs$$
$$= (C_{n+1} - C_n)(V_{R,n+1} - V_0) + q_n V_s$$

where q_n and q_{n+1} are the initial and final sample concentrations, respectively, in the stationary phase and C_n and C_{n+1} are the initial and final sample concentrations in the mobile phase, respectively. V_s is the volume of adsorbent in the column, V_0 is the holdup volume, and $V_{R,n+1}$ is the retention volume of the breakthrough curve. The retention volume is calculated from the area over the breakthrough curve:

$$V_R = \int_0^\infty \frac{(C_{n+1} - C)dV}{C_{n+1} - C_n}$$

The retention volume defined by the area method always gives the theoretically correct result for the amount adsorbed. In practice, it is, however, often easier and better to use the retention volume from half-height [i.e., at the concentration $(C_{n+1} + C_n)/2$] or the retention volume derived from the inflection point of the breakthrough curve. The reason for this is that the calculation of the area incorporates signal noise and it is very dependent on the integration limits. This is often a problem, especially when the mass transfer is slow, because, in this case, the plateau concentration C_{n+1} is only reached slowly and, therefore, systematic errors in the calculated area occur. It has been shown that the use of the retention volumes derived from the inflection point or the half-height gives satisfactory results. The halfheight method is,

Encyclopedia of Chromatography DOI: 10.1081/E-ECHR-120039994

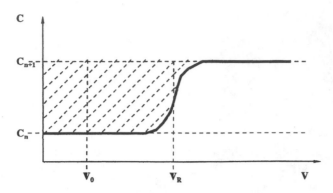

Fig. 1 Example of a frontal chromatography experiment; breakthrough curve of a single component.

however, easier to use and slightly more accurate than the inflection-point method.[5]

It has to be noted that the half-height and inflection-point methods do not give reliable results if the isotherm is concave upward and ascending concentration steps are performed. The same is true for a convex upward isotherm and descending concentration steps. The reason for this is that, in these cases, a diffuse breakthrough profile is obtained and, consequently, errors are made in the accurate determination of the retention volumes when they are derived from the half-height or the inflection point. The diffuse profile can, however, be used for the determination of isotherms by the frontal analysis by characteristic points method (FACP).

MODES OF FRONTAL ANALYSIS

There are two possibilities for performing a frontal chromatography experiment for the purpose of the determination of equilibrium isotherms. The step-series method uses a series of steps starting from $C_n = 0$ to C_{n+1}. After each experiment, the column has to be reequilibrated and a new step injection with a different end concentration C_{n+1} can be performed. In the staircase method, a series of steps is performed in a single run with concentration steps from 0 to C_1, C_1 to C_2, ..., C_n to C_{n+1}. The column does not have to be reequilibrated after each step and, therefore, the staircase method is faster than the step-series method. Both modes of frontal analysis give very accurate isotherm results.

Determination of Multicomponent Isotherms by Frontal Analysis

It is possible to extend the frontal chromatography method for the measurement of binary and multicomponent isotherms. In this case, the profiles are characterized by successive elution of several steep fronts. The use of these profiles for the determination of competitive isotherms in the binary case has been developed by Jacobsen et al.[8]

Combination of Frontal Analysis with Chromatographic Models

Frontal chromatography can be used in combination with chromatographic models to study mass-transfer and dispersion processes (e.g., the equilibrium dispersive or the transport model of chromatography.[7])

CONSTANT PATTERN, SELF-SHARPENING EFFECT, SHOCK-LAYER THEORY

Frontal chromatography generally requires the adsorption isotherm to be convex upward if the step injection is performed with ascending concentration (i.e., $C_{n+1} > C_n$) because, in this case, the profile of a breakthrough curve tends asymptotically toward a limit. After this constant profile has been reached, the profile migrates along the column without changing its shape. This state is called constant pattern.[9] This phenomenon arises because the self-sharpening effect associated with a convex isotherm is balanced by the dispersive effect of axial dispersion and a finite rate of mass-transfer kinetics. If the equilibrium adsorption isotherm is linear or concave upward, no constant pattern behavior is observed and the breakthrough curve spreads constantly during its migration through the column. This case is unfavorable. If the adsorption isotherm is concave upward, then a descending concentration step (i.e., $C_{n+1} < C_n$) leads to the formation of a constant pattern.

A very detailed study of the combined effects of axial dispersion and mass-transfer resistance under a constant pattern behavior has been conducted by Rhee and Amundson.[10] They used the *shock-layer* theory. The shock layer is defined as a zone of a breakthrough curve where a specific concentration change occurs (i.e., a concentration change from 10% to 90%). The study of the shock-layer thickness is a new approach to the study of column performance in nonlinear chromatography. The optimum velocity for minimum shock-layer thickness (SLT) can be quite different from the optimum velocity for the height equivalent to a theoretical plate (HETP).[9]

INSTRUMENTATION

There are many possibilities for performing frontal chromatography experiments. In general, standard chromatographic equipment can be used. The preparation of a series of solutions of known concentration can be easily accomplished by using a chromatograph with a gradient delivery system applied as a mobile-phase mixer. If this system is not available, then the solutions have to be prepared manually. Two pumps can be used to perform the step injections or a single pump with a gradient delivery system. An injector having a sufficient large loop can also be used. Even a single pump without gradient delivery system can be used. In this case, the step injection has to be made by manually switching the solvent inlet line to the prepared sample reservoir. The choice of the system is dependent on the application. For fast and accurate measurements of adsorption isotherms, a multisolvent gradient system with two pumps and a high-pressure mixer is a very good choice.

REFERENCES

1. Parcher, J. Adv. Chromatogr. **1978**, *16*, 151.
2. Antia, F.; Horváth, Cs. Operational modes of chromatographic separation processes. Ber. Bunsenges. Phys. Chem. **1989**, *93*, 968.
3. Lee, A.; Aliao; Horváth, Cs. Tandem separation schemes for preparative HPLC of proteins. J. Chromatogr. **1988**, *443*, 31.
4. James, D.; Phillips, C. The chromatography of gases and vapours. Part III. The determination of adsorption isotherms. J. Chem. Soc. **1954**, *1066*.
5. Shay, G.; Szekely, G. Gas adsorption measurements in flow systems. Acta Chim. Hung. **1954**, *5*, 167.
6. Guan, H.; Stanley, B.; Guiochon, G. Theoretical study of the accuracy and precision of the measurement of single component isotherms by the elution by characteristic points (ECP) method. J. Chromatogr. A. **1994**, *659*, 27.
7. Sajonz, P. *Ph.D. Thesis*; University Saarbrücken: Germany, 1996.
8. Jacobsen, J.; Frenz, J.; Horváth, Cs. Measurement of competitive adsorption isotherms by frontal chromatography. Ind. Eng. Chem. Res. **1987**, *26*, 43.
9. Guiochon, G.; Golshan-Shirazi, S.; Katti, A. *Fundamentals of Preparative and Nonlinear Chromatography*; Academic Press: Boston, 1994.
10. Rhee, H.; Amundson, N. A study of the shock layer in nonequilibrium exchange systems. Chem. Engng. Sci. **1972**, *27*, 199.

Fronting of Chromatographic Peaks: Causes

Ioannis N. Papadoyannis
Anastasia Zotou
Aristotle University of Thessaloniki, Thessaloniki, Greece

INTRODUCTION

Peaks with strange shapes represent one of the most vexing problems that can arise in a chromatographic laboratory. Fronting of peaks is a condition in which the front of a peak is less steep than the rear relative to the baseline. This condition results from nonideal equilibria in the chromatographic process.

DISCUSSION

Fronting peaks, as well as tailing or other misshaped peaks, can be hard to quantitate. Some data systems have difficulty in measuring peak size accurately. As a result, the precision and/or reliability of assay methods involving fronting or other misshaped peaks is often poor when compared to good chromatography. There are a number of different causes of peak fronting, and discovering why peaks are thus misshaped and then fixing the problem can be a difficult undertaking. Fortunately, there is a systematic approach based on logical analysis plus practical fixes that have now been documented in numerous laboratories. Fronting peaks are less commonly encountered in LC, but they are readily distinguished from other peak-shape problems. Fronting peaks are the opposite of tailing peaks. Whereas tailing peaks suggest that sample retention decreases with increasing sample size or concentration, fronting peaks suggest the opposite: retention increases with larger samples. In both cases, a decrease in sample size may eliminate peak distortion. However, this is often not practical, because some minimum sample size is required for good detectability. In the case of tailing peaks, it is believed that peak distortion often arises because large samples use up some part of the stationary phase. However, the cause of fronting peaks is seldom fully understood.

Ion-pair chromatography (IPC) is more susceptible to peak fronting than other modes in LC. *Column temperature* problems can cause fronting peaks in IPC. Fig. 1 shows the separation of an antibiotic amine at ambient temperature. Repeating the separation at 45°C eliminated the fronting problem. Some studies have shown peak fronting in IPC that can be corrected by operating at a higher column temperature, whereas some other separations are best carried out at lower temperatures. The reason for this peculiar peak-shape behavior is unclear, but it may be related to the presence of reagent micelles in the mobile phase for some experimental IPC conditions. Generally, it is good practice to run ion-pair separations under thermostatted conditions, because relative retention tends to vary with temperature in IPC. Usually, narrower bands and better separation results when temperatures of 40–50°C are used for IPC.

The use of a *sample solvent other than the mobile phase* is another cause of fronting peaks in IPC. In this case, the sample should only be injected as a solution in

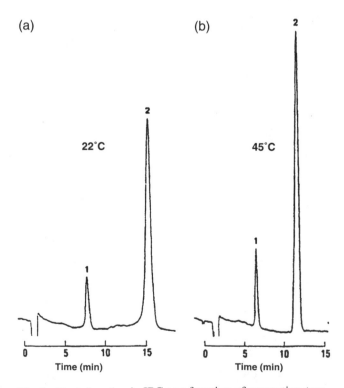

Fig. 1 Peak fronting in IPC as a function of separation temperature. Column: Zorbax C_8 mobile phase: 10 mM sodium dodecyl sulfate and 150 mM ammonium phosphate in 33% acetonitrile; pH: 6.0; flow rate: 2.0 mL/min; temperature: (a) = 22°C and (b) = 45°C. Peaks: 1 = lincomycin B; 2 = lincomycin A. (From Ref. 1 with permission from Elsevier Science.)

Encyclopedia of Chromatography DOI: 10.1081/E-ECHR-120039995

the mobile phase. No more than 25–50 µl of sample should be injected, if possible.

Silanol effects can adversely alter peak shape in IPC, just as in reversed-phase separations. Therefore, when separating basic (cationic) compounds, the column and mobile phase should be chosen bearing this in mind. When ion-pair reagents are used, however, silanol effects are often less important. The reason is that an anionic (acidic) reagent confers an additional negative charge on the column packing and this reduces the relative importance of sample retention by ion exchange with silanol groups. Similarly, cationic (basic) reagents are quite effective at blocking silanols because of the strong interaction between reagent and ionized silanol groups.

Still another cause of peak fronting is for the case of *anionic (acidic) sample molecules separated with higher-pH mobile phases*. For silica-based packings, the packing has an increasingly negative charge as the pH increases, and this results in the repulsion of anionic sample molecules from the pores of the packing. With larger sample sizes, however, this effect is overcome by the corresponding increase in ionic strength, caused by the sample. A remedy for this problem is to increase the ionic strength of the mobile phase, by increasing the mobile-phase buffer concentration to the range of 25–100 mM. It should be mentioned here that ionic or ionizable samples should never be separated with unbuffered mobile phases.

Finally, *column voids* and *blocked frits* can also cause peak fronting.

SUGGESTED FURTHER READING

Asmus, P.A.; Landis, J.B.; Vila, C.L. Liquid chromatographic determination of lincomycin in fermentation beers. J. Chromatogr. **1983**, *264* (2), 241.

Bidlingmeyer, B.A. *Practical HPLC Methodology and Applications*; John Wiley & Sons: New York, 1992; 20.

Dolan, J.W.; Snyder, L.R. *Troubleshooting LC Systems*; Humana Press: Totowa, NJ, 1989; 400–401.

Sadek, P.C.; Carr, P.W.; Bowers, L.D. Evaluation of several void-volume markers for reversed-phase HPLC. LC, Liq. Chromatogr. HPLC Mag. **1985**, *3*, 590.

GC/MS Systems

Raymond P. W. Scott
Scientific Detectors Ltd., Banbury, Oxfordshire, U.K.

INTRODUCTION

Despite the speed and accuracy of contemporary analytical techniques, the use of more than one, separately and in sequence, is still very time-consuming. To reduce the analysis time, many techniques are operated concurrently, so that two or more analytical procedures can be carried out simultaneously. The tandem use of two different instruments can increase the analytical efficiency, but due to unpredictable interactions between one technique and the other, the combination can be quite difficult in practice. These difficulties become exacerbated if optimum performance is required from both instruments. The mass spectrometer was a natural choice for the early tandem systems to be developed with the gas chromatograph, as it could easily accept samples present as a vapor in a permanent gas.

BACKGROUND INFORMATION

The first GC/MS system was reported by Holmes and Morrell in 1957, only 4 years after the first description of GC by James and Martin in 1953. The column eluent was split and passed directly to the mass spectrometer. Initially, only packed GC columns were available and thus the major problem encountered was the disposal of the relatively high flow of carrier gas from the chromatograph ($\sim$25 mL/min or more). These high flow rates were in direct conflict with the relatively low pumping rate of the MS vacuum system. This problem was solved either by the use of an eluent split system or by employing a vapor concentrator. A number of concentrating devices were developed (e.g., the jet concentrator invented by Ryhage and the helium diffuser developed by Biemann).

The jet concentrator consisted of a succession of jets that were aligned in series but separated from each other by carefully adjusted gaps. The helium diffused away in the gap between the jets and was removed by appropriate vacuum pumps. In contrast, the solute vapor, having greater momentum, continued into the next jet and, finally, into the mass spectrometer. The concentration factor was about an order of magnitude and the sample recovery could be in excess of 25%.

The Biemann concentrator consisted of a heated glass jacket surrounding a sintered glass tube. The eluent from the chromatograph passed directly through the sintered glass tube and the helium diffused radially through the porous walls and was continuously pumped away. The helium stream enriched with solute vapor passed into the mass spectrometer. Solute concentration and sample recovery were similar to the Ryhage device, but the apparatus was bulkier although somewhat easier to operate. An alternative system employed a length of porous polytetrafluorethylene (PTFE) tube, as opposed to one of sintered glass, but otherwise functioned in the same manner.

The introduction of the open-tubular columns eliminated the need for concentrating devices as the mass spectrometer pumping system could cope with the entire column eluent. Consequently, the column eluent could be passed directly into the mass spectrometer and the total sample can enter the ionization source. The first mass spectrometer used in a GC/MS tandem system was a rapid-scanning magnetic sector instrument that easily provided a resolution of one mass unit. Contemporary mass spectrometers have vastly improved resolution and the most advanced system (involving the triple quadrupole mass spectrometer) gives high in-line sensitivity, selectivity, and resolution.

IONIZATION TECHNIQUES FOR GC/MS

There are a number of ionization processes that are used, probably the most important being electron-impact ionization. Electron-impact ionization is a harsh method of ionization and produces a range of molecular fragments that can help to elucidate the structure of the molecule. Nevertheless, although molecular ions are usually produced that are important for structure elucidation, sometimes only small fragments of the molecule are observed, with no molecular ion invoking the use of alternative ionizing procedures. A diagram showing the configuration of an electron-impact ion source is shown in Fig. 1. Electrons, generated by a heated filament, pass across the ion source to an anode trap. The sample vapor is introduced in the center of the source and the solute molecules drift, by diffusion, into the path of the electron beam. Collision

Encyclopedia of Chromatography DOI: 10.1081/E-ECHR-120039996

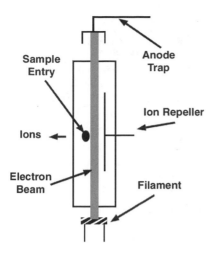

Fig. 1 An electron-impact ionization source: (a) Reagent gas methane; (b) reagent gas isobutane.

with the electrons produce molecular ions and ionized molecular fragments, the size of which is determined by the energy of the electrons. The electrons are generated by thermal emission from a heated tungsten or rhenium filament and accelerated by an appropriate potential to the anode trap. The magnitude of the collection potential may range from 5 to 100 V, depending on the electrode geometry and the ionization potential of the substances being ionized. The ions that are produced are driven by a potential applied to the ion-repeller electrode into the accelerating region of the mass spectrometer.

Unfortunately, with electron-impact ionization, there is a frequent absence of a molecular ion in the mass spectrum, which makes identification uncertain and complicates structure elucidation. One solution is to employ chemical ionization. If an excess of an appropriate reagent gas is fed into an electron-impact source, an entirely different type of ionization takes place. As the reagent gas is in excess, the reagent molecules are preferentially ionized and the reagent ions then collide with the sample molecules and produce sample + reagent ions or, in some cases, protonated ions. In this type of ionization, very little fragmentation takes place and parent ions + a proton or + a molecule of the reagent gas are produced. Little modification to the normal electron impact source is required and an additional conduit to supply the reagent gas is all that is necessary.

Chemical ionization was first observed by Munson and Field, who introduced it as an ionization procedure in 1966. A common reagent gas is methane and the partial pressure of the reagent gas is arranged to be about two orders of magnitude greater than that of the sample. The process is gentle and the energy of the most reactive reagent ions never exceeds 5 eV. Consequently, there is little fragmentation, and the

most abundant ion usually has a m/z value close to that of the singly-charged molecular ion. The spectrum produced depends strongly on the nature of the reagent ion; thus, different structural information can be obtained by choosing different reagent gases. This adds another degree of freedom in the operation of the mass spectrometer. Using methane as the reagent ion, the following reagent ions can be produced:

$$CH_4 \rightarrow CH_4^+, CH_3, CH_2^+$$
$$CH_4^+ + CH_4 \rightarrow CH_5^+ + CH_3$$
$$CH_3^+ + CH_4 \rightarrow C_2H_4^+ + H_2$$

Other reactions can occur that are not useful for ionization but, in general, these are in the minority. The interaction of positively charged ions with the uncharged sample molecules can also occur in a number of ways, and the four most common are as follows:

1. Proton transfer between the sample molecule and the reagent ion

$$M + BH^+ \rightarrow MH^+ + B$$

2. Exchange of charge between the sample molecule and the reagent ion

$$M + X^+ \rightarrow M^+ + X$$

3. Simple addition of the sample molecule to the reagent ion

$$M + X^+ \rightarrow MX^+$$

4. Anion extraction

$$AB + X^+ \rightarrow B^+ + AX$$

As an example, ions, which are formed when methane is used as the reagent gas, will react with a sample molecule largely by proton transfer; that is,

$$M + CH_5^+ \rightarrow MX^+ + CH_4$$

Some reagent gases produce more reactive ions than others and will produce more fragmentation. For example, methane produces more aggressive reagent ions than isobutane. Consequently, whereas methane ions produce a number of fragments by protonation, isobutane, by a similar protonation process, will produce almost exclusively the protonated molecular ion. This is shown in the mass spectra of methyl stearate in Fig. 2. Spectrum (a) was produced using methane as the reagent gas and exhibits fragments other than the protonated parent ion. In contrast, spectrum b obtained with butane as the reagent gas, exhibits the

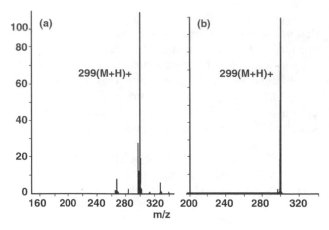

Fig. 2 Mass spectrum of methyl stearate produced by chemical ionization.

protonated molecular ion only. Continuous use of a chemical ionization source causes significant source contamination, which impairs the performance of the spectrometer and thus the source requires cleaning by baking-out fairly frequently. Retention data on two-phase systems coupled with matching electron-impact mass spectra or confirmation of the molecular weight from chemical ionization spectra are usually sufficient to establish the identity of a solute.

The inductively coupled plasma (ICP) source is used largely for specific element identification and evolved from the ICP atomic emission spectrometer; it is probably more commonly employed in LC/MS than GC/MS. In GC/MS, the ICP ion source is used in the assay of organometallic materials and in metal speciation analyses. The ICP ion source is very similar to the volatilizing unit of the ICP atomic emission spectrometer, and a diagram of the device is shown in Fig. 3. The argon plasma is an electrodeless discharge, often initiated by a Tesla coil spark, and

maintained by radio-frequency (rf) energy, inductively coupled to the inside of the torch by an external coil, wrapped around the torch stem. The plasma is maintained at atmospheric pressure and at an average temperature of about 8000 K. The ICP torch consists of three concentric tubes made from fused silica. The center tube carries the nebulizing gas, or the column eluent, from the gas chromatograph. Argon is used as the carrier gas, and the next tube carries an auxiliary supply of argon to help maintain the plasma and also to prevent the hot plasma from reaching the tip of the sample inlet tube. The outer tube also carries another supply of argon at a very high flow rate that cools the two inner tubes and prevents them from melting at the plasma temperature. The coupling coil consists of two to four turns of water cooled copper tubing, situated a few millimeters behind the mouth of the torch. The rf generator produces about 1300 W of rf at 27 or 40 MHz, which induces a fluctuating magnetic field along the axis of the torch. Temperature in the induction region of the torch can reach 10,000 K, but in the ionizing region, close to the mouth of the sample tube, the temperature is 7000–9000 K.

The sample atoms account for less than of the total number of atoms present in the plasma region; thus, there is little or no self-quenching. At the plasma temperature, over 50% of most elements are ionized. The ions, once formed, pass through the apertures in the apex of two cones. The first has an aperture about 1 mm inner diameter (i.d.) and ions pass through it to the second skimmer cone. The space in front of the first cone is evacuated by a high-vacuum pump. The region between the first cone and the second skimmer cone is evacuated by a mechanical pump to about 2 mbar and, as the sample expands into this region, a supersonic jet is formed. This jet of gas and ions flows through a slightly smaller orifice into the apex of the second cone. The emerging ions are extracted by negatively charged electrodes (−100 to −600 V) into the focusing region of the spectrometer, and then into the mass analyzer.

The ICP ion source has the advantages that the sample is introduced at atmospheric pressure, the degree of ionization is relatively uniform for all elements, and singlycharged ions are the principal ion product. Furthermore, sample dissociation is extremely efficient and few, if any, molecular fragments of the original sample remain to pass into the mass spectrometer. High ion populations of trace components in the sample are produced, making the system extremely sensitive. Nevertheless, there are some disadvantages: the high gas temperature and pressure evoke an interface design that is not very efficient and only about 1% of the ions that pass the sample orifice pass through the skimmer orifice. Furthermore, some molecular ion formation does occur in the plasma, the most troublesome being molecular ions formed with oxygen. These

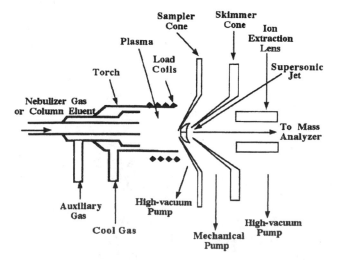

Fig. 3 ICP mass spectrometer ion source.

can only be reduced by adjusting the position of the cones, so that only those portions of the plasma where the oxygen population is low are sampled.

Although the detection limit of an ICP/MS is about 1 part in a trillion, as already stated, the device is rather inefficient in the transport of the ions from the plasma to the analyzer. Only about 1% pass through the sample and skimming cones and only about 10^{-6} ions will eventually reach the detector. One reason for ion loss is the diverging nature of the beam, but a second is due to space-charge effects, which, in simple terms, is the mutual repulsion of the positive ions away from each other. Mutual ion repulsion could also be responsible for some nonspectroscopic interelement interference (i.e., matrix effects). The heavier ions having greater momentum suffer less dispersion than the lighter elements, thus causing a preferential loss of the lighter elements.

MASS SPECTROMETERS FOR MS/GC TANDEM OPERATION

The most common mass spectrometer used in GC/MS systems is the quadrupole mass spectrometer, either as a single quadrupole or as a triple quadrupole, which can also provide MS/MS spectra. A diagram of a quadrupole mass spectrometer is shown in Fig. 4. The operation of the quadrupole mass spectrometer is quite different from that of the sector instrument. The instrument consists of four rods which must be precisely straight and parallel and so arranged that the beam of ions is directed axially between them. Theoretically, the rods should have a hyperbolic cross section, but in practice, less expensive cylindrical rods are nearly as satisfactory. A voltage comprising a DC component (U) and a rf component ($V_0 \cos wt$) is applied between adjacent rods, opposite rods being electrically connected. Ions are accelerated into the center, between the rods, by a potential ranging from 10 to 20 V. Once inside the quadrupole, the ions oscillate in the x and y dimensions induced by the high-frequency electric field. The mass range is scanned by changing U and V_0 while keeping the ratio U/V_0 constant. The quadrupole mass spectrometer is compact, rugged, and easy to operate, but its mass range does not extend to very high values. However, under certain circumstances, multiply-charged ions can be generated and identified by the mass spectrometer. This, in effect, increases the mass range of the device proportionally to the number of charges on the ion.

The quadrupole mass spectrometer can also be constructed to provide MS/MS spectra by combining three quadrupole units in series. A diagram of a triple quadrupole mass spectrometer is shown in Fig. 5. The sample enters the ion source and is usually fragmented by either an electron-impact or chemical ionization process. In the first analyzer, the various charged fragments are separated in the usual way, which then pass into the second quadrupole section, sometimes called the collision cell. The first quadrupole behaves as a straightforward mass spectrometer. Instead of the ions passing to a sensor, the ions pass into a second mass spectrometer and a specific ion can be selected for further study. In the center quadrupole section, the selected ion is further fragmented by collision ionization and the new fragments pass into the third quadrupole, which functions as a second analyzer. The second analyzer resolves the new fragments into their individual masses producing the mass spectrum. Thus, the exclusive mass spectrum of a particular molecular or fragment ion can be obtained from the myriad of ions that may be produced from the sample in the first analyzer. This is an extremely powerful analytical system that can handle exceedingly complex mixtures and very involved molecular structures.

Another form of the quadrupole mass spectrometer is the ion trap detector, which has been designed more

GC/MS Systems

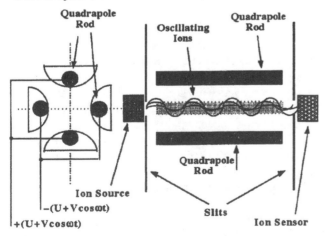

Fig. 4   Quadrupole mass spectrometer.

GC/MS Systems

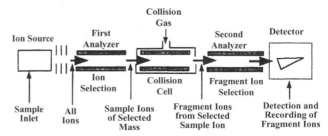

Fig. 5   Triple quadrupole mass spectrometer.

specifically as a chromatography detector than for use as a tandem instrument. The electrode orientation of the quadrupole ion trap mass spectrometer is shown in Fig. 6. The ion trap mass spectrometer has an electrode arrangement that consists of three cylindrically symmetrical electrodes comprised of two end caps and a ring. The device is small, the opposite internal electrode faces being only 2 cm apart. Each electrode has accurately machined hyperbolic internal faces. An rf voltage together with an additional DC voltage is applied to the ring, and the end caps are grounded. The rf voltage causes rapid reversals of field direction, so any ions are alternately accelerated and decelerated in the axial direction and vice versa in the radial direction. At a given voltage, ions of a specific mass range are held oscillating in the trap. Initially, the electron beam is used to produce ions, and after a given time, the beam is turned off. All the ions, except those selected by the magnitude of the applied rf voltage, are lost to the walls of the trap, and the remainder continue oscillating in the trap. The potential of the applied rf voltage is then increased, and the ions sequentially assume unstable trajectories and leave the trap via the aperture to the sensor. The ions exit the trap in order of their increasing m/z values. The first ion trap mass spectrometers were not very efficient, but it was found that the introduction of traces of helium to the ion trap significantly improved the quality of the spectra. The improvement appeared to result from ion-helium collisions that reduced the energy of the ions and allow them to concentrate in the center of the trap. The spectra produced are quite satisfactory for solute identification by comparison with reference spectra. However, the spectrum produced for a given substance will probably differ considerably from that produced by the normal quadrupole mass spectrometer.

The time-of-flight mass spectrometer was invented many years ago, but the performance of the modern version is greatly improved. A diagram of the time-of-flight mass spectrometer is shown in Fig. 7. In a

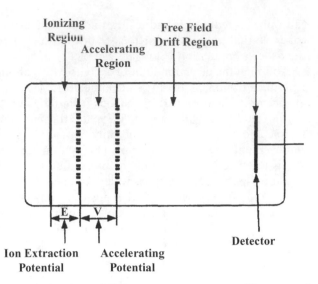

Fig. 7 The time-of-flight mass spectrometer. (Courtesy of VG Organic Inc.)

time-of-flight mass spectrometer, the following relationship holds:

$$t = \left(\frac{m}{2zeV}\right)^{1/2} L$$

where t is the time taken for the ion to travel a distance L, V is the accelerating voltage applied to the ion, and L is the distance traveled by the ion to the ion sensor.

The mass of the ion is directly proportional to the square of the transit time to the sensor. The sample is volatilized into the space between the first and second electrodes and a microsecond burst of electrons is allowed to produce ions. An extraction voltage is then applied for another short time period, which, as those further from the second electrode will experience a greater force than those closer to the second electrode, will focus the ions. After focusing, the accelerating potential (V) is applied for about 100 ns so that all the ions in the source are accelerated almost simultaneously. The ions then pass through the third electrode into the drift zone and are then collected by the sensor electrode. The particular advantage of the time-of-flight mass spectrometer is that it is directly compatible with surface desorption procedures. Consequently, it can be employed with laser-desorption and plasmadesorption techniques. An excellent discussion on general organic mass spectrometry is given in *Practical Organic Mass Spectrometry* edited by Chapman.[1]

The combination of the gas chromatograph with the single quadrupole mass spectrometer or with the triple quadrupole mass spectrometer are the most commonly used tandem systems. They are used extensively in forensic chemistry, in pollution monitoring and control, and in metabolism studies. The quadrupole mass spectrometers provide both high sensitivity and

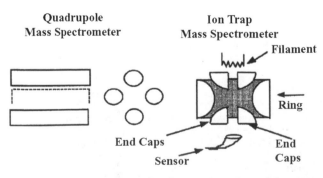

Fig. 6 Pole arrangement for the quadrupole and ion trap mass spectrometers.

good mass spectrometric resolution. They can be readily used with open-tubular columns, and an example of the use of the single quadrupole monitoring a separation from an open-tubular column is shown in Fig. 8. The column was 30 m long with a 0.25-mm i.d. and carried a 0.5-mm film of stationary phase. A 1-mL sample was used and the column was programmed from 50°C to 300°C at 10°C/min.

An elegant example of the use of GC/MS in the analysis of pesticides in river water is given by Vreuls et al.[2] A 1-mL sample was collected in an LC sample loop and the internal standard added. The sample was then displaced through a short column 1 cm long with a 2-mm i.d. packed with 10-mm particles of a proprietary PLRP-S adsorbent (styrene–divinylbenzene copolymer) by a stream of pure water. The extraction column was then dried with nitrogen and the adsorbed materials displaced into a gas chromatograph with 180 mL of ethyl acetate. The sample was passed through a short retention gap column and then to a retaining column. The GC oven was maintained at 70°C so that the ethyl acetate passed through the retaining column and was vented to waste. The solutes of interest were held in the retaining column at this temperature during the removal of the ethyl acetate. The temperature was then increased and the residual material separated on an analytical column using an appropriate temperature program. The eluents from the analytical column passed to a quadrupole mass spectrometer. An example of the chromatograms and spectra obtained are shown in Fig. 9. Fig. 9a shows the total ion current chromatogram from a sample of Rhine River water containing 200 ppt of the herbicides atrazine and simazine. The pertinent peaks are shown enlarged in the inset. Fig. 9b shows a section of the same chromatogram presented in the selected ion mode. It is seen that the herbicide peaks are clearly

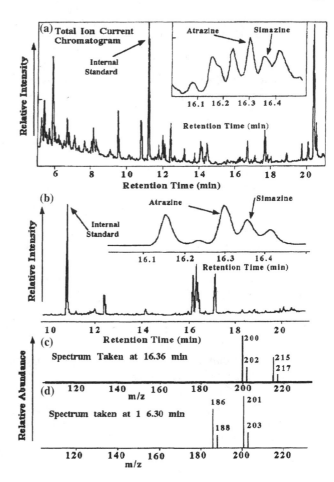

Fig. 9 Chromatogram and spectra from a sample of river water containing 200 ppt of atrazine and simazine. (From Ref. 2.)

and unambiguously revealed. In Figs. 9c and 9d, the individual mass spectra of atrazine (eluted at 16.30 min) and simazine (eluted at 16.36 min) are shown. The spectra are clear and more than adequate to confirm the identity of the two herbicides.

REFERENCES

1. Chapman, J.R., Ed.; *Practical Organic Mass Spectrometry*; John Wiley & Sons: New York, 1994.
2. Vreuls, J.J.; Bulterman, A.-J.; Ghijsen, R.T.; Brinkman, U.Th. Analyst **1992**, *117*, 1701.

SUGGESTED FURTHER READING

Message, G.M. *Practical Aspects of GC/MS*; John Wiley & Sons: New York, 1984.
Scott, R.P.W. *Tandem Techniques*; John Wiley & Sons: New York, 1984.

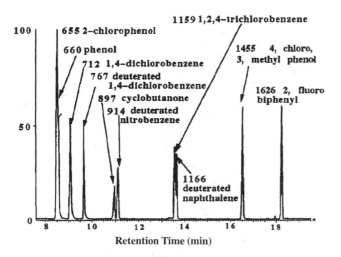

Fig. 8 A separation from an open-tubular column monitored by a single quadrupole mass spectrometer.

GC System Instrumentation

Mochammad Yuwono
Gunawan Indrayanto
Laboratory of Pharmaceutical Biotechnology, Airlangga University,
Jl. Dharmawangsa dalam, Surabaya, Indonesia

INTRODUCTION

GC was first described by Martin and James in 1952. It has become one of the most frequently used separation techniques for the analysis of gases and volatile liquids and solids. An important breakthrough in GC was the introduction of the open tubular column by Golay in 1958 and the adoption of fused silica capillary columns by Dandeneau and Zerenner in 1979.

Today, using of the capillary columns can solve many kinds of analytical problems, such as isomer separation and analysis of complex mixtures of natural products and biologicals.

The gas chromatograph involves volatilization of the sample in a heated inlet port (injector), separation of the component mixtures in a column, and detection of each component by a detector.

GC SYSTEM INSTRUMENTATION

GC, first described by James and Martin[1] in 1952, has become one of the most frequently used separation technique for the analysis of gases, volatile liquids, and solids. The major breakthrough of GC was the introduction of the open tubular column by Golay and Desty[2] in 1958 and the adoption of fused silica capillary columns by Dandeneau and Zerenner[3] in 1979. Today, using of the capillary columns can solve many analytical problems, such as isomeric separation and analysis of complex mixtures of natural products and biological samples.

The basic principle of a gas chromatograph involves volatilization of the sample in a heated inlet port (injector), separation of the component mixtures in a column, and detection of each component by a detector. Although the basic components remain the same, some improvements in gas chromatograph appeared in the commercial marketplace. GC with electronic integrators and computer-based data processing systems became common in 1970s, whereas in the 1980s, a computer was introduced to control all GC parameters automatically, such as column temperature, flow rates, inlet pressure, and sample injection, and to evaluate the data obtained. The automated equipment can be operated unattended overnight.[4,5]

Combinations of highly efficient separation columns, with specific or selective detectors, such as electron capture detector (ECD), GC/MS, and GC-Fourier transform infrared (FTIR) detector, make GC a more favorable technique. Multidimensional GC systems, which contain at least two columns operated in series, have also proved to be a powerful tool in the analytical chemistry of complex mixtures.

The dramatic advance in GC instrumentation is the introduction of portable gas chromatographs, which have been developed during 1990s to provide a field-based analysis. Recently, the micro high-speed GC portable has also appeared to carry out the analysis up to 10 times faster than conventional laboratory GCs.[6–8]

The GC system (Fig. 1) consists of a carrier gas supply system, an inlet to deliver sample to a column, the column where the separations occur, an oven as a thermostat for the column, a detector to register the presence of a chemical in the column effluent, and a data system to record, display, and evaluate the chromatogram.

CARRIER GAS

The carrier gas that is used as the mobile phase transfers the sample from the injector, through the column, and into the detector. For a laboratory gas chromatograph, the carrier gas is usually obtained from a commercial pressurized gas cylinder equipped with a two-stage regulator for coarse and fine flow control. In most instruments, provision is made for secondary fine-tuning of pressure and gas flow. In 1990, electronic pressure control was developed, which allows operation at constant pressure, constant flow, and pressure-programming modes. The gas flow can be read electronically on the instrument panel or measured using a soap-bubble flow meter at the outlet of the column. The carrier gas flow is directed through a sieve trap, or a series of traps to remove the moisture, organic matter, and oxygen, and then through frits to

Encyclopedia of Chromatography DOI: 10.1081/E-ECHR-120040660

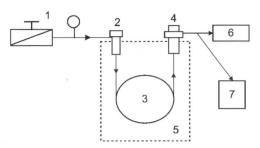

Fig. 1 Gas chromatography system schematic.

filter off any particulate matter.[9] It is also suggested to use copper tubing for the connection of the gas cylinder to the gas chromatographs. Polymer tubing should be avoided because the oxygen from the atmosphere can often permeate the tubing walls. The oxygen in the gas stream may cause degradation of some column stationary phases at elevated operating temperatures, thereby producing unstable baselines with the electron-capture detector and shortening filament lifetime for the thermal conductivity detector. To minimize contamination, high purity carrier gases are used, combined with additional chemical and or catalytic gas purifying devices. The gas chromatograph may have thermostatically controlled pneumatics to prevent drift, in which pressure regulators, flow controllers, and additional gas purifying traps and filters are housed. The carrier gas must be inert so that it reacts neither with the sample nor with the stationary phase at the operating temperature.[9] The choice of carrier gas requires consideration of the detector used, the separation problem to be solved, and the purity of the gases available. A further consideration in the selection of carrier gas is its availability and its cost. In practice, the choice of carrier gas will determine the efficiency of the GC system because the height equivalent to a theoretical plate depends on solute diffusivity in the carrier. The influence of the mobile-phase velocity on column efficiency and practical consequences of the carrier gas selection in capillary GC have been described in previous publications.[9,10] Normally, a compromise between inertness, efficiency, and operating cost make nitrogen or helium the most common GC carrier gases. The carrier gas flow can be determined by either linear velocity, expressed in cm sec^{-1}, or volumetric flow rate, expressed in mL min^{-1}. The linear velocity is independent of the column diameter, whereas the flow rate is dependent on the column diameter. For capillary columns, makeup gas is added at the column exit to obtain a total gas flow of 30–40 mL min^{-1} into the detector; it can be the same gas as the carrier gas or a different gas, depending on the type of detector being used.

SAMPLE INLET SYSTEMS

Sample introduction into the gas chromatograph is the first stage in the chromatographic process. It is of primary importance, especially in capillary GC, because its efficiency is reflected in the overall efficiency of the separation procedure and the quantitative results. The basic prerequisite of the sample injection system is that the sample should be introduced into the column as a narrow band, ideally with maintenance of constant pressure and flow. The specially designed inlet should be hot enough to flash-evaporate the sample and large enough in volume to allow the sample vapor to expand without blowing back through the septum. Care must be taken not to overheat the injector because the injection cell or the sample may decompose. The sample must be gaseous or an easily vaporized liquid or solid. Most organic compounds may be introduced onto the column in the form of a liquid sample, either as the neat compound or, in the case of a solid, as a solution. When the dilution of the solid sample would be undesirable, the solid may be encapsulated in glass capillaries and mechanically pushed into the heated injection block and crushed.[10,11]

For injecting gases and vapors, gas-tight syringes with Teflon-tipped plungers and syringe barrels are available. Many analysts favor using gas syringes for gas samples; however, the introduction of accurately measured volumes of gases remains a problem. In the alternative method, gas samples can be introduced onto a column using rotary gas switching valves, which generally consist of a rotating polymeric core, encased in a stainless-steel body. For repetitive or periodic injection of a large number of the same or different samples, auto samplers may be used.[10,12] The sample volume for analytical work depends on the dimensions of the column and on the sensitivity of the detector. For a packed column, sample size ranges from tenths of 1 μL up to 20 μL. Capillary columns need much less sample (0.01–1 μL). The most commonly used silicone-rubber septa may contain impurities that may bleed into the column above a certain temperature, resulting in unsteady baseline and ghost peaks. Recently, various kinds of septa have become available which can be used at very high temperatures.[13]

Packed and Open Tubular Column Inlet

Because of the variety of columns and samples that can be analyzed by GC, several injection techniques have been developed. The packed inlet system is designed mainly for packed and wide-bore columns. However, an adapter can be used to enable capillary columns to be used. When injection is carried out in the on-column mode, glass wool can be used for packing the injector. For capillary GC, split technique is most common, which is used for high concentration samples. This technique allows injection of samples virtually independent of the selection of solvent, at any column temperature, with little risk of band broadening or disturbing solvent effects. The splitless technique, on the other hand, is used for trace level analysis. The so-called cold injection techniques (on-column, temperature programmed vaporization, cooled needle split) have also been recently developed.[4,14,15]

Pyrolysis GC

Pyrolysis involves the thermal decomposition, degradation, or cracking of a large molecule into smaller fragments. Pyrolysis GC is an excellent technique for identifying certain types of compounds which cannot be analyzed by derivatization, e.g., polymers. The pyrolysis temperature is typically between 400°C and 1000°C. A number of analytical pyrolyzers have been introduced and are commercially available. The devices consist of platinum resistively heated and Curie point pyrolyzers. The carrier gas is directed through the system, and the platinum wire is heated to a certain temperature. The material decomposes, and the fragmentation products are analyzed.[16,17]

Headspace Analysis

Headspace analysis is an excellent technique for gas chromatography to analyze volatile samples in which the matrix is of no interest. It is readily applied to many analytical problems, such as monitoring of volatiles in soil and water, determination of monomers in polymers, aromas in food and beverages, etc. A variety of headspace auto samplers are commercially available, based on the principle of static or dynamic headspace. In static headspace, the sample is transferred to a headspace vial that is sealed and placed in a thermostat to drive the desirable component into the headspace sampling. An aliquot of the vapor phase is introduced into the GC system via a gas-tight syringe or a sample loop of a gas-sampling valve. Static headspace implies that the sample is taken from a single-phase equilibrium. To increase the detectability, dynamic headspace analysis has been developed. Driving the headspace out of the vial via an inert gas continuously displaces the phase equilibrium. A detailed discussion of headspace GC is reported in the previous work.[18]

Solid-Phase Microextraction

Solid-phase microextraction, first reported by Belardi and Pawliszyn in 1989, is an alternative sampling technique. The method has the advantages of convenience and simplicity, and it does not release environmentally polluting organic solvents into the atmosphere. The method is based on the extraction of analytes directly from liquid samples or from headspace of the samples onto a polymer- or adsorbent-coated fused silica fiber. After equilibration, the fiber is then removed and injected onto the gas chromatograph.[19–22]

Purge-and-Trap Methods

Purge and trap samplers have been developed for analysis of nonpolar and medium-polarity pollutants in water samples. The commercially available systems are all based on the same principle. Helium is purged through the sample that is contained in a sealed system, and the volatiles are swept continuously through an adsorbent trap where they are concentrated. After a selected time, purging is stopped, the carrier gas is directed through the trap via a six-way valve, and the trap is heated rapidly to desorb the solutes.[4,9]

OVEN

The column is ordinarily housed in a thermostatically controlled oven, which is equipped with fans to ensure a uniform temperature. The column temperature should not be affected by changes in the detector, injector, and ambient temperatures. Temperature fluctuations in column ovens can decrease the accuracy of the measured retention times and may also cause the peak splitting effect. For conventional ovens, the oven wall is well insulated using a wire coil of high thermal capacity, which is able to radiate heat into the inner volume of the oven. The characteristics of a more efficient method can accurately control the temperature of a column and allow the operator to change the temperature conveniently and rapidly for temperature programming. It is designed by suspending the column in an insulated air oven through which the air circulated at high velocity by means of fans or pumps. Most commercial instruments employ this design and allow for the adjustment and control of temperature between 50°C and 450°C. Subambient temperature operation would normally require a cryogenic cooling system using liquid nitrogen or carbon dioxide.[4,9,10]

DETECTOR

The detector in a gas chromatograph senses the differences in the composition of the effluent gases from the column and converts the column's separation process into an electrical signal, which is recorded. There are many detectors that can be used in GC, and each detector gives a different type of selectivity. An excellent discussion and review on developments of GC detectors has been published.[23]

Detectors may be classified on the basis of selectivity. A universal detector responds to all compounds in the mobile phase except carrier gas. A selective detector responds only to a related group of substances, and a specific detector responds to a single chemical compound. Most common GC detectors fall into the selective designation. Examples include flame ionization detector (FID), ECD, flame photometric detector (FPD), and thermoionic ionization detector. The common GC detector that has a truly universal response is the thermal conductivity detector (TCD). Mass spectrometer is another commercial detector with either universal or quasi-universal response capabilities.

Detectors can also be grouped into concentration-dependent detectors and mass-flow-dependent detectors. Detectors whose responses are related to the concentration of solute in the detector cell, and do not destroy the sample, are called concentration-dependent detectors, whereas detectors whose response is related to the rate at which solute molecules enter the detector are called mass-flow-dependent detectors. Typical concentration-dependent detectors are TCD and GC/FTIR. Important mass-flow-dependent detectors are the FID, thermoionic detector for N and P (N-, P-FID), flame photometric detector for S and P (FPD), ECD, and selected ion monitoring MS detector.

The FID is one of the most widely used GC detectors. The detection principle is based on the change in the electric conductivity of a hydrogen flame in an electric field when fed by organic compound(s). The resulting current is then directed into a high impedance operational amplifier for measurement. The FID is sensitive to all compounds which contain C–C or C–H linkages and considerably less sensitive up to insensitive to certain functional groups of organic compounds, such as alcohol, amine, carbonyl, and halogen. In addition, the detector is also insensitive toward non-combustible gases such as H_2O, CO_2, SO_2, and NO. A TCD, which was one of the earliest detectors for GC, is based on changes in the thermal conductivity of the gas stream caused by the presence of analyte molecules. This device is sometimes called a katharometer. Because the TCD reacts nonspecifically, it can be used universally for the detection of either organic or inorganic substances. In the ECD, the column effluent passes over a beta-emitter, such as nickel-63 or tritium. The electrons from emitter bombard the carrier gas (nitrogen), giving rise to ions and a burst of electrons. In the absence of an analyte, the ionization process

yields a constant standing current. However, this background current decreases in the presence of organic compounds that can capture electrons. The applications of the ECD illustrate the advantages of a highly sensitive special detector toward molecules that contain electronegative functional groups such as halogens, peroxide, quinines, or nitro groups.[22–24]

GC DATA SYSTEM

The GC data system performs the tasks of recording, handling, evaluation, and documentation of the chromatogram. In a modern gas chromatographic system, these can be performed by means of a computer with specialized software. Nowadays, software for calculating the quantitative results and for the method validation is available.[4]

CONCLUSIONS

Since the introduction of GC, the basic parts of a gas chromatograph have been unchanged in function and purpose, even though the improvement has been occurring in design and materials. One area of dramatic advance in GC instrumentation was the introduction of the open tubular columns. Consequently, most GC analyses in practice are performed in capillary columns that show the separation with high efficiencies and high resolution. The developments of GC instrumentation are occurring with sample handling techniques and refinements of detectors. The dramatic advance in GC instrumentation is the development of a small, high-speed, and portable gas chromatograph to provide a field-based analysis.

REFERENCES

1. James, A.T; Martin, A.J.P. Gas–liquid partition chromatography: the separation and micro-estimation of volatile fatty acids from formic acid to dodecanoic acid. Biochem. J. **1952**, *50*, 679.
2. Golay, M.J.E.; Desty, D. *Gas Chromatography*; Butterworths: London, 1958.
3. Dandeneau, R.D.; Zerenner, E.H. J. High Res. Chromatogr. **1979**, *2*, 351.
4. Schomburg, G. *Gas Chromatography, A Practical Course*; VCH Verlagsgesellschaft: Weinheimd, 1990.
5. Poole, C.F.; Poole, S.K. *Chromatography Today*; Elsevier: Amsterdam, 1991.

6. Eiceman, G.A.; Gardea-Torresdey, J.; Overton, E.; Carney, K.; Dorman, F. Gas chromatography. Anal. Chem. **2002**, *74*, 2771–2780.

7. See http://www.agilent.com/about/newsroom/pesrel/2002/30sep2002b.html.

8. See http://www.hnu.com/fpi/gc311.htm.

9. Sandra, J.F. Gas chromatography. In *Ullmann's Encyclopedia of Industrial Chemistry*; Wiley-VCH Verlag GmbH: Weinheim, 2002.

10. Ravindranath, B. *Principles and Practice of Chromatography*; Ellis Horwood Limited: Chichester, UK, 1989.

11. Sandra, P. *Sample Introduction in Capillary Gas Chromatography*; Hüthig Verlag: Heidelberg, 1985.

12. Grob, K.; Neukom, H.P., Jr. The influence of syringe needle on the precision and accuracy of vaporizing GC injections. J. High Res. Chrom. Comm. **1979**, *2*, 15–21.

13. Olsavicky, V.M. A comparison of high temperature septa for gas chromatography. J. Chromatogr. Sci. **1978**, *16*, 197–200.

14. Grob, K. *Classical Split and Splitless Injection in Capillary GC*; Hüthig Verlag: Heidelberg, 1986.

15. Grob, K. *On Column Injection in Capillary GC*; Hüthig Verlag: Heidelberg, 1987.

16. Wang, F.C.Y.; Burleson, A.D. Development of pyrolysis fast gas chromatography for analysis of synthetic polymers. J. Chromatogr. A. **1999**, *833* (1), 111–119.

17. Haken, J.K. Pyrolysis gas chromatography of synthetic polymers: a bibliography. J. Chromatogr. A. **1998**, *825* (2), 171–187.

18. Joffe, B.V.; Vitenberg, A.G. *Headspace Analysis and Related Methods in Gas Chromatography*; John Wiley: New York, 1984.

19. See http://gc.discussing.info/gs/r_hs-gs/micro-extraction.html (accessed 12/2/2002).

20. Scarlata, C.J.; Ebeler, S.E. Headspace solid-phase microextraction for the analysis of dimethyl sulfide in beer. J. Agric. Food Chem. **1999**, *47* (7), 2505–2508.

21. Mills, G.A.; Walker, V.; Mughal, H. Quantitative determination of trimethylamine in urine by solid-phase microextraction and gas chromatography mass spectrometry. J. Chromatogr. B. **1999**, *723* (1–2), 281–285.

22. Pinho, O.; Ferreira, I.M.P.L.V.O.; Ferreira, M.A. Solid-phase microextraction in combination with GC/MS for quantification of the major volatile free fatty acids in ewe cheese. Anal. Chem. **2002**, *74*, 5199–5204.

23. Buffington, R.; Wilson, M.K. *Detectors for Gas Chromatography—A Practical Primer*; Hewlett-Packard Corporation, 1987, Part No. 5958-9433.

24. Hill, H.H., McMinn, D.G., Eds.; *Detectors for Capillary Chromatography*; John Wiley & Sons: New York, 1992.

Golay Dispersion Equation for Open-Tubular Columns

Raymond P. W. Scott
Scientific Detectors Ltd., Banbury, Oxfordshire, U.K.

INTRODUCTION

The open-tubular column or capillary column is the one most commonly used in GC today. The equation that describes dispersion in open tubes was developed by Golay,[1] who employed a modified form of the rate theory, and is similar in form to that for packed columns. However, as there is no packing, there can be no multipath term and, thus, the equation only describes two types of dispersion. One function describes the longitudinal diffusion effect and two others describe the combined resistance to mass-transfer terms for the mobile and stationary phases.

DISCUSSION

The Golay equation takes the following form:

$$H = \frac{2D_m}{u} + \frac{f_1(k')r^2}{D_m}u + \frac{f_2(k')r^2}{K^2 D_s}u \tag{1}$$

where H is the height of a theoretical plate or the variance/unit length, D_m is the diffusivity of the solute in the mobile phase, D_s is the diffusivity of the solute in the stationary phase, r is the column radius, k' is the capacity ratio of the solute, K is the distribution coefficient of the solute, and u is the mobile-phase linear velocity.

Open-tubular columns behave in exactly the same way as packed columns with respect to pressure. The same mathematical arguments can be educed which results in the modified form of the equation shown in Eq. (2). As the column is geometrically simple, the respective functions of k' can also be explicitly developed.

$$H = \frac{2D_m}{u_0} + \frac{(1 + 6k' + 11k'2)r^2}{24(1 + k')^2 D_{m(0)}}u_0 \\ + \frac{2k'df^2}{3(1 + k')^2 D_s(\gamma + 1)}u_0 \tag{2}$$

where u_0 is the exit velocity of the mobile phase and $D_{m(0)}$ is the diffusivity of the solute measured at the

exit pressure. As the film is thin, $r \gg df$; then,

$$\frac{(1 + 6k' + 11k'2)r^2}{24(1 + k')^2 D_{m(0)}} \gg \frac{2k'df^2}{3(1 + k')^2 D_s(\gamma + 1)}$$

and, thus,

$$H = \frac{2D_{m(0)}}{u} + \frac{(1 + 6k' + 11k'2)r^2}{24(1 + k')^2 D_{m(0)}}u_0 \tag{3}$$

By differentiating Eq. (3) and equating it to zero, expressions can be obtained for u_{opt} and H_{min} in a manner similar to the method used for a packed column:

$$u_{0(\text{opt})} = 2\frac{D_{m(0)}}{r}\left(\frac{12(1 + k')^2}{1 + 6k' + 11k'2}\right)^{1/2} \tag{4}$$

$$H_{\text{min}} = \frac{r}{2}\left(\frac{1 + 6k' + 11k'2}{3(1 + k')^2}\right)^{1/2} \tag{5}$$

The approximate efficiency of a capillary column operated at its optimum velocity (assuming the inlet/outlet pressure ratio is small) can be simply calculated. If only the dead volume is considered (i.e., $k' = 0$), Eq. (3) reduces to

$$H = \frac{2D_m}{u} + \frac{1}{24}\frac{r^2}{D_m}u \tag{6}$$

Differentiating and equating to zero,

$$\frac{dH}{du} = -\frac{2D_m}{u^2} + \frac{1}{24}\frac{r^2}{D_m} = 0 \quad \text{or} \quad u = \frac{\sqrt{48}D_m}{r}$$

Substituting for u in Eq. (6) and simplifying,

$$H = \frac{2D_m r}{\sqrt{48}D_m} + \frac{1}{24}\frac{r^2}{D_m}\frac{\sqrt{48}D_m}{r} = 0.289r \\ + 0.289r = 0.577r$$

Encyclopedia of Chromatography DOI: 10.1081/E-ECHR-120039997

Thus, the efficiency of a capillary column of length (l) can be assessed as

$$n = \frac{l}{0.6r} \tag{7}$$

The column efficiency will be inversely proportional to the column radius and the analysis time will directly proportional to the column radius and inversely proportional to the diffusivity of the solute in the mobile phase.

REFERENCES

1. Golay, M.J.E. *Gas Chromatography, 1958*; Desty, D.H., Ed.; Butterworths: London, 1958; 36.

SUGGESTED FURTHER READING

Scott, R.P.W. *Introduction to Analytical Gas Chromatography*; Marcel Dekker, Inc.: New York, 1998.
Scott, R.P.W. *Techniques and Practice of Chromatography*; Marcel Dekker, Inc.: New York, 1996.

GPC/SEC: Effect of Experimental Conditions

Sadao Mori
PAC Research Institute, Mie University, Nagoya, Japan

INTRODUCTION

In order to calculate the molecular-weight averages of a polymer from the SEC chromatogram, the relationship between the molecular weight and the retention volume (called the "calibration curve") needs to be known, unless a molecular-weight-sensitive detector is used. The retention volume of a polymer changes with changing experimental conditions; therefore, when molecular-weight averages of the polymer are calculated using the calibration curve, care must be taken with the effect of experimental conditions.[1]

DISCUSSION

Sample concentration is one of the most important operating variables in SEC, because the retention volumes of polymers increase with increased concentration of the sample solution. The concentration dependence of the retention volume is a well-known phenomenon and the magnitude of the peak shift to higher retention volume is more pronounced for polymers with a higher molecular weights than for those with lower molecular weights. This phenomenon is almost improbable for polymers with a molecular weight lower than 10^4 and is observed ever at a low concentration, such as 0.01%, although the peak shift is smaller than that at a higher concentration.

In this sense, this concentration dependence of the retention volume should be called the "concentration effect," not "overload effect" or "viscosity effect." If a large volume of a sample solution is injected, an appreciable shift in retention volume is observed, even for low-molecular-weight polymers; this is called the "overload effect."

The retention volume increases with increasing concentration of the sample solution and the magnitude of the increase is related to the increasing molecular weight of the sample polymers.[2] The reason for the increase in retention volume with increasing polymer concentration is considered to result from the decrease in the hydrodynamic volume of the polymer molecules in the solution.

Molecular-weight averages calculated with calibration curves of varying concentrations may differ in value. As the influence of the sample concentration on the retention volume is based on the essential nature of the hydrodynamic volume of the polymer in solution, it is necessary to select experimental conditions that will reduce the errors produced by the concentration effect.

By rule of thumb, the preferred sample concentrations, if two SEC columns of 8 mm inner diameter (i.d.) × 25 cm in length are used, are as follows. The sample concentrations should be as low as possible and no more than 0.2%. For high-molecular-weight polymers, concentrations less than 0.1% are often required, and for low-molecular-weight polymers, concentrations of more than 0.2% are possible. The concentrations of polystyrene standards for calibration should be one-half of the unknown sample concentration. For polystyrene standards with a molecular weight over 10^6, it is preferable that they are one-eighth to one-tenth and for those with a molecular weights between 5×10^5 and 10^6, a quarter to one-fifth of the sample concentration.

The retention volume of a polymer sample increases as the injection volume increases.[3] In some cases, the increase in the retention volume from an injection volume increase from 0.1 to 0.25 mL was 0.65 mL, whereas that from 0.25 to 0.5 mL was only 0.05 mL, suggesting that a precise or constant injection is required even if the injection volume is as small as 0.1 or 0.05 mL. In view of the significant effect of the injection volume on the retention volume, it is important to use the same injection volume for the sample under examination as that used when constructing the calibration curve. The use of a loop injector is essential, and the same injection volume must be employed for all sample solutions including calibration standards, regardless of their molecular-weight values. The increase in the injection volume results in a decrease in the number of theoretical plates, due to band broadening, which means that the calculated values of the molecular-weight averages and distribution deviate from the true values (Fig. 1).

The retention volume in SEC increases with increasing flow rate.[3] This is attributed to nonequilibrium effects, because polymer diffusion between the intrapores and extrapores of gels is sufficiently slow that equilibrium cannot be attained at each point in the column. With a decreasing flow rate, the efficiency and the resolution are increased. Bimodal

Encyclopedia of Chromatography DOI: 10.1081/E-ECHR-120040000

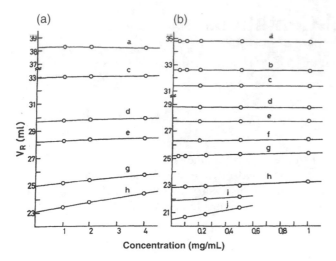

Fig. 1 Concentration dependence of retention volume for polystyrene in good solvents on polystyrene gel columns: (a) in toluene on microstyragel columns (3/8 in. × 1 ft × 4) (10^6, 10^5, 10^4, and 10^3 nominal porosity) at a flow rate 2 mL/min and injected volume 0.25 mL; (b) in Tetrahydrofuran on Shodex A 80 M columns (8 mm × 50 cm × 2) (mixed polystyrene gels of several nominal porosities) at a flow rate 1.5 mL/min and injected volume 0.25 mL. Molecular weight of polystyrene standards: (a) 2100; (b) 10,000; (c) 20,400; (d) 97,200; (e) 180,000; (f) 411,000; (g) 670,000; (h) 1,800,000; (i) 3,800,000; (j) 8,500,000.

distribution of a PS standard (NBS706) with a narrow molecular weight distribution was clearly observed at the lower flow rate.

Separation of molecules in SEC is governed, mainly, by the entropy change of the molecules between the mobile phase and the stationary phase, and the temperature independence of peak retention can be predicted. However, an increase in retention volume with increasing column temperature is often observed. A temperature difference of 10°C results in a 1% increase in the retention volume, which corresponds to a 10–15% change in molecular weight.[4]

Two main factors that cause retention-volume variations with column temperature are assumed: an expansion or a contraction of the mobile phase in the column and the secondary effects of the solute to the stationary phase. When the column temperature is 10°C higher than room temperature, the mobile phase (temperature of the mobile phase is supposed to be the same as room temperature in this case) will expand about 1% from when it entered the columns, resulting in an increase in the real flow rate in the column due to the expansion of the mobile phase and the decrease in the retention volume. The magnitude of the

retention-volume dependence on the solvent expansion is evaluated to be about one-half of the total change in the retention volume. The residual contribution to the change in retention volume is assumed to be that due to gel–solute interactions such as adsorption.

In order to obtain accurate and precise molecular-weight averages, the column temperature, as well as the difference of both temperatures, the solvent reservoir and the column oven, must be maintained.

Other factors affecting retention volume are the viscosity of the mobile phase, the sizes of gel pores, and the effective size of the solute molecules. Of these, the former two can be ignored, because they exhibit either no effect or only a small effect. The effective size of a solute molecule may also change with changing column temperature. The dependence of intrinsic viscosity on column temperature for PS in chloroform, tetrahydrofuran, and cyclohexane were tested.[5] The temperature dependence of intrinsic viscosity of PS solutions was observed over a range of temperatures. The intrinsic viscosity of PS in tetrahydrofuran is almost unchanged from 20°C up to 55°C, whereas the intrinsic viscosity in chloroform decreased from 30°C to 40°C. Cyclohexane is a theta solvent for PS at around 35°C and intrinsic viscosity in cyclohexane increased with increasing column temperature.

Because the hydrodynamic volume is proportional to the molecular size, the intrinsic viscosity can be used as a measure of the molecular size and optimum column temperatures and solvents must be those where no changes in intrinsic viscosity are observed.

REFERENCES

1. Mori, S.; Barth, H.G. *Size Exclusion Chromatography*; Springer-Verlag: New York, 1999; Chap. 5.
2. Mori, S. Effect of experimental conditions. In *Steric Exclusion Liquid Chromatography of Polymers*; Jančа, J., Ed.; Marcel Dekker, Inc.: New York, 1984.
3. Mori, S. High-speed gel permeation chromatography. A study of operational variables. J. Appl. Polym. Sci. **1977**, *21*, 1921.
4. Mori, S.; Suzuki, T. Effect of column temperatures on molecular weight determination by high performance size exclusion chromatography. Anal. Chem. **1980**, *52*, 1625.
5. Mori, S.; Suzuki, M. J. Liquid Chromatogr. **1984**, *7*, 1841.

GPC/SEC/HPLC Without Calibration: Multiangle Light Scattering Techniques

Philip J. Wyatt
Wyatt Technology Corporation, Santa Barbara, California, U.S.A.

INTRODUCTION

Traditional SEC or GPC as used to obtain molar masses and their distributions has been described elsewhere in this volume. The method suffers from three shortcomings:

1. The calibration standards generally differ from the unknown sample;
2. The results are sensitive to fluctuations in chromatography conditions (e.g., temperature, pump speed fluctuations, etc.); and
3. Calibration must be repeated frequently.

DISCUSSION

By adding a multiangle light-scattering[1] detector directly into the separation line, as shown schematically in Fig. 1, the eluting molar masses are determined *absolutely*, thus obviating the need for calibration and elimination of all of the three shortcomings listed. Fig. 1 illustrates also two most important elements associated with making quality light-scattering measurements: an in-line degasser and an in-line filter. The in-line degasser is essential to minimize dissolved gases and, thereby, prevent the production of bubbles during the measurement process. Scattering from such bubbles can overwhelm the signals from the solute molecules or particles. Perhaps even more importantly, the system requires that the mobile phase be dust-free. The filter illustrated is placed between the pump and the injector. Usually, this filter station is comprised of two holders, holding, respectively, a 0.20-μm filter followed by a 0.02- or 0.01-μm filter. Although providing for such pristine operating conditions may seem bothersome, it has been shown that so-called "dirty" solvents, although rarely affecting the refractive index detector (RID) signal, do actually contribute significantly to the degradation of HPLC and SEC columns as well as resulting in the more frequent need to rebuild pumps. The additional solvent cleanup effort is well worth it!

An "absolute" light-scattering (LS) measurement is one that is independent of calibration standards which have "known" molar masses to which the unknown is compared. A LS measurement requires the chromatographer to determine the fundamental properties of the solution (refractive index, dn/dc value) and the detector response (field of view, sensitivity, solid angle subtended at the scattering volume). In addition, other factors must be determined, such as the light wavelength and polarization, geometry of the scattering cell, refractive index of all regions through which the scattered and incident light will pass, and the ratio of the scattered light to the incident light. Generally, these determinations are made in conjunction with appropriate multiangle light-scattering (MALS) software. The importance of light scattering's independence of a set of reference molar masses to determine the molar mass of an unknown cannot be overemphasized.

THEORY

As described in detail in Refs.[1–3], the fundamental equation relating the quantities measured during a MALS detection and the quantities derived is, in the limit of "... vanishingly low concentrations...,",[2] given by

$$\frac{K^*c}{R(\theta)} \approx \frac{1}{MP(\theta)} + 2A_2c \qquad (1)$$

where $K^* = 4\pi^2(dn/dc)^2n_0^2(N_A\lambda_0^4)^{-1}$, M is the weightaverage molar mass, N_A is Avogadro's number, dn/dc is the refractive index increment, λ_0 is the vacuum wavelength, θ is the angle between the incident beam and the scattered light, and n_0 is the refractive index of the solvent. The refractive index increment, dn/dc, is measured off-line (or looked up in the literature) by means of a differential refractive index (DRI) operating at the same wavelength as the one used for the MALS measurements. It represents the incremental refractive index change dn of the solution (solvent plus solute) for an incremental change dc of the concentration in the limit of vanishingly small concentration. Most importantly, the

Encyclopedia of Chromatography DOI: 10.1081/E-ECHR-120040001

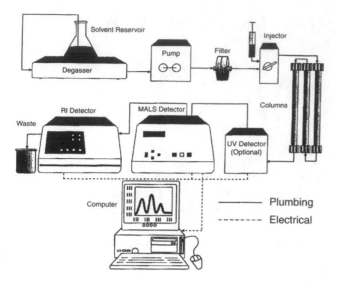

Fig. 1 Schematic diagram showing elements of traditional chromatograph with added MALS detector and dust- and bubble-reducing elements.

excess Rayleigh ratio, $R(\theta)$, and form factor $P(\theta)$ are defined respectively by

$$R(\theta) = \frac{f(\theta)_{\text{geom}}[I(\theta) - I_S(\theta)]}{I_0} \qquad (2)$$

$$P(\theta) = 1 - \alpha_1 \sin^2(\theta/2) + \alpha_2 \sin^2(\theta/2) - \cdots \qquad (3)$$

where

$$\alpha_1 = \frac{1}{3}\left(\frac{4\pi n_0}{\lambda_0}\right)\langle r_g{}^2 \rangle \qquad (4)$$

I_0 is the incident light intensity (ergs/cm^2 s), $f(\theta)_{\text{geom}}$ is a geometrical calibration constant that is a function of the solvent and scattering cell's refractive index and geometry, and $I(\theta)$ and $I_S(\theta)$ are the normalized intensities respectively of light scattered by the solution and by the solvent per solid angle. The mean square radius is given by Eq. (5), where the distances r_i are measured from the molecule's center of mass to the mass element m_i:

$$\langle r_g{}^2 \rangle = \frac{\sum_i r^2 m_i}{\sum_i m_i} = \frac{1}{M}\int r^2 dm \qquad (5)$$

For MALS measurement following GPC separation, the sample concentration at the LS detector is usually diluted sufficiently that the term $2A_2c$ often may be safely dropped from Eq. (1). In some applications involving very high molar masses, it is often worthwhile to perform an off-line determination of the second virial coefficient from a Zimm plot[1–2] to confirm its negligible effect on the derived molar mass of Eq. (1).

BASIC PRINCIPLES

In the limit of vanishingly small concentrations, and the extrapolation of Eq. (3) to very small angles, the two basic principles of light scattering are evident:

1. The amount of light scattered (in excess of that scattered by the mobile phase) at $\theta \approx 0$ is directly proportional to the product of the weightaverage molar mass and the concentration (ergo, measure the concentration and derive the mass!).
2. The angular variation of the scattered light at $\theta \approx 0$ is directly proportional to the molecule's mean square radius (i.e., size).

The successful application of absolute MALS measurements requires a sufficient number of resolved scattering angles to permit an accurate extrapolation to $\theta \approx 0$. Again, all required calculations are performed by the software. Whenever the mobile phase is changed, its corresponding refractive index must be entered into the software program, which should correct automatically for the resultant change of scattering geometry. Fig. 2 shows the normalized light-scattering signals at each scattering angle (detector) as a function of elution volume for a relatively broad sample. Also indicated is the corresponding concentration detector signal.

In conventional SEC measurements, it is necessary to calibrate the mass detector [DRI or ultraviolet (UV)] so that its response yields concentration directly. For example, a DRI detector, following calibration, should produce a response proportional to the refractive index change (Δn) detected. This is related to the concentration change Δc by the simple result $\Delta c = \Delta n/(dn/dc)$. Implicit in the use of a DRI detector, therefore, is that measurement of the concentration of the unknown requires that its differential refractive index, dn/dc, be measured, or otherwise determined.

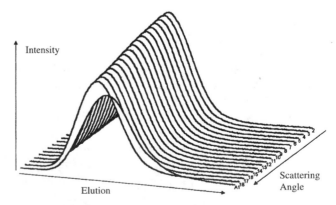

Fig. 2 Light-scattering and DRI signals from MALS setup shown in Fig. 1.

Combining SEC with MALS to produce absolute molar mass data without molecular calibration standards also requires prior calibration of the concentration detector as well as calibration of the MALS detector itself. The latter calibration involves the determination of all geometrical contributions such that the MALS detector measures the Rayleigh excess ratio at each scattering angle. This is most easily achieved by using a turbidity standard such as toluene. Details are found in Ref.[2]. Once the refractive index of the mobile phase is entered, the software[4] performs the required calibration.

DERIVED MASS, SIZE, AND CONFORMATION

The MALS detector produces the absolute molar mass and mean square $\langle r_g^2 \rangle$ radius at each eluting slice. The *root mean square* (rms) *radius* $r_g = \langle r_g^2 \rangle^{1/2}$ is often referred to by the misnomer "radius of gyration." There is a lower limit to its determination, which is generally about 8–10 nm. Below this value, MALS cannot generally produce a reliable value. Nevertheless, whenever both r_g and molar mass M are determined by MALS over a range of fractions present in an unknown sample, the sample's so-called conformation may be determined by plotting the logarithm of the rms radius versus the logarithm of the corresponding molar mass. A resultant slope of unity indicates a rod-like structure, a slope of 0.5–0.6 corresponds to a random coil, and a slope of 1/3 would indicate a sphere. Values below 1/3 generally suggest a highly branched molecular conformation.

Fig. 3 shows the MALS-derived molar mass and rms radius as a function of elution volume for a broad polystyrene sample. Measurements were made in

toluene at 690 nm. The value of dn/dc chosen was 0.11. From Fig. 3, it should be noted that the radius data begins to deteriorate around 10 nm, whereas the mass data extends to its detection limits. From the mass and radius data of Fig. 3, a conformation plot is easily generated with a slope of about 0.57 (i.e., corresponding to a random coil). These same data can also be used immediately to calculate the mass and size moments of the sample as well as its polydispersity as shown in the next section.

MASS AND SIZE MOMENTS

If we assume that the molecules in each slice, i, following separation by SEC, are monodisperse, the mass moments of each sample peak selected are calculated from the conventional definitions[3,5] by

$$M_n = \frac{\sum_i n_i M_i}{\sum_i n_i} = \frac{\sum_i c_i}{\sum_i c_i / M_i} \quad (6)$$

for the *number*-average molar mass, where n_i is the number of molecules of mass M_i in slice i and the summations are over all the slices present in the peak; the concentration c_i of the ith species, therefore, is proportional to $M_i n_i$;

$$M_w = \frac{\sum_i c_i M_i}{\sum_i c_i} = \frac{\sum_i n_i M_i^2}{\sum_i n_i M_i} \quad (7)$$

for the *weight*-average molar mass; and

$$M_z = \frac{\sum_i n_i M_i^3}{\sum_i n_i M_i^2} = \frac{\sum_i c_i M_i^2}{\sum_i c_i M_i} \quad (8)$$

for the z-average ("centrifuge") molar mass. Note how these "moments" are defined. In particular, the zaverage moment corresponds to "the next higher weighting" of both numerator and denominator by the factor Eqs. (6) and (7), of course, have a simple physical interpretation in terms of molecular numbers and concentration. From Eq. (8), it is a simple matter to write down expressions for the z + 1, z + 2, ... moments.

A similar set of expressions may be written down for the so-called size number and weight moments by replacing the mass terms M_i by the mean square radius values at each slice $\langle r_g^2 \rangle_i$ in Eqs. (6) and (7). A z-average term, on the other hand, takes on a more convoluted form.[5] For a random coil conformation under socalled theta conditions, the molar mass is directly proportional to $\langle r_g^2 \rangle$, and an expression that looks identical to Eq. (8) with one of the M_i of the numerator sum replaced by $\langle r_g^2 \rangle_i$ is obtained. However, in general, this "equivalence" is not the case

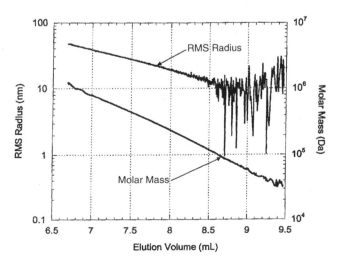

Fig. 3 Molar mass and rms radius generated from data of Fig. 2 as a function of elution volume.

and the "light-scattering" value LS is a better description, namely

$$\langle r_g^2 \rangle_{LS} = \frac{\sum_i c_i M_i \langle r_g^2 \rangle_i}{\sum_i M_i c_i} \tag{9}$$

Despite the non-random-coil-at-theta conditions, Eq. (9) is commonly referred to as the z-average mean square radius. The cross-term $M_i \langle r_g^2 \rangle_i c_i$ of Eq. (9) is a quantity measured directly by light scattering, at small $\sin^2(\theta/2)$, as clearly may be seen by expanding the term $1/P(\theta)$ in Eq. (1) using the expansion of $P(\theta)$ of Eq. (3).

POLYDISPERSITY

Within the peak selected, the sample polydispersity is simply the ratio of the weight to number average (viz. M_w/M_n) obtained from Eqs. (7) and (6), respectively.

DIFFERENTIAL MASS WEIGHT FRACTION DISTRIBUTION

The MALS measurements illustrated by Fig. 2 also may be used directly to calculate the differential mass weight fraction distribution, $x(M) = dW(M)/d(\log_{10}M)$ by using the measured $\log_{10}M$ as a function of the elution volume V. Thus, if the concentration detector's baseline subtracted response is $h(V)$, then $dW/dV = \pm h(V)/\int h(V) \, dV$, the integral representing the sum over all contributing concentrations to the peak. It is then easily shown[6] that

$$x(M) = -\frac{h(V)/\int h(V)dV}{d(\log_{10} M)/dV} \tag{10}$$

Note that for so-called "linear" column separations, the denominator $d(\log_{10}M)/dV$ is just a constant and, therefore, the differential weight fraction distribution will appear as a reflection (small mass first, from left to right) of the DRI signal. In general, column separations are not linear, so the DRI signal is not a good representation of the mass-elution distribution.

BRANCHING

The MALS measurements which eliminate the need for column calibration and all of its subsequent aberrations also permit the direct evaluation of branching phenomena in macromolecules because the basic quantitation of branching may only be achieved from such measurements as shown in the article by Zimm and Stockmayer.[7] Empirical approaches to quantitate branching, using such techniques as viscometry, have been shown to yield consistently erroneous results especially when long-chain branching becomes dominant.

REVERSED-PHASE AND OTHER SEPARATION TECHNIQUES

Because MALS determinations are independent of the separation mechanism, they may be applied to many types of HPLC. Reversed-phase separations are of particular significance because they cannot be calibrated, as sequential elutions do not occur in a monotonic or otherwise predictable manner. Again, as with all MALS chromatography measurements, all that is required is that the concentration and MALS's signals be available at each elution volume (slice).

Another separation technique of particular application for proteins, high-molar-mass molecules, and particles is the general class known as field-flow fractionation (FFF) in its various forms (cross-flow, sedimentation, thermal, and electrical). Once again, MALSdetection permits mass and size determinations in an absolute sense without calibration. For homogeneous particles of relatively simple structure, a concentration detector is not required to calculate size and differential size and mass fraction distributions. Capillary hydrodynamic fractionation (CHDF) is another particle separation technique that may be used successfully with MALS detection.

REFERENCES

1. Wyatt, P.J. Light scattering and the absolute characterization of macromolecules. Anal. Chim. Acta **1993**, *272*, 1.
2. Zimm, B.H. The scattering of light and the radial distribution function of high polymer solutions. J. Chem. Phys. **1948**, *16*, 1093, *16*, 1099 (1948)
3. Billingham, N.C. *Molar Mass Measurements in Polymer Science*; John Wiley & Sons: New York, 1977.
4. ASTRA® software Wyatt Technology: Santa Barbara, CA, 1999.
5. Wyatt, P.J. *Analytical and Preparative Separation Methods of Biomacromolecules*; Aboul-Enein, H.Y., Ed.; Marcel Dekker, Inc: New York, 1999.
6. Shortt, D.W. J. Liquid Chrom. **1993**, *16*, 3371 ± 3391.
7. Zimm, B.H.; Stockmayer, W.H. The dimensions of chain molecules containing branches and rings. J. Chem. Phys. **1949**, *17*, 1301.

SUGGESTED FURTHER READING

Huglin, M.B., Ed.; *Light scattering from Polymer Solutions*; Academic: London, 1972.

GPC/SEC: Introduction and Principles

Vaishali Soneji Lafita
Abbott Laboratories, Inc., Abbott Park, Illinois, U.S.A.

INTRODUCTION

The basic principle of chromatography involves the introduction of the sample into a stream of mobile phase that flows through a bed of a stationary phase. The sample molecules will distribute so that each spends some time in each phase. SEC is a liquid column chromatographic technique which separates molecules on the basis of their sizes or hydrodynamic volumes with respect to the average pore size of the packing. The stationary phase consists of small polymeric or silica-based particles that are porous and semirigid to rigid. Sample molecules that are smaller than the pore size can enter the stationary-phase particles and, therefore, have a longer path and longer retention time than larger molecules that cannot enter the pore structure. Very small molecules can enter virtually every pore they encounter and, therefore, elute last. The sizes, and sometimes the shapes, of the midsize molecules regulate the extent to which they can enter the pores. Larger molecules are excluded and, therefore, are rapidly carried through the system. The porosity of the packing material can be adjusted to exclude all molecules above a certain size. SEC is generally used to separate biological macromolecules and to determine molecular-weight distributions of polymers.

HISTORY

It is not obvious who was the first to use SEC. However, the first effective separation of polymers based on *gel filtration chromatography* (GFC) appears to be that reported by Porath and Flodin.[1] Porath and Flodin employed insoluble cross-linked polydextran gels, swollen in aqueous medium, to separate various water-soluble macromolecules. GFC generally employs aqueous solvents and hydrophilic column packings, which swell heavily in water. Moreover, at high flow rates and pressures, these lightly cross-linked soft gels have low mechanical stability and collapse. Therefore, GFC stationary phases are generally used with low flow rates to minimize high-back pressures. GFC is mainly used for biomolecule separations at low pressure.[2]

Moore described an improved separation technique relative to GFC and introduced the term GPC in 1964.[3] GPC performs the same separation as GFC, but it utilizes organic solvents and hydrophobic packings. Moore developed rigid polystyrene gels, cross-linked with divinylbenzene, for separating synthetic polymers soluble in organic media. These extensively cross-linked gels are mechanically stable enough to withstand high pressures and flow rates. The more rugged GPC quickly flourished in industrial laboratories where polymer characterization and quality control are of primary concern. Since its introduction in the 1960s, the understanding and utility of GPC has substantially evolved. GPC has been widely used for the determination of molecular weight (MW) and molecularweight distribution (MWD) for numerous synthetic polymers.[4]

Other names such as gel chromatography, exclusion chromatography, molecular sieve chromatography, gel exclusion chromatography, size separation chromatography, steric exclusion chromatography, and restricted diffusion chromatography have been utilized to reflect the principal mechanism for the separation. The fundamental mechanism of this chromatographic method is complex and certainly will not be readily incorporated into one term. Strong arguments have been made for many of the above-listed titles.[5] In an attempt to minimize the dispute over the proper name, the term SEC will be used in this entry, as it appears to be the most widely used.

MECHANISM

SEC is a LC technique in which a polymer sample, dissolved in a solvent, is injected into a packed column (or a series of packed columns) and flows through the column(s) and its concentration as a function of time is determined by a suitable detector. The column packing material differentiates SEC from other LC techniques where sample components primarily separate by differential adsorption and desorption. The SEC packing consists of a polymer, generally polystyrene, which is chemically cross-linked so that varying size pores are created. Several models are discussed by Barth et al.[6] to illustrate SEC separation theory. A rather simplified separation mechanism is described here. A polymer sample dissolved in the SEC mobile phase is injected in the chromatographic system. The column

Encyclopedia of Chromatography DOI: 10.1081/E-ECHR-120039998

eluent is monitored by a mass-sensitive detector, which responds to the weight concentration of polymer in the mobile phase. The most common detector for SEC is a differential refractometer. The raw data in SEC consists of a trace of detector response proportional to the amount of polymer in solution and the corresponding retention volume. A typical SEC sample chromatogram is depicted in Fig. 1. An SEC chromatogram generally is a broad peak representing the entire range of molecular weights in the sample. For synthetic polymers, this can extend from a few hundred mass units up to a million or more. The average molecular weight can be calculated in a number of ways. Both natural and synthetic polymers are molecules containing a distribution of molecular weights. The most commonly calculated molecular-weight averages using SEC are the weight-average molecular weight (M_w) and number-average molecular weight (M_n). These terms have been well defined by Cazes.[7]

The weight-average molecular weight is defined as

$$\bar{M}_w = \frac{\sum_{i=1}^{\infty} W_i M_i}{\sum_{i=1}^{\infty} W_i} \tag{1}$$

and the number-average molecular weight is defined as

$$\bar{M}_n = \frac{W}{\sum_{i=1}^{\infty} N_i} = \frac{\sum_{i=1}^{\infty} M_i N_i}{\sum_{i=1}^{\infty} N_i} \tag{2}$$

where W is the total weight of the polymer, W_i is the weight fraction of a given molecule i, N_i is the number of moles of each species i, and M_i is the molecular weight of each species i.

M_w is generally greater than or equal to M_n. The samples in which all of the molecules have a single molecular weight ($M_w = M_n$) are called monodisperse polymers. The degree of polydispersity (i.e., the ratio of M_w to M_n) describes the spread of the molecular-weight-distribution curve. The broader the SEC curve, the larger the polydispersity.

The detector response on the SEC chromatogram is proportional to the weight fraction of total polymer, and suitable calibration permits the translation of the retention volume axis into a logarithmic molecular-weight scale. Calibration of SEC is perhaps the most difficult aspect of the technique because polymer molecules are separated by size rather than by molecular weight. Size, in turn, is most directly proportional to the lengths of the polymer molecules in solution. A length, however, is proportional to molecular weight only within a single polymer type. An absolute SEC calibration would require the use of narrow molecular-weight range standards of the same polymer that is being analyzed. This is not always practical because a wide range of polymer types needs to be evaluated. SEC calibration is often achieved using the "universal calibration" technique, which assumes hydrodynamic volume is the sole determinate of retention time or volume.[8] A series of commercially available monodisperse molecular-weight polystyrenes are the most commonly used SEC calibration standards. If polystyrene standards are used to calibrate the analyses of any other type of polymer, the molecular weights obtained for a polymer sample are actually "polystyrene-equivalent" molecular weights. Numeric conversion factors are available for correlating "molecular weight per polystyrene length" to that of other polymers, but this approach only produces marginally better estimates of the absolute molecular weights. In addition, approaches such as these are usually invalid because the calibration curve for the polymer being analyzed does not often have the same shape as the curve generated with the polystyrene standards.

Size-exclusion chromatograms of narrow-distribution polystyrene standards along with a typical polystyrene calibration curve are shown in Fig. 2a, 2b. The peak retention volume and corresponding molecular weights produce a calibration curve. With a calibration curve, it is possible to determine M_w and M_n for a polymer. The SEC curve of a polymer sample is divided into vertical segments of equal retention volume. The height or area of each segment and the corresponding average molecular weight, calculated from the calibration curve, are then used for M_w and M_n calculations. There are several commercially available software packages

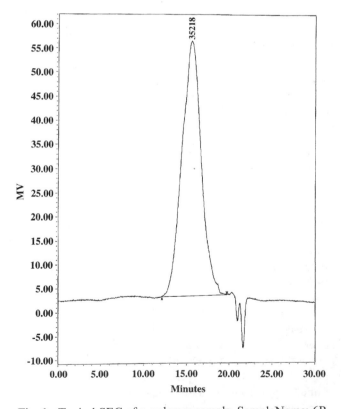

Fig. 1 Typical SEC of a polymer sample. SampleName: 6B Vial: 15 Inj: 1 Ch: 410 Type: Broad Unknown.

that simplify the calculation process for molecular-weight determinations.

APPLICATION

Until the mid-1960s, molecular-weight averages were determined only by techniques such as dilute solution viscosity, osmometry, and light scattering. Most of these techniques work best for polymers with a narrow MWD. None of these techniques, either alone or in combination, could readily identify the range of molecular weights in a given sample. SEC was introduced in the mid-1960s to determine MWDs and other properties of polymers. During the first two decades of SEC acceptance, the emphasis was on improving the fundamental aspects of chromatography, such as column technology, optimizing solvents, and the precision of analysis. Over the past 10 years, there has been an increasing demand for deriving more information from SEC, driven by the need to characterize, more fully, an increasingly complex array of new polymers. Significant developments in SEC detection systems include light scattering, viscometry, and matrix-assisted laser desorption ionization time-of-flight (MALDI–TOF) mass spectrometry and, most recently, nuclear magnetic resonance (NMR) detection in conjunction with SEC for determining MW and chemical composition of polymers. The use of SEC for measuring physiological properties of polymers, especially biopolymers, has become an important area of research.

Finally, SEC is merely a separation technique based on differences in hydrodynamic volumes of molecules. No direct measurement of molecular weight is made. SEC itself does not render absolute information on molecular weights and their distribution or on the structure of the polymers studied without the use of more specialized detectors (e.g., viscometry and light scattering). With these detectors, a "self-calibration" may be achieved for each polymer sample while it is being analyzed by SEC. However, it is possible to calibrate the elution time in relation to molecular weight

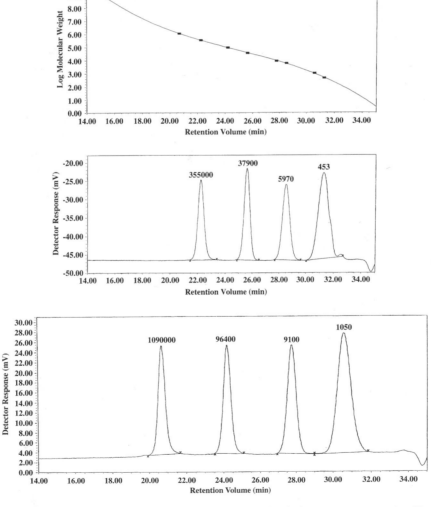

Fig. 2 Typical polystyrene narrow molecular-weight range standard chromatograms and calibration curve.

of known standards. With proper column calibration, or by the use of molecular-weight-sensitive detectors such as light scattering, viscometry, or mass spectrometry, MWD and average molecular weights can be obtained readily.[6] The combined use of concentration sensitive and molecular-weight-sensitive detectors has greatly improved the accuracy and precision of SEC measurements. Thus, SEC has become an essential technique that provides valuable molecular-weight information, which can be related to polymer physical properties, chemical resistance, and processability.

REFERENCES

1. Porath, J.; Flodin, P. Gel filtration: A method for desalting and group separation. Nature **1959**, *183*, 1657.
2. Danilov, A.V.; Vagenina, I.V.; Mustaeva, L.G.; Moshnikov, S.A.; Gorbunova, E.Y.; Cherskii, V.V.; Baru, M.B. Liquid chromatography on soft packing material, under axial compression: Size-exclusion chromatography of polypeptides J. Chromatogr. A. **1997**, *773*, 103.
3. Moore, J.C. Comments on "Gel permeation chromatography. I. A new method for molecular weight distribution of high polymers." J. Polym. Sci. Part A. **1964**, *2*, 835.
4. Lafita, V.S.; Tian, Y.; Stephens, D.; Deng, J.; Meisters, M.; Li, L.; Mattern, B.; Reiter, P. *Proc. Int. GPC Symp. 1998*; Waters Corp.: Milford, MA, 1998; 474–490.
5. Johnson, J.; Porter, R.; Cantow, M. J. Macromol. Chem. Part C. **1966**, *1*, 393.
6. Barth, H.G.; Boyes, B.E.; Jackson, C. Size exclusion chromatography and related separation techniques. Anal. Chem. **1998**, *70*, 251R.
7. Cazes, J. J. Chem. Educ. **1966**, *43*, A567.
8. Boyd, R.H.; Chance, R.R.; Ver Strate, G. Effective Dimensions of Oligomers in Size Exclusion Chromatography. A Molecular Dynamics Simulation Study. Macromolecules **1996**, *29* (4), 1182–1190.

GPC/SEC Viscometry from Multiangle Light Scattering

Philip J. Wyatt
Ron Myers
Wyatt Technology Corporation, Santa Barbara, California, U.S.A.

INTRODUCTION

Viscometric techniques have long been used in combination with GPC/SEC separations since the early discovery[1] that the elution of many classes of divers' polymers follows a so-called "universal calibration" curve. A plot of the logarithm of the hydrodynamic volume, $M[\eta]$, where M is the molar mass and $[\eta]$ the intrinsic (or "limiting") viscosity, against the elution volume V yields a common curve (differing for each mobile phase, operating temperature, and column set) along which polymers of greatly differing conformation appear to lie. Neglecting the fact that the errors of such fits can be quite large (the results are usually presented on a logarithmic scale), the concept of universal calibration (UC) allows one to estimate (from the UC curve) the molar mass of an eluting fraction by measuring only the intrinsic viscosity, $[\eta]$, and the corresponding elution volume (time). Key to the measurement of $[\eta]$ is the determination of the specific, η_{sp}, or relative, η_{rel}, viscosity and the concentration c, both in the limit as $c \to 0$. These viscosities are defined by

$$\eta_{sp} = \frac{\eta \pm \eta_0}{\eta_0} \tag{1}$$

and

$$\eta_{rel} = \frac{\eta}{\eta_0} \tag{2}$$

where η is the solution viscosity and η_0 is the viscosity of the pure solvent. Because $\eta_{rel} = \eta_{sp} + 1$, it is easily shown for η_{sp} small compared to unity that

$$\lim_{c \to 0} = \frac{\ln(\eta_{rel})}{c} = \lim_{c \to 0} \frac{\eta_{sp}}{c} = [\eta] \tag{3}$$

For the case of GPC/SEC elutions, the concentration c following separation is generally so small that Eq. (3) is assumed to be valid.

THE MARK–HOUWINK–SAKURADA EQUATION

Even without the use of a UC curve (one must be generated for each series of measurements), measurement of $[\eta_0]$ is believed by some to yield an intrinsic viscosity-weighted molar mass.[2] Most importantly, there is a historic interest in the relation of $[\eta]$ to molar mass and/or size. Indeed, the study and explanation of UC has occupied the theorists for some time and, accordingly, there are various formulations describing such relationships.[2] For linear polymers, the most popular empirical relationship between $[\eta]$ and molar 1mass is the Mark–Houwink–Sakurada (MHS) equation

$$[\eta] = KM^a \tag{4}$$

where K and a are the MHS coefficients. For many polymer–solvent combinations, a plot of $\log([\eta])$ versus $\log(M)$ is linear over a wide range of molar masses. In other words, both K and a are constant throughout the range. Thus, the equation may be used for such polymer–solvent combinations to determine molar mass by measuring $[\eta]$.

Unfortunately, for some solvent–polymer combinations, even for nearly ideal random coils such as polystyrene, the coefficients are not constant but vary with molar mass.

THE FLORY–FOX EQUATION

In the various theoretical attempts to explain the relation between $[\eta]$ and the molar mass M, a relation derived by Flory and Fox for random coil molecules is often applied to interpret viscometric measurements for even more general polymer structures. Although applicable to a broader range of polymers than the MHS equation, the Flory–Fox relation has its own shortcomings. Nevertheless, its frequent use and good correlation with experimental data over a wide range of polymer types confirms its potential for combination with light-scattering measurements to eliminate the need for separate viscometric determinations. In its most general form, the Flory–Fox equation is given by

$$M[\eta] = \Phi\left(\sqrt{6r_g}\right)^3 \tag{5}$$

Encyclopedia of Chromatography DOI: 10.1081/E-ECHR-120040002

where r_g is the root mean square radius (or "radius of gyration"). The excluded volume effect is taken into account by representing the Flory–Fox coefficient as $\Phi = \Phi_0(1 \pm 2.63\varepsilon + 2.86\varepsilon^2)$. The constant $\Phi_0 = 2.87 \times 10^{23}$ and ε is related to the MHS coefficient a by the relation $2a = 1 + 3\varepsilon$. Thus, ε ranges from 0 at the theta point to 0.2 for a good solvent. Eq. (5) is of particular interest because multiangle light-scattering (MALS) measurements [3] determine M and r_g directly. Thus, if a polymer–solvent combination is well characterized by Eq. (5), then this equation may be used directly to calculate the intrinsic viscosity without need for a viscometer.

VISCOMETRY WITHOUT A VISCOMETER

As we have seen earlier, $[\eta]$ may be calculated directly from the (absolute) MALS measurements of M and r_g using Eq. (5). For linear polymers spanning a relatively broad molecular range (an order of magnitude or more), the measurement of M and r_g permits the determination of the molecular conformation defined by

$$r_g = kM^\alpha \tag{6}$$

where k and α are constants generally calculated from the intercept and slope of the least-squares fitted plot of $\log(M)$ against $\log(r_g)$. Combining Eqs. (4) and (5), we obtain

$$KM^{\alpha-1} = \left(\Phi\sqrt{6r_g}\right)^3 \tag{7}$$

Solving for r_g and substituting into Eq. (6) yields

$$\frac{K^{1/3}M^{(a-1)/3}}{\Phi\sqrt{6}} = kM^\alpha \tag{8}$$

Therefore, we have the following relations between the coefficients:

$$a = 3\alpha - 1 \quad \text{and} \quad K = \Phi\left(\sqrt{6k}\right)^3 \tag{9}$$

Eq. (9) show that we can obtain the MHS coefficients a and K directly from a MALS measurement and a determination from such measurements of the molecular conformation parameters α and k. Note that when long-chain branching becomes significant and the molecular conformation becomes more compact such that $\alpha \rightarrow 1/3$, the MHS equation, Eq. (4), shows

that the intrinsic viscosity no longer varies with molar mass, but becomes constant. This condition also represents a failure of the Flory–Fox equation and the concepts associated with the use of intrinsic viscosity as a means (through UC, for example) to determine molar mass. For linear polymers for which MALS measurements yield values for r_g and M at each eluting slice, all of the important viscometric parameters may be derived directly from the Flory–Fox relation and the MHS equation, as has been shown. For more complex molecular structures or solvent–solute interactions, the MHS coefficients are no longer constants and the empirical theory itself begins to fail.

It is well known,[3] however, that the MALS measurements begin to fail in the determination of r_g once r_g falls below about 8–10 nm, even though the M values generated still remain precise. This lack of precision is due to the limitations of the laser "ruler" to resolve a size much below about one-twentieth of the incident wavelength. The trouble with empirical relations, such as the relation between intrinsic viscosity and molar mass, is that they too are often limited to regions where such concepts are applicable. For very small molar masses, the conformation of a polymer molecule may be poorly described by the same theory applied for the larger constituents of a sample. Although MALS conformation measurements may be extrapolated in r_g for the case of linear polymers, such extrapolations must be used with great caution. Similar remarks apply, of course, to the use of viscometric measurements for characterizing complex molecules whose conformations (and, therefore, MHS coefficients) are changing with M.

Fig. 1 presents the conformation plot for $\log(r_g)$ versus $\log(M)$ as obtained from a MALS measurement for the polystyrene broad linear standard NIST706

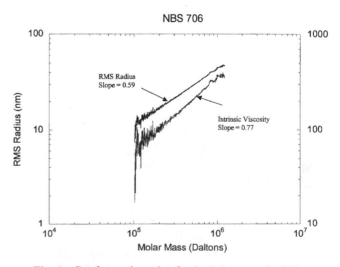

Fig. 1 Conformation plot for $\log(r_g)$ versus $\log(M)$.

in toluene. Superimposed thereon is a plot of the calculated $\log[\eta]$ as a function of $\log(M)$ for the same sample. From the latter plot, the MHS coefficients may be deduced by inspection of the slope and intercept to yield $a = 0.77$ and $K \approx 0.008$.

REFERENCES

1. Benoit, H.; Grubisic, Z.; Rempp, R. Reflections on "a universal calibration for gel permeation chromatography." J. Polym. Sci. B. **1967**, *5*, 753.

2. Kamide, K.; Saito, M. *Determination of Molecular Weight*; Cooper, A.R., Ed.; John Wiley & Sons: New York, 1989.

3. Wyatt, P.J. Light scattering and the absolute characterization of macromolecules. Anal. Chim. Acta. **1993**, *272*, 1.

SUGGESTED FURTHER READING

Billingham, N.C. *Molar Mass Measurements in Polymer Science*; John Wiley & Sons: New York, 1977.

Zimm, B.H. The Scattering of light and the radial distribution function of high polymer solutions. J. Chem. Phys. **1948**, *16*, 1093; *16*, 1099 (1948).

G

Gradient Development in TLC

Wojciech Markowski
Medical University, Lublin, Poland

INTRODUCTION

The main task of analytical TLC is separation of sample from matrix, separation of sample components, their identification, and the measurement of peak heights or areas for quantitative purposes. Finally, the peaks should be narrow and symmetrical. Two problems are related to the analysis: choice of suitable conditions of development, such that all components of the sample are eluted in optimal range of retention factor; and their separation, allowing for identification and quantitation.

GENERAL ELUTION PROBLEM

In practice, a problem that appears very often is components of widely differing retention properties being present in the sample.[1–3] For example, consider a model mixture composed of 15 components with capacity factors $k_0(j)$ forming a geometrical progression and exponentially dependent on the modifier concentration (molar or volume fraction φ), in accordance with the Snyder–Soczewiński model of adsorption (Fig. 1a–d).[4]

$$k_0(j) = \frac{25.6}{2^j}, \quad k(j,i) = \frac{k_0(j)}{c(i)^{m(j)}} \quad (1)$$

$$R_F(j,i) = \frac{1}{1 + k(j,i)} \quad (2)$$

where j is the code of the solute (1–15), i the number of elution step (elution fraction), $k(j,i)$ the capacity factor of solute j in the ith step, $R_F(j,i)$ the retardation factor of solute j in the fraction of eluent of concentration $\varphi(i)$, $\varphi(i)$ the concentration of modifier (mole/volume fraction) in the ith step, $v(i)$ the volume of eluent delivered in the ith step, and $m(j)$ the slopes of linear plots of $R_M(j,i) = R_M^0(j) - m(j)\log\varphi(i)$.

In Fig. 1 are presented chromatograms (simulated) to illustrate the "general elution problem" addressed to TLC. Consider a representative mixture for components of a wide range of polarity. If the elution conditions are suitable for weakly polar compounds, the strongly polar ones will remain at or near the start line. On the other hand, if the system is set up so that strongly polar components are separated, the weakly polar ones will accumulate at the mobile-phase front. It can be seen from the picture that no isocratic eluent can separate all components. A pure modifier of $\varphi = 1.0$ (100%) well separates solutes S3–S6, and the less polar solutes accumulate near the solvent front (Fig. 2a); for $\varphi = 0.5$ (50%), solutes S5–S8 are well separated, the remaining ones accumulating near either the start line or the front line (Fig. 2b); for $\varphi = 0.05$ (5.0%), solutes 1–11 accumulate near the start line; only solutes S12–S15 are separated in the optimal R_F window (Fig. 2c). Thus, only about one-third of the components can be satisfactorily separated by isocratic elution. This problem is named "the general elution problem" and is attacked in different ways in different modes of chromatography (Fig. 2d). Usually, some kind of change of parameters is done. This involves a stepwise or continuous change of chosen parameter. Both in column chromatography and in TLC, the isocratic mode is preferred unless the "general elution problem" is encountered. Its solution may consist of gradient elution (stepwise or continuous), gradation of stationary phase, development with a mixed eluent composed of solvents of different polarity (polyzonal TLC), or temperature programming.[3,5,6] When the strength of mobile phase delivered to the layer is programmed, this is called gradient development. One of advantages of gradient TLC is the feasibility of application of both simple and reversed gradients (decreasing modifier concentration) and a complex gradient—a combination of both types of gradient. The gradient of the mobile phase is both simple and practical in application. In a simple gradient, the eluent strength is increased from the beginning to the end of the development process. The reversed gradient can be applied in the case of multiple development (MD), and the eluent strength decreases with increase in number of developments.

DESCRIPTION OF THE MOBILE-PHASE GRADIENT

The gradient is defined by the variation of elution strength of a series of single solvents or a mixture of solvents, or by the variation of composition of the mobile phase, by the percentage content of the weaker

Encyclopedia of Chromatography DOI: 10.1081/E-ECHR-120040003

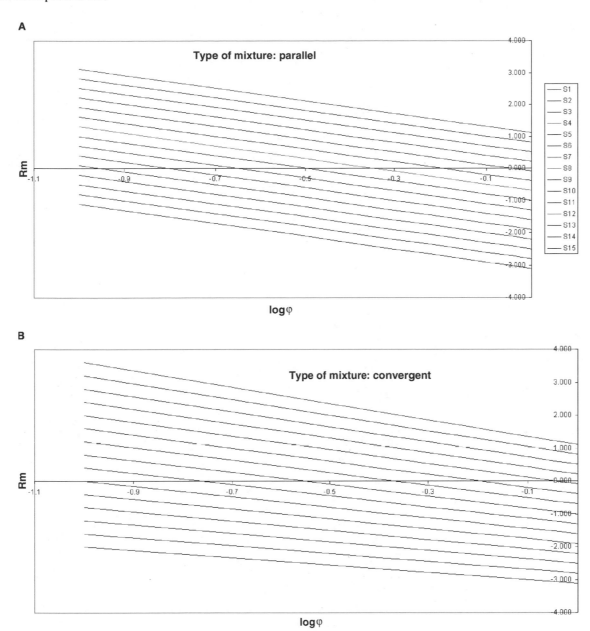

Fig. 1 Family of 15 $R_M(j, i) = R_M^0(j) - m(j) \log c(i)$ plots for model mixture of solutes S1–S15 according to adsorption model of Snyder–Soczewiński. (A) Family of parallel lines; (B) family of convergent lines; (C) family of divergent lines; (D) family of crossing lines. *(Continued next page.)* *(View this art in color at www.dekker.com.)*

component A and the stronger component B, called the modifier. The gradient is also characterized by its steepness, shape, and complexity. The steepness is defined by the concentration of the modifier of the first and last fractions of the eluent delivered to the adsorbent layer. When the differences in modifier concentrations between all steps are constant, the gradient is called linear. When these differences are large in the beginning and then decrease in the consecutive steps, it is known as a convex gradient program, and in the opposite case, as a concave gradient. The complexity of the gradient is related to the number of fractions

of mobile phase delivered to the layer and combination of three basic profiles and dimensions of the volume of fractions. Fig. 3 illustrates different profiles of stepwise simple gradient use in TLC.

MIGRATION OF SOLUTES UNDER MOBILE-PHASE GRADIENT CONDITIONS

The use of a stepwise gradient of the mobile phase is well described by the theoretical models.[4–8] In the ideal situation, the spots of solutes are overtaken by

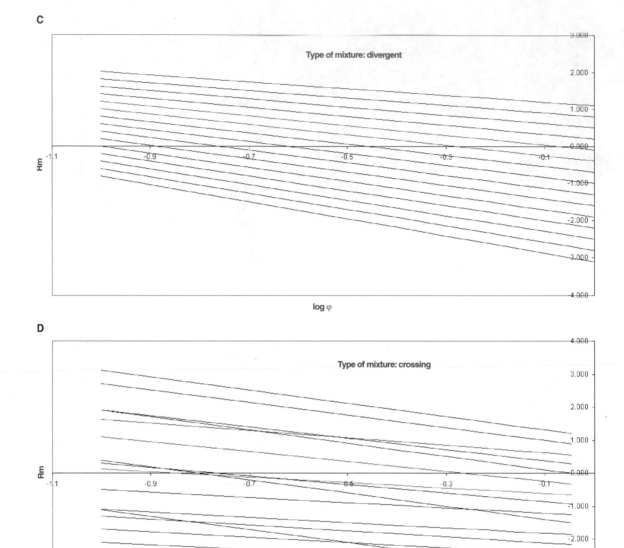

Fig. 1 *(Continued)*

consecutive fronts of increased modifier concentration, accelerating their migration so that even strongly retained solutes start to migrate. Depending on the polarities of the solutes, some migrate all the time in the first concentration zones [when the retardation factor of the solutes follow the condition $R_F(j,1) \geq 1 - v(1)$], or they are overtaken by the consecutive zones of higher concentration. It can be seen (Fig. 4) that both weakly polar (15–13) and strongly polar (1–5) compounds are well separated in the final chromatogram. The migration of the components under conditions of stepwise gradient with one void volume of mobile phase is describing by the following equations: In the case when solutes migrate only in the first

zone, the final position $R_P(j)$ is specified by

$$R_P(j) \;=\; R_F(j,1) \;=\; \frac{1}{1 + k(j,1)} \tag{3}$$

and for other solutes migrating through different zones of concentration, the final position is specified by

$$R_P(j) \;=\; \sum_{i=1}^{h-1} v(i)\,\frac{R_{F(j,i)}}{1 - R_{F(j,i)}}$$
$$+\; R_{F(j,h)}\left(1 - \sum_{i=1}^{h-1} v(i)\,\frac{1}{1 - R_{F(j,i)}}\right) \tag{4}$$

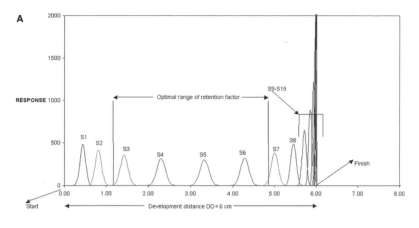

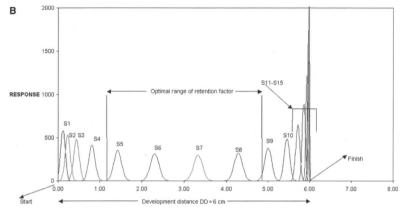

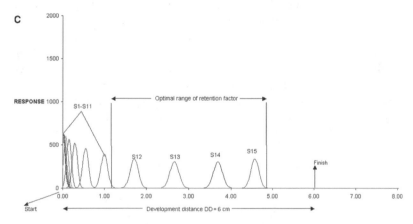

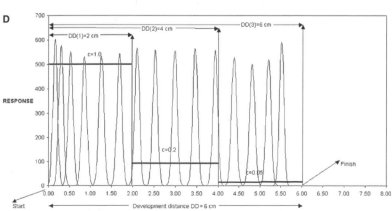

Fig. 2 Simulated chromatograms of hypothetical solutes S1–S15 (Fig. 1) for isocratic conditions of development. Concentration of modifier in volume fraction: (A) $\varphi = 1.0$; (B) $\varphi = 0.5$; (C) $\varphi = 0.05$. Other conditions of simulation: development distance $DD = 6$ cm; spreading of zone (according to Eqs. (8)–(10) and Belenkii model of spot broadening.[5] (D) Simulated chromatogram of hypothetical solutes S1–S15 for G IMD. Number of developments $n = 3$; increment of development distance $IDD = 2$ cm; total distance of development $DD = 6$ cm. Reverse gradient of mobile phase: $\varphi(1) = 1.0$; $\varphi(2) = 0.2$; $\varphi(3) = 0.05$. *(View this art in color at www.dekker.com.)*

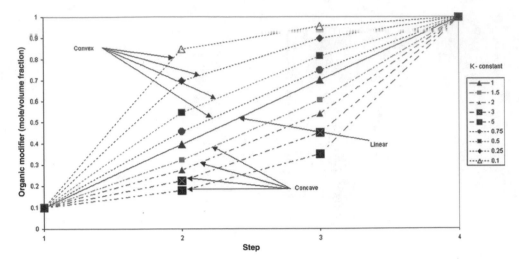

Fig. 3 Various stepwise gradient profiles used in simple gradient. Constant K is a measure of deviation from linearity by convex and concave profiles of gradient. *(View this art in color at www.dekker.com.)*

(For detailed derivation and discussion, see Ref.[7].) Both equations could be applied to formulate computer programs (in any programming language or using a spreadsheet) that calculate the final spot position for a given gradient program and retention–eluent composition relationships. The retention–eluent composition relationships are obtained from preliminary isocratic runs. The experimental results obtained in Refs.[9,10] are in good agreement with the theoretical values calculated from Eqs. (1) and (2) (average error 1.5% and 0.17%). The simulation of gradient development can help considerably in the selection of an optimal program for a given system of adsorbent/mobile phase.[11] In the model, it is assumed that the stagnant mobile phase in the pores of the adsorbent is rapidly displaced and demixing does not occur. In reality, demixing takes place (especially in the first fractions of low concentrations of modifier) and the exchange of the stagnant solvent in the pores with the mobile phase is slow; therefore, the boundaries of the concentration zones are not sharp but somewhat diffuse, and the solutes migrate in zones of intermediate properties. The demixing effect may cause deviation of the gradient profile from the planned one, especially for low concentration of modifier, and in consequence, the concentration fronts are delayed. For this reason, the distances of migration of solutes in lower concentration zones are smaller. The final relative position is lower in comparison to the position expected for ideal conditions. Because of the complexity of gradient development, only some rules of thumb can be given. In planning the gradient program, it is necessary to choose conditions under which the weakly retained components do not migrate with the front of the mobile phase and the strongly retained ones do not

remain on the start line. After the choice of adsorbent (the first choice is silica gel), the next step is selection of the eluent. After Ref.,[8] the following series of solvents with increasing elution strength can be applied to TLC on silica gel: heptane, trichloroethylene, dichloromethane, diisopropyl ether, ethyl acetate, and isopropanol. They can be used as one-component mobile phases or as components of multiple-component mobile phases. These solvents can be replaced by other solvents with similar strength but different selectivity belonging to the eight groups exploited in the Prisma model of optimization.[12] The conditions of development depend on the type of mixture of components. Inspection of many experiments carried out on the relationships between retention parameter (R_M) and $\log \varphi$ or φ permits one to distinguish the following groups of solutes (Fig. 1): parallel, crossing, convergent, divergent.[13] The parameters that influence the separation are number of steps, volumes of steps, and profile of gradient including difference in concentration between the first and last step. Additionally, with the same number of steps of gradient, the volumes of eluent fractions can be equal or different. In Fig. 4, there are examples showing the influence of profile (linear, concave) and different volumes of the steps on the separation of an illustrative mixture of 15 solutes (parallel). In Table 1 are presented the minimal resolution ($R_{S min}$—elementary criteria) and the multipeak criterion (MPC—product of the elementary criteria) values obtained for basic type of mixture and three programs of gradient (concave, linear, convex). These criteria were selected for estimation of the quality of chromatograms.[14] The $R_{S min}$ is given for the least resolved solute pair. The MPC is expressed as a percentage. When all compounds are equally

A

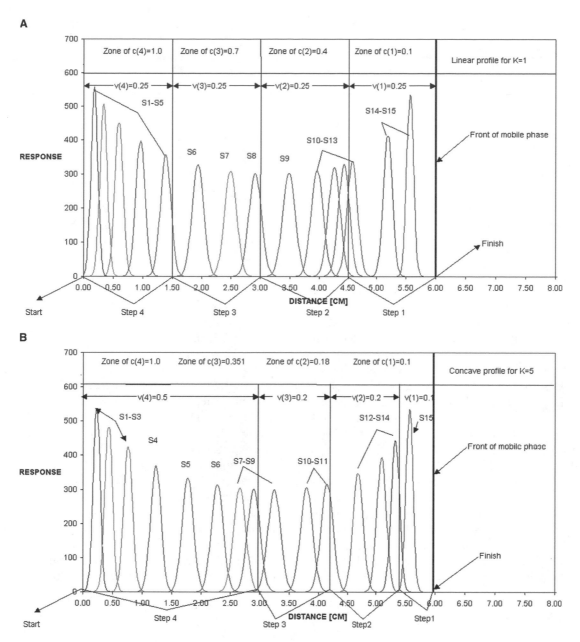

B

Fig. 4 Simulated chromatograms of hypothetical solutes S1–S15 for simple gradient development. (A) Linear profile for $K = 1$; (B) concave profile for $K = 5$. *(View this art in color at www.dekker.com.)*

spaced from each other and from the chosen boundaries, the function has a maximal value of 100%. It follows from the data presented in Table 1 that profiles can be ordered as follows: concave, linear, and convex. The application of the stepwise mobile-phase gradient greatly improves the separation of complex mixtures (e.g., plant extracts). In such cases, a gradient program with more steps of different concentrations is recommended; in many cases, the number of recognized spots is considerably increased in comparison to isocratic elution.[15] An example of practical application of a simple stepwise gradient is presented in Fig. 5, where separation of a quaternary alkaloid mixture is

reported.[16] The application of a stepwise gradient using organic eluents allows one to avoid more time consuming systems, like cellulose powder and aqueous eluents.[17]

GRADIENT MULTIPLE DEVELOPMENT (MD)

In all techniques of MD, the plate is repeatedly developed in the same direction, with intermittent removal of the mobile phase between consecutive developments. Three basic criteria can be used to classify the methods of MD.[18–20] They are distance

Table 1 Influence of gradient profile on selected criteria for estimation of quality of separation of different types of mixtures in stepwise gradient development

Profile of gradient	Parallel		Convergent		Divergent		Crossing	
	R_S[a]	MPC[b]	R_S	MPC	R_S	MPC	R_S	MPC
Linear	0.392	19.15	0.513	4.098	0.140	7.128	0.170	0.204
Convex	0.044	0.07	0.192	0.660	0.047	0.340	0.494	0.867
Concave	0.472	12.99	0.641	5.5	0.600	27.079	0.537	0.900

[a]Minimal values.
[b]Multipeak criterion in %.

of development, properties of the mobile phase used in the process of development, and automation of the development and drying processes—automatic multiple development (AMD). The simplest version of MD is when the distances of development are identical in each step (unidimensional multiple development—UMD) and the mobile phase in each step is identical as well. A variation of this technique, called incremental multiple development (IMD), consists of stepwise change of the development distance, which is shortest in the first step and is then increased, usually by a constant increment (equal distance or time); the last development step corresponds to the maximum development distance. If, in the process of MD, the solvent strength of the mobile phase is varied, the technique is called gradient multiple development (G UMD or G IMD). The change in the mobile phase may concern several or all steps. The MD technique gives the possibility to use, in sequence, systems of mobile phase with very different selectivity and of increasing or, in most cases, decreasing elution strength. The process of MD with any variation of distance and mobile-phase composition can be described by the model

and equations reported, modified to take into account the intermittent evaporation of solvents:[19]

$$SMD(j) = SMD(j, n - 1) + [DD(n) - SMD(j, n - 1)] \cdot R_F(j, n) \tag{5}$$

$$R_P(j) = \frac{SMD(j)}{DD(n)} \tag{6}$$

where $SMD(j)$ is the total migration distance of solute j after n steps, frequently expressed in millimeters, $SMD(j, n - 1)$ the sum of the migration distances of solute j after $n - 1$ steps, and $DD(n)$ the development distance in the last, nth step. $R_P(j)$ is the final position of the spot after n steps of gradient and is equal to the sum of distances traveled by a solute, divided by development distance corresponding to the distance between start and finish. Eqs. (5) and (6) are of typical recurrent type, in which the $(k - 1)$th value for the sum of migration distances traveled by a solute is needed to calculate the kth value. In the simplest case, e.g., two groups of solutes with strong differences

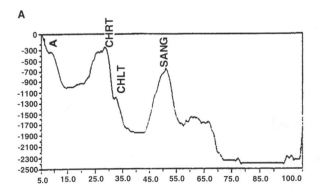
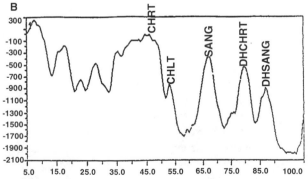

Fig. 5 (A) Densitogram obtained from micropreparative zonal chromatography of a mixture of quaternary alkaloids on silica plate with toluene/EtOAc/MeOH (83 : 15 : 2) as mobile phase. Detection by UV at $\lambda = 254$ nm. (B) Densitogram obtained from micropreparative zonal chromatography of a quaternary alkaloid mixture on silica plate. Gradient elution with T/EtOAc/MeOH, $n = 1$; 75 : 25 : 5, $n = 2$; 70 : 20 : 10, $n = 3$; 70 : 15 : 15, $n = 4$; EtOH/CHCl$_3$/AcOH (67 : 30 : 3) as mobile phases. Detection by UV at $\lambda = 254$ nm. (From Ref.[16]. Copyright Research Institute for Medicinal Plants, Hungary.)

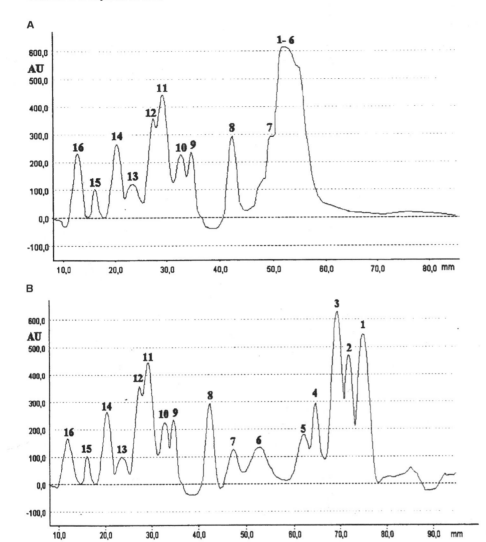

Fig. 6 Densitograms of flavonoids: (A) after first development with EtOAc/HCOOH/H_2O (85:15:0.5); (B) after second development with CH_2Cl_2/EtOAc/HCOOH (85:15:0.5). 1–7, Aglycones; 8–16, glycosides. (From Ref.[21]. Copyright Research Institute for Medicinal Plants, Hungary.)

in polarity, the version of a decreasing stepwise gradient with two steps can be applied. The layer is developed 2/3 of the distance with a polar eluent that separates the most polar components in the lower part of the chromatogram; the less polar components are accumulated in the front area. Their separation occurs in the second stage, when the layer is developed to the full distance with a less polar eluent (Fig. 6).[21,23] The polar components are not significantly affected by the mobile phase used in the second step and keep their positions from the first step. The incremental, multistep version of this technique, with programmed, automated development and evaporation steps, is called automated multiple development (AMD) and the method is considered to be the most effective and versatile TLC technique.[22]

Eqs. (5) and (6) could be applied to formulate computer programs that calculate the final values of $R_P(j)$ for a given gradient program and the retention–eluent composition relationships for selected systems. The systems very often applied in MD contain

thin layers of adsorbents with definite groups like diol, cyano, or amine.[24] The basic equation describing the retention of solutes in such systems has the form of Eq. (2) or the following formula:[25]

$$R_F(j, i) = \frac{1}{1 + k_0(j)/10^{m(j)\cdot c(i)}} \quad (7)$$

G IMD permits application of gradient profiles similar to those presented in Fig. 3. The gradient profiles begin with a high concentration of modifier and end with a low concentration. In MD, it is also possible to create segmented gradients. In the majority of practical applications, the number of systems used in the program of development can be more than two. The chromatograms obtained in MD (G IMD) show mostly evenly distributed peaks between the start and finish line. Fig. 7 shows a densitogram of chamomile extract obtained in MD with improved separation compared to the isocratic mode. The marked fractions on the densitogram were recovered from the plate

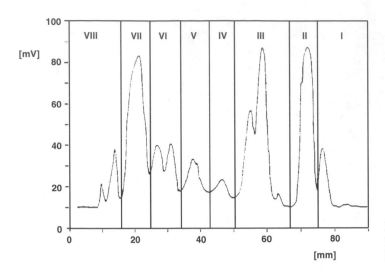

Fig. 7 Densitogram of chamomile extract obtained by MD. (From Ref.[26]. Copyright Research Institute for Medicinal Plants, Hungary.)

and further analyzed by GC/MS. Chromatograms obtained in this way are much simpler (compared to that of the extract), which simplifies their identification by GC/MS.[26]

The application of G IMD to the analysis of complex mixtures is based on the selection of mobile phases suitable to each group of solutes present in the mixture, e.g., polar, weakly polar, and/or nonpolar. In the next step, it is possible to design a gradient program. The profile of the gradient can be adjusted to specific properties of the components of particular groups. The number of steps may be varied depending on the number of components—more components require more steps.

Considering the distance of development, it is possible to develop chromatogram on the total distance with a mobile phase of rather low elution strength and, after removal of the mobile phase, to develop on a shortened distance with a very strong mobile phase. Fig. 8 illustrates the separation of two groups of alkaloids by this method.[27] The condition for the second step of development was that solutes to be separated in the second step still be on or near the start line after the first step. In this mode of MD, the compression effect related to multiple passing of the mobile phase is absent. In UMD, the application of a simple gradient is possible as well. It is used to fractionate complex mixtures by separating just a few solutes in each step. In this case, the plate has to be scanned after some steps and the results recorded. This mode cannot be applied when the picture of the final separation is required as a single chromatogram.

The AMD technique finds application in various fields, such as, for example, the determination of pesticides in water,[28,29] herbicides in plants,[30] and biogenic amines in fish meal,[31] the separation of gangliosides,[32] and the analysis of plant material by

coupling with other chromatographic (reverse-phase HPLC, HPLC/MS) and spectroscopic methods (UV, FTIR).[33–35]

MECHANISM OF COMPRESSION OF CHROMATOGRAPHIC ZONES

One of the advantages of gradient elution is the compression of zones. Each passage of the front of the mobile phase or of the concentration of the multiple-component mobile phase through a spot leads to compression of the spot in the direction of development. This is due to the fact that the front of the increased eluent concentration first reaches the lower edge of the spot, so that the solute molecules in this region start to move (MD) or accelerate their migration (gradient) earlier than the molecules in the farther parts of the spot. When the front of the mobile phase or of the concentration zone overtakes the whole spot, the compressed spot continues to migrate and gradually becomes more diffuse, as in isocratic elution. If the two mechanisms, compression and diffusion, become counterbalanced, the spot may migrate through considerable distances without any marked broadening (Fig. 9). The final width of the spot can be calculated from the equation

$$\sigma_x{}^2 = \sigma_{\text{spotting}}{}^2 + \sigma_{\text{chrom}}{}^2 + \sigma_{\text{inst}}{}^2 + (-\Delta)_{\text{compression}}$$
(8)

The last term of the equation is responsible for the compression effect. The contributions of particular terms of the equation can be estimated from the following equations. For spreading of the zone caused

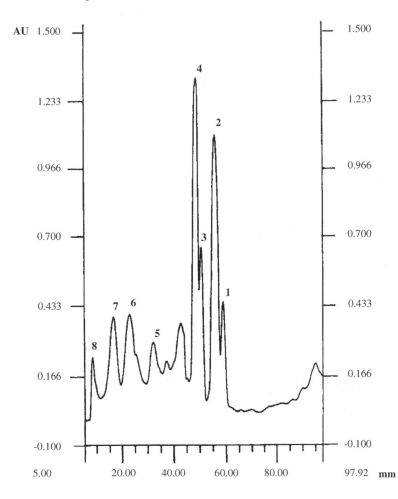

Fig. 8 Two-step gradient elution of the fraction of quaternary alkaloids of *Chelidonium majus* L. Stationary phase: Kieselgel Si 60. Mobile phase: first gradient step: T/EtOAc/MeOH (80:15:5, v/v), $DD(1) = 8$ cm; second gradient step: EtOH/CHCl$_3$/AcOH (67:30:3, v/v), $DD(2) = 4$ cm. Compounds: 1—chelirubine, 2—sanguinarine, 3—chelilutine, 4—chelerythrine, 5—corysamine, 6—berberine, 7—coptisine, 8—magnoflorine. (From Ref.[27]. Copyright Vieweg Verlag, Germany.)

by the process of chromatography,

$$\sigma_{\text{chrom}}^2 = \overline{H \cdot MD(j,i)} \qquad (9)$$

where $\overline{H}$ is the average height equivalent to a theoretical plate and $MD(j, i)$ is migration distance for solute j in the ith step. The values of this term depend on the efficiency of chromatographic layer and on the velocity of the mobile phase associated with migration distance $MD(j, i)$.

The compression effect can be calculated from the formula

$$\Delta = [0.25 \cdot w(j, i - 1) \cdot R_F(j, i)]^2 \qquad (10)$$

where $w(j, i-1)$ is the width of the zone at the end of the preceding step. The compression effect is most effective for zones with high R_F value. The contribution to zone broadening associated with the properties of the detection and recording devices are negligible. During the MD process, the migration distances of particular solutes are rather short; the contribution is not very

high and decreases with the progress of development. Additionally, the effect of compression keeps the spots narrow. The formation of more compact spots causes an increase in sensitivity of detection, in comparison to the spots after single development, and an increase of peak capacity.[36]

Incremental multiple development provides superior separation in comparison to multiple chromatography, in this case, by minimizing zone broadening and enhancing the zone center separation by migrations of the sample components over a longer distance while maintaining a mobile-phase flow rate range close to the best value for the separation. This variant can also be achieved by change of the point of delivery of the eluent to the layer.[37] Not all compounds are suitable for separation by MD. Compounds with significant vapor pressure may be lost during the repeated solvent evaporation steps. Certain solvents of low volatility and/or high polarity, such as acetic acid, triethylamine, dimethyl sulfoxide, and so forth are unsuitable as mobile phases because of the difficulty of removing them from the layer by vacuum evaporation between development steps. Water can be used, but the drying steps are then lengthy. The solvent residues remaining

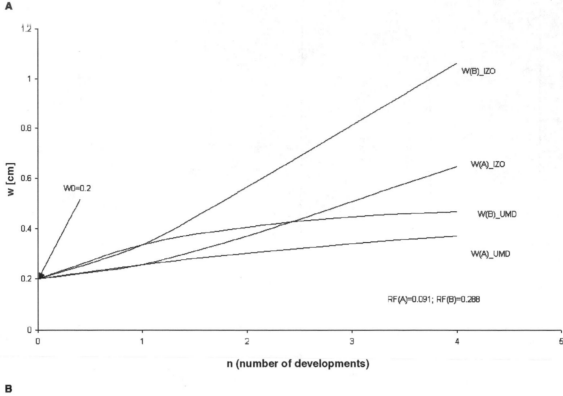

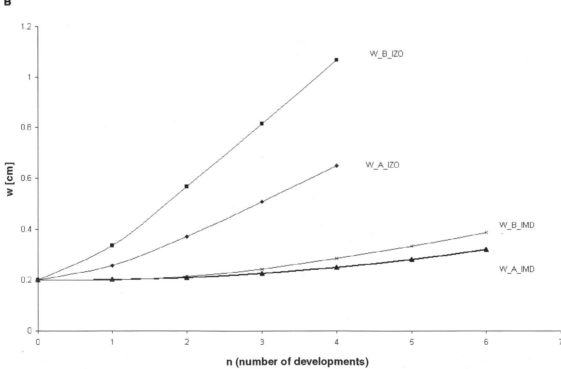

Fig. 9 Comparison of zone width in different modes of MD. (A) Zone width in a single isocratic development and unidimensional development. $R_F(A) = 0.091$, $R_F(B) = 0.288$, $\varphi = 0.05$. (B) Zone width in a single isocratic development and IMD. Increment of development distance $IDD = 1$ cm; number of developments $n = 6$; $\varphi = 0.5$. (C) Zone width in UMD and IMD. (D) Zone width in IMD and GIMD. Gradient: number of steps $n = 6$; concentration program: $\varphi(1) = 0.3$, $\varphi(2) = 0.2$, $\varphi(3) = 0.1$, $\varphi(4) = 0.05$, $\varphi(5) = 0.05$, $\varphi(6) = 0.05$; increment of development distance $IDD = 1$ cm. *(Continued next page.)* *(View this art in color at www.dekker.com.)*

C

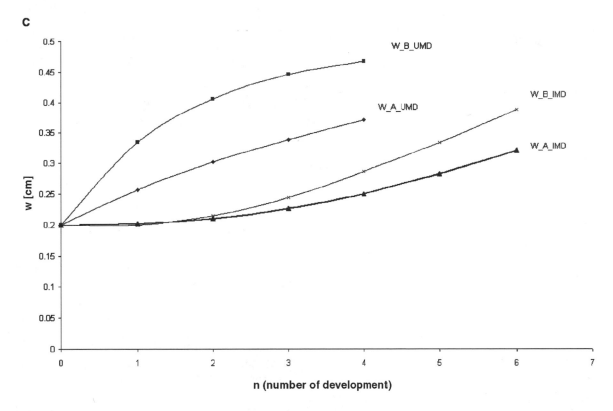

n (number of development)

D

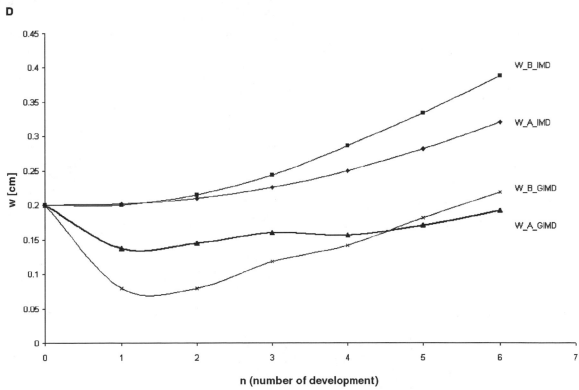

n (number of development)

Fig. 9 *(Continued)*

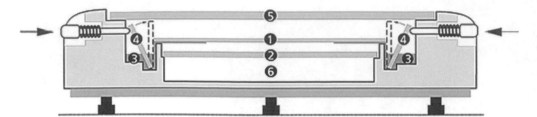

Fig. 10 Horizontal developing chamber (Camag). 1—HPTLC plate (layer facing down), 2—glass plate, 3—reservoir for developing solvent, 4—glass strip, 5—cover plate, 6—conditioning tray. (Photo courtesy of Camag.) *(View this art in color at www.dekker.com.)*

after the drying step can modify the selectivity of mobile phases used in later steps, resulting in irreproducible separations. Although precautions can be taken to minimize the production of artifact peaks in MD, the separation of light- and/or air-sensitive compounds is probably better handled by other techniques such as simple gradient development.

POLYZONAL TLC

Polyzonal TLC is the simplest method for formation of gradients in TLC. The main effect utilized is solvent demixing, which occurs in the case of the application of binary and polycomponent solvents, especially those of differentiated polarity. For an n-component mixed eluent, $n-1$ solvent fronts are formed, ordered in the sequence of polarity of the components. A gradient of eluent strength is thus formed along the layer; the solutes migrate in various zones, and the passage of fronts leads to compression of the TLC spots.[3,6]

EQUIPMENT FOR GRADIENT TLC

Depending on the type of gradient, various apparatuses are applied for its generation. Numerous gradient generators have been described.[3,5] The gradient of the mobile phase can be formed in some types of horizontal chambers (Fig. 10) (e.g., Camag, Muttenz, Switzerland; Chromdes, Lublin, Poland; Desaga, Wiesbaden, Germany). The generation of stepwise gradients is simple for sandwich chambers with distributors, which allow for complete absorption of the eluent fractions from the reservoir. For sandwich chambers, eluent fractions of increasing eluent strength (increasing concentrations of the polar modifier) are introduced under the distributor. After absorption of the preceding eluent fraction by the adsorbent layer, the next fraction of eluent of changed strength is delivered; the total volume of the eluent fractions corresponds to the development distance. Any gradient program, including continuous and multiple-component gradients, can be generated in this way.

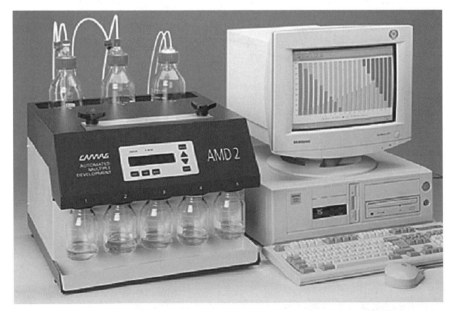

Fig. 11 Chamber for AMD (AMD2, Camag). AMD2 is controlled by winCats software through interface "EquiLink" or through keypad. Programmed gradient is displayed on screen of computer monitor or on screen of device. Dosage of solvents to generate gradient is created by piston pump following entered program. Solvent is completely removed from chamber, and plate with adsorbent is dried under vacuum. Solvent migration distances are monitored with a CCD sensor. (Photo courtesy of Camag.) *(View this art in color at www.dekker.com.)*

The process of MD can be fully automated (Fig. 11) (AMD2 chamber, Camag Scientific; TLC–MAT, Desaga). An apparatus comprises an N-type chamber with connections for adding and removing solvents and gas phases. AMD2 involves the use of a stepwise gradient of different mobile phases with decreasing strength in 10–30 successive developments increasing in length by about 1–5 mm. The initial solvent, which is the strongest, focuses the zones during the first short run, and the solvent is changed for each, or most, of the following cycles. The mobile phase is removed from the chamber, the plate dried and activated by vacuum evaporation, and the layer conditioned with a controlled atmosphere of vapors prior to the next development. High resolution and improved detection limits are achieved because zones are focused during each development stage. The widths of the separated zones are approximately constant at 1–3 mm, and the separation capacity for baseline-resolved peaks is 25–40. Zones migrate different distances according to their polarity. The reproducibility of values is 1–2% (CV) for multiple spots on the same plate or different plates from the same batch. A typical universal gradient for a silica gel layer involves 25 steps, with methanol, dichloromethane, or tert-butyl ether, and hexane as the component solvents.

A discontinuous gradient of the stationary phase can be obtained easily using an ordinary spreader. The trough is divided into separate chambers filled with suspensions of mixtures of adsorbents. The carrier plates are covered in the usual way.[3,5] Another method of formation of gradients of stationary-phase activity is the use of a Vario-KS chamber, which permits adsorption of various vapors on the adsorbent surface or control of the activity of the adsorbent.[6]

CONCLUSIONS

To sum up, the development process, applied as a simple gradient or combined with MD, can improve the separation power of planar chromatography in cases when transport of the mobile phase is controlled by capillary forces. The advantages of gradient development are as follows:

- More compact zones.
- Greater separation capacity.
- Greater sensitivity.

Other benefits of gradient follow:

- Possibility of separation of complex samples containing components with a wide spectrum of polarity or samples containing groups of solutes with different properties.
- More optimal use of solvents of different strength and selectivity.
- Possibility of coupling with HPLC and spectroscopic techniques.

In the case gradient development is carried out, the following conditions need to be fulfilled:

- Application of sample by an automatic device.
- Development with use of an automatic developing chamber.
- Densitometric evaluation and documentation with image processing.

Then the technique can be recognized as a very powerful modern instrumental system that offers reproducible and accurate quantitation for a wide variety of applications.

ARTICLE OF FURTHER INTEREST

Gradient Elution: Overview, p. 722.

REFERENCES

1. Jandera, P. On the way to a general theory of gradient elution. In *A Century of Separation, Science,* 1st Ed.; Issaq, H.J., Ed.; Marcel Dekker, Inc.: New York, 2002; 211–229.
2. Schoenmakers, P. Programmed analysis. In *Handbook of HPLC*; Katz, E., Eksteen, E., Schoenmakers, P., Miller, N., Eds.; Marcel Dekker, Inc.: New York, 1998; 193–232.
3. Gołkiewicz, W. Gradient development in thin-layer chromatography. In *Handbook of Thin-Layer Chromatography*; 3rd Ed.; Sherma, J., Fried, B., Eds.; Marcel Dekker, Inc.: New York, 1997; 135–154.
4. Soczewiński, E. Solvent composition effects in thin-layer systems of the type silica gel–electron donor solvent. Anal. Chem. **1969**, *41*, 179.
5. Geiss, F. Gradients. In *Fundamentals of Thin Layer Chromatography (Planar Chromatography)*; Huethig: Heidelberg, 1987; 388–397.
6. Snyder, L.R.; Saunders, D.L. Resolution in thin-layer chromatography with solvent or adsorbent programming. J. Chromatogr. **1969**, *44*, 1–13.
7. Soczewiński, E.; Markowski, W. Stepwise gradient development in thin-layer chromatography. III. A computer program for the simulation of

stepwise gradient elution. J. Chromatogr. **1986**, *370*, 63–73.

8. Soczewiński, E. Stepwise gradient development in thin-layer chromatography. Optimization of gradient program. J. Chromatogr. **1986**, *369*, 11–17.

9. Wang, Q.-S.; Yan, B.-W.; Zhang, Z.-C. Computer-assisted optimization of mobile phase composition in stepwise gradient HPTLC. J. Planar Chromatogr. **1994**, *7*, 229–232.

10. Markowski, W.; Soczewiński, E.; Matysik, G. A microcomputer program for the calculation of R_F values of solutes in stepwise gradient thin-layer chromatography. J. Liquid Chromatogr. **1987**, *10*, 1261–1267.

11. Markowski, W. Computer-assisted selection of the optimum gradient program in thin-layer chromatography. J. Chromatogr. **1989**, *485*, 517–532.

12. Nyiredy, Sz. Planar chromatographic method development using the Prisma optimization system and flow charts. J. Chromatogr. Sci. **2002**, *40*, 553–563.

13. Felinger, A.; Guiochon, G. Multicomponent interferences in overloaded gradient elution chromatography. J. Chromatogr. **1996**, *724*, 27–37.

14. Siouffi, A.-M. Some aspects of optimization in planar chromatography. J. Chromatogr. **1991**, *556*, 81–94.

15. Matysik, G.; Markowski, W.; Soczewiński, E.; Polak, B. Computer-aided optimization of stepwise gradient profiles in thin-layer chromatography. Chromatographia **1992**, *34*, 303–307.

16. Evgen'ev, M.I.; Evgen'ev, I.I.; Levinson, F.S. 4-Chloro-5,7-dinitrobenzofurazan and 7-chloro-4,6-dinitrobenzofuroxan—new spray reagents for the detection of amino compounds on thin-layer plates. J. Planar Chromatogr. **2000**, *13*, 199–209.

17. Matysik, G. Separation of DABS derivatives of amino acids by multiple gradient development (MGD) in thin-layer chromatography. Chromatographia **1996**, *43*, 301–303.

18. Markowski, W. Past, present and future of multiple development in planar chromatography. In *The Application of Chromatographic Methods in Phytochemical and Biomedical Analysis*; 4th International Symposium on Chromatography of Natural Products, Lublin–Kazimierz Dolny, Poland, June 14–17, 2004; Skubiszewski Medical University of Lublin: Lublin, 2004; L-25, 41.

19. Markowski, W. Computer-aided optimization of gradient multiple development thin-layer chromatography. Part II. Multi-stage development. J. Chromatogr. **1993**, *653*, 283–289.

20. Szabady, B. The different modes of development. In *Planar Chromatography. A Retrospective View for the Third Millennium*, 1st Ed.; Nyiredy, Sz., Ed.; Springer: Budapest, 2001; 88–102.

21. Soczewiński, E.; Wójciak-Kosior, M.; Matysik, G. Simultaneous separation of aglycones and glycosides of flavonoids by double-development TLC. J. Planar Chromatogr. **2004**, *17* (4), 261–263.

22. Poole, C.F.; Belay, M.T. Progress in automated multiple development. J. Planar Chromatogr. **1991**, *4* (9/10), 345–358.

23. Johansson, L.A. Chromatographic analysis of epicuticular plant waxes. Sver. Utsadesforen. Tidskr. **1985**, *95*, 129–136.

24. Lodi, G.; Betti, A.; Menziani, E.; Brandolini, V.; Tosi, B. Some aspects and examples of automated multiple development (AMD) gradient optimization. J. Planar Chromatogr. **1991**, *4* (3/4), 106–110.

25. Valko, K.; Snyder, L.R.; Glajch, J.L. Retention in reversed-phase liquid chromatography as a function of mobile-phase composition. J. Chromatogr. **1993**, *656*, 501–520.

26. Betti, A.; Lodi, G.; Fuzzati, N.; Coppi, S.; Benedetti, S. On the role of planar multiple development in a multidimensional approach to TLC–GC. J. Planar Chromatogr. **1991**, *4* (9/10), 360–364.

27. Gołkiewicz, W.; Gadzikowska, M. Isolation of some quaternary alkaloids from the extract of roots of *Chelidonium majus* L. by column and thin-layer chromatography. Chromatographia **1999**, *50* (1/2), 52–56.

28. de la Vigne, U.; Jaenchen, D. Determination of pesticides in water by HPTLC using automated multiple development (AMD). J. Planar Chromatogr. **1990**, *3* (1/2), 6–9.

29. Błądek, J.; Rostkowski, A.; Miszczak, M. Application of instrumental thin-layer chromatography and solid extraction to the analyses of pesticide residues in grossly contaminated samples of soil. J. Chromatogr. **1996**, *754*, 273–278.

30. Lautie, J.P.; Stankovic, V. Automated multiple development TLC of phenylurea herbicides in plants. J. Planar. Chromatogr. **1996**, *9* (3/4), 113–115.

31. Vega, M.H.; Saelzer, R.F.; Figueroa, C.E.; Rios, G.G.; Jaramillo, V.H. Use of AMD HPTLC for analysis of biogenic amines in fish meal. J. Planar Chromatogr. **1999**, *12* (1/2), 72–75.

32. Muthing, J.; Ziehr, H. Enhanced thin-layer chromatographic separation of G_{M1b}-type gangliosides by automated multiple development. J. Chromatogr. **1996**, *687*, 357–362.

33. Queckenberg, O.R.; Frahm, A.W. Chromatographic and spectroscopic coupling: a powerful

tool for the screening of wild Amaryllidaceae. J. Planar Chromatogr. **1993**, *6* (1/2), 55–61.

34. Galand, N.; Pothier, J.; Viel, C. Plant drug analysis by planar chromatography. J. Planar Chromatogr. **2002**, *40* (11/12), 585–597.

35. Kovar, K.-A.; Enßlin, H.K.; Frey, O.R.; Rienas, S.; Wolff, S.S. Applications of on-line coupling of thin layer chromatography and FTIR spectroscopy. J. Planar Chromatogr. **1991**, *4* (5/6), 246–250.

36. Essig, S.; Kovar, K.-A. The efficiency of thin-layer chromatographic systems: a comparison of separation numbers using addictive substances as an example. J. Planar Chromatogr. **1997**, *10* (3/4), 114–117.

37. Poole, S.K.; Poole, C.F. The influence of the solvent entry position on resolution in unidimensional multiple development thin layer chromatography. J. Planar Chromatogr. **1992**, *5*, 221–228.

G

Gradient Elution

J. E. Haky
Florida Atlantic University, Boca Raton, Florida, U.S.A.

INTRODUCTION

The term *gradient elution* refers to a systematic, programmed increase in the elution strength of the mobile phase during the chromatographic run. Of all the techniques used to provide quality separations among complex mixtures, gradient elution offers the greatest potential.[1] Basically, the composition of the mobile phase is varied throughout the separation so as to provide a continual increase in solvent strength and, thereby, a more convenient elution time and sharper peaks for all sample components.[2] What makes this method so useful is the ability to choose from a variety of different eluents. Although most instruments permit gradients to be automatically prepared from various concentrations of only a two-eluent mixture, sample mixtures of a wide range of polarities can be separated efficiently.

DISCUSSION

The process of mixing eluents is a sensitive one. When two solvents with a large difference in their elution strengths are used, even a small increase in the polar component produces a sharp rise in elution strength. Such an effect is undesirable because the components are almost always eluted at the beginning of the analysis and displacement effects may result from demixing of eluent mixtures.[1] According to Poole et al.,[3] the most frequently used gradients are binary solvent systems with a linear, convex, or concave increase in the percent volume fraction of the stronger solvent, as depicted in the following equations:

Linear gradient

$$\theta_B = \frac{t}{t_G} \tag{1}$$

Convex gradient

$$\theta_B = 1 - \left(1 - \frac{t}{t_G}\right)^n \tag{2}$$

Concave gradient

$$\theta_b = \left(\frac{t}{t_B}\right)^n \tag{3}$$

In these equations, θ_B is the volume fraction of the stronger eluting solvent, t is the time after the gradient begins, t_G is the total gradient time, and n is an integer controlling gradient steepness.

Complex gradients can be constructed by combining several gradient segments (i.e., rates of increase of strong solvent composition) to form the complete gradient program.[3] Linear gradients are most commonly used, with convex and concave gradients employed only when necessary to optimize more complex separations. In a linear-solvent-strength gradient, the logarithm of the capacity factor for each sample component, k', decreases linearly with time, according to Eq. (4):

$$\log k = \log k_0 - b\frac{[t]}{[t_m]} \tag{4}$$

In this equation, k_0 is the value of k determined isocratically in the starting solvent, b is the gradient steepness parameter, t is the time after the start of gradient and sample injection, and t_m is the column dead time. Ideally, this equation shows that a linear-solvent-strength gradient should result in equal resolution and bandwidths of all components. Unfortunately, this is not always possible. There are certain cases where linear-solvent-strength gradient is not the ideal method. In some cases, for example, b actually increases regularly with solute retention, which reduces the separation of late-eluting components. Such an effect is observed in the separation of polycyclic aromatic hydrocarbons.[3]

There are three things to consider when finding a suitable gradient for a separation: (a) the initial and final mobile-phase compositions, (b) the gradient shape, and (c) the gradient steepness.[3] A convex gradient leads to the elution of bands with a lower average capacity factor and a shorter total analysis time. In other words, the later-eluting bands appear wider and better resolved than the early eluting bands. A concave gradient resolves the early bands to a greater degree than the later bands.

Solvent selection is one of the most important facets of gradient elution. The choice of the first solvent influences the separation of the initial bands, whereas the strength of the final solvent influences the selectivity of the separation and the retention times and peak

Encyclopedia of Chromatography DOI: 10.1081/E-ECHR-120040005

shapes of later-eluting bands. If solvent B is too weak, the analysis time may become very long and the later-eluting bands might broaden excessively; thus, a stronger solvent B may be required.[3]

Abbott et al.[4] devised a method designed to predict the retention times in gradient elution under the assumption that the retention factor as determined under isocratic conditions is a log-linear function of solvent composition according to Eq. (5), where k_w is the retention factor obtained in water, φ_0 refers to the volume fraction of the organic component, and S refers to the solvent strength for which the values can be obtained as the negative slope of plots of $\log k$ versus volume fraction:

$$\log k = \log k_w - S\varphi_0 \qquad (5)$$

Engelhardt and Elgass[5] found that if the gradient volume is held constant and the initial and final compositions of the eluent are fixed, each component of a sample is eluted at a given solvent composition. Snyder et al.[6] derived a simple relationship between the elution time of a solute and the rate of change of solvent composition in gradient elution. Utilizing Eq. (6), they found that the elution time t_e is related to column dead time, t_0, and an experimental parameter b whereby k_0 is the retention factor that would be obtained in isocratic elution with mobile-phase composition used at the beginning of the gradient:[1]

$$t_e \left(\frac{t_0}{b} \right) \log(2.31 k_0 b + 1) + t_0 \qquad (6)$$

The parameter b is defined as

$$b = \frac{\Phi S t_0}{100} \qquad (7)$$

where Φ is the rate of increase in the concentration of the solvent component having eluent strength S and given as volume percent of organic solvent component per minute.[1]

Many technical problems can occur with gradient elution, some of which can be avoided through various methods. To begin, gradient elution relies upon the purity of the solvents used. The HPLC column can collect impurities, in the mobile phase, which may or may not elute as sharp peaks at a certain eluent composition. These can be mistaken for sample components. Such peaks are called "ghost peaks" and can result in inaccurate data. Water presents its own set of problems. Contaminated water can also result

in ghost peaks. Even deionization of water by ion exchangers can leach out organics from the resin.[1] For this reason, it is advisable to run a gradient first without injecting the sample and use commercially available, purified solvents, including the water, to determine if they result in the elution of ghost peaks.

Another thing to consider with gradient elution is changes in the eluent viscosity. When gradient elution with a hydro-organic mobile phase is used (e.g., methanol-water), systematic variations in the flow rate are expected under conditions of constant-pressure operation, and systematic variations in the operating pressure will be found when a constant flow rate is used.[1] The compressibility of the solvent is species-specific.

In summary, gradient elution is a powerful method for the separation and analysis of complex mixtures containing components with a wide variety of polarities and hydrophobicities. It can also be used to help establish an isocratic mobile phase for the analysis of simpler mixtures. In either case, utmost care must be taken in the selection and use of solvents of high purity and selectivity.

ACKNOWLEDGMENT

The author wishes to thank D.A. Teifer for technical assistance.

REFERENCES

1. Horvath, C. *High Performance Liquid Chromatography: Advances and Perspectives*; Academic Press: New York, 1980; Vol. 2.
2. Kirkland, J.J.; Glajch, J.L. Optimization of mobile phases for multisolvent gradient elution liquid chromatography. J. Chromatogr. **1983**, *255*, 27.
3. Poole, C.F.; Schuette, S.A. *Contemporary Practice of Chromatography*; Elsevier: Amsterdam, 1984.
4. Abbott, S.R.; Berg, J.R.; Achener, P.; Stevenson, R.L. Chromatographic reproducibility in high-performance liquid chromatographic gradient elution. J. Chromatogr. **1976**, *126*, 421.
5. Englehardt, H.; Elgass, H. Optimization of gradient elution: Separation of fatty acid phenacyl esters. J. Chromatogr. **1978**, *158*, 249.
6. Snyder, L.R.; Dolan, J.W.; Gant, J.R. Gradient elution in high-performance liquid chromatography: I. Theoretical basis for reversed-phase systems. J. Chromatogr. **1979**, *165*, 3.

Gradient Elution in Capillary Electrophoresis

Haleem J. Issaq
NCI-Frederick Cancer Research and Development Center, Frederick, Maryland, U.S.A.

INTRODUCTION

Gradient elution is routinely used in HPLC to achieve the complete resolution of a mixture which could not be resolved using isocratic elution. Unlike isocratic elution, where the mobile-phase composition remains constant throughout the experiment, in gradient elution the mobile-phase composition changes with time. The change could be continuous or stepwise, known as the *step-gradient*. In the continuous gradient, the analyst can pick one of three general shapes: linear, concave, or convex.

DISCUSSION

Gradient elution in HPLC is achieved using two pumps, two different solvent reservoirs, and a solvent mixer. In capillary electrophoresis (CE), electro-osmotic flow controls the flow of the mobile phase, which is, in most cases, an aqueous buffer and is used in place of a mechanical pump.

A manual step-gradient was used by Balchunas and Sepaniak[1] to separate a mixture of amines by MEKC. Stepwise gradients were produced by pipetting aliquots of a gradient solvent to the inlet reservoir which was filled with 2.5 mL of running buffer. A small magnetic stirring bar was used to ensure thorough mixing of the added gradient solvent with the starting mobile phase. The gradient elution solvent was manually added, in four 0.5 mL increments, spaced 5 min apart, 5 min after start of the experiment.

Bocek and his group[2] developed a method for controlling the composition of the operational electrolyte directly in the separation capillary in isotachophoresis (ITP) and capillary zone electrophoresis (CZE). The method is based on feeding the capillary with two different ionic species from two separate electrode chambers by simultaneous electromigration. The composition and pH of the electrolyte in the separation capillary is thus controlled by setting the ratio of two electric currents. This procedure can be used, in addition to generating the mobile-phase gradient, for generating pH gradients.[3,4] Sepaniak et al.[5–7] produced continuous gradients of different shapes (linear, concave, or convex) by using a negative-polarity configuration in which the inlet reservoir is at ground potential and the outlet reservoir at a very high negative potential. This configuration allows two syringe pumps to pump solutions into and out of the inlet reservoir. Tsuda[8] used a solvent-program delivery system, similar to that used in HPLC, to generate pH gradients in CZE. A pH gradient derived from temperature changes has also been reported.[9] Chang and Yeung[10] used two different techniques (i.e., the dynamic pH gradient and electro-osmotic flow gradient) to control selectivity in CZE. A dynamic pH gradient from pH 3.0 to 5.2 was generated by a HPLC gradient pump. An electro-osmotic flow gradient was produced by changing the reservoirs containing different concentrations of cetylammonium bromide for injection and running.

Capillary electrochromatography (CEC) is a separation technique which combines the advantages of micro-HPLC and CE. In CEC, the HPLC pump is replaced by electro-osmotic flow. Behnke and Bayer[11] developed a micro-bore system for gradient elution using 50- and 100-µm fused-silica capillaries, packed with 5 µm octadecyl reversed phase silica gel and voltage gradients, up to 30,000 V, across the length of the capillary. A modular CE system was combined with a gradient HPLC system to generate gradient CEC. Enhanced column efficiency and resolution were realized. Zare and his co-workers[12] used two high-voltage power supplies and a packed fused-silica capillary to generate an electro-osmotically driven gradient flow in an automated manner. The separation of 16 polycyclic aromatic hydrocarbons was resolved in the gradient mode; these compounds were not separated when the isocratic mode was employed. Others[13–16] used gradient elution in combination with CEC to resolve various mixtures.

Multiple, intersecting narrow channels can be formed on a glass chip to form a manifold of flow channels in which CE can be used to resolve a mixture of solutes in seconds. Harrison and co-workers[17] showed that judicious application of voltages to multiple channels within a manifold can be used to control the mixing of solutions and to direct the flow at the intersection of channels. The authors concluded that such a system, in which the applied voltages can be used to control the flow, can be used for sample dilution, pH adjustment, derivatization, complexation, or masking of interferences. Ramsey and co-workers[18]

Encyclopedia of Chromatography DOI: 10.1081/E-ECHR-120040006

used a microchip device with electrokinetically controlled solvent mixing for isocratic and gradient elution in MEKC. Isocratic and gradient conditions are controlled by proper setting of voltages applied to the buffer reservoirs of the microchip. The precision of such control was successfully tested for gradients of various shapes (linear, concave, or convex) by mixing pure buffer and buffer doped with a fluorescent dye. By making use of the electro-osmotic flow and employing computer control, very precise manipulation of the solvent was possible and allowed fast and efficient optimization of separation problems.

ACKNOWLEDGMENT

This project has been funded in whole or in part with federal funds from the National Cancer Institute, National Institutes of Health, under Contract No. NO1-CO-56000.

By acceptance of this article, the publisher or recipient acknowledges the right of the U.S. government to retain nonexclusive, royalty-free license to any copyright covering the article.

The content of this publication does not necessarily reflect the views of the Department of Health and Human Services, nor does the mention of trade names, commercial products, or organizations imply endorsement by the U.S. government.

REFERENCES

1. Balachunas, A.T.; Sepaniak, M.J. Gradient elution for micellar electrokinetic capillary chromatography. Anal. Chem. **1988**, *60*, 617.
2. Popsichal, J.; Deml, M.; Gebauer, P.; Bocek, P. Generation of operational electrolytes for isotachophoresis and capillary zone electrophoresis in a three-pole column. J. Chromatogr. **1989**, *470*, 43.
3. Bocek, P.; Deml, M.; Popsichal, J.; Sudor, J. Dynamic programming of pH—a new option in analytical capillary electrophoresis. J. Chromatogr. **1989**, *470*, 309.
4. Sustacek, V.; Foret, F.; Bocek, P. Simple method for generation of dymanic pH gradient in capillary zone electrophoresis. J. Chromatogr. **1989**, *480*, 271.
5. Sepaniak, M.J.; Swaile, D.F.; Powell, A.C. Instrumental developments in micellar electrokinetic capillary chromatography. J. Chromatogr. **1989**, *480*, 185.
6. Powell, A.C.; Sepaniak, M.J. J. Microcol. Separ. **1990**, *2*, 278.
7. Powell, A.C.; Sepaniak, M.J. Anal. Instrum. **1993**, *21*, 25.
8. Tsuda, T. pH gradient capillary zone electrophoresis using a solvent program delivery system. Anal. Chem. **1992**, *64*, 386.
9. Wang, C.W.; Yeung, E.S. Temperature programming in capillary zone electrophoresis. Anal. Chem. **1992**, *64*, 502.
10. Chang, H.-T.; Yeung, E.S. Optimization of selectivity in capillary zone electrophoresis via dynamic pH gradient and dynamic flow gradient. J. Chromatogr. **1992**, *608*, 65.
11. Behnke, B.; Bayer, E. Pressurized gradient electro-high-performance liquid chromatography. J. Chromatogr. **1994**, *680*, 93.
12. Yan, C.; Dadoo, R.; Zare, R.N.; Rakestraw, D.J.; Anex, D.S. Gradient elution in capillary electrochromatography. Anal. Chem. **1996**, *68*, 2726.
13. Schmeer, K.; Behnke, B.; Bayer, E. Capillary electrochromatography-electrospray mass spectrometry: a microanalysis technique. Anal. Chem. **1995**, *67*, 3656.
14. Taylor, M.R.; Teale, P.; Westwood, S.A.; Perrett, D. Analysis of corticosteroids in biofluids by capillary electrochromatography with gradient elution. Anal. Chem. **1997**, *69*, 2554.
15. Taylor, M.R.; Teale, P. Gradient capillary electrochromatography of drug mixtures with UV and electrospray ionisation mass spectrometric detection. J. Chromatogr. A. **1997**, *768*, 89.
16. Gfrorer, P.; Schewitz, J.; Psecker, K.; Tseng, L.-H.; Albert, K.; Bayer, E. Electrophoresis **1999**, *20*, 3.
17. Seller, K.; Fan, Z.H.; Fluri, K.; Harrison, J. Electroosmotic pumping and valveless control of fluid flow within a manifold of capillaries on a glass chip. Anal. Chem. **1994**, *66*, 3485.
18. Kutter, J.P.; Jacobson, S.J.; Ramsey, J.M. Integrated microchip device with electrokinetically controlled solvent mixing for isocratic and gradient elution in micellar electrokinetic chromatography. Anal. Chem. **1997**, *69*, 5165.

Gradient Elution: Overview

Ioannis N. Papadoyannis
Kalliopi A. Georga
Aristotle University of Thessaloniki, Thessaloniki, Greece

INTRODUCTION

Gradient elution is the elution method in which the mobile-phase composition changes during time. It may be considered as an analogy to the temperature programming in GC.

DISCUSSION

The main purpose of gradient elution is to move strongly retained components of a mixture faster, while having the least retained components well resolved. At the beginning of the analysis, the solvent used is appropriate to elute some of the components, but is "weak" in terms of its ability to remove other compounds from the column and separate them from one another.

Gradient elution operates on the principle that, under the initial mobile-phase conditions, many of the components have a k' (capacity factor) value of essentially infinity, in that these components are stopped in a narrow band near the head of the column. As the solvent composition is changed and its solvent strength is increased, sample components dissolve at a characteristic solvent strength and then migrate down the column, leaving the remaining components behind. Changes in the mobile-phase composition may be "continuous" with a predetermined set of conditions or may be done in "steps" of substantial solvent composition changes.

The typical gradients used in reversed-phase chromatography are linear or binary (i.e., involving two mobile phases). Convex and concave gradients are used occasionally for analytical purposes, particularly when dealing with multicomponent samples requiring extra resolution either at the beginning or at the end of the gradient (Fig. 1).

The concentration of the organic solvent is lower in the initial mobile phase (mobile phase A) than it is in the final mobile phase (mobile phase B). The gradient then, regardless of the absolute change in percent organic modifier, always proceeds from a condition of high polarity (high aqueous content, low concentration of organic modifier) to low polarity (higher concentration of organic modifier, lower aqueous content). Reversed-phase separations can be achieved using either a stepwise or a continuous gradient to elute sample components. Step gradients (i.e., a series of isocratic elutions at different percentages of B) are useful for applications such as desalting, but for separations requiring high resolution, a linear continuous gradient is required.

Step gradients are also ideal when performing process scale applications providing the desired resolution can be obtained; less complex instrumentation is required to generate step gradients. Additionally, step gradients can be generated more reproducibly than linear gradients.

Gradient shape (combination of linear gradient and isocratic conditions), gradient slope, and gradient volume are all important considerations in reversed-phase chromatography. Typically, when first performing a reversed-phase separation of a complex sample, a broad gradient is used for initial screening in order to determine the optimum gradient shape.

The ideal gradient shape and volume are empirically determined for a particular separation. Generally, the sample is chromatographed using a broad-range linear gradient to determine where the molecules of interest will elute. The initial conditions usually consist of mobile phase A containing 10% or less organic modifier and mobile phase B containing 90% or more organic modifier. The initial gradient runs from 0% B to 100% B over 10–30 column volumes. A blank gradient is usually run prior to injecting the sample, in order to detect any baseline disturbances resulting from the column or impurities originating in the mobile phase.

After the initial screening is completed, the gradient shape may be adjusted to optimize the separation of the desired components. This is usually accomplished by decreasing the gradient slope, where the desired components elute, and increasing it before and after. The choice of gradient slope will depend on how closely the contaminants elute to the target molecule. Generally, decreasing the gradient slope increases the resolution. However, the peak volume and retention time increase with decreasing gradient slope. Shallow gradients with short columns are generally optimal for high-molecular-weight bio-molecules.

Gradient slopes are generally reported as change in percent B per unit time (% B/min) or per unit volume

Encyclopedia of Chromatography DOI: 10.1081/E-ECHR-120040004

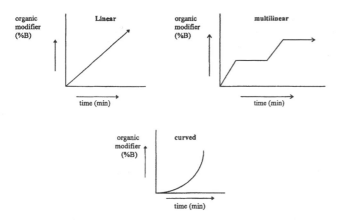

Fig. 1 Various gradient shapes.

(%B/mL). When programming a chromatography system in the time mode, it is important to remember that changes in flow rate will affect gradient slope and, therefore, resolution.

Resolution is also affected by the total gradient volume (gradient volume × flow rate). Although the optimum value must be determined empirically, a good rule of thumb is to begin with a gradient volume that is approximately 10–20 times the column volume. The slope can then be increased or decreased in order to optimize the resolution.

Except for optimizing gradient elution methods, which is a very important parameter in chromatographic analysis, another parameter of great significance is the mixing of the mobile-phase components.

There are two primary methods of mixing the mobile-phase components, known as "low-pressure mixing" and "high-pressure mixing." The first method employs electrically actuated solenoid valves located ahead of a single-solvent delivery system (pump). The precision of the gradient depends on the ability of the solenoid to reproducibly dispense solvents in segments of variable size (volume), depending on the composition desired. If reproducible retention times and stable detector baselines are to be obtained, these "segments of solvent plugs" must be well mixed into a homogenous mobile-phase stream before entering the chromatographic column.

The second method uses a separate solvent delivery device for each solvent, with each being capable of delivering smooth, precise flow rates of as low as a few microliters per minute. Gradients are formed by varying the delivery speeds and simply blending the concurrent solvent streams on the high-pressure side of the pumps.

Each method of gradient formation has advantages and disadvantages. The low-pressure gradient formation is preferred most of the time, because it uses only one pump, whereas the high-pressure method uses two pumps which might go wrong during use. Because

low-pressure gradient systems have only one pump, they are, of course, less expensive. These systems require extensive degassing and have a large lag time (delay volume) in starting the gradient, whereas, in high-pressure systems, degassing is desired but not essential.

Moreover, low-pressure systems often use three or four solvents. This multiple-solvent blending might also be useful for the optimization of both isocratic and gradient elution methods. This is an advantage that the high-pressure system also has; when not using a gradient system, the operator has two independent isocratic pumps.

Gradient elution is ideal for separating certain kinds of sample which cannot be easily handled by isocratic methods, because of their wide k' range. Nevertheless, there is a strong bias against the use of gradient elution in many laboratories.

Some of the reasons for not preferring gradient elution are as follows: Gradient equipment is not available in some laboratories, because of its higher cost, and gradient elution is more complicated and makes both method development and routine analysis more difficult; the most important issue is that it is not compatible with some HPLC detectors (e.g., refractive index detectors). Furthermore, gradient runs take longer, because of the need of column equilibration after each run. Baseline problems are more common with gradient elution and the solvent must be of high purity.

Although the disadvantages of gradient elution must be taken into serious consideration, many separations are only possible using gradient elution. The use of gradient elution for routine applications is suggested for the following kinds of sample:

1. Samples with a wide range of k'
2. Samples composed of large molecules >1000 in molecular weight
3. Samples containing late eluting interferences that can either foul the column or overlap subsequent chromatograms

Using gradient elution to develop HPLC methods has many advantages compared with using isocratic experiments. First, errors in solvent strength can be adjusted when changing from one solvent to another. Second, the ability to increase resolution during early exploratory runs is a distinct advantage when doing solvent mapping. Early bands often are severely overlapped in isocratic separations, so that it may not be clear how resolution is changing as separation conditions are varied. Gradient elution opens up the front of the chromatogram, allowing a better view of what is happening as conditions are varied (Fig. 2).

Third, using gradient elution runs during the initial stages of method development makes it easier to locate

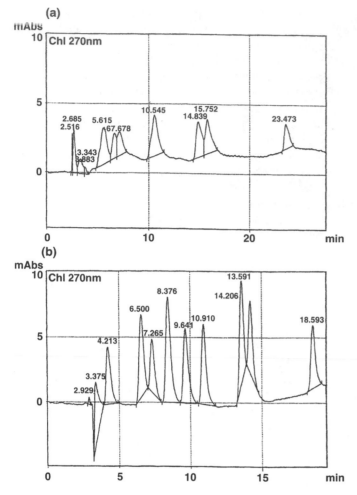

Fig. 2 HPLC analysis of eight methylxanthines: (a) with isocratic elution; (b) with gradient elution.

compounds that elute either very early or very late in the chromatogram. With isocratic separation, early-eluting compounds are often lost in the solvent front, whereas late-eluting compounds disappear into the baseline or overlap the next sample. Finally, gradient elution method development works for either gradient or isocratic elution.

In conclusion, gradient elution is not preferred as a quantitative technique because it is more complex than isocratic elution and, hence, more things can potentially go wrong. However, with proper control of operating parameters and good instrumentation, it is possible to obtain a separation with excellent quantitative results. This requires that the operator understand the hardware and determine that it is working correctly before attempting a separation. The ideal gradient system should be easy to operate, reproducible to provide consistent retention times, versatile to provide capability of generating various concave, convex, and linear gradient shapes, and convenient to provide a

rapid turnaround time to initial eluent conditions (equilibration) for fast throughput from analysis to analysis.

SUGGESTED FURTHER READING

Bidlingmeyer, B.A. *Practical HPLC Methodology and Applications*; John Wiley & Sons, Inc.: New York, 1992.

Papadoyannis, I.; Samanidou, V.; Georga, K. J. Liq. Chromatogr. **1996**, *19* (16), 2559.

Pharmacia Biotech. *Reversed Phase Chromatography*; Pharmacia Biotech: Uppsala, Sweden, 1996.

Snyder, L.R.; Glajch, J.L.; Kirkland, J.J. *Practical HPLC Method Development*; John Wiley & Sons, Inc.: New York, 1988.

Snyder, L.R.; Kirkland, J.J.; Glajch, J.L. *Practical HPLC Method Development*, 2nd Ed.; John Wiley & Sons, New York, 1997.

Gradient HPLC: Selection of a Gradient System

Pavel Jandera
Department of Analytical Chemistry, University of Pardubice, Pardubice CZ, Czech Republic

INTRODUCTION

Many HPLC analyses can be performed at constant, isocratic, operating conditions using isocratic elution. However, isocratic elution with a mobile phase of fixed composition often does not yield a successful separation of complex samples containing compounds that differ widely in retention characteristics. To keep the time of analysis within acceptable limits, the retention factors $k = (V_R/V_m - 1)$ of the most strongly retained sample components usually should be lower than 10 (V_R = retention volume, V_m = column holdup volume). For a satisfactory separation of both weakly and strongly retained sample compounds in a single run, operating conditions controlling retention, such as the composition or flow rate of the mobile phase, or the column temperature, should be varied during the chromatographic experiment. Flow programming in contemporary HPLC, using efficient small-particle columns, has only marginal effect on separation and is limited by the maximum instrumental pressure. Temperature programming is widely used in GC, but rarely in HPLC, because a large rise in temperature during the run is required to significantly reduce retention, and many columns, especially with stationary phases bonded onto a silica gel support, are not stable at temperatures higher than 60°C. Furthermore, only a few instruments that allow steep-enough temperature gradients are available because of a relatively slow response of the temperature inside the column to a change in the temperature setting in an air-heated thermostatted compartment, especially with conventional analytical columns of 2-mm inner diameter or larger.

For these reasons, solvent gradients are most frequently used in contemporary HPLC. In gradient elution, the composition of the mobile phase is changed during the chromatographic run, either stepwise or continuously, to increase the elution strength of the mobile phase, which allows to decrease the retention factors by two to three orders of magnitude in a single run. However, gradient elution requires more complicated equipment than isocratic HPLC, as two or more components of the mobile phase should be accurately mixed according to a preset time program; in addition, the selection of detectors is limited. Gradient runs generally take a longer time than isocratic elution because the column should be reequilibrated to initial gradient conditions after each run.

PRINCIPLES OF GRADIENT ELUTION

Mobile-phase gradients can be formed outside the separation column by pumping and mixing the liquid components according to a preset time program (external gradients), or can be generated inside the column as a consequence of changing the equilibrium between the components adsorbed on the stationary phase and in the solution, induced by incoming mobile phase (internal gradients). Therefore the second approach is suitable only for a limited number of separation cases and is much less frequently used. The retention of nonionic compounds in reverse-phase (RP) and normal-phase (NP) liquid chromatography depends mainly on their polarities and the polarities of the stationary and mobile phases; hence, external polarity (solvent strength) gradients, prepared by mixing solvents of different polarities, are suitable for their separation. Solvent strength gradients are often useful for the separation of ionic compounds; however, the mobile phase should contain buffers or other ionic additives so that ionic strength or pH gradients can be also used for separations in ion exchange or ion pair gradient chromatography of ionic compounds.

Two, three, or four mobile-phase components can be mixed to create binary, ternary, or quaternary gradients, respectively. Gradient elution either consists of a few consequent isocratic steps, or the composition of the mobile phase is changed continuously according to a preset program; this can be characterized by three parameters: 1) the initial concentration; 2) the steepness (slope); and 3) the shape (curvature) of the gradient, which all affect the elution time and the spacing of the peaks in the chromatogram. A linear gradient profile is used almost exclusively in practice and can

Encyclopedia of Chromatography DOI: 10.1081/E-ECHR-120041120

be described by Eq. (1):

$$\varphi = A + B't = A + \frac{\Delta\varphi}{t_G}t = A + \frac{B'}{F_M}V$$

$$= A + BV = A + \frac{\Delta\varphi}{V_G}V \qquad (1)$$

where A is the initial concentration φ of the strong solvent in the mobile phase at the start of the gradient, and B or B' is the steepness (slope) of the gradient (i.e., the increase in φ in the time unit, or in the volume unit of the mobile phase, respectively); V_G and t_G are the gradient volume and the gradient time during which the concentration φ is changed from the initial value A to the concentration $\varphi_G = A + \Delta\varphi$ at the end of the gradient; and $\Delta\varphi$ is the gradient range. Curved gradients are often substituted by multiple linear segmented gradients consisting of several subsequent linear gradient steps with different slopes B.

The theory of gradient elution chromatography allows prediction of the elution behavior of sample compounds by calculation from their isocratic retention data (or from two initial gradient experiments) in RP, NP, and ion exchange LC systems. Unlike isocratic conditions, the retention factors, change (decrease) during gradient elution and can be considered constant only in a very small (differential) volume of the mobile phase dV corresponding to migration along a differential part of the column holdup volume V_m, dV_m:

$$dV = k \cdot dV_m \qquad (2)$$

The differential equation Eq. (2) can be solved after introducing the dependence of k on the volume of the eluate passed through the column from the start of the gradient run V. Any dependence of k on V can be divided into two parts: 1) a dependence of k on the concentration of a strong eluting component in the mobile phase φ controlled by the thermodynamics of the distribution process (the retention equation); and 2) the parameters of Eq. (1) describing the gradient profile, adjusted by the operator.

RP Gradient LC

RP chromatography is, by far, the most widely used LC mode for the separation of complex mixtures based on different lipophilicities of sample compounds.[1] The effect of the volume fraction φ of the organic solvent in a binary aqueous–organic mobile phase on the retention factors k in RP chromatography can be very often described by a simple equation [Eq. (3)]:[1,2]

$$\log k = \log k_0 - m\varphi = a - m\varphi \qquad (3)$$

Here, k_0 is the retention factor of the sample solute extrapolated to pure water as the mobile phase and m characterizes the "solvent strength" (i.e., the change in log k per concentration unit of the organic solvent). Assuming the validity of Eq. (3), the retention volume V_R in RP gradient elution chromatography with linear gradients can be calculated from Eq. (4):[3]

$$V_R = \frac{1}{mB}\log\{2.31mB[V_m 10^{(a-mA)} - V_D] + 1\}$$
$$+ V_m + V_D \qquad (4)$$

V_D is the so-called gradient dwell volume [i.e., the volume of the mobile phase contained in the instrument parts (mixer, filter, and tubing) between the pump and the column]. In an ideal case, linear concentration gradients in RPLC correspond to linear solvent strength (LSS) gradients according to the model developed by Snyder and Dolan;[4] hence, Eq. (4) describes the retention data in LSS gradient elution.

NP (Adsorption) Gradient LC

In NP gradient LC on polar adsorbents, the concentration of one (or more) polar solvent(s) in a nonpolar solvent increases. A simple equation [Eq. (5)] can often adequately describe the experimental dependencies of the retention factors k of sample compounds on the volume fraction φ of a polar solvent B in a binary mobile phase comprised of two organic solvents with different polarities, if the sample solute is very strongly retained in the pure, less polar solvent:

$$k = k_0\varphi^{-m} \qquad (5)$$

k_0 and m in Eq. (5) depend on the nature of the solute and on the chromatographic system, but are independent of the concentration of the strong solvent Bν in the mobile phase. Assuming the validity of Eq. (5) in NP gradient chromatography with linear concentration gradients of a polar solvent B, the elution volume V_R of a sample solute can be calculated from Eq. (6):[5]

$$V_R = \frac{1}{B}[(m + 1)B(k_0 V_m - V_D A^m) + A^{(m+1)}]^{\frac{1}{m+1}}$$
$$- \frac{A}{B} + V_m + V_D \qquad (6)$$

Here, as in Eq. (4), V_m is the column holdup volume and V_D is the gradient dwell volume. In contrast to RP gradient elution, preferential adsorption of polar solvents from the mobile phase onto the surface of the polar adsorbent during a gradient run may lead to significant deviations of the actual gradient profile from the preset mobile-phase composition program and to a decrease in the reproducibility of the retention

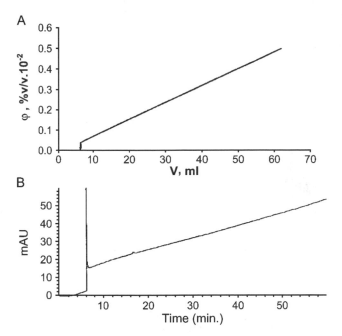

Fig. 1 (A) Calculated breakthrough curves in NP gradient elution HPLC. Simulated calculation using the experimental isotherm data and assuming $N = 5000$. Gradient dwell volume = 0.50 mL. (B) Record of the blank gradient detector trace showing the breakthrough of propan-2-ol at 6 min and a "ghost peak" of impurities displaced at the breakthrough volume. Column: silica gel Separon SGX (7.5 μm), 150 × 3.3 mm ID, 1 mL/min, 40°C. Gradient: 0 = 50% 2-propanol in 30 min (ν = concentration of propan-2-ol in the eluate; V = volume of the eluate from the start of the gradient).

data (Fig. 1). Furthermore, because of the strong preferential adsorption of polar solvents, column reequilibration times after the end of the gradient are often long in NPLC. This has been a reason for strong bias against the use of gradient elution in NP chromatography. To suppress these effects, which are most significant with gradients starting in pure nonpolar solvents, gradients should be started, rather, at 3% or more than at a zero concentration of the polar solvent, if possible, and nonlocalizing polar solvents should be used, such as dichloromethane, dioxane, or *tert*-butyl methyl ether.

Water is much more strongly adsorbed than polar organic solvents on polar adsorbents; hence, even trace water concentrations in the mobile phase decrease the adsorbent activity and very significantly affect the retention. As the distribution equilibrium of water and other polar solvents between the polar adsorbent and an organic mobile phase is strongly affected by temperature, it is very important to work with a thermostatted column. The reproducibility of the retention data in NP gradient LC can be considerably improved to the level comparable with RP gradient chromatography by keeping a constant temperature and

adsorbent activity, and by controlling the water content in the mobile phase best by using dehydrated solvents kept dry over activated molecular sieves and filtered before use.[5]

Ion Exchange Gradient LC

Because both adsorption on polar adsorbents and ion exchange are competing processes, Eqs. (5) and (6) can be used to describe, also, the effect of the ionic strength in the mobile phase on retention in isocratic and gradient ion exchange chromatography if φ has the meaning of the molar concentration of a salt (buffer) in the mobile phase. The dependence of the retention in gradient elution with pH gradients is less straightforward, as it is sometimes difficult to accomplish a linear change in pH during a gradient run.

EFFECTS OF GRADIENT PROFILE ON SEPARATION: COMPARISON WITH ISOCRATIC ELUTION

The gradient profile affects the retention in similar way as the concentration of the strong solvent in a binary mobile phase under isocratic conditions. This is illustrated in Fig. 2 for gradient elution RP chromatography separation of 10 homologous derivatives of *n*-alkylamines. At a constant gradient range (70%–100% methanol), the steepness of the gradient decreases as the gradient time increases from 10 to 40 min; the resolution improves, but the retention times increase (the top three chromatograms). The bottom three chromatograms show the effect of the gradient range on the separation at a constant steepness of the gradient (1% methanol/0.6 min)—as the initial concentration increases from 50% to 80% methanol, both retention and resolution decrease. The retention times of early eluting compounds are affected more significantly by the initial concentration of the strong solvent (methanol) than by the gradient steepness. The examples in Fig. 2 illustrate the importance of appropriate adjustment of both the gradient range and the initial gradient concentration to keep the analysis time short.

Equations (4) and (6) show that, for comparable retention times, a less steep gradient should be used to compensate for a higher parameter m in Eqs. (3) and (4), which usually increases with increasing size of the molecule. This behavior has the following practical consequences for gradient elution of high molecular compounds: 1) macromolecules may have so large an m that a very small change in the concentration of the strong solvent may change the retention from very strong to practically no retention, so that isocratic

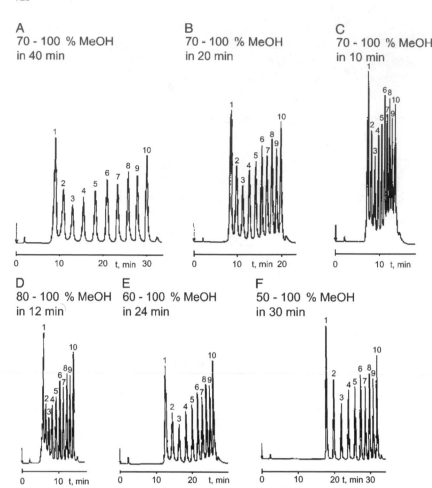

A
70 - 100 % MeOH
in 40 min

B
70 - 100 % MeOH
in 20 min

C
70 - 100 % MeOH
in 10 min

D
80 - 100 % MeOH
in 12 min

E
60 - 100 % MeOH
in 24 min

F
50 - 100 % MeOH
in 30 min

Fig. 2 RP gradient elution separation of 1,2-naphthoylenebenzimidazole alkylsulphonamides. Column: Lichrosorb RP-18, 10 μm (300 × 4mm id). Linear gradients of methanol in water with a constant gradient range but different gradient volumes (A–C), and with a constant gradient steepness (1.67% methanol/min) but different initial concentrations of methanol (D–F). Flow rate = 1 mL/min. The number of peaks agrees with the number of carbon atoms in alkyls.

separation of large molecules is difficult, if possible at all; 2) shallow gradients are usually required for separations of high molecular samples, so that the selection of a suitable combination of the gradient parameters A and B is more critical than for small molecules; and 3) samples with a broad range of molar masses may require a flatter gradient at the end of the chromatogram than at its start for regular band spacing (a convex gradient).

Under isocratic conditions, bandwidths increase for more strongly retained compounds, but the bandwidths in gradient elution chromatography are approximately constant for both early-eluting and late-eluting compounds. This is caused by increasing migration velocities of the bands along the column during gradient elution, so that all sample compounds eventually are eluted with very similar instantaneous retention factors k_e at the time they leave the column, which are approximately half the average retention factors (k^*) during the band migration along the column. The bandwidths decrease with steeper gradients (the three top chromatograms in Fig. 2). Because k_e values are usually significantly lower than the retention factors in isocratic LC, the peaks in gradient elution chromatography are generally narrower

and higher, improving the detector response and the sensitivity of determination. However, the beneficial effect of gradient elution on increasing sensitivity is often counterbalanced by an increased baseline drift and noise in comparison to isocratic HPLC. To avoid this inconvenience, high-purity solvents and mobile-phase additives are generally used in gradient elution HPLC.

The bandwidths w_g in gradient elution can be determined by introducing the appropriate instantaneous retention factor k_e at the elution of the peak maximum calculated using an appropriate gradient retention equation [e.g., Eq. (4) or Eq. (6)]:

$$W_g = \frac{4V_m(1 + k_e)}{\sqrt{N}} \qquad (7)$$

Combining the appropriate equations for the retention volumes of solutes 1 and 2 with adjacent bands and Eq. (7), for bandwidths, we can calculate the resolution in gradient HPLC as follows:

$$R_s = \frac{V_{R(2)} - V_{R(1)}}{w_g} \qquad (8)$$

Here, $V_{R(1)}, V_{R(2)}$ are the retention volumes of sample compounds with adjacent peaks; N is the number of theoretical plates determined under isocratic conditions; and V_m is the holdup volume of the column. It should be noted that the correct plate number value cannot be determined directly from a gradient elution chromatogram, as the retention factors k are continuously changing during the elution. To first approximation, additional gradient band compression resulting from a faster migration of the trailing edge of the band in a mobile phase with higher elution strength with respect to a slower migration of the leading edge can be neglected.

TRANSFER OF GRADIENT METHODS BETWEEN DIFFERENT INSTRUMENTS, COLUMNS, AND SEPARATION CONDITIONS

The transfer of gradient methods between various instruments and columns with different dimensions and efficiencies is less straightforward than with isocratic HPLC, as the gradient profile should be adapted to match changing column geometry and flow rate of the mobile phase to obtain predictable results. Because, in all equations for the gradient retention data, the product of the retention volume and of the gradient steepness parameter $V_R B$ is constant as long as the product $V_m B$ is kept constant, any change of the flow rate F, column length L, or diameter d_c at a constant gradient range [i.e., with constant concentrations of the stronger solvent at the start (A) and at the end (φ_G) of the gradient] can be compensated by appropriate change in the gradient time t_G to keep the ratio V_m/V_G constant.

This has several important practical consequences, as follows:

1. If the flow rate of the mobile phase increases from F_1 to F_2 by a factor $f = F_2/F_1 > 1$ and the gradient time t_G is kept constant, the gradient steepness parameter B decreases by the factor f, and the gradient volume V_G increases by the same factor. Hence, the retention volumes increase, too, so that the retention times do not decrease proportionally to the increased flow rate. To keep the gradient steepness (gradient volume) constant, the gradient time should decrease by the same factor f. For example, if the flow rate is increased from 1 to 2 mL/min, the gradient time should decrease from 20 to 10 min to maintain the same composition range of the mobile phase between the start and the end of the gradient, with the gradient volume $V_G = 20$ mL. Then, the retention times decrease proportionally to increasing flow rate, as in isocratic chromatography.

2. If a column with a larger inner diameter d_{c2} is used instead of the original column with the inner diameter d_{c1} at a constant column length (such as when upgrading an analytical method to semipreparative or preparative scale), the column holdup volume increases by the factor $f = (d_{c2}/d_{c1})^2$, but the retention volumes of the sample compounds increase less significantly and the separation may be impaired when the flow rate of the mobile phase does not change. To keep a constant product $V_m B$, the gradient steepness parameter B should decrease by increasing the flow rate by the same factor f. Then, the retention volumes increase f times, but the retention times do not change.

3. If a longer column (L_2) is used to increase the plate number with respect to the original column length (L_1) at a constant column inner diameter, flow rate of the mobile phase, and gradient time, the column holdup volume increases by the factor $f = L_2/L_1$, but the retention times and the retention volumes increase by less than f and the resolution may even decrease. This effect can be compensated by decreasing the gradient steepness (i.e., by increasing the gradient time by the factor f). The retention volumes and the retention times increase by the same factor and the expected increase in the column plate number and resolution is achieved.

TERNARY MOBILE-PHASE GRADIENTS

If the separation with binary gradients is unsatisfactory, ternary gradients can sometimes improve the selectivity by changing, simultaneously, the concentrations of two components with high elution strengths in a ternary mobile phase. For example, the early-eluting compounds show poor resolution with the gradients of methanol, but are better separated with gradients of acetonitrile in water, whereas the separation selectivity for the late-eluting compounds is better with a gradient of methanol than with gradients of acetonitrile in water. A ternary gradient with increasing concentration of methanol and simultaneously decreasing concentration of acetonitrile may improve the resolution of the sample.[6] Two specific types of ternary gradients are probably most useful in practice, as follows:

1. The "elution strength" (or "isoselective") ternary gradients, where the concentration ratio of two strong eluents is kept constant and the sum of the concentrations of the two eluents changes during the elution (so-called isoselective multisolvent gradient elution).

2. The "selectivity ternary gradients," where the sum of the concentrations of two strong eluents in the mobile phase is constant, but their concentration ratio changes during the elution.

OPTIMIZATION OF GRADIENT ELUTION

Gradient elution can be optimized using strategies common in isocratic HPLC. In RP gradient elution chromatography, the Dry-Lab G commercial software is probably the most popular tool for optimization of operating parameters.[7] Here, the retention data from two initial gradient runs are used to adjust, subsequently, the steepness and the range of the gradient, and, if necessary, other working parameters. This approach can be adapted to optimize segmented gradients. Structure-based commercial optimization software (e.g., Chromdream, Chromsword, or Eluex) predicts the retention based on additive contributions of the individual structural elements and, consequently, the optimum composition of the mobile phase is suggested. Such predictions are only approximate and do not take into account stereochemical and intramolecular interaction effects.

Some parameters may show synergistic effects on the separation. Appropriate selection of the concentration of the strong solvent in the mobile phase at the start of the gradient A is equally important as adjusting the gradient steepness B because each parameter influences, very significantly, the resolution and the time of analysis. The gradient steepness and the initial concentration of the strong solvent can be optimized simultaneously, using the simplex method,[8] or a simple strategy employing a preset concentration of the strong solvent φ_G at the end of the gradient and gradient volume V_G. Then, the steepness parameter B of the gradient depends on the initial concentration A, and the elution volumes V_R can be calculated as a function of a single parameter A:[6]

$$B = \frac{\varphi_G - A}{V_G} \qquad (9)$$

The differences between the retention volumes of compounds with adjacent peaks or corresponding resolution R_s can be plotted vs. the initial concentration of the strong solvent A in the form of a "window diagram" to select the optimum A that provides the desired resolution for all adjacent bands in the chromatogram in the shortest time. With optimized A, the corresponding gradient steepness parameter B can be calculated for the preset gradient volume V_G and final concentration φ_G using Eq. (9). An example of the "window diagram" for optimization of NP gradient elution chromatography is shown in

Fig. 3 (Top), and the corresponding optimized separation is shown in Fig. 3 (Bottom). In addition to the gradient steepness and initial concentration, the gradient shape can be adjusted for nonlinear gradients or segmented gradients with several subsequent linear steps with different gradient steepness.

The composition of mixed mobile phases for ternary or quaternary "isoselective gradient elution" can be optimized using "overlapping resolution mapping" strategy to adjust optimum separation selectivity based on seven or more initial experiments with solvent mixtures of approximately equal elution strengths. Based on the retention data from the initial experiments, either three-dimensional diagrams or contour "resolution maps" are constructed for all adjacent bands in a selectivity triangle space as a function of the concentration ratios of three solvents or two solvents and the pH of the mobile phase, from which the concentration ratio of the strong solvents that provides maximum resolution is selected.[9]

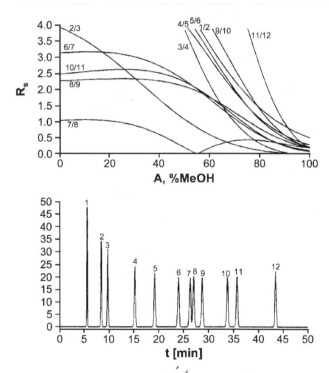

Fig. 3 Top: The resolution window diagram for RP gradient elution separation of phenylurea herbicides on a Separon SGX C18 7.5-μm column (150 × 3.3 mm ID) in dependence on the initial concentration of methanol in water at the start of the gradient A with optimum gradient volume V_G = 73 mL. Column plate number N = 5000; sample compounds: hydroxymetoxuron (1), desphenuron (2), phenuron (3), metoxuron (4), monuron (5), monolinuron (6), chlorotoluron (7), metobromuron (8), diuron (9), linuron (10), chlorobromuron (11), and neburon (12). Bottom: The separation with optimized binary gradient from 24% to 100% methanol in water in 73 min. Flow rate = 1 mL/min; T = 40°C.

INSTRUMENTAL ASPECTS OF GRADIENT ELUTION CHROMATOGRAPHY

Even though the most common UV and fluorimetric HPLC detectors can be used without problems in gradient elution if high-purity solvents are used as the mobile-phase components, some detectors are not compatible with gradient elution, such as the universal refractometric detector, which gives a response for almost all sample compounds, but also for the mobile-phase components. However, the only universal detector that can be used for gradient elution is the evaporative light-scattering (ELS) detector, which is less sensitive for UV-absorbing compounds than the UV detector. As the ELS detector gives response to the stray light on solid particles of analytes after evaporation of the solvent from the nebulized column effluent, its use is restricted to volatile mobile phases and nonvolatile analytes. Mass spectrometric detection is ideally suited for gradient elution HPLC, as it combines the features of universality and specific detection, including possibilities of on-line mass spectral analysis of each peak.

Electrochemical detectors are generally incompatible with gradient elution, except for the multichannel coulometric CoulArray detector, which is controlled by a software compensating for the gradient baseline drift during the elution. This detector allows highly sensitive and selective detection of oxidizable or reducible compounds in gradient HPLC.

In micro-HPLC with narrow bore or capillary columns, lower flow rates should be used at comparable linear mobile-phase velocities. The flow rate should decrease in proportion to the second power of the column inner diameter, so that micro-LC columns with 1 mm ID require flow rates in the range of 30–100 μL/min, whereas capillary columns with 0.3–0.5 mm ID require flow rates between 1 and 10 μL/min, and capillary or nano-LC columns with 0.075–0.1 mm ID require flow rates in the range of hundreds of nanoliters per minute. Special miniaturized pump systems are required to accurately deliver the mobile phase at very low flow rates in isocratic LC. It is technically much more difficult to keep very low flow rates constant during a gradient run than in isocratic LC, and to simultaneously change accurately the volume proportions of the mixed solvents according to a preset time program. In contemporary micro-LC and capillary LC practice, concentration gradients can be achieved using sophisticated LC pumping systems for the delivery of microliter-per-minute gradients that are either flow-split or sampled. High-precision microflow reciprocating pumps using precolumn flow splitting can be used for delivery of flow rates ranging down to 50 nL/min in micro-LC and capillary LC systems. If a precolumn flow splitter is used, only a small part of the mixed mobile phase from the pump flows through the column, whereas a larger part is diverted through a by-pass capillary. To avoid some problems connected with flow splitting, splitless systems with large inner volume syringe pumps for each solvent can be used, which deliver smooth flow. Mobile phase is not wasted and the systems are less affected by a change of the column backpressure in gradient runs where solvents with different viscosities are mixed.

The instrumental errors that can decrease the reproducibility of gradient elution data may originate from imperfect functioning of gradient pumps, especially when volatile or viscous solvents are mixed. These errors are usually most significant in the initial parts and the final parts of the gradient where the proportions of the solvents mixed are lower than 1 : 20 and rounding of the gradient is observed; this can reduce the retention times of the bands eluting near the start of the gradient and increase the retention times of bands eluting near the end of the gradient. It is usually more significant in the instruments with larger volumes between the gradient mixer and the column inlet, i.e., the "gradient dwell volume" V_D. The dwell volume may be as high as a few milliliters with some instruments and may differ from one instrument to another. It can be determined from a "blank" gradient. Much more important than contributing to the rounding of the gradient, dwell volume increases the retention times, as the sample bands migrate a certain distance along the column under isocratic conditions in a mobile phase with a low elution strength, before the front of the gradient gets to the actual position of the sample zone in the column. To avoid difficulties when a gradient HPLC method is transferred between the instruments with different V_D values and to improve the precision of predictive calculations of the gradient elution data, the correction for the gradient dwell volume should be accounted for in calculations, using appropriate equations such as Eq. (4) or Eq. (6).[5,7] The gradient delay due to the dwell volume can be relatively very significant in microcolumn gradient operation, especially in the system using precolumn flow splitting with 0.1 mm or lower ID capillaries.

When transferring gradient methods between instruments with different dwell volumes, these differences can be compensated for experimentally by programmed delay of the sample injection after the start of the gradient elution, or by insertion of a "mixing chamber"—an additional piece of tubing or a small precolumn packed with an inert material in front of the injector to obtain equal dwell volumes with different instruments. However, this approach contributes to the run time and may be impractical with narrow-diameter columns.

CONCLUSIONS

The elution with solvent gradients is the most efficient technique for improving the separation of complex samples by programmed change of retention during the HPLC separation run. An understanding of the theoretical principles of gradient elution is important for rational method development, optimization, and transfer between different instrumental systems and column geometries in RP, ion exchange, and NP modes. The gradient dwell volume of the system is the main instrumental factor complicating the method transfer and limiting rapid high-resolution in gradient micro-HPLC. The most important recent advances in the instrumentation for gradient elution HPLC have resulted in the development of sophisticated instrumentation for gradient micro-HPLC, and the availability of universal ELS detection for compounds that do not contain chromophores or fluorophores and of a sensitive multichannel coulometric detection for gradient HPLC of electroactive compounds. Gradient elution is well suited for HPLC/mass spectrometry applications.

REFERENCES

1. Snyder, L.R.; Dolan, J.W.; Gant, J.R. Gradient elution in high-performance liquid chromatography: I. Theoretical basis for reversed-phase systems. J. Chromatogr. **1979**, *165*, 3–30.
2. Jandera, P.; Churáček, J. Gradient elution in liquid chromatography: II. Retention characteristics (retention volume, bandwidth, resolution, plate number) in solvent-programmed chromatography—theoretical considerations. J. Chromatogr. **1974**, *91*, 223–235.
3. Jandera, P.; Churáček, J. Liquid chromatography with programmed composition of the mobile phase. Adv. Chromatogr. **1981**, *19*, 125–260.
4. Snyder, L.R.; Dolan, J.W. The linear-solvent-strength model of gradient elution. Adv. Chromatogr. **1998**, *38*, 115–187.
5. Jandera, P. Gradient elution in normal-phase high-performance liquid chromatographic systems. J. Chromatogr. A. **2002**, *965*, 239–261.
6. Jandera, P. Predictive calculation methods for optimization of gradient elution using binary and ternary gradients. J. Chromatogr. **1989**, *485*, 113–141.
7. Dolan, J.W.; Snyder, L.R. Maintaining fixed band spacing when changing column dimensions in gradient elution. J. Chromatogr. A. **1998**, *799*, 21–34.
8. Schoenmakers, P.J. *Optimisation of Chromatographic Selectivity*; Elsevier: Amsterdam, 1986.
9. Glajch, J.L.; Kirkland, J.J. Method development in high-performance liquid chromatography using retention mapping and experimental design techniques. J. Chromatogr. **1989**, *485*, 51–63.

SUGGESTED FURTHER READING

Jandera, P.; Churáček, J. *Gradient Elution in Column Liquid Chromatography, Theory and Practice*; Elsevier: Amsterdam, 1985.
Snyder, L.R.; Kirkland, J.J.; Glajch, J.L. *Practical HPLC Method Development*, 2nd Ed.; John Wiley & Sons: New York, 1997.

Headspace Sampling

Raymond P. W. Scott
Scientific Detectors Ltd., Banbury, Oxfordshire, U.K.

INTRODUCTION

Headspace sampling is usually employed to identify the volatile constituents of a complex matrix without actually taking a sample of the material itself. There are three variations of the technique: (a) static headspace sampling, (b) dynamic headspace sampling, and (c) purge and trapping.

DISCUSSION

The first technique, commonly used to monitor the condition of foodstuffs, particularly for detecting food deterioration (food deterioration is often accompanied by the characteristic generation of volatile products such as low-molecular-weight organic acids, alcohols, and ketones, etc.), involves first placing the sample in a flask or some other appropriate container and warming to about 40°C. Raising the temperature increases the distribution of the volatile substances of interest in the gas phase. A defined volume of the air above the material is withdrawn through an adsorption tube by means of a gas syringe. Graphitized carbon is often used as the adsorbing material, although other substances such as porous polymers can also be employed. Carbon adsorbents having relatively large surface areas ($\sim100\,m^2/g$) are used for adsorbing low-molecular-weight materials, whereas for large molecules, adsorbents of lower surface areas are used ($\sim5\,m^2/g$). After sampling, the adsorption trap is placed in an oven and connected to the chromatograph. The column is maintained at a low temperature (50°C or less) to allow the desorbed solutes to concentrate at the beginning of the column. The trap is then heated rapidly to about 300°C and a stream of carrier gas sweeps the desorbed solutes onto the column. When desorption is complete, the temperature of the column is programmed up to an appropriate temperature and the components of the headspace sample are separated and quantitatively assayed. The proportions of each component in the gas phase will not be the same as that in the sample, as they are modified by the distribution coefficient. Thus, analyses will be comparative or relative, but not absolute.

The second analytical procedure is somewhat similar, but a continuous stream of gas is passed over the sample and through the trap. This produces a much larger sample of the volatile substances of interest and, thus, can often detect trace materials. The adsorbed components are desorbed by heat in the same manner and passed directly onto a GC column. The results are still determined by the distribution coefficient of each solute between the sample matrix and the air and, thus, the quantitative results remain comparative or relative, but not absolute.

The third method (purge and trap) is used for liquids and, in particular, for testing for water pollution by volatile solvents. In this method, air or nitrogen is bubbled through the water sample and then through the adsorbent tube. In this way, the substances of interest can be completely leached from the water; the results will give the total quantity of each solute in the original water sample. Thus, with this method, the results can be actual and not relative or comparative. The solutes are desorbed by heat in exactly the same way as the previous two methods, but provision is usually made to remove the water that is also collected before developing the separation.

A good example of the use of headspace analysis is in the quality control of tobacco. Despite the health concern in the United States, tobacco is an extremely valuable export and its quality needs to be carefully monitored. Tobacco can be flue cured, air cured, fire cured, or sun cured, but the quality of the product can often be monitored by analyzing the vapors in the headspace above the tobacco.

The headspace over tobacco can be sampled and analyzed using a solid-phase micro-extraction (SPME) technique. The apparatus used for SPME is shown in Fig. 1. The basic extraction device consists of a length of fused-silica fiber, coated with a suitable polymeric adsorbent, which is attached to the steel plunger contained in a protective holder. The steps that are taken to sample a vapor are depicted in Fig. 1. The sample is first placed in a small headspace vial and allowed to come to equilibrium with the air in the vial.[1] The needle of the syringe containing the fiber is then made to pierce the cap, and the plunger pressed to expose the fiber to the headspace vapor. The fiber is left in contact with air above the sample for periods that can range from 3 to 60 min, depending on the nature of the sample.[2]

The fiber is then removed from the vial[3] and then passed through the septum of the injection system of

Encyclopedia of Chromatography DOI: 10.1081/E-ECHR-120040008

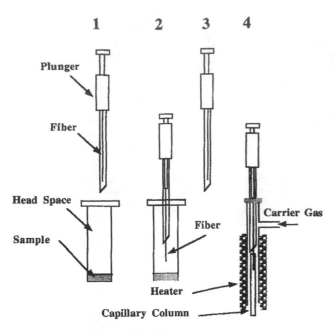

Fig. 1 The SPME apparatus.

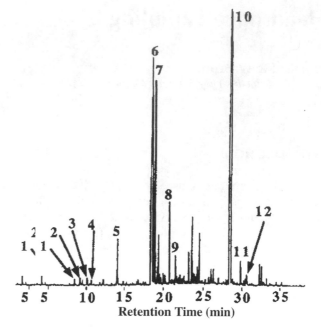

Fig. 2 A chromatogram of tobacco headspace. 1: Benzaldehyde; 2: 6-methyl-5-heptene-2-one; 3: phenylacetaldehyde; 4: ninanal; 5: menthol; 6: nicotine; 7: solanone; 8: geranyl acetone; 9: β-nicotyrine; 10: neophytadiene; 11: famesylacetone; 12: cembrene.

the gas chromatograph into the region surrounded by a heater (4). The plunger is again depressed and the fiber, now protruding into the heater, is rapidly heated to desorb the sample onto the GC column. In most cases, the column is kept cool so the components concentrate on the front of the column. When desorption is complete (a few seconds), the column can then be appropriately temperature programmed to separate the components of the sample. A chromatogram of the headspace sample, taken over tobacco, is shown in Fig. 2. The actual experimental details were as follows. One gram of tobacco (12% moisture) is placed in a 20-mL headspace vial and 3.0 mL of 3 M potassium chloride solution is added. The fiber is coated with polydimethyl siloxane (a highly dispersive adsorbent) as a 100-μm film. The vial is heated to 95°C and the fiber is left in contact with the headspace for 30 min. The sample is then desorbed from the fiber for 1 min at 259°C. The separation can be carried out on a column 30 cm long with a 250-μm inner diameter, carrying a 0.25-μm-thick film of 5% phenylmethylsiloxane. The stationary phase is predominantly dispersive, with a slight capability of polar interactions with strong polarizing solute groups by the polarized aromatic nuclei of the phenyl groups. Helium can be used

as the carrier gas, at 30 cm/s. The column is held isothermally at 40°C for 1 min, then programmed to 250°C at 6°C/min and held at 250°C for 2 min. It is seen that a clean separation of the components of the tobacco headspace is obtained and the resolution is quite adequate to compare tobaccos from different sources, tobaccos with different histories, and tobaccos of different quality.

REFERENCES

1. Grant, D.W. *Capillary Gas Chromatography*; Scott, R.P.W., Simpson, C.F., Katz, E.D., Eds.; John Wiley & Sons: Chichester, 1996.
2. Scott, R.P.W. *Introduction to Analytical Gas Chromatography*; Marcel Dekker, Inc.: New York, 1998.
3. Scott, R.P.W. *Techniques of Chromatography*; Marcel Dekker, Inc.: New York, 1995.

Helium Detector

Raymond P. W. Scott
Scientific Detectors Ltd., Banbury, Oxfordshire, U.K.

INTRODUCTION

The outer group of electrons in the noble gases is complete, and as a consequence, collisions between noble gas atoms and electrons are perfectly elastic. It follows that if a high potential is set up between two electrodes in a noble gas and ionization is initiated by a suitable radioactive source, electrons will be accelerated toward the anode and will not be impeded by energy absorbed from collisions with the noble gas atoms. However, if the potential of the anode is high enough, the electrons will develop sufficient kinetic energy that, on collision with a the noble gas atom, energy can be absorbed and a *metastable* atom can be produced. A metastable atom carries *no* charge, but adsorbs energy from collision with a high-energy electron by displacing an orbiting electron to an outer orbit.

DISCUSSION

Metastable helium atoms have an energy of 19.8 and 20.6 eV and thus can ionize and, consequently, detect all permanent gas molecules and, in fact, the molecules of all other volatile substances. A collision between a metastable atom and an organic molecule will result in the outer electron of the metastable atom collapsing back to its original orbit, followed by the expulsion of an electron from the organic molecule. The electrons produced by this process are collected at the anode and produce a large increase in anode current. However, when an ion is produced by collision between a metastable atom and an organic molecule, the electron, simultaneously produced, is also immediately accelerated toward the anode. This results in a further increase in metastable atoms and a consequent increase in the ionization of other organic molecules.

This cascade effect, unless controlled, results in an exponential increase in ion current. It is clear that the helium must be extremely pure or the production of metastable helium atoms would be quenched by traces of any other permanent gases that may be present.

Originally, a very complicated helium-purifying chain was necessary to ensure the helium detector's optimum operation. However, with high-purity helium becoming generally available, the helium detector is now a more practical system.

The metastable atoms that must be produced in the argon and helium detectors need not necessarily be generated from electrons induced by radioactive decay. Electrons can be generated by electric discharge or photometrically, which can then be accelerated in an inert gas atmosphere under an appropriate electrical potential to produce metastable atoms. This procedure is the basis of a highly sensitive helium detector that is depicted on the left-hand side of Fig. 1. The detector does not depend solely on metastable helium atoms for ionization and, for this reason, is called the helium discharge ionization detector (HDID).

The sensor consists of two cavities, one carrying a pair of electrodes across which a potential of about 550 V is applied. In the presence of helium, this potential initiates a gas discharge across the electrodes. The discharge gas passes into a second chamber that acts as the ionization chamber and any ions formed are collected by two plate electrodes having a potential difference of about 160 V. The column eluent enters the top of the ionization chamber and mixes with the helium from the discharge chamber and exits at the base of the ionization chamber.

In this particular detector, ionization probably occurs as a result of a number of processes. The electric discharge produces both electrons and photons. The electrons can be accelerated to produce metastable helium atoms which, in turn, can ionize the components in the column eluent. However, the photons generated in the discharge have, themselves, sufficient energy to ionize many eluent components and so ions will probably be produced by both mechanisms. It is possible that other ionization processes may also be involved, but the two mentioned are likely to account for the majority of the ions produced. The response of the detector is largely controlled by the collecting voltage and is very sensitive to traces of inert gases in the carrier gas. Peak reversal is often experienced at high collecting voltages, which may also indicate that some form of electron capturing may take place between the collecting electrodes. This peak reversal appears to be significantly reduced by the introduction of traces of neon in the helium carrier gas.

The helium discharge ionization detector has a high sensitivity toward the permanent gases and has been used very successfully for the analysis of trace components in ultrapure gases. It would appear that the

Encyclopedia of Chromatography DOI: 10.1081/E-ECHR-120040009

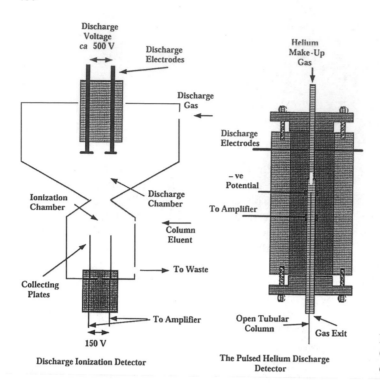

Fig. 1 The discharge ionization detector (courtesy of GOW-MAC Instruments) and the pulsed helium discharge detector (courtesy of Valco Instruments).

detector response is linear over at least two, and possibly three, orders of magnitude, with a response index probably lying between 0.97 and 1.03. In any event, any slight nonlinearity of the sensor can be corrected by an appropriate signal-modifying amplifier. The potential sensitivity of the detector to organic vapors appears to be about 1×10^{-13} g/mL.

THE PULSED HELIUM DISCHARGE DETECTOR

The pulsed helium discharge detector[1,2] is an extension of the helium detector, a diagram of which is shown on the right-hand side of Fig. 1. The detector has two sections: the upper section consisting of a tube 1.6 mm i.d. (where the discharge takes place) and the lower section, 3 mm i.d. (where reaction with metastable helium atoms and photons takes place). Helium makeup gas enters the top of the sensor and passes into the discharge section. The potential (about 20 V) applied across the discharge electrodes and for optimum performance is pulsed at about 3 kHz with a discharge pulse width of about 45 µs. The discharge produces electrons and high-energy photons (that can also produce electrons), and probably some metastable helium atoms. The photons and metastable helium atoms enter the reaction zone where they meet the eluent from the capillary column. The solute molecules are ionized and the electrons produced are collected at the lower electrode and measured by an appropriate high-impedance amplifier. The distance between the collecting electrodes is about 1.5 mm. The helium must

be 99.9995 pure, otherwise permanent gas impurities quench the production of metastable atoms. The base current ranges from 1×10^{-9} to 5×10^{-9} A, the noise level is about 1.2×10^{-13} A, and the ionization efficiency is about 0.07%. It is claimed to be about 10 times more sensitive than the flame ionization detector and to have a linear dynamic range of 10^5. The pulsed helium discharge detector appears to be an attractive alternative to the flame ionization detector and would eliminate the need for three different gas supplies. It does, however, require equipment to provide specially purified helium, which diminishes the advantage of using a single gas.

REFERENCES

1. Wentworth, W.E.; Vasnin, S.V.; Stearns, S.D.; Meyer, C.J. Pulsed discharge helium ionization detector. Chromatographia. **1992**, *34*, 219.
2. Wentworth, W.E.; Cai, H.; Stearns, S.D. Pulsed discharge helium ionization detector universal detector for inorganic and organic compounds at the low picogram level. J. Chromatogr. **1994**, *688*, 135.

SUGGESTED FURTHER READING

Scott, R.P.W. *Chromatographic Detectors*; Marcel Dekker, Inc.: New York, 1996.
Scott, R.P.W. *Introduction to Analytical Gas Chromatography*; Marcel Dekker, Inc.: New York, 1998.

Heterocyclic Bases: Analysis by LC

Monika Waksmundzka-Hajnos
Department of Inorganic Chemistry, Medical University, Lublin, Poland

INTRODUCTION

There are no absolute rules that formulate the influence of functional groups on pharmaceutical activities of compounds. However, it has been pointed out that an amine group or heterocyclic nitrogen atom possessing an aromatic ring causes an increase in a compound's biological activity. This is probably caused by the possibility of drug–receptor bonding, where electrostatic driving forces between ions present in the drug and receptor play the fundamental role. The presence of basic electron donor centers (amino group or heterocyclic nitrogen) makes possible their ionic interactions with acidic groups in proteins (–COOH), phospholipids, or nucleic acids (HPO_4^{2-}). Of course, other interactions, such as ion–dipole, dipole–dipole, induced dipole–dipole, H-bonds, and hydrophilic–hydrophobic specific interactions also play an important role in the fitting of drug molecules to the receptor. The chemical structures of drugs also play an important role in their solubility in body fluids and their transport through the lipid membranes of cells. Therefore, compounds possessing heterocyclic nitrogen are present in numerous drug groups. Examples of such drugs, their chemical structures, and pharmacological activities are presented in Table 1. Groups of alkaloids, which occur in plant organs with their pharmacological activities, are presented in Table 2.

STRUCTURE–CHROMATOGRAPHIC RETENTION RELATIONSHIPS

The relationship between the chemical structures of compounds and their chromatographic behavior has been considered by many scientists and was first reported by Martin[1] in partition chromatography, where the partition coefficient is regarded as an additive value. The hypothesis of the R_M additivity has raised worldwide discussion; hence, additivity rules were formulated. Deviation from the R_M additivity results from the complex character of the chromatographic process (change of a composition and volume proportions of phases), constitutional effects in molecules, because of reciprocal interactions of functional groups (internal hydrogen-bond effects, steric and electromeric effects) as well as ionization of substances.

Snyder[2] takes into consideration many factors influencing the value of adsorption energy of a molecule. Molecular planarity, steric hindrance, chemical interaction of adjacent functional groups (H-bonding), electronic interactions of some functional groups (induction and mesomeric effects), and simultaneous adsorption of two neighboring functional groups ("anchorage" effect of molecule on the adsorption sites especially on alumina surface) have been taken into account.

Retention in chromatographic systems can be connected with the properties of the chromatographed compounds. It should manifest itself in quantitative structure–retention relationships (QSRR) equations,[3] correlating retention parameters ($\log k$) with the properties of analytes and chromatographic system revealed by molecular descriptors: dipolarity/polarizability, ability to donate H-bonds, measure of analyte H-bond accepting potency, analyte molecular volume, and others.

ADSORPTION CHROMATOGRAPHY

Silica Gel

The retention–mobile phase composition relationships for heterocyclic bases in systems involving silica/binary eluent can be expressed as the linear plots of $\log k(R_M)$ vs. $\log X$ (X is the molar fraction of polar modifier in the eluent).[4] It proves a simple displacement model of retention of these substances in normal-phase systems. Slopes of $\log k$ vs. $\log X$ plots provide information about interactions of chromatographed substances with the adsorbent surface. Unit slopes, meaning single-point adsorption, are usually obtained for monofunctional heterocyclic bases (pyridine, quinoline, acridine without any functional groups). When unit slopes are obtained for bifunctional solutes, the complete delocalization of the weaker functional group takes place. It occurs for the apolar functional groups—for example, an alkyl chain. Sufficient adsorption energies of the two groups and suitable distance between the groups are two sufficient conditions of two-point adsorption for bifunctional heterocyclic bases (or other solutes). The experimental data indicate that for bases with two strongly polar

Encyclopedia of Chromatography DOI: 10.1081/E-ECHR-120041315

Table 1 Structure and pharmacological activity of some heterocyclic base derivatives

Drugs	Structure	Pharmacological activity
Benzodiazepine derivatives		Sedative, antiepileptic, hypnotic, anesthetic
1,4-Dihydropyridine derivatives		Ca-channel blocker, hypotensic
4-Aminoquinoline derivatives		Anti-inflammatory, antimalarial
8-Hydroxyquinoline derivatives		Antiseptic
Acridine derivatives		Antiseptic
Barbiturates		Hypnotic, sedative, anesthetic
Thiobarbiturates		Anesthetic
Hydantoin derivatives		Antiepileptic
Quinoline derivatives		Anesthetic
Phenothiazine derivatives		Sedative, spasmolytic, antihistaminic

(Continued)

Table 1 Structure and pharmacological activity of some heterocyclic base derivatives *(Continued)*

Drugs	Structure	Pharmacological activity
Butyrophenone derivatives		Sedative, psychotropic
5-Pyrazolone derivatives		Analgetic, antipyretic
3,5-Pyrazolidinedione derivatives		Analgetic, anti-inflammatory
Imidazole derivatives		Fungicidal
Imidazoline derivatives		Antihistaminic
Pyridine derivatives		Hypotensic, cardiac
Pyridine derivatives		Tuberculostatic
Monobactams		Antibiotic
Penicillins		Antibiotic
Cephalosporins		Antibiotic
Piperazine derivatives		Antidepressive

(Continued)

Table 1 Structure and pharmacological activity of some heterocyclic base derivatives *(Continued)*

Drugs	Structure	Pharmacological activity
Morpholine derivatives	R–N⟨ ⟩O	Analgetic, β-adrenolytic, antiepileptic
Piperidine derivatives	R₁–N⟨ ⟩–R₂	Neuroleptic, antidepressive, antihistaminic, psychotropic

functional groups, e.g., 4-aminopyridine, 5-hydroxy-quinoline, 5-aminoquinoline, two-point adsorption on silica occurs.[4] When the functional groups in the solute molecule differ significantly in their adsorption energies, the localization of the stronger group and the resulting delocalization of the weaker group may cause a decrease of the slope, for example, for quinoline derivatives with a methoxy group in the meta or para position.

The slope of R_M vs. $\log c$ plots depends also on solvent strength. For the polar component of an eluent with low solvent strength (diethyl ether, methyl ethyl ketone), the slopes of methoxy and acetyl derivatives are greater than for stronger ones. The use of diethyl-amine as the eluent modifier leads to single-point adsorption of almost all heterocyclic bases investigated because the weaker group is unable to compete with the solvent for surface hydroxyl groups. Only basic groups such as $-NH_2$ in 4-aminopyridine, 3-amino-pyridine, and 5-aminoquinoline can compete with diethylamine for surface hydroxyls, and the slopes for these solutes are approximately equal to 2.0.

Ortho substituted heterocyclic bases behave, on a silica surface, like monofunctional solutes and, in most cases, have slopes near unity [for example 2-acetylpyr-idine, 1-(pyridyl-2′)-ethan-1-ol], whereas the slope values for analogous para isomers are much higher. Also, 8-substituted derivatives of quinoline behave like ortho isomers (see 8-hydroxyquinoline). This is caused by the ortho effect—H-bond interactions of two neigh-boring polar groups (internally H-bonded groups), which cause their weaker adsorption and single-point interactions with surface silanols [compare slopes of 5-hydroxyquinoline and 8-hydroxyquinoline, 2-acetyl-pyridine and 4-acetylpyridine, 1-(pyridyl-2′) -ethan-1-ol and 1-(pyridyl-4′)-ethan-1-ol].[4]

Polar modifier influences separation selectivity of nitrogen bases on silica only to a small degree. Matyska and Soczewiński[5] compared separation selectivity of quinoline bases in equieluotropic eluent systems consisting of *n*-heptane and various modifiers (ethyl methyl ketone, ethyl acetate, diethyl ether, diiso-propyl ether) and with chloroform and the same modi-fiers. They concluded that retention and separation

selectivity of investigated nitrogen bases is, in most cases, similar. However, selectivity of separation of quinoline bases with two polar groups (hydroxy, amino derivatives) strongly depends on the modifier used as the eluent component.

Petrowitz[1] compared adsorptive properties of alkyl- and halogen-derivatives of different solutes, and also heterocyclic bases. For example, methyl deri-vatives of pyridine behave in a different way on silica, depending on the group position. Thus, α- and γ-methylpyridine are strongly retained on silica because of hyperconjugation and formation of a double bond with increase of basicity of heterocyclic nitrogen, which does not occur in the β-isomer. It is also mentioned that in homologous series of pyridine alkyl-derivatives, a decrease of adsorption ability with the increase of the alkyl chain length is observed. When a pyridine or qui-noline molecule has two functional groups (e.g., amino group in 2-position and methyl group in various posi-tions), the adsorption affinity depends on steric hin-drance of heterocyclic nitrogen by the neighboring methyl group. When retention behavior of quinolines with polar—hydroxy or amino group in the 8-position is observed, the methyl group in the 2-position signifi-cantly reduces the adsorption ability of molecules.

Retention behavior of halogenopyridines depends on the molecular mass of the halogen. The reduction of adsorption ability with an increase of halogen mole-cular mass is observed. The position of the halogen atom also has great meaning—a halogen neighboring the polar group or heterocyclic atom influences more adsorption ability and causes its decrease.

Alumina

Previous systematic investigations of organic compounds' adsorption from various chemical groups indicated analogies in the adsorption ability of silica and alumina.[6] However, alumina has heterogeneous surface active sites—electron donor oxygen atoms and electron acceptor aluminum ions. Numerous pub-lications[7] have drawn attention to certain regularities in the adsorption onto alumina of the substituted

Table 2 Alkaloids and their pharmacological activities

Alkaloid type	Group of alkaloids	Main active alkaloids	Biological activity
Pyridine, piperidine	Nicotine and anabazine	Nicotine, anabazine, nornicotine	Synapotolytic, toxic
	Cortex granati	Isopelletierine and derivatives	Toxic, anthelmintic
	Lobeline group	Lobeline, isolobeline, sedamine	Synapsotropic, analeptic, secretolytic
	Arecoline	Arecoline	Anthelmintic, purgative
	Conium	Coniine, coniidine	Toxic, poisonous
	Lycopodium	Lycopodine and derivatives	Toxic, poisonous
	Quinolizidine	Lupanine, sparteine, cytisine, lobeline	Toxic, antiarrythmic
Tropane	Tropine	Tropine, hyoscyamine, atropine, scopolamine	Paraspasmolytic
	Ecgonine	Ecgonine, cocaine	Anesthetic
Isoquinoline	Benzylisoquinoline	Papaverine	Spasmolytic
	Aporphine	Glaucine, magnoflorine, boldine	Spasmolytic, hypotensic, choleretic
	Protoberberine	Berberine, narcotine, palmatine	Antibacterial, choleretic, analgetic, antiarrythmic
	Benzophenanthridine	Chelidonine, sanguinarine, chelerythrine	Choleretic, anesthetic, spasmolytic
	Protopine	Protopine, cryptopine	Antiarrythmic
	Morphinan	Morphine, codeine, thebaine	Narcotic, analgetic, spasmolytic
	Emetine	Emetine, cephaline	Antimicrobial (protozoa)
Indole	Indole alkaloids, *Catharanthus roseus*	Vincristine, vinblastine, vindesine	Anticancer, cytostatic
	β-carboline	Harman, harmine	Hallucinogenic, antiparkinsonian
	Yohimbine	Yohimbine, serpentine	Spasmolytic, hypotensic
	Reserpine	Reserpine, rescinnamine	Psychotropic, anxiolytic, antiarrythmic
	Eburamine	Vincamine	Hypotensic
	Strychnos	Strychnine, brucine	Analeptic, toxic, convulsant
	Ibogaine	Ajmaline, ibogaine	Antiarrythmic, psychotropic
	Secale cornutum	Ergotamine, ergocriptine, ergozine	α-Adrenolytic, spasmolytic, hypotensic
Purine		Caffeine, theophylline, theobromine	Analeptic, diuretic, spasmolytic
Steroidal	*Fritillaria* isosteroidal	Cevanine, jervine veratramine	Antitussive, tracheal, and bronchial relaxative
	Veratrum album	Protoveratrine A and B, veratramine, germine, veratrine, protoverine	Hypotensic, insecticidal
	Solanum	Solanine, tomatine, tomatidine, solanidine	Toxic, antifungal
Quinoline	*Cinchona*	Cinchonine, quinidine, quinine, cinchonidine	Antipyretic, antimalarial, antiarrythmic
Pyrrolizidine		Retronecine, heliotridine	Hepatotoxic, cancerogenic, cytotoxic

pyridines and related aza aromatics. The contribution of the nitrogen atom in these adsorbates to total adsorption energy is markedly sensitive to the steric environment about the nitrogen atom. Klemm, Klopfstein, and Kelly stated[8] that the interaction between nitrogen and adsorbent is the result of charge-transfer complex formation. The presence of strong electron donor, as well as electron acceptor centers, favors especially considerable adsorption of molecules having, in adjacent positions, functional groups able to simultaneously interact with the surface aluminum ions produced by the formation of the chelate complex.[8] The same substances interact weakly with the silica surface. Retention behavior of hydroxyquinoline derivatives is quite different on alumina compared to silica.[1] 8-Hydroxyquinoline is strongly retained on the alumina surface due to anchorage of adjacent polar groups by the formation of a chelate with Al^{3+} ions, whereas, on a silica surface, it is weakly adsorbed. Rupture of an internal H-bond by methylation of the OH group leads to the increase of adsorption of methoxy derivative on silica, unlike alumina, on which 8-hydroxyquinoline is more strongly retained due to the anchorage effect than 8-methoxyquinoline (Table 3).[1] From the slopes of R_M vs. $\log X$ plots obtained for alumina in binary eluents, it is visible that in the case of the covering of a heterocyclic nitrogen by a methyl group or by a condensed ring, flat adsorption of the molecule is possible.

Magnesium Silicate—Florisil

Linear dependencies of R_M vs. $\log c$ of heterocyclic bases on Florisil layers indicate the displacement model of retention. It has been confirmed[1] that quinolines (methyl or benzo derivatives) are more strongly adsorbed onto Florisil than onto silica in nonaqueous n-heptane and polar modifier eluent systems. In spite of this, separation selectivity is better when Florisil is used (methylquinolines can be separated from dimethylquinolines and from quinoline). The influence

Table 3 R_F values of hydroxy and methoxy derivatives of quinoline on alumina and silica

Substance	R_F values in chromatographic systems			
	10% MeOH + B		40% AcOEt + Cx	
	SiO_2	Al_2O_3	SiO_2	Al_2O_3
8-Hydroxyquinoline	0.60	0.04	0.50	0.0
5-Hydroxyquinoline	0.16	0.29	0.14	0.15
8-Methoxyquinoline	0.34	0.78	0.06	0.28
5-Methoxyquinoline	0.48	0.82	0.26	0.66

Abbreviations: MeOH—methanol, B—benzene, AcOEt—ethyl acetate, Cx—cyclohexane.

of the neighboring methyl group in 8-methylquinoline on the decrease of adsorption affinity on Florisil is observed, as in the case of silica. The heterocyclic bases with a second electron acceptor group (6-nitroquinoline, 2-chloro-3-nitropyridine, 2-chloro-5-nitropyridine) are adsorbed more strongly onto Florisil active centers (OH groups and Mg^{2+} ions) than onto silica.[9]

Table 4 shows ΔR_M values for substituted quinolines in isoeluotropic eluent systems on different adsorbents.[10] The isoeluotropic series obtained for quinoline on various adsorbents is developed with solvents from different selectivity groups (Table 4). It is clearly seen that the highest ΔR_M values were obtained for quinoline bases on Florisil with dichloromethane, 2-propanol, and other modifiers, which proves highest selectivity of separation in this system.

Polar-Bonded Stationary Phases

Polar-bonded stationary phases, such as cyanopropyl, diol, or aminopropyl, bonded to a silica matrix, have moderate polarity and can be used in normal- and reversed-phase (RP) systems. The retention behavior of heterocyclic bases was also examined using these adsorbents by determination of R_M (log k) values of solutes by the use of eluents with various modifier concentrations.[1] It was statistically found that the Snyder–Soczewiński equation and Scott theory describe the retention of quinolines on polar-bonded stationary phases in normal-phase systems sufficiently well. It seems that results are consistent with a displacement model. The dispersive interactions between solute molecules and the polar component of an eluent seem also to have an important role.[1] Similarly, the retention–eluent composition relationships for quinolines on such layers in RP systems using aqueous eluents can be represented by a semilogarithmic equation.[1]

The selectivity of separation of quinoline bases using normal-phase systems and polar-bonded stationary phases was compared by log k_1 – log k_2 correlations.[11] The values of regression coefficients for all correlation lines are relatively low. This results from different selectivities and mechanism of separation of the heterocyclic bases in the investigated systems. Fig. 1 is a graphical comparison of quinoline separation selectivity by the use of various systems as log k spectrum. It is seen that all polar-bonded stationary phases can be used for the separation of quinoline derivatives, especially with 2-propanol or tetrahydrofuran as eluent modifiers.

Application of Polar Adsorbents in Separation of Heterocyclic Bases

Polar adsorbents, especially silica, are widely used for the separation of alkaloids and basic drugs, such as

Table 4 ΔR_M ($R_{M(QX)} - R_{M(Q)}$) values for substituted quinolines in isoeluotropic eluent systems[a]

Silica functional group	5% iPrOH	40% DX	50% THF	60% AcOEt	80% EtMeCO	50% Me$_2$CO	100% iPr$_2$O	100% DCM
CH$_3$	−0.05	−0.02	−0.02	0.04	0.00	0.02	0.07	0.12
CH$_3$O	0.04	0.13	0.09	0.14	0.04	0.07	0.27	0.41
NO$_2$	0.15	0.19	0.11	0.14	−0.04	0.00	0.37	0.12
C$_9$H$_6$N	−0.35	−0.20	−0.24	−0.56	−0.46	−0.15	−0.65	−0.33
C$_6$H$_4$	−0.23	−0.07	−0.15	−0.31	−0.24	−0.09	−0.26	0.27
Alumina functional group	**5% iPrOH**	**15% DX**	**20% THF**	**15% AcOEt**	**20% EtMeCO**	**20% Me$_2$CO**	**90% iPr$_2$O**	**90% DCM**
CH$_3$	−0.03	−0.02	−0.04	−0.03	−0.04	−0.08	0.04	0.04
CH$_3$O	0.05	0.23	0.15	0.19	0.13	0.08	0.28	0.21
NO$_2$	0.05	0.33	0.30	0.32	0.17	0.00	0.45	0.04
C$_9$H$_6$N	−0.20	−0.15	−0.23	−0.22	−0.29	−0.33	−0.32	−0.46
C$_6$H$_4$	−0.01	−0.02	0.00	−0.08	−0.07	−0.15	−0.04	0.00
Florisil functional group	**10% iPrOH**	**20% DX**	**30% THF**	**40% AcOEt**	**30% EtMeCO**	**30% Me$_2$CO**	**100% iPr$_2$O**	**100% DCM**
CH$_3$	−0.02	0.00	0.00	0.06	−0.02	−0.06	—	0.21
CH$_3$O	0.60	0.47	0.40	0.33	0.28	0.08	0.66	0.57
NO$_2$	0.70	0.59	0.32	0.19	0.09	−0.17	0.78	0.10
C$_9$H$_6$N	−0.26	−0.17	−0.21	−0.41	−0.39	−0.37	0.07	−0.69
C$_6$H$_4$	0.13	0.38	0.09	−0.04	0.00	−0.13	0.43	0.78

Abbreviations: iPrOH—2-propanol, DX—dioxane, THF—tetrahydrofuran, EtMeCo—ethylmethyl ketone, iPr$_2$O—diisopropyl ether, DCM—dichloromethane.
[a]The eluent strength of polar modifier—n-heptane binary mixtures was selected to give retention factor k, of 1 for quinoline. (From Ref.[10].)

barbiturates, benzodiazepines, and other pyridine and quinoline derivatives. Because of strong interactions of basic nitrogen with surface silanols, solvents with high eluent strength are used as mobile phases. In a review[12] describing TLC analysis of benzodiazepines, there are more than 40 papers cited where the use of silica layers is reported. Mixtures of highly polar solvents—alcohols (MeOH, EtOH), chloroform, and mostly ammonium aqueous solutions or ethylenediamine as ionization suppressing agents are used as eluents. Moreover, basic components of an eluent can block the surface acidic silanols of silica. In a few cases, the use of a polyamide with various eluents and alkyl-bonded phases with aqueous eluents is reported. The use of similar chromatographic systems, e.g., silica/chloroform + alcohol (MeOH, EtOH, BuOH), is reported for analytical TLC of polyhydroxy–chromone and flavonoid alkaloids.[13] Cellulose, with multicomponent aqueous eluent, has also been used. There were no reports of HPLC of flavonoid alkaloids until 2002. For the separation of polar chromone alkaloids, the use of silica with aqueous eluents containing ammonia is also reported.[13] Normal-phase systems can also be used for the isolation of chromone and flavonoid alkaloids using silica columns or preparative silica layers, mostly with gradient elution with chloroform + methanol. Because of the detection difficulty of polyhydroxy alkaloids (pyrrolidine, piperidine, pyrrolizidine, indolizidine, and nortropane classes) resulting from their lack of suitable chromophores for spectroscopic detection, analytical TLC is used for purity determination and detection in plant extracts and in pharmacokinetic studies.[14] Preparative planar chromatography has also been a separation method of choice for isolation of individual polyhydroxy alkaloids from mixtures. Silica gel, with combinations of chloroform, methanol, and aqueous ammonium, has been widely used for TLC in analytical, as well as on preparative, scale. Cinchona–quinoline alkaloids, mostly analyzed in RP systems, are also separated by normal-phase chromatography, mainly using bare silica columns or layers with mobile phases containing solvents such as chloroform, acetone, or

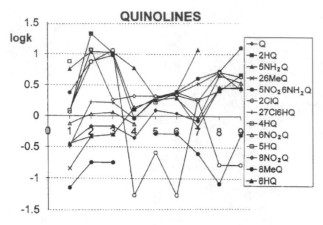

Fig. 1 Graphical comparison of $\log k$ values for quinoline bases in the following chromatographic systems. Diol phase: (1) 15% iPrOH, (2) 20% THF, (3) 20% DX; aminopropyl phase: (4) 10% iPrOH, (5) 10% THF, (6) 10% DX; cyanopropyl phase: (7) 10% iPrOH, (8) 5% THF, (9) 5% DX. All modifiers are dissolved in n-heptane. Q—quinoline, 2HQ—2-hydroxyquinoline, 5NH$_2$Q—5-aminoquinoline, 26MeQ—2,6-dimethylquinoline, 5NO$_2$6NH$_2$Q—5-nitro-6-aminoquinoline, 2ClQ—2-chloroquinoline, 27Cl6HQ—2,7-dichloro-6-hydroxyquinoline, 4HQ—4-hydroxyquinoline, 6NO$_2$Q—6-nitroquinoline, 5HQ—5-hydroxyquinoline, 8NO$_2$Q—8-nitroquinoline, 8MeQ—8-methylquinoline, 8HQ—8-hydroxyquinoline. (From Ref.[11].)

ethyl acetate with alcohols as polarity adjusters and ammonia or diethylamine as ionization suppressors and silanol blockers.[15] Fig. 2 presents an example of the separation of isoquinoline alkaloids by use of two-dimensional TLC (2D-TLC) on a silica plate.

RETENTION OF IONIZABLE WEAK BASES IN RPLC

HPLC separation of ionic samples is more complicated[16] than separation of nonionic compounds. For regular ionic samples, three HPLC methods: reversed-phase, ion-pair, or ion-exchange chromatography (IEC) can be chosen. Because of its simplicity, reversed-phase chromatography (RPC) is usually the best starting point. If RPC separation proves inadequate, the addition of an ion-pairing reagent to the mobile phase or application of IEC can be considered.

Separation Selectivity as a Function of pH and Mobile-Phase Composition

From the theory[17] for RP retention of ionic (e.g., basic) compounds as a function of pH, it can be assumed that a given solute (for example, a heterocyclic base) exists in ionized (+) and nonionized forms and

its capacity factor k is given by:

$$k = k^0(1 - F^+) + k_1 F^+ \tag{1}$$

where k_0 and k_1 refer to k values for nonionized and ionic forms and F^+ is the fraction of ionized solute molecules for the case of a basic solute:

$$F^+ = 1/\{1 + (K_a/[H^+])\} \tag{2}$$

The dependencies of $\log k$ (R_M) as a function of pH for selected alkaloids are given in Fig. 3.

The potential errors in the use of Eqs. (1) and (2) result from the following facts:[17] retention of solutes, especially protonated bases, by processes other than solvophobic interactions, e.g., with exposed silanols or metal contaminants;[18] change in K_a values as a function of ionic strength; solvophobic effect of ionic strength on solute retention; ion-pair interaction of sample ions with ionized buffer species; change in the sorption properties of the stationary phase (C$_8$ or C$_{18}$) as a result of changing ionization of silanols; a change in buffer type, when more than one buffer type is needed to cover a given pH range.

It is maintained[17] that computer simulations based upon the theoretical model [Eqs. (1) and (2)] are able to predict, accurately, retention and resolution of basic solutes as a function of pH. Predicted retention times and α values were significantly more difficult for the case of basic, rather than acidic, solutes, due to silanol effects (more significant for basic solutes).

As retention factors (k) can decrease by a factor of 10 or more for an ionized vs. a nonionized compound, it is often necessary to combine pH optimization with variation of solvent strength (%B, φ) in order to maintain a reasonable k range for the effective separation (1 < k < 20).[17]

$$k = f([H^+], \varphi) \tag{3}$$

Because the capacity factor of a weak base is the average of capacity factors of individual forms BH$^+$ and B:

$$k = k_0\left(\frac{[BH^+]}{[BH^+] + [B]}\right) + k_1\left(\frac{[B]}{[BH^+] + [B]}\right) \tag{4}$$

it can be transformed to the equation:

$$k = \frac{k_0 + k_1[H^+]/K_a}{1 + [H^+]/K_a} \tag{5}$$

However, the problem is more complex because the acidity constant K_a as well as the retention factor

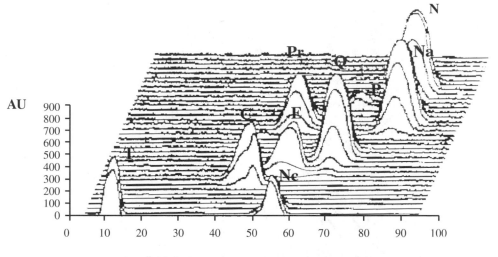

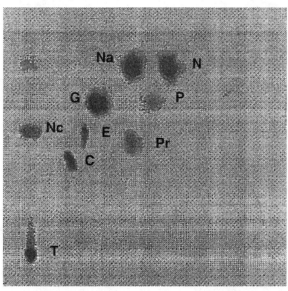

Fig. 2 Densitogram and videoscan from 2D-TLC of isoquinoline alkaloids separated on silica layer by use of aqueous methanol (8%) with 1% ammonia as the first direction eluent and multicomponent nonaqueous eluent with 0.1 M DEA as the second direction eluent. N—noscapine, Na—narcotine, Nc—narceine, G—glaucine, E—emetine, C—codeine, P—papaverine, Pr—protopine, T—tubocurarine. (From Petruczynik, A.; Waksmundzka-Hajnos, M; Hajnos, M.L. *J. Planar Chromatogr.* **2005**, *in press.*)

of the protonated form k_0 and ionized form k_1 vary with the concentration of modifier in the aqueous mobile phase (φ), although the general form of Eq. (5) is maintained. This problem of the retention factor as the combined function of pH and modifier concentration in aqueous mobile phase has been analyzed by several researchers.[1]

Marques and Schoenmakers[19] took two approaches for weak acids; nevertheless, ionized base can be counted as cationic acid BH$^+$:

1. At constant pH, they describe k_0 and k_1 as a function of concentration of organic modifier in the mobile phase—φ, taking, from previous

papers, the dependence of acidity constant as a function of φ:[20]

$$k = \delta + \frac{k_0(\varphi) + k_1(\varphi)[\text{H}^+]/K_a(\varphi)}{1 + [\text{H}^+]/K_a(\varphi)} \qquad (6)$$

2. The second approach starts with:

$$\ln k = A + B\varphi + C\varphi^2 \qquad (7)$$

where A is ln k for 0% modifier (methanol), which should be sigmoidal function of [H$^+$],

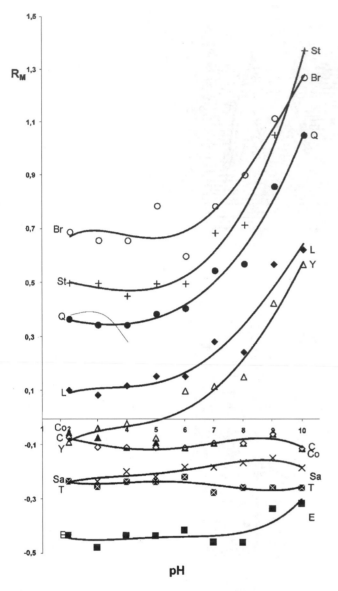

Fig. 3 Dependence of R_M vs. pH of mobile phase for investigated alkaloids. System: C18W/MeOH/water (8 : 2) buffered with phosphate buffer 0.01 M/L. E—emetine, T—theophylline, Sa—santonine, Co—colchicine, C—caffeine, Y—yohimbine, L—lobeline, Q—quinine, Br—brucine, St—strychnine. (From Petruczynik, A.; Waksmundzka-Hajnos, M; Hajnos, M.L. J. Chromatogr. Sci. **2005**, *in press.*.)

so that

$$A = \ln\left[\frac{k_0^w + k_1^w[H^+]/K_a^w}{1 + [H^+]/K_a^w}\right] \qquad (8)$$

where k_0^w is the capacity factor of B, k_1^w is the capacity factor of BH^+, and K_a^w is the acidity constant, all in pure water.

The first approach [Eq. (5)] is realized assuming k_0, k_1, and K_a as different functions of mobile-phase composition: linear, quadratic, cubic (for K_a), and $\delta = 0$ or $\delta \neq 0$ (δ is the constant shift parameter). All the models (class 1 models) were verified experimentally. The model approaching $\ln k_0$, $\ln k_1$, and $\ln K_a$ as quadratic function of φ and $\delta = 0$ is, in the authors' opinion, the best compromise between precision and practicality.

The second approach is realized assuming k_0 as a sigmoidal function of $[H^+]$, the B parameter as quadratic, cubic, or sigmoidal function of $[H^+]$, and $C = 0$ or C as a linear function of $[H^+]$ and $\delta = 0$ or $\delta \neq 0$. All models were verified experimentally (class 2 models).

Models approaching k_0 as a sigmoidal function of $[H^+]$, B as cubic function of $[H^+]$, $C = 0$ and $\delta \neq 0$ or k_0 as sigmoidal function of $[H^+]$, B as quadratic, and C as a linear function of $[H^+]$ and $\delta \neq 0$ are adequate for practical purposes.

Retention as a Function of Ionic Strength of Eluent

The pH of the mobile phase is a major factor in the separation of ionizable compounds. As mentioned earlier, the most widely used model[17] considers the retention factor as an average of k_0 and k_1 according to the mole fraction of the neutral and ionic forms. The mole fraction depends on pK_a and pH of mobile phase. The pH of the mobile phase is taken to be the same as that of the aqueous fraction and this implies a false assumption. Even when pH is measured after mixing the buffer with the organic modifier, the potentiometric system, calibrated with aqueous standards, does not measure the true pH of the mobile phase.[21]

The second problem is that[19] pH should be taken from the activity of hydrogen ions, and the effect of activity coefficients γ can be neglected in water, but when the percentage of the organic modifier in the mobile phase increases, the activity coefficients decrease and cannot be neglected. Similarly, for the dissociation constant, the concentration should be changed by activities. From the Debye–Hückel definition, an activity coefficient depends on the ionic strength I of the solution.

The pH scale of any amphiprotic solvent is limited by zero and pK_{ap} values (K_{ap} is the autoprotolysis constant of medium); it differs in a mixed solvent, for example, methanol–water, where different proton-transfer equilibria occur. Ionizable solutes dissolved in these mixtures are differently solvated, show different dissociation constants, and the pH scale of the medium changes with mobile-phase composition.

Because the retention of ionic solutes depends on K_a, pH, and solvent strength, it depends on the activity coefficients of ions in the medium and, therefore, on its ionic strength.

Application of RP-HPLC Systems for Heterocyclic Base Analysis

Optimization in RP separation, and controlling the selectivity of basic samples, can be performed similarly as for nonionic compounds by the variation of the solvent strength (%B) to obtain a satisfactory k range ($1 < k < 10$) or by change of the column type (C8, C18, phenyl, cyano). In applications of RP systems for the analysis of ionic compounds, the choice of suitable buffer is very important. Several properties such as buffer capacity, UV absorbance, and also solubility, stability, and interactions with the sample and chromatographic system, should be taken into account.

For RP separations when silica-based columns are used, the pH range of the mobile phase should be between 2 and 8. Therefore, for chromatographic analysis of heterocyclic bases, the following buffers may be used: phosphate buffer, (2.1–3.1, 6.2–8.2, and 11.3–13.3), acetate buffer (3.8–5.8), citrate buffer (2.1–6.4), carbonate buffer (3.8–5.8), formate buffer (2.8–4.8), and ammonia buffer (8.2–10.2) can be used.

Ionic samples, especially basic compounds, can interact with underivatized free silanols of silica-based alkyl-bonded columns. It appears that retention occurs by an ion-exchange process that involves protonated bases and ionized silanols. This case leads to increased retention, band tailing, and column-to-column irreproducibility. It is generally desirable to minimize these silanol interactions by an appropriate choice of experimental conditions. Silanol interactions can be reduced by selecting a column that is designed for basic samples with a reduced number of very acidic silanols that favor the retention process. The first method to reduce the silanol effect is the use of a low pH mobile phase ($2.0 < pH < 3.5$) to minimize the concentration of ionized silanols because, in this case, ionization of silanols is largely suppressed, giving rise to better peak shapes. The silanol effect can be further reduced by using a higher buffer concentration ($>10\,mM$) and the choice of buffer cations that are strongly held by the silanols ($Na^+ < K^+ < NH_4^+ <$ triethylammonium$^+$ < dimethyloctylammonium$^+$) and, therefore, block sample retention by ionized silanols. Successful analysis of bases can be obtained even with classical RP-HPLC silicas by incorporation of amines into the mobile phases, which then compete with the analytes for column silanol sites. Working at high pH (e.g., pH > 7.0) for the separation of basic compounds is also recommended. Weak bases (e.g., pyridines) may be nonionized at higher pH values and, thus, they may eliminate ionic interactions with acidic silanols. The problem, however, is with poor stability of silica-based columns, which are less stable for pHs over 6 and cannot be used at all for pH greater than 8. Only densely bonded alkyl end capped columns can be routinely used up to at least pH 11 when organic buffers and temperature under 40°C are used.

Reversed-phase HPLC is the method recommended for the screening of plant material, in which chromone alkaloids can be present.[13] Mostly C-18 columns were

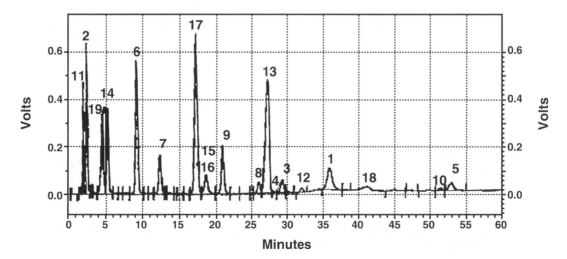

Fig. 4 HPLC chromatogram of the separation of alkaloids' standard mixture. Phenyl column, acetonitrile gradient 15–50%, acetate buffer at pH 3.5 + 0.05 M/L DEA. 1—berberine, 2—boldine, 3—chelidonine, 4—chelilutine, 5—chelerythrine, 6—codeine, 7—dionine, 8—emetine, 9—glaucine, 10—homochelidonine, 11—laudanosine, 12—noscapine, 13—narcotine, 14—narceine, 15—papaverine, 16—paracodine, 17—protopine, M. 18—sanguinarine, 19—tubocurarine. (From Petruczynik, A.; Waksmundzka-Hajnos, M. *in preparation*.)

applied with aqueous mobile phases containing high concentrations of buffer at pH 4.5–6.5, modified with methanol or acetonitrile. Polyhydroxy alkaloids of the pyrrolidine, piperidine, pyrrolizidine, indolizidine, and tropane classes, because of their lack of a suitable chromophore, have rarely been analyzed by conventional HPLC with UV detection. The detection problem could be surmounted by derivatization of the hydroxyl groups or by the use of other detection methods.[14] Amino columns, eluted with acetonitrile–water, or C-18 columns with buffered aqueous methanol with MS detection, were used for these purposes. Analysis of *Cinchona* alkaloids can also be performed by RP-HPLC, which is still the first choice for their separation.[15] Mostly, C-18 columns with aqueous mobile phases modified with methanol, acetonitrile, or tetrahydrofuran at acidic pH, have been used. The eluents with competitive amines to mask silanol effects were also used for the separations of *Cinchona* alkaloids.[16] Isosteroidal alkaloids, the main bioactive ingredient of *Fritillaria* species, do not display strong UV absorption and cannot be analyzed by conventional HPLC/UV. Systems with C18 columns and aqueous methanol containing amines were used when evaporative light-scattering detection was applied. Fig. 4 presents separation of isoquinoline alkaloid standards on a phenyl stationary phase with an eluent containing diethyl amine (DEA).

A review describing methods of measurement of benzodiazepines in biological samples[1,22] also reports a number of examples of the use of HPLC with alkyl-bonded phases and aqueous eluents. Mainly, C-18 columns eluted with aqueous acetonitrile or methanol at low pH and UV detection, were used in benzodiazepine analyses.

RETENTION OF HETEROCYCLIC BASES IN RP ION-PAIR LC

Ion-pair and RP-HPLC share several features. The columns and mobile phases used for these separations are generally similar, differing mainly in the addition of an ion-pairing reagent to the mobile phase for ion-pair chromatography (IPC). If RPC method development is unable to provide an adequate separation due to poor band spacing, IPC provides an important additional selectivity option.[1]

Parameters Influencing the Retention and Selectivity in IP Systems

For the analysis of basic compounds, anionic ion-pairing reagents, such as sulfonic acids, alkyl sulfonates, and other acids such as *bis*-(2-ethylhexyl)-ortho-phosphoric acid (HDEHP), have been

employed. When the concentration of the ion-pairing reagent gradually increases, then a distinct increase in retention of the analytes is observed, and in a limited range of concentrations, a linear relationship of $\log k$ and log of concentration of counterion is obtained[23,24] at the moment of approaching the saturation of surface concentration of hydrophobic counterions. Further increase of concentration does not lead to significant changes in retention even a decrease of retention is sometimes observed[1] (Fig. 5). The change of type and concentration of the counterion often causes variation in selectivity of separation.[25]

Additionally, the retention and selectivity in IP–RP systems can be controlled by the change of type and concentration of the organic modifier in the aqueous mobile phase[25] and the pH of the mobile phase,[1] which should be selected to obtain maximal ionization of solute molecules and ion-pairing reagent molecules for the

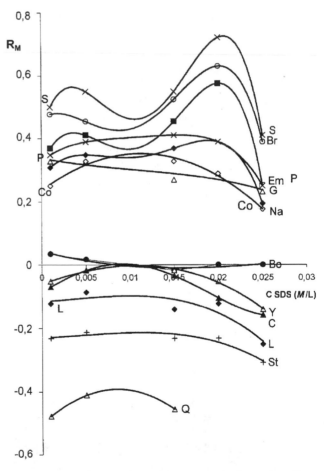

Fig. 5 Dependence of R_M vs. concentration of sodium dodecyl sulfate in mobile phase for investigated alkaloids. System: C18W/MeOH/water (8 : 2) buffered with phosphate buffer 0.01 M/L at pH 3. Em—emetine, S—santonine, Co—colchicine, C—caffeine, Y—yohimbine, L—lobeline, Q—quinine, Br—brucine, St—strychnine, P—papaverine, G—glaucine, Bo—boldine. (From Petruczynik, A.; Waksmundzka-Hajnos, M; Hajnos, M.L. J. Chromatogr. Sci. **2005**, *in press*.)

possibility of forming an ion pair. For the basic solutes, analyses of the pH range 7.0–7.5 are often applied.

The stationary phase (the length of the alkyl chains bonded to the silica support) also influences the retention of hydrophobic ion pairs.

Similarly, as for RP-HPLC separations, selectivity can be additionally varied by solvent type (methanol, tetrahydrofuran, acetonitrile), buffer concentration, and temperature.[1]

RP–IP Systems in Analysis of Heterocyclic Bases

The comparison of separation of *Chelidonium majus* L. alkaloids on a cyanopropyl column, with a buffered aqueous mobile phase, without (a) and with IP reagent (b) is presented in Fig. 6.

Some special problems for RP-IP-HPLC, such as positive and negative artifactual peaks appearing in a

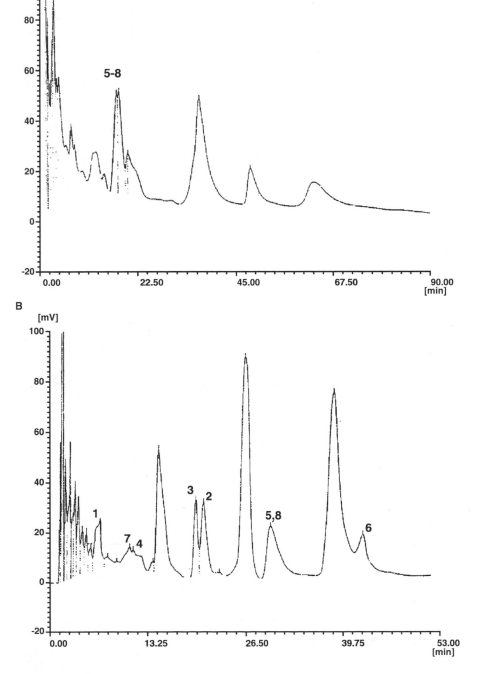

Fig. 6. Chromatogram of separation of *Chelidonium majus* L. extract in system: cyanopropyl–silica/20% MeCN + phosphate buffer pH 7.8 aqueous solution (A) and cyanopropyl–silica/20% MeCN + phosphate buffer pH 5.6 + 0.001 M octane-1-sulfonic acid sodium salt aqueous solution (B). 1—allocryptopine, 2—berberine, 3—chelerythrine, 4—chelidonine, 5—chelilutine, 6—chelirubine, 7—homochelidonine, 8—protopine. (From Petruczynik, A.; Gadzikowska, M.; Waksmundzka-Hajnos, M. Acta Pol. Pharm. Drug Res. **2002**, *59*, 61–64.).

blank run, occur and can interfere in the development of the method and its routine use. Another problem in IP separations is slow column equilibration, which is slower when an ion-pair reagent is more hydrophobic. The slow equilibration of the column with many ion-pair reagents can create problems if a gradient elution is used under these conditions.

CHROMATOGRAPHY OF WEAK BASES USING ION-EXCHANGE SYSTEMS

Today, IEC is used infrequently in comparison with other chromatographic methods. In most cases, IPC is more convenient because of its higher column efficiency, more stable and reproducible columns, and easier control over selectivity and resolution. There are, however, cases for using IEC instead of RP- or IP-HPLC, especially when organic ions have poor UV absorbance and need other detection (conductivity or MS). Then, completely volatile components of mobile phase are required. In such cases, IEC with volatile buffers fulfil this requirement, whereas ion-pair reagents are not sufficiently volatile in most cases; also, when compounds are isolated or purified by HPLC separation, the removal of mobile phase is necessary. When multistep separation is required, the aqueous buffer–salt mobile phase used for ion-exchange allows direct injection of a sample fraction onto an RP column for the next step of separation. This may be difficult with IP systems.

For the separation of weak bases, cation-exchange columns are used, which have negatively charged groups (e.g., sulfonic or carboxylic) attached to the stationary phase. Two kinds of cation-exchange columns can be used: weak cation exchanger (WCX) or strong cation exchanger (SCX). The retention of basic compounds X^+ on such stationary phases (R^-) can be manifested by the equilibrium of an ion exchange:

$$X^+ + R^-K^+ \leftrightarrow X^+R^- + K^+ \qquad (9)$$

where K^+ plays a role of counterion in mobile phase.

The increase of salt or buffer concentration in the mobile phase results in a decrease of the retention of sample compounds. Varying pH is usually a way to change the selectivity in IEC separations. Another way to change retention in IEC systems is the use of different counterions (displacers). Sometimes, the addition of organic modifiers, such as methanol or acetonitrile, is applied in IEC. This causes decreased retention of ionizable compounds.

There are several examples of the use of IEC systems for purification, isolation, and separation of heterocyclic bases. Based on ion-exchange SPE, a first and reliable procedure for the extraction of food tetrahydro-β-carboline, is accomplished on strong cation-exchange (SCX–benzenesulfonic acid cartridges) columns.[1] Water and/or alcoholic extracts of polyhydroxy alkaloids containing many other polar constituents are also purified by the use of resins.[14] The acidified aqueous solution is applied to the column and unretained neutral or acidic substances are eluted with water. The alkaloids, which are bound to the resin, accompanied by any other nonalkaloidal basic compounds, are then displaced with dilute ammonium hydroxide. Thus, an extension of the ion-exchange purification process—column chromatography—can be used for isolation of alkaloids on a preparative scale.[14] Also, with HPLC analysis, cation-exchange columns can be used; especially when UV detection is impossible because of lack of suitable chromophores, amperometric detection for the analysis of polyhydroxy alkaloids is applied. In such cases, a cation-exchange column, Dionex CS3, eluted with hydrochloric acid, was used for the separation and detection.[14] In the case of MS detection, the use of a separation process with a cation-exchange column, with the elution using volatile eluents, is also preferred.[14] Ion chromatography using a Dionex cation-exchange column, with the aqueous HCl as eluent, was applied to the analysis of theobromine and theophylline in foods and pharmaceutical preparations.[1]

CONCLUSIONS

It has been shown that there are many approaches to the separation of heterocyclic bases by chromatographic techniques. Normal-phase adsorption, reversed-phase, ion-pairing, and ion-exchange chromatographic methods have been reported extensively in the cited literature.

REFERENCES

1. Waksmundzka-Hajnos, M. Retention behaviour of heterocyclic bases. Research trends. Trends Heterocycl Chem. 2003, 9, 129–166.
2. Snyder, L.R. R_F values in thin-layer chromatography on alumina and silica. Adv. Chromatogr. 1967, 4, 3–11.
3. Kaliszan, R. Quantitative Structure Chromatographic Retention Relationships; John Wiley & Sons: New York, 1987.
4. Gołkiewicz, W.; Soczewiński, E. A simple molecular model of adsorption chromatography. VI. R_M—composition relationships of solutes with

two functional groups. Chromatographia **1972**, *5*, 594–601.

5. Matyska, M.; Soczewiński, E. Computer-aided optimization of liquid solid systems in TLC. Comparison of selectivity of various silica–diluent + modifier systems. J. Planar Chromatogr. **1990**, *3*, 264–268.

6. Wawrzynowicz, T.; Kuczmierczyk, J. A comparison of adsorption of organic compounds of different molecular structure on silica and alumina from nonaqueous solvents. Chem. Anal. (Warsaw) **1985**, *30*, 63–75.

7. Snyder, L.R. Adsorption from solution. III. Derivatives of pyridine, aniline and pyrrole on alumina. J. Phys. Chem. **1963**, *67*, 2344–2353.

8. Klemm, L.H.; Klopfstein, C.E.; Kelly, H.P. Thin layer chromatography of azines and of aromatic nitrogen heterocycles on alumina. J. Chromatogr. **1966**, *23*, 428–435.

9. Waksmundzka-Hajnos, M. Comparison of adsorption properties of Florisil and silica in HPLC. II. Retention behaviour of bi- and trifunctional model solutes. J. Chromatogr. **1992**, *623*, 15–23.

10. Waksmundzka-Hajnos, M.; Hawrył, A. Comparison of retention of phenols, aniline derivatives and quinoline bases in normal-phase TLC with binary isoeluotropic eluents. J. Planar Chromatogr. **1998**, *11*, 283–294.

11. Waksmundzka-Hajnos, M.; Petruczynik, A.; Hawrył, A. Comparison of chromatographic properties of cyanopropyl-, diol- and aminopropyl–polar bonded stationary phases by the retention of model compounds in normal-phase liquid chromatography systems. J. Chromatogr. A. **2001**, *919*, 39–50.

12. Klimes, J.; Kastner, P. Thin layer chromatography of benzodiazepines. J. Planar Chromatogr. **1993**, *6*, 168–180.

13. Houghton, P.J. Chromatography of chromone and flavonoid alkaloids. J. Chromatogr. A. **2002**, *967*, 75–84.

14. Molyneux, R.J.; Garden, D.R.; James, L.F.; Colegate, S.M. Polyhydroxy alkaloids: chromatographic analysis. J. Chromatogr. A. **2002**, *967*, 57–74.

15. McCalley, D.V. Analysis of the *cinchona* alkaloids by high performance liquid chromatography and other separation techniques. J. Chromatogr. A. **2002**, *967*, 1–19.

16. Snyder, R.L. Role of the solvent in liquid solid chromatography—a review. Anal. Chem. **1974**, *46*, 1384–1393.

17. Lewis, J.A.; Lommen, D.C.; Raddatz, W.D.; Dolan, J.W.; Snyder, L.R.; Molnar, I. Computer simulation for the prediction of separation as a function of pH for reversed-phase high-performance liquid chromatography. J. Chromatogr. **1992**, *592*, 183–195.

18. Scholten, A.B.; Claessens, H.A.; de Haan, J.W.; Cramers, C.A. Chromatographic activity of residual silanols of alkylsilane derivatized silica surface. J. Chromatogr. A. **1997**, *759*, 37–46.

19. Marques, R.M.L.; Schoenmakers, P.J. Modeling retention in reversed phase liquid chromatography as a function of pH and solvent composition. J. Chromatogr. **1992**, *592*, 157.

20. Rorabacher, D.B.; MacKellar, W.J.; Shu, F.R.; Bonavita, S.M. Solvent effects on protonation constants. Ammonia, acetate, polyamine and polyaminocarboxylate ligands in methanol–water mixtures. Anal. Chem. **1971**, *43*, 561–573.

21. Roses, M.; Bosch, E. Influence of mobile phases and acid–base equilibria on the chromatographic behaviour of protolytic compounds. J. Chromatogr. A. **2002**, *982*, 1–30.

22. Drummer, O.H. Methods for measurements of benzodiazepines in biological samples. J. Chromatogr. B. **1998**, *713*, 201–225.

23. Bidlingmeyer, B.A. J. Chromatogr. Sci. **1980**, *18*, 525.

24. Low, K.G.C.; Bartha, A.; Billiet, H.A.H.; de Galan, L. Systematic procedure for the determination of the nature of the solute prior to the selection of the mobile phase parameters for optimization of reversed-phase ion-pair chromatographic separations. J. Chromatogr. **1989**, *478*, 21–38.

25. Bieganowska, M.L.; Petruczynik, A. Thin-layer reversed phase chromatography of some alkaloids in ion-association systems. Part II. Chem. Anal. (Warsaw) **1994**, *39*, 445–454.

High-Speed SEC Methods

Peter Kilz
PSS Polymer Standards Service GmbH, Mainz, Germany

INTRODUCTION

SEC is the established method to determine macro-molecular properties in solution. It is the only technique that allows efficient measurement of property distributions for a wide range of applications. Recent trends in industrial laboratories and research institutes have been focused on increasing the analytical throughput in order to increase productivity. Quality control and combinatorial chemistry demand the optimization of high-throughput methods. Increased analytical throughput can also save time and resources (e.g., instrumentation) in production-related fields. In combinatorial research, high-throughput analytical techniques are a bare necessity, because of the huge numbers of samples being synthesized;[1,2] and references therein] In either situation, the slowest step in the process will determine the overall turnaround time. The importance of high-speed analytical techniques becomes obvious when research companies synthesize over 500 targets per day, but only about 100 samples can be analyzed. The potential of new synthetic methods and in-line production control cannot be fully utilized until the typical SEC run times of 40 min are substantially reduced.

METHODS FOR FAST SEC ANALYSES

There have been several approaches to overcome the traditionally slow SEC separations, which are caused by the diffusion processes in SEC columns. Most of them are column-related (see "High-Speed SEC Columns," "Small Particle Technology," and "Smaller SEC Column Dimensions"); one utilizes the column void volume (cf. "Overlaid Injections"), while another replaces separation with simplified sample preparation (see "Flow Injection Analysis"). Cloning existing methods and instrumentation is also reviewed with respect to the potential time gain (see "Cloning of SEC Systems"). Benefits and limitations of each method are summarized in Table 1.

High-Speed SEC Columns

The pore volume of the column packing has been shown to be one of the major factors influencing peak resolution in SEC. True high-speed separations, with good resolution, requires special high-speed columns, which allow fast flow rates, possess high separation volumes, and allow solutes to easily access the pores.[3] PSS GmbH is currently the only vendor of high-speed columns for SEC. Their high-speed columns replace conventional columns one to one, which allows for a trouble-free method transfer from an existing conventional application to a high-speed application. High-speed SEC can be performed in about 1 min, cutting down analysis time by about 10%, with similar resolution on existing instrumentation.[4] Fig. 1 shows a comparison of an SEC separation of polystyrene standards in THF on a conventional column and on a high-speed column, analyzed on the same instrument.

Precision and accuracy of high-speed separations have been investigated for various applications. Both the accuracy of molar mass results and the reproducibility have been comparable to results from conventional columns.[3]

Fig. 2 shows the overlay of 10 out of 60 repeats of a commercial polycarbonate sample analyzed in tetrahydrofurane (THF). They overlap almost perfectly. Each run took about 2.5 min, and the total run time for 60 repeats was about 2 hr.

The overall time savings can even be larger when taking the complete analytical process into account. The total run time of an instrument consists of the preparation and equilibration time, the time needed for running the calibration standards, and the run times for the unknown samples. If 10 individual standards are used for calibration and 10 samples are run, the total run time on a conventional system will be about 2 days. The same work carried out on a high-speed system will only require about 3 hr, and can be easily performed in a single day.[4]

The cost-saving aspects of high-throughput SEC techniques can be substantial and have been evaluated for different scenarios.[5]

Polyolefins, other synthetic polymers, and water-soluble macromolecules have been investigated in high-speed SEC systems. High-speed SEC can be a major time saver in two-dimensional chromatography applications, which require about 10 hr analysis time for cross-fractionation.[6] This can be reduced by a factor of 10, to about 1 hr, which makes it much more

Encyclopedia of Chromatography DOI: 10.1081/E-ECHR-120014223

Table 1 Synopsis of methods for increased SEC throughput

Approach	Advantages	Disadvantages	Beneficial for ...
Instrument cloning	No method change Easy to implement No additional training	High investment cost High maintenance Higher operating cost More people More space Limited throughput gain	Sample increase of up to 3×
High-speed column	No method change Uses existing equipment 1 : 1 application transfer No additional training Minimizes investment (column only) SEC separations in 1 min Time gain ca. 10× No additional shear High efficiency Runs with conventional software	No eluent savings	QC/QA Increased throughput (10×) Use with exiting methods
FIA	Uses existing equipment Saves eluent	No separation Limited time gain Not applicable for copolymers/blends Requires molar mass sensitive detectors Only primary information (conc., Mw, IV) Needs method change Needs special software	Samples difficult to separate Utilize existing instruments
Overlaid injections	No method change Uses existing equipment No additional training Low cost	Needs overlaid injection-ready software	QC/QA known samples
Small columns	Uses existing equipment Minimizes investment Saves eluent Runs with current software	Limited time savings Needs method adaption Optimization of: injection volumes detection systems Shear degradation Low efficiency	Low-resolution applications Low time-saving requirements Single detector applications
	Needs training Limited throughput increase		

interesting for many laboratories. Details on these and additional high-speed applications can be found in Ref.[4].

Small Particle Technology

Reducing particle size of the SEC column packings reduces the time requirements in SEC because of the increased mass transfer and resultant separation efficiency. Hence, columns can become smaller in dimensions while maintaining resolution. This approach has been used for many years. Column bank lengths dropped from several meters to now typically 60 cm with current SEC column particle sizes of 5 μm as compared with about 100 μm in the early 1960s. During the same period, time requirements dropped from about 6 hr to less than 1 hr.

Unfortunately, this approach is very limited now because of the high shear rates in columns packed

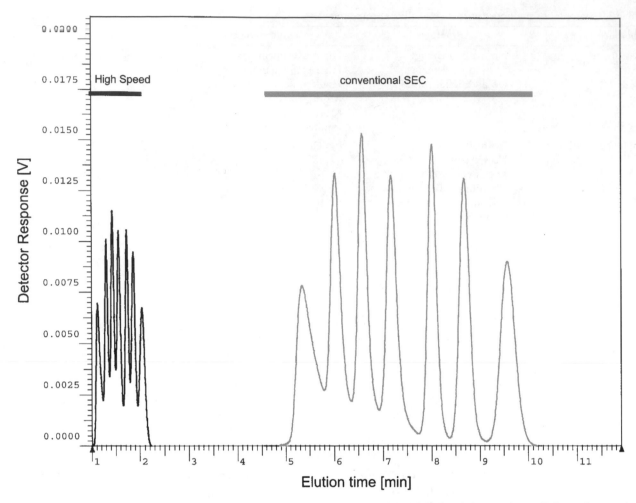

Fig. 1 Chromatogram of conventional SEC column (right part) compared to high-speed SEC column (left part); tested on identical instrument with polystyrene standards, in THF.

with small particles (less than 5 μm), which can cause polymer degradation.

Smaller SEC Column Dimensions

The reduction of column dimensions can, in theory, substantially reduce the time requirements of the separation. However, several limitations predicted by chromatographic theory have to be considered.[7,8] A study of the influence of column dimensions on fast SEC separations has been published in Ref.[4]. It has been found difficult to optimize and transfer existing methods and, in many cases, new equipment had to be purchased.

Overlaid Injections

This approach has also been used when SEC separations required hours; it can cut down analysis time

by a factor of 2. It utilizes the fact that about 50% of the SEC elution time is needed to transport the solutes through the interstitial volume of the columns. This allows us to inject another sample before the current one is already totally eluted. The optimum injection interval, Δt_{min}, can be calculated from the separation properties of the instrument:

$$\Delta t_{min} = (V_t - V_0)/F$$

where V_t is the total penetration volume of column(s), V_0 is the total exclusion volume of column(s), and F is the volumetric flow rate.

The required parameters are easily determined from a molar mass calibration curve.

Today, this method can be combined with appropriate software to automate data acquisition and data processing. It is easy to use, requires no additional investment, and no method modifications are necessary.

H

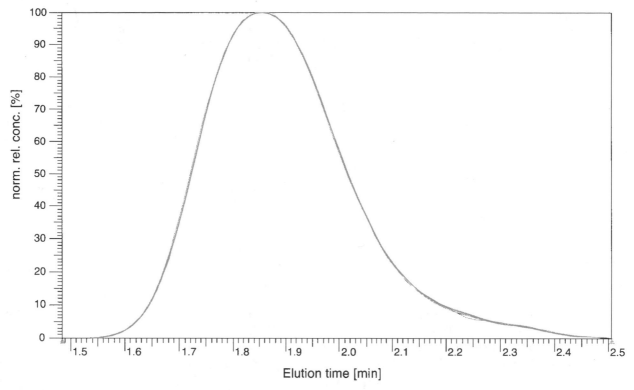

Fig. 2 Overlay of 10 out of 60 repeats of a commercial polycarbonate analysis, in THF, on PSS SDV 5 μm high speed 10^3, 10^5 Å column; measured Mw = (29,610 ± 150) g/mol (nominal sample molar mass by producer: 30,000 g/mol).

Flow Injection Analysis (FIA)

Another approach to cut down on analysis time is to avoid separation and inject samples directly into detector cells. FIA has received some attention recently and is, therefore, mentioned in this review. Because it does not rely on any separation, advantages and limitations will be summarized only.

This method uses the HPLC equipment for sample handling and requires molar mass sensitive detectors (such as light scattering and/or viscometry) to obtain a mean property values from each detector (Mw and/or IV, respectively). The FIA result from a concentration detector yields polymer content in a sample, which can also be determined with other well-established methods. The FIA approach requires expensive and well-maintained equipment, and will not save much time or solvent; furthermore, no distribution information is available.

Cloning of SEC Systems

The number of processed samples can be increased proportionally by increasing the number identical systems. The time and analytical requirements for each sample are not changed, but the number of samples per hour can be increased. Because no change in analytical methods is necessary, cloning SEC instruments and methods is straightforward and can be carried out in most environments.

This approach, however, is clearly limited by the availability of important resources such as laboratory space, operators, instrumentation, and software licenses. Cloning systems can become very costly; time and effort for instrument maintenance and operation increases proportionally.

True parallelization of analytical processes has, so far, not been very successful. In such set-ups, only the separation module (in general the column) is set up in parallel, while solvent delivery, injection, detection, and data processing are multiplexed. These systems will no longer be as simple in operation and maintenance as the cloned systems.

CONCLUSIONS

Time requirements of SEC experiments can be reduced substantially by using high-speed SEC columns. The availability of high-speed columns allows an increase in SEC separations by a factor of 10 and run times of 1 min are possible. Precision and accuracy of results are comparable with existing methods. Existing methods and instrumentation can still be used with high-speed columns.

The time gain of high-speed columns can open up SEC methodology for

a) Monitoring and controlling processes on-line;

b) Using SEC methods routinely in QC labs;

c) Allowing high-throughput screening for new materials design;

d) Being a useful tool in combinatorial chemistry; and

e) Studying monitoring time-critical processes.

REFERENCES

1. Nielson, R.B.; Safir, A.L.; Petro, M.; Lee, T.S.; Huefner, P. Polym. Mater. Sci. Eng. **1999**, *80*, 92.
2. Brocchini, S.; James, K.; Tangpasuthadol, V.; Kohn, J. A combinatorial approach for polymer design. J. Am. Chem. Soc. **1997**, *119*, 4553.
3. Kilz, P.; Reinhold, G.; Dauwe, C. *Proceedings of the International GPC Symposium 2000; Las Vegas, NV*; Waters Corp.: Milford, MA, 2001 (CD-ROM).
4. Kilz, P. Methods and columns for high speed SEC separations. In *Handbook for Size Exclusion Chromatography and Related Techniques*; Wu, C.-S., Ed.; Marcel Dekker: New York, 2002, *in press*.
5. Reinhold, G.; Hofe, T. GIT Fachz. Lab. **2000**, *44*, 556.
6. Kilz, P.; Pasch, H. Coupled LC techniques in molecular characterization. In *Encyclopedia of Analytical Chemistry*; Meyers, R.A., Ed.; John Wiley & Sons: New York, 2000; Vol. 9, 7495–7543.
7. Giddings, J.C.; Kucera, E.; Russell, C.P.; Myers, M.N. Statistical theory for the equilibrium distribution of rigid molecules in inert porous networks. Exclusion chromatography. J. Phys. Chem. **1968**, *72*, 4397.
8. Glockner, G. *Liquid Chromatography of Polymers*; Hüthig: Heidelberg, 1982.

High-Temperature High-Resolution GC

Fernando M. Lanças
J. J. S. Moreira
Laboratorio de CromatografiaInstituto de Quimica de São Carlos, Universidade de São Paulo, São Carlos/SP, Brazil

INTRODUCTION

Gas chromatography (GC), in its early days, used packed columns with chemically inert solid supports coated with stationary phases. These columns presented low efficiency due to the wide range of particle sizes used, causing inhomogeneity in the packed bed and, consequently, high instability due to a poor deactivation and thermal instability at high-temperature operations.[1] This characteristic limited the use of the GC to only volatile and low-mass molecular compounds. The later development of columns with a stationary phase coated on the inner wall of the capillary provided a more inert environment. In this form, columns with higher thermal stability and more efficiency (higher N) were produced, allowing the analysis of semivolatile and medium molecular mass compounds. This technique was named high-resolution gas chromatography (HRGC).[1] The possibility of using thermally stable, highly efficient columns, stimulated scientists to search for new stationary phases and chemical manufacturing processes to produce capillary columns with high thermal stabilities, capable of operating at higher temperatures[2] (to 360°C).

Lipsky and McMurray[3] suggested, in their pioneering work on high-temperature high-resolution gas chromatography (HT-HRGC), the use of column temperatures equal to, or higher than, 360°C. However, other column temperature values have also been reported for this technique.[4]

The thermal stability of the high-temperature capillary columns allowed the analysis of higher molecular masses (more than 600 Das) and nonvolatile compounds never before directly analyzed by gas chromatography.[2]

INSTRUMENTATION FOR HT-HRGC

The instrumentation used for HT-HRGC is the same as used for conventional GC, with only minor modifications.

Columns

The columns utilized in HT-HRGC are short (usually equal to, or shorter than, 10 m) coated with thin films (~0.1 μm or less) and having an inner diameter (i.d.) around 0.2 mm.[5]

A smaller inner diameter (e.g., 0.1 mm) can also be used, but with the inconvenience of limiting the work to more diluted samples in order to avoid column overload. On the other hand, this type of column permits carrier gas speeds higher than with columns of inner diameters in the range 0.2–0.3 mm. Columns with inner diameters equal to 0.1 mm exhibit fewer plates with the increment of the carrier gas speed, in contrast to the columns with equivalent characteristics, but of 0.3 mm i.d.[5] The increase of the carrier gas speed in smaller-i.d. columns performs an analysis in a shorter time, without undermining the efficiency of separation.[6]

Capillary columns, to be suitable to HT-HRGC, must be extremely robust and must be coated with a thin film of the stationary phase with the purpose of reducing the retention of the less volatile compounds and preventing stationary-phase bleed at high temperatures.[7]

Using such proper columns, elution of substances with carbon numbers in excess of n-C$_{130}$ has been reported, at column temperatures of up to 430°C.[8]

Tubing Material for HT-HRGC Columns

There are four major types of materials being utilized to prepare columns for high-temperature capillary columns:[2]

1. Glass (borosilicate)
2. Polyimide-clad fused silica
3. Aluminum-clad fused silica
4. Metal-clad fused silica

Columns of aluminum-clad fused silica,[2,4] and metal-clad fused silica support temperatures up to 500°C, representing an advantage in comparison with borosilicate glass columns, with a temperature limit to 450°C, and columns of polyimide-clad fused silica for high temperature,[2,9] limited temperature to 400–420°C. On the other hand, aluminum-clad fused silica columns present leakage, principally in the connections, after a short time of use.[2,9] Polyimide-clad fused-silica capillaries, after prolonged exposure to

Encyclopedia of Chromatography DOI: 10.1081/E-ECHR-120040010

temperatures above 380°C, tend to break spontaneously at many points, thus losing the polyimide coating.[9] Borosilicate columns are inexpensive, being an alternative to fused silica for high-temperature applications. However, these columns have been reported to leak when coupled with retention gap and to mass spectrometry detectors.[2] An important alternative for HT-HRGC are HT metal-clad fused-silica columns which resist temperatures above 500°C for long-term exposure.[9]

Stationary Phases

The first results on HT-HRGC[3,10] were published in 1983, dealing with stationary-phase immobilization (polysiloxane –OH terminated). Due to the column instability, when submitted to high temperature, stationary-phase loss was common at that time. These works can be considered to be the precursor of high-temperature gas chromatography, because the phase immobilization process developed resulted in a series of OH-terminated polysiloxane phases compatible with the inner surfaces of borosilicate glass and fused-silica tubing. These phases are thermally stable and capable of withstanding elevated temperatures[11] used in HT-HRGC. After this report, many other articles dealing with the ideal stationary phase for high-temperature gas chromatography appeared. Nonpolar stationary phases of the carborane–siloxane-type bonded phase (temperature range >480°C) and siloxane–silarylene copolymers suitable for HT-GC were developed[7] around 1988.

A medium-polarity stationary phase based on fluoralkyl–phenyl substitution, which is thermally stable up to 400°C, was reported,[12] and a CH_3O-terminated polydimethyl siloxane, diphenyl-substituted stationary phase made possible the analysis of complex high-molecular-mass mixtures such as free-base porphyrins and triglycerides using narrow-bore capillary columns.[5] Since these developments, a variety of stationary phases for analysis of specific analytes by HT-HRGC were found.[2]

Sample Introduction

The sampling and elution of such high-molecular-weight materials requires careful attention in order to avoid quantitative sample losses during the sample introduction step. In general, "cold" injection techniques are required for accurate nondiscriminative sample transfer into the column. Cold on-column and programmed temperature (PT) split/splitless injection have been used with success for a large number of HT-HRGC analyses. In certain cases, however, significant losses of compounds above n-C_{60} have been observed with PT splitless injection.[13] This effect was identified as a time-based discrimination process caused by purging the PT inlet too soon after injection, resulting in incomplete sample vaporization.[14]

Actually, same articles show the possibility of use split injection[8] in HT-HRGC analyses of substances up to C_{78}. However, volatile materials from the septum accumulate at the head of the column during the cool-down portion of the temperature program. When the columns are reheated to analyze the next sample, these accumulated volatiles are eluted, producing peaks, a baseline rise, or both. This difficulty can be solved using commercial septa already available for HT-HRGC, which exhibit very low bleed levels.

DETECTORS

High-temperature high-resolution GC is a technique similar to conventional GC; however, it presents high column bleeding due to the high temperature to which the column is submitted. Selective detectors, when used in HT-HRGC, require special attention. As an example, the electron-capture detector (ECD) is a very sensitive detector and should not be used in HT-HRGC because of its ability to detect column bleeding. This fact limits the detectors used to a few, such as the flame ionization detector (FID), alkali-flame ionization detector (AFID), and mass spectrometry detector (MS). In HT-HRGC, these detectors usually need small adjustments; for example, the MS detector requires a special interface when used for HT-HRGC.[2]

HT-HRGC APPLICATION

High-temperature high-resolution GC has opened to many scientists the opportunity to analyze compounds of high molecular mass (600 Da or more) with similar efficiency to conventional high-resolution gas chromatography (HRGC). Actually, HT-HRGC has been applied to the analyses of compounds from several different areas.[15–18] As a general rule, this will avoid the time-consuming and usually expensive step of derivatization. In natural products, underivatized triterpenic compounds found in medicinal plants can be analyzed by this technique. The HT-HRGC analysis of triterpenes in aqueous alcoholic extracts of *Maytenus ilicifolia* and *M. aquifolium* leaves clearly allows the detection of the presence of friedelan-3-ol and friedelin and, therefore, allows distinguishing between the two varieties;[15] this differentiation is very important in pharmacological studies, because they present different biological activities.

Cyclopeptidic alkaloids (molecular mass ~600 Da), a class of important alkaloids which present biological activity,were analyzed by HT-HRGC without derivatization.[16] Fig. 1 illustrates the separation of cyclopeptidic alkaloids in the chloroform fraction. The following selected compounds were identified: (1) Franganine, (2) Miriantine-A, (3) Discarine-C, and (4) Discarine-D.

Triacylglycerides from animal and vegetable sources have been separated and identified by HT-HRGC and high-temperature gas chromatography coupled to mass spectrometry (HT-HRGC/MS). Fig. 2 shows the chromatographic profile of palm oil (*Elaeis guineensis* L.) by HT-HRGC, and the triacylglyceride compounds identification.[17]

The HT-HRGC/MS technique was also used as an important tool to identify and quantify cholesterol present in the total lipid extracts of archeological bones and teeth, constituents of a new source of paleodietary information.[19]

High-Temperature High-Resolution GC

1 Franganine

2 Miriantine-A

3 Discarine-C

4 Discarine-D

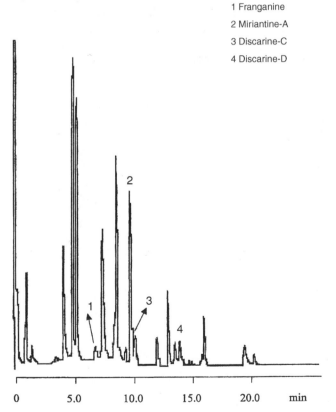

Fig. 1 Analysis of underivatized cyclopeptidic alkaloids in chloroform extract using HT-HRGC. Condition: fused-silica capillary column (6 m × 0.25 mm × 0.08 μm) coated with a LM-5 (5% phenyl, 95% polymethylsiloxane immobilized bonded phase) stationary phase. Temperature condition: column at 200°C (1 min), increased by 4°C/min, then 300°C (5 min); inlet: 250°C; FID detector: 310°C.

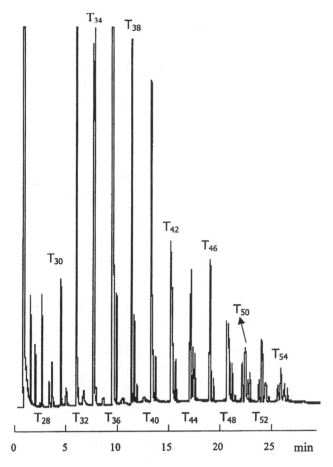

Fig. 2 Chromatogram of underivatized Palmist Oil (*Elaesis guineensis* L.) triacylglyceridic fraction using HT-HRGC. Condition: fused-silica capillary column (25 m × 0.25 mm × 0.1 μm) with the stationary phase OV-17-OH (50% phenyl, 50% methylpolysiloxane immobilized phase).Temperature condition: column at 350°C isothermic; injector: 360°C; FID detector: 380°C. T is the number of the underivatized triacilglyceride (e.g., T_{50} means a triacylglyceride having 50 carbon atoms).

The detection of vanadium, nickel, and porphyrins in crude oils were analyzed by high-temperature gas chromatography–atomic emission spectroscopy (HT-GC–AES), presenting characteristic metal distributions of oils from different sources.[18] Other related applications of HT-HRGC, including the analysis of α, β, and γ cyclodextrins, antioxidants, and oligosaccharides.[2]

Considering that HT-HRGC is still a young separation technique and that it presents several attractive features, including the analysis of higher-molecular-weight compounds within short analysis times, without the necessity of sample derivatization, we can envisage a bright future for this technique, with many new applications being developed in the near future.

REFERENCES

1. Fowlis, I.A. *Gas Chromatography*, 2nd Ed.; John Wiley & Sons: New York, 1994; 1–11.

2. Blum, W.; Aichholtz, R. *Hochtemperatur Gas-Chromatographie*; Hüthing: Germany, 1991; 26–114.

3. Lipsky, S.R.; McMurray, W.J. Role of surface groups in affecting the chromatographic performance of certain types of fused-silica glass capillary columns: II. Deactivation by esterification with alcohols and deactivation with specially prepared high-molecular-weight stationary phases. J. Chromatogr. **1983**, *279*, 59.

4. Lanças, F.M.; Galhiane, M.S. J. High Resolut. Chromatogr. Chromatogr. Commun. **1990**, *13*, 654.

5. Damasceno, L.M.P.; Cardoso, J.N.; Coelho, R.B. J. High Resolut. Chromatogr. Chromatogr. Commun. **1992**, *15*, 256.

6. Grob, K.; Tschuor, R. J. High Resolut. Chromatogr. Chromatogr. Commun. **1990**, *13*, 193.

7. Hubball, J. LC/GC **1990**, *8*, 12.

8. Hinshaw, J.V.; Ettre, L.S. J. High Resolut. Chromatogr. Chromatogr. Commun. **1989**, *12*, 251.

9. Blum, W.; Damasceno, L. J. High Resolut. Chromatogr. Chromatogr. Commun. **1987**, *10*, 472.

10. Verzele, M.; David, F.; van Roelenbosch, M.; Diricks, G.; Sandra, P. J. Chromatogr. **1983**, *270*, 99.

11. Lipsky, S.R.; Duffy, M.L. J. High Resolut. Chromatogr. Chromatogr. Commun. **1986**, *9*, 376.

12. Aichholz, R.; Lorbeer, E. J. Microcol. Separ. **1996**, *8*, 553.

13. Trestianu, S.; Zilioli, G.; Sironi, A.; Saravelle, C.; Munari, F.; Galli, M.; Gaspar, G.; Colin, J.; Jovelin, J.L. J. High Resolut. Chromatogr. Chromatogr. Commun. **1985**, *8*, 771.

14. Hinshaw, J.V. J. Chromatogr. Sci. **1987**, *25*, 49.

15. Lanças, F.M.; Vilegas, J.H.Y.; Antoniosi Filho, N.R. Chromatographia **1995**, *40*, 341.

16. Lanças, F.M.; Moreira, J.J.S. High temperature gas chromatography (HT-GC) analysis of underivatized cyclopeptidic alkaloids. In *Proc. of the 23rd Int. Symp. Capill. Chromatogr.*; 2000.

17. Antoniosi Filho, N.R. Analysis of the vegetable oils and fats using high resolution gas chromatography and computational methods. In *Ph.D. thesis*; University of São Paulo, Institute of Chemistry at São Carlos: Brazil, 1995; 140–152.

18. Zeng, Y.; Uden, P.C. J. High Resolut. Chromatogr. Chromatogr. Commun. **1994**, *17*, 223.

19. Stott, A.W.; Evershed, R.P. Analysis of cholesterol preserved in archaeological bones and teeth. Anal. Chem. **1996**, *68*, 4402.

Histidine in Body Fluids: Specific Determination by HPLC

Toshiaki Miura
College of Medical Technology, Hokkaido University, Sapporo, Hokkaido, Japan

Naohiro Tateda
Kiichi Matsuhisa
Asahikawa National College of Technology, Asahikawa, Hokkaido, Japan

INTRODUCTION

Amino acids in biological samples have been principally determined by HPLC with pre- or postcolumn chemical derivatization selective for a primary amino group. Although HPLC methods are applicable to the assay of all commonly encountered amino acids in biological samples, they are time-consuming and inadequate for the assay of a large number of samples when a specific amino acid is required to be assayed. In such cases, a rapid assay can be achieved by the use of a chemical derivatization that is selective for the individual amino acid, which renders the HPLC separation conditions to be very simple. As an example of such a case, this paper describes a rapid HPLC method for the determination of histidine in body fluids. The method is based on the separation by a reversed-phase, ion-pair chromatography followed by the selective postcolumn detection of histidine with fluorescence derivatization using *ortho*-phthalaldehyde (OPA).

SELECTIVE FLUORESCENCE DETECTION OF HISTIDINE WITH OPA

OPA has been known to give a fluorescent adduct with most primary amines in the presence of a thiol compound, but only with several biogenic amines such as histidine, histamine, and glutathione in the absence of a thiol compound in a neutral or alkaline medium. In the case of histidine, it gradually reacts with OPA alone in an alkaline medium, to give a relatively stable fluorescent adduct showing excitation and emission maxima at 360 and 440 nm, respectively.[1] Håkanson et al. optimized these reaction conditions and showed that the fluorescence intensity due to histidine reached a maximum 10 min after initiation of the reaction at pH 11.2–11.5, at 40°C. This fluorescence reaction is relatively selective for histidine and has been used in a batch method for the assay of histidine.[1]

On the other hand, we revealed the mechanistic pathway of the OPA-induced fluorescence reaction of histidine, as shown in Fig. 1.[2] In addition, we found that the fluorescent adduct of histidine rapidly forms in a neutral medium, although its stability is low.[3] These findings led us to optimize this fluorescence reaction for a postcolumn detection of histidine in its HPLC determination. Under the optimized conditions (for 30 sec at pH 7 and at 40°C), no significant fluorescence was observed with other biological substances, except for histamine and glutathione. The relative fluorescence intensities of histamine and glutathione were 14.4% and 11.8% of that given by histidine on a molar base, respectively.[3] Such high selectivity of this fluorescence reaction was reasonably explained by the fact that both the primary amino group and imidazole ring of histidine participate in the formation of the fluorescent adduct (Fig. 1).

Because the reaction temperature markedly influences the rates of formation and degradation of the fluorescent adduct, its precise control is an essential factor for the reproducibility of the postcolumn fluorescence detection. Therefore preheating of both the eluent and OPA reagent to a constant temperature of 40°C is required before their mixing, and these was achieved by insertion of preheater tubes for both the eluent and OPA reagent into the line. As described in the section "HPLC System and Conditions," the preheater tubes, as well as columns, resistor tube, and the reactor tube were placed in a column oven maintained at 40°C.

HPLC CONDITIONS FOR SEPARATION OF HISTIDINE

As described above, histamine and glutathione also show significant fluorescence in the postcolumn detection with OPA. The levels of glutathione are comparable or higher than those of histidine in many biological samples, such as liver, kidney, and blood (mainly in the erythrocytes). On the other hand, most biological samples normally contain histamine at markedly lower levels than histidine; in particular, the level of histamine in human serum or plasma is 10,000-fold lower than that of histidine. These facts

Encyclopedia of Chromatography DOI: 10.1081/E-ECHR-120025297

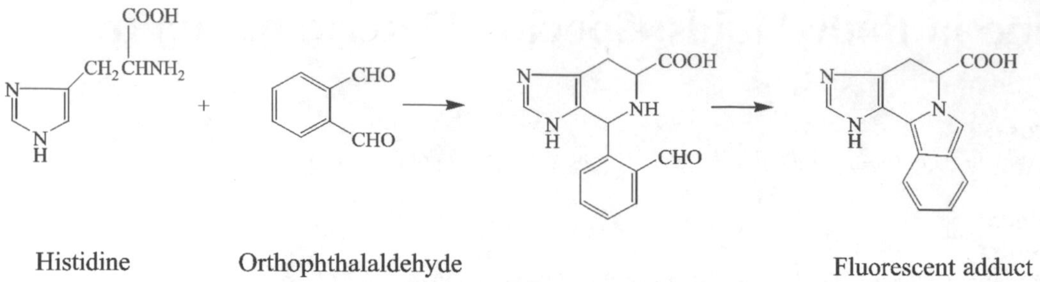

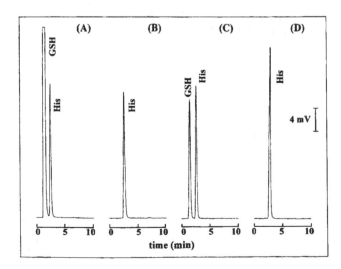

Fig. 1 Mechanistic pathway for the formation of fluorescent adduct in the reaction of histidine with *ortho*-phthalaldehyde.

indicate that the interfering biological substance is limited to glutathione in the HPLC method in the postcolumn fluorescence detection using OPA. Thus HPLC separation conditions had only to separate histidine from glutathione, which was easily achieved by a reversed-phase ion-pair chromatography on an ODS short column with a 5:95 (v/v) mixture of methanol and sodium phosphate buffer (35 mM, pH 6.2) containing 5.3 mM sodium octanesulfonate, at a flow rate of 0.5 mL/min and at 40°C. Under these conditions, histidine and glutathione were eluted at 2.7 and 1.4 min, respectively (Fig. 2A).

DETERMINATION OF HISTIDINE IN BODY FLUIDS

HPLC System and Conditions

The HPLC system comprised an L-6000 pump (Hitachi, Tokyo, Japan) and an LC-9A pump (Shimadzu, Kyoto, Japan) for deliveries of an eluent and the OPA reagent, a DGU-12A degasser (Shimadzu), a Rheodyne Model 7725i sample injector (Rheodyne, Cotati, CA, USA), a CTO-10A column oven (Shimadzu), an F-1050 fluorescence detector equipped with a 12-μL square flow cell, and a D-2500 data processor (Hitachi). Separation was performed at 40°C with a Develosil ODS UG-3 column (30 × 4.6 mm i.d., 3 μm; Nomura Chemical, Seto, Japan) as an analytical column, which was protected by a guard-pak cartridge column (Develosil ODS UG-5, 10 × 4.0 mm i.d., 5 μm), and with a 1:19 (v/v) mixture of methanol and sodium phosphate buffer (35 mM, pH 6.2) containing 5.3 mM sodium octanesulfonate as an eluent. The OPA reagent was a 15:1 (v/v) mixture of 50 mM sodium phosphate buffer (pH 8.0) and 50 mM OPA in methanol. Both the eluent and OPA reagent were filtered through a 0.45 μm membrane filter (Millipore, Bedford, MA, USA) before use. The eluent was delivered to the column at a flow rate of 0.5 mL/min through a preheater tube (stainless-steel tube, 10 m × 0.8 mm i.d.). Ten microliters of the sample solution was introduced to the column. The eluate from the column was added with OPA reagent delivered at a

flow rate of 0.5 mL/min to a mixing T-joint attached to the column through a preheater tube (stainless-steel tube, 10 m × 0.8 mm i.d.) and a resistor polytetrafluoroethylene (PTFE) tube (20 m × 0.25 mm i.d.). The mixture was passed through a reactor tube (coiled PTFE tube, 2.5 m × 0.5 mm i.d., coil diameter of 20 mm) and the generated fluorescence was detected at 435 nm with an excitation wavelength of 365 nm. All columns, preheater, resistor, and reactor tube were placed in the column oven which was maintained at 40°C.

Sample Preparation

Because of high selectivity of the postcolumn fluorescence detection with OPA, no sample pretreatment other than deproteinization was required for the assay of histidine in body fluids such as human serum, blood, and urine as follows:

Human serum was mixed with an equal volume of 6% (w/v) perchloric acid and was vortexed

Fig. 2 Typical HPLC chromatograms of histidine. (A) Standard histidine and glutathione. Injected amounts: histidine (His), 5 pmol; glutathione (GSH), 500 pmol. (B) Human serum. (C) Human blood. (D) Human urine. (See the section "HPLC System and Conditions" for chromatographic conditions.)

several times. The mixture was centrifuged at $10,000 \times g$ for 10 min at 4°C, then the supernatant was diluted 10-fold with water and was filtered through the 0.45-μm membrane filter. A portion of the filtrate was further diluted 10-fold with 0.01 M HCl for HPLC analysis.

Heparinized human blood (1.0 mL) was mixed with water (0.9 mL) and 60% (w/v) perchloric acid (0.1 mL), vortexed, and then centrifuged at 4°C and $10,000 \times g$ for 10 min. The supernatant (400 μL) was transferred to an Ultrafree-MC centrifugal filter unit (Durapore type, 0.22 μm) (Millipore) and centrifuged at 4°C and $10,000 \times g$ for 1 min. A portion of the filtrate was diluted 200-fold with 0.01 M HCl and injected onto the HPLC column.

Human urine was mixed with an equal volume of 6% (w/v) perchloric acid and was then filtered through the membrane filter. The filtrate was diluted 1000-fold with 0.01 M HCl and injected onto the HPLC column.

Evaluation of the Present HPLC Method

Fig. 2A shows the chromatogram of a 1:100 mixture of standard histidine and glutathione. The peak due to histidine was observed at 2.7 min with no interference from 100-fold excess of glutathione. The HPLC method gave a linear calibration curve ($r = 1.000$) over the range of 0.25–1000 pmol per injection (10 μL) with the coefficient of variation of 0.9% at 2 pmol ($n = 10$) and with the detection limit ($S/N = 8$) of 25 fmol.

Fig. 2B and D shows the typical chromatograms of deproteinized human serum and urine, respectively, which contain less glutathione than histidine. The high selectivity of the postcolumn detection made the chromatograms quite simple, where the peak due to histidine appeared as a sole peak. On the other hand, both glutathione and histidine were detected in human blood, which contains glutathione at a higher level than histidine (Fig. 2C).

Recoveries of the present HPLC method were tested by using a pooled human serum, blood, or urine, to which were added various amounts of histidine prior to the sample preparation. The mean recovery values were in the range of 101–104%. The values of histidine in human sera, blood, and urine, determined by the HPLC method, were 85.6 ± 15.0 μM ($n = 47$, mean $\pm$ SD), 95.3 μM ($n = 2$, 96.8 and 93.8 μM), and 1.13 ± 0.48 mmol/mg of creatinine ($n = 10$, mean $\pm$ S.D.), respectively, which were in good agreement with their reported values. The coefficients of the

day-to-day variation obtained with a pooled human serum, blood, or urine were below 1.0%.

CONCLUSIONS

Because of the high selectivity and sensitivity of the postcolumn fluorescence detection of histidine with OPA, the present HPLC method is applicable to a specific and rapid assay of histidine in human serum, blood, and urine after simple pretreatment. A recent paper demonstrated that the postcolumn detection with OPA was applicable to the simultaneous assays of histidine and its major metabolites (cis- and trans-urocanic acids) in human stratum corneum.[4] The postcolumn detection system was also applicable to the flow injection analysis (FIA) method for the assay of histidine in serum and urine. The FIA method enabled us to determine histidine in blood after pretreatment of the sample with N-ethylmaleimide (masking reagent of glutathione).[5] These methods are useful in the diagnosis of histidinanemia, one of hereditary metabolic disorders characterized by a virtual deficiency of histidine ammonia-lyase.

REFERENCES

1. Håkanson, R.; Rönnberg, A.L.; Sjölund, K. Improved fluorometric assay of histidine and peptides having NH_2-terminal histidine using o-phthalaldehyde. Anal. Biochem. **1974**, *59*, 98–109.
2. Yoshimura, T.; Kamataki, T.; Miura, T. Difference between histidine and histamine in the mechanistic pathway of the fluorescence reaction with ortho-phthalaldehyde. Anal. Biochem. **1990**, *188*, 132–135.
3. Tateda, N.; Matsuhisa, K.; Hasebe, K.; Kitajima, N.; Miura, T. High-performance liquid chromatographic method for rapid and highly sensitive determination of histidine using postcolumn fluorescence detection with o-phthaldialdehyde. J. Chromatogr. B. **1998**, *718*, 235–241.
4. Tateda, N.; Matsuhisa, K.; Hasebe, K.; Miura, T. Simultaneous determination of urocanic acid isomers and histidine in human stratum corneum by high-performance liquid chromatography. Anal. Sci. **2001**, *17*, 775–778.
5. Tateda, N.; Matsuhisa, K.; Hasebe, K.; Miura, T. Sensitive and specific determination of histidine in human serum, urine and stratum corneum by a flow injection method based on fluorescence derivatization with o-phthalaldehyde. J. Liq. Chromatogr. & Relat. Technol. **2001**, *24*, 3181–3196.

HPLC Column Maintenance

Sarah Chen
*Analytical Research Department, Merck Research Laboratories,
Rahway, New Jersey, U.S.A.*

INTRODUCTION

The column is arguably the most important component in HPLC separations. The availability of a stable, high-performance column is essential for developing a rugged, reproducible analytical method. Performance of columns from different vendors can vary widely. Separation selectivity, resolution, and efficiency depend on the type and quality of the column. Proper column maintenance is the key to ensure optimum column performance as well as an extended column lifetime. It ensures stability of column plate number, band symmetry, retention, and resolution. The major issues related to column performance and maintenance are discussed here.

COLUMN CONFIGURATION

The HPLC column usually consists of a length of stainless steel tubing packed with porous particles for separation. These particles are sealed in the tubing by an HPLC column end fitting at each end, containing porous frits to retain the packing particles. Typically, 2- and 0.5-μm pore size stainless steel frits are used for 5- and 3-μm particles, respectively.[1] Many problems arising from stainless steel columns can be traced to the inlet stainless steel frit, which has a higher surface area than the column walls, leading to possible sample adsorption. High backpressure, poor peak shapes, and low sample yields are indications of possible frit problems.

COLUMN PACKING MATERIALS

Silica-based packings are the most popular HPLC column packing materials because of their favorable physical and chemical properties.[2,3] The silica particles have high mechanical strength, as well as narrow pore size and particle size distribution. The surface of silica can be chemically modified with a large variety of bonding molecules having different functionalities. Silica-based packings are compatible with water and all organic solvents, and exhibit no swelling with change in solvents; this is in contrast to most polymer-based stationary phases.

Columns packed with porous, polymeric particles such as divinylbenzene cross-linked polystyrene, substituted methacrylates, and polyvinyl alcohols can also be used for HPLC method development[4] as can modified alumina and zirconia stationary phases.[5,6]

COLUMN MAINTENANCE

Proper column maintenance is very important to ensure optimal performance and extended column lifetimes. There are common procedures that apply to all columns, e.g., avoiding mechanical or thermal shock. There are procedures that are column specific, such as avoiding chloride-containing mobile phases to prevent "halide cracking" if the column tubing and frits are made of stainless steel (especially at low pH). Nonetheless, columns made with stainless steel tubing and packed with silica-based stationary phases are the most commonly used in HPLC. Thus problems associated with these columns and how to prevent such problems by proper column maintenance will be discussed here.

How to Ensure Retention and Resolution Reproducibility of HPLC Columns

Reproducible retention and resolution are very important when developing routine methods. Changes in resolution and retention can be a function of the column quality, its operation, instrumental effects, or variations in separation conditions. The first important step in maintaining retention reproducibility is through selection of a good-quality column with a less acidic and highly purified support. Choosing a favorable mobile-phase condition (pH, buffer type and concentration, additives, etc.) that can eliminate surface silanol interactions when separating basic compounds is also very important for column retention reproducibility. There should be minimal variation in laboratory temperature or column temperature for retention and

Encyclopedia of Chromatography DOI: 10.1081/E-ECHR-120038615

resolution reproducibility. Proper laboratory instrumentation and column storage conditions cannot be neglected either.

Poor retention reproducibility and tailing peaks often occur in poorly buffered mobile phases, because of an inappropriately selected buffer, too low a buffer concentration, or a pH out of the effective range of a buffer. Increasing buffer concentration can minimize some of the problems. However, the buffer concentration must not be too high; otherwise, the buffer may not be miscible with the organic portion of the mobile phase. Other factors such as tailing peaks, high backpressure, or loss of stationary phase will also result in poor retention reproducibility and will be discussed later.

How to Avoid Band Tailing

Band tailing causes inferior resolution and reduced precision. Thus conditions resulting in tailing or asymmetric peaks should be avoided. Peak asymmetry or band tailing can arise from several sources: partially plugged column frits, void(s) in the column, buildup of sample components and impurities on the column inlet following multiple sample injections, sample overload, solvent mismatch with reference to the sample, chemical or nonspecific interactions (e.g., silanol effects), contamination by heavy metals, and excess void volume in the HPLC system.

Tailing peaks are common with heavily used columns. During use, columns can develop severe band tailing or even a split peak for a single component. Such effects usually arise from the presence of a void in the inlet of the column and/or a dirty or partially plugged inlet frit. The cause of the void can be either a poorly packed column in which the packing settles during use or dissolution of silica packing at excessively high pH. Excessive system backpressure or pressure surges, caused by poorly operating pumps or sample injection valves, can also cause column voiding. The void can be eliminated by the addition of a new packing material at the inlet end of the column to fill the void. This can be done by carefully removing the end fittings from the column inlet. Packing material is added in a slurry form to the void column volume. For best results, packing material of the same type as the column packing should be used. Old frits should be replaced with new ones of the same type. However, sometimes it is difficult to achieve the initial column efficiency after such procedures.

The presence of strongly retained materials in real-world samples can result in peaks that are eluted long after the normal run time is over. These peaks can cause three kinds of problems in later runs. If the peaks are large, they can sometimes show up in a subsequent run as very broad peaks. If the peaks are very small, they can be hidden under a peak of interest and cause peak distortion. It is also possible that peaks eluting late in a run can be small enough or so strongly retained that they appear only as a minor baseline hump. The development of broader tailing peaks during column use may also indicate the buildup of strongly retained sample components (garbage) on the column. Purging the column with a strong solvent can eliminate this buildup. For reversed-phase columns, a 20-column volume purge (about 50 mL for a 250 × 4.6-mm i.d. column) with 100% acetonitrile is often adequate. In case a stronger solvent is needed, a mixture of 96% dichloromethane and 4% methanol with 0.1% ammonium hydroxide is often effective. Because dichloromethane is not miscible with aqueous mobile phases, it is necessary to flush the reversed-phase column with acetonitrile prior to and after the use of dichloromethane. Methanol is used for normal phase columns. Sometimes it may be sufficient to flush the column once a day. If strongly retained materials are known to exist in the sample, it is a good idea to flush the column with a strong solvent at the end of each run sequence so that any strongly retained materials are flushed out before the next analysis.

How to Avoid High Backpressure

High backpressure is one of the most commonly encountered problems when performing HPLC analysis. Normal column backpressure is observed after a new column has been installed and equilibrated with the mobile phase. Unfortunately, this pressure often will increase with time of use because of particles collecting on the column inlet or outlet frit. These particles can be sample impurities, mobile phase contaminants, or materials from the injector or autosampler rotary seal. Unfavorable buffer conditions, such as high pH, can dissolve silica particles. Following the breakdown of particles, resultant small particles can clog the frit at the outlet of the column. The presence of small particles in the system can result in increased backpressure, split peaks, tailing, and, eventually, overpressure shutdown of the HPLC system. Most often, plugged frits can be eliminated by back flushing the column with a strong solvent. If this does not alleviate the problem, the plugged inlet frit can be replaced with a new frit without disturbing the packing. When replacing the inlet frit, addition of packing material is often needed if a void is noticed at the column inlet. To reduce backpressure problems, samples should be cleaned up before injection. The sample treatment may include filtering the samples through a submicron membrane filter to remove particulates or using solid-phase extraction techniques to remove

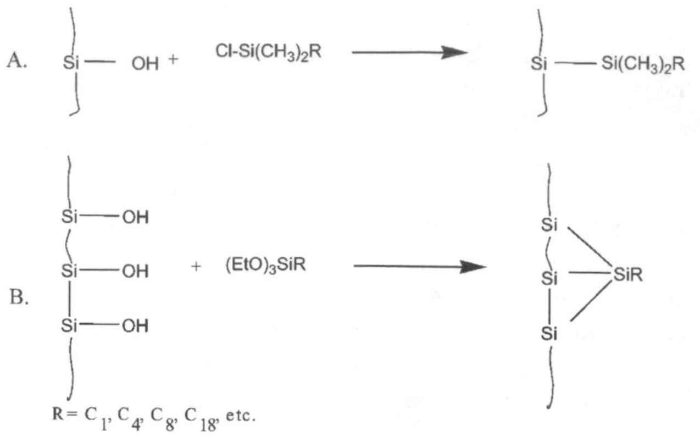

Fig. 1 Schematic diagram of monomeric (A) and polymeric (B) stationary phases for reversed phase HPLC.

highly retained sample components or matrix components. Only HPLC-grade or superior-grade solvents should be used to prepare the mobile phase, and buffer solutions should be filtered; alternatively, prefilters may be installed at the buffer reservoir. Rotary valve seals should be changed during routine maintenance procedures. Along with these preventive measures, it is advisable to use column prefilters, e.g., a guard column protection system. Particles then build up on the inexpensive, replaceable frit in the prefilter instead of in the permanent frit at the head of the analytical column. It is best to choose a guard column containing the same type of stationary phase as is contained in the analytical column. The length of a guard cartridge is usually 1 or 2 cm, with typical diameters ranging from 2.0 to 4.6 mm.

How to Prevent Loss of Stationary Phases

Column lifetime can be reduced significantly by the loss of the stationary phase during the separations. Stationary/mobile-phase combinations that lead to a rapid loss of bonded phase should be avoided. Column manufacturers' recommendations should be followed when using an HPLC column for separations. Commonly, reversed-phase columns with short-chain silane groups are the least stable, as the silane groups can be easily hydrolyzed with aggressive mobile phases (e.g., pH < 2.0). Reversed-phase columns with longer alkyl groups, such as C_8 or C_{18} (Fig. 1), are usually considered relatively stable because of the inaccessibility of the surface Si–O–R group. Polymeric C_8 or C_{18} is considered to be more stable than its monomeric counterparts. However, over long periods at very low or very high pH values, these columns can also lose bonded molecules. Use of sterically protected silane stationary phases will provide additional stability in aggressive low-pH environments.

The stability of the bonded organic ligand on a reversed-phase column also depends on the type and acidity of the silica used as the support. Packings made with fully hydroxylated silicas with a homogeneous distribution of silanol groups show superior stability. Higher bonded-phase stability apparently can occur for columns made with highly purified silica supports having a lower surface acidity.

Loss of stationary phase from silica-based columns is accelerated at higher temperatures. Temperatures above ~40°C should be used with caution when operating at intermediate or high pH values with phosphate buffers. Operation at pH < 3 and elevated temperature can degrade the bonded stationary phase more rapidly and cause retention reproducibility instabilities. However, there are specially end-capped columns available commercially that can withstand higher temperatures with significantly less hydrolysis.

One Hundred Percent Aqueous Mobile Phase with Reversed-Phase Column

Unless specified by the manufacturer, 100% aqueous mobile phase should be avoided with a reversed-phase column, such as C_{18} or C_8. Most reversed-phase columns exhibit decreased and poorly reproducible retention under more than 98% aqueous conditions. This problem has been attributed to the ligand collapse or incomplete wetting of the stationary phase. It has been reported that when a C_{18} column is washed with water, the bonded phase collapses or is incompletely wetted.[7] Subsequent flushing with mobile phase removes the wash solvent, but the stationary phase remained in the collapsed configuration; this has caused a change in retention and selectivity. It has been suggested that washing a reversed-phase column with water should be avoided. Furthermore, if the organic content of a mobile phase is too low, the stationary phase will tend to collapse onto itself in a low energy conformation. This collapse/incomplete wetting could lead to abnormal chromatographic behavior and generally undesirable results. To avoid this phenomenon, workers have sometimes used embedded polar phases, including amide or carbamate groups[8–11] which presumably do not undergo phase collapse in 100% water and can withstand mobile phases with high aqueous content. If 100% aqueous mobile phase is needed for a reversed-phase separation, columns with embedded polar groups should be used.

Other Factors in Column Maintenance

Storing a column filled with 100% organic solvent, such as acetonitrile, preserves the performance and lifetime of bonded-phase columns. Storage with buffered solutions (particularly those containing high concentrations of water and alcohols) should be avoided. When buffers are used, columns should be flushed with 15 to 20 column volumes of the same aqueous/organic mobile phase without buffer before converting to 100% organic for storage. Columns should be capped tightly during storage to prevent the packed bed from completely drying. Bacterial growth often occurs in buffers and aqueous mobile phases contained in columns that are prepared and stored at ambient temperature for more than a day. Particulates from this source can plug the column inlet and reduce column life significantly. As a result, mobile phases that are free of organic solvents should be discarded at the end of each day. Alternatively, 20% of organic modifier in the mobile phase retards bacterial growth. The organic modifier also assists in the mobile-phase degassing process.

CONCLUSIONS

In general, mechanical and thermal shock should be avoided to prevent disturbing the column packing bed. Pressure surges in the system should also be avoided. Choosing a suitable buffer system is very critical to maintaining retention and resolution reproducibility and preventing the loss of stationary phase. Samples containing particulates should be filtered before injection. The use of guard columns can reduce the "garbage" build up at the column inlet and reduce the risk of high backpressure. It is a good practice to flush the column frequently with a strong solvent after heavy use and to store it in an appropriate solvent.

REFERENCES

1. Snyder, L.R.; Kirkland, J.J.; Glajch, J.L. *Practical HPLC Method Development*, 2nd Ed.; John Wiley & Sons, Inc.: New York, 1997.
2. Iler, R.K. *The Chemistry of Silica*; Wiley: New York, 1979.
3. Unger, K.K. Porous silica. J. Chromatogr. Libr. **1979**, *16*, Elsevier, Amsterdam.
4. Tanaka, N.; Araki, M. Polymer-based packing materials for reversed-phase liquid chromatography. Adv. Chromatogr. **1989**, *30*, 81.
5. Pesek, J.J.; Sandoval, J.E.; Su, M. J. Chromatogr. **1990**, *630*, 91.
6. Sun, L.; Annen, M.J.; Lorenzano-Porras, F.; Carr, P.W.; McCormick, A.V. Synthesis of porous zirconia spheres for HPLC by polymerization-induced colloid aggregation (PICA). J. Colloid Interface Sci. **1993**, *163*, 91.
7. Wolcott, R.G.; Dolan, J.W. LC-GC **1999**, *17* (4), 316.
8. O'Gara, J.E.; Alden, B.A.; Walter, T.H.; Petersen, J.S.; Niederlander, C.L.; Neue, U.D. Simple preparation of a C8 HPLC stationary phase with an internal polar functional group. Anal. Chem. **1995**, *67* (20), 3809.
9. Cqajkwaka, T.; Hrabovsky, I.; Buszewski, B.; Gilpin, R.K.; Jaronieic, M. Comparison of the retention of organic acids on alkyl and alkylamide chemically bonded phases. J. Chromatogr. A. **1995**, *691*, 217.
10. Ascah, T.L.; Kallury, K.M.R.; Szafravski, C.A.; Corman, S.K.; Liu, F. J. Liq. Chromatogr. & Relat. Technol. **1996**, *19*, 3409.
11. Czajkowaka, T.; Jaronieic, M. Selectivity of alkylamide bonded-phases with respect to organic acids under reversed-phase conditions. J. Chromatogr. A. **1997**, *762*, 147.

H

HPLC Instrumentation: Troubleshooting

I. N. Papadoyannis
V. F. Samanidou
Laboratory of Analytical Chemistry, Chemistry Department, Aristotle University of Thessaloniki, Thessaloniki, Greece

INTRODUCTION

Despite the advances in technology and instrumentation, problems still arise when practicing HPLC, which cause headaches to chromatographers. During several stages of analysis, such as method development or routine operation, a variety of separation artifacts may be noticed. Pressure abnormalities, sample recovery, poor reproducibility, loss of resolution, instability, leaks, etc. are common problems.

General problems can be detected by smell, sight, or sound, although major symptoms in the LC system show up as changes in the chromatogram, such as irregular peak shapes, extra peaks, negative peaks, varying retention times, and many others. It is well known that, if a picture is worth a thousand words, then a chromatogram, to a chromatographer, is equally valuable. Any chromatographer who has injected many samples into an LC system has occasionally confronted more than one of the abovementioned problems. Some problems can be corrected by changes in the equipment, whereas others require modification of the assay procedure. However, many of the common LC problems can be prevented with routine preventive maintenance.

Guides for troubleshooting HPLC instrumentation provide analysts and laboratory technicians with a readily available, very useful aid for solving operational problems of equipment and techniques. In order that the chromatographer effectively solves an arising problem in HPLC, he/she should be aware of the role of operating parameters, as these are indicators of system performance. Step-by-step troubleshooting protocols for each system component should be followed to isolate the problem and its cause.

OVERVIEW

This article covers problem identification and procedures for solving them, as well as practices to maintain HPLC systems in good operating condition. It also guides users of HPLC equipment to investigate the source of a malfunction through each system component, from sample preparation to detection and integration.

PROBLEMS: CAUSES AND SOLUTIONS

In an HPLC system, problems can arise from many sources. Malfunction can be allocated to various points. Chromatographers should use not only their experience to locate problems but also all their senses (obviously, except taste) to identify LC problems. For example, a leak can be noticed by smell before it is actually seen. A strange noise indicates some kind of malfunction and a "hot" smell indicates an overheating module. Most problems, however, are identified by sight, and they can mainly be observed as changes in the chromatogram. As soon as the problem has been defined, actions should be taken to correct the malfunctioning component. The incident should be recorded, in a log book kept for this purpose, to help with further failure problems at a later time.

The HPLC user should know or learn what to look for and what to do to prevent HPLC problems and, finally, what can and should be done before calling a service technician. User's manuals, manufacturer's advice, books, articles in scientific journals, computer programs, and network sites can be used as resources for troubleshooting. A general rule is that one can know that a problem exists only when one knows how the system operates when it is working well. This means that keeping detailed records of system performance (log book) is very helpful.

Dolan (refer to "Suggested Additional Reading") has very effactually set forth five rules of thumb for HPLC troubleshooting.

1. Do not change more than one thing at a time.
2. A problem is considered as problem when it occurs more than once.
3. A questionable system component should be substituted/replaced with one that is known to be working properly. Known good parts should be put back into service while all failed parts should be thrown away.
4. An experienced chromatographer should try to anticipate what will fail next.
5. Good records of maintenance and troubleshooting actions should be kept.

Encyclopedia of Chromatography DOI: 10.1081/E-ECHR-120022508

In HPLC instrumentation troubleshooting, problems can be classified as follows. Major HPLC problems are discussed under this paragraph, while an extended summary of problem causes and remedy actions are tabulated in the respective tables.

Problems with the Chromatogram

Peak shape

Broad peaks, ghost peaks, pseudo peaks, negative peaks, peak doubling, peak fronting, peak tailing, spikes, no peaks. The major causes and their solutions are tabulated in Table 1.

Variable retention times

Retention time inconsistency (changing, increasing, decreasing), change in separation (loss of resolution). Table 2 summarizes their causes and solutions.

Baseline

Short-term noise, long-term noise, and drift. The causes of the baseline problems and their remedies are discussed in Table 3.

Pressure abnormalities

These include increased pressure, decreased pressure, unstable/fluctuating pressure, and high backpressure. Table 4 lists the major causes and their solutions.

Leaks

Leaks at various points, such as the column, fittings, the detector, the injection valve, or the pump. Table 5 summarizes their causes and solutions.

Change in quantitation

Including imprecision, change in selectivity, change in peak height–lack of sensitivity, poor sample recovery. Table 6 summarizes their causes and solutions.

Fig. 1 illustrates some of the changes that appear in the chromatogram as a result of the various problems.

Changes in the Chromatogram

Baseline

Baselines in chromatograms are not always smooth. On the contrary, a baseline may have spikes, noise, and other disturbances, indicating existing problems.

Indeed, by magnifying the baseline, chromatographers can obtain information to recognize the problem and correct it. For example, trace contaminants in water, buffer, and reagents may cause peaks when using gradient elution.

Very often baseline problems are related to detector problems. Many detectors are available for HPLC systems. The most common are fixed and variable wavelength ultraviolet spectrophotometers, refractive index, and conductivity detectors. Electrochemical and fluorescence detectors are less frequently used, as they are more selective. Detector problems fall into two categories: electrical and mechanical/optical. The instrument manufacturer should correct electrical problems. Mechanical or optical problems can usually be traced to the flow cell; however, improvements in detector cell technology have made them more durable and easier to use. Detector-related problems include leaks, air bubbles, and cell contamination. These usually produce spikes or baseline noise on the chromatograms or decreased sensitivity. Some cells, especially those used in refractive index detectors, are sensitive to flow and pressure variations. Flow rates or backpressures that exceed the manufacturer's recommendation will break the cell window. Old or defective source lamps, as well as incorrect detector rise time, gain, or attenuation settings will reduce sensitivity and peak height. Faulty or reversed cable connections can also be the source of problems.

Electronic noise from fluorescent lights and other common sources is often called 60-cycle noise because it coincides with the 60-Hz frequency of the alternating current servicing the laboratory.

To isolate the origin of the problem due to the detector, the chromatographer may perform the "dry cell test": by disconnecting the detector from the column and then blowing the cell dry with dry nitrogen. Under these conditions no drift should be observed.

Ghost peaks

These peaks appear in a chromatogram due to contamination of mobile phase, injector, column, or strongly retained compounds from previous sample injections. When an autosampler is used, the problem is often referred to as "carryover." Flushing the injector and the column with strong solvent will remove interfering or late eluting compounds. To correct carryover problems, the chromatographer should change injection size, check wash solvent, increase wash solvent volume, adjust its pH value, use a portion of organic solvent, change a needle seal, injection loop, and check fitting assembly.

Ghost peaks in gradient runs can be avoided by increasing the equilibration time between analyses.

Table 1 Peak shape problems and remedies

Peak shape

Fronting

1. Column overloaded. → Decrease sample amount or dilute sample.

2. Sample solvent incompatible with mobile phase. → Adjust solvent.

3. Nonresolved peak from another component. → Improve resolution by altering mobile or stationary phase.

4. Wrong pH value of mobile phase. → Adjust pH.

5. Channeling in column. → Replace or repack column.

Tailing

1. Secondary retention effects; residual silanol interactions. → Use ion pair reagent, or competing base or acid modifier. Triethylamine for basic compounds, acetate for acidic compounds.

2. Wrong pH value of mobile phase. → Adjust pH. For basic compounds lower pH usually provides more symmetric peaks.

3. Wrong stationary phase. → Change column.

4. Void at column inlet. → Repack top of column with stationary phase.

5. Wrong injection solvent. → Dissolve sample in mobile phase.

6. Interference in sample. → Check column performance with standards. Change mobile phase or stationary phase. Check selectivity.

7. Chelating solutes—trace metals in base silica. → Use high purity silica-based column with low trace-metal content, add EDTA or chelating compound to mobile phase; use polymeric column.

8. Unswept dead volume. → Minimize number of connections; ensure injector rotor seal is tight: ensure all compression fittings are correctly scaled.

9. Silica-based column-degradation at high temperature. → Reduce temperature to less than 50°C.

Double (split) peaks

1. Column voided. → Repack top of column with stationary phase; replace column.

2. Partially blocked frit. → Clean or replace the plugged frit. Install an in-line filter between pump and injector to remove solids from mobile phase or between injector and column to filter particulates from sample.

3. If only one peak is split co-eluting interfering components. → Use sample cleanup.

4. Sample solvent incompatible with mobile phase. → Inject samples in mobile phase.

5. Blocked frit. → Replace or clean frit, install 0.5-um porosity in-line filter between pump and injector to eliminate mobile-phase contaminants or between injector and column to eliminate sample contaminants.

6. Co-elution of interfering compound from previous injection. → Use sample cleanup; adjust selectivity by changing mobile or stationary phase. Flush column with strong solvent at end of ran; end gradient at higher solvent concentration.

7. Column overloaded. Sample volume too large. → Use higher-capacity stationary phase, increase column diameter, decrease sample amount.

8. Column void or channeling. → Replace column, or, if possible, open top endfitting and clean and fill void with glass beads or same column packing; repack column.

9. Injection solvent too strong. → Use weaker injection solvent or stronger mobile phase.

Broad peaks (all)

1. Large injection volume; detector operating outside linear dynamic range. → Injection of smaller sample volume or diluted sample (1 : 10).

2. High viscosity of mobile phase. → Change mobile phase, or increase column temperature. Change to lower viscosity solvent.

3. Poor column efficiency due to column void or column contaminated/worn out. → Repack top of column; replace column.

4. Incorrect detector settings. → Check settings and adjust.

5. Low mobile phase flow rate. → Increase flow rate.

(Continued)

Table 1 Peak shape problems and remedies *(Continued)*

Peak shape

  6. Tubing too long or too wide; large extra column volume. → Use right tubing, shorten path. Use low- or zero-dead-volume endfittings and connectors; use smallest possible diameter of connecting tubing (< 0.10 in. i.d.); connect tubing with matched fittings.

  7. Leaks between column and detector. → Check for leaks.

  8. Guard column contaminated/worn out. → Replace guard column.

  9. Retention times too long. → Use gradient elution or a stronger mobile phase for isocratic elution.

10. Too large volume of detector cell. → Use smaller cell volume.

11. Slow detector time constant → Adjust time constant to match peak width.

Broad peaks (some)

1. Late eluted peak from previous run. → Flush the column with a strong eluent after each run or end gradient at a higher concentration of strong solvent.

2. High molecular weight sample. → Optimize sample clean up.

Negative peaks

1. Highly UV absorbing mobile phase. → Change detection wavelength taking into account the UV cutoff of mobile phase solvents.

2. Refractive index of mobile phase is very different from RI of sample. → Change eluent.

3. Recorder connections. → Check polarity.

Ghost peaks

1. Dirty mobile phase. → Use HPLC grade solvents.

2. Carryover. Retained compound from previous injection. → Flush column with strong solvent to remove late eluting compounds. End gradient at higher solvent concentration.

3. Contamination of injector. → Flush injector.

4. Contamination of column. → Flush column with strong solvent.

5. Unknown interferences in sample. → Use sample cleanup or prefractionation before injection.

Spikes

1. Bubbles in mobile phase. → Degas mobile phase. Sparge it with helium (3–5 psi) during use; ensure that all fittings are tight; store column tightly capped.

2. Bubbles in detector. → Use back-pressure regulator at detector outlet.

Extra column dispersion

1. Wrong tubing dimensions. → Use short, small internal diameter (narrower) tubing between injector and column and between column and detector.

2. Detector overloaded. Outside linear dynamic range. → Use a low volume detector cell.

3. Large sample volume. → Inject small sample volumes.

No peaks

1. Detector off. → Check detector.

2. No flow. Pump off. → Start pump.

3. No sample. Sample deteriorated. → Check injector. Check sample stability.

4. Wrong settings on recorder or detector. → Check attenuation, gain, and detector wavelength.

5. Flow interrupted. → Check reservoirs, loop, degassing of mobile phase, and compatibility of mobile phase components.

6. Leaks. → Check fittings and pump for leaks and pump seals.

7. Air trapped in the system. → Prime pump.

Table 2 Retention time inconsistency

Variable retention times

1. Leaks. → Check for loose fittings, pump leaks, seals.

2. Change in mobile phase composition. → Prepare new. Ensure that gradient system is delivering correct composition. Prevent evaporation.

3. Air trapped in pump. → Prime pump. Degas mobile phase—Sparge it with helium (3–5 psi) during use.

4. Overloading. → Dilute sample.

5. Sample dissolved in a solvent that is incompatible with the mobile phase. → Dissolve sample in the mobile phase.

6. Temperature fluctuations. → Use column oven.

7. Isocratic elution: Insufficient equilibration time. → Pass 10–15 column volumes of mobile phase through column for equilibration.

8. Gradient elution: Insufficient column regeneration time. → Increase equilibrating time.

Loss of resolution

1. Leak. Pump flow problems. → Check for leaks.

2. Obstructed guard or analytical column. → Replace guard column. Reverse analytical column and flush disconnected from the detector. Change inlet frit. Replace the column.

3. Improperly prepared mobile phase; contaminated mobile phase. → Prepare fresh mobile phase. Check pump-proportioning valve for malfunction.

4. Sample overloaded. → Dilute sample and reinject.

5. Extra column dead volume. → Check system plumbing and all connections for dead volume.

6. Injector problem. → Leaking injection valve or a damaged or blocked needle has to be corrected.

7. Temperature fluctuations. → Use column oven.

No peaks and negative peaks

If no peaks are observed, then the chromatographer should check the detector, the connections, the flow (leaks, pump function, air bubbles), the sample (for its stability), and settings on the detector (e.g., wrong wavelength), or integration.

Negative peaks are due to wrong polarity of recorder, or absorbance or refractive index of mobile phase higher than that of solute.

Peak tailing and peak fronting

Peak Tailing (Peak Asymmetry Factor >1.2). This is attributed to the wrong pH value, wrong column, wrong sample solvent (mobile phase is better to be used), void volumes at column inlet (the column may need repacking), as well as to active sites within the column which can be solved with the use of a competing basic or acidic modifier.

If only some of the peaks tail, secondary retention effects, such as residual silanol interactions, may take place. Another possibility is that a small peak is eluting on the tail of a larger peak. If all peaks tail, this may be due to a bad column or build up of contamination on the column inlet frit.

Peak Fronting (Peak Asymmetry Factor <0.9). This indicates that a small band is eluting before a large band, a wrong pH value of the mobile phase is used, an overloaded column, a void volume at the inlet, or that the sample solvent is incompatible with the mobile phase.

Double peaks, rounded peaks, and broad peaks

A void volume in the column, or a partially blocked frit can possibly cause double or split peaks. In case that only one peak is a doublet, then co-eluting compounds may be present.

Rounded peaks are attributed to high concentrations (the detector response being outside the linear dynamic range), wrong sample solvent, or too high a setting of the detector or integrator time constant.

Additionally, these peak-related problems may be attributed to column overload, too long or too wide tubing, column contamination, low flow rate, etc.

If all peaks are broadened, possible causes include a large sample volume injected, or a viscous mobile phase, or a column that has lost its efficiency, possibly due to the presence of a column void. If only some peaks are broadened, then a peak from a previous run may be eluted late, or a high molecular mass sample, e.g., a protein or a polymer is present.

Table 3 Baseline

Baseline

Regular noise

1. Air bubbles. → Prime pump. Degas solvent. Sparge mobile phase with helium during use.

2. Pump pulsations. → Use a pulse dampener.

3. Incomplete mixing. Malfunctioning proportioning valves. → Ensure complete mixing. Clean or replace the proportioning valve; partially remix solvents.

4. Other electronic equipment on the same line. → Check electronic equipment in line. Correct as necessary.

5. Leaks. → Check pump for leaks, salt build-up. Check fittings and pump seals.

6. Continuous-detector lamp problem or dirty flow cell. → Replace UV lamp (each should last 2000 hr); clean and flush flow-cell.

Irregular noise

1. Leaks. → Check fittings, pump seals, and pump for leaks.

2. Electronics. → Locate problem. Get servicing. Isolate detector and recorder electronically. Use a voltage stabilizer for the LC system or use an independent electrical circuit for the chromatography equipment.

3. Insufficient grounding. → Establish sufficient grounding.

4. Flow cell contamination. → Clean detector cell.

5. Detector lamp failing. → Replace detector lamp.

6. Mobile phase mixer inadequate or malfunctioning. → Repair or replace the mixer or mix off-line in case of isocratic elution.

7. Air bubbles in detector. → Install backpressure regulator after detector.

8. Occasional sharp spikes—external electrical interference. → Use voltage stabilizer for LC system; use independent electrical circuit.

9. Periodic-pump pulses. → Service or replace pulse damper; purge air from pump; clean or replace check valves.

10. Random-contamination buildup. → Flush column with strong solvent; clean up sample; use HPLC grade solvent.

Drift

1. Strongly retained materials. → Flush column with strong solvent.

2. Default mixing. → Check mixer. Check flow rate and composition.

3. Air in the detector cell. → Clean cell. Use backpressure regulator at detector outlet.

4. Contamination of mobile phase. → Flush column with strong solvent; use HPLC grade solvents; clean-up sample.

5. Fluctuation of column temperature. → Use column oven.

6. Gradient elution. A. Positive direction. Absorbance of mobile phase B. → Add UV absorbing compound to mobile phase A.
   Negative direction. Absorbance of mobile phase A. → Add UV absorbing compound to mobile phase B.

7. Temperature at RI detector unstable. → Control changes in room temperature. Insulate column, use column oven, cover RI detector keeping it out of air currents.

8. Mobile phase not in equilibrium with column. → Allow more time for column equilibration.

System peaks

A system peak is a peak that originates from the chromatographic system itself, i.e., mobile phase and column, and not from the sample. Its appearance and size are sensitive to the sample composition, but its origins are generally the mobile phase components.

When a mobile phase is introduced to a column, its components undergo distribution until equilibrium is attained. Injection of a sample different from the mobile phase causes a small equilibrium perturbation at the column head. The equilibrium of each component of the mobile phase can be disturbed and, thereby, manifested by one system peak for each mobile phase additive, using the appropriate detection conditions.

System peaks are most often recognized as a pair of peaks, one positive and the other negative, which represent enrichment and depletion zones eluted from the column. They may vary in retention time and size, depending on the sample matrix, injection volume, mobile phase composition, and the stationary phase. Dissolving samples in the mobile phase is the best

Table 4 Pressure abnormalities

Increased pressure

1. Blocked flow lines: Pump. Injector. Tubing. → Locate obstruction, by systematic disconnection of system components. Replace or clean blocked components.

2. Obstructed column or guard column, from particulate buildup at top of column. → Replace guard column. Reverse analytical column and flush disconnected from the detector. Change inlet frit. Replace the column. Filter sample.

3. Salt precipitation. → Ensure mobile phase compatibility with buffer concentration.

4. High viscosity of mobile phase. → Use solvent of lower viscosity or increase temperature.

5. Microbial growth in the column. → Store column with at least 25% organic solvent. Add 0.02% sodium azide to aqueous mobile phases, or use a mobile phase with at least 10% organic solvent.

Decreased pressure

1. Leaks in the system: Fittings not tight. → Check all connection for leaks. Tighten or replace fittings. Replace or clean check valves.

2. Piston seal(s) worn. → Replace piston seal(s).

3. Air trapped in pump. → Prime pump.

4. Mobile phase interrupted. Insufficient flow from pump. → Check reservoirs, loop, degassing of mobile phase, and compatibility of mobile phase components.

Unstable-fluctuating pressure

1. Air bubbles in pump. → Degas mobile phase—Sparge it with helium (3–5 psi) during use. Prime pump.

2. Leaks in pump check valve or seals. → Replace or clean check valves; replace pump seals.

High backpressure

1. Plugged frit, pre-filter, guard column; plugged inlet frit. → Backflush column/cartridge. Replace frit, pre-filter, guard column; replace endfitting or frit assembly.

2. Irreversibly retained contaminants on the column head. → Column cleaning/regeneration.

3. Precipitation of buffer. → Flush with water at low flow rate.

4. Precipitation or aggregation of proteins in column, particulate matter trapped by the top and/or bottom filter build-up of lipids, DNA, or other macromolecules nonspecifically bound to the column microbial contamination. → Clean column, following column instructions with appropriate solvents, change top and/or bottom filter.

5. Column blocked with irreversibly adsorbed sample. → Improve sample cleanup; use guard column; reverse-flush column with strong solvent to dissolve blockage.

6. Column particle size too small (for example, 3 μm). → Use larger particle size (for example, 5 μm).

7. Microbial growth on column. → Use at least 10% organic modifier in mobile phase; use fresh buffer daily; add 0.02% sodium azide to aqueous mobile phase: store column in at least 25% organic solvent without buffer.

8. Mobile phase viscosity too high. → Use lower viscosity solvents or higher temperature. Replace frit or guard column.

9. Plugged frit in in-line filter or guard column; plugged inlet frit. → Replace end-fitting or frit assembly.

10. Polymeric columns—solvent change causes swelling of packing. → Use correct solvent with column; change to proper solvent composition! Consult manufacturer's solvent-compatibility chart; use a column with a higher percentage of cross-linking.

11. Salt precipitation (especially in reversed-phase chromatography with high concentration of organic solvent in mobile phase). → Ensure mobile phase compatibility with buffer concentration; decrease ionic strength and water-organic solvent ratio; premix mobile phase.

12. When injector disconnected from column-blockage in injector. → Clean injector or replace rotor.

---

way to minimize system peak effects. System peaks are especially important in ion-pairing chromatography and in ion chromatography. In the latter, they are pH dependent and they often interfere with sample components.

Pressure abnormalities

Lower pressure than anticipated is observed due to leaks, insufficient flow from the pump, air bubbles, and worn pump seals. Fluctuating pressure is

Table 5 Leaks

Leaks
Injector leaks
Rotor seal failure → Replace rotor seal.
Blocked loop → Clean or replace loop.
Loose injection-port seal → Tighten.
Waste line siphoning → Keep waste line above surface waste.
Waste line blockage → Replace waste line.
Column leaks.
Loose endfitting → Tighten endfitting.
Improper frit thickness → Use proper frit.
Detector leaks
Cell gasket failure → Prevent excessive backpressure; replace gasket.
Cracked cell window → Replace window.
Leaky fittings → Tighten or replace.
Blocked waste line → Replace waste line.
Leaky fittings
Loose fitting → Tighten.
Stripped fitting → Replace.
Overtightened fitting → Loosen and retighten; replace.
Dirty fitting → Disassemble and clean; replace.
Mismatched parts → Use parts from the same brand so that they match.
Leaks at pump
Loose check valves or fittings → Tighten.
Mixer seal failure → Replace mixer seal; replace mixer.
Pump seal failure → Repair or replace.
Pressure transducer failure → Repair or replace.
Pulse damper failure → Replace pulse damper.
Proportioning valve failure → Check diaphragms, replace if leaky; check for fitting damage, replace.
Purge valve → Tighten valve; replace purge valve.

attributed to air trapped in the pump, or leaking pump check valves or seals. Higher pressure than anticipated is due to blocked flow lines, particulate build up at the head of the column, or buffer salt precipitation. To locate blockage, components should be systematically disconnected, starting from the detector-end to column-end. Once the blocked component is located, it must be either cleaned or replaced. Back flushing of column will help to remove particulates at the top of column, thus reducing pressure. If buffers are used, their compatibility with the mobile phase should be checked to avoid precipitation within the system.

Change on separation (loss of resolution) and changes in height

Changes in separation, implying a loss of resolution, are attributed to leaks or an obstructed column. Change in height originates from sample deterioration,

leaks, nonreproducible sample volumes, and low detector response. A fresh sample should be checked, as well as detector settings and operating conditions.

Leaks

Leaks can be detected visually or by smell before even a pressure decrease is noticed. They can take place in different positions in the LC system, such as the column, the pump, the injection valve, or the detector. In case of leaks at the column or fittings, a leaky fitting should be tightened or replaced. The detector seal should be replaced if there is a leak at the detector.

A worn or scratched valve rotor in the injection valve should be replaced to prevent leaks in the injection valve.

In case of pump seal failure, pump seals should be replaced, or the piston should be checked for scratches and should be replaced if necessary.

Table 6 Imprecision

Change in selectivity

1. Not enough sample is injected. → Increase amount of injected sample.

2. Sample loop of injector is underfilled. → Overfill loop with sample.

3. Sample is lost during sample preparation. → Optimize sample preparation.

4. Autosampler line is blocked. → Clean blockage.

5. Detector attenuation is set too high. → Reduce detector attenuation.

6. Peaks are outside detector's linear range. → Dilute or enrich sample to reach linear range of detector.

7. Column is worn out. → Replace column.

8. Column temperature is altered. → Use column oven to maintain constant temperature.

Change in height–lack of sensitivity

1. Sample deterioration. → Use fresh sample.

2. Leak. → Check for pump leaks and fittings.

3. Nonreproducible sample volume. → Ensure loop is completely filled. Check autosampler. Check flow and clear any blockages.

4. Low detector response. → Check detector settings and operating conditions.

5. Detector attenuation is set too high. → Reduce detector attenuation.

6. Sample is lost during preparation. → Optimize sample preparation. Use internal standard during sample preparation.

7. Peaks are outside the linear range of the detector. → Dilute of enrich the sample until concentration is within the linear range of the detector.

8. First few sample injections—sample adsorption in injector sample loop or column. → Condition loop and column with concentrated sample.

9. Injector sample loop is underfilled. → Overfill loop with sample.

10. Not enough sample is injected. → Increase amount of sample injected.

11. Peaks are outside detector's linear range. → Dilute or concentrate sample to bring detector response into linear range.

Imprecision

1. Operator dependence during sample processing and clean-up. → Check all steps for errors.

2. Sample injection. → Check autosampler; fill loop completely.

3. Detection. → Clean flow-cell. Improve signal-to-noise ratio.

4. Separation. → Improve resolution.

5. Data processing and calibration. → Use internal standard. Calibrate frequently.

Poor sample recovery

1. Absorption or adsorption of proteins. → Reduce nonspecific interactions by changing HPLC mode; add protein-solubilizing agent, strong acid or base (with polymeric columns only), or detergent such as SDS to mobile phase.

2. Adsorption or chemisorption on column packing or on different hardware components. → Increase mobile phase strength; add competing base (for basic compounds) or use base-deactivated packing; ensure no reactive groups are present; use inert tubing and flow-path components, e.g., PEEK.

3. Irreversible adsorption on active-sites (less than 90% yield). → For basic compounds use end-capped, base-deactivated, sterically protected, high coverage, or polymeric reversed-phase. → For acidic compounds use endcapped or polymeric packing; acidify mobile phase.

Sample introduction

Problems with the sample introduction may arise both in manual injection and during autosampling. Table 7 summarizes the problems related to sample introduction.

In the case of autosamplers, although they are considered as time saving devices, their function is associated with some common problems. For example, needle depth adjustment is very critical when there is not enough sample available; a needle blockage may occur from a septum. Another common problem

H

Problem No. 1: No Peaks/Very Small Peaks

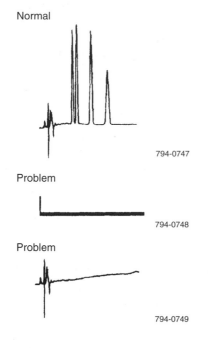

Normal

794-0747

Problem

794-0748

Problem

794-0749

1. Detector lamp off.
2. Loose/broken wire between detector and integrator or recorder.
3. No mobile phase flow.
4. No sample/deteriorated sample/wrong sample.

5. Settings too high on detector or recorder.

1. Turn lamp on.
2. Check electrical connections and cables.
3. See "No Flow" (Problem No. 2).
4. Be sure automatic sampler vials have sufficient liquid and no air bubbles in the sample. Evaluate system performance with fresh standard to confirm sample as source of problem.
5. Check attenuation or gain settings. Check lamp status. Auto-zero if necessary.

Problem No. 2: No Flow

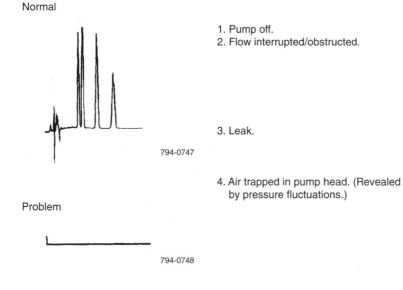

Normal

794-0747

Problem

794-0748

1. Pump off.
2. Flow interrupted/obstructed.

3. Leak.

4. Air trapped in pump head. (Revealed by pressure fluctuations.)

1. Start pump.
2. Check mobile phase level in reservoir(s). Check flow throughout system. Examine sample loop for obstruction or air lock. Make sure mobile phase components are miscible and mobile phase is properly degassed.
3. Check system for loose fittings. Check pump for leaks, salt buildup, unusual noises. Change pump seals if necessary.
4. Disconnect tubing at guard column (if present) or analytical column inlet. Check for flow. Purge pump at high flow rate (e.g., 5–10 mL/min), prime system if necessary. (Prime each pump head separately.) If system has check valve, loosen valve to allow air to escape. If problem persists, flush system with 100% methanol or isopropanol. If problem still persists, contact system manufacturer.

Fig. 1 Typical changes that appear in the chromatogram as a result of various problems in HPLC. (Reprinted with permission of Supelco, Bellefonte, PA, from Bulletin 826D, 1999.) (*Continued next page.*)

Problem No. 3: No Pressure/Pressure Lower Than Usual

Normal

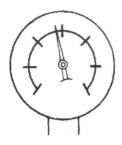

794-0750

Problem

794-0751

1. Leak.

2. Mobile phase flow interrupted/ obstructed.

3. Ait trapped in pump head. (Revealed by pressure fluctuations.)

4. Leak at column inlet end fitting.

5. Air trapped elsewhere in system.

6. Worn pump seal causing leaks around pump head.
7. Faulty check valve.
8. Faulty pump seals.

1. Check system for loose fittings. Check pump for leaks, salt buildup, unusual noises. Change pump seals if necessary.

2. Check mobile phase level in reservoir(s). Check flow throughout system. Examine sample loop for obstruction or air lock. Make sure mobile phase components are miscible and mobile phase is properly degassed.

3. Disconnect tubing at guard column (if present) or analytical column inlet. Check for flow. Purge pump at high flow rate (e.g., 10 mL/min), prime system if necessary. (Prime each pump head separately.) If system has check valve, loosen valve to allow air to escape.

4. Reconnect column and pump solvent at double the flow rate. If pressure is still low, check for leaks at inlet fitting or column end fitting.

5. Disconnect guard and analytical column and purge system. Reconnect column(s). If problem persists, flush system with 100% methanol or isopropanol.

6. Replace seal. If problem persists, replace piston and seal.
7. Rebuild or replace valve.
8. Replace seals.

Problem No. 4: Pressure Higher Than Usual

Normal

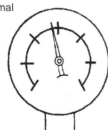

794-0750

Problem

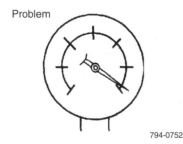

794-0752

1. Problem in pump, injector, in-line filter, or tubing.

2. Obstructed guard column or analytical column.

1. Remove guard column and analytical column from system. Replace with unions and 0.010″ ID or larger tubing to reconnect injector to detector. Run pump at 2–5 mL/min. If pressure is minimal, see Cause 2. If not, isolate cause by systematically eliminating system components, starting with detector, then in-line filter, and working back to pump. Replace filter in pump if present.

2. Remove guard column (if present) and check pressure. Replace guard column if necessary. If analytical column is obstructed, reverse and flush the column, while disconnected from the detector (page 14). If problem persists, column may be clogged with strongly retained contaminants. Use appropriate restoration procedure (Table 2, page 14). If problem still persists, change inlet frit (page 16) or replace column.

Fig. 1 *(Continued next page.)*

H

Problem No. 5: Variable Retention Times

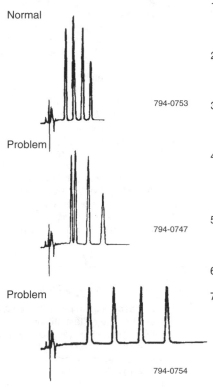

Normal

Problem

794-0753

794-0747

Problem

794-0754

1. Leak.

2. Change in mobile phase composition. (Small changes can lead to large changes in retention times.)

3. Air trapped in pump. (Retention times increase and decrease at random times.)

4. Column temperature fluctuations (especially evident in ion exchange systems).

5. Column overloading. (Retention times usually decrease as mass of solute injected on column exceeds column capacity.)

6. Sample solvent incompatible with mobile phase.

7. Column problem. (Not a common cause of erratic retention. As a column ages, retention times *gradually* decrease.)

1. Check system for loose fittings. Check pump for leaks, salt buildup, unusual noises. Change pump seals if necessary.

2. Check make-up of mobile phase. If mobile phase is machine mixed using proportioning values, hand mix and supply from one reservoir.

3. Purge air from pump head or check valves. Change pump seals if necessary. Be sure mobile phase is degassed.

4. Use reliable column oven. (Note: higher column temperatures increase column efficiency. For optimum results, heat eluant before introducing it onto column.)

5. Inject smaller volume (e.g., 10µL vs. 100µl) or inject the same volume after 1:10 or 1:100 dilutions of sample.

6. Adjust solvent. Whenever possible, inject samples in mobile phase.

7. Substitute new column of same type to confirm column as cause. Discard old column if restoration procedures fail (see page 14).

Problem No. 6: Loss of Resolution

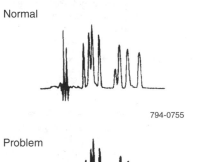

Normal

794-0755

Problem

794-0756

1. Mobile phase contaminated/ deteriorated (causing retention times and/or selectivity to change).

2. Obstructed guard or analytical column.

1. Prepare fresh mobile phase (page 2).

2. Remove guard column (if present) and attempt analysis. Replace guard column if necessary. If analytical column is obstructed, reverse and flush (page 14). If problem persists, column may be clogged with strongly retained contaminants. Use appropriate restoration procedure (Table 2, page 14). If problem still persists, change inlet frit (page 16) or replace column.

Fig. 1 *(Continued next page.)*

related to the autosampler is carryover that causes the appearance of a peak in a blank injection following injection of a high concentration of sample. Sample stability is a figure of merit that certainly has to be evaluated to avoid imprecision and lack of sensitivity.

PREVENTIVE MAINTENANCE

Many of the problems that the chromatographer encounters can be avoided if preventive maintenance is performed in routine operation, in every step of

Problem No. 7: Split Peaks

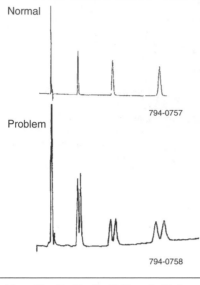

Normal

794-0757

Problem

794-0758

1. Contamination on guard or analytical column inlet.

2. Partially blocked frit.
3. Small (uneven) void at column inlet.

4. Sample solvent incompatible with mobile phase.

1. Remove guard column (if present) and attempt analysis. Replace guard column if necessary. If analytical column is obstructed, reverse and flush (page 14). If problem persists, column may be clogged with strongly retained contaminants. Use appropriate restoration procedure (Table 2, page 14). If problem still persists, inlet frit is probably (partially) plugged. Change frit (page 16) or replace column.

2. Replace frit (see above)
3. Repack top of column with pellicular particles of same bonded phase functionality. Continue using the column in reverse flow direction.

4. Adjust solvent. Whenever possible, inject samples in mobile phase.

Problem No. 8: Peaks Tail on Initial and Later Injections

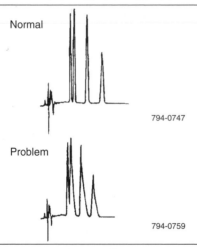

Normal

794-0747

Problem

794-0759

1. Sample reacting with active sites.

2. Wrong mobile phase pH.

3. Wrong column type.

4. Small (uneven) void at column inlet.
5. Wrong injection solvent.

1. First check column performance with standard column test mixture. If results for test mix are good, add ion pair reagent or competing base or acid modifier (page 2).

2. Adjust pH. For basic compounds, lower pH usually provides more symmetric peaks.

3. Try another column type (e.g., deactivated column for basic compounds).

4. See Problem No. 7.
5. Peaks can tail when sample is injected in stronger solvent than mobile phase. Dissolve sample in mobile phase.

Problem No. 9: Tailing Peaks

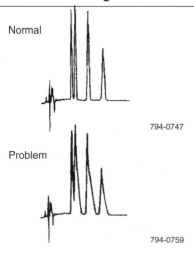

Normal

794-0747

Problem

794-0759

1. Guard or analytical column contaminated/worn out.

2. Mobile phase contaminated/deteriorated.
3. Interfering components in sample.

1. Remove guard column (if present) and attempt analysis. Replace guard column if necessary. If analytical column is source of problem, use appropriate restoration procedure (Table 2, page 14). If problem persists, replace column.

2. Check make-up of mobile phase (page 2).
3. Check column performance with standards.

Fig. 1 *(Continued next page.)*

H

Problem No. 10: Fronting Peaks

Normal

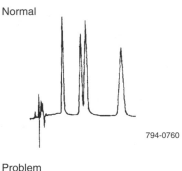

794-0760

Problem

794-0761

1. Column overloaded.

2. Sample solvent incompatible with mobile phase.

3. Shoulder or gradual baseline rise before a main peak may be another sample component.

1. Inject smaller volume (e.g., 10µL vs. 100µL). Dilute the sample 1:10 or 1:100 fold in case of mass overload.

2. Adjust solvent. Whenever possible, inject samples in mobile phase. Flush polar bonded phase column with 50 column volumes HPLC grade ethyl acetate at 2–3 times the standard flow rate, then with intermediate polarity solvent prior to analysis.

3. Increase efficiency or change selectivity of system to improve resolution. Try another column type if necessary (e.g., switch from nonpolar C18 to polar cyano phase).

Problem No. 11: Rounded Peaks

Normal

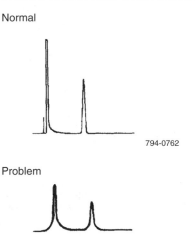

794-0762

Problem

794-0763

1. Detector operating outside linear dynamic range.
2. Recorder gain set too low.
3. Column overloaded.

4. Sample-column interaction.

5. Detector and/or recorder time constants are set too high.

1. Reduce sample volume and/or concentration.
2. Adjust gain.
3. Inject smaller volume (e.g., 10µL vs. 100µL) or 1:10 or 1:100 dilution of sample.
4. Change buffer strength, pH, or mobile phase composition. If necessary, raise column temperature or change column type. (Analysis of solute structure may help predict interaction.)
5. Reduce settings to lowest values or values at which no further improvements are seen.

Fig. 1 *(Continued next page.)*

chromatographic analysis, from sample preparation to the final step of an analysis.

Precautions regarding the use of the analytical columns, as well as the operation of each module, e.g., the pump, the detector, etc., may help keep away or postpone most of the abovementioned problems related to HPLC instrumentation.

Preventive Maintenance During Sample Preparation

As already mentioned, many problems concerning increased pressure and backpressure are due to plugged frits or system clogging from a sample matrix particulates or salt precipitation.

Preventive actions include the following:

1. Sample filtration to ensure it contains no solids.
2. Sample dissolution in the mobile phase or solvent weaker than the mobile phase.
3. Use of reduced sample volumes, whenever possible.

Mobile phase inlet filters, pre-injector and pre-column filters, saturator columns, and guard columns greatly reduce problems associated with complex separations.

Problem No. 12: Baseline Drift

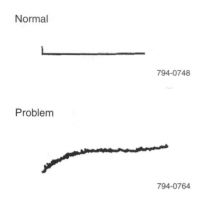

Normal

794-0748

Problem

794-0764

1. Column temperature fluctuation. (Even small changes cause cyclic baseline rise and fall. Most often affects refractive index and conductivity detectors, UV detectors at high sensitivity or in indirect photometric mode.)
2. Nonhomogeneous mobile phase. (Drift usually to higher absorbance, rather than cyclic pattern from temperature fluctuation.)
3. Contaminant or air buildup in detector cell.

4. Plugged outlet line after detector, (High pressure cracks cell window, producing noisy baseline.)
5. Mobile phase mixing problem or change in flow rate.

6. Slow column equilibration, especially when changing mobile phase.

7. Mobile phase contaminated, deteriorated, or not prepared from high quality chemicals.
8. Strongly retained materials in sample (high k') can elute as very broad peaks and appear to be a rising baseline. (Gradient analyses can aggravate problem.)
9. Detector (UV) not set at absorbance maximum but at slope of curve.

1. Control column and mobile phase temperature, use heat exchanger before detector.

2. Use HPLC grade solvents, high purity salts, and additives. Degas mobile phase before use, sparge with helium during use.
3. Flush cell with methanol or other strong solvent. If necessary, clean cell with 1 N HNO_3 (never with HCl and never use nitric acid with PEEK tubing or fittings.)
4. Unplug or replace line. Refer to detector manual to replace window.
5. Correct composition/flow rate. To avoid problem, routinely monitor composition and flow rate.
6. Flush column with intermediate strength solvent, run 10–20 column volumes of new mobile phase through column before analysis.
7. Check make-up of mobile phase (page 2).
8. Use guard column. If necessary, flush column with strong solvent between injections or periodically during analysis.
9. Change wavelength to UV absorbance maximum.

Problem No. 13: Baseline Noise (regular)

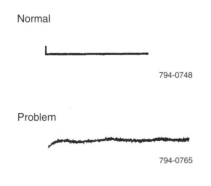

Normal

794-0748

Problem

794-0765

1. Air in mobile phase, detector cell, or pump.

2. Pump pulsations.

3. Incomplete mobile phase mixing.

4. Temperature effect (column at high temperature, detector unheated).
5. Other electronic equipment on same line.

6. Leak.

1. Degas mobile phase. Flush system to remove air from detector cell or pump.
2. Incorporate pulse damper into system.
3. Mix mobile phase by hand or use less viscous solvent.
4. Reduce differential or add heat exchanger.
5. Isolate LC, detector, recorder to determine if source of problem is external. Correct as necessary.
6. Check system for loose fittings. Check pump for leaks, salt buildup, unusual noises. Change pump seals if necessary.

Fig. 1 *(Continued next page.)*

Moreover, all solvents should be filtered through 0.2-μm filters, while particulates from samples can be removed by filtration through 0.45- or 0.2-μm syringe filters.

Bacterial Growth Precautions

Bacteria that grow in water reservoirs can yield by-products such as metabolites and dead bacteria, thus

Problem No. 14: Baseline Noise (irregular)

Normal

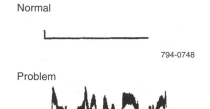

794-0748

Problem

794-0766

1. Leak.	1. Check system for loose fittings. Check pump for leaks, salt buildup, unusual noises. Change pump seals if necessary.
2. Mobile phase contaminated, deteriorated, or prepared from low quality materials.	2. Check make-up of mobile phase. (page 2).
3. Detector/recorder electronics.	3. Isolate detector and recorder electronically. Refer to instruction manual to correct problem.
4. Air trapped in system.	4. Flush system with strong solvent.
5. Air bubbles in detector.	5. Purge detector. Install back pressure regulator after detector. Check the instrument manual, particularly for RI detectors (excessive backpressure can cause the flow cell to crack).
6. Detector cell contaminated. (Even small amounts of contaminants can cause noise.)	6. Clean cell.
7. Weak detector lamp.	7. Replace lamp.
8. Column leaking silica or packing material.	8. Replace column and clean system.

Problem No. 15: Broad Peaks

Normal

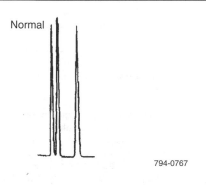

794-0767

Problem

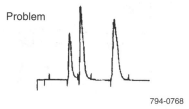

794-0768

1. Mobile phase composition changed.	1. Prepare new mobile phase.
2. Mobile phase flow rate too low.	2. Adjust flow rate.
3. Leak (especially between column and detector).	3. Check system for loose fittings. Check pump for leaks, salt buildup, and unusual noises. Change pump seals if necessary.
4. Detector settings incorrect.	4. Adjust settings.
5. Extra-column effects: a. Column overloaded	5. a. Inject smaller volume (e.g., 10μL vs. 100μL) or 1:10 and 1:100 dilutions of sample.
b. Detector response time or cell volume too large.	b. Reduce response time or use smaller cell.
c. Tubing between column and detector too long or ID too large.	c. Use as short a piece of 0.007–0.010" ID tubing as practical.
d. Recorder response time too high.	d. Reduce response time.
6. Buffer concentration too low.	6. Increase concentration.
7. Guard column contaminated/worn out.	7. Replace guard column.
8. Column contaminated/worn out.	8. Replace column with new one of same type. If new column does not provide narrow peaks, flush old column (Table 2, page 14), then retest.
9. Void at column inlet.	9. Replace column or open inlet end and fill void (page 16).
10. Peak represents two or more poorly resolved compounds.	10. Change column type to improve separation.
11. Column temperature too low.	11. Increase temperature. Do not exceed 75°C unless higher temperatures are acceptable to column manufacturer.

Fig. 1 *(Continued next page.)*

producing ghost peaks. Adding organic solvent to the aqueous part of the mobile phase, at a percentage of >20% or sodium azide 0.04% will prevent bacterial growth.

Sources of Contamination

Numerous sources of contamination should be taken into account: air particles from the laboratory

Problem No. 16: Change in Peak Height (one or more peaks)

Normal

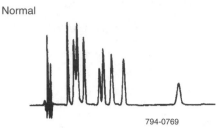

794-0769

Problem

794-0770

| 1. One or more sample components deteriorated or column activity changed. | 1. Use fresh sample or standard to confirm sample as source of problem. If some or all peaks are still smaller than expected, replace column. If new column improves analysis, try to restore the old column, following appropriate procedure (Table 2, page 14). If performance does not improve, discard old column. |

2. Leak especially between injection port and column inlet. (Retention also would change.)

2. Check system for loose fittings. Check pump for leaks, salt buildup, unusual noises. Change pump seals if necessary.

3. Inconsistent sample Volume.

3. Be sure samples are consistent. For fixed volume sample loop, use 2–3 times loop volume to ensure loop is completely filled. Be sure automatic sampler vials contain sufficient sample and no air bubbles. Check syringe-type injectors for air in systems with wash or flushing step, be sure wash solution does not precipitate sample components.

4. Detector or recorder setting changed.
5. Weak detector lamp.
6. Contamination in detector cell.

4. Check settings.
5. Replace lamp.
6. Clean cell.

Problem No. 17: Change in Selectivity

Normal

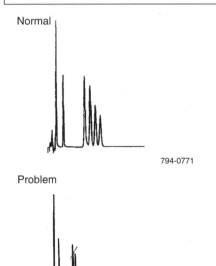

794-0771

Problem

794-0772

1. Increase of decrease solvent ionic strength, pH, or additive concentration (especially affects ionic solutes).
2. Column changed, new column has different selectivity from that of old column.

1. Check make-up of mobile phase (page 2).

2. Confirm identity of column packing. For reproducible analyses, use same column type. Establish whether change took place gradually. If so, bonded phase may have stripped. Column activity may have changed, or column may be contaminated.

3. Sample injected in incorrect solvent or excessive amount (100–200μL) of strong solvent.
4. Column temperature change.

3. Adjust solvent. Whenever possible, inject sample in mobile phase.

4. Adjust temperature. If needed, use column oven to maintain constant temperature.

Fig. 1   *(Continued next page.)*

environment, phthalates from plastic stoppers, plasticizers from plastic containers, detergents and cleaning agents from sample containers and glassware, stabilizing agents and additives from solvents, reagents and chemicals solvent impurities, purified water, and microorganisms. The use of reagent blanks (sample matrix is water in this case) and matrix blanks may help to monitor an analysis on a day-to-day basis.

The compatibility of an organic solvent with a buffer in the mobile phase must be checked as the buffer

Problem No. 18: Negative Peak(s)

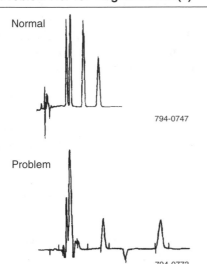

Normal

794-0747

Problem

794-0773

1. Recorder leads reversed.
2. Refractive index of solute less than that of mobile phase (RI detector).

3. Sample solvent and mobile phase differ greatly in composition (vacancy peaks).
4. Mobile phase more absorptive than sample components to UV wavelength.

1. Check polarity.
2. Use mobile phase with lower refractive index, or reverse recorder leads.
3. Adjust or change sample solvent. Dilute sample in mobile phase whenever possible.
4. a. Change polarity when using indirect UV detection, or
   b. Change UV wavelength or use mobile phase that does not adsorb chosen wavelength.

Problem No. 19: Ghost Peak

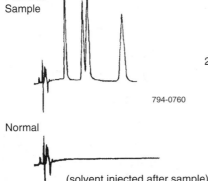

Previous
Sample

794-0760

Normal

(solvent injected after sample)
794-0774

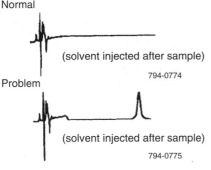

Problem

(solvent injected after sample)
794-0775

1. Contamination in injector or column.

2. Late eluting peak (usually broad) present in sample.

1. Flush injector between analyses (a good routine practice). If necessary, run strong solvent through column to remove late eluters. Include final wash step in gradient analyses, to remove strongly retained compounds.
2. a. Check sample preparation.
   b. Include (step) gradient to quickly elute component.

Fig. 1 *(Continued)*

salts can be easily precipitated when using on-line mixing, thus causing frit blockage, check valve malfunction, and other problems.

Good Column Practice: Column Protection

The most common problem associated with analytical columns is column deterioration. Deterioration may appear as poor peak shapes, split peaks, shoulders, loss of resolution, decreased retention times, and high backpressure. These symptoms indicate contaminants that have accumulated on the frit or column inlet, or there are voids, channels, or a depression in the packing bed. Deterioration is more evident in higher efficiency columns. For example, a column with 3-μm packing is more susceptible to plugging than one with 5- or 10-μm packing. Proper column protection and sample preparation are essential to prolong a column's life and obtain its best performance.

Filters and guard columns prevent particles and strongly retained compounds from accumulating on

Table 7   Problems related to sample introduction

Manual injector problems

Damaged rotor seal → Rebuild or replace valve.

Rotor too tight → Adjust rotor tension.

Valve misaligned → Adjust alignment.

Blocked loop → Replace loop.

Dirty syringe → Clean or replace syringe.

Blocked lines → Clean or replace lines.

Autoinjector problems

No air pressure → Supply proper pressure.

Rotor too tight → Adjust.

Valve misaligned → Adjust alignment.

Blockage → Clean or replace blocked portion.

Jammed mechanism → See service manual.

Faulty controller → Repair or replace controller.

Carryover problems →
1. Replace blank.
2. Change injection size.
3. Check fitting assembly and washing mechanism.
4. Use fresh wash solvent.
5. Increase wash volume.
6. Use more organic solvent in wash.
7. Adjust wash pH.
8. Change injection solvent.
9. Change needle seal, injection loop.
10. Replace valve.

the analytical column. Silica particles in a saturator column dissolve in high pH mobile phases, protecting the silica-based packing in the analytical column. The useful life of these disposable products depends on mobile phase composition, sample purity, pH value within the recommended range, etc. As these devices become contaminated or plugged with particles, pressure increases and peaks broaden or split.

Keeping records of column backpressure and important chromatographic parameters [number of theoretical plates (N), peak asymmetry factor (As), retention factor (k'), resolution factor (Rs)] helps to monitor the required column performance, while storage in the appropriate organic solvent extends column lifetime. Table 8 presents the preventive actions for column protection.

Lamp Failure

Many detectors track the number of hours the lamp is ignited. Although the lamp life may vary, the detector's meter reading can be a helpful guide for troubleshooting. Lamps can sometimes operate for more than 2000 h. To distinguish a lamp problem from air bubbles, one should stop the mobile phase flow. A lamp problem will persist when the flow is stopped, whereas, if the problem is due to the presence of a bubble, the baseline remains steady on- or off-scale. If the bubble stops in the flow cell, it causes a dramatic baseline shift, usually off-scale.

Solvent degassing and backpressure regulators after the detector minimize bubble formation. It is a helpful practice to perform blank runs every day to provide reference data that can be consulted at a later date.

Spikes in chromatograms can come from many sources, such as aging detector lamps or bubbles in the flow cell; both can be easily corrected. External electrical noise sources, such as ovens, refrigerators, cellular telephones, and fluorescent lights, and other possible noise sources such as system electronics or from external electronic sources, and laboratory power feed may be beyond a chromatographer's control.

Having an unused spare detector lamp available makes checking the problems attributed to the lamp performance an easy task.

Preventing and Solving Common Hardware Problems

Special care can be taken to avoid high backpressure in the LC system. Preventive actions to this end, as well as general preventive maintenance practices that can generally help reduce the failure rate, are summarized in Table 8.

Preventing leaks

Leaks are a common problem in HPLC analysis. Their occurrence can be minimized by avoiding interchanging hardware and fittings from different manufacturers. Incompatible fittings can be forced to fit initially, but repeated connections may eventually leak. If interchanging is unavoidable, the appropriate adapters should be used and all connections should be checked for leaks before proceeding.

Highly concentrated salts ($>0.2\,M$) and caustic mobile phases can reduce pump seal efficiency. The lifetime of injector rotor seals also depends on mobile phase conditions, e.g., operation at high pH. In some cases, prolonged use of ion pair reagents has a lubricating effect on pump pistons that may produce small leaks at the piston seal. Some seals do not perform well with certain solvents.

Table 8 Preventive maintenance for HPLC instrumentation

Preventive maintenance for HPLC columns	Preventive maintenance to avoid high backpressures	General preventive maintenance
Filter solvents before use.	Use HPLC or analytical grade buffers, freshly prepared, filtered and degassed before use.	Filtering mobile phase prevents often replacing inlet frits and check valves.
Use in-line filters for all columns, and guard columns for dirty samples; pretreat dirty samples, remove particulates.	Filter or centrifuge the sample to remove particulate matter. Turbid samples should not be injected onto column.	Degassing mobile phase prevents bubble formation.
Check samples for compatibility with mobile phase.	Set the maximum backpressure of the pumps at or slightly below the value suggested in the column instructions so that in case of increased pressure, the system will turn off before any damage occurs.	Inlet frits prevent check valve failure.
Avoid extreme column temperatures. Keep column temperature below 60°C.	Use the recommended flow rates for the column.	Rotor seal wear should not be over-tightened.
Flush column frequently/daily with strong solvent. Use stronger solvent protocol for dirty samples.	Ethanol, glycerol, high salt, urea, and cause increases in back pressure—reduce flow rate accordingly.	Sample filtration prevents injector function, frit blockage.
Keep the mobile phase pH between 3 and 7. If operating outside of this pH range use a precolumn.	When performing the separation at low temperatures, e.g., at 4°C, the recommended flow rate should be reduced by 50%.	In line filters or guard column prevent frit blockage.
Use fresh buffer solutions and aqueous mobile phases or treat them with sodium azide.	Wash the chromatography column thoroughly at the end of each separation.	Use of guard column or pre-column help avoiding void at top of column.
To prepare column for storage purge column of buffers and leave in appropriate solvent. Cap tightly.	Clean columns when needed, following the column instructions.	The use of restrictor after cell prevents bubble formation in cell.
Prevent microbial growth when storing columns by using 50–100 % organic/water mixtures or adding azide to gel filtration columns.	Maintain a log of each chromatographic run, including buffer composition, flow rate, observed backpressure (before and after sample application), sample composition, binding, and elution conditions.	Flushing buffer from LC prevents corrosive abrasive damage.
Avoid physically mishandling columns: banging, dropping, or over-tightening fittings.	Monitor column efficiency via regular column testing (i.e., acetone tests, function tests).	Keeping spare parts in lab can reduce waiting time intervals.

Instrument manufacturers' specifications should be consulted before using a pump under adverse conditions. To replace seals, refer to the maintenance section of the manufacturer's pump manual.

Unclogging the Column Frit

A clogged column frit is another common HPLC problem. To minimize this problem from the start, the use of a pre-column filter and/or guard column is recommended. To clean the clogged inlet frit, the column must be disconnected and reversed. Then it should be connected to the pump (but not to the detector), and solvent should be pumped through it at twice the standard flow rate. About 5–10 column volumes of solvent should be sufficient to remove small amounts of particulate material from the inlet frit. The performance of the cleaned column has to be evaluated by using a standard test mixture. It should be noted, however, that some columns are designated by the manufacturer as not to be used in a reverse flow mode. Of course, if the plugging prevents the column from being used anyway, then reverse flow treatment is an acceptable action.

Filling a Void/Replacing a Frit at the Column Inlet

Sometimes, neither solvent flushing nor restoration procedures restore a column's performance. If the column is proved to be the problem source, a void in the packing or a persistent obstruction on the inlet frit may exist. In this case, replacing the frit and/or topping the column with slurry of the sorbent material in a volatile solvent, e.g., acetone, to fill the void, may help.

Mobile Phase

Contamination of mobile phase reservoirs can often become a possible source of problems such as blocked frits, irregular pump performance, extra peaks, or noise in the chromatogram. Mobile phase degassing by helium sparging, sonication, and vacuum or heating (mostly in case of electrochemical detection) prior to, or during, use prevents air bubble formation.

Manually mixed mobile phases can be adequately degassed before pumping. However, if mixing takes place in the LC equipment, solvents must be simultaneously degassed. Besides bubble formation, oxygen is the primary problem interfering with detector response. In UV detectors, oxygen can cause a significant baseline rise. In fluorescence detectors, oxygen can adversely affect sensitivity by quenching sample fluorescence. Also, oxygen-free mobile phase is required for electrochemical detectors in the reductive mode.

Degassing is not required only in some cases of normal phase LC or nonaqueous SEC due to not significantly different solubility of air in the solvents used (e.g., hexane, toluene, dichloromethane, etc.).

Another case where degassing is not necessary is when high-pressure mixing is performed, where mobile phase components are mixed after the pump, at the high pressure side of the pump, so that gas is kept dissolved in solution by the high pressure.

Keeping Accurate Records

Most problems do not occur suddenly; rather, they usually develop gradually. Accurate record keeping, then, is of paramount importance in detecting and solving many gradually developing problems. When using a new column for the first time, it should be evaluated initially and at regular intervals thereafter. By keeping a written history of column efficiency, mobile phases used, lamp current, pump performance, etc., the chromatographer can monitor a system's performance.

Records also help prevent mistakes, such as introducing water into a silica column, or precipitating buffer in the system by adding too much organic solvent. Many analysts occasionally modify their HPLC systems for a variety of reasons. Reliable records are the best way to ensure that a modification does not introduce problems. For problems relating to pumps, detectors, automatic samplers, and data systems, instrument manuals provide suitable troubleshooting guides.

Referring to the maintenance and troubleshooting sections of an instrument's manual is highly recommended. Many individuals consult manuals only after a catastrophic failure and then only when all other problem-solving approaches have been exhausted. Modern HPLC systems often have self-diagnostic capabilities that help isolate the problem area within the instrument.

CONCLUSIONS

The common problems in HPLC concern pressure (high, low, or unstable, or none), leaks, quantitation (detection problems, injection problems, sample problems or data-system problems), chromatogram (peak shape), and hardware.

Troubleshooting HPLC instrumentation guides provide a systematic approach to isolating and correcting common HPLC problems. Referring to the maintenance and troubleshooting sections of instrument manuals is strongly recommended. Modern HPLC systems often have simple self-diagnostic capabilities that help isolate the problem area within the instrument.

It is good practice to run quality control samples (samples spiked at known levels) randomly among samples of unknown assay levels.

Performing a system suitability test each day provides a good assay reference.

Preventive maintenance helps the chromatographer to reduce instrument downtime, allowing for more efficiency and cost effectiveness in the HPLC laboratory.

SUGGESTED ADDITIONAL READING

Cooley, L.; Dolan, J. Reproducibility and carryover-A case study. LC GC Eur. **2001**, *11*, 209–214.

Dolan, J. Communicating with baseline. LC GC Eur. **2001**, *9*, 530–534.

Dolan, J. Attacking carryover problems. LC GC Eur. **2001**, *11*, 664–668.

Dolan, J. *Merck LC Troubleshooting Chrombook*; Merck: Rahway, NJ, USA.

Dolan, J. Problem isolation: Three more things. LC GC Int. **1993**, *6* (1), 14–17.

Dolan, J.; Snyder, L.R. *Troubleshooting LC Systems*; Humana Press: Clifton, NJ, USA.

Frasca, V. Troubleshooting high back pressure. Sci. Tools Pharm. Biotech. **1997**, *2*, 3.

HPLC Troubleshooting Technical Notes, 1st Ed.; Phenomenex Corp.: Torrance, CA, USA, 2 Feb, 1993.

http://kerouac.pharm.uky.edu/asrg/hplc/troubleshooting.html.

McDowall, R.D. Where did that peak come from?. LC GC Int. **1997**, *6*, 358–359.

Nelson, M.; Dolan, J. UV detection noise. LC GC Int. **1995**, 64–70.

Supelco HPLC troubleshooting guide. Bulletin **1999**, *826D*.

http://www.chromatography.co.uk/TECHNIQS/HPLC/trouble1.html.

http://www.chromtech.com.

http://www.dq.fct.unl.pt/QOF/hplcts.html.

http://www.fortunecity.de/lindenpark/lilienthal/8/trouble.html.

http://www.hplc1.com/shodex/english/dd.htm.

http://www.metachem.com/tech/troubleshoot.

http://www.rheodyne.com/tsguide/tsg.html.

H

Hybrid Micellar Mobile Phases

M. C. García-Alvarez-Coque
J. R. Torres-Lapasio
Departamento de Química Analítica, Universidad de Valencia, Valencia, Burjassot, Spain

INTRODUCTION

The first report on the analytical use of an aqueous solution of a surfactant, above its critical micellar concentration (CMC), as mobile phase in reversed-phase liquid chromatography (RPLC) was published in 1980.[1] The technique, named micellar liquid chromatography (MLC), is an interesting example of the modification of the chromatographic behavior taking advantage of secondary equilibria to vary both retention and selectivity.

Most MLC procedures use micelles of the anionic surfactant sodium dodecyl sulfate (SDS). Other useful surfactants include the cationic cetyltrimethylammonium bromide or chloride (CTAB or CTAC) and the nonionic Brij-35.[2] The separations are usually carried out in C_{18} or C_8 columns.

Inside the column, solutes are affected by the presence of micelles in the mobile phase and by the nature of the alkyl-bonded stationary phase, which is coated with monomers of surfactant (Fig. 1). As a consequence, at least two partition equilibria can affect the retention behavior. In the mobile phase, solutes can remain in the bulk water, be associated to the free surfactant monomers or micelle surface, be inserted into the micelle palisade layer, or penetrate into the micelle core. The surface of the surfactant-modified stationary phase is micelle-like and can give rise to similar interactions with the solutes, which are mainly hydrophobic in nature. With ionic surfactants, the charged heads of the surfactant in micelles and monomers adsorbed on the stationary phase are in contact with the polar solution, producing additional electrostatic interactions with charged solutes. Finally, the association of solutes with the nonmodified bonded stationary phase and free silanol groups still exists.

The most serious limitations of pure micellar solutions are their weak elution strength and poor efficiencies. As early as 1983,[3] the addition of a small percentage of 1-propanol was found to enhance the efficiencies and decrease the asymmetries of chromatographic peaks. Later, the term "hybrid micellar mobile phases" was given to the ternary eluents of water/organic solvent/micelles. Although 1-propanol is still the most frequently used additive, other alcohols (methanol, ethanol, 1-butanol, and 1-pentanol) and

organic solvents common in conventional RPLC (acetonitrile and tetrahydrofuran) have also been used. It should be noted that micellar solutions increase the solubility of butanol and pentanol in water to reach concentration levels considered useful in chromatography.

The concentration of organic solvent should be low enough to make the existence of micelles possible. Such maximal amount depends on the type of surfactant and organic solvent, and is usually unknown. For SDS, the maximal volume fractions of acetonitrile, propanol, butanol, and pentanol that seem to guarantee the presence of micelles are 20%, 15%, 10%, and 7% (v/v), respectively. However, analytical reports where authors claim the use of hybrid micellar mobile phases and these maximal values are exceeded—micelles do not exist—are not unusual. In such conditions, the system bears closer resemblance to an aqueous–organic system, although the surfactant monomers still affect the retention and efficiencies.

NATURE OF THE MOBILE AND STATIONARY PHASES

The presence of a small amount of an organic solvent in the micellar mobile phase produces changes in the micellization process which depend on the nature of the additive. A progressive increase in the CMC of SDS (8.2×10^{-3} M in water at 25°C) is observed by adding methanol and acetonitrile; the reverse is observed when ethanol, propanol, butanol, and pentanol are added. As an example, the CMC of SDS in the presence of 4% (v/v) organic solvent is 8.7×10^{-3} M (methanol), 9.2×10^{-3} M (acetonitrile), 7.4×10^{-3} M (ethanol), 5.9×10^{-3} M (propanol), 2.7×10^{-3} M (butanol), and 2.0×10^{-3} M (pentanol).[4]

Methanol, which has the shortest carbon chain, is more polar and soluble than other alcohols. SDS monomers are more easily solvated in an aqueous–methanol medium. This inhibits them from interacting and forming micelles. A similar behavior is expected for acetonitrile. Ethanol and propanol, which are also miscible with water, remain outside the micelles, dissolved in the bulk liquid, but interact with the micelle surface. Repulsion among the ionic heads of surfactant

Encyclopedia of Chromatography DOI: 10.1081/E-ECHR-120013363

Fig. 1 Solute–micelle and solute–stationary phase interactions in hybrid micellar mobile phases (see text for meaning of equilibrium constants).

monomers is reduced in the presence of these two alcohols, thus aiding the formation of micelles.

As the length of the alcohol alkyl chain increases, its affinity for the SDS micelle is enhanced. Butanol and pentanol are inserted in the intermonomer spaces of the micelle palisade (aligned with the surfactant molecules), the polar hydroxyl group orientated toward the Stern layer and the alkyl chains located in the nonpolar micelle core. A swollen mixed micelle is thus formed. Such micelles are geometrically hindered to allocate additional surfactant monomers, which is translated into a CMC reduction. However, above 4% butanol and 1.5% pentanol, the decay rate in the CMC values changes. An explanation for this behavior is that above a given concentration, the amount of alcohol entering the palisade is not significant, and the excess is solubilized in the core of the swollen micelle. Further additions of alcohol lead to a dramatic change in the mobile phase microstructure, yielding a microemulsion.

Organic solvents also induce changes in the properties of surfactant-coated stationary phases, such as polarity, surface area, or pore volume. Several studies have demonstrated that n-alcohols interpenetrate the C_{18}-bonded alkyl chains to form a single monolayer, structurally similar to an opened micelle (i.e., the hydroxyl group orientated toward the aqueous phase). The competition between organic solvent and surfactant molecules for the active sites on the column explains the reduction of adsorbed surfactant at increasing concentration of organic solvent in the mobile phase. For ionic surfactants, this reduction is linear with slopes that depend on the strength of

the solvent (methanol < ethanol < propanol < butanol < pentanol). Substitution of alcohol by some surfactant molecules on the stationary phase increases its effective polarity and decreases the retention time. With hybrid micellar mobile phases, the stationary phase resembles a solvated phase in aqueous–organic systems more, rather than in pure micellar systems.[5]

Solute retention varies with the concentration of propanol, butanol, and pentanol in the mobile phase in the same way as CMC does. This means that the collateral effects which change the CMC in an organic–micellar system are, at least partially, those that induce shorter retention with hybrid mobile phases: the modification of bulk water and micelle. As noted, another important factor that affects retention is the modification of the structure of the stationary phase. The analogous effects on both microenvironments (micelle and stationary phase) are evident in the parallel variation of solute–micelle and solute–stationary phase partition coefficients, as the concentration of organic solvent changes.[6]

In the hybrid system, solute partition equilibria are significantly displaced away from the micelle and the stationary phase toward the bulk aqueous–organic phase, which is more nonpolar (Fig. 1). However, as long as the integrity of micelles is maintained, the addition of organic solvent to micellar mobile phases will not create an aqueous–organic system. The chromatographic behavior of alkyl homologous series has been used to demonstrate that the separation mechanism in hybrid MLC is more similar to pure MLC than to conventional RPLC. In aqueous–organic systems, the retention of homologue compounds is described as follows:

$$\log\,k \;=\; \log\,\alpha(CH_2)n_C \;+\; \log\,\beta \qquad (1)$$

where n_C is the number of carbon atoms in the homologue, $\alpha(CH_2)$ is the nonspecific selectivity of a methylene group, and β is the contribution of the series functional group to the retention. In contrast to this behavior, a linear relationship between k and n_C is observed with either pure or hybrid micellar mobile phases.

RETENTION BEHAVIOR

In pure micellar mobile phases, retention is described by a hyperbolic relationship.[7]

$$k \;=\; \frac{\phi P_{AS}}{1 \;+\; \upsilon(P_{AM}\;-\;1)[M]} \;=\; \frac{K_{AS}}{1\;+\;K_{AM}[M]} \qquad (2)$$

where k is the retention factor, ϕ is the phase ratio,

P_{AS} and P_{AM} are the solute–stationary phase and solute–micelle partition coefficients, v is the specific volume of surfactant monomers, and $[M]$ is the concentration of surfactant forming micelles (total surfactant concentration minus the CMC). The expression with K_{AS} and K_{AM} is often used for simplicity.

The same model is valid for hybrid micellar mobile phases at fixed concentration of organic solvent, although both constants, K_{AS} and K_{AM}, decrease when the modifier concentration increases, especially for nonpolar solutes. An extended model, including the effect of changes in organic solvent concentration, has also been proposed:[7,8]

$$k = \frac{K_{AS}\frac{1 + K_{SD}\varphi}{1 + K_{AD}\varphi}}{1 + K_{AM}\frac{1 + K_{MD}\varphi}{1 + K_{AD}\varphi}[M]} \qquad (3)$$

where φ is the volume fraction of organic solvent; K_{AS} and K_{AM} are the partition constants in pure micellar eluents; and K_{AD}, K_{MD}, and K_{SD} measure the relative variations in solute concentration in bulk water, micelle, and stationary phase, respectively, in the presence of organic solvent, taking the pure micellar solution as reference. These constants increase with the organic solvent strength. For polar and moderately polar solutes eluted with propanol, K_{SD} can be disregarded and Eq. (3) is simplified as follows:

$$\frac{1}{k} = c_0 + c_1[M] + c_2\varphi + c_{12}[M]\varphi \qquad (4)$$

Chromatographic optimizations are usually performed at a preselected pH. However, a simultaneous consideration of the three factors (i.e., surfactant, organic solvent, and pH) expands the separation capability for some problems. The retention can be predicted from (see below) where K_{AS}, K_{AM}, K_{AD}, K_{MD}, and K_{SD} are the equilibrium constants associated to the basic species; and K_{HAS}, K_{HAM}, K_{HAD}, K_{HMD}, and K_{HSD} correspond to the acidic species. K_H is the protonation constant in the aqueous–organic bulk solvent, and $[H]$ is the proton concentration.

Eqs. (2–5) give accurate predictions of the retention with several types of surfactant (anionic, cationic, and nonionic) and organic solvent (alcohols and acetonitrile), and solutes of different polarity and charge, with errors usually below 3–5%.

ELUTION STRENGTH

One of the most serious problems of pure micellar eluents is their weak elution strengths. Shorter retention times are obtained by increasing the surfactant concentration, but the chromatographic efficiency usually deteriorates. The use of ultrawide pore or shorter chain-length-bonded stationary phases constitutes other alternatives, but the most practical solution seems to be the addition of a small amount of an organic solvent to the mobile phase. This strategy may result in adequate elution strengths, together with improved chromatographic efficiencies and selectivities, which will produce a favorable effect on both the resolution and the length of analysis time. However, the addition of an organic solvent is unsuitable for low retained solutes because the retention level may fall below the optimal retention range. In other instances, adequate retention times are achieved, but the efficiencies remain low.

The elution strength of the organic solvent in MLC has been described according to Eq. (3).[9] However, a simple logarithmic relationship (similar to that used in conventional RPLC) is customary:[10]

$$\log k = \log k_0 - S\varphi \qquad (6)$$

where S is the sensitivity of solute retention to changes in the volume fraction of organic solvent, and $\log k_0$ is the retention in a pure aqueous micellar solution. A similar equation has been used for the surfactant:

$$\log k = \log k_0' - S'[M] \qquad (7)$$

If Eq. (2) is rewritten in the logarithmic form:

$$\log k = \log K_{AS} - \log(1 + K_{AM}[M]) \qquad (8)$$

and linear relationships are assumed between $\log K_{AS}$ and $\log(1 + K_{AM}[M])$ with φ, the following results:

$$S = S_s - S_m \qquad (9)$$

S_s and S_m represent the sensitivity of variations in solute partitioning from bulk solvent into the stationary phase and micelles, respectively, with changes in φ. The negative sign in Eq. (9) reflects the competing

$$k = \frac{K_{AS}\dfrac{1 + K_{SD}\varphi}{1 + K_{AD}\varphi} + K_{HAS}\dfrac{1 + K_{HSD}\varphi}{1 + K_{HAD}\varphi}K_H[H]}{\left(1 + K_{AM}\dfrac{1 + K_{MD}\varphi}{1 + K_{AD}\varphi}[M]\right) + \left(1 + K_{HAM}\dfrac{1 + K_{HMD}\varphi}{1 + K_{HAD}\varphi}[M]K_H[H]\right)} \qquad (5)$$

nature of the two partitioning equilibria. In the absence of micelles, $S_m = 0$ and $S = S_s$, which represents the solvent strength parameter in conventional aqueous–organic RPLC. Eq. (9) also shows that the elution strength in hybrid micellar systems will be generally smaller. The magnitude of the reduction depends on the interaction degree of solutes and organic solvents with micelles. Also, the range of elution strength values for solutes of diverse polarity is narrower than in conventional RPLC.

Longer-length alcohols, which are more hydrophobic, are able to shorten the retention in a larger extent. Retention times are smaller for micellar solutions containing greater amounts of additive, but even relatively low amounts can produce dramatic effects. These are attenuated as the surfactant concentration is increased. Finally, decrease in retention times is more intense for more hydrophobic solutes. Elution strength is usually greater in alcohols than in surfactants. However, changes in the retention of positively charged solutes— such as catecholamines and β-blockers, eluted with SDS micellar mobile phases—are larger when the concentration of surfactant is varied due to the high affinity of these compounds for the negatively charged micelles.

The elution strength of hybrid micellar mobile phases was measured for a number of organic additives (alcohols, alkane diols, alkanes, alkylnitriles, and dipolar aprotic solvents, such as dimethyl sulfoxide and dioxane) added to micellar SDS, CTAC, and Brij-35.[11] Benzene and 2-ethylanthraquinone were used as probe compounds. The presence of alcohols, alkane diols, alkylnitriles, and dipolar aprotic solvents produced a diminution of the retention times, reaching remarkable levels for the most hydrophobic compound (2-ethylanthraquinone). The observed elution strength order roughly paralleled the octanol–water partition coefficients of the additives, $P_{o/w}{}^a$ (Fig. 2), or their ability to bind to micelles, $K_{AM}{}^a$. In contrast, alkanes (pentane, hexane, and cyclohexane) had relatively little effect on the retention.

Selection of the most suitable organic solvent should consider the polarity of analytes and their association with the surfactant. Solute log $P_{o/w}$ values can be used, in most cases, as a guide to make this decision.[12] Thus, with SDS as surfactant, a low concentration of propanol (~1% v/v) is useful to separate compounds with log $P_{o/w} < 1$ as amino acids. A greater concentration of this solvent (~5% to 7%) is needed for compounds in the range $1 < \log P_{o/w} < 2$ as diuretics and sulfonamides. Pentanol (~2% to 6%) is more convenient for low polar compounds with log $P_{o/w} > 3$ as steroids. For basic compounds, such as phenethylamines with $0 < \log P_{o/w} < 2$ and β-blockers with $1 < \log P_{o/w} < 3$, propanol is too weak due to the strong electrostatic interaction

between the positively charged solutes and the anionic surfactant adsorbed on the stationary phase. In this case, a high concentration of propanol (~15%), or preferably, a moderate concentration of butanol (<10%) should be used.

When complex mixtures of compounds in a wide range of polarities are eluted with hybrid micellar mobile phases at increasing volume fraction of organic solvent, the retention times of late-eluting compounds are reduced in a larger extent than earlier peaks. The effect is similar to that obtained with gradient elution in conventional RPLC but using isocratic elution and low amounts of organic solvent.

A

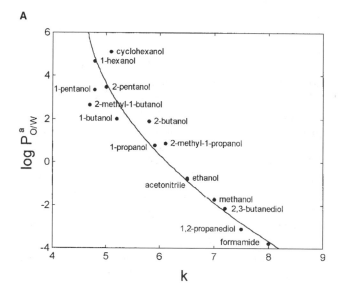

B

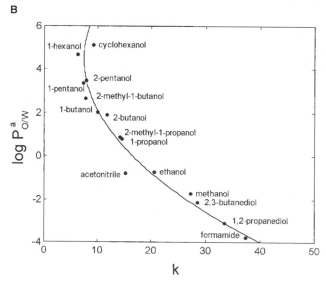

Fig. 2 Correlation between octanol–water partition coefficients of the organic solvents (log $P_{o/w}{}^a$), and the retention factors of: (A) benzene and (B) 2-ethylanthraquinone in hybrid SDS micellar mobile phases. The concentration of surfactant and organic solvent was 0.285 M and 5% (v/v), respectively. (From Ref. [11].)

EFFICIENCY

Low efficiencies, observed especially for highly hydrophobic solutes eluted with pure micellar mobile phases, hindered the initial development of MLC. The addition of a small amount of organic solvent at least partially remedied this situation. Efficiency enhancements—dramatic in some cases—have been reported for solutes eluted with SDS mobile phases in the presence of alcohols, alkane diols, alkylnitriles, and dipolar aprotic solvents. Concomitant with the enhanced efficiencies, improvements in peak asymmetry are observed. The reason for these effects is an increased solute mass transfer between micelles/stationary phase and aqueous phase due to the greater solute–micelle exchange rate constants, lower stationary phase viscosity, and smaller amount of adsorbed surfactant.[5,11]

Plots of $P_{o/w}{}^a$ or $K_{AM}{}^a$ vs. plate counts for benzene and 2-ethylanthraquinone eluted with micellar SDS, in the presence of several alkanols and alkane diols, show an initial steep increase in efficiency, after which an approximately constant value is reached (Fig. 3). Among the alcohols, maximal efficiency for benzene and 2-ethylanthraquinone is obtained with the butanols and the pentanols, with enhancement factors of ~2.5 and ~25 (compared to pure SDS), respectively. However, final efficiencies for the latter compound are much lower compared to that for benzene. Dipolar aprotic modifiers (acetonitrile or dimethylsulfoxide) appear to be somewhat more effective in enhancing efficiencies than alcohols with comparable $P_{o/w}{}^a$. Some recent work has shown the advantage of using acetonitrile as additive in MLC for the analysis of sulfonamides, tetracyclines, and the most polar steroids.

Chromatographic efficiency seems to be linked to the additive-to-surfactant concentration ratio in the micellar mobile phase. The plate numbers increase with this ratio but reach a maximum level (e.g., at pentanol/SDS = 6 and acetonitrile/CTAC = 12).[5] The organic solvent/surfactant ratio affects the exchange rates of the solute between micelle/stationary and aqueous phases. It also controls the extent of the surfactant coverage and the fluidity of the organic layer on the stationary phase.

Satisfactory results are obtained with compounds of different nature. Of particular interest is the case of basic compounds, such as phenethylamines and β-blockers, which experience large efficiency enhancements in SDS systems.[12] This makes the use of special columns less necessary. The surfactant layer adsorbed on the column prevents the interaction of basic compounds with free silanol groups, which accounts for the low efficiencies observed with conventional columns in aqueous–organic RPLC. For acidic compounds such as sulfonamides, the efficiencies are

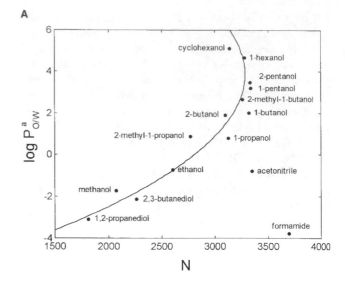

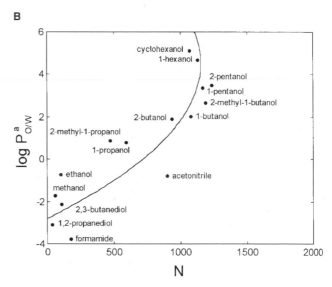

Fig. 3 Correlation between octanol–water partition coefficients of the organic solvents and the efficiencies (plate counts) of: (A) benzene and (B) 2-ethylanthraquinone in hybrid SDS micellar mobile phases. See details in Fig. 2.

comparable with both MLC and conventional RPLC, but for low polar compounds such as steroids, efficiencies are comparably poorer in MLC. However, in this technique, low polar steroids are eluted at sufficiently short retention times via a small amount of a strong organic solvent. Meanwhile, in conventional RPLC, a high amount of organic solvent is needed to decrease the retention time to practical values.

SELECTIVITY

The rate of change in retention at varying surfactant and organic solvent concentration depends on the solute charge and polarity, as well as on the nature of both modifiers. The existence of different intermolecular forces governing the retention, the magnitude of which is altered by each modifier, explains this behavior. The more hydrophobic the solute, the more intense is the effect of the organic solvent on the elution strength (i.e., for a given increment in the concentration of an organic solvent, the partition coefficients for more hydrophobic solutes decrease more than for hydrophilic solutes). Hence, the selectivity changes. Several examples have been published where a hybrid mobile phase was able to achieve an acceptable separation—impossible with pure micellar eluents—within much shorter analysis time.

The presence of micelles has a great influence on the chromatographic selectivity of organic solvents. As a result, the solvent classification established by Snyder in conventional RPLC, based on selectivity, does not seem to be valid in MLC with hybrid mobile phases.[10] Thus, for example, according to Snyder, 1-propanol and 1-butanol belong to the same selectivity group and, consequently, although the elution strength is greater when using butanol, both alcohols yield similar chromatographic selectivities in aqueous–organic mobile phases. Because in MLC organic solvents not only associate with micelles but also compete with them to interact with the solutes, the selectivities for 1-propanol and 1-butanol vary. The different impact of micelles of different surfactants on the selectivity is another factor that should be considered.

In conventional RPLC, a systematic decrease in selectivity usually occurs when the volume fraction of organic solvent is increased.[10] In contrast, in the presence of micelles, the selectivity may increase, decrease, or remain unchanged with the addition of both surfactant and organic solvent. Although the elution strength increases with the concentration of both micelle and organic solvent, their effect on the selectivity can be quite different, even opposite.

The behavior of solutes depends on the relative change of their apparent solute–micelle association constants and partitioning constants into the stationary phase. These usually decrease with an increase in volume fraction of modifier. The magnitude of the diminution is, however, not equal for various solutes. Improving the resolution and simultaneously reducing the separation time are the two more important goals in most optimization strategies. Simultaneous enhancements in selectivity and elution strength can lead to better separations in shorter analysis times.

Method development in MLC requires first the selection of a surfactant/organic solvent system. The second step concerns the optimization of the selectivity. The separation can be improved by varying only one factor, or modifying one after optimizing the other (i.e., proton, surfactant, and organic solvent concentration). However, in operations like this, the best

separation conditions can be easily missed. Reliable optimal conditions can be obtained only when all factors are simultaneously taken into account. This requires the use of an interpretive optimization strategy (i.e., based on the description of the retention behavior and peak shape of solutes). In this task, the product of free peak areas or purities has proved to be the best optimization criterion.[13] An interactive computer program is available to obtain the best separation conditions.[14]

The experimental domain that should be examined to obtain the best separation should cover a region of concentrations of surfactant and organic solvent as wide as possible, with some restrictions. The lower concentration of surfactant should be well above the CMC (e.g., 0.05 M for SDS and 0.04 M for CTAB). Surfactant concentrations exceeding 0.20 M are not convenient due to the high viscosity of the mobile phase and degradation of the efficiencies. Polarity of the solutes should be considered for the choice of the right organic solvent (nature and concentration range). This concentration must be low enough to guarantee the existence of micelles. Typically, resolution diagrams are complex, with several local maxima.

The optimization methodologies currently developed allow the separation of complex mixtures and the comparison of chromatographic performance in different situations (i.e., different organic solvents or columns). Analyses are often possible via a single organic solvent, but a mixture of solutes showing a wide range of polarities, such as steroids with log $P_{o/w} = 3-8$, may require two mobile phases, each with a particular organic solvent. In this example, acetonitrile permits the appropriate elution of the least retained steroids with higher efficiencies, whereas pentanol is used to elute highly nonpolar steroids with lower retention times.

CONCLUSIONS

Pure micellar mobile phases are certainly attractive, considering the increasing restriction in the use of organic solvents in laboratories. For this reason, hybrid micellar mobile phases were first belittled because some of MLC's appeal was considered to be lost. However, most reported analytical procedures in MLC utilize these eluents. The main reason is that in most cases, retention of solutes with pure micellar mobile phases is too high, which necessitates the addition of an organic solvent to achieve adequate retention times. Peak shape and symmetry are also improved. Procedures that use the hybrid eluents still have the advantage of requiring significantly smaller amounts of organic solvent with respect to conventional RPLC. In MLC, the organic solvent is also highly retained in micellar solution, which reduces the risk of evaporation. The mobile phases can therefore be kept stable for a long time. The toxicity, flammability, environmental impact, and cost of RPLC are consequently reduced. Finally, sample preparation is expedited due to the solubilization capability of micellar media, avoiding laborious steps to separate the matrix, previously performed to sample injection. All these features have allowed the development of multiple applications that are highly competitive against conventional RPLC.

ACKNOWLEDGMENTS

This work was supported by Project BQU2001-3047 (Ministerio de Ciencia y Tecnología, MCYT, Spain and FEDER funds). JRTL thanks the MCYT for a Ramón y Cajal position.

REFERENCES

1. Armstrong, D.W.; Henry, S.J. Use of an aqueous mobile phase for separation of phenols and PAHs via HPLC. J. Liq. Chromatogr. **1980**, *3*, 657–662.
2. Berthod, A.; García-Alvarez-Coque, C. *Micellar Liquid Chromatography*; Cazes, J., Ed.; Marcel Dekker, Inc.: New York, 2000.
3. Dorsey, J.G.; DeEchegaray, M.T.; Landy, J.S. Efficiency enhancement in micellar liquid chromatography. Anal. Chem. **1983**, *55*, 924–928.
4. Lopez-Grio, S.; Baeza-Baeza, J.J.; García-Alvarez-Coque, M.C. Influence of the addition of modifiers on solute–micelle interaction in hybrid micellar liquid chromatography. Chromatographia **1998**, *48*, 655–663.
5. Berthod, A. Causes and remediation of reduced efficiency in micellar liquid chromatography. J. Chromatogr. A. **1997**, *780*, 191–206.
6. Marina, M.L.; García, M.A. Evaluation of distribution coefficients in micellar liquid chromatography. J. Chromatogr. A. **1997**, *780*, 103–116.
7. García-Alvarez-Coque, M.C.; Torres-Lapasio, J.R.; Baeza-Baeza, J.J. Modelling of retention behaviour of solutes in micellar liquid chromatography. A review. J. Chromatogr. A. **1997**, *780*, 129–148.
8. Lopez-Grio, S.; Baeza-Baeza, J.J.; García-Alvarez-Coque, M.C. Modelling of the elution behaviour in hybrid micellar eluents with different organic modifiers. Anal. Chim. Acta **1999**, *381*, 275–285.
9. Lopez-Grio, S.; Baeza-Baeza, J.J.; García-Alvarez-Coque, M.C. Evaluation of the elution strength of organic modifier, and surfactant in micellar mobile phases. J. Liq. Chromatogr. & Relat. Technol. **2001**, *24*, 2765–2783.
10. Kord, A.S.; Khaledi, M.G. Controlling solvent strength and selectivity in MLC: role of organic modifiers and micelles. Anal. Chem. **1992**, *64*, 1894–1900.
11. Lopez-Grio, S.; García-Alvarez-Coque, M.C.; Hinze, W.L.; Quina, F.H.; Berthod, A. Effect of a variety of organic additives on retention, and efficiency in micellar liquid chromatography. Anal. Chem. **2000**, *72*, 4826–4835.
12. Caballero, R.D.; Ruiz-Angel, M.J.; Simo-Alfonso, E.; García-Alvarez-Coque, M.C. Micellar liquid chromatography: a suitable technique for screening analysis. J. Chromatogr. A. **2002**, *947*, 31–45. *in press.*
13. Carda-Broch, S.; Torres-Lapasio, J.R.; García-Alvarez-Coque, M.C. Evaluation of several global resolution functions for liquid chromatography. Anal. Chim. Acta **1999**, *396*, 61–74.
14. Torres-Lapasio, J.R. *Michrom Software*; Cazes, J., Ed.; Marcel Dekker, Inc.: New York, 2000.

Hydrodynamic Equilibrium in Countercurrent Chromatography

Petr S. Fedotov
Boris Ya. Spivakov
Vernadsky Institute of Geochemistry and Analytical Chemistry, Russian Academy of Sciences, Moscow, Russia

INTRODUCTION

In all cases, countercurrent chromatography (CCC) utilizes a hydrodynamic behavior of two immiscible liquid phases through a tubular column space which is free of a solid support matrix. The most versatile form of CCC, called the hydrodynamic equilibrium system, applies a rotating coil in an acceleration field (either in the unit gravity or in the centrifuge force field). Two immiscible liquid phases confined in such a coil distribute themselves along the length of the coil to form various patterns of hydrodynamic equilibrium.[1]

DISCUSSION

According to the hypothesis proposed by Ito,[2] the multitude of hydrodynamic phenomena observed in the rotating coils can be attributed to the following types of liquid distribution.

1. The basic hydrodynamic equilibrium (the two liquid phases are evenly distributed from one end of the coil, called the head, and any excess of either phase is accumulated at the other end, called the tail). Here, the tail-head relationship of the rotating coil is defined by the direction of the Archimedean screw force which drives all objects toward the head of the coil.
2. The unilateral hydrodynamic equilibrium [the two solvent phases are unilaterally distributed along the length of the coil, one phase (head phase) entirely occupying the head side and the other phase (tail phase) the tail side of the coil]. The head phase can be the lighter or the heavier phase and also can be the aqueous or the nonaqueous phase, depending on the physical properties of the liquid system and the applied experimental conditions. This type of equilibrium may also be called bilateral, indicating the distribution of the one phase on the head side and the other phase on the tail side.[3]

To illustrate the process of establishing the hydrodynamic equilibrium, it is worthwhile to begin with the distribution of two immiscible solvent phases in the "closed" coil, simply rotated around the horizontal axis in the unit gravitational field. The coil is filled with equal volumes of the lighter and heavier phase and then sealed at both ends. At a slow rotation of 10–20 revolutions per minute (rpm), two liquid phases are evenly distributed in the coil (basic hydrodynamic equilibrium) due to the Archimedean screw force. As the rotational speed increases, the heavier phase quickly occupies more space on the head side of the coil and, at the critical speed range of 60–100 rpm, the two phases are completely separated along the length of the coil, with the heavier phase on the head side and the lighter phase on the tail side (unilateral hydrodynamic equilibrium).

After this critical speed range, the amount of the heavier phase on the head side decreases sharply, reaching substantially below the 50% level at about 160 rpm. Further increase of the rotational speed again distributes the two phases fairly evenly throughout the coil, apparently due to the strong radial centrifugal force field produced by the rotation of the coil. The phase distribution described can be observed in many solvent systems [chloroform-acetic acid-water (2/2/1), hexane-methanol, *n*-butanol-water, etc.], glass coils [10–20 mm inner diameter (i.d.)] with different helical diameters (5–20 cm) being applied.

As a first approximation, the complex hydrodynamic phenomenon taking place in the rotating coil may be explained by the interplay between two force components acting on the fluid. The tangential force component (F_t) generates the Archimedean screw effect to move two phases toward the head of the coil, and the radial force component (F_r) which acts against the Archimedean force. The critical speed range is the most interesting. An increase of the rotational speed up

Encyclopedia of Chromatography DOI: 10.1081/E-ECHR-120040012

to 60–100 rpm alters the balance of the hydrodynamic equilibrium by an enhanced radial centrifugal force field that increases the net force field acting at the bottom of the coil and decreases that acting at the top. Under this asymmetrical force distribution, the movement of the heavier phase toward the head is accelerated, whereas the movement of the lighter phase toward the head is retarded. This results in a unilateral hydrodynamic phase distribution in the rotating coil.

The hydrodynamic equilibrium condition may be used for performing CCC as follows. First, the coil is completely filled with the stationary phase, either the lighter or the heavier phase, and the other phase is introduced from the head end of the coil while the coil is rotated around its axis. Then, the two liquid phases establish equilibrium in each turn of the coil and the mobile phase finally emerges from the tail end of the coil, leaving some amount of the stationary phase permanently in the coil. Solutes locally introduced at the head of the coil are subjected to a partition process between two phases and eluted in order of their partition coefficients. In general, higher retention of the stationary phase significantly improves the peak resolution. Consequently, the unilateral hydrodynamic equilibrium condition provides a great advantage in performing CCC, because the system permits retention of a large amount of stationary phase in the coil if the lighter phase is eluted in a normal mode (head-to-tail direction) or the heavier phase in a reversed mode (tail-to-head direction).

In general, the retention of the stationary phase in the coil rotated in the unit gravity field entirely relies on relatively weak Archimedean screw force. In this situation, application of a high flow rate of the mobile phase would cause a depletion of the stationary phase from the column. This problem can be solved by the utilization of synchronous planetary centrifuges, free of rotary seals, which enable one to increase the rotational speed and, consequently, enhance the Archimedean screw force. The seal-free principle can be applied to various types of synchronous planetary motion. In all cases, the holder revolves around the centrifuge axis and simultaneously rotates about its own axis at the same angular velocity ω.

When the coil is mounted coaxially around the holder, which revolves around the central axis of the device and counterrotates about its own axis, two axis being parallel (Type I), two solvent phases are distributed along the length of the coil according to the basic hydrodynamic equilibrium. It does not favor the stationary-phase retention. Another, similar planetary motion, except that the holder revolves around the central axis of the centrifuge and rotates about its own axis in the same direction (Type J), produces, regardless of the rotational speed, a totally different phase-distribution pattern which is typical for the unilateral hydrodynamic equilibrium. The unilateral distribution can also be attained in the coaxially mounted coils in cross-axis planetary centrifuges.[4] It is important to note that all the planetary motions providing the unilateral distribution form an asymmetrical centrifuge force field that closely resembles that observed in the coil rotating at the critical speed in the unit gravity.

The unilateral hydrodynamic equilibrium conditions provide the basis for high-speed CCC (HSCCC, $\omega = 800$ rpm or more) which has mainly gained acceptance for CCC separations. The stroboscopic observation on two-phase flow through the running spiral column of a Type J system reveals the following pattern. When the lower phase (chloroform) is eluted through the stationary lighter phase (water) from the head toward the tail of the spiral column, a large volume of the stationary phase is retained in the column and the spiral column is divided into two distinct zones: the mixing zone in about one-fourth of the area near the center of the centrifuge and the settling zone showing a linear interface between the two phases in the rest of the area. The mixing zone is always fixed at the vicinity of the central axis of the centrifuge while the spiral column undergoes the planetary motion. In other words, the mixing zone in each loop is traveling through the spiral column toward the head at a rate equal to the column rotation. Consequently, at any portion of the column, the two liquid phases are subjected to a typical partition process of repetitive mixing and settling at a high frequency, over 13 times per second at 800 rpm of column revolution, while the mobile phase is being continuously pumped through the stationary phase.[3]

At a first approximation, the hydrodynamic phenomenon observed also may be explained by the interplay between two force components acting on the fluid. At the distal portion of the coil, both the strong radial force field and the reduced relative flow of the two phases establish a clear and stable interface between the two liquid phases. At the proximal portion of the spiral column, where the strength of the radial-force component is minimized, the effect of the Archimedean screw force becomes visualized as agitation at the interface caused by the relative movement of two liquid layers.[2]

It should be noted that the centrifuge force field acts on the fluid in the rotated coil in parallel with other forces of different nature[5]:

F_A, buoyancy force due to the difference between the stationary and mobile phases

F_i, inertial force caused by coil motion, comprises components of centrifugal force field

F_η, viscosity force due to the overflow of the stationary phase along the coil tube walls

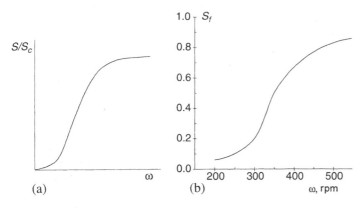

Fig. 1 (a) Theoretical ω_dependence of S/S_c (b) Experimental ω dependence of S_f for the n_decane–water system. Planetary centrifuge of Type J; $\beta = 0.37$; flow rate = 1 mL/min.

F_γ, interfacial tension force

F_W, adhesion force

F_h, hydraulic resistance force caused by moving of two immiscible phases relative to each other

HYDRODYNAMIC EQUILIBRIUM IN CCC

The following balance of these forces of a different nature is considered:

$$F_i = F_A + F_\eta + F_\gamma + F_W + F_h$$

From this, the basic equation of the stationary-phase retention process can be derived, a number of assumptions and complex theoretical treatments being required. Taking as example the planetary centrifuge of Type J, the average cross-sectional area of a stationary-phase layer has been estimated for hydrophobic liquid systems, which are characterized by high values of interfacial tension γ, low values of viscosity η, and low hydrodynamic equilibrium settling times:

$$\left(\sqrt{\frac{S_c}{S}} - 1 \right)(S_c - S) \approx \frac{v_m^{1/2} r^{1/2} \eta_s}{\rho_m \Delta \rho R \omega^{3/2}}$$

where S and S_c are the cross-sectional areas of the stationary-phase layer and the spiral column, respectively, v_m is the linear speed of the mobile phase flow, η_s is the viscosity of the stationary phase, and r and R are rotation and revolution radii, respectively; ρ_m is the density of the mobile phase and $\Delta \rho$ is the density difference between two phases. After a few assumptions, it can be rewritten as

$$\frac{S}{S_c} \approx 1 - k_1 \frac{\beta^{1/4}}{\omega^{3/4} R^{1/4}} \approx 1 - k_2 \frac{1}{\omega^{3/4}}$$

where $\beta = r/R$, is a proportional coefficient characterizing peculiarities of the liquid system (it is dependent on the interfacial tension, viscosity of the stationary

phase, and density difference between two phases); $k_2 = k_1 (r^{1/4}/R^{1/2})$.

The ratio of the cross-sectional area of the stationary-phase layer to that of the coil tube (S/S_c) governs the volume of the stationary phase retained in the column. The theoretical dependence of S/S_c on the rotation speed ω and the experimental dependencies of the S_f value (ratio of the volume of the stationary phase retained in the column to the total column volume) on ω for n-decane-water and chloroform-water liquid systems are in good agreement (Fig. 1).

Hence, an approach based on considering the balance of forces of a different nature acting on the fluid in the rotating coil may give some correlation among the peculiarities of the liquid system, operation conditions, design parameters of the planetary centrifuge, and the stationary-phase retention. However, any rigorous mathematical model describing the complex hydrodynamic equilibrium of two liquid phases in the rotating coiled column has not been yet elaborated. This issue remains open.

REFERENCES

1. Conway, W.D. *Countercurrent Chromatography. Apparatus, Theory and Application*; VCH: New York, 1990.
2. Ito, Y. J. Liquid Chromatogr. **1992**, *15*, 2639.
3. Ito, Y. Principle, apparatus, and methodology of highspeed countercurrent chromatography. In *High-Speed Countercurrent Chromatography*; Ito, Y., Conway, W.D., Eds.; John Wiley & Sons: New York, 1996; 3–44.
4. Menet, J.-M.; Shimomiya, K.; Ito, Y. Studies on new cross-axis coil planet centrifuge for performing counter-current chromatography: III. Speculations on the hydrodynamic mechanism in stationary phase retention. J. Chromatogr. **1993**, *644*, 239.
5. Fedotov, P.S.; Kronrod, V.A.; Maryutina, T.A.; Spivakov, B.Ya. J. Liquid Chromatogr. & Related Technol. **1996**, *19*, 3237.

Hydrophilic Vitamins: Analysis by TLC

Fumio Watanabe
Emi Miyamoto
Department of Health Science, Kochi Women's University, Kochi, Japan

INTRODUCTION

The benefit of using TLC for the identification of unknown vitamins and related compounds by comparing R_f values of the unknown compounds with authentic vitamins is beyond doubt. The quantification of the separated vitamins can be performed by the use of modern densitometry. TLC [or high-performance thin-layer chromatography (HPTLC)], as a powerful separation and analytic tool, is used particularly with pharmaceutical preparations and food products. Because amounts of most hydrophilic vitamins are low, or very low, in tissues or body fluids, bioautography or derivatization is used before densitometry.

In this section, we summarize the recent advance of TLC analysis for hydrophilic vitamins.

THIAMINE (VITAMIN B₁)

To investigate thiamine metabolism in mammals, thiamine (R_f values: 0.16, 0.04, and 0.03), urinary excretion of thiamine metabolites [thiochirome (R_f values: 0.31, 0.28, and 0.33), thiazole (R_f values: 0.85, 0.79, and 0.81), and 2-methyl-4-amino-pyrimidinecarboxylic acid (R_f values: 0.42, 0.21, and 0.26)], and related compounds [pyrimidinesulfonic acid (R_f values: 0.48, 0.39, and 0.46), α-hydroxyethylthiamine (R_f values: 0.23, 0.09, and 0.06), N'-methylnicotinamide (R_f values: 0.31, 0.06, and 0.05)] were analyzed and identified by TLC on silica gel with acetonitrile–water (40 : 10 vol/vol) adjusted to a pH of 2.54, 4.03, and 7.85 with formic acid as solvents, respectively.[1] Although N'-methylnicotinamide and thiochrome could not be separated in single-phase chromatography at pH 2.54, a second phase at right angle, with a pH 4.03 solvent, separated these quite clearly without affecting the resolution of the other compounds.[1]

The quantitative analysis of thiamine hydrochloride (vitamin B₁), using HPTLC on silica gel plates with two different mobile phases, was elaborated.[2] After TLC separation, vitamin B₁ was derivatized by the use of *tert*-butyl hypochlorite or potassium hexacyanoferrate (III)–sodium hydroxide as reagents. The *tert*-butyl hypochlorite reagent formed yellow-fluorescing derivatives with a limit of detection of less than 3 ng per chromatogram zone. The potassium hexacyanoferrate (III)–sodium hydroxide reagent led to a bluish-fluorescing derivative with a limit of detection of 500 ng per chromatogram zone.

RIBOFLAVIN (VITAMIN B₂)

TLC on silica gel 60 plates was used in various TLC solvent systems for both determination and identification of flavin derivatives in baker's yeast[3] and foods (plain yogurt and bioyogurt, raw egg white, and egg powder).[4,5] The R_f values of two unknown compounds found in plain yogurt were identical to those of 7α-hydroxyriboflavin (R_f values: 0.32 and 0.21) and riboflavin-β-galactoside (R_f values: 0.14 and 0.10), but not to those of other flavin compounds [flavin adenine dinucleotide or FAD (R_f values: 0 and 0), flavine mononucleotide or FMN (R_f values: 0 and 0.05), 10-hydroxyethylflavin (R_f values: 0.71 and 0.40), riboflavin (R_f values: 0.55 and 0.32), and 10-formylmethylflavin (R_f values: 0.86 and 0.76)] by TLC on silica gel with chloroform–methanol–ethyl acetate (5 : 5 : 2) and 1-butanol–benzyl alcohol–glacial acetic acid (8 : 4 : 3) as solvents, respectively.[4]

7α-Hydroxyriboflavin was identified in blood plasma from humans, following oral administration of riboflavin supplements, by fluorescence after TLC [benzene–1-butanol–methanol–water (1 : 2 : 1 : 1 vol/vol)] and by its spectrum.[6]

PYRIDOXINE (VITAMIN B₆)

TLC of vitamin B₆ compounds, on various layers in different solvents, was studied.[7] The R_f values of pyridoxine, pyridoxal, pyridoxamine, pyridoxal ethyl acetate, 4-pyridoxic acid, 4-pyridoxic acid lactone, pyridoxine phosphate, pyridoxal phosphate, and pyridoxamine phosphate were 0.62, 0.68, 0.12, 0.54, 0.91, 0.91, 0.95, 0.95, and 0.86, respectively, by TLC on silica gel HF₂₅₄ with 0.2% NH₄OH in water as solvent. When adsorbents containing fluorescent

Encyclopedia of Chromatography DOI: 10.1081/E-ECHR-120028858

indicators are used, all forms and derivatives of vitamin B_6 can be detected through fluorescence, or through quenching of indicator fluorescence in ultraviolet (UV) light (254 nm).

When radioactive pyridoxine hydrochloride was orally supplemented to evaluate vitamin B_6 metabolism in adult domestic cats, two unknown radioactive compounds (compounds X and Y) were excreted in the urine.[8] The R_f values of compound X (R_f values: 0.95, 0.83, 0.2, 0.5, and 0.62) and compound Y (R_f values: 0.35, 0.20, 0.22, 0.32, and 0.25) were identical to those of pyridoxine-3-sulfate and N-methylpyridoxine, respectively, but not to those of pyridoxine (R_f values: 0.73, 0.83, 0.78, 0.52, and 0.62) in various solvent systems [0.5% ammonium hydroxide, 95% ethanol, chloroform–methanol (3:1 vol/vol), isoamyl alcohol–acetone–triethylamine–water (24:18:8:6 vol/vol), and 2-butanol–1.5 N ammonium hydroxide (3:1 vol/vol), respectively] by TLC on silica gel plates.

COBALAMIN (VITAMIN B_{12})

Usual dietary sources of vitamin B_{12} are animal food products (meat, milk, eggs, and shellfish), but not plant food products. To evaluate whether foods contain true vitamin B_{12} or inactive corrinoids, vitamin B_{12} compounds were purified and characterized using TLC on silica gel.[9] The R_f values of the unknown vitamin B_{12} compound, purified from an algal health food (Spirulina tablets) were identical to those of pseudo-vitamin B_{12} (R_f values: 0.14 and 0.42), but not to those of vitamin B_{12} (or 5,6-dimethylbenzimidazolyl cyanocobamide) (R_f values: 0.23 and 0.56), benzimidazolyl cyanocobamide (R_f values: 0.18 and 0.52), 5-dydroxy-benzimidazolyl cyanocobamide (R_f values: 0.20 and 0.47), and p-cresolyl cyanocobamide (R_f values: 0.38 and 0.62) by TLC on silica gel 60 with 1-butanol–2-propanol–water (10:7:10 vol/vol) and 2-propanol–NH_4OH (28%)–water (7:1:2 vol/vol) as solvents, respectively. The results indicate that an inactive vitamin B_{12} compound (pseudovitamin B_{12}) is predominant in Spirulina tablets.[10]

Because amounts of vitamin B_{12} are very low in foods, tissues, and body fluids, bioautography is used before densitometry. A selected strain of *Escherichia coli* is used as a microorganism for the bioautography. Growth spots are enhanced by the addition of 2,3,5-triphenyltetrazolium chloride, which is converted to the red-colored formazan by *E. coli* growth. Determination of vitamin B_{12} in human plasma and erythrocytes was accomplished by one-dimensional bioautography.[11] A sensitive two-dimensional bioautography was also developed to investigate B_{12} metabolism in health and a wide range of diseases.[12]

NICOTINIC ACID AND NICOTINAMIDE

Nicotinic acid and nicotinamide and their derivatives were analyzed by TLC on MN 300G cellulose plates in various solvent systems (K. Shibata, personal communications, October 16, 2001). The R_f values of nicotinamide adenine dinucleotide phosphate or $NADP^+$ (R_f values: 0.03, 0.50, and 0.70), nicotinamide adenine dinucleotide or NAD^+ (R_f values: 0.13, 0.61, and 0.58), nicotinic acid adenine dinucleotide (R_f values: 0.15, 0.52, and 0.57), nicotinamide mononucleotide (R_f values: 0.11, 0.63, and 0.73), nicotinic acid mononucleotide (R_f values: 0.13, 0.47, and 0.75), nicotinamide (R_f values: 0.87, 0.88, and 0.45), and nicotinic acid (R_f values: 0.77, 0.82, and 0.55) are shown in various solvent systems [1 M ammonium acetate–95% ethanol (3:7), pH 5.0; 2-butyric acid–ammonia–water (66:1.7:33), and 600 g of ammonium sulfate in 0.1 M sodium phosphate–2% 1-propanol (pH 6.8), respectively]. The detection is performed by illumination under short-wavelength (257.3 nm) UV light. Urinary metabolites of the vitamin could be analyzed by TLC.[13]

PANTOTHENIC ACID

A rapid, simple, and specific TLC method has been developed for the estimation of panthenol and pantothenic acid in pharmaceutical preparations containing other vitamins, amino acids, syrups, enzymes, etc.[14] The vitamin was extracted with ethanol (from tablets and capsules) or benzyl alcohol (from liquid oral preparations) and isolated from other ingredients by TLC on silica gel 60 plates with 2-propanol–water (85:15 vol/vol) as a solvent. β-Alanine (panthothenate) or β-alanol (panthenol) was liberated by heating for 20 min at 160°C. The liberated amines were visualized with the ninhydrin reaction and estimated by spectrodensitometry at 490 nm. Recoveries for panthenol and pantothenic acid were 99.8 ± 2.25% and 100.2 ± 1.7%, respectively.

BIOTIN

Unidentified biotin metabolites were analyzed and identified in urine from healthy adults by TLC.[15,16] Three unknown biotin metabolites were identified as biotin sulfone (R_f values: 0.49 and 0.17), bisnorbiotin methyl ketone (R_f values: 0.78 and 0.29), and tetra-norbiotin-l-sulfoxide (R_f values: 0.22 and 0.01) by derivatization with p-demethylaminocinnamaldehyde after TLC on microcellulose with 1-butanol–acetic acid–water (4:1:1) and 1-butanol as solvents, respectively.[15]

FOLIC ACID

Folates [pteroylmonoglutamates (PteGlu)] and related compounds were separated by TLC on cellulose powder (MN300 UV$_{254}$) with 3.0% (wt/vol) NH$_4$Cl and 0.5% (vol/vol) 2-mercaptoethanol as solvents.[17] The R_f values of PteGlu (R_f value: 0.24), H$_2$-PteGlu (R_f value: 0.1), 5,10-CH=H$_4$-PteGlu (R_f value: 0.32), H$_4$-PteGlu (R_f value: 0.56), 5-CHO-H$_2$-PteGlu (R_f value: 0.72), 5-HCNH-H$_4$-PteGlu (R_f value: 0.72), 5,10-CH$_2$-H$_4$-PteGlu (R_f value: 0.75), 5-CH$_3$-H$_4$-PteGlu (R_f value: 0.8), 10-CHO-H$_4$-PteGlu (R_f value: 0.82), 5-CH$_3$-H$_2$-PteGlu (R_f value: 0.87), 10-CHO-PteGlu (R_f value: 0.7), and 10-CHO-H$_2$-PteGlu (R_f value: 0.73) were shown in this TLC system, which is applied to evaluate the transport and metabolism of reduced folates in blood.

A TLC densitometric method could be applied to evaluate the purity of folic acid preparations for the final purpose of determination of the N-(4-aminobenzol)-L-glutamic acid content as an impurity.[18] The separation was performed in 1-propanol–NH$_4$OH (25%)–ethanol (2:2:1) and toluene–methanol–glacial acetic acid–acetone (14:4:1:1) as solvents. The silica gel plates developed were scanned at 278 nm.

ASCORBIC ACID (VITAMIN C)

TLC has been widely used to determine ascorbic acid concentrations in foods,[19–21] pharmaceutical preparations,[21–23] and biological materials.[21,24–25] Isomers of ascorbic acid and their oxidation product, dehydroascorbic acid, were separated by TLC on sodium borate-impregnated silica gel and cellulose plates.[21] This TLC method has been adapted to separate and identify ascorbic acid and dehydroascorbic acid in fresh orange and lime juices, pharmaceutical preparations (ascorbic acid), and guinea pig tissues (liver, kidney, and eye lens) and fluids (plasma and urine).

The components of an analgesic mixture (paracetamol, ascorbic acid, caffeine, and phenylephrine hydrochloride) were separated by HPTLC on silica gel plates with methylene chloride–ethyacetate–ethanol–formic acid (3.5:2:4:0.5 vol/vol) as the mobile phase.[22] The plates were scanned at 264 nm for ascorbic acid (R_f value: 0.53), 254 nm for paracetamol (R_f value: 0.87), and 274 nm for phenylephrine hydrochloride (R_f value: 0.22) and caffeine (R_f value: 0.69).

Ascorbic acid and dipyrone (metamizole) are sometimes combined in pharmaceutical dosage forms to relive pain and fever. Simultaneous determination of ascorbic acid and dipyrone was done by TLC on silica gel using water–methanol (95:5 vol/vol) as the solvent.[23] The developed plates were directly scanned at 260 nm. The R_f values for ascorbic acid and dipyrone were 0.92 and 0.65, respectively.

MULTIVITAMIN COMPLEX

Vitamin B$_1$, vitamin B$_2$, and nicotinic acid, all of which frequently occur together in foods, were separated by TLC and fluorimetrically determined by using a commercially available fiber optic-based instrument.[26] A fluorescent tracer (fluoresceinamine, isomer II) was used to label the nicotinic acid. Vitamin B$_1$ was converted to fluorescent thiochrome by oxidizing with potassium ferricyanide solution in aqueous sodium hydroxide. These vitamins were separated by HPTLC on silica gel using methanol–water (70:30 vol/vol) as mobile phase. Under these conditions, the R_f values of the vitamin B$_1$, vitamin B$_2$, and nicotinic acid derivatives were 0.73, 0.86, and 0.91, respectively.

A vitamin B complex (vitamin B$_1$, vitamin B$_2$, vitamin B$_6$, vitamin B$_{12}$, and folic acid) was also separated into its components using TLC plates impregnated with different transition metal ions.[27] CuSO$_4$ at 0.4% impregnation in all the employed solvent systems resulted in the simultaneous resolution of constituents of the vitamin B complex with appreciable differences in R_f values.

Water-soluble vitamins (vitamin B$_1$, vitamin B$_6$, vitamin B$_{12}$, and vitamin C) in "Kombucha" drink (a curative liquor) were separated by TLC on silica gel plates with water as the solvent.[28] The plates were visually examined under UV light at 254- and 366-nm wavelengths. The four vitamins were identified and determined by comparing the R_f values with the reference values (vitamin B$_1$, 0.21; vitamin B$_6$, 0.73; vitamin B$_{12}$, 0.34; and vitamin C, 0.96).

An overpressured layer chromatographic procedure, with photodensitometric detection for the simultaneous determination of water-soluble vitamins in multivitamin pharmaceutical preparations, was developed and evaluated.[29] HPTLC on silica gel plates with 1-butanol–pyridine–water (50:35:15 vol/vol) as mobile phase was used. The quantitation was carried out without derivatization [vitamin B$_2$ (R_f value: 0.30), vitamin B$_6$ (R_f value: 0.64), folic acid (R_f value: 0.37), nicotinamide (R_f value: 0.80), and vitamin C (R_f value: 1.02)] or after spraying ninhydrin reagent [calcium panthothenate (R_f value: 0.72)] or 4-demethyl-aminocinnamaldehyde [vitamin B$_{12}$ (R_f value: 1.84) and biotin (R_f value: 0)].

CONCLUSIONS

TLC is used as a powerful separation tool particularly for the analysis of pharmaceutical preparations and food products. The separated vitamins can be quantified by the use of densitometry. In biological materials (tissues and body fluids), which contain only trace amounts of vitamins, bioautography or derivatization is used before densitometry.

TLC offers great advantages (simplicity, flexibility, speed, and relative low expense) for the separation and analysis of hydrophilic vitamins.

REFERENCES

1. Ziporin, Z.Z.; Waring, P.P. Thin-layer chromatography for the separation of thiamine, N'-methylnicotinamide, and related compounds. Methods Enzymol. **1970**, *18A*, 86–87.

2. Funk, W.; Derr, P. Characterization and quantitative HPTLC determination of vitamin B_1 (thiamine hydrochloride) in a pharmaceutical product. J. Planar Chromatogr. **1990**, *3*, 149–152.

3. Gliszczynska, A.; Koziolowa, A. Chromatographic determination of flavin derivatives in baker's yeast. J. Chromatogr. A. **1998**, *822*, 59–66.

4. Gliszczynska-Swiglo, A.; Koziolowa, A. Chromatographic determination of riboflavin and its derivatives in food. J. Chromatogr. A. **2000**, *881*, 285–297.

5. Gliszcznska, A.; Koziolowa, A. Chromatographic identification of a new flavin derivative in plain yogurt. J. Agric. Food Chem. **1999**, *47*, 3197–3201.

6. Zempleni, J.; Galloway, J.R.; McCormick, D.B. The identification and kinetics of 7α-hydroxyriboflavin (7-hydroxymethylriboflavin) in blood plasma from humans following oral administration of riboflavin supplements. Int. J. Vitam. Nutr. Res. **1996**, *66*, 151–157.

7. Ahrens, H.; Korytnyk, W. Pyridoxine chemistry: XXI. Thin-layer chromatography and thin-layer electrophoresis of compounds in the vitamin B_6 group. Anal. Biochem. **1969**, *30*, 413–420.

8. Coburn, S.P.; Mahuren, J.D. Identification of pyridoxine 3-sulfate, pyridoxal 3-sulfate, and N-methylpyridoxine as major urinary metabolites of vitamin B_6 in domestic cats. J. Biol. Chem. **1987**, *262*, 2642–2644.

9. Watanabe, F.; Miyamoto, E. TLC separation and analysis of vitamin B_{12} and related compounds in food. J. Liq. Chromatogr. & Relat. Technol. **2002**, *25*, 1561–1577.

10. Watanabe, F.; Katsura, H.; Takenaka, S.; Fujita, T.; Abe, K.; Tamura, Y.; Nakatsuka, T.; Nakano, Y. Pseudovitamin B_{12} is the predominant cobamide of an algal health food, Spirulina tablets. J. Agric. Food Chem. **1999**, *47*, 4736–4741.

11. Gimsing, P.; Nexo, E.; Hippe, E. Determination of cobalamins in biological material: II. The cobalamins in human plasma and erythrocytes after desalting on nonpolar adsorbent material, and separation by one-dimensional thin-layer chromatography. Anal. Biochem. **1983**, *129*, 296–304.

12. Linnell, J.C.; Hoffbrand, A.V.; Peters, T.J.; Matthews, D.M. Chromatographic and bioautographic estimation of plasma cobalamins in various disturbances of vitamin B_{12} metabolism. Clin. Sci. **1971**, *40*, 1–16.

13. Shibata, K.; Taguchi, H. Nicotinic acid and nicotinamide. In *Modern Chromatographic Analysis of Vitamins*, 3rd Ed.; De Leenheer, A.P., Lambert, W.E., Van Bocxlaer, J.F., Eds.; Marcel Dekker, Inc.: New York, 2000; 325–364.

14. Nag, S.S.; Das, S. Identification and quantitation of panthenol and pantothenic acid in pharmaceutical preparations by thin-layer chromatography and densitometry. J. AOAC Int. **1992**, *75*, 898–901.

15. Zempleni, J.; McCormick, B.; Mock, D.M. Identification of biotin sulfone, bisnorbiotin methyl ketone, and tetranorbiotin-*l*-sulfoxide in human urine. Am. J. Clin. Nutr. **1997**, *65*, 508–511.

16. Zempleni, J.; Mock, D.M. Advanced analysis of biotin metabolites in body fluids allows a more accurate measurement of biotin bioavailability and metabolism in humans. J. Nutr. **1999**, *129*, 494S–497S.

17. Brown, J.P.; Davidson, G.E.; Scott, J.M. Thin-layer chromatography of peteroglutamates and related compounds. Application to transport and metabolism of reduced folates in blood. J. Chromatogr. **1973**, *79*, 195–207.

18. Krzek, J.; Kwiecien, A. Densitometric determination of impurities in drugs: Part IV. Determination of N-(4-aminobenzoyl)-L-glutamic acid in preparations of folic acid. J. Pharm. Biomed. Anal. **1999**, *21*, 451–457.

19. Beljaars, P.R.; Horrock, W.V.S.; Rondags, T.M.M. Assay of L(+)-ascorbic acid in buttermilk by densitometric transmittance measurement of the dehydroascorbic acid. J. Assoc. Off. Anal. Chem. **1974**, *57*, 65–69.

20. Okamura, M. Distribution of ascorbic acid analogs and associated glycorides in mushrooms. J. Nutr. Sci. Vitaminol. **1994**, *40*, 81–94.

21. Roomi, M.W.; Tsao, C.S. Thin-layer chromatographic separation of isomers of ascorbic acid and dehydroascorbic acid as sodium borate complexes on silica gel and cellulose plates. J. Agric. Food Chem. **1998**, *46*, 1406–1409.

22. El-Sadek, M.; El-Shanawany, A.; Aboul Khier, A. Determination of the components of analgesic

mixture using high-performance thin-layer chromatography. Analyst 1990, *115*, 1181–1184.

23. Aburjai, T.; Amro, B.I.; Aiedeh, K.; Abuirjeie, M.; Al-Khalil, S. Second derivative ultraviolet spectrophotometry and HPTLC for the simultaneous determination of vitamin C and dipyrone. Pharmazie 2000, *55*, 751–754.

24. DiMattio, J. A comparative study of ascorbic acid entry into aqueous and vitreous tumors of the rat and guinea pig. Invest. Ophthalmol. Vis. Sci. 1989, *30*, 2320–2331.

25. Chatterjee, I.B.; Banerjee, A. Estimation of dehydroascorbic acid in blood of diabetic patients. Anal. Biochem. 1979, *98*, 368–374.

26. Diaz, A.N.; Paniaqua, A.G.; Sanchez, F.G. Thin-layer chromatography and fibre-optic fluorimetric quantitation of thiamine, riboflavin and niacin. J. Chromatogr. A. 1993, *655*, 39–43.

27. Bhushan, R.; Parshad, V. Improved separation of vitamin B complex and folic acid using some new solvent systems and impregnated TLC. J. Liq. Chromatogr. & Relat. Technol. 1999, *22*, 1607–1623.

28. Bauer-Petrovska, B.; Petrushevska-Tozi, L. Mineral and water soluble vitamin content in the Kombucha drink. Int. J. Food Sci. Technol. 2000, *35*, 201–205.

29. Postaire, E.; Cisse, M.; Le Hoang, M.D.; Pradeau, D. Simultaneous determination of water-soluble vitamins by over-pressure layer chromatography and photodensitometric detection. J. Pharm. Sci. 1991, *80*, 368–370.

Hydrophobic Interaction Chromatography

Karen M. Gooding
Eli Lilly and Company, Indianapolis, Indiana, U.S.A.

INTRODUCTION

Hydrophobic interaction chromatography (HIC) is a mode of separation in which molecules in a high-salt environment interact hydrophobically with a nonpolar bonded phase. HIC has been predominantly used to analyze proteins, nucleic acids, and other biological macromolecules by a hydrophobic mechanism when maintenance of the three-dimensional structure is a primary concern.[1–4] The main applications of HIC have been in the area of protein purification because the recovery is frequently quantitative in terms of both mass and biological activity.

In HIC, a high-salt environment causes the association of hydrophobic patches on the surface of an analyte with the nonpolar ligands of the bonded phase. Elution is generally effected by an "inverse" gradient to lower salt concentration. This is considered "inverse" because it is the opposite of gradients used for ion-exchange chromatography. Effective salts for HIC are those which are "antichaotropic"; that is, they promote the ordering of water molecules at interfaces. Because interaction is only with the surface of a macromolecule such as a protein, the number of amino acids involved in the chromatography is relatively small, and changes in surface structure can cause differential binding and, hence, separation.

Reversed-phase chromatography (RPC) and HIC are both based on interactions between hydrophobic moieties, but the operational aspects of the techniques render selectivities totally different. The physical properties and selectivities of the two methods are contrasted in Fig. 1. The bonded phase of HIC supports consists of a hydrophilic matrix into which hydrophobic chains are inserted, generally in low density. This can be contrasted with the higher-density organosilane chemistry used in RPC. The chromatograms illustrate that both the selectivity and the number of peaks obtained for a protein mixture vary between the two modes. Cytochrome c is not retained at all by HIC and myoglobin is split into two peaks by RPC. A primary reason for the vast difference is the mobile-phase environment for each method. The organic solvents and generally acidic conditions used in RPC cause denaturation of most proteins and even splitting into subunits, whereas the high-salt concentrations at neutral pH used in HIC result in stabilization of globular or three-dimensional structures for biological macromolecules.

The hydrophobic amino acid residues of globular proteins are generally folded inside the structure or located in a few patches on the surface. As a protein is denatured, the buried amino acids are exposed, yielding more sites for hydrophobic binding. The hydrophobic interaction system thus encounters primarily surface amino acids— far fewer hydrophobic residues than the reversed phase.

SUPPORTS

Bonded phases for HIC consist of a hydrophilic polymeric layer into which hydrophobic ligands are inserted. The hydrophilic layer totally covers the silica or polymer matrix, providing a wettable and noninteractive surface which is neutral to the protein. In HIC, even short ligands cause substantial binding and there is a definite relationship between ligand chain length and retention, contrary to the minimal effect of chain length observed in the RPC of proteins and peptides. The ligand chains are postulated either to interact with hydrophobic surface patches on the proteins or to be inserted into their hydrophobic pockets; it is the latter interaction which is strengthened by and related to chain length. The strength of the binding causes some proteins to bind irreversibly if the ligand is too long; therefore, most ligands are either aromatic or 1–3 carbon alkyl chains.

Because HIC supports are designed for macromolecules, they either possess pore diameters of at least 300 Å to allow inclusion or are nonporous. Both silica and polymer matrices are used because the hydrophilic polymeric coating minimizes or eliminates most matrix—based effects. The absolute retention and selectivity of an HIC support may be affected by the specific composition of the bonded phase, as well as the ligand. For example, protein mixtures have shown distinct selectivity on different HIC columns which have propyl functional groups.[5]

OPERATION

Mobile Phase

In HIC, the concept of weak and strong solvents is different than in other modes because the weak

Encyclopedia of Chromatography DOI: 10.1081/E-ECHR-120040013

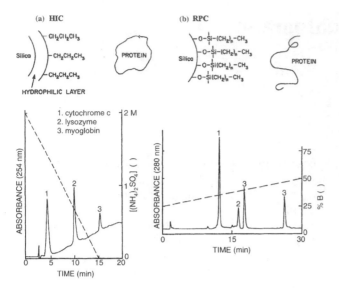

Fig. 1 (a) SynChropak Propyl; 15 min gradient from 2 M– 0 M (NH$_4$)$_2$SO$_4$ in 0.1 M potassium phosphate, pH 6.8. (b) SynChropak RPP (C$_{18}$); 30-min gradient from 25% to 50% ACN with 0.1% TFA. (Used with permission of Eichrom.)

solvent, or the one which promotes binding, is that containing high-salt concentration. The strong solvent, or one which causes elution, is that with low-salt concentration.

Salt

The most important variable in HIC retention, other than the ligand chain, is the composition of the salt used to promote binding. The effectiveness is based on the molal surface tension increment, which is parallel to the Hofmeister salting-out series for precipitation of proteins. The strength of HIC binding for some commonly used salts is K$_3$citrate > Na$_2$SO$_4$ > (NH$_4$)$_2$SO$_4$ > Na$_2$HPO$_4$ > NaCl.

Although potassium citrate and sodium sulfate cause stronger retention, ammonium sulfate is probably the most popular choice for HIC. Besides being effective for retention, it is highly soluble, stabilizing for enzymes, and resistant to microbial growth. Ammonium sulfate is available in high purity because of its use for salt fractionation. Sodium sulfate is less soluble and may precipitate under conditions of high concentration. The initial concentration of salt must be at a level high enough to cause binding of all the proteins to the bonded phase to avoid variable retention of early eluting peaks, which may also be broad.[6] Most proteins will bind when 2 M ammonium sulfate is used. In HIC, the concentration of antichaotropic salt is proportional to log k, as has been shown for conalbumin in four different salts.[7] The exact relationship varies for each salt, as well as for the specific protein.

pH

In HIC, the mobile phase should be buffered to provide control of ionization because amino acids which are not ionized are more hydrophobic than those which are charged. The effect of pH on hydrophobicity produces some variation of retention with pH; however, it is not directly related to the pI of the analyte because only surface amino acids interact with the ligands. In a study of the effect of pH on retention by HIC for a series of lysozymes from different bird species, those containing histidine residues in the hydrophobic contact region exhibited deviation for pH values of 6–8, which is near the pK of histidine (pK = 6).[7]

Additives

Because HIC is based on surface-tension phenomena, changing those characteristics by the addition of surfactants affects retention. In a study of the effects of surfactants on retention of proteins by HIC, the addition of CHAPS {3-[(3-cholamidopropyl) dimethylammonio]-1-propane sulfonate} to the mobile phase resulted in shortened retention, improvement of peak shape, and a change in peak order for enolase and bovine pancreatic trypsin inhibitor.[8] The effects were dependent on the concentration of the surfactant. Surfactants can usually be washed easily from hydrophobic interaction columns because the bonded phases are neither highly hydrophobic nor ionic.

The hydrophobic basis of HIC means that alcohols may reduce interaction with supports; however, disruption of protein conformation may also occur. Because of the high salt concentrations used in HIC, organic solvents should only be added after compatibility with the mobile phase has been tested to ensure that precipitation will not take place. Generally, no more 2 Hydrophobic Interaction Chromatography than 10% organic is added. Other additives that increase the stability of a given protein can often be included in the mobile phase for HIC without adversely changing the separation.

Flow Rate and Gradient

Almost all HIC separations are performed in the gradient mode because proteins bind with multipoint interactions. The flow rate and gradient have an effect on retention in HIC because HIC follows the linear solvent strength model.[9] The time of the gradient is another determinant in improving resolution in that longer gradients provide increased resolution. Generally, a 20–60-min gradient from 2 M–0 M ammonium sulfate in 0.02 M buffer at neutral pH, with a moderate flow rate (1 mL/min for a 4.6-mm inner diameter), will provide a satisfactory starting point for an HIC analysis.[1]

Temperature

Hydrophobic interaction chromatography is different than other modes of chromatography in that it is an entropy-driven process, characterized by increased retention with increased temperature. This is a major benefit when subambient temperatures must be used to preserve the structure and biological activity of labile proteins. Retention is usually decreased rather than increased as temperatures are lowered. In one study, the retention of lysozyme was relatively unchanged throughout a temperature range 0–45°C, whereas bovine serum albumin exhibited two peaks which changed in proportion with temperature, as well as increased in retention.[10] Some of the increase in retention with elevated temperatures, in this or other studies, can be attributed to protein unfolding and the increased exposure of internal hydrophobic residues, especially when peak broadening also occurs.

Loading

Loading capacities for proteins on HIC columns are quite high because proteins retain their globular forms during the procedure.[1,4] High loading is generally accompanied by high recoveries of biological activity. Dynamic and absolute loading capacities of HIC supports are in the range of 10 mg/mL and 30 mg/mL, respectively. Loading is also related to the relative sizes of the pore diameter and the solute, with 300 Å giving maximum capacity for many proteins.

APPLICATIONS

The primary application for HIC has been in protein analysis and purification due to the good selectivity and preservation of biological activity.[1–4] Because of the major differences in selectivity, HIC can be used as an orthogonal technique to RPC, as well as to ionexchange and size-exclusion chromatography.

Although the best HPLC method for peptide analysis is RPC, HIC offers a different selectivity for those peptides possessing three-dimensional conformations under high-salt conditions. When the separations of peptide mixtures by HIC and RPC have been compared, peaks were generally narrower on RPC due to the organic mobile phase. In a study of calcitonin variants, it was seen that peptides with certain amino acid substitutions could not be resolved by RPC, but were separated by HIC.[11] The main utility of HIC for peptide separations seems to lie in applications

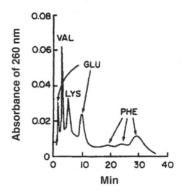

Fig. 2 Column: Polyol HIC, 100 nm; mobile phase: 0.7 M disodium hydrogen phosphate, pH 6.3. (Reprinted from El Rassi and Horvath, J. Chromatogr. 326, **1985**, 79.)

for extremely hydrophilic or hydrophobic peptides, or those with three-dimensional structures stable in high salt.

The separation of nucleic acids, particularly t-RNA, has been another useful application of HIC for biological macromolecules. The tertiary structure of t-RNA has made analysis under the gentle conditions of HIC very feasible.[12] Fig. 2 shows an example of the purification of t-RNA molecules specific for different amino acids on a 100-nm polyol HIC column. Separation of t-RNA molecules has also been accomplished successfully by using HIC conditions on supports with alkylamino ligands, which are functionally similar to those traditionally used to separate nucleic acids.[1]

CONCLUSIONS

Hydrophobic interaction chromatography is a mode of chromatography particularly effective for the analysis of proteins and other macromolecules. The hydrophobic interactions are primarily with nonpolar groups on the surface of the analytes due to maintenance of the tertiary structure. High loading and recovery of both mass and biological activity are achieved.

REFERENCES

1. Cunico, R.L.; Gooding, K.M. Hydrophobic interaction chromatography. In *Basic HPLC and CE of Biomolecules*; Bay Bioanalytical Laboratories: Richmond, CA, 1998.
2. Shansky, R.E.; Wu, S.-L.; Figueroa, A.; Karger, B.L. Hydrophobic interaction chromatography

of proteins. In *HPLC of Biological Macromolecules*; Gooding, K.M., Regnier, F.E., Eds.; Marcel Dekker, Inc.: New York, 1990; 95.

3. Aguilar, M.I.; Hearn, M.T.W. Reversed-phase and hydrophobic-interaction chromatography of proteins. In *HPLC of Proteins, Peptides and Polynucleotides*; Hearn, M.T.W., Ed.; VCH: New York, 1991; 247.

4. Ingraham, R.H. Hydrophobic interaction chromatography of proteins. In *High-Performance Liquid Chromatography of Peptides and Proteins*; Mant, C.T., Hodges, R.S., Eds.; CRC Press: Boca Raton, FL, 1991; 425.

5. Alpert, A.J. High-performance hydrophobic-interaction chromatography of proteins on a series of poly(alkyl aspart-amide)-silicas. J. Chromatogr. **1986**, *359*, 85.

6. Kato, Y.; Kitamura, T.; Nakatani, S.; Hashimoto, T. High-performance hydrophobic interaction chromatography of proteins on a pellicular support based on hydrophilic resin. J. Chromatogr. **1989**, *483*, 401.

7. Fausnaugh, I I ; Regnier, F.E. Solute and mobile phase contributions to retention in hydrophobic interaction chromatography of proteins. J. Chromatogr. **1986**, *359*, 131.

8. Wetlaufer, D.B.; Koenigbauer, M.R. Surfactant-mediated protein hydrophobic-interaction chromatography. J. Chromatogr. **1986**, *359*, 55.

9. Snyder, L.R. Gradient elution separation of large biomolecules. In *HPLC of Biological Macromolecules*; Gooding, K.M., Regnier, F.E., Eds.; Marcel Dekker, Inc: NewYork, 1990; 95.

10. Goheen, S.C.; Engelhorn, S.C. Hydrophobic interaction high-performance liquid chromatography of proteins. J. Chromatogr. **1984**, *317*, 55.

11. Heinitz, M.L.; Flanigin, E.; Orlowski, R.C.; Regnier, F.E. Correlation of calcitonin structure with chromatographic retention in high-performance liquid chromatography. J. Chromatogr. **1988**, *443*, 229.

12. ElRassi, Z.; Horvath, Cs. High-performance liquid chromatography of tRNAs on novel stationary phases. J. Chromatogr. **1985**, *326*, 79.

Hydroxy Compounds: Derivatization for GC Analysis

Igor G. Zenkevich
Chemical Research Institute, St. Petersburg State University, St. Petersburg, Russia

INTRODUCTION

The hydroxyl group is one of the most propagated functional groups in organic compounds. Important biogenic substances (carbohydrates, phenolic acids, flavones, etc.) belong to the class of hydroxy compounds. One of the principal directions of the metabolism of different ecotoxicants and drugs in vivo is their hydroxylation followed by formation of conjugates with carbohydrates or amino acids. For example, the bio-oxidation of widespread environmental pollutants—polychlorinated biphenyls (PCBs)—leads to hydroxy-PCBs.[1] The determination of OH compounds has been one of the important problems of GC analysis during the more than half a century that this method has been in existence.

The main classes of hydroxy compounds are aliphatic alcohols with OH groups attached to sp^3 hybridized carbon atoms and phenols, in which the OH groups are located in the aromatic systems attached to sp^2 carbon atoms. Some carbonyl compounds (preferably having few C=O groups, like 2,4-alkanediones, ketocarboxylic acids and their esters, 1,2- and 1,3-cycloalkanediones, etc.) exist in equilibrium with their corresponding enols in accordance with the general scheme $-CH_2-CO- \rightleftharpoons -CH=C(OH)-$, which means the presence of hydroxyl groups in these molecules as well. Another group of hydroxy compounds—namely, the carboxylic acids, having the structural fragment COOH—are discussed separately.

A simple rule for the prediction of the possibility of GC analysis of organic compounds is based on the reference data of their boiling points. If any compound can be distilled without decomposition at pressures from atmospheric to 0.01–0.1 Torr, it can be subjected to GC analysis at least on standard nonpolar polydimethylsiloxane stationary phases. In accordance with this rule, most of the monofunctional hydroxy compounds (alcohols, phenols) and their S-analogs (thiols, thiophenols, etc.) can be analyzed directly. The confirmation of chromatographic properties of any analyte should be not only verbal (at the binary "yes/no" level) but also based on their GC retention indices (RIs) as the most objective criteria, for example:

Compound	T_b, °C	$RI_{nonpolar}$
1-Tetradecanol	290.8	1664 ± 12
2,6-Di-*tert*-butyl-4-methyl phenol	265	1491 ± 10
1-Decanethiol	239.2	1320 ± 7
2-Methylbenzenethiol	194	1061 ± 11

All RIs with standard deviations are randomized interlaboratory data.

The chemical properties of hydroxy compounds depend on the presence of active hydrogen atoms in the molecule. The pK_a values for aliphatic alcohols are comparable to that of water (≈ 16), but phenols having weak acidic properties are characterized by $pK_a \approx 9$–10. Enols of carbonyl compounds are also usually weak acids. An increase in the number of polar functional groups in the molecules leads to an increase in the strength of intermolecular interactions. This is manifested in the rise in melting and boiling points of compounds, which can increase the temperature limits of their thermal stability. For example, some aliphatic diols and triols have boiling points at atmospheric pressure and, hence, are volatile enough for GC analysis. Similar compounds with four or more hydroxyl groups have no boiling points at atmospheric pressure, and this means that their GC analysis is impossible. The same restrictions are valid for the series of polyfunctional phenols:

Compound (number of OH groups)	T_b, °C	$RI_{nonpolar}$
Glycerol (3)	290.5	1196 ± 28
meso-Erythritol (4)	329–331	≈1320
Xylitol (5)	None	None
Hydroquinone (2)	287	1338 ± 14
Pyrogallol (3)	309	1420–1550
1,2,4-Benzenetriol (3)	None	None

Even the simplest bifunctional compounds of these classes being analyzed on nonpolar phases show broad nonsymmetrical peaks on chromatograms. This leads to low detection limits and bad reproducibility of RIs

Encyclopedia of Chromatography DOI: 10.1081/E-ECHR-120039948

(the position of the peaks' maxima strongly depends on the quantities of analytes) compared with nonpolar compounds. The general method to avoid all of the above mentioned problems is based on the conversion of hydroxy compounds to thermally stable and less polar volatile derivatives. This task is a most important precursor to derivatization. This chemical treatment may be used not only for nonvolatile compounds but also for the so-called semivolatile substances. The less polar products typically yield narrower chromatographic peaks, which provide better signal-to-noise ratio and, hence, lower detection limits. Nonpolar derivatives have much better interlaboratory reproducibility of RIs compared with initially polar compounds.

METHODS OF DERIVATIZATION OF HYDROXY COMPOUNDS

The principal methods of hydroxy compound derivatization may be classified in accordance with the following types of chemical reactions:

- Silylation: $R(OH)_n + nXSi(CH_3)_3$
  $\rightarrow R(OTMS)_n + nXH$
- Acylation: $R(OH)_n + nR'COX + nB$
  $\rightarrow R(OCOR')_n + nBH^+X^-$
- Alkylation: $R(OH)_n + nR'X$
  $\rightarrow R(OR')_n + nXH$

where the generally accepted codification of structural fragment X is as the carrier of the target silyl, acyl, or alkyl chemical functions.

The experimental details of these derivatization reactions are presented in some well-known specialized texts.[2–5] The first two reactions mentioned are sensitive to the presence of water in the samples (typical silylation and acylation reagents react with water more quickly than with the target organic compounds) and usually cannot proceed in aqueous media. The analysis in these conditions typically requires special sample preparation (drying before derivatization).

The large group of silylation reactions [trimethylsilyl (TMS) derivatives are most widely used] implies the replacement of active hydrogen atoms in molecules of analytes by silyl groups donated by different O-, N- or C-silylating reagents. The relative order of OH compound reactivity in general is prim–OH>sec-OH>tert-OH≈Ar–OH>R–SH. The first part of Table 1 includes physicochemical and GC constants of numerous reagents listed in the approximate order of increasing silyl donor strength. The second part of this table presents compounds later introduced into analytical practice, with still unestimated relative silylation activity, as well as a few non-TMS reagents. Besides

the individual chemicals, their different combinations are known as the so-called silylating mixtures, e.g., hexamethyldisilazane (HMDS) + trimethyl-chlorosilane (TMCS) or HMDS + dimethyl formamide (DMFA). The triple mixture bis-TMS-acetamide/TMS-imidazole/ TMCS (BSA/TMSI/TMCS) (1 : 1 : 1 v/v/v) seems to be the most active currently known silylating agent. It may be used for the derivatization of all types of substances with active hydrogen atoms, including enols of carbonyl compounds and even aci-forms of aliphatic nitro compounds.

Each reagent in Table 1 is characterized not only by its own retention index (RI), but also by the RI values of the principal by-products of the reaction. This permits us to predict the possible overlapping of their chromatographic peaks with signals of derivatives of target analytes.

The list of recommended silylating compounds is constantly changing. Some older ones have been excluded and replaced by more effective reagents. For example, in one book,[2] N-trimethylsilyl-N-phenyl acetamide, $C_6H_5N(COCH_3)SiMe_3$ [the Ph analog of N-methyl-TMS-acetamide (MSA)] was discussed as the recommended silylating reagent. However, no new examples of its application have been recently published. The principal reason for its being excluded from analytical practice is the inconvenient RIs of both this reagent itself (1493 ± 28, estimated value for nonpolar polydimethylsiloxane phases) and the by-product of the reaction (N-phenyl acetamide, 1362 ± 11). This window of GC RIs may include peaks of target derivatives and, hence, it should be free from overlap with the initial reagents and by-products.

Standard mass spectra of TMS derivatives of aliphatic hydroxy compounds indicate no peaks of molecular ions $M^{+\cdot}$, but in all cases, ions $[M–CH_3]^+$ are reliably registered. The same derivatives of phenols and enols of carbonyl compounds with π–p conjugated systems C=C–O in the molecules indicate the signals of $M^{+\cdot}$ of high intensities. Typical base peaks in mass spectra of O-TMS derivatives are $[Si(CH_3)_3]^+$ (m/z 73) and $[Si(CH_3)_2OH]^+$ (m/z 75).

The principal disadvantage of TMS derivatives is their easy postreaction hydrolysis. Most of these compounds cannot exist in aqueous media. Another type of derivatives, namely dimethyl-tert-butylsilyl ethers, have approximately 10^3 times lower hydrolysis constants owing to sterical hindrance of Si–O bonds by tert-butyl groups. Unfortunately, their mass spectra are not very informative for the elucidation of structures of unknown analytes; the base peaks for most of them belong to noncharacteristic ions $[C_4H_9]^+$ with m/z 57 and $[M–C_4H_9]^+$. Other trialkylsilyl derivatives are recommended for the derivatization of hydroxy compounds, but in practice, they are less available and used rarely.[6]

Table 1 Physicochemical and gas chromatographic properties of some silylating reagents

Reagent (abbreviation)	MW	T_b, °C (P)	d_4^{20}	n_D^{20}	$RI_{nonpolar}$	By-products $(RI_{nonpolar})$[a]
Most widely used reagents (*in order of increasing silyl donor strength*)						
Hexamethyldisilazane (HMDS)	161	126	0.774	1.408	817 ± 29	NH_3 (ND)[b]
Trimethylchlorosilane (TMCS)	108	57.7	0.858	1.338	560 ± 8	HCl (ND)
N-Methyl-N-trimethylsilyl acetamide (MSA)	145	159–161	0.904	1.439	947 ± 14	$CH_3CONHCH_3$ (816 ± 22)
N-Trimethylsilyl diethylamine (TMSDEA)	145	125–126	0.767	1.411	817 ± 11[c]	$(C_2H_5)_2NH$ (548 ± 8)
N-Trimethylsilyl dimethylamine (TMSDMA)	117	84	0.732	1.397	660 ± 4[c]	$(CH_3)_2NH$ (425 ± 16)
N-Methyl-N-trimethylsilyl trifluoroacetamide (MSTFA)	199	130–132	1.079	1.380	826 ± 3[c]	$CF_3CONHCH_3$ (540)
N,O-bis-Trimethylsilyl acetamide (BSA)	203	71–73 (35)	0.832	1.418	1008[c]	CH_3CONH_2 (711 ± 19)
N,O-bis-Trimethylsilyl trifluoroacetamide (BSTFA)	257	145–147	0.974	1.384	887[c]	CF_3CONH_2 (675 ± 11)
N-Trimethylsilyl imidazole (TMSI)	140	222–223	0.957	1.476	1176 ± 18[c]	Imidazole (1072 ± 17)
Later proposed and special reagents (*in order of increasing molecular weights*)						
2-(Trimethylsilyloxy)propene (IPOTMS)	130	—	0.780	1.395	675 ± 12	Acetone (472 ± 12)
Chloromethyldimethyl chlorosilane (CMDCS)	142	114	1.086	1.437	755 ± 8[c]	HCl (ND)
N-Trimethylsilyl pyrrolidine (TMSP)	143	139–140	0.821	1.433	862 ± 5[c]	Pyrrolidine (686 ± 10)
Dimethyl-tert-butyl chlorosilane (DMTBCS, TBDMS-Cl)	150	125	—	—	729 ± 11[c]	HCl (ND)
Ethyl(trimethylsilyl)acetate (ETSA)	160	156–159	0.876	1.415	930 ± 5[c]	$CH_3CO_2C_2H_5$ (602 ± 9)
bis-Trimethylsilyl methylamine (BSMA)	175	144–147	0.799	1.421	903 ± 18[c]	CH_3NH_2 (348 ± 12)
Trimethylsilyl trifluoroacetate (TMSTFA)	186	88–90	1.076	1.336	674 ± 5[c]	CF_3CO_2H (744 ± 6)
Bromomethyldimethyl chlorosilane (BMDCS)	186	—	—	—	842 ± 14[c]	HCl (ND)
bis-Trimethylsilyl formamide (BSFA)	189	158	0.885	1.437	948 ± 14[c]	$HCONH_2$ (637 ± 6)
bis-Trimethylsilyl urea (BSU)	204	—	—	—	1237 ± 11	$CO(NH_2)_2$ (ND)
N,O-bis-Trimethylsilyl carbamic acid	205	77–78 (mp)[d]	—	—	—	H_2NCO_2H, NH_3, CO_2 (ND)
Trimethylsilyl trifluoromethanesulfonate	222	77 (80)	1.228	1.360	—	CF_3SO_2OH
Dimethyl-tert-butylsilyl trifluoroacetamide (MTBSTFA)	241	168–170	1.023	1.402	996 ± 13[c]	CF_3CONH_2 (675 ± 11)
Dimethylpentafluorophenyl chlorosilane (in mixture with dimethylpentafluorophenylsilyl amine, 1:1 v/v)	260	88–90 (10)	1.384	1.447	—	HCl (ND)
N-Methyl-N-trimethylsilyl heptafluorobutanamide (MSHFBA)	299	148	1.254	1.353	906 ± 11[c]	$C_3F_7CONH_2$ (750 ± 18)

[a]Common hydrolysis by-products for all reagents are trimethylsilanol $(CH_3)_3SiOH$ (RI 584 ± 8) or dimethyl-tert-butyl silanol (RI 753 ± 18; estimated RI value).

[b]ND: by-product is not detected after formation of nonvolatile salts with bases (HCl) or acids (NH_3).

[c]Estimated RI values.

[d]Melting point.

Some halogenated silylating reagents are used in the synthesis of derivatives for GC analysis with selective (element-specific) detectors, for example $ClCH_2SiMe_2Cl$, $BrCH_2SiMe_2Cl$, $C_6F_5SiMe_2Cl$ in the case of electron capture detection. For the use of NP(nitrogen/phosphorus)-selective flame photometric detectors, the conversion of hydroxy compounds into 2-cyanoethyldimethylsilyl derivatives using $(C_2H_5)_2N$–$Si(CH_3)_2$–CH_2CH_2CN (2-cyanoethyldimethylsilyl diethylamine) as reagent is recommended.[7]

Polyfunctional hydroxy compounds, in accordance with general principles of derivatization, require the modification of all functional groups in their molecules. For instance, within the important group of hydroxy carbonyl compounds, the steroids, all carbonyls prior to silylation should be converted to alkoxyimino fragments $>C=O \rightarrow >C=N-OR$ or hydroxyimino fragments (with following silylation) $>C=O \rightarrow >C=N-OH \rightarrow >C=N-O-SiMe_3$ (the same methods are used for carbohydrates). Besides that, there is a single-step mode of their derivatization by treatment with most active silylation reagents that leads to per-TMS derivatives of enols. For selective derivatization of hydroxy and carbonyl groups of different reactivity in steroids, the maximal variety of reagents has been proposed.[2–5] These complex organic compounds with large retention parameters require fewer restrictions on the chemical origin of reagents. For example, one of them, namely N-methoxy-N,O-bis-TMS carbamate (BSMOC) [CH_3-ON(TMS)CO_2TMS], has an RI on standard nonpolar phases of 1143 $\pm$ 3 and has never been used

Reagent (abbreviation) for derivatization of OH compounds	Derivative	ΔMW
Acetic anhydride/[(H^+) or bases (pyridine, trialkylamies)]	R(Ar)OCOCH$_3$	42
Trifluoroacetic anhydride (TFAA), bis-trifluoroacetyl methylamine (MBTFA), or trifluoroacetylimidazole (TFAI)	R(Ar)OCOCF$_3$	96
Pentafluoropropionic anhydride (PFPA)	R(Ar)OCOC$_2$F$_5$	146
Heptafluorobutyric anhydride (HFBA)	R(Ar)OCOC$_3$F$_7$	196
Pentafluorobenzoyl chloride	R(Ar)OCOC$_6$F$_5$	194
Chloroacetic anhydride	R(Ar)OCOCH$_2$Cl	76
Dichloroacetic anhydride	R(Ar)OCOCHCl$_2$	110
Trichloroacetic anhydride or trichloroacetyl chloride	R(Ar)OCOCCl$_3$	144
N-Trifluoroacetyl-L-prolyl chloride (N-TFA-L-Pro-Cl)[a]	N-TFA-L-prolyl esters	193
Diethyl chlorophosphate	R(Ar)OP(O)(OC$_2$H$_5$)$_2$	136

(Continued)

Reagent (abbreviation) for derivatization of OH compounds	Derivative	ΔMW
2-Chloro-1,3,2-dioxaphospholane	2-Alkoxy-1,3,2-dioxaphospholanes	90
NaNO$_2$/(H$^+$) (special derivatization of simplest C$_1$–C$_5$ aliphatic alcohols into volatile alkyl nitrites for head-space analysis)[9]	RONO	29

[a]Used for synthesis of diastereomeric derivatives of enantiomeric alcohols.

for derivatization of organic compounds other than steroids.[8]

A second group of hydroxy compound derivatization reactions includes acylation of OH groups with the formation of esters. The most important are listed in the previous table.

It is recommended that these reactions be conducted in the presence of bases without active hydrogen atoms (pyridine, triethylamine, etc.). These basic media are necessary for the conversion of acidic by-products into nonvolatile salts to protect the acid-sensitive analytes from decomposition and avoid the appearance of extra peaks of by-products on the chromatograms. Exceptions are indicated by "reagent/(H$^+$)." Phenols can be converted into Na-salts before acylation.

A third group of derivatization reactions of hydroxy compounds for GC analysis includes formation of their alkyl or substituted benzyl ethers:

Reagent (abbreviation)	Derivative	ΔMW
Diazomethane (as diethyl ether solutions in the presence of HBF$_4$)	R(Ar)OCH$_3$	14
Methyl iodide/dimethylformamide (DMFA), K$_2$CO$_3$	R(Ar)OCH$_3$	14
Pentafluorobenzyl bromide (PFB-Br)	R(Ar)OCH$_2$C$_6$F$_5$	180
3,5-bis-(Trifluoromethylbenzyl) dimethylanilinium fluoride (BTBDMA-F) (only for phenols during GC injection)[10]	ArOCH$_2$C$_6$H$_3$-3,5-(CF$_3$)$_2$	226

Methylation by diazomethane is a simple method of derivatization of relatively acidic compounds like phenols (pK_a 9–10) or carboxylic acids (pK_a 4.4 $\pm$ 0.2). The application of this reagent for methylation of aliphatic alcohols needs additional acid catalysis. Methyl iodide is the most convenient reagent for the synthesis of permethylated derivatives of polyols (including carbohydrates) and phenols. Dimethyl sulfate (CH_3O)$_2$SO$_2$ can be used in basic aqueous media for the methylation of phenols, but the yields of methyl ethers in this case are not enough for quantitative determinations of initial compounds by GC.

Some special derivatization methods, which lead to the formation of cyclic products, were recommended for glycols (triols, tetrols, etc. including carbohydrates), amino alcohols, and polyfunctional aromatic hydroxy compounds (aminophenols, substituted salicylic acids). An appropriate location of two functional groups [(OH)$_2$, (OH) + (NH$_2$) or (OH) + (COOH)] in 1,2-(*vic*), 1,3-, or *ortho* (in aromatic series) positions is necessary for their realization:

Initial compound	Reagent(s)	Product	ΔMW
1,2-Diols	Acetone/(H$^+$)	2,2-Dimethyl-1,3-dioxolanes	40
	Methane- or butaneboronic acid: CH$_3$B(OH)$_2$ or C$_4$H$_9$B(OH)$_2$	2-Methyl- or 2-butyl-1,3,2-dioxaborolanes (Me or Bu boronates)	24 (Me), 66 (Bu)
2-Amino alcohols	Methane- or butaneboronic acid	2-Methyl or 2-butyl 1,3,2-oxazaborolanes	24 (Me), 66 (Bu)
Salicylic acids	Di-*tert*-butyl dichlorosilane	Di-*tert*-butyl silylene derivatives[11]	72

The choice of chemical reactions that can be used for the derivatization of hydroxy compounds in aqueous samples (when it is impossible to remove the large excess of water) is relatively small compared with their total number. For example, such hydrophilic and polar compounds as monoethanolamine (HOCH$_2$CH$_2$NH$_2$) (MEA) can be preconcentrated from the air of industrial areas only as a mixture with water (owing to air humidity) and require derivatization prior to GC analysis. All silylation and acylation reactions cannot be used in this case. An appropriate method is based on the interaction of MEA with benzaldehyde. This reaction proceeds in aqueous solutions and leads to the corresponding Schiff base HOCH$_2$CH$_2$N=CHC$_6$H$_5$, which can exist in the tautomeric cyclic form, namely 2-phenyl-oxazolidine.[12]

The number of proposed methods for alcoholic and phenolic S-analog derivatization in GC analysis is significantly smaller than those for alcohols. The objective reason of this is the lower frequency of their determinations in real analytical practice. In accordance with general recommendations, thiols and thiophenols may be converted into TFA (PFP, HFB) esters or PFB ethers (S-TMS derivatives seems not as stable as O-TMS ethers). In addition to the optimization of chromatographic parameters, the derivatization of these compounds is necessary to prevent their oxidation by atmospheric oxygen before analysis.

CONCLUSIONS

The simplest monofunctional hydroxy compounds are stable and volatile enough for their direct GC analysis. However, any polyfunctional compounds of this class, owing to their high polarity and low volatility, usually cannot in practice be analyzed in native form. Hence, their derivatization by various reagents is not only strongly recommended, but at present a generally accepted and routine procedure of sample preparation for GC analysis.

Besides the optimization of retention parameters and peak shaping, the function of derivatization is the protection of analytes from chemical transformations, which is most important for S-analogs of hydroxy compounds.

ARTICLES OF FURTHER INTEREST

Acids: Derivatization for GC Analysis, p. 3.
Amines, Amino Acids, Amides, and Imides: Derivatization for GC Analysis, p. 57.
Derivatization of Analytes in Chromatography: General Aspects, p. 427.
Kovats' Retention Index System, p. 901.

REFERENCES

1. Zenkevich, I.G.; Moeder, M.; Koeller, G.; Schrader, S. Using new structurally related additive schemes in the precalculation of GC retention indices of polychlorinated hydroxy biphenyls on HP-5 stationary phase. J. Chromatogr. A. **2004**, *1025* (2), 227–236.

2. Blau, K., King, G.S., Eds.; *Handbook of Derivatives for Chromatography*; John Wiley & Sons: Chinchester, U.K., 1978; 576.

3. Knapp, D.R. *Handbook of Analytical Derivatization Reactions*; John Wiley & Sons: New York, 1979; 741.

4. Drozd, J. *Chemical Derivatization in Gas Chromatography*. In *Journal of Chromatography Library*; Elsevier: Amsterdam, 1981; Vol. 19, 232.

5. Blau, K., Halket, J.M., Eds.; *Handbook of Derivatives for Chromatography*, 2nd Ed.; John Wiley & Sons: New York, 1993; 369.

6. Poole, C.F.; Zlatkis, A. Trialkylsilyl ether derivatives (other than TMS) for gas chromatography and mass spectrometry. J. Chromatogr. Sci. **1979**, *17* (3), 115–123.

7. Bertrand, M.J.; Stefanidis, S.; Sarrasin, B. 2-Cyanoethyldimethyl(diethyl)aminosilane, a silylating reagent for selective GC analysis using a nitrogen–phosphorus detector. J. Chromatogr. **1986**, *351*, 47–56.

8. Szederkenyi, F.; Ambrus, G.; Horvat, G.; Ilkov, E. *N*-Methoxy *N,O-bis*-TMS carbamate as a derivatization reagents for the gas chromatography of sitosterol degradation products. J. Chromatogr. **1988**, *446*, 253–257.

9. Ioffe, B.V.; Vitenberg, A.G. *Head-Space Analysis and Related Methods in Gas Chromatography*; Wiley-Interscience Publishers: New York, 1984; 276.

10. Amijee, M.; Cheung, J.; Wells, R.J. Development of 3,5-*bis*-(trifluoromethyl)benzyl-dimethylphenyl-lammonium fluoride, an efficient new on-column derivatization reagent. J. Chromatogr. A. **1996**, *738*, 57–72.

11. Brooks, C.J.W.; Cole, W.T. Cyclic di-*tert*.-butyl silylene derivatives of substituted salicylic acids and related compounds. A study by gas chromatography–mass spectrometry. J. Chromatogr. **1988**, *441*, 13–29.

12. Zenkevich, I.G.; Chupalov, A.A. Gas chromatographic determination of monoethanolamine in air of industrial areas. Zh. Anal. Khim. (Russ.) **1996**, *51* (6), 642–646.

Immobilized Antibodies: Affinity Chromatography

Monica J. S. Nadler
Beth Israel Deaconess Medical Center and Harvard Medical School, Boston, Massachusetts, U.S.A.

Tim Nadler
Applied Biosystems, Framingham, Massachusetts, U.S.A.

INTRODUCTION

Antibodies are serum proteins that are generated by the immune system which bind specifically to introduced antigens. The high degree of specificity of the antibody–antigen interaction plays a central role in an immune response, directing the removal of antigens in concert with complement lysis (humoral immunity). Importantly, this high degree of specific binding has been exploited as an analytical tool: Antigens can be detected, quantified, and purified from sources in which they are in low abundance with numerous contaminants. Examples include enzyme-linked immunosorbent assays (ELISAs), Ouchterlony assays, and Western blots. Antibodies that are specifically immobilized on high-performance chromatographic media offer a means of both detection and purification that is unparalleled in specificity, versatility, and speed.

We will focus, here, on the use of immobilized antibodies for analytical affinity chromatography, which offers a number of advantages over standard partition chromatography. The first advantage is the specificity imparted by the antibody itself, which allows an antigen to be completely separated from any contaminants. During a chromatographic run with an antibody affinity column, all of the contaminants wash through the column unbound, and the bound antigen is subsequently eluted, resulting in only two peaks generated in the chromatogram (contaminants in the flowthrough step and antigen in the elution step). With antibodies which are immobilized on high-speed media such as perfusive media,[1–2] typical analytical chromatograms can be generated in less than 5 min and columns can last for hundreds of analyses. In Fig. 1, an example of 5 consecutive analytical affinity chromatography assays are shown, followed by the results of the last 5 assays of a set of 5000. Note that, here, the cycle time for loading, washing out the unbound material, eluting the bound material, and reequilibration of the affinity column is only 0.1 min (6 s). Also note that the calibration curve has changed little between the first analysis and after 5000 analyses, demonstrating both the durability and reproducibility

of this analytical technique. Although many soft-gel media are also available for antibody immobilization, these media do not withstand high linear velocity and, therefore, are not suited for high-performance affinity chromatography.

Affinity chromatography using immobilized antibodies offers several advantages over conventional chromatographic assay development. First, assay development can be very rapid because specificity is an inherent property of antibody and solvent mobile-phase selection is limited to a capture buffer and an elution buffer, which is often the same from one antibody to the next. Therefore, there is less "column scouting" for appropriate conditions. In addition, the assays are fast (see above) and chromatograms yield only two peaks instead of multiple peaks. Furthermore, the two peaks in the affinity chromatogram indicate both antigen concentration (from the eluted peak) and purity (from the ratio of the eluted peak to the total peak area). Thus, affinity chromatography with immobilized antibodies allows both fast assay development and rapid analysis times.

The limitations of immobilized antibody affinity chromatography are few. First, plentiful amounts of antibody, usually milligram quantities, are required to get reasonable ligand density on a useful amount of chromatographic media. Also, it is optimal if the antibody is antigen affinity purified, so that when it is immobilized, no other contaminating proteins with competing specificity dilute the antibody's concentration. Finally, the antibody must be amenable to affinity chromatography such that it is not irreversibly denatured by the immobilization process and can withstand many cycles of antigen capture and elution. Both monoclonal and polyclonal antibodies have been used successfully.

IMMOBILIZATION CHEMISTRIES

Many different chemistries can be used to immobilize antibodies onto chromatographic media and only a few will be discussed. In most cases, the

Encyclopedia of Chromatography DOI: 10.1081/E-ECHR-120039872

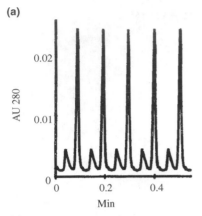

(a)

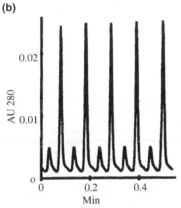

(b)

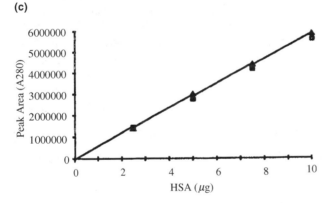

(c)

Fig. 1 Examples of affinity chromatography with an epoxy-immobilized polyclonal anti-human serum albumin (HSA) antibody in a 2.1-mm-inner diameter × 30-mm POROS CO column run at 5 mL/min (8000 cm/h) using phosphate-buffered saline for loading and 12 mM HCl with 150 mM NaCl for elution. The sample was 10 μg HSA at 1 mg/mL. Part (a) shows the first five analyses of a relatively pure sample of HSA, where the first small peak is the unbound contaminant and the larger peak is the elution of the HSA from the affinity column; (b) shows the results of the last 5 analyses from a set of 5000; and (c) shows the calibration curve before (squares) the 5000 analyses and after (triangles).

chromatographic media is coated with the active chemistry, which will then react with the antibody. These include amine reactive chemistries such as

epoxide-, aldehyde-, and cyanogen bromide (CNBr)-activated media, carboxyl reactive chemistries such as carbodiimides, aldehyde-reactive chemistries such as amino and hydrazide, and thiol-reactive chemistries such as iodoacetyl and reduce thiol media. Although there are several antibody isotypes (IgA, IgE, IgG, IgM), the most common antibody immobilized for affinity chromatography is IgG, which is composed of four polypeptide chains (two heavy and two light) which are disulfide linked to form a Y-shaped structure capable of binding two antigens. For best results, it is also important to antigen affinity purify the antibody prior to immobilization to yield optimum binding capacity and a wider dynamic range for analytical work. Also note that the antibody may be digested with pepsin or papain to separate the constant region from the antigenbinding domains, which may then be immobilized.

Antibodies are very often immobilized through their amino groups either through the N-terminal amines or the epsilon amino groups of lysine. Reactions with epoxide-activated media are performed under alkaline conditions and lead to extremely stable linkages between the chromatographic support and the antibody. Similarly, immobilization using an aldehyde-activated media first proceeds through a Schiff base intermediate which must then be reduced (often by sodium cyanoborohydride) to yield a very stable carbon–nitrogen bond linking the antibody to the media. N-Hydroxy-succinimide-activated media also couples via primary amines and leads to a stable linkage in a single-step reaction. The major advantage of these chemistries is that they are extremely stable due to the formation of covalent bonds to the media. Although less stable but easy to use is CNBr-activated media, which also immobilizes antibodies through their primary amines.

Antibodies can also be immobilized through their carboxyl groups by first treating them with a carbodiimide such as EDC (1-ethyl-3-[3-dimethlaminopropyl]-carbodiimide) followed by immobilization on an amine-activated chromatographic resin. It is important to note that EDC does not add a linker chain between the antibody and the media, but simply facilitates the formation of an amide bond between the antibody's carboxyl and the amine on the media. Coupling through sulfhydryls on free cysteines can be accomplished with thiol-activated media by formation of disulfide bonds between the media and the antibody. However, this coupling is not stable to reducing conditions and a more stable iodoacetyl-activated media is often preferred because the resulting carbon–sulfur bond is more stable. Free cysteines can be generated in the antibody by use of mild reducing agents (e.g., 2-mercaptoethylamine), which can selectively reduce disulfide bonds in the hinge region of the antibody.

Alternatively, antibodies may also be immobilized through their carbohydrate moieties. One method involves oxidation of the carbohydrate with sodium periodate to generate two aldehydes in the place of vicinyl hydroxyls. These aldehydes may then be coupled either directly to hydrazide-activated media or through amine-activated media with the addition of sodium cyanoborohydride to reduce the Schiff base. The primary advantages of these chemistries is to offer alternative linkages to the antibody beyond primary amines.

In addition, antibodies may also be coupled to other previously immobilized proteins. For example, the antibody may be first captured on protein A or protein G media and then cross-linked to the immobilized protein A or G with reagents such as glutaraldehyde or dimethyl pimelimidate. The advantage here is that the antibody need not be pure prior to coupling because the protein A or protein G will selectively bind only antibody and none of the other serum proteins. The disadvantage is that free protein A or protein G will still be available to cross-react with any free antibody in samples to be analyzed, which will only be problematic with serum-based samples. Antibody coupling does not need to be covalent to be effective. For example, biotinylated antibodies can be coupled to immobilized streptavidin. The avidin–biotin interaction is extremely strong and will not break under normal antigen elution conditions. The advantage of this immobilization protocol is that many different biotinylation reagents are available in a wide range of chemistries and linker chain lengths. Once biotinylated and free biotin are removed, the antibody is simply injected onto the streptavidin column and it is ready for use. Immobilization can be accomplished through hydrophobic interaction by simply injecting the antibody onto a reversed-phase column and then blocking with an appropriate protein solution such as albumin, gelatin, or milk. This is analogous to techniques used to coat ELISA plates and perform Western blots, and although this noncovalent coupling is not stable to organic solvents and detergents, it can last for hundreds of analyses under the normal aqueous analysis conditions. The advantage of this immobilization

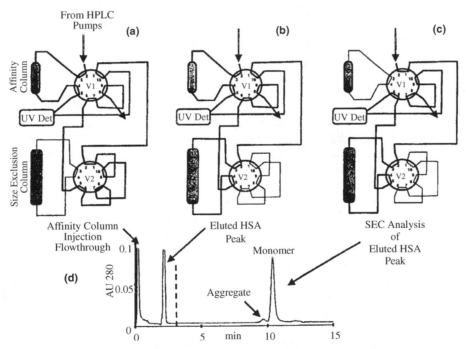

Fig. 2 Example of a multidimensional LC analysis for albumin aggregates using immobilized antibody affinity chromatography with SEC. Part (a) shows the flow path during the loading of the sample to capture the albumin monomer and aggregates while allowing all other proteins to elute to waste. Part (b) shows the transfer of the albumin and its aggregates to the size-exclusion column. Part (c) shows the flow path used to elute the size-exclusion column to separate the aggregate and monomer. Part (d) shows the UV trace from this analysis. Note that in this plumbing configuration, the albumin passes through the detector twice, once as it is transferred from the affinity to the size-exclusion column and again as the albumin elutes from the size exclusion column. The affinity column is a 2.1-mm-inner diameter (i.d.) × 30 mm POROS XL column to which anti-human serum albumin has been covalently cross-linked, run at 1 mL/min, loaded in PBS, and eluted with 12 mM HCl. The size-exclusion column is a 7.5-mm-i.d. × 300-mm Ultrasphere OG run at 1 mL/min with 100 mM potassium phosphate with 100 mM sodium phosphate, pH 7.0. The sample was 100 μg heat-treated albumin.

is that it can be done very quickly (in several minutes) by simply injecting an antibody first and then a blocking agent.

OPERATION

A wide range of buffers can be used for loading the sample and eluting the bound antigen; however, for best analytical performance, a buffer system that has low a low ultraviolet (UV) cutoff and rapid reequilibration properties is desirable. One of the better examples is phosphate-buffered saline (PBS) for loading and 12 mM HCl with 150 mM NaCl. The NaCl is not required in the elution buffer but helps to minimize baseline disturbances due to the refractive index change between the PBS loading buffer and the elution buffer because both will contain about 150 mM NaCl. UV detection is well suited for these assays and wavelengths at 214 or 280 nm are commonly used.

For analytical work, large binding capacities are not required, but increased capacity does increase the dynamic range of the analysis. However, the dynamic range can be increased by injecting a smaller volume of sample onto the column at the expense of sensitivity at the low end of the calibration curve. Likewise, sensitivity can be increased by injecting more sample volume.

APPLICATION EXAMPLES

The most obvious way to use immobilized antibodies for analytical affinity chromatography is to simply use it in a traditional single-column method to determine an antigen's concentration and/or purity. However, there are a number of ways this technique can be advanced to more sophisticated analyses. For example, instead of immobilizing an antibody, the antigen may be immobilized to quantify the antibody as has been done with the Lewis Y antigen.[3] However, the analysis is still a single-column method.

Immobilized antibodies have also been used extensively in multidimensional liquid chromatography (MDLC) analyses. As shown in Fig. 2, an affinity column with immobilized anti-HSA is used to capture all of the human serum albumin in a sample, allowing all of the other components to flow through to waste. Then, the affinity chromatography column is eluted directly into a size-exclusion column where albumin monomers and aggregates are separated and quantified. In this example, neither mode of chromatography would be sufficient by itself. The affinity media does not distinguish between monomer and aggregate, and the size-exclusion column would not be able to discriminate between albumin and the other coeluting proteins in the sample. Other MDLC applications employing immobilized antibodies include an acetylcholine esterase assay utilizing SEC[4], combinations of immobilized antibodies with reversed-phase analysis,[5–7] protein variant determination using immobilized antibodies to select hemoglobin from a biological sample followed by oncolumn proteolytic digestion, and LC/MS peptide mapping.[8]

There are many more examples of immobilized antibodies used for affinity chromatography which are not mentioned here, but it was the goal of this section to present some of the capabilities of this technique for analytical chromatographic applications.

REFERENCES

1. Afeyan, N.B.; Gordon, N.F.; Mazsaroff, I.; Varady, L.; Fulton, S.P.; Yang, Y.B.; Regnier, F.E. Flow-through particles for the high-performance liquid chromatographic separation of biomolecules: Perfusion chromatography. J. Chromatogr. **1990**, *519* (1), 1.
2. Afeyan, N.B.; Gordon, N.F.; Regnier, F.E. Automated real-time immunoassay of biomolecules. Nature **1992**, *358* (6387), 603.
3. Schenerman, M.A.; Collins, T.J. Determination of a monoclonal antibody binding activity using immunodetection. Anal. Biochem. **1994**, *217* (2), 241.
4. Vanderlaan, M.; Lotti, R.; Siek, G.; King, D.; Goldstein, M. Perfusion immunoassay for acetylcholinesterase: analyte detection based on intrinsic activity. J. Chromatogr. A. **1995**, *711* (1), 23.
5. Cho, B.Y.; Zou, H.; Strong, R.; Fisher, D.H.; Nappier, J.; Krull, I.S. Immunochromatographic analysis of bovine growth hormone releasing factor involving reversed-phase highperformance liquid chromatography-immunodetection. J. Chromatogr. A. **1996**, *743* (1), 181.
6. Battersby, J.E.; Vanderlaan, M.; Jones, A.J. Purification and quantitation of tumor necrosis factor receptor immunoadhesin using a combination of immunoaffinity and reversed-phase chromatography. J. Chromatogr. B. **1999**, *728* (1), 21.
7. Holtzapple, C.K.; Buckley, S.A.; Stanker, L.H. Determination of four fluoroquinolones in milk by on-line immunoaffinity capture coupled with reversed-phase liquid chromatography. J. AOAC Int. **1999**, *82* (3), 607.
8. Hsieh, Y.L.; Wang, H.; Elicone, C.; Mark, J.; Martin, S.A.; Regnier, F. Automated analytical system for the examination of protein primary structure. Anal. Chem. **1996**, *68* (3), 455.

Immobilized Metal Affinity Chromatography

Roy A. Musil
Althea Technologies, Inc., San Diego, California, U.S.A.

INTRODUCTION

The foundations for immobilized metal affinity chromatography (IMAC) were first laid in 1961 when Helferich introduced "ligand-exchange chromatography".[1] The modern-day usage of this technique and its practical applications as a purification tool did not emerge, however, until 1975 and the seminal work by Porath et al.[2].

Among the many new protein purification approaches introduced in recent years, IMAC stands out for its ease of use and widespread applicability. This highly versatile and efficient technique is based on the interaction between biological molecules and covalently bound chelating ligands immobilized on a chromatographic support. Indeed, because the popularization of the Qiagen Qiaexpress® bacterial expression and one-step purification system,[3] the use of IMAC has become nearly ubiquitous as tool for molecular biologists.

DISCUSSION

The principle behind IMAC lies in the fact that many transition metal ions [i.e., Ni(II) and Cu(II)] can coordinate to the amino acids histidine, cysteine, and tryptophan via electron-donor groups on the amino acid side chains.

An IMAC column may be loaded with a given metal–ion by perfusing the column with a metal–ion solution until equilibrium is reached between the metal chelated to the stationary phase and the metal ion in solution. The solid support (typically agarose, cross-linked dextran, or silica) is covalently linked to a metal-chelating ligand. The two most common ligands are iminodiacetic acid (IDA) and nitrilotriacetic acid (NTA).[4] All major chromatography suppliers now offer their own brands of IMAC supports, with IDA typically the ligand of choice. The IDA residue is very suitable as an immobilized chelating agent because a bidentate chelating moiety remains free after immobilization, to which a metal ion can be coordinated. The NTA ligand contains an additional chelating site for metal ions, which can minimize metal leakage on the column. Free coordination sites of the metal ion are then used to bind different proteins and peptides.

Pearson systematized metal ions into three categories according to their reactivity toward nucleophiles: hard, intermediate, and soft.[5] Hard metal ions, such as Fe(III), prefer oxygen, whereas soft metal ions prefer sulfur. Intermediate types of ions such as Cu(II), Zn(II), Ni(II), and Co(II) coordinate nitrogen but also oxygen and sulfur. All the metals mentioned have been successfully employed for use in IMAC.[6] The immobilized metal–ion adsorbents may be prepared by charging the chelating gels with a slightly acid solution of the metal salt (pH 3–5). Charging the gel under acidic conditions is essential in the case of Fe(III) to avoid the formation of ferric hydroxide particles in solution. The use of colored Ni(II) or Cu(II) ions facilitates checking of leakage and the possible presence of metal ions bound to the protein eluate.

In his pioneering contribution, Porath postulated that the histidine, cysteine, and tryptophan residues of a protein were most likely to form stable coordination bonds with chelated metal ions at near neutral pH.[2] To date, an analysis of several protein models[7] lend full to his original theory. Having said that, histidine, by far and away, plays the most prominent role in IMAC binding. In a very real sense, IMAC has subtly become synonymous as a histidine affinity technique. The absence of a histidine residue on a protein surface correlates with the lack of retention of that protein on any IDA–metal column. The presence of even a single histidine on a protein surface, available for coordination, results in retention of that protein on an IDA–Cu(II) column. Also, a protein needs to display at least two histidine residues on its surface to be retained on an IDA–Ni(II) column. Thus, beyond its role as a purification technique, IMAC has been used as a tool to probe the surface topography of proteins.[6]

The Qiaexpress® system is based on the selectivity of Ni–NTA for proteins with an affinity tag of six consecutive histidine residues: the 6x His tag. The 6x His tag is much smaller than such affinity tags as glutathione S-transferase, protein A, and maltose-binding protein and is uncharged at physiological pH. It has been shown to rarely contribute to a protein's immunogenicity, interfere with protein structure, function, or affect secretion from its expression system.[3]

As in any chromatography technique, one can break down the separation process to its two most fundamental aspects: adsorption and elution. On a more

Encyclopedia of Chromatography DOI: 10.1081/E-ECHR-120040014

practical level, the execution of an IMAC experiment involves five discrete steps which can be readily automated: column equilibration (charging of the gel), sample loading, removal of unbound material (washing), elution, and regeneration.

Adsorption of a protein to an IMAC column has to be performed at a pH at which an electron-donor group(s) on the protein's surface is at least partially unprotonated. Because the pK_a value of histidine groups (which supply the strongest metal interactions) lies in the neutral range, the binding of protein samples to the column should normally occur at a pH value of approximately 7. However, the actual pK_a value of an individual amino acid varies strongly depending on the neighboring amino acid value. Various experiments show that depending on the protein structure, pK_a the value of an amino acid can deviate from the theoretical value up to one pH unit.[4] Therefore, an application buffer of pH 8 often achieves improved binding. In order to eliminate any nonspecific electrostatic interactions, it is common to include salt in the equilibrating buffer. Typically, sodium chloride is used in concentrations between 0.1 M and 1.0 M.[8]

The buffer itself should not effectively compete with a protein for coordination to the metal ligand. Sodium phosphate or sodium acetate are recommended buffers (depending on the pH choice) and the presence of EDTA or sodium citrate should be avoided. The presence of detergents (Triton X-100, Tween-20, urea, etc.) in the buffer does not normally affect the adsorption of proteins.[4]

Elution of proteins can be achieved by one of three methods: protonation, ligand exchange, or column stripping. Protonation is the most common method and probably the simplest. The pH is reduced by either a linear gradient or step-gradient in the range of pH 8 to 3 or 4, reflecting the titration of the histidyl residues. Most proteins elute between pH 6 and 4. Again, sodium phosphate or sodium acetate are the buffers of choice. Competitive elution with ammonium chloride (0 M–2.0 M), imidazole (0 M–0.5 M) or its analogs histidine (0 M–0.05 M) and histamine yield similar selectivity.[8] Competitive elution with a linear gradient or step-gradient is best run at a constant nearly neutral pH. The final method of elution is to use chelating agents such as EDTA or EGTA (0.05 M solutions) which will strip the metal ions from the gel and cause the proteins to elute. Unless the protein of interest is the only one still bound on the column, this method will result only in recovery and not in purification or resolution.[3] Another undesirable feature of this protocol is that the eluate will contain a high concentration of free metal ion. Most resin manufacturers recommend that to maintain reproducibility and consistency, IMAC columns should be stripped of their metal ligands after each use and subsequently recharged with

metal before the next run.[6] Recently, Fe(III)–IMAC has found specific application in the separation of phosphorylated macromolecules and other biological substances.[9] Unlike Cu(II)–IDA complexes which have no formal charge, the metal–ligand complex Fe(III)–IDA has a net positive charge. In terms of use, the highest protein capacity is reached at low pH (<6) rather than at or above neutrality and at low ionic strength rather than at high salt concentrations. Electrostatic interactions for Fe(III) complexes play an important in protein binding.[2] However, Fe(III)–IMAC systems do not interact with phosphoproteins in the same way as ordinary ionexchange resins. Fe(III)–IMAC can be employed to resolve proteins with a wide range of isoelectric points (pI 4–11) something that is not generally possible in a simple, single ion-exchange chromatographic step.

Immobilized metal affinity chromatography has been shown to be effective for isolating proteins from crude mixtures, as well as for selective separations of closely related proteins.[2] With respect to separation efficiency, IMAC compares well with biospecific affinity chromatography and the immobilized metalion complexes are much more robust than antibodies or enzymes. These factors make IMAC particularly well suited for scale-up to process scale chromatography. The main scale-up points to be aware of are the degree to which the column is metal saturated, the chelating agent content of the sample, and the potential of leached metal (or its interactions) within the product eluate.

Leakage of metals from the column during elution can be the most significant problem due to their toxicity, but there are several ways to avoid this pitfall. Some references suggest the precaution of underloading IMAC columns (by as much as 20%) with the metal ion to begin with.[8] Another precaution is to add EDTA with imidazole, histidine, or histamine to the column fractions. EDTA competitively blocks formation of coordination complexes between protein carboxyl clusters and divalent metal cations, whereas the imidazolium groups block histidyl–metal complexation. For best reproducibility and general ease of use, a two-column format is preferred for process scale. A second scavenging column with 5% of the volume of the metal saturated purification column is simply placed in line.

Besides accommodating raw feedstreams, the relative independence of protein binding from salt concentration offers a great deal of flexibility for process sequencing. IMAC can follow virtually any other technique without the requirement for buffer exchange. The main exception is hydrophilic interaction chromatography (HIC) with ammonium sulfate. IMAC can also be used as the initial capture step enabling purification up to a 1000-fold[4] and subsequent preequilibration for downstream low-ionic-strength methods. Although high salt loading improves IMAC-binding specificity,[5] its concentration can be reduced after the major contaminants are washed through the column. Even with sodium chloride concentrations of up to 0.1 M, just a minor dilution can allow for the following charge-based chromatography method. This flexibility reveals IMAC as a valuable tool for streamlining the overall process design.

REFERENCES

1. Helferich, F. Ligand exchange: a novel separation technique. Nature **1961**, *189*, 1001.
2. Porath, J.; Carlsson, J.; Olsson, I.; Belfrage, G. Metal chelate affinity chromatography, a new approach to protein fractionation. Nature **1975**, *258*, 598.
3. *The Qiaexpressionist*, 2nd Ed.; Qiagen Inc.: Chatsworth, CA, 1992.
4. Porath, J. Immobilized metal ion affinity chromatography. Protein Express. Purif. **1992**, *3*, 263.
5. Sulkowski, E. Purification of proteins by IMAC. Trends Biotechnol. **1985**, *3*, 170.
6. Sulkowski, E. The saga of IMAC and MIT. BioEssays **1989**, *10*, 170.
7. Johnson, R.D.; Arnold, F.H. Multipoint binding and heterogenity in immobilized metal affinity chromatography. Biotechnol. Bioeng. **1995**, *48*, 437.
8. Porath, J.; Olin, B. Immobilized metal ion affinity adsorption and immobilized metal ion affinity chromatography of biomaterials. Serum protein affinities for gel-immobilized iron and nickel ions. Biochemistry **1983**, *22*, 1621.
9. Holmes, L.D.; Schiller, M.R. J. Liquid Chromatogr. & Relat. Technol. **1997**, *20*, 123.

Immunoaffinity Chromatography

David S. Hage

Chemistry Department, University of Nebraska, Lincoln, Nebraska, U.S.A.

INTRODUCTION

Immunoaffinity chromatography (IAC) refers to any chromatographic method in which the stationary phase consists of antibodies or antibody-related binding agents. *Antibodies*, or *immunoglobulins*, are a diverse class of glycoproteins that are produced by the body in response to a foreign agent, or antigen. The high selectivity of antibodies in their interactions with other molecules and the ability to produce antibodies against a wide range of substances have made IAC a popular purification tool for the isolation of hormones, peptides, enzymes, proteins, receptors, viruses, and subcellular components. The high selectivity of IAC has also made it appealing as a means of developing a variety of specific analytical methods. This review examines the basic principles of IAC and discusses the various formats in which it can be used.

HISTORY OF IMMUNOAFFINITY CHROMATOGRAPHY

The use of an immobilized ligand for the purification of antibodies was first reported in 1935, when antigens adsorbed on kaolin and charcoal were employed for the isolation of antibodies associated with syphilis and tuberculosis.[1] In 1936, Landsteiner and van der Scheer[2] began to immobilize targets for antibodies by using a diazo coupling method and chicken erythrocyte stroma as the support material. However, the first modern use of IAC is generally credited to Campbell, Luescher, and Lerman,[3] who coupled serum albumin to *p*-aminobenzylcellulose in 1951 for antibody purification. There are now thousands of methods based on IAC that have been reported for the isolation of chemicals and biochemicals,[4–8] with a growing number of applications also appearing for chemical analysis.[5,8–11]

ANTIBODY STRUCTURE

The key component of any IAC method is the antibody preparation used as the stationary phase. The basic structure of a typical antibody (i.e., immunoglobulin G or an IgG-class antibody) consists of four polypeptides that are linked by disulfide bonds to form a Y- or T-shaped structure (Fig. 1). The two upper arms of this structure are called the *Fab fragments* and contain two identical antigen binding regions. The lower stem region is known as the *Fc fragment*, and has a structure which is highly conserved between antibodies that belong to the same class. Other classes of antibodies (e.g., IgA, IgM, IgD, and IgE) have the same basic structure as IgG but may contain multiple units that are crosslinked through the presence of additional peptide chains. The amino acid composition within the Fab fragments is highly variable from one type of antibody to the next. It is this variability that allows the body to produce antibodies with a variety of affinities and binding specificities for foreign agents.

Typical antigens in nature include bacteria, viruses, and foreign proteins from animals or plants. All of these agents are fairly large compared to the binding sites on an antibody. As a result, these antigens usually have many different locations on their surfaces to which an antibody can bind; each of these locations is called an *epitope*. Smaller antigens (i.e., those with molecular masses below several thousand daltons) are too small to produce an immune response by themselves. However, these can be made to give rise to antibody production if they are coupled to a larger species, such as a carrier protein. The agent that is coupled to the carrier protein is then called a *hapten*.

ANTIBODY PRODUCTION

One way to produce antibodies to a given compound is to inject the corresponding antigen or hapten–carrier conjugate into a suitable laboratory animal, such as a mouse or rabbit. Samples of the animal's blood are then taken at specified intervals to collect any antibodies produced to the foreign agent. This method results in a heterogeneous mixture of antibodies that bind with a variety of strengths and to various epitopes on the antigen or hapten–carrier conjugate. These antibodies are called *polyclonal antibodies*, since they are produced by several different immune system cell lines within the body. Techniques have also been developed that allow for the isolation of single antibody-producing cells and the subsequent hybridization of these with cancer cells to produce new cell lines that

Encyclopedia of Chromatography DOI: 10.1081/E-ECHR-120040015

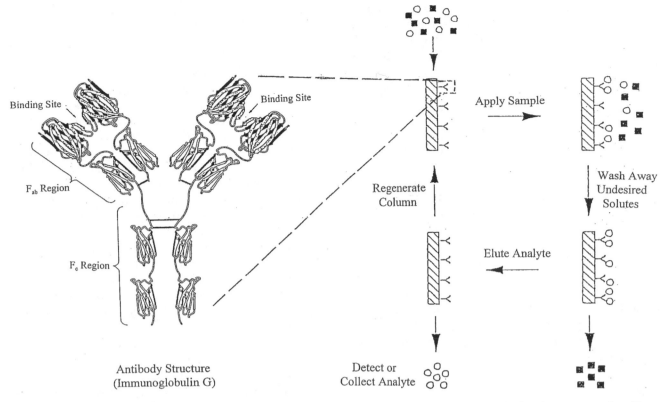

Fig. 1 Basic structure of an antibody and an example of its use in the on–off mode of immunoaffinity chromatography. (From Ref.[11].)

are stable and relatively easy to grow over long periods of time. These combined immune system/cancer cells are known as *hybridomas*, and their product is a single type of well-defined antibody called a *monoclonal antibody*. Both polyclonal and monoclonal antibodies have been used in IAC methods.[8,9]

IMMUNOAFFINITY SUPPORTS

The support material is another important item to consider in the development of a successful IAC method. In the past, most IAC applications have been based on low-performance supports (i.e., nonrigid media that can be operated under gravity or in the presence of peristaltic flow or a small applied vacuum). The supports used in this situation have typically been carbohydrate-related materials, like agarose and cellulose, or synthetic organic supports, such as acrylamide-based polymers. However, IAC can also be used in HPLC if more rigid, pressure-resistant, and higher efficiency materials are employed. Some examples of HPLC supports that have been used for IAC include derivatized glass, silica, polystyrene-based perfusion media, and azalactone beads. When these types of supports are used in IAC, the resulting method is often referred to as *high-performance immunoaffinity*

chromatography (HPIAC). Of these two approaches, low-performance IAC is the method most often used for the purification of solutes or in sample pretreatment prior to analysis by other techniques. High-performance immunoaffinity chromatography can also be used for sample pretreatment or compound isolation, but it is more commonly employed as an analytical tool for the measurement of specific chemicals in complex mixtures.[8]

ANTIBODY IMMOBILIZATION

One common approach for antibody immobilization involves direct, covalent attachment between the support and free amine groups on the antibodies. Examples include reductive amination (i.e., the Schiff base method) or the reaction of antibodies with supports that have been activated with reagents such as carbonyldiimidazole or *N*-hydroxysuccinimide. Antibodies or antibody fragments can also be immobilized through more site-selective methods. For instance, free sulfhydryl groups that are generated during the production of Fab fragments can be used to couple these fragments to thiol-activated supports. Another example involves the mild oxidation of the carbohydrate residues, which occur in the Fc region of antibodies, followed by the reaction of these oxidized residues with amine- or hydrazide-activated materials. The main advantage of site-selective immobilization is that it produces immobilized antibodies or antibody fragments that have fairly well-defined points of attachment and greater accessibility of their binding regions to analytes, giving rise to higher binding activities than are obtained by more general coupling methods.

Noncovalent immobilization can also be used for the site-selective coupling of antibodies to supports. This often involves absorbing the antibody to a secondary ligand such as protein A or protein G, which both bind to the Fc region of many antibody classes. This binding is quite strong under physiological conditions but can be easily disrupted by decreasing the pH of the surrounding solution. This method is useful when high antibody activity is needed or when it is desirable to have frequent replacement of antibodies in the IAC column. Other secondary ligands that can be used to adsorb antibodies to affinity columns are avidin or streptavidin, which bind to antibodies that have been labeled with biotin groups.[8]

APPLICATION AND ELUTION CONDITIONS

The mobile phase used in IAC is another factor to consider when using this method. The application buffer used during sample injection should facilitate quick and efficient binding of the analyte to immobilized antibodies. This mobile phase is usually selected so that it mimics the natural surroundings of the antibody (i.e., physiological pH and ionic strength). The association equilibrium constants for antibody–antigen interactions under such conditions are often in the range of 10^6–10^{12} M^{-1}. This results in extremely strong binding between the analytes and the immunoaffinity column during sample application.

Although it is possible to use isocratic elution for IAC columns that contain low affinity antibodies (i.e., those with association constants below 10^6 M^{-1}), this is not practical for higher affinity antibodies. The only way solutes can be quickly eluted from these antibodies is to change the column conditions to lower the effective strength of the antibody–analyte interaction. This is done by applying an elution buffer to the column. Usually, an acidic buffer (pH 1–3) or one that contains a chaotropic agent (e.g., sodium thiocyanate) is used for analyte elution, but occasionally a competing agent, an organic modifier, a temperature change, or a denaturing agent is employed. The elution buffer is typically applied in a step gradient; however, more gradual linear or nonlinear gradients can also be used.[8,9]

TRADITIONAL IMMUNOAFFINITY CHROMATOGRAPHY

There are a variety of formats in which IAC can be performed. The simplest format (shown in Fig. 1) is the *on–off mode*. In this technique, the sample is first injected onto the IAC column in the presence of the application buffer. As the analyte is being retained, other compounds in the sample pass through nonretained and are washed from the column. After these nonretained solutes have been removed, the elution buffer is applied. The analyte is then collected or detected as it elutes from the column. Afterwards, the initial application buffer is reapplied and the antibodies are allowed to regenerate before the next sample is injected.

This particular format is the one most commonly used in IAC for the purification of compounds. This has been used with a broad array of compounds, ranging from proteins and glycoproteins to carbohydrates, lipids, bacteria, viral particles, drugs, and environmental agents.[4–8]

Immunoaffinity chromatography has been especially popular in the isolation of specific antibodies or antigens, where an appropriate ligand (i.e., either an antibody or purified antigen) is used to isolate its counterpart from a sample. This results in the isolation of an immunologically specific agent, regardless of its chemical or biochemical nature. The on–off mode is also used in analytical applications that involve analytes, which are labeled, or that occur at sufficiently high levels to allow their direct detection as they elute from the IAC column. In this application, the on–off mode of IAC is sometimes referred to as the *direct detection mode*.[9]

IMMUNOEXTRACTION METHODS

Another set of IAC methods are those that involve *immunoextraction*.[9–11] This refers to the use of IAC for the removal of a specific solute or group of solutes from a sample prior to determination by a second analytical method. Off-line immunoextraction is the easiest way for combining IAC with techniques such as GC or HPLC. In this method, antibodies are typically immobilized onto a low-performance support that is packed into a small disposable syringe or solid-phase extraction cartridge. After sample application and the washing away of undesired sample components, an elution buffer is applied and the analyte is collected. In most situations, the collected fraction is dried down and reconstituted in a solvent more suited for analysis (e.g., a volatile solvent for compound quantitation by GC). This approach has been used to analyze substances in samples ranging from plasma and urine to food, water, and soil extracts.

The relative ease with which IAC can be directly coupled to an HPLC system makes on-line immuno-extraction appealing as a means for automating and reducing the time required for sample pretreatment in HPLC. While IAC has been directly coupled with both size exclusion and ion-exchange chromatography, the vast majority of on-line immunoextraction has involved coupling IAC with reversed-phase LC (RPLC).[9] An example of such a system is shown in Fig. 2.[1,2] Part of the reason for the popularity of this combination is the widespread use of reversed-phase HPLC in routine chemical separations. Another reason is the fact that the elution buffer for an IAC column is an aqueous solvent with little or no organic modi-fier, making this a weak mobile phase for reversed-phase columns. On-line immunoextraction coupled with reversed-phase HPLC has been used to quantitate compounds in such samples as food extracts, bodily fluids, enzyme digests, cell extracts, and environmental samples. Immunoaffinity chromatography has also been used on-line with capillary electrophoresis and mass spectrometry.[13,14] In addition, there has been work in which on-line immunoextraction has been used with GC.[15]

CHROMATOGRAPHIC IMMUNOASSAYS

Another important technique in IAC is the use of immobilized antibody columns to perform *chroma-tographic* (or *flow-injection*) *immunoassays*.[9,11,16] One way this can be performed is through a competi-tive binding format (Fig. 3). The simplest approach to a competitive binding scheme is to mix the sample and a labeled analyte analog (the label) and apply these simultaneously to the IAC column; this is known as a *simultaneous injection competitive binding immuno-assay*. Alternatively, if the sample is first applied to

the IAC column and followed later by a separate injec-tion of the label, the technique is called a *sequential injection competitive binding immunoassay*. In both these formats, an indirect measure of the analyte is obtained by examining the amount of label that elutes in either the nonretained or retained IAC fractions. An alternative format is the *displacement competitive binding immunoassay*. Here, the IAC column is first saturated with the labeled analog, followed by applica-tion of sample to the column. As the analyte travels through the column, it is able to bind to any antibody sites that are momentarily unoccupied by the label as these undergo local dissociation and reassociation. This results in displacement of the label from the column, with the degree of this displacement being directly proportional to the amount of applied analyte.

A *sandwich immunoassay* in IAC involves the use of two different antibodies that each bind to the ana-lyte of interest. The first is attached to a solid-phase support and is used for extraction of the analyte from samples. The second contains an easily measured label and is added in solution to the analyte either before or after extraction. This label allows a substance to be quantitated by providing a signal proportional to the amount of analyte in the IAC column.

In a *one-site immunometric assay*, the sample is first incubated with a known excess of labeled antibodies (or Fab fragments) specific for the analyte of interest. After this binding has occurred, the mixture is applied to a column that contains an immobilized analog of the analyte; this is done to extract any antibodies not bound to the analyte. Those antibodies that are bound to the analyte will pass through the column in the nonretained peak. Detection is performed by either looking at the nonretained labeled antibodies or by monitoring the amount of excess antibodies that later dissociate from the column during the elution step.

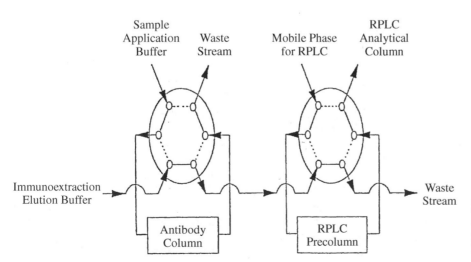

Sample Application Buffer   Waste Stream   Mobile Phase for RPLC   RPLC Analytical Column

Immunoextraction Elution Buffer   Waste Stream

Antibody Column   RPLC Precolumn

Fig. 2 Typical scheme for the on-line coupling of immunoextraction with reversed-phase LC (RPLC). (From Ref.[12].)

Step 1: Inject sample and label onto antibody column

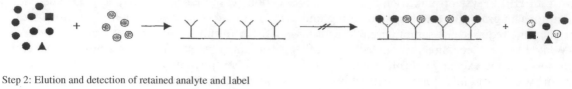

Step 2: Elution and detection of retained analyte and label

● = Analyte 1   ⊛ = Label   ■,▲ = Nonretained sample elements

Fig. 3 Typical scheme for a chromatographic competitive binding immunoassay. This particular example is for the simultaneous injection format. (From Ref.[16].)

A variety of detection schemes have been used with chromatographic immunoassays. This has involved the use of chemicals tags that act as chromophores, fluro-phores, chemiluminescent agents, or electrochemically active species. Enzyme labels like β-galactosidase, alkaline phosphatase, and horseradish peroxidase have also been employed. In this latter case, the enzyme is used to generate products for detection by absorbance, fluorescence, chemiluminescence, electro-chemical, or thermometric measurements. Liposomes, radiolabels, and bioassays have also been used for detection in chromatographic immunoassays.[11,16]

POSTCOLUMN IMMUNODETECTION

The technique of *postcolumn immunodetection* in-volves the use of an IAC column attached to the exit of an analytical HPLC system. The IAC column in this approach serves to collect and retain a specific analyte from the HPLC column eluent for later detec-tion.[9,17] The direct detection mode of IAC is the sim-plest approach for postcolumn immunodetection if the analyte of interest is capable of generating a sufficient signal for detection. For instance, this approach has been used to measure acetylcholinesterase by cap-turing it by IAC after elution from a size exclusion chromatography, with a chromogenic substrate later being applied to the IAC column for detection of this enzyme.[18]

One item that must be considered in such work is the need to adjust the eluent of the HPLC analytical column to a pH, ionic strength, and polarity appro-priate for an IAC application buffer.[19] This item is especially important when using immunodetection for RPLC, where an appreciable amount of organic modifier may be present in the mobile phase leaving the analytical column. One solution to this problem is to combine the analytical column eluent with a dilution buffer prior to sample application onto the

IAC column. Immunodetection by the on–off mode also requires that the eluting analyte be present in a conformation that is recognized by antibodies in the IAC column.

Chromatographic sandwich immunoassays have been used for postcolumn immunodetection;[19] how-ever, the one-site immunometric assay is a more common approach.[19,17,20] This involves taking the analytical HPLC column eluent and combining it with a solution of labeled antibodies or Fab fragments that will bind the analyte of interest. The column eluent and antibody or Fab mixture is then allowed to react in a mixing coil and passed through an immunodetection column that contains an immobilized analog of the analyte. The antibodies or Fab fragments bound to the analyte will pass through this column and to the detector, where they will provide a signal proportional to the amount of bound analyte. If desired, the immu-nodetection column can later be washed with an eluting solvent to dissociate the retained antibodies or Fab fragments; but a sufficiently high binding capa-city is generally used so that a reasonably large amount of analytical column eluent can be analyzed before the immunodetection column must be regenerated.

CONCLUSIONS

Immunoaffinity chromatography is a powerful separa-tion method for the selective binding of targets from samples. This method makes use of the antibodies as specific ligands within affinity columns, with different antibodies being used for different target compounds. There are several ways in which antibodies can be used in affinity columns. The first of these is the on–off mode, which can be used for the purification of com-pounds by immunoaffinity columns or for the direct detection of such agents. It is also possible to use antibodies for the indirect detection of compounds; this is accomplished through chromatographic immunoassays in which a labeled antibody or labeled analog of the target is used to determine the amount of a given chemical in a sample. Antibodies can also be used with affinity columns for the postcolumn detection of analytes as they elute from other types of chromatographic columns. These formats have been used in numerous applications that span from clinical and pharmaceutical testing to environmental work and food analysis.

ARTICLE OF FURTHER INTEREST

Affinity Chromatography: An Overview, p. 33.

REFERENCES

1. d'Allesandro, G.; Sofia, F. The adsorption of antibodies from the sera of syphilitics and tuberculosis patients. Z. Immun. **1935**, *84*, 237–250.
2. Landsteiner, K.; van der Scheer, J. Cross reactions of immune sera to azoproteins. J. Exp. Med. **1936**, *63*, 325–339.
3. Campbell, D.H.; Luescher, E.; Lerman, L.S. Immunologic adsorbents I. Isolation of antibody by means of a cellulose–protein antigen. Proc. Natl. Acad. Sci. U.S.A. **1951**, *37*, 575–578.
4. Calton, G.J. Immunosorbent separations. Meth. Enzymol. **1984**, *104*, 381–387.
5. Phillips, T.M. High performance immunoaffinity chromatography. An introduction. LC Mag. **1985**, *3*, 962–972.
6. Ehle, H.; Horn, A. Immunoaffinity chromatography of enzymes. Bioseparation **1990**, *1*, 97–110.
7. Nakajima, M.; Yamaguchi, I. Purification of plant hormones by immunoaffinity chromatography. Kagaku To Seibutsu. **1991**, *29*, 270–275.
8. Hage, D.S.; Phillips, T.M. Immunoaffinity chromatography. In *Handbook of Affinity Chromatography*; Hage, D.S., Ed.; Marcel Dekker: New York, 2005 (Chapter 6).
9. Hage, D.S. Survey of recent advances in analytical applications of immunoaffinity chromatography. J. Chromatogr. B. **1998**, *715*, 3–28.
10. de Frutos, M.; Regnier, F.E. Tandem chromatographic–immunological analyses. Anal. Chem. **1993**, *65*, 17A–25A.
11. Hage, D.S.; Nelson, M.A. Chromatographic immunoassays. Anal. Chem. **2001**, *73*, 198A–205A.
12. Hage, D.S.; Rollag, J.G.; Thomas, D.H. Analysis of atrazine and its degradation products in water by tandem high-performance immunoaffinity chromatography and reversed-phase liquid chromatography. In *Immunochemical Technology for Environmental Applications*; Aga, D.S., Thurman, E.M., Eds.; ACS Press: Washington, D.C, 1997 (Chapter 10).
13. Heegaard, N.H.H.; Schou, C. Affinity ligands in capillary electrophoresis. In *Handbook of Affinity Chromatography*; Hage, D.S., Ed.; Marcel Dekker: New York, 2005 (Chapter 26).
14. Briscoe, C.J.; Clarke, W.; Hage, D.S. Affinity mass spectrometry. In *Handbook of Affinity Chromatography*; Hage, D.S., Ed.; Marcel Dekker: New York, 2005 (Chapter 27).
15. Farjam, A.; Vreuls, J.J.; Cuppen, W.J.G.M.; Brinkman, U.A.T.; de Jong, G.J. Direct introduction of large-volume urine samples into an on-line immunoaffinity sample pretreatment-capillary gas chromatography system. Anal. Chem. **1991**, *63*, 2481–2487.
16. Moser, A.C.; Hage, D.S. Chromatographic immunoassays. In *Handbook of Affinity Chromatography*; Hage, D.S., Ed.; Marcel Dekker: New York, 2005 (Chapter 29).
17. Irth, H.; Oosterkamp, A.J.; Tjaden, U.R.; van der Greef, J. Strategies for online coupling of immunoassays to HPLC. Trends Anal. Chem. **1995**, *14*, 355–361.
18. Vanderlaan, M.; Lotti, R.; Siek, G.; King, D.; Goldstein, M. Perfusion immunoassay for acetylcholinesterase: analyte detection based on intrinsic activity. J. Chromatogr. A. **1995**, *711*, 23–31.
19. Cho, B.Y.; Zou, H.; Strong, R.; Fisher, D.H.; Nappier, J.; Krull, I.S. Immunochromatographic analysis of bovine growth hormone releasing factor involving reversed-phase high-performance liquid chromatography-immunodetection. J. Chromatogr. A. **1996**, *743*, 181–194.
20. Irth, H.; Oosterkamp, A.J.; van der Welle, W.; Tjaden, U.R.; van der Greef, J. Online immunochemical detection in liquid chromatography using fluorescein-labeled antibodies. J. Chromatogr. **1993**, *633*, 65–72.

Immunodetection

E. S. M. Lutz
AstraZeneca R&D Mölndal, Mölndal, Sweden

INTRODUCTION

Monitoring a LC effluent by means of an immunoassay provides sensitive and selective detection in combination with the separation of cross-reactive compounds.[1,2] When implementing the immunoassay as a postcolumn reaction detection system after LC, it is frequently referred to as immunodetection.[3,4] Automation and assay speed are the main advantages of immunodetection over off-line coupling of immunoassays to LC by means of fraction collection.[5,6]

The typical setup of immunodetection is illustrated in Fig. 1.[5,6] The column effluent is mixed with labeled antibodies which will bind selectively to the analytes while passing through a reaction coil. This binding is based on the affinity between analyte and antibody and is characterized by the association and dissociation rate constants of the affinity reaction. Whereas the association rate constant (k_{+1}) is diffusion controlled, the dissociation rate constant (k_{-1}) depends on the interactions between the antibody and its antigen. Generally, k_{+1} lies in the range of 10^7–10^8 L/ml s, whereas k_{-1} is comparably slow (10^3–10^5 s^{-1}). The volume of the reaction coil and the flow rates used determine the reaction time during which the labeled antibodies can bind to analyte molecules. Typical reaction times lie in the range of a few minutes and thus allow the fast association reaction to take place, whereas the dissociation reaction can practically be neglected. Quantification of the analyte concentration is then possible by distinguishing labeled antibody which has bound to analyte from free antibody. For that purpose, the free and the bound antibody needs to be separated (e.g., by means of an affinity column which traps free antibody), whereas the analyte–antibody complex passes the affinity column unretained for detection in a conventional flow through the detector. Using this setup, both analyte recognition and quantification occurs through the labeled antibody.

Alternatively, it is possible to use untreated antibodies in combination with a labeled antigen; see Fig. 2.[5,6] Under these circumstances, a two-step reaction is performed after the analytical separation. First, the column effluent is mixed with antibodies to allow the recognition of analyte(s). In the second step, the labeled antigen is added to saturate the fraction of free antibodies and allow quantification. When binding of labeled antigen to the antibody causes a change in detection properties, the reaction mixture can be monitored directly for quantification (homogeneous assay). However, generally the labeled antigen which has reacted with the antibody needs to be separated from the free labeled antigen to allow quantification. Again, affinity columns can be used for this purpose. Other forms of separating free and bound labeled antigens comprise restricted access columns, free-flow electrophoresis, and cross-flow filtration.

REAGENTS

So far, primarily antibodies and their Fab fragments have been implemented in immunodetection for analyte recognition. Antibodies can be raised against virtually any compound of interest; accordingly, their implementation into detection for LC provides a general approach. Antibody affinity and selectivity can be modulated by appropriate design of the hapten, by adequate screening of the antibodies, and by site-directed mutagenesis. Because the chemical structure and properties of antibodies against different antigens is comparatively homogenous, immobilization, stabilization, calibration, and storage procedures can be standardized.

The approach of implementing a biological assay as a postcolumn reaction detection system after LC can not only be applied to antibodybased assays (immunoassays) but also to assays employing other affinity interactions with high association and low dissociation rate constants, such as receptors. Information obtained from such a detection system not only provides quantitative results but also indicates the biological activity of the detected compound.

Requirements with respect to the label used to mark one of the immunoreagents are comparable to those in other postcolumn reaction detection systems.[4] The label should preferably allow sensitive and rapid detection and be nontoxic, stable, and commercially available. So far, mainly fluorescence labels have been employed (e.g., fluorescein), although, in principle, also liposomes, time-resolved fluorescence, and electrochemical or enzymatic labels are feasible. On the other hand, labels providing a slow response, including radioactive isotopes and glow-type chemiluminescence,

Encyclopedia of Chromatography DOI: 10.1081/E-ECHR-120040016

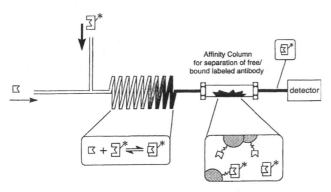

Fig. 1 Scheme of the immunodetection system employing labeled antibodies (Ɔ*) and an affinity column for separating labelled antibodies which have reacted with an analyte (Ɔ) from free-labeled antibodies.

are less suitable for immunodetection. When attaching the label to the immunoreagent, care has to be taken not to affect the affinity reaction between the antibody and its antigens and thus deteriorate assay performance.

A concern in recent research involving immunodetection has been availability, quality, and cost of reagents, especially of antibody and receptor preparations. In the future, this concern may be overcome with novel cloning techniques providing possibilities to drastically reduce the cost of producing proteins as well as to develop proteins for specific applications.

INTERFACING LC/IMMUNODETECTION

The attractiveness of immunodetection consists in its on-line coupling to a separation step, such as liquid chromatography. Parameters to consider are band broadening caused by the postcolumn reaction and interference of the LC mobile phase with the immunoreaction.[4,7]

In conventional immunoassays with long incubation times, the environment in which the affinity reaction is taking place needs to be strictly controlled with respect to, for example, pH, salt, and organic modifier content. In contrast, immunodetection takes

place within a few minutes, entailing less stringent requirements with respect to reaction conditions. Nevertheless, the mobile phase needs to be consistent with the affinity reaction; that is, the mobile phase should not denature the immunoreagents or compete with the analyte for the available binding sites. Mobile-phase compatibility has mainly been evaluated with reversed-phase LC, as it is a frequently used analytical separation technique and constitutes the greatest challenges in interfacing to biological assays. The crucial consideration is the organic-modifier content in the reversed- phase LC mobile phase. Investigations have shown that up to 15–25% (v/v) of organic modifier can be used in immunodetection without affecting the antibody–antigen interaction.[5] These results are in concurrence with immunodetection systems which have been coupled to reversedphase LC. At higher concentrations of organic modifier, the affinity reaction can be hampered seriously, which typically is overcome by dilution of the column effluent.[8,9]

However, many interesting analytes (e.g., peptides and proteins) are commonly separated by means of a gradient. The challenge of coupling immunodetection to gradient LC is twofold: On the one hand, the affinity interaction will be affected; on the other hand, the detection properties of the label will vary. For example, using a gradient of organic modifier affects the conformation of the antibody and, thus, its affinity characteristics, as well as the detection properties of a fluorescence label. When acceptance of an increasing baseline[10,11] is out of the question, additional interfacing between the separation step and the immunodetection is required. One approach is to introduce a buffer-exchange step after the separation (e.g., with on-line dialysis[12] or an ion-exchange column[7]). However, this will introduce extra band broadening as well as affect robustness with yet another part in the system.

APPLICATIONS

Feasibility of LC/immunodetection has been shown for quantitative analysis and for screening for biological

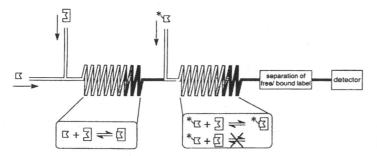

Fig. 2 Scheme of the immunodetection system employing untreated antibodies (Ɔ) and a labeled antigen (*Ɔ).

Table 1 Review of immunodetection applications

Application area	Compound	Matrix	Immunodetection type	Detection limit	Refs.
Protein bioanalysis	Granulocyte colony-stimulating factor (GCSF)	Plasma	Fluorescence-labeled antibodies, affinity chromatography	0.6 nmol/L	[10]
	Growth-hormone-releasing factor (GHRF)	Plasma	Fluorescence and enzyme-labeled antibodies, affinity chromatography, interface for dilution between LC and immunodetection for compatibility	0.2 ng/mL	[9]
	Urokinase	Plasma	Fluorescence-labeled receptor, affinity column	40 nmol/L	[11]
	Interleukine 4	?	Fluorescence-labeled antibodies, affinity chromatography	2 fmol	[16]
Drug bioanalysis	Diogoxin	Plasma	Fluorescence-labeled antibodies, affinity chromatography	0.2 nmol/L	[8]
Biomarker analysis	Sulfodipeptide Leukotrienes	Urine, human cell culture extract	Untreated antibodies and fluorescence-labeled ligand, reversed-phase restricted-access chromatography	0.4 nmol/L	[15]

activity, as summarized in Table 1. For analytical purposes, immunodetection in combination with LC is most promising in those cases when conventional immunoassays or conventional LC methods by themselves do not suffice for accurate analytical determinations. Being an approach offering high selectivity and sensitivity, applications are directed toward measurement of trace levels of compounds which lack appropriate detection properties in complex matrices. This is illustrated, for example, by measuring endogenous levels of the protein granulocyte colony-stimulating factor (GCSF) in biological matrices.[10] Affinity chromatography for sample preparation introduces high selectivity into the system but does not provide a means to improve detection properties of the protein. By combining the affinity chromatography for sample preparation with an analytical separation by means of reversed-phase LC and immunodetection, levels of GCSF were determined in plasma.

In other bioanalytical applications, the emphasis lies more on overcoming cross-reactivity. For example, the heart glycoside digoxin is cross-reactive with several of its metabolites as well as with plasma constituents, thus hampering approaches solely based on immunoassays. By treating plasma samples with solid-phase extraction on a restricted access column, coupled on-line to reversed-phase LC/immunodetection, digoxin and two of its cross-reactive metabolites were analyzed in patients treated with the heart glycoside.[8]

A key to successful drug development is the identification of new lead compounds. Lead compounds can be identified through receptor assays, where the receptor– ligand interaction reflects the biomolecular mechanism associated with a disorder. By implementing the receptor interactions into postcolumn reaction detection systems, biologically active compounds can be separated and detected with high sensitivity and selectivity. This concept has been described for the analysis of estrogens using a recombinant steroid-binding domain of the human estrogen receptor for analyte recognition and coumestrol, a fluorescent estrogen, as reporter molecule.[13] Prior to detection samples were treated online and automated by reversed-phase solid-phase extraction and reversed-phase LC. Selectivity of this system is demonstrated for analysis in urine samples, which shows the feasibility of using this method in the determination of the abuse of steroid hormones in performance doping or cattle breeding. When performing both LC/immunodetection and mass spectrometry, information on biological activity is combined with structure elucidation.[11]

Recently, also direct recognition of active ligands attached to bead surfaces has been achieved with immunodetection.[14] This provides a rapid and automated screening tool which is compatible with solidphase bound compounds originating from solid-phase chemistry in combinatorial chemistry. However, this approach has so far only been published for a model system.

CONCLUSIONS

Immunodetection coupled on-line to LC as a tool for quantitative analysis has been developed for model compounds as well as been applied in relevant applications. The approach is particularly appealing for trace analysis in complicated matrices and for identifying ligands for certain receptors in drug discovery. However, each application still requires a fair amount of method development and optimization, and obtaining the desired, pure immunoreagents still is a concern, although advances in recombinant protein production are providing us with an increased choice and availability of affinity reagents at reduced cost.

In parallel with miniaturization of LC, immunodetection will be downscaled, thus lowering reagent consumption and, consequently, cost. However, increased challenges with respect to band broadening in a postcolumn reaction detection system and nonspecific binding to capillary walls will need to be addressed. The trend toward miniaturization will simultaneously give the opportunity to couple immunodetection to other analytical separation methods, such as capillary electrophoresis. Combination of immunodetection with mass spectrometry enables the combination of information of biological activity with structure elucidation. Other developments in immunodetection concern the separation of free and bound labels, enabling the implementation of suspended materials in detection LC, including, for example, suspended membrane receptors, whole cells, and molecularly imprinted polymers serving as artificial receptors. Consequently, immunodetection potentially conquer new application areas, such as the evaluation of absorption profiles of drugs and the investigation of drug metabolism on a cellular level.

REFERENCES

1. Mattiasson, B.; Nilsson, M.; Berdén, P.; Håkansson, H. Flow-ELISA: binding assays for process control. Tr.A. C. **1990**, *9*, 317.
2. De Frutos, M.; Regnier, F.E. Tandem chromatographic immunological analyses. Anal. Chem. **1992**, *65*, 17.
3. Hage, D.S. Survey of recent advances in analytical applications of immunoaffinity chromatography. J. Chromatogr. B. **1998**, *715*, 3.

4. Krull, I.S.; Cho, B.-Y.; Strong, R.; Vanderlaan, M. LC–GC Int. **May 1997**, *278*.

5. Irth, H.; Oosterkamp, A.J. Tr.A. C. **1995**, *14*, 355.

6. Lutz, E.S.M.; Oosterkamp, A.J.; Irth, H. Chim. Oggi **1997**, *15*, 11.

7. Shahdeo, K.; March, C.; Karnes, H.T. Postcolumn immunodetection following conditioning of the HPLC mobile phase by on-line ion-exchange extraction. Anal. Chem. **1997**, *69*, 4278.

8. Oosterkamp, A.J.; Irth, H.; Beth, M.; Unger, K.K; Tjaden, U.R.; van der Greef, J. Bioanalysis of digoxin and its metabolites using direct serum injection combined with liquid chromatography and on-line immunochemical detection. J. Chromatogr. B. **1994**, *653*, 55.

9. Cho, B.-Y.; Zou, H.; Strong, R.; Fisher, D.H.; Nappier, J.; Krull, I.S. Immunochromatographic analysis of bovine growth hormone releasing factor involving reversed-phase high-performance liquid chromatography-immunodetection. J. Chromatogr. A. **1996**, *743*, 181.

10. Miller, K.J.; Herman, A.C. Affinity chromatography with immunochemical detection applied to the analysis of human methionyl granulocyte colony stimulating factor in serum. Anal. Chem. **1996**, *68*, 3077.

11. Oosterkamp, A.J.; van der Hoeven, R.; Glässgen, W.; König, B.; Tjaden, U.R.; van der Greef, J.; Irth, H. Gradient reversed-phase liquid chromatography coupled on-line to receptor-affinity detection based on the urokinase receptor. J. Chromatogr. B. **1998**, *715*, 331.

12. Kaufmann, M.; Schwarz, T.; Batholmes, P. Continuous buffer exchange of column chromatographic eluates using a hollow-fibre membrane module. J. Chromatogr. A. **1993**, *639*, 33.

13. Oosterkamp, A.J.; Villaverde Herraiz, M.T.; Irth, H.; Tjaden, U.R.; van der Greef, J. Reversed-phase liquid chromatography coupled on-line to receptor affinity detection based on the human estrogen receptor. Anal. Chem. **1996**, *68*, 1201.

14. Lutz, E.S.M.; Irth, H.; Tjaden, U.R.; van der Greef, J. Biochemical detection for direct bead surface analysis. Anal. Chem. **1997**, *69*, 4878.

15. Oosterkamp, A.J.; Irth, H.; Heintz, L.; Marko-Varga, G.; Tjaden, U.R.; van der Greef, J. Simultaneous determination of cross-reactive leukotrienes in biological matrices using on-line liquid chromatography immunochemical detection. Anal. Chem. **1996**, *68*, 4101.

16. Schenk, T.; Irth, H.; Heintz, L.; Marko-Varga, G.; Tjaden, U.R.; van der Greef, J. submitted.

Industrial Applications of Countercurrent Chromatography

Alain Berthod
Laboratoire des Sciences Analytiques, CNRS UMR5180, Université de Lyon I, CPE, Villeurbanne, France

Serge Alex
Centre d'Etudes des Procédés Chimiques du Québec, Montreal, Quebec, Canada

INTRODUCTION

Industries produce material products and goods. The chemical industry produces millions of metric tons of simple chemicals such as soda, ethylene, sulfuric acid, and urea, and relatively lesser quantities of fine and/or complex chemicals such as chiral drugs, catalysts, antibiotics, and delicate perfumes.

Countercurrent chromatography (CCC) is useful in the production of the latter class of chemicals. This entry explains the role that CCC can play in industrial processes, revealing concepts and ideas rather than detailing examples that can be found elsewhere. At the moment, only a handful of chemical companies use CCC in commercial processes. They are apparently very successful with the technique, because they purchase more CCC systems, thereby CCC becoming part of the production process. The problem is that these companies do not make or want their competitors to know that CCC works.

THE LIQUID STATIONARY PHASE

It is the liquid nature of the stationary phase in CCC that renders it useful, for three reasons:

1. The solutes can access the whole volume of the stationary phase, not just the surface of a solid stationary phase as in most other chromatographic techniques. Large amounts of substance can be processed in a single run. Countercurrent chromatography is truly a preparative technique.
2. The retention mechanism is very simple. The only physicochemical parameter responsible for solute retention is the liquid–liquid distribution coefficient, D, also called partition coefficient. The retention volume is

$$V_R = V_M + DV_S = V_C + (D - 1)V_S \quad (1)$$

where V_M and V_S are the volumes of the mobile and stationary phases, respectively, inside the CCC system. The sum of these two volumes is the CCC system volume, V_C.

3. It is possible to switch the roles of the phases during a run. The mobile phase becomes the stationary phase and vice versa.

The combination of these three points produces the following advantages:

- A study and optimization of a separation of a complex mixture can be accomplished with a low-volume CCC system, injecting small quantities of the sample. The biphasic liquid system composition is *optimized rapidly* because the volumes are reduced.
- If the retention volumes of some constituents of the sample are too large, the dual mode is used. The retained constituents are eluted in the reversed mode. It is absolutely certain that *no part of the sample can be trapped* inside the CCC apparatus.

Once the separation is optimized, the distribution coefficient of each constituent is then calculated [Eq. (1)]. The very same liquid system is used in a large-volume CCC system. So, the *retention volumes can be predicted* because the distribution coefficients, which depend entirely upon the liquid–liquid system, are the same as for the small-volume system. The scaling-up is straightforward.[1]

LARGE-SCALE SEPARATION OR PURIFICATION

Classical Use of the Technique

In a recent work, the fractionation of a tannin sample was studied. The separation was optimized with a 150-ml CCC apparatus. The butanol–ethyl acetate–water (pH 2.8) system (3.5 : 46.5 : 50% v/v) was found to be efficient for the separation. The distribution coefficients of the 12 peaks were calculated. Subsequently, it was possible to fractionate 26 g of the tannin sample in one run with the same liquid

Encyclopedia of Chromatography DOI: 10.1081/E-ECHR-120040017

833

system and a 2-L CCC system.[1] The dual-mode approach was used for this separation. The high loading capability of large-volume CCC systems is commonly used in industry. Large amounts (gram-to-kilogram scale) of natural products with high added values are separated by CCC. Alkaloids, antibiotics, enzymes, macrolides, peptides, rare fatty acids, saponins, tannins, taxoids and/or precursors of Taxol®, and other fine chemicals have been isolated, separated, and/or purified by preparative CCC.[1]

What makes CCC preferred to classical prep LC are as follows:

1. An original or unique selectivity obtained with a subtle polarity difference between the two liquid phases.
2. The possibility of injecting heavily concentrated or even polyphasic samples (e.g., fermentation broths).
3. The gentle interactions during the separation process that preserve delicate molecules (c.g., proteins) from denaturation.

Displacement Chromatography

In displacement chromatography, the sample to be purified is injected in a large volume, or even continuously, into the CCC machine. The sample components have different affinities for the stationary phase in an exclusive way: The component with a higher affinity for the stationary phase displaces another one with a lower affinity. Bands of pure components form. This method of using CCC offers the maximum throughput capability.[1] Displacement CCC can be done using the solute ionization capability (pH zone refining) or complexation capability (ion exchange CCC).

pH Zone Refining

pH zone refining is a form of displacement chromatography for ionizable compounds. It sorts compounds by their ionization constants, K_a, using a stationary phase with a different pH from the mobile phase.[2] For example, the stationary phase is an acidic organic phase (e.g., methyl *tert*-butyl ether (MTBE) with 1% acetic acid) and the mobile phase is an aqueous basic buffer (e.g., 0.015 M NH₃ solution, pH = 10). The acid in the stationary phase is considered as the retainer. The base in the mobile phase is the displacer.

Up to 60% of the CCC system volume of a mixture of organic acids can be injected into the apparatus (in the ammonium salt form at pH = 10).[2] The injected organic salts are protonated by the acidic stationary MTBE. The protonated molecular forms stay in the organic phase, and the ionized basic forms prefer the aqueous phase. The weaker acid is protonated first, so it is most retained. Bands of pure organic acids form in the order of decreasing pK_a values.

Complexation CCC

Complexation can be used to separate metallic ions on an industrial scale. Fig. 1 illustrates the process in the case of the separation of nickel and cobalt ions.[3] A complexing agent (e.g., diethyl hexyl phosphoric acid) is added to a heptane stationary phase. A large volume (up to 20 times the CCC machine volume V_C) of the ionic solution is injected. The nickel ions are displaced in the aqueous phase. The cobalt ions can be collected in the stationary phase. More than two ions can

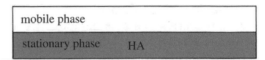

Step 1: aqueous mobile phase, heptane + HA stationary phase

Step 2: the CCC machine works like a deionizer, bands form

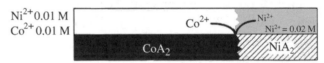

Step 3: only nickel ions are eluted more concentrated

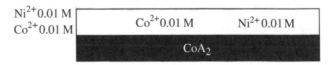

Step 4: cobalt ions are in the liquid stationary phase

Fig. 1 Removal and separation of cobalt and nickel ions by CCC. Stationary phase: heptane + diethyl hexyl phosphoric acid (HA 0.5 M); mobile phase: aqueous solution of cobalt and nickel acetate (0.01 M each). Step 1: The CCC machine is equilibrated with water. Step 2: The ionic solution is introduced into the machine, the ions are extracted into the stationary phase, and the cobalt complex displaces the nickel one, and is less stable. Step 3: The stationary phase is saturated in nickel ions. The greenish effluent leaving the machine contains only nickel ions two times more concentrated than the entering solution. Cobalt ions are still extracted, displacing nickel ions. Step 4: End of the process—the stationary phase is saturated in cobalt ions. The machine is stopped, the dark blue stationary phase is collected, and the cobalt ions are recovered by an acid wash. (Adapted from Ref.[3].)

be separated in bands of increasing complexation constant order.[3] Because no ions can stay trapped inside the CCC machine, it could be a very potent tool in the separation of radionuclides in the processing of nuclear wastes.

Extraction

Countercurrent chromatography can be used to extract and concentrate, in a low volume of stationary phase, a component present in large volumes of mobile phase. It was shown that a 60-ml CCC instrument was able to extract 285 mg of a nonionic surfactant contained in 20 L of water (at 16.5 ppm or mg/L) and to concentrate it into 30 ml of ethyl acetate (at 9500 ppm or 9.5 g/L).[4]

CONTINUOUS PLUG-FLOW REACTOR

The use of a CCC system as an original and powerful plug-flow liquid–liquid reactor for a biphasic catalytic reaction was also demonstrated.[5] Benzaldehyde (BZA) can be reduced to benzyl alcohol (BZOH) by sodium formate in the aqueous phase, at room temperature, when a ruthenium–phosphine complex is used. Benzaldehyde is located in the cyclohexane mobile phase. Sodium formate, the complex, and BZOH are located in the aqueous stationary phase. A 79% conversion of BZA to BZOH was obtained at 30°C.[5] Even more interesting is injecting a 200-μL plug of BZA into the CCC apparatus with the aqueous phase containing the catalyst and sodium formate (5 M); the result shown in Fig. 2 was obtained. The elution band profile of the BZOH formed should allow one to model the kinetic behavior of the catalyst complex used. This study needs numerous batch experiments with different experimental conditions using each time, some amount of an expensive and rare new catalyst. This can be done in one

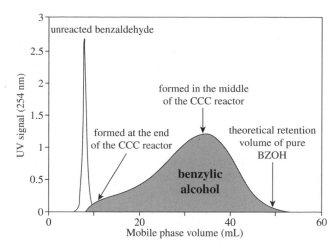

Fig. 2 Fast evaluation of a new catalyst capability: reduction of BZA to benzyl alcohol (BZOH) by sodium formate catalyzed by a ruthenium–triphenylphosphine trisulfonated sodium complex. Twenty micromoles of BZA was injected. Seventeen micromoles of BZOH was formed (85% conversion). Conditions: A 59-ml CCC machine, 52 ml of 5 M sodium formate aqueous solution, 7 ml of cyclohexane at 1.5 ml/min in the ascending tail to head direction, 750 rpm. (Adapted from Ref.[5].)

run, for one temperature, with the CCC reactor, saving time and catalyst.

INDUSTRIAL CCC EQUIPMENT

In the 20th century, only the now-redundant Sanki Engineering Ltd., a Japanese company, marketed industrial CCC machines based on the hydrostatic scheme. Recently, new large-scale CCC machines based on the hydrodynamic scheme (coiled open tubes) were developed in the U.K., while modern and reliable hydrostatic preparative CCC machines (channels and ducts) were developed in France (Table 1). These 21st century machines are much more efficient

Table 1 Large-scale CCC machines, production in g to kg day^{-1} m^{-3}

Manufacturer	Sanki[a]	Brunel–Dynex[b]	Armen[c]	Partus[d]
Model and CCC type	LLI hydrostatic	Maxi hydrodynamic	CPC hydrostatic	Partitron 25 hydrostatic
Volume (L)	5–30	5	25	25
Flow rate (L/min)	0.05–0.7	0.1–2	0.1–2	0.2–10
Rotor rotation (rpm)	1000	50–850	1500	1500
Pressure (kg/cm^2)	40	5	70	150

[a]Sanki no longer exists. EverSeiko Corporation, 4F Takagi Building, 4-36-17, Ikebukuro, Toshima-ku, Tokyo 71-0014, Japan, can be contacted.
[b]Brunel Institute of Bioengineering, Dynamic Extraction Ltd, Uxbridge, UB8 3PH, U.K.
[c]Armen Instrument, Z.I de Kermelin, 16, rue Ampère, F-56890 Saint Ave, France. Tel.: + 33-297-618-400; fax: + 33-297-618-500; e-mail: contact@armen-instrument.com.
[d]Partus Technologies, 2 Allée Caquot, F-51100 Reims, France. Tel./fax: + 33-326-918-620; e-mail: rodolphe.margraff@wanadoo.fr.

and reliable, and, upon industrial will, they will definitively prove the CCC capability in terms of purity of the produced compounds, solvent saving and throughput.

CONCLUSIONS

At the moment, CCC is scarcely used in industry. The reasons are that the technique is not well known because only few small companies market good and reliable CCC systems. The capabilities of a liquid stationary phase in industrial environment are very large. The CCC technique could be of great help in many industrial processes including classical ones such as extraction, purification, and separation of fragile compounds, as well as novel ways as in use of a CCC system as a powerful liquid–liquid reactor.

ARTICLES OF FURTHER INTEREST

Coil Planet Centrifuges, p. 349.
Octanol–Water Distribution Constants by Countercurrent Chromatography, p. 1131.

REFERENCES

1. Berthod, A.; Billardello, B. Countercurrent chromatography: fundamentally a preparative tool. Adv. Chromatogr. **1999**, *40*, 503–538.
2. Ito, Y.; Shinomiya, K.; Fales, H.M.; Weisz, A. *Modern Countercurrent Chromatography*; Conway, W.D., Petroski, R.J., Eds.; ACS Symposium Series No. 368; American Chemical Society: Washington, D.C., 1995; 156–183.
3. Berthod, A.; Xiang, J.; Alex, S.; Collet-Gonnet, C. Chromatographie à contre courant et micelles inverses pour la séparation et l' extraction de cations métalliques. Can. J. Chem. **1996**, *74*, 277–286.
4. Berthod, A. The support-free liquid stationary phase. In *Countercurrent Chromatography*; Comprehensive Analytical Chemistry; Elsevier: Amsterdam, 2002; Vol. XXXVIII.
5. Berthod, A.; Talabardon, K.; Caravieilhes, S.; De Bellefon, C. Original use of the liquid nature of the stationary phase in counter-current chromatography: II. A liquid–liquid reactor for catalytic reactions. J. Chromatogr. **1998**, *828*, 523–530.

Influence of Organic Solvents on pK_a

Ernst Kenndler
Institute for Analytical Chemistry, University of Vienna, Vienna, Austria

INTRODUCTION

The influence of solvents on the ionization equilibrium is related to their electrostatic and their solvation properties. The value of the ionization constant of an analyte is closely determined, in practice, by the pH scale in the particular solvent. It is clear that it is most desirable to have a universal scale which is able to describe acidity (and basicity) in a way that is generally valid for all solvents. It is, in principle, not the definition of an acidity scale in theory which complicates the problem; it is the difficulty of approximating the measured values in practice to the specifications of the definition. The pH scale, as is common in water, is applicable only to some organic solvents (i.e., mainly those for which the solvated proton activity is compatible with the Brønsted theory of acidity). The applicability of an analog to the pH scale in water decreases with decreasing relative permittivity of the solvents and with their increasing aprotic character.

A scale that would enable us to compare the acidity in all solvents could be based on the transfer activity coefficient on the proton (see the entry Capillary Electrophoresis in Nonaqueous Media). The effect of the solvent of any species can be expressed in the same way as for the proton by this concept and applied to all particles involved in the thermodynamic equilibrium.

MEDIUM EFFECT AND IONIZATION CONSTANTS OF WEAK ACIDS AND BASES

The ionization of a weak neutral acid, HA, is described according to

$$HA = H^+ + A^-$$

For the weak base, B, for formal reasons, it is favourably expressed for its conjugated cation acid:

$$HB^+, \text{ by } HB^+ = H^+ + B$$

The corresponding changes of the ionization constants of these weak acids can be expressed by the particular transfer activity coefficients, $_m\gamma_i$, on the single species, i, according to

$$\Delta pK_{a,HA} = {}_s pK_{a,HA} - {}_w pK_{a,HA} = \log\left(\frac{{}_m\gamma_{H^+}\,{}_m\gamma_{A^-}}{{}_m\gamma_{HA}}\right) \tag{1}$$

$$\Delta pK_{a,HB^+} = {}_s pK_{a,HB^+} - {}_w pK_{a,HB^+} = \log\left(\frac{{}_m\gamma_{H^+}\,{}_m\gamma_{B}}{{}_m\gamma_{HB^-}}\right) \tag{2}$$

S and W indicate organic solvent and water, respectively. The transfer activity coefficients, $_m\gamma_i$, is the ratio of the activity coefficients in the particular solvents: $_m\gamma_i = {}_w\gamma_i/{}_s\gamma_i$.

Eqs. (1) and (2) enable the interpretation of the changes in pK_a values in the different organic solvents, compared to water. One must take into account not only the stabilization of the proton (this is given by the mutual basicity of the solvents) but also the different ability to stabilize the other individual particles. Besides the neutral particles HA and B, oppositely charged ions H^+ and A^- take part in the equilibrium in case of the neutral acids, and equally charged HB^+ and B^+ ions in case of the cation acid. This occurrence leads to a different change of the pK_a values for these two different types of weak electrolytes, depending on the ability of the solvent to stabilize anions or cations.

Nearly all organic solvents stabilize anions worse than water. For this reason, the pK_a values of neutral acids are larger in organic solvents (it is obvious that strongly basic solvents like amines may level out this effect). In lower alcohols, for example, the pK_a values increase by several units. In acetonitrile, which has generally even less cation stabilization ability in addition, the pK_a values may increase by 16 units. The effect of many solvents on weak bases (expressed by the pK_a of the corresponding cation acid) is much lower. pK_a values change only by few units. An exception is acetonitrile, due to the reason mentioned earlier.

It should be noted that traces of water have a great influence on the shift of pK_a values in all solvents, because there is a steep change of the pK_a with increasing water content of the organic solvent. It should be also pointed out that the pK_a of the silanol groups at

Encyclopedia of Chromatography DOI: 10.1081/E-ECHR-120040018

Table 1 Ionization constants of neutral and cation acids of type HA and HB+, respectively, in water, amphiprotic, and dipolar aprotic solvents

Acid	pK_a						
	W	**MeOH**	**EtOH**	**t-BuOH**	**ACN**	**DMSO**	**DMF**
Acetic	4.73	9.7	10.3	14.2	22.3		13.3
Chloro acetic	2.81	7.8	8.3	12.2	18.8		10.1
Benzoic	4.21	9.4	10.1	15.1	20.7	11.0	12.3
3,4-Dimethyl benzoic	4.4	9.7		15.4	21.2	11.4	13.0
3-Bromo benzoic	3.81	8.8	9.4	13.5	20.3	9.7	11.3
4-Nitro benzoic	3.45	8.3	8.9	12.0	18.7	9.0	10.6
2,4,6-Trinitro phenol	0.3	3.7	4.1	4.8	11.0		
Ammonium	9.2				16.5	10.5	
Ethylammonium	10.6				18.4	11.0	
Triethylammonium	10.7	10.9			18.5	9.0	
Anilinium	4.6		5.7		10.6	3.6	
Pyridinium	5.2	5.2			12.3	3.4	

the surface of the commonly used fused-silica material is affected by the choice of the solvent, too. It is also shifted to higher values (in water the pK_a is around 5–6).

SUGGESTED FURTHER READING

Bates, R.G. Medium effect and pH in nonaqueous and mixed solvents. In *Determination of pH, Theory and Practice*; John Wiley & Sons: New York, 1973; 211–253.

Kenndler, E. Organic solvents in capillary electrophoresis. In *Capillary Electrophoresis Technology*; Guzman, N.A., Ed.; Marcel Dekker, Inc.: New York, 1993; 161–186.

Kolthoff, I.M.; Chantooni, M.K. General introduction to acid-base equilibria in nonaqueous organic solvents. In *Treatise on Analytical Chemistry*; Kolthoff, I.M., Elving, P.J., Eds.; John Wiley & Sons: New York, 1979; 239–301.

Sarmini, K.; Kenndler, E. Ionization constants of weak acids and bases in organic solvents. J. Biophys. Biochem. Methods **1999**, *38*, 123–137.

Injection Techniques for Capillary Electrophoresis

Robert Weinberger
CE Technologies Inc., Chappaqua, New York, U.S.A.

INTRODUCTION

In LC, a loop containing a defined volume is used to introduce the sample into the flowing mobile phase. Injection in high-performance capillary electrophoresis (HPCE) differs in two ways: (a) the injection volume is not as well defined and (b) injection is performed with the electric field turned off. Both of these features can contribute to quantitative errors of analysis. In addition, the length of the injection plug must be kept quite small to maintain the efficiency of the electrophoretic process.[1] The use of stacking electrolytes permits large injections to be made.[2] This is necessary to achieve acceptable limits of detection.

There are two modes of injection in capillary electrophoresis (CE): hydrodynamic injection and electrokinetic injection. In hydrodynamic injection, pressure or vacuum are placed on the inlet sample vial or the outlet waste vial, respectively. For electrokinetic injection, the voltage is activated for a short time with the capillary and electrode immersed in the sample.

The general process of performing an injection and run is as follows:

1. The capillary is rinsed with 0.1 N sodium hydroxide or 0.1N phosphoric acid for 1–2 min.
2. A second rinse with background electrolyte (BGE) is performed for 2–3 min.
3. The inlet side of the capillary is immersed in the sample.
4. Injection is performed for 1–30 s.
5. The voltage is ramped up (15 s) to the designated value and the separation is performed.
6. The process repeats for the next sample.

VOLUMETRIC CONSTRAINTS ON INJECTION SIZE

Because the entire internal volume of a 50-cm × 50-μm-inner diameter (i.d.) capillary is only 981 nL, the injection volume must be kept quite small. The contribution to band broadening (variance) from a plug injection is given by

$$\sigma_{inj}^{2} = \frac{l_{inj}^{2}}{12} \tag{1}$$

where l is the length of the injection plug. To calculate the band broadening from the injection process, the diffusion-limiting case can be considered using the Einstein equation:

$$\sigma_{diff}^{2} = 2D_{m}t \tag{2}$$

Because the squares of the variances are additive, the contributions to band broadening from injection and diffusion can be inserted into the theoretical plate equation:

$$N = \left(\frac{L_{d}}{\sigma_{tot}}\right)^{2} \tag{3}$$

For a 50-cm capillary and a solute migration time of 600 s, the impact of the injection size for a small molecule $(D_{m} = 10^{-5}\, cm^{2}/s)$ and large molecule $(D_{m} = 10^{-6}\, cm^{2}/s)$ is shown in Fig. 1.

As illustrated in Fig. 1, injection of 1% (0.5 cm) of the capillary volume with sample produces a 92% loss of efficiency for a large molecule and an 8% loss of efficiency for a small molecule. Because diffusion is a limiting cause of efficiency, the large molecule provides a higher number of theoretical plates. The more efficient the separation process, the more difficult it is to maintain that inherent efficiency.

This model assumes that the sample is dissolved in BGE. Through the use of a low-ionic-strength solution as the sample diluent, sample stacking permits large-volume injections to be made.

In a well-controlled separation, injection can be the greatest source of band broadening.[1] This is one of the reasons that micromachined systems may become important for high-resolution DNA separations. In this case, it is possible to inject minute amounts of sample and use shortened separation channels.[3] Sensitivity does not suffer because laser-induced detection is employed.

HYDRODYNAMIC INJECTION

The volume of material injected per unit time (V_{t}, nL/s) is determined by the Poiseuille equation.

$$V_{t} = \frac{\Delta P D^{4} \pi}{128 \eta L} \tag{4}$$

where ΔP equals the pressure drop, D is the capillary internal diameter, η is the viscosity, and L is the length

Encyclopedia of Chromatography DOI: 10.1081/E-ECHR-120040019

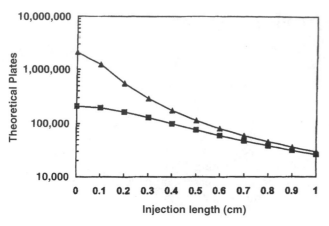

Fig. 1 Effect of the injection zone length on the number of theoretical plates for a small molecule and large molecule (■, $D_m = 10^{-5}$ cm^2/s) and large molecule (▲, $D_m = 10^{-6}$ cm^2/s) as solved by Eq. (3). Conditions: capillary length = 50 cm to detector; migration time = 600 s.

of the capillary. On some instruments, the pressure is generated by raising the capillary inlet side (siphoning).

The problems generated using an open-ended injection system as shown by the Poiselle equation dictate that changes in the experimental conditions will result in variations of the amount of material injected. Internal standards are best used to compensate for some of the experimental variables.

Pressure-driven systems are preferred compared to vacuum-driven systems for two reasons: (a) Generation of pressures over 1 atm is important when viscous polymer networks are used for size separations and (b) interface to the mass spectrometer is simpler.

ELECTROKINETIC INJECTION

The quantity (Q) of a solute injected is given by

$$Q = (\mu_{ep} + \mu_{eo})\pi r^2 ECt \qquad (5)$$

where μ_{ep} and μ_{eo} are the electrophoretic and electro-osmotic mobilities, respectively, r is the capillary radius, E is the field strength, t is the time of injection, and C is the concentration of each solute.

As illustrated in Fig. 2, solutes with high mobility are preferably injected compared to those with low mobility.[4] Note the smaller peak heights for lithium and arginine compared to rubidium when electrokinetic injection is employed. Solutes that have identical mobility in free solution show no such bias (e.g., oligonucleotides and DNA fragments). This is fortunate because it is often necessary to use electrokinetic injection with gel-filled capillaries or when high-viscosity polymer networks are employed.

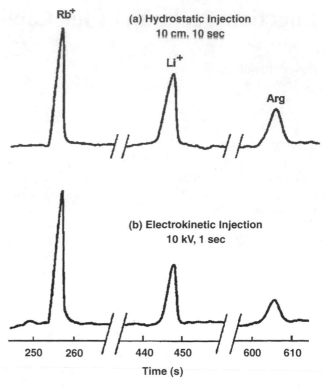

Fig. 2 Hydrostatic versus electrokinetic injection. Buffer: 20 mM MES adjusted with histidine to pH 6.0; solutes: Rb$^+$, Li$^+$, and arginine, 5×10^{-5} M injection: (top) hydrostatic, $\Delta h = 10$ cm, $t = 10$ s; (bottom) electrokinetic, 1 s at 10 kV; detection: conductivity. (Reprinted with permission from Anal. Chem. **1988**, *60*, 375.)

The problem with electrokinetic injection is that the field strength at the point of injection is inversely proportional to the sample conductivity. Calibration curves for ionic solutes show negative deviations from linearity because of this. Internal standards are necessary unless it is certain that the sample conductivities are identical. Low-conductivity samples are preferable because they stack.

The advantage of electrokinetic injection is that extreme trace enrichment is possible.[5] If the electro-osmotic flow approaches zero, it is possible to inject only solute ions, omitting the sample diluent.

"SHORT-END" INJECTION

The section of capillary between the outlet vial and the detector can be used for high-speed separations if sufficient selectivity is designed into the separation.[6] The process is as follows. (1) Equilibrate the capillary in BGE as usual. (2) Place the sample at the capillary outlet. (3) Inject by pressuring the outlet vial or with electrokinetic injection using negative polarity (inlet-side negative). (4) Set the power supply to negative polarity and perform the usual voltage ramp. Because the capillary length is short, the injection should also be kept small and stacking buffers should be used. Be sure to set the detector time constant to 10–20% of the peak width to minimize that form of band broadening. Depending on the instrument, the short-end of the capillary usually ranges from 6 to 10 cm.

INJECTION ARTIFACTS, PROBLEMS, AND SOLUTIONS

No Injection

A plugged capillary is the usual culprit. Cut a few millimeters of the inlet or pressurize the outlet with a syringe to unplug it. When plugged, the observed current is usually zero. No injection can also occur if an empty or incorrect sample vial is used, if an incorrect vial is called for in the method, if the vial cap is missing or badly leaking, or if the external pressure source (if required) is not activated. It is possible that the capillary is broken. Breaks usually occur at the detection window. Check that the voltage polarity is correctly set.

Peak Tailing

Peak tailing can result from a poorly cut capillary inlet. If the capillary is not cut squarely, a concentration gradient can occur upon injection.[7]

Peak Splitting

Artifactual-injection-related peak splitting can occur under certain conditions. When the sample diluent contains organic solvents and micellar electrokinetic capillary electrophoresis or cyclodextrin containing electrolytes are employed, splitting can occur due to the distribution of the solute between two phases moving at different speeds at the point of injection.[8] The problem is solved by dissolving the solute in aqueous media. If the sample is insoluble in totally aqueous solvents, 6 M urea can be added both to the BGE and the sample diluent.

A fracture near the capillary inlet can also cause peak splitting. The break can occur if the capillary hits a vial wall or seal. The polyimide coating keeps the cracked portion intact. During injection, the sample moves into the capillary from both the open end of the tube and through the crack. The split peak is usually smaller than the main component and always has a migration time that is a little shorter. This is confirmed by examining the capillary inlet. If fractured, a small piece often detaches and the peak splitting is resolved.

REFERENCES

1. Huang, X.; Coleman, W.F.; Zare, R.N. Analysis of factors causing peak broadening in capillary zone electrophoresis. J. Chromatogr. **1989**, *480*, 95.
2. Burgi, D.; Chien, R.-L. Optimization in sample stacking for high-performance capillary electrophoresis. Anal. Chem. **1991**, *63*, 2042.
3. Jacobson, S.C.; Hergenroder, R.; Koutny, L.B.; Warmack, R.J.; Ramsey, M.J. Effects of injection schemes and column geometry on the performance of microchip electrophoresis devices. Anal. Chem. **1994**, *66*, 1107.
4. Huang, X.; Gordon, M.J.; Zare, R.N. Bias in quantitative capillary zone electrophoresis caused by electrokinetic sample injection. Anal. Chem. **1988**, *60*, 375.
5. Zhang, C.-X.; Thormann, W. Head-column field-amplified sample stacking in binary system capillary electrophoresis. 2. Optimization with a preinjection plug and application to micellar electrokinetic chromatography. Anal. Chem. **1998**, *70*, 540.
6. Euerby, M.R.; Johnson, C.M.; Cikalo, M.; Bartle, K.D. Short-end injection—rapid analysis capillary electrochromatography. Chromatographia **1998**, *47*, 135.
7. Guttman, A.; Schwartz, H.E. Artifacts related to sample introduction in capillary gel electrophoresis affecting separation performance and quantitation. Anal. Chem. **1995**, *67*, 2279.
8. Weinberger, R. Am. Lab. **1997**, *29*, 24.

Inorganic Elements: Analysis by CCC

E. Kitazume
Faculty of Humanities and Social Sciences, Iwate University, Morioka, Japan

INTRODUCTION

Countercurrent chromatography (CCC) has been applied to preconcentration and separation of inorganic elements since the end of the 1980s. Since the early 1990s, certain inorganic elements, including rare earths, have been separated by high-speed countercurrent chromatography (HSCCC). In addition, preconcentration and separation of inorganics from geological samples have been studied. Many of the features of HSCCC have convinced us that this method can be successfully used in separation of inorganic elements and inorganic analytical chemistry. However, the liquid systems for inorganics are somewhat complicated as compared with those for the separation of organics because they usually contain significant amounts of an extracting reagent which influences kinetic properties and viscosities of the two-phase system.

To achieve high sensitivity for analyzing trace inorganic elements in a sample solution using an atomic absorption spectrometry (AAS) or ICP atomic emission spectrometry (ICP/AES), conventional preconcentration methods such as evaporation, ion exchange, solvent extraction techniques, etc. have been used. However, there are several problems in their methodologies for the determination of ultratrace elements; for example, peak broadening for the ion exchange technique, low enrichment factor for solvent extraction techniques, etc. are encountered. It is difficult to directly work with under 0.5-mL concentrated sample solution by conventional methods. If there were effective methods for concentrating traces into 0.1 mL or less volume solution, absolute detection limits for trace analysis using techniques such as AAS, ICP/AES, and ICP-mass spectrometry (ICP/MS) would be greatly decreased, as well as eliminating their matrix effects.

pH Peak Focusing

pH-peak-focusing countercurrent chromatography (pH-PFCCC) is a unique technique which is based on neutralization between mobile and stationary phases.[1,2] It has been applied to the separation and enrichment of organic compounds such as indole auxins, bromoacetyl thyroxine and its analogs, dinitrophenyl amino acids, transretinoic acid, and diazepam.

Neutralization is initiated at the mobile phase front, but advances through the column at a slower pace, forming a sharp border between basic and acidic zones. Trace impurities in the sample solution are concentrated at this narrow pH boundary in the column. This has great potential for on-line enrichment and subsequent analysis of trace inorganic elements by interfacing HSCCC with analytical instruments such as nonflame atomic absorption spectrometry (NFAAS), ICP/AES, and ICP/MS.

Recently, the feasibility of a HSCCC centrifuge in enriching several metallic elements was demonstrated using pH-PFCCC. Under optimum conditions, an excellent enrichment factor of over 100 is achieved with good recovery of Ca, Cd, Mg, Mn, Pb, and Zn at a concentration of several parts per billion by on-line detection using a direct-current plasma atomic emission spectrometer (DCP-AES) as the detector.[3] In addition, many metal ions were efficiently enriched into an eluent of 100 μL or less by HSCCC, resulting in a greater than 100-fold increase in peak intensities for a 10-mL sample solution. Preconcentrating the sample solution, substantially improving detection limits, facilitates conventional trace determination of inorganic elements by instrumental analysis; however, usually, metals could not be separated from each other. If the pH border can be sufficiently broad, and the order of the elements in the column is maintained until the final stage of the elution, each concentrated metal ion may be separated mutually, as in displacement chromatography.

Under appropriate experimental conditions with pH-PFCCC, major matrix elements such as Ca and Mg in enriched tap water were found to be separated from trace elements.[4] Trace Cd, which will appear between chromatographic bands of Mg and Ca when using di(2-ethylhexyl) phosphoric acid (DEHPA) as an extracting reagent, was well determined without interference by alkali metals.

EXTRACTION REAGENT FOR SEPARATION OF INORGANIC ELEMENTS

The existence of extracting reagent in the mobile phase is an essential factor in separation and enrichment of inorganic elements. It complicates the determination of several important factors, e.g., distribution

Encyclopedia of Chromatography DOI: 10.1081/E-ECHR-120027337

coefficients, peak resolution, and separation efficiency. Some basic researches revealed that kinetic properties of specific systems used in HSCCC affect the separation efficiency. Moreover, mass-transfer rates into organic stationary phases are significantly responsible for separating mode, i.e., stepwise or isocratic elution. In addition, the values of the distribution coefficients, determined by the batch extraction measurements in the systems, are sometimes considerably different from those of the dynamic distribution coefficients calculated from elution curves plotted from experimental data. Further theoretical and basic investigations are necessarily concerned with extraction kinetics, as well as hydrodynamic behavior of the two phases in the HSCCC column.

In Table 1, typical extracting reagents used for separation and enrichment of inorganic elements are summarized. Organophosphorus extractants are often used because of their solubility properties. Di(2-ethylhexyl) phosphoric acid is commonly applied to industrial separations because of its high extractability and high separation factors between many inorganic elements, especially for rare earth elements. Other metal ions are extracted as well as the trivalent metal ions.

ENRICHMENT ANALYSIS IN THE EFFLUENT

Large-Scale Enrichment Followed by Conventional Elution

Enrichment of the desired trace elements prior to their determination cannot simply overcome such problems as interference, toxic or radioactive samples, etc., but can also provide highly sensitive determination of trace elements.

The parts-per-billion level of metal ions in 500 mL of the mobile phase was continuously concentrated into small volumes of the stationary phase retained in the column. Concentrated metal ions were simultaneously eluted with nitric acid and determined by the emission intensity with a direct current plasma atomic emission spectrometer. The recoveries of Ca, Cd, Mg, Mn, Pb,

and Zn ranged over 88% at the concentration, of each, of 10 ppb in 500 mL of the sample solution. Versatility of this method was further demonstrated in determination of trace metals in tap water and deionized water.

Rare earth elements have been enriched into a stationary phase composed of toluene including 2-ethylhexylphosphonic acid mono-2-ethylhexyl ester (EHPA) from 1 L of aqueous solution and eluted with a stepwise pH gradient. As many elements remained in the column head because of their high partition coefficients to the stationary phase, they can be eluted with mobile phase and also separated mutually.

A large-scale enrichment technique is very useful for determining extremely low level concentrations of metals in solution when a large amount of sample such as natural water is available.

Enrichment Using pH-Zone Refining Technique as Preconcentration Method of Inorganic Analysis for Subsequent Determination

Even if sufficient sample size, in volume, may not be available, enrichment techniques that concentrate trace metals in microliter samples are sometimes quite useful because modern instrumental detection systems such as AAS, ICP/AES, ICP/MS, etc. do not need a large sample size. Moreover, if trace metals that have been separated from their major substances can be concentrated in an extremely small area of the polytetrafluoroethylene (PTFE) tube in HSCCC, this would be an ideal flow-injection analysis system for determination of inorganics. From this point of view, the recently developed pH-zone refining technique has great potential for enrichment, especially for instrumental inorganic trace analysis.

In the pH-zone refining technique, a basic organic solution containing a complex-forming reagent, such as DEHPA, as a stationary phase is used. After sample solution is introduced into the column, metal ions stay close to the sharp pH border region in the small-bore PTFE tube. Then, the trace inorganic ions in the sample are moved by the acid effluent (diluted

Table 1 Typical extracting reagent for separation and enrichment of inorganic elements using HSCCC

Extracting reagent	Two-phase system	Inorganic element
Di(2-ethylhexyl) phosphoric acid (DEHPA)	HCl, organic acid–heptane	Rare earth, heavy metals
2-Ethylhexylphosphonic acid mono-2-ethylhexylester (EHPA)	Carboxylic acid–toluene	Rare earth
Dinonyltin dichloride	HCl, HNO$_3$–Methylisobutylketone (MIBK)	Orthophosphate and pyrophosphate
Cobalt dicarbolide	HNO$_3$–nitrobenzen	Cs and Sr
Tetraoctylethylenediamine (TOEDA)	HCl, HNO$_3$, organic acid–chloroform	Alkali, alkaline-earth, rare earth, heavy metals, Hf, Zr, Nb, Ta

hydrochloric or nitric acid, etc.) to the tail of the column while concentrating in the sharp-moving pH "interface," and finally eluted as small fractions containing concentrated inorganic ions. This enrichment method for trace organic impurities has been called "pH-peak-focusing countercurrent chromatography (pH-PFCCC)."

The concentration procedure is modeled in Figs. 1–3. In Fig. 1, sample ion is concentrated into the column head just after the concentration procedure is started. After the stationary phase has been introduced into the system, the HSCCC is started at an appropriate rotational rate, followed by the sample and the mobile-phase pumps. For example, monovalent metal or inorganic ions (M^+) are concentrated into the stationary phase as MR in Fig. 1. The dark-shaded zone shows an organic stationary phase, which includes organophosphorus extractants (HR) and an alkali such as ammonia. The bright-shaded zone shows a sample phase with the pH adjusted by ammonia. When the column rotation is started, as the organic stationary phase is lighter than the mobile phase (diluted acid solution, HCl in Fig. 1), it moves to the head of the column by an Archimedean screw effect (ASE). The driving force, based on ASE, increases with the rotational speed of the column. Therefore the stationary phase can be retained in the column by selecting an appropriate rotational rate and pump rate, even if the mobile phase is introduced from the head direction (left side in Fig. 1) into the column. The retention ratio of the stationary phase to the whole column varies from 20% to 70%, but is stable when all conditions including pump and rotational speed are constant. So the position of the mobile phase is stable in a column while in operation. Inorganic ions (M^+) form complexes with ligand ions (R^-) and are mainly concentrated on the column head, shown as MR in Fig. 1. A proton is transferred into the mobile phase when the inorganic ion extracted into the stationary phase.

Fig. 2 shows a displacement procedure as well as the concentration procedure in the HSCCC column

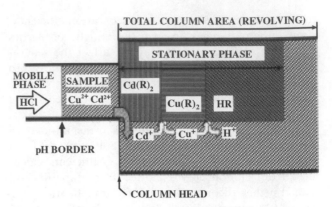

Fig. 2 Metal displacement process in stationary phase.

after most ions in the sample are extracted on the top of the stationary phase. If sample contains two kinds of divalent metal ions (Cu and Cd) in a large amount of solution, each metal might be arranged by difference of its affinity or partition rate to the stationary phase. As Cd ion is usually more extractable than Cu ion, Cd can displace Cu at the end (left) of the Cu band in the stationary phase. However, the bandwidth of Cd is increasing until the entire sample was introduced into the column.

Metal-focusing process after the sample injection is shown in Fig. 3. After all ions in the sample are extracted in the stationary phase, mobile phase is introduced in the column. Ammonia in the stationary phase begins to be neutralized with hydrochloric acid in the mobile phase. The neutralization area, where reaction between the acid and the base has just finished, is shown as pH border in Fig. 3. The pH border proceeds from left to right (the same direction as the mobile phase); however, its flow rate is lower than that of the mobile phase because of the delay based on the neutralization between the acid in the mobile phase and the base in the stationary phase. Therefore the flow rate of the pH border may be controlled by adjusting the concentration of base and acid in each phase.

As a result, Cd zone will be enriched despite increasing its retention time of chromatogram. On the other

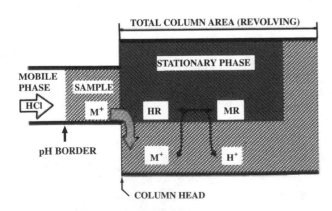

Fig. 1 Metal extraction into stationary phase.

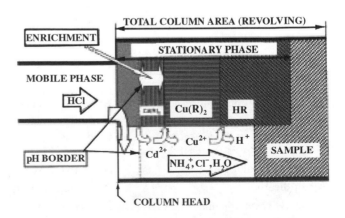

Fig. 3 Metal enrichment process after sample injection.

hand, as the border between Cd and Cu also moves by proceeding elution of Cd, Cu zone also moves to tail with enrichment.

If the concentration of hydrochloric acid just after the sample is very low or in case of using gradient elution mode, pH border shown as dotted line in Fig. 3 will move more slowly.

If the speed of pH border would be relatively so high compared with the movement of Cu zone, pH border will catch up the border between Cu zone and reagent zone shown as HR in Fig. 3. Then, both peaks could not be separated well in such quick movement of pH border.

In both concentration mechanisms described above, there is little diffusion process observed as in usual elution procedures with ion exchange or other chromatographic separation method, such as conventional HSCCC and HPLC. If there is no basic compound such as ammonia in the stationary phase, ions move to the tail with a different flow rate as a function of distribution ratio between the stationary phase and the mobile phase. Many ions can be separated from each other in that case, but because there is no pH border in the column, concentration is not effected, but only separation with diffusion.

Fig. 4 shows the typical concentration results for a 10-ppm solution of cadmium, magnesium, and zinc. The injected sample solution contained 50 μg of each in 5 mL of 0.1 M tartaric acid solution, adjusted to pH to 9.25. The mobile phase was pumped at a flow rate of 0.05 mL/min. Rotational speed was 950 rpm. The eluent was collected every 2 min (0.1-mL fractions). The fractions were diluted 1:10 with water, and the emission intensity for each element was measured by plasma atomic emission spectrometer. The emission intensities for each element were increased 20-fold compared with the original sample solution. The results of this study demonstrated the high-performance capabilities of the pH-zone refining technique. Trace elements in the sample solution could be

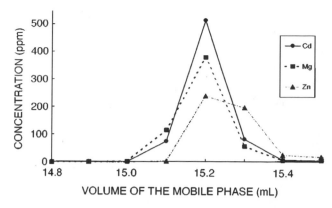

Fig. 4 Typical concentration results for 10-ppm solution of cadmium, manganese, and zinc. Apparatus: HSCCC centrifuge with 10.0-cm revolution radius; column: one multilayer coil; sample: 5 mL of each 10-ppm solution (pH 9.25) in 0.1 M tartaric acid; mobile phase: 0.1 M HCl saturated with ether; stationary phase: 6 mL of 0.2 M DEHPA and 0.18 M ammonia in ether; column: 0.5 mm i.d. × 32 m; flow rate: 0.05 mL/min; rotational speed: 950 rpm.

successfully concentrated into a small volume, almost under 0.1 mL with an enormous level of enrichment.

On the other hand, if the speed of pH border would be relatively slow as in Fig. 3, so that the pH border cannot catch up the border between the Cu zone and the reagent zone until the border passes through the stationary phase, both peaks would be separated because the Cu zone will be eluted before coming close to the enriched Cd zone. So the order of the elements will be maintained until the final stage of elution in the tail of the column. Whereas the use of large-bore column may accelerate the peak appearance because of the longitudinally short distance of the stationary phase, each metal may be separated well compared with the case of using a small-bore column. As well as enrichment, chromatographic separation would be essential for exact measurement. The speed of the pH

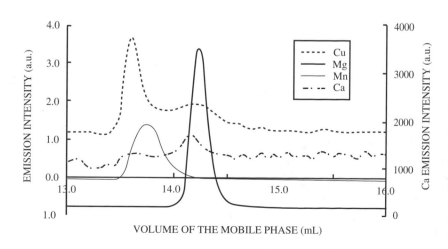

Fig. 5 Enrichment profiles of Mg, Cu, Mn, and Ca in 10-mL tap water. Experimental conditions: column, one monolayer coil, 0.5 mm i.d. × 10 m (2 mL); sample, 10 mL of tap water (pH 8.0) in 0.1 M tartaric acid; mobile phase, 0.1 M HCl including 0.1 M tartaric acid; stationary phase, 0.22 M DEHPA and 0.20 M ammonia in heptane; flow rate, 0.1 mL/min at enrichment stage and 1.0 mL/min at detection stage; rotational speed, 1200 rpm; Sf, 7.5%.

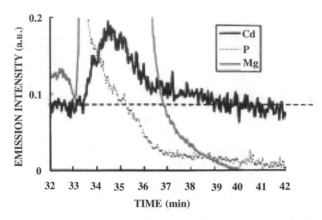

Fig. 6 Enrichment profiles of Cd and Mg in 30-mL tap water. Experimental conditions: column, one monolayer coil, 1.6 mm i.d. × 0.86 m (2 mL); sample, 30 mL of tap water (pH 7.1) in 0.1 M tartaric acid; mobile phase, 0.1 M HCl including 0.1 M tartaric acid; stationary phase, 2 mL of 0.22 M DEHPA and 0.20 M ammonia in heptane; flow rate, 1.0 mL/min; rotational speed, 800 rpm; Sf, 20%.

border can be controlled by choosing appropriate experimental conditions, such as bore size of a column and molar ratio between acid and base.

When a small-bore column, such as 0.5 mm i.d., was used, all peaks appeared simultaneously if the volume of the retained stationary phase was over 15%.[3] However, if the retention volume of the stationary phase was less than 10% in a small-bore column, peak separation was observed as shown in Fig. 5.[4] Each peak was detected by plasma atomic emission spectrometry. This peak profile shows enrichment profiles with separation of Mg, Cu, Mn, and Ca in tap water. The intensity of Ca is shown in the right axis, 3 orders higher than that of the other elements, while the intensity of Mn is amplified 10 times. The spectral interference of Ca to the signal of Cu is observed. This separation phenomenon is considered to be quite useful for exact determination of trace metals.

On the other hand, in the case of using a large bore column, such as 1.6 mm i.d., with the same volume of stationary phase, peak separation was observed even if there was sufficient retention volume of more than 15%. This phenomenon may be explained by shorter longitude of the stationary phase in a large bore column, compared with a small-bore column. Fig. 6 shows the signals of Mg, phosporus, and Cd for 30 mL of tap water. The large peak of the Cd emission line may be the result of spectral interference caused by the scaled-out peak of Mg. On the shoulder of the Cd peak, a small peak was observed. This may be the real emission signal of Cd. If using a large peak on the determination of Cd,

the result was 3 ppb. On the other hand, the result determined by the small peak was about 1 ppb. Using the pH-PFCCC system, precision results for the determination of Cd or other elements which are influenced by matrices elements could be obtained.

CONCLUSIONS

In contrast to HPLC, the unique feature of CCC is that there is no solid support in the column. As the distribution abilities, including the capacity of the stationary phase, are easy to control, CCC can be applied to the separation, enrichment, and purification of inorganics over a wide range of concentration. In particular, enrichment of trace elements using pH-peak-focusing countercurrent chromatography will be an ideal preconcentration method for subsequent inorganic determination of modern instrumental analytical methods. Countercurrent chromatography can be combined directly with the flow injection technique, and it shows great potential for preconcentration of selected desired trace inorganic elements prior to their final detection and quantitation. On-line enrichment and subsequent analysis, based on a simple and high-performance enrichment system for inorganics, may take the place of conventional sample preparation using a beaker and separatory funnel in future investigations in this field.

REFERENCES

1. Ito, Y., Conway, W.D., Eds.; *High-Speed Counter-current Chromatography*; John Wiley & Sons: New York, 1996.
2. Ito, Y.; Shibusawa, Y.; Fales, H.M.; Cahnmann, H.J. Studies on an abnormally sharpened elution peak observed in countercurrent chromatography. J. Chromatogr. A. **1992**, *625*, 177.
3. Kitazume, E.; Higashiyama, T.; Sato, N.; Kanetomo, M.; Tajima, T.; Kobayashi, S. On-line microextraction of metal traces for subsequent determination by plasma atomic emission spectrometry using pH peak focusing countercurrent chromatography. Anal. Chem. **1999**, *71*, 5515.
4. Kitazume, E.; Takatsuka, T.; Sato, N.; Ito, Y. Mutual metal separation system with enrichment using pH-peak focusing countercurrent chromatography. J. Liq. Chromatogr. & Relat. Technol. **2003**, *27*, 427.

Instrumentation of Countercurrent Chromatography

Yoichiro Ito
*National Heart, Lung, and Blood Institute, National Institutes of Health,
Bethesda, Maryland, U.S.A.*

INTRODUCTION

Countercurrent chromatography (CCC) is a support-free liquid–liquid partition system where solutes are partitioned between the mobile and stationary liquid phases in an open column space. The instrumentation, therefore, requires a unique approach for achieving both retention of the stationary phase and high partition efficiency in the absence of a solid support. The variety of existing CCC systems may be divided into two classes,[1] i.e., hydrostatic and hydrodynamic equilibrium systems. The principle of each system can be illustrated by a simple coil, as shown in Fig. 1.

TWO BASIC CCC SYSTEMS

The basic hydrostatic equilibrium system (Fig. 1, left) utilizes a stationary coil. The mobile phase is introduced at the inlet of the coil, which has been filled with the stationary phase. The mobile phase then displaces the stationary phase completely on one side of the coil (dead space), but only partially displaces it on the other side due to the effect of gravity. This process continues until the mobile phase elutes from the coil. Once this hydrostatic equilibrium state is established throughout the column, the mobile phase only displaces the same phase while leaving the stationary phase permanently in the coil. Consequently, the solutes locally introduced at the inlet of the coil are subjected to a continuous partition process between two phases at each helical turn and separated according to their partition coefficients in the absence of a solid support.

The basic hydrodynamic equilibrium system (Fig. 1, right) uses a rotating coil, which generates an Archimedean screw effect, whereby all objects of different densities in the coil are driven toward one end, conventionally called "head," the other end being the tail. The mobile phase introduced through the head of the coil is mixed with the stationary phase to establish hydrodynamic equilibrium, with a portion of the stationary phase retained in each turn of the coil. This process continues until the mobile phase elutes from the tail of the coil. After hydrodynamic equilibrium is established throughout the coil, the mobile phase displaces only the same phase while leaving the other phase stationary in the coil. Consequently, solutes introduced locally at the head of the coil are subjected to an efficient partition process between the two phases and separated according to their partition coefficients.

Each basic system has its specific advantages as well as disadvantages. The hydrostatic system provides stable retention of the stationary phase but has relatively low partition efficiency due to the limited degree of mixing. The hydrodynamic system, on the other hand, has a high partition efficiency in a short elution time, but the retention of the stationary phase tends to become unstable due to violent mixing, often resulting in emulsification and extensive carryover of the stationary phase.

DEVELOPMENT OF HYDROSTATIC CCC SYSTEMS

In the early 1970s, the hydrostatic system was quickly developed into several efficient CCC schemes, as shown in Fig. 2.[2] The development has been done by utilizing unit gravity (Fig. 2, top) or centrifugal force (Fig. 2, bottom).

In droplet CCC, which utilizes unit gravity, one side of the coil (Fig. 1, left), entirely occupied by the mobile phase, is reduced to a fine flow tube, while the other side of the coil is replaced by a straight tubular column. The column is first filled with the stationary phase and the mobile phase is introduced into the column in a proper direction so that it forms a string of droplets in the stationary phase by the effect of gravity. The system necessitates the formation of droplets, which limits the choice of solvent system. In order to allow more universal application of solvent systems, a locular column was devised by inserting centrally perforated disks into the tube at regular intervals to form a number of compartments called "locules." The locular column is held at an angle and rotated along its axis to mix the two phases in each locule. As in droplet CCC, the lower phase is eluted from the upper end of the locular column and the upper phase from the lower end for better retention of the stationary phase.

In the toroidal coil CCC (helix CCC) system, operated under a centrifugal force, the dimensions of the

Encyclopedia of Chromatography DOI: 10.1081/E-ECHR-120040020

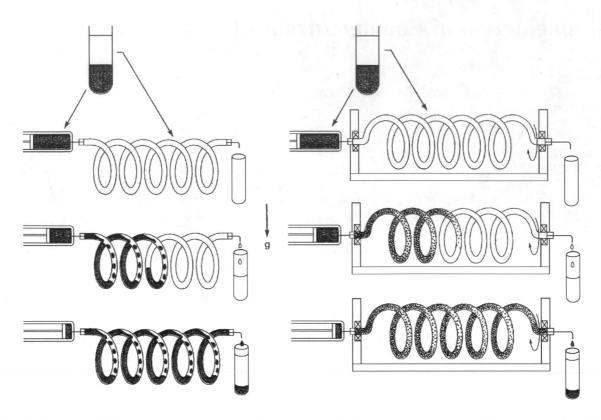

Fig. 1 Two basic CCC systems: hydrostatic equilibrium system (left) and hydrodynamic equilibrium system (right).

coil are reduced (Fig. 2, lower left). The coil is mounted around the periphery of a centrifuge bowl so that the stable, radially acting centrifugal force field retains the stationary phase, either upper or lower, on one side of the coil, as in the basic hydrostatic system (Fig. 1, left). The effective column capacity and retention of the stationary phase can be increased by replacing

the coil with a locular column arrangement (centrifugal partition chromatography).

INSTRUMENTATION OF HYDROSTATIC CCC SYSTEMS

Helix CCC (Toroidal Coil CCC)

Fig. 3A shows the design of the original helix CCC centrifuge rotor. A long helical column (typically 0.2–0.3 mm I.D. × 40 m length coiled onto a 0.85 mm O.D. tube making ca. 8000 turns) is accommodated around the periphery of the rotor. The mobile phase is introduced into the coil from a rotating syringe mounted at the center of the rotor, while the effluent from the outlet of the coil is collected through a rotary seal at the upper end of the syringe plunger. The system has a high partition efficiency of several thousand theoretical plates.[3]

The above original design has been improved using a seal-free centrifuge system based on nonplanetary motion (Fig. 5, bottom middle). Fig. 3B shows a cross-sectional view of the seal-free toroidal coil centrifuge. A long helical tube is accommodated around the periphery of the column holder. The seal-free (non-planetary) motion of the column is achieved by a set of four miter gears: When the motor drives the gear

Droplet CCC

Rotation Locular CCC

Basic HSES

Toroidal Coil CCC

Centrifugal Partition Chromatography

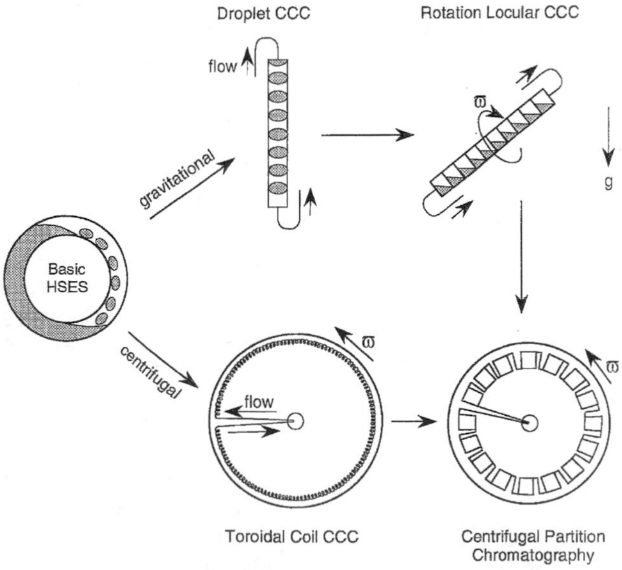

Fig. 2 Development of hydrostatic CCC systems.

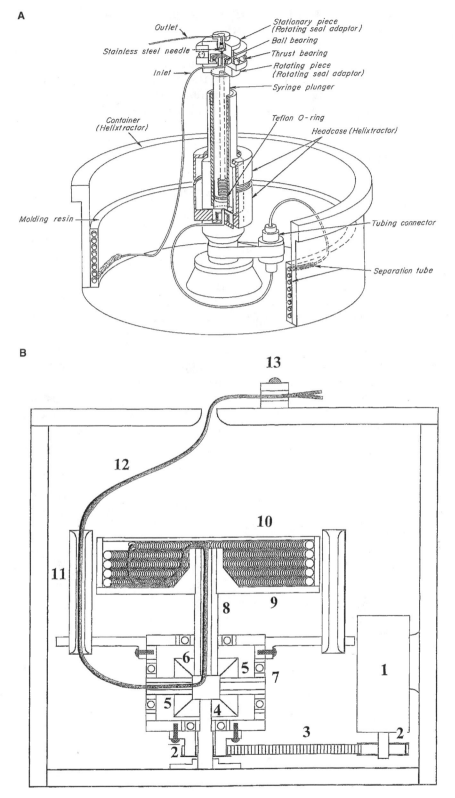

A

Outlet
Stationary piece (Rotating seal adaptor)
Stainless steel needle
Ball bearing
Thrust bearing
Inlet
Rotating piece (Rotating seal adaptor)
Syringe plunger
Container (Helixtractor)
Teflon O-ring
Headcase (Helixtractor)
Molding resin
Tubing connector
Separation tube

B

13
12
10
11
8 9
6 5 7
5 4
2 3 2
1

Fig. 3 Design of helix CCC (toroidal coil CCC) apparatus. (A) Schematic drawing of centrifuge head of original helix CCC apparatus; (B) cross-sectional view of advanced design of helix CCC equipped with seal-free flow-through device. 1, Motor; 2, toothed pulleys; 3, toothed belt; 4, stationary miter gear; 5, horizontal idler miter gear; 6, inverted upper miter gear mounted at bottom of column holder shaft (8); 7, gear box; 9, column holder; 10, coiled separation column; 11, hollow tube support; 12, flow tubes; 13, clamps.

box with a pair of toothed pulleys and a toothed belt, a pair of idler miter gears, engaged with the stationary miter gear at the bottom, rotate about their own axis at the same speed in a revolving gear box. This motion is further conveyed to the top miter gear, which is directly mounted at the lower end of the column holder shaft. Consequently, the column holder rotates at double the speed of an ordinary flow-through centrifuge system but without the need for the conventional rotary seal device. A pair of flow tubes from the coiled column pass through the holder shaft downward, through the horizontal shaft of an idler miter gear, and then through the vertical tube support (left) to exit the system at the center of the centrifuge cover, where they are tightly fixed with a pair of clamps, as indicated in the diagram. Because of the near-symmetrical arrangement of the design, the system is well balanced and the column can be rotated at a high speed up to 2000 rpm, while the elution can be performed through its seal-free flow-through system to eliminate complications, such as leakage of solvent and cross-contamination of solutes, that are often caused by the use of a conventional rotary seal device.[4]

Centrifugal Partition Chromatograph

Fig. 4 schematically shows a design of the centrifugal partition chromatographic system (see Fig. 2). The separation disk has a series of partition chambers connected by narrow ducts. The disk is rotated up to 2000 rpm. The flow-through system is made by a pair of rotary seals, which can maintain leak-free elution up to 60 bar (ca. 800 psi), although the seals should be kept clean. The system is computerized and the operation is programmed as in an HPLC system.[5]

DEVELOPMENT OF HYDRODYNAMIC CCC SYSTEMS

The performance of the basic hydrodynamic CCC system (Fig. 1, right) is remarkably improved by rotating the coil in a centrifugal force field, i.e., by applying a planetary motion to the coil. During the 1970s, a series of flow-through centrifuge schemes has been developed for performing CCC. In these centrifuge systems, the use of the conventional rotary seal devise is eliminated, since it leads to various complications such as leakage, clogging, and cross-contamination. These seal-less flow-through centrifuge schemes are divided into three classes: synchronous, nonplanetary, and nonsynchronous, according to the mode of planetary motion (Fig. 5).

In type I synchronous planetary motion (Fig. 5, upper left), a vertical holder revolves around the central axis of the centrifuge while it counter-rotates around its own axis at the same angular velocity. This counter-rotation of the holder unwinds the twist of the tube bundle caused by revolution, thus eliminating the need for the rotary seal. This principle works well for the rest of the synchronous schemes with tilted (types I-L and I-X), horizontal (types L and X), inversely tilted (types J-L and J-X), and even inverted orientation (type J) of the holder. When a

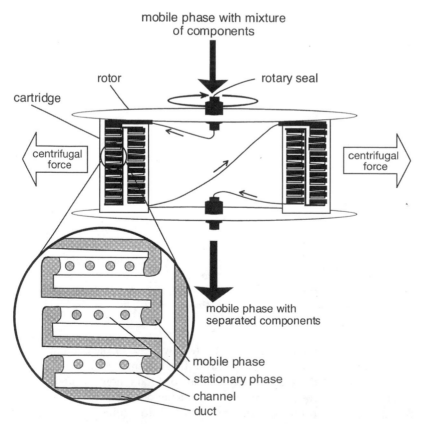

Fig. 4 Design of centrifugal partition chromatograph.

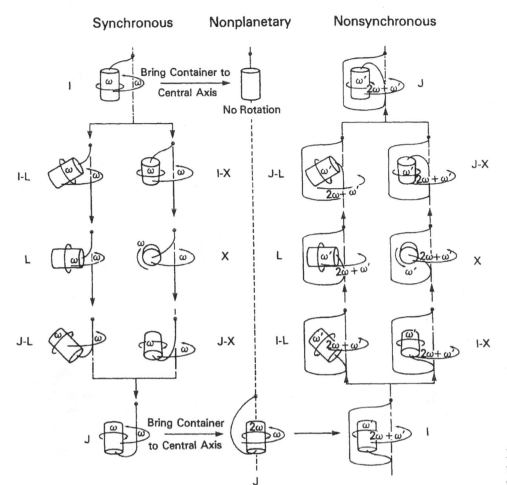

Fig. 5 Series of seal-less flow-through centrifuge systems for performing CCC.

holder of type I is moved to the center of the centrifuge, the counter-rotation of the holder cancels out the revolution effect, resulting in no rotation (Fig. 5, upper center). In contrast, when this shift is applied to type J planetary motion, the rotation of the holder is added to the revolution, resulting in the rotation of the holder at doubled speed, while the tube bundle revolves around the holder to unwind the twisting (Fig. 5, bottom center). This nonplanetary scheme is a transitional form to nonsynchronous planetary motion. On the basis of the nonplanetary scheme, the holder is again shifted toward the periphery to undergo a synchronous planetary motion. Since the net revolution speed of the coil is the sum of the nonplanetary and synchronous planetary motions, the ratio of the rotation and revolution becomes freely adjustable.

Several useful CCC systems have been developed from these centrifuge schemes. The nonplanetary scheme has been used for toroidal coil CCC,[4,6] centrifugal precipitation chromatography,[7,8] and online apheresis in blood banks.[9,10] The nonsynchronous scheme has been applied to the partitioning of cells

with polymer phase systems and also to cell elutriation with physiological solutions.[11,12] The type J synchronous scheme has been further developed into a highly efficient CCC system called high-speed CCC.[13,14]

Development of High-Speed Countercurrent Chromatography (HSCCC)

The development of HSCCC was initiated by the discovery that when type J planetary motion is applied to an end-closed coil coaxially mounted on a holder (Fig. 6A), the two solvent phases are completely separated in such a way that one phase occupies the head side and the other phase the tail side of the coil.

This bilateral hydrodynamic distribution can be utilized for performing CCC, as illustrated in Fig. 6B, where each of the coils is schematically shown as a straight tube to indicate the overall distribution of the two phases. The top coil shows the bilateral distribution of the two phases as mentioned above, with the white phase occupying the head side and the black

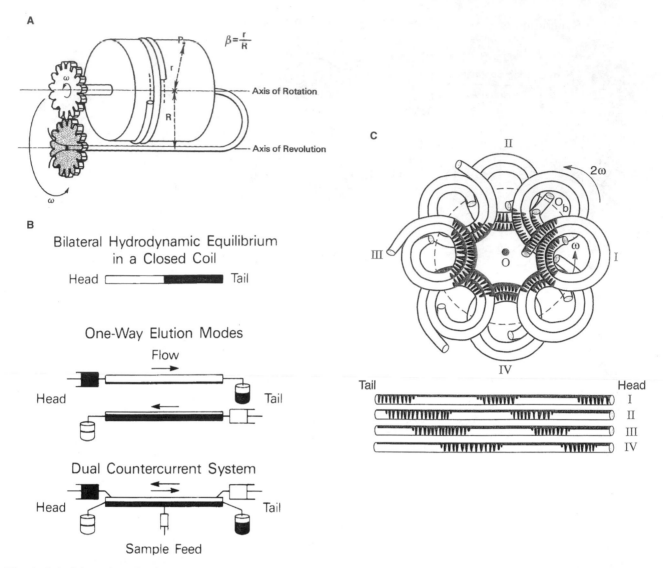

Fig. 6 Principle and mechanism of high-speed CCC. (A) Coaxial coil orientation on holder of type J coil planet centrifuge; (B) mechanism of high-speed CCC; (C) distribution of mixing and settling zones in spiral column undergoing type J synchronous planetary motion.

phase the tail side. This hydrodynamic distribution of the two phases can be utilized for performing CCC. In the middle diagram, the upper coil is filled with the white phase and the black phase is introduced from the head end. The mobile black phase then rapidly travels through the coil, leaving a large volume of the white phase stationary in the coil. Similarly, the lower coil is filled with the black phase and the white phase is introduced from the tail end. The mobile white phase then travels through the coil, leaving a large volume of the black phase stationary in the coil. In either case, solutes locally injected at the inlet of the coil are efficiently partitioned between the two phases and quickly eluted from the coil in the order of their partition coefficients, thus yielding high partition efficiency in a short elution time.

The present system also permits simultaneous introduction of the two phases through the respective terminals, as illustrated in the bottom coil. This dual countercurrent operation requires an additional flow tube at each terminal to collect the effluent, and if desired, a sample injection port is created at the middle portion of the coil. This system has been effectively applied to foam CCC[15,16] and dual CCC.[17]

The hydrodynamic motion of the two solvent phases in the rotating spiral column has been observed under stroboscopic illumination (Fig. 6C). As shown in the upper diagram, the spiral column is divided into two areas, a mixing zone near the center of the centrifuge and a settling zone in the rest of the area. The lower diagram shows the motion of the mixing zone by stretching the spiral column from position I to IV.

A

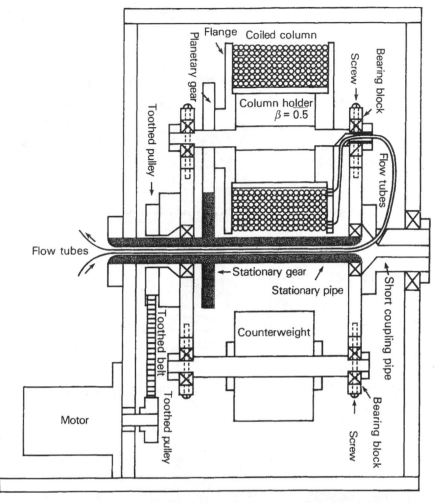

B

Fig. 7 Design of high-speed CCC apparatus. (A) Original multilayer coil planet centrifuge: cross-sectional view through center of apparatus; (B) photograph of most advanced prototype of high-speed CCC centrifuge equipped with a set of three multilayer coils connected in series.

It demonstrates that the mixing zones travel through the spiral column at a rate of one round per revolution. This implies high efficiency of this system: Solutes present in any portion in the column are subjected to an efficient partition cycle of repeating mixing and settling at an enormously high frequency (13 cycles/sec at 800 rpm of column revolution).

INSTRUMENTATION OF HYDRODYNAMIC CCC SYSTEMS

Type J HSCCC

Fig. 7A shows a cross-sectional view of the original design of the HSCCC centrifuge.[18] The rotary frame holds a large multilayer coil holder and a counter-weight mass symmetrically to balance the centrifuge system. Twist-free type J synchronous planetary motion is provided by coupling a pair of identical gears, the planetary gear mounted on the column holder flange and the stationary sun gear (shaded) on the centrifuge axis. The flow tubes from the separation column first pass through the center of the holder and then, forming an arch, enter the central stationary pipe (shaded) via a side hole made in the short rotary shaft (right). These tubes can maintain their integrity for many runs when protected with a short segment of Tygon tubing to avoid direct contact with metal parts.

Later, the above original design was improved by eliminating the counterweight mass and arranging two to three identical columns symmetrically around the rotary frame. Fig. 7B shows the most advanced form of high-speed CCC centrifuge, equipped with a set of three multilayer coil separation columns. All three columns are serially connected with flow tubes through a counter-rotation hollow pipe to prevent twisting.

This type J high-speed CCC system equipped with the multilayer coil separation column can separate a variety of natural and synthetic products with high partition efficiency using organic–aqueous two-phase solvent systems. The system, however, fails to retain a satisfactory amount of stationary phase in low interfacial aqueous–aqueous polymer phase systems due to intensive emulsification.

Cross-Axis Coil Planet Centrifuge (X-Axis CPC)

In this CCC system, the column holder revolves around the vertical axis of the centrifuge while it rotates about its horizontal axis at the same angular velocity.[19] This second HSCCC system is based on a hybrid between type L and X synchronous systems (Fig. 5), and it leads to bilateral hydrodynamic distribution of the two phases in an end-closed coaxial multilayer coil as in the type J HSCCC (Fig. 6B). However, in contrast to type J synchronous planetary motion, the centrifugal vectors fluctuate in a three-dimensional space where one component steadily acts across the diameter of the tube to stabilize the retention of the stationary phase. This stabilizing effect becomes greater as the hybrid approaches the type L synchronous system, while the phase-mixing effect is reduced. The optimum column position for separating proteins with a polyethylene glycol (PEG) and potassium phosphate system is at around $L/X = 1.5$ (Fig. 8), where X is the distance from the axis of the holder to the central axis of the centrifuge and L the length of column shift from the center along the rotary shaft. For highly viscous and very low interfacial tension

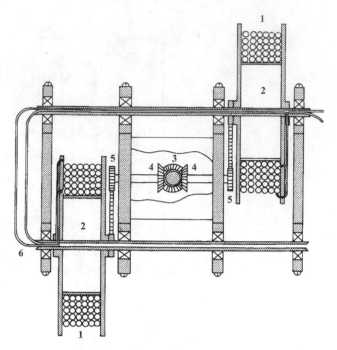

Fig. 8 Cross-sectional view through central horizontal plane of $1.5L/X$ cross-axis coil planet centrifuge. 1, Multilayer coil separation column; 2, column holder; 3, stationary miter gear; 4, horizontal miter gears each equipped with toothed pulley (5) at peripheral end; 6, flow tubes.

polymer phase systems such as dextran/PEG, $L/X = 3$ provides satisfactory retention of the stationary phase. Fig. 8 shows a cross-sectional view through the horizontal plane of the X-axis CPC ($L/X = 1.5$). A pair of column holders is mounted symmetrically around the rotary frame. When the motor (not shown) drives the rotary frame around the centrifuge axis, a pair of horizontal miter gears engaged with a stationary sun gear (center) rotates about their own axes on the rotating frame. This motion is further conveyed to each column holder by coupling with a pair of pulleys, one at the end of the gear shaft and the other mounted on the flange of the column holder, with a toothed belt. The flow tubes leading from each holder, as indicated in the diagram, are not twisted when they are supported at the center of the centrifuge cover. These tubes maintain their integrity for many runs when lubricated with grease and protected with a short sheath of Tygon tubing to prevent direct contact with metal parts. This apparatus has been successfully used for the purification of various kinds of proteins with PEG/potassium phosphate systems.

REFERENCES

1. Ito, Y. Countercurrent chromatography (mini-review). J. Biophys. Biochem. Meth. **1980**, *3*, 77–87.

2. Ito, Y. Recent advances in countercurrent chromatography (review). J. Chromatogr. **1991**, *538*, 3–25.

3. Ito, Y.; Bowman, R.L. Countercurrent chromatography: liquid–liquid partition chromatography without solid support. Science **1970**, *167*, 281–283.

4. Matsuda, K.; Matsuda, S.; Ito, Y. Toroidal coil counter-current chromatography. Achievement of high resolution by optimizing flow-rate, rotation speed, sample volume and tube length. J. Chromatogr. A. **1998**, *808*, 95–104.

5. Marchal, L.; Foucault, A.P.; Patissier, G.; Rosant, J.-M.; Legrand, J. Centrifugal partition chromatography: an engineering approach. In *Countercurrent Chromatography: The Support-Free Liquid Stationary Phase*; Berthod, A., Ed.; Elsevier: New York, 2002; 115–157 (Chapter 5).

6. Ito, Y.; Bowman, R.L. Countercurrent chromatography with flow-through centrifuge without rotating seals. Anal. Biochem. **1978**, *85*, 614–617.

7. Ito, Y. Centrifugal precipitation chromatography applied to fractionation of proteins with ammonium sulfate. J. Liq. Chromatogr. & Rel. Technol. **1999**, *22*, 2825–2836.

8. Ito, Y. Centrifugal precipitation chromatography: principle, apparatus and optimization of key parameters for protein fractionation by ammonium sulfate precipitation. Anal. Biochem. **2000**, *277* (1), 143–153.

9. Ito, Y.; Suaudeau, J.; Bowman, R.L. New flow-through centrifuge without rotating seals applied to plasmapheresis. Science **1975**, *189*, 999–1000.

10. Ito, Y. Sealless continuous flow centrifuge. In *Apheresis: Principles and Practice*; McLeod, B., Price, T.H., Drew, M.J., Eds.; AABB Press: Bethesda, MD, 1997; 9–13.

11. Ito, Y.; Blamblett, G.T.; Bhatnagar, R.; Huberman, M.; Leive, L.; Cullinane, L.M.; Groves, W. Improved non-synchronous flow-through coil planet centrifuge without rotating seals. Principle and application. Sep. Sci. Technol. **1983**, *18*, 33–48.

12. Okada, T.; Metcalf, D.D.; Ito, Y. Purification of mast cells with an improved nonsynchronous flow-through coil planet centrifuge. Int. Arch. Allergy Immunol. **1996**, *109*, 376–382.

13. Ito, Y. High-speed countercurrent chromatography. CRC Crit. Rev. Anal. Chem. **1986**, *17*, 65–143.

14. Ito, Y., Conway, W.D., Eds.; *High-peed Countercurrent Chromatography*; Wiley Interscience: New York, 1996.

15. Ito, Y. Foam countercurrent chromatography based on dual countercurrent system. J. Liq. Chromatogr. **1985**, *8*, 2131–2152.

16. Oka, H. Foam countercurrent chromatography. In *High-Speed Countercurrent Chromatography*; Ito, Y., Conway, W.D., Eds.; Wiley Interscience: New York, 1996; 107–120 (Chapter 5).

17. Lee, Y.W. Dual countercurrent chromatography. In *High-Speed Countercurrent Chromatography*; Ito, Y., Conway, W.D., Eds.; Wiley Interscience: New York, 1996; 93–104 (Chapter 5).

18. Ito, Y.; Sandlin, J.L.; Bowers, W.G. High-speed preparative countercurrent chromatography (CCC) with a coil planet centrifuge. J. Chromatogr. **1982**, *244*, 247–257.

19. Ito, Y.; Menet, J.-M. Coil planet centrifuges for high-speed countercurrent chromatography. In *Countercurrent Chromatography*; Menet, J.-M., Thiébaut, D., Eds.; Marcel Dekker: New York, 1999; 87–119 (Chapter 3).

Intrinsic Viscosity of Polymers: Determination by GPC

Yefim Brun
Waters Corporation, Milford, Massachusetts, U.S.A.

INTRODUCTION

The intrinsic viscosity is a widely used measure of molecular weight, M, and size (dimensions) of macromolecules in dilute solution. Important information about macromolecular architecture and conformations can be obtained from the molecular-weight dependence of intrinsic viscosity for a homologous series of polymers. SEC provides a unique opportunity to measure this dependence in a single chromatographic run. Another striking coincidence is that the dimensions of a macromolecule associated with its frictional properties in dilute solution (i.e., with the intrinsic viscosity) determine the elution time (volume) in size-exclusion separation. This allows one to use the intrinsic viscosity measurement as a crucial intermediate step in the determination of molecular weights and molecular-weight distributions of polymers. From the above, it might be assumed that on-line intrinsic viscosity measurements represent an important aspect of contemporary GPC.

SOLUTION VISCOSITY

There are several dilute solution viscosity quantities used in the determination of the intrinsic viscosity. The size of macromolecules in solution is associated with an increase in viscosity of the solvent brought about by the presence of these molecules. Relative viscosity is a dimensionless quantity representing a solution/solvent viscosity ratio, $\eta_r = \eta/\eta_0$, where η and η_0 are the solution and solvent viscosities, respectively. The specific viscosity $\eta_{sp} = \eta_r - 1$ is the fractional increase in viscosity between the solution and solvent. The effect of the concentration can be normalized by division of η_{sp} by the concentration C, expressed in grams per deciliter (g/dL). This concentration-normalized viscosity is termed the reduced specific viscosity or η_{sp}/C (the IUPAC preferred term is viscosity number). Another related term, the inherent viscosity, is expressed as $\eta_{inh} = (\ln \eta_r)/C$ (the IUPAC preferred term is logarithmic viscosity number).

In order to relate viscosity to molecular weight, the value of reduced (or inherent) viscosity is extrapolated to zero concentration. This parameter is called the intrinsic viscosity, $[\eta]$, and is usually expressed in deciliters per gram (dL/g) (the IUPAC preferred term is limiting viscosity number, mL/g):

$$[\eta] = \lim(\eta_{sp}/C) = \lim \eta_{inh}(C \to 0) \qquad (1)$$

Practically, the limit in Eq. (1) is achieved when the concentration is so low that the frictional interactions between an individual macromolecule and a solvent are not affected by the presence of other macromolecules in the same solution. Under these conditions, which are typical for the GPC/SEC experiments, the difference between the reduced and intrinsic viscosities is negligible.

CALCULATION OF INTRINSIC VISCOSITY IN GPC

The opportunity to measure the dilute polymer solution viscosity in GPC came with the continuous capillary-type viscometers (single capillary or differential multicapillary detectors) coupled to the traditional chromatographic system before or after a concentration detector in series (see the entry Viscometric Detection in GPC/SEC). Because liquid continuously flows through the capillary tube, the detected pressure drop across the capillary provides the measure for the fluid viscosity according to the Poiseuille's equation for laminar flow of incompressible liquids.[1] Most commercial on-line viscometers provide either relative or specific viscosities measured continuously across the entire polymer peak. These measurements produce a viscometry elution profile (chromatogram). Combined with a concentration-detector chromatogram (the concentration versus retention volume elution curve), this profile allows one to calculate the instantaneous intrinsic viscosity $[\eta]_i$ of a polymer solution at each data point i (time slice) of a polymer distribution. Thus, if the differential refractometer is used as a concentration detector, then for each sample slice i,

$$[\eta]_i \approx \frac{\eta_{sp,i}}{C_i} = \frac{\nu \eta_{sp,i}}{\Delta n_i} \qquad (2)$$

Encyclopedia of Chromatography DOI: 10.1081/E-ECHR-120040021

where Δn is the refractive index change due to the polymer in solution, detected by the refractometer and $\nu = dn/dc$ is the refractive index increment of the polymer. The quantity calculated from Eq. (2) is often designated as "observed" intrinsic viscosity, $[\eta]_{obs}$.

As can be seen from Eq. (2), the observed intrinsic viscosity for each slice is proportional to the ratio of two detectors' responses. It follows that detector noise, which is an irreducible component of the measurement process, introduces noise in the intrinsic viscosity that depends on this ratio. However, two detectors have different sensitivities at the tails of polymer distribution: the concentration detector is less sensitive to the high-molecular-weight end and the viscometer is less sensitive to the opposite end. Thus, the noise increases dramatically on both tails of the distribution, where the ratio (2) does not produce physically meaningful values. For example, the logarithm of intrinsic viscosity computed from the slice ratios (2) sometimes does not increase monotonically with molecular weight (i.e., with decreasing elution volume V) even for the flexible coillike polymers (curves 2 in Fig. 1).

Fitting a smooth, multivariate model to a time series of noisy data is an effective way to produce a more precise estimate of the measured quantity at each sample time. Typically, the logarithm of intrinsic viscosity is modeled as a low-order polynomial in elution volume V using a least-squares fitting to the experimental data.[2–4] The intrinsic viscosity calibration curve ($\log[\eta]_{fit}$ versus V) obtained this way depends on properties of the polymer as well as that of the

chromatographic system (e.g., columns). It can be used for diagnostic information concerning the GPC–viscometry system[5] and also to refine such system parameters as interdetector volume (see the entry Interdetector Delay Volume) and band broadening (see the entry Axial Dispersion Correction in GPC/SEC).

APPLICATION OF INTRINSIC VISCOSITY IN GPC TO POLYMER CHARACTERIZATION

Molecular Weight and Molecular-Weight Distribution Determination

The most important feature that has been added to conventional GPC by the viscometer detector through the intrinsic viscosity calculation is the ability to determine the "absolute" MWD without any additional assumptions about the polymer chemical structure. This goal is accomplished by applying the universal calibration concept, which establishes the hydrodynamic volume $H = [\eta]M$ as a universal parameter governing the size-exclusion separation. The MWD can be determined in three steps. First, the set of narrow polydispersity polymers with known molecular weights (narrow standards) covering the entire region of column size-exclusion separation is selected to construct the universal (or hydrodynamic volume) calibration curve, $\log H$ versus elution volume V (see the entry Calibration of GPC/SEC with Universal Calibration Techniques). This curve is then used to calculate the molecular-weight calibration curve via the relationship $\log M = \log H - \log[\eta]_{fit}$, where these quantities are obtained from the (smooth) hydrodynamic and intrinsic viscosity calibration curves. Finally, the MWD is constructed by plotting the concentration C as a function of $\log M$ across the polymer distribution (curves 1 in Fig. 1). Different statistical moments of this distribution (i.e., average molecular weights, including viscosity-average molecular weight M_v) can be calculated by appropriate summation over the slice data and compared with the values obtained by bulk measurements.[6]

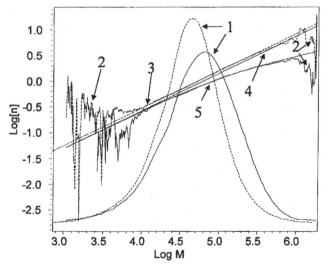

Fig. 1 Molecular-weight-distribution (MWD) and viscosity-law plots for NIST PE1475 high-density polyethylene (dashed lines) and NIST PE1476 low-density polyethylene (solid lines). The curves are (1) MWD, (2) observed viscosity, (3) fitted viscosity for linear polyethylene, (4) extrapolated fitted viscosity for polyethylene with short-chain branches only, and (5) fitted viscosity for branched polyethylene.

Intrinsic Viscosity Distribution

The intrinsic viscosity is a fundamental property of the polymer sample in solution, and thus the intrinsic viscosity distribution (IVD) (C versus $\log[\eta]$) with associated statistical moments may be used to characterize polymers without converting this distribution into a MWD.[7] The IVD can be determined in GPC–viscometry directly, without resorting to universal calibration. This distribution depends not only on the polymer sample itself but also on the solvent and the temperature, and hence does not possess the versatility of the MWD. Nevertheless, the IVD measurement in GPC–viscometry is much less sensitive to experimental conditions than any calibration curve and, hence, can be successfully used in industry (e.g., for quality control of polymers in production).

Size and Molecular Structure of Polymers

The GPC–viscometry with universal calibration provides the unique opportunity to measure the intrinsic viscosity as a function of molecular weight (viscosity law, $\log[\eta]_{fit}$ versus $\log M$) across the polymer distribution (curves 3 and 4 in Fig. 1). This dependence is an important source of information about the macromolecule architecture and conformations in a dilute solution. Thus, the Mark–Houwink equation usually describes this law for linear polymers: $\log[\eta] = \log K + \alpha \log M$ (see the entry Mark–Houwink Relationship). The value of the exponent α is affected by the macromolecule conformations: Flexible coils have the values between 0.5 and 0.8, the higher values are typical for stiff anisotropic ("rod"-like) molecules, and much lower (even negative) values are associated with dense spherical conformations.

The determination of the viscosity law in GPC–viscometry is even more important for branched polymers. Branches reduce the sizes of a macromolecule, including its hydrodynamic volume H. This size reduction is reflected by the changes in the shape and position of the viscosity law plot for a branched polymer. Short-chain branches usually do not change the linearity and slope of the Mark–Houwink plot and just decrease the value of parameter K, whereas the long-chain branches cause bending of the corresponding plot.

These features of the viscosity-law plots for branched polymers are demonstrated in Fig. 1 with two NIST (National Institute of Standards and Technology, U.S.A.) polyethylene standards as examples: high-density linear polyethylene PE1475 and low-density branched polyethylene PE1476. This last one contains both short- and long-chain branches. Dashed straight line 3 represents the Mark–Houwink plot for linear polyethylene, parallel solid line 4 takes into account the short-chain branches, and polyethylene with both types of branches (PE1476) is described by solid curve 5 (see the entry Long-Chain Polymer Branching, Determination by GPC/SEC for further discussion)

For further information on the intrinsic viscosity determination in GPC, including the use of the light-scattering detector, see Ref.[8] and the entry GPC/SEC Viscometry from Multiangle Light Scattering.

REFERENCES

1. Mays, J.W.; Hadjichristidis, N. Polymer characterization using dilute solution viscometry. In *Modern Methods of Polymer Characterization*; Barth, H.G., Mays, J.W., Eds.; John Wiley & Sons: New York, 1991; 227–269.
2. Kuo, C.-Y.; Provder, T.; Koehler, M.E.; Kah, A.F. Use of a viscometric detector for size exclusion chromatography. In *Detection and Data Analysis in Size Exclusion Chromatography*; Provder, T., Ed.; American Chemical Society: Washington, D.C., 1987; 130–154.
3. Lew, R.; Cheung, P.; Balke, S.T.; Mourey, T.H. SEC–viscomter detector systems. I. Calibration and determination of Mark-Houwink constants. J. Appl. Polym. Sci. **1993**, *47*, 1685–1700.
4. Brun, Y.; Nielson, R.; Gorenstein, M.; Hay, N. New results in polymer characterization using multidetector GPC. In *Proceedings, International GPC Symposium*; 1998; 48–67.
5. Balke, S.T.; Cheung, P.; Lew, R.; Mourey, T.H. Single-capillary viscometer used for accurate determination of molecular-weights and Mark-Houwink constants. J. Liquid Chromatogr. **1990**, *13*, 2929–2955.
6. Lesec, J. Problems encountered in the determination of average molecular weights by GPC viscometry. In *Liquid Chromatography of Polymers and Related Materials II*; Chromatographic Science Series; Cazes, J., Delamare, X., Eds.; Marcel Dekker, Inc.: New York, 1980; Vol. 13, 1–17.
7. Yau, W.W.; Rementer, S.W. J. Liquid Chromatogr. **1990**, *13*, 627–675.
8. Jackson, C.; Barth, H.G. Molecular weight-sensitive detectors for size exclusion chromatography. In *Handbook of Size Exclusion Chromatography*; Chromatographic Science Series; Wu, C., Ed.; Marcel Dekker, Inc.: New York, 1995; Vol. 69, 103–145.

Ion Chromatography Principles: Suppressed and Nonsuppressed

Ioannis N. Papadoyannis
Victoria F. Samanidou
Aristotle University of Thessaloniki, Thessaloniki, Greece

INTRODUCTION

Ion chromatography (IC) is a mode of HPLC in which ionic analyte species are separated on cationic or anionic sites of the stationary phase. The separation mechanisms can be broadly compared to ion exchange, using fixed-site exchange resins of various composition and ion-interaction methods, using a variety of columns as substrates to support dynamically exchanged or permanently bonded ionic groups. Alternative approaches of minor significance also exist. The mobile phase is an aqueous buffer solution. The rate of migration of the ion (inorganic ions and organic acids and bases) through the column is directly dependent on the type and concentration of eluent ions. Retention is based on the affinity of different ions for the ion-exchange sites and on the competition between eluent buffer ions and analyte ions, which is dependent on the ionic strength of the buffer and can be adjusted by altering the pH of the mobile phase or the concentration of any organic modifier in it.

DISCUSSION

Ion chromatography operates at pressures ranging from several hundred to several thousand pounds per square inch. In most cases, the same chromatographic components (pumps, injectors, etc.) can be employed in both HPLC and IC. Most of the chromatographic principles developed in HPLC stand for IC also, with possible minor modifications. Injection volumes in ion chromatography are generally somewhat larger than those normally in HPLC, typically in the range 50–100 µL, in contrast to HPLC, where 5–20 µL are injected.

It was in 1975 when Small and his co-workers introduced the high-pressure operation mode of ion chromatography. In their original paper, they described a novel system for the chromatographic determination of inorganic ions, in which a resin was used for the separation and a second ion-exchange column was combined to chemically suppress the background conductance of the eluent, thus improving detection limits for eluted ions. Since 1975, IC has grown rapidly and ion chromatographic methods for ions are currently among the best available and have been applied to a wide range of inorganic species. This can be attributed to concurrent advances in separation technology and detection methods.

Detection techniques can be subdivided into three broad categories:

1. Electrochemical detection (using conductivity, amperometry, or potentiometry).
2. Spectroscopic detection (using ultraviolet/visible (UV/vis) absorbance, refractive index, fluorescence, atomic absorption or atomic emission).
3. Techniques based on postcolumn reactions.

Conductivity detectors provide the advantage of universal detection, as all ions are electrochemically conducting. Thus, the majority of ion chromatography detectors rely upon conductivity measurements.

The principle of conductivity-detector operation is the differential measurement of conductance of the eluent, prior to and during elution of the analyte ion. The detector response depends on analyte concentration, the degree of ionization of both eluent and analyte (governed by the eluent pH), and limiting equivalent conductances of the eluent cation and of the eluent and analyte anions (where an anion-exchange system is considered). If the eluent and analyte are fully ionized, the signal is proportional to the analyte concentration and to the difference (positive or negative) in limiting equivalent conductances which determines sensitivity.

Conductivity detection provides a sensitive measure of ion concentrations in solution, but its measurement is hampered by high conductivity of the eluent, as ion exchange requires a competing electrolyte to displace the analytes from the column. In order to eliminate background conductivity and thus to improve the analyte signal, H, Smith et al. proposed the use of a second ion-exchange column. In this way, two different ion chromatography techniques are distinguished: eluent suppressed and nonsuppressed (also called single-column ion chromatography) using different

Encyclopedia of Chromatography DOI: 10.1081/E-ECHR-120040022

packing materials and different eluents, leading to specific advantages and disadvantages for each technique.

ELUENT SUPPRESSED ION CHROMATOGRAPHY

Various schemes have been devised to improve the signal-to-noise ratio (S/N) by decreasing the background signal of the eluent/displacer or increasing the conductance of the analyte, or both.

The principle of conductivity suppression is the reduction of background conductivity by converting the eluent to a less conductive medium (H_2O) through acid–base neutralization while the analyte ions' conductivity is increased, by converting them to a more conductive medium: Anions are converted to their acid forms and cations to their hydroxide forms. These reactions lead to higher S/N ratios, thus significantly improving baseline stability and detection limits.

Suppressor devices include packed column suppressors, hollow-fiber membrane suppressors, micromembrane suppressors, suspension postcolumn reaction suppressors, autoregenerated electrochemical suppressors, and so forth.

The packed column suppressor, originally introduced by Small et al., suffers from a number of drawbacks, such as time shifts due to Donnan exclusion effects, band broadening (due to a large dead volume and high dispersion), and oxidation of nitrite, which is easily oxidized to nityrate, due to the formation of nitrous acid in the suppressor. Because of these limitations, they were only practical for isocratic elution. However, the main disadvantage of the method is the necessity for periodical regeneration of the suppressor (also called stripper) to restore its ion-exchange capacity.

For anion analysis, the regenerant must supply a source of hydrogen ions to convert the eluent anions to a less conductive form. The most common regenerant is dilute sulfuric acid, whereas for cation analysis, the most common regenerant is hydroxide (sodium, potassium, or tetramethylammonium hydroxide).

The preferred eluents for anions are dilute carbonate–bicarbonate mixture, sodium hydroxide and, for common alkali metals and simple amines, dilute mineral acids (HCl, HNO_3, $BaCl_2$, $AgNO_3$, amino acids, alkyl and aryl sulfonic acids). The most common choice is HCl, but in the case of divalent ions, an eluent of much higher affinity for the ion-exchange resin, such as $AgNO_3$, must be used.

Typical neutralization reactions for chemical suppressors are as follows:

Anion-exchange chromatography:

Eluent reaction: $NaOH + resin–SO_3^-H^+ \rightarrow$
$Resin–SO_3^-Na^+ + H_2O$

Analyte reaction: $NaX + resin–SO_3^-H^+ \rightarrow$
$Resin–SO_3^-Na^+ + HX$, where
X = anions (Cl^-, Br^-, NO_2^-, etc.).

Cation-exchange chromatography:

Eluent reaction: $HCl + resin–NR_3^+OH^- \rightarrow$
$Resin–NR_3^+Cl^- + H_2O$

Analyte reaction: $MCl + resin–NR_3^+OH^- \rightarrow$
$Resin–NR_3^+Cl^- + MOH$, where
M = cations (Na^+, K^+, etc.).

The reaction, in the case of bicarbonate, yields the largely undissociated carbonic acid that does not contribute significantly to the conductivity.

Without chemical suppression, the contribution to the total measured conductivity from the eluent is many orders of magnitude higher than that from the analyte, leading to low sensitivity (Fig. 1).

Some of the drawbacks that packed column suppressors have were eliminated when hollow-fiber membrane suppressors were introduced in 1981. These were found to be even more convenient and efficient, with low dead volume and high capacity, and they are dynamically regenerated. Eluent passes through the core of the fiber and regenerant washes the outside. However, they have also limited suppression capacity and are restricted only to isocratic operation.

Micromembrane suppressors introduced in 1985 use thin, flat ion-exchange membranes to enhance

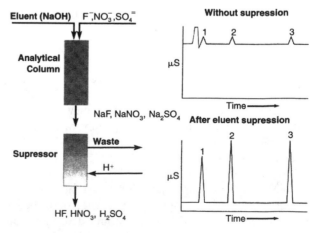

Fig. 1 The effect of the background conductivity suppression on the monitored signal of the analyte anions, after separation by means of ion chromatography. Peaks: 1 = fluoride, 2 = nitrate, 3 = sulfate.

ion transport while maintaining a very low dead volume, providing a high suppression capacity, with low dispersion.

Later, electrochemically regenerated suppression modules were introduced, where an electrochemical process is used to regenerate a solid-phase chemical suppressor for continuous reagent-free operation. Self-regenerating suppressors are similar to micromembrane suppressors, except that regenerant hydronium and hydroxide ions are produced, *in situ*, by electrolysis of water supplied by recycle or an external source. This is achieved by incorporating electrodes inside the regenerant chambers; thus, external acid or base supply are unnecessary. The two electrolysis reactions taking place are

$$\text{Anode: } 2H_2O \rightarrow 4H^+ + O_2 + 4e$$

$$\text{Cathode: } 2H_2O + 2e \rightarrow H_2 + 2OH^-$$

Another technique of improving the S/N ratio is the one that uses postcolumn addition of a solid-phase reagent (SPR), which is a colloidal suspension of ultra-fine ion-exchange particles. The SPR reacts with the analyte to increase its conductivity. Additionally, the SPR has a low electrophoretic mobility and, hence, conductance. This technique avoids the dead time due to suppressor column and also eliminates the regeneration cycle.

NONSUPPRESSED SINGLE-COLUMN ION CHROMATOGRAPHY

Another approach of ion chromatography is the non-suppressed single column, in which no suppressor device is used. In this case, the only method for improving the sensitivity is to maximize the difference between mobile-phase conductivity and analyte conductivity.

Nonsuppressed single-column ion chromatography (SCIC) was introduced in 1979 by Gjerde and coworkers, based on a two-principal innovation:

1. The use of a special anion-exchange resin of very low capacity (0.007–0.007 mEq/g).
2. The adoption of an eluent having a very low conductivity, which can be passed directly through the conductometric detector. Typical eluents used are benzoate, phthalate, or other aromatic acid salts, with low limiting equivalent conductances (leading to direct detection) or potassium hydroxide eluent, with high conductivity for anions or dilute nitric acid for cations, leading to indirect detection mode (decrease of conductivity as the analyte is eluting).

The major limitation of nonsuppressed conductivity detection is that gradient systems cannot be used; thus, the background conductivity remains constant.

Virtually every type of HPLC detector can be combined with SCIC: refractive index, UV absorbance (direct and indirect), electrochemical, and so forth.

A typical nonsuppressed SCIC separation obtained with a low-capacity resin-based strong anion exchanger (PRP-X100 Hamilton) used as the analytical column is illustrated in Fig. 2 for the simultaneous determination of eight inorganic anions (F^-, Br^-, NO_2^-, Cl^-, NO_3^-, PO_4^{3-}, SO_4^{2-}, CO_3^{2-}), with conductometric detection, using a mixture of 2.0 mM sodium benzoate and 2.5 mM p-hydroxybenzoic acid (pH 9.0 adjusted with 1N NaOH) as eluent, with the organic modifier methanol 8% v/v, at a flow rate of 0.7 mL/min. The detection limits (S/N = 3) were 100 µg/L for carbonate and 50 µg/L for the rest of

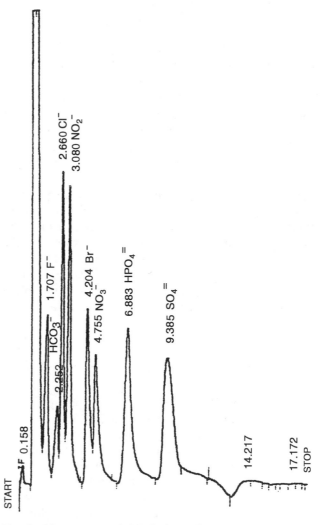

Fig. 2 Nonsuppressed SCIC determination of eight inorganic anions.

the cited anions, when 50 μL of the samples were injected onto the analytical column.

COLUMNS

Two types of packing materials are commonly used for ion chromatography: silica-based and polymer-based ion exchangers. The polymer-based ion exchangers typically contain a PSDVB (polystyrenedivinylbenzene) backbone, lightly sulfonated (cation exchanger) or lightly aminated (anion exchanger), whereas the silica-based ion exchangers use a porous silica bead, chemically prepared to form the anion or cation exchanger. The resins have the advantage that they can be used over the entire pH range, whereas silica based materials can be used in a narrow working pH range (2–6.5).

Detection limits for ions vary with the sensitivity of the detector, with the volume of sample injected, and with the identity, concentration, and pH of the eluent, as well as with chromatographic factors, such as column efficiency and so forth.

COMPARISON OF ESIC AND SCIC

The main advantages of eluent suppressed ion chromatography (ESIC) are that a wide range of eluents and columns can be used, the wide dynamic range, and the higher sensitivity; the main disadvantage is the periodical necessity for suppressor-column regeneration.

On the other hand, SCIC is rapid, sensitive, with easy sample preparation, and simple instrumentation; however, it requires a significant difference in conductance between eluent and analyte ions and the temperature stability is crucial. The answer to the question of which IC technique is most efficient is dependent on several considerations, such as the nature of sample analytes, the concentration of the solute ions, the sensitivity required, the equipment available, and so forth.

APPLICATIONS

Ion chromatography, suppressed and nonsuppressed, can be applied both to anion and cation analysis. The current situation is that the methods for anion determination have far outnumbered those for cation analysis, for the reason that there are available methods for the latter, which are rapid and sensitive (e.g., AAS, ICP, ASV). It is difficult to mention all the ionic species detectable by this analytical technique. Practically, any compound that can be converted to an ionic form is amenable to analysis by IC. Among the inorganic ions determined are (F^-, Br^-, NO_2^-, Cl^-, NO_3^-, PO_4^{3-}, SO_4^{2-}, CO_3^{2-}, CrO_4^{2-}, I^-, IO_3^-, $C_2O_4^{2-}$, BrO_3^-, SCN^-, Na^+, K^+, Mg^{2+}, Ca^{2+}, NH_4^+, at the ppm or ppb levels, in drinking water, food samples, food additives, beverages, environmental samples (soil extracts, rain water, surface water or groundwater), cosmetics, pharmaceuticals, biomedical, plating bath analysis, biological fluids, industrial process products, wastewater, and so forth. IC is also capable of speciation analysis of polyvalent anions or transition metal ions with multiple oxidation states, at levels lower than those possible with ICP or AAS. Organic species of biological and biochemical interest can also be determined.

SUGGESTED FURTHER READING

Dasgupta, P. Ion chromatography: the state of the art. Anal. Chem. **1992**, *64* (15), 775A–783A.

Gierde, D.; Fritz, J.; Schmuckler, G. Anion chromatography with low-conductivity eluents. J. Chromatogr. **1979**, *186*, 509–519.

Gierde, D.; Schmuckler, G.; Fritz, J. Anion chromatography with low-conductivity eluents. II. J. Chromatogr. **1980**, *187*, 35–45.

Haddad, P.; Heckenberg, A. Determination of inorganic anions by high-performance liquid chromatography. J. Chromatogr. **1984**, *300*, 357–394.

Henderson, I.; Saari-Nordhaus, R.; Anderson, J., Jr. Sample preparation for ion chromatography by solid-phase extraction. J. Chromatogr. **1991**, *546*, 61–71.

Henshall, A.; Rabin, S.; Statler, J.; Stilian, J. Int. Chromatogr. Lab. **1993**, *12*, 7–14.

Papadoyannis, I.; Samanidou, V.; Moutsis, K. J. Liquid Chromatogr. **1998**, *21* (3), 361–379.

Papadoyannis, I.; Samanidou, V.; Zotou, A. Highly selective simultaneous determination of eight inorganic anions by single column high pressure anion chromatography in drinking water. J. Liquid Chromatogr. **1995**, *18* (7), 1383–1403.

Pietrzyk, D.; Iskandarani, Z.; Schmitt, G. J. Liquid Chromatogr. **1986**, *9* (12), 2633–2659.

Saari-Nordhaus, R.; Anderson, J., Jr. Int. Chromatogr. Lab. **1994**, *18*, 4–10.

Schmuckler, G. Recent developments in ion chromatography. J. Chromatogr. **1984**, *313*, 47–57.

Small, H.; Stevens, T.; Bauman, W. Novel ion exchange chromatographic method using conductimetric detection. Anal. Chem. **1975**, *47* (11), 1801–1809.

Tarter, J. *Ion Chromatography*; Chromatographic Science Series; Marcel Dekker, Inc.: New York, 1987; Vol. 37.

Walker, T.; Akbari, N.; Ho, T. J. Liquid Chromatogr. **1991**, *14* (4), 619–641.

Ion-Exchange Buffers

J. E. Haky
Florida Atlantic University, Boca Raton, Florida, U.S.A.

INTRODUCTION

Ion-exchange chromatography is a separation method based on the exchanging of ions in a solution with ions of the same charge present in a porous insoluble solid. The method is used for the deionization of water.[1,2] It is often employed for the separation and identification of the rare earth and transuranium elements.[2] Additionally, ion-exchange chromatography is also used in clinical laboratories for the automated separation and analysis of amino acids and other physiologically important amines used for pharmaceutical purposes.[3]

DISCUSSION

In ion-exchange chromatography, ions are separated on the basis of their differences in relative affinity for ionic functional groups on the stationary phase. Anionic and cationic functional groups are covalently attached to the stationary phase, usually resins, which are amorphous particles of organic material.[1–3] Sulfonated styrene-based polymers are the most widely used cation-exchange resin, and similar polymers containing quaternary ammonium groups are the most widely used anion exchangers.[4] Oppositely charged solute ions are attracted to ionic functional groups on the stationary phase by electrostatic forces. Retention is based on the attraction between solute ions and charged sites bound to the stationary phase.[4,5]

Due to the desirable solvent and ionizing properties of water, most ion-exchange chromatographic separations are carried out in aqueous media. Once the selection of the column type has been made, the resolution of components in the sample can be optimized by adjusting ionic strength, temperature, flow rate, and, most importantly, the pH and concentration of buffer or organic modifier in the mobile phase.

Solvent strength, which is defined as the ability of the solvent to elute a given solute from the stationary phase, increases with increased ionic strength of the mobile phase. Selectivity is generally not affected by changes in ionic strength, except for samples containing solutes with different valence charges. With increased temperature, the rate of solute exchange between the stationary phases and mobile phases increases, and the viscosity of the mobile phase

decreases, resulting in increased solvent strength. Solvent strength also increases with the volume percent of organic modifier for hydrophobic solutes. However, most ion-exchange chromatography is performed in totally aqueous mobile phases, due to the hydrophilic nature of most ionic solutes. Flow rates of the mobile phase can change resolution in ion-exchange chromatography, but the effects are often minimal.[5]

Increases in mobile-phase pH cause decreases in solute retention in cation-exchange chromatography and increases in retention in anion-exchange chromatography. Separation selectivity can also be greatly influenced by small changes in pH. In ion-exchange chromatography with aqueous mobile phases, buffers are used to maintain the pH in the mobile phase. A buffered solution can resist the changes in pH when an acid or base is added or when dilution is occurring. The pH of a buffer is given by the Henderson–Hasselbalch equation:

$$ \mathrm{pH} = \mathrm{p}K_a + 109\frac{[\mathrm{A}^-]}{[\mathrm{HA}]} \tag{1} $$

where $\mathrm{p}K_a$ refers to the acid dissociation constant of the species in the denominator, HA, and A refers to the conjugate base of the acid HA. Buffer capacity, the measure of how well a solution resists changes in pH when a strong acid or base is added, increases as the concentration of the buffer increases. However, the pH of a buffer solution is virtually independent of dilution. When the $\mathrm{pH} = \mathrm{p}K_a$ the maximum buffer capacity is met and a good working range of the buffer is approximately when the $\mathrm{pH} = \mathrm{p}K_a = 1 \pm 1$.[1] A buffer is very easy to make. For example, to prepare 1.00 L of buffer containing 0.100 M tris(hydroxymethyl) aminomethane hydrochloride at pH of 7.4, simply weigh out 0.100 mol of its hydrochloride salt and dissolve it in a beaker containing about 900 mL of water. Then, add a base (e.g., NaOH), until the pH is exactly 7.4. Then, quantitatively transfer the solution to a volumetric flask. Finally, dilute to the volumetric mark and mix.[1]

By increasing the buffer concentration, the concentration of the counterions are increased in the mobile phase and stronger competition is provided between the sample components and the counterions for the exchangeable ionic centers, resulting in reduced solute

Encyclopedia of Chromatography DOI: 10.1081/E-ECHR-120040023

Table 1 Typical buffers for ion-exchange chromatography

Buffer salt	pH Range
Ammonia	8.2–10.2
Ammonium acetate	8.6–9.8
Ammonium phosphate	2.2–6.5
Citric acid	2.0–6.0
Disodium hydrogen citrate	2.6–6.5
Potassium dihydrogen phosphate	2.0–8.0/9.0–13
Potassium hydrogen phthalate	2.2–6.5
Sodium acetate	4.2–5.4
Sodium borate	8.0–9.8
Sodium dihydrogen phosphate	2.0–6.0/8.0–12
Sodium formate	3.0–4.4
Sodium perchlorate	8.0–9.8
Sodium nitrate	8.0–10.0
Triethanolamine	6.7–8.7

retention.[5] As stated earlier, selectivity and retention can also be adjusted by changing the pH of the mobile phase. This occurs because such a change in pH modifies the character of both the ionexchange medium and the acid–base equilibrium as well as the degree of ionization of the sample.[3] A pH gradient in which the pH of the mobile phase is changed during the chromatographic analysis can also be used to control the solvent strength and retention of ionic solutes. Such gradients can also be used to control selectivity.[6]

The working pH range for a separation can be estimated from the pK_a values of the sample components. If such pK_a values are not available, they can often be estimated by considering the number and types of functional groups present and the molecular structures of the components in the sample.[1] In order to ensure that solutes are ionized and retained by the ion exchanger, the optimum buffer pH of the mobile phase should be 1 or 2 pH units above the pK_a of acids and 1 or 2 pH units below the pK_a of bases.[3]

Two criteria should be met when choosing the components of the buffer. First, the buffer must be able to maintain the operating pH for the separation to be performed. Second, the exchangeable buffer counterion must yield the desired eluent strength.[3] Some common buffer salts used in ion-exchange chromatography and their usable pH ranges are summarized in Table 1. Examples of their use includes the chromatography of amino acids, polymeric, cation exchanger using various combinations of citrate and borate buffer.[3] Additionally, carbohydrates can be separated by anion-exchange chromatography using an aqueous solution of sodium hydroxide–sodium acetate as the eluent.[3] Being weak acids, the ion-exchange behavior of such compounds is significantly affected by the pH of the mobile phase. Similar separations of ionizable compounds through the use of ion-exchange chromatography with these and other buffers have been reported.[1–7]

ACKNOWLEDGMENT

The author wishes to thank H. Seegulum for technical assistance.

REFERENCES

1. Harris, D.C. *Quantitative Chemical Analysis,*, 5th Ed.; W. H. Freeman: New York, 1998; 755–766.
2. Walton, H.F. *Ion-Exchange Chromatography*; Hutchinson and Ross: Dowden, U.K., 1976.
3. Poole, C.F.; Poole, S.K. *Chromatography Today*; Elsevier: New York, 1991; 422–439.
4. Small, H. *Ion Chromatography*; Plenum: New York.
5. Gjerde, D.T.; Fritz, J.S. *Ion Chromatography*, 2nd Ed.; Huthig: New York, 1987.
6. Snyder, L.R.; Kirkland, J.J. *Introduction to Modern Liquid Chromatography*, 2nd Ed.; John Wiley & Sons: New York, 1979; 410–452.
7. Rieman, W.; Walton, H.F. *Ion Exchange in Analytical Chemistry*; Pergamon Press: New York, 1976.

Ion Exchange: Mechanism and Factors Affecting Separation

Karen M. Gooding
Eli Lilly and Company, Indianapolis, Indiana, U.S.A.

INTRODUCTION

Ion-exchange chromatography (IEC) is a technique in which ionic solutes bind to charged functional groups on the bonded phase. The power and versatility of IEC as an analytical and preparative technique is due in large part to the ability to drastically change the selectivity through manipulation of the mobile phase. Although it is obvious that the pH determines the charge on the support and the analytes, the nature of the salt is an equally important parameter. The constituent ions of the salt associate with the support functional groups and/or those of the solute, yielding distinct ionic interactions. Mobile-phase additives, temperature, and gradient conditions also contribute to the separation in IEC.

MOBILE PHASE

pH

Adjustment of the pH is a critical factor in IEC because the pH dictates the charge of both the solutes and the ion exchanger, thus controlling their affinity for one another or their ability to release from a bound state. The essential nature of pH in the process necessitates its exact control; therefore, any mobile phase used for IEC should contain an effective buffer (0.02 M–0.1 M) within its optimum pH range. Some common buffers which cover much of the range of pH used in IEC are phosphate, citrate, acetate, and tris(hydroxymethyl) aminomethane (Tris).[1–2] The pH should be selected to yield ionization of the functional groups on the support as well as those on the analytes. For molecules with a single charge, the pH should be at least two units from the pK in the direction of ionization. The guideline for zwitterions is that the pH be at least two units from the isoelectric point (pI). A pH near neutrality is often effective for complex mixtures of diverse substances. Even carbohydrates, whose hydroxyl groups do not ionize until the pH is greater than 12, can be separated by IEC when the pH is adjusted to a high enough value with low concentrations of base as the mobile phase.[3]

The choice of pH for IEC of proteins or other macromolecules is not as simplistic as it is for small molecules. Although using the pI as a guide frequently yields an adequate separation, the pI encompasses all the charged groups in the molecule, whereas, because of their defined tertiary structures, only the surface amino acids of proteins are actually involved in the binding. Under denaturing conditions, more amino acids are likely to be exposed to the bonded phase.

Salt Concentration

IEC is a very predictable technique because the mechanism is well defined. The capacity factor (k) for the binding of an ionic solute to an ion-exchange functional group in IEC is directly related to the concentration (c) of salt in the mobile phase:

$$\log k = \log K_0 + Z_c \log\left(\frac{1}{c}\right)$$

where K_0 is the distribution coefficient and Z_c is an experimentally determined parameter that reflects the apparent number of ionic charges associated with the process of a specific solute with a specific surface.[4] For isocratic separations of simple molecules with up to several charges, the analysis time can be optimized along with resolution by adjustment of the salt concentration. For more complex analytes or mixtures, salt gradients are often necessary to achieve acceptable separations. Generally, a gradient from 0 M to 1 M salt in a buffer at a suitable pH will yield a preliminary separation.

An opposite mechanism to ion exchange occurs when the ionic strength is too low. Ion exclusion is a phenomenon in which a charged analyte is repelled by the like charges within a pore. This is very likely to occur if water is used alone as the mobile phase with ion-exchange supports or with other modes of silica-based columns. Adding buffer and salt usually eliminates the problem.

Salt Composition

Elution with increased concentrations of salt is the most common and readily controlled method of achieving displacement of molecules which are strongly bound by an ion exchanger. The salt counterions

Encyclopedia of Chromatography DOI: 10.1081/E-ECHR-120040024

competitively displace solute ions from the charged sites on the stationary phase. Smaller, more highly charged ions are most effective at this displacement. Specifically, the strength of displacement for cations

$$Mg^{2+} > Ca^{2+} > NH_4^+ > Na^+ > K^+$$

is and for anions, it is

$$SO_4^{2-} > HPO_4^{2-} > Cl > CH_3COO^-$$

The strength of the ions for displacement is not necessarily related to optimum selectivity or resolution. Selectivity is dictated by the effect of the salt on both the solute and the bonded phase. Besides displacing the solute from the support, either of the ions of the salt can complex with the ion-exchange functional group or the solute, alter the tertiary structure of the solute, or enhance hydrophobic properties. It is this combination of effects which results in selectivity. For example, when a mixture of proteins was run on a polyethyleneimine (PEI) weak anion-exchange

column with gradients formed with 1.0 N salt, substitution of sodium acetate for sodium phosphate produced not only longer retention but also much better resolution of the proteins.[5] Sodium phosphate produced narrower peaks with less tailing, but the peaks had only slight differences in retention. In this case, the short retention was proven to be due to a special affinity of phosphate for PEI, which did not occur with anion-exchange supports having quaternary (Q) or diethylaminoethanol (DEAE) functional groups. The salt effects on selectivity encompass anions and cations in both anion-exchange and cation-exchange chromatography, as illustrated in Fig. 1, implying that the selectivity occurs because of ionic interactions with the functional groups of both the support and the solute. In the case of adenosine 5-diphosphate (ADP), divalent ions like calcium can bridge between the oxygens in the phosphate and thus reduce the ionic properties. Phosphate salts reduce the retention of ADP on PEI supports due to the phosphate–PEI affinity discussed earlier. Another example of ion-based selectivity is the excellent resolution obtained for sugars when a calcium salt is used with a

Effect of salt on Retention in IEC

Anion-Exchange

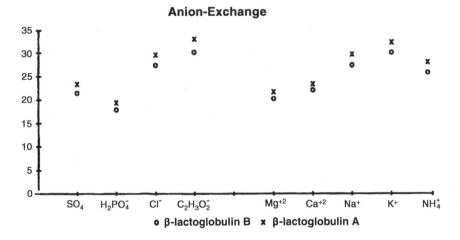

o β-lactoglobulin B   x β-lactoglobulin A

Cation-Exchange

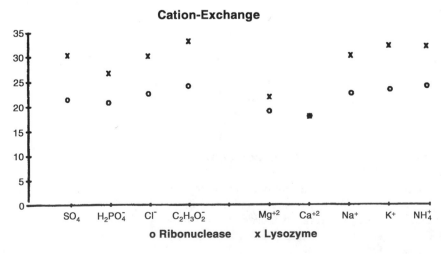

o Ribonuclease   x Lysozyme

Fig. 1 Anion-exchange chromatography (AEX): SynChropak AX300 (polyethyleneimine, 300 Å, 6 μm); cation-exchange chromatography (CEX): SynChropak CM300 (carboxymethyl, 300 Å, 6 μm); 30-min gradient (0–1N) of sodium or chloride salts in 0.02 M Tris, pH 7. (Reprinted with permission of MICRA Scientific.)

cation-exchange resin. This ability to change selectivity so dramatically by varying the salt significantly broadens the utility of IEC.

The only restrictions on the choice of salt are those involving analyte solubility or stability. Volatile salts such as ammonium acetate even allow IEC to be interfaced with mass spectrometry or evaporative lightscattering detection. It is very important that a given salt be totally stripped from a support before changing to other ions to avoid mixed ion effects. An acid such as trifluoroacetic acid is often effective as a bridge/washing solvent for this purpose.

SURFACTANTS AND ORGANIC SOLVENTS

Secondary separation which may be present in IEC is generally size exclusion or hydrophobicity. Size exclusion will occur if macromolecules are larger than the pores in a support. Hydrophobic interactions are most often observed under conditions of high salt for solutes with significant nonpolar characteristics, such as certain peptides. The hydrophobicity of an ion-exchange support is due to either the matrix or the cross-linking agents which were employed in the synthesis of the bonded phase. Any hydrophobic interactions are fundamentally undesirable and can be minimized by the addition of 1–10% of an organic solvent, such as methanol, ethanol, or acetonitrile, to the running buffer. The solubility of the salt in the organic mobile phase should always be verified to avoid precipitation.

Nonionic detergents may also reduce hydrophobic interactions with a column. These detergents, such as CHAPS or urea, can also be added to ion-exchange mobile phases to aid in the solubilization of membrane or other insoluble proteins. Such detergents are easy to equilibrate and remove from ion-exchange columns; however, ionic detergents should be avoided because of their very strong binding to the column or the solutes.

FLOW RATE AND GRADIENT

Small molecules can often be effectively separated isocratically by IEC; however, due to multipoint interactions, isocratic IEC of proteins and most biological macromolecules is not usually feasible, yielding no resolution and extreme tailing. Such complex molecules are generally separated by gradient elution.

As a salt gradient proceeds to higher levels in IEC, molecules elute at a specific salt concentration, generally without binding from secondary effects. The relationship of gradient conditions to elution (k^*) can be described by

$$k^* = 0.87 t_G \frac{F}{V_M} \left(\log \frac{C_2}{C_1} \right) Z$$

where C_1 and C_2 are the total salt concentrations (salt plus buffer) at the beginning and the end of the gradient, respectively, Z is the effective charge on the solute molecule, F is the flow rate; V_M is the total mobile-phase volume, and t_G is the gradient time.[6] The Z number will vary with solute and pH. An initial ion-exchange protocol of a 20–30-min linear gradient from 0 M–1 M salt in a buffer at a suitable pH will usually yield a separation which can be later optimized, if necessary. For shortest analysis times, a gradient should begin at the highest salt concentration where the analytes are bound and it should end at the lowest ionic strength that causes elution. The pH gradients may also be used to elicit elution during IEC, although this has been a less popular strategy than salt gradients. Ion-exchange columns can be effectively

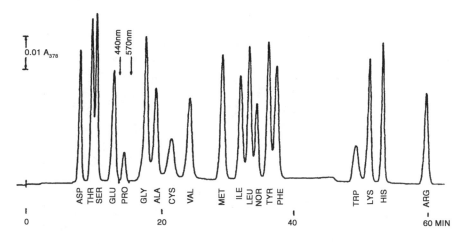

Amino Acid Analysis by Cation-Exchange Chromatography

Fig. 2 Column: Micropak AA (sulfonated polystyrene); solvent A: $0.2\,M$ sodium citrate, pH 3.25; solvent B: $1\,M$ sodium citrate, pH 7.40. Gradient: 5 min 100% A; 100–75% A in 20 min; 75–70% A in 5 min; 70–35% A in 5 min; 10 min 35%; 35–0% A in 1 min. $T = 50°C$ for 25 min, then 90°C. Detection after ninhydrin postcolumn reaction. (Reprinted from Amino acid analysis with ninhydrin postcolumn derivatization, *LC at Work*, Varian Associates, with permission.)

washed with a mobile phase of higher ionic strength than the upper gradient limit or with low pH. For gradients, intermediate flow rates of 1mL/min for a 4.6-mm-inner diameter column are usually satisfactory.

TEMPERATURE

The use of elevated temperature in IEC reduces the mobile-phase diffusion coefficient and concomitantly decreases band spreading. Most mobile phases in IEC are composed of water with salts and thus produce efficiencies which are less than those obtained in modes using organic solvents. Because increased temperatures decrease retention, they may permit the use of lower salt concentrations. Elevated temperatures have been especially effective in amino acid analyses by cation-exchange chromatography, as illustrated in Fig. 2.

CONCLUSIONS

The effectiveness of IEC as a method for separating charged species is enhanced by the ability of many operational factors to change the selectivity and resolution. Salt concentration, salt composition, and pH are the most important operational parameters which strengthen the versatility of the technique.

REFERENCES

1. Cunico, R.L.; Gooding, K.M.; Wehr, T. Ion-exchange chromatography. In *Basic HPLC and CE of Biomolecules*; Bay Bioanalytical Laboratories: Richmond, VA, 1998.
2. *Ion-Exchange Chromatography, Principles and Methods*; Pharmacia Biotech: Sweden, 1998.
3. Townsend, R.R. High-pH anion exchange chromatography of recombinant glycoprotein glycans. In *High Performance Liquid Chromatography: Principles and Methods in Biotechnology*; Katz, E.D., Ed.; John Wiley & Sons: New York, 1996.
4. Aguilar, M.I.; Hodder, A.N.; Hearn, M.T.W. HPIEC of proteins. In *HPLC of Proteins, Peptides and Polynucleotides*; Hearn, M.T.W., Ed.; VCH: New York, 1991; 199.
5. Nowlan, M.P.; Gooding, K.M. HPIEC of proteins. In *High-Performance Liquid Chromatography of Peptides and Proteins*; Mant, C.T., Hodges, R.S., Eds.; CRC Press: Boca Raton, FL, 1991.
6. Snyder, L.R. Gradient elution separation of large biomolecules. In *HPLC of Biological Macromolecules: Methods and Applications*; Gooding, K.M., Regnier, F.E., Eds.; Marcel Dekker, Inc.: New York, 1990.

Ion-Exchange Stationary Phases

Karen M. Gooding
Eli Lilly and Company, Indianapolis, Indiana, U.S.A.

INTRODUCTION

In ion-exchange chromatography (IEC), molecules bind by the reversible attraction of electrostatic charges located on the outer surface of a solute molecule with dense clusters of groups with an opposite charge on an ion-exchange support. To maintain electrical neutrality, the charges on both the analytes and the matrix are associated with ions of opposite charge, termed counterions, which are either provided by preequilibration with the mobile phase or during manufacturing. Because a solute must displace the counterions on the matrix during attachment, the technique is termed "ion exchange." If the support possesses a positive charge, it is used for anion-exchange chromatography, whereas if it carries a negative charge, it is for cation exchange. Generally, the molecule of interest will have a charge that is opposite (positive or negative) to that on the support and the same as the competitively displaced counterions.

There are several major variables which distinguish ion-exchange packings and determine their utility for specific classes of solutes and for analytical or preparative applications. Those variables are as follows:

1. Structure of the bonded phase, including the chemistry of the functional group, its pK, and the properties of the spacer arm and/or bonded phase layer
2. Charge density and related nominal capacity
3. Properties of the support matrix, including composition and pore diameter

BONDED PHASE

Functional Groups

The functional groups of an ion-exchange bonded phase are ionizable under specific pH conditions. The extent of their charge dependence on pH is the basis for distinguishing two types of ion exchangers—strong and weak. These designations do not refer to the strength of binding or to the capacity of the gel, but simply to the pK of the ionizable ligand group, similar to the designations for acids and bases. The structures and approximate pK and pH ranges of some typical strong and weak ion-exchange groups are shown in Table 1.[1–5]

Generally, strong ion-exchange groups retain their charge over a wide range of pH, with binding capacity dropping off at the extremes. For example, quaternary ammonium (Q) resins are strong anion-exchange groups which are effective throughout the pH range of about 2–12. Similarly, sulfonyl groups are strong cation-exchange groups that remain negatively charged until acidic pH levels are used. Strong ion-exchange groups can be considered to possess a permanent positive or negative charge.

The diminished ionization of weak ion-exchange groups near neutral pH result in less predictable separations if operation in this range is necessary for analyte stability, as in the case of many proteins. In these cases, the use of a strong ion exchanger allows the pH of the mobile phase to be manipulated to protonate or deprotonate the analytes without changing the ionic properties of the packing. For example, certain amino acids are most highly charged at pH less than 4, where a weak cation-exchange support would not be fully charged, but a strong cation-exchange group would.

Because weak ion-exchange groups are not fully charged in certain pH ranges, column equilibration may require more mobile phase or time under those conditions. Conversely, highly bound molecules may release more easily from supports which are not totally ionized. Clearly, careful consideration of the titration curves for an ion-exchange support is an essential aspect of designing appropriate conditions for a separation. A complete description of the charged group of an ion exchanger is necessary to understand its pH characteristics because they are dependent on the exact chemical composition of the bonded phase and the matrix. Convenient descriptions such as "strong," "S," "stable weak ion-exchange," and so forth do not sufficiently describe the ionic characteristics of the packing. The exact pK and functional pH range are also affected by the chemistry of the remainder of the bonded phase and of the matrix. For example, a silica matrix may ion-pair with cationic functional groups or a polymeric layer with amines may ion-pair with anionic functional groups. The actual titration curves, pK, and/or pH range for a given support should always be consulted.

Encyclopedia of Chromatography DOI: 10.1081/E-ECHR-120040025

Table 1 Properties of ion-exchange groups

	Functional group	Type	pK	pH Range (approximate)
Anion exchange				
DEAE (diethylaminoethyl)	$-O-CH_2-CH_2-N^+H(CH_2CH_3)_2$	Weak	15–9	2–9
PEI (polyethyleneimine)	$(-NHCH_2CH_2)_n-N(CH_2CH_2-)_{n'}CH_2CH_2NH_2$	Weak	15–9	2–9
Q (quaternary ammonium)	$-CHOH-CH_2-N^+(CH_3)_3$	Strong	>13	2–12
Cation exchange				
CM (carboxymethyl)	$-O-CH_2-COO^-$	Weak	14–6	6–10
SP (sulfopropyl)	$-CH_2-CH_2-CH_2SO_3^-$	Strong	<1	4–13
S (sulfonate)	$-R-CH_2SO_3^-$ (R may be methyl with hydroxyl or amide groups)	Strong	<1	3–11

Hydrophobic Spacer Arms

Ion-exchange functional groups are chemically bonded to the support, often through a polymeric layer which totally covers the matrix. The chemical nature of this coupling chemistry and its spatial characteristics can affect the chromatographic properties. Hydrophobic linkages may impart a nonpolar aspect to the separations. Spacer arms make the functional groups more accessible by distancing them from the support surface. Tentacle IEC bonded phases are a spacer design incorporating a hydrophilic ligand arm.[6]

Charge Density

The number of charges, as measured by titration, defines the nominal capacity of a support. the charge density of an ion-exchange support is determined by the number of ionic groups divided by the surface area or the volume. Typical values range from 3 to 370 µEq/mL of support. The lower values are generally found in nonporous supports. High loading capacities are associated with IEC, especially for porous supports. Weak ion-exchange groups only have maximum capacity in the pH range where they maintain charge —pH less than 9 for DEAE supports and pH greater than 6 for CM.

Counterions

In certain cases, ion-exchange columns are preequilibrated with distinct counterions by the manufacturer. These ions, such as calcium for amino acids, impart a specific selectivity (see the entry Ion-Exchange, Mechanism and Factors Affecting Separation). Alternatively, a layer of counterions is applied by the user by conditioning a column with the salt of interest. An intermediate step of washing with a weak acid may accelerate the equilibration process.

MATRIX

Composition

Ion-exchange supports based on derivatized cellulose and agarose have been popular since the 1960s, particularly for protein analysis. For HPLC, less compressible supports, such as silica and cross-linked polymers, are most commonly used.

Carbohydrate Matrix

Carbohydrate supports such as dextran or agarose are very hydrophilic and easily derivatized with ionic functional groups. They have been very popular for analysis and purification of biological molecules like proteins. One major drawback to these supports is that their volume changes with mobile-phase composition. This has been alleviated in part by higher cross-linking.

Silica Matrix

In silica-based ion exchangers, the silica is bonded through a polymeric layer to a charged ligand group. Operating pH is generally limited to pH 2–8 due to the silica backbone. Although some small-pore silica-based ion exchangers have been synthesized with silane bonding, large-pore supports (≥ 300 Å) designed for protein analysis have polymeric layers containing ionic functional groups which are very stable and even protect the silica matrix from erosion. Silica columns have several advantages:

1. High mechanical stability
2. Minimal shrinkage or swelling with changes in counterions
3. Stability to organic modifiers (with the restriction of salt solubility)
4. High capacity
5. Good mass transfer
6. Large variety of particle and pore sizes

Polymeric Matrix

Polymeric matrices are also widely available for IEC. Polystyrene cross-linked with divinylbenzene (PSDVB) is one such polymer, typically available with pore diameters of at least 1000 Å. The repetitive structure of polystyrene permits reproducible coupling of both strong and weak ion-exchange groups; cross-linking adds the rigidity required for high-pressure applications. These polymeric supports have most of the same advantages as silica for IEC. Methacrylate copolymers, which are also used as matrices in IEC, are more hydrophilic than PSDVB.

Pellicular Matrix

A third group of ion-exchange supports are pellicular, consisting of a solid inert core made of PSDVB agglomerated with 350 nm functionalized latex. The quaternary amine groups are closely and uniformly bound on the microbeads, improving flow and reducing nonspecific retention. These pellicular supports are primarily used for carbohydrate analysis.[7]

Pore Diameter

Pore diameter is a major determinant in ion-exchange capacity because as the pore diameter decreases, there is a tremendous increase in surface area. Nominal loading capacity is directly related to the surface area and the ligand density; consequently, matrices with the smallest pores exhibit the highest ion-exchange capacities for small, totally included solutes.

The ion-exchange capacities of picric acid correlate with surface area. For example, that of a 100-Å pore was seen to be 1415 µmol/g, whereas that of a 300-Å pore was only 656 µmol/g.[8] The capacities for macromolecules such as proteins do not relate directly to surface area because they are excluded by size from portions of small pores and are effectively prevented from reaching all the reactive exchange sites.[5,8] Consequently, larger pores exhibit maximum capacity for macromolecules. For example, a 300-Å pore exhibited maximum capacities of 98 and 130 mg/g for ovalbumin (45,000 MW) and bovine serum albumin (65,000 MW) respectively, because they were able to permeate and bind to the optimum available surface area.[5,8]

REFERENCES

1. Cunico, R.L.; Gooding, K.M.; Wehr, T. Ion-exchange chromatography. In *Basic HPLC and CE of Biomolecules*; Bay Bioanalytical Laboratories: Richmond, VA, 1998.
2. *Ion-Exchange Chromatography, Principles and Methods*; Pharmacia Biotech.: Sweden.
3. Katz, E.D., Ed.; *High Performance Liquid Chromatography: Principles and Methods in Biotechnology*; John Wiley & Sons: New York, 1996.
4. Aguilar, M.I.; Hodder, A.N.; Hearn, M.T.W. HPIEC of proteins. In *HPLC of Proteins, Peptides and Polynucleotides*; Hearn, M.T.W., Ed.; VCH: New York, 1991; 199.
5. Mant, C.T., Hodges, R.S., Eds.; *High-erformance Liquid Chromatography of Peptides and Proteins*; CRC Press: Boca Raton, FL, 1991.
6. Muller, W. J. Chromatogr. **1990**, *510*, 133.
7. Analysis of Carbohydrates by HPAE-PAD, Technical Note 20, Dionex. 1993.
8. Vanecek, G.; Regnier, F.E. Variables in the high-performance anion-exchange chromatography of proteins. Anal. Biochem. **1980**, *109*, 345.
9. Nowlan, M.P.; Gooding, K.M. HPIEC of proteins. In *High-Performance Liquid Chromatography of Peptides and Proteins*; Mant, C.T., Hodges, R.S., Eds.; CRC Press: Boca Raton, FL, 1991; 203.

Ion-Exclusion Chromatography

Ioannis N. Papadoyannis
Victoria F. Samanidou
Aristotle University of Thessaloniki, Thessaloniki, Greece

INTRODUCTION

Ion exclusion is the term that describes the mechanism by which ion-exchange resins are used for the fractionation of neutral and ionic species. Ionic compounds are rejected by the resin, due to Donnan exclusion, and they are eluted in the void volume of the column. Nonionic or weakly ionic substances penetrate into the pores of the packing, they are retained and, thus, separation is achieved, as they partition between the liquid inside and outside the resin particles.

Ion-exclusion chromatography is a mode of HPLC and, thus, the same equipment can be used, with the proper eluent, column, and detection technique. The technique is mostly used for the analysis of organic acids, sugars, alcohols, phenols, and organic bases. It provides a convenient way to separate molecular acids from highly ionized substances. Ionized acids pass rapidly through the column while molecular acids are held up to varying degrees. A conductivity detector is commonly used. Carboxylic acids can be separated by using water, a dilute mineral acid, or a dilute benzoic or succinic acid as eluent.

DISCUSSION

As neutral species, rather than ions, are being separated, ion-exclusion chromatography cannot be considered as a form of ion chromatography; although ionexchange polymers are used, ion-exchange mechanisms are not involved.

Anions, most commonly simple carboxylic acids (e.g., tartaric, malic, citric, lactic, acetic, succinic, formic, propionic, butyric, etc.), are separated on cation-exchange resins in acidic form. Salts of weak acids can also be analyzed, as they are converted to the corresponding acid by the hydrogen ions in the exchanger. Cations (weak bases and their salts) are separated on anion-exchange resins in the hydroxide form.

In order to understand the mechanism of ion-exclusion chromatography, the behavior of the resin, in an aquatic medium, must be taken into account. In this case, three parts can be distinguished:

1. The resin network
2. The liquid inside the resin particles
3. The liquid between the resin particles

The first acts as a semipermeable membrane between the stationary liquid phase within the resin and the mobile liquid phase between the resin beads.

Ionic groups are fixed on the resin and movement of ions across the membrane takes place as predicted by Donnan theory. The ion-exclusion mechanism involves interaction between partially ionized species and fully ionized polymer matrix. Electrostatic repulsive forces, between strong electrolytes (e.g., chloride, in the case of using HCl as eluent) and the ionic groups fixed on the resin (e.g., sulfonate), prevent them from entering into the resin, due to high ionic concentration inside the resin. Because ionized analytes are not retained, they are excluded from the polymer, migrate rapidly through the column, and are eluted at the column void volume. Partially ionized and neutral species (e.g., the undissociated forms of the analyte acids), as they penetrate into the pores of the resin, are distributed between the mobile phase in the column and the immobilized liquid in the pores of the packing. Separation is accomplished by differences in acid strength, size, and hydrophobicity.

The degree of retardation increases with the decrease of the ionization degree and, additionally, depends on polar attractions between analyte and fixed functional groups and on different van der Waals forces between an analyte and the hydrocarbon part of the resin. Elution order is related to pK_a values for ionic species and to the molecular size for neutral compounds.

Members of a homologous series, such as formic, acetic, and propionic acids, elute in the order of increasing pK_a (decreasing acid strength). Dibasic acids elute sooner than monobasic acids of the same carbon number. Isoacids elute earlier than normal acids. Double bonds retard elution, whereas keto groups increase elution rate.

Microporous polystyrene divinylbenzene resins are used, operating at pressures sometimes exceeding 3000 psi; unlike silica-based packings, they are stable from pH 0 to 14. For ion-exclusion separation of organic acids and weakly acidic compounds, strongly acidic, high-capacity, sulfonated styrene divinylbenzene in the hydrogen form are used. For organic bases, separation columns are packed with strongly basic copolymer with a quaternary ammonium

Encyclopedia of Chromatography DOI: 10.1081/E-ECHR-120040026

functional group. The degree of cross-linking (the percentage of divinylbenzene in the copolymer) affects the retention of weakly ionized species; the lower the degree of cross-linking, the longer the retention time of acid, either strong or weak. This is due to the fact that as cross-linking decreases, ions more readily penetrate the resin, where they are held up.

As aforementioned, a large number of organic and weak inorganic acids can be eluted from the hydrogen form of cation-exchange resin using water as the eluent. However, the addition of mineral acid to the water eluent suppresses the ionization of strong and moderately strong organic acids, allowing them to partition into the resin phase and, thus, improve selectivity, as retention times on the resin are increased. The addition of inorganic salts, such as $(NH_4)_2SO_4$ or organic modifiers (acetonitrile, isopropanol, ethanol, methanol), to the eluent may improve separation. Acetonitrile, for example, decreases the retention time of relatively nonpolar compounds.

Ion-exclusion chromatography can couple to ion chromatography to improve the chromatographic resolution of inorganic anions and organic acids in complex matrices. The dual system can be either in the order IEC/IC or IC/IEC.

Various detection systems can be used in ion-exclusion chromatography, among them ultraviolet (UV)/vis spectrophotometry, conductivity, electrochemistry, fluorometry, refractive index (RI) measurement, are the most common techniques. Additionally, combined detection systems (e.g., UV/amperometry, UV/RI) may be used, leading to enhanced selectivity.

Ultraviolet detection is useful, especially when water or sulfuric acid, which do not absorb in the UV region, are used. Detection for most nonaromatic carboxylic acids is accomplished at 210 nm.

Conductivity detection is preferred when water is used as eluent; then ionizable analytes are readily detected. However, in the case where HCl is used as the eluent, the analytical column is followed by a suppressor column, packed with a cation-exchange resin in the silver form. The hydrogen ions of the eluent are exchanged for silver ions, which then precipitate chloride ions, thus removing the ions contributed by the eluent and enhancing the analyte's signal.

Electrochemical detectors (coulometric and amperometric) are used when the analytes are electrochemically active or capable of being coupled to an electrochemical reaction.

Refractive index monitors are used in food analysis, for detecting carbohydrates, alcohols, and other substances with weak or no UV absorption.

With the combination of RI and UV, simultaneous detection of organic acids, carbohydrates, and

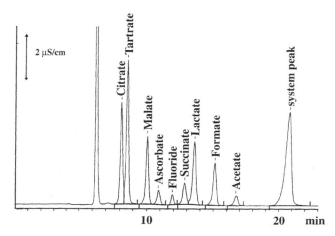

Fig. 1 Determination of organic acids and fluoride using ion-exclusion chromatography with direct conductivity detection, using mmol/L H_2SO_4, and 10% acetone as eluent. (From Metrohm Ltd., with permission.)

alcohols with one sample injection can be achieved. Postcolumn reactions can be used for fluorometric detection of amino acids, with excellent sensitivity and selectivity.

Ion-exclusion chromatography finds numerous applications for identification and determination of acidic species in complex matrix materials, such as dairy products, coffee, wine, beer, fruit juice, and other commercial products which can be quickly analyzed with minimal sample preparation before injection (usually only filtration, dilution, or centrifugation). Organic acid determination is also of great importance in biomedical research (e.g., physiological samples, in which most of the Krebs cycle acids (tricarboxylic acid cycle) are present).

Organic acids can be detected in the parts per billion range. With preconcentration, this limit can be further decreased. A typical ion-exclusion chromatogram of organic acids separation is presented in Fig. 1.

SUGGESTED FURTHER READING

Gierde, D.; Fritz, J. *Ion Chromatography*, 2nd Ed.; Alfred Huethig Verlag: New York, 1987.

Gierde, D.; Mehra, H. *Advances in Ion Chromatography*; Jandik, P., Cassidy, R., Eds.; Century International: Franklin, MA, 1989; Vol. 1.

Haddad, P.; Jackson, P. *Ion Chromatography, Principles and Application*; Elsevier: Amsterdam, 1990.

Kaine, L.; Crowe, J.; Wolnic, K. Forensic applications of coupling non-suppressed ion-exchange chromatography with ion-exclusion chromatography. J. Chromatogr. **1992**, *602*, 141–247.

Metrohm IC Application Note No. O-5, Application Notes, Metrohm, Herisau, (1996).

Small, H. *Ion Chromatography*; Plenum Press: New York, 1989.

Tanaka, K.; Fritz, J. Determination of bicarbonate by ion-exclusion chromatography with ion-exchange enhancement of conductivity detection. Anal. Chem. **1987**, *59*, 708–712.

Tarter, J. *Ion Chromatography*; Chromatographic Science Series; Marcel Dekker, Inc.: New York, 1987; Vol. 37.

Togami, D.; Treat-Clemons, L.; Hometchko, D. Int. Lab. **1990**, *2*, 29–33.

Ion-Interaction Chromatography

Teresa Cecchi
Università degli Studi di Camerino, Camerino, Italy

INTRODUCTION

Under reversed-phase (RP) HPLC conditions, ionic compounds are weakly retained. On the contrary, when an ion-interaction reagent (IIR), which is a large lipophilic ion, is added to the mobile phase, ionized species of opposite charge are separated on RP columns with adequate retention. This is the chromatographic approach of RP ion-interaction chromatography (IIC). It has become a widely used separation mode in analytical HPLC because it provides a useful and flexible alternative to ion-exchange chromatography. Better selectivity, enhanced resolution, and retention are usually gained by this separation strategy.

According to the qualitative retention model of Bidlingmeyer, the lipophilic IIR, flowing under isocratic conditions, dynamically adsorbs onto the alkyl-bonded apolar surface of the stationary phase, forming a primary charged ion layer. The corresponding counterions are found in the diffuse outer region to form an electrical double layer. This charged stationary phase can then more strongly retain analyte ions of the opposite charge.

Unlike conventional ion exchange, IIC can be used to separate nonionic and ionic or ionizable compounds in the same sample, because retention of an analyte involves its transfer trough the electrical double layer and depends on both electrostatic interactions and adsorptive (RP) effects.

In recent's, many examples of applications of IIC have been reported. They essentially concern the separation of organic and inorganic ions in the environmental, pharmaceutical, food, and clinical fields.

RETENTION MECHANISM

The larger number of names (e.g., ion-pair chromatography, dynamic ion-exchange chromatography, hetaeric chromatography, soap chromatography) which have been given to the IIC mode sheds light on the uncertainty concerning the retention mechanism. A majority of the proposed models are stoichiometric. They suggest that the oppositely charged analyte and IIR form a complex, according to a clear reaction scheme, either in the mobile-phase (ion-pair model) or at the stationary-phase surface (dynamic ion-exchange model). According to the first theory, the uncharged ion pair between oppositely charged analyte and IIR, which is formed in the mobile phase, is then more strongly retained by the stationary phase. The second theory presumes that solute ions undergo an ion-exchange process, at exchange sites dynamically generated by the adsorption of the IIR at the stationary phase. Knox and Hartwick demonstrated that both models lead to identical retention equation.

These models, although of practical and intuitive value, are not well founded in physical chemistry. The pioneeristic, even if qualitative, work of Bidlingmayer demonstrated that IIRs adsorb onto the stationary phase. It follows that stoichiometric equilibrium constants, which depend on the change in free energy of adsorption of the analyte, cannot be considered constant if the IIR concentration in the mobile phase increases, because the stationary phase surface properties (including its charge density) are modified. The multibody interactions and long-term forces involved in IIC can better be described by a thermodynamic approach.

A quantitative nonstoichiometric model was developed by Ståhlberg and coworkers. The model applies the Gouy–Chapman electrostatic theory to describe the interactions between charged species and it does not assume the formation of any chemical complexes: The adsorption of the IIR onto the stationary phase establishes a certain electrostatic surface potential, because its counterion has a lower adsorption tendency. An electrical double layer develops and a difference in electrostatic potential is created between the electroneutral bulk of the mobile phase and the net charged surface. The intuitive view of the effect of the IIR on retention is an electrostatic repulsion or attraction of the analyte to the charged stationary-phase surface, according to the analyte and IIR charge status. However, the adsorptive (RP) effects are also considered, to evaluate the total free energy of adsorption of the solute: The latter is partitioned into a "chemical" and an electrostatic free energy. This is a first approximation: The "chemical" part depends on the concentration of the IIR, as it determines a dynamic modification

Encyclopedia of Chromatography DOI: 10.1081/E-ECHR-120040027

of the stationary-phase properties. This electrostatic theory of IIC has been implemented by taking into account the competition between IIR and analyte for a limited surface area, and the different surface area requirements of analyte and IIR (multisite occupancy model). However, the main drawback of this powerful electrostatic theory is the complex algebraic form of the resulting equations; hence, a series of approximations has to be made to obtain a relationship between the analyte capacity factor and mobile-phase concentration of IIR, which is of interest for practical work.

Cantwell and co-workers proposed a surface adsorption, diffuse-layer ion-exchange double-layer model in which they underlined the role of the diffuse part of the double layer by assigning a stoichiometric constant for the exchange of ions.

Stranahan and Deming proposed a thermodynamic model for IIC in which the distribution a sample between the mobile and the stationary phase is discussed in terms of chemical potentials in both phases.

Additional peaks relative to the number of components injected are often obtained in IIC. These so-called "system peaks" confirm the proposed mechanism of dynamic functionalization of the stationary phase. They can be explained by taking into account that IIR ions are locally adsorbed onto (desorbed from) the stationary phase by injection of adsorbophilic solute ion of the opposite charge (of same charge). This change in the eluent composition, created by the sample injection, migrates along the column and give a signal if at least one of the eluent components can be detected. The same rationale provides the explanation for the indirect ultraviolet (UV) visualization (or amplification) of otherwise non-UV-absorbing samples, when a UV-absorbing lipophilic ion is added to the eluent.

INFLUENCE OF EXPERIMENTAL PARAMETERS ON RETENTION

The optimization of separations performed with IIC and the rationalization of analytes retention behavior are not easy tasks because they are influenced by many interdependent factors. This allows a fine modulation of their effects to achieve tailor-made separations.

Experimental design can be very helpful, and a number of chemometric optimization methods are present in the literature. Neural network models provided a good prediction power and a great versatility, without the need to develop any equations.

The following presents the effect of varying some individual factors on analyte retention.

ION-INTERACTION REAGENT

Type

The hydrophobic character of the IIR increases with increasing its chain length. More lipophilic reagents have higher adsorption constants, hence the effect of increasing chain length is qualitatively similar to the effect of increasing IIR concentration (see below) with regard to the degree of stationary-phase coverage. The use of multiply-charged IIRs allows the chromatographer to obtain larger changes of analyte retention. If chiral compounds are used as the IIR, the separation of the enantiomeric forms of the analyte may be achieved. The most popular IIRs are listed in Table 1.

Concentration

If the eluent concentration of the IIR increases, the amount of the adsorbed IIR also increases, according to its adsorption isotherm. This induces a higher surface potential on the stationary phase but also adsorption competes between analyte and IIR for the available stationary phase sites. Therefore, the following hold:

1. If the charge status of analyte and the IIR is the same, a decrease in retention is observed because of electrostatic repulsion between solute and charged stationary phase, and because of adsorption competition.
2. If the charge status of analyte and the IIR is the opposite, an increase in retention is expected because of electrostatic attraction between solute and charged stationary phase. A parabola-like dependence of analyte capacity factors on IIR concentration is observed if the investigated concentration range is broad. For narrower ranges, a linear increase may hold. Some authors have emphasized that analyte retention passes through a maximum because if the ionic strength is not kept constant when increasing the IIR concentration, there is a competition between analyte ion and IIR counterion; this competition counteracts the retention increase.

Table 1 Commonly used ion-interaction reagents

Cationic IIRs	Anionic IIRs
Tetramethylammonium	Butanesulfonate
Tetraethylammonium	Pentanesulfonate
Tetrabutylammonium	Hexanesulfonate
Cetyltrimethylammonium	Octanesulfonate
Octylammonium (from octylamine)	Dodecanesulfonate

However, a foldover of the plot may still occur even if the ionic strength is kept constant, because there is a critical value of the IIR concentration at which the positive effect of the electrostatic attraction is balanced by the negative effect of adsorption competition for the available stationary-phase surface area.

3. If the analyte is uncharged, a very weak decrease in retention is usually observed, primarily because of adsorption competition for the stationary phase.

Increasing the IIR above its critical micelle concentration leads into the field of micellar chromatography in which analyte may partition between the mobile phase and both the stationary phase and the micelle.

MOBILE-PHASE COMPOSITION

Organic-Modifier Concentration

In IIC, the logarithm of the analyte capacity factor is described as a linear function of the organic-modifier concentration in the mobile phase. When the sample ion is in the same charge status as the IIR, the slope of the linear relationship, if compared to the original RP slope, becomes steeper (the contrary is observed for oppositely charged combinations). This can be explained by taking into account that the organic modifier, through desorption effects, decreases the retention of ionic solutes via the simultaneous decrease of the free energy of adsorption of both the analyte and IIR.

Ionic Strength

An increase in salt concentration in the bulk mobile phase provides those counterions which are able to reduce, according to the Gouy–Chapman electrostatic theory, the electrostatic stationary-phase surface potential. Hence, the adsorption of the IIR may increase, even if its concentration in the eluent is the same, because of lower electrostatic "self"-repulsion. However, the net effect is a reduced surface potential: The ion interactions decrease, and analyte retention may be modulated.

From an intuitive point of view, the inorganic ions are eluting agents because they limit the interaction of oppositely charged analyte and IIR, via a competing equilibrium for adsorbed lipophilic ions. This view gives the rationale for the use of mobile-phase additives, such as sodium carbonate, to avoid the unnecessarily high resolution which may be obtained between analytes of different charge.

It has to be emphasized that the nature of the electrolyte ions influences the surface potential value because the effective surface charge concentration is reduced if slight hydrophobic, adsorbophilic electrolytic counterions are included in the eluent.

The influence of moderate increase of ionic strength on the "chemical" part of the free energy relative to the analyte transfer from the mobile to the stationary phase has been usually neglected.

MOBILE-PHASE pH

The eluent pH value affects the degree of ionization of the species involved in ion interaction. Hence, the greatest retention is obtained for completely dissociated species. This is the opposite of what is observed in RP chromatography.

Unexpected pH dependencies were explained by (a) competition between negative analyte ions and OH^- ions for interaction with the electrical double layer and (b) a mixed retention mechanism in which RP partition or interaction with unreacted silanols from the stationary-phase base may play a significant role.

RP STATIONARY PHASE

A number of different packings were used in IIC, including the newly developed graphitized carbon column, which has excellent chemical and physical resistance. The use of polymeric material has the drawback of poor physical resistance. However, a wider pH range is investigable and the affinity for certain IIR is higher, by comparison with the silica-based RP columns. However, discordant results are present in literature reports with regard to the chromatographic efficiency.

With regard to the silica-based RP stationary phase, unreacted residual silanol groups may play a significant role in IIC because it was shown that they are ion-exchange sites not only for analyte cations but also for alkylammonium IIR. The higher retentions that were noticed for the silica-based stationary phase if compared to end-capped or polymer-based packings supports this.

The reproducibility of results obtained with silica-based RP, of the same declared characteristics but from different manufacturers, was sometimes poor, probably because of the properties of the silica used and the different reaction conditions in the alkylation of the support.

Stationary phases with higher hydrophobicities and adsorption capacities show increased retention of both solute and IIR. Hence, an increased capacity factor value should be expected, even if anomalies can be due to direct competition of solute and IIR for the available stationary phase.

TEMPERATURE

Temperature control is very important for obtaining reproducible separations. Indeed, the adsorption of the IIR onto the stationary phase follows an adsorption isotherm; hence, an increase of the column temperature leads to a decreased amount of the adsorbed IIR, even if its concentration in the mobile phase is constant. This, in turn, determines a decreased absolute surface potential and a modification of the solutes' capacity factors. Usually, a temperature increase results in an improved resolution and faster separation, even if a reversal of the elution sequence of the components of a mixture may sometimes be observed, because of the interplay of electrostatic and RP interaction which are characterized by different enthalpies.

SUGGESTED FURTHER READING

Bartha, A.; Ståhlberg, J. J. Chromatogr. A. **1994**, *668*, 255–284.

Bidlingmeyer, B.A. J. Chromatogr. Sci. **1980**, *18*, 525–539.

Chen, J.C.; Weber, S.G.; Glavina, L.L.; Cantwell, F.F. Electrical double-layer models of ion-modified (ion-pair) reversed-phase liquid chromatography. J. Chromatogr. **1993**, *656*, 549–576.

Gennaro, M.C. Adv. Chromatogr. **1995**, *35*, 343–381.

Knox, J.H.; Hartwick, R.A. Mechanism of ion-pair liquid chromatography of amines, neutrals, zwitterions and acids using anionic hetaerons. J. Chromatogr. **1981**, *204*, 3–21.

Okamoto, T.; Isozaki, A.; Nagashima, H. Studies on elution conditions for the determination of anions by supressed ion-interaction chromatography using a graphitized carbon column. J. Chromatogr. A. **1998**, *800*, 239–245.

Pietrzy, D. J. Chromatogr. Sci. **1998**, *78*, 413–462.

Sacchero, G.; Bruzzoniti, M.C.; Sarzanini, C.; Mentasti, E.; Metting, H.J.; Coenegracht, P.M.J. Comparison of prediction power between theoretical and neural-network models in ion-interaction chromatography. J. Chromatogr. A. **1998**, *799*, 35–45.

Stranahan, J.; Deming, S.N. Thermodynamic model for reversed-phase ion-pair liquid chromatography. Anal. Chem. **1982**, *54*, 2251–2256.

Weiss, J. *Ion Chromatography*, 2nd Ed.; VCH: Weinheim, 1995; 239–289.

Ion-Interaction Chromatography: Comprehensive Thermodynamic Approach

Teresa Cecchi
Dipartimento Scienze Chimiche, Università degli Studi di Camerino, Camerino,
Italy Chemistry Department, ITIS Montani, Fermo (AP), Italy

INTRODUCTION

Reverse-phase (RP) chromatography is, by far, the most widely used separation mode in HPLC. A wide variety of mobile-phase additives that may also modify the surface of the packing material are used in optimization procedures. Ion interaction chromatography (IIC) is an RP technique that involves the use of ion interaction reagents (IIRs) that are large lipophilic ions; they are retained much more than their nonadsorbophilic counterions. An electrostatic potential difference between the stationary phase and the bulk mobile phase develops and influences ionic solute retention. When IIR is used, conventional RP columns are able to retain oppositely charged analytes with adequate retention. Improved resolution and selectivity broaden the scope of separation of organic and inorganic ions.

IIC analyses are challenging because of the high number of easily tunable but interdependent mobile-phase variables. Retention models are often sought because there is a need for a detailed understanding of the underlying phenomena that govern solute distribution between the two phases. The retention equations of a theoretical model can also be advantageously used during method development, involving the setup of custom modes in computer-supported software tools.

ION INTERACTION INTERPRETATION

To obtain a simple interpretation of the experimental findings in IIC, theoretical chromatographers first adopted a stoichiometric strategy that pioneered this separation mode. Unfortunately, the reaction schemes of stoichiometric models in both the mobile phase (ion pair model)[1] and stationary phase (dynamic ion exchange model)[2] lack a firm foundation in physical chemistry[3] because they are not able to account for the stationary-phase modification that results from the addition of the IIR to the eluent, and they fail to properly describe experimental results, as pointed out by Bidlingmeyer et al.[4] Key insights on these retention

models were also provided by Knox and Hartwick.[5] The contrast between the simplicity and practicality of stoichiometric relationships, and the complexities of the thermodynamics solutes undergo is an important underlying theme of meditation for a model maker. Later, electrostatic theories[3,6] tried to pattern chromatographic findings according to electrostatic interactions between charged IIRs and an analyte, thereby disregarding chemical equilibria in the IIC system, but some predictions of theirs are at variance with experimental evidence.[7–9] Conversely, the recently developed extended thermodynamic approach to IIC[7–18] considers that a major contribution to the distribution of charged solutes between phases is the electrostatic potential difference between them, but also capitalizes on the importance of complex formation at a thermodynamic level, and not a stoichiometric level, to take into account the stationary-phase electrostatic potential that results from IIR adsorption. In the following, we will discuss in detail how this retention model may help the chromatographer to make educated guesses and to simplify the complex task of a successful optimization in the parameter space of IIC separations.

THEORY

Influence of IIR Concentration on the Retention Behavior of Charged, Neutral, and Zwitterionic Analytes

Amphiphilic IIR ions (H) dynamically adsorb onto the stationary phase, forming a primary charged ion layer, and counterions in the diffuse outer region form an electrical double layer. The adsorption isotherm of an IIR can be described by the Freundlich equation:

$$[LH] = a[H]^b \tag{1}$$

where a and b are constants; [H] and [LH] are, respectively, the mobile-phase and the stationary-phase concentrations of the IIR. The use of the Freundlich

Encyclopedia of Chromatography DOI: 10.1081/ECHR-120040673

adsorption isotherm is not empirical[19] because it is related to the potential modified Langmuir adsorption isotherm, which holds for the adsorption of ions that have a substantial hydrophobic moiety. Because IIR counterions have a negligible adsorbophilic aptitude, an electrical potential difference Ψ° develops between the surface and the bulk solution. This potential is given by the rigorous Gouy–Chapman theory equation: if the ionic strength is high enough and the Debye length is low, we are allowed to use a planar surface geometry;[3] we have the following relationship:[3,7]

$$
\Psi = \frac{2RT}{F} \ln \left\{ \frac{[LH]|z_H|F}{\left(8\varepsilon_0\varepsilon_r RT \sum_i c_{0i}\right)^{\frac{1}{2}}} \right.
$$

$$
\left. + \left[\frac{([LH]z_H F)^2}{8\varepsilon_0\varepsilon_r RT \sum_i c_{0i}} + 1 \right]^{\frac{1}{2}} \right\} \tag{2}
$$

where F is the Faraday constant, R is the gas constant, T is the absolute temperature, z_H is the charge of the IIR, ε_r is the dielectric constant of the medium, ε_0 is the vacuum permittivity, and $\sum c_{0i}$ is the mobile-phase concentration of singly charged electrolytes. For the sake of simplicity, we will indicate the following:

$$
f = \frac{|z_H|F}{\left(8\varepsilon_0\varepsilon_r RT \sum_i c_{0i}\right)^{\frac{1}{2}}} \tag{3}
$$

where f (m^2/mol) is a constant, which can be evaluated from experimental conditions.

The complex equation Eq. (2) can be advantageously approximated, for high surface potential, by the following very simple expression:[10,12,14,19]

$$
\Psi = \alpha + \beta \ln[LH] \tag{4}
$$

where α and β are constants[10] that depend on experimental conditions.

For surface potential below 25 mV, Eq. (2) can be linearized[3,11,15] and approximated by the following:

$$
\Psi = \frac{z_H[LH]F}{\kappa\varepsilon_0\varepsilon_r} \tag{5}
$$

where κ is the inverse Debye length.

It has been demonstrated[7] that the course of analyte retention on the mobile-phase and stationary-phase concentrations of the IIR can be described,

respectively, by the following two expressions

$$
k = \phi[L]_T \frac{K_{LE}\dfrac{\gamma_L\gamma_E}{\gamma_{LE}}\exp(-z_E F\Psi/RT)}{\left(1 + K_{EH}\dfrac{\gamma_E\gamma_H}{\gamma_{EH}}[H]\right)} \\[2pt]
\times \frac{+ K_{EHL}\dfrac{\gamma_E\gamma_H\gamma_L}{\gamma_{EHL}}[H]}{\left(1 + K_{LH}\dfrac{\gamma_L\gamma_H}{\gamma_{LH}}\exp(-z_H F\Psi/RT)[H]\right)} \tag{6}
$$

$$
k = \phi \frac{K_{LE}\dfrac{\gamma_L\gamma_E}{\gamma_{LE}}\exp(-z_E F\Psi/RT) + \dfrac{K_{EHL}}{a^{1/b}}\dfrac{\gamma_E\gamma_H\gamma_L}{\gamma_{EHL}}[LH]^{1/b}}{\left(1 + \dfrac{K_{EH}}{a^{1/b}}\dfrac{\gamma_E\gamma_H}{\gamma_{EH}}[LH]^{1/b}\right)} \\[2pt]
\times ([L]_T - [LH]) \tag{7}
$$

where ϕ is the phase ratio of the column; K_{EL}, K_{EHL}, K_{EH}, and K_{HL}, are the thermodynamic equilibrium constants for adsorption of the analyte E onto the stationary phase, ion pair formation in the stationary phase, ion pair formation in the eluent, and adsorption of the IIR onto the stationary phase, respectively; γ represents the activity coefficient for each species; z_E is the charge of the analyte; and $[L]_T$ estimates the total ligand surface concentration.

The first term in the numerator of Eqs. (6) and (7) describes the electrostatic interaction between the analyte and the charged stationary phase. This interaction will be attractive (repulsive) if the analyte is oppositely (similarly) charged to the IIR; hence, its retention increases (decreases) with increasing IIR concentration. The second term in the numerator of Eqs. (6) and (7) accounts for ion pair formation at the stationary phase that results in a retention increase. The left factor of the denominator of Eq. (6) and the denominator of Eq. (7) accounts for ion pair formation in the mobile phase that tends to reduce retention because the analyte is withdrawn from the stationary phase toward the eluent. Both terms concerning ion pair formation are missing if the analyte and IIR are similarly charged, or if the analyte is neutral. The right-hand factor of the denominator of Eq. (6) and the right-hand factor of Eq. (7) describe adsorption competition between the analyte and the IIR. From Eq. (6), it is clear that Ψ° always runs counter to further adsorption of the IIR because Ψ° is of the same sign as z_H. From Eq. (7), it is evident that adsorption competition, if it applies, shows only if [LH] is high, compared to $[L]_T$.

Eluent pH is an important optimization parameter because it influences the analyte ionization, which, in turn, controls the magnitude of electrostatic interactions.

When Eqs. (1)–(3) are substituted into Eqs. (6) and (7), the following two expressions are obtained:

$$k = \frac{c_1\left(a[H]^b f + \left((a[H]^b f)^2 + 1\right)^{\frac{1}{2}}\right)^{(\pm 2|z_E|)} + lc_2[H]}{(1 + c_3[H])\left(1 + c_4[H]\left(a[H]^b f + \left((a[H]^b f)^2, + 1\right)^{\frac{1}{2}}\right)^{(-2|z_H|)}\right)}$$

(8)

$$k = \frac{d_1\left([LH]f + \left(([LH]f)^2 + 1\right)^{\frac{1}{2}}\right)^{\pm 2|z_E|} + d_2[LH]^{1/b}}{(1 + d_3[LH]^{1/b})} \times (d_4 - [LH])$$

(9)

In the exponent of the first term of the numerator, the plus sign applies for oppositely charged analytes and IIRs, whereas the minus sign applies for similarly charged analytes and IIRs, as expected according to the electrostatic behavior; Eqs. (8) and (9) are also able to take into account, via the magnitude of z_E in the exponents, that the electrostatic interaction is stronger for multiply charged analytes.

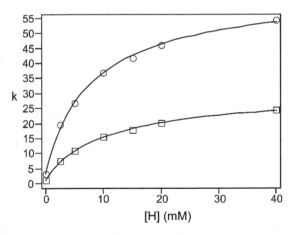

Fig. 1 Dependence of k on adrenaline (squares) and L-tyrosine hydrazide (circles), on mobile-phase concentration of 1-hexanesulfonate. Column: Synergi Hydro-RP (Phenomenex) 150 × 4.6 mm ID, particle size 4 µm, and bonded phase coverage 4.05 µmol/m². Eluent: phosphate buffer 37.10 mM KH$_2$PO$_4$ and 4.29 mM Na$_2$HPO$_4$ calculated to provide a pH of 6.0. After addition of the desired amount of sodium 1-hexanesulfonate, NaCl was added so that the total sodium concentration was 50 mM (constant ionic strength). Experimental data were fitted by Eq. (8).

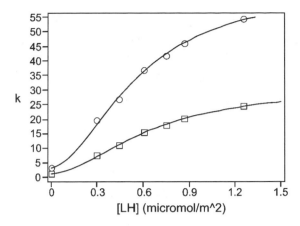

Fig. 2 Dependence of k on adrenaline (squares) and L-tyrosine hydrazide (circles), on stationary-phase concentration of 1-hexanesulfonate. Conditions as in Fig. 1. Experimental data were fitted by Eq. (9).

When Eqs. (8) and (9) are fitted to experimental data, excellent results are obtained, as shown in Figs. 1 and 2 for analytes charged oppositely to the IIRs; in Figs. 3 and 4 for analytes charged similarly to the IIRs; and in Fig. 5 for a doubly charged analyte, as detailed in the captions. Tables 1 and 2 detail parameter estimates and their standard deviations, correlation coefficients, and sum of square errors obtained from the best fit of these experimental data by Eqs. (8) and (9), respectively. Fitting parameters $c_1 - c_4$ and $d_1 - d_4$ are not simple adjustable constants, but have a clear physical meaning,[7] and this allows the parameter estimates to be commented on. As expected, $c_2 - c_3$ and $d_2 - d_3$, which are related to ion pair equilibrium constants, decrease with decreasing analyte lipophilicity; d_4, which represents the total ligand concentration, compares well with the bonded phase coverage of the

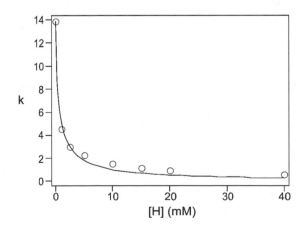

Fig. 3 Dependence of k on sodium p-toluenesulfonate, on mobile-phase concentration of 1-hexanesulfonate. Conditions as in Fig. 1. Experimental data were fitted by Eq. (8). (From Ref.[18].)

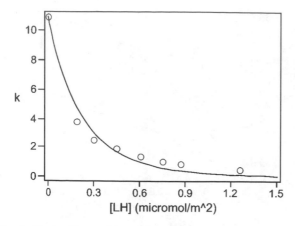

Fig. 4 Dependence of k sodium salicylate on the stationary-phase concentration of 1-hexanesulfonate. Conditions as in Fig. 1. Experimental data were fitted by Eq. (9). (From Ref.[18].)

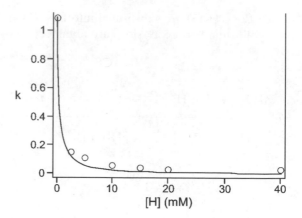

Fig. 5 Dependence of k on 2,6-naphthalenedisulfonate, on mobile-phase concentration of 1-hexanesulfonate. Conditions as in Fig. 1. Experimental data were fitted by Eq. (8). (From Ref.[18].)

column ($4.05\,\mu mol/m^2$), as expected; from the estimated c_4, which represents the equilibrium constant for the IIR adsorption, we obtain an averaged $\Delta G° = -15.6\,kJ/mol$, which is a very reasonable value for the standard free energy of adsorption of hexanesulfonate.[19,20] In Table 1, c_1 is missing for all analytes except 2,6-naphthalenedisulphonate because it was not considered an optimization parameter, since it represents k_0 that is analyte retention without the IIR in the mobile phase and it can be obtained from experimental results. When c_1 was estimated by the model (as for the doubly charged 2,6-naphthalendisulphonate, for which other fitting parameters do not apply because electrostatic repulsion entirely models its retention behavior), the percent error is very low (0.6%).

When Eq. (4) is used to obtain the surface potential that has to be substituted into Eq. (6), we obtain[10] the following expression (a parallel expression can also be obtained from Eq. (7) to model retention data as

a function of the stationary-phase concentration of the IIR):

$$k = \frac{c_1[H]^{z_E/z_H(b-1)} + c_2[H]}{(1 + c_3[H])(1 + c_4[H]^b)} \quad (10)$$

For a detailed description of $c_1 - c_4$ in Eq. (10), the reader is referred to Ref.[10] Fig. 6 details how this equation properly describes the retention of neutral analytes in IIC;[12] in this case the above expression can be simplified because z_E, c_2, and c_3 are all zero and only adsorption competitions model the retention decrease with increasing IIR concentration.

The surface potential is easily predicted to alter the retention behavior of zwitterions.[13–15,21] The molecular electrical dipole is subjected to a torque moment that arranges it parallel to the lines of the nonhomogeneous electrical field, with the head oppositely charged to the electrostatic surface potential facing

Table 1 Summary of parameter estimates and their standard deviation, correlation coefficient (r), and sum of square errors (SSE) for the best fit of experimental retention data (some of which are shown in Figs. 1, 3, and 5) as a function of the mobile-phase concentration of the IIR by Eq. (8)

Analyte	c_1	c_2 (mM^{-1})	c_3 (mM^{-1})	c_4 (mM^{-1})	r Eq. (8)	SSE Eq. (8)
L-Tyrosine hydrazide	—	4.627 ± 0.295	0.111 ± 0.006	—	0.999	2.531
Adrenaline	—	1.368 ± 0.084	0.090 ± 0.004	—	0.999	0.321
Octopamine		0.989 ± 0.087	0.085 ± 0.005	—	0.999	0.380
Sodium p-toluenesulfonate	—	—	—	0.629 ± 0.144	0.995	1.337
Sodium salicylate	—	—	—	0.488 ± 0.140	0.994	1.064
2,6-Naphthalenedisulfonate	1.097 ± 0.035	—	—	—	0.996	0.007

Conditions as in Fig. 1.
(From Ref.[18].)

Table 2 Summary of parameter estimates and their standard deviation, correlation coefficient (r), and sum of square errors (SSE) for the best fit of experimental retention data (some of which are shown in Figs. 2 and 4) as a function of the stationary-phase concentration of the IIR by Eq. (9)

Analyte	d_1	d_2 (m^2/μmol)$^{(1/b)}$	d_3 (m^2/μmol)$^{(1/b)}$	d_4 (μmol/m^2)	r Eq. (9)	SSE Eq. (9)
L-Tyrosine hydrazide	—	101.829 ± 10.444	3.011 ± 0.229	—	0.999	4.587
Adrenaline	—	35.116 ± 2.971	2.350 ± 0.146	—	0.999	0.606
Octopamine	—	25.396 ± 2.300	2.204 ± 0.144	—	0.999	0.408
Sodium p-toluenesulfonate	—	—	—	1.311 ± 0.393	0.987	3.380
Sodium salicylate	—	—	—	1.616 ± 0.595	0.987	2.085
2,6-Naphthalenedisulfonate	1.097 ± 0.036	—	—	—	0.996	0.008

Conditions as in Fig. 1.
(From Ref.[18].)

the stationary phase. Hence, the electrical force is always attractive and it pushes the dipole toward the interphase, where the field is stronger. This force was used to calculate the electrostatic contribution to the electrochemical potential of the analyte, and to the thermodynamic equilibrium constant for its adsorption. It was demonstrated that, from Eq. (6), the following relationship is quantitatively able to model zwitterions retention and can be advantageously used in life science chromatography:

$$k = \frac{c_1\left(a[H]^b f + \left((a[H]^b f)^2 + 1\right)^{\frac{1}{2}}\right)^{2c_2\kappa/F}}{1 + c_4[H]\left(a[H]^b f + \left((a[H]^b f)^2 + 1\right)^{\frac{1}{2}}\right)^{-2|z_H|}} \tag{11}$$

In Eq. (11), c_1 and c_4 are the already discussed parameters, whereas c_2 is related to the molecular dipole: the higher it is, the stronger is the retention increase on IIR addition. A parallel expression can also be obtained from Eq. (7) to model retention data as a function of the stationary-phase concentration of the IIR.[13] A fractional charge approach to the IIC of zwitterions was also recently put forward.

Influence of Organic Modifier Concentration

A bivariate treatment of the simultaneous effects of IIR mobile-phase concentration and organic modifier percentage in the eluent on analyte retention gives the following relationship[16] where φ is the percentage (% vol/vol) of methanol in the mobile phase; $c_{1_0} - c_{4_0}$ are the already discussed parameters $c_1 - c_4$ when the organic modifier is not present in the eluent; a_0 and b_0 are a and b when the organic modifier is not present in the eluent; and h is a parameter that accounts for the eluent ionic strength. If the latter is not constant and the dependence of h on total ionic concentration is explicitly introduced in Eq. (12), (see below) allows a multivariate approach to IIC; m_1, m_2, m_3, m_4, m_5, and m_6 are parameters that depend on experimental conditions: m_1 depends on the analyte characteristic, m_2, m_3, and m_6 depend on the peculiarity of IIR, whereas m_4 and m_5 depend on the nature of both the analyte and the IIR. For analytes oppositely charged to the IIR, the active fitting parameters are m_1, m_4, m_5, c_{2_0}, and c_{3_0} because m_2, m_3, a_0, and b_0 are readily obtained from the fitting of the Freundlich constants as functions of the organic modifier concentration in the eluent. The bivariate nonlinear regression of the retention of a typical analyte oppositely charged to the IIR gives parameter estimates that were used to graphically present Eq. (12) in Fig. 7: it is rewarding to observe that retention decreases with increasing organic modifier concentration and increases with increasing IIR concentration. This increase is steeper at low organic modifier percentages because the organic modifier reduces analyte retention both directly (it decreases the analyte free energy of adsorption) and indirectly (it decreases the IIR free energy of

$$k = \frac{c_{1_0}e^{-m_1\varphi}\left(\frac{a_0 e^{-m_2\varphi}[H]^{(b_0+m_3\varphi)}}{\left(h(\varepsilon_{H_2O}-(\varepsilon_{H_2O}-\varepsilon_{MeOH})\varphi)\right)^{\frac{1}{2}}} + \left(\left(\frac{a_0 e^{-m_2\varphi}[H]^{(b_0+m_3\varphi)}}{\left(h(\varepsilon_{H_2O}-(\varepsilon_{H_2O}-\varepsilon_{MeOH})\varphi)\right)^{\frac{1}{2}}}\right)^2 + 1\right)^{\frac{1}{2}}\right)^{(\pm 2|z_E|)} + c_{2_0}e^{-m_4\varphi}[H]}{(1 + c_{3_0}e^{-m_5\varphi}[H])\left(1 + c_{4_0}e^{-m_6\varphi}[H]\left(\frac{a_0 e^{-m_2\varphi}[H]^{(b_0+m_3\varphi)}}{\left(h(\varepsilon_{H_2O}-(\varepsilon_{H_2O}-\varepsilon_{MeOH})\varphi)\right)^{\frac{1}{2}}} + \left(\left(\frac{a_0 e^{-m_2\varphi}[H]^{(b_0+m_3\varphi)}}{\left(h(\varepsilon_{H_2O}-(\varepsilon_{H_2O}-\varepsilon_{MeOH})\varphi)\right)^{\frac{1}{2}}}\right)^2 + 1\right)^{\frac{1}{2}}\right)^{(-2|z_H|)}\right)} \tag{12}$$

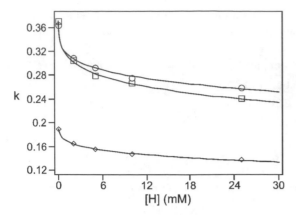

Fig. 6 Dependence of k on dimethylformamide (squares), dimethylchetone (circles), and dimethylsulfoxide (rhombs), on mobile-phase concentration of tetrabutylammonium bromide. Column: Res Elut 5 C_{18} (Varian), 25 cm × 4.6 mm ID, 5 μm. Eluent: 81.6 mM phosphate buffer, pH 7.2—methanol, 85:15, vol/vol containing tetrabutylammonium bromide. Experimental data were fitted by Eq. (10) with all z_E, c_2, and c_3 equal to zero. (From Ref.[12].)

adsorption: a lower surface concentration of the IIR results in a lower electrostatic attraction of the analyte and hence in a lower retention).

Noteworthy, in the absence of the IIR, Eq. (12) reduces to one which describes the influence of the organic modifier on RP-HPLC retention. A relationship similar to Eq. (12) can be obtained to describe the simultaneous effects of the IIR stationary-phase

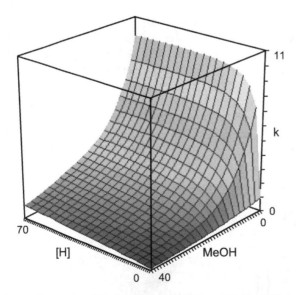

Fig. 7 Retention behavior of a typical analyte in IIC as a function of IIR mobile-phase concentration and organic modifier percentage in the eluent, according to Eq. (12).

concentration and organic modifier percentage in the eluent on analyte retention.[16]

Influence of Ionic Strength

In IIC method development, when a compensatory electrolyte is not added to the eluent, the ionic strength is not constant; hence, the dependence of f Eq. (3) on the ionic concentration must be explicitly addressed.[17] At a fixed IIR surface concentration, the electrostatic potential decreases with increasing electrolyte concentration because counterions in the diffuse layer shield the surface charge. However, with increasing ionic strength, there is an increased surface concentration of the IIR because counterions lower the self-repulsion forces between the similarly charged adsorbed IIR ions. Yet, this Donnan effect is not able to compensate the decreased surface potential due to a higher concentration of counterions in the diffuse layer. The strong interplay between these issues can be easily taken into account using the adsorption isotherm of the IIR obtained when its counterion concentration is not kept constant in the eluent.[17] The retention of a solute oppositely (similarly) charged to the IIR is predicted to decrease (increase) with increasing ionic strength because of the lower net electrostatic attraction (repulsion).

CONCLUSIONS

The distinguishing features that set this retention model apart are the following:[7–18,21]

1. It is general. From two equations, quantitative predictions can be made for the retention behavior of charged, multiply charged, neutral, and zwitterionic solutes in IIC as a function of both the mobile-phase and the stationary-phase concentrations of the IIR. Retention equations can also quantitatively take into account the influence of the organic modifier concentration and the ionic strength; in the absence of IIR, they reduce to the well-known relationships of RP-HPLC.
2. It reduces to the relationships of stoichiometric or electrostatic approaches in IIC, respectively, if the surface potential or ion pairing equilibria are disregarded.
3. It is able to rationalize experimental behaviors that cannot be explained by other outstanding electrostatic retention models: 1) different theoretical curves when k is plotted as a function of the stationary-phase concentration of the IIR, for different IIRs; 2) dependence of

the ratio of the retention of two different analytes on the IIR concentration; 3) dependence of the k/k_0 ratio on the analyte nature if the experimental conditions are the same; and 4) better agreement between electrostatic retention model predictions and experimental findings for analytes similarly charged as the IIR.

4. It is able to explain why the electrostatic approach is sometimes at variance with experimental evidence (this happens if experimental data underscore and involve complex formation).

5. Adjustable parameters have a clear physical meaning and their estimates are reliable because they compare well with literature estimates and with direct experimental measurements.

6. It is able to rationalize, in a quantitatively unprecedented way, with very low percent errors, experimental evidence qualitatively or semiquantitatively explained by other models.

As a concluding remark, it has to be emphasized that every acceptable retention theory must be consistent with fundamental physics, as well as describe experimental findings. The complex multiplicity of phenomena involved in an IIC system requires a complex description of the thermodynamics solutes have undergone: this description is epistemologically acceptable if the model is well founded in physical chemistry and if it is able to describe experimental data better than previous models.

REFERENCES

1. Horvath, C.; Melander, W.; Molnar, I.; Molnar, P. Enhancement of retention by ion-pair formation in liquid chromatography with nonpolar stationary phases. Anal. Chem. **1977**, *49* (14), 2295–2305.

2. Kissinger, P.T. Comments on reverse-phase ion-pair partition chromatography. Anal. Chem. **1977**, *49* (6), 883.

3. Bartha, A.; Stahlberg, J. Electrostatic retention model of reversed-phase ion-pair chromatography. J. Chromatogr. A. **1994**, *668*, 255–284.

4. Bidlingmeyer, B.A.; Deming, S.N.; Price, W.P., Jr.; Sachok, B.; Petrusek, M. Retention mechanism for reversed-phase ion-pair liquid chromatography. J. Chromatogr. **1979**, *186*, 419–434.

5. Knox, J.H.; Hartwick, R.A. Mechanism of ion-pair liquid chromatography of amines, neutrals, zwitterions and acids using anionic heterons. J. Chromatogr. **1981**, *204*, 3–21.

6. Cantwell, F.F. Retention model for ion-pair chromatography based on double-layer ionic adsorption and exchange. J. Pharm. Biomed. Anal. **1984**, *2* (2), 153–164.

7. Cecchi, T.; Pucciarelli, F.; Passamonti, P. Extended thermodynamic approach to ion-interaction chromatography. Anal. Chem. **2001**, *73* (11), 2632–2639.

8. Cecchi, T. Extended thermodynamic approach to ion-interaction chromatography: a thorough comparison with the electrostatic approach and further quantitative validation. J. Chromatogr. A. **2002**, *958* (1–2), 51–58.

9. Cecchi, T.; Pucciarelli, F.; Passamonti, P. Ion interaction chromatography of neutral molecules. Chromatographia **2001**, *53* (1–2), 27–34.

10. Cecchi, T.; Pucciarelli, F.; Passamonti, P. An extended thermodynamic approach to ion-interaction chromatography for high surface potential: use of a potential approximation to obtain a simplified retention equation. Chromatographia **2001**, *54* (9–10), 589–593.

11. Cecchi, T.; Pucciarelli, F.; Passamonti, P. Extended thermodynamic approach to ion-interaction chromatography for low surface potential: use of a linearized potential expression. J. Liq. Chromatogr. & Relat. Technol. **2001**, *24* (17), 2551–2557.

12. Cecchi, T.; Pucciarelli, F.; Passamonti, P. Ion interaction chromatography of neutral molecules: A potential approximation to obtain a simplified retention equation. J. Liq. Chromatogr. & Relat. Technol. **2001**, *24* (3), 291–302.

13. Cecchi, T.; Pucciarelli, F.; Passamonti, P.; Cecchi, P. The dipole approach to ion interaction chromatography of zwitterions. Chromatographia **2001**, *54* (1–2), 38–44.

14. Cecchi, T.; Cecchi, P. The dipole approach to ion interaction chromatography of zwitterions: use of a potential approximation to obtain a simplified retention equation. Chromatographia **2002**, *55* (5–6), 279–282.

15. Cecchi, T.; Cecchi, P. The dipole approach to ion interaction chromatography of zwitterions: use of the linearized potential expression for low surface potential. J. Liq. Chromatogr. & Relat. Technol. **2002**, *25* (3), 415–420.

16. Cecchi, T.; Pucciarelli, F.; Passamonti, P. Extended thermodynamic approach to ion-interaction chromatography. The influence of the organic modifier concentration. Chromatographia **2003**, *58* (7–8), 411–419.

17. Cecchi, T.; Pucciarelli, F.; Passamonti, P. Extended thermodynamic approach to ion-interaction

chromatography. A mono- and bivariate strategy to model the influence of ionic strength. J. Sep. Sci. **2004**, *27*, *in press*.

18. Cecchi, T.; Pucciarelli, F.; Passamonti, P. Extended thermodynamic approach to ion-interaction chromatograph: effect of the electrical charge of the solute ion. J. Liq. Chromatogr. & Relat. Technol. **2004**, *27* (1), 1–15.

19. Davies, J.T.; Rideal, E.K. Adsorption at liquid interfaces. In *Interfacial Phenomena*; Academic Press: New York, 1961; 154–216.

20. Rosen, M.J. Adsorption of surface-active agents at interfaces: the electrical double layer. In *Surfactant and Interfacial Phenomena*, 2nd Ed.; John Wiley & Sons: New York, 1978; 33–106.

21. Cecchi, T.; Pucciarelli, F.; Passamonti, P. Ion-Interaction Chromatography of Zwitterions. The Fractional Charge Approach to Model the Influence of the Mobile Phase Concentration of the Ion-Interaction Reagent The Analyst. 2004; 129, 1037–1046 (article B404721D available: DOI 10.1039/b404721d http://www.rsc.org/is/journals/current/analyst/anlpub.htm).

Ion-Pairing Techniques

Ioannis N. Papadoyannis
Anastasia Zotou
Aristotle University of Thessaloniki, Thessaloniki, Greece

INTRODUCTION

Ion-pair chromatography (IPC) is of relatively recent origin, being first applied in the mid-1970s. Much of the development work in both theory and practice was performed by Schill and co-workers. At various times, IPC has also been called extraction chromatography, chromatography with a liquid ion exchanger, soap chromatography, paired-ion chromatography and ion-pair partition chromatography.

BACKGROUND INFORMATION

When solute ions (A^-) are added to a chromatographic system containing pairing ions (B^+) and associated counter ions (C^-), the degree of retention of (A^-) depends on the following equilibrium:

$$A_{aq}^- + B_{aq}^+ = AB_{org} \qquad (1)$$

with an extraction constant

$$E_{AB} = \frac{[AB_{org}]}{[A_{aq}^-][B_{aq}^+]} \qquad (2)$$

In the simplest case of IPC, it can be assumed that the sample and counterions are soluble only in the aqueous mobile phase and the ion pair formed is soluble only in the organic stationary phase.

Assuming that the concentration of the pairing ion in the aqueous phase is high compared to that of the solute ion, the *distribution coefficient* of A^-, D_A^-, is given by

$$D_A^- = \frac{[AB_{org}]}{[A_{aq}^-]} = E_{AB}[B_{aq}^+] \qquad (3)$$

The *capacity factor k'* is related to E_{AB} as follows (in the reversed-phase mode):

$$k' = \frac{V_S}{V_m}\left(\frac{[AB_{org}]}{[A_{aq}^-]}\right) \qquad (4)$$

$$= \frac{V_S}{V_m}(E_{AB}[B_{aq}^+]) \qquad (5)$$

or

$$k' = D_A^- \frac{V_S}{V_m} \qquad (6)$$

Because the capacity factor k' is proportional to $1/D$ in normal-phase chromatography and to D in reversed-phase chromatography, it follows that k' in IPC is inversely proportional to the pairing ion concentration in the normal-phase situation but directly proportional in the reversed-phase case.

When the pairing ion is very hydrophobic, B^+ will be extracted into the organic phase with its normal counterion C^-, according to

$$C_{aq}^- + B_{aq}^+ = CB_{org}(\text{ion pair}) \qquad (7)$$

Subtracting it from Eq. (1) gives

$$A_{aq}^- + CB_{org} = AB_{org} + C_{aq}^- \qquad (8)$$

This is very similar to ion-exchange chromatography with an equilibrium constant:

$$K_{IE} = \frac{[AB_{org}][C_{aq}^-]}{[CB_{org}][A_{aq}^-]} \qquad (9)$$

This gives

$$D_A^- = \frac{[AB_{org}]}{[A_{aq}^-]} = K_{IE}\frac{[CB_{org}]}{[C_{aq}^-]} \qquad (10)$$

from which it follows that k' is inversely proportional to the concentration of the counterion in the aqueous phase.

The latter situation is usual in the reversed-phase mode, where the hydrophobic ion is adsorbed onto the bonded hydrocarbon of the packing material. Thus, we can distinguish three different techniques:

1. Normal-phase IPC, where the support is coated with an aqueous stationary phase containing the pairing ion and the ion pairs are partitioned between the stationary phase and an organic mobile phase.

Encyclopedia of Chromatography DOI: 10.1081/E-ECHR-120040028

2. Reversed-phase IPC, where the liquid stationary phase is organic and the pairing ion is introduced in the aqueous mobile phase.
3. Reversed-phase IPC, using a chemically bonded stationary phase and a hydrophobic pairing ion in the aqueous mobile phase.

The use of bonded-phase partition systems is generally preferred over mechanically held stationary phases; this gives advantage to technique 3.

NORMAL-PHASE ION-PAIR CHROMATOGRAPHY

The support is loaded with the aqueous stationary phase containing the pairing ion by one of the following three methods:

1. The stationary phase or a concentrated solution of the stationary phase in acetone is pumped through the packed column bed. The excess is then removed by passing eluent or hexane, followed by eluent saturated with stationary phase, until equilibrium is reached. Equilibrium is normally achieved when stable k' values are obtained for a series of representative solutes. This usually requires passage of several hundred milliliters of eluent.
2. The stationary phase can be loaded onto the column in several large plugs (0.1–1.0 mL) using a stopped-flow technique and equilibrium is achieved in the same way as previously.
3. The eluent, which has been preequilibrated with the stationary phase, is pumped through the column until stable k' values are obtained. The stationary phase is adsorbed onto the support surface, but at equilibrium, the pores of the support are not as completely filled as they are in the columns obtained by the first two methods. The equilibration can be a very time-consuming procedure, but the columns thus obtained are stable and reproducible.

Because the columns are in an equilibrium situation, it is obvious that gradient elution is not possible. In the normal-phase situation, the k' value of a solute is inversely proportional to the pairing ion concentration. Because the pairing ion is in the stationary phase, this concentration is not readily changed, and for this reason, retention is normally controlled by modification of the eluent. Hydrocarbon or chlorinated hydrocarbon solvents are usually employed with a small percentage of alcohol as a modifier. Varying the concentration or nature of the alcohol can produce the required changes in retention or selectivity.

Very high efficiencies have not usually been achieved with normal-phase IPC; the advantage of the reversed-phase mode, where the pairing ion concentration can be easily altered, has led to almost total takeover in the ion-pair field. The normal phase has two advantages compared to the reversed-phase mode:

1. The use of ultraviolet (UV) absorbing or fluorescent ions to enhance or enable the detection of nonabsorbing solutes.
2. The possibility of varying selectivity by varying the organic-phase composition.

REVERSED-PHASE ION-PAIR CHROMATOGRAPHY

Most often, pentanol or butyronitrile is used as the stationary phase loaded onto a hydrophobic support such as silanized silica.

The equilibration time depends on the hydrophobicity of the support (the coating of pentanol on a hydrocarbon-bonded silica takes no longer than 2 h at a flow rate of approximately 1 mL/min).

The retention of analytes can be regulated by varying the following factors:

1. The capacity factor increases with the hydrophobicity of the pairing ion. For hydrophilic solutes, hydrophobic pairing ions are chosen, and vice versa.
2. The capacity factor increases linearly with pairing ion concentration. Alternatively, gradient elution can be performed by decreasing the pairing ion concentration.
3. The choice of the organic phase affects the selectivity of the system.

REVERSED-PHASE ION-PAIR CHROMATOGRAPHY USING A CHEMICALLY BONDED STATIONARY PHASE

This is, by far, the most commonly used form of reversed-phase IPC. This technique has been also called *soap chromatography* although, in soap chromatography, the use of detergents as counterions is introduced. Here, the columns (with C_8 or C_{18} packings) are prepared by equilibrating the stationary bonded phase with the mobile phase containing the pairing ion. The ion-pair reagent is attracted to the stationary phase because of its hydrophobic alkyl group and the charge carried by the reagent thereby attaches to the stationary phase.

The surface of a C_8 or C_{18} column packing is shown in Fig. 1 as a rectangle covered by sorbed molecules of

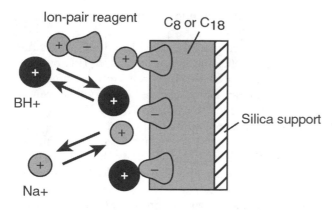

Fig. 1 Pictorial representation of IPC retention of a protonated base (BH⁺); Na⁺ is the mobile-phase cation; the IPC reagent is hexane sulfonate. (From L. R. Snyder, J. J. Kirkland, and J. L. Glajch, *Practical HPLC Method Development*, 2nd ed., 1997, by permission of John Wiley & Sons, Inc.)

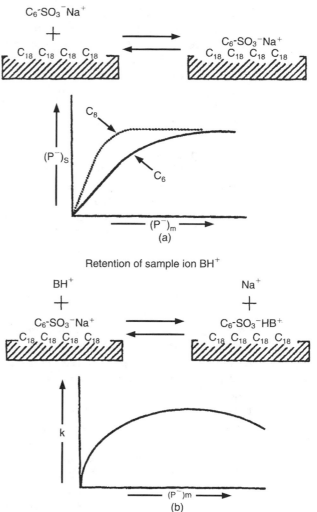

Fig. 2 Effect of ion-pair reagent concentration on separation. (a) Sorption of the ion-pair reagent as a function of concentration for reagents of different hydrophobicity (C_6- and C_8-sulfonates); (b) retention as a function of reagent concentration. (From L. R. Snyder, J. J. Kirkland, and J. L. Glajch, *Practical HPLC Method Development*, 2nd ed., 1997, by permission of John Wiley & Sons.)

a negative ion-pair reagent (e.g., hexane sulfonate, C_6-SO_3^-). The negative charge on the stationary phase is balanced by the positive ions (Na^+) from the reagent and/or buffer. A positively charged sample ion (protonated base BH^+) can exchange with a Na^+ ion as shown (arrows), resulting in the retention of the sample ion by an ion-exchange process.

For each ion-pair reagent, the column uptake increases for a higher reagent concentration in the mobile phase, but then levels off as the column becomes saturated with the reagent. The more hydrophobic reagents are retained more strongly and saturate the column at a lower mobile-phase reagent concentration (10^{-5} M), but equilibration may take several hours. Less hydrophobic ion-pair reagents are added at a slightly higher concentration (10^{-4} M–10^{-3} M) and equilibrium is reached much faster (1–2 h at a flow rate of 1 mL/min). This is shown in Fig. 2a, where the concentration of reagent in the stationary phase $(P^-)_S$ is plotted versus the concentration of reagent in the mobile phase $(P^-)_m$ for two reagents of different hydrophobicity.

The change in sample retention, as the ion-pair reagent concentration increases, is shown in Fig. 2b for a hydrophilic sample compound BH^+. Once the column becomes saturated with the reagent, the sample retention levels off. Because IPC retention involves an ion-exchange process, further increases in reagent concentration lead to an increase in the counterion concentration (Na^+), which competes with the retention of the sample ion on the column.

In practice, when very hydrophobic pairing ions are used, the columns are irreversibly altered, because the ions can never be completely removed. Once equilibrium is reached, the columns are stable and can be used for several months. The columns should be stored in the mobile phase because of the lengthy equilibration times. Only if the column is not used for an extended period of time should one consider storing the column in an organic solvent.

DESIGN OF AN ION-PAIR SEPARATION

Unless there is a specific reason to choose a normal-phase system, IPC should be carried on in the reversed-phase mode, using chemically bonded stationary phases.

The best counterion and pH depend on the kind of sample to be separated. Most ion-pair reagents used today are either alkyl sulfonates or tetraalkyl ammonium salts, either of which allow UV detection above 210 nm. The IPC aqueous phase must be adequately buffered with respect to both pH and concentration of the counterion.

Inadequate buffering of the aqueous phase is a source of band tailing in IPC. Conventional buffers, such as citrate and phosphate, have been used and, in some cases, the counterion itself is an adequate buffer. For separations at low pH, 0.1 M–0.2 M solutions of a strong acid provide adequate buffering. Inadequate buffering of the aqueous phase is a source of band tailing in IPC.

In reversed-phase IPC, maximum k' values are obtained at intermediate values of pH, where the sample compounds are completely ionized and ion-pair formation is at a maximum. As the pH of the mobile phase is lowered, sample anions A^- begin to form the un-ionized acids HA, leading to a smaller number of sample ion pairs in the stationary phase. Acids are usually separated at a pH of 7–9, whereas bases are separated at a pH of 1–6.

In reversed-phase systems, the solvent strength is readily varied by changing the counterion or its concentration. When all sample ions are fully ionized, a change in solvent strength via a change in counterion concentration leads to minimal changes in separation selectivity. The concentration of the counterion is usually 0.005 M–0.05 M, except for perchlorate (0.5 M–1 M) or the detergents used in soap chromatography (e.g., 1 wt% of counterion). Buffer concentrations are similar to those used in ion-exchange chromatography (0.001 M–0.5 M).

An increase in the alkyl chain length of the counterion increases retention in reversed-phase IPC by up to 2.5 times per added $-CH_2-$ group in the counterion.

Apart from an increase in the counterion concentration, an increase in ionic strength of the aqueous phase generally reduces the formation of ion pairs, as a result of the competition of secondary ions in forming ion pairs with the counterion. One study showed a twofold to threefold change in k' for each doubling of ionic strength.

For reproducible separations by IPC, it is important to thermostat the column. Temperature effects in IPC are more important than in some other methods.

SUGGESTED FURTHER READING

Bidlingmeyer, B.A. *Practical HPLC Methodology and Applications*; John Wiley & Sons: New York, 1992; 157–165.

Gilbert, M.T. *High Performance Liquid Chromatography*; IOP Publishing, Wright: Bristol, U.K, 1987; 227–253.

Snyder, L.R.; Kirkland, J.J. *Introduction to Modern Liquid Chromatography*, 2nd Ed.; John Wiley & Sons: New York, 1979; 454–482.

Snyder, L.R.; Kirkland, J.J.; Glajch, J.L. *Practical HPLC Method Development*, 2nd Ed.; John Wiley & Sons: New York, 1997; 317–341.

Su, S.C.; Hartkopf, A.V.; Karger, B.L. J. Chromatogr. **1976**, *199*, 523.

Isocratic HPLC: Selection of a System

Pavel Jandera
*Department of Analytical Chemistry, University of Pardubice,
Pardubice, Czech Republic*

INTRODUCTION

Any new sample type delivered for HPLC analysis requires an adequate separation method. Previous experience with similar samples is very useful and many methods can be looked up in the literature; however, it is often necessary to modify earlier established methods to suit laboratory equipment or sample matrices, or to improve sample throughput in the laboratory. In many cases, a desired separation can be achieved with a few experiments, but some separation problems are more difficult and their solution may require considerable experimental effort. Nowadays, effects of experimental conditions on HPLC separations are well understood, and this knowledge can be used for effective method development. First, the objective of the separation should always be kept in mind. Any information available on the sample is very helpful in HPLC method development: the matrix, the approximate number of sample components, their chemical structures, concentrations, solubilities, and other properties provide clues for the selection of sample pretreatment approach (if necessary), suitable detection conditions, and a separation system. Most often, variable-wavelength or diode array ultraviolet (UV) absorbance detectors are used, but mass spectrometry (MS), coupled to HPLC, is becoming increasingly popular because it offers valuable structural information for unknown samples. Other detection techniques are used less frequently—mainly fluorimetric and electrochemical detection for sensitive and selective environmental, biological, and food analyses, or light-scattering detection for analysis of compounds that do not absorb in the UV region and neither can be oxidized nor reduced. For adequate separation, suitable chromatographic mode, stationary phase, mobile phase, flow rate, column dimension, and temperature should be selected. Many analyses can be performed at constant operating conditions using the so-called isocratic elution.

SELECTION OF A COLUMN AND PACKING MATERIAL

Many HPLC separations are performed with conventional analytical columns, which are 10–25 cm long

and 3–4.6 mm in diameter. The column plate number, the pressure drop across the column, and the separation time at a constant flow rate are directly proportional to the column length. With short (2–6 mm), "high-speed" columns of the same diameter, simple separations can be accomplished in 1–3 min, so that the productivity of the laboratory is considerably increased and solvent consumption per analysis is reduced.

The flow rate and consumption of the mobile phase at a constant flow rate, and the pressure drop increase with the second power of the column diameter. Separations on "microbore" columns (15–5 cm long and 1–2 mm ID) need even less mobile phase and allow high mass sensitivity of detection, which is useful for analyses of small sample amounts with mass spectrometric detection. Separations on high-speed microbore, and especially capillary HPLC columns of 0.1–0.5 mm internal diameter, are subject to more significant extra-column contributions of the injector, detector, and connecting tubing to band broadening in comparison to conventional analytical columns, so that miniaturized HPLC requires specially designed low-volume injectors and detectors, often at the cost of decreased sensitivity of detection. Hence, microbore and capillary HPLC columns have been, so far, more frequently used for HPLC/MS trace analysis than for other routine quantitative analytical applications.

HPLC columns contain, usually, spherical particle packings, which are carefully sorted to fractions with narrow size distribution to provide high separation efficiency. Totally porous packing materials most frequently used for separations of small molecules in contemporary HPLC have pore sizes of 7–12 nm and specific surface area of 150–400 m^2/g, but wide-pore particles with pore sizes of 15–100 nm and relatively low specific surface area of 10–150 m^2/g, or nonporous materials are used for separations of macromolecules. Perfusion materials, designed especially for the separation and isolation of biopolymers, contain very broad pores (400–800 nm) throughout the whole particle, which are interconnected by smaller pores. Column efficiency and flow resistance increase with small particles, and a high pressure has to be used to maintain required flow rate and to keep an acceptable time of analysis. However, the maximum operating pressure

Encyclopedia of Chromatography DOI: 10.1081/E-ECHR-120041123

is 30–40 MPa, with common instrumentation for HPLC. Hence, short columns should be used with small-diameter particles. Five-micrometer particles are most often used in conventional analytical columns, and particles of 3–4 µm (less often, 1–2 µm) are common in short, "high-speed" columns for rapid, simple separations.

Instead of packed columns, monolithic (continuous bed), analytical, or capillary columns in the form of a rod with flow-through pores offer high porosity and improved permeability. Silica-based monolithic columns are generally prepared by gelation of a silica sol to a continuous sol–gel network, onto which a C_{18} or another stationary phase is subsequently chemically bonded. Such columns provide comparable efficiency and sample capacity as conventional columns packed with 5-µm particle materials, but have three to five times lower flow resistance, thereby allowing higher flow rates and fast HPLC analyses. Rigid polyacrylamide, polyacrylate, polymethacrylate, or polystyrene monolithic columns are prepared by in situ polymerization.

SELECTION OF HPLC SEPARATION MODE

The first step in HPLC method development consists of selecting an appropriate separation mode. Many neutral compounds can be separated either by reverse-phase (RP) or normal-phase (NP) chromatography. An RP system is usually the best first choice because it is likely to result in a satisfactory separation of a great variety of nonpolar, polar, and even ionic compounds. Lipophilic samples often can be separated either by nonaqueous RP chromatography or NP chromatography. Weak acids or bases can be analyzed by RP chromatography with buffered mobile phases, strong acids, or strong bases by ion pair or ion exchange chromatography (IEC). Special chiral columns or chiral selector additives to the mobile phase can be used for separation of optical isomers (enantiomers).

Macromolecules are usually separated and characterized by SEC on columns packed with inert materials (gels) characterized by controlled pore distributions, based on different accessibilities of the pores for molecules of different sizes, with the larger molecules eluting first. For some lower polymers with molecular masses in the range $10^3–10^4$ Da, "interactive" (i.e., RP-HPLC or NP-HPLC) modes provide better selectivity of separation than SEC. Many ionizable biopolymers such as peptides, proteins, oligonucleotides, and nucleic acids can be separated by IEC or RP chromatography on wide-pore packing materials, with mobile phases containing trifluoroacetic acid or triethylammonium acetate as ion pair reagents.

RPLC

The stationary phase in RP chromatography—usually an alkyl immobilized on an inorganic support—is less polar than the aqueous–organic mobile phase. Nonpolar samples are more strongly retained than the polar ones, and the retention increases with increasing polarity of the mobile phase, so that very lipophilic samples may require nonaqueous mobile phases.

Silica gel-based materials for RP chromatography with nonpolar (most often C_8 or C_{18} alkyls) or moderately polar stationary phases covalently bonded via Si–O–Si–C bonds are prepared by chemical modification of the silanol (Si–OH) groups on the silica gel surface by chloro-silane or alkoxy-silane reagents, and are relatively stable to hydrolysis. The retention in RP increases with increasing surface coverage and length of the bonded alkyl chains, so that C_{18} phases show greater retention than C_8-bonded phases.

Monofunctional silane reagents yield efficient stationary phases with flexible "furlike" or "brush-like" structure of the chains bonded on the silica surface. When bifunctional or trifunctional silanes are used for modification, Cl or alkoxy groups are introduced into the stationary phase, which are subject to hydrolysis and react with excess molecules of reagents to form a polymerized spongelike bonded phase structure. Stationary phases prepared in that way usually show stronger retention but lower separation efficiency (plate number) than monomerically bonded stationary phases.

Bonded phases prepared by modification of silica gel, in turn prepared by gelation of soluble silicates, are not stable in mobile phases with pH > 8, where the silica gel slowly dissolves. However, chemically bonded groups on silica support can hydrolyze at pH < 3, causing stationary phase "bleeding." The hydrolysis is enhanced at higher temperatures, so that many bonded phases are not stable at temperatures higher than 60°C. New materials prepared using high-purity silica particles made by aggregation of silica sols are stable up to pH 9–10 and do not contain metal contaminants that can form chelates, which can cause tailing of polar compounds.

Rather bulky silanization reagents can chemically modify no more than 50% of the original silanol groups. The residual silanol groups may interact with polar solutes, especially basic solutes, often causing strong and irreversible retention and poor separation with tailing or distorted peaks. Some residual silanol groups can be removed by a subsequent "end-capping" reaction with small-molecule trimethylchlorosilane or hexamethyldisilazane reagents. Another approach relies on using diisopropyl chlorosilane or diisobutyl chlorosilane reagents in a single silanization step to provide steric shielding of residual silanols.

Stationary phases prepared by modification of the silica gel surface with bidentate silanes containing C_{18} or C_8 alkyls and two reactive groups separated by a $-CH_2-CH_2-$ or a $-CH_2-CH_2-CH_2-$ bridging group show high bonding density and improved stability over a broad pH range. Finally, hybrid stationary phases with methyl groups incorporated into the silica gel structure contain lower concentrations of silanol groups and show improved pH stability. Stationary phases with chemically bonded, branched hydrocarbons, perfluoroalkanes, polyethylene glycol, cholesterol, or alkylaryl groups show different separation selectivities, which can be useful for specific separations. For example, chemically bonded phenyl groups show preferential retention of aromatic compounds and increased shape selectivity for planar and rigid rodlike molecules. Incorporating amide or carbamate groups into the alkyl-bonded phases improves retention behavior in highly aqueous mobile phases.

Materials with inorganic or porous hydrophobic or (less frequently) hydrophilic organic polymer matrices and graphitized carbon are stable over a broad pH range from 0 to 12–14; hence, they are useful for separations of basic compounds. RP phases on aluminium and zirconium oxide supports exhibit hardness and mass transfer properties comparable to silica, and can be prepared by forming a cross-linked polystyrene, polybutadiene, or alkylated polymethylsiloxane layer on the support surface to which alkyls are attached. The inorganic surface, encapsulated by a nonpolar stationary phase, does not come into contact with the mobile phase or with the analyte, so these materials can be used in the pH range 1–14.

Porous graphitized carbon adsorbents with sufficient hardness, and well-defined and stable pore structures without micropores are now available with increased affinity for aromatic and polar substances, allowing difficult separations of some hydrophilic or isomeric compounds.

The mobile phase in RP chromatography contains water and one or more organic solvents, most frequently acetonitrile, methanol, tetrahyrofuran, or propanol. By the choice of the organic solvent, selective polar interactions (dipole–dipole, proton–donor, or proton–acceptor) with analytes can be either enhanced or suppressed, and the selectivity of separation can be adjusted. Binary mobile phases are usually well suited for the separation of a variety of samples, but ternary or, less often, quaternary mobile phases may offer improved selectivity for some difficult separations. The retention times t_R are controlled by the concentration of the organic solvent in the aqueous–organic mobile phase. Equation (1) is widely used to describe the effect of the volume fraction of methanol or acetonitrile φ on the retention factors $k = t_R/t_0 - 1$:[1]

$$k = k_0 10^{-m\varphi} \tag{1}$$

where t_0 is the column dead (holdup) time. The constants m and k_0 in Eq. (1) increase as the polarity of the solute decreases, or as its size increases. Furthermore, the constant m increases with decreasing polarity of the organic solvent.

NPLC

In NP chromatography (also called adsorption or liquid–solid chromatography), the stationary phase is more polar than the mobile phase. The retention increases as the polarity of the mobile phase decreases, and polar analytes are more strongly retained than nonpolar ones (i.e., the opposite of RP chromatography; Fig. 1). The column packing is either an inorganic adsorbent (silica gel or, less often, aluminium oxide), or a moderately polar bonded phase [cyanopropyl – $(CH_2)_3-CN$, diol $-(CH_2)_3-O-CH_2-CHOH-CH_2-OH$, or aminopropyl $-(CH_2)_3-NH_2$] chemically bonded

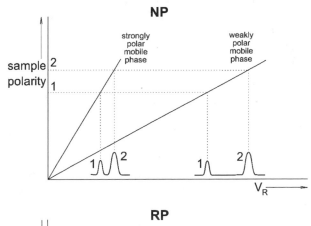

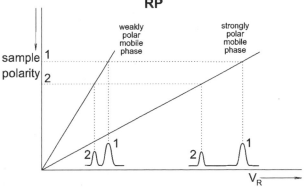

Fig. 1 Schematic diagram of the effects of sample and mobile phase polarities on retention in NPLC and RPLC. V_R = retention volume.

onto silica gel, and the mobile phase usually is a mixture of a nonpolar solvent and one or more strongly or moderately polar solvents. NP behavior can be sometimes observed also in nonaqueous RP liquid chromatography (NARPLC), probably because of the activity of polar residual silanol groups.

RP chromatography generally offers a better selectivity than NPLC for the separation of molecules differing in hydrophobic parts of the molecules, but there are some practical reasons for selecting NP chromatography methods in specific cases, as follows:

1. A lower organic mobile phase viscosity offers a lower pressure drop across the column than in aqueous–organic mobile phases used in RPLC at a comparable flow rate.
2. HPLC columns are usually more stable and have longer lifetimes in organic solvents than in aqueous–organic mobile phases.
3. Many samples are more soluble or less prone to decompose in organic mobile phases than in aqueous mobile phases, and do not cause injection problems in NPLC, as is occasionally observed in RPLC.
4. Unlike RP chromatography, NP chromatography enables the direct injection of samples extracted into a nonpolar solvent.
5. NPLC is usually better-suited for the separation of positional isomers or stereoisomers than RPLC.

Very large changes in separation selectivity are possible by changing either the mobile phase or the stationary phases in NPLC. Proton donor–acceptor interactions cause strong retention of basic compounds on silica gel in nonaqueous mobile phases, whereas acidic compounds show increased affinities to aminopropyl columns. The elution strength is proportional to the polarity of the mobile phase. Great changes in the selectivity of NP chromatography separations can be achieved by selecting solvents with appropriate types of selective polar interactions. With some simplification, the dependence of the retention factor k on the volume fraction φ of the polar solvent B in binary mobile phases can be described by Eq. (2):[2]

$$k = k_0 \varphi^{-m} \tag{2}$$

The constants k_0 and m depend on the nature of the solute and on the chromatographic system, but are independent of the concentration φ, and k_0 is the retention factor in pure solvent B. The parameter m theoretically corresponds to the number of molecules of the strong solvent B necessary to displace one adsorbed sample molecule.

LC Separation of Ionic Compounds

Ionized compounds are usually much less retained than noncharged compounds; their separation in RP chromatography is usually possible only with ionic additives to the mobile phase. Basic compounds can interact with residual silanols in alkyl silica-bonded phases ionized to SiO^- anions; consequently, strong retention and tailing peaks are observed. Alkylamine additives to the mobile phase sometimes improve peak shape by blocking the silanol groups. Repulsive interactions with the negatively charged residual $-SiO^-$ groups in the alkyl silica-bonded phases may cause ionic exclusion and poor separation of strong acids, which elute close to the column holdup time, often as asymmetrical peaks.

In mobile phases containing 10–50 mM phosphate or acetate buffers of pH 2–8.5, the ionization of weak acids (at pH < 7) or bases (at pH > 7) can be more or less suppressed to improve separation and peak symmetry. By adjusting the pH in the range ± 1.5 U around the pK_a, differences in the degree of ionization of the individual sample components can often be utilized to control the separation selectivity. The retention is usually adjusted by the addition of up to 30–40% acetonitrile, methanol, or tetrahydrofuran to the mobile phase.

Strong acids and strong bases are completely ionized over a broad pH range and their chromatographic behavior is usually only little affected by adding a buffer to the mobile phase. Such compounds can be separated by RP ion pair or IEC. In ion pair chromatography (IPC), ion pair reagents whose molecules contain a strongly acidic or strongly basic group and a bulky hydrocarbon part are added to the mobile phase. Basic substances can usually be separated using C_6–C_8 alkanesulphonates, and acidic substances can be separated using tetralkylammonium salts. Ion pair additives significantly increase retention and improve peak symmetry through the formation of neutral ionic-associated species, called ion pairs, with increased affinity to a nonpolar stationary phase. The retention in IPC can be controlled by the type and concentration of the ion pair reagent or of the organic solvent in the mobile phase. Increasing the number and size of alkyls in the reagent molecules enhance retention in the reagent concentration range between 10^{-4} and 10^{-2} mol/L.

Nowadays, IEC is used mainly for the separation of small inorganic ions or ionic biopolymers such as oligonucleotides, nucleic acids, peptides, and proteins, rather than in the analysis of small organic ions, for which RP chromatography and IPC usually offer higher efficiency and better resolution. IEC columns are packed with fine particles of ion exchangers, which contain charged ion exchange groups covalently attached to a solid matrix (either an organic cross-linked styrene–divinylbenzene or ethyleneglycol–methacrylate copolymer), or inorganic support to which a functional group is chemically bonded via a spacer: a propyl, or phenylpropyl moiety. Strong cation exchangers contain $-SO_3^-$ sulphonate groups and strong anion exchangers $-N(CH_3)_3^+$ contain quaternary ammonium groups, completely ionized over a broad pH range (pH $= 2$–12). Weak cation exchangers contain carboxylic or phosphonic acid groups, which are ionized only in alkaline solutions, and tertiary or secondary amino groups (e.g., diethyl aminoethyl) in weak anion exchangers are ionized only in acidic mobile phases. Ion exchange separations require aqueous or aqueous–organic mobile phases containing counterions ($10^{-2} - 10^{-1}$ mol/L salts, buffers, ionized acids, or bases), which compete with the sample ions for the ion exchange groups. The retention in IEC decreases with increasing concentration of counterions in the mobile phase and with decreasing ion exchange capacity of the column (1–5 mEq/g with organic polymer ion exchangers and 0.3–1 mEq/g with silica-based ion exchangers). Weak acids are usually separated by anion exchange chromatography at pH > 6 and weak bases are separated by cation exchange chromatography at pH < 6, and their retention increases with increasing ionization. Varying the pH of the mobile phase can adjust the separation selectivity, whereas the retention is controlled by the ionic strength.

STRATEGIES FOR SELECTING AND OPTIMIZING ISOCRATIC SEPARATION CONDITIONS

Once a suitable HPLC separation mode has been selected, experimental conditions can be adjusted using either an empirical method or a systematic method development approach. The separation of two sample compounds is conveniently characterized by resolution R_S:

$$R_S \cong \frac{\sqrt{N}}{4}(r_{1,2} - 1)\frac{k}{1 + k}$$
$$= \text{Efficiency} \times \text{Selectivity}$$
$$\times \text{Capacity} \tag{3}$$

Here, N is the column efficiency expressed in terms of plate number; $r_{1,2} = k_2/k_1$ is the separation factor, which characterizes the selectivity of separation; and k is the average retention factor of the two sample compounds 1 and 2 as a measure of capacity contribution to the resolution. The three terms contributing to the resolution depend on many experimental

conditions, which can be adjusted either simultaneously or in subsequent steps.

For accurate quantitative analysis, the resolution usually should be 1–1.5. However, too high a resolution may result in excessive analysis time. The chromatogram in Fig. 2A schematically illustrates adequate separation of a three-component sample. If the separation is not satisfactory, it can be improved according to the following strategy:

1. Poorly resolved peaks appearing close to the column holdup volume, such as in the example in Fig. 2B, show that the retention is too low and should be increased; this is best done by decreasing the elution strength of the mobile phase, whereas the elution strength should be increased if the retention volumes are too large (Fig. 2C).
2. If the retention times are adequate and partial separation of the bands is apparent but the bands are relatively broad (Fig. 2D), the resolution can be possibly improved by increasing the efficiency (i.e., the plate number of the column).
3. If the bands are narrow, but are not well separated from each other, such as in Fig. 2E, the selectivity should be improved:

    a) By changing the components of a binary mobile phase;

    b) By using ternary or more complex mobile phases or mobile phase additives inducing specific interactions with sample components; or

    c) By using another HPLC column with a different stationary phase.

4. If the resolution of early eluted bands is unsatisfactory and the separation time is long, the sample separation is usually improved by temperature or solvent gradients.

Control of Separation Efficiency

The efficiency contribution to the resolution (i.e., the column plate number, N) increases with decreasing particle size of the column packing, with increasing column length, and, to a lesser extent, with decreasing flow rate of the mobile phase. In this case, improved separation often is traded off for an increase in the pressure drop across the column, or for an increase in the run time. Further, column efficiency often improves as the temperature increases because of decreasing viscosity of the mobile phase and increasing diffusion coefficient and mass transfer.

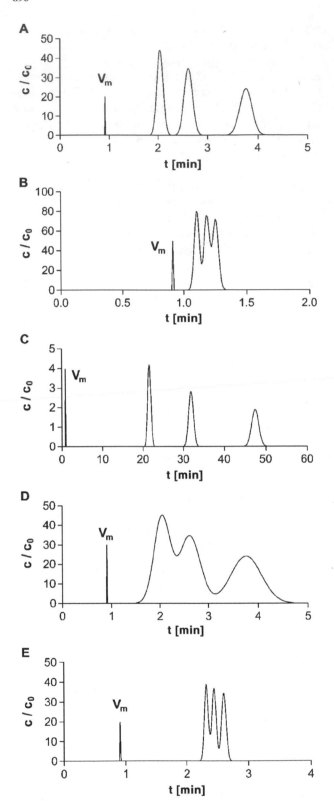

Fig. 2 Examples of chromatographic separation of a three-component sample. (A) Satisfactory separation. (B) Unsatisfactory separation, too low retention. (C) Good resolution, but too long time of separation. (D) Unsatisfactory separation, too low column efficiency. (E) Unsatisfactory separation, good retention and column efficiency, but too low separation selectivity.

The chemistry of a stationary phase and the composition of the mobile phase often have only minor effects on the separation efficiency in the absence of strong interactions with adsorption centers, and when the porosity of the packing material and the viscosity of the mobile phase do not change very significantly.

Control of Retention and Separation Selectivity

The retention and the selectivity of separation depend primarily on the chemistry of the stationary phase and the mobile phase and, to a lesser extent, on temperature. The retention usually decreases by 1–2% when the column temperature increases by 1°C. A change in retention is often accompanied by a change in separation selectivity $r_{1,2}$ so that temperature regulation can be used for optimizing the resolution. For example, increased temperature usually affects, favorably, the separation selectivity of ionic compounds. The regulation of temperature is convenient and simple, but is usually less effective for improving HPLC separations than for varying the composition of the mobile phase. Furthermore, many HPLC columns are not stable at temperatures above 60°C.

For a successful HPLC separation, the appropriate selection of the mobile phase is equally important as the correct choice of the separation column. In NPLC, the elution strength increases, whereas in RPLC, it decreases with increasing solvent polarity. Single-component mobile phases do not allow a fine adjustment of the elution strength, as there is only a limited selection of solvents compatible with UV and other common detection techniques, so that mixed mobile phases composed of solvents with different elution strengths should be used. Increasing the concentration of the strong solvent in a mixed mobile phase speeds up the elution. The retention of weak acids and weak bases in RPLC increases when the pH of the mobile phase is adjusted to suppress their ionization. A change in the retention, induced by an increase or decrease in the concentration of the strong solvent, is often accompanied by a change in the separation selectivity. The effects of the mobile phase on separation can be predicted, and the separation can be optimized using either a commercial software such as Dry Lab I, or simple predictive calculations employing, for example, Eq. (1) or Eq. (2).

Ternary and more complex mobile phases contain at least two different strong solvents with different predominant selective polar contributions (dipole–dipole, proton–donor, and proton–acceptor) in a weak solvent. Fine selectivity tuning is often possible by appropriate selection of the concentration ratios of the strong solvents.[3] In RP chromatography, acetonitrile with dipole–dipole properties, tetrahydrofuran

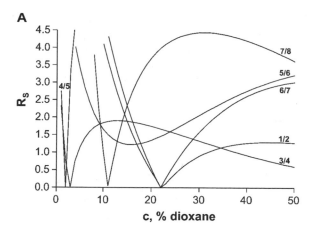

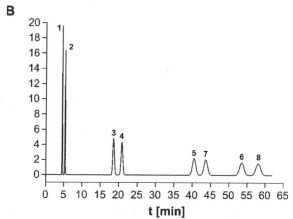

Fig. 3 (A) The window diagram (the dependence of the resolution on the concentration of dioxane in *n*-heptane as the mobile phase) for a mixture of eight phenylurea herbicides on a Separon SGX 7.5-μm silica gel column (150 × 3.3 mm ID). (B) The separation with an optimized concentration of 13% dioxane in the mobile phase for maximum resolution. Column plate number $N = 5000$, $T = 40°C$, flowrate 1 mL/min. Sample compounds: neburon (1), chlorobromuron (2), 3-chloro-4-methylphenylurea (3), desphenuron (4), isoproturon (5), diuron (6), metoxuron (7), and deschlorometoxuron (8).

parameter (such as the concentration of the strong solvent in a binary mobile phase, pH, temperature, etc.) to predict the resolution as a function of the optimized parameter using empirical or simple model-based calculations. Then, plots are constructed (the "window diagrams")[4] in which the range of the optimized parameter is searched for the value that provides the desired resolution for all adjacent bands in the chromatogram in the shortest time. Two examples of window diagrams (Figs. 3 and 4) illustrate the approach for optimization of binary mobile phases in NPLC and the impact of the selection of a strong solvent on the time of optimized separation.

Some parameters may show synergistic effects on the separation. In this case, simultaneous optimization of two or more parameters at a time can provide better results than their sequential optimization. Simple methods can be used for sequential multiparameter optimization in HPLC. Alternatively, the composition of mixed mobile phases can be optimized using an

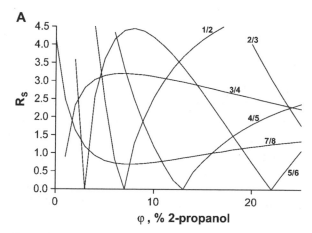

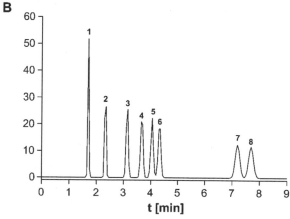

Fig. 4 (A) The window diagram (the dependence of the resolution on the concentration) of 2-propanol in *n*-heptane as the mobile phase. (B) The separation with an optimized concentration of 19% 2-propanol in the mobile phase for maximum resolution. Compounds and column are as in Fig. 3.

with proton acceptor properties, and methanol with both proton donor and proton acceptor properties are used as strong solvents in water; in NP chromatography, a nonlocalizing solvent (dichloromethane), a basic localizing solvent (methyl-*t*-butyl ether), and a nonbasic localizing solvent (acetonitrile or ethyl acetate) are mixed and diluted with hexane or heptane to adjust the elution strength. For any three-component or four-component mobile phase, the proportions of the individual selective contributions to the polarity are proportional to the concentration ratios of the three strong solvents, whereas the elution strength is controlled by the concentration of water in RPLC and alkane in NPLC.

Single-parameter optimization employs several experiments at preselected values of the optimized "overlapping resolution mapping" strategy to adjust the optimum separation selectivity based on seven or more initial experiments with solvent mixtures of approximately equal elution strength. Based on the retention data from the initial experiments, either three-dimensional diagrams or contour "resolution maps" are constructed for all adjacent bands in the selectivity triangle space as a function of the concentration ratios of three solvents or two solvents and the pH of the mobile phase, from which the composition of the mobile phase that provides maximum resolution is selected.

Structure-based commercial optimization software (e.g., Chromdream, Chromsword, or Eluex) incorporate some features of the "expert system," as the retention is predicted based on the additive contributions of the individual structural elements and, consequently, the optimum composition of the mobile phase is suggested. Such predictions are only approximate and do not take into account stereochemical and intramolecular interaction effects.

CONCLUSIONS

Suitable detection conditions, separation mode, appropriate column dimensions, and packing materials (stationary phase) in isocratic LC should be selected, keeping in mind the objective of separation, using the information on sample properties, such as solubility, polarity, presence of specific functional groups in sample compounds, sample amount and concentration, etc. For the control of separation selectivity, resolution, and time of analysis, appropriate selection of the mobile phase is equally as important as the choice of the stationary phase, whereas the variation in flow rate and temperature usually has a lesser effect on separation. Trial-and-error optimization of separation is facilitated by a knowledge of the principles of

retention mechanisms. Isocratic separations can be optimized using single-parameter or multiparameter strategies, using either commercial software or simple computer-aided predictive calculations.

REFERENCES

1. Snyder, L.R.; Dolan, J.W.; Gant, J.R. Gradient elution in high-performance liquid chromatography: I. Theoretical basis for reversed-phase systems. J. Chromatogr. **1979**, *165*, 3–30.
2. Jandera, P.; Churáček, J. Gradient elution in liquid chromatography: I. The influence of the composition of the mobile phase on the capacity ratio (retention volume, band width, and resolution) in isocratic elution—Theoretical considerations. J. Chromatogr. **1974**, *91*, 207–221.
3. Glajch, J.L.; Kirkland, J.J.; Snyder, L.R. Practical optimization of solvent selectivity in liquid–solid chromatography using a mixture–design statistical technique. J. Chromatogr. **1982**, *238*, 269–280.
4. Price, W.P.; Deming, S.N. Optimized separation of scopoletin and umbelliferone and cis–trans isomers of ferulic and *p*-coumaric acids by reverse-phase high-performance liquid chromatography. Anal. Chim. Acta **1979**, *108*, 227–231.

SUGGESTED FURTHER READING

Neue, U.D. *HPLC Columns: Theory, Technology and Practice*; Wiley-VCH: New York, 1997.

Schoenmakers, P.J. *Optimisation of Chromatographic Selectivity*; Elsevier: Amsterdam, 1986.

Snyder, L.R.; Kirkland, J.J.; Glajch, J.L. *Practical HPLC Method Development*, 2nd Ed.; John Wiley & Sons: New York, 1997.

Index